Physiology in Health and Disease

Published on behalf of The American Physiological Society by Springer

Physiology in Health and Disease

This book series is published on behalf of the American Physiological Society (APS) by Springer. Access to APS books published with Springer is free to APS members.
APS publishes three book series in partnership with Springer: *Physiology in Health and Disease* (formerly *Clinical Physiology*), *Methods in Physiology*, and *Perspectives in Physiology* (formerly *People and Ideas*), as well as general titles.

More information about this series at http://www.springer.com/series/11780

Kirk L. Hamilton • Daniel C. Devor
Editors

Ion Channels and Transporters of Epithelia in Health and Disease

Springer

American Physiological Society

Editors
Kirk L. Hamilton
Department of Physiology
University of Otago
Dunedin, Otago
New Zealand

Daniel C. Devor
Department of Cell Biology
University of Pittsburgh
Pittsburgh, Pennsylvania
USA

Physiology in Health and Disease
ISBN 978-1-4939-3364-8 ISBN 978-1-4939-3366-2 (eBook)
DOI 10.1007/978-1-4939-3366-2

Library of Congress Control Number: 2015957986

Springer New York Heidelberg Dordrecht London

Printed on acid-free paper

Springer Science+Business Media LLC New York is part of Springer Science+Business Media (www.springer.com)

Preface

Ion channels and transporters play critical roles both in the homeostasis of normal function of the human body and during the disease process. Indeed, as of 2005, 16 % of all Food and Drug Administration-approved drugs targeted ion channel and transporters, highlighting their importance in the disease process. Further, the Human Genome Project provided a wealth of genetic information that has since been utilized, and will again in the future, to describe the molecular pathophysiology of many human diseases. Over the recent years, our understanding of the pathophysiology of many diseases has been realized. The next great "step" is a combined scientific effort in basic, clinical, and pharmaceutical sciences to advance treatments of molecular diseases.

A number of unique ion channels and transporters are located within epithelial tissues of various organs including the kidney, intestine, pancreas, and respiratory tract and all play crucial roles in various transport processes responsible for maintaining homeostasis. Ultimately, understanding the fundamentals of ion channels and transporters, in terms of function, modeling, regulation, molecular biology, trafficking, structure, and pharmacology, will shed light on the importance of ion channels and transporters in basic physiology and pathophysiology of human diseases.

This book contains chapters written by notable world-leading scientists and clinicians in their respective research fields. The book consists of four parts. The first part of the book is entitled "Basic Epithelial Ion Transport Principles and Function" (Chaps. 1–8) and spans the broad fundamentals of chloride, sodium, potassium, and bicarbonate transepithelial ion transport, the most recent developments in cell volume regulation, the mathematical modeling of these processes, the mechanisms by which these membrane proteins are correctly sorted to the apical and basolateral membranes, and protein folding of ion channels and transporters. The chapters in Part I provide the foundation of the molecular "participants" and epithelial cell models that play key roles in transepithelial ion transport function of epithelia detailed throughout the rest of this volume.

The second part is entitled "Epithelial Ion Channels and Transporters" and contains seventeen chapters (Chaps. 9–25) in which authors have concentrated their discussion on a particular ion channel or transporter ranging from chloride channels to the Na^+/K^+-ATPase, for example. Generally, the authors have initially provided a broad perspective of the physiology/biology of a particular ion channel or transporter in epithelial tissues, followed by a focused in-depth discussion of the latest physiology, cell biology, and molecular biology of the ion channel/transporter, and then finish their discussion on aspects of pathophysiology and disease.

It will be appreciated following the discussion of the various ion channels and transporters that many of these transport proteins are potential pharmacological targets for possible treatment of disease. Therefore, the third part is entitled "Pharmacology of Potassium Channels" that consists of two chapters (Chaps. 26 and 27) that provide the latest developments on the pharmacology of calcium-activated potassium channels and small-molecule pharmacology of inward-rectified potassium channels. It should be noted, however, that pharmacological information about various ion channels and transporters is also provided in some of the chapters found within Part II of this volume.

Finally, the last part in the book is entitled "Diseases in Epithelia" and consists of two chapters (Chaps. 28 and 29). These chapters are designed to bridge the basic cellular models and epithelial transport functions discussed throughout this volume with a compelling clinical perspective, from bench to bedside. In these chapters, Dr. Whitcomb discusses the role of ion channels and transporters in pancreatic disease, while Dr. Ameen and her colleagues similarly provide insight into the secretory diarrheas.

Our utmost goal, with this book, was to provide a comprehensive and authoritative volume that encapsulates the most recent research findings in the basic physiology of ion channels and transporters of molecular diseases from the laboratory bench top to the bedside. Additionally, we hope that the book will be very exciting and useful to a range of readers from students to research scientists providing a wealth of up-to-date research information in the field of epithelial ion channels and transporters in health and disease.

The undertaking of a book of this scale would always be a "mountain" of work. We want to give our heartfelt thanks to all of our authors who have taken time from their very busy work and nonwork schedules to provide excellent chapters, which provided depth of knowledge, informative description, and coverage of basic physiology and pathophysiology of the topic of their particular chapters.

We want to thank Dr. Dee Silverthorn who planted the "seed" that developed into this volume, which stemmed from a featured topic session entitled "Ion Channels in Health and Disease" held during the Experimental Biology meetings in Boston in April 2013 (chaired by KLH). We thank the members of the American Physiology Society (APS) Book Committee who had faith in us to pursue such an exciting book.

As with any book, this volume would not have been possible without the excellent partnership between the APS and Springer-Verlag and the publishing team at Heidelberg, Germany (Britta Mueller, Springer Editor, and Jutta

Lindenborn, Project Coordinator). We wish to thank Portia Wong, our Developmental Editor at Springer+Business Media (San Mateo, CA), and her team who assisted with the early stages of the publishing process that greatly added to this contribution. Finally, a special thanks to Shanthi Ramamoorthy (Production Editor, Books) and Ramya Prakash (Project Manager) of Publishing – Springer, SPi Content Solutions – SPi Global and their production team who assisted us throughout the final stages of the publication of our book.

Finally, we want to thank our mentors Douglas C. Eaton and the late Dale J. Benos for KLH, Michael E. Duffey and Raymond A. Frizzell for DCD, and our colleagues who guided us over the years to be able to undertake this volume.

Dunedin, New Zealand — Kirk L. Hamilton
Pittsburgh, PA — Daniel C. Devor
June 2015

Contents

About the Editors

Kirk L. Hamilton was born in Baltimore, Maryland, in 1953. He gained his undergraduate (biology/chemistry) and M.Sc. (ecology) degrees from the University of Texas at Arlington. He obtained his Ph.D. at Utah State University under the tutelage of Dr. James A. Gessaman, where he studied incubation physiology of barn owls. His first postdoctoral position was at the University of Texas Medical Branch in Galveston, Texas, under the mentorship of Dr. Douglas C. Eaton where he studied epithelial ion transport, specifically, the epithelial sodium channel (ENaC). He had further postdoctoral training at the University of Alabama, Birmingham, under the supervision of the late Dr. Dale J. Benos where he further studied ENaC and nonspecific cation channels. He took his first academic post in the Department of Biology at Xavier University of Louisiana in New Orleans (1990–1994). Since arriving in the Department of Physiology at the University of Otago in 1994 where he is currently senior lecturer (top step above the bar, UK system), he has focused his research on the molecular physiology and trafficking of potassium channels (specifically KCa3.1). He has published more than 50 papers and book chapters. Dr. Devor and he have been collaborators since 1999. When he is not working, he enjoys playing guitar (blues and jazz) and volleyball. Kirk is married to Judith Rodda, a recent Ph.D. graduate in spatial ecology. They have two children, Nathan (b. 1995) and Emma (b. 1998).

Daniel C. Devor was born in Vandercook Lake, Michigan, in 1961. His education took him through Southampton College of Long Island University, where he studied marine biology, before entering SUNY Buffalo for his Ph.D., under the guidance of Dr. Michael E. Duffey, where he studied the role of basolateral potassium channels in regulating transepithelial ion transport. He subsequently did his postdoctoral work at the University of Alabama, Birmingham, under the mentorship of Dr. Raymond A. Frizzell, where he studied both apical CFTR and basolateral KCa3.1 in intestinal and airway epithelia. He joined the University of Pittsburgh faculty in 1995 where he is currently a professor of cell biology. During this time, he has continued to study the regulation, gating, and trafficking of KCa3.1 as well as the related family member, KCa2.3, publishing more than 50 papers on these topics. When not in the lab, he enjoys photography and growing exotic plants. Dan is married to Catherine Seluga, an elementary school teacher. They have three children, Caitlin (b. 1990), Emily (b. 1993), and Daniel (b. 1997).

Part I
Basic Epithelial Ion Transport Principles and Function

Chapter 1
Fundamentals of Epithelial Cl^- Transport

Bruce D. Schultz and Daniel C. Devor

Abstract While epithelial solute transport predates recorded history, our understanding of epithelial function has risen from the most basic level only recently. This chapter provides a historical perspective of epithelial electrophysiology and an initial foundation for much of the information contained in this volume. Epithelial cell models are presented in their contemporary contexts to demonstrate the philosophical breakthroughs that they heralded along with the novel techniques that made them possible. The text touches on the roles that unique physiological systems such as eel gill, frog skin, rabbit intestine, and cultured cell lines have contributed to our understanding. Two examples of diseases associated with abnormal Cl^- transport, cholera diarrhea and cystic fibrosis, are discussed, and underlying mechanisms that contribute to the pathology are identified. A hypothetical cell model with the minimal complement of transport proteins that are required for Cl^- secretion (Na^+/K^+-ATPase, $Na^+/K^+/2Cl^-$ cotransporter, K^+ channel, Cl^- channel) along with their required localization to the mucosal (apical) or serosal (basolateral) membrane is presented. Selected examples of these transport mechanisms are presented and discussed in the light of their discoveries, biophysical characteristics, pharmacology, genetic identities, and their molecular partners. A recently published comprehensive cell model is presented as the climax of the chapter that sets the stage for distinct components that are presented in greater detail throughout this volume.

Keywords Cl^- secretion • CFTR • K^+ channels • Na^+/K^+-ATPase • TMEM channels • Epithelia • Cotransporters

---

B.D. Schultz (✉)
Department of Anatomy and Physiology, Kansas State University, 1620 Denison Avenue, 127 Coles Hall, Manhattan, KS 66506, USA
e-mail: bschultz@vet.ksu.edu

D.C. Devor (✉)
Department of Cell Biology, School of Medicine, University of Pittsburgh, S331 BST 3500 Terrace St, Pittsburgh, PA 15261, USA
e-mail: dd2@pitt.edu

K.L. Hamilton, D.C. Devor (eds.), *Ion Channels and Transporters of Epithelia in Health and Disease*, Physiology in Health and Disease,
DOI 10.1007/978-1-4939-3366-2_1

1.1 Introduction

Fluid and electrolyte balance is critical for life whether viewed at the level of an individual cell, tissue, organ, or individual. The physiology or pathology associated with fluid intake or fluid loss has been appreciated since antiquity, although not always well managed to achieve proper health—the contribution of active solute transport to fluid movement simply was not appreciated. Although Charles Darwin (1839) did not attribute solute transport functions to any particular tissues, he wrote that Galapagos tortoises captured near lowland lagoons had urinary bladders that were distended with dilute fluid that was said to decrease in volume gradually and to become less pure (i.e., solutes became more concentrated) as animals were captured in highlands more distant from the lagoons. As early as the mid-nineteenth century, oral rehydration therapy was promoted as a treatment for cholera diarrhea along with the admonition that the "... water for drinking must be pure" (Johnson 1866) and especially free from sewage, providing a premonition to the role of bacteria and bacterial toxins in the regulation of epithelial solute and fluid transport (Snow 1849, 1855). Notable physiologists of the nineteenth century such as Ludwig Thiry, Luis Vella, Rudolf Heidenhain and Ivan Pavlov studied epithelial fluid secretion and its regulation without appreciating the contributions of solute transport to fluid flux. Cellular mechanisms that account for solute (e.g., ions, sugars, amino acids, etc.) transport were described initially in the mid-twentieth century, and these mechanisms are a focus of ongoing research. Additional mechanisms continue to be discovered. New arrangements of well-defined and novel mechanisms, which account for solute and fluid flux in distinct tissues or situations, continue to be described. Their identities, orientations, distributions, regulation and functions are the topics addressed in this volume.

All cells live in an extracellular fluid milieu and they maintain an intracellular environment that allows for cytosolic function. Within the vertebrate body, epithelial cells separate body compartments that can contain variable volumes and diverse solute compositions. The generation and maintenance of distinct fluid compositions in these extracellular body compartments are critical to the overall health of the individual and epithelial cells are responsible for moving solutes selectively from one compartment to another. Specialization is observed throughout the body. Gastric secretions are acid, while pancreatic secretions are alkaline and have substantial buffering capacity. Airway surface fluid must be maintained with an appropriate thickness (i.e., volume), viscosity, and pH to allow for mucociliary clearance. Cells lining the inner ear generate the endolymph that is required for hearing (Wangemann 2002, 2006). Mammary epithelia generate milk for a period following gestation, although there are wide variations in milk composition over time and across species. Abnormal epithelial function is associated with both male and female infertility (Denning et al. 1968; Gervais et al. 1996; Johannesson et al. 1998; Hoglund et al. 2006). The examples are nearly endless. Tightly regulated epithelial transport and barrier functions contribute to virtually all aspects of bodily function.

This overview of epithelial Cl^- transport will present information from a historical perspective and will focus on concepts and techniques that have played critical roles in developing the knowledge base. Often in this history, breakthroughs resulted from the development of novel techniques or novel applications of proven techniques. Examination of background material has not been exhaustive, although care is taken to identify many notable contributions. A portion of the pumps, transporters, and ion channels that are described in detail in this volume will be introduced and simple arrangements of these components that can account for observed phenomena will be shown.

1.2 Active Cl^- Transport

Research to examine the movement of Cl^- across an epithelium was initiated by August Krogh and his colleagues at the University of Copenhagen in the 1930s. Prior to this time, the characterization of epithelial transport was limited by analytical techniques that, by current standards, have very low precision. For example, gastric and pancreatic secretion characteristics were studied with the use of anionic or cationic dyes as surrogate markers (Ingraham and Visscher 1935). George de Hevesy, however, initiated the use of isotopes to study basic questions in physiology and metabolism (Hevesy and Hogfer 1934; Hahn et al. 1937, 1939). Krogh, who worked closely with de Hevesy, was quick to appreciate the advantages of these new tools and actively promoted the use of isotopes (which they obtained from Niels Bohr) to characterize metabolic pathways (Hevesy et al. 1935; Krogh 1937b). Work with frogs provided strong evidence that Cl^- could be taken up from dilute bathing solutions along with Na^+ and K^+ and could be concentrated to a 10,000-fold gradient across the skin (Krogh 1937a). Ca^{2+} uptake was not detectable with this preparation and in the presence of $CaCl_2$, the uptake of Cl^- was associated with bath alkalinization, which provided initial evidence for Cl^-/base exchange. In retrospect, these observations are quite exciting, but at the time the author stated that “The present investigation raises the question ... regarding the biological significance of this power [i.e., Cl^- uptake]” (Krogh 1937a). Nonetheless, Krogh extended his observations across phyla and reported active anion and cation transport across the skin or gills of frogs, goldfish, crayfish, and dragonfly larva (Krogh 1939). Ion transport was energy dependent and could be poisoned by cyanide. Importantly, the results showed that the anion uptake mechanism was selective for Cl^- or Br^-, but virtually excluded I^-, NO_3^-, and SCN^- (Krogh 1939). After comparing the uptake characteristics across this diverse selection of animals, Krogh concluded that cells accounting for transport are unknown, but reside in frog skin, in fish gills, and likely in the intestinal wall. He further speculated that these processes are likely regulated by nerves.

Hans Ussing, a student and colleague of August Krogh in Copenhagen, provided the next major breakthrough in the field of epithelial ion transport. Through the 1940s, isotopic tracers were used to characterize the distribution and turnover of ions in the body and the movement of ions into and out of cells. At the same time, measurements were being made to determine the electromotive forces across cell membranes in various physiological and nonphysiological buffers. Eventually, experiments were initiated to examine the movement of ions across isolated epithelial layers, specifically frog skin, and it was determined that some components of this transport could be modeled as passive diffusion while other aspects of transport required energy utilization to move ions against their electrochemical gradients. Importantly, Ussing built on earlier reports (Francis 1933) and introduced a means to separate the active transport of ions across epithelial cells from the passive movement driven by the naturally occurring transepithelial electrical potential—it was known that the isolated frog skin generated a transepithelial electrical potential, which was serosa positive (Ussing 1949a). Ussing's technique, the "Ussing chamber," coupled a standard flux chamber with external electrical circuitry to nullify the transepithelial voltage by the injection of electrical current (Ussing and Zerahn 1951). The system has great sensitivity, being able to resolve active transport of 1 μA on a minute-to-minute basis or 0.62 nanoequivalents/min. However, the system is blind to electroneutral processes and to charge carriers (e.g., Na^+ absorption cannot be differentiated from K^+ absorption or from Cl^- secretion). Thus, less sensitive complimentary systems such as radiotracer flux must be used to define the charge carrier(s). Figure 1.1 shows that the frog skin transmural potential was attributed to net active Na^+ absorption and could contribute a driving force for the net flux of other ions (Ussing and Zerahn 1951). Ussing appreciated that there were significant mucosal-to-serosal and serosal-to-mucosal fluxes with a smaller net absorptive flux. Na^+ could be absorbed from the external environment, accumulate in the epithelial cells, and then be transferred to the basolateral or blood side of the epithelium, showing quite clearly that both the apical and the basal cell membranes play roles, and likely unique roles, in ion transport (i.e., the two-membrane model) (Koefoed-Johnsen and Ussing 1958). Importantly, in their initial studies, the authors fully accounted for short-circuit current, that is, the current required to maintain the transmural voltage at zero, with the absorption of

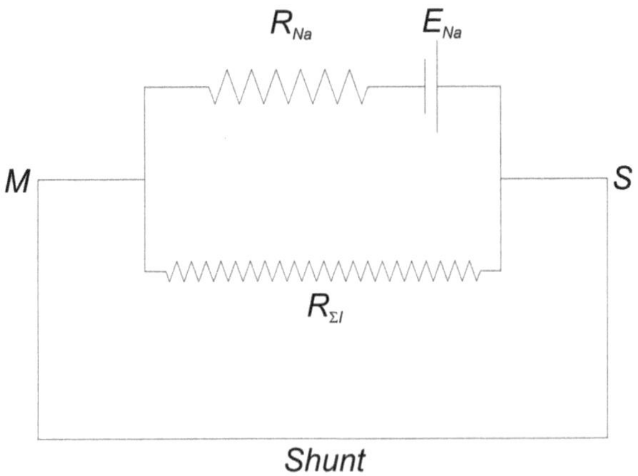

Fig. 1.1 Equivalent circuit representing short-circuited frog skin. E_{Na}: electromotive force of the sodium-transporting mechanisms. R_{Na}: resistance to the sodium current. $R_{\Sigma I}$: resistance to passive ions. *Shunt* represents the net effect of the applied electromotive force between mucosa (*M*) and serosa (*S*) [Modified from Ussing and Zerahn (1951)]

Na^+. Only in the presence of adrenalin did the authors express a concern that the active transport of other ions might be implicated in the response (Ussing and Zerahn 1951). Indeed, in follow-up papers, the authors concluded that there was no net Cl^- flux in basal conditions, but that adrenalin stimulated net Cl^- secretion (Koefoed-Johnsen et al. 1952a, b). At the same time, Adrian Hogben, another of Ussing's colleagues, reported active Cl^- secretion by frog gastric epithelium (Hogben 1951). Additional evidence for active gastric Cl^- secretion was provided in a follow-up paper that also suggested the possibility of Cl^-/HCO_3^- exchange (Hogben 1955). Perhaps one of the most important conclusions to come from early studies of frog skin is an initial cell model that attributes distinct functions to the mucosal (facing the environment) and serosal (facing the blood) membranes. Figure 1.2 depicts a cell in which Na^+ enters passively across the apical membrane moving down its electrochemical gradient, whereas K^+ passively exits the cell across the basal membrane. In this model, it is critical that the apical membrane is virtually impermeant to K^+ and that the basal membrane is virtually impermeant to Na^+. The authors invoked a Na^+/K^+ pump in the basal membrane to complete transepithelial Na^+ transport. That a Na^+/K^+ pump(s) exhibited stoichiometric linkage had been postulated for cephalopod nerves (Hodgkin and Keynes 1955), and it had been reported quite recently, at that time, that the enzyme derived from crab nerves required the presence of both Na^+ and K^+ for Mg^{2+}-dependent ATP hydrolysis to proceed (Skou 1957). This minimal cell model (Fig. 1.2) accounted for vectorial active transepithelial ion movement and set the stage for subsequent critical experiments to establish a role for anion secretion in health and disease. For a more thorough treatment of the fundamentals of Na^+ absorption across epithelia, the reader is directed to Chap. 2 in this volume.

Initially, Ussing concluded that there was not a reason to suppose that Cl^- was actively transported by the epithelium—evidence supported the conclusion that Cl^- moved down the electrical potential established by active Na^+ transport, at least in basal conditions (Ussing 1949b). However, work by Hogben (1951, 1955) and other

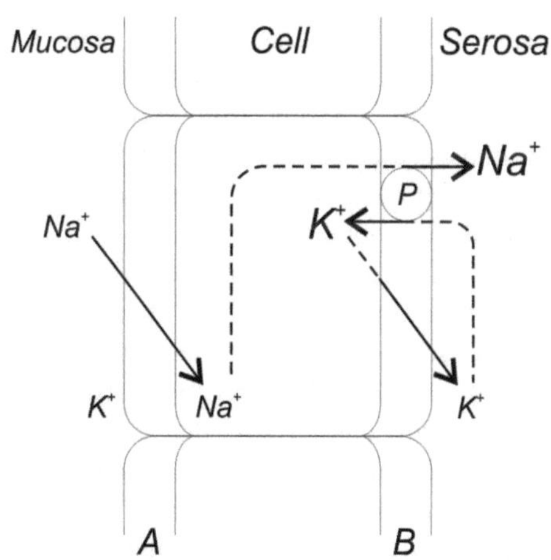

Fig. 1.2 A cell model illustrating the initial hypothesis that the frog skin potential resulted from the passive permeabilities of the apical (A) and basal (B) cell membranes to Na^+ and K^+, respectively, along with the Na^+/K^+ pump (P) in the basal membrane [Modified from Koefoed-Johnsen and Ussing (1958)]

subsequent reports provided substantial evidence for both Cl^- absorption and Cl^- secretion by epithelia, although underlying mechanisms for active transport in either direction were not defined (Koefoed-Johnsen et al. 1952b; Curran and Solomon 1957; Zadunaisky et al. 1963; Zadunaisky 1966; Finn et al. 1967). Experiments were conducted to test for solvent drag that might accompany hydrostatically driven fluid flux, a commonly hypothesized mechanism to account for solute flux. Hogben, however, was able to rule out this possibility and build a solid case for active electrogenic Cl^- secretion. Although rudimentary by today's standards, two important mechanisms were postulated to contribute to active transport in addition to Na^+/K^+ ATPase. The first is "exchange diffusion," which was introduced initially to account for Na^+/H^+ exchange across gastric mucosa (Teorell 1939). As stated above, exchange diffusion was implicated in the uptake of Cl^- across frog skin and to account for gastric secretion (Krogh 1937a; Ussing 1949b; Hogben 1955). A second salient concept, cell membrane pores that allow the movement of solutes and/or solvents down electrochemical gradients, was introduced in the 1930s (Ingraham and Visscher 1935; Curran and Solomon 1957). These fundamental mechanisms—channels, exchangers, and ATP-dependent pumps—are now well recognized, and myriad hypothetical arrangements of these (along with cotransporters that were described later) have been proposed to account for net flux across epithelia. However, cell models incorporating these components were not postulated until much later.

Although a variety of other techniques continued to be used, the Ussing-style flux chamber was adopted rapidly and was modified for the evaluation of tissues from a number of sources. In general, initial experiments focused on freshly excised epithelial sheets such as frog skin, toad bladder (Crabbe 1961), and rabbit ileum (Schultz et al. 1964). In each of these cases, Na^+ absorption could be demonstrated readily, but anion transport proved to be more elusive. Using open circuit conditions and other approaches, Cl^- absorption could be observed, but the possibility that this was secondary to active Na^+ transport, with Cl^- moving as a passive countervalent ion, could not be ruled out. By incorporating measurements of transepithelial electrical potential, rat ileal perfusion with radiotracer fluxes was used to conclude that both Na^+ and Cl^- were actively absorbed (Curran and Solomon 1957) in much the same way that Krogh had characterized uptake across the skin of a living frog two decades earlier (Krogh 1937a). Subsequently, Zadunaisky et al. (1963) provided compelling evidence for Cl^- absorption in Ussing chamber studies. Short-circuit current across frog skin in standard Ringer solution did not agree with measured $^{24}Na^+$ flux (i.e., there was a residual current), but in Cl^--free Ringer solution, the values for $^{24}Na^+$ flux and short-circuit current concurred. Subsequent $^{36}Cl^-$ flux experiments were used to account for the initial discrepancy observed in standard Ringer solution and thus to demonstrate active Cl^- absorption. The underlying reason for the inequality between Na^+ flux and short-circuit current was that the concurrent absorptive flux of Na^+ and Cl^- is electroneutral and therefore "invisible" to the Ussing chamber voltage-clamp system. Importantly, ouabain, a glycoside ATPase inhibitor, reduced both Na^+ and Cl^- absorption, which suggested that these were active transport processes.

Although it was not appreciated at the time, the outcomes revealed the activity of dual transport pathways. In addition to a ouabain-sensitive electrogenic Na^+ absorption mechanism, which had been observed in a number of tissues, the new results suggested that an additional process was expressed, either stoichiometric NaCl absorption or concurrent electrogenic Cl^- absorption. In this era, authors often speculated about the activity of a "Cl^- pump."

Other experiments examining basal transport properties revealed evidence for epithelial Na^+ and/or Cl^- secretion. Specialized tissues such as avian salt glands (Schmidt-Nielsen et al. 1958), spiny dogfish rectal glands (Burger and Hess 1960), and tear or nasal glands from loggerhead turtles and marine iguanas (Schmidt-Nielsen and Fange 1958) were reported to secrete fluids with Na^+ and Cl^- concentrations ranging from 500 to nearly 1000 mEq/l. Since these concentrations are substantially greater than plasma, one would conclude that an energy-consuming ion transport process must be involved for at least one of these ions. Since the lumen in an actively secreting gull nasal gland was positive relative to the interstitium, it was concluded that the active component of secretion was likely Na^+ and that Cl^- moved passively as a countervalent ion (Thesleff and Schmidt-Nielsen 1962). Such a conclusion was consistent with most previous reports that failed to gain evidence for active Cl^- transport in basal conditions. A distinct contrast, however, was shown when an Ussing-style flux chamber was developed to study frog cornea, which exhibited an aqueous humor-positive spontaneous potential. Secretory Cl^- fluxes were of the same magnitude as the short-circuit current. Furthermore, the short-circuit current was absent when Cl^- was removed, the magnitude of the current was a saturating function of the Cl^- concentration in the bathing medium, and no net Na^+ flux was detected in these studies (Zadunaisky 1966). Thus, unequivocal evidence for active Cl^- secretion by an unstimulated epithelium was obtained.

The stage was set to characterize underlying mechanisms to account for active epithelial Cl^- transport. However, many obstacles remained—some conceptual and some practical. Ussing's early work introduced the "two-membrane" model that attributed distinct functions to the mucosal (apical) and serosal (basal) cell membranes (Fig. 1.2) (Ussing 1949a; Ussing and Zerahn 1951). Yet, throughout the 1960s and sometimes beyond, conceptual models to describe ion transport were reduced to a single barrier, which confounded activities occurring at the serosal and mucosal membranes—a complete cell model to account for Cl^- movement would not be seen routinely until the 1970s (Linzell and Peaker 1971c; Schultz and Frizzell 1972; Wright 1972). Some experiments had been conducted to assess "sidedness" of buffer composition, but the concept of separate cell "loading" and "exit" mechanisms was seldom seen. Among the more pressing of the practical issues was to identify simple systems that expressed only one predominant transport process, which would allow it to be studied in relative isolation. Many of the early studies examining Cl^- transport were conducted in "basal" or unstimulated conditions and/or in complex tissues, which prevented the detection of some processes. This, however, was soon to change.

1.3 Cl^- Transport Regulation

The first indications that active Cl^- transport could be regulated acutely came from early Ussing chamber studies. In frog skin, where electrogenic Na^+ absorption predominates, Koefoed-Johnsen et al. (1952a; 1952b) reported adrenalin-induced Cl^- secretion, supporting Krogh's prediction that ion transport was likely regulated by nerves (Krogh 1939). The authors concluded that mucus gland cells likely accounted for this phenomenon. This conclusion was built on earlier reports suggesting that subpopulations of cells in complex tissues could have specialized functions. For example, Keys and coworkers reported that specialized Cl^--secreting cells at the base of eel gill leaflets were responsible for Cl^- secretion in a marine environment, but not in freshwater (Keys 1931a; Keys and Willmer 1932). An in vitro system to assess eel gill permeability was developed with the goal of defining acute or chronic changes that occurred when animals transitioned between seawater and freshwater and that showed high energy consumption with substantial Cl^- secretion in the marine, but not in the freshwater environment (Keys 1931b). Histological appearance was used to implicate this cell population (Keys and Willmer 1932). Although not focused on Cl^- secretion, Wright and coworkers (1940) (reviewed by Florey et al. (1941)) reported that sympathetic stimulation inhibited and that cholinergic agonists stimulated feline intestinal fluid secretion and that atropine and some anesthetics such as barbiturates blocked these effects, which again highlighted the role of nerves in the regulation of epithelial functions. These observations reinforced the thought that epithelial cells were responsible for electrolyte and fluid transport and demonstrated that transport processes were induced, modified, or regulated by extracellular signals or conditions.

1.3.1 Cholera

The global cholera pandemic of 1961–1970 fueled substantial interest in the control of epithelial secretion and ultimately provided compelling evidence for the role of epithelial ion transport in the disease process (Pierce et al. 1971a). Asiatic cholera was first described by Westerners in India in the late eighteenth century and first received general attention following the pandemic of 1817–1824, which reached from India to China and to the shores of the Mediterranean. In what proved to be one of the great epidemiological studies of all time, John Snow described the role played by contaminated water or food in the transmission of cholera (Snow 1849, 1855). Snow reported the rapid onset of explosive watery diarrhea that might be preceded by malaise and abdominal cramps, but typically was not accompanied by fever or other outward signs of illness. Cholera-induced intestinal fluid secretion is impressive by any standard. As a localized epidemic occurred in Bangkok in 1958, Watten et al. (1959) documented 24 h "rice-water" stool volumes for adult patients of up seventeen liters! Cholera diarrhea is a classic example of pathological

Cl^- and/or HCO_3^-secretion. Importantly, delineation of cholera etiology provided both the impetus and the framework to define mechanisms regulating Cl^- secretion.

Electrophysiological studies examining intestinal anion transport initially were focused primarily on basal or unstimulated conditions, and net Cl^- transport was difficult to detect in these conditions (Clarkson and Toole 1964; Schultz et al. 1964). It was known that nerve stimulation induced feline intestinal fluid secretion, but the underlying mechanism(s) was unknown (Wright et al. 1940; Florey et al. 1941). Likewise, inoculation with either *Vibrio cholera* or media containing cholera toxin was known to induce intestinal fluid accumulation or diarrhea in rabbits, although fluid accumulation was not observed consistently in other species (De and Chatterje 1953; Dutta and Habbu 1955). Competing hypotheses included that cholera toxin inhibited the Na^+/K^+-ATPase or that it caused an increase in villus capillary permeability (De and Chatterje 1953; Finkelstein et al. 1966; Pierce et al. 1971b). Ultimately, rabbit ileal loop preparations (Burrows and Musteikis 1966; Leitch and Burrows 1968) and canine Thiry-Vella intestinal loops (Carpenter et al. 1968; Pierce et al. 1971a) were used to demonstrate that luminal inoculation led to fluid secretion. Seminal observations were reported by Field, Greenough, and coworkers, who showed that the dialyzed filtrate from a *Vibrio cholera* culture induced an increase in short-circuit current that was accompanied by an increase in Cl^- secretion across isolated rabbit ileum (Field et al. 1969; Greenough et al. 1969). The results showed quite clearly that cholera toxin caused intestinal Cl^- secretion and suggested that fluid secretion was osmotically driven. The authors built on an earlier report, which showed that exogenous cAMP or theophylline, a phosphodiesterase inhibitor that was used to block the breakdown of endogenously produced cAMP, caused a sustained increase in short-circuit current (Field et al. 1968). Exposure to cholera toxin substantially reduced the effect of theophylline, suggesting that a similar mechanism may be involved in both pathways. At the same time, work with canine Thiry-Vella loops showed that exogenous cAMP, prostaglandin E1, theophylline, or cholera toxin could be used to stimulate fluid secretion of 0.5 to 1 ml cm^{-1} h^{-1}, the authors speculating that the key common mediator was cytosolic cAMP. Subsequently, Sharp and coworkers (Chen et al. 1971; Sharp and Hynie 1971; Sharp et al. 1971), using rabbit and human intestinal tissue homogenates, and Kimberg et al. (1971), using guinea pig and rabbit intestinal tissues, showed that cholera toxin caused a substantial and prolonged increase in adenylyl cyclase activity while having no detectable effect on phosphodiesterase. Importantly, it was shown that prostaglandins produced their effects when exposed to the serosal side of the cells, whereas cholera toxin was effective only from the mucosal aspect, but that both were mediated by cytosolic adenylyl cyclase activity. Additional work would be required to fully describe the links between extracellular signaling molecules and the modulation of epithelial anion secretion. However, this flurry of reports in the late 1960s and early 1970s demonstrated that a variety of extracellular signaling molecules including vasopressin, prostaglandins and microbial toxins could be used to upregulate anion secretion, and it appeared that cAMP was a common cytosolic agent mediating the responses. The reader is directed to Chap. 29 for a more thorough description of the molecular mechanisms underlying the secretory diarrheas.

1.3.2 Crypts Are the Site of Intestinal Fluid Secretion

The studies outlined above demonstrate that, given appropriate stimulation, particular intestinal segments are capable of secreting vast quantities of fluids and electrolytes; however, these studies did not identify the cell type(s) that contribute to electrolyte secretion, which was a long-standing question. Florey et al. (1941) pointed to histological publications from the nineteenth century and indicated that “It has not, for instance, been settled whether the crypts secrete and the villi absorb. . ..” Early evidence that the crypts were the site of fluid secretion was obtained by showing that they dilated in response to pilocarpine injection into human subjects (Trier 1964). Isaacs et al. (1976) observed a greater density of nerve endings surrounding the crypts compared to the surface cells, suggesting they were the cells responding to neurotransmitters. Stronger evidence was provided by a number of studies showing that selectively damaging either surface or crypt cells with Na_2SO_4 or cycloheximide, respectively, resulted in an inhibition of cholera toxin- or acetylcholine-induced short-circuit current and Cl^- secretion only when the crypts were damaged (Serebro et al. 1969; Roggin et al. 1972; Browning et al. 1978). Ultimately, Welsh et al. (1982b) demonstrated that the crypts were the site of fluid secretion by (i) directly observing the secretion of fluid using phase contrast microscopy and (ii) using electrophysiological methods to directly demonstrate a change in the crypt cells’ potential profile.

1.4 Initial Cell Models for Cl^- Transport

The development of hypothetical cell models to account for ion fluxes was moved forward by the growing appreciation that the nature of particular tissues, or that the physiological or pathophysiological state of these tissues, impacts ion transport. Early studies with the Ussing-style flux chamber demonstrated net Cl^- secretion by frog gastric mucosa (Hogben 1955) and by frog skin that was exposed to adrenalin (Koefoed-Johnsen et al. 1952b). However, these observations predate the two-membrane epithelial model depicted in Fig. 1.2, and no attempt was made to place Cl^- secretion in the context of a cell system, *per se*. Throughout the period from the 1950s through the early 1970s, a number of observations were made that ultimately were essential in developing a cell model to account for Cl^- secretion. Krogh (1937a) and Hogben (1955) suggested the likelihood of Cl^-/HCO_3^- exchange, which was supported by later observations (Hubel 1967, 1969) (reviewed by Keynes (1969)). Curran and Solomon (1957) suggested the presence of membrane pores—although the concept of ion channels had not yet appeared in this literature. Pharmacological agents began to be used to characterize overall transport processes. Ouabain became widely used to test for dependence on Na^+/K^+-ATPase activity (Bonting 1966). Ethacrynic acid, furosemide, and other sulfonamides began to be used as in vitro diagnostic tools, although the underlying mechanisms of action were, at best, speculative. Assessments with these agents tended to focus on changes in Na^+ transport that were attributed to Na^+/K^+-ATPase inhibition (Hook and Williamson 1965; Walker et al. 1967; Bank 1968), although this mechanism was soon ruled out (Bourgoignie et al. 1969). It was also learned that, in some tissues, Cl^- secretion was totally or partially dependent on the presence of Na^+, although Na^+ was not actively transported, which foreshadowed the discovery of electroneutral cotransporters that harnessed the electrochemical potential of one solute, typically Na^+, to drive the uptake of another solute (Zadunaisky 1966; Field et al. 1968) (see below and see Chaps. 10–13). There was wide speculation regarding the expression of a "Cl^- pump(s)" as highlighted in the exhaustive review by Keynes (1969). The stage was set to build a cell diagram with components that could account for active Cl^- transport, but additional electrical driving forces still had to be defined.

Membrane potential and concentration gradients play critical roles in the movement of ions across mucosal or serosal cell membranes and transepithelial electrochemical potentials drive ions through the paracellular pathway. Numerous laboratories impaled epithelial cells with microelectrodes to measure both transmembrane and transepithelial potentials in various conditions while also determining cytosolic ion concentrations (Giebisch 1958; Edmonds and Nielsen 1968; Shoemaker et al. 1970; Rose and Schultz 1971). It was noted that the predominant monovalent ions, Na^+, K^+, and Cl^-, were not distributed across cell membranes in accordance with measured membrane potentials. Typically, there was a substantial inwardly directed electrochemical gradient for Na^+, while both K^+ and Cl^- exhibited outwardly directed electrochemical gradients of lesser magnitudes (for reviews, see Schultz and Frizzell (1972) and Frizzell and Duffey (1980)). Rose and Schultz (1971) proposed an electrical equivalent circuit that showed electrical coupling between the mucosal and serosal membranes and implicated a conductive pathway for Na^+ in the mucosal membrane. Serosal Na^+/K^+-ATPase activity could account for the outward movement of Na^+ and the inward movement of K^+ against their electrochemical gradients and for the distribution of these cations across the cell membrane. A critical aspect of this electrical model was to incorporate a "shunt" resistance representing the paracellular pathway. The model was particularly robust because the inclusion of the shunt resistance allowed the model to account for differences in serosal membrane potential between "tight" epithelia, such as frog skin, and "leaky" epithelia, such as rabbit ileum, as was demonstrated nicely by mathematical derivation in a subsequent letter (Schultz 1972).

In a series of papers and an elegant review, Linzell and Peaker (1971a, b, c, d) determined serosal, cytosolic, and mucosal ion composition for guinea pig mammary epithelium and the electrical potential across the mucosal and serosal membranes. Based on these observations, ion transport mechanisms were predicted, as shown in Fig. 1.3. The authors speculated that the mucosal membrane was freely

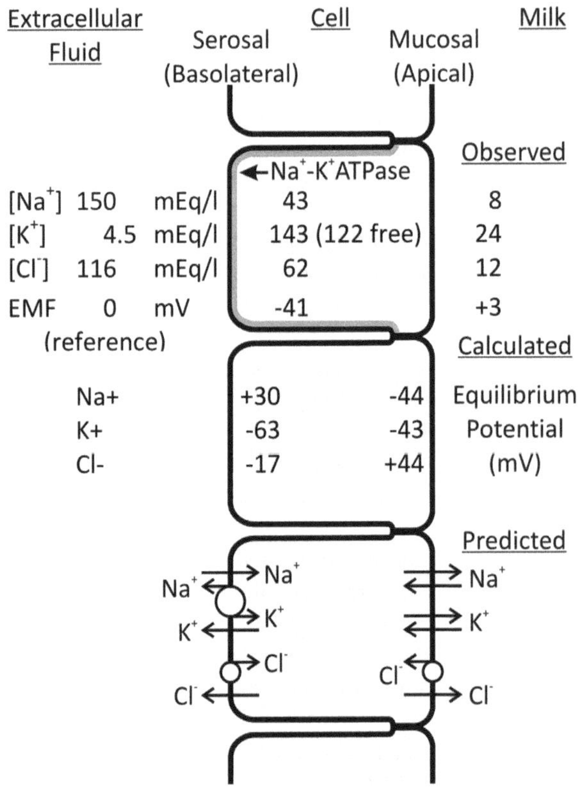

Fig. 1.3 Distribution of ions in mammary tissue of guinea pig. *Top*: measured concentrations in extracellular fluid and milk and calculated intracellular concentrations from an analysis of whole tissue and extracellular and milk spaces (Linzell and Peaker 1971b) and measured membrane potentials (Evans et al. 1971). Distribution of Na^+-K^+-ATPase is taken from Kimura (1969). *Middle*: equilibrium potentials for each ion across apical and basal membranes calculated from Nernst equation. *Bottom*: scheme that could account for observed distribution of ions and membrane potentials. Active transport mechanisms probably exist for Na^+, K^+, and possibly for Cl^- on basolateral membrane, but possibly for Cl^- only on apical membrane [Modified from Linzell and Peaker (1971c)]

permeable to both Na^+ and K^+—both cations were distributed such that their Nernst potential was equal to the measured potential across this membrane. The authors surmised that activity of the serosal Na^+/K^+-ATPase could account for the concentration of these cations in the cytosol and by extension in the milk. Cl^-, however, was not distributed according to its Nernst potential at either membrane and mechanisms had to be envisioned that could account for Cl^- entry into the cell against an electrochemical gradient. The authors speculated, as others had in reference to other epithelia (Zadunaisky and Lande 1971), that an active pump for Cl^- uptake might be present in either or both membranes. The model was both novel and groundbreaking because it was able to use biophysical principles and known or predicted transporters to account for differences in mucosal and serosal fluid compositions. A schematic model to account for Cl^- secretion by choroid plexus was published a year later (Wright 1972). This model included new mechanisms such as a putative Na^+/H^+ antiporter in the serosal membrane working in concert with the carbonic acid reaction to account for HCO_3^- entry into the cell. An ion transporter working in conjunction with the carbonic acid reaction was not new, *per se*. Turnberg et al. (1970) previously published a model for human ileal mucosa that linked Na^+/H^+ exchange and Cl^-/HCO_3^- exchange to account for electroneutral uptake of Na^+, Cl^-, and/or HCO_3^- from various ion-substituted buffers. The cell model proposed by Wright (1972) made a philosophical breakthrough by including a subset of known transport mechanisms and by placing the Na^+/K^+-ATPase in the mucosal membrane, demonstrating that components could be rearranged to account for differences in overall transport function between tissues. These models, however, included fundamental shortcomings. The authors included arrows across the serosal and/or mucosal membranes to indicate the movement of each ion, Na^+, K^+, and Cl^-, down (or sometimes against!) their respective electrochemical gradients. However, no permeation pathway is depicted and certainly no regulated pathway is envisioned by their presentations. Furthermore, these models did not extend to address one more recurring theme in epithelial anion transport, the movement of HCO_3^-. Krogh (1937a) showed that Cl^- uptake through frog skin was associated with bath alkalinization, suggesting Cl^-/HCO_3^- exchange, a mechanism that was suggested also for frog gastric mucosa (Hogben 1955). Cholera diarrhea is alkaline (Watten et al. 1959; Leitch and Burrows 1968), and Turnberg et al. (1970) suggested that a Cl^-/HCO_3^- exchange mechanism was expressed in small intestine. Nonetheless, the unveiling of these cell models to account for Cl^- transport was a substantial step forward.

1.5 Cl^- Conductances

Ion permeability pathways across cell membranes were undefined in early schematic models to account for ion transport. That cell membranes exhibited “conductances” to certain ions was long known (Cole and Curtis 1941), and it was demonstrated subsequently that these conductive pathways exhibited selectivity for one ion over another, at least in excitable cells (Hodgkin and Huxley 1952a, b; Hodgkin et al. 1952). It was speculated that epithelial cells, too, included membrane pores that allowed the permeation of solutes (Ingraham and Visscher 1935; Curran and Solomon 1957). However, the structure and mechanics of these pores remained virtually undefined. Ultimately, three distinct electrophysiology-based techniques were developed that contributed substantially to the discovery, kinetic description and localization of permeation pathways through mucosal and/or serosal membranes.

Fluctuation analysis or noise analysis, which was enabled substantially by the development of computer algorithms (Cooley and Tukey 1965), was used initially to describe K^+-selective conductances in nodes of Ranvier (Derksen and Verveen 1966; Verveen et al. 1967). Lindemann and Van Driessche (Lindemann and Van Driessche 1977, 1978; Van Driessche and Zeiske 1980) used a similar approach, albeit with stationary kinetics, to demonstrate the presence of both K^+- and Na^+-conductive pathways across cell membranes in frog skin. The authors demonstrated that amiloride could be used to block Na^+ channels and introduce a new component in the power spectrum that shifted in a concentration-dependent manner (see Chaps. 2 and 18 in this volume). Amiloride was without effect on the K^+ channels identified in this system. An exhaustive review (Van Driessche and Zeiske 1985) demonstrated that this technique was embraced widely to define cation channels in a variety of epithelia. Not until 20 years later was the technique used to describe epithelial Cl^- channels in the mucosal membrane of an immortalized colonic epithelial cell line, T84 cells (Dharmsathaphorn et al. 1984), and the modulation of their activity by novel diarylsulfonylureas (Singh et al. 2004), as will be described further in subsequent paragraphs.

The membrane patch voltage-clamp (i.e., patch-clamp) technique provided a second independent method to define the biophysical characteristics of ion channels. In their Nobel prize winning work, Neher and Sakmann (1976) made an electrically tight seal between a small aperture (i.e., ~10 μm^2) glass pipette and a frog muscle cell membrane to record stochastic deflections in the current records that were consistent with the inward flow of cations. Critical criteria that the authors required to define a membrane channel were that: (1) the amplitude was dependent on membrane potential; (2) the channel amplitude and kinetic rates should be consistent with predictions from fluctuation analysis; and (3) the pharmacology should be consistent with other assay systems (Neher and Sakmann 1976). Surprisingly, the authors did not articulate a requirement for ion selectivity. Nonetheless, the authors reported a cholinergically stimulated 22 pS channel (at 8 °C), which met the fundamental requirements that they had set. The patch-clamp technique was enhanced greatly by a subsequent report from the same laboratory (Hamill et al. 1981). Smaller pipettes (~1 μm^2) that were fire polished and electrically isolated with Sylgard were used a single time to achieve a gigaohm seal with a cell membrane. The authors showed well-resolved openings and closings of single channels in membranes of cells derived from different tissues, demonstrated voltage dependence of channel amplitude, and showed that current could be carried through a particular class of channels by different ions (e.g., Cs^+ and Na^+). Importantly, the report introduced four distinct recording configurations: cell-attached, whole-cell, excised inside-out, and excised outside-out patches, each with its own applications, strengths, and weaknesses. The technique was embraced rapidly for epithelial ion transport studies. Petersen and colleagues (Maruyama and Peterson 1982; Maruyama et al. 1983) reported Ca^{2+}- and voltage-regulated cation channels in pancreatic and salivary acinar cells. Apical membrane Cl^- channels were first reported by Nelson et al. (1984) in A6 cells, which were originally derived from *Xenopus* kidney (Perkins and Handler 1981). The authors reported a very large conductance (360 pS), stilbene-sensitive channel that was ~10-fold selective for Cl^- over Na^+ and remained active over a narrow voltage range (±20 mV).

The perfused duct was a third powerful technique developed to define epithelial ion transport and ultimately this technique played a key role in identifying an apical Cl^- conductance. In 1966, Maurice "Mo" Burg and coworkers (Burg et al. 1966) at the National Institutes of Health reported that they had developed a method to dissect viable rabbit nephron segments and, using concentric pipettes, to perfuse them while measuring net fluid loss from the lumen. A companion paper (Grantham and Burg 1966) showed that the technique could be used to test for effects of vasopressin and cAMP on water and urea permeability. The technique was enhanced to include electrical measurements and cable theory was introduced as a means to assess the outcomes (Burg et al. 1968; Lutz et al. 1973). In subsequent studies, the authors reported collecting duct transepithelial resistance of nearly 1 $k\Omega$ cm^2 (Helman et al. 1971), approximately 40 times greater than the resistance reported for proximal convoluted tubules (Giebisch et al. 1964). The technique was used extensively to define ion transport characteristics in various nephron segments, including active Cl^- absorption in the thick ascending limb of Henle's loop, and to localize the action of furosemide (Burg et al. 1973), a sulfonamide diuretic that was later shown to selectively inhibit a $Na^+/K^+/2Cl^-$ cotransporter (Geck et al. 1980) (see Chap. 11). The assay system was used to screen antracene-9-carboxylate derivatives, which led to the discovery of diphenylamine-2′-carboxylate (DPC) as a Cl^- conductance blocker. DPC immediately was characterized further using the patch-clamp technique on the apical membrane of cells derived from shark (*Squalus acanthias*) rectal gland tubule, which showed that it blocked inwardly directed current carried by Cl^-. Thus, a first pharmacological tool that selectively targeted epithelial Cl^- channels was identified. The study of carboxylates was expanded to screen more than 200 derivatives in this assay system, which led to the identification of 5-nitro-2-(3-phenylpropylamino)-benzoate (NPPB) as a high-affinity (half maximal inhibitory concentration = 80 nM), selective epithelial Cl^- channel blocker (Wangemann et al. 1986). Greger (1985) provided an extensive review focused on the thick ascending limb, which included a comprehensive model to account for Cl^- reabsorption from the nephron lumen (Fig. 1.4). The model is similar to that described by others to account for Cl^- secretion by airway gland cells and shark rectal glands (Nadel and Davis 1978; Frizzell et al. 1981; Welsh et al. 1982a). Importantly, this schematic model accounts for both the entry and exit of all pertinent ions (i.e., no ion is either accumulated or depleted by the depicted mechanisms running in concert), it accounts for the energy source that is required to move ions against their electrochemical gradients, and it shows pharmacological agents that can be used to selectively block each component—the inhibition of any one component will disable the entire system. The ouabain-sensitive ATPase pumps Na^+ out of and K^+ into the cell with a 3:2 stoichiometry, which reduces cytosolic Na^+ concentration and makes the cytosol electronegative relative to the bath (see Chap. 10). The Na^+ electrochemical gradient is harnessed by an electrically silent $Na^+/K^+/2Cl^-$ cotransporter, which is inhibited by

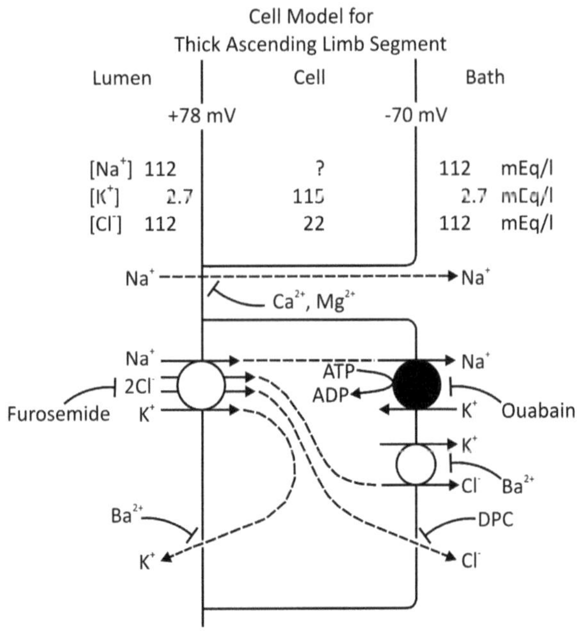

Fig. 1.4 Cell model for thick ascending limb segment. ----, diffusion; carrier; active pump; DPC, diphenylamine-2′-carboxylate [Modified from Greger (1985)]

furosemide, to accumulate Cl^- in the cytosol above its equilibrium potential (see Chap. 11). Both Cl^- and K^+ exit the cell via facilitated diffusion through regulated ion-selective channels. DPC can be used to block Cl^- exit, and Ba^{2+} was shown to block K^+ channels. Pharmacological agents such as these have proven to be immeasurably important when defining contributions of particular components in complex tissues. The identification of additional selective blockers remains a high priority (see Chaps. 24, 26, and 27).

The perfused duct system was used to make the seminal observation that cystic fibrosis (CF), the most common life-shortening recessive genetic disease in Caucasians, is characterized by the apparent absence of an epithelial Cl^--conductive pathway. Cystic fibrosis of the pancreas, first described in the medical literature by Dorothy Andersen (1938), is characterized by steatorrhea, failure to thrive, pancreatic insufficiency, meconium ileus, intestinal obstruction, recurrent pulmonary infections, male infertility and elevated concentrations of salt in sweat. Using sweat ducts derived from CF patients and from healthy controls, Quinton and coworkers (Quinton 1983, 1986; Quinton and Bijman 1983; Bijman and Quinton 1984) showed that patient ducts exhibited a greater lumen-negative transepithelial potential when compared to their unaffected counterparts and that perfusion with solutions having reduced Cl^- concentrations caused transepithelial hyperpolarization in healthy ducts, but was without effect in patient ducts (Quinton 1983). The authors concluded that the fundamental defect in CF was an abnormally low epithelial Cl^- permeability. This conclusion was supported by contemporary observations using freshly excised nasal epithelium and cultured nasal or tracheal epithelial cells, which showed normal expression of amiloride-sensitive Na^+ absorption, but relative impermeability to Cl^- (Knowles et al. 1983; Widdicombe et al. 1985; Yankaskas et al. 1985).

1.5.1 Apical Cl^- Conductances

The stage was set to determine the molecular or biophysical identity of the Cl^- conductance that was apparently absent in CF epithelia. The possibilities included that a Cl^- channel was absent or nonfunctional or that a critical regulatory component was genetically inactivated (Frizzell et al. 1986). Both Welsh (1986) and Frizzell et al. (1986) employed airway epithelial cells with cell-attached and excised inside-out patch-clamp configurations to show the presence of an apical outwardly rectifying Cl^- channel (ORCC) that was observed frequently in cell-attached recordings from healthy individuals, but was not observed in cell-attached patches from patients. Following excision, an outwardly rectifying channel was observed in both CF and healthy tissues. While there were some similarities in the channels defined by these authors, there were substantial differences regarding the biophysical characteristics: e.g., channel size as defined by slope conductance at the reversal potential (20 or 50 pS) and sensitivity to Ca^{2+}. Additionally, Frizzell et al. (1986) reported that the channel displayed subconductance states, while Welsh (1986) reported partial block by anthracene-9-carboxylate and a $Cl^-:Na^+$ selectivity ratio of 6.7:1. These reports set the tone for CF research for several years and numerous other reports describing ORCCs in epithelial apical membranes were published, especially focusing on the activation of these channels by protein kinases and the lack of kinase-dependent activation in CF cells (Hwang et al. 1989; Kunzelmann et al. 1989; Li et al. 1989) (reviewed by Guggino (1993)). As was shown previously with ouabain for Na^+/K^+-ATPase, furosemide for the $Na^+/K^+/2Cl^-$ cotransporter, and amiloride to block a Na^+ conductance, effective and selective blockers can provide compelling evidence for contributions of distinct transporters in different and/or complex ion transport systems. However, few Cl^- channel blockers were available. It was reported that the ORCC was inhibited by NPPB (Kunzelmann et al. 1989) and by 4,4′-dinitrostilbene-2,2′-disulfonic acid (DNDS) (Bridges et al. 1989), although others had reported previously that a related stilbene, 4-acetamido-4′-isothiocyanotostilbene-2,2′-sulfonic acid (SITS) was without effect on Cl^- uptake by normal and CF airway epithelia (Stutts et al. 1985). Bridges and coworkers (Singh et al. 1995) employed voltage-clamped synthetic lipid bilayers with membrane vesicles isolated from rat colon epithelia to develop para-sulfonated calixarenes as high-affinity selective blockers of the

ORCC. While the discovery of the ORCC provided much excitement, it was determined ultimately that ORCC was likely not the primary determinant of the anion transport defect in CF.

Anion conductances with distinct biophysical characteristics were observed in colonic and pancreatic epithelial cells. Gray et al. (1989) reported that the dominant channel type observed in cell-attached pancreatic cell membrane patches was a 4–7 pS channel that was Cl^- selective and activated by forskolin. The small conductance channel was not active in excised membrane patches, but the ORCC was observed frequently in this configuration. The authors suggested that the predominant small conductance channel might work in concert with a Cl^-/HCO_3^- exchanger to account for high concentrations of HCO_3^- in pancreatic secretions (Gray et al. 1989). Frizzell and coworkers (Worrell et al. 1989; Cliff and Frizzell 1990) employed the whole-cell membrane patch to identify three distinct types of Cl^- conductances in T84 cells. Hypoosmotic shifts were used to induce cell swelling that activated both a K^+ conductance and an outwardly rectifying Cl^- conductance in the whole-cell configuration with time-dependent activation at negative potentials and time-dependent inactivation at positive potentials. The authors speculated that this Cl^- conductance played a major role in colonic cell volume regulation (Worrell et al. 1989). Forskolin or cAMP was used to activate a Cl^- conductance, without an effect on K^+ conductance, that was linear and showed no time-dependent activation or inactivation (Cliff and Frizzell 1990). Ca^{2+} ionophores, ionomycin and A23187, activated a Cl^- conductance that exhibited an outward rectification, time-dependent inactivation at negative potentials and time-dependent activation at positive potentials (Cliff and Frizzell 1990). Taken together, the observations from the Frizzell laboratory suggest that there are at least three classes of Cl^- channels in T84 cells that have distinctly different activation mechanisms and biophysical fingerprints. The outcomes, however, do not inform the reader regarding the identity or localization (apical or basolateral) of the corresponding Cl^- channels. An often-stated goal for the treatment of CF is to activate an alternative Cl^- channel that can circumvent the disease pathology. These outcomes suggested the potential for this treatment scheme.

1.5.2 CFTR

The cystic fibrosis transmembrane conductance regulator, CFTR, provided the first opportunity to link a molecular identity (i.e., a gene) with a particular Cl^- conductance. The gene coding for the protein that is defective in CF tissues was identified in 1989 and named CFTR (Kerem et al. 1989; Riordan et al. 1989; Rommens et al. 1989). The sequences of approximately one half million base pairs on the long arm of human chromosome seven were evaluated to identify a span of approximately 250 KB that contain the coding regions for CFTR (Rommens et al. 1989). Twenty-four exons were identified that coded for a protein that includes 1480 amino acids (Riordan et al. 1989). Haplotype analysis suggested that a single three-base deletion accounted for approximately 70 % of disease-associated alleles and that the remaining 30 % included a variety of different mutations (Kerem et al. 1989) (see Chap. 24). Sequence analysis suggested that the resulting protein included two membrane-spanning domains with six membrane-spanning segments in each, two nucleotide-binding domains, and a large regulatory domain that contained numerous charged residues and sixteen consensus sites for phosphorylation by either PKC or PKA (Riordan et al. 1989). Access to the cDNA coding for CFTR provided an unprecedented opportunity to define the molecular basis of CF and to develop therapies targeted to circumvent or cure the disease.

Biophysical characteristics of CFTR were examined closely using a variety of artificial expression systems. Within a year after the sequence of CFTR was determined, the protein was expressed in CFPAC-1 cells (Drumm et al. 1990), a cell line derived from a pancreatic tumor in a CF patient (Schoumacher et al. 1990). Whole-cell patch-clamp recordings revealed a forskolin- or cAMP-stimulated conductance that had high Cl^-:Na^+ selectivity and a linear current–voltage relationship (Drumm et al. 1990), similar to that reported by Cliff and Frizzell (1990). Similar observations were made when CFTR was expressed in HeLa, CHO, or NIH 3T3 cells (Anderson et al. 1991c) and when expressed with the baculovirus expression system in Sf9 cells (Kartner et al. 1991), which were derived from *Spodoptera frugiperda* ovary (Vaughn et al. 1977). Cell-attached recordings following CFTR expression in either Sf9 cells or *Xenopus* oocytes revealed a small (7–10 pS) linear Cl^- channel that was activated by cAMP (Bear et al. 1991; Kartner et al. 1991) and exhibited substantial similarities to channels that had been reported in pancreatic duct cells (Gray et al. 1989). The most compelling evidence that CFTR was a Cl^- channel and not a Cl^- channel regulator came from work by Anderson et al. (1991b) showing that mutations in the putative transmembrane segments were associated with differences in whole-cell anion selectivity when expressed in HeLa cells. The evidence was compelling that CFTR coded for the small linear channel and that mutations in this channel accounted for CF-associated Cl^- transport deficiencies. The relationship of CFTR to other ion channels and the potential to harness alternative Cl^- channels remained undetermined. These topics are covered in detail elsewhere in this volume (see Chaps. 16 and 24). Potential alternative anion channels are also covered in detail in this volume (Chaps. 15 and 17).

1.5.2.1 CFTR Pharmacology and Potential Therapeutic Applications

Access to CFTR expression systems provided unprecedented opportunities to define physiological and pharmacological regulators of the CFTR anion channel (see Chaps. 16 and 24). The deduced protein structure suggested that kinases and nucleotides could affect protein function (Riordan et al. 1989), and this was certainly shown to be the case by a variety of reports. Welsh and colleagues (Anderson et al. 1991a; Anderson and Welsh 1992) provided initial compelling evidence that nucleotides played a critical role in channel gating. Venglarik and

Schultz (Venglarik et al. 1994; Schultz et al. 1995b) employed the excised membrane patch configuration and fluctuation analysis to derive the underlying rate constants for the interaction of nucleotides with the CFTR channel and to determine the relationship with channel gating. Sheppard and Welsh (1992) reported that glibenclamide and tolbutamide, ATP-dependent K^+ channel blockers, inhibited the CFTR-associated conductance in whole-cell recordings. Excised membrane patches were used in conjunction with fluctuation analysis to demonstrate that these sulfonamides directly interacted with the CFTR channel (Schultz et al. 1996; Venglarik et al. 1996) and that structurally related compounds, diarylsulfonylureas, which have no detectable effect on K^+ channels, inhibit CFTR, as well (Schultz et al. 1995a). This structure was exploited through medicinal chemistry to produce analogs that blocked CFTR in excised membrane patches and in intact epithelial monolayers (Singh et al. 2004). DASU-01 was used as a probe in fluctuation analysis, and based on the outcomes, estimates of unitary current, channel number and open probability could be made for CFTR channels in a native cell environment (Singh et al. 2004). DASU-02 has been used in combination with other selective channel blockers to demonstrate that CFTR plays a critical role in intestinal secretion induced by *E. coli* heat-stable and heat-labile toxins and by extension to cholera toxin (O'Donnell et al. 2000; Veilleux et al. 2008), making it a potential lead compound for the treatment of secretory diarrhea. Alternatively, Cuppoletti et al. (2014) employed DASU-02 to demonstrate a contrast between CFTR-mediated whole-cell current and current that was stimulated by lubiprostone and inhibited by methadone, which the authors attribute to ClC-2 Cl^- channels. CLC-2, anoctamin 1 (also known as T_{MEM}16A), and CFTR channels, their pharmacologies and the general topic of secretory diarrheas are handled in detail in Chaps. 15–17, 24, 26, and 29 in this volume.

1.6 Apical Cl^-/HCO_3^- Exchangers

As indicated in previous paragraphs, a functional linkage between Cl^- and HCO_3^- transport has been observed or postulated frequently. Among the examples presented were frog skin (Krogh 1937a) and frog gastric mucosa (Hogben 1955), while other contemporary examples could be presented (e.g., mosquito gut (Keynes 1969), rabbit small intestine (Schultz and Frizzell 1972), rabbit colon (Frizzell et al. 1976), human ileum (Turnberg et al. 1970)). Research concerning HCO_3^- transporters, and especially concerning anion exchangers, has distinct challenges associated with it. First, the carbonic acid reaction can be catalyzed both in the cytosol and extracellular fluid by carbonic anhydrases, which can produce or consume HCO_3^- in either compartment. Second, CO_2, which is another component of the carbonic acid reaction, is volatile, is present in large quantities in the environment, and can cross cell membranes more easily than HCO_3^-. Thus, it is challenging to "clamp" the HCO_3^- concentration at a particular level or to measure the production or disappearance of HCO_3^- during an experiment. Finally, many of the HCO_3^- transport mechanisms are electroneutral, which greatly reduces the utility of electrophysiological techniques. Nonetheless, useful techniques have been developed and two major classes of HCO_3^- transporters have been identified.

The route to the discovery of anion exchangers was initiated in 1971 when proteins from human erythrocyte ghosts were resolved on polyacrylamide gels (Fairbanks et al. 1971). Six major protein bands were observed, and "Band 3" (as it became known) accounted for more than 25 % of the total intensity on the gel. Calculations suggested that the Band 3 protein was present in two- to fivefold molar excess when compared to the intensity of other bands (Fairbanks et al. 1971). Further research showed that Band 3 allowed for anion permeability across the membrane and that permeation could be inhibited by a number of compounds including pyridoxal phosphate and disulfonic stilbenes (Zaki et al. 1975; Rothstein et al. 1976). Band 3, which later became known as anion exchanger 1 (AE1), was the first transport protein to be cloned and expressed (Kopito and Lodish 1985). Subsequently, a number of related proteins were identified by homology cloning and a gene family, SLC4, containing ten members has been described (Romero et al. 2004; Alper 2006). The gene family includes not only classical anion exchangers (SLC4A1-4) but also electrogenic and electroneutral Na^+/HCO_3^- cotransporters (SLC4A4-8). The latter proteins play important roles in systemic acid–base balance by mediating Na^+-coupled HCO_3^- recovery in the kidney and in loading pancreatic cells for HCO_3^- secretion that ultimately buffers the proximal small intestine. The reader is directed to Chap. 13 for a more extensive description of SLC4 HCO_3^- transporters.

A second family of HCO_3^- transporters, the SLC26 gene family, has been described more recently, although certain family members are associated with medical conditions that were described more than 50 years ago. The founding member of the gene family, Sat-1 (SLC26A1), is a sulfate transporter that was cloned from rat hepatocytes (Bissig et al. 1994). Like many of the anion transporters that have been described, SLC26A1 is inhibited by disulfonic stilbenes. SLC26A2 is a sulfate transporter that is predominantly expressed in cartilage and in the intestine. Naturally occurring mutations are associated with diastrophic dysplasia, which was first described by Lamy and Maroteaux in 1960 (Everett and Green 1999). The gene was cloned in 1994 (Hastbacka et al. 1994). SLC26A3 is a Cl^-/HCO_3^- exchanger that is known also as DRA (downregulated in adenoma), which is highly expressed in normal colonic epithelium, but expressed at lower levels in polyps and in adenocarcinomas of the colon (Schweinfest et al. 1993). The same gene has been associated with Cl^- losing diarrhea (CLD), which was first described in 1945 as congenital alkalosis with diarrhea (Darrow 1945; Gamble et al. 1945). More recently, Hoglund and colleagues (Hihnala et al. 2006; Hoglund et al. 2006; Kujala et al. 2007) reported that mutations in SLC26A3 are associated also with male infertility. Mutations in SLC26A4 are associated with deafness and goiter, a condition first described by Vaughan Pendred (1896). The gene associated with Pendred syndrome was speculated initially to be a sulfate transporter based on sequence homology (Everett et al. 1997), but it is now known to function as a Cl^-/HCO_3^- exchanger. Recently, it was reported by Wangemann and colleagues

(Choi et al. 2011; Li et al. 2013) that SLC26A4 expression is required during fetal development in the endolymphatic sac, but not in the cochlea or vestibular sensory organs, for hearing and balance to develop, and that it is not required during postnatal life for hearing to be maintained. SLC26A6 is an anion exchanger that is expressed in many tissues and at higher levels in kidney, where it may have a role in formate or oxalate transport. SLC26A6 null mice appear to be healthy, but have some renal transport abnormalities (Wang et al. 2005). There has been speculation that there are physical or functional interactions between CFTR and SLC26A6 or SLC26A3 that can account for CFTR-dependent HCO_3^- secretion, especially in the pancreas where segregated expression was suggested (Lee et al. 1999; Ko et al. 2004; Singh et al. 2010). More recently, it was shown that both SLC26A3 and SLC26A6 are expressed at high levels in or near the mucosal membrane of all epithelial cells lining porcine vas deferens with no detectable gradient in expression over the length of the duct (Pierucci-Alves et al. 2015). Although protein localization was not assessed, there were no detectable differences in the expression of either SLC26A3 or SLC26A6 in the vas deferens of pigs lacking CFTR or expressing the most common mutant form of CFTR, ΔF508 (Pierucci-Alves et al. 2011). Obviously, additional work is required to demine the roles that these transport proteins play in normal physiology of different tissues. Excellent reviews of this transporter family have been published (Everett and Green 1999; Mount and Romero 2004), and the reader is directed to Chap. 14 for a complete discussion focused on SLC26 anion transporters.

1.7 Evidence for a $Na^+/K^+/2Cl^-$ Cotransporter

As pointed out by Tidball (1961), Cl^- typically is secreted against an electrochemical potential gradient such that the process must include an active energy-consuming process. This active step could occur at either the basolateral membrane where Cl^- uptake occurs, or Cl^- could be actively extruded across the apical membrane. Measurements of intracellular Cl^- from numerous laboratories demonstrated that Cl^- is accumulated above electrochemical equilibrium in the cell (Reuss et al. 1983; Welsh 1983b; Greger and Schlatter 1984b; Shorofsky et al. 1984), indicative of Cl^- being actively transported into the cell across the basolateral membrane. Whether this active step involved a pump or a cotransport process remained an open question, however.

This question began to be addressed when Al-Bazzaz and Al-Awqati (1979) demonstrated that Cl^- secretion from dog trachea could be blocked by simply removing Na^+ from the serosal solution, being consistent with Na^+-coupled entry of Cl^- in to the cell. More direct evidence for Na^+-coupled Cl^- entry was provided by measuring intracellular Cl^- with ion-selective microelectrodes in the presence and absence of bath Na^+ (Welsh 1983b). Initially, intracellular Cl^- was found to be accumulated above electrochemical equilibrium. However, when Na^+ was removed from the bath, intracellular Cl^- activity declined to a value very near that expected for an equilibrium distribution (Welsh 1983b). Identical results were reported by Reuss et al. (1983) in bullfrog corneal epithelium. By carrying out Na^+ and Cl^- influx studies in dog trachea, Widdicombe et al. (1983) provided additional evidence for coupled Na^+-Cl^- cotransport. That is, these authors demonstrated that removal of Na^+ reduced Cl^- influx, while removal of Cl^- reduced Na^+ influx into the cells.

Demonstrating a dependence on K^+ in the uptake of Cl^- across the basolateral membrane proved to be more difficult. This stems from the fact that simply removing K^+ from the serosal solution will inhibit the Na^+/K^+-ATPase, thereby confounding the results. Greger and Schlatter (1984b) proposed that 2 Cl^- ions were transported for each Na^+ in the spiny dogfish (*Squalus acanthias*) rectal gland, based on the Hill coefficients for Na^+-dependent Cl^- transport and Cl^--dependent Na^+ transport. This notion was extended by showing that furosemide, a known blocker of the cotransporter (Geck et al. 1980), caused the concentration of Cl^- to drop in the cell twice as fast as Na^+ (Greger and Schlatter 1984a). These authors also demonstrated that removal of serosal K^+ caused a drop in intracellular Cl^- that was not due to an effect on the Na^+/K^+-ATPase and, therefore, proposed a $1Na^+$:1 K^+:$2Cl^-$ stoichiometry for the carrier (Greger and Schlatter 1984b). This stoichiometry was similarly proposed for this carrier in the MDCK cell line when it was found that the uptake of radiolabeled Na^+ only occurred in the presence of both Cl^- and K^+, while the uptake of radiolabeled Rb^+ (a tracer typically used in place of K^+) required both Na^+ and Cl^- (McRoberts et al. 1982). Finally, O'Grady et al. (1986), working in winter flounder intestine, carried out simultaneous NaCl and RbCl influx measurements and demonstrated a ratio of 2.2 for Cl^-:Na^+ and 1.8 for Cl^-:Rb^+, confirming a $1Na^+$:1 K^+:$2Cl^-$ stoichiometry.

One of the consequences of proposing a $1Na^+$:1 K^+:$2Cl^-$ stoichiometry for the cotransporter is that it must be electroneutral and therefore insensitive to membrane voltage. Early evidence that this was the case came from duck red blood cells (Haas et al. 1982). In these studies, the duck red cells were exposed to the K^+ ionophore valinomycin, thereby increasing membrane K^+ permeability such that the membrane voltage was dominated by the K^+ reversal potential (E_K). Thus, the uptake of Na^+ could be studied while bath K^+, and therefore membrane voltage, was varied. Under these conditions, the uptake of Na^+ was not affected by changes in membrane voltage, demonstrating the electroneutrality of the cotransporter (Haas et al. 1982). Similar experiments were carried out on MDCK cells using valinomycin with a similar lack of dependence on membrane voltage (McRoberts et al. 1982). Finally, electrophysiological methods in canine tracheal epithelium confirmed that the cotransporter is electroneutral by confirming that the potential across the basolateral membrane was unaffected when the cotransporter was blocked by either Cl^- removal (Welsh 1983b) or furosemide (Welsh 1983a).

As noted above, Cl^- entry into the cell requires that it moves against an electrochemical potential. Thus, the coupling of Cl^- entry to Na^+ and K^+, such that this process is electroneutral, is an energetically efficient way of accumulating Cl^- above electrochemical equilibrium. Indeed, using values for intracellular

Na^+, K^+, and Cl^-, a net driving force of 40–70 mV can be calculated, favoring Cl^- entry. This indicates that there is sufficient energy in the established ion gradients to drive Cl^- in to the cell above electrochemical equilibrium using a $1Na^+$:1 K^+:$2Cl^-$ transport process.

A complete account of the regulation, cell biology, and molecular biology of the $Na^+/K^+/2Cl^-$ cotransporters is discussed in Chap. 11 of this volume, and the reader is directed there for additional details.

1.8 Evidence for a Basolateral Membrane K^+ Channel

As detailed above, it was recognized early on that both Na^+ absorption and Cl^- secretion were associated with the exit of K^+ across the basolateral membrane of epithelia. Indeed, as can be appreciated from the above discussion, activation of an apical Cl^- conductance alone would induce an initial Cl^- secretory response, followed by a return to steady state. This is because Cl^- is initially above its electrochemical equilibrium, due to the $Na^+/K^+/2Cl^-$ cotransporter, such that Cl^- can flow down its electrochemical potential gradient upon opening of a Cl^- channel exit pathway in the apical membrane. Subsequently, however, activation of the Cl^- conductance alone would result in the apical membrane depolarizing until it approaches the reversal potential for Cl^- (E_{Cl}), after which there would be no net driving force favoring Cl^- exit from the cell, resulting in an attenuation of Cl^- secretion. Clearly then, an additional conductive pathway, capable of hyperpolarizing the cell such that Cl^- remains above electrochemical equilibrium, is required, the aforementioned K^+ exit pathway.

Early electrophysiological studies demonstrated, in both canine trachea (Welsh et al. 1982a) and spiny dogfish rectal gland (Greger and Schlatter 1984b), that cAMP-mediated agonists induced activation of a basolateral membrane K^+ conductance. Using microelectrodes, Welsh et al. (1982a) showed that epinephrine caused an initial depolarization of membrane potential, due to the activation of an apical Cl^- conductance, followed approximately 20 s later by a repolarization that was associated with a decrease in the basolateral membrane fractional resistance, consistent with activation of a basolateral K^+ conductive pathway. Further studies in both canine trachea (Smith and Frizzell 1984) and shark rectal gland (Greger and Schlatter 1984a) preparations demonstrated that cAMP-mediated agonist-induced Cl^- secretion could be inhibited by either raising serosal K^+ or by adding the nonspecific K^+ channel blocker Ba^{2+} to the serosal bath. Both of these maneuvers would depolarize the cell, decreasing the electrochemical driving force for Cl^- exit across the apical membrane. These early studies confirmed a role for a basolateral membrane K^+ conductance in the Cl^- secretory response.

While activation of a basolateral membrane K^+ conductance hyperpolarizes the cell and thus maintains the electrochemical driving force for Cl^- exit, activation of this conductive pathway plays an additional important role that can sometimes be overlooked. That is, early investigators noted that when cells are stimulated to secrete Cl^-, there is an increase in intracellular Na^+ (Greger et al. 1984; Shorofsky et al. 1986). This is due to Na^+ entering on the $Na^+/K^+/2Cl^-$ cotransporter with Cl^- as detailed above. This Na^+ is removed from the cell via the Na^+/K^+-ATPase, which results in a potential increase in intracellular K^+. However, it has been shown that intracellular K^+ does not change during agonist stimulation (Greger and Schlatter 1984b; Smith and Frizzell 1984). Thus, K^+ entry via both the Na^+/K^+-ATPase and the $Na^+/K^+/2Cl^-$ cotransporter must be equaled by K^+ exit across the basolateral membrane. Indeed, in the absence of an increase in K^+ permeability, there would be a requirement for the driving force for K^+ exit to increase. Since potential across the basolateral membrane is either not changed during stimulation (Smith and Frizzell 1984) or slightly decreased (Greger et al. 1984), this would require an increase in cellular K^+ to levels 100–600 mM above normal (Welsh et al. 1982a; Smith and Frizzell 1984), which, as stated above, does not happen. Although a role for a basolateral membrane K^+ conductive pathway in the Cl^- secretory response was confirmed by these studies, the identities of these K^+ channels would remain elusive for many more years until selective blockers and molecular cloning techniques could be applied.

1.8.1 Identification of the cAMP- and Ca^{2+}-Activated Basolateral Membrane K^+ Channels

The original studies of Frizzell and colleagues (Welsh et al. 1982a, 1983; Heintze et al. 1983; Smith and Frizzell 1984), Greger and colleagues (Greger and Schlatter 1984a, b; Greger et al. 1984), and Shorofsky et al. (1983) clearly suggested that cAMP was regulating a basolateral membrane K^+ conductance. Additional microelectrode studies in locust hindgut similarly showed cAMP activation of a basolateral K^+ channel (Hanrahan and Phillips 1984), and patch-clamp studies confirmed activation of a basolateral K^+ channel in *Necturus* oxyntic cells (Ueda et al. 1987). Dharmsathaphorn and colleagues (Weymer et al. 1985; Mandel et al. 1986) demonstrated that a colonic epithelial cell line, T84, also exhibited a cAMP-activated K^+ conductance that could be blocked by Ba^{2+}. In addition to cAMP, Ca^{2+}-mediated agonists have been shown to stimulate transepithelial Cl^- secretion via a similar model. Frizzell (1977) demonstrated that a Ca^{2+} ionophore, A23187, could induce Cl^- secretion from rabbit colon. Similarly, Bolton and Field (1977) demonstrated that A23187 increased serosa-to-mucosa Cl^- flux, resulting in Cl^- secretion in rabbit ileal mucosa. These authors further demonstrated that carbachol stimulated Cl^- secretion in the absence of any detectable change in cytosolic cAMP (Bolton and Field 1977). In addition, Zimmerman et al. (1982) demonstrated that bethanechol induced Cl^- secretion from rat intestine in a Ca^{2+}-dependent, cAMP-independent manner. Finally, using the T84 cell line, Dharmsathaphorn and Pandol (1986) demonstrated that carbachol induced Cl^- secretion by increasing intracellular Ca^{2+} with no detectable change in cAMP, identical to what had been described in ex vivo tissue. Indeed, these authors went on to show that carbachol stimulated K^+ efflux across the basolateral membrane, indicative of an increase in basolateral K^+ conductance as shown for cAMP-mediated agonists. While these data confirm that both cAMP- and Ca^{2+}-mediated agonists activate a basolateral membrane K^+ conductance, in the absence of selective blockers and molecular techniques, little additional evidence was forthcoming to differentiate among these conductances.

McRoberts et al. (1985) initially demonstrated in T84 cells that the cAMP- and Ca^{2+}-activated basolateral membrane K^+ conductances were differentially inhibited by Ba^{2+}, consistent with these two second messengers activating unique K^+ channels. Unfortunately, there were no blockers available at that time that could allow the investigator to conclusively identify Ca^{2+}- and cAMP-dependent K^+ channels. A first breakthrough in this line of investigation came when McCann et al. (1990) and Devor and Frizzell (1993) characterized the Ca^{2+}-activated basolateral membrane K^+ channel in airway and colonic epithelia, respectively, subsequently demonstrating that it was blocked with high affinity by charybdotoxin, but was insensitive to block by the Maxi-K channel blockers, paxilline and iberiotoxin (Devor et al. 1996b). A second critical breakthrough in the pharmacology of these channels came when Lohrmann et al. (1995) characterized the 1,4-sulfonylaminochromanoles as a novel class of inhibitors of the cAMP-mediated K^+ channel in secretory epithelia, with 293B being the most potent compound described. Third, Devor and coworkers (Devor et al. 1996a, b) described the first class of openers of the Ca^{2+}-activated K^+ channels, 1-ethyl-2-benzimidazolinone (1-EBIO). These authors demonstrated that 1-EBIO activated a charybdotoxin-sensitive basolateral K^+ conductance that was insensitive to 293B, whereas cAMP-mediated Cl^- secretion was insensitive to charybdotoxin, but blocked by 293B. Finally, Rufo et al. (1996) characterized clotrimazole as a blocker of Ca^{2+}-mediated Cl^- secretion and that this effect was due to block of a basolateral K^+ conductance. Devor et al. (1997) subsequently confirmed that this was due to a direct inhibition of the charybdotoxin-sensitive basolateral K^+ channel (see Chap. 20 for additional details). The use of these pharmacological tools unequivocally demonstrated that cAMP- and Ca^{2+}-mediated agonists activate unique basolateral membrane K^+ channels and played a pivotal role in confirming their molecular identities.

1.8.2 KCa3.1 Is the Ca^{2+}-Activated Basolateral Membrane K^+ Channel

The molecular identity of the Ca^{2+}-activated, basolateral membrane K^+ channel was revealed in 1997 when Joiner et al. (1997) and Ishii et al. (1997) nearly simultaneously reported the cloning of this channel. This channel was the fourth member of the KCNN gene family that contains the SK1-3 family members previously reported by Kohler et al. (1996). SK1-3 encode the small conductance, Ca^{2+}-activated, and apamin-sensitive K^+ channels expressed primarily in nerve and smooth muscle (Adelman et al. 2012). Given the ~40 % similarity in amino acid sequence, Joiner et al. (1997) referred to the novel clone as SK4. However, given the intermediate conductance of the channel, based on historical precedence, Ishii et al. (1997) referred to this channel as IK1. Based on the standard nomenclature derived from the International Union of Pharmacology (IUPHAR; (Gutman et al. 2003)), this channel is now referred to as KCa3.1. These channels were posited to have 6 transmembrane domains (S1–S6), with the reentrant pore loop between S5 and S6 and cytosolic N- and C-termini. Joiner et al. (1997) and Ishii et al. (1997) demonstrated that KCa3.1 was activated by Ca^{2+} and was blocked by charybdotoxin and clotrimazole while being insensitive to blockers of Maxi-K and small conductance K^+ channels, similar to what had been reported for the endogenously expressed basolateral membrane, Ca^{2+}-activated K^+ channel of epithelia. Further studies confirmed KCa3.1 was activated by 1-EBIO (Pedersen et al. 1999; Syme et al. 2000; Jensen et al. 2001), consistent with data in both T84 cells (Devor et al. 1996a, b) and murine tracheal epithelia (Devor et al. 1996a) and human bronchial epithelia (Devor et al. 2000). Subsequent to the cloning of KCa3.1 and the realization that this channel displayed identical biophysical and pharmacological characteristics to the basolateral membrane Ca^{2+}-activated K^+ conductance of secretory epithelia, Warth et al. (1999) and Gerlach et al. (2000) confirmed by RT-PCR and Northern blot, respectively, that this was the channel expressed in T84 cells. Finally, Flores et al. (2007) carried out studies on KCa3.1 knockout mice and demonstrated that Ca^{2+}-mediated Cl^- secretion was completely eliminated in both distal colon and small intestinal epithelium, which resulted in a marked reduction in water content in the stools. In total, these studies unequivocally demonstrate that KCa3.1 is the basolateral membrane K^+ channel activated by Ca^{2+}-mediated agonists as a means of maintaining transepithelial Cl^- secretion across intestinal epithelia. For a complete account of KCa3.1 in epithelial transport, the reader is directed to Chap. 20 of this volume.

1.8.3 KCNQ1 (Kv7.1)/KCNE3 (Mirp2) Is the cAMP-Activated Basolateral Membrane K^+ Channel

The demonstration that 293B blocked cAMP-mediated transepithelial Cl^- secretion with high affinity by inhibiting the basolateral membrane K^+ conductance, with no effect on the apical membrane Cl^- conductance, was a critical turning point in the identification of this K^+ channel (Lohrmann et al. 1995). The next step came when it was discovered that expression of minK (for minimal K^+ channel, now called KCNE1), a protein first cloned in 1988 (Takumi et al. 1988), in *Xenopus laevis* oocytes resulted in a slowly activating, voltage-dependent K^+ current that was blocked by 293B (reviewed in Suessbrich and Busch (1999)). However, it was difficult to imagine how a protein of 130 amino acids and a single transmembrane domain could itself form a K^+ channel. In addition, KCNE1 is expressed at very low levels in colonic epithelia, the 293B-sensitive current does not display the same biophysical characteristics of *Xenopus laevis*-expressed KCNE1, and expression of KCNE1 in mammalian expression systems failed to induce a 293B-sensitive current. This conundrum was solved with the discovery that KCNE1 co-assembles with KvLQT1 (Kv7.1, KCNQ1) in *Xenopus* oocytes to form this slowly activating current (Barhanin et al. 1996; Sanguinetti et al. 1996). KvLQT1 was identified as the cardiac K^+ channel responsible for long QT syndrome (type 1; LQT1) (Wang et al. 1996b) and was subsequently shown to be expressed in various tissues, including the intestine. KCNQ1, being a member of the Kv superfamily of voltage-gated K^+ channels, has an architecture of 6 transmembrane domains (S1–S6), with the reentrant pore loop lying between S5 and S6 and cytosolic N- and C-termini. Co-expression of KCNE1 and KCNQ1 resulted in a dramatically increased macroscopic current that displayed slowed activation and almost no inactivation kinetics (Splawski et al. 1997) and was blocked by various chromanol derivatives (including 293B) with a similar profile to that described for the basolateral K^+ conductance in distal colon (Suessbrich et al. 1996). These results suggested that KCNQ1 was the basolateral membrane K^+ conductance responsible for maintaining a favorable electrochemical driving force for transepithelial Cl^- secretion (Bleich et al. 1997).

While the above results argued for KCNQ1/KCNE1 channels being the basolateral membrane, cAMP-activated K^+ conductance in epithelia, an additional puzzle remained to be solved. That is, KCNQ1/KCNE1 channels are slowly activating and voltage dependent, whereas the cAMP-activated basolateral membrane K^+ conductance is voltage independent and constitutively active. A possible solution to this puzzle came with the cloning of KCNE3, another small, single transmembrane domain subunit, which, when co-expressed with KCNQ1 in *Xenopus* oocytes, resulted in a near-instantaneous activation of the channel complex resulting in a constitutively active, 293B-blockable K^+ conductance similar to what has been described in colonic epithelia (Schroeder et al. 2000). These results suggest that KCNQ1 may be obligatorily associated with KCNE3 in certain epithelia resulting in a linear current–voltage relationship for this K^+ channel as described in colonic epithelia (Schroeder et al. 2000). However, expression of KCNQ1/KCNE3 in mammalian cells does not reproduce these biophysical characteristics (Mazhari et al. 2002; Jespersen et al. 2004). Thus, while there is clear evidence demonstrating that KCNQ1 is the basolateral membrane K^+ channel α-subunit activated by cAMP in secretory epithelia, there may be additional components to the channel complex required to produce the biophysical properties observed endogenously in epithelia. As the KCNQ1/KCNE complex is not further discussed in this volume, we will briefly review the gating, trafficking, and regulation of this channel complex below.

1.8.3.1 How Do KCNE β-Subunits Alter KCNQ1 Gating?

A significant question, based on the above discussion, is how do KCNE β-subunits alter KCNQ1 gating. Indeed, while the conductance of KCNQ1 has been reported to be between 0.7 and 7 pS, it is known that the KCNE β-subunits increase the single-channel conductance between four- and sevenfold (Romey et al. 1997; Pusch 1998; Sesti and Goldstein 1998; Yang and Sigworth 1998), explaining the increased macroscopic current when KCNQ1/KCNEx are co-expressed. Several investigators have shown that either the pore region or voltage-sensing domain of KCNQ1 may physically interact with KCNE proteins (Melman et al. 2004; Panaghie et al. 2006; Nakajo and Kubo 2007; Shamgar et al. 2008; Chung et al. 2009), which would allow for an altered channel gating/conductance. Indeed, a triplet of amino acids in the transmembrane domain of KCNE1/3 has been shown to be responsible for the modulation of KCNQ1 currents (Angelo et al. 2002; Melman et al. 2004). However, it is still debated whether 2 or 4 KCNE β-subunits associate with the tetrameric KCNQ1 channel to induce these biophysical changes (Wang et al. 1996a, 1998).

1.8.3.2 Basolateral Targeting and Trafficking of KCNQ1

As detailed above, KCNQ1 is a basolateral membrane, cAMP-activated K^+ channel in secretory epithelia. It is now known that KCNQ1 is targeted to the basolateral membrane via a tyrosine-based motif in the N-terminus and that association with KCNE β-subunits does not alter this trafficking pattern (Jespersen et al. 2004). However, more recent evidence has shown that KCNE1 can redirect KCNQ1 to the apical membrane following mutation of this critical basolateral sorting motif (David et al. 2013). Further, these authors provided several lines of evidence to suggest that KCNQ1 and KCNE1 associate early in the biosynthetic pathway, but that this occurs in a post-ER compartment (David et al. 2013). It has also been shown that maintaining KCNQ1 in the basolateral membrane of polarized MDCK cells requires the activity of phosphoinositide 3-kinase (PI3K), i.e., inhibition of PI3K induces KCNQ1 endocytosis. This appears to be due to phosphorylation of serum- and glucocorticoid-inducible kinase 1 (SGK1) by PI3K as inhibition of SGK1 similarly induced KCNQ1 endocytosis. Finally, knockdown of the ubiquitin ligase Nedd4-2 abrogated the PI3K inhibition effects, whereas the endocytosis of a KCNQ1 mutant that does not bind Nedd4-2 was unaffected by PI3K or SGK1 inhibition. This led Andersen et al. (2013) to propose a model whereby SGK1 phosphorylation of Nedd4-2 eliminates the Nedd4-2-dependent endocytosis of KCNQ1. These authors also demonstrated that during the polarization of MDCK cells, KCNQ1 is endocytosed. They proposed a model whereby protein kinase C (PKC) phosphorylates the LKB1 kinase, which subsequently activates AMP-activated protein kinase (AMPK), which phosphorylates Nedd4-2, which then associates with and ubiquitylates KCNQ1, ultimately resulting in its endocytosis (Andersen et al. 2012).

1.8.3.3 cAMP-Dependent Regulation of KCNQ1

As detailed above, basolateral membrane KCNQ1 is activated by increasing cAMP/PKA. It was demonstrated that the cAMP-/PKA-dependent regulation of KCNQ1 occurs via the A-kinase anchoring protein (AKAP), Yotiao (AKAP9), which associates with the C-terminus of KCNQ1 via a leucine zipper (Potet et al. 2001). Through this association, Yotiao recruits PKA, protein phosphatase 1, and adenylyl cyclase to the channel (Potet et al. 2001; Li et al. 2012). Following agonist activation, PKA phosphorylates Ser27 in the N-terminus of KCNQ1 resulting in both increased current and faster activation kinetics, although the exact mechanism by which phosphorylation of the N-terminus alters channel gating is not yet clear. Importantly, this regulation of KCNQ1 appears to depend upon KCNE1 as disease-causing mutations in both KCNQ1 and KCNE1 can disrupt this regulation (Marx et al. 2002; Kurokawa et al. 2004; Dvir et al. 2014). Interestingly, Harvey and colleagues (Condliffe et al. 2001; O'Mahony et al. 2007) have shown that 17β-estradiol inhibits transepithelial Cl^- secretion in colonic epithelia in a gender-specific manner—inhibiting only in female rats. The authors demonstrated that this was due to a complete inhibition of basolateral KCNQ1, where it associates with KCNE3, and that it occurred via PKCδ- and PKA-dependent signaling pathways. It was further shown that both PKCδ and the C1 form of PKA (PKAC1) associated with KCNQ1 following 17β-estradiol activation. As expected, this association only occurred in colonic crypts from female rats and not in male rats. Further, the association of PKAC1 could be inhibited by blocking PKCδ activity suggesting the PKA-dependent effects were downstream of PKCδ (O'Mahony et al. 2007). Reenstra (1993) previously demonstrated that phorbol ester-dependent activation of PKC reduced transepithelial Cl^- secretion in the T84 colonic cell line; consistent with the notion that 17β-estradiol-dependent activation of PKCδ inhibits Cl^- secretion in rat colonic crypts. However, it is unclear how PKAC1 association with KCNQ1 in colonic crypts results in channel inhibition. Given that the regulation of KCNQ1 depends upon association with its β-subunit as well as accessory proteins such as Yotiao, there may well be differences depending on the tissue studied as well as the route of second messenger activation. Indeed, Kurokawa et al. (2009) demonstrated that co-expression of either KCNE1 or KCNE2 with KCNQ1 confers a response to PKA, whereas co-expression with KCNE3 does not. Chimeras between KCNE1 and KCNE3 identified the intracellular C-terminus of

KCNE1 as being the portion of the protein capable of conferring this PKA-dependent activation of KCNQ1.

For a more extensive review of the gating and regulation of KCNQ1/KCNE, the reader is directed to several excellent reviews (Jespersen et al. 2005; Zaydman and Cui 2014; Liin et al. 2015; Nakajo and Kubo 2015).

1.9 Conclusion: An Extensive Cell Model

At this juncture it seems appropriate to respond to the statement published in 1937 by Nobel laureate August Krogh regarding epithelial Cl^- uptake, "The present investigation raises the question … regarding the biological significance of this power" (Krogh 1937a). There is now little doubt regarding the significance of Cl^- transport in light of the many pathologies that can be attributed to changes in the expression or activity of individual proteins such as CFTR, SLC26A3 or SLC26A4. Pavlov, Krogh, Ussing, Burg, Neher, Sackman, Mullis, Agre, Venter, Collins and others have provided groundbreaking techniques and insights that allowed generations of scientists to build a progressively better understanding of epithelial function. There is no longer a question regarding the importance of Cl^- transport. Rather, one necessarily asks, "What epithelial mechanisms are required to accomplish Cl^- transport in health and which molecular targets can be modulated to remedy disease?" Minimal models can account for secretion with as few as four components: Na^+/K^+-ATPase, $Na^+/K^+/2Cl^-$ cotransporter, and a K^+ channel in the serosal membrane and a Cl^- channel in the mucosal membrane. This "classical model" is presented in a cell near the top of Fig. 1.5 as depicted by Frizzell and Hanrahan (2012) in an excellent recent review. Much of this chapter shows that unique transporters fulfill these general roles in different cells and tissues. Additionally, the components can be paired with different partners or arranged in different configurations to account for differences in transport that are observed across tissues (e.g., see Fig. 1.4). The lower cells in Fig. 1.5 present a few recent additions to the classical model including distinct Cl^- and K^+ channels, Na^+/H^+ exchangers (e.g., NHE), anion exchangers (e.g., AE2), Na^+/HCO_3^- cotransporters (e.g., NBC1), and an H^+-ATPase. The diagram is not exhaustive, but provides a new, fuller framework as a foundation for ongoing studies that are designed to better understand the mechanisms of ion transport. Many of the chapters in this volume describe these components in great detail and build on this simple model with the addition of other transport proteins. The authors characterize the structures, functions and regulation of these epithelial transporters in health and disease.

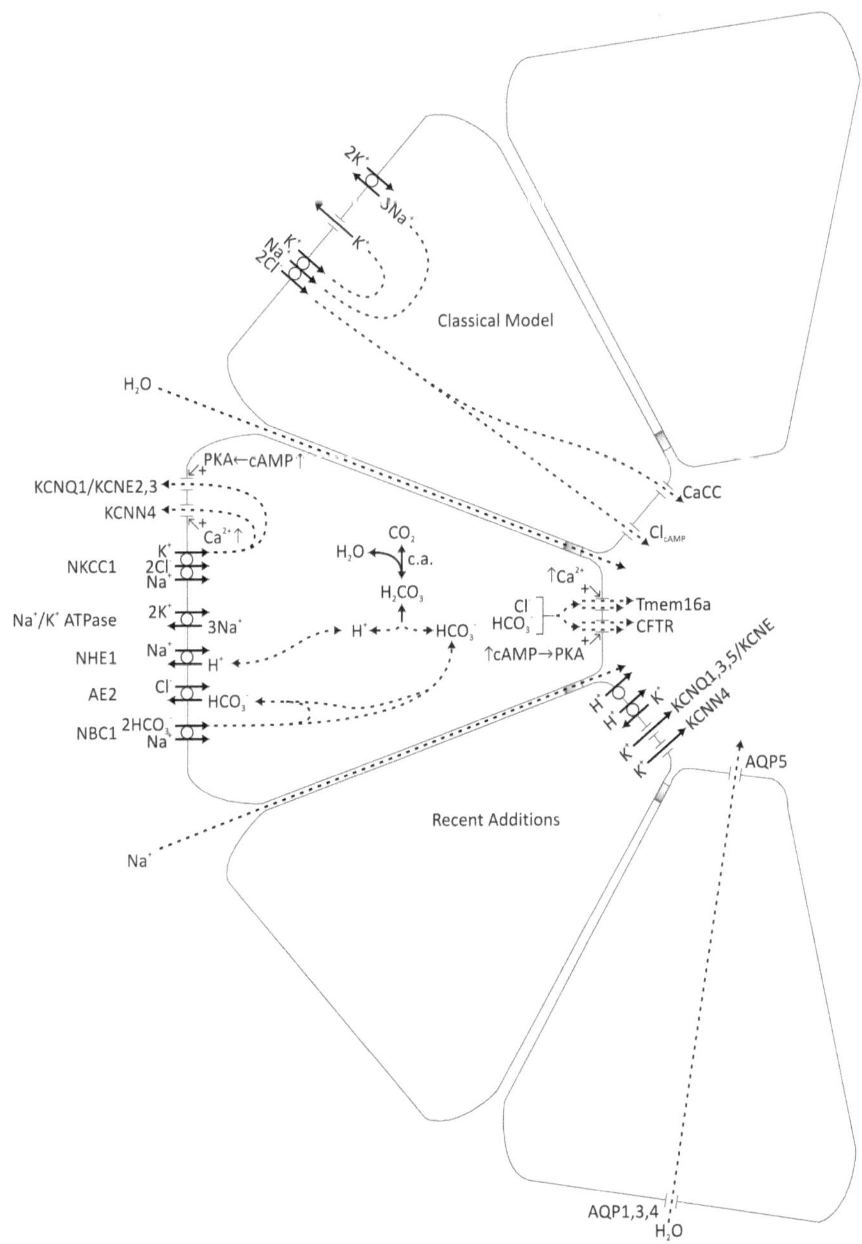

Fig. 1.5 Schematic of the transport pathways in secretory epithelia. As shown in the "classical model," Cl^- entry across the basolateral membrane is driven primarily via Na^+ cotransport. Apical Cl^- efflux is mediated by cAMP- and Ca^{2+}-activated anion conductances. The scheme labeled "Recent additions" identifies the numerous transporters and channels proposed to mediate ion and fluid secretion in various epithelia. Accordingly, basolateral Cl^- loading may occur via both NKCC1 and AE2. Cl^- and HCO_3^- exit involves apical CFTR, Tmem16A, and probably other Cl^- channels/transporters, such as members of the SLC26A family (not shown). NBC1-mediated HCO_3^- entry during cAMP stimulation maintains intracellular pH and generates electrogenic

References

Adelman JP, Maylie J, Sah P (2012) Small-conductance Ca^{2+}-activated K^+ channels: form and function. Annu Rev Physiol 74:245–269. doi:10.1146/annurev-physiol-020911-153336

Al-Bazzaz FJ, Al-Awqati Q (1979) Interaction between sodium and chloride transport in canine tracheal mucosa. J Appl Physiol Respir Environ Exerc Physiol 46:111–119

Alper SL (2006) Molecular physiology of SLC4 anion exchangers. Exp Physiol 91:153–161. doi:10.1113/expphysiol.2005.031765

Andersen DH (1938) Cystic fibrosis of the pancreas and its relation to celiac disease: a clinical and pathological study. Am J Dis Child 56:344–399

Andersen MN, Krzystanek K, Jespersen T, Olesen SP, Rasmussen HB (2012) AMP-activated protein kinase downregulates Kv7.1 cell surface expression. Traffic 13:143–156. doi:10.1111/j.1600-0854.2011.01295.x

Andersen MN, Krzystanek K, Petersen F, Bomholtz SH, Olesen SP, Abriel H, Jespersen T, Rasmussen HB (2013) A phosphoinositide 3-kinase (PI3K)-serum- and glucocorticoid-inducible kinase 1 (SGK1) pathway promotes Kv7.1 channel surface expression by inhibiting Nedd4-2 protein. J Biol Chem 288:36841–36854. doi:10.1074/jbc.M113.525931

Anderson MP, Welsh MJ (1992) Regulation by ATP and ADP of CFTR chloride channels that contain mutant nucleotide-binding domains. Science 257:1701–1704

Anderson MP, Berger HA, Rich DP, Gregory RJ, Smith AE, Welsh MJ (1991a) Nucleoside triphosphates are required to open the CFTR chloride channel. Cell 67:775–784

Anderson MP, Gregory RJ, Thompson S, Souza DW, Paul S, Mulligan RC, Smith AE, Welsh MJ (1991b) Demonstration that CFTR is a chloride channel by alteration of its anion selectivity. Science 253:202–205

Anderson MP, Rich DP, Gregory RJ, Smith AE, Welsh MJ (1991c) Generation of cAMP-activated chloride currents by expression of CFTR. Science 251:679–682

Angelo K, Jespersen T, Grunnet M, Nielsen MS, Klaerke DA, Olesen SP (2002) KCNE5 induces time- and voltage-dependent modulation of the KCNQ1 current. Biophys J 83:1997–2006. doi:10.1016/S0006-3495(02)73961-1

Bank N (1968) Physiological basis of diuretic action. Annu Rev Med 19:103–118. doi:10.1146/annurev.me.19.020168.000535

Barhanin J, Lesage F, Guillemare E, Fink M, Lazdunski M, Romey G (1996) K(V)LQT1 and IsK (minK) proteins associate to form the I(Ks) cardiac potassium current. Nature 384:78–80. doi:10.1038/384078a0

Bear CE, Duguay F, Naismith AL, Kartner N, Hanrahan JW, Riordan JR (1991) Cl^- channel activity in *Xenopus* oocytes expressing the cystic fibrosis gene. J Biol Chem 266:19142–19145

Bijman J, Quinton PM (1984) Influence of abnormal Cl^- impermeability on sweating in cystic fibrosis. Am J Physiol Cell Physiol 247:C3–C9

Bissig M, Hagenbuch B, Stieger B, Koller T, Meier PJ (1994) Functional expression cloning of the canalicular sulfate transport system of rat hepatocytes. J Biol Chem 269:3017–3021

Fig. 1.5 (continued) transepithelial HCO_3^- secretion. NHE1 maintains intracellular pH during stimulation by Ca^{2+} secretagogues when cells are hyperpolarized and net Cl^- secretion is favored. There is evidence for apical H^+ secretion mediated by vacuolar proton pumps, H^+/K^+-ATPase, and also apical K^+ channels, which contribute to apical hyperpolarization and thus enhance the driving force for anion secretion. Most transepithelial Na^+ flux occurs passively through the paracellular pathway driven by the lumen-negative voltage arising from electrogenic anion secretion. In both schemes, regulated apical anion conductance is rate limiting at low/medium levels of anion secretion, but basolateral K^+ conductance becomes increasingly limiting at higher secretory rates, explaining the additive nature of secretagogues that operate via different second messenger pathways [Modified from Frizzell and Hanrahan (2012)]

Bleich M, Briel M, Busch AE, Lang HJ, Gerlach U, Gogelein H, Greger R, Kunzelmann K (1997) KVLQT channels are inhibited by the K^+ channel blocker 293B. Pflugers Arch 434:499–501
Bolton JE, Field M (1977) Ca ionophore-stimulated ion secretion in rabbit ileal mucosa: relation to actions of cyclic 3′,5′-AMP and carbamylcholine. J Membr Biol 35:159–173
Bonting SL (1966) Studies on sodium-potassium-activated adenosinetriphosphatase. XV. The rectal gland of the elasmobranchs. Comp Biochem Physiol 17:953–966
Bourgoignie J, Klahr S, Yates J, Guerra L, Bricker NS (1969) Characteristics of ATPase system of turtle bladder epithelium. Am J Physiol 217:1496–1503
Bridges RJ, Worrell RT, Frizzell RA, Benos DJ (1989) Stilbene disulfonate blockade of colonic secretory Cl^- channels in planar lipid bilayers. Am J Physiol Cell Physiol 256:C902–C912
Browning JG, Hardcastle J, Hardcastle PT, Redfern JS (1978) Localization of the effect of acetylcholine in regulating intestinal ion transport. J Physiol 281:15–27
Burg M, Grantham J, Abramow M, Orloff J (1966) Preparation and study of fragments of single rabbit nephrons. Am J Physiol 210:1293–1298
Burg MB, Issaacson L, Grantham J, Orloff J (1968) Electrical properties of isolated perfused rabbit renal tubules. Am J Physiol 215:788–794
Burg M, Stoner L, Cardinal J, Green N (1973) Furosemide effect on isolated perfused tubules. Am J Physiol 225:119–124
Burger JW, Hess WN (1960) Function of the rectal gland in the spiny dogfish. Science 131:670–671
Burrows W, Musteikis GM (1966) Cholera infection and toxin in the rabbit ileal loop. J Infect Dis 116:183–190
Carpenter CC, Sack RB, Feeley JC, Steenberg RW (1968) Site and characteristics of electrolyte loss and effect of intraluminal glucose in experimental canine cholera. J Clin Invest 47:1210–1220. doi:10.1172/JCI105810
Chen LC, Rohde JE, Sharp GW (1971) Intestinal adenyl-cyclase activity in human cholera. Lancet 1:939–941
Choi BY, Kim HM, Ito T, Lee KY, Li X, Monahan K, Wen Y, Wilson E, Kurima K, Saunders TL, Petralia RS, Wangemann P, Friedman TB, Griffith AJ (2011) Mouse model of enlarged vestibular aqueducts defines temporal requirement of Slc26a4 expression for hearing acquisition. J Clin Invest 121:4516–4525. doi:10.1172/JCI59353
Chung DY, Chan PJ, Bankston JR, Yang L, Liu G, Marx SO, Karlin A, Kass RS (2009) Location of KCNE1 relative to KCNQ1 in the I(KS) potassium channel by disulfide cross-linking of substituted cysteines. Proc Natl Acad Sci USA 106:743–748. doi:10.1073/pnas.0811897106
Clarkson TW, Toole SR (1964) Measurement of short-circuit current and ion transport across the ileum. Am J Physiol 206:658–668
Cliff WH, Frizzell RA (1990) Separate Cl^- conductances activated by cAMP and Ca^{2+} in Cl^- secreting epithelial cells. Proc Natl Acad Sci USA 87:4956–4960
Cole KS, Curtis HJ (1941) Membrane potential of the squid giant axon during current flow. J Gen Physiol 24:551–563
Condliffe SB, Doolan CM, Harvey BJ (2001) 17beta-oestradiol acutely regulates Cl^- secretion in rat distal colonic epithelium. J Physiol 530:47–54
Cooley JW, Tukey JW (1965) An algorithm for the machine calculation of complex Fourier series. Math Comput 19:297–301
Crabbe J (1961) Stimulation of active sodium transport by the isolated toad bladder with aldosterone in vitro. J Clin Invest 40:2103–2110. doi:10.1172/JCI104436
Cuppoletti J, Chakrabarti J, Tewari KP, Malinowska DH (2014) Differentiation between human ClC-2 and CFTR Cl^- channels with pharmacological agents. Am J Physiol Cell Physiol 307: C479–C492. doi:10.1152/ajpcell.00077.2014
Curran PF, Solomon AK (1957) Ion and water fluxes in the ileum of rats. J Gen Physiol 41:143–168
Darrow CW (1945) Congenital alkalosis with diarrhea. J Pediatr 26:519–532. doi:10.1016/S0022-3476(45)80079-3

Darwin C (1839) Journal of researches into the geology and natural history of the various countries visited by H.M.S. Beagle [under the command of Captain Fitzroy, R.N. from 1832 to 1836]. Colburn, London

David JP, Andersen MN, Olesen SP, Rasmussen HB, Schmitt N (2013) Trafficking of the I_{Ks} -complex in MDCK cells: site of subunit assembly and determinants of polarized localization. Traffic 14:399–411. doi:10.1111/tra.12042

De SN, Chatterje DN (1953) An experimental study of the mechanism of action of *Vibrio cholerae* on the intestinal mucous membrane. J Pathol Bacteriol 66:559–562

Denning CR, Sommers SC, Quigley HJ Jr (1968) Infertility in male patients with cystic fibrosis. Pediatrics 41:7–17

Derksen HE, Verveen AA (1966) Fluctuations of resting neural membrane potential. Science 151:1388–1389

Devor DC, Frizzell RA (1993) Calcium-mediated agonists activate an inwardly rectified K^+ channel in colonic secretory cells. Am J Physiol Cell Physiol 265:C1271–C1280

Devor DC, Singh AK, Bridges RJ, Frizzell RA (1996a) Modulation of Cl^- secretion by benzimidazolones. II. Coordinate regulation of apical G_{Cl} and basolateral G_K. Am J Physiol Lung Cell Mol Physiol 271:L785–L795

Devor DC, Singh AK, Frizzell RA, Bridges RJ (1996b) Modulation of Cl^- secretion by benzimidazolones. I. Direct activation of a Ca^{2+}-dependent K^+ channel. Am J Physiol Lung Cell Mol Physiol 271:L775–L784

Devor DC, Singh AK, Gerlach AC, Frizzell RA, Bridges RJ (1997) Inhibition of intestinal Cl^- secretion by clotrimazole: direct effect on basolateral membrane K^+ channels. Am J Physiol Cell Physiol 273:C531–C540

Devor DC, Bridges RJ, Pilewski JM (2000) Pharmacological modulation of ion transport across wild-type and ΔF508 CFTR-expressing human bronchial epithelia. Am J Physiol Cell Physiol 279:C461–C479

Dharmsathaphorn K, Pandol SJ (1986) Mechanism of chloride secretion induced by carbachol in a colonic epithelial cell line. J Clin Invest 77:348–354. doi:10.1172/JCI112311

Dharmsathaphorn K, McRoberts JA, Mandel KG, Tisdale LD, Masui H (1984) A human colonic tumor cell line that maintains vectorial electrolyte transport. Am J Physiol Gastrointest Liver Physiol 246:G204–G208

Drumm ML, Pope HA, Cliff WH, Rommens JM, Marvin SA, Tsui L-C, Collins FS, Frizzell RA, Wilson JM (1990) Correction of the cystic fibrosis defect in vitro by retrovirus-mediated gene transfer. Cell 62:1227–1233

Dutta NK, Habbu MK (1955) Experimental cholera in infant rabbits: a method for chemotherapeutic investigation. Br J Pharmacol Chemother 10:153–159

Dvir M, Strulovich R, Sachyani D, Ben-Tal Cohen I, Haitin Y, Dessauer C, Pongs O, Kass R, Hirsch JA, Attali B (2014) Long QT mutations at the interface between KCNQ1 helix C and KCNE1 disrupt I(KS) regulation by PKA and PIP(2). J Cell Sci 127:3943–3955. doi:10.1242/jcs.147033

Edmonds CJ, Nielsen OE (1968) Transmembrane electrical potential differences and ionic composition of mucosal cells of rat colon. Acta Physiol Scand 72:338–349. doi:10.1111/j.1748-1716.1968.tb03856.x

Evans MH, Linzell JL, Peaker M (1971) Membrane potentials in the mammary gland of the lactating rat. J Physiol 213:49P–50P

Everett LA, Green ED (1999) A family of mammalian anion transporters and their involvement in human genetic diseases. Hum Mol Genet 8:1883–1891

Everett LA, Glaser B, Beck JC, Idol JR, Buchs A, Heyman M, Adawi F, Hazani E, Nassir E, Baxevanis AD, Sheffield VC, Green ED (1997) Pendred syndrome is caused by mutations in a putative sulphate transporter gene (PDS). Nat Genet 17:411–422. doi:10.1038/ng1297-411

Fairbanks G, Steck TL, Wallach DF (1971) Electrophoretic analysis of the major polypeptides of the human erythrocyte membrane. Biochemistry 10:2606–2617

Field M, Plotkin GR, Silen W (1968) Effects of vasopressin, theophylline and cyclic adenosine monophosphate on short-circuit current across isolated rabbit ileal mucosa. Nature 217:469–471

Field M, Fromm D, Wallace CK, Greenough WBI (1969) Stimulation of active chloride secretion in small intestine by cholera exotoxin. J Clin Invest 48:24a

Finkelstein RA, Atthasampunna P, Chulasamaya M, Charunmethee P (1966) Pathogenesis of experimental cholera: biologic ativities of purified procholeragen A. J Immunol 96:440–449

Finn AL, Handler JS, Orloff J (1967) Active chloride transport in the isolated toad bladder. Am J Physiol 213:179–184

Flores CA, Melvin JE, Figueroa CD, Sepulveda FV (2007) Abolition of Ca^{2+}-mediated intestinal anion secretion and increased stool dehydration in mice lacking the intermediate conductance Ca^{2+}-dependent K^+ channel Kcnn4. J Physiol 583:705–717. doi:10.1113/jphysiol.2007.134387

Florey HW, Wright RD, Jennings MA (1941) The secretions of the intestine. Physiol Rev 21:36–69

Francis WL (1933) Output of electrical energy by frog-skin. Nature 131:805

Frizzell RA (1977) Active chloride secretion by rabbit colon: calcium-dependent stimulation by ionophore A23187. J Membr Biol 35:175–187

Frizzell RA, Duffey ME (1980) Chloride activities in epithelia. Fed Proc 39:2860–2864

Frizzell RA, Hanrahan JW (2012) Physiology of epithelial chloride and fluid secretion. Cold Spring Harb Perspect Med 2:a009563. doi:10.1101/cshperspect.a009563

Frizzell RA, Koch MJ, Schultz SG (1976) Ion transport by rabbit colon. I. Active and passive components. J Membr Biol 27:297–316

Frizzell RA, Welsh MJ, Smith PL (1981) Electrophysiology of chloride-secreting epithelia. Soc Gen Physiol Ser 36:137–145

Frizzell RA, Rechkemmer G, Shoemaker RL (1986) Altered regulation of airway epithelial cell chloride channels in cystic fibrosis. Science 233:558–560

Gamble JL, Fahey KR, Appleton J, MacLachlan E (1945) Congenital alkalosis with diarrhea. J Pediatr 26:509–518. doi:10.1016/S0022-3476(45)80078-1

Geck P, Pietrzyk C, Burckhardt BC, Pfeiffer B, Heinz E (1980) Electrically silent cotransport on Na^+, K^+ and Cl^- in Ehrlich cells. Biochim Biophys Acta 600:432–447

Gerlach AC, Gangopadhyay NN, Devor DC (2000) Kinase-dependent regulation of the intermediate conductance, calcium-dependent potassium channel, hIK1. J Biol Chem 275:585–598

Gervais R, Dumur V, Letombe B, Larde A, Rigot JM, Roussel P, Lafitte JJ (1996) Hypofertility with thick cervical mucus: another mild form of cystic fibrosis? JAMA 276:1638

Giebisch G (1958) Electrical potential measurements on single nephrons of Necturus. J Cell Physiol 51:221–239

Giebisch G, Klose RM, Malnic G, Sullivan WJ, Windhager EE (1964) Sodium movement across single perfused proximal tubules of rat kidneys. J Gen Physiol 47:1175–1194

Grantham JJ, Burg MB (1966) Effect of vasopressin and cyclic AMP on permeability of isolated collecting tubules. Am J Physiol 211:255–259

Gray MA, Harris A, Coleman L, Greenwell JR, Argent BE (1989) Two types of chloride channel on duct cells cultured from human fetal pancreas. Am J Physiol Cell Physiol 257:C240–C251

Greenough WBI, Pierce NF, Al Awqati Q, Carpenter CCJ (1969) Stimulation of gut electrolyte secretion by prostaglandins, theophylline, and cholera exotoxin. J Clin Invest 48:32a

Greger R (1985) Ion transport mechanisms in thick ascending limb of Henle's loop of mammalian nephron. Physiol Rev 65:760–797

Greger R, Schlatter E (1984a) Mechanism of NaCl secretion in rectal gland tubules of spiny dogfish (*Squalus acanthias*). II. Effects of inhibitors. Pflugers Arch 402:364–375

Greger R, Schlatter E (1984b) Mechanism of NaCl secretion in the rectal gland of spiny dogfish (*Squalus acanthias*). I. Experiments in isolated in vitro perfused rectal gland tubules. Pflugers Arch 402:63–75

Greger R, Schlatter E, Wang F, Forrest JN Jr (1984) Mechanism of NaCl secretion in rectal gland tubules of spiny dogfish (*Squalus acanthias*). III. Effects of stimulation of secretion by cyclic AMP. Pflugers Arch 402:376–384

Guggino WB (1993) Outwardly rectifying chloride channels and CF: a divorce and remarriage. J Bioenerg Biomembr 25:27–35

Gutman GA, Chandy KG, Adelman JP, Aiyar J, Bayliss DA, Clapham DE, Covarriubias M, Desir GV, Furuichi K, Ganetzky B, Garcia ML, Grissmer S, Jan LY, Karschin A, Kim D, Kuperschmidt S, Kurachi Y, Lazdunski M, Lesage F, Lester HA, McKinnon D, Nichols CG, O'Kelly I, Robbins J, Robertson GA, Rudy B, Sanguinetti M, Seino S, Stuehmer W, Tamkun MM, Vandenberg CA, Wei A, Wulff H, Wymore RS (2003) International union of pharmacology. XLI. Compendium of voltage-gated ion channels: potassium channels. Pharmacol Rev 55:583–586. doi:10.1124/pr.55.4.9

Haas M, Schmidt WF 3rd, McManus TJ (1982) Catecholamine-stimulated ion transport in duck red cells. Gradient effects in electrically neutral [Na + K + 2Cl] Co-transport. J Gen Physiol 80:125–147

Hahn LA, Hevesy GC, Lundsgaard EC (1937) The circulation of phosphorus in the body revealed by application of radioactive phosphorus as indicator. Biochem J 31:1705–1709

Hahn LA, Hevesy GC, Rebbe OH (1939) Do the potassium ions inside the muscle cells and blood corpuscles exchange with those present in the plasma? Biochem J 33:1549–1558

Hamill OP, Marty A, Neher E, Sakmann B, Sigworth FJ (1981) Improved patch-clamp techniques for high-resolution current recording from cells and cell-free membrane patches. Pflugers Arch 391:85–100

Hanrahan JW, Phillips JE (1984) KCl transport across an insect epithelium: II. Electrochemical potentials and electrophysiology. J Membr Biol 80:27–47

Hastbacka J, de la Chapelle A, Mahtani MM, Clines G, Reeve-Daly MP, Daly M, Hamilton BA, Kusumi K, Trivedi B, Weaver A, Coloma A, Lovett M, Buckler A, Kaitila I, Lander ES (1994) The diastrophic dysplasia gene encodes a novel sulfate transporter: positional cloning by fine-structure linkage disequilibrium mapping. Cell 78:1073–1087

Heintze K, Stewart CP, Frizzell RA (1983) Sodium-dependent chloride secretion across rabbit descending colon. Am J Physiol Gastrointest Liver Physiol 244:G357–G365

Helman SI, Grantham JJ, Burg MB (1971) Effect of vasopressin on electrical resistance of renal cortical collecting tubules. Am J Physiol 220:1825–1832

Hevesy G, Hogfer E (1934) Elimination of water from the human body. Nature 134:879

Hevesy GV, Hofer E, Krogh A (1935) The permeability of the skin of frogs to water as determined by D_2O and H_2O. Skand Arch Physiol 72:199–214

Hihnala S, Kujala M, Toppari J, Kere J, Holmberg C, Hoglund P (2006) Expression of SLC26A3, CFTR and NHE3 in the human male reproductive tract: role in male subfertility caused by congenital chloride diarrhoea. Mol Hum Reprod 12:107–111. doi:10.1093/molehr/gal009

Hodgkin AL, Huxley AF (1952a) The components of membrane conductance in the giant axon of *Loligo*. J Physiol 116:473–496

Hodgkin AL, Huxley AF (1952b) Currents carried by sodium and potassium ions through the membrane of the giant axon of *Loligo*. J Physiol 116:449–472

Hodgkin AL, Keynes RD (1955) Active transport of cations in giant axons from *Sepia* and *Loligo*. J Physiol 128:28–60

Hodgkin AL, Huxley AF, Katz B (1952) Measurement of current–voltage relations in the membrane of the giant axon of *Loligo*. J Physiol 116:424–448

Hogben CA (1951) The chloride transport system of the gastric mucosa. Proc Natl Acad Sci USA 37:393–395

Hogben CA (1955) Active transport of chloride by isolated frog gastric epithelium: origin of the gastric mucosal potential. Am J Physiol 180:641–649

Hoglund P, Hihnala S, Kujala M, Tiitinen A, Dunkel L, Holmberg C (2006) Disruption of the SLC26A3-mediated anion transport is associated with male subfertility. Fertil Steril 85:232–235

Hook JB, Williamson HE (1965) Lack of correlation between natriuretic activity and inhibition of renal NaK-activated ATPase. Proc Soc Exp Biol Med 120:358–360
Hubel KA (1967) Bicarbonate secretion in rat ileum and its dependence on intraluminal chloride. Am J Physiol 213:1409–1413
Hubel KA (1969) Effect of luminal chloride concentration on bicarbonate secretion in rat ileum. Am J Physiol 217:40–45
Hwang TC, Lu L, Zeitlin PL, Gruenert DC, Huganir R, Guggino WB (1989) Cl^- channels in CF: lack of activation by protein kinase C and cAMP-dependent protein kinase. Science 244:1351–1353
Ingraham RC, Visscher MB (1935) Studies on the elimination of dyes in the gastric and pancreatic secretions, and inferences there from concerning the mechanisms of secretion of acid and base. J Gen Physiol 18:695–716
Isaacs PE, Corbett CL, Riley AK, Hawker PC, Turnberg LA (1976) In vitro behavior of human intestinal mucosa. The influence of acetyl choline on ion transport. J Clin Invest 58:535–542. doi:10.1172/JCI108498
Ishii TM, Silvia C, Hirschberg B, Bond CT, Adelman JP, Maylie J (1997) A human intermediate conductance calcium-activated potassium channel. Proc Natl Acad Sci USA 94:11651–11656
Jensen BS, Strobaek D, Olesen SP, Christophersen P (2001) The Ca^{2+}-activated K^+ channel of intermediate conductance: a molecular target for novel treatments? Curr Drug Targets 2:401–422
Jespersen T, Rasmussen HB, Grunnet M, Jensen HS, Angelo K, Dupuis DS, Vogel LK, Jorgensen NK, Klaerke DA, Olesen SP (2004) Basolateral localisation of KCNQ1 potassium channels in MDCK cells: molecular identification of an N-terminal targeting motif. J Cell Sci 117:4517–4526. doi:10.1242/jcs.01318
Jespersen T, Grunnet M, Olesen SP (2005) The KCNQ1 potassium channel: from gene to physiological function. Physiology (Bethesda) 20:408–416. doi:10.1152/physiol.00031.2005
Johannesson M, Landgren BM, Csemiczky G, Hjelte L, Gottlieb C (1998) Female patients with cystic fibrosis suffer from reproductive endocrinological disorders despite good clinical status. Hum Reprod 13:2092–2097
Johnson G (1866) Rules for the treatment of epidemic diarrhoea and cholera. Br Med J 2:63–65
Joiner WJ, Wang LY, Tang MD, Kaczmarek LK (1997) hSK4, a member of a novel subfamily of calcium-activated potassium channels. Proc Natl Acad Sci USA 94:11013–11018
Kartner N, Hanrahan JW, Jensen TJ, Naismith AL, Sun SZ, Ackerley CA, Reyes EF, Tsui LC, Rommens JM, Bear CE, Riordan R (1991) Expression of the cystic fibrosis gene in non-epithelial invertebrate cells produces a regulated anion conductance. Cell 64:681–691
Kerem B, Rommens JM, Buchanan JA, Markiewicz D, Cox TK, Chakravarti A, Buchwald M, Tsui LC (1989) Identification of the cystic fibrosis gene: genetic analysis. Science 245:1073–1080
Keynes RD (1969) From frog skin to sheep rumen: a survey of transport of salts and water across multicellular structures. Q Rev Biophys 2:177–281
Keys AB (1931a) Chloride and water secretion and absorption by the gills of the eel. Z vergl Physiol 15:364–388. doi:10.1007/bf00339115
Keys AB (1931b) The heart-gill preparation of the eel and its perfusion for the study of a natural membrane in situ. Z vergl Physiol 15:352–363. doi:10.1007/bf00339114
Keys A, Willmer EN (1932) "Chloride secreting cells" in the gills of fishes, with special reference to the common eel. J Physiol 76(368–378):362
Kimberg DV, Field M, Johnson J, Henderson A, Gershon E (1971) Stimulation of intestinal mucosal adenyl cyclase by cholera enterotoxin and prostaglandins. J Clin Invest 50:1218–1230. doi:10.1172/JCI106599
Kimura T (1969) Electron microscopic study of the mechanism of secretion of milk. Nihon Sanka Fujinka Gakkai Zasshi 21:301–308
Knowles MR, Stutts MJ, Spock A, Fischer N, Gatzy JT, Boucher RC (1983) Abnormal ion permeation through cystic fibrosis respiratory epithelium. Science 221:1067–1070

Ko SB, Zeng W, Dorwart MR, Luo X, Kim KH, Millen L, Goto H, Naruse S, Soyombo A, Thomas PJ, Muallem S (2004) Gating of CFTR by the STAS domain of SLC26 transporters. Nat Cell Biol 6:343–350

Koefoed-Johnsen V, Ussing HH (1958) The nature of the frog skin potential. Acta Physiol Scand 42:298–308. doi:10.1111/j.1748-1716.1958.tb01563.x

Koefoed-Johnsen V, Levi H, Ussing HH (1952a) The mode of passage of chloride ions through the isolated frog skin. Acta Physiol Scand 25:150–163. doi:10.1111/j.1748-1716.1952.tb00866.x

Koefoed-Johnsen V, Ussing HH, Zerahn K (1952b) The origin of the short-circuit current in the adrenaline stimulated frog skin. Acta Physiol Scand 27:38–48. doi:10.1111/j.1748-1716.1953.tb00922.x

Kohler M, Hirschberg B, Bond CT, Kinzie JM, Marrion NV, Maylie J, Adelman JP (1996) Small-conductance, calcium-activated potassium channels from mammalian brain. Science 273:1709–1714

Kopito RR, Lodish HF (1985) Primary structure and transmembrane orientation of the murine anion exchange protein. Nature 316:234–238

Krogh A (1937a) Osmotic regulation in the frog (*R. esculenta*) by active absorption of chloride ion. Skand Arch Physiol 76:60–74

Krogh A (1937b) The use of isotopes as indicators in biological research. Science 85:187–191. doi:10.1126/science.85.2199.187

Krogh A (1939) The active uptake of ions into cells and organisms. Proc Natl Acad Sci USA 25:275–277

Kujala M, Hihnala S, Tienari J, Kaunisto K, Hastbacka J, Holmberg C, Kere J, Hoglund P (2007) Expression of ion transport-associated proteins in human efferent and epididymal ducts. Reproduction 133:775–784, doi:133/4/775 [pii]10.1530/rep.1.00964

Kunzelmann K, Pavenstadt H, Greger R (1989) Properties and regulation of chloride channels in cystic fibrosis and normal airway cells. Pflugers Arch 415:172–182

Kurokawa J, Motoike HK, Rao J, Kass RS (2004) Regulatory actions of the A-kinase anchoring protein Yotiao on a heart potassium channel downstream of PKA phosphorylation. Proc Natl Acad Sci USA 101:16374–16378. doi:10.1073/pnas.0405583101

Kurokawa J, Bankston JR, Kaihara A, Chen L, Furukawa T, Kass RS (2009) KCNE variants reveal a critical role of the beta subunit carboxyl terminus in PKA-dependent regulation of the I_{Ks} potassium channel. Channels (Austin) 3:16–24

Lee MG, Wigley WC, Zeng W, Noel LE, Marino CR, Thomas PJ, Muallem S (1999) Regulation of Cl^-/HCO_3^- exchange by cystic fibrosis transmembrane conductance regulator expressed in NIH 3T3 and HEK 293 cells. J Biol Chem 274:3414–3421

Leitch GJ, Burrows W (1968) Experimental cholera in the rabbit ligated intestine: ion and water accumulation in the duodenum, ileum and colon. J Infect Dis 118:349–359

Li M, McCann JD, Anderson MP, Clancy JP, Liedtke CM, Nairn AC, Greengard P, Welsch MJ (1989) Regulation of chloride channels by protein kinase C in normal and cystic fibrosis airway epithelia. Science 244:1353–1356

Li Y, Chen L, Kass RS, Dessauer CW (2012) The A-kinase anchoring protein Yotiao facilitates complex formation between adenylyl cyclase type 9 and the I_{Ks} potassium channel in heart. J Biol Chem 287:29815–29824. doi:10.1074/jbc.M112.380568

Li X, Sanneman JD, Harbidge DG, Zhou F, Ito T, Nelson R, Picard N, Chambrey R, Eladari D, Miesner T, Griffith AJ, Marcus DC, Wangemann P (2013) SLC26A4 targeted to the endolymphatic sac rescues hearing and balance in Slc26a4 mutant mice. PLoS Genet 9(7), e1003641. doi:10.1371/journal.pgen.1003641

Liin SI, Barro-Soria R, Larsson HP (2015) The KCNQ1 channel - remarkable flexibility in gating allows for functional versatility. J Physiol. doi:10.1113/jphysiol.2014.287607

Lindemann B, Van Driessche W (1977) Sodium-specific membrane channels of frog skin are pores: current fluctuations reveal high turnover. Science 195:292–294

Lindemann B, Van Driessche W (1978) The mechanism of Na-uptake through Na-selective channels in the epithelium of frog skin. In: Hoffman JF (ed) Membrane transport processes, vol 1, Membrane transport processes. Raven, New York, pp 155–178

Linzell JL, Peaker M (1971a) The effects of oxytocin and milk removal on milk secretion in the goat. J Physiol Lond 216:717–734

Linzell JL, Peaker M (1971b) Intracellular concentrations of sodium, potassium and chloride in the lactating mammary gland and their relation to the secretory mechanism. J Physiol Lond 216:683–700

Linzell JL, Peaker M (1971c) Mechanism of milk secretion. Physiol Rev 51:564–597

Linzell JL, Peaker M (1971d) The permeability of mammary ducts. J Physiol Lond 216:701–716

Lohrmann E, Burhoff I, Nitschke RB, Lang HJ, Mania D, Englert HC, Hropot M, Warth R, Rohm W, Bleich M, Greger R (1995) A new class of inhibitors of cAMP-mediated Cl^- secretion in rabbit colon, acting by the reduction of cAMP-activated K^+ conductance. Pflugers Arch 429:517–530

Lutz MD, Cardinal J, Burg MB (1973) Electrical resistance of renal proximal tubule perfused in vitro. Am J Physiol 225:729–734

Mandel KG, Dharmsathaphorn K, McRoberts JA (1986) Characterization of a cyclic AMP-activated Cl-transport pathway in the apical membrane of a human colonic epithelial cell line. J Biol Chem 261:704–712

Maruyama Y, Peterson OH (1982) Single-channel currents in isolated patches of plasma membrane from basal surface of pancreatic acini. Nature 299:159–161

Maruyama Y, Gallacher DV, Petersen OH (1983) Voltage and Ca^{2+}-activated K^+ channel in basolateral acinar cell membranes of mammalian salivary glands. Nature 302:827–829

Marx SO, Kurokawa J, Reiken S, Motoike H, D'Armiento J, Marks AR, Kass RS (2002) Requirement of a macromolecular signaling complex for beta adrenergic receptor modulation of the KCNQ1-KCNE1 potassium channel. Science 295:496–499. doi:10.1126/science.1066843

Mazhari R, Nuss HB, Armoundas AA, Winslow RL, Marban E (2002) Ectopic expression of KCNE3 accelerates cardiac repolarization and abbreviates the QT interval. J Clin Invest 109:1083–1090. doi:10.1172/JCI15062

McCann JD, Matsuda J, Garcia M, Kaczorowski G, Welsh MJ (1990) Basolateral K^+ channels in airway epithelia. I. Regulation by Ca^{2+} and block by charybdotoxin. Am J Physiol Lung Cell Mol Physiol 258:L334–L342

McRoberts JA, Erlinger S, Rindler MJ, Saier MH Jr (1982) Furosemide-sensitive salt transport in the Madin-Darby canine kidney cell line. Evidence for the cotransport of Na^+, K^+, and Cl^-. J Biol Chem 257:2260–2266

McRoberts JA, Beuerlein G, Dharmsathaphorn K (1985) Cyclic AMP and Ca^{2+}-activated K^+ transport in a human colonic epithelial cell line. J Biol Chem 260:14163–14172

Melman YF, Um SY, Krumerman A, Kagan A, McDonald TV (2004) KCNE1 binds to the KCNQ1 pore to regulate potassium channel activity. Neuron 42:927–937. doi:10.1016/j.neuron.2004.06.001

Mount DB, Romero MF (2004) The SLC26 gene family of multifunctional anion exchangers. Pflugers Arch 447:710–721

Nadel JA, Davis B (1978) Regulation of Na^+ and Cl^- transport and mucous gland secretion in airway epithelium. Ciba Found Symp 54:133–147

Nakajo K, Kubo Y (2007) KCNE1 and KCNE3 stabilize and/or slow voltage sensing S4 segment of KCNQ1 channel. J Gen Physiol 130:269–281. doi:10.1085/jgp.200709805

Nakajo K, Kubo Y (2015) KCNQ1 channel modulation by KCNE proteins via the voltage-sensing domain. J Physiol. doi:10.1113/jphysiol.2014.287672

Neher E, Sakmann B (1976) Single-channel currents recorded from membrane of denervated frog muscle fibres. Nature 260:799–802

Nelson DJ, Tang JM, Palmer LG (1984) Single-channel recordings of apical membrane chloride conductance in A6 epithelial cells. J Membr Biol 80:81–89

O'Donnell EK, Sedlacek RL, Singh AK, Schultz BD (2000) Inhibition of enterotoxin-induced porcine colonic secretion by diarylsulfonylureas in vitro. Am J Physiol Gastrointest Liver Physiol 279:G1104–G1112

O'Grady SM, Musch MW, Field M (1986) Stoichiometry and ion affinities of the Na-K-Cl cotransport system in the intestine of the winter flounder (*Pseudopleuronectes americanus*). J Membr Biol 91:33–41

O'Mahony F, Alzamora R, Betts V, LaPaix F, Carter D, Irnaten M, Harvey BJ (2007) Female gender-specific inhibition of KCNQ1 channels and chloride secretion by 17beta-estradiol in rat distal colonic crypts. J Biol Chem 282:24563–24573. doi:10.1074/jbc.M611682200

Panaghie G, Tai KK, Abbott GW (2006) Interaction of KCNE subunits with the KCNQ1 K^+ channel pore. J Physiol 570:455–467. doi:10.1113/jphysiol.2005.100644

Pedersen KA, Schroder RL, Skaaning-Jensen B, Strobaek D, Olesen SP, Christophersen P (1999) Activation of the human intermediate-conductance Ca^{2+}-activated K^+ channel by 1-ethyl-2-benzimidazolinone is strongly Ca^{2+}-dependent. Biochim Biophys Acta 1420:231–240

Pendred V (1896) Deaf-mutism and goitre. Lancet 2:532. doi:10.1016/S0140-6736(01)74403-0

Perkins FM, Handler JS (1981) Transport properties of toad kidney epithelia in culture. Am J Physiol Cell Physiol 241:C154–C159

Pierce NF, Carpenter CC Jr, Elliott HL, Greenough WB 3rd (1971a) Effects of prostaglandins, theophylline, and cholera exotoxin upon transmucosal water and electrolyte movement in the canine jejunum. Gastroenterology 60:22–32

Pierce NF, Greenough WB 3rd, Carpenter CC Jr (1971b) *Vibrio cholerae* enterotoxin and its mode of action. Bacteriol Rev 35:1–13

Pierucci-Alves F, Akoyev V, Stewart JC 3rd, Wang LH, Janardhan KS, Schultz BD (2011) Swine models of cystic fibrosis reveal male reproductive tract phenotype at birth. Biol Reprod 85:442–451. doi:10.1095/biolreprod.111.090860

Pierucci-Alves F, Akoyev V, Schultz BD (2015) Bicarbonate exchangers SLC26A3 and SLC26A6 are localized at the apical membrane of porcine vas deferens epithelium. Physiol Rep 3(4): e12380, doi:10.14814/phy2.12380

Potet F, Scott JD, Mohammad-Panah R, Escande D, Baro I (2001) AKAP proteins anchor cAMP-dependent protein kinase to KvLQT1/IsK channel complex. Am J Physiol Heart Circ Physiol 280:H2038–H2045

Pusch M (1998) Increase of the single-channel conductance of KvLQT1 potassium channels induced by the association with minK. Pflugers Arch 437:172–174

Quinton PM (1983) Chloride impermeability in cystic fibrosis. Nature 301:421–422

Quinton PM (1986) Missing Cl conductance in cystic fibrosis. Am J Physiol Cell Physiol 251: C649–C652

Quinton PM, Bijman J (1983) Higher bioelectric potentials due to decreased chloride absorption in the sweat glands of patients with cystic fibrosis. N Engl J Med 308:1185–1189. doi:10.1056/NEJM198305193082002

Reenstra WW (1993) Inhibition of cAMP- and Ca-dependent Cl^- secretion by phorbol esters: inhibition of basolateral K^+ channels. Am J Physiol Cell Physiol 264:C161–C168

Reuss L, Reinach P, Weinman SA, Grady TP (1983) Intracellular ion activities and Cl-transport mechanisms in bullfrog corneal epithelium. Am J Physiol Cell Physiol 244:C336–C347

Riordan JR, Rommens JM, Kerem B, Alon N, Rozmahel R, Grzelczak Z, Zielenski J, Lok S, Plavsic N, Chou JL, Drumm ML, Iannuzzi MC, Collins FS, Tsui LC (1989) Identification of the cystic fibrosis gene: cloning and characterization of complementary DNA. Science 245:1066–1073

Roggin GM, Banwell JG, Yardley JH, Hendrix TR (1972) Unimpaired response of rabbit jejunum to cholera toxin after selective damage to villus epithelium. Gastroenterology 63:981–989

Romero MF, Fulton CM, Boron WF (2004) The SLC4 family of HCO_3^- transporters. Pflugers Arch 447:495–509. doi:10.1007/s00424-003-1180-2

Romey G, Attali B, Chouabe C, Abitbol I, Guillemare E, Barhanin J, Lazdunski M (1997) Molecular mechanism and functional significance of the MinK control of the KvLQT1 channel activity. J Biol Chem 272:16713–16716

Rommens JM, Iannuzzi MC, Kerem B, Drumm ML, Melmer G, Dean M, Rozmahel R, Cole JL, Kennedy D, Hidaka N, Zsiga M, Buchwald M, Riordan JR, Tsui LC, Collins FS (1989) Identification of the cystic fibrosis gene: chromosome walking and jumping. Science 245:1059–1065

Rose RC, Schultz SG (1971) Studies on the electrical potential profile across rabbit ileum. Effects of sugars and amino acids on transmural and transmucosal electrical potential differences. J Gen Physiol 57:639–663

Rothstein A, Cabantchik ZI, Knauf P (1976) Mechanism of anion transport in red blood cells: role of membrane proteins. Fed Proc 35:3–10

Rufo PA, Jiang L, Moe SJ, Brugnara C, Alper SL, Lencer WI (1996) The antifungal antibiotic, clotrimazole, inhibits Cl^- secretion by polarized monolayers of human colonic epithelial cells. J Clin Invest 98:2066–2075. doi:10.1172/JCI119012

Sanguinetti MC, Curran ME, Zou A, Shen J, Spector PS, Atkinson DL, Keating MT (1996) Coassembly of K_VLQT1 and minK (IsK) proteins to form cardiac I_{Ks} potassium channel. Nature 384:80–83. doi:10.1038/384080a0

Schmidt-Nielsen K, Fange R (1958) Salt glands in marine reptiles. Nature 182:783–784

Schmidt-Nielsen K, Jorgensen CB, Osaki H (1958) Extrarenal salt excretion in birds. Am J Physiol 193:101–107

Schoumacher RA, Ram J, Iannuzzi MC, Bradbury NA, Wallace RW, Hon CT, Kelly DR, Schmid SM, Gelder FB, Rado TA, Frizzell RA (1990) A cystic fibrosis pancreatic adenocarcinoma cell line. Proc Natl Acad Sci USA 87:4012–4016

Schroeder BC, Waldegger S, Fehr S, Bleich M, Warth R, Greger R, Jentsch TJ (2000) A constitutively open potassium channel formed by KCNQ1 and KCNE3. Nature 403:196–199. doi:10.1038/35003200

Schultz SG (1972) Electrical potential differences and electromotive forces in epithelial tissues. J Gen Physiol 59:794–798

Schultz SG, Frizzell RA (1972) An overview of intestinal absorptive and secretory processes. Gastroenterology 63:161–170

Schultz SG, Zalusky R, Gass AE Jr (1964) Ion transport in isolated rabbit ileum. 3. Chloride fluxes. J Gen Physiol 48:375–378

Schultz BD, Singh AK, Aguilar-Bryan L, Frizzell RA, Bridges RJ (1995a) LY295501; a sulfonylurea that blocks CFTR Cl^- channels, but does not alter pancreatic β-cell function. Ped Pulm Suppl 12:200

Schultz BD, Venglarik CJ, Bridges RJ, Frizzell RA (1995b) Regulation of CFTR Cl^- channel gating by ADP and ATP analogues. J Gen Physiol 105:329–361

Schultz BD, DeRoos AD, Venglarik CJ, Singh AK, Frizzell RA, Bridges RJ (1996) Glibenclamide blockade of CFTR chloride channels. Am J Physiol Lung Cell Mol Physiol 271:L192–L200

Schweinfest CW, Henderson KW, Suster S, Kondoh N, Papas TS (1993) Identification of a colon mucosa gene that is down-regulated in colon adenomas and adenocarcinomas. Proc Natl Acad Sci USA 90:4166–4170

Serebro HA, Iber FL, Yardley JH, Hendrix TR (1969) Inhibition of cholera toxin action in the rabbit by cycloheximide. Gastroenterology 56:506–511

Sesti F, Goldstein SA (1998) Single-channel characteristics of wild-type I_{Ks} channels and channels formed with two minK mutants that cause long QT syndrome. J Gen Physiol 112:651–663

Shamgar L, Haitin Y, Yisharel I, Malka E, Schottelndreier H, Peretz A, Paas Y, Attali B (2008) KCNE1 constrains the voltage sensor of Kv7.1 K^+ channels. PLoS ONE 3(4), e1943. doi:10.1371/journal.pone.0001943

Sharp GW, Hynie S (1971) Stimulation of intestinal adenyl cyclase by cholera toxin. Nature 229:266–269

Sharp GW, Hynie S, Lipson LC, Parkinson DK (1971) Action of cholera toxin to stimulate adenyl cyclase. Trans Assoc Am Physicians 84:200–211
Sheppard DN, Welsh MJ (1992) Effect of ATP-sensitive K^+ channel regulators on cystic fibrosis transmembrane conductance regulator chloride currents. J Gen Physiol 100:573–591
Shoemaker RL, Makhlouf GM, Sachs G (1970) Action of cholinergic drugs on *Necturus* gastric mucosa. Am J Physiol 219:1056–1060
Shorofsky SR, Field M, Fozzard HA (1983) Electrophysiology of Cl secretion in canine trachea. J Membr Biol 72:105–115
Shorofsky SR, Field M, Fozzard HA (1984) Mechanism of Cl secretion in canine trachea: changes in intracellular chloride activity with secretion. J Membr Biol 81:1–8
Shorofsky SR, Field M, Fozzard HA (1986) Changes in intracellular sodium with chloride secretion in dog tracheal epithelium. Am J Physiol Cell Physiol 250:C646–C650
Singh AK, Venglarik CJ, Bridges RJ (1995) Development of chloride channel modulators. Kidney Int 48:985–993
Singh AK, Schultz BD, van Driessche W, Bridges RJ (2004) Transepithelial fluctuation analysis of chloride secretion. J Cyst Fibros 3(Suppl 2):127–132, doi:S1569199304000815 [pii]10.1016/j.jcf.2004.05.027
Singh AK, Riederer B, Chen M, Xiao F, Krabbenhoft A, Engelhardt R, Nylander O, Soleimani M, Seidler U (2010) The switch of intestinal Slc26 exchangers from anion absorptive to HCO_3^- secretory mode is dependent on CFTR anion channel function. Am J Physiol Cell Physiol 298: C1057–C1065, doi:ajpcell.00454.2009 [pii]10.1152/ajpcell.00454.2009
Skou JC (1957) The influence of some cations on an adenosine triphosphatase from peripheral nerves. Biochim Biophys Acta 23:394–401
Smith PL, Frizzell RA (1984) Chloride secretion by canine tracheal epithelium: IV. Basolateral membrane K permeability parallels secretion rate. J Membr Biol 77:187–199
Snow J (1849) On the mode of the communication of cholera. John Churchill, London
Snow J (1855) On the mode of the communication of cholera (second edition), 2nd edn. John Churchill, London
Splawski I, Timothy KW, Vincent GM, Atkinson DL, Keating MT (1997) Molecular basis of the long-QT syndrome associated with deafness. N Engl J Med 336:1562–1567. doi:10.1056/NEJM199705293362204
Stutts MJ, Cotton CU, Yankaskas JR, Cheng E, Knowles MR, Gatzy JT, Boucher RC (1985) Chloride uptake into cultured airway epithelial cells from cystic fibrosis patients and normal individuals. Proc Natl Acad Sci USA 82:6677–6681
Suessbrich H, Busch AE (1999) The I_{Ks} channel: coassembly of I_{sK} (minK) and KvLQT1 proteins. Rev Physiol Biochem Pharmacol 137:191–226
Suessbrich H, Bleich M, Ecke D, Rizzo M, Waldegger S, Lang F, Szabo I, Lang HJ, Kunzelmann K, Greger R, Busch AE (1996) Specific blockade of slowly activating I_{sK} channels by chromanols - impact on the role of I_{sK} channels in epithelia. FEBS Lett 396:271–275
Syme CA, Gerlach AC, Singh AK, Devor DC (2000) Pharmacological activation of cloned intermediate- and small-conductance Ca^{2+}-activated K^+ channels. Am J Physiol Cell Physiol 278:C570–C581
Takumi T, Ohkubo H, Nakanishi S (1988) Cloning of a membrane protein that induces a slow voltage-gated potassium current. Science 242(4881):1042–1045
Teorell T (1939) On the permeability of the stomach mucosa for acids and sme other substances. J Gen Physiol 23:263–274
Thesleff S, Schmidt-Nielsen K (1962) An electrophysiological study of the salt gland of the herring gull. Am J Physiol 202:597–600
Tidball CS (1961) Active chloride transport during intestinal secretion. Am J Physiol 200:309–312
Trier JS (1964) Studies on small intestinal crypt epithelium. I. Evidence for the mechanisms of secretory activity by undifferentiated crypt cells of the human small intestine. Gastroenterology 47:480–495

Turnberg LA, Bieberdorf FA, Morawski SG, Fordtran JS (1970) Interrelationships of chloride, bicarbonate, sodium, and hydrogen transport in the human ileum. J Clin Invest 49:557–567. doi:10.1172/JCI106266

Ueda S, Loo DD, Sachs G (1987) Regulation of K^+ channels in the basolateral membrane of *Necturus* oxyntic cells. J Membr Biol 97:31–41

Ussing HH (1949a) The active ion transport through the isolated frog skin in the light of tracer studies. Acta Physiol Scand 17:1–37. doi:10.1111/j.1748-1716.1949.tb00550.x

Ussing HH (1949b) Transport of ions across cellular membranes. Physiol Rev 29:127–155

Ussing HH, Zerahn K (1951) Active transport of sodium as the source of electric current in the short-circuited isolated frog skin. Acta Physiol Scand 23:110–127. doi:10.1111/j.1748-1716.1951.tb00800.x

Van Driessche W, Zeiske W (1980) Spontaneous fluctuations of potassium channels in the apical membrane of frog skin. J Physiol 299:101–116

Van Driessche W, Zeiske W (1985) Ionic channels in epithelial cell membranes. Physiol Rev 65:833–903

Vaughn JL, Goodwin RH, Tompkins GJ, McCawley P (1977) The establishment of two cell lines from the insect *Spodoptera frugiperda* (Lepidoptera; Noctuidae). In Vitro 13:213–217

Veilleux S, Holt N, Schultz BD, Dubreuil JD (2008) *Escherichia coli* EAST1 toxin toxicity of variants 17–2 and O 42. Comp Immunol Microbiol Infect Dis 31:567–578, doi:S0147-9571(07)00107-5 [pii]10.1016/j.cimid.2007.10.003

Venglarik CJ, Schultz BD, Frizzell RA, Bridges RJ (1994) ATP alters current fluctuations of cystic fibrosis transmembrane conductance regulator: evidence for a three-state activation mechanism. J Gen Physiol 104:123–146

Venglarik CJ, Schultz BD, de Roos AD, Singh AK, Bridges RJ (1996) Tolbutamide causes open channel blockade of cystic fibrosis transmembrane conductance regulator Cl^- channels. Biophys J 70:2696–2703

Verveen AA, Derksen HE, Schick KL (1967) Voltage fluctuations of neural membrane. Nature 216:588–589

Walker WG, Cooke CR, Iber FL, Lesch M, Caranasos GJ (1967) Topics in clinical medicine. A symposium. Uses and complications of diuretic therapy. Johns Hopkins Med J 121:194–216

Wang KW, Tai KK, Goldstein SA (1996a) MinK residues line a potassium channel pore. Neuron 16:571–577

Wang Q, Curran ME, Splawski I, Burn TC, Millholland JM, VanRaay TJ, Shen J, Timothy KW, Vincent GM, de Jager T, Schwartz PJ, Toubin JA, Moss AJ, Atkinson DL, Landes GM, Connors TD, Keating MT (1996b) Positional cloning of a novel potassium channel gene: KVLQT1 mutations cause cardiac arrhythmias. Nat Genet 12:17–23. doi:10.1038/ng0196-17

Wang W, Xia J, Kass RS (1998) MinK-KvLQT1 fusion proteins, evidence for multiple stoichiometries of the assembled I_{sK} channel. J Biol Chem 273:34069–34074

Wang Z, Wang T, Petrovic S, Tuo B, Riederer B, Barone S, Lorenz JN, Seidler U, Aronson PS, Soleimani M (2005) Renal and intestinal transport defects in Slc26a6-null mice. Am J Physiol Cell Physiol 288:C957–C965

Wangemann P (2002) Adrenergic and muscarinic control of cochlear endolymph production. Adv Otorhinolaryngol 59:42–50

Wangemann P (2006) Supporting sensory transduction: cochlear fluid homeostasis and the endocochlear potential. J Physiol 576:11–21. doi:10.1113/jphysiol.2006.112888

Wangemann P, Wittner M, Di Stefano A, Englert HC, Lang HJ, Schlatter E, Greger R (1986) Cl^-channel blockers in the thick ascending limb of the loop of Henle. Structure activity relationship. Pflugers Arch 407(Suppl 2):S128–S141

Warth R, Hamm K, Bleich M, Kunzelmann K, von Hahn T, Schreiber R, Ullrich E, Mengel M, Trautmann N, Kindle P, Schwab A, Greger R (1999) Molecular and functional characterization of the small Ca^{2+}-regulated K^+ channel (rSK4) of colonic crypts. Pflugers Arch 438(4):437–444

Watten RH, Morgan FM, Yachai Na S, Vanikiati B, Phillips RA (1959) Water and electrolyte studies in cholera. J Clin Invest 38:1879–1889. doi:10.1172/JCI103965
Welsh MJ (1983a) Inhibition of chloride secretion by furosemide in canine tracheal epithelium. J Membr Biol 71:219–226
Welsh MJ (1983b) Intracellular chloride activities in canine tracheal epithelium. Direct evidence for sodium-coupled intracellular chloride accumulation in a chloride-secreting epithelium. J Clin Invest 71:1392–1401
Welsh MJ (1986) An apical-membrane chloride channel in human tracheal epithelium. Science 232:1648–1650
Welsh MJ, Smith PL, Frizzell RA (1982a) Chloride secretion by canine tracheal epithelium: II. The cellular electrical potential profile. J Membr Biol 70:227–238
Welsh MJ, Smith PL, Fromm M, Frizzell RA (1982b) Crypts are the site of intestinal fluid and electrolyte secretion. Science 218:1219–1221
Welsh MJ, Smith PL, Frizzell RA (1983) Chloride secretion by canine tracheal epithelium: III. Membrane resistances and electromotive forces. J Membr Biol 71:209–218
Weymer A, Huott P, Liu W, McRoberts JA, Dharmsathaphorn K (1985) Chloride secretory mechanism induced by prostaglandin E1 in a colonic epithelial cell line. J Clin Invest 7:1828–1836. doi:10.1172/JCI112175
Widdicombe JH, Nathanson IT, Highland E (1983) Effects of "loop" diuretics on ion transport by dog tracheal epithelium. Am J Physiol Cell Physiol 245:C388–C396
Widdicombe JH, Welsh MJ, Finkbeiner WE (1985) Cystic fibrosis decreases the apical membrane chloride permeability of monolayers cultured from cells of tracheal epithelium. Proc Natl Acad Sci USA 82:6167–6171
Worrell RT, Butt AG, Cliff WH, Frizzell RA (1989) A volume-sensitive chloride conductance in human colonic cell line T84. Am J Physiol Cell Physiol 256:C1111–C1119
Wright EM (1972) Mechanisms of ion transport across the choroid plexus. J Physiol 226:545–571
Wright RD, Jennings MA, Florey HW, Lium R (1940) The influence of nerves and drugs on secretion by the small intestine and an investigation of the enzymes in intestinal juice. Exp Physiol 30:73–120
Yang Y, Sigworth FJ (1998) Single-channel properties of I_{Ks} potassium channels. J Gen Physiol 112:665–678
Yankaskas JR, Cotton CU, Knowles MR, Gatzy JT, Boucher RC (1985) Culture of human nasal epithelial cells on collagen matrix supports. A comparison of bioelectric properties of normal and cystic fibrosis epithelia. Am Rev Respir Dis 132:1281–1287
Zadunaisky JA (1966) Active transport of chloride in frog cornea. Am J Physiol 211:506–512
Zadunaisky JA, Lande MA (1971) Active chloride transport and control of corneal transparency. Am J Physiol 221:1837–1844
Zadunaisky JA, Candia OA, Chiarandini DJ (1963) The origin of the short-circuit current in the isolated skin of the South American frog *Leptodactylus Ocellatus*. J Gen Physiol 47:393–402
Zaki L, Fasold H, Schuhmann B, Passow H (1975) Chemical modification of membrane proteins in relation to inhibition of anion exchange in human red blood cells. J Cell Physiol 86:471–494. doi:10.1002/jcp.1040860305
Zaydman MA, Cui J (2014) PIP2 regulation of KCNQ channels: biophysical and molecular mechanisms for lipid modulation of voltage-dependent gating. Front Physiol 5:195. doi:10.3389/fphys.2014.00195
Zimmerman TW, Dobbins JW, Binder HJ (1982) Mechanism of cholinergic regulation of electrolyte transport in rat colon in vitro. Am J Physiol Gastrointest Liver Physiol 242:G116–G123

Chapter 2
Fundamentals of Epithelial Na^+ Absorption

Alexander Staruschenko, Daria V. Ilatovskaya, and Kenneth R. Hallows

Abstract The maintenance of electrolyte balance is essential for the control of many functions in the human body. Na^+, K^+, and Cl^- are key electrolytes that contribute to a variety of processes ranging from the maintenance of cellular membrane potential to the regulation of cell volume and extracellular fluid volume. Fundamentals of epithelial Cl^- and K^+ transport are discussed in the preceding and following chapters and will be only briefly touched upon here. Na^+ absorption occurs across the epithelial barriers of many organs, including the lung, gastrointestinal tract, exocrine glands, and kidney. Na^+ is the primary determinant of blood volume, and a number of physiological mechanisms that control blood pressure mediate their effects by adjusting Na^+ balance in the kidney. This chapter describes classical fundamentals of epithelial Na^+ absorption and highlights some recent mechanisms involved in physiological regulation of Na^+ transport in specific epithelia with a particular emphasis on the kidney.

Keywords Kidney • Hypertension • Na^+/K^+ pump • NHE • SGLT • NKCC • NCC • ENaC • NBC • NPT

2.1 Introduction

In this book, specific chapters deal with all aspects of transport of the individual solutes and electrolytes. In the present chapter, we focus on general principles of epithelial Na^+ absorption and mechanisms controlling this process. As recently summarized in an excellent review by Kotchen and colleagues, current levels of salt

A. Staruschenko (✉) • D.V. Ilatovskaya
Department of Physiology, Medical College of Wisconsin, 8701 Watertown Plank Rd, Milwaukee, WI 53226, USA
e-mail: staruschenko@mcw.edu

K.R. Hallows (✉)
Division of Nephrology and Hypertension, Department of Medicine, Keck School of Medicine of USC, 2020 Zonal Av., IRD 806, Los Angeles, CA 90033, USA
e-mail: hallows@usc.edu

K.L. Hamilton, D.C. Devor (eds.), *Ion Channels and Transporters of Epithelia in Health and Disease*, Physiology in Health and Disease,
DOI 10.1007/978-1-4939-3366-2_2

consumption in modern society generally exceed total body salt needs and are associated with adverse clinical outcomes. High salt intake is associated with hypertension and increased rates of cardiovascular disease. Experimental studies continue to provide information about mechanisms for these adverse effects of salt (Kotchen et al. 2013). Thus, despite the critical role of NaCl in the maintenance of multiple mechanisms in the human body, many individuals, especially those on Western diets, need to limit salt consumption. When we eat too much salt, several organs, especially the kidneys, act to help us excrete the excess salt load. Sodium absorption in the kidney is precisely regulated and controlled by numerous mechanisms, many of which are reviewed in this chapter. Multiple studies have focused on transport processes in the kidney. As a consequence, much of our current knowledge of the principles of epithelial transport has been derived from studies on renal epithelia. Thus, we provide here the fundamentals of epithelial Na^+ absorption, focusing primarily on the mechanisms involved in the transport of this important ion in the kidney. In addition, modulation of transepithelial Na^+ reabsorption in the lung plays a key role in determining airway surface liquid volume and airway clearance, which is dysregulated in certain diseases like cystic fibrosis. The following section of this chapter will briefly describe sodium transport in the lung and other organs.

Although our understanding of the specific details and mechanisms of function of epithelial Na^+ absorption has greatly expanded recently, the main concepts remain remarkably similar to those described 50–60 years ago. Several excellent review articles and book chapters have described these mechanisms, and we briefly summarize them here. In addition, we recommend to the reader additional renal physiology textbooks, including those edited by Schrier, Brenner, Boron, and Guyton/Hall.

Na^+ is the principal extracellular cation and is present within cells at a much lower concentration than it is in the extracellular space. Typically, Na^+ concentrations are 10–20 mM and 135–145 mM in the intracellular and extracellular compartments in humans, respectively. The glomeruli filter approximately 25,000 mEq of Na^+ per day. As sodium is freely filtered by glomeruli, the kidneys must reabsorb the vast majority of Na^+ as the filtrate flows along the nephron. Na^+ absorption in the kidney is co-regulated with the transport of other ions (including chloride, phosphate, bicarbonate, potassium, protons, calcium, and magnesium), neutral solutes (e.g., glucose and amino acids), and water via directly or functionally coupled transport processes at both the apical and basolateral membranes. This is accomplished by the integrated function of consecutive sodium transporters and channels along the nephron.

2.2 General Concepts of Sodium Absorption in Epithelia

2.2.1 Basic Principles of Sodium Transport

Epithelial sodium (Na^+) absorption represents in general reabsorption of sodium ions from the tubules back into the blood. In this case, it is called reabsorption rather than absorption because sodium returns to the circulation after it had been secreted the renal tubules. The classical concept of transepithelial NaCl absorption, which was formulated more than 50 years ago by Koefoed-Johnsen and Ussing for the frog skin (Koefoed-Johnsen and Ussing 1958) is still valid in general for sodium reabsorption in the kidney. Palmer and Andersen briefly discussed the historical background and wide-ranging impact and implications of this seminal work (Palmer and Andersen 2008).

In the kidney, Na^+, along with other ions such as chloride (Cl^-) and bicarbonate (HCO_3^-), is highly reabsorbed. However, the rate of reabsorption depends on body needs and is precisely controlled to maintain homeostatic balance. To be reabsorbed, Na^+ must be transported across the tubular epithelial membranes in processes that will be discussed in this chapter. Reabsorption of Na^+ across the tubular epithelium into the interstitium includes active transport, passive diffusion, or both. The rate of total transport across a cellular membrane (dM/dt) is the sum of passive (PT) and active (carrier-mediated; CM) transport and can be described by using the Michaelis–Menten Eq. (2.1) (Sugano et al. 2010):

$$\frac{\mathrm{d}M}{\mathrm{d}t} = A \times (P_{PT} + P_{CM}) \times C = A \times P_{PT} \times C + \frac{V_{\max} \times C}{K_m + C} \tag{2.1}$$

where A is the surface area of a membrane (length^2), P is the permeability (length/time), C is the concentration of a permeant species (amount/length^3), K_m is the Michaelis constant (amount/length^3), and $V_{\max}$ is the maximum carrier-mediated transport (amount/time) (Sugano et al. 2010).

As shown in Fig. 2.1, there are two basic mechanisms: (1) the transcellular route whereby sodium is transported through the cell across both the apical and basolateral plasma membranes and (2) the paracellular route, where sodium ions traverse the spaces between cell–cell junctions. After Na^+ is reabsorbed across the tubular epithelial cells into the interstitial fluid, sodium is then transported through peritubular capillary walls into the blood by ultrafiltration. This process is mediated by Starling forces (Starling 1896; Levick and Michel 2010), a combination of hydrostatic pressure and colloid osmotic forces, which will not be covered further in this chapter.

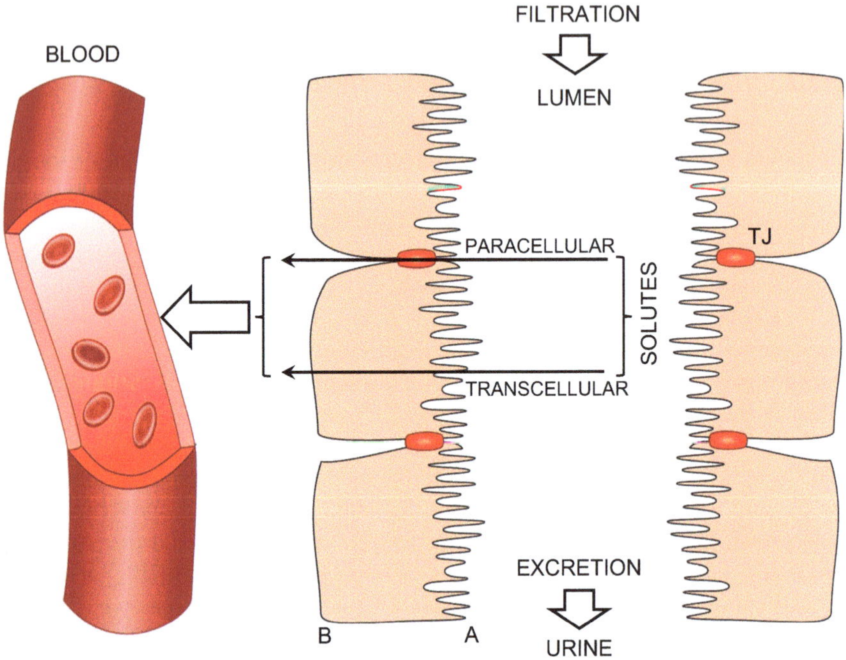

Fig. 2.1 General concept of active and passive Na^+ transport through the cell. Reabsorption of filtered Na^+ from the tubular lumen across the epithelial cells back into the blood. Na^+ is transported either by active transport through the cells (transcellular pathway) or between the cells (paracellular pathway) by diffusion. Tight junctions (TJ) are shown

2.2.2 *Cytosolic Diffusion*

As described in Einstein's theory of Brownian motion, molecular diffusion (or diffusion) consists of the random movements of a molecule with respect to adjacent molecules and occurs as the consequence of thermal energy. Since the diffusional movement of an individual molecule is random, a concentration gradient is required for any net transfer of molecules to occur across a membrane. Thus, the concentration gradient represents a major driving force for net transport. In addition, the movements of charged solutes (ions) are governed by electrical gradients created by potential differences between compartments, and thus net ion fluxes are dictated by their electrochemical gradients across membranes.

2.2.3 *Maintenance of Membrane Potential*

In addition to its critical role in the maintenance of electrolyte homeostasis, Na^+ concentrations play an important role in the maintenance of cellular membrane potential, as originally formulated by Goldman, Hodgkin, and Katz (Goldman 1943; Hodgkin and Katz 1949). Studies of various segments of the nephron have revealed heterogeneity of the electrical properties along the segments. The electrophysiological properties, such as transepithelial voltage and resistance (V_t and R_T, respectively), vary widely. Na^+, along with K^+ and Cl^-, contributes to the maintenance of concentration and charge differences across cell membranes. The intracellular and extracellular concentration differences between K^+ and Na^+ create an electrochemical gradient across the plasma membrane known as the cellular membrane potential. In kidney epithelial cells, the cellular membrane potential is maintained mostly by ionic gradients generated by the Na^+/K^+-ATPase and governed by the relative permeabilities of the major permeant ions, K^+, Cl^-, and Na^+, through the ensemble of channels and transporters present in each cell type. As tight control of the cellular membrane potential is recognized as being critical in other organs such as brain, heart, and muscle, the maintenance of membrane potential in renal epithelial cells is also important for the coordinated transport of solutes and overall kidney function.

For charged solutes, the driving force for transport is the sum of the chemical and electrical potential gradients. The Nernst Eq. (2.2) describes the equilibrium condition for a membrane permeable only to a single ionic species:

$$V_m = V_2 - V_1 = \frac{RT}{ZF} \ln \frac{C_o}{C_i} \tag{2.2}$$

where R is the gas constant, T is the absolute temperature, Z is the valence of the solute, F is the Faraday constant, and C and V are concentration and electrical potential terms, respectively. At equilibrium, then, the voltage (V_m) across an ideally selective membrane is defined by the concentrations of the permeant ion on the inside and outside of the membrane, C_i and C_o, respectively.

For systems containing more than one permeant ion, i.e., almost all real physiological systems, the equilibrium membrane potential (V_m) can be described by the Goldman–Hodgkin–Katz (GHK) Eq. (2.3; equation is shown for Na^+, K^+, and Cl^-):

$$V_m = \frac{RT}{F} \ln \left[\frac{(P_{Na}C_{Na}^o + P_K C_K^o + P_{Cl}C_{Cl}^i)}{(P_{Na}C_{Na}^i + P_K C_K^i + P_{Cl}C_{Cl}^o)} \right]. \tag{2.3}$$

where P_x is the permeability of the respective solutes, such as Na^+, K^+, and Cl^-. Thus, in a polarized epithelial system containing multiple charged solutes, one can estimate the transmembrane voltages at both membranes as a function of the relative transmembrane concentrations and permeabilities of each solute on both sides of the apical and basolateral membranes.

2.2.4 Mechanisms of Na^+ Transport Across the Plasma Membrane

Na^+ absorption in the kidney includes a series of transport mechanisms. Active transcellular transport is the main mechanism for Na^+ absorption along the nephron. However, passive paracellular transport of Na^+ at least partially mediates sodium reabsorption in the kidney. Thus, in the proximal convoluted tubules (PCT) and thick ascending limb (TAL) of Henle's loop, sodium reabsorption takes both transcellular and paracellular routes (Breiderhoff et al. 2012; Muto et al. 2012). Passive paracellular transport is the predominant route for passive ion flows in the PCT where the paracellular resistance is low. In the case of active transport, Na^+ ions have to cross both the apical and basolateral membranes (or vice versa) in series. Due to hydrophobic properties of the membrane lipids, electrolytes such as Na^+ cannot cross the membrane freely and have to interact with specific transport proteins, which will be discussed in corresponding sections. Many of these transport proteins are being identified on a molecular basis, and our knowledge about these transport mechanisms has significantly improved in recent years.

2.2.4.1 Active Transcellular Transport

The coupling of solute transport to an energy source can take two forms: primary active transport, where solute transport is coupled directly to an energy-yielding reaction, and secondary active transport, where solute movement against its electrochemical gradient is energized by the movement of another solute down its own gradient.

The most common example of primary active transport is the transport of Na^+ and K^+ by the Na^+/K^+-ATPase (also often referred to as the sodium pump or Na^+/K^+ pump), which is an electrogenic transmembrane ATPase mediating the exchange of 3 Na^+ ions outward for 2 K^+ ions inward at the expense of ATP hydrolysis with each cycle. Therefore, the pump is electrogenic. The Na^+/K^+ pump is localized at the basolateral membrane of many epithelial cells in different segments of the kidney, but it is especially abundant in the TAL of Henle's loop, cells of the distal convoluted tubule (DCT), connecting tubule (CNT) cells, and principal cells of the collecting duct (CD). In general, segments with high rates of active Na^+ transport have high Na^+/K^+ pump activity (Feraille and Doucet 2001). Na^+/K^+ pump functions as a heterodimeric protein complex comprised of catalytic $\boldsymbol{\alpha}$- and auxiliary $\boldsymbol{\beta}$-subunits. The Na^+/K^+ pump is responsible for maintaining the intracellular Na^+ activity at a low level, which provides the energy for the Na^+-coupled transport of many other solutes into the cell. More specific details on the structure and function of the Na^+/K^+ pump are provided in Chap. 11.

In contrast to primary active transport, in secondary active transport, also known as coupled transport, stored energy is used to transport molecules across a membrane instead of direct coupling to ATP hydrolysis. Na^+, because of its steep inward electrochemical gradient maintained by the Na^+/K^+ pump, often participates in the transport of other solutes. Depending on the direction of the movement of coupled solutes (same or opposite), secondary active transporters are classified either as cotransporters (or symporters) or exchangers (or antiporters). Robert K. Crane first proposed flux coupling in 1960 suggesting a model for the sodium–glucose cotransporter (SGLT; described in Sect. 2.3.2.1) (Crane 1960). Multiple cotransporters and exchangers involved in Na^+ absorption are expressed in epithelial cells. Specific transporters mediating Na^+ influx in exchange for efflux of other solutes are discussed below in sections describing Na^+ transport in defined segments.

There are multiple electrogenic Na^+-driven cotransporters expressed in epithelial cells in the kidney, such as Na^+-coupled HCO_3^- transporters (Parker and Boron 2013). In addition to electrogenic transporters, Na^+ entry is also mediated by electroneutral (i.e., when no net charge is transported across the membrane) Na^+/H^+ exchangers (e.g., NHE3) and Na^+-coupled cotransporters (e.g., NCC and NKCC). Shown on Fig. 2.2 are examples of both primary and secondary active

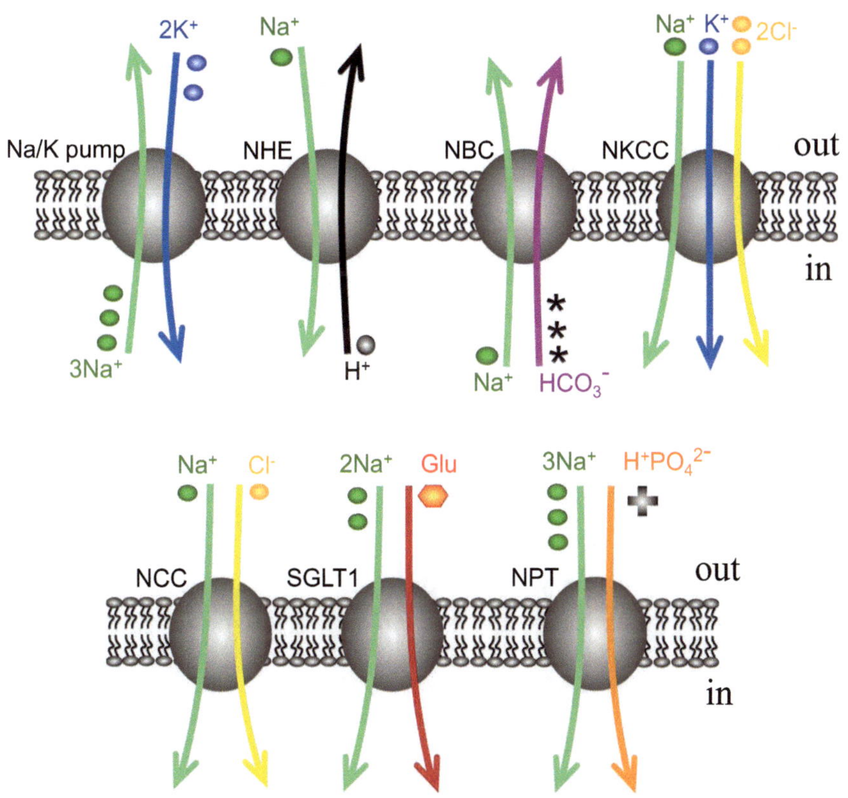

Fig. 2.2 Ion movement in the main Na^+ transporters involved in epithelial sodium absorption. Shown are primary transport pathways for Na^+/K^+-ATPase, Na^+/H^+ exchanger (NHE), Na–HCO_3 cotransporter (NBC), Na^+–K^+–Cl^- cotransporter (NKCC), Na^+–Cl^- cotransporter (NCC), sodium–glucose cotransporter (SGLT), and Na^+–phosphate cotransporter (NPT). *Arrows* demonstrate direction of ion movement inside or outside of the cell

transporters mediating epithelial Na^+ absorption. As seen from Fig. 2.2, ion/substrate coupling stoichiometry of the transporters varies among transporters. For instance, the coupling stoichiometry of the SGLT1 cotransporter is 2 Na^+ ions to 1 glucose molecule per transport cycle (interestingly, SGLT1 stoichiometry is different from SGLT2, which is 1 Na^+:1 glucose), but for NKCC 1 Na^+ ion is transported in the same direction as 1 K^+ and 2 Cl^- ions. Importantly, due to their various nature and function, they have different topologies in plasma membranes. Figure 2.3 illustrates some examples of the topologies of various transporters.

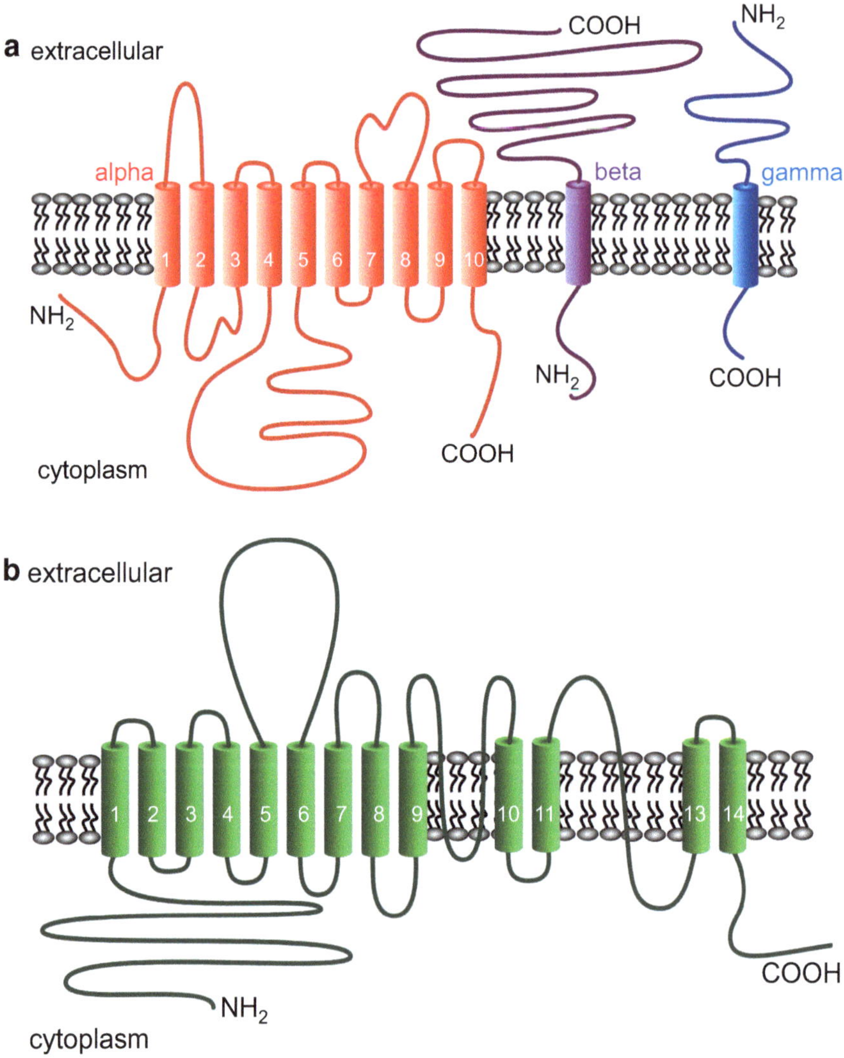

Fig. 2.3 (continued)

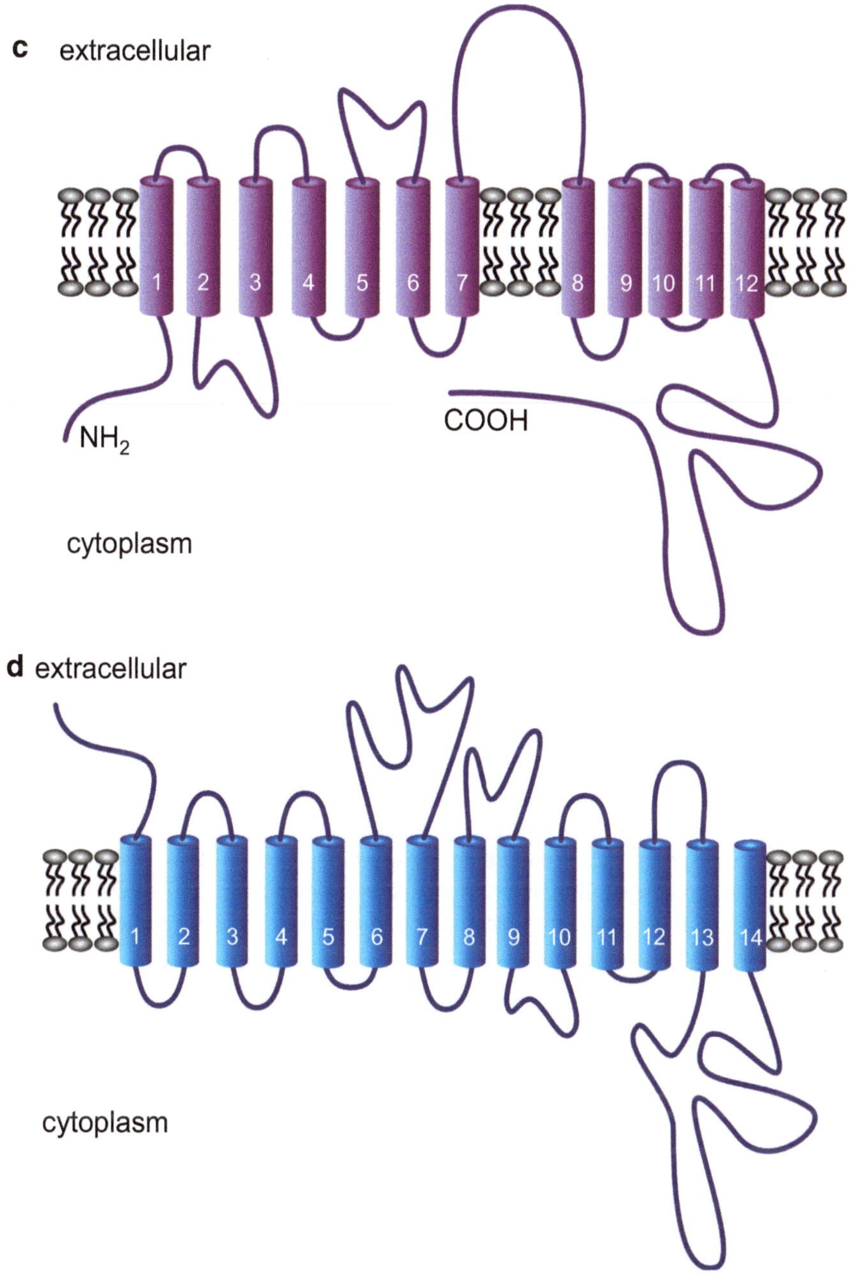

Fig. 2.3 Membrane topology of several Na^+ transporters involved in epithelial sodium absorption. (**a**) Organization of the ***α***- and ***β***-subunits of Na^+/K^+-ATPases in the plasma membrane. The *γ*-subunit, which is found associated with some isoforms of Na^+/K^+-ATPase, is also shown. (**b**) Presumed topology of the electrogenic Na^+–HCO_3^- cotransporter. The model shows the extended cytosolic amino- and carboxy-terminal domains (NH_2 and C-termini) linked via a transmembrane

2.2.4.2 Passive Paracellular Transport

The paracellular pathway in renal tubular epithelia such as the PCT, which reabsorbs the largest fraction of filtered NaCl, is important for the transport of electrolytes and water (Krug et al. 2012; Muto et al. 2012). The gatekeeper of the paracellular pathway is the tight junction, which is located at apical cell–cell interactions of adjacent epithelial cells (see Fig. 2.1). The tight junction separates the apical domain from the basolateral domain and provides a barrier to paracellular movement of water and ions. Claudins are key integral proteins that provide the barrier function and permit selective paracellular transport (Yu 2015). Claudin-2 is the main claudin responsible for reabsorption of Na^+ (Kiuchi-Saishin et al. 2002). In addition to the claudin family members, there are several other tight junction transmembrane proteins that might directly influence the adhesive barrier, including occludins, junctional adhesion molecules, and tricellulin (Anderson and Van Itallie 2009). Interestingly, approximately 30 % of the Na^+ that is transported from lumen to blood by the transcellular pathway diffuses back to the urine by the paracellular pathway in the proximal tubule.

2.2.5 *Methods of Na^+ Transport Measurement*

To a large degree, advancement in our understanding of the physiology of epithelial transport of sodium and other electrolytes has been afforded by the development of key measurement methods and tools. Stockand and colleagues have recently provided a clear overview and description of traditional and contemporary in vivo and ex vivo tools used to study renal tubule transport physiology (Stockand et al. 2012). These include isotopic flux measurements, micropuncture, and microperfusion techniques and the patch-clamp technique performed ex vivo in various configurations on isolated kidney tubules.

Fig. 2.3 (continued) domain that includes 14 transmembrane spans, one of which is thought to be an extended region rather than an α-helix [modified from Parker and Boron (2013)]. (**c**) The topology proposed for Na^+-coupled chloride cotransporters (NKCC) [modified from Gamba (2005)]. (**d**) Secondary structure of Na^+–glucose cotransporter SGLT. Shown are 14 transmembrane helices with both the NH_2 and COOH-termini facing the extracellular solution [modified from Turk et al. (1996), Wright et al. (2011)]

2.3 Sodium Homeostasis and Its Role in the Kidney

2.3.1 Role of Sodium Reabsorption in the Passive Diffusion of Water, Urea, and Other Solutes

Na^+ reabsorption is critical for the transport of other solutes and water homeostasis. When sodium is transported out of the kidney tubule, its concentration decreases inside the tubule while increasing in the interstitial space. Various active and passive transport processes are involved in the maintenance of water and solute homeostasis in different nephron segments. Shown in Fig. 2.4 are basic mechanisms by which water, chloride, and urea reabsorption are coupled with sodium transport in the kidney. The generation of transmembrane concentration differences of sodium and other solutes induces osmotic water flow from the lumen to the renal interstitium. Certain segments of the nephron, such as the proximal tubule, are highly permeable to water and small ions via the tight junctions in these segments. In addition, AQP1 was shown to be abundant in the apical and basolateral membranes of the proximal tubules and descending thin limbs of Henle's loop, where this water channel provides the transcellular pathway for water following small osmotic gradients at the apical and basolateral surfaces of the cell (Nielsen et al. 1993; Agre 2000). In contrast to the proximal tubules, less leaky epithelia such as the loop of Henle, distal tubules, and collecting duct prevent the osmotic flow of water across the tight junctions of the plasma membrane. However, vasopressin, also known as arginine vasopressin (AVP) or antidiuretic hormone (ADH), greatly increases water permeability in the collecting tubules, as discussed later (see Sect. 2.3.2.4). Therefore, changes in sodium reabsorption significantly influence the reabsorption of water and many other solutes.

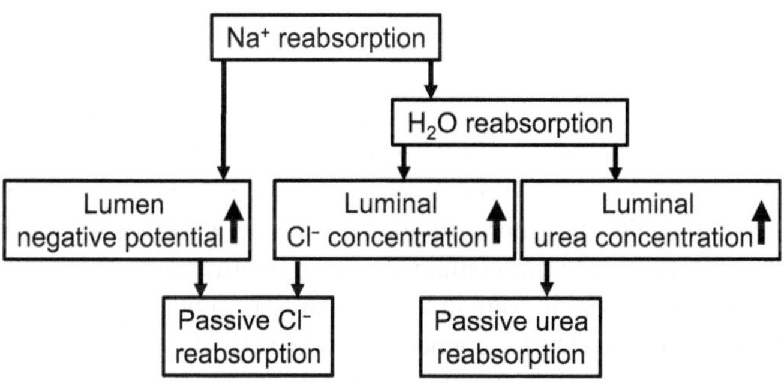

Fig. 2.4 Mechanisms of coupling of sodium and water, chloride, and urea reabsorption. When sodium is reabsorbed, anions such as chloride and bicarbonate are transported along with it; as Na^+ is positively charged, its transport from the lumen creates a lumen-negative potential, which promotes the diffusion of chloride through the paracellular pathway. Furthermore, transmembrane concentration differences of sodium facilitates water reabsorption from the lumen to the renal interstitium, which in turn can promote additional reabsorption of chloride ions and urea due to the increased concentration of these solutes present in the tubular lumen resulting from the water loss

In addition to water, sodium reabsorption plays an important role in the transport of other solutes. When sodium is reabsorbed across epithelial cells, anions such as chloride and bicarbonate are transported along with sodium through coupled transporters. As Na^+ is positively charged, its transport through channels from the lumen creates a lumen-negative potential, which promotes the diffusion of anions such as chloride through the paracellular pathway in certain nephron segments (e.g., PCT). Osmotic water flow from the lumen to the interstitium also promotes the additional reabsorption of chloride ions due to the increased chloride concentration present in the tubular lumen resulting from the water loss. Thus, sodium reabsorption is coupled to the passive reabsorption of chloride by changes in the electrical potential and chloride concentration gradient. Moreover, sodium and chloride transport are tightly linked since these ions can also be reabsorbed by secondary active transport through coupled cotransporters (e.g., NCC and NKCC, as discussed below).

Urea is mainly transported by UT-A and UT-B urea transporters (Klein et al. 2011, 2012). Besides, urea is also passively reabsorbed from the tubule. As described above, water is reabsorbed by osmosis from the lumen. Thus, urea concentration in the lumen increases, consequently, and urea is also reabsorbed by created concentration gradient. However, urea permeates the tubule to a much lesser extent than water or chloride. In addition to these facilitated urea transporters, sodium-dependent, secondary active urea transport mechanisms have been characterized. Thus, it was shown that removing sodium from the perfusate completely inhibited net urea reabsorption, demonstrating that active urea transport is dependent upon the presence of sodium in the tubule lumen (Isozaki et al. 1994; Sands et al. 1996). However, the active urea transporters have not yet been cloned (Klein et al. 2011).

2.3.2 Sodium Absorption in Different Nephron Segments

Precise regulation of Na^+ absorption by the kidneys relies on sequential actions of the various nephron segments, each with highly specialized transport capabilities. Each human kidney contains about one million nephrons capable of forming urine (Hoy et al. 2005). Figure 2.5 provides an overview of consecutive segments of the nephron and corresponding Na^+ transport proteins along the nephron. Every nephron is comprised of a renal corpuscle containing the glomerulus and Bowman's capsule, a proximal tubule (proximal convoluted and straight tubules), a loop of Henle, a distal convoluted tubule (DCT), connecting tubule (CNT), and the collecting duct system, which includes the initial collecting tubule (ICT), the cortical collecting duct (CCD), the outer medullary collecting duct (OMCD), and the inner medullary collecting duct (IMCD). All of the components of the nephron, including the collecting duct system, which collects urine from several nephrons, are functionally interconnected. In general, the absolute rates of Na^+ reabsorption are greatest in the proximal tubule and fall as the tubular fluid proceeds from proximal to distal segments. The proximal tubule reabsorbs the majority of the filtered Na^+ load (up

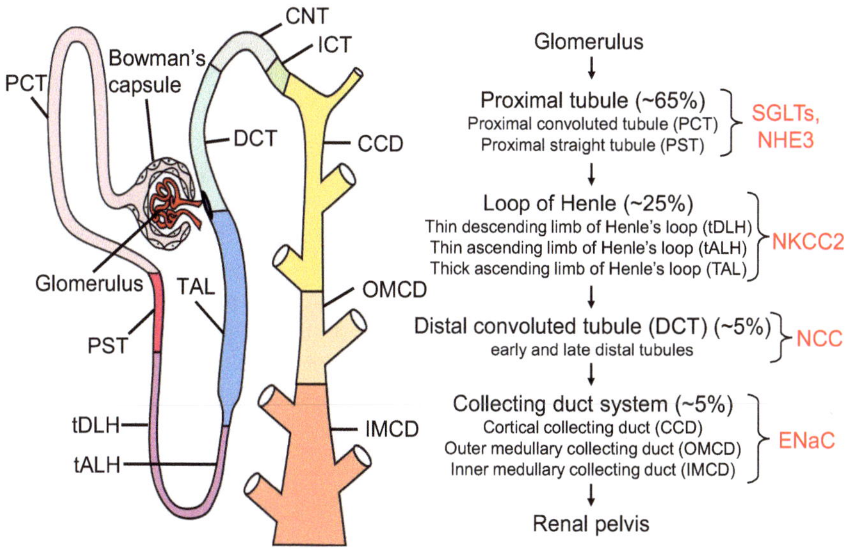

Fig. 2.5 Structure and consecutive segments of the nephron. Proximal convoluted tubule (PCT), proximal straight tubule (PST), thin descending limb of Henle's loop (tDLH), thin ascending limb of Henle's loop (tALH), thick ascending limb of Henle's loop (TAL), distal convoluted tubule (DCT), connecting tubule (CNT), initial collecting tubule (ICT), cortical collecting duct (CCD), outer medullary collecting duct (OMCD), and inner medullary collecting duct (IMCD) segments of nephron are shown. Approximate percentage of sodium reabsorption in different segments and major sodium transporters mediating these processes are also depicted [modified from Staruschenko (2012)]

to 60–70 %). However, as discussed below, the proximal tubule has a rather limited ability to alter Na^+ transport. In contrast, the DCT and collecting ducts reabsorb only a minor fraction (approximately 5 %) of the filtered Na^+, but are finely regulated by different physiological factors.

2.3.2.1 Proximal Tubule

The proximal tubule (PT) is the major site for Na^+ absorption in the nephron. There is considerable heterogeneity of both morphologic and functional characteristics along the proximal tubule. Typically, the proximal tubule consists of convoluted and straight tubules (PCT and PST, respectively). Based on the ultrastructure, it can be further divided into S1, S2, and S3 segments. The first two segments are located in the kidney cortex, and the straight segments (S3) descend into the outer medulla. The net rates of the Na^+ transport in the late proximal segment are, in general, lower than in the initial proximal convoluted tubule (Jacobson 1982). Importantly, the absorption of sodium in the proximal tubule provides the driving force for absorption of other solutes, such as bicarbonate, glucose, phosphate, and amino acids (Parker and Boron 2013; Skelton et al. 2010). Under most circumstances, fluid at any given point along the proximal tubule has virtually the same Na^+ concentration and osmolality as plasma (Ullrich et al. 1963). The isosmotic nature of proximal tubule fluid absorption derives from the high water permeability of this segment (Andreoli et al. 1978), which effectively clamps the osmolality of the tubular fluid at that of plasma. Although Na^+ transport in the proximal tubule occurs in the absence of large electrical or chemical gradients, the bulk of Na^+ absorption in the proximal tubule involves active transport. Earlier studies demonstrated that Na^+ can be reabsorbed in this tubule segment against both concentration (Ullrich et al. 1963; Giebisch et al. 1964) and electrical (Barratt et al. 1974) gradients. A significant amount of Na^+ transport in the proximal tubules also occurs passively (Schafer et al. 1975). For example, in the late convoluted and straight proximal tubules, Na^+ diffuses passively out of the tubule driven by the lumen-positive electrical potential difference in those segments. This potential difference derives from a Cl^- concentration gradient across the tubule wall. Shown in Fig. 2.6 is a simplified scheme of Na^+ reabsorption in proximal tubules.

NHE3 NHE3 is a Na^+/H^+ exchanger involved in large amounts of neutral Na^+ absorption in the PCT. NHE3 (in human is encoded by the *SLC9A3* gene) is one of the nine isoforms of the mammalian Na^+/H^+ exchanger (NHE) gene family (Donowitz and Li 2007). NHE3 is localized at the apical plasma membrane of the proximal tubules and is responsible for the majority of sodium absorption in the kidney. NHE3 performs an electroneutral exchange of one extracellular Na^+ ion with one extracellular proton (see Fig. 2.2). NHE3 is functionally linked to a Cl^-/HCO_3^- exchanger and takes part in neutral NaCl absorption (Soleimani 2013).

The activity of NHE3 is regulated allosterically by the intracellular pH, stimulated by cellular acidification. NHE3-mediated neutral NaCl absorption is highly regulated by dietary conditions and the endocrine environment under both physiological and pathophysiological conditions. NHE3 cycles between the apical plasma membrane and the recycling compartments under basal conditions. The mechanisms involved in acute regulation of NHE3 include changes in plasma membrane expression and turnover, which can be modulated by altering both endocytosis and exocytosis rates, and changes in the cellular half-life of NHE3 (for review, see Donowitz and Li (2007)).

SGLT There are two classes of glucose transporters involved in glucose homeostasis. Sodium–glucose cotransporters or symporters (SGLTs) are the active transporters driven by the inward Na^+ gradient maintained by the Na^+/K^+ pump. Transport of glucose via facilitated diffusion is mediated by GLUT transporters (uniporters). SGLTs are exquisitely selective for Na^+ as the cation required for cotransport with glucose. One interpretation of the cation selectivity data is that cation binding initiates a change in the conformation of the sugar binding site (Wright et al. 2011). Interestingly, SGLT1 (in humans is encoded by the *SLC5A1* gene) and SGLT3 (*SLC5A4*) have two strongly interacting Na^+ binding sites, while SGLT2 (*SLC5A2*) only has one Na^+ binding site (Díez-Sampedro et al. 2001; Hummel et al. 2011). As demonstrated in Fig. 2.2, SGLT1 has a 2:1 Na^+ to glucose transport stoichiometry.

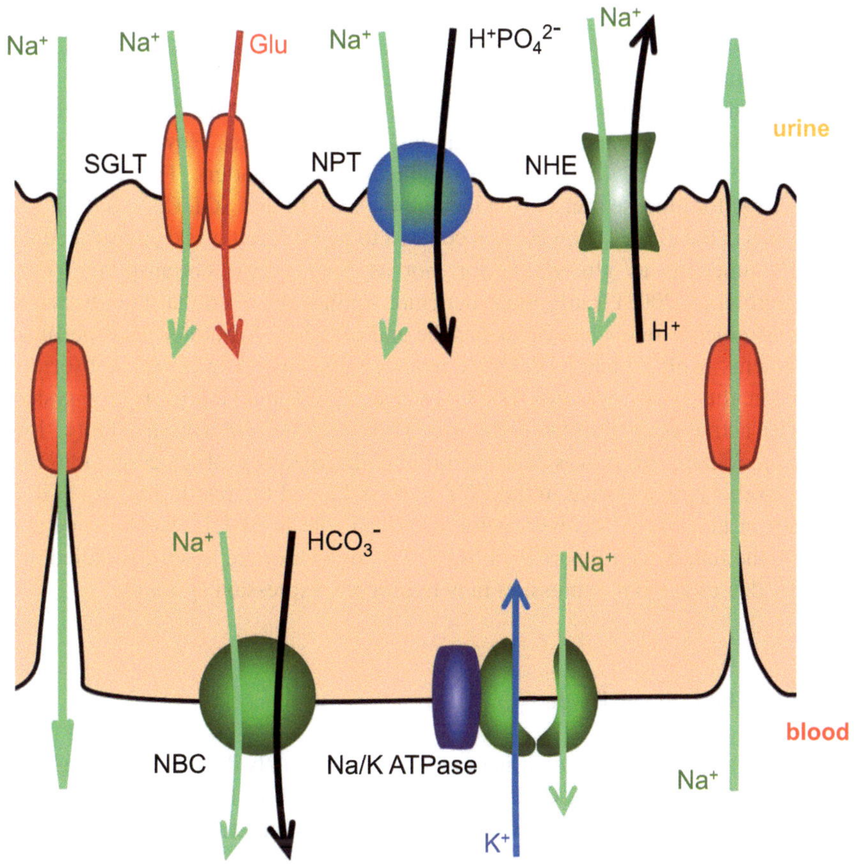

Fig. 2.6 Simplified scheme of Na^+ reabsorption in proximal tubules. In the proximal tubules, sodium can be reabsorbed against both concentration and electrical gradients; however, a significant amount of Na^+ transport also occurs passively. The majority of sodium ions are reabsorbed via the Na^+/H^+ exchanger (NHE); sodium–glucose cotransporter (SGLT) cotransports Na^+ and glucose into the cells, and so does the Na^+–phosphate cotransporter (NPT) along with phosphate ions. Importantly, the absorption of sodium in the proximal tubule provides the driving force for absorption of other solutes, such as bicarbonate, glucose, phosphate, and amino acids. On the basolateral side, sodium is removed from the cells with the help of Na^+/HCO_3^-cotransporter (NBC) and Na^+/K^+-ATPase

SGLT1 cotransports Na^+ and glucose in the PCT. Activity of this transporter depolarizes the plasma membrane, which can serve as a signal for additional mechanisms. Interestingly, in the absence of glucose, SGLT1 can still conduct Na^+ currents. The first evidence that SGLT1 worked as uniporter was reported in 1990 (Umbach et al. 1990), soon after SGLT1 was cloned in 1987 (Hediger et al. 1987). Thus, it was shown that SGLT inhibitor phlorizin blocked a current in the absence of glucose. This current accounted for approximately 8 % of the total Na^+–glucose cotransporter current and was not observed in control cells (Umbach et al.

1990). Moreover, this current was saturated with increasing external Na^+ concentration (Loo et al. 1999). In addition, SGLT1 transports water and urea along with Na^+ and glucose (Loo et al. 1999; Wright et al. 2011).

There are 12 members of the human SGLT (SLC5) gene family, including cotransporters for sugars, anions, vitamins, and fatty acids. However, only four members of the SLC5 family are Na^+–glucose cotransporters. Two of these are responsible for glucose transport in the kidney: SGLT1 and SGLT2. SGLT3 is also expressed in the kidney. However, in contrast to SGLT1/2, it was demonstrated that SGLT3 functions as glucose sensor, not as Na^+–glucose cotransporter (Díez-Sampedro et al. 2003). Early micropuncture studies identified that the glomerular filtrate is glucose-free when it reaches the end of the PCT. Further microperfusion experiments with isolated nephrons revealed that active glucose reabsorption occurs in the early segments of PCT via a low-affinity high-capacity system, whereas active reabsorption in the late PST occurred by a high-affinity, low-capacity system. As it was later identified, the majority of glucose (and Na^+, respectively) is transported in the PCT by SGLT2 and the rest in PST by SGLT1 (Wright et al. 2011). However, while a critical role of SGLT2 has been demonstrated, the functional properties of SGLT2 in the kidney are less well studied than SGLT1 due to its poor expression in heterologous expression systems.

2.3.2.2 The Loop of Henle

The loop of Henle consists of structurally and functionally distinct thin limbs (thin ascending limb of Henle (tALH), thin descending limb of Henle (tDLH)) and the thick ascending limb (TAL) (refer to Fig. 2.5). Compared to the proximal tubule, the loop of Henle reabsorbs a smaller amount of filtered Na^+ (approximately 25 % of the total load (Bennett et al. 1968)). The key difference between the two segments is that Henle's loop taken as a whole reabsorbs more salt than water, whereas in the proximal tubule, this occurs in essentially equal proportions. Interestingly, the reabsorption of salt and water is performed in different places along the loop. The descending limb reabsorbs very little Na^+ or Cl^-, but it allows transport of water; both tALH and TAL are impermeable to water, but significantly reabsorb sodium and chloride (which is why they are often referred to as "diluting segments"). As a result, the fluid leaving the loop is hypo-osmotic relative to plasma, indicating that more salt than water has been reabsorbed. Ultimately, the loop of Henle is a nephron segment that allows water and salt excretion to be regulated independently.

Thin Ascending and Thin Descending Loops of Henle Intense water extraction via aquaporin water channels (Nielsen et al. 2002; Verkman 2006; Halperin et al. 2008) and near absence of sodium reabsorption in the tDLH concentrate tubular fluid (Kokko 1970; Kokko and Rector 1972). High interstitial levels of NaCl and urea provide additional osmotic energy for water reabsorption by tDLH. Therefore, when the fluid exits the tDLH, it possesses a favorable gradient of luminal sodium, which facilitates further passive sodium reabsorption in the water-impermeable tALH (Imai and Kokko 1974). The formation of dilute urine by the loop of Henle begins at the tALH. Transepithelial movement of sodium from the lumen occurs via the paracellular pathway, whereas chloride diffuses transcellularly via ClC-K1 chloride channels, which are selectively localized to the tALH and appear to be the major mediators of chloride movement there (Reeves et al. 2001). At large, the osmolarity of the fluid leaving the tALH is decreased due to a fall in NaCl content, which occurs almost entirely via passive and paracellular pathways.

Thick Ascending Loop of Henle As the luminal fluid moves into the TAL tubules, the reabsorbing properties of the epithelium change. In contrast to the tDLH and tALH, the major mechanisms of sodium and chloride reabsorption in the TAL are active. As with the tALH, the TAL consists of a tight epithelium, which is impermeable to water (Hebert 1986).

As shown in Fig. 2.7, the dominant step of luminal sodium and chloride uptake in this segment occurs via the sodium–potassium–chloride cotransporter (Na^+–K^+–$2Cl^-$; NKCC2 isoform; see below) (Greger 1981; Koenig et al. 1983; Markadieu and Delpire 2014). ClC-K1 (hClC-Ka in human) localized at the basolateral membrane allows passive Cl^- transepithelial resorption in the TAL (Fahlke and Fischer 2010; Imbrici et al. 2014). An accessory ClC-K channel subunit, barttin, encoded by the gene product of *BSND*, is required to form functional channel (Birkenhager et al. 2001; Estevez et al. 2001). It was reported that ClC-K1 is upregulated in water-restricted animals (Fushimi et al. 1993; Uchida et al. 1993). Further studies revealed that the ClC-K1 knockout mice can maintain salt and water balance under normal conditions but are unable to concentrate urine under water restriction; deletion of ClC-K1 results in diabetes insipidus (Matsumura et al. 1999). The electroneutral process of Na^+–K^+–$2Cl^-$ cotransport is driven largely by the favorable electrochemical gradient for Na^+ entry, which is maintained by the continuous operation of the Na^+/K^+ pump in the basolateral membrane. Na^+–K^+–$2Cl^-$ transport is selectively inhibited by so-called loop diuretics (e.g., furosemide and bumetanide), as discussed further in Sect. 2.3.5.1 (Russell 2000). The apical membrane of TAL cells also expresses a Na^+/H^+ antiporter isoform (major isoform being NHE3 (Amemiya et al. 1995; Borensztein et al. 1995; Laghmani et al. 1997; Capasso et al. 2002)), which provides an alternative mechanism for the uptake of Na^+ and regulates intracellular pH.

Interestingly, NKCC2 requires equal amounts of Na^+ and K^+ to enter the cell. However, as there is much less K^+ than Na^+ in the lumen, the tubular fluid would seem to get depleted of K^+ long before enough Na^+ was reabsorbed. This issue is resolved by K^+ recycling via the apical K^+-conductive pathway. K^+ leaks back into the lumen through apical potassium channels (mostly through ROMK channels, but involvement of other potassium channels has also been reported) (Guggino et al. 1987; Wang et al. 1990; Wang 1994, 2012; Giebisch 2001; Hebert et al. 2005), and is again available for Na^+–K^+–$2Cl^-$ cotransport. Furthermore, K^+ channels actually

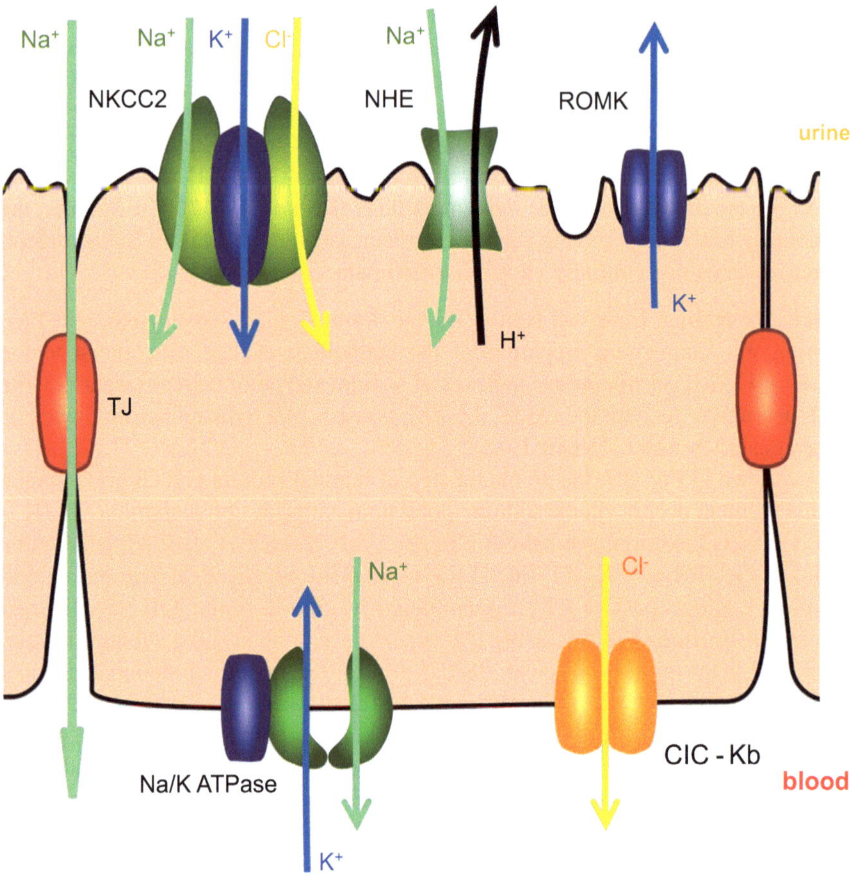

Fig. 2.7 Simplified scheme of Na^+ reabsorption in the loop of Henle. Transepithelial movement of sodium from the lumen in the tDLH and tALH occurs via the paracellular pathway, whereas chloride diffuses transcellularly via ClC-K1 chloride channels. In contrast to the tDLH and tALH, the major mechanisms of sodium and chloride reabsorption in the TAL are active. The dominant step of luminal sodium and chloride uptake in this segment occurs via the sodium–potassium–chloride symporter (NKCC2 isoform); an alternative mechanism for the uptake of Na^+ from the lumen is provided by the apical the Na^+/H^+ antiporter (NHE). ClC-K1 channels and Na^+/K^+-ATPase localized at the basolateral membrane allow Cl^- and Na^+ transepithelial resorption, respectively. K^+ enters the cell as a result of Na^+/K^+-ATPase function and then leaks back into the urine through apical potassium (ROMK) channels

dominate the apical membrane conductance of the TAL, creating a lumen-positive transepithelial voltage, which provides the driving force for the paracellular movement of Na^+. This passive diffusion of Na^+ accounts for as much as 50 % of total Na^+ reabsorption in this segment. The Cl^- that enters TAL epithelial cells apically via NKCC2 exits via ClC-Kb Cl^- channels on the basolateral membrane (Fig. 2.7).

NKCC2 The uptake of Na^+ in the loop of Henle is coupled to that of Cl^- and K^+. The coupling ratio is fixed at 2 Cl^- per 1 Na^+ and 1 K^+ (see Fig. 2.2). The renal-specific apical form of this cotransporter encoded by the *SLC12A1* gene was denoted NKCC2 (also known as BSC1 or bumetanide-sensitive cotransporter), based on high homology with the basolateral Na^+–K^+–2Cl^- cotransporter NKCC1 (Gamba 2005). After its initial cloning from rat and rabbit (Gamba et al. 1994; Payne and Forbush 1994), NKCC2 was identified in the mouse exclusively in the kidney (Igarashi et al. 1995) and later in humans (Simon et al. 1996a).

NKCC2 belongs to a family of electroneutral cation–chloride cotransporters, which are all members of the amino acid polyamine cotransporter (APC) superfamily. As resolved from similarity to other crystallized APC family members, NKCC2 contains 12 transmembrane domains (see Fig. 2.3c) with a large extracellular loop between membrane segments 7 and 8; the *SLC12A1* gene encodes the protein with molecular weight of 115–120 kDa (approximately 1100 a.a.) (Gamba et al. 1994). There have been at least six different isoforms of NKCC2 identified in the mouse kidney (Payne and Forbush 1994; Igarashi et al. 1995; Yang et al. 1996; Simon et al. 1996a; Mount et al. 1999; Plata et al. 2001; Gamba 2001; Ares et al. 2011); such molecular diversity results from alternative splicing variants. Interestingly, NKCC2 isoforms are differentially expressed along the TAL. In situ RNA hybridization and RT-PCR studies have revealed that isoform A is expressed in the renal cortex and medulla outer stripe, F in the outer medulla with higher density in the inner stripe, and B in the renal cortex. In the macula densa, both isoforms A and B were found (Castrop and Schnermann 2008). Furthermore, A, B and F isoforms have different transport capabilities; A and B possess higher affinity to Na^+, K^+, and Cl^-, than the F isoform (Plata et al. 2002).

The role of apical NKCC2-mediated Na^+–K^+–2Cl^- cotransport in the TAL was initially demonstrated by mutations in *SLC12A1*, which are associated with the Bartter syndrome (Simon et al. 1996a), a disorder characterized by hypokalemia, metabolic alkalosis, and hyperaldosteronism (see Table 2.1). NKCC2 knockout mice reproduce a clinical phenotype characteristic of this illness (clinical features included severe renal failure, high plasma potassium, and metabolic acidosis) (Takahashi et al. 2000). Furthermore, NKCC2 has been shown to play a role in the cardiovascular diseases, including hypertension. This is particularly important as this cotransporter is the main target of loop diuretics (e.g., furosemide and bumetanide), which are potent pharmacological agents currently in wide clinical use (Gamba 2005). Inhibition of NKCC2 results in reduced salt reabsorption in the TAL, which increases salt delivery to the distal nephron and produces substantial diuresis and natriuresis. It has been shown that NKCC2 along the TAL is activated in the Milan hypertensive rats, and this significantly contributes to the increase in systemic blood pressure in these rats (Carmosino et al. 2011). Several rare functionally impairing mutations in human NKCC2 were associated with lower blood pressure (Monette et al. 2011). In Dahl salt-sensitive (SS) rats, high salt intake increases apical NKCC2 expression and phosphorylation in the TAL as compared with salt-resistant animals, and this may contribute to enhanced NaCl reabsorption in SS rats during high salt intake (Haque et al. 2011; Ares et al. 2012). In general,

Table 2.1 Inherited disorders of renal sodium transport

Disorder	Affected gene product	OMIM #	Nephron distribution	Functional consequences	Clinical features
Hypotensive disorders					
Bartter, type 1 (antenatal)	NKCC2 (*SLC12A1*)	600839	TAL	Decreased NKCC cotransport	Hypotension, hypokalemia, metabolic alkalosis
Bartter, type 2 (antenatal)	ROMK (*KCNJ1*)	600359	TAL, CCD	Decreased apical K^+ recycling	Same as above
Bartter, type 3 (infantile)	ClC-Kb (*CLCNKB*)	602023	TAL	Decreased basolateral Cl^- flux	Hypokalemia, alkalosis, hyperreninemia
Bartter, type 4	Barttin (*BSND*)	606412	tAL, TAL, inner ear	Decreased basolateral Cl^- flux	As above, with sensorineural deafness
Bartter, type 5	CaSR (*CASR*)	601199	PT, TAL, parathyroid, thyroid, brain	Activating mutation of Ca^{2+}-sensing receptor with decreased K^+ recycling and NaCl reabsorption	Hypokalemia, alkalosis, hyperreninemia, hypocalcemia
Gitelman	NCCT (*SCL12A3*)	600968	DCT	Decreased NaCl cotransport	Hypokalemia, metabolic alkalosis, hypocalciuria
Pseudohypoaldosteronism type 1 (autosomal recessive)	ENaC subunits ($\boldsymbol{\alpha}$, $\boldsymbol{\beta}$, $\boldsymbol{\gamma}$) (*SCNN1A*, *SCNN1B*, *SCNN1G*)	264350	Collecting duct	Decreased amiloride-sensitive sodium transport	Hypotension, hyperkalemia, metabolic acidosis
Pseudohypoaldosteronism type 1 (autosomal dominant)	Mineralocorticoid receptor (*NR3C2*)	177735	Collecting duct	Decreased response to mineralocorticoids	Hypotension, hyperkalemia (less severe than recessive)
Hypertensive disorders					
Liddle	$\boldsymbol{\beta}$- or $\boldsymbol{\gamma}$-ENaC (*SCNN1B*, *SCNNIG*)	177200	Collecting duct	Increased cell surface expression and activity of ENaC channels	Hypertension, hypokalemia, metabolic alkalosis, responsive to amiloride

Pseudohypoaldosteronism type 2 (Gordon)	WNK1 or WNK4 (*WNK1*, *WNK4*)	145260	DCT, collecting duct	Increased NCC cotransport, increased paracellular Cl^- permeability	Hypertension, hyperkalemia, metabolic acidosis, responsive to thiazide diuretics
Apparent mineralocorticoid excess	11*β*-HSD2 (*HSD11B2*)	218030	DCT, collecting duct	Decreased oxidation of glucocorticoids causing activation of mineralocorticoid receptor	Hypertension, hypokalemia, metabolic alkalosis, responsive to dexamethasone
Other disorders					
Renal glucosuria	SGLT2 (*SLC5A2*)	233100	Proximal tubule	Decreased sodium-coupled glucose absorption	Decreased threshold for glucosuria
Proximal RTA	NBC1 (*SLC4A4*)	603345	Proximal tubule	Decreased sodium-coupled HCO^-_3 transport across basolateral membrane	Metabolic acidosis, ocular and dental abnormalities, growth and mental retardation

Updated references available at http://www.ncbi.nlm.nih.gov/entrez/query.fcgi?db=OMIM. Data provided are based on (Subramanya et al. 2013)
TAL thick ascending limb, *tAL* thin ascending limb, *DCT* distal convoluted tubule, *OMIM* # accession number for Online Mendelian Inheritance in Man.

NKCC2 constitutes the major salt transporting pathway in mammalian TAL, and its proper function is fundamental not only for salt reabsorption but for the kidney's ability to produce urine that is more dilute or concentrated than plasma, which is an essential for survival of the land mammals (Gamba 2005). Further details on structure and function of NKCC are provided in Chap. 12.

2.3.2.3 Distal Convoluted Tubule

The distal convoluted tubule (DCT) is the initial segment of the aldosterone-sensitive distal nephron (ASDN), which consists of DCT, the connecting tubule (CNT), and the collecting ducts (CD). DCT can be further divided into two functionally different segments, denominated "early" and "late" DCT (Reilly and Ellison 2000) by their sensitivity to aldosterone. Distal tubules reabsorb much smaller fractions of filtered Na^+ load than PCT or the loop of Henle (5–10 %) (Hierholzer and Wiederholt 1976), and this process occurs almost exclusively by the transcellular pathway. NaCl concentrations generated along the loop of Henle cannot be maintained by the DCT, and within the initial 20 % of the DCT, salt concentration almost doubles (Schnermann et al. 1982). Later luminal Na^+ concentration decreases to a value of approximately 30 mM at the end of DCT (Khuri et al. 1975). This occurs to a large extent via active Na^+ transport and because the transepithelial voltage becomes progressively more negative moving from the early to late DCT (Wright 1971; Barratt et al. 1975; Hayslett et al. 1977; Allen and Barratt 1985). The absorption of Na^+ in the DCT is dependent on the load. Higher tubular flow rates result in the increased delivery of Na^+ to the normally unsaturated later parts of the distal tubule (Khuri et al. 1975; Shimizu et al. 1989).

The major step of apical Na^+ intake in the early DCT is via the Na^+–Cl^- cotransporter (NCC) (Velazquez et al. 1987), which is characteristically different from the NKCC2 in the TAL, independent of K^+ and highly sensitive to a different class of drugs—thiazide diuretics (see Sect. 2.3.5.2 and Fig. 2.8). As mentioned above, lumen transepithelial electrical potential is more negative in the late rather than early DCT, and this is probably caused by differences in the Na^+ transporters; the late DCT, as distinct from the early DCT, also has an amiloride-sensitive pathway of Na^+ reabsorption (Reilly and Ellison 2000), which is mediated by the epithelial Na^+ channel (ENaC; see Sect. 2.3.2.4) (Costanzo 1984; Ciampolillo et al. 1996; Schmitt et al. 1999; Loffing et al. 2000, 2001; Campean et al. 2001). However, species differences have been reported for ENaC expression in the late DCT. ENaC has been found in the DCT of mouse and rat renal tissues (Schmitt et al. 1999; Loffing et al. 2000) but not in human or rabbit kidney (Velazquez et al. 2001; Biner et al. 2002).

The removal of Na^+ from the basolateral side of the DCT cells is controlled by the very intense activity of the Na^+/K^+ pump in both early and late segments (Doucet 1988). Basolateral K^+ is then removed via the channels which belong to the K_{ir} family (Ookata et al. 2000; Lourdel et al. 2002; Hamilton and Devor 2012; Zhang et al. 2013; Zaika et al. 2013), likely Kir4.1 homomer (known as KCNJ10) or

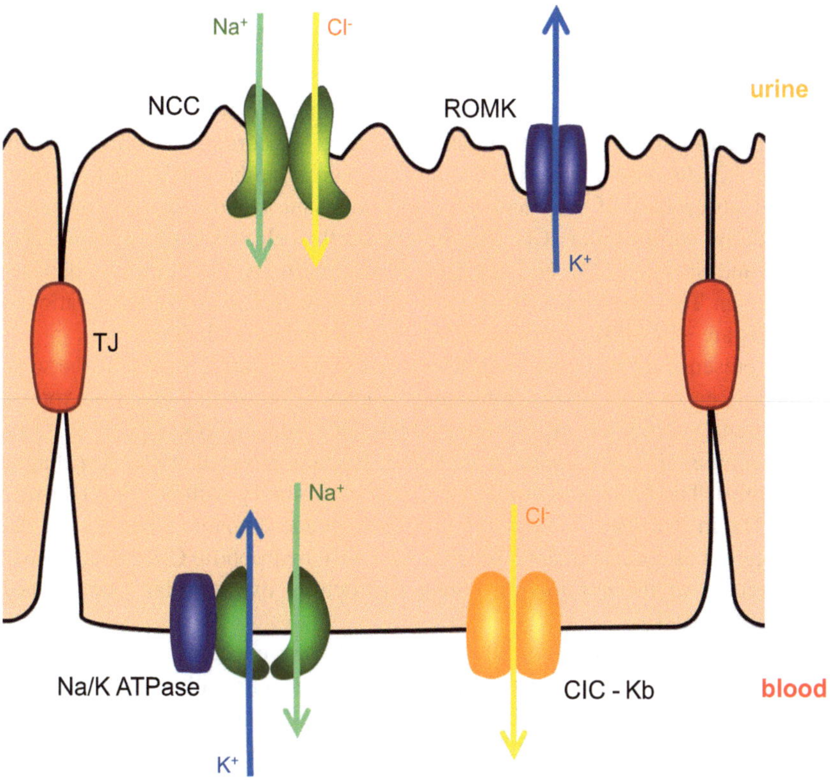

Fig. 2.8 Simplified scheme of Na^+ reabsorption in the distal convoluted tubules. The major step of apical Na^+ intake in the early DCT is via the Na^+–Cl^- cotransporter (NCC). The removal of Na^+ from the basolateral side of the DCT cells is controlled by the very intense activity of the Na^+/K^+ pump; apically, K^+ is secreted via ROMK channels. Similar to the TAL, basolateral Cl^- efflux is mediated via the ClC-K channels, which significantly ensure the maintenance of the gradient for the entry of Na^+ and mitigate against intracellular Cl^- concentration increases resulting from apical NaCl entry via Na^+–Cl^- cotransport

the Kir4.1/Kir5.1 (KCNJ10/16) heteromer, which help maintain the activity of the Na^+/K^+ pump and, thus, salt reabsorption capacity (Bockenhauer et al. 2009; Reichold et al. 2010; Paulais et al. 2011). Apically, K^+ is secreted via ROMK channels (Xu et al. 1997; Wang and Giebisch 2009; Wang et al. 2010). Similar to the TAL, basolateral Cl^- efflux is mediated via the ClC-K (mostly through ClC-K2/ClC-Kb) channels (Fahlke and Fischer 2010). These chloride channels significantly ensure the maintenance of the gradient for the entry of Na^+ and mitigate against intracellular Cl^- concentration increases resulting from apical NaCl entry via Na^+–Cl^- cotransport (Estevez et al. 2001; Lourdel et al. 2003). The apical membranes and tight junctions of DCT have almost no water permeability.

NCC The Na^+–Cl^- cotransporter (NCC, or *SLC12A3*) is a cation-coupled Cl^- cotransporter that, like NKCC2, belongs to the SLC12 family (Gamba 2005). NCC is dose dependently inhibited by thiazide diuretics (e.g., chlorthalidone, hydrochlorothiazide, bendroflumethiazide, and metolazone) rather than loop diuretics (Stokes 1984; Costanzo 1985; Ellison et al. 1987; Gamba et al. 1994) and is thus also known as the thiazide-sensitive cotransporter (TSC). Moreover, it was also shown to be insensitive to barium, acetazolamide, furosemide, amiloride, 4,4′-diisothiocyanostilbene-2,2′-disulfonic acid (DIDS), and ouabain (Gamba 2005). NCC mediates coupled absorption of Na^+ and Cl^-; this mechanism provides the means for maintaining intracellular Cl^- levels above the electrochemical equilibrium (Boron and Sackin 1983). Na^+ and Cl^- absorption in DCT is interdependent, and mammalian TSC has high affinity for both ions (half-maximal concentrations of both ions are ~10 mM) (Velazquez et al. 1984; Monroy et al. 2000).

The presence of NCC was first discovered in the urinary bladder of the winter flounder (Renfro 1975, 1977). Subsequent studies demonstrated that it is expressed in the renal cortex of rat, mouse, rabbit, and humans (Gamba et al. 1994; Mastroianni et al. 1996; Simon et al. 1996c; Bostanjoglo et al. 1998; Kunchaparty et al. 1999). The basic structure of NCC shares similarity with NKCC2 (shown in Fig. 2.3c), in that the protein has 12 transmembrane domains flanked by a short amino-terminal domain and a long predominantly hydrophilic C-terminal domain, both located in the cytoplasm; a long hydrophilic extracellular loop connects segments 7 and 8.

Loss-of-function mutations in the *SLC12A3* gene encoding NCC lead to the development of the Gitelman syndrome, which is clinically characterized by hypokalemia, hypomagnesemia, metabolic alkalosis, arterial hypotension, and hypocalciuria (see Table 2.1). On the other hand, increased NCC activity (caused by a gain-of-function mutation) results in the opposite symptoms including hyperkalemic metabolic acidosis, arterial hypertension, and hypercalciuria, known as Gordon syndrome, pseudohypoaldosteronism type II, or familial hyperkalemic hypertension (see Table 2.1) (Gamba 2005). New roles for NCC in sodium and potassium handling and blood pressure regulation have been unraveled. For instance, the recent discoveries that NCC is activated by angiotensin II but inhibited by dietary potassium shed light on how the kidney handles sodium during hypovolemia (accompanied with high angiotensin II) and hyperkalemia (Moes et al. 2014). In addition, much of the complex molecular machinery controlling the transporter's activity has been recently revealed. For example, modification of multiple kinases or ubiquitin ligases (including WNKs, SGK1, SPAK, Nedd4-2, Cullin-3, Kelch-like 3, and others), alteration of intracellular calcium signaling, changes in hormonal status (aldosterone, vasopressin, insulin, angiotensin II), and circadian rhythms all modulate the transporter's activity and may contribute to disease states such as hypo- or hypertension, and have detrimental effects on the ability of the kidneys to excrete sodium, potassium, calcium, and protons (Chiga et al. 2008; Vallon et al. 2009; Mutig et al. 2010; Gamba 2012; Chavez-Canales et al. 2013; Lee et al. 2013; Rieg et al. 2013; Saritas et al. 2013; Eladari et al. 2014; Gailly et al. 2014; Lagnaz et al. 2014; Moes et al. 2014; Richards et al. 2014; Terker et al. 2014). Further details on the structure, function, and regulation of NCC are provided in Chap. 13.

2.3.2.4 Connecting Tubule and Collecting Duct

Heterogenic vectorial transport in this segment is tightly controlled. This segment of the nephron is the most sensitive to hormones, which oversee the fine control of plasma Na^+ and K^+ concentrations. A variety of hormones (e.g., aldosterone, angiotensin II, vasopressin, atrial natriuretic peptide, and insulin) are involved in this tight regulation of CNT and CCD ion transport function. The late DCT, CNT, and CCD are often referred to as the aldosterone-sensitive distal nephron (ASDN). Aldosterone, which is the final element of the renin–angiotensin–aldosterone system (RAAS), increases the reabsorption of Na^+ and water and the secretion of K^+ ions.

The apical entry of Na^+ in the CNT and the CCD is regulated by the epithelial Na^+ channel (ENaC). Within the scope of this overview, only some aspects of ENaC regulation will be provided. Different aspects of ENaC regulation are highlighted in Chap. 19 and are subjects of several excellent recent reviews (Loffing and Korbmacher 2009; Bhalla and Hallows 2008; Butterworth et al. 2009; Soundararajan et al. 2010; Rossier 2014).

As described in Sect. 2.2.4.1, there are different transporters and exchangers involved in the maintenance of sodium homeostasis in the kidney. Interestingly, ENaC is the only "classical" channel mediating sodium reabsorption in the renal tubules. Considering that ENaC is responsible for the fine-tuning of sodium reabsorption in the last nephron segment, the role of this channel in sodium reabsorption in the kidney in critical and unique. While other transport proteins are also regulated by different factors, control of ENaC plays a key role in the final Na^+ composition of urine. Evolutionarily, the ENaC/degenerin (Deg) superfamily members appeared in Metazoan ancestors (Studer et al. 2011), along with the Na^+/K^+ pump **α**- and **β**-subunits. The emergence of the ENaC/Deg superfamily and the complete form of the active Na^+/K^+ pump afforded more strict control of salt balance with changes in the environment. Like other genes involved in membrane transport and osmolarity regulation, the hypothesis that these genes helped regulate the transition to multicellularity is appealing. Evolution of ENaC and the Na^+/K^+ pump as limiting factors of aldosterone action on Na^+ transport has been recently discussed by Rossier and colleagues (Studer et al. 2011).

Since ENaC is a channel that generally conducts Na^+ ions in only one direction, it is possible to apply the patch-clamp electrophysiological approach to directly measure inward ENaC-mediated Na^+ flux in native cells (Mironova et al. 2013; Stockand et al. 2012) and in overexpression mammalian systems (Staruschenko et al. 2006). Expression of ENaC subunits in *Xenopus* oocytes is also successfully used to assess ENaC function (Malik et al. 2005; Zhou et al. 2013; Chen et al. 2014; Krappitz et al. 2014). In combination with contemporary molecular genetics and biochemical tools, electrophysiology enables us to identify the function and specific mechanisms of action of renal ENaC channels. This has led to an explosive growth in investigation and subsequent understanding of many physiological and pathophysiological processes performed by ENaC in the aldosterone-sensitive distal nephron. Single-channel data provide the most illustrative picture of channel-mediated Na^+ transport since transients from closed to open states represent flux of Na^+ ions through ENaC when channel is positioned in the open state. A more thorough description of ENaC structure, function, and regulation can be found in Chap. 19.

2.3.3 Physiological Regulation of Na^+ Absorption

The tight regulation of transcellular Na^+ concentrations is so important that multiple mechanisms work in concert to control them. The expression and activity of Na^+ channels and transporters are regulated by specific hormones and different extra- and intracellular regulatory mechanisms. Recently, significant progress has been made in our understanding of the cellular and molecular mechanisms responsible for Na^+ absorption in the kidney. Multiple laboratories have employed electrophysiological, biochemical, microscopical, molecular, and genetic methods to study ion channels and transporters mediating sodium transport in both normal and pathological conditions. The development and application of various tools to study proteins mediating renal function have revealed some of the intriguing physiological functions of the kidney. Research advances naturally resulted in the cloning of multiple genes that are involved in water and electrolyte transport in different renal tubular segments. Moreover, new knowledge about the function of particular segments of the kidney has been accrued as a result of the development and application of gene deletion techniques, both conventional and cell-specific gene knockouts (Rubera et al. 2009; Lang and Shumilina 2013; Wen et al. 2014).

Importantly, the same mechanisms controlling Na^+ absorption are implicated in the maintenance of other ions and solutes. Thus, multiple studies have demonstrated that the same mechanisms controlling activation of ENaC result in the downregulation of ROMK K^+ channels and vice versa (Frindt et al. 2011). Similar findings are reported for integrated control of sodium, chloride, bicarbonate, phosphate, and other ions. Recent studies have also revealed cross talk between ENaC and Na^+/K^+ pump that may play a role in the correlation between Na^+ delivery and reabsorption independently of hormonal influence (Wang et al. 2014).

2.3.4 Tubuloglomerular Feedback (TGF) Mechanisms

Changes in sodium reabsorption and glomerular filtration are closely coordinated to avoid fluctuations in urinary excretion. Autoregulation of renal blood flow and thus glomerular filtration rate (GFR) maintains constant delivery of ions etc to the distal nephron where fine control of their transport by hormones such as aldosterone and arginine vasopressin (AVP) governs water and electrolyte balance. The TGF mechanism operates at the single-nephron level to maintain distal delivery within the narrow limits of the reabsorptive capacity of distal tubules (Braam et al. 1993). The mechanism for autoregulation during an acute increase in blood pressure is a reduction in afferent arteriolar radius that normalizes flow to the glomerulus. This arteriolar constriction is driven both by a myogenic mechanism and TGF, a response that senses NaCl delivery and transport by the apical NKCC cotransporter in the macula densa (MD) (McDonough et al. 2003). NKCC was proposed to be essential for the sensing step since intraluminal application of furosemide abolished the TGF-mediated reduction in single-nephron GFR (Wright and Schnermann 1974; Franco et al. 1988). MD is a region formed by about 20 cells in the tubular epithelium between the TAL and DCT where it establishes contact with its parent glomerulus and afferent arteriole. MD cell volume increases when increments are isosmotic and shrinks if osmolality increases. The MD cells are different from TAL and DCT epithelial cells both anatomically and functionally. For instance, in contrast to the water-impermeable TAL and DCT segments, the apical membrane of MD cells is permeable to water (Komlosi et al. 2006). Furthermore, MD cells have lower Na^+/K^+ pump activity than the adjacent tubular epithelial cells (Schnermann and Marver 1986; Komlosi et al. 2009).

TGF stabilizes nephron function through negative feedback by establishing an inverse relationship between the tubular NaCl load and the GFR of the same nephron. Although the molecular mechanisms of TGF signaling are still not completely uncovered, it is generally accepted that purinergic signaling via ATP or adenosine is at least partially responsible for the tubular signal-dependent regulation of glomerular hemodynamics (Braam et al. 1993; Peti-Peterdi 2006; Komlosi et al. 2009; Schnermann 2011). The sensor function is also coupled with other intracellular signaling events including changes in Cl^- concentrations, intracellular calcium, pH, membrane depolarization, cell volume, etc. Furthermore, in addition to purines, MD produces and releases prostaglandin E_2 and nitric oxide (Komlosi et al. 2009). MD also plays a critical role in renin secretion, which is the major component of the renin–angiotensin–aldosterone system (RAAS). Products of the RAAS formed downstream from renin act to conserve salt by causing glomerular vasoconstriction and by increasing reabsorption of salt and water from the proximal and distal nephrons.

In addition to the MD contacts with afferent arteriole, it was previously reported that the superficial nephrons of the renal cortex also come into close proximity to the corresponding afferent arteriole through the CNT (Barajas et al. 1986; Dorup et al. 1992; Capasso 2007). Following studies by Ren et al. provided evidence of the functional connection between the CNT and afferent arteriole (Ren et al. 2007). This cross talk was called "connecting tubule glomerular feedback" (CTGF) to differentiate it from TGF. Inhibition of ENaC with amiloride blocked the dilatation induced by CTGF, but inhibition of NCC with hydrochlorothiazide failed to prevent renal afferent arterioles dilatation (Ren et al. 2007). Thus, in addition to its critical role in collecting ducts, ENaC is important for sodium reabsorption in upstream CNT segment (Rubera et al. 2003; Nesterov et al. 2012). Recent studies provided some further details about specific mechanisms involved in CTGF (Ren et al. 2013, 2014; Wang et al. 2013). The CTGF response has not been widely studied and this mechanism needs to be more fully explored. However, these studies emphasize the role of the connecting tubules and collecting ducts, where the final control of urinary electrolytes takes place.

2.3.5 Pharmacological Control of Na^+ Absorption

2.3.5.1 Loop Diuretics

As suggested earlier, the apical membranes of TAL epithelial cells are hyperpolarized because of high ROMK K^+ channel activity. Thus, the transmembrane potential of these apical membranes is strongly dependent on the equilibrium potential for K^+ (V_K). In contrast, the basolateral membrane has many Cl^- channels, so the basolateral transmembrane potential is less negative than V_K; (i.e., the Cl^- conductance depolarizes the basolateral membrane). As a result of hyperpolarization of the luminal membrane and depolarization of the basolateral membrane, there exists a transepithelial potential difference of ~10 mV, with the lumen positive with respect to the interstitial space. The lumen-positive potential provides an important driving force for the paracellular flux of Na^+, Ca^{2+}, and Mg^{2+} into the interstitial space. Inhibitors of Na^+–K^+–$2Cl^-$ cotransport thus also attenuate Ca^{2+} and Mg^{2+} reabsorption in the TAL by abolishing the transepithelial potential difference. Mutations in genes encoding the Na^+–K^+–$2Cl^-$ cotransporter, the apical K^+ channel, the basolateral Cl^- channel, the Barttin subunit of this Cl^- channel, or the calcium-sensing receptor on the basolateral membrane of these cells can all be causes of Bartter syndrome (inherited hypokalemic alkalosis with salt wasting and hypotension; see Table 2.1) (Simon et al. 1996a, b, c; Lifton et al. 2001).

Na^+–K^+–$2Cl^-$ cotransport inhibitors are chemically diverse. The drugs currently available in this group are furosemide and bumetanide, which contain a sulfonamide moiety; ethacrynic acid, which is a phenoxyacetic acid derivative; and torsemide, which is a sulfonylurea (Wargo and Banta 2009). These drugs inhibit NKCC2 at the apical membrane of the TAL cells of Henle's loop, so these diuretics are often called "loop diuretics." Diuretics acting at sites distal to the TAL have lower efficacy because a much smaller percentage of the filtered Na^+ reaches these distal sites. In contrast, NKCC inhibitors are highly efficacious because ~25 % of the filtered Na^+ normally is reabsorbed by the TAL and nephron segments distal to the TAL do not possess the reabsorptive capacity to rescue the flood of unreabsorbed salt from the TAL.

2.3.5.2 Thiazide Diuretics

Figure 2.8 illustrates the current model of electrolyte transport in the late DCT. As with other nephron segments, transport is powered by a Na^+/K^+ pump in the basolateral membrane. Energy in the electrochemical gradient for Na^+ is harnessed by the NCC cotransporter in the luminal membrane and moves Cl^- into the epithelial cell against its electrochemical gradient. Cl^- then exits the basolateral membrane passively via a Cl^- channel. Thiazide diuretics inhibit the Na^+–Cl^- cotransporter. Mutations in the Na^+–Cl^- cotransporter cause a form of inherited hypokalemic alkalosis (Gitelman syndrome; see Table 2.1) (Gamba 2005, 2012).

Because the first inhibitors of the NCC were 1,2,4-benzothiazide-1,1-dioxides, this class of diuretics became known as thiazide diuretics. Later, drugs that are pharmacologically similar to thiazide diuretics, but are not chemically thiazides, were developed and are called thiazide-like diuretics. There are many thiazide and thiazide-like diuretics available in the USA, including hydrochlorothiazide, chlorothiazide, metolazone, and chlorthalidone. These drugs (especially hydrochlorothiazide) are often the first-line treatments for patients with hypertension. Moreover, because some resistance to the effects of high-dose loop diuretics may occur over time or in patients with chronic kidney disease or heart failure due to compensatory upregulation of DCT NCC expression and function, thiazide diuretics (esp. metolazone or chlorothiazide) are often used in this setting to potentiate diuresis in these patients (Fliser et al. 1994; Jentzer et al. 2010).

2.3.5.3 Amiloride and Its Analogues

Principal cells in the late DCT, CNT, and CD regulate ENaC expression in their apical membranes, which provides a conductive pathway for Na^+ entry into the cell down the electrochemical gradient created by the basolateral Na^+ pump (Staruschenko 2012). The higher permeability of the luminal membrane for Na^+ depolarizes the luminal membrane but not the basolateral membrane, creating a lumen-negative transepithelial potential difference (TEPD). This transepithelial voltage provides an important driving force for the secretion of K^+ into the lumen via K^+ channels (primarily ROMK and the Ca^{2+}-activated K^+ channel (BK); see Fig. 2.9) in the luminal membrane. Loop and thiazide diuretics increase distal Na^+ delivery, a situation that is often associated with increased K^+ and H^+ excretion. The increased luminal Na^+ concentration in the distal nephron induced by such diuretics augments depolarization of the apical membrane and thereby enhances the lumen-negative TEPD, which facilitates K^+ excretion. In addition to principal cells, the CD also contains type A intercalated cells (Al-Awqati 2013) that mediate H^+ secretion into the tubular lumen via the apical vacuolar H^+-ATPase (V-ATPase or proton pump), and this pump is aided by partial depolarization of the apical membrane. However, increased distal Na^+ delivery is not the only mechanism by which diuretics increase K^+ and H^+ excretion. Activation of the renin–angiotensin–aldosterone system by diuretics also contributes to diuretic-induced K^+ and H^+ excretion, as discussed in the next section on mineralocorticoid antagonists.

Considerable evidence indicates that amiloride blocks ENaC in the luminal membrane of principal cells in the late distal tubule and CD (Warnock et al. 2014). Importantly, amiloride blocks ENaC in the low nM concentration range. Used in higher concentration, it is also able to inhibit other sodium channels and

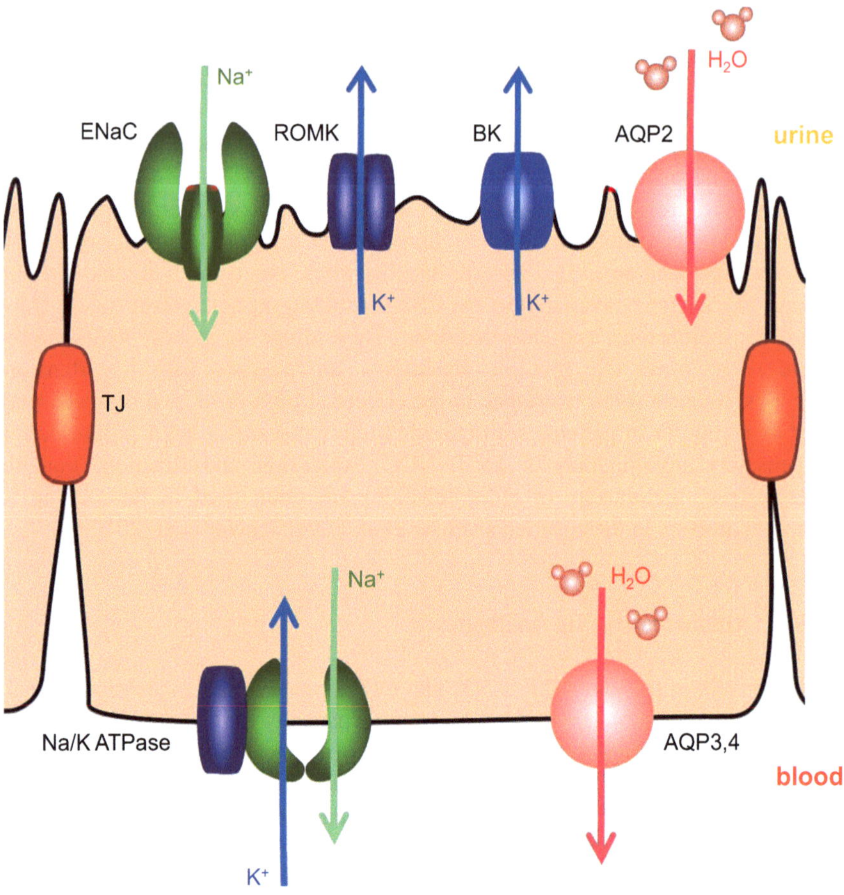

Fig. 2.9 Simplified scheme of Na^+ reabsorption in the connecting tubules/cortical collecting duct. The apical entry of Na^+ in the CNT and the CCD is regulated by the epithelial Na^+ channel (ENaC); water follows the concentration gradient created by sodium transport and is taken up from the urine via AQP2 (aquaporin 2) channels. Basolaterally, sodium is reabsorbed via the Na^+/K^+-ATPase, and water is removed by aquaporins 3 and 4 (AQP3 and 4). Secretion of K^+ into the lumen occurs primarily via ROMK and the Ca^{2+}-activated K^+ channel (BK)

transporters, such as NHE (Kleyman and Cragoe 1988; Teiwes and Toto 2007). Appropriate modification of amiloride has produced analogues that are several hundredfold more active than amiloride against specific transporters (Kleyman and Cragoe 1988). Liddle syndrome is an autosomal dominant form of low-renin, volume-expanded hypertension that results from C-terminal truncation mutations in the *β*- or *γ*-ENaC subunits, which prevents the channel's retrieval from the apical membrane and subsequent degradation, thus leading to increased basal ENaC expression and activity at the apical membrane (see Table 2.1) (Shimkets et al. 1994; Hansson et al. 1995a, b; Lifton et al. 2001; Ronzaud and Staub 2014).

Triamterene and amiloride are the only two drugs of this class in clinical use. Amiloride is a pyrazinoylguanidine derivative, and triamterene is a pteridine (Tamargo et al. 2014). Thus, both drugs are organic bases and are transported by the organic base secretory mechanism in the proximal tubule (McKinney 1984). Both drugs cause small increases in NaCl excretion and are often employed for their antikaliuretic actions to offset the effects of other diuretics that increase K^+ excretion and may cause hypokalemia. Consequently, triamterene and amiloride, along with spironolactone (see next section), are often classified as potassium-sparing diuretics. These drugs are also often used in the setting of ongoing lithium treatment to limit the potential damage to CD epithelial cells caused by lithium entry via ENaC, which over time may result in nephrogenic diabetes insipidus (Kortenoeven et al. 2009; Kishore and Ecelbarger 2013).

2.3.5.4 Mineralocorticoid Receptor (MR) Antagonists

Mineralocorticoids induce salt and water retention and increase K^+ and H^+ excretion by binding to specific MRs. Currently, two MR antagonists are available in the USA: spironolactone and eplerenone. These MR antagonists are also called K^+-sparing diuretics or aldosterone antagonists. Epithelial cells in the late distal tubule and CD contain cytosolic MRs that have high affinity for aldosterone. Aldosterone enters the epithelial cell from the bloodstream via the basolateral membrane and binds to MRs. The MR–aldosterone complex then translocates to the nucleus where it binds to specific hormone-responsive elements of DNA and thereby regulates the expression of multiple gene products called aldosterone-induced proteins (AIPs), which notably include the α-ENaC subunit, the serum and glucocorticoid-regulated kinase (SGK1) and the glucocorticoid-induced leucine zipper protein (GILZ). A number of different proposed effects of AIPs have been described, including increased cellular protein expression and localization of Na^+ pumps and Na^+ channels at the plasma membrane, changes in the permeability of tight junctions, and increased activity of enzymes in the mitochondria that are involved in ATP production (Law and Edelman 1978; Yu 2015). The net effect of AIPs is to increase Na^+ conductance of the luminal membrane and sodium pump activity of the basolateral membrane. Consequently, transepithelial NaCl transport is enhanced, and the lumen-negative transepithelial voltage is increased. The latter effect increases the driving force for secretion of K^+ and H^+ into the tubular lumen.

2.3.5.5 SGLT Inhibitors

SGLT2 expressed in the PT has become one of the major targets for regulation blood glucose levels in diabetes (Oliva and Bakris 2014; De Nicola et al. 2014; Vallon et al. 2011, 2014). The classic competitive inhibitor of SGLTs is phlorizin (Ehrenkranz et al. 2005). Phlorizin has a higher affinity for hSGLT2 (K_i 40 nM) than hSGLT1 (200 nM). Several pharmaceutical companies have attempted to modify the phlorizin structure to further enhance selectivity for SGLT2 over SGLT1 and develop oral SGLT2 inhibitors. Examples of these new FDA-approved SGLT2 inhibitors include dapagliflozin and canagliflozin (Riser Taylor and Harris 2013).

2.4 Sodium Balance and Its Role in Other Organs

2.4.1 Sodium Absorption in the Lung

Alveoli The alveoli of the lung promote the exchange of oxygen and CO_2 between the air spaces and the blood and contain two main cell types, termed alveolar type I and type II cells. Efficient gas exchange requires the presence of a thin liquid layer on apical surface of the alveolar epithelium, which is finely regulated by a balance between passive secretion of fluid from the vasculature by a paracellular route governed by permeability of claudins and active reabsorption governed by the presence of apical Na^+ channels (reviewed in Eaton et al. (2009)). Although more is known about the surfactant-secreting type II cells, both alveolar cell types appear to contain a variety of ion transport proteins, including two forms of amiloride-sensitive channels. One form has the electrophysiological signature of ENaC expressed in the kidney and colon, as it is highly selective for Na^+ over other cations and has a low single-channel conductance of ~5 pS and strong sensitivity to amiloride. The other form is relatively nonselective for Na^+ over K^+, has a higher single-channel conductance (~20 pS) and a weaker sensitivity to amiloride, and may contain primarily α-ENaC subunits rather than all three subunits. Na^+ entering at the apical membrane exits at the basolateral membrane via the Na^+/K^+ pump. ENaC in the airways is highly regulated by agonists interacting with G protein-coupled receptors (e.g., purinergic and adrenergic agents), circulating hormones (e. g., glucocorticoids), chemokines and inflammatory mediators (e.g., TNF-α and interleukins), and reactive oxygen and nitrogen species. These regulatory factors may differentially modulate the function of highly selective ENaC channels versus nonselective cation channels (Eaton et al. 2009). Further details about the role of ENaC and other cation channels in the lung are provided in Chap. 19.

Airways Under normal conditions in the conducting airways, a balance is maintained between fluid reabsorption and fluid secretion to preserve airway surface liquid (ASL) height, cilia function, and thus normal mucociliary clearance. This balance appears to be regulated by ASL height itself and signaling factors such as secreted nucleotides, although the mechanisms that underlie this absorption–secretion coupling are still under active investigation. Apical membrane ENaC and possibly cyclic nucleotide-gated cation channels contribute to active transcellular Na^+ reabsorption, which occurs in concert with basolateral Na^+/K^+ pump activity and paracellular Cl^- reabsorption (Hollenhorst et al. 2011). This active salt absorption is coupled with passive transcellular water flow mediated by aquaporin water

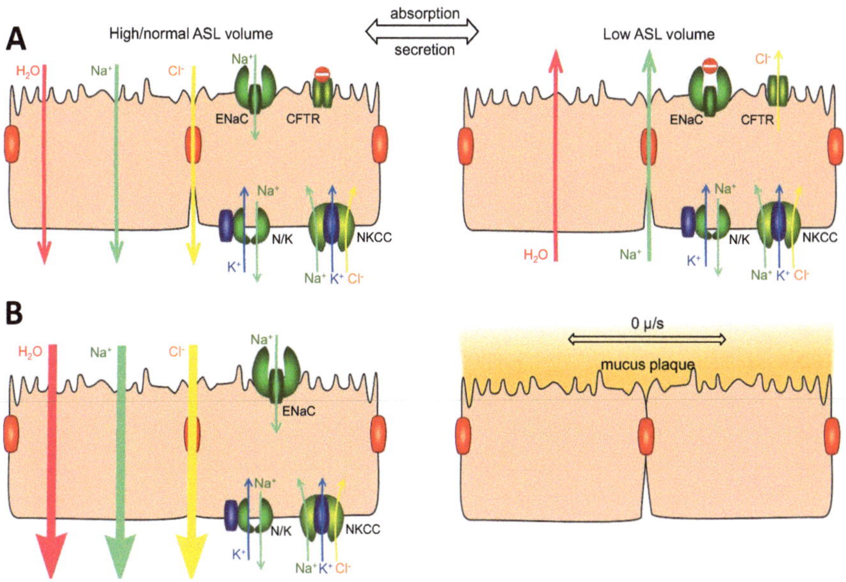

Fig. 2.10 Balance is maintained between airway surface liquid absorption and secretion mediated by transcellular and paracellular Na^+, Cl^-, and water transport pathways under normal conditions (**a**), but this balance gets disturbed in patients with cystic fibrosis, who have excessive ENaC-mediated Na^+ (and thus fluid) absorption (**b**). See text for further details

channels (see Fig. 2.10) (Donaldson and Boucher 2003). On the other hand, fluid secretion in the conducting airways occurs via transcellular Cl^- secretion predominantly through apical CFTR channels, but also through Ca^{2+}-activated Cl^- channels like TMEM16A (Rock et al. 2009), coupled with basolateral NKCC-mediated Cl^- entry, paracellular Na^+ secretion (with ENaC inhibited), and transcellular water flow from the basolateral to apical compartment (Fig. 2.10).

Disease Correlations The importance of ENaC in the regulation of alveolar and airway surface liquid volume is evidenced by the fact that loss-of-function ENaC mutations underlie pseudohypoaldosteronism type I, which is associated with neonatal respiratory distress syndrome caused by pulmonary edema (Keszler and Sivasubramanian 1983; Malagon-Rogers 1999). Hypoxia decreases ENaC expression in nongenetic forms of this disorder among premature infants. In addition to inducing increased surfactant production, dexamethasone may improve this condition by counteracting hypoxia-induced decreases in ENaC-mediated sodium reabsorption. Decreased ENaC activity in the lung is also seen in patients susceptible to high-altitude pulmonary edema (Scherrer et al. 1999). A potential therapeutic role for ENaC modulation in the acute respiratory distress syndrome has also been explored (Matthay et al. 2005).

Perhaps the most thoroughly studied role for ENaC in pulmonary disease is in patients with cystic fibrosis (CF), caused by mutations in the CFTR Cl^- channel (see Chap. 17). In patients with CF lung disease, the normal balance between fluid secretion and reabsorption in the airway gets disrupted due to a loss of CFTR function. As CFTR normally inhibits ENaC activity (by unclear mechanisms), a constitutive gain of ENaC functional activity appears to ensue in the airways of patients with CF, which might be a major contributing factor in the pathogenesis of CF airway disease, causing low ASL volume, mucous plugging, infections and inflammation, and chronic bronchiectasis with lung parenchymal destruction (Fig. 2.10) (Donaldson and Boucher 2007). Of note, a ***β***-ENaC transgenic lung mouse model revealed that overexpression of ENaC appears to be sufficient to recapitulate the CF lung phenotype of decreased airway surface liquid volume and increased airway inflammation in mice (Mall et al. 2004). However, this view that low ASL volume, per se, underlies much of the pathogenesis of CF lung disease has been challenged, and the actual situation that exists in human CF lung disease is still far from settled (Song et al. 2009).

2.4.2 Sodium Absorption in the Gastrointestinal and Endocrine Systems

Small Intestine There are regional differences in the mechanisms of Na^+ transport and salt absorption vs. secretion in the gastrointestinal tract. Electroneutral NaCl absorption predominates in the small intestine and proximal colon and occurs at the apical membrane via parallel Na^+/H^+ exchange (mediated by NHE2 and NHE3) and Cl^-/HCO_3^- exchange (Kato and Romero 2011). These transport proteins are regulated by intracellular pH and intracellular $[HCO_3^-]$ directly, as H^+ and HCO_3^- are substrates, and indirectly through pH-dependent allosteric effects. In addition, there is nutrient-coupled Na^+ absorption, which occurs via SGLTs for Na^+–glucose cotransport, along with several Na^+–amino acid cotransporters and Na^+-coupled solute carriers. This salt and fluid absorption is balanced by electrogenic Cl^- secretion mediated by apical CFTR, which is stimulated under conditions that increase cellular cAMP, cGMP, or Ca^{2+} levels. These same signaling mediators concomitantly inhibit electroneutral NaCl transport. Once Na^+ and Cl^- are apically absorbed, the combination of a basolateral Cl^- channel (ClC-2), the Na^+/K^+-ATPase, and the Kir7.1 K^+ channel, which allows K^+ pumped in by the Na^+/K^+ pump to recycle back out of the cell, affords the net transfer of NaCl across the basolateral membrane (Kato and Romero 2011).

Functional interactions among these transporters, along with a network of kinase signaling cascades, coordinate the regulation of salt absorption from the small intestine. The endocrine, autonomic, and immune systems can regulate epithelial NaCl transport function via the enteric nervous system and gene expression of the various transport proteins (Kato and Romero 2011). Glucocorticoids stimulate apical membrane NHE3 via SGK1 and also stimulate SGLT1 transporters (Grahammer et al. 2006). The mineralocorticoid aldosterone also weakly stimulates electroneutral NaCl absorption in the small intestine.

Colon In the colon there is electroneutral NaCl reabsorption under basal conditions, but aldosterone is a key stimulator of electroneutral NaCl absorption in the proximal colon and ENaC-mediated electrogenic Na^+ reabsorption in the distal colon, where it works to enhance Na^+ (salt) absorption from these segments of the gut (Harvey et al. 2008). Thus, in the distal colon, aldosterone causes a switch from electroneutral NaCl absorption to stimulated electrogenic Na^+ absorption by inducing expression of apical ENaC and basolateral Na^+/K^+-ATPase (Kunzelmann and Mall 2002). In parallel, net K^+ absorption is converted to net K^+ secretion by induction of apical K^+ channels (Sweiry and Binder 1989). ENaC expression at the apical membrane is generally more near the tips of the villi, whereas CFTR Cl^- channel expression is more pronounced at the apical membrane near the crypts of the villi. Like in the small intestine, CFTR is activated during secretion.

Disease Correlations Whereas enhanced apical Cl^- secretion via CFTR in the small intestine plays a key role in several secretory diarrheal diseases, such as cholera, where cholera toxin-induced cAMP formation promotes PKA-dependent CFTR-mediated fluid secretion, inhibition of sodium and chloride absorption in the colon is important for diarrheal disorders such as ulcerative colitis (Sandle 2005). Recent studies indicate reduced expression/activity of apical Na^+ channels and basolateral Na^+/K^+-ATPase, leading to a loss of electrogenic Na^+ absorption in the distal colon and rectum (Greig et al. 2004). There is also likely to be a decrease in electroneutral NaCl cotransport, which is present throughout the colon. Preliminary work on basolateral K^+ channel abundance and activity in colonic epithelial cells suggests that whole-cell K^+ conductance is decreased in ulcerative colitis, leading to epithelial cell depolarization and further limitation of Na^+ absorption. In addition, there is a marked reduction in colonic epithelial resistance, which reflects a decrease in the integrity of intercellular tight junctions and the presence of apoptotic foci (Sandle 2005).

2.5 Sodium Transport in Epithelia and Human Diseases

In the kidney, the primary role of renal Na^+ transport is the control of extracellular fluid volume. Thus, a number of genetic disorders are characterized by either hypo- or hypertension. Many Na^+ transport proteins have been linked to specific genetic disorders in the kidney (Table 2.1). However, mutations in Na^+ channels and transporters may cause disease manifestations in other organs such as lung and colon (described above) and in other tissues. Importantly, Na^+ channels and transporters are highly critical for human diseases not only due to certain mutations identified in those proteins. Mutations in multiple genes causing human diseases are also involved in signaling mechanisms, where Na^+ transport proteins are the final end points of those signaling pathways. Thus, ion channels and transporters might be attractive targets for various pharmaceutical drugs even when diseases are mediated by mutations in other proteins.

2.6 Final Conclusions

Epithelial Na^+ reabsorption in the kidneys and other tissues plays key roles in regulating total body salt and water balance, airway function, and fluid and electrolyte reabsorption from the gut, glands, and other organs. In addition to focusing on individual Na^+ channels and other transporters, future work will be required to provide us with detailed information about the integration between various components of sodium absorption and their tight regulation.

Acknowledgments Research from the authors' laboratories is supported by NIH grants DK075048 (to K.R.H.) and HL108880 (to A.S.) and a Ben J. Lipps Research Fellowship from the ASN (to D.I).

References

Agre P (2000) Homer W. Smith award lecture. Aquaporin water channels in kidney. J Am Soc Nephrol 11(4):764–777

Al-Awqati Q (2013) Cell biology of the intercalated cell in the kidney. FEBS Lett 587(13):1911–1914

Allen GG, Barratt LJ (1985) Origin of positive transepithelial potential difference in early distal segments of rat kidney. Kidney Int 27(4):622–629

Amemiya M, Loffing J, Lotscher M, Kaissling B, Alpern RJ, Moe OW (1995) Expression of NHE-3 in the apical membrane of rat renal proximal tubule and thick ascending limb. Kidney Int 48 (4):1206–1215

Anderson JM, Van Itallie CM (2009) Physiology and function of the tight junction. Cold Spring Harb Perspect Biol 1(2):a002584

Andreoli TE, Schafer JA, Troutman SL (1978) Perfusion rate-dependence of transepithelial osmosis in isolated proximal convoluted tubules: estimation of the hydraulic conductance. Kidney Int 14(3):263–269

Ares GR, Caceres PS, Ortiz PA (2011) Molecular regulation of NKCC2 in the thick ascending limb. Am J Physiol Renal Physiol 301(6):F1143–F1159

Ares GR, Haque MZ, Delpire E, Ortiz PA (2012) Hyperphosphorylation of Na-K-2Cl cotransporter in thick ascending limbs of Dahl salt-sensitive rats. Hypertension 60(6):1464–1470

Barajas L, Powers K, Carretero O, Scicli AG, Inagami T (1986) Immunocytochemical localization of renin and kallikrein in the rat renal cortex. Kidney Int 29(5):965–970

Barratt LJ, Rector FC Jr, Kokko JP, Seldin DW (1974) Factors governing the transepithelial potential difference across the proximal tubule of the rat kidney. J Clin Invest 53(2):454–464

Barratt LJ, Rector FC Jr, Kokko JP, Tisher CC, Seldin DW (1975) Transepithelial potential difference profile of the distal tubule of the rat kidney. Kidney Int 8(6):368–375

Bennett CM, Brenner BM, Berliner RW (1968) Micropuncture study of nephron function in the rhesus monkey. J Clin Invest 47(1):203–216

Bhalla V, Hallows KR (2008) Mechanisms of ENaC regulation and clinical implications. J Am Soc Nephrol 19(10):1845–1854

Biner HL, Arpin-Bott MP, Loffing J, Wang X, Knepper M, Hebert SC, Kaissling B (2002) Human cortical distal nephron: distribution of electrolyte and water transport pathways. J Am Soc Nephrol 13(4):836–847

Birkenhager R, Otto E, Schurmann MJ, Vollmer M, Ruf EM, Maier-Lutz I, Beekmann F, Fekete A, Omran H, Feldmann D, Milford DV, Jeck N, Konrad M, Landau D, Knoers NV, Antignac C, Sudbrak R, Kispert A, Hildebrandt F (2001) Mutation of BSND causes Bartter syndrome with sensorineural deafness and kidney failure. Nat Genet 29(3):310–314

Bockenhauer D, Feather S, Stanescu HC, Bandulik S, Zdebik AA, Reichold M, Tobin J, Lieberer E, Sterner C, Landoure G, Arora R, Sirimanna T, Thompson D, Cross JH, Van't Hoff W, Al Masri MO, Tullus K, Yeung S, Anikster Y, Klootwijk E, Hubank M, Dillon MJ, Heitzmann D, Arcos-Burgos M, Knepper MA, Dobbie A, Gahl WA, Warth R, Sheridan E, Kleta R (2009) Epilepsy, ataxia, sensorineural deafness, tubulopathy, and KCNJ10 mutations. N Engl J Med 360(19):1960–1970

Borensztein P, Froissart M, Laghmani K, Bichara M, Paillard M (1995) RT-PCR analysis of Na$^+$/H$^+$ exchanger mRNAs in rat medullary thick ascending limb. Am J Physiol Renal Physiol 268 (6 Pt 2):F1224–F1228

Boron WF, Sackin H (1983) Measurement of intracellular ionic composition and activities in renal tubules. Annu Rev Physiol 45:483–496

Bostanjoglo M, Reeves WB, Reilly RF, Velazquez H, Robertson N, Litwack G, Morsing P, Dorup J, Bachmann S, Ellison DH (1998) 11Beta-hydroxysteroid dehydrogenase, mineralocorticoid receptor, and thiazide-sensitive Na-Cl cotransporter expression by distal tubules. J Am Soc Nephrol 9(8):1347–1358

Braam B, Mitchell KD, Koomans HA, Navar LG (1993) Relevance of the tubuloglomerular feedback mechanism in pathophysiology. J Am Soc Nephrol 4(6):1257–1274

Breiderhoff T, Himmerkus N, Stuiver M, Mutig K, Will C, Meij IC, Bachmann S, Bleich M, Willnow TE, Müller D (2012) Deletion of claudin-10 (Cldn10) in the thick ascending limb impairs paracellular sodium permeability and leads to hypermagnesemia and nephrocalcinosis. Proc Natl Acad Sci USA 109(35):14241–14246

Butterworth MB, Edinger RS, Frizzell RA, Johnson JP (2009) Regulation of the epithelial sodium channel by membrane trafficking. Am J Physiol Renal Physiol 296(1):F10–F24

Campean V, Kricke J, Ellison D, Luft FC, Bachmann S (2001) Localization of thiazide-sensitive Na$^+$-Cl$^-$ cotransport and associated gene products in mouse DCT. Am J Physiol Renal Physiol 281(6):F1028–F1035

Capasso G (2007) A new cross-talk pathway between the renal tubule and its own glomerulus. Kidney Int 71(11):1087–1089

Capasso G, Rizzo M, Pica A, Di Maio FS, Moe OW, Alpern RJ, De Santo NG (2002) Bicarbonate reabsorption and NHE-3 expression: abundance and activity are increased in Henle's loop of remnant rats. Kidney Int 62(6):2126–2135

Carmosino M, Rizzo F, Ferrari P, Torielli L, Ferrandi M, Bianchi G, Svelto M, Valenti G (2011) NKCC2 is activated in Milan hypertensive rats contributing to the maintenance of salt-sensitive hypertension. Pflugers Arch 462(2):281–291

Castrop H, Schnermann J (2008) Isoforms of renal Na-K-2Cl cotransporter NKCC2: expression and functional significance. Am J Physiol Renal Physiol 295(4):F859–F866

Chavez-Canales M, Arroyo JP, Ko B, Vazquez N, Bautista R, Castaneda-Bueno M, Bobadilla NA, Hoover RS, Gamba G (2013) Insulin increases the functional activity of the renal NaCl cotransporter. J Hypertens 31(2):303–311

Chen J, Kleyman TR, Sheng S (2014) Deletion of alpha-subunit exon 11 of the epithelial Na$^+$ channel reveals a regulatory module. Am J Physiol Renal Physiol 306(5):F561–F567

Chiga M, Rai T, Yang SS, Ohta A, Takizawa T, Sasaki S, Uchida S (2008) Dietary salt regulates the phosphorylation of OSR1/SPAK kinases and the sodium chloride cotransporter through aldosterone. Kidney Int 74(11):1403–1409

Ciampolillo F, McCoy DE, Green RB, Karlson KH, Dagenais A, Molday RS, Stanton BA (1996) Cell-specific expression of amiloride-sensitive, Na$^+$-conducting ion channels in the kidney. Am J Physiol Cell Physiol 271(4 Pt 1):C1303–C1315

Costanzo LS (1984) Comparison of calcium and sodium transport in early and late rat distal tubules: effect of amiloride. Am J Physiol Renal Physiol 246(6 Pt 2):F937–F945

Costanzo LS (1985) Localization of diuretic action in microperfused rat distal tubules: Ca and Na transport. Am J Physiol 248(4 Pt 2):F527–F535
Crane RK (1960) Intestinal absorption of sugars. Physiol Rev 40:789–825
De Nicola L, Gabbai FB, Liberti ME, Sagliocca A, Conte G, Minutolo R (2014) Sodium/glucose cotransporter 2 inhibitors and prevention of diabetic nephropathy: targeting the renal tubule in diabetes. Am J Kidney Dis 64(1):16–24
Díez-Sampedro A, Eskandari S, Wright EM, Hirayama BA (2001) Na^+-to-sugar stoichiometry of SGLT3. Am J Physiol Renal Physiol 280(2):F278–F282
Díez-Sampedro A, Hirayama BA, Osswald C, Gorboulev V, Baumgarten K, Volk C, Wright EM, Koepsell H (2003) A glucose sensor hiding in a family of transporters. Proc Natl Acad Sci U S A 100(20):11753–11758
Donaldson SH, Boucher RC (2003) Update on pathogenesis of cystic fibrosis lung disease. Curr Opin Pulm Med 9(6):486–491
Donaldson SH, Boucher RC (2007) Sodium channels and cystic fibrosis. Chest 132(5):1631–1636
Donowitz M, Li X (2007) Regulatory binding partners and complexes of NHE3. Physiol Rev 87 (3):825–872
Dorup J, Morsing P, Rasch R (1992) Tubule-tubule and tubule-arteriole contacts in rat kidney distal nephrons. A morphologic study based on computer-assisted three-dimensional reconstructions. Lab Invest 67(6):761–769
Doucet A (1988) Function and control of Na-K-ATPase in single nephron segments of the mammalian kidney. Kidney Int 34(6):749–760
Eaton DC, Helms MN, Koval M, Bao HF, Jain L (2009) The contribution of epithelial sodium channels to alveolar function in health and disease. Annu Rev Physiol 71:403–423
Ehrenkranz JR, Lewis NG, Kahn CR, Roth J (2005) Phlorizin: a review. Diabetes Metab Res Rev 21(1):31–38
Eladari D, Chambrey R, Picard N, Hadchouel J (2014) Electroneutral absorption of NaCl by the aldosterone-sensitive distal nephron: implication for normal electrolytes homeostasis and blood pressure regulation. Cell Mol Life Sci 71(15):2879–2895
Ellison DH, Velazquez H, Wright FS (1987) Thiazide-sensitive sodium chloride cotransport in early distal tubule. Am J Physiol Renal Physiol 253(3 Pt 2):F546–F554
Estevez R, Boettger T, Stein V, Birkenhager R, Otto E, Hildebrandt F, Jentsch TJ (2001) Barttin is a Cl^- channel beta-subunit crucial for renal Cl^- reabsorption and inner ear K^+ secretion. Nature 414(6863):558–561
Fahlke C, Fischer M (2010) Physiology and pathophysiology of ClC-K/barttin channels. Front Physiol 1:155
Feraille E, Doucet A (2001) Sodium-potassium-adenosinetriphosphatase-dependent sodium transport in the kidney: hormonal control. Physiol Rev 81(1):345–418
Fliser D, Schroter M, Neubeck M, Ritz E (1994) Coadministration of thiazides increases the efficacy of loop diuretics even in patients with advanced renal failure. Kidney Int 46(2):482–488
Franco M, Bell PD, Navar LG (1988) Evaluation of prostaglandins as mediators of tubuloglomerular feedback. Am J Physiol Renal Physiol 254(5 Pt 2):F642–F649
Frindt G, Houde V, Palmer LG (2011) Conservation of Na^+ versus K^+ by the rat cortical collecting duct. Am J Physiol Renal Physiol 301(1):F14–F20
Fushimi K, Uchida S, Hara Y, Hirata Y, Marumo F, Sasaki S (1993) Cloning and expression of apical membrane water channel of rat kidney collecting tubule. Nature 361(6412):549–552
Gailly P, Szutkowska M, Olinger E, Debaix H, Seghers F, Janas S, Vallon V, Devuyst O (2014) P2Y receptor activation inhibits the expression of the sodium-chloride cotransporter NCC in distal convoluted tubule cells. Pflugers Arch 466(11):2035–2047
Gamba G (2001) Alternative splicing and diversity of renal transporters. Am J Physiol Renal Physiol 281(5):F781–F794
Gamba G (2005) Molecular physiology and pathophysiology of electroneutral cation-chloride cotransporters. Physiol Rev 85(2):423–493

Gamba G (2012) Regulation of the renal Na^+-Cl^- cotransporter by phosphorylation and ubiquitylation. Am J Physiol Renal Physiol 303(12):F1573–F1583
Gamba G, Miyanoshita A, Lombardi M, Lytton J, Lee WS, Hediger MA, Hebert SC (1994) Molecular cloning, primary structure, and characterization of two members of the mammalian electroneutral sodium-(potassium)-chloride cotransporter family expressed in kidney. J Biol Chem 269(26):17713–17722
Giebisch G (2001) Renal potassium channels: function, regulation, and structure. Kidney Int 60 (2):436–445
Giebisch G, Klose RM, Malnic G, Sullivan WJ, Windhager EE (1964) Sodium movement across single perfused proximal tubules of rat kidneys. J Gen Physiol 47:1175–1194
Goldman DE (1943) Potential, impedance, and rectification in membranes. J Gen Physiol 27 (1):37–60
Grahammer F, Henke G, Sandu C, Rexhepaj R, Hussain A, Friedrich B, Risler T, Metzger M, Just L, Skutella T, Wulff P, Kuhl D, Lang F (2006) Intestinal function of gene-targeted mice lacking serum- and glucocorticoid-inducible kinase 1. Am J Physiol Gastrointest Liver Physiol 290(6):G1114–G1123
Greger R (1981) Chloride reabsorption in the rabbit cortical thick ascending limb of the loop of Henle. A sodium dependent process. Pflugers Arch 390(1):38–43
Greig ER, Boot-Handford RP, Mani V, Sandle GI (2004) Decreased expression of apical Na^+ channels and basolateral Na^+, K^+-ATPase in ulcerative colitis. J Pathol 204(1):84–92
Guggino SE, Guggino WB, Green N, Sacktor B (1987) Ca^{2+}-activated K^+ channels in cultured medullary thick ascending limb cells. Am J Physiol Cell Physiol 252(2 Pt 1):C121–C127
Halperin ML, Kamel KS, Oh MS (2008) Mechanisms to concentrate the urine: an opinion. Curr Opin Nephrol Hypertens 17(4):416–422
Hamilton KL, Devor DC (2012) Basolateral membrane K^+ channels in renal epithelial cells. Am J Physiol Renal Physiol 302(9):F1069–F1081
Hansson JH, Nelson-Williams C, Suzuki H, Schild L, Shimkets R, Lu Y, Canessa C, Iwasaki T, Rossier B, Lifton RP (1995a) Hypertension caused by a truncated epithelial sodium channel gamma subunit: genetic heterogeneity of Liddle syndrome. Nat Genet 11(1):76–82
Hansson JH, Schild L, Lu Y, Wilson TA, Gautschi I, Shimkets R, Nelson-Williams C, Rossier BC, Lifton RP (1995b) A de novo missense mutation of the beta subunit of the epithelial sodium channel causes hypertension and Liddle syndrome, identifying a proline-rich segment critical for regulation of channel activity. Proc Natl Acad Sci U S A 92(25):11495–11499
Haque MZ, Ares GR, Caceres PS, Ortiz PA (2011) High salt differentially regulates surface NKCC2 expression in thick ascending limbs of Dahl salt-sensitive and salt-resistant rats. Am J Physiol Renal Physiol 300(5):F1096–F1104
Harvey BJ, Alzamora R, Stubbs AK, Irnaten M, McEneaney V, Thomas W (2008) Rapid responses to aldosterone in the kidney and colon. J Steroid Biochem Mol Biol 108(3–5):310–317
Hayslett JP, Boulpaep EL, Kashgarian M, Giebisch GH (1977) Electrical characteristics of the mammalian distal tubule: comparison of Ling-Gerard and macroelectrodes. Kidney Int 12 (5):324–331
Hebert SC (1986) Hypertonic cell volume regulation in mouse thick limbs. II. Na^+-H^+ and Cl^--HCO_3^- exchange in basolateral membranes. Am J Physiol Cell Physiol 250(6 Pt 1):C920–C931
Hebert SC, Desir G, Giebisch G, Wang W (2005) Molecular diversity and regulation of renal potassium channels. Physiol Rev 85(1):319–371
Hediger MA, Coady MJ, Ikeda TS, Wright EM (1987) Expression cloning and cDNA sequencing of the Na^+/glucose co-transporter. Nature 330(6146):379–381
Hierholzer K, Wiederholt M (1976) Some aspects of distal tubular solute and water transport. Kidney Int 9(2):198–213
Hodgkin AL, Katz B (1949) The effect of sodium ions on the electrical activity of giant axon of the squid. J Physiol 108(1):37–77

Hollenhorst MI, Richter K, Fronius M (2011) Ion transport by pulmonary epithelia. J Biomed Biotechnol 2011:174306

Hoy WE, Hughson MD, Bertram JF, Douglas-Denton R, Amann K (2005) Nephron number, hypertension, renal disease, and renal failure. J Am Soc Nephrol 16(9):2557–2564

Hummel CS, Lu C, Loo DD, Hirayama BA, Voss AA, Wright EM (2011) Glucose transport by human renal Na^+/D-glucose cotransporters SGLT1 and SGLT2. Am J Physiol Cell Physiol 300 (1):C14–C21

Igarashi P, Vanden Heuvel GB, Payne JA, Forbush B 3rd (1995) Cloning, embryonic expression, and alternative splicing of a murine kidney-specific Na-K-Cl cotransporter. Am J Physiol Renal Physiol 269(3 Pt 2):F405–F418

Imai M, Kokko JP (1974) Sodium chloride, urea, and water transport in the thin ascending limb of Henle. Generation of osmotic gradients by passive diffusion of solutes. J Clin Invest 53 (2):393–402

Imbrici P, Liantonio A, Gradogna A, Pusch M, Camerino DC (2014) Targeting kidney CLC-K channels: pharmacological profile in a human cell line versus Xenopus oocytes. Biochim Biophys Acta 1838(10):2484–2491

Isozaki T, Lea JP, Tumlin JA, Sands JM (1994) Sodium-dependent net urea transport in rat initial inner medullary collecting ducts. J Clin Invest 94(4):1513–1517

Jacobson HR (1982) Transport characteristics of in vitro perfused proximal convoluted tubules. Kidney Int 22(5):425–433

Jentzer JC, DeWald TA, Hernandez AF (2010) Combination of loop diuretics with thiazide-type diuretics in heart failure. J Am Coll Cardiol 56(19):1527–1534

Kato A, Romero MF (2011) Regulation of electroneutral NaCl absorption by the small intestine. Annu Rev Physiol 73:261–281

Keszler M, Sivasubramanian KN (1983) Pseudohypoaldosteronism. Am J Dis Child 137(8):738–740

Khuri RN, Strieder N, Wiederholt M, Giebisch G (1975) Effects of graded solute diuresis on renal tubular sodium transport in the rat. Am J Physiol 228(4):1262–1268

Kishore BK, Ecelbarger CM (2013) Lithium: a versatile tool for understanding renal physiology. Am J Physiol Renal Physiol 304(9):F1139–F1149

Kiuchi-Saishin Y, Gotoh S, Furuse M, Takasuga A, Tano Y, Tsukita S (2002) Differential expression patterns of claudins, tight junction membrane proteins, in mouse nephron segments. J Am Soc Nephrol 13(4):875–886

Klein JD, Blount MA, Sands JM (2011) Urea transport in the kidney. Compr Physiol 1(2):699–729

Klein JD, Blount MA, Sands JM (2012) Molecular mechanisms of urea transport in health and disease. Pflugers Arch 464(6):561–572

Kleyman TR, Cragoe EJ Jr (1988) Amiloride and its analogs as tools in the study of ion transport. J Membr Biol 105(1):1–21

Koefoed-Johnsen V, Ussing HH (1958) The nature of the frog skin potential. Acta Physiol Scand 42(3–4):298–308

Koenig B, Ricapito S, Kinne R (1983) Chloride transport in the thick ascending limb of Henle's loop: potassium dependence and stoichiometry of the NaCl cotransport system in plasma membrane vesicles. Pflugers Arch 399(3):173–179

Kokko JP (1970) Sodium chloride and water transport in the descending limb of Henle. J Clin Invest 49(10):1838–1846

Kokko JP, Rector FC Jr (1972) Countercurrent multiplication system without active transport in inner medulla. Kidney Int 2(4):214–223

Komlosi P, Fintha A, Bell PD (2006) Unraveling the relationship between macula densa cell volume and luminal solute concentration/osmolality. Kidney Int 70(5):865–871

Komlosi P, Bell PD, Zhang ZR (2009) Tubuloglomerular feedback mechanisms in nephron segments beyond the macula densa. Curr Opin Nephrol Hypertens 18(1):57–62

Kortenoeven ML, Li Y, Shaw S, Gaeggeler HP, Rossier BC, Wetzels JF, Deen PM (2009) Amiloride blocks lithium entry through the sodium channel thereby attenuating the resultant nephrogenic diabetes insipidus. Kidney Int 76(1):44–53

Kotchen TA, Cowley AW Jr, Frohlich ED (2013) Salt in health and disease-a delicate balance. N Engl J Med 368(13):1229–1237

Krappitz M, Korbmacher C, Haerteis S (2014) Demonstration of proteolytic activation of the epithelial sodium channel (ENaC) by combining current measurements with detection of cleavage fragments. J Vis Exp 5:89. doi:10.3791/51582

Krug SM, Gunzel D, Conrad MP, Lee IF, Amasheh S, Fromm M, Yu AS (2012) Charge-selective claudin channels. Ann N Y Acad Sci 1257:20–28

Kunchaparty S, Palcso M, Berkman J, Velazquez H, Desir GV, Bernstein P, Reilly RF, Ellison DH (1999) Defective processing and expression of thiazide-sensitive Na-Cl cotransporter as a cause of Gitelman's syndrome. Am J Physiol Renal Physiol 277(4 Pt 2):F643–F649

Kunzelmann K, Mall M (2002) Electrolyte transport in the mammalian colon: mechanisms and implications for disease. Physiol Rev 82(1):245–289

Laghmani K, Borensztein P, Ambuhl P, Froissart M, Bichara M, Moe OW, Alpern RJ, Paillard M (1997) Chronic metabolic acidosis enhances NHE-3 protein abundance and transport activity in the rat thick ascending limb by increasing NHE-3 mRNA. J Clin Invest 99(1):24–30

Lagnaz D, Arroyo JP, Chavez-Canales M, Vazquez N, Rizzo F, Spirli A, Debonneville A, Staub O, Gamba G (2014) WNK3 abrogates the NEDD4-2-mediated inhibition of the renal Na^+:Cl^- cotransporter. Am J Physiol Renal Physiol 307(3):F275–F286

Lang F, Shumilina E (2013) Regulation of ion channels by the serum- and glucocorticoid-inducible kinase SGK1. FASEB J 27(1):3–12

Law PY, Edelman IS (1978) Induction of citrate synthase by aldosterone in the rat kidney. J Membr Biol 41(1):41–64

Lee DH, Maunsbach AB, Riquier-Brison AD, Nguyen MT, Fenton RA, Bachmann S, Yu AS, McDonough AA (2013) Effects of ACE inhibition and ANG II stimulation on renal Na-Cl cotransporter distribution, phosphorylation, and membrane complex properties. Am J Physiol Cell Physiol 304(2):C147–C163

Levick JR, Michel CC (2010) Microvascular fluid exchange and the revised Starling principle. Cardiovasc Res 87(2):198–210

Lifton RP, Gharavi AG, Geller DS (2001) Molecular mechanisms of human hypertension. Cell 104(4):545–556

Loffing J, Korbmacher C (2009) Regulated sodium transport in the renal connecting tubule (CNT) via the epithelial sodium channel (ENaC). Pflugers Arch 458(1):111–135

Loffing J, Pietri L, Aregger F, Bloch-Faure M, Ziegler U, Meneton P, Rossier BC, Kaissling B (2000) Differential subcellular localization of ENaC subunits in mouse kidney in response to high- and low-Na diets. Am J Physiol Renal Physiol 279(2):F252–F258

Loffing J, Loffing-Cueni D, Valderrabano V, Klausli L, Hebert SC, Rossier BC, Hoenderop JG, Bindels RJ, Kaissling B (2001) Distribution of transcellular calcium and sodium transport pathways along mouse distal nephron. Am J Physiol Renal Physiol 281(6):F1021–F1027

Loo DDF, Hirayama BA, Meinild A-K, Chandy G, Zeuthen T, Wright EM (1999) Passive water and ion transport by cotransporters. J Physiol 518(1):195–202

Lourdel S, Paulais M, Cluzeaud F, Bens M, Tanemoto M, Kurachi Y, Vandewalle A, Teulon J (2002) An inward rectifier K^+ channel at the basolateral membrane of the mouse distal convoluted tubule: similarities with Kir4-Kir5.1 heteromeric channels. J Physiol 538(Pt 2):391–404

Lourdel S, Paulais M, Marvao P, Nissant A, Teulon J (2003) A chloride channel at the basolateral membrane of the distal-convoluted tubule: a candidate ClC-K channel. J Gen Physiol 121 (4):287–300

Malagon-Rogers M (1999) A patient with pseudohypoaldosteronism type 1 and respiratory distress syndrome. Pediatr Nephrol 13(6):484–486

Malik B, Yue Q, Yue G, Chen XJ, Price SR, Mitch WE, Eaton DC (2005) Role of Nedd4-2 and polyubiquitination in epithelial sodium channel degradation in untransfected renal A6 cells expressing endogenous ENaC subunits. Am J Physiol Renal Physiol 289(1):F107–F116

Mall M, Grubb BR, Harkema JR, O'Neal WK, Boucher RC (2004) Increased airway epithelial Na^+ absorption produces cystic fibrosis-like lung disease in mice. Nat Med 10(5):487–493

Markadieu N, Delpire E (2014) Physiology and pathophysiology of SLC12A1/2 transporters. Pflugers Arch 466(1):91–105

Mastroianni N, De Fusco M, Zollo M, Arrigo G, Zuffardi O, Bettinelli A, Ballabio A, Casari G (1996) Molecular cloning, expression pattern, and chromosomal localization of the human Na-Cl thiazide-sensitive cotransporter (SLC12A3). Genomics 35(3):486–493

Matsumura Y, Uchida S, Kondo Y, Miyazaki H, Ko SB, Hayama A, Morimoto T, Liu W, Arisawa M, Sasaki S, Marumo F (1999) Overt nephrogenic diabetes insipidus in mice lacking the CLC-K1 chloride channel. Nat Genet 21(1):95–98

Matthay MA, Robriquet L, Fang X (2005) Alveolar epithelium: role in lung fluid balance and acute lung injury. Proc Am Thorac Soc 2(3):206–213

McDonough AA, Leong PKK, Yang LE (2003) Mechanisms of Pressure Natriuresis. Ann N Y Acad Sci 986(1):669–677

McKinney TD (1984) Further studies of organic base secretion by rabbit proximal tubules. Am J Physiol Renal Physiol 246(3 Pt 2):F282–F289

Mironova E, Bugay V, Pochynyuk O, Staruschenko A, Stockand J (2013) Recording ion channels in isolated, split-opened tubules. Methods Mol Biol 998:341–353

Moes AD, van der Lubbe N, Zietse R, Loffing J, Hoorn EJ (2014) The sodium chloride cotransporter SLC12A3: new roles in sodium, potassium, and blood pressure regulation. Pflugers Arch 466(1):107–118

Monette MY, Rinehart J, Lifton RP, Forbush B (2011) Rare mutations in the human Na-K-Cl cotransporter (NKCC2) associated with lower blood pressure exhibit impaired processing and transport function. Am J Physiol Renal Physiol 300(4):F840–F847

Monroy A, Plata C, Hebert SC, Gamba G (2000) Characterization of the thiazide-sensitive Na^+-Cl^- cotransporter: a new model for ions and diuretics interaction. Am J Physiol Renal Physiol 279(1):F161–F169

Mount DB, Baekgaard A, Hall AE, Plata C, Xu J, Beier DR, Gamba G, Hebert SC (1999) Isoforms of the Na-K-2Cl cotransporter in murine TAL I. Molecular characterization and intrarenal localization. Am J Physiol Renal Physiol 276(3 Pt 2):F347–F358

Mutig K, Saritas T, Uchida S, Kahl T, Borowski T, Paliege A, Bohlick A, Bleich M, Shan Q, Bachmann S (2010) Short-term stimulation of the thiazide-sensitive Na^+-Cl^- cotransporter by vasopressin involves phosphorylation and membrane translocation. Am J Physiol Renal Physiol 298(3):F502–F509

Muto S, Furuse M, Kusano E (2012) Claudins and renal salt transport. Clin Exp Nephrol 16(1):61–67

Nesterov V, Dahlmann A, Krueger B, Bertog M, Loffing J, Korbmacher C (2012) Aldosterone-dependent and -independent regulation of the epithelial sodium channel (ENaC) in mouse distal nephron. Am J Physiol Renal Physiol 303(9):F1289–F1299

Nielsen S, Smith BL, Christensen EI, Knepper MA, Agre P (1993) CHIP28 water channels are localized in constitutively water-permeable segments of the nephron. J Cell Biol 120(2):371–383

Nielsen S, Frokiaer J, Marples D, Kwon TH, Agre P, Knepper MA (2002) Aquaporins in the kidney: from molecules to medicine. Physiol Rev 82(1):205–244

Oliva RV, Bakris GL (2014) Blood pressure effects of sodium-glucose co-transport 2 (SGLT2) inhibitors. J Am Soc Hypertens 8(5):330–339

Ookata K, Tojo A, Suzuki Y, Nakamura N, Kimura K, Wilcox CS, Hirose S (2000) Localization of inward rectifier potassium channel Kir7.1 in the basolateral membrane of distal nephron and collecting duct. J Am Soc Nephrol 11(11):1987–1994

Palmer LG, Andersen OS (2008) The two-membrane model of epithelial transport: Koefoed-Johnsen and Ussing (1958). J Gen Physiol 132(6):607–612

Parker MD, Boron WF (2013) The divergence, actions, roles, and relatives of sodium-coupled bicarbonate transporters. Physiol Rev 93(2):803–959

Paulais M, Bloch-Faure M, Picard N, Jacques T, Ramakrishnan SK, Keck M, Sohet F, Eladari D, Houillier P, Lourdel S, Teulon J, Tucker SJ (2011) Renal phenotype in mice lacking the Kir5.1 (Kcnj16) K^+ channel subunit contrasts with that observed in SeSAME/EAST syndrome. Proc Natl Acad Sci U S A 108(25):10361–10366

Payne JA, Forbush B 3rd (1994) Alternatively spliced isoforms of the putative renal Na-K-Cl cotransporter are differentially distributed within the rabbit kidney. Proc Natl Acad Sci U S A 91(10):4544–4548

Peti-Peterdi J (2006) Calcium wave of tubuloglomerular feedback. Am J Physiol Renal Physiol 291(2):F473–F480

Plata C, Meade P, Hall A, Welch RC, Vazquez N, Hebert SC, Gamba G (2001) Alternatively spliced isoform of apical Na^+-K^+-Cl^- cotransporter gene encodes a furosemide-sensitive Na^+-Cl^- cotransporter. Am J Physiol Renal Physiol 280(4):F574–F582

Plata C, Meade P, Vazquez N, Hebert SC, Gamba G (2002) Functional properties of the apical Na^+-K^+-2Cl- cotransporter isoforms. J Biol Chem 277(13):11004–11012

Reeves WB, Winters CJ, Andreoli TE (2001) Chloride channels in the loop of Henle. Annu Rev Physiol 63:631–645

Reichold M, Zdebik AA, Lieberer E, Rapedius M, Schmidt K, Bandulik S, Sterner C, Tegtmeier I, Penton D, Baukrowitz T, Hulton SA, Witzgall R, Ben-Zeev B, Howie AJ, Kleta R, Bockenhauer D, Warth R (2010) KCNJ10 gene mutations causing EAST syndrome (epilepsy, ataxia, sensorineural deafness, and tubulopathy) disrupt channel function. Proc Natl Acad Sci U S A 107(32):14490–14495

Reilly RF, Ellison DH (2000) Mammalian distal tubule: physiology, pathophysiology, and molecular anatomy. Physiol Rev 80(1):277–313

Ren Y, Garvin JL, Liu R, Carretero OA (2007) Crosstalk between the connecting tubule and the afferent arteriole regulates renal microcirculation. Kidney Int 71(11):1116–1121

Ren Y, D'Ambrosio MA, Garvin JL, Wang H, Carretero OA (2013) Prostaglandin E2 mediates connecting tubule glomerular feedback. Hypertension 62(6):1123–1128

Ren Y, D'Ambrosio MA, Garvin JL, Leung P, Kutskill K, Wang H, Peterson EL, Carretero OA (2014) Aldosterone sensitizes connecting tubule glomerular feedback via the aldosterone receptor GPR30. Am J Physiol Renal Physiol 307(4):F427–F434

Renfro JL (1975) Water and ion transport by the urinary bladder of the teleost Pseudopleuronectes americanus. Am J Physiol 228(1):52–61

Renfro JL (1977) Interdependence of active Na^+ and Cl^- transport by the isolated urinary bladder of the teleost, Pseudopleuronectes americanus. J Exp Zool 199(3):383–390

Richards J, Ko B, All S, Cheng KY, Hoover RS, Gumz ML (2014) A role for the circadian clock protein Per1 in the regulation of the NaCl co-transporter (NCC) and the with-no-lysine kinase (WNK) cascade in mouse distal convoluted tubule cells. J Biol Chem 289(17):11791–11806

Rieg T, Tang T, Uchida S, Hammond HK, Fenton RA, Vallon V (2013) Adenylyl cyclase 6 enhances NKCC2 expression and mediates vasopressin-induced phosphorylation of NKCC2 and NCC. Am J Pathol 182(1):96–106

Riser Taylor S, Harris KB (2013) The clinical efficacy and safety of sodium glucose cotransporter-2 inhibitors in adults with type 2 diabetes mellitus. Pharmacotherapy 33(9):984–999

Rock JR, O'Neal WK, Gabriel SE, Randell SH, Harfe BD, Boucher RC, Grubb BR (2009) Transmembrane protein 16A (TMEM16A) is a Ca^{2+}-regulated Cl^- secretory channel in mouse airways. J Biol Chem 284(22):14875–14880

Ronzaud C, Staub O (2014) Ubiquitylation and control of renal Na^+ balance and blood pressure. Physiology (Bethesda) 29(1):16–26

Rossier BC (2014) Epithelial sodium channel (ENaC) and the control of blood pressure. Curr Opin Pharmacol 15:33–46

Rubera I, Loffing J, Palmer LG, Frindt G, Fowler-Jaeger N, Sauter D, Carroll T, McMahon A, Hummler E, Rossier BC (2003) Collecting duct-specific gene inactivation of αENaC in the mouse kidney does not impair sodium and potassium balance. J Clin Invest 112(4):554–565
Rubera I, Hummler E, Beermann F (2009) Transgenic mice and their impact on kidney research. Pflugers Arch 458(1):211–222
Russell JM (2000) Sodium-potassium-chloride cotransport. Physiol Rev 80(1):211–276
Sandle GI (2005) Pathogenesis of diarrhea in ulcerative colitis: new views on an old problem. J Clin Gastroenterol 39(4 Suppl 2):S49–S52
Sands JM, Martial S, Isozaki T (1996) Active urea transport in the rat inner medullary collecting duct: functional characterization and initial expression cloning. Kidney Int 49(6):1611–1614
Saritas T, Borschewski A, McCormick JA, Paliege A, Dathe C, Uchida S, Terker A, Himmerkus N, Bleich M, Demaretz S, Laghmani K, Delpire E, Ellison DH, Bachmann S, Mutig K (2013) SPAK differentially mediates vasopressin effects on sodium cotransporters. J Am Soc Nephrol 24(3):407–418
Schafer JA, Patlak CS, Andreoli TE (1975) A component of fluid absorption linked to passive ion flows in the superficial pars recta. J Gen Physiol 66(4):445–471
Scherrer U, Sartori C, Lepori M, Allemann Y, Duplain H, Trueb L, Nicod P (1999) High-altitude pulmonary edema: from exaggerated pulmonary hypertension to a defect in transepithelial sodium transport. Adv Exp Med Biol 474:93–107
Schmitt R, Ellison DH, Farman N, Rossier BC, Reilly RF, Reeves WB, Oberbaumer I, Tapp R, Bachmann S (1999) Developmental expression of sodium entry pathways in rat nephron. Am J Physiol Renal Physiol 276(3 Pt 2):F367–F381
Schnermann J (2011) Maintained tubuloglomerular feedback responses during acute inhibition of P2 purinergic receptors in mice. Am J Physiol Renal Physiol 300(2):F339–F344
Schnermann J, Marver D (1986) ATPase activity in macula densa cells of the rabbit kidney. Pflugers Arch 407(1):82–86
Schnermann J, Briggs J, Schubert G (1982) In situ studies of the distal convoluted tubule in the rat. I. Evidence for NaCl secretion. Am J Physiol Renal Physiol 243(2):F160–F166
Shimizu T, Yoshitomi K, Taniguchi J, Imai M (1989) Effect of high NaCl intake on Na^+ and K^+ transport in the rabbit distal convoluted tubule. Pflugers Arch 414(5):500–508
Shimkets RA, Warnock DG, Bositis CM, Nelson-Williams C, Hansson JH, Schambelan M, Gill JR Jr, Ulick S, Milora RV, Findling JW, Canessa CM, Rossier BC, Lifton RP (1994) Liddle's syndrome: heritable human hypertension caused by mutations in the beta subunit of the epithelial sodium channel. Cell 79(3):407–414
Simon DB, Karet FE, Hamdan JM, DiPietro A, Sanjad SA, Lifton RP (1996a) Bartter's syndrome, hypokalaemic alkalosis with hypercalciuria, is caused by mutations in the Na-K-2Cl cotransporter NKCC2. Nat Genet 13(2):183–188
Simon DB, Karet FE, Rodriguez-Soriano J, Hamdan JH, DiPietro A, Trachtman H, Sanjad SA, Lifton RP (1996b) Genetic heterogeneity of Bartter's syndrome revealed by mutations in the K^+ channel, ROMK. Nat Genet 14(2):152–156
Simon DB, Nelson-Williams C, Bia MJ, Ellison D, Karet FE, Molina AM, Vaara I, Iwata F, Cushner HM, Koolen M, Gainza FJ, Gitleman HJ, Lifton RP (1996c) Gitelman's variant of Bartter's syndrome, inherited hypokalaemic alkalosis, is caused by mutations in the thiazide-sensitive Na-Cl cotransporter. Nat Genet 12(1):24–30
Skelton LA, Boron WF, Zhou Y (2010) Acid–base transport by the renal proximal tubule. J Nephrol 23(Suppl 16):S4–S18
Soleimani M (2013) SLC26 Cl^-/HCO_3^- exchangers in the kidney: roles in health and disease. Kidney Int 84(4):657–666
Song Y, Namkung W, Nielson DW, Lee JW, Finkbeiner WE, Verkman AS (2009) Airway surface liquid depth measured in ex vivo fragments of pig and human trachea: dependence on Na^+ and Cl^- channel function. Am J Physiol Lung Cell Mol Physiol 297(6):L1131–L1140
Soundararajan R, Pearce D, Hughey RP, Kleyman TR (2010) Role of epithelial sodium channels and their regulators in hypertension. J Biol Chem 285(40):30363–30369

Starling EH (1896) On the absorption of fluids from the connective tissue spaces. J Physiol 19 (4):312–326
Staruschenko A (2012) Regulation of transport in the connecting tubule and cortical collecting duct. Compr Physiol 2(2):1541–1584
Staruschenko A, Booth RE, Pochynyuk O, Stockand JD, Tong Q (2006) Functional reconstitution of the human epithelial Na^+ channel in a mammalian expression system. Methods Mol Biol 337:3–13
Stockand JD, Vallon V, Ortiz P (2012) In vivo and ex vivo analysis of tubule function. Compr Physiol 2(4):2495–2525
Stokes JB (1984) Sodium chloride absorption by the urinary bladder of the winter flounder. A thiazide-sensitive, electrically neutral transport system. J Clin Invest 74(1):7–16
Studer RA, Person E, Robinson-Rechavi M, Rossier BC (2011) Evolution of the epithelial sodium channel and the sodium pump as limiting factors of aldosterone action on sodium transport. Physiol Genomics 43(13):844–854
Subramanya AR, Pastor-Soler NM, Reeves WB, Hallows KR (2013) Tubular sodium transport. In: Falk RJ, Neilson EG, Schrier RW, Coffman TM, Molitoris BA (eds) Schrier's diseases of the kidney, Lippincott Wilkins and Williams, Philadelphia, pp 159–193
Sugano K, Kansy M, Artursson P, Avdeef A, Bendels S, Di L, Ecker GF, Faller B, Fischer H, Gerebtzoff G, Lennernaes H, Senner F (2010) Coexistence of passive and carrier-mediated processes in drug transport. Nat Rev Drug Discov 9(8):597–614
Sweiry JH, Binder HJ (1989) Characterization of aldosterone-induced potassium secretion in rat distal colon. J Clin Invest 83(3):844–851
Takahashi N, Chernavvsky DR, Gomez RA, Igarashi P, Gitelman HJ, Smithies O (2000) Uncompensated polyuria in a mouse model of Bartter's syndrome. Proc Natl Acad Sci U S A 97(10):5434–5439
Tamargo J, Solini A, Ruilope LM (2014) Comparison of agents that affect aldosterone action. Semin Nephrol 34(3):285–306
Teiwes J, Toto RD (2007) Epithelial sodium channel inhibition in cardiovascular disease. A potential role for amiloride. Am J Hypertens 20(1):109–117
Terker AS, Yang CL, McCormick JA, Meermeier NP, Rogers SL, Grossmann S, Trompf K, Delpire E, Loffing J, Ellison DH (2014) Sympathetic stimulation of thiazide-sensitive sodium chloride cotransport in the generation of salt-sensitive hypertension. Hypertension 64(1):178–184
Turk E, Kerner CJ, Lostao MP, Wright EM (1996) Membrane topology of the human Na^+/glucose cotransporter SGLT1. J Biol Chem 271(4):1925–1934
Uchida S, Sasaki S, Furukawa T, Hiraoka M, Imai T, Hirata Y, Marumo F (1993) Molecular cloning of a chloride channel that is regulated by dehydration and expressed predominantly in kidney medulla. J Biol Chem 268(6):3821–3824
Ullrich KJ, Schmidt-Nielson B, O'Dell R, Pehling G, Gottschalk CW, Lassiter WE, Mylle M (1963) Micropuncture study of composition of proximal and distal tubular fluid in rat kidney. Am J Physiol 204:527–531
Umbach JA, Coady MJ, Wright EM (1990) Intestinal Na^+/glucose cotransporter expressed in Xenopus oocytes is electrogenic. Biophys J 57(6):1217–1224
Vallon V, Schroth J, Lang F, Kuhl D, Uchida S (2009) Expression and phosphorylation of the Na^+-Cl^- cotransporter NCC in vivo is regulated by dietary salt, potassium, and SGK1. Am J Physiol Renal Physiol 297(3):F704–F712
Vallon V, Platt KA, Cunard R, Schroth J, Whaley J, Thomson SC, Koepsell H, Rieg T (2011) SGLT2 mediates glucose reabsorption in the early proximal tubule. J Am Soc Nephrol 22 (1):104–112
Vallon V, Gerasimova M, Rose MA, Masuda T, Satriano J, Mayoux E, Koepsell H, Thomson SC, Rieg T (2014) SGLT2 inhibitor empagliflozin reduces renal growth and albuminuria in proportion to hyperglycemia and prevents glomerular hyperfiltration in diabetic Akita mice. Am J Physiol Renal Physiol 306(2):F194–F204

Velazquez H, Good DW, Wright FS (1984) Mutual dependence of sodium and chloride absorption by renal distal tubule. Am J Physiol Renal Physiol 247(6 Pt 2):F904–F911
Velazquez H, Ellison DH, Wright FS (1987) Chloride-dependent potassium secretion in early and late renal distal tubules. Am J Physiol Renal Physiol 253(3 Pt 2):F555–F562
Velazquez H, Silva T, Andujar E, Desir GV, Ellison DH, Greger R (2001) The distal convoluted tubule of rabbit kidney does not express a functional sodium channel. Am J Physiol Renal Physiol 280(3):F530–F539
Verkman AS (2006) Roles of aquaporins in kidney revealed by transgenic mice. Semin Nephrol 26 (3):200–208
Wang WH (1994) Two types of K^+ channel in thick ascending limb of rat kidney. Am J Physiol Renal Physiol 267(4 Pt 2):F599–F605
Wang T (2012) Renal outer medullary potassium channel knockout models reveal thick ascending limb function and dysfunction. Clin Exp Nephrol 16(1):49–54
Wang WH, Giebisch G (2009) Regulation of potassium (K) handling in the renal collecting duct. Pflugers Arch 458(1):157–168
Wang WH, White S, Geibel J, Giebisch G (1990) A potassium channel in the apical membrane of rabbit thick ascending limb of Henle's loop. Am J Physiol Renal Physiol 258(2 Pt 2):F244 F253
Wang WH, Yue P, Sun P, Lin DH (2010) Regulation and function of potassium channels in aldosterone-sensitive distal nephron. Curr Opin Nephrol Hypertens 19(5):463–470
Wang H, D'Ambrosio MA, Garvin JL, Ren Y, Carretero OA (2013) Connecting tubule glomerular feedback in hypertension. Hypertension 62(4):738–745
Wang Y-B, Leroy V, Maunsbach AB, Doucet A, Hasler U, Dizin E, Ernandez T, de Seigneux S, Martin P-Y, Féraille E (2014) Sodium transport is modulated by p38 kinase–dependent cross-talk between ENaC and Na, K-ATPase in collecting duct principal cells. J Am Soc Nephrol 25 (2):250–259
Wargo KA, Banta WM (2009) A comprehensive review of the loop diuretics: should furosemide be first line? Ann Pharmacother 43(11):1836–1847
Warnock DG, Kusche-Vihrog K, Tarjus A, Sheng S, Oberleithner H, Kleyman TR, Jaisser F (2014) Blood pressure and amiloride-sensitive sodium channels in vascular and renal cells. Nat Rev Nephrol 10(3):146–157
Wen D, Cornelius RJ, Sansom SC (2014) Interacting influence of diuretics and diet on BK channel-regulated K homeostasis. Curr Opin Pharmacol 15:28–32
Wright FS (1971) Increasing magnitude of electrical potential along the renal distal tubule. Am J Physiol 220(3):624–638
Wright FS, Schnermann J (1974) Interference with feedback control of glomerular filtration rate by furosemide, triflocin, and cyanide. J Clin Invest 53(6):1695–1708
Wright EM, Loo DDF, Hirayama BA (2011) Biology of human sodium glucose transporters. Physiol Rev 91(2):733–794
Xu JZ, Hall AE, Peterson LN, Bienkowski MJ, Eessalu TE, Hebert SC (1997) Localization of the ROMK protein on apical membranes of rat kidney nephron segments. Am J Physiol Renal Physiol 273(5 Pt 2):F739–F748
Yang T, Huang YG, Singh I, Schnermann J, Briggs JP (1996) Localization of bumetanide- and thiazide-sensitive Na-K-Cl cotransporters along the rat nephron. Am J Physiol Renal Physiol 271(4 Pt 2):F931–F939
Yu AS (2015) Claudins and the kidney. J Am Soc Nephrol 26(1):11–9
Zaika OL, Mamenko M, Palygin O, Boukelmoune N, Staruschenko A, Pochynyuk O (2013) Direct inhibition of basolateral Kir4.1/5.1 and Kir4.1 channels in the cortical collecting duct by dopamine. Am J Physiol Renal Physiol 305(9):F1277–F1287
Zhang C, Wang L, Thomas S, Wang K, Lin DH, Rinehart J, Wang WH (2013) Src family protein tyrosine kinase regulates the basolateral K channel in the distal convoluted tubule (DCT) by phosphorylation of KCNJ10 protein. J Biol Chem 288(36):26135–26146
Zhou R, Tomkovicz VR, Butler PL, Ochoa LA, Peterson ZJ, Snyder PM (2013) Ubiquitin-specific peptidase 8 (USP8) regulates endosomal trafficking of the epithelial Na^+ channel. J Biol Chem 288(8):5389–5397

Chapter 3
Physiologic Influences of Transepithelial K^+ Secretion

Dan R. Halm

Abstract Cellular ionic balance relies on ion channels and coupled transporters to maintain and use the transmembrane electrochemical gradients of the cations Na^+ and K^+. High intracellular K^+ concentration provides a ready reserve within the body allowing epithelia to secrete K^+ into the fluid covering the apical membrane in the service of numerous physiologic activities. A major role for transepithelial K^+ secretion concerns the balance of total body K^+ such that excretion of excess K^+ in the diet safeguards against disturbances to cellular balance. Accomplishing this transepithelial flow involves two archetypical cellular mechanisms, Na^+ absorption and Cl^- secretion. Ion channels for K^+, Na^+, and Cl^-, as well as cotransporters, exchangers, and pumps contribute to produce transepithelial flow by coupling electrochemical gradients such that K^+ flow enters across the basolateral membrane and exits through the apical membrane. Beyond excretion, transepithelial K^+ secretion serves to create the high K^+ concentration of endolymph in the inner ear that supports sensation of sound and body orientation. For several epithelia such as those in airways and gastric mucosa, the elevated K^+ concentration of apical fluid may occur largely as a consequence of supporting the secretion of other ions such as Cl^-or H^+. Less well-appreciated consequences of K^+ secretion may result as in saliva and colonic luminal fluid where a high K^+ concentration likely influences interactions with the resident microbiome. Independent control of K^+ secretion also allows for specific adjustments in rate that serve the physiology of organs large and small.

Keywords Airway fluid • Ammonium transport • Gastric acid • Inner ear • Microbiome • Nutrient absorption • Olfaction • Pancreatic acini • Renal excretion • Saliva • Sweat

D.R. Halm (✉)
Department of Neuroscience, Cell Biology and Physiology, Wright State University Boonshoft School of Medicine, 3640 Colonel Glenn Hwy, Dayton, OH 45435, USA
e-mail: dan.halm@wright.edu

K.L. Hamilton, D.C. Devor (eds.), *Ion Channels and Transporters of Epithelia in Health and Disease*, Physiology in Health and Disease,
DOI 10.1007/978-1-4939-3366-2_3

3.1 Introduction

Potassium and sodium comprise the major cations in body fluids and thereby command a pervasive influence on cellular ionic balance, from establishing the electrical potential difference across the cell's plasma membrane to nutrient uptake that supports cellular metabolism (Williams and Dawson 2006). The high Na^+ and low K^+ concentrations of extracellular fluid likely mimic the relative abundance of these ions in the ocean during early stages of development for this eukaryotic cellular balance (Knauth 2005; Pinti 2005). Sequestering K^+ inside cells and excluding Na^+ to the extracellular space became a cornerstone of ionic balance that permits maintenance of near constant cellular volume (Chap. 4). Constancy of these extracellular concentrations, maintained by the kidneys for vertebrate animals (Malnic et al 2013), permits all cells in the body to operate under reliable environmental conditions.

Key components of this cellular process for maintaining ionic and osmotic balance are the Na^+/K^+ pump (Chap. 10) and K^+ channels (Chaps. 19–21, 26, 27). These proteinaceous nanomachines operating in the plasma membrane guide flow of Na^+ and K^+ into and out of cells. The combined action of these two types of nanomachines lowers Na^+ and raises K^+ concentrations intracellularly such that a plasma membrane electrical potential difference (ψ_m) develops that is negative inside the cell compared with the extracellular fluid. As simple as this system may seem, it underlies the orientation of transmembrane flow for most other small solutes a cell encounters.

Using membrane nanomachines with specific solute coupling stoichiometries allows the inward orientation of the Na^+ electrochemical gradient to influence the flow of various solutes (Chaps. 11–13) such that the intracellular concentrations of these solutes exceed the expectations governed by the electrochemical energy of that solute alone. Similarly, the dominance of K^+ channels in contributing to plasma membrane conductance (G_m) sets the polarity of ψ_m, thereby influencing flow through all the types of ion channels present (Chaps. 9, 15–24).

Epithelial cells comprise the barrier that defines the inside of the body. By bounding the extracellular space, epithelia permit a stable internal environment via selective entry and exit of substances into and from the body. These absorptive and excretory events occur across epithelia located in specialized organs including the gastrointestinal tract, kidneys, and lungs. The intricate tubular geometry of these epithelia leads to a functional distinction between excretion and secretion. In the broadest biologic sense, secretion means the exit from a cell, but in the context of epithelial cells, the focus is on exit across the apical membrane and thereby out of the body. The clearest phrasing for this physiologic focus is to consider transepithelial secretion the net movement from extracellular fluid into the cell across the basolateral membrane and then out across the apical membrane. As for the distinction between excretion and transepithelial secretion, accomplishing excretion requires that no further opportunity exists to retrieve the secreted K^+. Given the variation in epithelial transport along renal tubules, for example, the simple transepithelial secretion of K^+ at a proximal site could be reversed by transepithelial absorption from the tubular fluid at more distal sites. Ultimate renal excretion of K^+ (or any other solute) therefore occurs when urine passes out of the body. This type of serial processing of tubular fluid K^+ not only allows renal K^+ excretion to be highly regulated but also illustrates that transepithelial K^+ secretion can lead to high $[K^+]$ on epithelial surfaces that supports a range of physiologic functions beyond the simple need for excretion of excess K^+.

Any transepithelial K^+ secretion leads to increased $[K^+]$ in the aqueous fluid covering the apical membrane with the volume of this space determining how rapidly that increase occurs. These exo-epithelial apical compartments often are the lumens of small bore tubules (5–30 μm diameter) such as in the kidney, salivary gland, or intestinal crypts of Lieberkühn but also can be a thin film as on the airway or skin surface. Again in the case of excretion, the higher the concentration attained, the smaller the volume of water required to eliminate the solute, with the only limitation being the requirements for handling other solutes at the same time. The renal tubule uses an elaborate set of regulatory schemes to produce appropriate excretion of multiple solutes nearly simultaneously.

Many occurrences of K^+ secretion operate primarily to support the absorption/secretion of other ions without leading to eventual K^+ excretion. These situations can be seen as extensions of the regulated cellular mechanisms that lead to K^+ excretion from the kidney (see Sect. 3.3.1), but instead use K^+ as a recycling cofactor to promote another epithelial action (see Sect. 3.3.3). The stomach provides an example for this type of balance in the gastric parietal cell where apical K^+ channels support acid secretion via an H^+/K^+ pump without creating an impact on the ultimate gastrointestinal excretion of K^+. Transepithelial K^+ secretion also can act primarily to make K^+ the dominant cation within an extracellular compartment, instead of Na^+ (see Sect. 3.3.4). The endolymph of the inner ear provides the most dramatic example of changing the dominant cation, where the high luminal $[K^+]$ serves in the sensation of sound and body orientation (see Sect. 3.3.2). These contributions of K^+ secretion to the physiology of numerous organs require precise coupling of transmembrane electrochemical gradients.

3.2 Pathways for Transepithelial K^+ Secretion

Transepithelial transport of solute drives fluid flow across epithelia as well as altering the composition of the aqueous fluid covering the apical membrane. Electrogenic Na^+ absorption and Cl^- secretion (Chaps. 1 and 2) typify the transport mechanisms responsible for flow via the cellular pathway across epithelia. Paracellular flow in comparison occurs through the tight junctions and along the lateral spaces between epithelial cells (Frömter and Diamond 1972; Furuse 2010). Selectivity for this shunt pathway is governed primarily by characteristics of tight junction proteins. Together these parallel cellular and paracellular routes combine to determine the result of epithelial actions.

Insertion of ion channels, cotransporters, exchangers, and pumps into cell membranes creates the pathways for the solute flow observed across epithelia, but the electrochemical energy gradients of these solutes set the direction for the flow. Chemical gradients (Eq. 3.1a) developed by pump (ATPase) activity constitute the ultimate electromotive forces (*EMF* or simply *E*) for ions, and the combination of all the conductive membrane components determines the size and orientation of ψ_m (Sackin and Palmer 2013; Williams and Dawson 2006). The familiar driving force for ions flowing though channels (Eq. 3.1b) indicates whether that ion type will flow into or out of the cell. Conveniently, the same formalism allows a calculation of the driving forces for other transporters as well by adding or subtracting "$\Delta\mu_j/F$" terms (Eq. 3.1b) as appropriate for cotransport, exchange, or pump activity. A determination of the direction for transepithelial flow whether absorptive or secretory, therefore, results from the cellular control of ion composition and ψ_m.

$$\Delta^{21}\mu_j = \left(\mu_{j2} - \mu_{j1}\right) = RT\ln\left(\frac{C_{j2}}{C_{j1}}\right) + z_jF(\psi_2 - \psi_1) \tag{3.1a}$$

$$\begin{aligned} \frac{\Delta^{21}\mu_j}{F} &= \frac{RT}{F}\ln\left(\frac{C_{j2}}{C_{j1}}\right) + z_j(\psi_2 - \psi_1) = z_j(\psi_m - E_j); \\ E_j &= -\frac{RT}{z_jF}\ln\left(\frac{C_{j2}}{C_{j1}}\right) \end{aligned} \tag{3.1b}$$

Epithelial ion flow can be analyzed using the schematic approach of an equivalent circuit to represent the transport elements (Chap. 6, Lewis et al. 1996; Schultz et al. 1981; Thompson 1986). The conductance (*G*) of each membrane is the sum of the individual conductances from all the ion channels and other electrogenic transporters; and each membrane *EMF* is the weighted average of the E_j's (concentration gradients as in Eq. 3.1b) for all of the conductors present (Fig. 3.1). This circuit illustrates the influence of these E_j's on the transepithelial electrical potential difference (ψ_t) as well as on the apical (ψ_a) and basolateral (ψ_b) membranes. During steady-state transepithelial flow, cell influx clearly needs to equal efflux which is accomplished through adjustments to conductances and driving forces. As driving force decreases or increases, a commensurate increase or decrease in conductance is required to maintain the flow rate steady across the membrane. Any epithelial cell using a secretory mechanism with apical membrane K^+ channels maintains the size and orientation of ψ_a such that K^+ exits into the lumen, $z_K(\psi_a - E_K^a) > 0$.

The dependence of electrogenic K^+ secretion on ψ_a allows flow of other ions to influence K^+ exit through the apical membrane into the lumen. Equation 3.2 provides this relationship of ψ_a to conductive ion flow, derived from the equivalent circuit in Fig. 3.1 (Halm and Frizzell 1991). Similar relations can be derived for ψ_b and ψ_t. For the case that conductance through the paracellular route (G_{pr}) is much smaller than via the cellular route (G_{cr}), the E_j's of the apical membrane alone determine ψ_a (Eq. 3.2b). This situation raises a crucial point, that the most conductive ions exert the largest influence on determining the size and orientation of ψ_m.

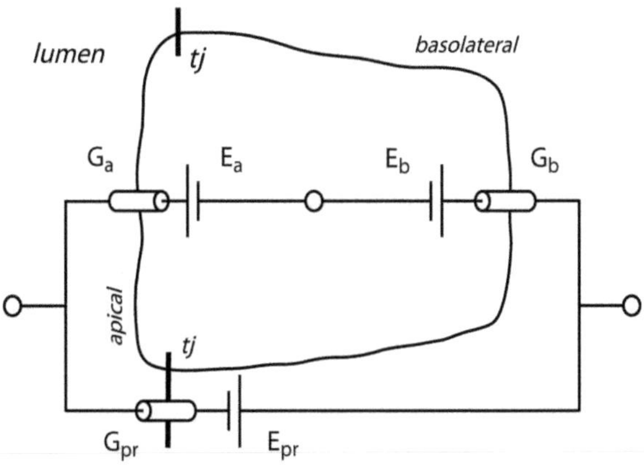

Fig. 3.1 Epithelial equivalent circuit. The pathways for ion flow are indicated by conductors (*G*) and batteries (*E*) at the apical (a) and basolateral (b) membranes as well as in the paracellular route (pr). The apical and basolateral conductors represent channels or other electrogenic transporters, and the conductor for the paracellular route represents a combination of tight junction proteins (tj) and lateral space conductivity. The distributed nature of the lateral space conductance along the lateral membrane (Clausen et al 1979; Frömter and Diamond 1972; Sackin and Palmer 2013) is lumped together as part of this simplified G_{pr}. The *EMF's* (*E*) are represented as batteries for the conductive ions, as in Eqs.3.1b and 3.2a, 3.2b, 3.2c

As long as the E_j's for some of the conductive ions are more positive than E_K^a, then $z_K(\psi_a - E_K^a)$ will be positive and K^+ will exit into the lumen. If only K^+ channels open, then ψ_a equals E_K and K^+ exit stops after a very brief period of K^+ flow that charges the membrane capacitance. This stoppage of flow results directly from the impediment imposed by the physical constraint of macroscopic electroneutrality, where microscopic charge flow generates a ψ_m opposing further ion flow.

$$\psi_a = \left(\frac{G_{pr}}{G_{cr} + G_{pr}}\right)\left\{\left(\frac{G_a}{G_a + G_b} + \frac{G_{cr}}{G_{pr}}\right)E_a + \left(\frac{G_b}{G_a + G_b}\right)\quad(E_b - E_{pr})\right\}; \qquad (3.2a)$$

$$G_{cr} = G_a\left(1 + \frac{G_a}{G_b}\right)^{-1}$$

$$G_{pr} << G_{cr} \rightarrow \psi_a \approx E_a = \sum\left(\frac{G_j^a}{G_a}\right)E_j^a \qquad (3.2b)$$

$$G_{pr} >> G_{cr} \rightarrow \psi_a \approx \left(\frac{G_a}{G_a + G_b}\right)E_a + \left(\frac{G_b}{G_a + G_b}\right)E_b = \sum\left(\frac{G_j}{G_{cell}}\right)E_j; \qquad (3.2c)$$

$$G_{cell} = \sum G_j$$

Epithelial ion flows within various organs illustrate the many inventive strategies evolved for promoting counter charge flow to deal with this electroneutrality stricture. For epithelia with even modest G_{pr}, the E_j's of the basolateral membrane and paracellular pathway also contribute to ψ_a, via the electrical shunting of G_{pr}. In this more typical situation (Eq. 3.2a), the largest driving force for apical K^+ exit occurs when basolateral channels open for ions with E_j's more positive than E_K^a (Dawson and Richards 1990; Li et al. 2003). This combining of specific transporter types in the apical and basolateral membranes harnesses the available electrochemical energy (generated by ion pumps) to support the ultimate goal of transepithelial fluid movement and/or alterations in apical fluid composition.

The influence of the paracellular pathway on transcellular ion flow has several features. As just noted, the electrical connection between the lumen and the interstitial compartment allows basolateral membrane E_j's to alter ψ_a and thereby the driving forces of electrogenic solute flow across the apical membrane. Since active transcellular ion flow often produces a nonzero ψ_t, net ion flows occur through the G_{pr} driven by ψ_t (Eq. 3.2a). These net flows lead to changes in luminal composition such that the concentrations of the ions (solutes) relative to one another differ over time, further altering transepithelial driving forces (E_{pr}). Any resulting increase or decrease in the total number of solutes in the luminal space also leads to an osmotic gradient across the epithelium that drives water flow. In this way, paracellular features complement cellular events allowing epithelial actions to serve body demands for absorption or secretion of solutes and water.

$$\psi_t = \left(\frac{G_{cr}}{G_{cr} + G_{pr}}\right)(-E_a + E_b) + \left(\frac{G_{pr}}{G_{cr} + G_{pr}}\right)E_{pr} \tag{3.3}$$

Additionally, similar shunting paths occur if more than one cell type exists in proximity within an epithelial region. The principal and intercalated cells of the renal distal tubule and collecting duct provide an example (Kriz and Kaissling 2013; Malnic et al. 2013). Differences in the ion selectivity between the cell types alter driving forces as the expanded number of equivalent circuit components contribute to ψ_a and ψ_b, as well as ψ_t. The shunting action may occur via active transepithelial ion flow as for intercalated cells (H^+ and K^+ secretion) or more simply as a transcellular conductive route (G_{Cl}^a and G_{Cl}^b) as for mitochondrial-rich cells in amphibian skin and other epithelia (Fan et al. 2012; Jakab et al. 2013; Larsen 2011).

3.2.1 Cellular Mechanisms for Transepithelial K^+ Flow

The cellular model proposed by Koefoed-Johnsen and Ussing (1958) for electrogenic Na^+ absorption formalized the concepts necessary to describe transepithelial flow via the cellular pathway, by including the balance of influx and efflux that maintains cellular homeostasis while accomplishing movement across an epithelial cell layer (Chap. 2). Subsequent studies of epithelial solute transport built directly on these ideas. The concept of Na^+-coupled glucose transport (Crane 1965; Wright et al. 2011) fits with this cellular mechanism such that Na^+-dependent glucose absorption contributes to the osmotic gradient driving water absorption in the small intestine and renal proximal tubule. Based on these insights, fluid secretion was envisioned to occur by a simple switching of the membrane localization for these transport elements such that solute and fluid flow would be guided into the lumen rather than into the body. The cellular model that best fits observations in many epithelia relocates only the Na^+-dependent uptake elements to the basolateral membrane and includes apical membrane Cl^- channels (Ernst and Mills 1977; Field 1993; Frizzell et al. 1979; Silva et al. 1977), which leads directly to electrogenic Cl^- secretion (Chap. 1). K^+ plays a central role in these epithelial archetypes for absorption and secretion.

Electrogenic Na^+ Absorption Producing K^+ secretion coupled to Na^+ absorption only requires the additional presence of apical membrane K^+ channels in the Koefoed-Johnsen-Ussing model (Fig. 3.2). The first proposal of this coupling was made to account for the Na^+ dependence of renal K^+ excretion occurring in cells of the distal tubule and collecting duct (Berliner 1961; Giebisch et al. 1967). Opening of K^+ channels in both apical and basolateral membranes allows for a variable rate of K^+ exit into the lumen as the K^+ taken up by the Na^+/K^+ pump exits from the cell according to the size of the G_K and driving forces at apical and basolateral membranes. When all K^+ exit occurs through apical channels, the Na^+/K^+ coupling stoichiometry of the Na^+/K^+ pump sets an upper limit on the maximal rate for K^+ secretion. The opposing nature of the Na^+ and K^+ concentration gradients at the apical membrane, created by the Na^+/K^+ pump, assures that Na^+ enters and K^+ exits, since ψ_a resides between the value of E^a_{Na} and E^a_K, such that $z_{Na}(\psi_a - E^a_{Na}) < 0$ and $z_K(\psi_a - E^a_K) > 0$. When cellular regulation dramatically decreases G^a_K relative to G^a_{Na} thereby reducing K^+ secretion, the rate of Na^+ absorption becomes more dependent upon the electrical properties of the paracellular pathway and basolateral membrane to maintain an inward driving force for Na^+ (Eq. 3.2a). The critical dependence of K^+ secretion on the rate of Na^+ absorption becomes most apparent as ongoing Na^+ absorption depletes Na^+ in the apical fluid layer leading to decreased E^a_{Na} and limited pump turnover thereby attenuating K^+ secretion.

Electrogenic Cl^- Secretion Opening apical membrane K^+ channels in a cell already capable of Cl^- secretion produces a coupling of K^+ secretion with Cl^-secretion (Fig. 3.3). This proposed combination of ion secretions accounts for the dependence of electrogenic K^+ secretion in mammalian colon on Na^+ uptake across the basolateral membrane (Frizzell et al. 1983; Halm and Frizzell 1986). As noted for the Na^+ absorbing cell, the presence of K^+ channels in the apical and basolateral membranes readily sets up a regulatory scheme for K^+ secretion to occur via adjustments in the relative conductance at the two membranes (Li and Halm 2002). Activating both Cl^- and K^+ channels in the apical membrane nearly assures exit into the lumen for both ions since ψ_a will reside between E^a_{Cl} and E^a_K, with $z_{Cl}(\psi_a - E^a_{Cl})$ and $z_K(\psi_a - E^a_K)$ positive. Low rates of K^+ secretion associated with few apical K^+ channels open make Cl^- secretion more dependent on the properties of

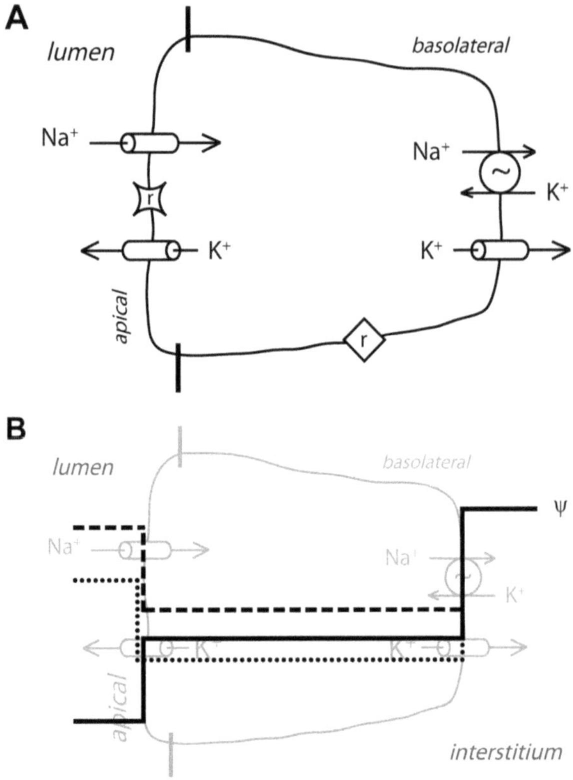

Fig. 3.2 Electrogenic Na^+ absorption and K^+ secretion. The Koefoed-Johnsen and Ussing model for Na^+ absorption (Koefoed-Johnsen and Ussing 1958) includes apical membrane Na^+ channels together with Na^+/K^+ pumps and K^+ channels in the basolateral membrane that constitute the minimal set of transporters required for absorption. (**a**) Addition of apical membrane K^+ channels to this cellular mechanism produces K^+ secretion dependent on the rate of Na^+ absorption. Variation in opening of basolateral K^+ channels adjusts the K^+ secretory rate. Receptors (r) in the basolateral and apical membranes transduce extracellular signals into activation or inhibition of transepithelial Na^+ and K^+ flow. (**b**) The E_j's of the conductive ions determine the electrical potential profile across this epithelium (ψ, referenced to the interstitial space). For the idealized case of $G_{pr} = 0$ and $G_K^a = 0$ (Eq. 3.2), a staircase-type profile occurs (*solid line*) with $\psi_a = E_{Na}^a$ and $\psi_b = E_K^b$. A well-type profile occurs with increases in G_{pr} such that E_K^b contributes to ψ_a as well as E_{Na}^a contributing to ψ_b (*dashed line*) or during increases in G_K^a that alter the polarization of ψ_a via E_K^a. Mineralocorticoid stimulation hyperpolarizes ψ_b (*dotted line*), apparently via the action of Na^+/K^+ pumps (Koeppen and Giebisch 1985) (see Sect. 3.3.1)

the paracellular pathway and basolateral membrane (Eq. 3.2a). Higher rates of K^+ secretion produced by opening apical K^+ channels readily enhances the driving force for apical Cl^- exit by moving ψ_a away from E_{Cl}^a. This possibility of adjusting G_K^a to maximize the rate of Cl^- secretion (Cook and Young 1989; Palk et al. 2010) allows a separate regulation of G_{pr} to optimize paracellular flow, with the consequence of sizable K^+ secretion leading to an increase in luminal $[K^+]$.

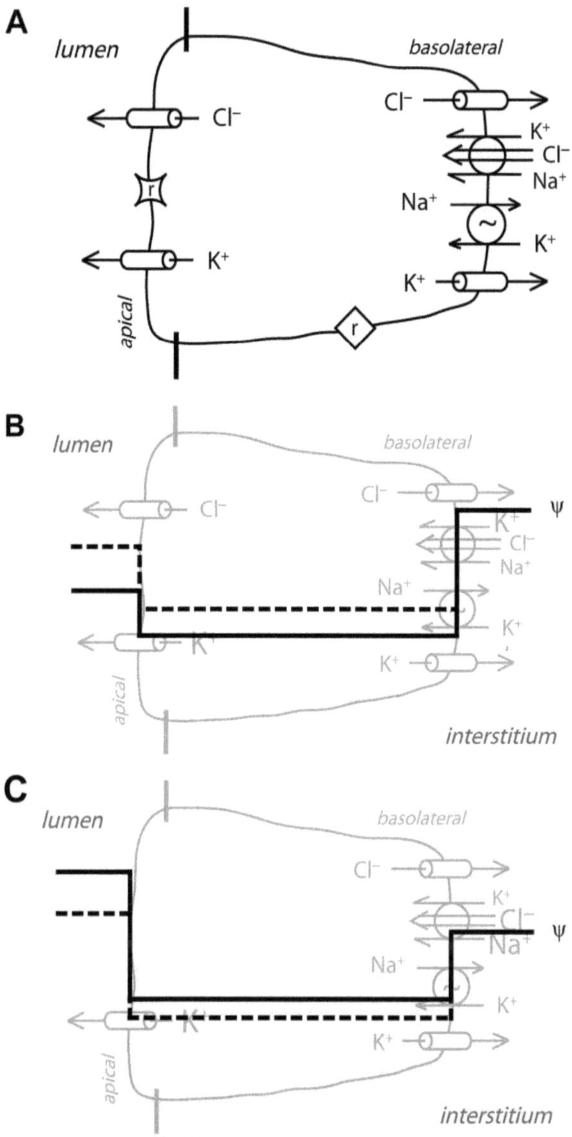

Fig. 3.3 Electrogenic Cl^- and K^+secretion. The cellular model for Cl^- secretion includes apical membrane Cl^- channels together with Na^+/K^+ pumps, Na^+-K^+-$2Cl^-$ cotransporters, and K^+ channels in the basolateral membrane. (**a**) Apical membrane K^+ channels in this cellular mechanism produce K^+ secretion dependent on the rate of Cl^- secretion. Opening basolateral Cl^- channels allows the K^+ secretory rate to exceed the rate of Cl^- secretion. Receptors (r) in the basolateral and apical membranes transform extracellular signals into activation or inhibition of transepithelial Cl^- and K^+ flow. (**b**) The E_j's of the conductive ions determine the electrical potential profile across this epithelium (ψ, referenced to the interstitial space). For the idealized case of $G_{pr} = 0$ and $G_K^a = 0$ (Eq. 3.2), a well-type profile occurs (*solid line*) with $\psi_a = E_{Cl}^a$ and $\psi_b = E_K^b$. Increasing G_{pr} and G_K^a alters the profile (*dashed line*) with E_K^b and E_K^a hyperpolarizing ψ_a as well as E_{Cl}^adepolarizing ψ_b. (**c**) For the case of $G_{Cl}^a = 0$ with Cl^- secretion thereby absent, a

An extra feature of this cellular mechanism for K^+ and Cl^- secretion emerges from a need for a basolateral membrane Cl^- exit path during K^+ secretion with low Cl^- secretory rates. In mammalian colon, renal intercalated cells, and the inner ear, K^+ secretion often occurs with negligible Cl^- secretion (Fig. 3.3c), and basolateral membrane Cl^- channels appear to account for the Cl^- exit that balances Cl^- entry via Na^+-K^+-2Cl^- cotransporters (He et al. 2011; Li et al. 2003; Liu et al. 2011; Marcus and Wangemann 2009). In the absence of Cl^- secretion, apical K^+ exit depends completely on the depolarizing influence of E^b_{Cl} (and possibly E_{pr}) on ψ_a so that $z_K(\psi_a - E^a_K) > 0$. Whether these basolateral Cl^- channels open during Cl^- secretion remains to be fully explored, but Cl^- channel inhibitors acting on the basolateral side lead to increased Cl^- secretion, suggesting a general applicability of a cellular mechanism that includes Cl^- channels opening in both apical and basolateral locations (Duta et al. 2006; Fischer et al. 2007; He et al. 2011). Adjustments of G^b_{Cl} also may allow optimization of thc driving force for basolateral K^+ exit, $z_K(\psi_b - E^b_K)$, without resorting to large increases in G_{pr}. With this insight, both Cl^- and K^+ enter the cell via Na^+-dependent uptake and exit across either the apical or basolateral membrane according to regulatory influences that govern the open status of the respective channels.

3.2.2 Paracellular K^+ Flow

Epithelia absorbing Na^+ or secreting Cl^- by these electrogenic cellular mechanisms develop a lumen negative ψ_t (Eq. 3.2a) with an immediate consequence that ions are driven electrically through the paracellular pathway. Over time the presence of any G_{pr} for K^+ leads inexorably to a higher luminal [K^+] due to this transepithelial driving force acting on K^+, $-z_K(\psi_t - E^{pr}_K)$. Given sufficient time, the [K^+] of the fluid residing over the apical membrane reaches a value determined by ψ_t ($E^{pr}_K = \psi_t$). For $\psi_t = -40$ mV, the luminal [K^+] would rise to 22 mM (assuming interstitial K^+ of 5 mM). A paracellular route for transepithelial K^+ secretion can be distinguished experimentally from the cellular mechanisms (Figs. 3.2 and 3.3) by maintaining $\psi_t = 0$ and the ionic composition of the fluid on the two sides of the epithelium identical (all $E^{pr}_j = 0$), generally in an ex vivo or in vitro setting (Sackin and Palmer 2013). Without a transepithelial driving force, net paracellular K^+ flow ceases but transcellular flow continues. Electrogenic K^+ secretion in the absence of Na^+ absorption or Cl^-secretion (Fig. 3.3c) leads to increased luminal [K^+] along with a lumen positive ψ_t, such that both E^{pr}_K and ψ_t act to drive K^+ out of the lumen via the paracellular pathway. The net amount of K^+ secreted into the lumen depends on

---

Fig. 3.3 (continued) well-type profile occurs (*solid line*) with $\psi_a = E^a_K$. Increasing G_{pr} to a value greater than zero alters the profile (*dashed line*) with E^b_{Cl} depolarizing ψ_a away from E^a_K

whether the fluid remains stationary allowing the luminal [K^+] to attain equilibrium with ψ_t or flows away to a location distant from the site of secretion.

3.2.3 Transport Proteins Supporting K^+ Secretion

The specific transport proteins found in K^+ secretory epithelia provide the pathways for flow and contribute to producing appropriate driving forces for maintaining directionality of movement. Both Na^+ absorptive and Cl^- secretory epithelia require Na^+/K^+ pumps (Chap. 10) in the basolateral membrane to establish the electrochemical gradients needed for transepithelial flow. After K^+ uptake across the basolateral membrane via Na^+/K^+ pumps (*ATP1A*, Na/K-ATPase, EC3.6.3.9), as well as Na^+-K^+-$2Cl^-$ cotransporters (*SLC12A2*, NKCC1, Chap. 11) in the Cl^- secretory scheme, K^+ exit across the apical membrane generally occurs through K^+ channels.

Seventy genes encode the family of K^+ channel proteins, many of which undergo alternative splicing such that the number of different functional channel types is much larger (González et al. 2012; Gutman et al. 2003; Yu and Catterall 2004). Although all conduct K^+, differences in control of channel opening allow each K^+ channel type to mesh with the specific signaling cascades used in various cells. For cells absorbing Na^+ (Fig. 3.2), $K_{ir}1.1$ (*KCNJ1*, RomK) and $K_{Ca}1.1$ (*KCNMA1*, BK, Maxi-K) commonly contribute to the G_K^a permitting simultaneous K^+ secretion (Nakamoto et al. 2008; Wang and Huang 2013). K^+ secretory cells using a Cl^- secretory type mechanism (Fig. 3.3) often have $K_{Ca}1.1$, $K_{Ca}3.1$ (*KCNN4*, IK), and $K_V7.1$ (*KCNQ1*, KVLQT) in the apical membrane (Barmeyer et al. 2010; Estilo et al. 2008; Grahammer et al. 2001; Joiner et al. 2003; Marcus and Wangemann 2009; Sørensen et al. 2010; Zhang et al. 2012). Auxiliary β-subunits further increase K^+ channel diversity by altering gating kinetics and membrane localization, particularly for $K_V7.1$ and $K_{Ca}1.1$ (Pongs and Schwarz 2010). In gastric parietal cells, *KCNE2* enhances $K_V7.1$ targeting to the apical membrane while also minimizing voltage-dependence gating and conferring acid activation, all of which promotes gastric acid secretion (Abbott 2012). The β-subunits for $K_{Ca}1.1$ alter Ca^{2+} sensitivity and voltage dependence, with $K_{Ca}\beta1$ (*KCNMB1*) present in the Na^+-absorbing principal cells and $K_{Ca}\beta4$ (*KCNMB4*) in the acid-secreting intercalated cells of the kidney (Grimm et al. 2007; Holtzclaw et al. 2011). Other K^+ channel types have been observed in the apical membrane (Hayashi and Novak 2013; Heitzmann and Warth 2008; O'Grady and Lee 2005), but $K_{ir}1.1$, $K_{Ca}1.1$, $K_{Ca}3.1$, and $K_V7.1$ (Chaps. 19–21) may represent a group particularly suited to promote the apical K^+ exit needed for K^+ secretion.

Additional transporter proteins often reside in the apical membrane along with K^+ channels, including other ion channel types, cotransporters, exchangers, and pumps. The epithelial Na^+ channel ENaC (*SCNN1*) supports Na^+ absorption (Fig. 3.2; Chaps. 2 and 18), and, along with CFTR (*ABCC7*) various Cl^- channels such as the Ca^{2+}-activated TMEM16A (*ANO1*) contribute to Cl^- secretion (Fig. 3.3; Chaps. 1, 14–17; Huang et al. 2012; Woodward and Guggino 2013). Na^+ uptake across the apical membrane also occurs via Na^+-K^+-$2Cl^-$ cotransporters (*SLC12A1*, NKCC2) and Na^+-Cl^- cotransporters (*SLC12A3*, NCC) in cells of the renal thick ascending limb of Henle's loop (TALH) and distal tubule (Chaps. 11 and 12, Gamba 2013), as well as the fish intestine (Frizzell et al. 1984; Musch et al. 1982) and urinary bladder (Dawson and Frizzell 1989; Gamba et al. 1993; Stokes et al. 1984). The presence of either Na^+-K^+-$2Cl^-$ cotransporters or Na^+-Cl^- cotransporters illustrates that the net rate and direction of K^+ transport depend on the combined action of all influx and efflux pathways (Fig. 3.4). Coupling of apical K^+ influx with acid secretion occurs via H^+/K^+ pump (H/K-ATPase, EC3.6.3.10), most dramatically (*ATP4A*, gHKA) in gastric parietal cells (Heitzmann and Warth 2007; Kopic et al. 2010) (Fig. 3.5) but also (*ATP12A*, cHKA) in the distal colon and the kidney intercalated cell (Crowson and Shull 1992; Gumz et al. 2010; Suzuki and Kaneko 1987). Intercalated cells also operate electrogenic H^+ pumps (*ATP6V*, V-ATPase, EC3.6.3.14) in the apical membrane together with $K_{Ca}1.1$ (Wagner et al. 2004). In addition to K^+ channels, K^+ exit across the apical membrane into the lumen could occur via any transporter with an appropriate driving force (Eq. 1). Renal distal tubule cells employ K^+-Cl^- cotransporters (*SLC12A3*, KCC) for that purpose, $-(z_K E_K^a + z_{Cl} E_{Cl}^a) > 0$, together with apical membrane Na^+-Cl^- cotransporters supplying Na^+ influx for Na^+/K^+-ATPase turnover (Ellison et al. 1987; Malnic et al. 2013). In combination these apical transporters influence the fluid covering the apical membrane by decreasing apical fluid pH or by contributing to fluid absorption while K^+ secretion increases the $[K^+]$ of the remaining fluid.

Basolateral membrane transporters must complement apical K^+ exit by providing appropriate pathways and driving forces for K^+ influx into the K^+ secretory epithelial cell (Bachmann et al. 2011). The Na^+/K^+-ATPase provides the primary energy for this K^+ influx, and in cells using the Cl^- secretory mechanism Na^+-K^+-$2Cl^-$ cotransporters provide Na^+ influx for maintaining Na^+/K^+-ATPase turnover as well as contributing to K^+ influx, $-(z_{Na} E_{Na}^b + z_K E_K^b + 2 z_{Cl} E_{Cl}^b) < 0$. K^+ channels and Cl^- channels make up the other key components of basolateral ion transport, but the molecular identity of the channel types remains to be fully defined. In Na^+ absorbing cells (Fig. 3.2; Chap. 2), K^+ channels provide a basolateral exit path such that the K^+ secretory rate can be lower than defined by the Na^+ absorption rate and the Na^+/K^+ coupling stoichiometry of the Na^+/K^+ pump. Basolateral Cl^- channels provide a necessary exit pathway in K^+ secretory cells relying on basolateral Na^+-K^+-$2Cl^-$ cotransporters for maintaining Na^+/K^+ pump turnover and also for those that absorb Cl^-. Cells of the stria marginalia in the inner ear use CLC-Ka and CLC-Kb along with the β-subunit barttin (*CLCNKA*, *CLCNKB*, and *BSND*) to support primary electrogenic K^+ secretion (Hibino and Kurachi 2006; Marcus and Wangemann 2009). Colonic epithelial cells likely employ TMEM16A and CLC-2 (Chaps. 15 and 17) to support K^+ secretion in Cl^- secretory and absorptive cells, respectively (Catalán et al. 2002; He et al. 2011; Zdebik et al. 2004). The range of biophysical behavior exhibited by Cl^- currents in colonic epithelial cells suggests

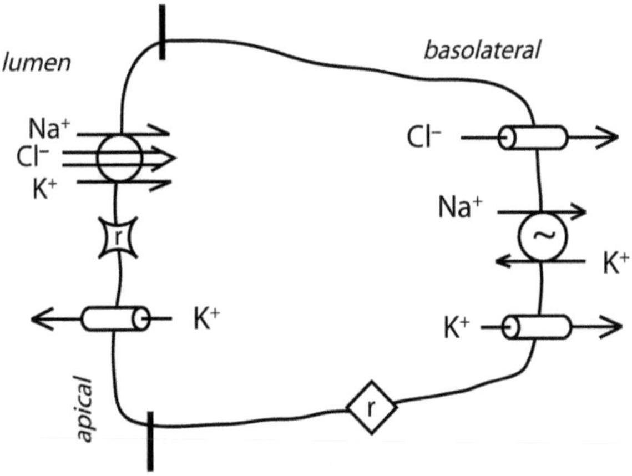

Fig. 3.4 Electroneutral Na^+Cl^- absorption and K^+ secretion. The cellular model for Na^+Cl^- absorption in the thick ascending limb of Henle's loop of the kidney (TALH) and in the fish intestine includes apical membrane Na^+-K^+-$2Cl^-$ cotransporters and K^+ channels together with basolateral membrane Na^+/K^+ pumps, Cl^- channels, and K^+ channels. Apical membrane K^+ channels in this cellular mechanism produce K^+ secretion dependent on the rate of Na^+Cl^- absorption. Basolateral and apical membrane receptors (r) activate or inhibit transepithelial Cl^-, Na^+, and K^+ flow (see Sect. 3.3.3)

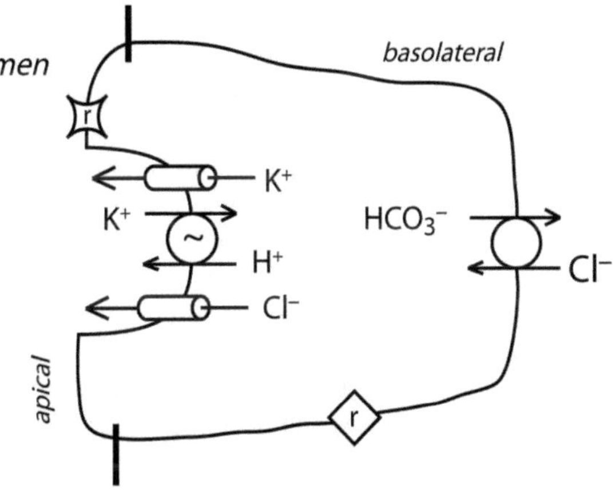

Fig. 3.5 Gastric acid secretion. The cellular model for acid secretion in gastric parietal cells includes apical domain H^+/K^+ pumps, Cl^- channels, and K^+ channels that reside in tubulo-vesicles prior to stimulated fusion with the apical membrane. Together with basolateral membrane Cl^-/HCO_3^- exchangers, these transporters suffice to generate activated secretory rates for H^+ and Cl^-. Basolateral and apical membrane receptors (r) activate or inhibit transepithelial H^+ and Cl^- flow

that other Cl^- channel types likely contribute (Li et al. 2003; Mignen et al. 2000). Renal tubule cells use CLC-Kb in Cl^- absorbing segments, TALH and distal tubule, and in intercalated cells (Gamba et al. 2013; Stauber et al. 2012). The combined action of these basolateral transport proteins assures continued K^+ secretion during stimulation by physiologic agonists.

3.3 Physiologic Contributions of Transepithelial K^+ Secretion

3.3.1 Potassium Excretion

Excretion constitutes a paramount motivation for transepithelial K^+ secretion. Maintaining cell volume and electrical activity requires sufficient K^+ for it to be the major intracellular cation. Since dietary intake of either animal or plant components includes cell contents, K^+ in the diet generally exceeds that needed to maintain balance of total body composition. Excretion of any excess K^+ occurs primarily via the kidneys (~90 %), with the remainder via the gut (Agarwal et al. 1994; Malnic et al. 2013; Wrong et al. 1981). The amount excreted is the product of the $[K^+]$ and the volume of fluid ejected from the body proper. While the typical $[K^+]$ in urine (20–70 mM) and fecal water (~75 mM) are similar for humans, daily urine volume ranges from 0.5 L to 18 L (typical ~1.5 L) and fecal water remains below 0.2 L. Any increase in fecal water loss constitutes diarrhea with attendant pH disturbances and dehydration, because unlike the kidney, colonic transport that adjusts luminal fluid pH is not governed by requirements for maintaining systemic pH. Even though higher gut loss of K^+ can be beneficial during renal failure (Sandle and Hunter 2010), this limit on fecal volume restricts the possibility of large K^+ losses via physiologic compensation by the gut. In contrast to terrestrial mammals, seawater teleost fish produce a low urine volume which aids in water conservation but also limits renal K^+ excretion. The apparent route of K^+ excretion proceeds via Cl^- secreting cells of the gill that include apical K^+ channels ($K_{ir}1.1$), similar to Fig. 3.3 (Furukawa et al. 2012). Salt glands of desert iguanid lizards also secrete a fluid with high $[K^+]$ that relieves the kidneys of producing urine to manage excess K^+ (Hazard 2001; Shuttleworth et al. 1987). Total body K^+ balance ultimately is achieved by regulating the increases and decreases in activity of K^+ transporting epithelial cells.

Cellular K^+ Secretory Mechanisms Potassium secretion in both the kidney and gut occurs in cells using the Na^+ absorptive type (Fig. 3.2) and Cl^- secretory type (Fig. 3.3) mechanisms. Positioning these cells in distal locations allows a definitive influence on the final excreted fluid. In the kidney these cells reside in the distal tubule and collecting duct, with principal cells absorbing Na^+ and intercalated cells using Na^+ entry via Na^+-K^+-$2Cl^-$ cotransporters in the basolateral membrane (Liu et al. 2011; Malnic et al. 2013). Although transport localization in the colon remains unsettled (Grotjohann et al. 1998; Halm and Rick 1992; Köckerling and Fromm 1993; Köckerling et al. 1993; Rechkemmer et al. 1996; Sørensen et al. 2011; Zhang et al. 2012), Na^+ absorbing cells likely reside in the surface epithelium and

Cl^- secretory cells in the crypts of Lieberkühn. The presence of these distinct K^+ secretory cell types in proximity at distal locations suggests a physiologic advantage for coupling K^+ secretion to different cellular driving forces. At the least, establishing both mechanisms provides for flexible control of luminal [K^+] in combination with other fluid and solute transport requirements.

Dependence on Na^+ Absorption Coupling K^+ secretion to Na^+ absorption allows the requirements for K^+ excretion to work with the need for Na^+ conservation in the typical paleo-environment of Na^+ scarcity, with a low [Na^+] in the final urine and fecal fluid combining with higher [K^+]. As noted earlier, apical membrane K^+ channels ($K_{ir}1.1$ and $K_{Ca}1.1$) aid in maintaining a sizable driving force for apical Na^+ entry via Na^+ channels (ENaC) that sustains Na^+ absorption without resorting to the higher G_{pr} required to harness E_K^b for setting a driving force to assure apical Na^+ entry (Eq. 3.2a). This lower G_{pr} aids in limiting paracellular backflux of actively transported ions, thereby preserving the ion concentrations of the fluid destined for excretion.

Mineralocorticoid activation increases Na^+ absorption and K^+ secretion in the distal segments of the renal tubules and in the distal colon, increasing Na^+ retention and K^+ excretion. For electrogenic Na^+ absorption to substantially exceed K^+ secretion, paracellular electrical shunting needs to occur to maintain the driving force for basolateral K^+ exit by depolarizing ψ_b via the influence of E_{Na}^a such that $z_K(\psi_b - E_K^b) > 0$. Sustained mineralocorticoid stimulation over many days leads to a renal K^+ secretory rate near to or above 2/3 the rate of Na^+ absorption (O'Neil and Helman 1977; Schwartz and Burg 1978; Stokes et al. 1981). Since the coupling ratio of the Na^+/K^+ pump ($3Na^+/2\,K^+$) sets the limit on the direct dependence of K^+ secretion upon Na^+ absorption, basolateral K^+ exit via G_K^b may cease in this condition, $z_K(\psi_b - E_K^b) \approx 0$, such that all of the K^+ taken in by the Na^+/K^+ pump exits across the apical membrane. Strict interpretation of these results must consider that the lumen negative ψ_t drives Na^+ and K^+ passively into the renal tubule lumen such that the ratio of K^+ secretion to Na^+ absorption increases. The ~3-fold decrease in G_{pr} during this mineralocorticoid stimulated state (Koeppen and Giebisch 1985) would reduce paracellular flow of Na^+ and K^+ supporting the concept of negligible K^+ exit via G_K^b. Additionally, intercalated cells contribute to K^+ secretion without directly altering Na^+ absorption.

The series arrangement of G_K^b and G_K^a suggests another possible mechanism for coupling K^+ secretion to Na^+ absorption (Fig. 3.2). Mineralocorticoid activation of cortical collecting duct principal cells hyperpolarizes ψ_b to around -103 mV, seemingly larger than E_K^b such that $z_K(\psi_b - E_K^b) < 0$ and K^+ enters principal cells via G_K^b (Koeppen and Giebisch 1985; Malnic et al. 2013; Sansom and O'Neil 1985; Sansom and O'Neil 1986). The concurrent decrease in G_{pr} reduces the depolarizing influence of E_{Na}^a on ψ_b, thereby favoring this basolateral K^+ entry. With apical K^+ exit maintained by the depolarizing influence of E_{Na}^a on ψ_a, a conductive transcellular K^+ secretory shunt would be established, increasing the rate of K^+ secretion to a value above 2/3 the rate of Na^+ absorption. The hyperpolarization of

ψ_b to values more negative than E_K^b was proposed to occur via high turnover of Na^+/ K^+ pumps (with an electrogenic $3Na^+/2$ K^+ coupling ratio) that supports the observed increase of Na^+ flow, thus producing an apparent reversal of conductive K^+ flow across the basolateral membrane. The large *EMF* of the Na^+/K^+ pump, more negative than −120 mV (Rakowski and Paxson 1988; Rakowski et al. 1989), could support such an action if the effective conductance of the Na^+/K^+ pumps becomes comparable to the other components of G_b (mostly G_K^b). Complicating this interpretation would be any changes to $[K^+]$ resulting from the high Na^+/K^+ pump turnover. A local depletion of interstitial $[K^+]$ due to this high pump activity, from the typical 5 mM to 3 mM, would increase E_K^b and likely stop secretory flow through this transcellular K^+ shunt, $z_K(\psi_b - E_K^b) \geq 0$.

Mammalian distal colon exhibits K^+ secretion in vivo that relies on ENaC-dependent electrogenic Na^+ absorption (Edmonds 1981; Wrong et al. 1981), but during ex vivo measurements with the transepithelial electrochemical gradients nullified by short circuiting and identical bath composition, K^+ secretion generally fails to depend on apical Na^+ entry via ENaC (Frizzell et al. 1976; McCabe et al. 1984; Rechkemmer and Halm 1989). A demonstration of K^+ secretion sensitivity to the ENaC inhibitor amiloride that may be consistent with coupling to Na^+ absorption has been accomplished, but only under the unusual condition of low $[Na^+]$ on the basolateral side of the colonic mucosa (Sweiry and Binder 1989). Colon from turtle exhibits amiloride-sensitive transcellular K^+ secretion during current clamping that maintains ψ_t near the spontaneous level (Halm and Dawson 1984). Voltage-clamping $\psi_t$to zero eliminates this transcellular K^+ secretion, supporting the concept of voltage dependent G_K^athat limits K^+ secretion via this colonic cell type to conditions with lumen negative ψ_t, as occurs during high rates of electrogenic Na^+ absorption.

Dependence on Basolateral Membrane Na^+-K^+-$2Cl^-$ Cotransporter Secreting K^+ by a mechanism independent of Na^+ absorption frees the epithelium from changes in luminal Na^+. Turning to basolateral uptake of Na^+ via Na^+-K^+-$2Cl^-$ cotransporters to provide for continued turnover of Na^+/K^+ pumps establishes a steady driving force for basolateral K^+ influx (Fig. 3.3). Secretagogue activation of K^+ secretion in the colon occurs together with Cl^-secretion that ranges from small and transient to large and sustained (Halm and Frizzell 1986; Hosoda et al. 2002; Rechkemmer et al. 1996; Zhang et al. 2009b). Several of these secretagogues, adrenergic, cholinergic, peptidergic, and prostanoid, activate electrogenic K^+ secretion without accompanying sustained Cl^- secretion (Carew and Thorn 2000; Halm and Halm 2001; Liao et al. 2005; Traynor and O'Grady 1996; Zhang et al. 2009b) supporting the concept that Cl^- exits conductively via basolateral Cl^- channels (Halm and Frizzell 1986; He et al. 2011; Li et al. 2003). Renal intercalated cells also use this secretory mechanism with Na^+/K^+ pumps and Na^+-K^+-$2Cl^-$ cotransporters operating together with conductive Cl^- exit only across the basolateral membrane (Liu et al. 2011; Sabolić et al. 1999). In addition, $K_V1.3$ contributes to G_K^aalong with $K_{Ca}1.1$ in intercalated cells (Carrisoza-Gaytán et al. 2010). A consequence of secreting K^+ via apical K^+ channels without either Na^+ channels or Cl^- channels in the apical membrane (Fig. 3.3c) is a need to maintain sufficient G_{pr} so that basolateral E_js, in this case E^b_{Cl}, depolarize ψ_a away from E^a_K such that $z_K(\psi_a - E^a_K) > 0$. The relatively high luminal [K^+] in the renal tubule and colon decreases the magnitude of E^a_K constraining the required size for E^b_{Cl}, which needs to be more positive than E^a_K in order to sustain K^+ secretion. The proximity of cells using this K^+-alone secretory mechanism (Fig. 3.3c) to the Na^+ absorptive type cells (Fig. 3.2) in the kidney and the colon leads to interactions that alter the rates of all the conductively transported ions. The lumen positive influence on the orientation of ψ_t promoted by the K^+-alone secretory cell enhances the driving force for Na^+ absorption, and the Na^+-absorptive cells similarly generate a lumen negative influence on ψ_t that contributes to apical K^+ exit from these K^+-alone secretory cells.

The K^+-alone secretory cells of the kidney and colon concurrently extrude H^+ via apical membrane pumps. Renal intercalated cells use electrogenic H^+ pumps (*ATP6V*, V-ATPase) to support systemic acid/base balance, which also contributes a lumen positive influence on ψ_t that enhances Na^+ absorption in neighboring principal cells (Andersen et al. 1985; Malnic et al. 2013; Wagner et al. 2004). The colon mucosa during ex vivo measurements exhibits net K^+ secretion in the proximal region and net K^+ absorption in the distal region (Foster et al. 1984; Rechkemmer and von Engelhardt 1993; Sullivan and Smith 1986). Stimulation by secretagogues leads toward net K^+ secretion in the distal region (Halm and Frizzell 1986; Rechkemmer et al. 1996). These K^+ secretory cells in the distal colon secrete H^+ using H^+/K^+ pumps (*ATP12A*, cHKA; Shao et al. 2010; Suzuki and Kaneko 1987) accounting for the observed net K^+ absorption. Activity of these apical H^+/K^+ pumps influences conductive K^+ exit across the apical membrane such that complete inhibition of these apical pumps limits K^+ secretion (Rechkemmer et al. 1996). Genetic deletion of *ATP12A* (cHKA) results in higher fecal K^+ excretion and leads to more severe hypokalemia during a K^+ restricted diet (Meneton et al. 1998). Renal compensation during dietary K^+ restriction lowers K^+ excretion from the kidney likely via the action of both gHKA and cHKA (Crambert 2014; Gumz et al. 2010; Malnic et al. 2013). Switching to H^+/K^+ pumps for renal acid secretion during K^+ restriction has the consequence of reducing the enhancement of Na^+ absorption produced via the lumen positive influence on ψ_t generated by electrogenic H^+ pumps (V-ATPase).

Interactions with Ammonium Systemic acid/base balance imposes further conditions on K^+ excretion, since the kidney generates ammonium (NH_4^+) to support acid excretion (Weiner and Verlander 2013), and the enteric microbiome produces NH_4^+ while fermenting the remnants of the diet (Davila et al. 2013; Wrong et al. 1981). The charge and size similarity of K^+ and NH_4^+ makes NH_4^+ an important physiologic competitor leading to potential influences on K^+ transport, with Na^+-K^+-$2Cl^-$ cotransporters (Kinne et al. 1986), Na^+/K^+ pumps (Halm and Dawson 1983; Skou and Esmann 1992; Wall and Koger 1994), H^+/K^+ pumps (Codina et al. 1999; Cougnon et al. 1999; Swarts et al. 2005), and the K^+ channels, $K_{ir}1.1$ (Yang et al. 2012), $K_{Ca}1.1$ (Eisenman et al. 1986), $K_{Ca}3.1$ (Christophersen 1991), and K_V1

(Heginbotham and MacKinnon 1993), having significant NH_4^+ permeability via interaction at K^+ binding sites. This multitude of cellular transport routes constitutes a cause of toxicity as increasing plasma [NH_4^+] leads to disruption of cellular physiology.

Ammonium excretion proceeds via selective transport proteins (Weiner and Verlander 2014; Worrell et al. 2008), but the presence of K^+ transporters complicates the task of trapping NH_4^+ in the lumen (Nakamura et al. 1999; Weiner and Verlander 2013). Apical membrane K^+ channels in particular could dissipate this NH_4^+ gradient, since the cell negative ψ_a contributes a substantial driving force for NH_4^+ entry from the lumen into the epithelial cells of renal tubules and colon, $z_{NH4}(\psi_a - E^a_{NH4}) < 0$. The physiologic range of ψ_a and the opposing orientation for E^a_K and E^a_{NH4} maintain the driving forces for K^+ exit and NH_4^+ entry. However, several features of K^+ channels act to limit apical NH_4^+ uptake, including ionic competition for binding sites and a reduced probability of opening during NH_4^+ passage (Bennekou and Christophersen 1990; Eisenman et al. 1986; Heginbotham and MacKinnon 1993; Yang et al. 2012). The characteristics of $K_{ir}1.1$ (Yang et al. 2012), conducting K^+ in renal TALH cells and principal cells, illustrate how the multisite permeation pathway in K^+ channels likely blunts NH_4^+ entry. Competition with Mg^{2+} for an extracellular site limits NH_4^+ access to the conduction path, and stronger K^+ binding within the selectivity filter blocks inward progress of NH_4^+ at physiologic ψ_a.

3.3.2 Epithelial K^+ Gradients Supporting Sensory Physiology

The inner ear senses sound waves (cochlea) and body orientation (vestibular labyrinth) using transepithelial electrochemical gradients to transduce these physical forces into electrical signals for nerve conduction (Hibino and Kurachi 2006; Marcus and Wangemann 2009). The sensory cells reside in the epithelia bounding the endolymph that has a [K^+] of ~150 mM. The vestibular dark cell epithelium secretes K^+ into the endolymph using the K^+-alone secretory mechanism (Fig. 3.3c), with $K_V7.1$/*KCNE1* as the apical K^+ channel and CLC-K/barttin as the basolateral Cl^- channel. High endolymph [K^+] reduces E^a_K to low levels requiring a depolarized ψ_a in the secreting cells so that K^+ exits from the cell into the endolymph, $z_K(\psi_a - E^a_K) > 0$. This condition implies a high intracellular [Cl^-] that produces a low E^b_{Cl} for setting this driving force. The high endolymph [K^+] depolarizes vestibular hair cells during activation of sensory mechanoelectrical transducer channels.

In the cochlea, the combination of two epithelia in series comprises the stria vascularis that maintains the high endolymph [K^+] in this region, generating an endocochlear potential of +80 to +100 mV (Hibino and Kurachi 2006; Marcus and Wangemann 2009; Wangemann 2006). The marginal cell epithelium of the stria vascularis secretes K^+ directly into the endolymph using essentially the same transport proteins as in the vestibular dark cell epithelium. A syncytium of intermediate and basal cells together with fibrocytes forms the other stria vascular epithelium that secretes K^+, by the K^+-alone mechanism, into the intra-strial space facing the basolateral membrane of the marginal cell epithelium. The apical K^+ channel ($K_{ir}4.1$, *KCNJ10*) for this syncytial epithelium resides only in the intermediate cells. Generation of the stria vascularis electrochemical gradient occurs with the syncytial epithelium producing a ψ_t of +90 mV and an intra-strial $[K^+]$ of ~2 mM. The marginal epithelium secretes from this intra-strial space into the endolymph with a ψ_t near 0 mV. At the apical membrane of the syncytial epithelium, the low intra-strial $[K^+]$ increases E_K^a, thereby hyperpolarizing ψ_a. A low E_{Cl}^b aids in maintaining the large ψ_t and the outward K^+ electrochemical gradient at the apical membrane. This tandem epithelial arrangement creates the endocochlear potential that depolarizes hair cells in the adjacent organ of Corti resulting in sensory nerve activity and hearing.

Terrestrial vertebrates sense olfactory cues via the main olfactory epithelium and the vomeronasal organ (Lucero 2013; Tirindelli et al. 2009). Detection of odorants and pheromones occurs through receptors on olfactory sensory neurons projecting into the apical surface fluid. This exo-epithelial fluid on the olfactory epithelium and the vomeronasal organ exhibits a $[K^+]$ of ~70 mM (Kim et al. 2012; Reuter et al. 1998). Similar to the inner ear, the resulting low value for E_K^a, near −25 mV, contributes to depolarization of the sensory neurons, likely modulating signal generation and the sensory response. The transport character of the columnar (support) cells in the olfactory epithelium resembles the tracheal mucosa with capacities for electrogenic Na^+ absorption using ENaC and Cl^- secretion via CFTR and/or TMEM16A (Dauner et al. 2012; Grubb et al. 2009; Merigo et al. 2011). These columnar epithelial cells open apical membrane $K_{Ca}1.1$ channels during purinergic activation, generating a lumen positive ψ_t consistent with K^+ secretion that likely maintains the high exo-epithelial $[K^+]$ (Grubb et al. 2009; Hegg et al. 2009; Vogalis et al. 2005). Coordination between these epithelial sensory neurons and columnar cells serves to match olfactory sensitivity to the external odorant environment (Lucero 2013).

3.3.3 Transport Cofactor

Many epithelia generate K^+ secretion using apical membrane K^+ channels that results in only modest increases of the apical fluid $[K^+]$. In these cases the K^+ secretion often serves as part of a K^+ circulation that promotes transepithelial flow of other ions and solutes. Generally, retrieval of the secreted K^+ occurs by absorptive pathways in the secretory cell (parietal cell, TALH cell) or by fluid absorption at a distant site. The primary example of this second type of retrieval consists of gut handling of saliva and airway fluid. Removing the consequence of excretion allows these epithelia to focus K^+ secretion on enhancing other transport activities.

Gastric Acid Secretion Parietal cells secrete H^+ at high rates leading to a luminal pH as low as 1 to 2 and [K^+] of 10–20 mM (Heitzmann and Warth 2007; Kopic et al. 2010). The secretory mechanism involves H^+/K^+ pumps and K^+ channels (Fig. 3.5) trafficked into the apical membrane by stimulated generation of a canaliculus from vesicles residing in the apical pole of the parietal cell (Forte and Zhu 2010). During lower rates of secretion, the transport mechanism resembles the K^+Cl^- secretory cell (Fig. 3.3) found in the distal colon that includes apical H^+/K^+ pumps (Halm and Frizzell 1986; Suzuki and Kaneko 1987). The $K_V7.1$/*KCNE2* K^+ channel provides the dominant contribution to G_K^a, with $K_{ir}1.1$ and $K_{ir}4.1$ (*KCNJ10*) also involved (Lambrecht et al. 2005; Song et al. 2011; Vucic et al. 2014). Localization studies support close proximity of $K_V7.1$/*KCNE2* and $K_{ir}4.1$ with H^+/K^+ pumps in activated canalicular membranes, but $K_V7.1$/*KCNE2* and H^+/K^+ pumps reside in separate resting vesicle populations (Heitzmann et al. 2004; Nguyen et al. 2013; Song et al. 2011).

Fluid secretion driven by apical exit of K^+ and Cl^- provides flow that propels secreted acid out of the canaliculus and gastric pit (Heitzmann et al. 2004). In addition to providing K^+ in the apical fluid for turnover of H^+/K^+ pumps, a large G_K^a moves ψ_a toward E_K^a which aids in maintaining an outward Cl^- driving force at the apical membrane, $z_{Cl}(\psi_a - E_{Cl}^a) > 0$, for the continued Cl^- secretion needed to support acid secretion. Since the 0.08 mM K_m^K of the gastric H^+/K^+ pump (Shao et al. 2010) more than assures maximal operation at the typical luminal [K^+], the luminal level of K^+ may derive more as a consequence of the G_K^a needed for maintaining adequate rates of Cl^- secretion.

Pancreatic Acinar Enzyme Release Digestion of nutrients in the intestinal lumen requires enzymes released from pancreatic acinar cells. These digestive enzymes reside in zymogen granules within the apical pole of acinar cells such that release of the stored contents into the lumen involves fusion of the granule membrane with the apical membrane followed by fluid flow that propels the contents out (Thévenod 2002). Vesicular Cl^- channels and K^+ channels promote ion secretion that drives water flow via aquaporins into to the widening fusion pit. The presence of these channels in the vesicular membrane focuses the osmotic gradient and water flow at the points of release (Fig. 3.6). Since acinar stimulation primarily activates Cl^- channels in the apical membrane (Heitzmann and Warth 2008), the K^+ secretion from the fusion events likely contributes only a minor amount to the ionic composition of the secreted acinar fluid.

Na^+Cl^- Absorption The thick ascending limb of Henle's loop (TALH) in the kidney absorbs Na^+ and Cl^- using Na^+-K^+-$2Cl^-$ cotransporters (NKCC2) as the apical step in the cellular mechanism (Fig. 3.4). During transit of tubular fluid along the TALH, absorption decreases [Na^+], [Cl^-], and [K^+], with [K^+] dropping from ~8 mM to ~1 mM at the distal end (Vallon et al. 1997). Longitudinal differences in splice variants of NKCC2 promote absorption in the face of these decreasing substrate concentrations, with the variant in the distal half of TALH exhibiting a ~4-fold higher affinity for K^+ than the variant in the proximal half (Giménez et al.

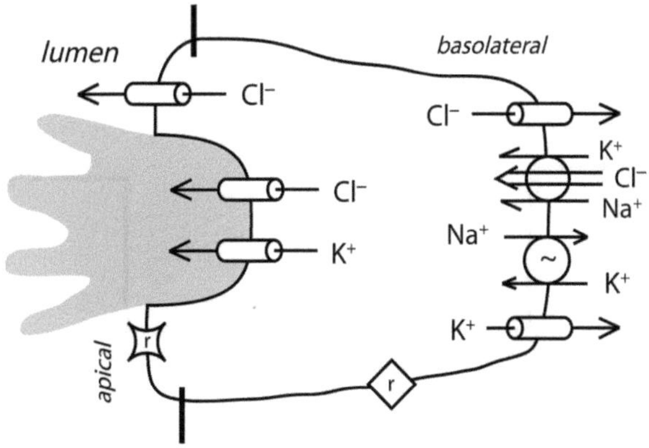

Fig. 3.6 Exocytotic apical release. The cellular model for enzyme release from pancreatic acinar cells includes vesicular Cl^- channels and K^+ channels together with basolateral membrane transporters supporting Cl^- secretion (Fig. 3.3). A local osmotic gradient drives water flow into granules fused with the apical membrane. Basolateral and apical membrane receptors (r) activate or inhibit acinar secretions

2002). K^+ secretion via apical $K_{ir}1.1$ assists in maintaining Na^+ and Cl^- absorption by adding K^+ to the tubular fluid such that luminal [K^+] permits uptake via Na^+-K^+-$2Cl^-$ cotransporters. The presence of NH_4^+ in the tubular fluid as part of acid/base management likely contributes to Na^+ and Cl^- absorption (Weinstein 2010), since Na^+-K^+-$2Cl^-$ cotransporters also accept NH_4^+ at the K^+ transport site. In addition, the lumen positive ψ_t generated by K^+ secretion drives absorption of Na^+ as well as Mg^{2+} and Ca^{2+} via the paracellular pathway.

Intestines of fish absorb fluid driven by Na^+ and Cl^- absorption using a cellular mechanism similar to the TALH (Fig. 3.4). Marine fish ingest seawater during feeding and to replace water lost to the environment. Since intestinal water uptake requires Na^+Cl^- absorption, the gills excrete the excess Na^+ and Cl^-. Progressive fluid absorption leads to a luminal composition of low [Na^+], [Cl^-], and [K^+] dominated by Mg^{2+} and SO_4^{2-} from the ingested seawater, with Ca^{2+} precipitating as $CaCO_3$ (Whittamore et al. 2010). As in TALH, intestinal K^+ secretion maintains luminal [K^+] for continued uptake via Na^+-K^+-$2Cl^-$ cotransporters and supports a lumen positive ψ_t for paracellular Na^+ absorption (Frizzell et al. 1984). The G_K^aresponsible for K^+ secretion also creates a ψ_a supporting Na^+ coupled nutrient absorption (Fig. 3.7; Halm et al. 1985). Apical uptake of glucose and some amino acids occurs via Na^+ coupled transporters using the Na^+ electrochemical gradient, $z_{Na}(\psi_a – E_{Na}^a)$, that drives luminal nutrient concentrations to low levels (Bröer 2008; Musch et al. 1987; Thompson and Kleinzeller 1985; Wright et al. 2011). As luminal [Na^+] drops due to absorption, the Na^+ driving force for this nutrient uptake also apparently decreases. Assuming a near constant intracellular [Na^+] (Hudson and Schultz 1984), E_{Na}^a would decrease by ~43 mV. In addition, G_K^a increases in response to lower luminal [Na^+], possibly via reduced ion uptake by Na^+-K^+-$2Cl^-$ cotransporters, such that ψ_a hyperpolarizes by ~42 mV, to −106 mV (Halm

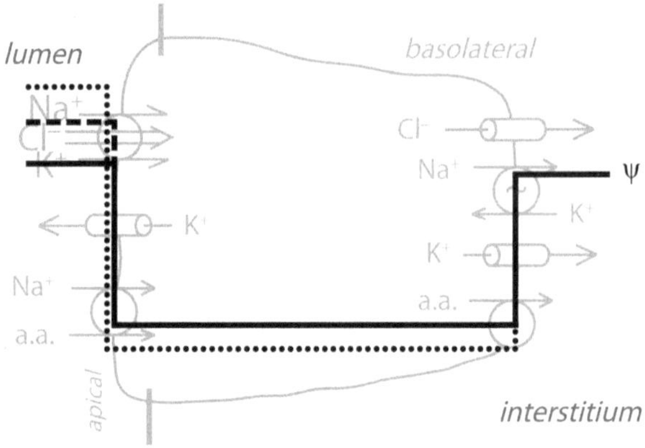

Fig. 3.7 Intestinal nutrient absorption. The cellular model for Na^+Cl^- absorption in fish intestine includes apical membrane Na^+-K^+-$2Cl^-$ cotransporters, K^+ channels, and Na^+-coupled amino-acid transporters together with basolateral membrane Na^+/K^+ pumps, Cl^- channels, and amino-acid transporters (Halm et al. 1985). At high luminal [Na^+] and [Cl^-], ψ_t is small (*solid line*); decreases in luminal [Na^+] and [Cl^-] during absorption hyperpolarize ψ_a (*dashed line*); associated decreases in luminal [K^+] to 1 mM further hyperpolarize ψ_a (*dotted line*)

et al. 1985), which compensates for the decreased E^a_{Na} such that $z_{Na}(\psi_a - E^a_{Na})$ remains ~120 mV inward. The presence of electrogenic K^+ secretion in these intestinal cells contributes to the driving forces for Na^+-K^+-$2Cl^-$ cotransporters, paracellular Na^+ absorption, as well as nutrient absorption.

Cl^- Secretion The driving force for apical Cl^- exit during Cl^- secretion (Fig. 3.3) requires hyperpolarization of ψ_a away from E^a_{Cl} so that $z_{Cl}(\psi_a - E^a_{Cl}) > 0$. Opening basolateral K^+ channels provides this hyperpolarization (Frizzell and Hanrahan 2012), but only in conjunction with a sizable G_{pr} that electrically couples apical and basolateral membranes (Eq. 3.2) producing an added complication of likely nonselective paracellular flow of ions that may partially dissipate the active transcellular Cl^- flow. Simulations of salivary acinar Cl^- secretion indicate that modest activation of G^a_K increases Cl^- secretion by 10–25 % (Cook and Young 1989; Palk et al. 2010), lessening the dependence on G_{pr}. Supporting this concept of E^a_K contributing to the driving force for apical Cl^- exit, stimulation of parotid acinar cells activates apical K^+ currents consistent with the presence of $K_{Ca}1.1$ and $K_{Ca}3.1$ (Almassy et al. 2012). As this primary acinar secretion flows along into the salivary ducts, these epithelial cells alter the ion composition by a mechanism similar to the Na^+ absorptive model (Fig. 3.2). Submandibular ducts secrete K^+ using apical $K_{Ca}1.1$ leading to a saliva [K^+] of ~35 mM (Mangos et al. 1973; Nakamoto et al. 2008). In these ductal cells, the G^a_K producing K^+secretion also assists in setting the driving force for apical Na^+ entry.

Airway epithelia also secrete Cl^- such that the resulting apical fluid supports muco-ciliary clearance. An associated K^+ secretion, only ~3 % of the Cl^- secretory rate (Clarke et al. 1997; Welsh et al. 1983), leads to apical fluid [K^+] of 15–30 mM

(Joris et al. 1993; Knowles et al. 1997; Namkung et al. 2009). The $K_{Ca}1.1$ K^+ channel (Manzanares et al. 2011), and possibly Kv7 (Moser et al. 2008; Namkung et al. 2009), supports secretagogue activated K^+ secretion with ciliary beating in this apical fluid layer. Since the physiologic secretory stimulus via ATP produces transient activation, the augmentation of Cl^- secretion and thereby fluid secretion likely occur as short bursts dispersed to locally activated ATP release sites along the airway.

Just as salivary glands exhibit region-specific transport, type-2 cells of alveoli absorb Na^+ electrogenically with G^a_{Cl} providing a pathway for electrically coupled transcellular Cl^- absorption (O'Grady and Lee 2003). The observed localization of several K_V type K^+ channels in the apical membrane (Lee et al. 2003) supports the possibility for K^+ secretion (Nielson 1986) that also alters Cl^- flow via this transcellular route. In the absence of activated G^a_K, Na^+ and Cl^- channels determine ψ_a such that both ions enter the cell, $z_{Na}(\psi_a - E^a_{Na}) < 0$ and $z_{Cl}(\psi_a - E^a_{Cl}) < 0$. Opening apical K^+ channels increases apical fluid $[K^+]$ while hyperpolarizing ψ_a toward E^a_K such that the Cl^- driving force may reach zero or reverse direction, $z_{Cl}(\psi_a - E^a_{Cl}) \geq 0$. Inhibition of these $K_V$channels by hypoxia (O'Grady and Lee 2005) would enhance apical Cl^- entry via G^a_{Cl}, thereby contributing to maximal thinning of alveolar surface liquid that promotes gas exchange.

Several epithelia in reproductive organs produce luminal fluid with elevated $[K^+]$ consistent with K^+ secretion (Casslén and Nilsson 1984; Iritani et al. 1974; Wales et al. 1966). The endometrial gland epithelium secretes K^+ using $K_{Ca}2.1$ and $K_{Ca}2.3$ at a rate ~25 % that of Cl^- secretion (Palmer et al. 2008), providing another example for the concept that opening apical K^+ channels may enhance Cl^- secretion. The estradiol stimulated switching of expression from $K_{Ca}2.1$ to $K_{Ca}2.3$ may relate to estrus cycle-dependent increases in luminal $[K^+]$ (Casslén and Nilsson 1984).

Sweat glands and mammary glands elaborate a fluid onto the skin surface by producing a primary fluid secretion in the secretory coil followed by ductal modification (Quinton et al. 1999; Sato 2001; Shennan and Peaker 2000). Human eccrine sweat $[K^+]$ is 7–8 mM via actions of these secretory coil and duct cells (Bijman and Quinton 1984; Sato and Sato 1990). Mammary epithelial cell lines recapitulate milk production by exhibiting Cl^- secretion (Blaug et al. 2003) and Na^+ absorption (Palmer et al. 2011). K^+ secretion accompanies Na^+ absorption via apical $K_{Ca}3.1$ activated by P2Y-purinergic receptors (Palmer et al. 2011), supporting a process in which ductal modification elevates $[K^+]$ to 15–25 mM and lowers $[Na^+]$ to ~8 mM (Shennan and Peaker 2000). Frog skin produces a surface liquid layer by a similar process driven via exocrine gland Cl^- secretion followed by electrogenic Na^+ absorption across the skin surface epithelium (Larsen 2011; Larsen and Ramløv 2013; Watlington and Huf 1971). Recirculation of Na^+ allows evaporative water loss without major Na^+ loss. A Ca^{2+}-activated K^+ channel consistent with $K_{Ca}1.1$ opens during Cl^- secretion in these exocrine glands (Sørensen et al. 2001) leading to $[K^+]$ in this fluid of 15–20 mM (Larsen and Ramløv 2013). In contrast, rat and mouse paw sweat glands produce sweat with $[K^+]$ of 140–190 mM and $[Na^+]$ of 60–

10 mM, with a primary fluid from the secretory coil of 120–160 mM [K^+] and 60–35 mM [Na^+] (Brusilow et al. 1968; Quatrale and Laden 1968; Sato and Sato 1978; Sato et al. 1994). Compared with the inner ear that produces endolymph with a high [K^+], secretory coil cells secrete K^+ with a ψ_a ranging from −50 mV to −40 mV so that apical exit of K^+ via channels would require unreasonably high intracellular [K^+] to assure $z_K(\psi_a - E_K^a) > 0$ (Sato 1980). Instead, K^+ secretion apparently relies on an apical transporter that couples K^+ flow to a large outwardly directed energy gradient, perhaps apical exit of lactate, unless K^+ secretion occurs in a minority cell type with a ψ_a near 0 mV.

3.3.4 Apical Fluid Composition

Several epithelia produce an apical fluid with K^+ as the dominant cation. In the cases that G_K^a aids in potentiating Cl^- secretion, luminal [K^+] generally remains lower than [Na^+]. Aside from the excretory action by the kidney and the high endolymph [K^+] of the inner ear that serves sensory transduction, the utility of the other high [K^+] exo-epithelial fluids remains more obscure. In particular, the colonic lumen exhibits a high [K^+] space from which nutrient and fluid absorption occur. Although colonic K^+ secretion has been suggested to operate in support of Cl^- secretion (Nanda Kumar et al. 2010), many physiologic secretagogues stimulate K^+ secretion largely independent of Cl^- secretion (Fig. 3.3c; see Sect. 3.3.1). The K^+ secreted into the colonic lumen may serve instead primarily as an ionic osmolyte for maintaining luminal volume in the face of continual absorption of Na^+ together with short-chain fatty acids. This augmentation of luminal fluid volume provides space for the large community of resident bacteria to ferment poorly digestible meal components into short-chain fatty acids as well as other metabolites (Davila et al. 2013; Preidis and Versalovic 2009). Using K^+ rather than Na^+ for sustaining this microbiome space allows the final excretion of gut fluid to be more nearly devoid of Na^+, as well as contributing a modest amount to total body K^+ excretion.

The presence of K^+ in these exo-epithelial fluids also may confer other advantages, such as maintaining the epithelial barrier against infection. Saliva suppresses microbial growth using secreted proteins and specific ionic composition (Carpenter 2013; van't Hof et al. 2014; Viejo-Díaz et al. 2004; Wang and Germaine 1993). By secreting K^+ and reabsorbing Na^+, salivary ducts generate saliva that restricts bacterial use of Na^+-dependent transporters and diminishes bacterial K^+ gradients, E_K^{bac}. Similarly, the bioenergetics enforced by a low Na^+ and high K^+ environment may contribute to stability of the colonic microbiome and resistance to overpopulation by virulent strains (Epstein and Schultz 1965; Häse and Barquera 2001; Häse et al. 2001). The high [K^+] sweat of rodent paws also may serve an antimicrobial role, as well as possibly enhancing sensation of pheromones by adding K^+ to inhaled aerosols (see Sect. 3.3.2). In particular, exo-epithelial fluids high in K^+

may limit bacterial growth and virulence bioenergetically, thereby easing the burden on other components of immune system surveillance.

3.4 Signaling Pathways for Transepithelial K^+ Secretion

Regulation of K^+ secretion often follows the control of Na^+ absorption and Cl^- secretion (Chaps. 1 and 2), since these cellular mechanisms (Figs. 3.2 and 3.3) depend on basolateral Na^+/K^+ pumps that provide K^+ uptake and since opening either Na^+ or Cl^- channels in the apical membrane generally enhances the driving force for K^+ exit via apical channels (see Sect. 3.2.1). The control mechanisms that maintain the steady-state balance of K^+ uptake and exit during K^+ secretion remain largely unknown, but responses to physiologic stimuli provide insights into epithelial regulation. Stimulation of electrogenic Na^+ absorption by aldosterone acts via increasing the number of Na^+ channels (ENaC) in the apical membrane by slowing endocytosis (Butterworth et al. 2009). Direct control of apical K^+ channel activity also occurs in the renal principal and intercalated cells responsible for K^+ excretion via cellular signaling pathways leading to channel phosphorylation and interaction with membrane lipids (Satlin et al. 2006; Wang and Huang 2013; Welling 2013). Mirroring the regulation of ENaC, $K_{ir}1.1$ activity depends on the kinetics of membrane insertion and endocytosis, thus providing major control points for adjusting G_K^a. Increases in tubular fluid flow activate $K_{Ca}1.1$ via increases in cytosolic Ca^{2+} produced by mechanical deformation of the primary cilium and paracrine activation of purinergic receptors. A gut-derived factor also stimulates renal K^+ excretion, as well as extrarenal K^+ sequestration, such that plasma $[K^+]$ remains nearly constant during ingestion of K^+ containing meals (Lee et al. 2007; Oh et al. 2011; Youn 2013). Acting through inhibition of distal tubule Na^+-Cl^- cotransporters (NCC) via rapid dephosphorylation, this gut-derived factor leads to increased fluid flow and Na^+ delivery to the collecting duct that activates electrogenic Na^+ absorption as well as K^+ secretion (Sorensen et al. 2013).

Colonic K^+ secretion often occurs without activation of Cl^- secretion indicating the presence of signaling pathways allowing independent control of apical K^+ and Cl^- channels. Since both K^+ secretion and Cl^- secretion use cAMP production as a key signaling event, activating K^+ secretion without Cl^- secretion requires separation between these signaling cascades either within distinct cell types or via functional compartmentalization within a common secretory cell. Support for compartmentalization comes from β-adrenergic activation that produces transient Cl^- secretion and sustained K^+ secretion (Halm et al. 2010; Zhang et al. 2009a; Zhang et al. 2009b). The transient Cl^- secretion results from cAMP production via a transmembrane adenylyl cyclase activated by β2-adrenergic receptors. In contrast, activation of sustained K^+ secretion proceeds via soluble adenylyl cyclase activated by a β1/β2-adrenergic receptor complex. Phosphodiesterases likely contribute to the functional compartmentalization by restricting cAMP spread within the cell. These secretory cells also respond to aldosterone with K^+ secretion independent of Na^+ absorption and Cl^- secretion (Halm and Halm 1994; Rechkemmer and Halm 1989). The activation requires gene transcription and protein translation, occurring with a delay of ~15 min and half time of ~40 min, suggesting a mechanism critically dependent on a signaling protein with a relatively short half-life. Paracrine signaling also activates apical K^+channels in Cl^- secretory cells, as illustrated by stimulation of K^+ secretion in the airway and colon by release of ATP into the apical fluid (Clarke et al. 1997; Manzanares et al. 2011; Sausbier et al. 2006; Zhang et al. 2009b). This range of control for colonic K^+ secretion allows for an adaptable maintenance of luminal [K^+] that serves both excretion and the enteric microbiome.

3.5 Summary

Transepithelial K^+ secretion contributes to a wide range of physiologic actions by increasing the [K^+] of exo-epithelial spaces. This K^+ secretion generally occurs in cells absorbing Na^+ or secreting Cl^- such that the electrochemical gradients for these ions couple to maintain the rate and orientation of transepithelial flow. Several epithelia, notably from the inner ear, kidney, and colon, modify the Cl^- secretory mechanism to produce K^+ secretion without transcellular Cl^- flow, by redirecting Cl^- taken up via Na^+-K^+-$2Cl^-$ cotransporters back across the basolateral membrane through Cl^- channels. Renal and colonic K^+ secretion together produce K^+ excretion into a fluid with high [NH_4^+], which competes for passage through many K^+ transport proteins including the apical K^+ channels. Sensation in the inner ear depends crucially on the high [K^+] of endolymph, while olfaction likely modulates sensory signals via high apical fluid [K^+]. Apical K^+ channels involved in secretion also contribute to maintaining Cl^- exit in several Cl^- secretory epithelia as well as to intestinal nutrient uptake. The elevated [K^+] in fluid on several apical surfaces, including saliva and sweat as well as airway and colonic fluid, likely influences the activity of the resident microbiome living in these exo-epithelial fluids. The physiologic impact of K^+ secretion clearly ranges from the macroscopic issues of total body balance attended to by excretion to microscopic manipulations that adjust transport for sensation and cellular metabolism.

References

Abbott GW (2012) KCNE2 and the K^+ channel: the tail wagging the dog. Channels (Austin) 6: 1–10

Agarwal R, Afzalpurkar R, Fordtran JS (1994) Pathophysiology of potassium absorption and secretion by the human intestine. Gastroenterology 107:548–571

Almassy J, Won JH, Begenisich TB et al (2012) Apical Ca^{2+}-activated potassium channels in mouse parotid acinar cells. J Gen Physiol 139:121–133
Andersen OS, Silveira JE, Steinmetz PR (1985) Intrinsic characteristics of the proton pump in the luminal membrane of a tight urinary epithelium. The relation between transport rate and $\Delta\mu H$. J Gen Physiol 86:215–234
Bachmann O, Juric M, Seidler U et al (2011) Basolateral ion transporters involved in colonic epithelial electrolyte absorption, anion secretion and cellular homeostasis. Acta Physiol (Oxf) 201:33–46
Barmeyer C, Rahner C, Yang Y et al (2010) Cloning and identification of tissue-specific expression of KCNN4 splice variants in rat colon. Am J Physiol Cell Physiol 299:C251–C263
Bennekou P, Christophersen P (1990) The gating of human red cell Ca^{2+}-activated K^+-channels is strongly affected by the permeant cation species. Biochim Biophys Acta 1030:183–187
Berliner RW (1961) Renal mechanisms for potassium excretion, vol 55, The Harvey lectures, series. Academic, New York, pp 141–171
Bijman J, Quinton PM (1984) Influence of abnormal Cl^- impermeability on sweating in cystic fibrosis. Am J Physiol Cell Physiol 247:C3–C9
Blaug S, Rymer J, Jalickee S et al (2003) P2 purinoceptors regulate calcium-activated chloride and fluid transport in 31EG4 mammary epithelia. Am J Physiol Cell Physiol 284:C897–C909
Bröer S (2008) Amino acid transport across mammalian intestinal and renal epithelia. Physiol Rev 88:249–286
Brusilow SW, Ikai K, Gordes E (1968) Comparative physiological aspects of solute secretion by the eccrine sweat gland of the rat. Proc Soc Exp Biol Med 129:731–732
Butterworth MB, Edinger RS, Frizzell RA et al (2009) Regulation of the epithelial sodium channel by membrane trafficking. Am J Physiol Renal Physiol 296:F10–F24
Carew MA, Thorn P (2000) Carbachol-stimulated Cl^- secretion in mouse colon: evidence of a role for autocrine prostaglandin-E_2 release. Exp Physiol 85:67–72
Carpenter GH (2013) The secretion, components, and properties of saliva. Annu Rev Food Sci Technol 4:267–276
Carrisoza-Gaytán R, Salvador C, Satlin LM et al (2010) Potassium secretion by voltage-gated potassium channel Kv1.3 in the rat kidney. Am J Physiol Renal Physiol 299:F255–F264
Casslén B, Nilsson B (1984) Human uterine fluid, examined in undiluted samples for osmolarity and the concentrations of inorganic ions, albumin, glucose, and urea. Am J Obstet Gynecol 150:877–881
Catalán M, Cornejo I, Figueroa CD et al (2002) CLC-2 in guinea pig colon: mRNA, immunolabeling, and functional evidence for surface epithelium localization. Am J Physiol Gastrointest Liver Physiol 283:G1004–G1013
Christophersen P (1991) Ca^{2+}-activated K^+ channel from human erythrocyte membranes: single channel rectification and selectivity. J Membr Biol 119:75–83
Clarke LL, Chinet T, Boucher RC (1997) Extracellular ATP stimulates K^+ secretion across cultured human airway epithelium. Am J Physiol Lung Cell Mol Physiol 272:L1084–L1091
Clausen C, Lewis SA, Diamond JM (1979) Impedance analysis of a tight epithelium using a distributed resistance model. Biophys J 26:291–317
Codina J, Pressley TA, DuBose TD Jr (1999) The colonic H+, K+-ATPase functions as a Na+-dependent K+(NH4+)-ATPase in apical membranes from rat distal colon. J Biol Chem 274: 19693–19698
Cook DI, Young JA (1989) Effect of K^+ channels in the apical plasma membrane on epithelial secretion based on secondary active Cl^- transport. J Membr Biol 110:139–146
Cougnon M, Bouyer P, Jaisser F et al (1999) Ammonium transport by the colonic H^+-K^+-ATPase expressed in Xenopus oocytes. Am J Physiol Cell Physiol 277:C280–C287
Crambert G (2014) H-K-ATPase type 2: relevance for renal physiology and beyond. Am J Physiol Renal Physiol 306:F693–F700
Crane RK (1965) Na^+-dependent transport in the intestine and other animal tissues. Fed Proc 24: 1000–1006

Crowson MS, Shull GE (1992) Isolation and characterization of a cDNA encoding the putative distal colon H, K-ATPase. Similarity of deduced amino acid sequence to gastric H+, K+-ATPase and Na+, K+-ATPase and mRNA expression in distal colon, kidney, and uterus. J Biol Chem 267:13740–13748
Dauner K, Lissmann J, Jeridi S et al (2012) Expression patterns of anoctamin 1 and anoctamin 2 chloride channels in the mammalian nose. Cell Tissue Res 347:327–341
Davila AM, Blachier F, Gotteland M et al (2013) Intestinal luminal nitrogen metabolism: role of the gut microbiota and consequences for the host. Pharmacol Res 68:95–107
Dawson DC, Frizzell RA (1989) Mechanism of active K^+ secretion by flounder urinary bladder. Pflugers Arch 414:393–400
Dawson DC, Richards NW (1990) Basolateral K^+ conductance: role in regulation of Na^+Cl^- absorption and secretion. Am J Physiol Cell Physiol 259:C181–C195
Duta V, Duta F, Puttagunta L et al (2006) Regulation of basolateral Cl^- channels in airway epithelial cells: the role of nitric oxide. J Membr Biol 213:165–174
Edmonds CJ (1981) Amiloride sensitivity of the transepithelial electrical potential and of sodium and potassium transport in rat distal colon in vivo. J Physiol 313:547–559
Eisenman G, Latorre R, Miller C (1986) Multi-ion conduction and selectivity in the high-conductance Ca^{2+}-activated K^+ channel from skeletal muscle. Biophys J 50:1025–1034
Ellison DH, Velázquez H, Wright FS (1987) Mechanisms of sodium, potassium and chloride transport by the renal distal tubule. Miner Electrolyte Metab 13:422–432
Epstein W, Schultz SG (1965) Cation transport in *Escherichia coli*: V. Regulation of cation content. J Gen Physiol 49:221–234
Ernst SA, Mills JW (1977) Basolateral plasma membrane localization of ouabain-sensitive sodium transport sites in the secretory epithelium of the avian salt gland. J Cell Biol 75:74–94
Estilo G, Liu W, Pastor-Soler N et al (2008) Effect of aldosterone on BK channel expression in mammalian cortical collecting duct. Am J Physiol Renal Physiol 295:F780–F788
Fan S, Harfoot N, Bartolo RC et al (2012) CFTR is restricted to a small population of high expresser cells that provide a forskolin-sensitive transepithelial Cl^- conductance in the proximal colon of the possum, Trichosurus vulpecula. J Exp Biol 215:1218–1230
Field M (1993) Intestinal electrolyte secretion. History of a paradigm. Arch Surg 128:273–278
Fischer H, Illek B, Finkbeiner WE et al (2007) Basolateral Cl^- channels in primary airway epithelial cultures. Am J Physiol Lung Cell Mol Physiol 292:L1432–L1443
Forte JG, Zhu L (2010) Apical recycling of the gastric parietal cell H+, K+-ATPase. Annu Rev Physiol 72:273–296
Foster ES, Hayslett JP, Binder HJ (1984) Mechanism of active potassium absorption and secretion in the rat colon. Am J Physiol Gastrointest Liver Physiol 246:G611–G617
Frizzell RA, Hanrahan JW (2012) Physiology of epithelial chloride and fluid secretion. Cold Spring Harb Perspect Med 2(6):a009563. doi:10.1101/cshperspect.a009563
Frizzell RA, Koch MJ, Schultz SG (1976) Ion transport by rabbit colon. I. Active and passive components. J Membr Biol 27:297–316
Frizzell RA, Field M, Schultz SG (1979) Sodium-coupled chloride transport by epithelial tissues. Am J Physiol Renal Physiol 236:F1–F8
Frizzell RA, Halm DR, Krasny EJ Jr (1983) Active secretion of chloride and potassium across the large intestine. In: Skadhauge E, Heintze K (eds) Intestinal absorption and secretion, Falk symposium 36. MTP Press Ltd, Boston, pp 313–324
Frizzell RA, Halm DR, Musch MW et al (1984) Potassium transport by flounder intestinal mucosa. Am J Physiol Renal Physiol 246:F946–F951
Frömter E, Diamond J (1972) Route of passive ion permeation in epithelia. Nat New Biol 235: 9–13
Furukawa F, Watanabe S, Kimura S et al (2012) Potassium excretion through ROMK potassium channel expressed in gill mitochondrion-rich cells of Mozambique tilapia. Am J Physiol Regul Integr Comp Physiol 302:R568–R576

Furuse M (2010) Claudins, tight junctions, and the paracellular barrier. In: Yu ASL (ed) Claudins, vol 65, Current topics in membranes. Elsevier, Boston, pp 1–19

Gamba G (2013) Physiology and pathophysiology of the Na^+Cl^- co-transporters in the kidney. In: Alpern RJ, Moe OW, Caplan M (eds) Seldin and Giebisch's the kidney: physiology and pathophysiology, 5th edn. Elsevier, Boston, pp 1047–1080

Gamba G, Saltzberg SN, Lombardi M et al (1993) Primary structure and functional expression of a cDNA encoding the thiazide-sensitive, electroneutral sodium-chloride cotransporter. Proc Natl Acad Sci USA 90:2749–2753

Gamba G, Wang W, Schild L (2013) Sodium chloride transport in the loop of Henle, distal convoluted tubule, and collecting duct. In: Alpern RJ, Moe OW, Caplan M (eds) Seldin and Giebisch's the kidney: physiology and pathophysiology, 5th edn. Elsevier, Boston, pp 1143–1179

Giebisch G, Klose RM, Malnic G (1967) Renal tubular potassium transport. Bull Schweiz Akad Med Wiss 23:287–312

Giménez I, Isenring P, Forbush B (2002) Spatially distributed alternative splice variants of the renal Na^+-K^+-Cl^- cotransporter exhibit dramatically different affinities for the transported ions. J Biol Chem 277:8767–8770

González C, Baez-Nieto D, Valencia I et al (2012) K^+ channels: function-structural overview. Compr Physiol 2:2087–2149

Grahammer F, Herling AW, Lang HJ et al (2001) The cardiac K^+ channel KCNQ1 is essential for gastric acid secretion. Gastroenterology 120:1363–1371

Grimm PR, Foutz RM, Brenner R et al (2007) Identification and localization of BK-β subunits in the distal nephron of the mouse kidney. Am J Physiol Renal Physiol 293:F350–F359

Grotjohann I, Gitter AH, Köckerling A et al (1998) Localization of cAMP- and aldosterone-induced K^+ secretion in rat distal colon by conductance scanning. J Physiol 507:561–570

Grubb BR, Rogers TD, Boucher RC et al (2009) Ion transport across CF and normal murine olfactory and ciliated epithelium. Am J Physiol Cell Physiol 296:C1301–C1309

Gumz ML, Lynch IJ, Greenlee MM et al (2010) The renal H^+-K^+-ATPases: physiology, regulation, and structure. Am J Physiol Renal Physiol 298:F12–F21

Gutman GA, Chandy KG, Adelman JP et al (2003) International union of Pharmacology. XLI. Compendium of voltage-gated ion channels: potassium channels. Pharm Rev 55:583–586

Halm DR, Dawson DC (1983) Cation activation of the basolateral sodium-potassium pump in turtle colon. J Gen Physiol 82:315–329

Halm DR, Dawson DC (1984) Control of potassium transport by turtle colon: role of membrane potential. Am J Physiol Cell Physiol 247:C26–C32

Halm DR, Frizzell RA (1986) Active K^+ transport across rabbit distal colon: relation to Na^+ absorption and Cl^- secretion. Am J Physiol Cell Physiol 251:C252–C267

Halm DR, Frizzell RA (1991) Ion transport across the large intestine. In: Field M, Frizzell RA (eds) Intestinal absorption and secretion. Schultz SG (ed) The gastrointestinal system. Rauner BB (ed) Handbook of physiology, Sect 6, vol IV, Chap 8. American Physiological Society, Bethesda, pp 257–273

Halm DR, Halm ST (1994) Aldosterone stimulates K^+ secretion prior to onset of Na^+ absorption in guinea pig distal colon. Am J Physiol Cell Physiol 266:C552–C558

Halm DR, Halm ST (2001) Prostanoids stimulate K^+ secretion and Cl^- secretion in guinea pig distal colon via distinct pathways. Am J Physiol Gastrointest Liver Physiol 281:G984–G996

Halm DR, Rick R (1992) Secretion of K^+ and Cl^- across colonic epithelium: cellular localization using electron microprobe analysis. Am J Physiol Cell Physiol 262:C1392–C1402

Halm DR, Krasny EJ Jr, Frizzell RA (1985) Electrophysiology of flounder intestinal mucosa. II. Relation of the electrical potential profile to coupled NaCl absorption. J Gen Physiol 85: 865–883

Halm ST, Zhang J, Halm DR (2010) β-adrenergic activation of K^+ and Cl^- secretion in guinea pig distal colonic epithelium proceeds via separate cAMP signaling pathways. Am J Physiol Gastrointest Liver Physiol 299:G81–G95

Häse CC, Barquera B (2001) Role of sodium bioenergetics in *Vibrio cholerae*. Biochim Biophys Acta 1505:169–178
Häse CC, Fedorova ND, Galperin MY et al (2001) Sodium ion cycle in bacterial pathogens: evidence from cross-genome comparisons. Microbiol Mol Biol Rev 65:353–370
Hayashi M, Novak I (2013) Molecular basis of potassium channels in pancreatic duct epithelial cells. Channels (Austin) 7:432–441
Hazard LC (2001) Ion secretion by salt glands of desert iguanas (*Dipsosaurus dorsalis*). Physiol Biochem Zool 74:22–31
He Q, Halm ST, Zhang J et al (2011) Activation of the basolateral membrane Cl^- conductance essential for electrogenic K^+ secretion suppresses electrogenic Cl^- secretion. Exp Physiol 96: 305–316
Hegg CC, Irwin M, Lucero MT (2009) Calcium store-mediated signaling in sustentacular cells of the mouse olfactory epithelium. Glia 57:634–644
Heginbotham L, MacKinnon R (1993) Conduction properties of the cloned Shaker K^+ channel. Biophys J 65:2089–2096
Heitzmann D, Warth R (2007) No potassium, no acid: K^+ channels and gastric acid secretion. Physiology (Bethesda) 22:335–341
Heitzmann D, Warth R (2008) Physiology and pathophysiology of potassium channels in gastrointestinal epithelia. Physiol Rev 88:1119–1182
Heitzmann D, Grahammer F, von Hahn T et al (2004) Heteromeric KCNE2/KCNQ1 potassium channels in the luminal membrane of gastric parietal cells. J Physiol 561:547–557
Hibino H, Kurachi Y (2006) Molecular and physiological bases of the K^+ circulation in the mammalian inner ear. Physiology (Bethesda) 21:336–345
Holtzclaw JD, Grimm PR, Sansom SC (2011) Role of BK channels in hypertension and potassium secretion. Curr Opin Nephrol Hypertens 20:512–517
Hosoda Y, Karaki S, Shimoda Y et al (2002) Substance P-evoked Cl^- secretion in guinea pig distal colonic epithelia: interaction with PGE_2. Am J Physiol Gastrointest Liver Physiol 283: G347–G356
Huang F, Wong X, Jan LY (2012) International union of basic and clinical pharmacology. LXXXV: calcium-activated chloride channels. Pharmacol Rev 64:1–15
Hudson RL, Schultz SG (1984) Sodium-coupled sugar transport: effects on intracellular sodium activities and sodium-pump activity. Science 224:1237–1239
Iritani A, Sato E, Nishikawa Y (1974) Secretion rates and chemical composition of oviduct and uterine fluids in sows. J Anim Sci 39:582–588
Jakab RL, Collaco AM, Ameen NA (2013) Characterization of CFTR high expresser cells in the intestine. Am J Physiol Gastrointest Liver Physiol 305:G453–G465
Joiner WJ, Basavappa S, Vidyasagar S et al (2003) Active K^+ secretion through multiple K_{Ca}-type channels and regulation by IK_{Ca} channels in rat proximal colon. Am J Physiol Gastrointest Liver Physiol 285:G185–G196
Joris L, Dab I, Quinton PM (1993) Elemental composition of human airway surface fluid in healthy and diseased airways. Am Rev Respir Dis 148:1633–1637
Kim S, Ma L, Jensen KL et al (2012) Paradoxical contribution of SK3 and GIRK channels to the activation of mouse vomeronasal organ. Nat Neurosci 15:1236–1244
Kinne R, Kinne-Saffran E, Schütz H et al (1986) Ammonium transport in medullary thick ascending limb of rabbit kidney: involvement of the Na+, K+, Cl – -cotransporter. J Membr Biol 94: 279–284
Knauth LP (2005) Temperature and salinity history of the Precambrian ocean: implications for the course of microbial evolution. Palaeogeog Palaeoclim Palaeoecol 219:53–69
Knowles MR, Robinson JM, Wood RE et al (1997) Ion composition of airway surface liquid of patients with cystic fibrosis as compared with normal and disease-control subjects. J Clin Invest 100:2588–2595
Köckerling A, Fromm M (1993) Origin of cAMP-dependent Cl^- secretion from both crypts and surface epithelia of rat intestine. Am J Physiol Cell Physiol 264:C1294–C1301

Köckerling A, Sorgenfrei D, Fromm M (1993) Electrogenic Na^+ absorption of rat distal colon is confined to surface epithelium: a voltage-scanning study. Am J Physiol Cell Physiol 264: C1285–C1293
Koefoed-Johnsen V, Ussing HH (1958) The nature of the frog skin potential. Acta Physiol Scand 42:298–308
Koeppen BM, Giebisch GH (1985) Mineralocorticoid regulation of sodium and potassium transport by the cortical collecting duct. In: Graves S (ed) Regulation and development of membrane transport processes, vol 39, Society of general physiologists series. Wiley, New York, pp 89–104
Kopic S, Murek M, Geibel JP (2010) Revisiting the parietal cell. Am J Physiol Cell Physiol 298: C1–C10
Kriz W, Kaissling B (2013) Structural organization of the mammalian kidney. In: Alpern RJ, Moe OW, Caplan M (eds) Seldin and Giebisch's the kidney: physiology and pathophysiology, 5th edn. Elsevier, Boston, pp 595–691
Lambrecht NW, Yakubov I, Scott D et al (2005) Identification of the K^+ efflux channel coupled to the gastric H^+-K^+-ATPase during acid secretion. Physiol Genomics 21:81–91
Larsen EH (2011) Reconciling the Krogh and Ussing interpretations of epithelial chloride transport - presenting a novel hypothesis for the physiological significance of the passive cellular chloride uptake. Acta Physiol (Oxf) 202:435–464
Larsen EH, Ramløv H (2013) Role of cutaneous surface fluid in frog osmoregulation. Comp Biochem Physiol A Mol Integr Physiol 165:365–370
Lee SY, Maniak PJ, Ingbar DH et al (2003) Adult alveolar epithelial cells express multiple subtypes of voltage-gated K^+ channels that are located in apical membrane. Am J Physiol Cell Physiol 284:C1614–C1624
Lee FN, Oh G, McDonough AA et al (2007) Evidence for gut factor in K^+ homeostasis. Am J Physiol Renal Physiol 293:F541–F547
Lewis SA, Clausen C, Wills NK (1996) Impedance analysis of epithelia. In: Wills NK, Reuss L, Lewis SA (eds) Epithelial transport: a guide to methods and experimental analysis. Chapman & Hall, New York, pp 118–145
Li Y, Halm DR (2002) Secretory modulation of basolateral membrane inwardly rectified K^+ channel in guinea pig distal colonic crypts. Am J Physiol Cell Physiol 282:C719–C735
Li Y, Halm ST, Halm DR (2003) Secretory activation of basolateral membrane Cl^- channels in guinea pig distal colonic crypts. Am J Physiol Cell Physiol 284:C918–C933
Liao T, Wang L, Halm ST et al (2005) The K^+ channel K_VLQT (Kcnq1) located in the basolateral membrane of distal colonic epithelium is not essential for activating Cl^- secretion. Am J Physiol Cell Physiol 289:C564–C575
Liu W, Schreck C, Coleman RA et al (2011) Role of NKCC in BK channel-mediated net K^+ secretion in the CCD. Am J Physiol Renal Physiol 301:F1088–F1097
Lucero MT (2013) Peripheral modulation of smell: fact or fiction? Semin Cell Dev Biol 24:58–70
Malnic G, Giebisch G, Muto S et al (2013) Regulation of K^+ excretion. In: Alpern RJ, Moe OW, Caplan M (eds) Seldin and Giebisch's the kidney: physiology and pathophysiology, 5th edn. Elsevier, Boston, pp 1659–1715
Mangos JA, McSherry NR, Nousia-Arvanitakis S et al (1973) Secretion and transductal fluxes of ions in exocrine glands of the mouse. Am J Physiol 225:18–24
Manzanares D, Gonzalez C, Ivonnet P et al (2011) Functional apical large conductance, Ca^{2+}-activated, and voltage-dependent K^+ channels are required for maintenance of airway surface liquid volume. J Biol Chem 286:19830–19839
Marcus DC, Wangemann P (2009) Cochlear and vestibular function and dysfunction. In: Alvarez-Leefmans FJ, Delpire E (eds) Physiology and pathology of chloride transporters and channels in the nervous system – from molecules to diseases. Elsevier, Boston, pp 425–437
McCabe RD, Smith PL, Sullivan LP (1984) Ion transport by rabbit descending colon: mechanisms of transepithelial potassium transport. Am J Physiol Gastrointest Liver Physiol 246: G594–G602

Meneton P, Schultheis PJ, Greeb J et al (1998) Increased sensitivity to K+ deprivation in colonic H+, K+-ATPase-deficient mice. J Clin Invest 101:536–542

Merigo F, Mucignat-Caretta C, Cristofoletti M et al (2011) Epithelial membrane transporters expression in the developing to adult mouse vomeronasal organ and olfactory mucosa. Dev Neurobiol 71:854–869

Mignen O, Egee S, Liberge M et al (2000) Basolateral outward rectifier chloride channel in isolated crypts of mouse colon. Am J Physiol Gastrointest Liver Physiol 279:G277–G287

Moser SL, Harron SA, Crack J et al (2008) Multiple KCNQ potassium channel subtypes mediate basal anion secretion from the human airway epithelial cell line Calu-3. J Membr Biol 221:153–163

Musch MW, Orellana SA, Kimberg LS et al (1982) Na^+-K^+-Cl^- co-transport in the intestine of a marine teleost. Nature 300:351–353

Musch MW, McConnell FM, Goldstein L et al (1987) Tyrosine transport in winter flounder intestine: interaction with Na^+-K^+-$2Cl^-$ cotransport. Am J Physiol Regul Integr Comp Physiol 253:R264–R269

Nakamoto T, Romanenko VG, Takahashi A et al (2008) Apical maxi-K ($K_{Ca}1.1$) channels mediate K^+ secretion by the mouse submandibular exocrine gland. Am J Physiol Cell Physiol 294: C810–C819

Nakamura S, Amlal H, Galla JH et al (1999) NH_4^+ secretion in inner medullary collecting duct in potassium deprivation: role of colonic H^+-K^+-ATPase. Kidney Int 56:2160–2167

Namkung W, Song Y, Mills AD et al (2009) In situ measurement of airway surface liquid [K^+] using a ratioable K^+-sensitive fluorescent dye. J Biol Chem 284:15916–15926

Nanda Kumar NS, Singh SK, Rajendran VM (2010) Mucosal potassium efflux mediated via Kcnn4 channels provides the driving force for electrogenic anion secretion in colon. Am J Physiol Gastrointest Liver Physiol 299:G707–G714

Nguyen N, Kozer-Gorevich N, Gliddon BL et al (2013) Independent trafficking of the KCNQ1 K^+ channel and H^+-K^+-ATPase in gastric parietal cells from mice. Am J Physiol Gastrointest Liver Physiol 304:G157–G166

Nielson DW (1986) Electrolyte composition of pulmonary alveolar subphase in anesthetized rabbits. J Appl Physiol 60:972–979

O'Grady SM, Lee SY (2003) Chloride and potassium channel function in alveolar epithelial cells. Am J Physiol Lung Cell Mol Physiol 284:L689–L700

O'Grady SM, Lee SY (2005) Molecular diversity and function of voltage-gated (Kv) potassium channels in epithelial cells. Int J Biochem Cell Biol 37:1578–1594

Oh KS, Oh YT, Kim SW et al (2011) Gut sensing of dietary K^+ intake increases renal K^+ excretion. Am J Physiol Regul Integr Comp Physiol 301:R421–R429

O'Neil RG, Helman SI (1977) Transport characteristics of renal collecting tubules: influences of DOCA and diet. Am J Physiol Renal Physiol 233:F544–F558

Palk L, Sneyd J, Shuttleworth TJ et al (2010) A dynamic model of saliva secretion. J Theor Biol 266:625–640

Palmer ML, Schiller KR, O'Grady SM (2008) Apical SK potassium channels and Ca^{2+}-dependent anion secretion in endometrial epithelial cells. J Physiol 586:717–726

Palmer ML, Peitzman ER, Maniak PJ et al (2011) $K_{Ca}3.1$ channels facilitate K^+ secretion or Na^+ absorption depending on apical or basolateral P2Y receptor stimulation. J Physiol 589: 3483–3494

Pinti DL (2005) The origin and evolution of the oceans. In: Gargaud M, Barbier B, Martin H, Reisse J (eds) Lectures in astrobiology, vol 1, Advances in astrobiology and biogeophysics. Springer, Berlin, pp 83–112

Pongs O, Schwarz JR (2010) Ancillary subunits associated with voltage-dependent K^+ channels. Physiol Rev 90:755–796

Preidis GA, Versalovic J (2009) Targeting the human microbiome with antibiotics, probiotics, and prebiotics: gastroenterology enters the metagenomics era. Gastroenterology 136:2015–2031

Quatrale RP, Laden K (1968) Solute and water secretion by the eccrine sweat glands of the rat. J Invest Dermatol 51:502–504

Quinton PM, Elder HY, Jenkinson DM et al (1999) Structure and function of human sweat glands. In: Laden K (ed) Antiperspirants and deodorants, 2nd edn. Marcel Dekker, New York, pp 17–57

Rakowski RF, Paxson CL (1988) Voltage dependence of Na^+/K^+ pump current in *Xenopus* oocytes. J Membr Biol 106:173–182

Rakowski RF, Gadsby DC, De Weer P (1989) Stoichiometry and voltage dependence of the sodium pump in voltage-clamped, internally dialyzed squid giant axon. J Gen Physiol 93: 903–941

Rechkemmer G, Halm DR (1989) Aldosterone stimulates K^+ secretion across mammalian colon independent of Na^+ absorption. Proc Natl Acad Sci USA 86:397–401

Rechkemmer G, von Engelhardt W (1993) Absorption and secretion of electrolytes and short-chain fatty acids in the guinea pig large intestine. In: Clauss W (ed) Ion transport in vertebrate colon, vol 16, Advances in comparative and environmental physiology. Springer, New York, pp 139–167

Rechkemmer G, Frizzell RA, Halm DR (1996) Active K^+ transport across guinea pig distal colon: action of secretagogues. J Physiol 493:485–502

Reuter D, Zierold K, Schröder WH et al (1998) A depolarizing chloride current contributes to chemoelectrical transduction in olfactory sensory neurons in situ. J Neurosci 18:6623–6630

Sabolić I, Herak-Kramberger CM, Breton S et al (1999) Na^+/K^+-ATPase in intercalated cells along the rat nephron revealed by antigen retrieval. J Am Soc Nephrol 10:913–922

Sackin H, Palmer LG (2013) Electrophysiological analysis of transepithelial transport. In: Alpern RJ, Moe OW, Caplan M (eds) Seldin and Giebisch's the kidney: physiology and pathophysiology, 5th edn. Elsevier, Boston, pp 177–216

Sandle GI, Hunter M (2010) Apical potassium (BK) channels and enhanced potassium secretion in human colon. Q J Med 103:85–89

Sansom SC, O'Neil RG (1985) Mineralocorticoid regulation of apical cell membrane Na^+ and K^+ transport of the cortical collecting duct. Am J Physiol Renal Physiol 248:F858–F868

Sansom SC, O'Neil RG (1986) Effects of mineralocorticoids on transport properties of cortical collecting duct basolateral membrane. Am J Physiol Renal Physiol 251:F743–F757

Satlin LM, Carattino MD, Liu W et al (2006) Regulation of cation transport in the distal nephron by mechanical forces. Am J Physiol Renal Physiol 291:F923–F931

Sato K (1980) Electrochemical driving forces for K^+ secretion by rat paw eccrine sweat gland. Am J Physiol Cell Physiol 239:C90–C97

Sato K (2001) The mechanism of eccrine sweat secretion. In: Gisolfi CV, Lamb DR, Nadel ER (eds) Exercise, heat, and thermoregulation, vol 6, Perspectives in exercise science and sports medicine. Cooper Publishing Group, Traverse City MI, pp 85–117

Sato F, Sato K (1978) Secretion of a potassium-rich fluid by the secretory coil of the rat paw eccrine sweat gland. J Physiol 274:37–50

Sato K, Sato F (1990) Na^+, K^+, H^+, Cl^-, and Ca^{2+} concentrations in cystic fibrosis eccrine sweat in vivo and in vitro. J Lab Clin Med 115:504–511

Sato K, Cavallin S, Sato KT et al (1994) Secretion of ions and pharmacological responsiveness in the mouse paw sweat gland. Clin Sci (Lond) 86:133–139

Sausbier M, Matos JE, Sausbier U et al (2006) Distal colonic K^+ secretion occurs via BK channels. J Am Soc Nephrol 17:1275–1282

Schultz SG, Thompson SM, Suzuki Y (1981) Equivalent electrical circuit models and the study of Na transport across epithelia: nonsteady-state current–voltage relations. Fed Proc 40: 2443–2449

Schwartz GJ, Burg MB (1978) Mineralocorticoid effects on cation transport by cortical collecting tubules in vitro. Am J Physiol Renal Physiol 235:F576–F585

Shao J, Gumz ML, Cain BD et al (2010) Pharmacological profiles of the murine gastric and colonic H, K-ATPases. Biochim Biophys Acta 1800:906–911

Shennan DB, Peaker M (2000) Transport of milk constituents by the mammary gland. Physiol Rev 80:925–951
Shuttleworth TJ, Thompson JL, Dantzler WH (1987) Potassium secretion by nasal salt glands of desert lizard *Sauromalus obesus*. Am J Physiol Regul Integr Comp Physiol 253:R83–R90
Silva P, Stoff J, Field M et al (1977) Mechanism of active chloride secretion by shark rectal gland: role of Na^+-K^+-ATPase in chloride transport. Am J Physiol Renal Physiol 233:F298–F306
Skou JC, Esmann M (1992) The Na+, K+-ATPase. J Bioenerg Biomembr 24:249–261
Song P, Groos S, Riederer B et al (2011) Kir4.1 channel expression is essential for parietal cell control of acid secretion. J Biol Chem 286:14120–14128
Sorensen MV, Grossmann S, Roesinger M et al (2013) Rapid dephosphorylation of the renal sodium chloride cotransporter in response to oral potassium intake in mice. Kidney Int 83:811–824
Sørensen JB, Nielsen MS, Gudme CN et al (2001) Maxi K^+ channels co-localised with CFTR in the apical membrane of an exocrine gland acinus: possible involvement in secretion. Pflugers Arch 442:1–11
Sørensen MV, Sausbier M, Ruth P et al (2010) Adrenaline-induced colonic K^+ secretion is mediated by $K_{Ca}1.1$ (BK) channels. J Physiol 588:1763 1777
Sørensen MV, Strandsby AB, Larsen CK et al (2011) The secretory $K_{Ca}1.1$ channel localises to crypts of distal mouse colon: functional and molecular evidence. Pflugers Arch 462:745–752
Stauber T, Weinert S, Jentsch TJ (2012) Cell biology and physiology of CLC chloride channels and transporters. Compr Physiol 2:1701–1744
Stokes JB, Lee I, Williams A (1981) Potassium secretion by cortical collecting tubule: relation to sodium absorption, luminal sodium concentration, and transepithelial voltage. Am J Physiol Renal Physiol 241:F395–F402
Stokes JB, Lee I, D'Amico M (1984) Sodium chloride absorption by the urinary bladder of the winter flounder. A thiazide-sensitive, electrically neutral transport system. J Clin Invest 74: 7–16
Sullivan SK, Smith PL (1986) Active potassium secretion by rabbit proximal colon. Am J Physiol Gastrointest Liver Physiol 250:G475–G483
Suzuki Y, Kaneko K (1987) Acid secretion in isolated guinea pig colon. Am J Physiol Gastrointest Liver Physiol 253:G155–G164
Swarts HG, Koenderink JB, Willems PH et al (2005) The non-gastric H, K-ATPase is oligomycin-sensitive and can function as an H+, NH4+-ATPase. J Biol Chem 280:33115–33122
Sweiry JH, Binder HJ (1989) Characterization of aldosterone-induced potassium secretion in rat distal colon. J Clin Invest 83:844–851
Thévenod F (2002) Ion channels in secretory granules of the pancreas and their role in exocytosis and release of secretory proteins. Am J Physiol Cell Physiol 283:C651–C672
Thompson SM (1986) Relations between chord and slope conductances and equivalent electromotive forces. Am J Physiol Cell Physiol 250:C333–C339
Thompson KA, Kleinzeller A (1985) Glucose transport in intestinal epithelia of winter flounder. Am J Physiol Regul Integr Comp Physiol 248:R573–R577
Tirindelli R, Dibattista M, Pifferi S et al (2009) From pheromones to behavior. Physiol Rev 89:921–956
Traynor TR, O'Grady SM (1996) Regulation of colonic ion transport by GRP. I. GRP stimulates transepithelial K^+ and Na^+ secretion. Am J Physiol Cell Physiol 270:C848–C858
Vallon V, Osswald H, Blantz RC et al (1997) Potential role of luminal potassium in tubuloglomerular feedback. J Am Soc Nephrol 8:1831–1837
van't Hof W, Veerman EC, Nieuw Amerongen AV et al (2014) Antimicrobial defense systems in saliva. Monogr Oral Sci 24:40–51
Viejo-Díaz M, Andrés MT, Fierro JF (2004) Modulation of in vitro fungicidal activity of human lactoferrin against *Candida albicans* by extracellular cation concentration and target cell metabolic activity. Antimicrob Agents Chemother 48:1242–1248

Vogalis F, Hegg CC, Lucero MT (2005) Electrical coupling in sustentacular cells of the mouse olfactory epithelium. J Neurophysiol 94:1001–1012
Vucic E, Alfadda T, MacGregor GG et al (2014) $K_{ir}1.1$ (ROMK) and $K_V7.1$ (KCNQ1/K_VLQT1) are essential for normal gastric acid secretion: importance of functional $K_{ir}1.1$. Pflugers Arch 46(7):1457–1468
Wagner CA, Finberg KE, Breton S et al (2004) Renal vacuolar H^+-ATPase. Physiol Rev 84: 1263–1314
Wales RG, Wallace JC, White IG (1966) Composition of bull epididymal and testicular fluid. J Reprod Fertil 12:139–144
Wall SM, Koger LM (1994) NH_4^+ transport mediated by Na^+-K^+-ATPase in rat inner medullary collecting duct. Am J Physiol Renal Physiol 267:F660–F670
Wang YB, Germaine GR (1993) Effects of pH, potassium, magnesium, and bacterial growth phase on lysozyme inhibition of glucose fermentation by *Streptococcus mutans 10449*. J Dent Res 72:907–911
Wang WH, Huang CL (2013) The molecular biology of renal K^+ channels. In: Alpern RJ, Moe OW, Caplan M (eds) Seldin and Giebisch's the kidney: physiology and pathophysiology, 5th edn. Elsevier, Boston, pp 1601–1627
Wangemann P (2006) Supporting sensory transduction: cochlear fluid homeostasis and the endocochlear potential. J Physiol 576:11–21
Watlington CO, Huf EG (1971) β-adrenergic stimulation of frog skin mucous glands: non-specific inhibition by adrenergic blocking agents. Comp Gen Pharmacol 2:295–305
Weiner ID, Verlander JW (2013) Renal ammonia metabolism and transport. Compr Physiol 3: 201–220
Weiner ID, Verlander JW (2014) Ammonia transport in the kidney by *Rhesus* glycoproteins. Am J Physiol Renal Physiol 306:F1107–F1120
Weinstein AM (2010) A mathematical model of rat ascending Henle limb. I. Cotransporter function. Am J Physiol Renal Physiol 298:F512–F524
Welling PA (2013) Regulation of renal potassium secretion: molecular mechanisms. Semin Nephrol 33:215–228
Welsh MJ, Karp P, Ruppert TR (1983) Evidence for basolateral membrane potassium conductance in canine tracheal epithelium. Am J Physiol Cell Physiol 244:C377–C384
Whittamore JM, Cooper CA, Wilson RW (2010) HCO_3^- secretion and $CaCO_3$ precipitation play major roles in intestinal water absorption in marine teleost fish in vivo. Am J Physiol Regul Integr Comp Physiol 298:R877–R886
Williams JA, Dawson DC (2006) Cell structure and function. In: Mulholland MW, Lillemoe KD, Doherty GM, Maier RV, Upchurch GR Jr (eds) Greenfield's surgery: scientific principles and practice, 4th edn. Lippincott Williams & Wilkins, Philadelphia, pp 2–42
Woodward OM, Guggino WB (2013) Anion channels. In: Alpern RJ, Moe OW, Caplan M (eds) Seldin and Giebisch's the kidney: physiology and pathophysiology, 5th edn. Elsevier, Boston, pp 1019–1045
Worrell RT, Merk L, Matthews JB (2008) Ammonium transport in the colonic crypt cell line, T84: role for Rhesus glycoproteins and NKCC1. Am J Physiol Gastrointest Liver Physiol 294: G429–G440
Wright EM, Loo DD, Hirayama BA (2011) Biology of human sodium glucose transporters. Physiol Rev 91:733–794
Wrong OM, Edmonds CJ, Chadwick VS (eds) (1981) The large intestine: it's role in mammalian nutrition and homeostasis. MTP Press, Lancaster, UK, p 217
Yang L, Edvinsson J, Sackin H et al (2012) Ion selectivity and current saturation in inward-rectifier K^+ channels. J Gen Physiol 139:145–157
Youn JH (2013) Gut sensing of potassium intake and its role in potassium homeostasis. Semin Nephrol 33:248–256
Yu FH, Catterall WA (2004) The VGL-Chanome: a protein superfamily specialized for electrical signaling and ionic homeostasis. Sci STKE 253:re15

Zdebik AA, Cuffe JE, Bertog M et al (2004) Additional disruption of the CLC-2 Cl^- channel does not exacerbate the cystic fibrosis phenotype of cystic fibrosis transmembrane conductance regulator mouse models. J Biol Chem 279:22276–22283

Zhang J, Halm ST, Halm DR (2009a) Adrenergic activation of electrogenic K^+ secretion in guinea pig distal colonic epithelium: involvement of β1- and β2-adrenergic receptors. Am J Physiol Gastrointest Liver Physiol 297:G269–G277

Zhang J, Halm ST, Halm DR (2009b) Adrenergic activation of electrogenic K^+ secretion in guinea pig distal colonic epithelium: desensitization via the Y2-neuropeptide receptor. Am J Physiol Gastrointest Liver Physiol 297:G278–G291

Zhang J, Halm ST, Halm DR (2012) Role of the BK channel ($K_{Ca}1.1$) during activation of electrogenic K^+ secretion in guinea pig distal colon. Am J Physiol Gastrointest Liver Physiol 303: G1322–G1334

Chapter 4
Volume Regulation in Epithelia

Erik Hviid Larsen and Else Kay Hoffmann

Abstract We review studies on regulatory volume decrease (RVD) and regulatory volume increase (RVI) of major ion and water transporting vertebrate epithelia. The rate of RVD and RVI is faster in cells of high osmotic permeability like amphibian gallbladder and mammalian proximal tubule as compared to amphibian skin and mammalian cortical collecting tubule of low and intermediate osmotic permeability. Crosstalk between entrance and exit mechanisms interferes with volume regulation both at aniso-osmotic and iso-osmotic volume perturbations. It has been proposed that cell volume regulation is an intrinsic function of iso-osmotic fluid transport that depends on Na^+ recirculation. The causative relationship is discussed for a fluid-absorbing and a fluid-secreting epithelium of which the Na^+ recirculation mechanisms have been identified.

A large number of transporters and ion channels involved in cell volume regulation are cloned. The volume-regulated anion channel (VRAC) exhibiting specific electrophysiological characteristics seems exclusive to serve cell volume regulation. This is contrary to K^+ channels as well as cotransporters and exchange mechanisms that may serve both transepithelial transport and cell volume regulation. In the same cell, these functions may be maintained by different ion pathways that are separately regulated. RVD is often preceded by increase in cytosolic free Ca^{2+}, probably via influx through TRP channels, but Ca^{2+} release from intracellular stores has also been observed. Cell volume regulation is associated with specific ATP release mechanisms and involves tyrosine kinases, mitogen-activated protein kinases, WNKs and Ste20-related kinases that are modulated by osmotic stress and cell volume perturbations.

Keywords Regulatory volume decrease (RVD) • Regulatory volume increase (RVI) • Aniso-osmotic and iso-osmotic volume regulation • Crosstalk • Absorbing epithelia • Exocrine glands • Heterocellular epithelia • Osmotic

---

E.H. Larsen
Molecular Integrative Physiology, Department of Biology, University of Copenhagen, Universitetsparken 13, 2100 Copenhagen Ø, Denmark
e-mail: ehlarsen@bio.ku.dk

E.K. Hoffmann (✉)
Cell and Developmental Biology, Department of Biology, University of Copenhagen, Universitetsparken 13, 2100 Copenhagen Ø, Denmark
e-mail: ekhoffmann@bio.ku.dk

K.L. Hamilton, D.C. Devor (eds.), *Ion Channels and Transporters of Epithelia in Health and Disease*, Physiology in Health and Disease,
DOI 10.1007/978-1-4939-3366-2_4

permeability • Volume regulation and iso-osmotic transport • VRAC • K^+ channel subfamilies • NKCC • NHE • TRP channels • Osmo-signalling • Osmosensing

4.1 Introduction

Extracellular volume and electrolyte homeostasis depend on water flows and solute fluxes through epithelia. High resistance or 'tight' epithelia, which have developed paracellular junctions of low ion permeability, have the capacity to maintain large transepithelial gradients of solute concentrations and osmolality. The transported fluid may be hypo- or hyperosmotic to the body fluids. Both extracellular water volume and electrolyte concentrations are regulated by these epithelia. Distal tubules of the vertebrate kidney, amphibian epidermis and urinary bladder, vertebrate colon and gills of teleost fish are examples of tight epithelia. Low resistance or 'leaky' epithelia of significant paracellular ion conductance have capacity for transporting large volumes of iso-osmotic fluid between body compartments at transepithelial osmotic equilibrium. Kidney proximal tubules, small intestine, gallbladder and the acinus epithelium of exocrine glands belong to this class. This chapter is concerned with the question: How do epithelia take care of their own volume and ion concentrations while serving whole-body fluid homeostasis? The pertinent issues concern (1) osmotic water permeability of the epithelial cell membranes, (2) cell volume response to osmotic challenges, (3) cell volume as a signal of iso-osmotic transport, (4) molecular identity of channels and transporters, (5) nature of the sensing mechanisms of cell volume and/or cytoplasmic osmolality, and (6) design of coordinating signalling pathways.

4.2 Concepts in Cell Volume and Electrolyte Homeostasis

4.2.1 *Application of the van't Hoff Law*

Animal body cells are in osmotic equilibrium with the extracellular fluid. A change in solute concentration of the extracellular fluid, ΔC, changes its osmotic pressure by $\Delta\pi$ according to the van't Hoff law:

$$\Delta\pi = \phi RT\Delta C \tag{4.1}$$

where ϕ is the osmotic coefficient, which is 0.93 for a 150 mM NaCl solution, and R and T have their usual meanings. With few exceptions, the water permeability of the plasma membrane is orders of magnitudes larger than the permeabilities of common solutes implying that their reflection coefficients would be near unity. Thus, according to Eq. (4.1), cell volume $V^{(c)}$ would respond to the above change in extracellular concentration such that

$$\Delta V^{(c)}\Delta\pi = \phi RT\sum m^{(c)} \tag{4.2}$$

where $\sum m^{(c)}$ is the total amount of intracellular osmolytes. If a fraction of $V^{(c)}$ is occupied by an osmotically inactive non-solvent volume v', the osmotically active solution of the cell occupies a volume that is equal to $V^{(c)} - v'$. Let $V_0^{(c)}$ be the volume at osmotic equilibrium of the cell in an iso-osmotic solution of osmotic pressure $\pi_0^{(c)}$ and $V^{(c)}$ be the volume at equilibrium in a hypo- or hyperosmotic solution with the osmotic pressure $\pi^{(c)}$; then,

$$\pi_0^{(c)}\left(V_0^{(c)} - v'\right) = \pi^{(c)}\left(V^{(c)} - v'\right) \tag{4.3}$$

Equation (4.3) can be rearranged to give

$$V^{(c)}/V_0^{(c)} = \left[\left(V_0^{(c)} - v'\right)/V_0^{(c)}\right]\left(\pi_0^{(c)}/\pi^{(c)} - 1\right) + 1 \tag{4.4}$$

The above relationship between $V^{(c)}/V_0^{(c)}$ and $\left(\pi_0^{(c)}/\pi^{(c)} - 1\right)$ depicts a straight line of slope:

$$\alpha = \left(V_0^{(c)} - v'\right)/V_0^{(c)} \tag{4.5}$$

Equation (4.4) applies if the cell does not eliminate or accumulate osmolytes during transition from one steady state to the other, which provides a simple way of calculating v' by Eq. (4.5) after having obtained α by curve fitting of Eq. (4.4) to experimental data. Many types of animal cells regulate their volume by eliminating osmolytes when swelled or accumulating osmolytes when shrunken. If the rate of volume regulation is slow relative to the speed of solution shifts and volume recording, the above Eqs. (4.4)–(4.5) would still be useful when applied to 'instantaneous' changes. Subsequent change in $V^{(c)}$ as a function of time would then be governed by cell volume regulatory processes, which sets the stage for their experimental investigation. Epithelial studies have reported $v'/V_0^{(c)}$ ratios that range from 0.13 in rabbit kidney cortical collecting duct (Strange and Spring 1987b) to 0.41 in isolated epithelial cells of mouse choroid plexus (Hughes et al. 2010).

4.2.2 *Cell Water Homeostasis Depends on Metabolic Energy*

A cell of non-permeant negatively charged intracellular macroions and diffusible small anions kept electroneutral by K^+, which is in osmotic equilibrium with a surrounding NaCl solution, constitutes a Gibbs-Donnan system that will tend to swell. This is because Na^+ and Cl^- by electrodiffusion enter the cell together with an osmotically equivalent amount of water, a process that cannot proceed to equilibrium before the cell bursts (see Sten-Knudsen 2002 for comprehensive mathematical treatment). This fatal situation is prevented to develop by the working of the Na^+/K^+ pump, which powers the back flux of Na^+ (and Cl^-) to the extracellular fluid fuelled by ATP hydrolysis. In the classical 'pump-leak' description of the cell's osmotic equilibrium with the extracellular fluid, Cl^- was assumed to be in thermodynamic equilibrium across the plasma membrane (Leaf 1959; Tosteson and Hoffman 1960; Ussing 1960). This applies to a number of body cells and as discussed below to some epithelial cells. In most cells, however, Cl^- is kept above thermodynamic equilibrium by inwardly directed cotransport with Na^+ that enters the cell along this ion's electrochemical potential gradient. Independent of the energetics of the Cl^- distribution, the intracellular Cl^- pool and the pools of other intracellular small osmolytes are most important determinants of the volume of intracellular water. In cells with Cl^- being above thermodynamic equilibrium, the water volume associated with the Cl^- pool is under dynamic control by a balance between passive flows of Cl^- and K^+ through specific channels and oppositely directed cotransport of these two ions with Na^+ (Fig. 4.1A). Thus, an osmotically swelled cell may restore its volume by leaking K^+ and Cl^- to the extracellular fluid either through specific channels or by cotransport, termed *regulatory volume decrease* (RVD), whereas the volume of shrunken cells would be restored by uptake of Cl^- and K^+ coupled to cotransport with Na^+, which is termed *regulatory volume increase* (RVI). As indicated in Fig. 4.1A, a set of Na^+/H^+- and Cl^-/HCO_3^- exchange mechanisms may also serve this latter mentioned function. In addition, the intracellular pools of other small osmolytes like amino acids and sugars contribute to cell volume regulation by similar dynamic balance between solute accumulation (uptake or synthesis) and solute elimination (release or degradation). This framework was developed for a small number of cell types, duck erythrocytes (Kregenow 1971), Ehrlich ascites tumour cells (Hoffmann 1978; Hoffmann et al. 1979; Hoffmann et al. 1983), amphiuma red blood cells (Cala 1980), human lymphocytes (Grinstein et al. 1982), lymphoid leukaemia cells (Gelfand et al. 1984) and Chinese hamster ovary cells (Sarkadi et al. 1984), but is now accepted as general scheme of cell volume regulation (see reviews by Choe and Strange 2009, Hoffmann et al. 2009, Hoffmann and Pedersen 2011 and Lang 2013).

4.2.3 Iso-osmotic and Aniso-osmotic Cell Volume Regulation

The volume regulatory machinery may be activated by changes in intracellular amount of osmolytes, termed *iso-osmotic cell volume regulation*, or by changes in extracellular osmolality, i.e., *aniso-osmotic cell volume regulation*. Iso-osmotic cell volume changes are brought about by upsetting the balance between entry and exit of osmolytes, by synthesis/metabolism of organic osmolytes and by turnover of intracellular macromolecules that leads to change in amount of constituent monomers. In turn, this drives water into or out of the cell until osmotic equilibrium is

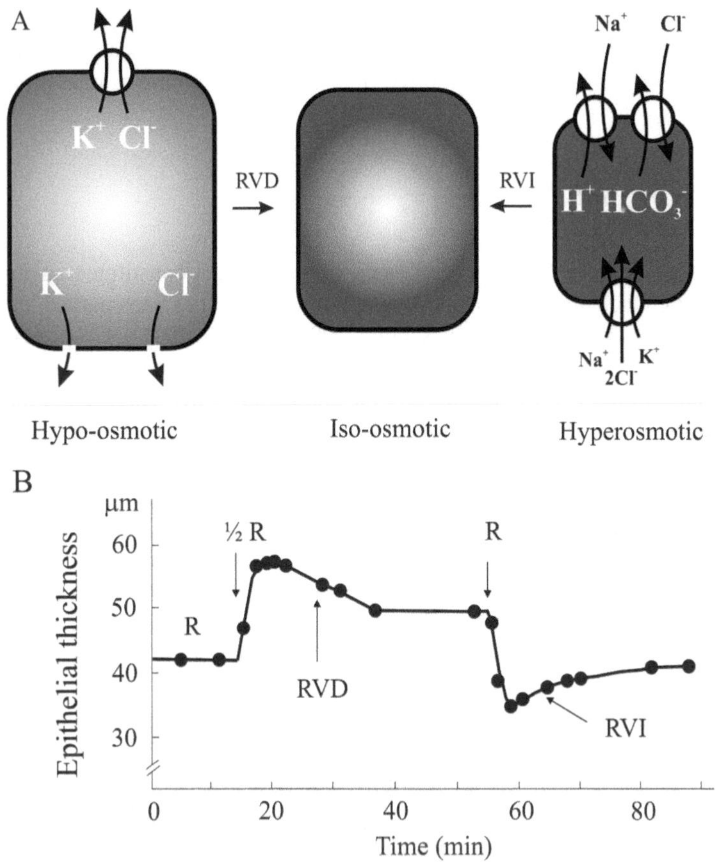

Fig. 4.1 Ion mechanisms of cell volume regulation. (**A**) The osmotically swollen cell may eliminate K^+ and Cl^- either through ion-selective channels or by coupled transport. By eliminating an osmotically equivalent amount of water, cell volume returns towards the iso-osmotic volume (RVD). The osmotically shrunken cell may regain volume by water uptake associated with uptake of K^+ and Cl^- that are coupled to and driven by uptake of Na^+ or mediated by a set of Na^+/H^+ and Cl^-/HCO_3^- exchange mechanisms (RVI). The transporters' localization to epithelial membrane domain depends on cell type and is here randomly indicated for convenience. (**B**) Several years before the above concepts were introduced (see text for references), cell volume regulation was disclosed in the first real-time study of volume responses of an epithelium (frog skin) to external osmotic perturbations. The preparation was exposed to a 20-times diluted Ringer's solution on the outside (see text for details), while the inside bath was either Ringer's solution (R) or Ringer's solution diluted 50 % by distilled water (indicated by ½ R) [The figure is redrawn from MacRobbie and Ussing (1961)]

re-established. This activates volume regulation for bringing cell water volume back to its normal iso-osmotic state. In mammals, the kidneys maintain appropriate extracellular ion and water homeostasis. Therefore, body cells are more often challenged by iso-osmotic than by aniso-osmotic cell volume perturbations (see Chamberlin and Strange 1989). Perturbation of cell volume by change in extracellular osmolality is a common challenge to transporting epithelia. This applies to epithelia that generate or maintain transcellular osmotic gradients and probably also to low- resistance fluid-transporting epithelia that seem to depend on cell volume regulation for securing the transported fluid being iso-osmotic to interstitial fluid.

4.2.4 'Crosstalk'

Coordinated activity of solute entry and exit pathways constitutes an important homeostatic mechanism of transporting epithelia, which has been named 'crosstalk' (Schultz 1981; Diamond 1982; Schultz 1992; Harvey 1995). Examples are (*i*) activity of the Na^+/K^+ pump versus that of apical Na^+ channels and basolateral K^+ channels of tight epithelia (MacRobbie and Ussing 1961; Davis and Finn 1982), (*ii*) coordinated activity of apical Na^+ channels and basolateral K^+ channels in frog skin (Harvey et al. 1988), (*iii*) coupling of apical rheogenic alanine/2 Na^+ uptake and activity of K^+ channels of intestinal mucosa (Gunter-Smith et al. 1982), (*iv*) Na^+ absorption and H^+ secretion by α-type mitochondrion-rich (MR) cells (Harvey 1992), (*v*) apical Cl^- channel activation and multiple basolateral Na^+ transport pathways of secretory acinar cells (Robertson and Foskett 1994) and (*vii*) coupled activity of apical Cl^- channels of mitochondria cells and apical Na^+ channels of principal cells of amphibian skin (Larsen 2011). Different types of signals govern these regulations: intracellular pH (Harvey et al. 1988; Harvey and Ehrenfeld 1988), intracellular free $[Ca^{2+}]$ (Grinstein and Erlij 1978; Taylor and Windhager 1979; Chase and Alawqati 1983; Tinel et al. 2000), ATP (Tsuchiya et al. 1992; Welling 1995; Urbach et al. 1996), membrane-bound G-proteins (Cantiello et al. 1989), sub-membrane actin filaments (Wang et al. 2005), concomitant fall in intracellular Cl^- concentration and rise in intracellular Ca^{2+} concentration (Robertson and Foskett 1994) and membrane potential (Larsen and Harvey 1994; Larsen 2011). Volume regulation and crosstalk operate together in securing homeostasis in cell volume, electrolyte concentrations and driving forces of transmembrane solute fluxes, which otherwise would be submitted to large variations depending on the functional activity of the epithelium. As discussed in this chapter, this may result in surprising responses to experimentally induced cell volume perturbations.

4.3 Osmotic Permeability of Epithelial Cell Membranes

Studies on water content of tissue slices corrected for interstitial water by an impermeable extracellular marker proved the significance of energy metabolism and active Na^+ transport for regulation of epithelial cell volume (Macknight and Leaf 1977). There were problems in following up these studies with sufficient time and space resolution as it turned out that the osmotic permeability (hydraulic conductance) of normal functioning epithelia is very high. Added to this, some epithelia are heterocellular, which requires visual identification of cell types of specific functions to be studied individually. In the body, polarized epithelia are often exposed to solutions of different compositions on the two sides, which impose a further challenge to in vitro studies. Epithelial cell volume regulation is therefore studied in an Ussing chamber for flat-type preparations (MacRobbie and Ussing 1961, Spring and Hope 1978) and in a Burg microperfusion set-up for tubular preparations (Burg and Knepper 1986), both of which permit unilateral perturbation of bathing solutions. For measuring the osmotic permeability of the cell membranes (P_f) and recording 'instantaneous' volume responses to rapid external osmotic perturbations, unstirred layer artefacts must be eliminated, which has been accomplished by several laboratories (Schafer et al. 1978; Spring and Hope 1979; Carpi-Medina et al. 1983; Welling et al. 1983b; Carpi-Medina et al. 1984; Cotton et al. 1989). The most advanced technology applies quantitative DIC microscopy by online computer-controlled fast optical sectioning, storing of video images of section frames and offline computation of cell volume as function of time (Marsh et al. 1985; Strange and Spring 1986; Guggino et al. 1990). Another powerful technology relies on measuring the concentration of a non-permeable cytosolic probe molecule for calculating time-dependent changes in cell water volume in response to unilateral manipulation of bath composition (Reuss 1985; Ford et al. 2005). A quite different approach, only aiming at P_f, applies stopped flow and light scattering techniques for following volume changes of vesicles of isolated apical or basolateral membrane, respectively (van Heeswijk and van Os 1986). Examples of P_f values corrected for membrane area amplification by apical microvilli or basolateral infoldings are listed in Table 4.1 chosen here for illustrating the range covered by epithelial cell membranes. Except for the small intestine, P_f of leaky epithelia is significantly larger than P_f of tight epithelia. It is also seen that antidiuretic hormone (ADH) stimulates apical P_f of tight epithelia with no effect on the basolateral membrane (MacRobbie and Ussing 1961; Strange and Spring 1987b). The osmotic permeability is associated with the expression of aquaporin water channels (Preston et al. 1992; Agre et al. 1993; Nielsen and Agre 1995; Nielsen et al. 2002; Nielsen et al. 2013), which is dealt with in detail in Chap. 25 of the present book. While P_f values listed in Table 4.1 are a measure of the density of water channels of the respective membrane domains, the membranes' total osmotic permeability may be significantly larger due to the gain in membrane area by luminal microvilli and basolateral infoldings. For example, the total osmotic water permeability of the S_2 segment of rabbit proximal tubule is 1260–4500 $\mu m \cdot s^{-1}$ for the luminal membrane and 1400–5000 $\mu m \cdot s^{-1}$ for the basolateral membrane (Gonzalez et al. 1982; Schafer 1990). Likewise, taking into account area gains in intact cortical collecting tubules, the ratio of the total membrane osmotic permeability of apical and basolateral membrane would be 1:7 after hormone stimulation rather than about 1:1 as indicated in Table 4.1 (Strange and Spring 1987b).

Table 4.1 Examples of osmotic permeabilities (P_f) of luminal and basolateral plasma membranes of transporting epithelia; areas are corrected for apical microvilli and basolateral infoldings, respectively

Epithelium	Luminal membrane P_f (μm·s^{-1})	Basolateral membrane P_f (μm·s^{-1})	Method [Ref.]
Rana temporaria skin, control[a] + ADH[a]	≤1 3.8 ± 1.2	24 ± 1 23.5 ± 1.7	Optical [1]
Necturus gallbladder	545	347	Optical [2]
Necturus gallbladder	660	440	TMA^+ as volume marker [3]
Rat small intestine	60 ± 0.4	60 ± 15	Stopped flow, light scattering [4]
Rat proximal tubule, isolated plasma membrane vesicles	604 ± 20	557 ± 137	Stopped flow, light scattering [4]
Rabbit proximal tubule, S_2 segment	70–250	100–375	Optical [5]
Amphiuma kidney, diluting segment	1.0 ± 0.3	2.5 ± 0.3	Optical [6]
Rabbit cortical collecting tubule (PC), control + ADH	19.8 ± 1.9[b] 92.2 ± 8.6[b]	68.2 ± 3.5[c] 74.1 ± 5.8[c]	Optical [7]
Rabbit cortical collecting tubule (IC), control + ADH	25.0 ± 2.3[b] 86.2 ± 6.0[b]	70.5 ± 4.6[c] 61.2 ± 9.6[c]	Optical [7]
WT-RCCD1 cells[d]	10 ± 0.6		BCECF as volume marker [8]
AQP2-RCCD1 cells[e]	123 ± 7		BCECF as volume marker [8]

PC principal cells, *IC* intercalated cells
[a]Normalized for apical membrane area
[b]At a 100 mosmol/kg osmotic gradient
[c]At a 100 mosmol/kg osmotic gradient. Tubule lumen filled with oil shortly before basolateral solution shift
[d]Clonal cell line of rat cortical collecting tubule, wild type
[e]RCCD1 stably transfected with AQP2
[1] MacRobbie and Ussing (1961), [2] Persson and Spring (1982), [3] Cotton et al. (1989), [4] van Heeswijk and van Os (1986), [5] Gonzáles et al. (1984), Carpi-Medina and Whittembury (1988), [6] Guggino et al. (1985), [7] Strange and Spring (1987b), [8] Ford et al. (2005)

4.4 Cell Volume Response to Osmotic Challenges in Extrarenal Epithelia

4.4.1 Amphibian Skin

Amphibian skin epithelium is heterocellular with Na^+ transporting principal cells and mitochondrion-rich (MR) cells that are excluded from the functional syncytium of principal cells. Populations of MR cells are differentiated for passive and active Cl^- uptake and acid/base regulation, respectively, recently reviewed by Larsen et al. (2014).

4.4.1.1 Principal Cells

The first real-time study of epithelial cell volume aimed at testing the two-membrane theory of epithelial ion transport (MacRobbie and Ussing 1961). It was indicated that ADH selectively increases apical P_f (see Table 4.1). While a number of observations were in good agreement with the theory, the study also raised problems that were not solved until about 20 years later. Thus, with sulphate as major anion, the epithelium behaved as expected for a simple osmometer of a relatively large non-osmotic volume (Eq. 4.3). However, as shown in Fig. 4.1B after substitution with chloride on the inside, the acute osmotic volume expansion in hypo-osmotic solution was followed by a return of the cell volume towards its initial value. When re-exposed to normal osmolality, the epithelium shrank initially, as expected, but then gained fluid for returning towards the original iso-osmotic volume. These unexpected late responses were disclosed long before the concept of cell volume regulation was introduced (see Sect. 4.2.2). It was troublesome, in addition, that pump inhibition by ouabain did not swell the cells as predicted for a Gibbs-Donnan system. Because of inhibition of the apical Na^+- and the basolateral K^+, permeability volume stayed the same for about 30 min. Even more surprising, this was often followed by cell shrinkage (see Fig. 4.2A). The ouabain-induced cell shrinkage found its logical explanation once it was appreciated that similar to many non-epithelial cells, the Cl^- pool in frog skin is maintained above thermodynamic equilibrium by Na^+, K^+ and $2Cl^-$ cotransport in parallel with dynamic Cl^- and K^+ leak permeabilities (Ussing 1982; Giraldez and Ferreira 1984; Harvey and Kernan 1984; Dörge et al. 1985; Ussing 1985). Thus, computer-assisted mathematical analysis showed that the frog skin model with a NKCC1 transport mechanism in the serosal membrane has the capacity to reproduce the above-mentioned perplex time course of epithelial cell volume following ouabain inhibition of the Na^+/K^+ pump. In particular, on the way towards Gibbs-Donnan equilibrium, transient cell shrinkage results from Cl^- pool depletion associated with decreased Cl^- uptake via the serosal cotransporter caused by passive increase in $[Na^+]_c$ (Larsen 1991). See also Sect. 4.5.1 on kidney proximal tubule.

4.4.1.2 Mitochondrion-Rich Cells

When exposed on the outside to Ringer's solution, apical Cl^- channels of MR cells can be activated by membrane depolarization, which mediates passive Cl^- uptake together with active Na^+ uptake by principal cells, reviewed by Larsen (2011). All evidence considered indicates that this is an adaptation to the life on land where the

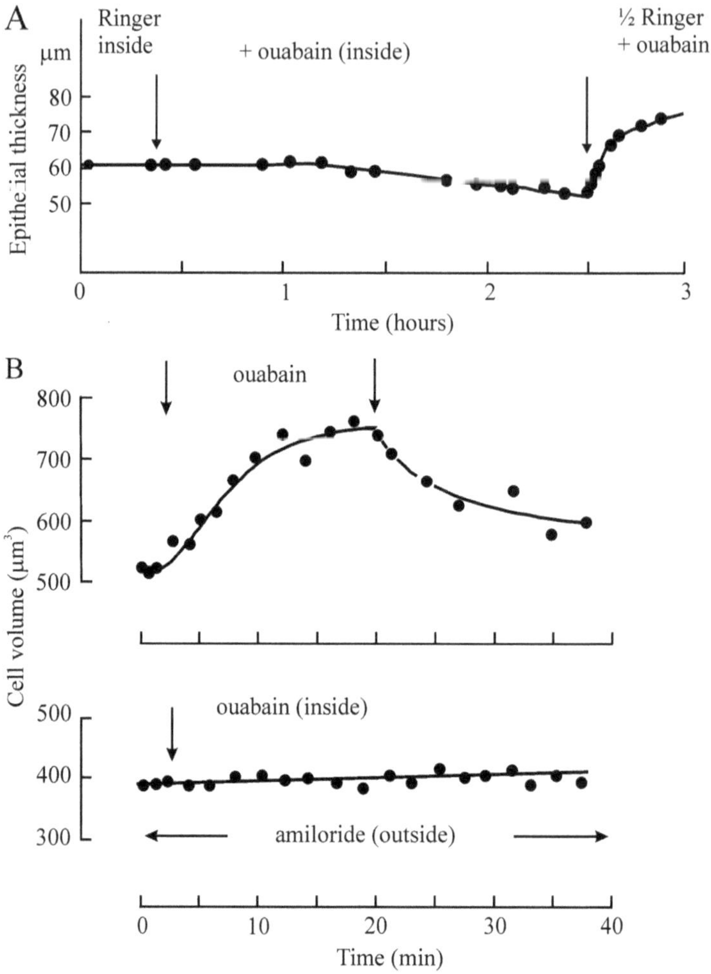

Fig. 4.2 Response of anuran epidermal cells to ouabain. (**A**) Principal cells of frog skin; the immediate response to block the Na^+/K^+ pump is not swelling. Rather, the volume stayed constant for about ½ h. This is due to inhibition of the apical and basolateral cation conductances, later denoted crosstalk. Thereafter, the cell volume decreased. See text for references and further discussion. Ringer's solution on the outside diluted 20 times by distilled water (see text for details) and Ringer's solution or a 50 % hypo-osmotic Ringer's solution (indicated by ½ R) on the inside as indicated [Redrawn from MacRobbie and Ussing (1961)]. (**B**) The mitochondrion-rich (MR) cell bathed with Ringer's solution on both sides swelled with ouabain on the inside, which was very significantly reduced by blocking the apical Na^+ entrance mechanism by amiloride. Redrawn from Larsen et al. (1987)

skin is covered by subepidermal gland secretions from which ions are reabsorbed during evaporative water loss (Larsen and Ramløv 2013; Larsen et al. 2014). Serosal ouabain evoked relatively fast cell swelling of toad skin MR cells, which was reversible upon wash with fresh Ringer's solution and could be prevented by amiloride in apical bath (Fig. 4.2B). In bilateral Ringer's solutions, $[Cl^-]_c$ was passively distributed at a concentration of 19.8 ± 1.7 mM. The cells swelled in parallel with voltage activation of apical Cl^- channels due to uptake of Cl^- across the apical membrane and K^+ uptake across the membrane facing the lateral intercellular space. Upon bilateral changes in external osmolality, the volume responded in agreement with Eq. 4.4 with $v'/V_0^{(c)} = 0.21$. Neither RVD nor RVI was observed within 5–10 min following solution shifts (Larsen et al. 1987). Frog skin exposed on the outside to a Ringer's solution that is diluted about 20 times by distilled water ($[Na^+] = [Cl^-] = 5.5$ mM) leads to closure of the apical Cl^- channels of MR cells (reviewed by Larsen 2011). This was shown to result in a significant osmotic increase in MR cell volume and in a cytosolic osmolality calculated to be about 166 mosmol/kg (Spring and Ussing 1986). Applying this protocol, Spring and Ussing (1986) found that subsequent dilution of the serosal Ringer's solution by 50 % with distilled water resulted in acute further osmotic volume expansion. This was followed by a complete RVD suggested to be caused by release of K^+, Cl^- and unidentified intracellular negatively charged diffusible osmolytes (Spring and Ussing 1986).

4.4.2 *Gallbladder*

Gallbladder epithelium modifies the bile secreted by the liver cells by iso-osmotic reabsorption of Na^+, Cl^- and HCO_3^-. The epithelium is formed by a monolayer of regular slender cells of an apical membrane with short 0.4–5-μm microvilli and a flat basal membrane resting on the basal lamina (Tormey and Diamond 1969). The studies by Kenneth Spring and his co-workers used *Necturus maculosus* gallbladder with epithelial cells of simple geometry (Spring and Hope 1978, 1979). Thus, Persson and Spring (1982) measured for the first time individual membrane permeabilities of a leaky epithelium (Table 4.1). The cells effectively regulate the volume in hyperosmotic (RVI) and hypo-osmotic (RVD) solutions of an initial volume response in agreement with that of a perfect osmometer with $v'/V_0^{(c)} = 0.18$ (see Fig. 4.3A). With both Cl^-- and K^+ activity being well above electrochemical equilibrium at control conditions, RVD was caused by transient activation of basolateral electrogenic KCl transport (Larson and Spring 1984). RVI depended on parallel operation of apical Na^+/H^+ and Cl^-/HCO_3^- exchange mechanisms (Fisher et al. 1981; Ericson and Spring 1982a; Ericson and Spring 1982b) regardless of the sidedness of the hyperosmotic challenge (Marsh and Spring 1985). Mucosal bumetanide at low concentrations inhibited apical NaCl uptake (Larson and Spring 1983) and transcellular fluid flow (Davis and Finn 1985) indicating a role of apical NKCC2 in both NaCl uptake and cell volume regulation. The relative importance of the above apical ion uptake mechanisms is cAMP dependent (Garvin and Spring 1992). Serosal ouabain caused acute swelling (see Fig. 4.3B) at a rate depending

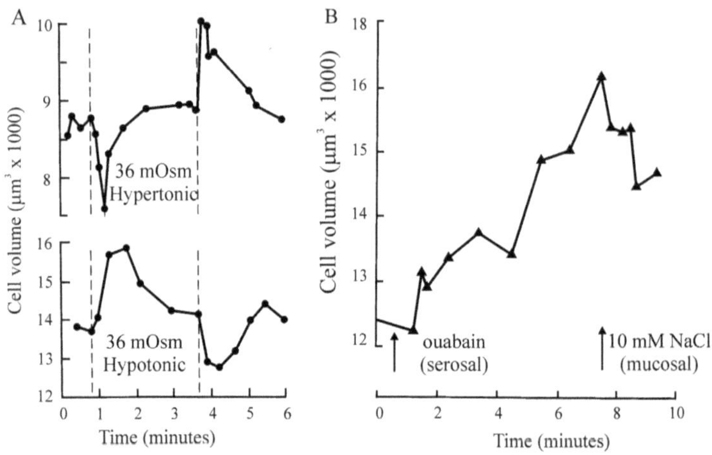

Fig. 4.3 *Necturus* gallbladder. (**A**) In response to addition or removal of mannitol, cell lost or gained water. Subsequent RVI or RVD proceeded at a significant rate. The osmotic challenges shown at the graphs were applied on the mucosal side of the epithelium. RVI was observed also with hyperosmotic mucosal bath achieved by adding NaCl or urea (not shown). (**B**) Serosal ouabain caused rapid swelling of the cell bathed with Ringer + 36 mM mannitol on both sides. At the right arrow, 90 % of mucosal NaCl was replaced by an osmotically equivalent amount of mannitol, which reduced mucosal [NaCl] to 10 mM [Redrawn from Persson and Spring (1982)]

upon mucosal [NaCl], which was abolished if all NaCl was replaced by mannitol (Ericson and Spring 1982a), or in K^+-free mucosal bath (Davis and Finn 1985).

4.4.3 Small Intestine

Intestinal homeostatic mechanisms are important both because the osmolality of intestinal fluid varies with water and food intake and because the pools of cellular osmolytes and metabolites depend on whether the intestine is in a resting or an absorbing/secreting state. Villus cells are mainly engaged in absorption of NaCl, nutrients and water, while crypts are mainly engaged in fluid secretion. The absorptive surface of the large columnar cells is increased by elaborated microvilli. The serosal membrane is flat, but the area of lateral membranes is increased by interdigitations with neighbouring cells. Absorption of nutrients is linked to Na^+ transport at the apical membrane, which also contains Na^+/H^+ exchangers, Cl^-/HCO_3^- exchangers and CFTR Cl^- channels (Xia et al. 2014). K^+ channels, Cl^- channels and NKCC1 are expressed at the basolateral membrane. Na^+/K^+ pumps at the lateral membranes create hyperosmotic lateral spaces that govern fluid uptake by osmosis. In human small intestine, this amounts to ~6.5 L/day. Stimulation by cAMP, cGMP, Ca^{2+} or enterotoxins inhibits the apical electroneutral exchange mechanisms and stimulates the apical CFTR channel, whereby the epithelium switches from absorption to secretion. Human small intestine secretes about 1 L of fluid per day. Discussions of intestinal fluid exchanges and intestinal ion transporters and their functions are given by Kato and Romero (2011).

4.4.3.1 Intestinal Crypt Cells

Isolated guinea pig intestinal crypts regulate cell volume in both hypo-osmotic (O'Brien et al. 1991) and hyperosmotic solutions (O'Brien et al. 1993). The technique applied prevented analysis of the sidedness of the osmotic membrane responses. The RVD was almost complete within 20 min and effectively inhibited by K^+ channel blockers (Ba^{2+}, quinine) and a Cl^- channel blocker (9ACA). The RVI led to complete volume recovery in about 20 min (Fig. 4.4A) and was abolished by bumetanide inhibition of the NKCC1 cotransporter (Fig. 4.4B). RVI was decoupled with the added osmolyte being D-mannitol (Fig. 4.4C), probably due to lack of appropriate ion gradients for NKCC1. Secretagogues (vasoactive intestinal peptide and carbachol) initiated cell shrinkage with no spontaneous RVI, but the volume loss was reversible upon removal of the hormone (Fig. 4.4D). Interestingly, volume regulation of intestinal crypts in hypo-osmotic solutions is impaired in cystic fibrosis mice (Valverde et al. 1995). This is not due to lack of a Cl^- exit channel, but by absence of a K^+ pathway, because the K^+-selective valinomycin ion pore restored full RVD (Valverde et al. 2000). In human intestine 407 cells, RVD is associated with release of organic osmolytes through a non-identified PKC-activated pathway (Tomassen et al. 2004).

4.4.3.2 Intestinal Villus Cells

Schultz and co-workers discovered crosstalk between apical nutrient uptake and basolateral K^+ channels, the activation of which restores the apical membrane potential and the driving force of the apical rheogenic cotransport (Grasset et al. 1983; Gunter-Smith et al. 1982; Gunter-Smith et al. 1982). The addition of D-glucose or L-alanine, but not L-glucose or D-alanine, caused immediate and fast swelling of isolated villus enterocytes followed by RVD, which was associated with loss of K^+ and Cl^- through volume-activated ion channels (MacLeod and Hamilton 1991). Like intestinal crypts, RVI of hyperosmotically shrunken cells from jejunum villi also depended on NKCC2 cotransport, which was inhibited by bumetanide (Macleod and Hamilton 1990). Ouabain-induced swelling proceeded slowly for levelling off already at about 110 % of the iso-osmotic cell volume (Macleod and Hamilton 1990). The mechanism preventing further development of the Gibbs-Donnan response was not investigated.

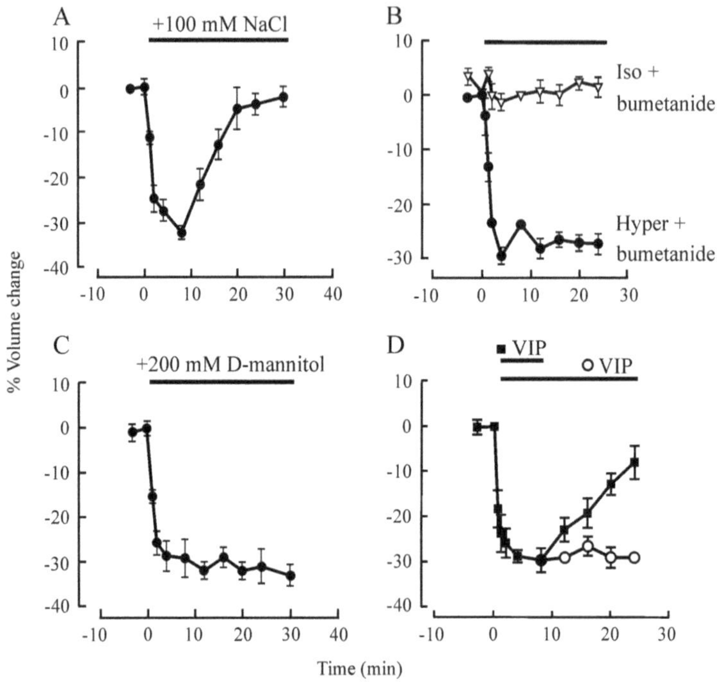

Fig. 4.4 Intact guinea pig intestinal crypts at 30 °C. (**A**) Well-developed RVI in response to hyperosmotic challenge achieved with 100 mM NaCl in bath. (**B**) RVI inhibited by bumetanide, hyperosmolality obtained by adding 100 mM NaCl to bath. (**C**) Cell shrinkage with 200 mM D-mannitol added to bath; RVI decoupled. (**D**) Crypts stimulated to fluid secretion by vasoactive intestinal peptide (VIP) exhibit iso-osmotic shrinkage and remain shrunken until VIP is being removed [Redrawn from O'Brien et al. (1993)]

4.4.4 Upper Airways

The upper airways from nasal cavity to bronchi are lined by ciliated columnar cells covered by an airway surface layer (ASL) of a near iso-osmotic periciliary liquid (PCL) with mucous on top. Ciliary beating moves mucus and periciliary fluid towards the throat where it is swallowed, which is important for normal function of the pulmonary host defence; dysfunction may lead to severe pathological conditions like pulmonary obstruction and cystic fibrosis (see Chap. 16). ASL is renewed by flow from distal areas and by submucosal gland secretion. The surface of the small airways is relatively very large, and the height of ASL is the same in distal and proximal regions. Because the volume flows relative to epithelial surface area are about the same, it follows that during passage from distal to proximal airways water and ions are absorbed probably at iso-osmotic proportions. This is energized by Na^+ uptake via apical ENaC and basolateral Na^+/K^+ pumps (Willumsen and Boucher 1991; Boucher 1994; Grubb et al. 1997; Boucher 1999, 2003; Tarran et al. 2001). The apical membrane also harbours cAMP-activated

CFTR and Ca^{2+}-activated Cl^- channels. A basolateral NKCC1 cotransporter maintains $[Cl^-]_c$ above thermodynamic equilibrium across the basolateral membrane (Willumsen et al. 1989). In the absorbing state, uptake of Cl^- is passive probably through paracellular pathways. Chloride secretion is driven by the NKCC1 cotransporter at the serosal membrane and downhill movement of Cl^- through the apical membrane's Cl^- channels (Willumsen et al. 1989; Willumsen and Boucher 1991). Thus, PCL volume of a depth ≤ 7 μm is balanced by ENaC-activated fluid absorption and CFTR-stimulated fluid secretion regulated by not yet fully understood complex interactions between ATP and ADP of ASL (Lazarowski et al. 2004; Tarran et al. 2006; Higgins et al. 2014). The volume sensing mechanism is unknown, but previous studies indirectly indicated that osmotically induced epithelial volume changes may provide an early signal. Thus, cell shrinkage was seen only in response to mucosal hyperosmolality, and the cells remained shrunken until physiological saline again was applied on the mucosal surface as expected for an epithelial cell that serves sensing of ASL osmolality (Willumsen et al. 1994). RVD is developed in human tracheal and bronchial cells and is defect in tracheal cells from cystic fibrosis (CF) patients. This was caused by altered regulation of an intermediate conductance, Ca^{2+}-regulated K^+ channel. The recording technique did not allow for analysis of the sidedness of the osmotic membrane response (Vazquez et al. 2001; Fernandez-Fernandez et al. 2002). The swelling-activated entry of Ca^2 via the TRPV4 channel was abolished in cells of CF patients, which is compatible with the hypothesis that TRPV4 is a mechano- and osmoreceptor (see Sect. 4.8.2) and is of significance for maintenance of mechanosensitivity and iso-osmotic equilibrium (Arniges et al. 2004) (see Sect. 4.9 for discussion).

4.4.5 Exocrine Glands

The acinar epithelium of exocrine glands generates an iso-osmotic primary secretion of $[K^+]$ higher than that of plasma (Petersen and Gallacher 1988; Petersen 1993), which may be modified by the epithelium lining the excretory duct (Novak and Greger 1988; Novak 2011). Carbachol-induced increase in acinar $[Ca^{2+}]_c$ results in Cl^- and K^+ loss and iso-osmotic cell shrinkage (Nauntofte and Dissing 1988; Takemura et al. 1991; Petersen 1992; Foskett et al. 1994; Speake et al. 1998). This leads to active secretion of Cl^- (Cook et al. 1994; Nielsen and Nielsen 1999) driven by increased activity of the NKCC1 cotransporter (Mills 1985; Foskett et al. 1994) and/or the NHE1 mechanism (Manganel and Turner 1991). Stimulation by cAMP-linked agonists (prostaglandin E_2, isoproterenol) leads to ion secretion through co-expressed apical (luminal) CFTR Cl^- channels and maxi-K^+ channels (Sørensen et al. 2001). Stimulation by Ca^{2+}-linked agonists (carbachol) activates Ca^{2+}-dependent apical anoctamin-1 (TMEM16A) Cl^- channels (Romanenko et al. 2010). Experiments on perfused shark rectal gland showed that it is the transient cell shrinkage that triggers cotransport by NKCC1 (Greger et al. 1999).

In single isolated parotid cells, large amplitude oscillations of volume and intracellular Na^+, K^+ and Cl^- concentrations driven by oscillations in intracellular $[Ca^{2+}]$ have been demonstrated by Foskett and his co-workers (Wong and Foskett 1991; Foskett et al. 1994). Frog skin acinar cells in nystatin-permeabilized whole-cell patch-clamp configuration did not exhibit membrane potential oscillations neither during Ca^{2+}- nor cAMP-coupled secretion (Sørensen and Larsen 1999). This indicates that oscillations in $[K^+]_c$ (and cell volume) are not associated with continued secretion by an intact acinar epithelium. The studies discussed above, however, have shown that an iso-osmotic cell volume decrease precedes the secretory response similar to the secretion-stimulated intestinal crypt epithelium (see Fig. 4.4D).

Acinar cells isolated from rat lacrimal glands exhibit aniso-osmotic volume regulation. The cells lost volume in hyperosmotic solutions, but an RVI was only observed under certain conditions depending on temperature and buffer (Douglas and Brown 1996). Thus, in HEPES-buffered solutions at 37 °C, RVI was dependent on ion uptake via a NKCC1 mechanism. In contrast, at 20 °C and in HCO_3^--buffered solutions, inhibitor studies indicated that RVI required ion uptake via Cl^-/HCO_3^- and Na^+/H^+ exchangers. RVI was also supported by uptake of a mixture of neutral amino acids via system-A Na^+-coupled amino acid cotransport. RVD was observed in hypo-osmotic solutions probably controlled by channel-mediated inflow of Ca^{2+} (Speake et al. 1998). Isolated parotid cells also exhibit classical RVD (Foskett et al. 1994). Since the above studies were performed on isolated cells, the sidedness of the above osmotic membrane responses could not be assessed.

4.4.6 Teleost Gill and Opercular Epithelium

The euryhaline killifish *Fundulus heteroclitus* tolerates salinities ranging from freshwater (FW) to seawater (SW) and adapts quickly to changes in salinity (reviewed by Grosell in Larsen et al. 2014). The killifish opercular epithelium has a high density of chloride-secreting (MR) cells and is an excellent model of an osmosensing Cl^--secreting epithelium (Foskett 1982; Foskett et al. 1983; Marshall 2011). The MR cell has a ouabain-inhibitable Na^+/K^+ ATPase and a bumetanide-sensitive NKCC1 cotransporter in the basolateral membrane with a CFTR anion channel in the apical membrane (Evans et al. 2005; Marshall and Grosell 2005). There is an increase of $\sim$70 mosmol/kg in plasma osmolality during the transition of the killifish to high salinity (Zadunaisky et al. 1995) that initiates rapid increase in salt secretion across the epithelium. This seems to involve a hyperosmotic (shrinkage) activation of the basolateral NKCC1 cotransporter and a suite of the protein kinases PKC, MLCK, p38 MAPK, JUNK, oxidative stress-responsive kinase 1 (OSR1) and Ste20-related proline/alanine-rich kinase (SPAK) (Hoffmann et al. 2002; Marshall et al. 2005) (see Fig. 4.5). A major decrease in plasma osmolality accompanies freshwater transition of the killifish, especially in the first

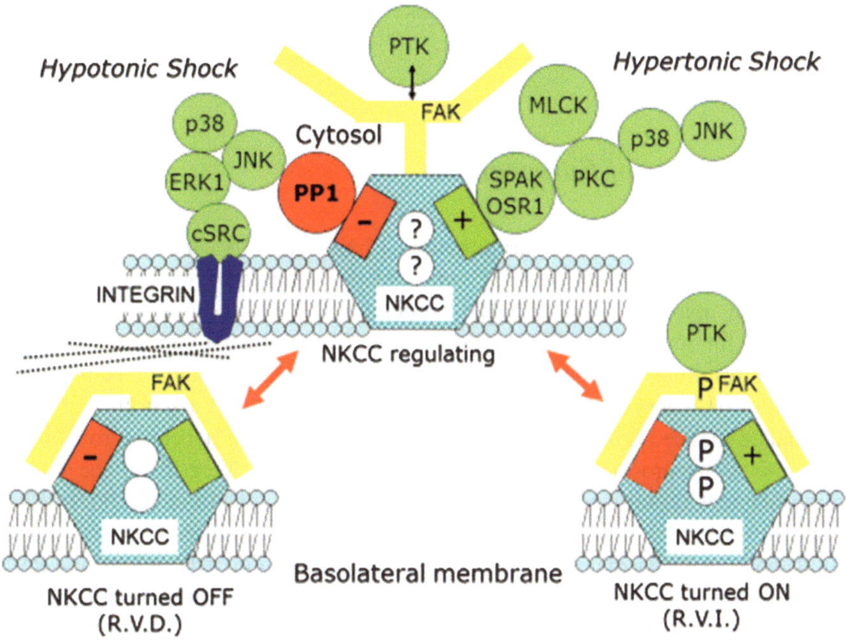

Fig. 4.5 Hypothetical model of the osmotic regulation of NKCC in the basolateral membrane of killifish opercular epithelium. *Right-hand side*. JNK, PKC, MLCK and P38 MAPK kinases near the basolateral membrane are all activated by hyperosmotic stimuli and by NKCC activation associated with regulatory volume increase (RVI). Kinases co-localized with NKCC are also included (FAK, OSR1 and SPAK). Because SPAK and OSR1 in other systems co-immunoprecipitate with NKCC, they are placed nearest to NKCC. The order of the other kinases in the cascade is unknown. *Left-hand side*. The inhibitory cascade terminating with PP1 shown in other systems to co-immunoprecipitate with NKCC. Known players are JNK and P38 MAPK. Shown are also integrin, ERK and cSRC because they mediate hypo-osmotic stimuli in other cell types. The model proposes three conditions: an unstable regulating phase that results in NKCC becoming phosphorylated and stimulated during RVI (*right*) or dephosphorylated and inactivated during RVD (*left*). The reason why FAK is depicted as occluding the phosphorylation sites of NKCC is because of the unusual, dual effect of the PTK inhibitor genistein on the chloride current: when the chloride current is high, it is decreased by genistein; when the chloride current is low, it is the increased by genistein [From Hoffmann, Schettino and Marshall, Comparative Biochemistry and Physiology A. 148, 29–43, 2007. With permission]

12 h after transfer (Marshall et al. 2000). This results in hypo-osmotically (cell swelling) controlled deactivation of the NKCC1 cotransporter involving Ca^{2+} and an unidentified protein phosphatase. The mitogen-activated protein kinase (MAPK) p38 has an inhibitory role in the hypo-osmotic deactivation of NKCC1 (Marshall et al. 2005). Finally, the focal adhesion kinase (FAK), which co-localizes with NKCC1 at the basolateral membrane, has an important role in setting the level of NKCC1 activity. It is likely that chloride cells respond to hypo-osmotic shock using integrin β1 as an osmosensor coupled to dephosphorylation of FAK that leads to

NKCC1 deactivation (Marshall et al. 2008). Sensing mechanisms are further dealt with in Sect. 4.8.

4.4.7 *Intestine of European Eel*

In eel intestinal epithelium transepithelial Cl^- absorption is via an apical NKCC2 cotransporter and a basolateral Cl^- channel. K^+ is returned to intestinal lumen through an apical K^+ channel. The model for the eel intestine is of the type described for mammalian TAL (Sect. 4.5.3). Downstream migratory eels drink seawater resulting in hyperosmotic exposure of the apical side of the intestinal epithelium (Lionetto et al. 2001), reviewed by Grosell in Larsen et al. (2014). Shrinkage-induced stimulation of NKCC2 and RVI play a central role in the response to hyperosmotic challenge, which depends on integrity of the F-actin microfilament system and involves PKC and MLCK (Lionetto et al. 2002). When the eel moves back to freshwater, the intestine becomes exposed to hypo-osmotic fluids both on the basolateral and the apical side resulting in cell swelling and RVD (recorded by morphometric measurement of the epithelium height) associated with swelling activation of several anion and K^+ conductive pathways in apical and basolateral membranes (Lionetto et al. 2005), reviewed by Hoffmann et al. (2007). The calcium-activated large conductance (BK) K^+ channel in the apical membrane is a particularly important participant in this response (Lionetto et al. 2008) where it is activated under hypo-osmotic conditions.

4.5 Cell Volume Response to Osmotic Challenges in Renal Epithelia

Most segments of vertebrate nephron have evolved ability to volume-regulate when exposed to aniso-osmotic solutions. In the sections below, examples of principal mechanisms are discussed based on representative studies. Comprehensive reviews are given by Hebert (1987), Montroserafizadeh and Guggino (1990) and Lang (2013).

4.5.1 *Kidney Proximal Tubules*

The cells have apical microvilli and elaborated interdigitations of lateral plasma membranes at the base of the cells. As a result, the area of lateral membranes is significantly amplified and similar to the area of brushborder surface and intervening membranes (Welling and Welling 1988). The proximal tubule of human kidney reabsorbs ~120 L/day of a primary filtrate of 180 L/day as iso-osmotic fluid. This is energized by active translateral Na^+ transport via Na^+/K^+ pumps at the plasma membranes lining the intercellular space of complicated microanatomy, reviewed in detail by Weinstein (2008) and Zhuo and Li (2013). Water reabsorption is passive and secondary to solute reabsorption at iso-osmotic proportions. When challenged by hypo-osmotic peritubular solution, the cells of perfused rabbit proximal straight tubule swelled immediately but reversed to the iso-osmotic control volume in continued presence of the hypo-osmotic challenge (Fig. 4.6A). When returning to the control bath of 290 mosmol/kg, the cells shrank initially but reversed fast to the volume prior to the osmotic perturbation (Kirk et al. 1987B). In contrast, following a hyperosmotic peritubular challenge (390 mosmol/kg), the cells shrank as expected for a perfect osmometer but stayed shrunken for 15–20 min until the return to the iso-osmotic solution of 290 mosmol/kg (Fig. 4.6B). Reintroduction of iso-osmotic bath resulted in a transient volume increase that exceeded the control volume before return to the iso-osmotic control value (discussed below).

When luminal Na^+ was replaced by choline$^+$, the cells shrank reflecting the expected decreased inflow of Na^+ through the apical membrane and continued pumping of Na^+ out of the cells across the lateral membranes (Fig. 4.6C). Like the aniso-osmotic shrinkage in Fig. 4.6B, the iso-osmotic shrinkage shown in Fig. 4.6C did not elicit back-regulation of cell volume. Interestingly, the tubule cells regulate volume following iso-osmotic swelling obtained by replacing a fraction of NaCl of the luminal perfusion solution and the peritubular bath with an osmotically equivalent amount of urea (Fig. 4.6D). This indicates that RVD can be elicited by swelling per se in the absence of cytosolic dilution. Like in the experiment on cytosolic dilution at aniso-osmotic conditions (Fig. 4.6B), following return to the control solution, the cells swelled transiently over and above the iso-osmotic control volume. Kirk et al. (1987a) pointed out that such a cell volume response in excess of the steady-state iso-osmotic volume may reflect a period where the rate of solute entry across the apical membrane transiently is out of phase with solute exit across the basolateral membrane. The RVD is associated with loss of K^+ through channels in the peritubular membrane (Kirk et al. 1987a; Beck and Potts 1990; Schild et al. 1991).

The immediate response to pump inhibition by ouabain is not cell swelling, but cell shrinkage (see Fig. 4.7 (Kirk et al. 1987a)). The authors mention that the cells remained shrunken for at least 25 min. At this point the cells began swelling, and they swelled continually during 30–60 min until structural changes took place like blebbing and vacuolation. Thus, this would indicate initial transient loss of cellular osmolytes kept above thermodynamic equilibrium by the Na^+/K^+ pump that preceded the inevitable swelling. One such candidate would be Cl^- of a cellular activity that is 1.3–2 times higher than predicted for passive distribution (Cassola et al. 1983; Ishibashi et al. 1988). The apical uptake mechanisms comprise Na^+/H^+ and Cl^-/base exchangers (Alpern 1987) in parallel with V-type proton pumps, reviewed by Weinstein (2008). Their coordinated operation would result in cellular accumulation of Cl^- of which Na^+/H^+ exchange depends directly on

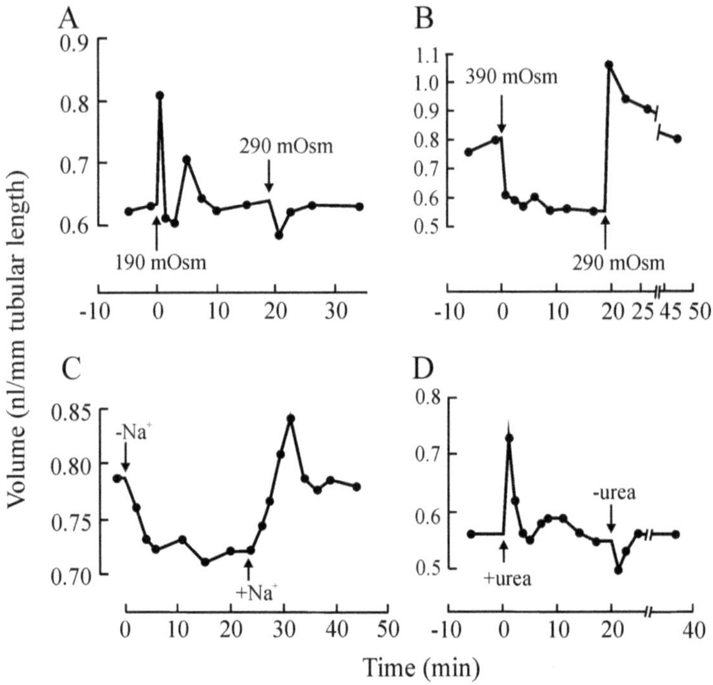

Fig. 4.6 Perfused rabbit proximal straight tubule at 38 °C. (**A**) Regulatory volume decrease following hypo-osmotic challenge by removal of NaCl from the luminal perfusion solution and peritubular bath. (**B**) Response of the tubule to a hyperosmotic challenge by adding NaCl to the luminal perfusion solution and peritubular bath. Note the absence of RVI. (**C**) Reversible iso-osmotic volume decrease obtained by exchanging luminal Na^+ for $choline^+$. Note the absence of RVI. (**D**) Regulatory volume decrease by iso-osmotic substitution of 50 mM NaCl for 100 mM urea in luminal perfusion solution and peritubular bath. The apparently 'heavily damped oscillations' after the second solution shifts in (**B**) and (**C**) were suggested to reflect periods where entrance and exit of osmolytes across the opposite membranes are out of phase. A similar explanation might apply to the transient volume attenuation during RVD in (**A**) and (**D**) [Redrawn from Kirk et al. (1987b)]

Na^+/K^+ pump activity. Thus, there is an explanation to hand for the ouabain-induced initial cell shrinkage that complies with the mechanism of frog skin discussed above (Fig. 4.2A). Furthermore, the experimental results shown in Fig. 4.7 demonstrate that pump inhibition by ouabain decouples RVD. The authors hypothesized that this is caused by inhibition of the basolateral K^+ permeability secondary to pump inhibition (crosstalk) as described in previous studies on (amphibian) proximal tubule (Matsumura et al. 1984; Messner et al. 1985).

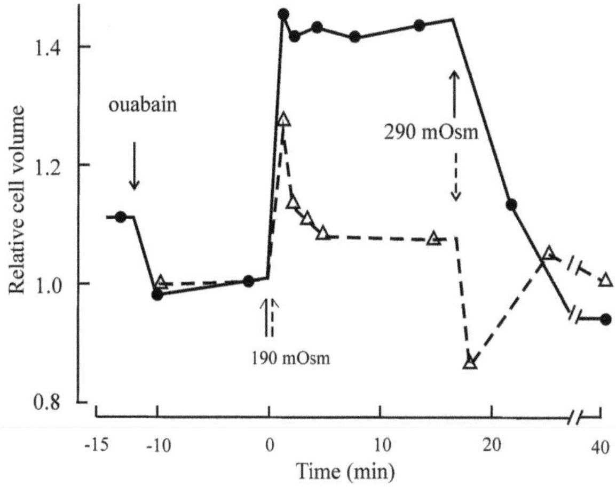

Fig. 4.7 Perfused rabbit proximal straight tubule at 38 °C. The immediate response to pump inhibition by ouabain is not cell swelling, but cell shrinkage. Furthermore, pump inhibition decouples RVD. Both of these non-trivial observations are discussed in the text [Redrawn from Kirk et al. (1987a)]

4.5.2 Cortical Collecting Tubule

Cortical collecting tubule (CCT) is heterocellular with principal (PC) and intercalated (IC) cells (Welling et al. 1983a). Principal cells reabsorb Na^+ through ENaC and secrete K^+ through ROMK and maxi-K^+ (BK) channels in the luminal membrane (Wang and Giebisch 2009). Of the two types of intercalated cells, the apical membrane of α-MR cells (type A ICs) is configured for acid secretion via apical H^+ pumps, K^+ reabsorption via apical H^+/K^+ exchangers and K^+ secretion through apical BK channels. The β-MR cell (type B IC) is configured for base secretion with H^+ pumps at the basolateral membrane and Cl^-/HCO_3^- exchangers at the apical membrane, reviewed by Breton and Brown (2013). The osmotic permeability of CCT is controlled by ADH (Table 4.1) that stimulates translocation of aquaporin-2 water channels from intracellular vesicles to the apical membrane (Nielsen et al. 1995). In situ, this results in reabsorption of water from tubular fluid. Thereby, this portion of the nephron participates in controlling the volume and the degree of dilution of the urine. A relatively large P_f of the basolateral membrane was reasoned to protect ADH-stimulated cells from swelling and cytoplasmic dilution (Strange and Spring 1987a; Strange 1988).

Both PCs and ICs exhibit typical RVD at 37 °C (see Fig. 4.8A, B), which were significantly reduced or fully inhibited at room temperature (Strange 1988). Inhibitor and ion substitution studies did not lead to clear answers about the release mechanisms for the efflux of osmolytes during RVD. RVI was observed only in one type of ICs, whereas PCs and the other IC type remained shrunken for 20–40 min following a serosal hyperosmolality (Strange and Spring 1987b). The RCCT-28A CCT cell line harbours a cell swelling-activated ~300-pS apical Cl^- channel that is activated by membrane stretch and disruption of F-actin (Schwiebert et al. 1994).

It is interesting to compare the time course of RVD in Fig. 4.8 with RVD of rabbit proximal tubule (Fig. 4.6A) and RVD of *Necturus* gallbladder (Fig. 4.3A).

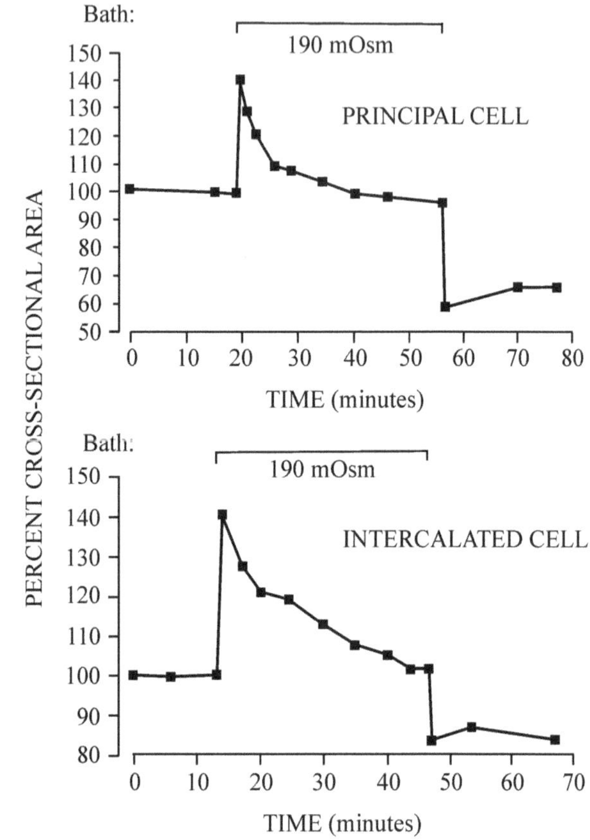

Fig. 4.8 Perfused rabbit renal cortical collecting tubule at 36–38 °C. During the control period, the tubules were exposed to bath and perfusion solutions of about 240 mosmol/kg. RVD elicited by a change in osmolality of peritubular bath to 190 mosmol/kg. The initial response to bath dilution indicates complete swelling in less than 1 s. For this reason, cross-sectional area of the mid-zone of the cell was used for estimating water exchange rates rather than the time-consuming whole-cell optical sectioning technique. *Panel above*, principal cell; *panel below*, intercalated cell [Redrawn from Strange (1988)]

They were all studied under conditions of practically eliminated unstirred layer effects and of about similar osmolality challenges, i.e. nearly similar 'instantaneous' driving forces for water. Yet RVD proceeds with a significantly slower rate in cells of smallest water permeability (see Sect. 4.3 above), indicating that expression levels of water channels, ion transporters and ion channels are positively correlated.

Ouabain caused swelling of PC, but not of IC. The swelling was immediate and fast and abolished or reduced by 95 % in Na^+-free luminal perfusion solution or application of luminal amiloride, respectively (Strange 1989). With an anion-selective microelectrode, $[Cl]_c$ in rat CCT was estimated to be near equilibrium (Schlatter et al. 1990), and subsequent studies have adopted the assumption of passive Cl^- distribution in principal cells of CCT (Staruschenko 2012). Thus, ouabain-induced increase in $[Na^+]_c$ instantly would be followed by an increase in $[Cl^-]_c$ and an associated cell swelling, which should continue until the cell bursts if the cell behaves in accordance with a nonregulated Gibbs-Donnan system. Indeed, cell swelling commenced with fast rate right after addition of ouabain. Surprisingly, however, rather than continued swelling, the cells exhibited spontaneous iso-osmotic RVD (see Fig. 4.9A). Thus, Strange (1989) stated that in CCT principal cells, pump inhibition seems to reduce apical P_{Na} (crosstalk). Following long-term exposure and removal of ouabain, cell volume rapidly decreased to a level that was significantly lower than that of the iso-osmotic control volume, and a second ouabain exposure had little or no effect on cell volume (Fig. 4.9B). This supported the hypothesis that pump inhibition resulted in a rapid inhibition of P_{Na}. It was more difficult to explain the late and complete RVD during continued presence of peritubular ouabain (Fig. 4.9A). Strange (1989) discussed the possibility that ouabain binding is transient upon prolonged exposure and that pumps become released for resuming active transport. Thus, it would be pumps that have lost ouabain sensitivity that energizes recovery of cell volume. In a subsequent investigation of this late RVD in the presence of ouabain, evidence was provided for K^+ loss via an apical K^+ pathway different from the conductance mediating the resting K^+ secretion (Strange 1990).

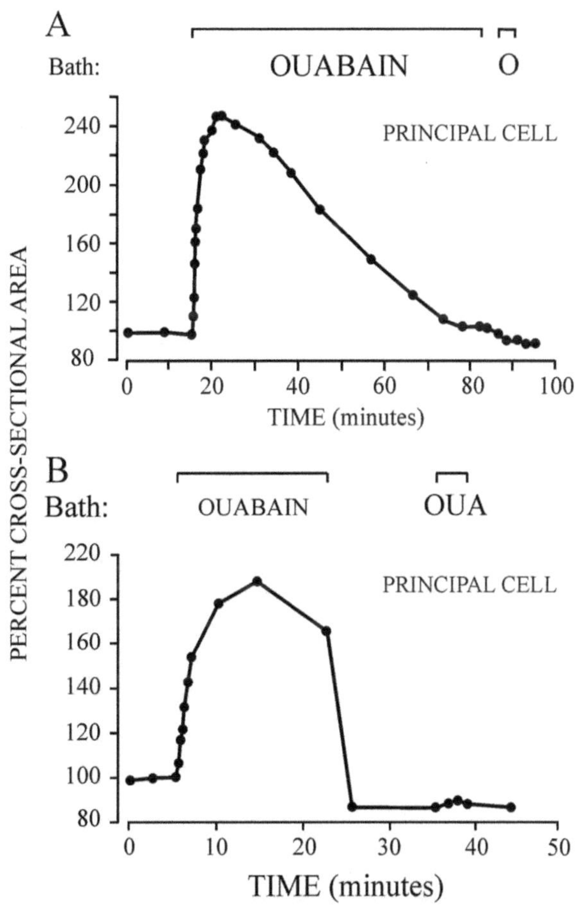

Fig. 4.9 Perfused rabbit renal cortical collecting tubule at 36–38 °C. (**A**) Complex response to long-term exposure of ouabain on principal cell volume. Pump inhibition induced immediate and fast cell swelling. However, this was followed by spontaneous iso-osmotic cell shrinkage. After wash at 80 min, a second addition of ouabain had practically no effect. (**B**) Effects of ouabain removal shortly after activation of iso-osmotic cell shrinkage of principal cell. The cell rapidly shrank to a volume significantly below the iso-osmotic volume prior to ouabain treatment. The second application of ouabain failed to initiate any significant cell swelling [Redrawn from Strange (1989)]

4.5.3 Medullary and Papillary Portions of Mammalian Nephron

By countercurrent multiplication, the loop of Henle has the capacity to generate an osmotic gradient in the interstitial space that increases linearly towards the tip of the papilla. Thus, the epithelial cells may be facing both hypo-osmotic and hyperosmotic conditions depending on whether the organism is in diuretic or antidiuretic state. Transport of NaCl across the thin descending limb of Henle (tDLH) is passive with Na^+/K^+ pumps serving maintenance of ionic gradients, which are important for RVD during hypo-osmotic volume regulation. Both Cl^- and HCO_3^- participate in RVD, which is associated with K^+ loss across the peritubular membrane (Lopes et al. 1988). In the absence of ADH, cells of mouse thick ascending limb (TAL) shrank and remained shrunken as a perfect osmometer when exposed to a hyperosmotic bath. In the presence of ADH, however, TAL cells exhibited perfect RVI (Hebert 1986a). During RVI in mouse thick ascending limb, intracellular Na^+ and Cl^- were accumulated via parallel operation of Na^+/H^+ and Cl^-/HCO_3^- exchange across the basolateral membrane (Hebert 1986b). In TAL of rabbit kidney, NaCl would accumulate via a basolateral NKCC1 of increased activity in hypertonic media (Eveloff and Calamia 1986). In the papillary region, the cells are capable of preventing shrinkage to hyperosmotic challenge by net uptake of NaCl and/or by genome-directed synthesis of intracellular organic osmolytes (Montroserafizadeh and Guggino 1990; Burg et al. 1997; Ferraris and Burg 2006; Jeon et al. 2006). Polyols (sorbitol, inositol), trimethylamines (betaine, glycerophosphocholine) and urea constitute the common organic osmolytes of inner medulla during antidiuresis that maintain cell volume and function in the very strong hyperosmotic environment (Bagnasco et al. 1986; Balaban and Burg 1987).

With an apical NKCC2 cotransporter, the early distal tubule of *Amphiuma* kidney is regarded as equivalent to mammalian TAL. Ouabain-induced cell swelling of perfused isolated tubules was prevented by inhibition of the cotransporter (Guggino et al. 1985). The cells did not show RVD in response to bath dilution. The cells just swelled as expected for a perfect osmometer of a non-solvent volume of 18 %. Independent $[Cl^-]_c$ measurements confirmed unchanged cellular Cl^- pool (Guggino et al. 1985). Cell swelling after addition of luminal furosemide, on the other hand, was followed by significant loss of cellular Cl^-. The authors pointed out that the unchanged Cl^- pool after aniso-osmotic bath dilution indicates crosstalk between entrance and exit pathways, so that the presumed increased ion loss through the peritubular membrane was compensated by a stimulated ion uptake via the apical cotransporter.

4.6 Epithelial Cell Volume as Signal for Regulating Iso-osmotic Transport

It is a distinct feature of both low-resistance (Schatzmann et al. 1958; Curran 1960; Diamond 1964) and high-resistance epithelia (Schafer 1993; Nielsen and Larsen 2007; Gaeggeler et al. 2011) that they are capable of transporting iso-osmotic fluid at transepithelial osmotic equilibrium. Experimental evidence shows that this type of transport is energized by Na^+/K^+ pumps and that the water flow in several cases is due to osmosis (Altenberg and Reuss 2013). In all investigated epithelia, the pumps are abundantly expressed at the membranes lining the lateral intercellular spaces (Stirling 1972; Mills and Ernst 1975; Mills et al. 1977; Ernst and Mills 1977; Mills and DiBona 1978; Grosell 2007), including lateral plasma membrane infoldings near the base of proximal tubule cells (Kashgarian et al. 1985; Maunsbach and Boulpaep 1991). Thus, at transepithelial osmotic equilibrium, the lateral intercellular spaces constitute intraepithelial compartments that couple a passive flow of water with the active flux of Na^+. Since the fluid of the osmotic coupling compartment is hyperosmotic relative to the solution of departure, truly iso-osmotic transport would be possible if a fraction of the pumped Na^+ is recirculated back into the coupling compartment (the lateral space) (Ussing and Nedergaard 1993; Nedergaard et al., 1999). The physical justification for this theory is reinforced by the fact that exit of ions from the above coupling compartments is due to convection-diffusion across an interface of relatively high ion diffusion coefficient. This implies that the osmolality of the fluid emerging from the lateral space would be larger than that of the lateral space; for discussion and mathematical treatment, see Larsen et al. (2000). Hoffmann and Ussing (1992) discussed that changes in cell volume are signals for regulating the recirculation flux and inferred that cell volume regulation is an integrated mechanism of iso-osmotic transport. Below, this hypothesis is discussed for an absorbing and a secreting epithelium.

In collagenase-isolated (denuded) toad skin epithelium, β-adrenergic receptor stimulation leads to activation of the Cl^- conductance of MR cells (Willumsen et al. 1992), probably by opening 8-pS Cl^- channels in the apical membrane (Sørensen and Larsen 1996). In addition, AQP2a isoform water channels are inserted into the apical membrane of principal cells (Bellantuono et al. 2008), and ENaC of principal cells is activated (Ogushi et al. 2010). As a result, the epithelium generates a transepithelial flow of iso-osmotic fluid (Nielsen and Larsen 2007) that is energized by Na^+/K^+ pumps at the lateral membranes (Larsen et al. 2009). In principal cells, $[Na^+]_c$ is governed by apical Na^+ uptake via ENaC, pumping of Na^+ into the lateral space and uptake ('recirculation') of Na^+ via the NKCC1 cotransporter at the serosal membrane (Rick et al. 1978; Dörge et al. 1985). Since the cotransport fluxes are stimulated in shrunken cells (Ussing 1982; Ussing 1985), this mechanism is a strong candidate for the cell volume-regulated recirculation pathway. Thus, if the transported fluid becomes hyperosmotic leading to cell shrinkage, or hypo-osmotic leading to cell swelling, the recirculation flux would be stimulated or reduced, respectively, whereby ion fluxes and water flow return to iso-osmotic proportions.

The Na^+ recirculation theory was originally developed for iso-osmotic transport in exocrine glands by Ussing and Eskesen (1989) and Ussing et al. (1996), who provided experimental evidence for significant recirculation of Na^+ in frog skin submucosal glands during adrenaline-stimulated secretion. Whole-cell patch-clamp studies of microdissected frog glands showed that both cAMP- and Ca^{2+}-dependent Cl^- channels are present in the acinar epithelium and are activated by physiological stimulation together with both K^+ and Na^+ channels (Sørensen and Larsen 1999). Single-channel recording from apical (luminal) membrane patches disclosed a CFTR Cl^- channel co-expressed with a maxi-K^+ channel and a 5-pS Na^+ channel (Sørensen and Larsen 1998; Sørensen et al. 1998; Sørensen et al. 2001). The Na^+ channel of the apical membrane shown in Fig. 4.10 is the obvious candidate for the Na^+ recirculation pathway. In Ehrlich ascites tumour cells (EATC), P_{Na} is a continuous function of cell volume in such a way that P_{Na} decreases upon cytoplasmic dilution and increases in response to cell shrinkage (Hoffmann 1978; Hoffmann et al. 2009). Applied to acinar cells, an increase in luminal fluid osmolality that leads to cell shrinkage will result in increased P_{Na} and thus increased Na^+ recirculation, i.e. an RVI response. In turn, this would restore iso-osmotic volume flow and cell volume. In contrast, if the luminal fluid becomes diluted, the acinar cells swell. The evoked RVD will then decrease P_{Na} whereby iso-osmotic transport and cell volume are being restored. If acinar cells function as sensors of luminal osmolality, it is expected that P_f of the apical membrane is much larger than that of the basolateral membrane. Immunofluorescence labelling of amphibian AQP-x5 (homologous to mammalian AQP5) verified exclusive expression in the apical membrane of subepidermal glands. AQP-h3BL was similarly immunolocalized in the basolateral membrane (Hillyard et al. 2009). The relative water permeability of the two membrane domains has not been studied.

4.7 Molecular Identity of Channels and Transporters Involved in Cell Volume Regulation

4.7.1 Chloride Channels

A key player in RVD is the volume-regulated anion channel (VRAC), which is expressed in all vertebrate cells investigated (Hoffmann et al. 2009). In most cell types, the K^+ conductance is much larger than the Cl^- conductance. Thus, a marked increase in Cl^- conductance is essential for volume regulation. The activation of a conductive Cl^- pathway by cell swelling was first demonstrated in EATC (Hoffmann 1978) and human lymphocytes (Grinstein et al. 1982). An electrophysiological study of a swelling-activated epithelial Cl^- current was first carried out on human intestinal cells (Hazama and Okada 1988). The VRAC current that is

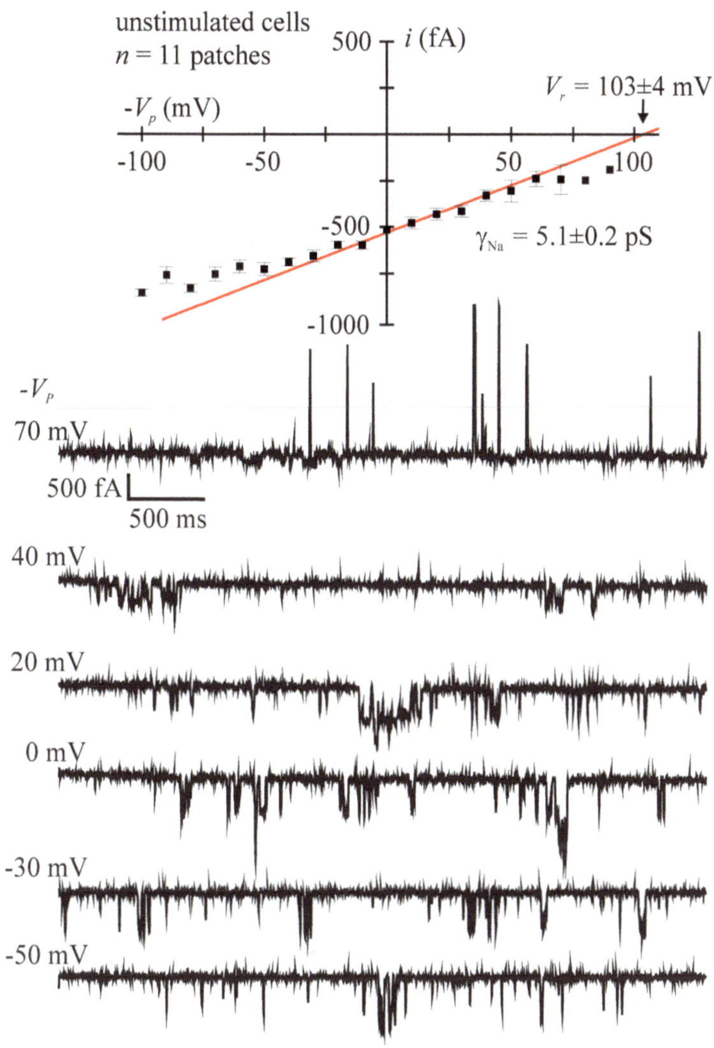

Fig. 4.10 The apical (luminal) Na^+ recirculation pathway of frog skin subepidermal gland epithelium. Na^+ channels in cell-attached patches on the apical membrane of microdissected frog skin mucous gland recorded with Ringer in the pipette. *Upper panel*: The *curve* was obtained by linear regression to points $-V_p > 0$ mV giving a single channel conductance $\gamma_{Na} = 5.1 \pm 0.2$ pS and a reversal potential $-V_r = 103 \pm 4$ mV. With a membrane potential of unstimulated acinar cells of $V_m = -69.5 \pm 0.7$ mV (Sørensen and Larsen, 1999), according to usual sign convention, the reversal potential would be $+33$ mV. *Lower panels*: Examples of single-channel recordings at indicated pipette potentials ($-V_p$) [Redrawn from Sørensen et al. (1998)]

reversibly activated by cell swelling (see Fig. 4.11A) has characteristic biophysical fingerprints (Worrell et al. 1989; Nilius et al. 1997; Okada 1997; Nilius and Droogmans 2003; Stutzin and Hoffmann 2006; Hoffmann et al. 2009; Okada et al. 2009), (*i*) slow deactivation at positive potentials (Fig. 4.11B, C), (*ii*) mild outward rectification in symmetrical configuration and excess GHK rectification in asymmetrical configuration (Fig. 4.11C) and (*iii*) Eisenman type I permeability sequence, $I^- > Br^- > Cl^- > F^-$ (Fig. 4.11D). Single-channel recordings have been performed in rat and mouse renal inner medullary collecting duct (Boese et al. 1996; Meyer and Korbmacher 1996) and human bronchial epithelium (Toczylowska-Maminska and Dolowy 2012). The channel exhibits conductances between 10 and 20 pS at negative potentials and between 40 and 80 pS at positive potentials, which qualitatively conforms to the instantaneous *I-V* relationship of excess GHK rectification in whole-cell configuration (Fig. 4.11C). The mechanism of VRAC activation is not fully described, although many details are known such as: (*i*) the current is activated in the absence of free intracellular Ca^{2+} but requires binding of intracellular ATP to the channel protein, whereas there is an open channel block by extracellular ATP (Eggermont et al. 2001; Okada et al. 2009);

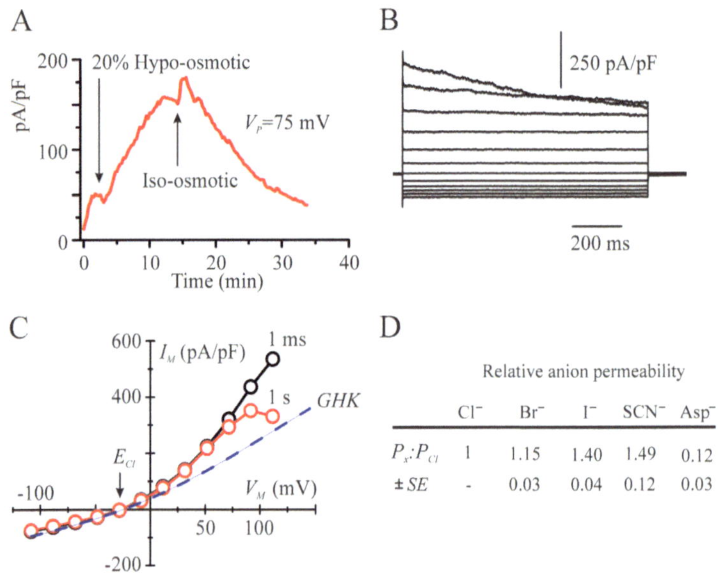

Fig. 4.11 Macroscopic phenotype of VRAC of human embryonic kidney cells at 22 °C studied by whole-cell patch-clamp technique with pipette solution of 40 mM Cl^-, 3 mM ATP and D-gluconic acid lactone and TRIS as major anion and cation, respectively. (**A**) Reversible channel activation by diluting [NaCl] of bath PBS to 20 % hypo-osmolality. V_P is pipette potential. (**B**) VRAC currents obtained from a holding potential of −20 mV by a step protocol with steps ranging from $V_P = -100$ to 120 mV with 20-mV increments and step length of 1 s. Note the slow inactivation of VRAC currents at hyperpolarizing membrane potentials. (**C**) Current–voltage relationships obtained from currents of panel B 1 ms (*black symbols*) and 1 s (*red symbols*) after voltage steps. V_M is membrane potential, which is V_P corrected for liquid junction potential. *Blue dashed line* indicated by *GHK* is fit of the GHK current equation to data points about the reversal potential with $P_{Cl} = 5.8 \cdot 10^{-12}$ cm³/s/pF. The deviation from the 1-ms relationship indicates rectification of VRAC currents in excess of simple electrodiffusion. (**D**) Relative anion permeabilities of VRAC characterized by $I^- > Br^- > Cl^-$ ($n = 3$). E. H. Larsen unpublished

(*ii*) reduced intracellular ionic strength triggers activation of VRAC (Voets et al. 1999) probably by affecting associated signalling protein molecules; (*iii*) the small G-protein RhoA and the Rho-Rho kinase pathway is involved in the activation of VRAC (Nilius et al. 1999; Pedersen et al. 2002a), presumably via modulation of F-actin polymerization (Klausen et al. 2006); (*iv*) phosphatidylinositol 3,4,5-trisphosphate (PIP3) stimulates VRAC (Yamamoto et al. 2008); (*v*) pharmacological evidence indicates a role for tyrosine kinases in regulation of VRAC (see Hoffmann et al. 2009); and (*vi*) reactive oxygen species (ROS), such a H_2O_2, that are released upon cell swelling stimulate VRAC (Browe and Baumgarten 2004; Shimizu et al. 2004).

The molecular identity of VRAC was for years unknown, and many proteins were wrongly suggested to be VRAC (reviewed by Pedersen et al. 2015). Recently, however, two groups both using a cell-based fluorescence assay for iodide fluxes through VRAC and a genome-wide RNAi screen have identified a four-transmembrane protein LRRC8A as an essential component of VRAC (Qiu et al. 2014; Voss et al. 2014). It was suggested that VRAC is formed by LRRC8 hexamers of LRRC8A and one or more of the other LRRC8 isoforms (Voss et al. 2014).

4.7.2 *Potassium Channels*

The K^+ channel picture is quite diverse, and so far members of nine different potassium subfamilies have been shown to be activated after cell swelling including several examples in epithelia. The cell volume-sensitive K^+ channels discussed below are indicated in red in Fig. 4.12. General discussions of the physiology and biophysics of epithelial K^+ channels are given in Chaps. 3 and 19–21 of this book, while Chaps. 26 and 27 contain extended discussions of their pharmacology.

4.7.2.1 Large Conductance or Maxi-(BK) K^+ Channel

This channel is expressed in the apical membrane of rat medullary thick ascending limb (mTAL) (Taniguchi and Guggino 1989), rat collecting tubule (Stoner and Morley 1995), rabbit kidney proximal tubule (Dube et al. 1990), basolateral and apical membrane of acinar cells in frog skin glands (Andersen et al. 1994; Sørensen et al. 2001), human bronchial epithelial cells (Fernandez-Fernandez et al. 2002) and eel intestinal epithelium (Lionetto et al. 2008). In several cases, it was activated at hypo-osmotic conditions. It has been discussed that BK channels themselves are stretch activated; however, examined for mTAL, it turned out not to be the case. Rather, stretch appears to elevate cytosolic Ca^{2+}, which in turn activates Ca^{2+}-controlled BK channels (Taniguchi and Guggino 1989). Similarly, in the majority of cells, activation of BK channels after cell swelling is secondary to an increase in

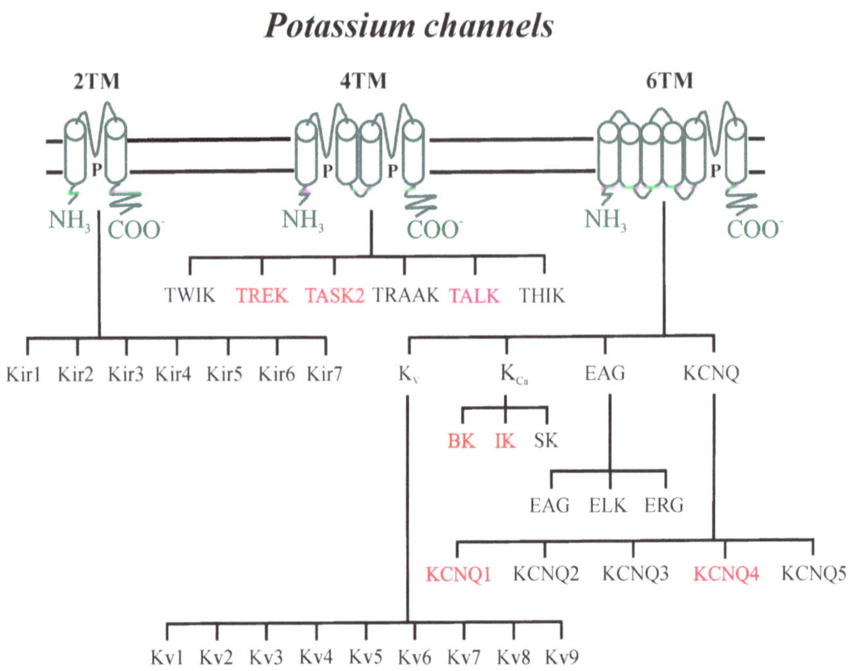

Fig. 4.12 K^+ channel families. Channels activated by cell swelling and discussed in Sect. 4.7.2 are given in *red colour* [Modified from Hoffmann et al. (2009)]

$[Ca^{2+}]_c$ (Calderone 2002), which in some cases is coupled to an autocrine/paracrine swelling-induced ATP release (Wehner 2006).

4.7.2.2 RVD-Mediating Intermediate Conductance (IK) Channels

IK potassium channels have been found in human parotid gland (Begenisich et al. 2004; Thompson and Begenisich 2006), human intestine (Wang et al. 2003), human embryonic kidney (HEK293) cells (Jorgensen et al. 2003), renal A6 cells (Urbach et al. 1999), mouse proximal tubule (Barriere et al. 2003a) and human lens epithelium (Lauf et al. 2008). Swelling activation is in some cells, e.g. intestine 407 cell line, dependent on an increase in $[Ca^{2+}]_c$, which is receptor mediated and coupled to an autocrine/paracrine swelling-induced release of ATP (Okada et al. 2001). When expressed in HEK293 cells, activation of *h*IK channels by cell swelling does, however, not depend on an increase in $[Ca^{2+}]_c$, but a certain level of $[Ca^{2+}]_c$ is required, and the swelling-induced activation of *h*IK channels depends upon integrity of the F-actin cytoskeleton (Jorgensen et al. 2003).

4.7.2.3 The Two-Pore Domain K^+ (K2P) Channel KCNK5

KCNK5 (*K2P5.1*) also named TASK-2 is a two-pore domain K^+ channel belonging to the TALK subgroup of the K2P family of proteins (Fig. 4.12) (see Cid et al. 2013). KCNK5 has a role in volume regulation in mouse renal proximal tubule showed on KCNK5-KO mice (Barriere et al. 2003b). Its role in volume regulation and the associated signalling pathways were studied in detail in EAT cells (Niemeyer et al. 2001a; Niemeyer et al. 2001b). There is a direct tyrosine kinase-dependent phosphorylation of KCNK5 during cell swelling important for channel activation (Kirkegaard et al. 2010). Recently, it was suggested that G-protein modulation of TASK-2 is part of the mechanism by which the channel responds to cell volume changes (Cid et al. 2013). Finally, L'Hoste et al. (2007) showed that hypo-osmotically induced cell swelling of a cell line from mouse proximal kidney tubule is followed by a rapid increase in extracellular pH due to an efflux of HCO_3^- that activates TASK-2 channels. Other volume-sensitive (or 'mechanically gated') members of the KCNK family are KCNK2 and KCNK4 (Maingret et al. 1999a; Maingret et al. 1999b, 2000).

4.7.2.4 KCNQ Channels

The apical KCNQ1 potassium channel was identified as an essential player in RVD of the human mammary epithelial cell line MCF-7 (vanTol et al. 2007), in cells from the inner ear (the KCNQ1 β-subunit KCNE1) (Wangemann et al. 1995) and in murine tracheal epithelial cells (the KCNQ1/KCNE1 complex) (Lock and Valverde 2000). Activation of KCNQ1 channels is implicated in secretion-associated shrinkage in rabbit parietal cells (Bachmann et al. 2007). KCNQ4 channels are also sensitive to changes in volume of non-epithelial cells (Grunnet et al. 2003; Hougaard et al. 2004), but whether such a mechanism plays a role in epithelia has to our knowledge not been studied.

4.7.3 *Na^+/H^+ Exchangers*

The Na^+/H^+ exchanger NHE1 (*SLC9A1*), of the *SLC9A* family, was initially described as a pH regulator but is also involved in cell volume regulation by being activated during cell shrinkage and uptake of sodium ions (Rotin and Grinstein 1989) together with chloride via a Cl^-/HCO_3^- exchanger as first suggested by Cala (1980). NHE1 is essentially ubiquitous and is localized to the basolateral membrane of many types of epithelial cells, e.g. ileus villus, intestinal crypts, rat alveolar epithelium and several nephron segments (Wakabayashi et al. 1997). NHE1 is both shrinkage activated (e.g. Orlowski and Grinstein 2011) and swelling inhibited (Elsing et al. 2007). While other plasma membrane NHE isoforms, including NHE2 and NHE4, are also activated by cell shrinkage (see below), their expression is more restricted, and NHE1 is by far the most important mechanism of NHE-mediated RVI, reviewed by Hoffmann et al. (2009). The activation of NHE1 by cell shrinkage involves an alkaline shift in H^+ sensitivity of the exchanger, e.g. Wakabayashi et al. (1997) and Hendus-Altenburger et al. (2014), and a change in the allosteric affinity constant of NHE1 in favour of the activated state that leaves V_{max} unchanged (Lacroix et al. 2008). Several molecular mechanisms for the activation of NHE1 by cell shrinkage have been described, and a valuable discussion can be found in Hendus-Altenburger et al. (2014). It has been demonstrated that NHE1 activation by cell shrinkage is a consequence of shrinkage per se rather than an increased osmolality (Krump et al. 1997). While this cannot fully account for its shrinkage sensitivity, NHE1 responds directly to mechanical stress applied to the plasma membrane or to changes in lipid packing and cholesterol concentration (Lacroix et al. 2008; Tekpli et al. 2008; Fuster et al. 2004). This suggests that NHE1 mechanosensitivity is associated with the ability to sense membrane differential packing (Pang et al. 2012) and/or is due to direct interaction between NHE1 and specific lipids (Wakabayashi et al. 1997). The precise mechanisms through which cell shrinkage eventually translates into activation of NHE1 remain to be elucidated. Post-translational mechanisms, directly or indirectly, play a role in congruence with known roles of protein kinases and phosphatases that modulate shrinkage-activated NHE1 activity (Pedersen et al. 2002b). Shrinkage activation of NHE1 was originally thought not to involve altered phosphorylation of NHE1 directly (Grinstein et al. 1992), but a recent mass spectrometry analysis of *Amphiuma tridactylum* NHE1 demonstrated that NHE1 is phosphorylated on multiple residues during osmotic cell shrinkage (Rigor et al. 2011). Post-translational regulation of NHE1 during cell shrinkage might also involve altered interaction of NHE1 with its multiple binding partners, most of which are regulated by phosphorylation (see Hendus-Altenburger et al. 2014).

In addition to NHE1, plasma membrane NHEs include NHE2–NHE5. NHE2 (*SLCA2*) and NHE3 (*SLCA3*) are apical isoforms restricted to various types of epithelial cells in the stomach/gastrointestinal tract, kidney and gallbladder (Wakabayashi et al. 1997). For NHE2, depending on cell type, shrinkage activation, shrinkage inhibition and lack of shrinkage effect have been reported. There is agreement, however, that NHE3 is inhibited by cell shrinkage and activated by cell swelling (see, e.g. Hendus-Altenburger et al. 2014). NHE4 is basolateral and found mainly in the stomach, at intermediate levels in the small intestine and colon and at low levels in the kidney (Orlowski and Grinstein 2004). Finally, NHE4 is potently shrinkage activated in a manner dependent on the cytoskeleton and is important for RVI in the kidney medulla (Bookstein et al. 1994), whereas NHE5 is inhibited by cell shrinkage (Attaphitaya et al. 2001).

4.7.4 Na^+-K^+-$2Cl^-$ Cotransporters (NKCC)

While this subsection focuses on the cotransporters' significance for epithelial volume regulation, more general accounts are presented in Chap. 11 of this book. NKCC1 is encoded by the *SLC12A2* gene, whereas NKCC2 is encoded by *SLC12A1*. Both NKCC1 and NKCC2 form homodimers (Moore-Hoon and Turner 2000). NKCC1 and NKCC2 are both activated by cell shrinkage and by a decrease in $[Cl^-]_c$ (Markadieu and Delpire 2014). The two stimuli are often working in opposite direction as cell shrinkage caused by an increase in external osmolality is accompanied by an increase in intracellular Cl^-. Thus, decrease in cell volume gives an activation signal and increase in intracellular Cl^- an inhibition signal. As a result, many cell types fail to exhibit RVI when exposed directly to a hyperosmotic solution, but perform RVI only when they are returned to iso-osmotic conditions after a hypo-osmotic pretreatment that results in Cl^- loss. This is exemplified in Fig. 4.6A, B above and reviewed by Hoffmann et al. (2009).

NKCC1 is widely expressed in vertebrates and is a central effector in cell volume regulation. It is located to the basolateral membrane of cells of secretory and absorptive functions, e.g. exocrine glands, airways, intestinal tract, inner medullary collecting duct and shark rectal gland (Hebert et al. 2004; Markadieu and Delpire 2014). In salivary glands, where NKCC1 is involved in the secretion of fluid, there is a severe impairment of salivation in NKCC1-deficient mice (Evans et al. 2000). In the small intestine, NKCC1 together with the Na^+/K^+ pump drives fluid secretion (Jakab et al. 2011) (see Sect. 4.4.3). In marginal cells of the inner ear, NKCC1 has been proposed as a component of the entry pathway for K^+ that is secreted into the endolymph thus playing a critical role in hearing (Flagella et al. 1999). NKCC2 exhibits a highly restricted pattern of expression. It is expressed in the apical membrane of the thick ascending limb of Henle (TALH) where it reabsorbs a large amount of NaCl (Ares et al. 2011). NKCC2 homologues are also found in non-mammalian vertebrates including shark kidney (Gagnon et al. 2002) and eel intestine (Lionetto and Schettino 2006).

Two families of protein kinases, the *w*ith *n*o lysine (*K*) kinases (WNKs) and the Ste20-related kinases named Ste20-related proline/alanine-rich kinase (SPAK) and oxidative stress-responsive kinase 1 (OSR1), play major roles in the shrinkage activation (see Sect. 4.9.4 for references). The regulation of NKCC1/NKCC2 by the kinases is complex and excellently described in a recent review by Gagnon and Delpire (2012). Beyond SPAK/OSR1, other ser/thr protein kinases are involved in shrinkage-induced NKCC1 activation, including PKC (Gagnon and Delpire 2012), MLCK (Klein and O'Neill 1995; Hoffmann and Pedersen 2007; Lionetto et al. 2002) and FAK (Hoffmann et al. 2007). PKC-δ acts upstream of SPAK and OSR1 in the hyperosmotically induced activation of NKCC1. Thus, SPAK kinase activity increased in airway epithelial cells after treatment with a PKC-δ activator and decreased after treatment with a PKC-δ inhibitor (see Gagnon and Delpire 2012). Furthermore, PKC and MLCK contribute to hyperosmotic activation of the apical NKCC2 in intestinal epithelial cells of European eel, which is strongly dependent on integrity of both F-actin and microtubules (Lionetto et al. 2002; Lionetto and Schettino 2006).

4.8 Putative Sensors of Cell Volume and Cell Volume Changes

Cell volume changes result in alterations in (1) macromolecular crowding, (2) cellular ionic strength or concentrations of specific ions or (3) mechanical/chemical changes in the lipid bilayer. All three mechanisms play a role in response to volume perturbations. None of them, however, have been specifically studied in epithelia and will not be dealt with here. The reader is referred to reviews by Burg (2000), Minton (2006) and Pedersen and Nilius (2007). For our purpose, we focus on the nature of 'osmosensors'. The mechanism of osmosensing is still incompletely understood in vertebrate cells as compared to osmosensing in bacteria and fungi (Bahn et al. 2007). Several membrane proteins, like integrins, growth factor receptors (GFRs), cytokine receptors, calcium-sensing receptors (CaRs) and transient receptor potential (TRP) channels, seem to play a role as sensors of cell volume changes. Schematically, Fig. 4.13 indicates how a membrane protein may sense changes in cell volume, which shall be detailed below.

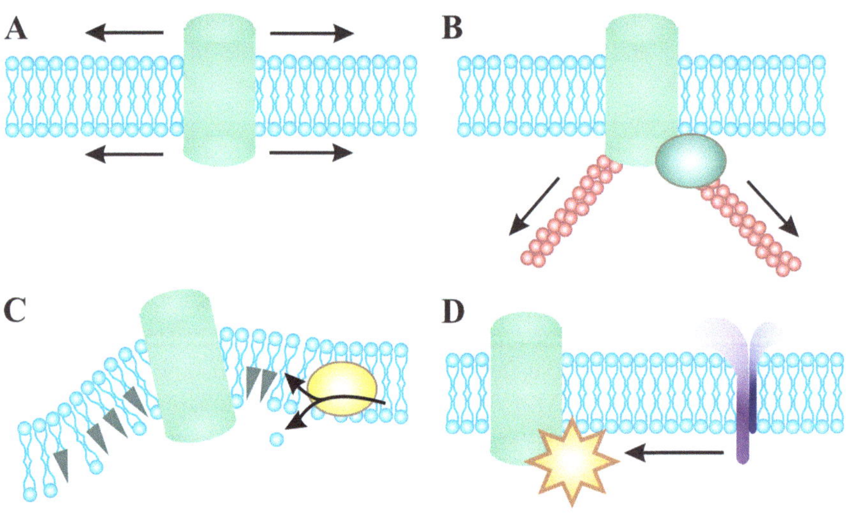

Fig. 4.13 Mechanisms by which a cell volume change may be transmitted to a membrane protein. In all panels, the volume-sensitive membrane protein is depicted as a *green cylinder*. (**A**) Stretch sensitivity of the protein. (**B**) Direct or indirect attachment to cytoskeleton. (**C**) Changes in membrane curvature either through changes in membrane composition or mechanically. (**D**) Integrins activated by cell volume perturbation interacting with the membrane protein. See text for further details [Redrawn from Pedersen et al. (2011)]

4.8.1 *Integrins and Other Receptors*

Integrins link the internal actin cytoskeleton to the extracellular matrix that may mediate mechano-transduction. Integrin heterodimers associate with cytoskeletal and signalling proteins in focal adhesions (FAs). Upon integrin activation, the focal adhesion kinase (FAK) is phosphorylated, which mediates a coordinating role in the signalling events (see Pedersen et al. 2011). Direct evidence for roles of integrins in osmo-/volume sensing in epithelia is so far rather limited. Integrin alpha1beta1 ($\alpha_1\beta_1$-integrin) is a player in regulation of renal medullary osmolyte concentration, and hyperosmotic conditions induce mRNA and protein expression of β_1-integrin in MDCK cells (see Burg et al. 2007). The increase in osmolyte accumulation induced by hyperosmotic treatment was strongly impaired in the kidney medulla of α1-integrin-null mice (Moeckel et al. 2006). Integrins contribute to the osmosensitivity of some ion transport proteins. Examples are the TRPV4-mediated Ca^{2+} influx in epithelial cells that occurs under hypo-osmotic stress and during activation by cell shrinkage of NKCC1 (see review by Hoffmann et al. 2009). Studies of cell shrinkage in kidney cells indicate that hyperosmotic conditions block signalling by growth factor receptor protein tyrosine kinases immediately downstream of Ras (Copp et al. 2005). The role of Ca^{2+} receptors (CaRs) has been studied as salinity sensors in fish (Nearing et al. 2002; Okada et al. 2001; Fiol and Kultz 2007).

4.8.2 *Transient Receptor Potential (TRP) Channels*

The role of TRP channels in cell volume regulation is probably to mediate an increase in $[Ca^{2+}]_c$, which stimulates RVD by activating Ca^{2+}-controlled K^+ and Cl^- channels. Thus, the increase in $[Ca^{2+}]_c$ after cell swelling especially in epithelia seems to depend on Ca^{2+} influx via TRP channels (Arniges et al. 2004; Numata et al. 2007; Strotmann et al. 2000; Liedtke and Kim 2005; Cohen 2005). All TRPV ('V' for vanilloid) channels are Ca^{2+} permeable with $P_{Ca}/P_{monovalent}$ between 1 and 10 for TRPV1–TRPV4 and $P_{Ca}/P_{monovalent} > 100$ for TRPV5 and TRPV6 (Owsianik et al. 2006). Several TRPs are sensitive to cell volume and/or to membrane stretch (Christensen and Corey 2007; Pedersen and Nilius 2007), and TRPC1, TRPC6, TRPV4 and TRPM7 are all candidate volume sensors, best documented for TRPV4 (Nilius et al. 2007). Complete description of osmotic activation of TRPV4 is still awaiting (Everaerts et al. 2010; Pochynyuk et al. 2013). 'With no lysine kinase 4' (WNK4) and aquaporin-5 are involved in the osmosensitivity (Fu et al. 2006; Liu et al. 2006), and recently hypo-osmotic TRPV4 activation was found to depend on phosphatidylinositol 4,5-diphosphate (PIP2) binding to the cytosolic N-terminus (Garcia-Elias et al. 2013). Finally, a series of highly conserved aromatic residues in transmembrane (TM) helices 5–6 have profound importance for TRPV4 activity in response to cell swelling (Klausen et al. 2014).

4.8.3 Phospholipases of the Phospholipase A2 (PLA2) Family

PLA2-mediated hydrolysis of phospholipids for releasing arachidonic acid initiates a swelling-induced signalling cascade that includes potent biological substances of the eicosanoid family, some of which are involved in volume regulation (Hoffmann et al. 2007). Thus, osmotic sensitivity of TRPV4 (Vriens et al. 2004) depends on activation of phospholipase A_2 and subsequent production of arachidonic acid metabolite 5′-6′-epoxyeicosatrienoic acid (EET). In Ehrlich cells, the arachidonic acid metabolite leukotriene D4 (LTD_4) is released upon hypo-osmotic stress and activates the TASK-2 channel (Hougaard et al. 2000). Several PLA2 isoforms may be stimulated during cell swelling. The swelling-induced arachidonic acid release and RVD thus involve Ca^{2+}-dependent PLA2 activity in rat inner medulla collecting duct cells (Tinel et al. 2000).

4.8.4 Cytoskeleton

The cytoskeleton, specifically F-actin, is important for sensing and transduction of cell volume perturbations (Ingber 2006). Reorganization of the cortical F-actin occurs rapidly after volume changes leading to net increase or net decrease in actin polymerization after shrinkage and swelling, respectively (Hoffmann et al. 2009). This change in cortical F-actin was found in non-adherent cells. In epithelial cells, cytosolic polymerized F-actin changes very little, and the change in cortical F-actin goes along with a change in 'stress fibres' (Di Ciano-Oliveira et al. 2006). Little is known about the significance of the intermediate filaments or microtubules. In shark rectal gland, hypo-osmotic swelling affects neither microtubules nor intermediate filaments (Henson et al. 1997). In opossum kidney cells, vimentin is reorganized into short fragments after hypo-osmotic treatment (Dartsch et al. 1994).

The cytoskeleton may influence volume regulation by several mechanisms: (*i*) the transport protein may interact with the cytoskeleton, (*ii*) the cytoskeleton may regulate signalling events that are controlling transport proteins, or (*iii*) dependent on the cytoskeleton, cell shrinkage may inhibit and cell swelling stimulate exocytosis (van der Wijk et al. 2003). In turn, this regulates transporters/channels through signal molecules like the swelling-induced release of ATP in intestine 407 cells (van der Wijk et al. 2003), or through insertion of transport proteins, for examples, see Hoffmann et al. (2009). The Rho family small G-proteins, Rho, Rac and Cdc42, are regulators of actin organization and sensitive to cell volume changes (Lewis et al. 2002; Hall 1998). They may be involved in cytoskeletal reorganizations during RVD and RVI. Hyperosmotic cell shrinkage in kidney epithelial cells increases Rho activity (Di Ciano-Oliveira et al. 2003). In intestine 407 cells, a p21Rho signalling cascade and actin filaments are involved in activation of VRAC (Tilly et al. 1996).

4.9 Signal Transduction in Response to Cell Volume Perturbations

4.9.1 Free Intracellular Ca^{2+} Concentration

Increase in cytosolic free Ca^{2+} is involved in RVD in many epithelial cells. In human intestine 407 culture cells, a biphasic rise in $[Ca^{2+}]_c$ depends on Ca^{2+} influx from extracellular fluid as well as Ca^{2+} release from intracellular stores (Hazama and Okada 1990), which act synergistically in activating K^+ and Cl^- channels. Also in rat, the inner medullary collecting duct is the channel activation dependent on Ca^{2+} release from intracellular stores (Tinel et al. 2000), but this is not the general mechanism, reviewed by Hoffmann et al. (2009). Importance of Ca^{2+} entry for RVD is reported for intestinal cells (MacLeod and Hamilton 1999), thick ascending loop of Henle (Montrose-Rafizadeh et al. 1991; Tinel et al. 2002), rabbit corneal epithelium (Wu et al. 1997) and mouse distal colon (Mignen et al. 1999). The influx of Ca^{2+} is probably predominantly via TRP channels, which play a crucial role in the response to mechanical and osmotic perturbations in a wide range of cell types (Pedersen and Nilius 2007) (see Sect. 4.8.2). In renal cortical collecting tubule, TRPV4 associates with AQP2 in hypo-osmotic activation of TRPV4 and subsequent Ca^{2+} regulation of cell volume (Galizia et al. 2012).

It was found that swelling-induced VRAC activation does not depend on raised intracellular $[Ca^{2+}]$ (see Sect. 4.7.1 and Hoffmann et al. 2009). However, as suggested by Okada's group, it is possible that activated VRAC molecules are located in the immediate vicinity of individual Ca^{2+} channels that generate so-called Ca^{2+} nanodomains of very high intracellular $[Ca^{2+}]$ (Akita et al. 2011).

4.9.2 Role of ATP Release

ATP is released during cell swelling in several epithelial cells, which results in stimulation of purinergic receptors and Ca^{2+} influx (Novak 2011). In many secretory epithelia, luminally applied UTP/ATP increases intracellular Ca^{2+} and stimulates Ca^{2+}-activated Cl^- channels (CaCC), most likely via P2Y2 receptors. In mouse colonic epithelium and salivary gland, the Ca^{2+}-activated Cl^- channel anoctamin-1 (and other anoctamins) is activated by cell swelling through such an autocrine mechanism that depends on ATP release, binding of ATP to purinergic P2Y2 receptors and increase in $[Ca^{2+}]_c$ (Almaca et al. 2009). A number of ATP release mechanisms have been proposed, including ATP release via anion channels (Sabirov and Okada 2005).

4.9.3 *Mitogen-Activated Protein Kinases (MAPKs)*

p38 MAPKs, c-Jun N-terminal kinases (JNK1, JNK2) and extracellular signal-regulated kinase 1/2 (ERK1/ERK2) are modulated by osmotic stress in many mammalian epithelial cells including the cells in medullary thick ascending limb (Roger et al. 1999) and human tracheal epithelial cells (Liedtke and Cole 2002). Cell swelling is generally associated with ERK1/ERK2 stimulation and cell shrinkage with stimulation of p38 MAPK and JNK. However, there are exceptions where opposite effects are seen which support the idea that the MAPKs are activated by cell volume disturbances in general rather than by swelling or shrinkage as such (see Hoffmann et al. 2009).

There is evidence for involvement of ERK, p38 MAPK and JNK in the control of RVI in certain epithelia (Liedtke and Cole 2002; Roger et al. 1999) and of RVD in others. In rabbit corneal epithelium, swelling-induced ERK and JNK stimulation precedes Cl^- and K^+ channel activations, whereas p38 activation occurs as a consequence of RVD (Pan et al. 2007). The most well-studied role of MAPK activation during osmotic stress is the long-time adaptation at the level of gene transcription regulation predominantly via p38 MAPK-mediated effects on the osmoprotective transcription factor TonEBP (tonicity-responsive enhancer binding protein), which leads to increased osmolyte transporter expression (Tsai et al. 2007). It is well established that MAPK activation by osmotic stress in the yeast is via the p38 MAPK orthologue HOG1, which is activated by small G-proteins and some scaffold proteins. The mechanism by which cell shrinkage leads to activation of the MAPKs in mammalian cells is still incompletely understood, but there might well be parallels to the yeast pathway. Thus, the small G-protein Cdc42 and/or Rac may be important upstream activators of p38 MAPK after cell shrinkage (Uhlik et al. 2003; Friis et al. 2005).

4.9.4 *With No Lysine Kinases (WNKs) and Ste20-Related Kinases*

The WNK has its name due to the lack of lysine in subdomain II (Huang et al. 2007). The best described role of the WNKs is in regulation of epithelial Na^+, K^+, Cl^-, HCO_3^- and Ca^{2+} transporters (Yang et al. 2003; Vitari et al. 2005; McCormick and Ellison 2011). Among these are the cation-chloride cotransporters, the renal outer medulla K^+ channel (ROMK1), ENaC, CFTR and TRPV4 and TRPV5 (see Peng and Warnock 2007). WNKs are ser/thr protein kinases that

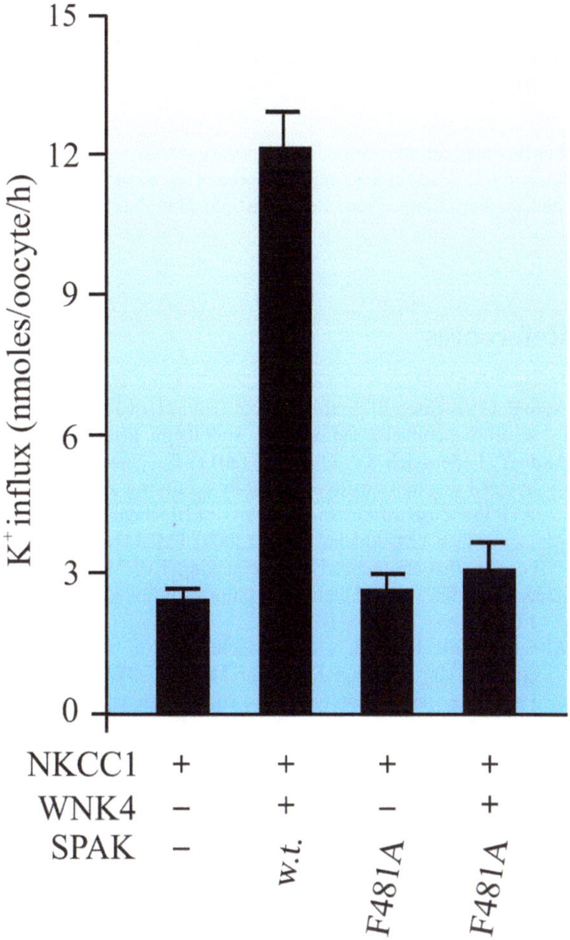

Fig. 4.14 Loss of NKCC1 function when expressed with the SPAK F481A mutant or in the absence of WNK4. Expressed in *Xenopus laevis* oocytes with the K^+ influx via NKCC1 monitored by $^{86}Rb^+$. Bars represent means ± SE ($n = 20$–25 oocytes) [Data from Delpire Laboratory, redrawn from Gagnon and Delpire (2012)]

include four isoforms in mammalian cells, named WNK1–WNK4 (Verissimo and Jordan 2001; Kahle et al. 2006; Peng and Warnock 2007). They are volume-sensitive kinases and are thus very important in secretory epithelia as they regulate NKCC1 during cell shrinkage (Park et al. 2012), which is the main focus below.

The WNKs do not generally act directly on targeted transporters, but through phosphorylation of SPAK and OSR1 (Vitari et al. 2005; Richardson and Alessi 2008; Delpire 2009; Delpire and Austin 2010). Figure 4.14 shows that mutations in the binding pocket of SPAK results in loss of WNK4-SPAK activation of NKCC1. A high expression level of SPAK is found in several Cl^--secreting epithelia like the gastric mucosa, sublingual gland and salivary gland. In the salivary gland, SPAK is localized to the basolateral membrane where NKCC1 is highly expressed. SPAK and OSR1 co-localize with NKCC2 in the thick ascending limb of Henle (Gagnon and Delpire 2012). The WNKs seem to have a scaffolding role in activation of

SPAK/OSR1 as demonstrated in a study of NKCC1 regulation (Piechotta et al. 2003). Thus, during cell shrinkage, WNK1 and WNK4 bind, phosphorylate and activate OSR1 and SPAK (Richardson and Alessi 2008; Delpire and Austin 2010).

Acknowledgements Stine F. Pedersen is acknowledged for suggestions to and critical reading of Sect. 4.7.3. Work in the authors' laboratories was supported by the Natural Science Foundation, the Carlsberg Foundation, the Augustinus Foundation and the Brødrene Hartmann Foundation.

References

Agre P, Christensen EI, Smith BL et al (1993) Distribution of the aquaporin CHIP in secretory and resorbtive epithelia and capillary endothelia. Proc Natl Acad Sci USA 90:7275–7279

Akita T, Fedorovich SV, Okada Y (2011) Ca^{2+} nanodomain-mediated component of swelling-induced volume-sensitive outwardly rectifying anion current triggered by autocrine action of ATP in mouse astrocytes. Cell Physiol Biochem 28:1181–1190

Almaca J, Tian YM, Aldehni F et al (2009) TMEM16 proteins produce volume-regulated chloride currents that are reduced in mice lacking TMEM16A. J Biol Chem 284:28571–28578

Alpern RJ (1987) Apical membrane chloride base-exchange in the rat proximal convoluted tubule. J Clin Invest 79:1026–1030

Altenberg GA, Reuss L (2013) Mechanisms of water transport across cell membranes and epithelia. In: Alpern RJ, Caplan M, Moe OW (eds) Seldin and Giebisch's The Kidney. Physiology and Pathophysiology, vol 1. Elsevier/Academic, Amsterdam, pp 95–120

Andersen HK, Urbach V, Van Kerkhove E et al (1994) Maxi K^+ channels in the basolateral membrane of the exocrine frog skin gland regulated by intracellular calcium and pH. Pflugers Arch 431:52–65

Ares GR, Caceres PS, Ortiz PA (2011) Molecular regulation of NKCC2 in the thick ascending limb. Am J Physiol Renal Physiol 301:F1143–F1159

Arniges M, Vazquez E, Fernandez-Fernandez JM et al (2004) Swelling-activated Ca^{2+} entry via TRPV4 channel is defective in cystic fibrosis airway epithelia. J. Biol Chem 279:54062–54068

Attaphitaya S, Nehrke K, Melvin JE (2001) Acute inhibition of brain-specific Na^+/H^+ exchanger isoform 5 by protein kinases A and C and cell shrinkage. Am J Physiol Cell Physiol 281: C1146–C1157

Bachmann O, Heinzmann A, Mack A et al (2007) Mechanisms of secretion-associated shrinkage and volume recovery in cultured rabbit parietal cells. Am J Physiol Gastrointest Liver Physiol 292:G711–G717

Bagnasco S, Balaban RS, Fales HM et al (1986) Predominant osmotically active organic solutes in rat and rabbit renal medullas. J Biol Chem 261:5872–5877

Bahn YS, Xue CY, Idnurm A et al (2007) Sensing the environment: lessons from fungi. Nat Rev Microbiol 5:57–69

Balaban RS, Burg MB (1987) Osmotically active organic solutes in the renal inner medulla. Kidney Int 31:562–564

Barriere H, Belfodil R, Rubera I et al (2003a) Role of TASK2 potassium channels regarding volume regulation in primary cultures of mouse proximal tubules. J Gen Physiol 122:177–190

Barriere H, Rubera I, Belfodil R et al (2003b) Swelling-activated chloride and potassium conductance in primary cultures of mouse proximal tubules. Implication of KCNE1 protein. J Membr Biol 193:153–170

Beck JS, Potts DJ (1990) Cell swelling, cotransport activation and potassium conductance in isolated perfused rabbit kidney proximal tubules. J Physiol 425:369–378

Begenisich T, Nakamoto T, Ovitt CE et al (2004) Physiological roles of the intermediate conductance, Ca^{2+}-activated potassium channel Kcnn4. J Biol Chem 279:47681–47687
Bellantuono V, Cassano G, Lippe C (2008) The adrenergic receptor substypes in frog (*Rana esculenta*) skin. Comp Biochem Physiol C 148:160–164
Boese SH, Kinne RK, Wehner F (1996) Single channel properties of swelling-activated anion conductance in rat inner medullary collecting duct cells. Am J Physiol Renal Physiol 271: F1224–F1233
Bookstein C, Musch MW, Depaoli A et al (1994) A unique sodium-hydrogen exchange isoform (Nhe-4) of the inner medulla of the rat-kidney is induced by hyperosmolarity. J Biol Chem 269:29704–29709
Boucher RC (1994) Human airway ion transport. Part One. Am J Resp Crit Care 150:271–281
Boucher RC (1999) Topical review: molecular insights into the physiology of the ‘thin film’ of airway surface liquid. J Physiol 516:631–638
Boucher RC (2003) Regulation of airway surface liquid volume by human airway epithelia. Pflugers Arch 445:495–498
Breton S, Brown D (2013) Regulation of luminal acidification by the V-ATPase. Physiology (Bethesda) 28:318–329
Browe DM, Baumgarten CM (2004) Angiotensin II (AT1) receptors and NADPH oxidase regulate Cl- current elicited by beta 1 integrin stretch in rabbit ventricular myocytes. J Gen Physiol 124:273–287
Burg MB (2000) Macromolecular crowding as a cell volume sensor. Cell Physiol Biochem 10:251–256
Burg MB, Knepper MA (1986) Single tubule perfusion techniques. Kidney Int 30:166–170
Burg MB, Kwon ED, Kultz D (1997) Regulation of gene expression by hypertonicity. Ann Rev Physiol 59:437–455
Burg MB, Ferraris JD, Dmitrieva NI (2007) Cellular response to hyperosmotic stresses. Physiol Rev 87:1441–1474
Cala PM (1980) Volume regulation by amphiuma red-blood-cells - Nature of the ion flux pathways. Fed Proc 39:379–379
Calderone V (2002) Large-conductance, Ca^{2+}-activated K^+ channels: function, pharmacology and drugs. Curr Med Chem 9:1385–1395
Cantiello HF, Patenaude CR, Ausiello DA (1989) G-protein subunit, alpha-I-3, activates a pertussis toxin-sensitive Na^+ channel from the epithelial-cell line, A6. J Biol Chem 264:20867–20870
Carpi-Medina P, Whittembury G (1988) Comparison of transcellular and transepithelial water osmotic permeabilities (P_{os}) in the isolated proximal straight tubule (PST) of the rabbit kidney. Pflugers Arch 412:66–74
Carpi-Medina P, Gonzáles E, Whittembury G (1983) Cell osmotic water permeability of isolated rabbit proximal convoluted tubules. Am J Physiol Renal Physiol 244:F554–F563
Carpi-Medina P, Lindemann B, Gonzáles E et al (1984) The continous measurement of tubular volume changes in response to step changes in contraluminal osmolarity. Pflugers Arch 400:343–348
Cassola AC, Mollenhauer M, Frömter E (1983) The intracellular chloride activity of rat kidney proximal tubular cells. Pflugers Arch 399:259–265
Chamberlin ME, Strange K (1989) Anisosmotic cell volume regulation: a comparative view. Am J Physiol 257:C159–C173
Chase HS, Alawqati Q (1983) Calcium reduces the sodium permeability of luminal membrane-vesicles from toad bladder - Studies using a fast-reaction apparatus. J Gen Physiol 81:643–665
Choe K, Strange K (2009) Volume regulation and osmosensing in animal cells. In: Evans DH (ed) Osmotic and Ionic Regulation. Cells and Animals. CRC Press, Taylor and Francis Group, Boca Raton, FL, pp 37–67
Christensen AP, Corey DP (2007) TRP channels in mechanosensation: direct or indirect activation? Nat Rev Neurosci 8:510–521

Cid LP, Roa-Rojas HA, Niemeyer MI et al (2013) TASK-2: a K2P K^+ channel with complex regulation and diverse physiological functions. Front Physiol 4(Article 198):1–9

Cohen DM (2005) TRPV4 and the mammalian kidney. Pflugers Arch 451:168–175

Cook DI, Van Lennep EW, Roberts ML et al (1994) Secretion by the major salivary glands. Physiol Gastrointest Tract 2:1061–1117

Copp J, Wiley S, Ward MW et al (2005) Hypertonic shock inhibits growth factor receptor signaling, induces caspase-3 activation, and causes reversible fragmentation of the mitochondrial network. Am J Physiol Cell Physiol 288:C403–C415

Cotton CU, Weinstein AM, Reuss L (1989) Osmotic water permeability of Necturus gallbladder epithelium. J Gen Physiol 93:649–679

Curran PF (1960) Na, Cl, and water transport by rat ileum *in vitro*. J Gen Physiol 43:1137–1148

Dartsch PC, Kolb HA, Beckmann M et al (1994) Morphological alterations and cytoskeletal reorganization in opossum kidney (Ok) cells during osmotic swelling and volume regulation. Histochemistry 102:69–75

Davis CW, Finn AL (1982) Sodium transport inhibition by amiloride reduces basolateral membrane potassium conductance in tight epithelia. Science 216:525–527

Davis CW, Finn AL (1985) Effects of mucosal sodium removal on cell volume in *Necturus* gallbladder epithelium. Am J Physiol Cell Physiol 249:C304–C312

Delpire E (2009) The mammalian family of sterile 20p-like protein kinases. Pflugers Arch 458:953–967

Delpire E, Austin TM (2010) Kinase regulation of Na^+-K^+-$2Cl^-$ cotransport in primary afferent neurons. J Physiol Lond 588:3365–3373

Di Ciano-Oliveira C, Sirokmany G, Szaszi K et al (2003) Hyperosmotic stress activates Rho: differential involvement in Rho kinase-dependent MLC phosphorylation and NKCC activation. Am J Physiol Cell Physiol 285:C555–C566

Di Ciano-Oliveira C, Thirone ACP, Szaszi K et al (2006) Osmotic stress and the cytoskeleton: the R(h)ole of Rho GTPases. Acta Physiol 187:257–272

Diamond JM (1964) Transport of salt and water in rabbit and Guinea pig Gallbladder. J Gen Physiol 48:1–14

Diamond JM (1982) Trans-cellular cross-talk between epithelial-cell membranes. Nature 300:683–685

Dörge A, Rick R, Beck F-X et al (1985) Cl transport across the basolateral membrane in frog skin epithelium. Pflugers Arch 405(Suppl 1):S8–S11

Douglas IJ, Brown PD (1996) Regulatory volume increase in rat lacrimal gland acinar cells. J Membr Biol 150:209–217

Dube L, Parent L, Sauve R (1990) Hypotonic shock activates a maxi K^+ channel in primary cultured proximal tubule cells. Am J Physiol Renal Physiol 259:F348–F356

Eggermont J, Trouet D, Carton I, Nilius B (2001) Cellular function and control of volume-regulated anion channels. Cell Biochem Biophys 35:263–274

Elsing C, Gosch I, Hennings JC et al (2007) Mechanisms of hypotonic inhibition of the sodium, proton exchanger type 1 (NHE1) in a biliary epithelial cell line (Mz-Cha-1). Acta Physiol 190:199–208

Ericson A, Spring KR (1982a) Coupled NaCl entry into *Necturus* gallbladder epithelial-cells. Am J Physiol Cell Physiol 243:C140–C145

Ericson AC, Spring KR (1982b) Volume regulation by *Necturus* gallbladder - Apical Na^+-H^+ and Cl^--HCO_3^- exchange. Am J Physiol Cell Physiol 243:C146–C150

Ernst SA, Mills JW (1977) Basolateral plasma membrane localization of ouabain-sensitive sodium transport sites in the secretory epithelium of the avian salt gland. J Cell Biol 75:74–94

Evans RL, Park K, Turner RJ, Watson GE et al (2000) Severe impairment of salivation in $Na^+/K^+/2Cl^-$ cotransporter (NKCC1)-deficient mice. J Biol Chem 275:26720–26726

Evans DH, Piermarini PM, Choe KP (2005) The multifunctional fish gill: dominant site of gas exchange, osmoregulation, acid–base regulation, and excretion of nitrogenous waste. Physiol Rev 85:97–177

Eveloff JL, Calamia J (1986) Effect of osmolarity on cation fluxes in medullary thick ascending limb cells. Am J Physiol Renal Physiol 250:F176–F180
Everaerts W, Nilius B, Owsianik G (2010) The vanilloid transient receptor potentialchannel TRPV4: from structure to disease. Progr Biophys Mol Biol 103:2–17
Fernandez-Fernandez JM, Nobles M, Currid A et al (2002) Maxi K^+ channel mediates regulatory volume decrease response in a human bronchial epithelial cell line. Am J Physiol Cell Physiol 283:C1705–C1714
Ferraris JD, Burg MB (2006) Tonicity-dependent regulation of osmoprotective genes in mammalian cells. Contrib Nephrol 152:125–141
Fiol DF, Kultz D (2007) Osmotic stress sensing and signaling in fishes. FEBS J 274:5790–5798
Fisher RS, Persson BE, Spring KR (1981) Epithelial cell volume regulation - bicarbonate dependence. Science 214:1357–1359
Flagella M, Clarke LL, Miller ML et al (1999) Mice lacking the basolateral Na-K-2Cl cotransporter have impaired epithelial chloride secretion and are profoundly deaf. J Biol Chem 274:26946–26955
Ford P, Rivarola V, Chara O et al (2005) Volume regulation in cortical collecting duct cells: role of AQP2. Biol Cell 97:687–697
Foskett JK (1982) The chloride cell: definitive identification as the salt-secretory cell in teleosts. Science 215:164–166
Foskett JK, Bern HA, Machen TE et al (1983) Chloride cells and the hormonal control of teleost fish osmoregulation. J Exp Biol 106:255–281
Foskett JK, Wong MMM, Sueaquan G et al (1994) Isosmotic modulation of cell-volume and intracellular ion activities during stimulation of single exocrine cells. J Exp Zool 268:104–110
Friis MB, Friborg CR, Schneider L et al (2005) Cell shrinkage as a signal to apoptosis in NIH 3T3 fibroblasts. J Physiol 567:427–443
Fu Y, Subramanya A, Rozansky D et al (2006) WNK kinases influence TRPV4 channel function and localization. Am J Physiol Renal Physiol 290:F1305–F1314
Fuster D, Moe OW, Hilgemann DW (2004) Lipid- and mechanosensitivities of sodium/hydrogen exchangers analyzed by electrical methods. Proc Natl Acad Sci USA 101:10482–10487
Gaeggeler HP, Guillod Y, Loffing-Cueni D et al (2011) Vasopressin-dependent coupling between sodium transport and water flow in a mouse cortical collecting duct cell line. Kidney Int 79:843–852
Gagnon KB, Delpire E (2012) Molecular physiology of Spak and Osr1: two Ste20-related protein kinases regulating ion transport. Physiol Rev 92:1577–1617
Gagnon E, Forbush B, Flemmer AW et al (2002) Functional and molecular characterization of the shark renal Na-K-Cl cotransporter: novel aspects. Am J Physiol Renal Physiol 283:F1046–F1055
Galizia L, Pizzoni A, Fernandez J et al (2012) Functional interaction between AQP2 and TRPV4 in renal cells. J Cell Biochem 113:580–589
Garcia-Elias A, Mrkonjic S, Pardo-Pastor C et al (2013) Phosphatidylinositol-4,5-biphosphate-dependent rearrangement of TRPV4 cytosolic tails enables channel activation by physiological stimuli. Proc Natl Acad Sci USA 110:9553–9558
Garvin JL, Spring KR (1992) Regulation of apical membrane ion transport in *Necturus* gallbladder. Am J Physiol Cell Physiol 263:C187–C193
Gelfand EW, Cheung RKK, Ha K et al (1984) Volume regulation in lymphoid leukemia-cells and assignment of cell lineage. New Eng J Med 311:939–944
Giraldez F, Ferreira KTG (1984) Intracellular chloride activity and membrane potential in stripped frog skin (*Rana temporaria*). Biochim Biophys Acta 769:625–628
Gonzáles E, Carpi-Medina P, Linares H et al (1984) Osmotic water permeability of the apical membrane of proximal straight tubular (PST) cells. Pflugers Arch 402:337–339
Gonzalez E, Carpimedina P, Whittembury G (1982) Cell osmotic water permeability of isolated rabbit proximal straight tubules. Am J Physiol Renal Physiol 242:F321–F330

Grasset E, Gunter-Smith P, Schultz SG (1983) Effects of Na-coupled alanine transport on intracellular K-activities and the K-conductance of the basolateral membranes of Necturus small intestine. J Membr Biol 71:89–94
Greger R, Heitzmann D, Hug MJ et al (1999) The Na+,2Cl−, K+ cotransporter in the rectal gland of Squalus acanthias is activated by cell shrinkage. Pflugers Arch 438:165–176
Grinstein S, Erlij D (1978) Intracellular calcium and regulation of sodium-transport in frog skin. Proc R Soc Ser B Biol 202:353–360
Grinstein S, Clarke CA, Rothstein A (1982) Increased anion permeability during volume regulation in human-lymphocytes. Philos T Roy Soc B 299:509–518
Grinstein S, Woodside M, Sardet C et al (1992) Activation of the Na^+/H^+ antiporter during cell-volume regulation - Evidence for a phosphorylation-independent mechanism. J Biol Chem 267:23823–23828
Grosell M (2007) Intestinal transport processes in marine fish osmoregulation. In: Baldisserotto B, Mancera JM, Kapoor BG (eds) Fish Osmoregulation. Science Publishers Inc., Enfield, New Hamshire, USA, pp 332–357
Grubb BR, Schiretz FR, Boucher RC (1997) Volume transport across tracheal and bronchial airway epithelia in a tubular culture system. Am J Physiol Cell Physiol 273:C21–C29
Grunnet M, Jespersen T, MacAulay N et al (2003) KCNQ1 channels sense small changes in cell volume. J Physiol 549:419–427
Guggino WB, Oberleithner H, Giebisch G (1985) Relationship between cell volume and ion transport in the early distal tubule of the Amphiuma kidney. J Gen Physiol 86:31–58
Guggino WB, Markakis D, Amzel LM (1990) Measurements of volume and shape changes in isolated tubules. Method Enzymol 191:371–379
Gunter-Smith PJ, Grasset E, Schultz SG (1982) Sodium-coupled amino-acid and sugar transport by Necturus small intestine - an equivalent electrical circuit analysis of a rheogenic cotransport system. J Membr Biol 66:25–39
Hall A (1998) Rho GTPases and the actin cytoskeleton. Science 279:509–514
Harvey BJ (1992) Energization of sodium absorption by the H^+-ATPase pump in mitochondria-rich cells of frog skin. J Exp Biol 172:289–309
Harvey BJ (1995) Cross-talk between sodium and potassium channels in tight epithelia. Kidney Intl 48:1191–1199
Harvey BJ, Ehrenfeld J (1988) Role of Na^+/H^+ exchange in the control of intracellular pH and cell membrane conductance in frog skin epithelium. J Gen Physiol 92:793–810
Harvey BJ, Kernan RP (1984) Intracellular ion activities in frog skin in relation to external sodium and effects of amiloride and/or ouabain. J Physiol 349:501–517
Harvey BJ, Thomas SR, Ehrenfeld J (1988) Intracellular pH controls cell membranes Na^+ and K^+ conductances and transport in frog skin epithelium. J Gen Physiol 92:767–791
Hazama A, Okada Y (1988) Electrophysiological evidence for independent activation of K^+ and Cl^- conductances during the regulatory volume decrease in cultured epithelial-cells. Comp Biochem Phys A 90:809–809
Hazama A, Okada Y (1990) Biphasic rises in cytosolic free Ca^{2+} in association with activation of K^+ and Cl^- conductance during the regulatory volume decrease in cultured human epithelial-cells. Pflugers Arch 416:710–714
Hebert SC (1986a) Hypertonic cell-volume regulation in mouse thick limbs. 1. ADH dependency and nephron heterogeneity. Am J Physiol Cell Physiol 250:C907–C919
Hebert SC (1986b) Hypertonic cell-volume regulation in mouse thick limbs. 2. Na^+-H^+ and Cl^--HCO_3^- exchange in basolateral membranes. Am J Physiol Cell Physiol 250:C920–C931
Hebert SC (1987) Volume regulation in renal epithelial cells. Sem Nephrol 7:48–60
Hebert SC, Mount DB, Gamba G (2004) Molecular physiology of cation-coupled Cl^- cotransport: the SLC12 family. Pflugers Arch 447:580–593
Hendus-Altenburger R, Kragelund BB, Pedersen SF (2014) Structural dynamics and regulation of the mammalian SLC9A family of Na^+/H^+ exchangers. In: Bevensee MO (ed) Exchangers, vol 73, Current topics in membranes. Academic/Elsevier, Burlington, pp 69–148, Chapter 2

Henson JH, Roesener CD, Gaetano CJ et al (1997) Confocal microscopic observation of cytoskeletal reorganizations in cultured shark rectal gland cells following treatment with hypotonic shock and high external K^+. J Exp Zool 279:415–424

Higgins G, Buchanan P, Perriere M et al (2014) Activation of P2RY11 and ATP release by lipoxin A4 restores the airway surface liquid Layer and epithelial repair in cystic fibrosis. Am J Resp Cell Mol Biol 51:178–190

Hillyard SD, Møbjerg N, Tanaka S et al (2009) Osmotic and Ion Regulation in Amphibians. In: Evans DH (ed) Osmotic and Ionic Regulation. Cells and Animals. Taylor & Francis Group, Boca Raton, Florida, pp 367–441

Hoffmann EK (1978) Regulation of cell volume by selective changes in the leak permeabilities of Ehrlich ascites tumore cells. In: Jørgensen CB, Skadhauge E (eds) Proceedings of Alfred Benzon Symposium XI. Osmotic and Volume Regulation. Munksgaard, Copenhagen, pp 397–417

Hoffmann EK, Pedersen SF (2007) Shrinkage insensitivity of NKCC1 in myosin II-depleted cytoplasts from Ehrlich ascites tumor cells. Am J Physiol Cell Physiol 292:C1854–C1866

Hoffmann EK, Pedersen SF (2011) Cell volume homeostatic mechanisms: effectors and signalling pathways. Acta Physiol 202:465–485

Hoffmann EK, Ussing HH (1992) Membrane mechanisms in volume regulation in vertebrane cells and epithelia. In: Giebisch G, Schafer JA, Ussing H, Kristensen P (eds) Membrane transport in biology, vol 5. Springer, Berlin, pp 317–399

Hoffmann EK, Simonsen LO, Sjøholm C (1979) Membrane potential, chloride exchange, and chloride conductance in ehrlich mouse ascites tumour cells. J Physiol 296:61–84

Hoffmann EK, Sjøholm C, Simonsen LO (1983) Na^+, Cl^- co-transport in Ehrlich ascites tumor cells activated during volume regulation (regulatory volume increase). J Membr Biol 76:269–280

Hoffmann EK, Hoffmann E, Lang F et al (2002) Control of Cl^- transport in the operculum epithelium of Fundulus heteroclitus: long- and short-term salinity adaptation. Biochim Biophys Acta 1566:129–139

Hoffmann EK, Schettino T, Marshall WS (2007) The role of volume-sensitive ion transport systems in regulation of epithelial transport. Comp Biochem Physiol A Mol Integr Physiol 148:29–43

Hoffmann EK, Lambert IH, Pedersen SF (2009) Physiology of cell volume regulation in vertebrates. Physiol Rev 89:193–277

Hougaard C, Niemeyer MI, Hoffmann EK et al (2000) K^+ currents activated by leukotriene D-4 or osmotic swelling in Ehrlich ascites tumour cells. Pflugers Arch 440:283–294

Hougaard C, Klaerke DA, Hoffmann EK et al (2004) Modulation of KCNQ4 channel activity by changes in cell volume. Biochim Biophys Acta 1660:1–6

Huang CL, Cha SK, Wang HR et al (2007) WNKs: protein kinases with a unique kinase domain. Exp Mol Med 39:565–573

Hughes ALH, Pakhomova A, Brown PD (2010) Regulatory volume increase in epithelial cells isolated from the mouse fourth ventricle choroid plexus involves Na^+-H^+ exchange but not Na^+-K^+-$2Cl^-$ cotransport. Brain Res 1323:1–10

Ingber DE (2006) Cellular mechanotransduction: putting all the pieces together again. FASEB J 20:811–827

Ishibashi K, Sasaki S, Yoshiyama N (1988) Intracellular chloride activity of rabbit proximal straight tubule perfused in vitro. Am J Physiol Renal Physiol 255:F49–F56

Jakab RL, Collaco AM, Ameen NA (2011) Physiological relevance of cell-specific distribution patterns of CFTR, NKCC1, NBCe1, and NHE3 along the crypt-villus axis in the intestine. Am J Physiol Gastrointest Liver Physiol 300:G82–G98

Jeon US, Kim J-A, Sheen MR, Kwon HM (2006) How tonicity regulates genes: story of TonEBP transcriptional activator. Acta Physiol 187:241–247

Jorgensen NK, Pedersen SF, Rasmussen HB et al (2003) Cell swelling activates cloned Ca^{2+}-activated K^+ channels: a role for the F-actin cytoskeleton. Biochim Biophys Acta 1615:115–125

Kahle KT, Rinehart J, Ring A, Gimenez I, Gamba G, Hebert SC, Lifton RP (2006) WNK protein kinases modulate cellular Cl^- flux by altering the phosphorylation state of the Na-K-Cl and K-Cl cotransporters. Physiology 21:326–335

Kashgarian M, Biemesderfer D, Caplan M et al (1985) Monoclonal antibody to Na, K-ATPase: immunocytochemical localization along nephron segments. Kidney Int 28:899–913

Kato A, Romero MF (2011) Regulation of electroneutral NaCl absorption by the small intestine. Annu Rev Physiol 73:261–281

Kirk KL, DiBona DR, Schafer JA (1987a) Regulatory volume decrease in perfused proximal nephron: evidence for a dumping of cell K^+. Am J Physiol Renal Physiol 252:F933–F942

Kirk KL, Schafer JA, DiBona DR (1987b) Cell volume regulation in rabbit proximal straight tubule perfused in vitro. Am J Physiol Renal Physiol 252:F922–F932

Kirkegaard SS, Lambert IH, Gammeltoft S et al (2010) Activation of the TASK-2 channel after cell swelling is dependent on tyrosine phosphorylation. Am J Physiol Cell Physiol 299:C844–C853

Klausen TK, Hougaard C, Hoffmann EK et al (2006) Cholesterol modulates the volume-regulated anion current in Ehrlich-Lettre ascites cells via effects on Rho and F-actin. Am J Physiol Cell Physiol 291:C757–C771

Klausen TK, Janssens A, Prenen J et al (2014) Single point mutations of aromatic residues in transmembrane helices 5 and-6 differentially affect TRPV4 activation by 4 alpha-PDD and hypotonicity: implications for the role of the pore region in regulating TRPV4 activity. Cell Calcium 55:38–47

Klein JD, O'Neill WC (1995) Volume-sensitive myosin phosphorylation in vascular endothelial cells: correlation with Na-K-2Cl cotransport. Am J Physiol Cell Physiol 269:C1524–C1531

Kregenow FM (1971) The response of duck erythrocytes to hypertonic media. Further evidence for a volume-controlling mechanism. J Gen Physiol 58:396–412

Krump E, Nikitas K, Grinstein S (1997) Induction of tyrosine phosphorylation and Na^+/H^+ exchanger activation during shrinkage of human neutrophils. J Biol Chem 272:17303–17311

L'Hoste S, Barriere H, Belfodil R et al (2007) Extracellular pH alkalinisation by Cl^-/HCO_3^- exchanger is crucial for TASK2 activation by hypotonic shock in proximal cell lines from mouse kidney. Am J Physiol Renal Physiol 292:F628–F638

Lacroix J, Poet M, Huc L et al (2008) Kinetic analysis of the regulation of the Na^+/H^+ exchanger NHE-1 by osmotic shocks. Biochemistry 47:13674–13685

Lang F (2013) Cell volume control. In: Alpern RJ, Moe OW, Caplan M (eds) Seldin and Giebisch's the kidney, vol 1, 5th edn. Academic, London, pp 121–141

Larsen EH (1991) Chloride transport by high-resistance heterocellular epithelia. Physiol Rev 71:235–283

Larsen EH (2011) Reconciling the Krogh and Ussing interpretations of epithelial chloride transport - presenting a novel hypothesis for the physiological significance of the passive cellular chloride uptake. Acta Physiol 202:435–464

Larsen EH, Harvey BJ (1994) Chloride currents of single mitochondria-rich cells of toad skin epithelium. J Physiol 478:7–15

Larsen EH, Ramløv H (2013) Role of cutaneous surface fluid in frog osmoregulation. Comp Biochem Physiol A Mol Integr Physiol 165:365–370

Larsen EH, Ussing HH, Spring KR (1987) Ion transport by mitochondria-rich cells in toad skin. J Membr Biol 99:25–40

Larsen EH, Sørensen JB, Sørensen JN (2000) A mathematical model of solute coupled water transport in toad intestine incorporating recirculation of the actively transported solute. J Gen Physiol 116:101–124

Larsen EH, Willumsen NJ, Møbjerg N et al (2009) The lateral intercellular space as osmotic coupling compartment. Acta Physiol 195:171–186

Larsen EH, Deaton LE, Onken H et al (2014) Osmoregulation and Excretion. Compr Physiol 4:405–573

Larson M, Spring KR (1983) Bumetanide inhibition of NaCl transport by *Necturus* Gallbladder. J Membr Biol 74:123–129

Larson M, Spring KR (1984) Volume regulation by *Necturus* gallbladder - basolateral KCl exit. J Membr Biol 81:219–232

Lauf PK, Misri S, Chimote AA, Adragna NC (2008) Apparent intermediate K conductance channel hyposmotic activation in human lens epithelial cells. Am J Physiol Cell Physol 294: C820–C832

Lazarowski ER, Tarran R, Grubb BR et al (2004) Nucleotide release provides a mechanism for airway surface liquid homeostasis. J Biol Chem 279:36855–36864

Leaf A (1959) Maintenance of concentration gradients and regulation of cell volume. Annal N Y Acad Sci 72:396–404

Lewis A, Di Ciano C, Rotstein OD et al (2002) Osmotic stress activates Rac and Cdc42 in neutrophils: role in hypertonicity-induced actin polymerization. Am J Physiol Cell Physiol 282:C271–C279

Liedtke CM, Cole TS (2002) Activation of NKCC1 by hyperosmotic stress in human tracheal epithelial cells involves PKC-delta and ERK. Biochim Biophys Acta 1589:77–88

Liedtke W, Kim C (2005) Functionality of the TRPV subfamily of TRP ion channels: add mechano-TRP and osmo-TRP to the lexicon! Cell Mol Life Sci 62:2985–3001

Lionetto MG, Schettino T (2006) The Na^+-K^+-$2Cl^-$ cotransporter and the osmotic stress response in a model salt transport epithelium. Acta Physiol 187:115–124

Lionetto MG, Giordano ME, Nicolardi G et al (2001) Hypertonicity stimulates Cl^- transport in the intestine of fresh water acclimated eel, *Anguilla anguilla*. Cell Physiol Biochem 11:41–54

Lionetto MG, Pedersen SF, Hoffmann EK et al (2002) Roles of the cytoskeleton and of protein phosphorylation events in the osmotic stress response in eel intestinal epithelium. Cell Physiol Biochem 12:163–163

Lionetto MG, Giordano ME, De Nuccio F et al (2005) Hypotonicity induced K^+ and anion conductive pathways activation in eel intestinal epithelium. J Exp Biol 208:749–760

Lionetto MG, Rizzello A, Giordano ME et al (2008) Molecular and functional expression of high conductance Ca^{2+} activated K^+ channels in the eel intestinal epithelium. Cell Physiol Biochem 21:373–384

Liu XB, Bandyopadhyay B, Nakamoto T et al (2006) A role for AQP5 in activation of TRPV4 by hypotonicity - Concerted involvement of AQP5 and TRPV4 in regulation of cell volume recovery. J Biol Chem 281:15485–15495

Lock H, Valverde MA (2000) Contribution of the IsK (MinK) potassium channel subunit to regulatory volume decrease in murine tracheal epithelial cells. J Biol Chem 275:34849–34852

Lopes AG, Amzel LM, Markakis D et al (1988) Cell volume regulation by the thin descending limb og Henle's loop. Proc Natl Acad Sci USA 85:2873–2877

Macknight ADC, Leaf A (1977) Regulation of cellular volume. Physiol Rev 57:510–573

Macleod RJ, Hamilton JR (1990) Regulatory volume increase in mammalian jejunal villus cells is due to bumetanide-sensitive NaKCl2 cotransport. Am J Physiol Gastrointest Liver Physiol 258: G665–G674

MacLeod RJ, Hamilton JR (1991) Volume regulation initiated by Na^+-nutrient cotransport in isolated mammalian villus enterocytes. Am J Physiol Gastrointest Liver Physiol 260:G26–G33

MacLeod RJ, Hamilton JR (1999) Ca^{2+}/calmodulin kinase II and decreases in intracellular pH are required to activate K^+ channels after substantial swelling in villus epithelial cells. J Membr Biol 172:59–66

MacRobbie EAC, Ussing HH (1961) Osmotic behaviour of the epithelial cells of frog skin. Acta Physiol Scand 53:348–365

Maingret F, Fosset M, Lesage F et al (1999a) TRAAK is a mammalian neuronal mechano-gated K^+ channel. J Biol Chem 274:1381–1387

Maingret F, Patel AJ, Lesage F et al (1999b) Mechano- or acid stimulation, two interactive modes of activation of the TREK-1 potassium channel. J Biol Chem 274:26691–26696
Maingret F, Patel AJ, Lesage F et al (2000) Lysophospholipids open the two-pore domain mechano-gated K^+ channels TREK-1 and TRAAK. J Biol Chem 275:10128–10133
Manganel M, Turner RJ (1991) Rapid secretagogue-induced activation of Na^+/H^+ exchange in rat parotid acinar-cells - Possible interrelationship between volume regulation and stimulus-secretion coupling. J Biol Chem 266:10182–10188
Markadieu N, Delpire E (2014) Physiology and pathophysiology of SLC12A1/2 transporters. Pflugers Arch 466:91–105
Marsh DJ, Spring KR (1985) Polarity of volume-regulatory increase by *Necturus* gallbladder epithelium. Am J Physiol Cell Physiol 249:C471–C475
Marsh DJ, Jensen PK, Spring KR (1985) Computer-based determination of size and shape in living cells. J Microsc 137:281–292
Marshall WS (2011) Mechanosensitive signalling in fish gill and other ion transporting epithelia. Acta Physiol 202:487–499
Marshall WS, Grosell M (2005) Ion transport, osmoregulation, and acid–base balance. In: Evans DH, Claiborne JB (eds) The physiology of fishes, 3rd edn. CRC, Boca Raton, pp 177–230
Marshall WS, Bryson SE, Luby T (2000) Control of epithelial Cl^- secretion by basolateral osmolality in the euryhaline teleost *Fundulus heteroclitus*. J Exp Biol 203:1897–1905
Marshall WS, Ossum CG, Hoffmann EK (2005) Hypotonic shock mediation by p38 MAPK, JNK, PKC, FAK, OSR1 and SPAK in osmosensing chloride secreting cells of killifish opercular epithelium. J Exp Biol 208:1063–1077
Marshall WS, Katoh F, Main HP et al (2008) Focal adhesion kinase and β1 integrin regulation of Na^+, K^+, 2 Cl^- cotransporter in osmosensing ion transporting cells of killifish, *Fundulus heteroclitus*. Comp Biochem Phys A 150:288–300
Matsumura Y, Cohen B, Guggino WB et al (1984) Regulation of the basolateral potassium conductance of the Necturus proximal tubule. J Membr Biol 79:153–161
Maunsbach AB, Boulpaep EL (1991) Immunoelectron microscope localization of Na, K-ATPase in transport pathways in proximal tubule epithelium. Micron Microsc Acta 22:55–56
McCormick JA, Ellison DH (2011) The WNKs: atypical protein kinases with pleiotropic actions. Physiol Rev 91:177–219
Messner G, Wang W, Paulmichl M et al (1985) Ouabain decreases apparent potassium-conductance in proximal tubules of the amphibian kidney. Pflugers Arch 404:131–137
Meyer K, Korbmacher C (1996) Cell swelling activates ATP-dependent voltage-gated chloride channels in M-1 mouse cortical collecting duct cells. J Gen Physiol 108:177–193
Mignen O, Le Gall C, Harvey BJ et al (1999) Volume regulation following hypotonic shock in isolated crypts of mouse distal colon. J Physiol 515:501–510
Mills JW (1985) Ion transport across the exocrine glands of the frog skin. Pflugers Arch 405(suppl 1):S44–S49
Mills JW, DiBona DR (1978) Distribution of Na^+-pump sites in the frog gallbladder. Nature 271:273–275
Mills JW, Ernst SA (1975) Localization of sodium pump sites in frog urinary bladder. Biochim Biophys Acta 375:268–273
Mills JW, Ernst SA, DiBona DR (1977) Localization of Na^+-pump sites in frog skin. J Cell Biol 73:88–110
Minton AP (2006) Macromolecular crowding. Curr Biol 16:R269–R271
Moeckel GW, Zhang L, Chen X (2006) Role of integrin alpha1beta1 in the regulation of renal medullary osmolyte concentration. Am J Physiol Renal Physiol 290:F223–F231
Montroserafizadeh C, Guggino WB (1990) Cell-volume regulation in the nephron. Ann Rev Physiol 52:761–772
Montrose-Rafizadeh C, Guggino WB, Montrose MH (1991) Cellular differentiation regulates expression of Cl^- transport and cystic fibrosis transmembrane conductance regulator mRNA in human intestinal cells. J Biol Chem 266:4495–4499

Moore-Hoon ML, Turner RJ (2000) The structural unit of the secretory Na^+-K^+-$2Cl^-$ cotransporter (NKCCl) is a homodimer. Biochemistry 39:3718–3724
Nauntofte B, Dissing S (1988) Cholinergic-induced electrolyte transport in rat parotid acini. Comp Biochem Phys A 90:739–746
Nearing J, Betka M, Quinn S et al (2002) Polyvalent cation receptor proteins (CaRs) are salinity sensors in fish. Proc Natl Acad Sci USA 99:9231–9236
Nedergaard S, Larsen EH, Ussing HH (1999) Sodium recirculation and isotonic transport in toad small intestine. J Membr Biol 168:241–251
Nielsen S, Agre P (1995) The aquaporin family of water channels in kidney. Kidney Int 48:1057–1068
Nielsen R, Larsen EH (2007) Beta-adrenergic activation of solute coupled water uptake by toad skin epithelium results in near-isosmotic transport. Comp Biochem Phys A 148:64–71
Nielsen MS, Nielsen R (1999) Effect of carbachol and prostaglandin E2 on chloride secretion and signal transduction in the exocrine glands of frog skin (Rana esculenta). Pflugers Arch 438:732–740
Nielsen S, Chou C-L, Marples D et al (1995) Vasopressin increases water permeability of kidney collecting duct by inducing translocation of aquaporin-CD water channels to plasma membrane. Proc Natl Acad Sci USA 92:1013–1017
Nielsen S, Frokiaer J, Marples D et al (2002) Aquaporins in the kidney: from molecules to medicine. Physiol Rev 82:205–244
Nielsen S, Kwon T-H, Dimke H et al (2013) Chapter 41 - Aquaporin water channels in mammalian kidney. In: Alpern RJ, Moe OW, Caplan M (eds) Seldin and Giebisch's the kidney, 5th edn. Academic, London, pp 1405–1439
Niemeyer MI, Cid LP, Barros LF et al (2001a) Modulation of the two-pore domain acid-sensitive K^+ channel TASK-2 (KCNK5) by changes in cell volume. J Biol Chem 276:43166–43174
Niemeyer MI, Cid LP, Sepulveda FV (2001b) K^+ conductance activated during regulatory volume decrease. The channels in Ehrlich cells and their possible molecular counterpart. Comp Biochem Phys A 130:565–575
Nilius B, Droogmans G (2003) Amazing chloride channels: an overview. Acta Physiol Scand 177:119–147
Nilius B, Eggermont J, Voets T et al (1997) Properties of volume-regulated anion channels in mammalian cells. Prog Biophys Mol Biol 68:69–119
Nilius B, Voets T, Prenen J et al (1999) Role of Rho and Rho kinase in the activation of volume-regulated anion channels in bovine endothelial cells. J Physiol 516:67–74
Nilius B, Owsianik G, Voets T et al (2007) Transient receptor potential cation channels in disease. Physiol Rev 87:165–217
Novak I (2011) Purinergic signalling in epithelial ion transport: regulation of secretion and absorption. Acta Physiol 202:501–522
Novak I, Greger R (1988) Properties of the luminal membrane of isolated perfused rat pancreatic ducts. Effect of cyclic AMP and blockers of chloride transport. Pflugers Arch 411:546–553
Numata T, Shimizu T, Okada Y (2007) TRPM7 is a stretch- and swelling-activated cation channel involved in volume regulation in human epithelial cells. Am J Physiol Cell Physiol 292:C460–C467
O'Brien JA, Walters RJ, Sepulveda FV (1991) Regulatory volume decrease in small intestinal crypts is inhibited by K^+ and Cl^- channel blockers. Biochim Biophys Acta 1070:501–504
O'Brien JA, Walters RJ, Valverde MA et al (1993) Regulatory volume increase after hypertonicity- or vasoactive-intestinal-peptide-induced cell-volume decrease in small-intestinal crypts is dependent on Na^+-K^+-$2Cl^-$ cotransport. Pflugers Arch 423:67–73
Ogushi Y, Kitagawa D, Hasegawa T et al (2010) Correlation between aquaporin and water permeability in response to vasotocin, hydrin and β-adrenergic effectors in the ventral pelvic skin of the tree frog Hyla japonica. J Exp Biol 213:288–294
Okada Y (1997) Volume expansion-sensing outward-rectifier Cl^- channel: fresh start to the molecular identity and volume sensor. Am J Physiol Cell Physiol 273:C755–C789

Okada Y, Maeno E, Shimizu T (2001) Receptor-mediated control of regulatory volume decrease (RVD) and apoptotic volume decrease (AVD). J Physiol 532:3–16
Okada Y, Sato K, Numata T (2009) Pathophysiology and puzzles of the volume-sensitive outwardly rectifying anion channel. J Physiol 587:2141–2149
Orlowski J, Grinstein S (2004) Diversity of the mammalian sodium/proton exchanger SLC9 gene family. Pflugers Arch 447:549–565
Orlowski J, Grinstein S (2011) Na^+/H^+ exchangers. Compr Physiol 1:2083–2100
Owsianik G, Talavera K, Voets T et al (2006) Permeation and selectivity of TRP channels. Ann Rev Physiol 68:685–717
Pan Z, Capo-Aponte JE, Zhang F et al (2007) Differential dependence of regulatory volume decrease behavior in rabbit corneal epithelial cells on MAPK superfamily activation. Exp Eye Res 84:978–990
Pang V, Counillon L, Lagadic-Gossmann D et al (2012) On the role of the difference in surface tensions involved in the allosteric regulation of NHE-1 induced by low to mild osmotic pressure, membrane tension and lipid asymmetry. Cell Biochem Biophys 63:47–57
Park S, Hong JH, Ohana E et al (2012) The WNK/SPAK and IRBIT/PP1 pathways in epithelial fluid and electrolyte transport. Physiology 27:291–299
Pedersen SF, Nilius B (2007) Transient receptor potential channels in mechanosensing and cell volume regulation. Method Enzymol 428:183–207
Pedersen SF, Beisner KH, Hougaard C et al (2002a) Rho family GTP binding proteins are involved in the regulatory volume decrease process in NIH3T3 mouse fibroblasts. J Physiol 541:779–796
Pedersen SF, Varming C, Christensen ST et al (2002b) Mechanisms of activation of NHE by cell shrinkage and by calyculin A in ehrlich ascites tumor cells. J Membr Biol 189:67–81
Pedersen SF, Kapus A, Hoffmann EK (2011) Osmosensory mechanisms in cellular and systemic volume regulation. J Am Soc Nephrol 22:1587–1597
Pedersen SF, Klausen TK, Nilius B (2015) The identification of a volume-regulated anion channel: an amazing Odyssey. Acta Physiol 213:868–881
Peng JB, Warnock DG (2007) WNK4-mediated regulation of renal ion transport proteins. Am J Physiol 293:F961–F973
Persson B-E, Spring KR (1982) Gallbladder epithelial cell hydraulic water permeablity and volume regulation. J Gen Physiol 79:481–505
Petersen OH (1992) Stimulus-secretion coupling - Cytoplasmic calcium signals and the control of ion channels in exocrine acinar-cells. J Physiol 448:1–51
Petersen OH (1993) Regulation of isotonic fluid secretion in exocrine acini. In: Ussing HH, Fischbarg J, Sten-Knudsen O, Larsen EH, Willumsen NJ (eds) Proceeding of Alfred Benzon Symposium 34. Isotonic Transport in Leaky Epithelia. Munksgaard, Copenhagen, pp 103–146
Petersen OH, Gallacher DV (1988) Electrophysiology of pancreatic and salivary acinar cells. Ann Rev Physiol 50:65–80
Piechotta K, Garbarini N, England R et al (2003) Characterization of the interaction of the stress kinase SPAK with the Na^+-K^+-$2Cl^-$ cotransporter in the nervous system - Evidence for a scaffolding role of the kinase. J Biol Chem 278:52848–52856
Pochynyuk O, Zaika O, O'Neil RG et al (2013) Novel insights into TRPV4 function in the kidney. Pflugers Arch 465:177–186
Preston GM, Carroll TP, Guggino WB et al (1992) Appearance of water channels in *Xenopus* oocytes expressing red cell CHIP28 protein. Science 256:385–387
Qiu ZZ, Dubin AE, Mathur J et al (2014) SWELL1, a plasma membrane protein, is an essential component of volume-regulated anion channel. Cell 157:447–458
Reuss L (1985) Changes in cell volume measured with an electrophysiologic technique. Proc Natl Acad Sci USA 82:6014–6018
Richardson C, Alessi DR (2008) The regulation of salt transport and blood pressure by the WNK-SPAK/OSR1 signalling pathway. J Cell Sci 121:3293–3304

Rick R, Dörge A, Von Arnim E et al (1978) Electron microprobe analysis of frog skin epithelium: evidence for a syncytial sodium transport compartment. J Membr Biol 39:313–331
Rigor RR, Damoc C, Phinney BS et al (2011) Phosphorylation and activation of the plasma membrane Na^+/H^+ exchanger (NHE1) during osmotic cell shrinkage. PLoS ONE 6(12)
Robertson MA, Foskett JK (1994) Na^+ transport pathways in secretory acinar-cells - membrane cross-talk mediated by $[Cl^-]_I$. Am J Physiol Cell Physiol 267:C146–C156
Roger F, Martin PY, Rousselot M et al (1999) Cell shrinkage triggers the activation of mitogen-activated protein kinases by hypertonicity in the rat kidney medullary thick ascending limb of the Henle's loop - Requirement of p38 kinase for the regulatory volume increase response. J Biol Chem 274:34103–34110
Romanenko VG, Catalan MA, Brown DA et al (2010) TMEM16A encodes the Ca^{2+}-activated Cl^- channel in mouse submandibular salivary gland acinar cells. J Biol Chem 285:12990–13001
Rotin D, Grinstein S (1989) Impaired cell-volume regulation in Na^+-H^+ exchange-deficient mutants. Am J Physiol Cell Physiol 257:C1158–C1165
Sabirov RS, Okada Y (2005) ATP release via anion channels. Purinerg Signal 1:311–328
Sarkadi B, Attisano L, Grinstein S, Buchwald M, Rothstein A (1984) Volume regulation of Chinese-Hamster ovary cells in anisoosmotic media. Biochim Biophys Acta 774:159–168
Schafer JA (1990) Transepithelial osmolality differences, hydraulic conductivities, and volume absorption in the proximal tubule. Ann Rev Physiol 52:709–726
Schafer JA (1993) The rat collecting duct as an isosmotic volume reabsorber. In: Ussing HH, Fischbarg J, Sten-Knudsen O, Larsen EH, Willumsen NJ (eds) Proceeding of Alfred Benzon Symposium 34. Isotonic Transport in Leaky Epithelia. Munksgaard, Copenhagen, pp 339–354
Schafer JA, Patlak CS, Troutman SL et al (1978) Volume absorption in the parts recta. II. Hydraulic conductivity coefficient. Am J Physiol Renal Physiol 234:F340–F348
Schatzmann HJ, Windhager EE, Solomon AK (1958) Single proximal tubules of the Necturus kidney. II. Effect of 2, 4-dinitro-phenol and ouabain on water reabsorption. Am J Physiol 195:570–574
Schild L, Aronson PS, Giebisch G (1991) Basolateral transport pathways for K^+ and Cl^- in rabbit proximal tubule:effects on cell volume. Am J Physiol Renal Physiol 260:F101–F109
Schlatter E, Greger R, Schafer JA (1990) Principal cells of cortical collecting ducts of the rat are not a route of transepithelial Cl^- transport. Pflugers Arch 417:317–323
Schultz SG (1981) Homocellular regulatory mechanisms in sodium-transporting epithelia: avoidance of extinction by "flush-through". Am J Physiol Renal Physiol 10:F579–F590
Schultz SG (1992) Membrane crosstalk in sodium-absorbing epithelial cells. In: Seldin DW, Giebisch G (eds) The kidney: physiology and pathophysiology. Raven, New York, pp 287–299
Schwiebert EM, Mills JW, Stanton BA (1994) Actin-based cytoskeleton regulates a chloride channel and cell volume in a renal cortical collecting duct cell line. J Biol Chem 269:7081–7089
Shimizu T, Numata T, Okada Y (2004) A role of reactive oxygen species in apoptotic activation of volume-sensitive Cl^- channel. Proc Natl Acad Sci USA 101:6770–6773
Sørensen JB, Larsen EH (1996) Heterogeneity of chloride channels in the apical membrane of isolated mitochondria-rich cells from toad skin. J Gen Physiol 108:421–433
Sørensen JB, Larsen EH (1998) Patch clamp on the luminal membrane of exocrine gland acini from frog skin (*Rana esculenta*) reveals the presence of cystic fibrosis transmembrane conductance regulator-like Cl^- channels activated by cyclic AMP. J Gen Physiol 112:19–31
Sørensen JB, Larsen EH (1999) Membrane potential and conductance of frog skin gland acinar cells in resting conditions and during stimulation with agonists of macroscopic secretion. Pflugers Arch 439:101–112
Sørensen JB, Nielsen MS, Nielsen R et al (1998) Luminal ion channels involved in isotonic secretion by Na^+-recirculation in exocrine gland-acini. Roy Dan Acad Sci Lett Biol Ser 49:179–191

Sørensen JB, Nielsen MS, Gudme CN et al (2001) Maxi K^+ channels co-localised with CFTR in the apical membrane of an exocrine gland acinus: possible involvement in secretion. Pflugers Arch 442:1–11

Speake T, Douglas IJ, Brown PD (1998) The role of calcium in the volume regulation of rat lacrimal acinar cells. J Membr Biol 164:283–291

Spring KR, Hope A (1978) Size and shape of the lateral intercellular spaces in a living epithelium. Science 200:54–57

Spring KR, Hope A (1979) Fluid transport and the dimensions of cell and interspaces of living *Necturus* gallbladder. J Gen Physiol 73:287–305

Spring KR, Ussing HH (1986) The volume of mitochondria-rich cells of frog skin epithelium. J Membr Biol 92:21–26

Staruschenko A (2012) Regulation of transport in the connecting tubule and cortical collecting duct. Compr Physiol 2:1541–1584

Sten-Knudsen O (2002) Biological membranes. Theory of transport, potentials and electric impulses. Cambridge University Press, Cambridge

Stirling CE (1972) Radioautographic localization of sodium pump sites in rabbit intestine. J Cell Biol 53:704–714

Stoner LC, Morley GE (1995) Effect of Basolateral or apical hyposmolarity on apical maxi-K channels of everted rat collecting tubule. Am J Physiol Renal Physiol 268:F569–F580

Strange K (1988) RVD in principal and intercalated cells of rabbit cortical collecting tubule. Am J Physiol Cell Physiol 255:C612–C621

Strange K (1989) Ouabain-induced cell swelling in rabbit cortical collecting tubule: NaCl transport by principal cells. J Membr Biol 107:249–261

Strange K (1990) Volume regulation following Na^+ pump inhibition in CCT principal cells: apical K^+ loss. Am J Physiol Renal Physiol 258:F732–F740

Strange K, Spring K (1986) Methods for imaging renal tubule cells. Kidney Int 30:192–200

Strange K, Spring KR (1987a) Absence of significant cellular dilution during ADH-stimulated water reabsorption. Science 235:1068–1070

Strange K, Spring KR (1987b) Cell membrane water permeability of rabbit cortical collecting duct. J Membr Biol 96:27–43

Strotmann R, Harteneck C, Nunnenmacher K et al (2000) OTRPC4, a nonselective cation channel that confers sensitivity to extracellular osmolarity. Nat Cell Biol 2:695–702

Stutzin A, Hoffmann EK (2006) Swelling-activated ion channels: functional regulation in cell-swelling, proliferation and apoptosis. Acta Physiol 187:27–42

Takemura T, Sato F, Suzuki Y et al (1991) Intracellular ion concentrations and cell volume during cholinergic stimulation of eccrine secretory coil cells. J Membr Biol 119:211–219

Taniguchi J, Guggino WB (1989) Membrane stretch - a physiological stimulator of Ca^{2+}-activated K^+ channels in thick ascending limb. Am J Physiol Renal Physiol 257:F347–F352

Tarran R, Grubb BR, Gatzy JT et al (2001) The relative roles of passive surface forces and active ion transport in the modulation of airway surface liquid volume and composition. J Gen Physiol 118:223–236

Tarran R, Trout L, Donaldson SH et al (2006) Soluble mediators, not cilia, determine airway surface liquid volume in normal and cystic fibrosis superficial airway epithelia. J Gen Physiol 127:591–604

Taylor A, Windhager EE (1979) Possible role of cytosolic calcium and Na-Ca exchange in regulation of transepithelial sodium transport. Am J Physiol Renal Physiol 236:F505–F512

Tekpli X, Huc L, Lacroix J et al (2008) Regulation of Na^+/H^+ exchanger 1 allosteric balance by its localization in cholesterol- and caveolin-rich membrane microdomains. J Cell Physiol 216:207–220

Thompson J, Begenisich T (2006) Membrane-delimited inhibition of maxi-K channel activity by the intermediate conductance Ca^{2+}-activated K channel. J Gen Physiol 127:159–169

Tilly BC, Edixhoven MJ, Tertoolen LGJ et al (1996) Activation of the osmo-sensitive chloride conductance involves p21(rho) and is accompanied by a transient reorganization of the F-actin cytoskeleton. Mol Biol Cell 7:1419–1427

Tinel H, Kinne-Saffran E, Kinne RKH (2000) Calcium signalling during RVD of kidney cells. Cell Physiol Biochem 10:297–302

Tinel H, Kinne-Saffran E, Kinne RKH (2002) Calcium-induced calcium release participates in cell volume regulation of rabbit TALH cells. Pflugers Arch 443:754–761

Toczylowska-Maminska R, Dolowy K (2012) Ion transporting proteins of human bronchial epithelium. J Cell Biochem 113:426–432

Tomassen SFB, Fekkes D, de Jonge HR et al (2004) Osmotic swelling-provoked release of organic osmolytes in human intestinal epithelial cells. Am J Physiol Cell Physiol 286:C1417–C1422

Tormey JM, Diamond JM (1969) The ultrastructure route of fluid transport in rabbit gall bladder. J Gen Physiol 50:2031–2060

Tosteson DC, Hoffman JF (1960) Regulation of cell volume by active cation transport in high and low potassium sheep red cells. J Gen Physiol 44:169–194

Tsai TT, Guttapalli A, Agrawal A et al (2007) MEK/ERK signaling controls osmoregulation of nucleus pulposus cells of the intervertebral disc by transactivation of TonEBP/OREBP. J Bone Miner Res 22:965–974

Tsuchiya K, Wang W, Giebisch G et al (1992) ATP is a coupling modulator of parallel Na, K-ATPase-K-channel activity in the renal proximal tubule. Proc Natl Acad Sci USA 89:6418–6422

Uhlik MT, Abell AN, Johnson NL et al (2003) Rac-MEKK3-MKK3 scaffolding for p38 MAPK activation during hyperosmotic shock. Nat Cell Biol 5:1104–1110

Urbach V, Van Kerkhove E, Maguire D et al (1996) Cross-talk between ATP-regulated K^+ channels and Na^+ transport via cellular metabolism in frog skin principal cells. J Physiol 491:99–109

Urbach V, Leguen I, O'Kelly I et al (1999) Mechanosensitive calcium entry and mobilization in renal A6 cells. J Membr Biol 168:29–37

Ussing HH (1960) Active and passive transport of the alkali metal ions. In: Ussing HH, Kruhøffer P, Thaysen JH, Thorn NA (eds) The Alkali Metal Ions in Biology. Springer Verlag, Berlin, Göttingen, Heidelberg, pp 45–143

Ussing HH (1982) Volume regulation of frog skin epithelium. Acta Physiol Scand 114:363–369

Ussing HH (1985) Volume regulation and basolateral co-transport of sodium, potassium, and chloride ion in frog skin epithelium. Pflugers Arch 405(suppl 1):S2–S7

Ussing HH, Eskesen K (1989) Mechanism of isotonic water transport in glands. Acta Physiol Scand 136:443–454

Ussing HH, Nedergaard S (1993) Recycling of electrolytes in small intesine of toad. In: Ussing HH, Fischbarg J, Sten-Knudsen O, Larsen EH, Willumsen NJ (eds) Proceeding of Alfred Benzon Symposium 34. Isotonic Transport in Leaky Epithelia. Munksgaard, Copenhagen, pp 26–36

Ussing HH, Lind F, Larsen EH (1996) Ion secretion and isotonic transport in frog skin glands. J Membr Biol 152:101–110

Valverde MA, O'Brien JA, Sepúlveda FV et al (1995) Impaired cell volume regulation in intestinal crypt epithelia of cystic fibrosis mice. Proc Natl Acad Sci USA 92:9038–9041

Valverde MA, Vazquez E, Munoz FJ et al (2000) Murine CFTR channel and its role in regulatory volume decrease of small intestine crypts. Cell Physiol Biochem 10:321–328

van der Wijk T, Tomassen SFB, Houtsmuller AB et al (2003) Increased vesicle recycling in response to osmotic cell swelling - Cause and consequence of hypotonicity-provoked ATP release. J Biol Chem 278:40020–40025

van Heeswijk MPE, van Os CH (1986) Osmotic water permeabilities of brush border and basolateral membrane vesicles from rat renal cortex and small intestine. J Membr Biol 92:183–193 vanTol BL, Missan S, Crack J et al (2007) Contribution of KCNQ1 to the regulatory volume decrease in the human mammary epithelial cell line MCF-7. Am J Physiol Cell Physiol 293: C1010–C1019

Vazquez E, Nobles M, Valverde MA (2001) Defective regulatory volume decrease in human cystic fibrosis tracheal cells because of altered regulation of intermediate conductance Ca^{2+}-dependent potassium channels. Proc Natl Acad Sci USA 98:5329–5334

Verissimo F, Jordan P (2001) WNK kinases, a novel protein kinase subfamily in multi-cellular organisms. Oncogene 20:5562–5569

Vitari AC, Deak M, Morrice NA et al (2005) The WNK1 and WNK4 protein kinases that are mutated in Gordon's hypertension syndrome phosphorylate and activate SPAK and OSR1 protein kinases. Biochem J 391:17–24

Voets T, Droogmans G, Raskin G et al (1999) Reduced intracellular ionic strength as the initial trigger for activation of endothelial volume-regulated anion channels. Proc Natl Acad Sci USA 96:5298–5303

Voss FK, Ullrich F, Münch J et al (2014) Identification of LRRC8 heteromers as an essential component of the volume-rRegulated anion channel VRAC. Science 344:634–638

Vriens J, Watanabe H, Janssens A et al (2004) Cell swelling, heat, and chemical agonists use distinct pathways for the activation of the cation channel TRPV4. Proc Natl Acad Sci USA 101:396–401

Wakabayashi S, Shigekawa M, Pouyssegur J (1997) Molecular physiology of vertebrate Na^+/H^+ exchangers. Physiol Rev 77:51–74

Wang W-H, Giebisch G (2009) Regulation of potassium (K) handling in the renal collecting duct. Pflug Arch 458:157–168

Wang J, Morishima S, Okada Y (2003) IK channels are involved in the regulatory volume decrease in human epithelial cells. Am J Physiol Cell Physiol 284:C77–C84

Wang GX, Dai YP, Bongalon S et al (2005) Hypotonic activation of volume-sensitive outwardly rectifying anion channels (VSOACs) requires coordinated remodeling of subcortical and perinuclear actin filaments. J Membr Biol 208:15–26

Wangemann P, Liu J, Shen Z et al (1995) Hypoosmotic challenge stimulates transepithelial K^+ secretion and activates apical I-Sk channel in vestibular dark cells. J Membr Biol 147:263–273

Wehner F (2006) Cell volume-regulated cation channels. Contrib Nephrol 152:25–53

Weinstein AM (2008) Sodium and chloride transport. Proximal nephron. In: Alpern RJ, Hebert SC (eds) Seldin and Giebisch's The Kidney. Physiology and Pathophysiology. Fourth edn. Academic Press, Elsevier, New York, pp 793–847

Welling PA (1995) Cross-talk and role of K_{ATP} channels in the proximal tubule. Kidney Int 48:1017–1023

Welling LW, Welling DJ (1988) Relationship between structure and function in renal proximal tubule. J Electron Microsc Tech 9:171–185

Welling LW, Evan AP, Welling DJ et al (1983a) Morphometric comparison of rabbit cortical connecting tubules and collecting ducts. Kidney Int 23:358–367

Welling LW, Welling DJ, Ochs TJ (1983b) Video measurement of basolateral membrane hydraulic conductivity in the proximal tubule. Am J Physiol Renal Physiol 245:F123–F129

Willumsen NJ, Boucher RC (1991) Sodium transport and intracellular sodium activity in cultured human nasal epithelium. Am J Physiol Cell Physiol 261:C319–C331

Willumsen NJ, Davis CW, Boucher RC (1989) Intracellular Cl^- activity and cellular Cl^- pathways in cultured human airway epithelium. Am J Physiol Cell Physiol 256:C1033–C1044

Willumsen NJ, Vestergaard L, Larsen EH (1992) Cyclic AMP-and β-agonist-activated chloride conductance of a toad skin epithelium. J Physiol 449:641–653

Willumsen NJ, Davis CW, Boucher RC (1994) Selective responce of human airway epithelia to luminal but nor serosal solution hypertonicity. Possible role for proximal airway epithelia as an osmolality transducer. J Clin Invest 94:779–787

Wong MMY, Foskett JK (1991) Oscillations of cytosolic sodium during calcium oscillations in exocrine acinar cells. Science 254:1014–1016

Worrell RT, Butt AG, Cliff WH et al (1989) Cell Physiology - a volume-sensitive chloride conductance in human colonic cell-line T84. Am J Physiol 256:C1111–C1119

Wu X, Yang H, Iserovich P, Fischbarg J et al (1997) Regulatory volume decrease by SV40-transformed rabbit corneal epithelial cells requires ryanodine-sensitive Ca^{2+}-induced Ca^{2+} release. J Membr Biol 158:127–136

Xia W, Yu Q, Riederer B et al (2014) The distinct roles of anion transporters Slc26a3 (DRA) and Slc26a6 (PAT-1) in fluid and electrolyte absorption in the murine small intestine. Pflugers Arch 466:1541–1556

Yamamoto S, Ichishima K, Ehara T (2008) Regulation of volume-regulated outwardly rectifying anion channels by phosphatidylinositol 3,4,5-trisphosphate in mouse ventricular cells. Biomed Res-Tokyo 29:307–315

Yang CL, Angell J, Mitchell R et al (2003) WNK kinases regulate thiazide-sensitive Na-Cl cotransport. J Clin Invest 111:1039–1045

Zadunaisky JA, Cardona S, Au L et al (1995) Chloride transport activation by plasma osmolarity during rapid adaptation to high salinity of *Fundulus heteroclitus*. J Membr Biol 143:207–217

Zhuo JL, Li XC (2013) Proximal nephron. Compr Physiol 3:1079–1123

Chapter 5
Fundamentals of Bicarbonate Secretion in Epithelia

Ivana Novak and Jeppe Praetorius

Abstract Certain epithelia secrete HCO_3^- to buffer luminal fluids and protect cells against acids. Dysregulation of HCO_3^- secretion can lead to conditions such as malabsorption, acid/base disturbances, cystic fibrosis, biliary cirrhosis and peptic and duodenal ulcers. In addition to transport of HCO_3^- across the epithelium, epithelial cells also need to maintain intracellular pH, despite significant HCO_3^- extrusion and sometimes even despite exposure to external acid. In this chapter, we will introduce the main plasma membrane acid/base transporters and describe their role in general cellular homeostasis. The same transporters are also used in building the molecular machinery for vectorial HCO_3^- transport, i.e. bicarbonate secretion. We will highlight HCO_3^--secreting epithelia by examples from the digestive system (pancreas, salivary glands, hepatobiliary system and duodenum), the renal collecting duct type-B intercalated cell, and the choroid plexus epithelium of the brain. We seek an integrative approach to understand the HCO_3^- secretion processes by combining historical perspectives with molecular and genetic studies as well as studies of selected regulatory systems.

Keywords Secretory epithelia • Bicarbonate secretion • Pancreas • Salivary glands • Duodenum

5.1 Introduction

5.1.1 Overview

A number of epithelia in our body secrete significant amount of HCO_3^-, which is often accompanied by fluid secretion. One of the important early observations was made on the pancreas, which secretes pancreatic juice rich in HCO_3^- and poor in

I. Novak (✉)
Molecular and Integrative Physiology, Department of Biology, University of Copenhagen, 2100 Copenhagen, Denmark
e-mail: inovak@bio.ku.dk

J. Praetorius
Department of Biomedicine, Aarhus University, Aarhus, Denmark

K.L. Hamilton, D.C. Devor (eds.), *Ion Channels and Transporters of Epithelia in Health and Disease*, Physiology in Health and Disease,
DOI 10.1007/978-1-4939-3366-2_5

Cl^-, a relation between two anions that together with later studies on isolated pancreatic duct epithelium became important steps for understanding general cellular models for HCO_3^- secretion (Fig. 5.1a, b). In general, the purpose of HCO_3^- secretion is to buffer and protect cells against acids produced and secreted by cells during digestive or metabolic processes and to solubilize proteins and other macromolecules. Dysregulation of these processes can lead to serious diseases such as cystic fibrosis, biliary cirrhosis and peptic and duodenal ulcers. In addition to transporting significant amounts of HCO_3^- from the interstitium to lumen, epithelia face another major challenge—they have to defend their intracellular pH (pH_i). This fact is a challenge to scientists, as it is often difficult to study and separate the transepithelial acid/base transport as opposed to transport across the single plasma cell membrane exerted for the purpose of pH_i regulation. In the first part of the chapter, we will introduce the main H^+/HCO_3^- transporters and describe their role in general cellular acid/base homeostasis. These "building blocks" will then be used to equip epithelial cells so that they can perform vectorial HCO_3^- transport, i.e. secretion. Other ion channels and transporters necessary for overall transepithelial HCO_3^- transport will be given in specific tissues/organs. We will focus on HCO_3^--secreting epithelia of the digestive system (pancreas, salivary glands, hepatobiliary system and duodenum), choroid plexus epithelium of the brain, and renal collecting ducts. Combining the historical perspectives with molecular and genetic studies in this chapter, we hope to mark more integrative approach that will help us to understand the challenges of HCO_3^- secretion.

5.1.2 Cellular Acid/Base Homeostasis

In secretory epithelial cells, and in other cells, the intracellular pH is maintained within the range 7.1–7.4, as most cellular processes have a pH optimum within this range (Boron and Boulpaep 2012). The balance between production, consumption, and transmembrane movement of acid/base equivalents determines the pH_i. The cellular buffering power is not regulating the steady-state pH_i, but determines the size and rate of the pH change inflicted by an acute acid or base challenge (Boron and Boulpaep 2012). The intrinsic buffering power is set by the cellular weak acid/base pairs such as phosphate, bicarbonate and anionic proteins. In addition to the intrinsic buffers, the open buffer system of CO_2/HCO_3^- enables very efficiently buffering of pH. Virtually all cells express plasma membrane ion transporters that contribute to cellular pH homeostasis. Some of these exploit the inward gradient for Na^+ to drive acid or base transport. Other transporters are dependent on, for example, the Cl^- gradient, the HCO_3^- gradient, electrical gradient or ATP hydrolysis to drive the transport. Most cells express acid/base transporters, depending on the function of the specific cell, and especially in HCO_3^--secreting epithelia, a great variety of such transporters are found.

Sodium-Hydrogen Exchangers (NHEs, *SLC9*) Most of the NHEs mediate the electroneutral exchange of intracellular H^+ for extracellular Na^+, given the typical

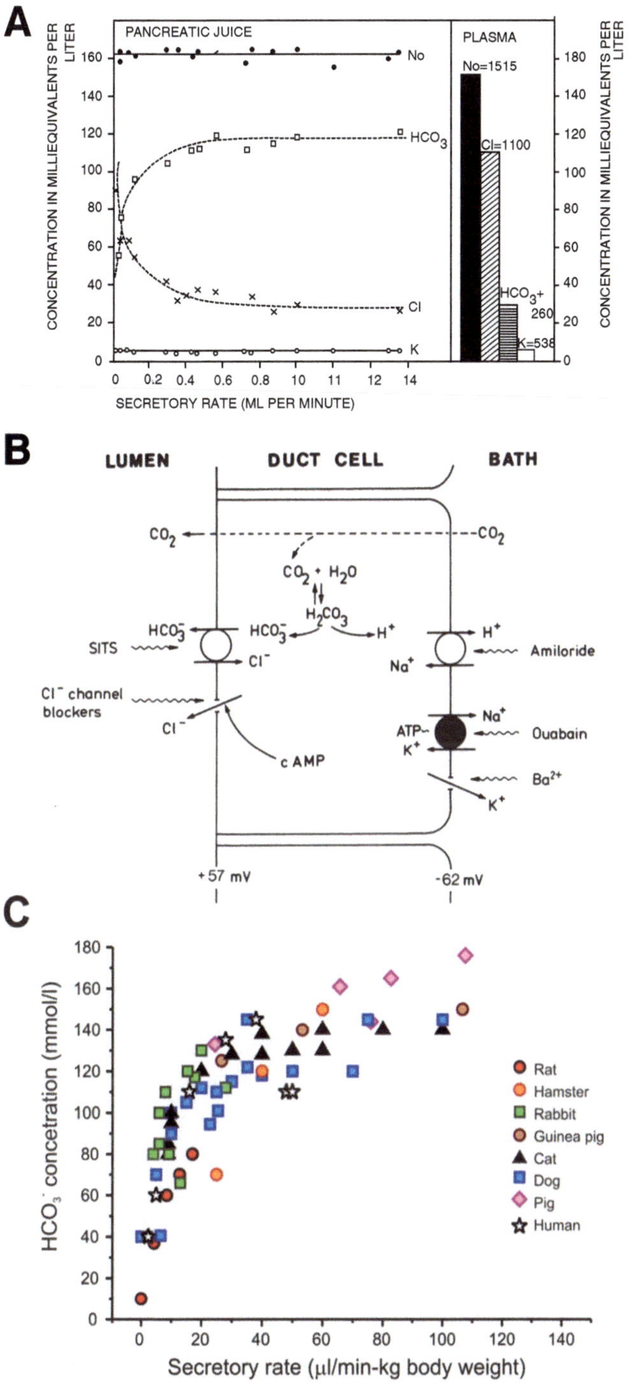

Fig. 5.1 (**a**) The classical electrolyte excretion curves showing the relationship between secretory rates and electrolyte concentrations in pancreatic juice collected from the dog pancreas stimulated with secretin [Reproduced with permission from Bro-Rasmussen et al. (1956)]. Similar excretory

ionic distribution and intracellular pH. Of the nine members of the NHE gene family, only NHE1 (*SLC9A1*) seems ubiquitously expressed and is therefore regarded as the central cellular acid extruder (Orlowski and Grinstein 2004). NHE1 has also been shown to be important for cell volume regulation, migration and proliferation mainly in non-epithelial and cultured cells (Pedersen 2006). NHE2 and NHE3 (*SLC9A2* and *A3*, respectively) are luminal proteins mainly found in Na^+-absorptive epithelia. Nevertheless, both can be found alongside potent HCO_3^- secretory machinery in the alkaline secretory cells of the stomach surface, duodenal villus cells and exocrine gland ducts. In these cases, NHE2 and NHE3 could potentially favour HCO_3^- absorption rather than secretion (Praetorius et al. 2000). The last plasma membrane *SLC9* member, NHE4 (*SLC9A4*), is a basolateral alternative to NHE1 in specialized cells of the kidney, stomach and liver. NHEs are inhibited by amiloride and its derivatives as well as cariporide with highest potency towards NHE1 (Scholz et al. 1995).

Sodium Bicarbonate Cotransporters (NBCs and NDCBEs, *SLC4*) The electrogenic Na^+-HCO_3^- cotransporter NBCe1 (*SLC4A4*) was the first Na^+-HCO_3^- cotransporter to be identified at the molecular level (Romero et al. 1997). NBCe1 transporter was structurally similar to the anion exchanger, AE1 (*SLC4A1*) (see below). NBCe1 mediates electrogenic Na^+-HCO_3^- cotransport with either 1:2 or 1:3 stoichiometry depending on the tissue and is localized to the basolateral surface in epithelial cells involved in vectorial HCO_3^- transport in the kidney, intestine and pancreatic ducts (Boron and Boulpaep 1983; Schmitt et al. 1999). The second electrogenic NBC, i.e. NBCe2 (*SLC4A5*), displays similar transport properties as NBCe1, also with varying Na^+:HCO_3^- stoichiometries (Pushkin et al. 2000; Sassani et al. 2002; Virkki et al. 2002). NBCe2 expression pattern is more controversial; NBCe2 is described in epithelial tissues such as the liver, testis, kidney, lung and the choroid plexus, where it is localized to the luminal membrane (Abuladze et al. 2004; Pushkin et al. 2000; Virkki et al. 2002; Bouzinova et al. 2005).

The Na^+-dependent exchange of Cl^- and HCO_3^- has been described in various tissues (Boron and Knakal 1992; Liu et al. 1990; Schlue and Deitmer 1988). Two Na^+-dependent Cl^- and HCO_3^- exchangers have been described, NDCBE1 (*SLC4A8*) and NCBE (*SLC4A10*) (Grichtchenko et al. 2001; Virkki et al. 2003; Wang et al. 2000). These transporters were characterized as electroneutral and DIDS sensitive and work with an apparent stoichiometry of $1Na^+$:$1Cl^-$:$2HCO_3^-$, where Na^+ and HCO_3^- are normally imported and Cl^- is extruded from the cells. The Cl^- dependence of NCBE has been challenged by compelling experiments in a study by Parker and colleagues

Fig. 5.1 (continued) curves were obtained for cat pancreas (Case et al. 1969). Similar excretory curves are expected for human pancreas. (**b**) The cellular model of ion transport in a pancreatic duct cell as established from electrophysiological studies of isolated perfused rat pancreatic ducts [Reproduced with permission from Novak and Greger (1988b)]. (**c**) The relation between secretory rates and HCO_3^- concentrations in pancreatic juice of various species. Secretion was stimulated with secretin and secretory rates were corrected for body weights [Reproduced with permission from Novak et al. (2011)]

(Parker et al. 2008). The only epithelial expression sites described thus far is the choroid plexus and connecting tubules for NCBE and hepatobiliary system for NDCBE1 (Grichtchenko et al. 2001; Wang et al. 2000; Strazzabosco et al. 1997; Banales et al. 2006b; Praetorius et al. 2004a). At these sites, the transporters may well take part in transepithelial movement of both Na^+ and HCO_3^-.

The electroneutral NBC, NBC3 or NBCn1 (Choi et al. 2000; Pushkin et al. 1999), also belongs to the *SLC4* gene family (*SLC4A7*). As the name indicates, the apparent Na^+:HCO_3^- stoichiometry is 1:1. This means that it is normally importing the two ions into cells. NBCn1 is expressed in the basolateral membranes in many epithelia including HCO_3^- secretory epithelia, such as the stomach surface cells, duodenum, colon and choroid plexus. Except for epithelial variants of NBCn1, NBCs and NDCBEs are inhibited by stilbene derivatives such as DIDS and SITS (Aalkjacr and Cragoe 1988; Boedtkjer et al. 2006; Bouzinova et al. 2005; Odgaard et al. 2004; Praetorius et al. 2001, 2004a). The drug S0859 seems to be a general inhibitor of Na^+-HCO_3^- cotransporter (Larsen et al. 2012). For further details on NBCs, see Chap. 14 of this volume.

Classical Anion Exchangers (AEs, *SLC4*) AE1-3 are Na^+-independent Cl^-/HCO_3^- exchangers which are electroneutral and all belong to the *SLC4* gene family (*SLC4A1-3*). AE1, or band-3 protein, was first demonstrated in red blood cells where it exports HCO_3^- (Lux et al. 1989). After entry into the red blood cells, CO_2 is hydrated and carbonic anhydrases accelerate the subsequent formation of HCO_3^- and H^+. H^+ is buffered by hemoglobin and HCO_3^- extrudes in exchange for Cl^- (the chloride shift) by AE1. Type-A intercalated cells of the renal collecting duct express an epithelial variant of AE1. This basolateral plasma membrane anion exchanger may play a supportive role for the apical acid secretion by extruding HCO_3^- to the blood side (Kollert-Jons et al. 1993). Another member, AE2, is expressed basolaterally in most epithelia, except for hepatobiliary system, and is involved in the protection of the cells against alkalization (Alper 2006). AE2 deletion in mice results in a severe phenotype with growth retardation, gastrointestinal dysplasia, biliary cirrhosis and death before weaning (Gawenis et al. 2004; Concepcion et al. 2013). AE3 is expressed mainly in excitable tissues, such as the brain and heart, but is also found in gastrointestinal enterocytes (Yannoukakos et al. 1994). Human AE3 point mutations have been associated with seizures, most likely as a consequence of the impaired neuronal pH_i regulation (Hentschke et al. 2006). AEs like NBCs and NDCBEs are inhibited by DIDS. For interested readers, see Chap. 15 for detailed information about SLCs.

Promiscuous Anion Exchangers A separate gene family of anion exchangers with a more promiscuous anion transport profile has taken a central position in transepithelial HCO_3^- movement. The *SLC26* genes give rise to 12 transporters, which are expressed in many different tissues and mediate very diverse functions (Alper and Sharma 2013; Mount and Romero 2004). Dysfunction of the intestinally expressed downregulated in adenoma (DRA, *SLC26A3*) produces congenital chlorodiarrhoea (Hoglund et al. 1996), which is caused by reduced luminal Cl^-/HCO_3^- exchange in the intestinal tract (Melvin et al. 1999). Pendrin, *SLC26A4*, is defective in the Pendred syndrome, in which patients suffer from impaired hearing and thyroid function. The symptoms result from dysfunctional thyroid Cl^-/I^- exchange, defective Cl^-/HCO_3^- exchange in the stria vascularis of the inner ear and in mice probably also decreased renal collecting duct HCO_3^- reabsorption (Masmoudi et al. 2000; Royaux et al. 2000, 2001). PAT-1 (or CFEX) is encoded by *SLC26A6* and also exchanges Cl^- with HCO_3^- and seems necessary for normal pancreatic and duodenal bicarbonate secretion (Ko et al. 2002b; Shcheynikov et al. 2006). Finally, deletion of *SLC26A9* has been shown to impair luminal alkalization in the gastric mucosa (Demitrack et al. 2010).

Vacuolar H^+-ATPase and H^+/K^+-ATPase Vacuolar H^+-ATPases, (V-ATPases), are expressed ubiquitously in the lysosomal system, but certain cells are known to express V-ATPases on the plasma membrane (Breton and Brown 2013). The V-ATPases are large transmembrane protein complexes consisting of several subunits and resemble the mitochondrial ATP synthases. Only complexes with certain specific composition of subunits can reside in the plasma membrane. The energy resulting from ATP hydrolysis is exploited to move H^+ across the membrane without counter ion transport. Hence, V-ATPases are electrogenic. Epithelial cells, such as renal intercalated cells and epididymis, use V-ATPase for transepithelial transport (Brown et al. 2009; Pastor-Soler et al. 2008). The V-ATPase is inhibited by bafilomycin and concanamycin (Huss and Wieczorek 2009). A separate group of P-type ATPases mediate exchange of H^+ and K^+, i.e. the H^+/K^+-ATPases (Fellenius et al. 1981). H^+/K^+-ATPases are classified into two families, gastric and non-gastric (also called colonic), and α-subunits are coded by *ATP4A* and *ATP12A (ATP1AL1)* genes, respectively (Modyanov et al. 1991, 1995; Sachs et al. 2007; Forte and Zhu 2010; Sangan et al. 2000). The pump complex consists of two α- and two β-subunits, whereby the gastric α-subunit assembles with gastric β-subunit (*ATP4B*), while non-gastric α-subunits can assemble with gastric or Na^+/K^+-ATPase β-subunits. The gastric H^+/K^+-ATPase is best known from the gastric corpus/fundus glands, where it mediates potent H^+ secretion for gastric acid, but is also expressed in the kidney and cochlea, and the non-gastric form is expressed in the colon, kidney, skin and placenta (Pestov et al. 1998). Some HCO_3^--secreting epithelia also express these pumps (see below). Proton pump inhibitors such as omeprazole are potent inhibitors of gastric H^+/K^+-ATPases, while high concentrations of potassium-competitive acid blockers and ouabain probably inhibit the non-gastric type (Grishin and Caplan 1998; Grishin et al. 1996; Swarts et al. 2005).

Carbonic Anhydrases HCO_3^- and H^+ are the major biologically relevant base and acid, respectively. The most potent cellular pH homeostasis and base secretion rely on a steady supply of these ion species. The hydration of CO_2 occurs spontaneously at sufficient rate, while the uncatalysed hydrolysis of H_2CO_3 is quite slow for biological purposes. Thus, the carbonic anhydrases, which catalyse the conversion of $CO_2 + H_2O$ to HCO_3^- and H^+, are of major importance both for pH homeostasis and bicarbonate secretion, and there are 14 different CA isoenzymes: CA I–III and VII are cytosolic; IV, IX and XII are membrane associated; V is mitochondrial; and VI is secreted (Supuran 2008). The canonical and ubiquitously expressed form is CAII. This enzyme has one of the fastest turnovers in mammalian biology (Maren 1962) and is a soluble cytosolic protein. In recent years, it has become evident that other forms of carbonic anhydrases are resident in the plasma membrane, either with extracellular enzyme activity (GPI anchored) or with cytosolic sub-membrane activity. Acetazolamide has been used to block carbonic anhydrases for decades and seems to block both cytosolic and membrane-associated enzyme forms. In any case, inhibition of HCO_3^- formation by this drug has profound impact on epithelial bicarbonate secretion in several tissues. Interestingly, some investigators have reported the physical as well as functional interaction between carbonic anhydrases and bicarbonate transporters, such as AE2 and NBCe1, while other researchers have negated these findings (McMurtrie et al. 2004; Boron 2010). It is believed that such interactions could facilitate transport by securing substrate to the HCO_3^- transporters (McMurtrie et al. 2004).

5.1.3 Vectorial Bicarbonate Transport

It is evident that epithelia that secrete bicarbonate need to employ one of the abovementioned mechanisms to extrude HCO_3^- from the cytosol into the luminal/apical compartment. Epithelia with potent bicarbonate extrusion are generally equipped with anion channels, with promiscuous anion exchangers or with an electrogenic Na^+-HCO_3^- cotransporter at the site of exit. However, to avoid cellular acidification, the cells must have just as effective means of getting rid of H^+ across the opposite plasma membrane to avoid damaging acidification and to sustain the production of new HCO_3. So, in the case of luminal HCO_3^- secretory epithelia, the cells must have sufficient acid extrusion mechanisms such as NHE1/4 or a H^+-ATPase. Alternatively, the luminal HCO_3^- secretion can be supported by basolateral HCO_3^- entry through any of the Na^+-driven HCO_3^- transporters.

In the following sections, selected epithelia with distinct acid/base transport characteristics will be described: pancreatic ducts, salivary glands, hepatobiliary system, duodenum, collecting duct and choroid plexus. While the first four organs have high to very high HCO_3^- output, the choroid plexus epithelium has an intermediate HCO_3^- output, and the terminal renal collecting ducts exemplify epithelia with little or no transepithelial HCO_3^- movement, where a subset of specialized cells mediate HCO_3^- secretion depending on the acid/base status. Thus, similarities and differences in molecular machinery for HCO_3^- transport between these tissues may help establish hypotheses regarding the functional roles of specific acid/base transporters and ion channels.

5.2 Pancreas

5.2.1 *The Prototype of a Bicarbonate Secretor Is a Complex Gland: Integrated Function and Morphology*

The pancreas and other exocrine glands are composed of at least two main types of epithelia—secretory acini/endpieces and excretory ducts. Thaysen and coworkers (Bro-Rasmussen et al. 1956) proposed the two-stage hypothesis of secretion for complex exocrine glands, and this can still be used as a starting point to understand their integrated function. Basically, it says that acini/endpieces secrete fluid similar to that in plasma in their electrolyte composition, and they secrete macromolecules such as enzymes. The ducts may modify this secretion, and in the pancreas, they do so by adding a secretion of their own (Fig. 5.1a). Pancreatic ducts are generally regarded as leaky epithelia expressing aquaporins, and they are able to secrete fluid that is HCO_3^- rich and alkaline (Bro-Rasmussen et al. 1956; Steward and Ishiguro 2009; Wilschanski and Novak 2013; Ishiguro et al. 1998; Fernandez-Salazar et al. 2004; Novak et al. 2011). In humans, the maximum HCO_3^- output in secretin-stimulated gland is about 500 μmol/h/g pancreas tissue weight. This output would be at least five times higher if it is assumed that it arises from pancreatic ducts contributing around 20 % to pancreas mass in humans.

The well-established function of pancreatic bicarbonate secretion is that it contributes to buffering of acid chyme entering the duodenum; the other contributors are the duodenal epithelium and bile duct epithelium (Ainsworth et al. 1992). Recently, it has been discussed whether the bicarbonate secretion has an additional function already within the pancreas (Hegyi et al. 2011; Wilschanski and Novak 2013; Novak et al. 2013). That is, there are some indications that the acinar secretion might be acidic and the function of adjoining ducts may be therefore to alkalinize this acinar secretion very early in its passage through the duct tree and thus prevent premature activation of digestive enzymes and maintain the balance in exo-/endocytosis in acini (Freedman and Scheele 1994; Freedman et al. 2001; Behrendorff et al. 2010; Hegyi and Petersen 2013). The third possible function for bicarbonate secretion is to solubilize mucins, and although this has not been proven for the pancreas (Quinton 2008, 2010), it has been shown that the very early key symptom in cystic fibrosis is reduced HCO_3^- secretion and mucoviscidosis in the pancreas (Andersen 1938; Kopelman et al. 1985, 1988).

Pancreatic juice collected from the pancreas stimulated with the main "secretagogue" in many species, secretin, has electrolyte composition that depends on secretory rates (Fig. 5.1a). Basically, at high secretory rates, pancreatic juice is rich in HCO_3^- and poor in Cl^-, and as secretion decreases, HCO_3^- falls and Cl^- increases in a mirror-like fashion. Na^+ concentrations do not change with flow rate and are plasma-like. K^+ concentrations are similar to or higher than in plasma.

Over many years, it has been regarded that some animals produce juice low in HCO_3^- concentrations (mice, rats, rabbits), while the pancreas of other species produce secretion with high HCO_3^- concentrations (man, dog, cat, pig, guinea pig).

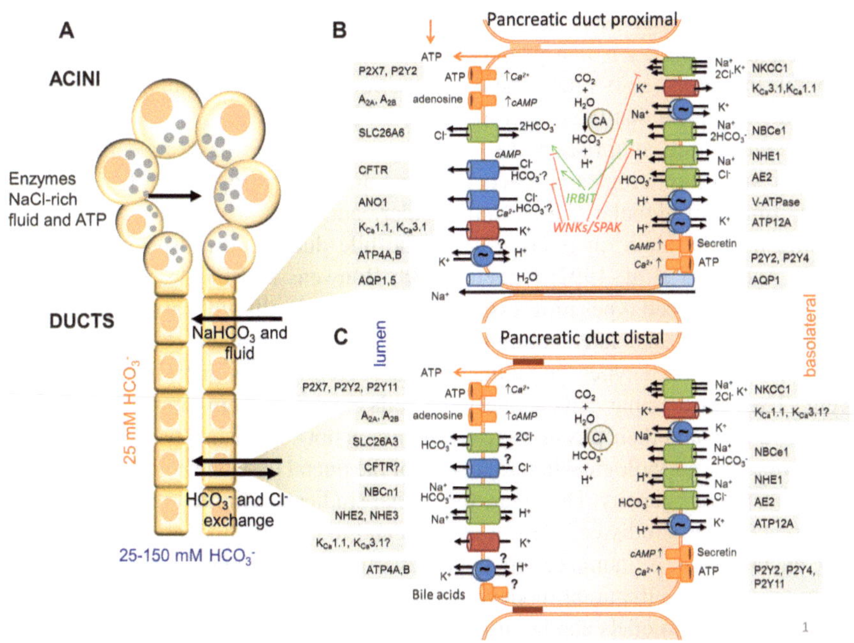

Fig. 5.2 (**a**) Schematic diagram showing simplified pancreas with acini, proximal and distal ducts and pancreatic juice with typical range of HCO_3^- concentrations. Insets show two types of cells with the cellular models for a HCO_3^--secreting cell (**b**) and a cell that is exchanging HCO_3^- for Cl^- (**c**). Interstitial/plasma HCO_3^- concentration is 25 mmol/l and pancreatic juice contains 25–150 mmol/l HCO_3^- and depends on the secretory rate and stimulation (see Fig. 5.1). Molecular identities of ion transporters, channels and receptors are discussed in the text; question marks indicate unclear identities, localization or functions. Intracellular signalling is simplified, and stimulatory (*green*) and inhibitory (*red*) pathways and other interactions between cAMP and Ca^{2+} signalling (*double-headed arrow*) are discussed in Sect. 6.2.4.3

Nevertheless, close analysis shows that HCO_3^- concentrations in pancreatic juice among different species may depend on the relative proportion of acinar to duct cells contributing to the final secretion. If corrected for this, it becomes apparent that secretion of all species, summarized in Fig. 5.1c, falls within one excretory curve, which implicates similar secretory mechanisms in all species. But why does HCO_3^- fall and Cl^- increase with the falling secretory rate (Fig. 5.1a)? These curves are often pictured but are rarely elaborated. One explanation is provided by the admixture hypothesis, which states that the final secretion is a mix of fluids with different compositions (acini and ducts) (Fig. 5.2a). Another theory is the exchange theory (implying exchange of HCO_3^- for Cl^-). This is most apparent at low secretory rates, that is, when secretion from acini and proximal ducts is low, distal ducts are not overridden by incoming secreted fluid, and thus, they use their full capacity to simply exchange HCO_3^- for Cl^-, a process referred to recently as "HCO_3^- salvaging". This HCO_3^-/Cl^- exchange was demonstrated many years ago on the cat main pancreatic duct (see below) that was perfused with various solutions and could carry out such an exchange (Case et al. 1969). Yet another so far theoretical possibility is that pancreatic ducts can also secrete H^+, a process most obvious in low secretory rates. These explanations might not be mutually exclusive, and they implicate that the ductal tree is heterogeneous.

The ductal tree comprises 5–20 % of the pancreas tissue mass, depending on the species, and morphologically, ducts are quite different—progressing from intercalated, small intralobular, larger intralobular, and inter-/extralobular, eventually joining into the main ducts and in turn the bile duct in some species (Kodama 1983; Ashizawa et al. 1997; Githens 1988; Bouwens and Pipeleers 1998; Gmyr et al. 2004). The cell types progress from the flat small cells with large long primary cilia to cuboidal and later columnar cells with short primary cilia. Large ducts contain several cell types, including mucus-secreting cells and single endocrine cells. The centro-acinar cells are very flat cells extending from intercalated cells to acinar lumen, and their physiological function is not established.

Given the morphological heterogeneity of the ductal tree, one could expect some functional heterogeneity. However, functional studies are limited to ducts that can be isolated or microdissected from animals and to culture models, mostly pancreatic ductal adenocarcinoma cell models. Nevertheless, combined with data acquired from transgenic cell/animal models, immunohistochemistry and many other techniques, coherent models can be proposed.

5.2.2 HCO_3^- and H^+ Transporters in Pancreatic Ducts

CFTR and Cl^-/HCO_3^- Exchangers The first studies of cellular mechanisms for pancreatic duct HCO_3^- transport showed, surprisingly, that secretin/cAMP-activated Cl^- channels on the luminal membranes of isolated rat pancreatic ducts (Fig. 5.1b) (Novak and Greger 1988b; Gray et al. 1988). Almost in parallel, the cystic fibrosis transmembrane conductance regulator, CFTR, was discovered (Riordan et al. 1989; Kerem et al. 1989), and it was shown to have properties of a Cl^- channel, also in the pancreatic ducts (Tabcharani et al. 1991; Gray et al. 1993) (Fig. 5.2b). Subsequently, CFTR was immunolocalized in human and rodent pancreas to intercalated and small intralobular ducts, which also express other key proteins in HCO_3^- secretion, aquaporins and carbonic anhydrases (Hyde et al. 1997; Marino et al. 1991; Kumpulainen and Jalovaara 1981; Burghardt et al. 2003). Since pancreatic HCO_3^- and fluid secretion is defective in cystic fibrosis, and since the underlying signatures are mutations in CFTR, the channel has been considered as the key element in the pancreatic duct secretion (Wilschanski and Novak 2013). Nevertheless, the question whether and how CFTR Cl^- channels could transport HCO_3^- has been a long-lasting challenge. Over the years, many patch-clamp studies of CFTR have shown that in physiological-like conditions, the permeability ratio $PHCO_3^-/PCl^-$ is 0.2–0.5, implying that secretion of Cl^- would dominate. Recent studies suggest that CFTR could become more HCO_3^- permeable if intracellular Cl^- was reduced (Ishiguro et al. 2009; Park et al. 2010, 2012).

Whether this HCO_3^- permeability of CFTR could be maintained in physiological conditions and whether there is sufficient electrochemical gradient remain to be settled. It is proposed that some cell volume/Cl^- regulatory mechanism could be contributing (see Sect. 5.2.4.3). For additional information about CFTR, see Chap. 17 of this volume.

The first proposal for HCO_3^- exit pathway was the Cl^-/HCO_3^- exchange mechanism coupled with luminal Cl^- channels, thus allowing Cl^- recirculation and net HCO_3^- secretion, though this would only account for 60–80 mM HCO_3^- in secretion (Novak and Greger 1988a, b). Through intensive efforts in molecular and cell biology, the following Cl^-/HCO_3^- exchangers were identified. The anion exchanger*SLC26A3*, also known as downregulated in adenoma (DRA), and *SLC26A6*, also known as putative anion exchanger-1 (PAT-1), were found expressed on the luminal membrane of large mouse and human pancreatic ducts (Greeley et al. 2001; Lohi et al. 2000). These two exchangers have different stoichiometries showing $Cl^-:HCO_3^-$ of 2:1 for *SLC26A3* and 1:2 for *SLC26A6*. It was proposed that *SLC26A3* was expressed in more distal ducts. *SLC26A6* was more proximal on the luminal membrane of intralobular ducts and rarely on larger ones (Ko et al. 2002b, 2004). Theoretically, the Cl^-/HCO_3^- exchange of 1:2 for *SLC26A6* would be thermodynamically more favourable for HCO_3^- secretion, while *SLC26A3* would favour HCO_3^- absorption (Fig. 5.2b, c). There is a functional coupling between *SLC26A6* and CFTR, and this involves the R domain of CFTR and sulphate transporter anti-sigma (STAS) domains of *SLC26A6* exchanger (Ko et al. 2004; Dorwart et al. 2008; Wang et al. 2006; Stewart et al. 2009). Nevertheless, studies using knockout strategy for *SLC26A* exchanger showed some interdependence between the two isoforms and varied effects on duct/pancreas secretion (Ishiguro et al. 2007; Song et al. 2012).

AE2 (*SLC4A2*), another anion exchanger from the *SLC4* family, was demonstrated, usually on the basolateral membranes, in a number of pH_i studies in pancreatic ducts (Stuenkel et al. 1988; Zhao et al. 1994; Rakonczay et al. 2006). However, immunohistochemical studies are not congruent as to which membrane the transporter is localized (Hyde et al. 1999; Roussa et al. 2001; Kulaksiz and Cetin 2002). Most likely, AE2 is more involved in pH_i regulation rather than transepithelial HCO_3^- transport.

Calcium-Activated Cl^- Channels In addition to cyclic AMP-regulated secretion, a number of studies show that agonists such as acetylcholine and extracellular nucleotides (see below) act via Ca^{2+} signalling to stimulate Ca^{2+}-activated Cl^- channels (CaCC) (see Chap. 17) and thus could support duct secretion (Gray et al. 1989, 1994; Pahl and Novak 1993; Hug et al. 1994; Winpenny et al. 1998; Szalmay et al. 2001; Pascua et al. 2009). The identity of CaCC channels has been difficult to pinpoint (see Duran et al. 2010). After suggestions of CCl-2 and bestrophins, the TMEM16/ANO family was discovered (Caputo et al. 2008; Schroeder et al. 2008; Yang et al. 2008). One of the members, TMEM16A/ANO1, is regarded a good candidate for CaCC in epithelia. In pancreatic ducts, recent studies on human duct cell lines shows that they express TMEM16A/ANO1, which targets to the luminal membrane upon stimulation and gives rise to the secretory potential in polarized duct epithelia (Wang et al. 2013; Wang and Novak 2013). There is also one immunohistological study on human pancreatic sections showing that the ANO1/DOG-1 antibody localizes to centro-acinar and small duct cells, and the channel is grossly overexpressed in pancreatic cancer and other gastrointestinal stromal tumours (Bergmann et al. 2011). This channel could be relevant for pancreatic HCO_3^- secretion. One study shows that at least in an expression system of HEK293 cells and mouse submandibular acinar cells, TMEM16A is directly modulated by calmodulin, and not with no lysine kinases (WNKs), and it increases HCO_3^- permeability (Jung et al. 2013). Nevertheless, it cannot be excluded that there are also other CaCC and/or that Ca^{2+} regulates also CFTR (see below). Further information about TMEM16 can be found in Chap. 17 of this volume.

NBCs, NHEs and Carbonic Anhydrases HCO_3^- transport across the luminal membrane, whatever the mechanism is, relies on provision of cellular HCO_3^-. One well-supported solution is the import of HCO_3^- across the basolateral membrane via a Na^+-coupled process, i.e. Na^+-HCO_3^- cotransporters, NBC. One NBC isoform was cloned from the pancreas, pNBC (NBCe1B, *SLC4A4*), and it transports 1 Na^+:2 HCO_3^-, and putative inhibitor H_2DIDS inhibits about 50 % of duct secretion (Abuladze et al. 1998; Ishiguro et al. 1998; Choi et al. 1999). The pNBC is found on the basolateral membranes of epithelial cells of a duct tree in the human pancreas, though in rat pancreas, acinar and duct labelling was occasionally on both membranes (Marino et al. 1999; Satoh et al. 2003). If pancreatic ducts were relying predominantly on this transporter, secretion would be highly dependent on provision of HCO_3^- from interstitium/plasma rather than endogenous CO_2 production, CA activity and H^+ extrusion mechanism. This was found to be the case for isolated cat pancreas and guinea pig ducts, though H_2DIDS inhibited about 50 % of secretions (Schulz 1971; Case et al. 1970; Ishiguro et al. 1998).

Another isoform, the electroneutral NBCn1 (NBC3, *SLC4A7*), is also expressed in the pancreas (Damkier et al. 2006), though one study shows that in mouse ducts, it interacts with CFTR, and it is inhibited by cAMP and therefore should be placed on the luminal membrane and possibly regulate HCO_3^- salvage (Park et al. 2002b) (Fig. 5.2c).

An alternative or additional solution for cellular HCO_3^- (and H^+) provision is carbonic anhydrase (CA)-catalysed hydration of CO_2, provided from metabolism and/or HCO_3^-/CO_2 buffer system. Isoforms CAII, IV, VI, IX and XII are expressed in human pancreas and cultured duct tumour cells (Kumpulainen and Jalovaara 1981; Nishimori et al. 1999; Nishimori and Onishi 2001). CAII and CAIV interact with H^+/HCO_3^- transporters; however, localization of the CA isoforms does not always match the predicted localization of the transporters in pancreatic ducts. For example, CAII is found intracellularly and on the luminal membrane (Alvarez et al. 2001), and it seems to interact with NHE1 and NBC3 (Li et al. 2002; Loiselle et al. 2003). CAIV is expressed in the luminal membrane of the ductal tree (centro-acinar cells and in intercalated, intralobular and interlobular ductal cells) (Fanjul et al. 2004; Mahieu et al. 1994), but it interacts with NBCe1 in expression studies

(Alvarez et al. 2003). CA IX and XII are expressed on the basolateral membranes of normal and pathological samples of pancreas (Kivela et al. 2000; Juhasz et al. 2003). Carbonic anhydrases have been somewhat neglected in pancreatic duct studies in recent years. Nevertheless, CAs are key enzymes in pancreatic duct function, as their inhibition leads to marked effects on pH_i and pancreatic secretion (Hollander and Birnbaum 1952; Case et al. 1979; Cheng et al. 1998; Steward et al. 2005; Rakonczay et al. 2006).

Intracellular H^+, generated from CA activity or metabolism, can be extruded out of the cell by a Na^+/H^+ exchanger (NHE). Such exchanger was proposed based on the observation that pancreatic duct secretion could be maintained efficiently without HCO_3^- by a number of weak lipid soluble acids, such as acetate (Schulz et al. 1971; Case et al. 1979). NHE, sensitive to amiloride and derivatives, has been detected in many studies monitoring secretion of whole pancreas in different species and later on in isolated pancreatic ducts (Wizemann and Schulz 1973; Veel et al. 1992; Novak and Greger 1988a; Ishiguro et al. 1998; de Ondarza and Hootman 1997; Fernandez-Salazar et al. 2004; Szucs et al. 2006). Nevertheless, amiloride-type inhibitors could decrease secretion by about 20–50 %. One of the NHE isoforms, the ubiquitous NHE1 (*SLC9A1*), is the major pH_i regulator. In functional studies, it was revealed that NHE contributed significantly to pH_i regulation in many duct preparations including pig, guinea pig, rat and mice ducts and human duct cell lines (Veel et al. 1992; Szucs et al. 2006; de Ondarza and Hootman 1997; Ishiguro et al. 2000; Novak and Christoffersen 2001; Lee et al. 2000; Demeter et al. 2009; Rakonczay et al. 2006; Olszewski et al. 2010). There is some molecular evidence for NHE1 expression in normal ducts, and localization appears to be on the basolateral membrane (Lee et al. 2000; Roussa et al. 2001); thus function in secretion (and pH_i regulation) could be supported. In addition, NHE2 and 3 isoforms are expressed on the luminal membrane of main ducts and are proposed to interact with CFTR via PDZ domains (Lee et al. 2000; Ahn et al. 2001; Marteau et al. 1995). These exchangers would then not support secretion, but conduct HCO_3^- salvage (or pH_i regulation) (Fig. 5.2c).

Proton Pumps The above models do provide a number of answers, but still we are left with the problem of how to explain high HCO_3^- concentrations and why inhibitors of NHE1, NBC and CA are relatively ineffective in blocking secretion (Fernandez-Salazar et al. 2004; Grotmol et al. 1986). The above transporters and ion channels rely on gradients that are created by the Na^+/K^+-ATPase. Another solution to create HCO_3^-/H^+ gradients would be to extrude H^+ via the V-ATPase. In one early study, V-ATPase on the basolateral membrane was proposed (Villanger et al. 1995), and V-ATPase was detected on basolateral membrane of intralobular ducts, although occasionally some cells had luminal staining (Roussa et al. 2001). A number of functional studies gave contradictory findings (Zhao et al. 1994; Ishiguro et al. 1996; de Ondarza and Hootman 1997; Cheng et al. 1998), perhaps depending on which parameters were measured. It seems that contribution of the pump to pH_i regulation is relatively small (compared to NHE1), but inhibition of secretion or short-circuit currents with V-ATPase blockers can be significant.

Recently, other types of H^+ pumps have been detected in pancreatic ducts. Both rodent (and human) ducts express the gastric-type H^+/K^+-ATPases (*ATP4A* and *ATP4B*) and non-gastric type H^+/K^+-ATPase (*ATP12A*) (Novak et al. 2011; Wang et al. 2015). Inhibition of these with proton pump inhibitors such as omeprazole and SCH28080 reduced pH_i recovery in response to acid loads, and more importantly, they reduced secretion in isolated pancreatic ducts (Novak et al. 2011; Giannuzzo et al. 2013; Wang et al. 2015). The immunohistochemical study showed that H^+/K^+-ATPases (mainly colonic type) are localized to the basolateral membrane and thus is consistent with HCO_3^- secretion model. However, some H^+/K^+-ATPases were also localized to the luminal membrane, especially the gastric form (Novak et al. 2011; Wang et al. 2015). At present, the function of these pumps in pancreatic ducts is unclear, similar to other HCO_3^--secreting epithelia such as the airway epithelia (Novak et al. 2013). It is speculated that these luminal pumps are creating a buffer zone protecting cells against bulk secretion which is pH > 8. In addition, luminal H^+/K^+ pumps in distal ducts would by virtue of H^+ secretion have more impact on pancreatic juice composition at low flow rates and minor at high flow rates, thus explaining excretory curves for HCO_3^- (Fig. 5.1). Furthermore, the luminal H^+/K^+ pumps would recirculate K^+ extruded by the luminal K^+ channels (Hayashi et al. 2012; Novak et al. 2013).

K^+ Channels In addition to HCO_3^-/H^+ transporters, it is now appreciated that K^+ channels are important for pancreatic duct secretion. They maintain the resting potential, and during stimulation, opening of K^+ channels and hyperpolarization of the membrane potential maintains the driving force for Cl^- or HCO_3^- exit (Novak and Greger 1988a, 1991). Conductance for K^+ (G_K) is present both on the basolateral and luminal membranes, and equivalent-circuit analysis has shown that the luminal K^+ channels contribute with at least with 10 % to the total conductance in stimulated duct (Novak and Greger 1988a, 1991). Modelling in salivary glands confirms that such ratio of luminal to basolateral K^+ channels would optimize secretion without destroying the transepithelial potential and transport (Almassy et al. 2012; Cook and Young 1989). Another function of luminal K^+ channels could be to contribute to secreted K^+, as pancreatic juice contains 4–8 mM K^+ (Sewell and Young 1975; Caflisch et al. 1979; Seow et al. 1991). The molecular identities and function of only some K^+ channels are known (see Hayashi and Novak 2013). The best studied until now are the Ca^{2+}-activated K^+ channels. The $K_{Ca}1.1$ channels (maxi-K, BK, coded by *KCNMA1*) are present in pancreatic ducts (Hede et al. 2005; Venglovecz et al. 2011) (Fig. 5.2b). Earlier patch-clamp studies indicate that these channels are located basolaterally (Gray et al. 1990; Hede et al. 1999). However, recent studies indicate that these channels are expressed on the luminal membrane and activated by low concentrations of bile acids (see below). Evidence for another K^+ channel activated by extracellular ATP via purinergic P2 receptors was provided in studies of rat and dog duct epithelia (Hug et al. 1994; Nguyen et al. 1998), and later the intermediate conductance,

$K_{Ca}3.1$ channel (IK, SK4, coded by *KCNN4*), was documented (Hede et al. 2005; Jung et al. 2006; Hayashi et al. 2012). Immunolocalization indicates that $K_{Ca}3.1$ is expressed on both luminal and basolateral membranes (Fig. 5.2b). The $K_{Ca}3.1$ channel activator EBIO enhanced secretion potential in Capan-1 monolayer indicating that these channels are important in pancreatic duct secretion (Hayashi et al. 2012; Wang et al. 2013). Recent studies on pancreatic ducts offer molecular identities of several other K^+ channels, including KVLQT1, HERG, and EAG2; Slick and Slack (Hayashi et al. 2012); and interestingly the pH sensor TASK-2 (Fong et al. 2003). Nevertheless, the function and regulation of these channels in pancreatic physiology needs to be studied.

Aquaporins and NKCC1 Taking that pancreatic ducts are secretory, water follows paracellularly and transcellulary via aquaporins (AQPs). AQP1 is expressed on centro-acinar cells and luminal and basolateral membrane of intercalated ducts, and AQP5 is expressed luminally, and labelling decreases in larger ducts in the human pancreas and is more distal in rodent pancreas (Burghardt et al. 2003, 2006). Notably, AQPs are co-expressed with CFTR in the same cells.

Upon stimulation of secretion, there would be a significant reduction in cell volume due to solute transport followed by osmotically obliged water. Subsequently, the cell volume would need to be reinstituted, and one of the most important transporters in that respect is the Na^+-K^+-$2Cl^-$ cotransporter (NKCC1, *SLC12A2*). This transporter is expressed in pancreatic ducts; however, it is not clear whether shrinkage activation of NKCC1 occurs in pancreatic ducts, or first following withdrawal of stimuli, as is the case in salivary acinar cells (see Sect. 5.3.2). Additionally, NKCC1 could provide cellular Cl^- for Cl^--driven fluid transport. In fact, diuretics such as bumetanide can inhibit duct/pancreas secretion, but the effect depends on the species (Fernandez-Salazar et al. 2004; Grotmol et al. 1986).

5.2.3 Integrating Ion Channels and Transporters to Pancreatic Ducts

Taking the above-described channels and ion transporters and placing them into one cell model becomes rather problematic—such cell would secrete and absorb at the same time. Therefore, it should be also considered where these transporters are localized within the pancreatic duct tree. Since there are only very few functional studies on native ducts from different regions of the ductal tree, we have to resort to taking into account heterogeneity in duct morphology and immunohistochemistry and interaction between channels/transporters in expression studies. Studies summarized in the preceding sections have indicated that small proximal ducts express CFTR, CA, *SLC26A*, AE2, NBC1e and AQP1. The larger, distal ducts express *SLC26A3* and possibly CFTR, NBCn1 and NHE3, as well. It is not yet clear whether AE2, K^+ channels, H^+ pumps and NKCC1 are differentially expressed.

For simplicity, if one inserts these transporters into two cells (Fig. 5.2b, c), it becomes apparent that the first cell has a potential to secrete, while the second has the possibility to exchange HCO_3^- for Cl^-. It cannot be excluded that these two models are two different states of one cell at different times or conditions. However, a very likely scenario is that one cell represents small proximal ducts that are secreting HCO_3^--rich fluid, and the other represents large interlobular/lobar ducts that are modifying incoming fluid but not contributing with a net fluid secretion. The simplest interpretation for these data obtained for large distal ducts is that they reabsorb HCO_3^- in exchange for Cl^-—thus giving the rise to the excretory curve (Fig. 5.1). The ion transport model for pancreatic acinar cells is reviewed elsewhere (Heitzmann and Warth 2008) and is similar to Cl^- secreting salivary acini (see Sect. 5.3).

5.2.4 *Regulation of Pancreatic Duct Secretion*

The classical HCO_3^--evoking secretagogue is secretin, though a number of other hormones and transmitters can also evoke and co-regulate HCO_3^- secretion. Even cholinergic stimulation and cholecystokinin (CCK) can evoke HCO_3^- secretion in some species, and they can potentiate the secretin effect on the volume of secretion (Hickson 1970; Holst 1993; Park et al. 1998; You et al. 1983; Evans et al. 1996; Chey and Chang 2001; Szalmay et al. 2001). Here, we will consider the novel paracrine and autocrine regulators of pancreatic ducts—those secreted by acini and ducts themselves (nucleotides) and those that are entering duct retrogradely (bile acids). Subsequently, we will consider novel interaction between signalling pathways and ion transporters and how they can in an integrative way affect pancreatic duct secretion.

5.2.4.1 Purinergic Signalling

It is now becoming accepted that purinergic signalling can contribute to integrating acinar and duct functions, in particular fine-tuning duct performance. Pancreatic acini release ATP, some of which is stored in zymogen granules and released upon hormonal and cholinergic stimulation (Sorensen and Novak 2001; Haanes and Novak 2010), and small amounts of ATP can be also detected in pancreatic juice, though most is hydrolysed by ectonucleotidases (Kordas et al. 2004; Yegutkin et al. 2006) (Fig. 5.2a). ATP is also most likely released from nerves as well as from acini and duct cells in response to cell volume changes and mechanical and chemical stress. In contrast to acini, pancreatic ducts express a number of functional purinergic (P2Y2, P2Y4, P2Y11, P2X4 and P2X7) and adenosine (A2A and A2B) receptors that regulate various epithelial transporters (see Novak 2008, 2011) (Fig. 5.2b, c). For example, luminal ATP (or UTP) can increase anion and fluid secretion, and this involves the regulation of TMEM16A/ANO1 and CFTR, as well as $K_{Ca}3.1$ and Cl^-/HCO_3^- exchange (Chan et al. 1996; Hug et al. 1994; Ishiguro et al. 1999; Namkung et al. 2003; Hede et al. 2005; Jung et al. 2006; Novak et al. 2010; Hayashi et al. 2012; Wang et al. 2013). In addition, luminal ATP stimulates P2X7 receptors and potentiates cholinergically evoked ductal secretion (Novak et al. 2010). Furthermore, ATP/UTP also potentiates cAMP-evoked mucin secretion (Jung et al. 2010). Ca^{2+} signalling and P2Y2 and P2X7 receptors in particular have been considered in these actions. Adenosine receptors via cAMP signalling regulate CFTR (Novak et al. 2008). From the basolateral side, ATP released by nerves and/or distended epithelium can also affect secretion, and some purinergic receptors are inhibitory to secretion (e.g. P2Y2 receptors inhibit $K_{Ca}1.1$ channels), while other P2 receptors, including P2Y11 receptors, may have positive effects on secretion (Hede et al. 1999, 2005; Ishiguro et al. 1999; Nguyen et al. 2001; Wang et al. 2013). A number of processes in purinergic signalling are pH sensitive, and it will be relevant to investigate those in the microenvironment of the duct epithelium (Novak et al. 2013). Due to the fact that nucleotides/nucleoside could stimulate a multitude of P2 and adenosine receptors acting via Ca^{2+} and cAMP signalling, interactions need to be considered. For further information about P2X receptors in epithelial, the reader is directed to Chap. 9.

5.2.4.2 Bile Acids

Systemic bile acids (primary or secondary) produced in the liver and by gut microbiota are becoming regarded as important physiological regulators of a wide range of cells. In the exocrine pancreas, though, bile acids (BA) may exert additional effects, as they can enter pancreatic duct tree by reflux following outflow obstruction by gallstones, and apparently affect both duct and acinar cells. Biliary acute pancreatitis (AP), or gallstone obstruction-associated AP, accounts for a significant percentage of clinical cases of AP, and animal and cellular models are important tools for understanding the development of this disease (Wan et al. 2012). Many studies have been carried out on acinar cells, and it has been shown that high concentrations (mM) of, for example, taurine-conjugated BA cause large increases in intracellular Ca^{2+}, activation of intracellular trypsinogen and necrosis (Voronina et al. 2002, 2004; Gerasimenko et al. 2006; Kim et al. 2002). These BA effects are mediated by TGR5/Gpbar1 receptor, which is expressed on the apical surface of pancreatic acini in mice (Perides et al. 2010).

Only a few studies show that BA also have effects on pancreatic ducts, but these may be important since pancreatic ducts are pivotal for the maintenance of physiological function of the whole pancreas. BA have bimodal effect on pancreatic ducts inducing pancreatic fluid hypersecretion in early stages of pancreatitis and hyposecretion during the onset of the disease (Czako et al. 1997). Studies on isolated ducts and duct epithelia show that this bimodal effect may be related to the concentration of BA used. At high (>mM) concentrations, BA have detrimental effect. For example, in guinea pig ducts, nonconjugated BA, chenodeoxycholic acid (CDCA) at 1 mM, caused large sustained increase in Ca^{2+}, inhibited

HCO_3^- transport, caused mitochondrial damage and increased permeability of duct cells and caused mitochondrial damage (Venglovecz et al. 2008; Maleth et al. 2011). In bovine duct cell monolayer, 5 mM taurodeoxycholic acid (TDCA) decreased epithelial resistance due to damage of the epithelial barrier (Alvarez et al. 1998). At lower concentrations, BA have positive effects. For example, in epithelium derived from the dog main duct, TDCA increased luminal G_{Cl} and basolateral G_K in Ca^{2+}-dependent manner (Okolo et al. 2002). In other studies on guinea pig ducts and CFPAC cells, it was shown that 0.1 mM CDCA increased Ca^{2+} via PLC and IP_3 (inositol 1,4,5-trisphosphate) and stimulated HCO_3^- transport (i.e. pH_i monitoring), though NBC, NHE, AE or CFTR or other Cl^- channels seem not to be primary targets (Venglovecz et al. 2008; Ignath et al. 2009). Recent studies on guinea pig ducts show that a low-dose CDCA activated maxi-K^+ channels on the luminal membrane and thereby could initiate the secretory machinery (Venglovecz et al. 2011). Thus, it is proposed at high concentrations, BA are damaging, but at low concentrations, BA would be able to promote duct secretion and thus wash out refluxed bile. However, it is not yet clear whether these physiological-like BA signals are mediated via TGR5/Gpbar1 receptors. Recent study shows that BA effects are mediated by ATP release and purinergic signalling (Kowal et al. 2015).

5.2.4.3 Synergistic Intracellular Signalling: Calcium, cAMP and Cell Volume

In pancreatic ducts, as in other biological systems, physiological regulation would involve the stimulation of several types of receptors and coordination of several signalling pathways to stimulate relevant ion transporters on both luminal and basolateral membranes to achieve transcellular secretion of ions/fluid, as well as balancing cell volume and pH_i changes. Utilizing synergism of signalling pathways would ensure maximum effect without running each pathway at a maximum capacity, which could be detrimental to cell survival, as exemplified by Ca^{2+}-mediated cellular toxicity (Berridge 2012). Here, we summarize evidence for the interaction of Ca^{2+} and cAMP signalling pathways and their effect on pancreatic duct ion transport.

First of all, some agonists, such as ubiquitous nucleotides/nucleosides, signal via multiple receptors, coupled with G_q, G_s and G_i proteins (P2Y and adenosine receptors) and ligand-gated ion channels (P2X receptors) (Burnstock 2007; Surprenant and North 2009; Bilbao et al. 2012). Pancreatic ducts express G-protein-coupled P2Y receptors, P2X receptors and various nucleotidases, such that ATP would have multiple effects via cAMP and Ca^{2+} signalling (Novak 2008, 2011) (Fig. 5.2).

Another synergistic mechanism occurs at the ion channel level, where Ca^{2+}-sensitive K^+ channels can alter the driving force for anion secretion through cAMP/PKA-regulated CFTR. Further potentially synergistic mechanism to increase secretory output is the parallel anion transport through CaCC channels and CFTR channels in pancreatic ducts. However, there is evidence that there is some interdependence between CFTR and CaCC, such that malfunctioning CFTR (CF models) downregulates expression or function of CaCC in parallel to CFTR (Gray et al. 1994; Winpenny et al. 1995; Pascua et al. 2009; Wang et al. 2013).

The central channel in pancreatic duct is CFTR, which is a part of signalling complex that includes scaffolds, adaptors and many regulatory enzymes associated with cAMP/PKA signalling (Frizzell and Hanrahan 2012). Recent studies show that there is cross talk between Ca^{2+} signalling and CFTR activation. There are Ca^{2+}-sensitive adenylate cyclases 1 and 8 (Namkung et al. 2010; Martin et al. 2009) and Ca^{2+}-dependent activation of tyrosine kinases (Src2/Pyk complex), both of which could eventually alter activity of CFTR, as shown for airway and intestinal epithelia (Billet and Hanrahan 2013; Billet et al. 2013). Another effect at the CFTR level would be priming of some PKC isoforms that enhance activity of CFTR (see Billet and Hanrahan 2013).

Furthermore, synergy between Ca^{2+} and cAMP signalling could be exerted by the third messenger IP_3 receptor-binding protein released with IP3 (IRBIT). Agonists that couple with G_s increase cAMP and via PKA phosphorylation of the IP_3 receptor, and receptors that activate G_q increase level of IP_3. Increased affinity of IP_3R to IP_3 facilitates the release of IRBIT from the apical pools, which then translocates and coordinates epithelial fluid and HCO_3^- secretion by stimulating NBCe1B and CFTR and *SLC26A6* (Yang et al. 2009, 2011). This type of synergy is well studied in pancreatic ducts using genetic modifications of *SLC26A6* and IRBIT (Park et al. 2013). Using cAMP/Ca^{2+} signalling agonist pairs such as forskolin/carbachol, secretin/carbachol, forskolin/carbachol—synergy in fluid secretion and HCO_3^- flux is revealed.

Lastly, the inhibitory pathways, which are downstream of Ca^{2+} and cAMP, should be considered. Cell signalling pathways involving volume- and low Cl^--sensitive with no lysine kinases (WNKs), acting via Ste20-like kinases, SPS-related proline/alanine-rich kinase (SPAK) and oxidative stress-responsive kinase (OSR1), may be key factors in secretory epithelia, since they regulate NKCC1 and other transporters (Kahle et al. 2006; McCormick and Ellison 2011). Basically these kinases are activated by hyperosmolarity (cell shrinkage) and low intracellular Cl^- and thus would restore cell volume. In relation to pancreas, WNK1 and 4 are expressed in lateral membranes of interlobular and main pancreatic ducts, and they inhibit NKCC1 and *SLC26A6* (Choate et al. 2003; Kahle et al. 2004). Other studies show that WNKs inhibit CFTR (Yang et al. 2007, 2011). For example, in mice ducts, it was shown that WNKs and SPAK reduced expression of CFTR and NBCe1 and duct secretion (Yang et al. 2011). IRBIT increases membrane surface expression of NBCe1B, CFTR and *SCL26A6* and thus overcomes antagonizing WNK/SPAK signalling, which otherwise reduces secretion (see above). It seemed somewhat surprising then that the WNK/OSR1/SPAK system stimulated by low intracellular Cl^- could change the permeability of CFTR in favour of HCO_3^-, i.e. $PHCO_3^-/PCl^-$ increased from 0.24 to 1.09, and at the same time inhibited *SLC26A6* and *A3* (Park et al. 2010, 2012). It is proposed that WNK signalling for distal ducts and IRBIT signalling for proximal ducts could be a part of the mechanisms underlying overall pancreatic duct function (see Lee et al. 2012; Park and Lee 2012).

The above section indicates that cell volume regulation, e.g. via WNK/OSR1/SPAK system, may be important for pancreatic ducts. Similarly, autocrine and paracrine signalling via volume-sensitive ATP release must be a key regulator in short- and long-term cell volume and ion transport in epithelia, including the pancreatic duct. Although cell volume regulation is a cornerstone in epithelial physiology and pathophysiology, we know very little about this process in pancreatic ducts (Pedersen et al. 2013b).

5.3 Salivary Glands

5.3.1 Salivary Glands: Heterogenous Structures and Functions

Saliva is a complex mixture of fluid containing amylase, lipase, glycoproteins (e.g. mucins, vitamin B12 binding haptocorrin), proline-rich proteins and proteins regulating calcium phosphate and hydroxyapatite formation (e.g. statherins, histatins and cystatins), growth factors (e.g. EGF), antibacterial agents (immunoglobulins, lysozyme, lactoferrin), water and electrolytes (including varied concentrations of HCO_3^-) and minerals. The major functions of the salivary glands are to protect the teeth and oral-oesophageal mucosa (by modulating re-/demineralization of teeth enamel, by protecting gingiva and by initiating antibacterial actions), initiate digestive processes, enhance taste perception and provide lubrication, and provide pH buffering capacity (bicarbonate, phosphate, proteins) (Humphrey and Williamson 2001; Matsuo 2000; Pedersen et al. 2002, 2013a). In some animals, salivary glands have additional functions, e.g. in grooming and evaporative cooling keeping oral cavity moist in panting, regulating salt homeostasis (e.g. in crocodiles). Major human salivary glands supply about 90 % of the whole saliva and comprise three pairs of exocrine glands: parotid glands (P), submandibular (submaxillary) (SM) glands and sublingual glands (SL). The rest of secretion is provided by hundreds of minor glands spread throughout the oral cavity (Pedersen et al. 2013a). The largest glands in human, parotid glands, contain serous acinar cells and secrete amylase-rich secretions. The submandibular glands contain seromucous and serous acini and produce mucous saliva. The sublingual glands are the smallest glands and contain prevalently mucous acini and produce mucin-rich viscous secretion. Secretions that originate in acini are conducted through short intercalated ducts (that may be secretory) to striated ducts, characterized by many mitochondria and folds on the basal membrane, and to excretory (extralobular) ducts leading to the main excretory ducts of Stensen (P) and Wharton (SM) or a series of excretory ducts (SL) (Pedersen et al. 2013a). In some glands/animals, e.g. male rodents, the striated ducts of SM contain granules and are referred to as granular ducts (Schneyer et al. 1972). Salivary glands are under control of both branches of autonomic nervous systems and higher brain centers and autocrine/

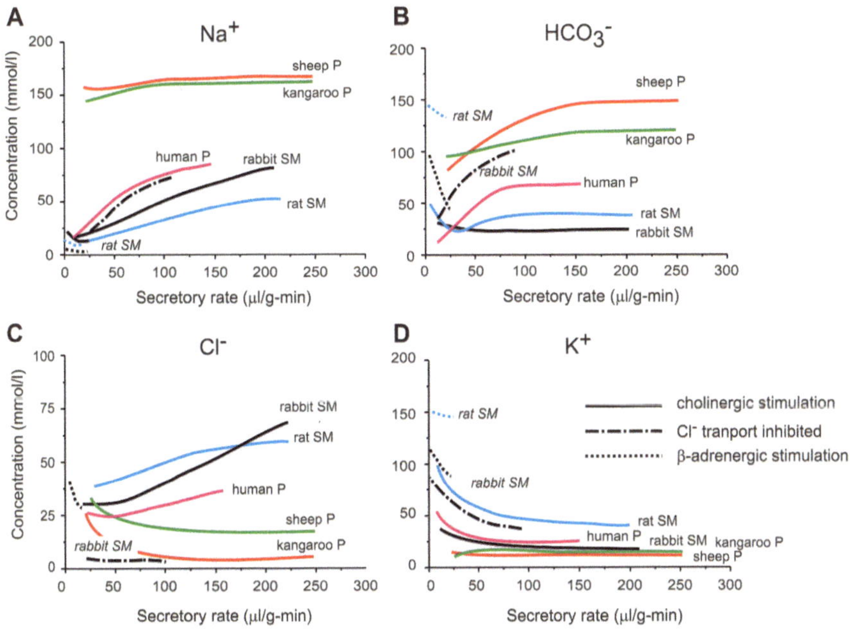

Fig. 5.3 The relation between secretory rates and electrolyte concentrations in saliva collected from salivary glands of various species. Preparations were in vivo glands or ex vivo perfused glands stimulated with cholinergic agonist, e.g. pilocarpine or carbachol (*full lines*). In some experiments on perfused glands, Cl^- transport was inhibited, e.g. with furosemide (*dot-dash line*). Some glands were stimulated with β-adrenergic agonist, isoproterenol (*dotted line*). Secretory rates were corrected for gland weights, and data were redrawn from publications on the rabbit submandibular gland (Case et al. 1980, 1984; Novak and Young 1986), the rat submandibular gland (Young and Martin 1971), the human parotid gland (Thaysen et al. 1954), the sheep parotid gland (Compton et al. 1980) and the kangaroo parotid gland (Beal 1984)

paracrine regulation (Schneyer et al. 1972; Garrett 1987; Pedersen et al. 2002, 2013a; Proctor and Carpenter 2014).

The major component of saliva is water (99 %) and electrolytes, and there are special relationships between those and secretory rates (see Fig. 5.3). Based on these relationships, Thaysen and coworkers proposed that saliva was formed in two stages—in acini and in ducts (Thaysen et al. 1954). Similar hypotheses were proposed for the pancreas and other exocrine glands, though importantly, the function of ducts differs among various gland types (Bro-Rasmussen et al. 1956; Schwartz and Thaysen 1956). In the following paragraphs, the simplest scenario valid for most salivary glands will be outlined (Fig. 5.4). In the first stage, salivary acini generate primary saliva that is isotonic plasma-like fluid that is high in Na^+ and Cl^- concentrations, and K^+ concentrations are slightly above the plasma (Schneyer et al. 1972; Young et al. 1980). The anion gap is most likely HCO_3^- secreted at about plasma concentrations (see Sect. 5.3.4 for exceptions). In the second stage, the ducts reabsorb Na^+ and Cl^- and partially compensate electrolytes by secreting some K^+ and HCO_3^-. The ductal epithelium is electrically tight and

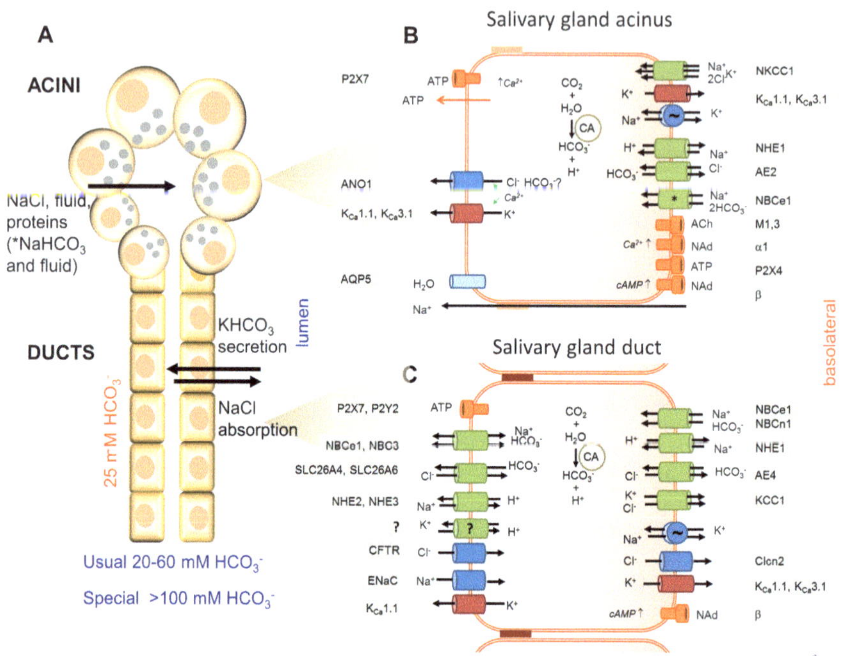

Fig. 5.4 (**a**) Schematic diagram showing simplified salivary gland with acini and excretory ducts and saliva with typical range of HCO_3^- concentrations for most cholinenergically stimulated salivary glands and special salivary glands and/or special circumstance (see Fig. 5.3). (**b**) *Insets* show the cellular models for acinar cell that can secrete Cl^- relatively independent of HCO_3^- and rely on NKCC1 and double exchange system (e.g. SM glands). Some acini rely primarily on HCO_3^- transport (e.g. parotid glands) and express NBCe1, marked with *. See also Fig. 5.3. (**c**) Cellular model of a duct cell that is absorbing Na^+ and Cl^- (via luminal ENaC-CFTR channels or double exchange system) and secreting K^+ and HCO_3^-

water impermeable (Young et al. 1980). In this respect, the salivary ducts are fundamentally different to the pancreatic ducts, the latter being a leaky and secretory epithelium (see Sect. 5.2.1). Due to hypertonic salt transport in salivary ducts, the final saliva in many gland types and species is hypotonic. Importantly though, tonicity and electrolyte patterns depend on the secretory rate (acini) and saturation of various transporters on the downstream ducts. Above-described processes would give rise to the simplest excretory curves (Fig. 5.3: rat, rabbit, human glands).

A number of studies were conducted to verify the two-stage theory, including micropuncture studies that involve sampling and analyzing fluid at or close to acini/intercalated ducts and in downstream ducts and studies on isolated salivary ducts. These are summarized in earlier reviews in this field (Martinez et al. 1966; Young and Schögel 1966; Young et al. 1980). In the current research on exocrine glands, many advanced techniques on cellular/genetic level are used, but it is still very valuable to take the integrative approach and return to the whole gland secretion and electrolyte patterns. Nonetheless, understanding of ion transport in salivary glands is particularly challenging as there are three different major glands (parotid, submandibular and sublingual), there are large interspecies and even male/female variations in structure and regulation, and these may reflect very varied salivary gland functions.

In the following section, we will describe the basic ion transport mechanisms in acini and ducts of most common experimental animals (rat, mouse, rabbit) stimulated with cholinergic stimulation that evokes largest fluid secretion rates (Figs. 5.3 and 5.4). Later (Sects. 5.3.4 and 5.3.5), we will consider other modes of stimulation, e.g. sympathetic, as well as specific glands/animals and experimental conditions that evoke saliva with unusual ion compositions. This holds in particular HCO_3^- secretion, which shows the most bewildering variety of excretion patterns (Fig. 5.3) and may originate in acini and/or ducts (Young et al. 1980; Novak 1993) (Fig. 5.3B). In humans, parotid glands can secrete up to 40–60 mM HCO_3^-, though mixed saliva from major and minor salivary glands rarely exceeds 20–25 mM HCO_3^- (Thaysen et al. 1954; Bardow et al. 2000). Human parotid glands have HCO_3^- output ranging from about 40 to 500 μmol/h/g gland weight. Rat and rabbit submandibular glands stimulated with cholinergic stimulus have similar HCO_3^- output ranging from about 40 to 400 μmol/h/g gland weight. In sheep and kangaroo parotid, the output ranges from around 100 to 1500 μmol/h/g gland weight (see Fig. 5.3).

5.3.2 Ion Channels and Transporters in Salivary Gland Acini

Since many salivary glands commonly studied (e.g. rodent P and SM) can secrete very efficiently without exogenous HCO_3^- and/or with CA inhibitors, the ion transport models for acini are based predominantly on the transport of Cl^-. Cl^- is transported across the basolateral membrane via loop diuretic-sensitive Na^+-Cl^- cotransporter, later identified as NKCC1 in several preparations including human parotid acini (Case et al. 1982, 1984; Martinez and Cassity 1983; Turner et al. 1986; Nauntofte and Poulsen 1986; Moore-Hoon and Turner 1998; Evans et al. 2000; Nakamoto et al. 2007) (Fig. 5.4B). An alternative mechanism for NaCl transport is the parallel transport via Na^+/H^+ and Cl^-/HCO_3^- exchangers, as proposed from studies of isolated glands (Novak and Young 1986; Turner and George 1988). Interestingly, the Cl^-/HCO_3^- exchanger, most likely AE2, is upregulated when NKCC1 is inhibited or genetically silenced (Evans et al. 2000). AE2 is expressed together with NHE1 on the basolateral membrane of acini (He et al. 1997; Lee et al. 1998; Park et al. 1999). In *NHE1−/−* mice, parotid acini express higher activity of AE2 and CAII, as determined by pH_i measurements, indicating increased Cl^- and HCO_3^- transport across the plasma membranes, though no data on salivary secretion are available (Gonzalez-Begne et al. 2007). There is also evidence for NBC1e expression on salivary gland acini (see Sect. 5.3.4), but in many glands, transport of Cl^- is sufficient to drive full fluid secretion.

Most effective agonists for salivary secretion, e.g. acetylcholine acting via muscarinic receptors, increase intracellular Ca^{2+} concentrations, and Ca^{2+} signalling has been a well-studied mode of stimulus-secretion coupling (Petersen and Tepikin 2008; Ambudkar 2012), though more recent studies show that similar to the pancreas, there is a synergism between Ca^{2+} and cAMP signalling (see Jung and Lee 2014; Ahuja et al. 2014). In salivary acini, Ca^{2+} signals are essential in regulating Cl^- efflux via the luminal channels. These CaCC properties are corresponding to recently identified Cl^- channel TMEM16A/ANO1, and submandibular glands from *TMEM16a−/−* mice produce lower amount of saliva (Romanenko et al. 2010). Some TMEM16A/ANO1 is also expressed on the luminal membrane of intercalated ducts, though another CaCC candidate bestrophin-2 may be relevant to duct function (Romanenko et al. 2010). The expression of TMEM16A/ANO1 is also found on the luminal membrane of human parotid acini and intercalated ducts (Chenevert et al. 2012) (Fig. 5.4B). Latest studies in HEK293 cells and SM gland acini indicate that TMEM16A/ANO1 anion selectivity is dynamically modulated by Ca^{2+}/calmodulin, possibly increasing $PHCO_3^-/PCl^-$ (Jung et al. 2013). Regarding CFTR, the protein is expressed on the luminal membrane of ducts, but there are contradicting reports regarding the expression of CFTR in SM of rodent acini. Nevertheless, since in mice ΔF508 mutation in CFTR or inhibition of CFTR had no significant effect on salivary secretion rate, other Cl^- channels must have rescued secretion (He et al. 1997; Zeng et al. 1997; Catalan et al. 2010).

One of the most marked effects in salivary acini is the loss of intracellular K^+ upon stimulation, as observed in initial studies on in vivo glands and isolated acini (Burgen 1956; Nauntofte 1992). Also many electrophysiological studies on plasma membrane potentials in acini reported hyperpolarizing "secretory potentials", which would be consistent with increased K^+ conductance, and Ca^{2+}-activated maxi-K^+ channels were characterized in patch-clamp studies (Imai 1965; Petersen and Poulsen 1967; Maruyama et al. 1983; Petersen and Gallacher 1988). Ca^{2+}-activated K^+ channels (BK – $K_{Ca}1.1$ and IK – $K_{Ca}3.1$) have been identified on the basolateral membrane (Wegman et al. 1992; Park et al. 2001a; Nehrke et al. 2003; Begenisich et al. 2004; Romanenko et al. 2007) (Fig. 5.4b). It is now well accepted that the basolateral K^+ channels serve for K^+ recirculation necessary for the operation of the Na^+/K^+ pump and thus secretion. However, micropuncture studies and analysis of fluid close to acini indicated that the primary secretion has K^+ concentrations higher than the plasma, e.g. up to 10 mM K^+ (Young and Schögel 1966; Mangos et al. 1966, 1973) (see Schneyer et al. 1972), indicating that there may be some secretory mechanisms for K^+ on the luminal membrane. Hence, it was proposed that K^+ channels are also present on the luminal membrane of salivary acini, and various mathematical models verified that luminal K^+ channels are necessary for creating the driving force for Cl^- exit and account for at least 10–20 % of total K^+ conductance in acinar cells (Cook and Young 1989; Palk et al. 2010). Recent studies on mouse parotid acini using spatially localized manipulation of Ca^{2+} and whole-cell patch clamp show that the very small area of the luminal membrane, e.g. approximately 3–8 % of overall plasma membrane (Poulsen and

Bundgaard 1994), expresses high density of $K_{Ca}1.1$ and $K_{Ca}3.1$ channels (Almassy et al. 2012). These channels exhibit some interdependence/interaction (Thompson and Begenisich 2006, 2009). In submandibular acini, it seems that the apical Ca^{2+} signals stimulate CaCC and only when signals spread to the basolateral membrane and/or the membrane is depolarized, then the basolateral K^+ channels are activated. It also seems that either is $K_{Ca}1.1$ or $K_{Ca}3.1$ can support full secretion in mouse submandibular gland; and only a double knockout of these K^+ channels reduces secretion significantly (Romanenko et al. 2007). Whether incongruence between parotid and submandibular acini is related to different Ca^{2+} signalling or patterns of K^+ channel expression is not clear yet. Nevertheless, the most significant K^+ secretion is contributed by the ducts (see Sect. 5.3.3) (Fig. 5.4c).

In normal salivary secretion, water transport, occurring by transcellular and paracellular routes and cell volume regulation, is dependent on the expression of water channels, aquaporins. The most important aquaporin in salivary gland acini is AQP5, as determined in knockout studies on mice (Ma et al. 1999; Krane et al. 2001; Murakami et al. 2006; Kawedia et al. 2007). Cell volume regulation is important in many cellular functions, including epithelial transport (Pedersen et al. 2013b). In secreting epithelia, physiological stimulus leads to the opening of luminal Cl^- channels and basolateral/luminal K^+ channels, and osmotically driven water movement leads to the shrinkage of secreting cells. Basolateral transporters and pH_i regulating mechanisms need to be activated to provide ions for luminal exit. Nevertheless, in the secretory state, these mechanisms are unable to maintain the cell volume of salivary secretory cells which remain shrunken, until the stimulus is terminated, after which, cell volume recovers (Dissing et al. 1990; Foskett 1990; Nakahari et al. 1990, 1991; Lee and Foskett 2010).

5.3.3 Ion Channels and Transporters in Salivary Gland Ducts

The cornerstone in salivary duct ion transport is NaCl absorption (and $KHCO_3$ secretion), and it is apparent that Na^+ and Cl^- excretion curves are following each other (Fig. 5.3a, c). One possible mechanism for NaCl absorption is the electroneutral model—luminal Na^+/H^+ and Cl^-/HCO_3^- exchangers. The alternative model is the parallel activity of epithelial Na^+ channels (ENaC) and CFTR (Fig. 5.4c). There is evidence for both systems, and it has been proposed that ducts of low-HCO_3^- secretors (mouse and rabbit SM) are dominated by Na^+ and Cl^- channels on the luminal membrane, while ducts of high HCO_3^- secretors (rat SM) are dominated by the double exchangers (Chaturapanich et al. 1997).

There is solid evidence for ENaC expression on the luminal membrane of salivary ducts (Fig. 5.4c), and ENaC is regulated by ubiquitin-protein ligase Nedd4 (Komwatana et al. 1996b; Dinudom et al. 1998, 2001; Cook et al. 2002). There are number of electrophysiological and inhibitors studies on isolated ducts and glands supporting the evidence for ENaC, e.g. low concentrations of amiloride lead to increased NaCl content in saliva (Bijman et al. 1981; Komwatana et al. 1996a). CFTR is expressed on the luminal membrane of salivary ducts, and its inhibition by specific inhibitors and CFTR knockout leads to decreased NaCl absorption and increased salt excretion in saliva, as seen in murine models of CF (Dinudom et al. 1995; Zeng et al. 1997; Catalan et al. 2010). If CFTR is to transport Cl^- from lumen to the cell, it requires markedly depolarized luminal membrane potential, which has not been measured, but quantitative modelling of salivary ion transporters strongly supports this model (Patterson et al. 2012). Exit pathway for Cl^- on the basolateral membrane is not clear, and proposals include a hyperpolarization-activated Cl^- channel (Clcn2), KCl cotransporter (KCC1) or Cl^-/HCO_3^- exchanger (AE4) (Romanenko et al. 2008; Roussa et al. 2002; Ko et al. 2002a). Note that the basolateral membrane needs to be more hyperpolarized than the luminal membrane to permit Cl^- exit out of the cell towards the interstitium.

The molecular basis for the alternative electroneutral NaCl transport model is more difficult to pinpoint. Salivary ducts express NHE2 and NHE3 on the luminal membrane, and their function is not clear, as knockout of either NHE isoform has no effect on salivary secretion in mice (Park et al. 2001b; Lee et al. 1998). Salivary ducts also express SLC26A4 and A6 (Shcheynikov et al. 2008), and although they differ in coupling $Cl^-:HCO_3^-$ (i.e. 1:1 vs. 1:2), they could ensure Cl^- influx into duct cells and HCO_3^- efflux, and it seems that either one can explain the excretory curves in a model simulation of pancreatic ducts (Patterson et al. 2012).

There are several other transporters that could contribute to duct HCO_3^- secretion (or HCO_3^- absorption). On the basolateral membrane, the means of HCO_3^- import into duct cells could be NBCe1 (e.g. guinea pig SM) or NBCn1 (e.g. rat SM ducts, human SM and P) (Li et al. 2006; Gresz et al. 2002). Alternatively, or in addition, NHE1 on the basolateral membrane together with CAII could be a part of HCO_3^-/H^+ system involved in secretion and/or pH_i regulation (Park et al. 1999). Interestingly, some NBC transporters, NBC3 and NBCe1, are also expressed on the luminal membrane of several types of salivary ducts, and their proposed function is to absorb (salvage) HCO_3^- (Park et al. 2002b; Li et al. 2006; Gresz et al. 2002). Presumably, HCO_3^- ductal absorption would occur if salivary acini were secreting primary fluid rich in HCO_3^- (see below).

Regarding H^+ pumps, immunohistochemical studies show a presence of V-ATPases in intracellular compartments in SM granular and main ducts in rats with normal acid/base balance, but in the rat parotid gland, the pumps are close to the luminal membrane of the striated and excretory ducts (Roussa et al. 1998; Roussa and Thevenod 1998). The pump has not been studied functionally, though heterogenic distribution may indicate that the parotid glands have special acid/base challenges. H^+/K^+ pump (not the putative passive H^+/K^+ exchanger; see Sect. 5.3.4) has not been seriously considered in salivary glands, except for one study where proton pump inhibitors (omeprazole and SCH28080) did not have effect on pH_i on striated ducts from the rat parotid in a given experimental condition (Paulais et al. 1994).

In many species, the collected saliva has K^+ concentrations several fold higher than the plasma, and it is inversely related to secretory rate, and often HCO_3^- excretion follows similar pattern (see exception 5.3.4). Therefore, it is not surprising that one of the original proposals to explain $KHCO_3$ secretion (Knauf et al. 1982; Paulais et al. 1994) was a K^+/H^+ exchanger working in parallel with Na^+/H^+ exchange and Cl^-/HCO_3^- exchange (Fig. 5.4c). Together these would perform a functional K^+/HCO_3^- cotransport on luminal membrane and provide Na^+ and Cl^- for the basolateral exit. Such K^+/H^+ transporter has not been cloned, and several studies show that K^+ and HCO_3^- transport is not very tightly coupled (Nakamoto et al. 2008; Chaturapanich et al. 1997). Therefore, other K^+ exit pathways have been considered, such as the luminal K^+ channels. Striated and excretory SM ducts express $K_{Ca}1.1$ channels, and knockout studies show that much decreased K^+ secretion in whole glands, indicating that indeed, these channels are important for ductal secretion (Nakamoto et al. 2008).

Using above transport components for salivary glands ducts, and experimental values obtained from various studies and modelling, it has been possible to reproduce excretory curves similar to Fig. 5.3 for major ions in saliva, including HCO_3^- concentrations 35–45 mM and K^+ of 20–60 mM (Patterson et al. 2012).

5.3.4 *Salivary Glands Can Secrete Very High Bicarbonate and/or Potassium: Where and When?*

Now that the basics of ion transport in salivary gland acini and ducts have been presented, we can consider three special circumstances in which some salivary glands secrete saliva with very high HCO_3^- concentrations, with/without accompanying K^+ (Fig. 5.3b, d), and try to resolve how this happens.

First of all, as introduced above, ducts can secrete HCO_3^- (and K^+), though without accompanying water fluxes and the lower the secretory rate of saliva is, the higher the concentrations of HCO_3^- and K^+ are (Fig. 5.3b, Fig. 5.3: hyperbolic curves). These patterns can be particularly dominant with some forms of stimulation (see Sect. 5.3.5).

Salivary glands of some animals have a large capacity to secrete HCO_3^-, and concentrations are almost as high as in the pancreas. This is the case for foregut fermenters such as sheep, cattle, camels and kangaroos, where usually parotid glands supply well-buffered saliva to stabilize pH of the fermenting digesta and saliva can contain 110–120 mM HCO_3^- (and also 20–60 mM phosphate), though K^+ is relatively low, 5–15 mM (Kay 1960; Young and van Lennep 1979; Beal 1984) (Fig. 5.3). The analysis of relationship between the salivary flow rate and electrolyte concentrations suggests that the primary secretion itself must be high in HCO_3^- concentration and that the duct contribution is relatively small. This conclusion is supported by micropuncture studies on sheep parotid, which showed that Cl^- concentrations were about 50 mM (Compton et al. 1980). By inferences then,

HCO_3^- must have been correspondingly high (due to very small volumes, H^+/HCO_3^- could not be measured) and that furosemide, and inhibitor of NKCC1, had relatively small effect on saliva secretion (Wright et al. 1986). Further studies showed that the sheep parotid acini exhibited acetylcholine-stimulated and Na^+-dependent HCO_3^- influx, and a Na^+-HCO_3^- cotransport was proposed (Poronnik et al. 1995; Steward et al. 1996). Subsequently, it was verified that bovine parotid acini express NBCe1B at high levels (on mRNA and protein levels) and show large electrogenic currents (Yamaguchi and Ishikawa 2005) (Fig. 5.4b). The transporter is localized on the basolateral membrane of acini and interacts with IRBIT (Yamaguchi and Ishikawa 2012).

The NBCe1 transporter is also expressed in other acini, such as rat and human parotid acini, and this is in accordance with observations that parotid glands can secrete saliva with relatively high HCO_3^- concentrations (Roussa et al. 1999; Park et al. 2002a; Bardow et al. 2000). Correspondingly, the excretory curve patterns are consistent with acinar origin of HCO_3^- secretion (Fig. 5.3b). In submandibular glands, acinar secretion is not dependent on HCO_3^--driven transport normally, as the Cl^--driven transport is dominating (Case et al. 1982, 1984). Nevertheless, if Cl^- transport is inhibited, these glands can also produce saliva with concentrations >100 mM HCO_3^- (although secretion rate is diminished) (Fig. 5.3b: rabbit SM). This indicates that, most likely, SM acini also have the machinery to transport and drive HCO_3^- transport (Novak and Young 1986).

5.3.5 Regulation of Salivary Gland Secretion

The main regulation of saliva secretion is via autonomic nervous system and little if any by hormones (Proctor and Carpenter 2014). Parasympathetic and sympathetic stimulation leads to the activation of muscarinic M1 and M3 receptors and $\alpha 1$ receptors on salivary gland cells, respectively, that via $G_{q/11}$ eventually leads to Ca^{2+} signalling. Furthermore, noradrenaline stimulates the cAMP/PKA signalling pathway via β receptors and G_s. Both parasympathetic and sympathetic stimulations result in approximately similar plasma-like primary secretion, as measured in, e.g. rat SM (Young and Martin 1971). However, parasympathetic stimulation causes 6–8 times higher fluid secretion compared to sympathetic one, the latter resulting in more concentrated protein secretion. Salivary ducts are also well innervated by autonomic nerves, and they modify secretion mainly by stimulating K^+ and HCO_3^- secretion (Young and Martin 1971; Martin and Young 1971). In particular, β-adrenergic receptors stimulation can result in disproportionately greater effect on K^+ and HCO_3^- concentrations in small volumes of the final saliva (Case et al. 1980) (Fig. 5.3b, d: dotted lines). Furthermore, sympathetic stimulation via β-adrenergic receptors (at low concentrations) can modulate saliva composition by the activation of CFTR, thereby increasing the rate of Na^+ absorption (Dinudom et al. 1995).

The effects of acetylcholine and noradrenaline are modulated by a number of non-adrenergic non-cholinergic (NCNA) co-transmitters, including VIP, substance P, neuropeptide Y, NO, ATP and others (Pedersen et al. 2013a; Proctor and Carpenter 2014). Under physiological conditions, it may be expected that multiple transmitters would collaborate to induce salivary fluid and protein secretion. The synergistic effects of cAMP/PKA and Ca^{2+} signalling have been demonstrated in many studies, and recent data indicate that IRBIT mediates synergism between these signalling pathways (Bruce et al. 2002; Ahuja et al. 2014; Jung and Lee 2014).

Extracellular nucleotides/nucleosides are probably NANC and also autocrine/paracrine signalling molecules that have important function in coordinating salivary gland functions including HCO_3^- secretion. The purinergic signalling has been pioneered in salivary glands (see Novak 2011). It is most likely that ATP is a co-transmitter with acetylcholine or noradrenaline, and a number of P2 receptors are expressed and functional on salivary gland acini. In particular, P2X4 and P2X7 receptors have been described in early studies, and it is clear that in physiological conditions, they do not form permeable pores, but rather remain cation channels/receptor that co-regulate secretion, for example, by aiding Ca^{2+} signalling and stimulating K^+ and Cl^- channels. Recent studies show that in the rat parotid, acini P2X4 receptors are located on the basolateral membrane, while P2X7 receptors are enriched on the luminal membrane, and they evoke spatially distinct Ca^{2+} signals and effects on protein exocytosis (Bhattacharya et al. 2012) and fluid secretion in mouse salivary glands (Nakamoto et al. 2009; Novak et al. 2010) (Fig. 5.4). Interestingly, the P2Y receptors are expressed transiently, e.g. during gland development (P2Y1) and during stress (P2Y2). Furthermore, ATP is most likely also stored in secretory granules, and it is secreted into duct lumen and ATP can be detected in whole saliva (Ishibashi et al. 2008; Novak et al. 2010). In addition, the duct epithelium also releases ATP spontaneously or in response to, for example, mechanical stimulation (Shitara et al. 2009; Ryu et al. 2010). Here it acts via P2Y2 receptors to increase Cl^- absorption by stimulating CFTR (Ishibashi et al. 2008). It is not clear how HCO_3^- transport is affected and how luminally expressed P2X7 receptors regulate salivary duct function.

5.4 Hepatobiliary System

5.4.1 Hepatobiliary System: Concerted Action of Several Types of Epithelial Cells

The hepatobiliary system shares many similarities with the pancreas. It secretes highly complex bile containing 95 % water with electrolytes, bile salts and bilirubin, lipids, proteins, enzymes, peptides and amino acids, nucleotides, heavy metals and vitamins (Boyer 2013). Most importantly, bile contains 15–55 mM HCO_3^-. The hepatobiliary system is composed of several types of epithelial cells:

hepatocytes, cholangiocytes and gallbladder epithelial cells, all of which contribute to formation of bile in several stages. Hepatocytes, the major liver cell population (65 %), secrete primary or canalicular bile, which is driven by bile salt-dependent transport and bile salt-independent transport of bicarbonate and reduced glutathione and osmotically obliged water flux. Canalicular bile is delivered to the intrahepatic bile ducts, which are lined with cholangiocytes, forming about 3–5 % of liver cell population, and these modify bile by secretion and absorption (Fig. 5.5a). The intrahepatic biliary duct system is the most important epithelium in the liver regarding the ability to secrete HCO_3^--containing fluid in response to secretin and other regulators. This secretion contributes to alkalization and fluidization of bile, which prevents protonation and absorption of weak lipophilic acids. In addition, distally it provides the buffer function in the duodenum—similar to the pancreatic duct system (Boyer 2013). In the last stage, gallbladder concentrates and stores bile but also modifies it by secretion in some species and lastly delivers it to the duodenum during feeding. Assuming that in a human the canalicular bile acid-independent flow and bile duct flow together are about 4 μl/min/kg body weight and final bile contains 15–55 mM HCO_3^- (Banales et al. 2006b; Boyer 2013), it can be approximated that HCO_3^- output is about 0.3–0.7 μmol/h/g liver weight. Considering the whole organ though, HCO_3^- output of 300–700 μmol/h is a significant output.

5.4.2 Canalicular Bile Salt-Independent Flow Generated by Hepatocytes

Canalicular bile fluid is secreted by hepatocytes and secretion has two components (see above). The bile salt-dependent fraction is driven by a series of organic anionic transporters, BA transporters and multidrug resistance transporters (MRP) (Esteller 2008). The bile salt-independent fraction is driven by transport of glutathione and HCO_3^-, each contributing about 50 % to this fraction (Banales et al. 2006b; Boyer 2013; Esteller 2008). Glutathione is secreted via organic anion transporter MPR2/ABCC2. Bicarbonate is secreted via a DIDS-sensitive Na^+-independent Cl^-/HCO_3^- exchanger (AE2) expressed on the luminal membrane (Fig. 5.5b) (Meier et al. 1985; Martinez-Anso et al. 1994). Apparently, CFTR is not expressed in hepatocytes, and it might be another Cl^- channel that together with AE2 coordinates or performs HCO_3^- secretion, and water follows via AQP8 (Banales et al. 2006b). HCO_3^- loaders on the basolateral membrane of hepatocytes are most likely NHE1 and also NHE2 and 4 (Moseley et al. 1986; Pizzonia et al. 1998). In addition, NHE3 is expressed on the canalicular membrane and more importantly on the apical membrane of cholangiocytes, where it most likely participates in fluid absorption (Mennone et al. 2001). Electrogenic Na^+-HCO_3^- cotransport was demonstrated in studies using plasma membrane vesicles and microelectrodes, and subsequently, NBCe1/*SLC4A4* and NBCe2/NBC4/*SLC4A5* were identified (Fitz

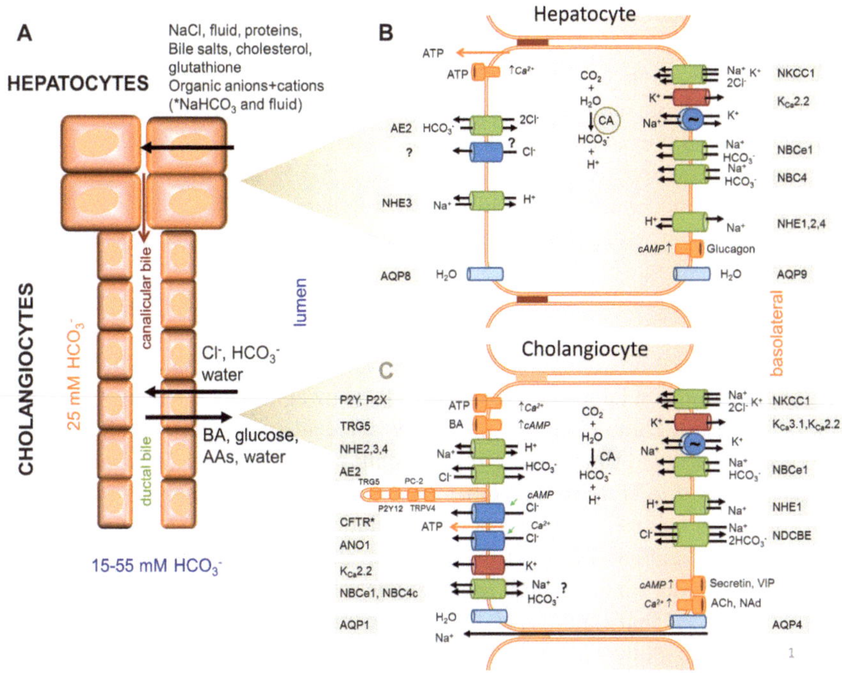

Fig. 5.5 (**a**) Schematic diagram showing simplified hepatobiliary system composed of hepatocytes and cholangiocytes. Hepatocytes secrete canalicular bile fluid driven by bile salt-dependent fraction and bile salt-independent fraction. Cholangiocytes secrete Cl^-, HCO_3^- and fluid and absorb bile acids, glucose, amino acids and water. (**b**) *Inset* shows cellular model for bile salt-independent HCO_3^--driven fluid secretion only. (**c**) *Inset* shows cellular model for ion transport and regulation in cholangiocytes with a primary cilium (expressing several receptors and channels). CFTR* indicates that CFTR is functional in large but not small cholangiocytes

et al. 1989; Renner et al. 1989). However, the expression of NBCe1 is low, while NBC4, especially NBC4c variant, is high in the liver (Pushkin et al. 2000; Abuladze et al. 2004).

5.4.3 *Intrahepatic Biliary Duct System: Ion Transport in Cholangiocytes*

Intrahepatic biliary duct system is a network of interconnecting bile ductules and ducts that extend from canals of Hering, partly lined with hepatocytes, to ductules (<15 μm) lined with cholangiocytes and then converge to large bile ducts (>15 μm) of increasing diameter to interlobular, septal and segmental ducts and then hepatic and common hepatic ducts (Tabibian et al. 2013). Along the duct system, cholangiocytes change in size, morphology, regulation and function. Although they account only for 3–5 % of liver population, they are responsible for about

30–40 % of bile volume, depending on species. The magnitude of ductal contribution of basal bile flow depends on species and can vary from 10 to 30 %. Primarily cholangiocytes secrete Cl^-, HCO_3^- and fluid, and they reabsorb bile acids, glucose and amino acids. Secretin stimulates HCO_3^- secretion, and this determines the pH and hydration of bile and thereby decreases protonation of glycine-conjugated bile acids and thus their passive reabsorption (see Sect. 5.4.5), and lastly it contributes to alkalization of duodenal contents (Boyer 2013; Beuers et al. 2012). In the following section, we will only focus on cholangiocyte ion channels and transporters involved in HCO_3^- transport.

Ductal bile secretion is a regulated process that is initiated by the transport of Cl^- across the luminal membrane into the ductal lumen and coupling to AE2, as verified in isolated bile duct unit (IBDU) preparations. CFTR has been identified and it is functional in large but not small IBDU (Fitz et al. 1993; Alpini et al. 1997; Cohn et al. 1993; Dutta et al. 2011). However, recent studies have challenged the role of CFTR as the primary route for Cl^- exit; rather they ascribed the protein a regulatory role in ATP release (see below). In response to mechanical stimulation (fluid flow and cell swelling) and ATP stimulation, cholangiocytes isolated from various species exhibit CaCC (Tabibian et al. 2013). The molecular identity of these channels was unknown until recent studies showed that TMEM16A/ANO1 is a good candidate (Dutta et al. 2011, 2013). It seems that CFTR is only expressed in larger bile ducts, while TMEM16A is expressed in both small and large bile ducts (Dutta et al. 2011). Interestingly, CaCC contribution to anion transport is several fold higher than CFTR.

Similar to pancreatic ducts, Cl^- channels are functionally coupled with Cl^-/HCO_3^- exchanger, and AE2 is indeed the main effector of both basal and stimulated Cl^-/HCO_3^- exchange (Banales et al. 2006a, b; Strazzabosco et al. 1997; Aranda et al. 2004; Concepcion et al. 2013). Interestingly, *AE2−/−* mice develop biochemical, histological and immunological alterations resembling primary biliary cirrhosis (Concepcion et al. 2013). The luminal membrane of cholangiocytes also expresses amiloride-insensitive NHE2/*SLC9A2* and amiloride-sensitive NHE3/*SCL9A3* and NHE4/*SLC9A4* (Banales et al. 2006b). The function of these is not clear, though NHE3 knockout studies showed that the exchanger is important in fluid absorption in resting duct epithelium and also interacts with CFTR (Mennone et al. 2001). The Cl^- secretory response is maintained by small conductance Ca^{2+}-activated K^+ channels, SK2 ($K_{Ca}2.2$, *KCNN2*) expressed on the luminal membrane (Feranchak et al. 2004).

NHE1 is expressed on the basolateral membrane of cholangiocytes, and it has a multitude of functions, including regulation of pH_i and cell volume and transepithelial HCO_3^- transport in rat and human bile ducts (Spirli et al. 1998; Banales et al. 2006b). In addition, another acid extruder, V-ATPase, was identified and functional in pig cholangiocytes (Villanger et al. 1993), though it seems not important in human cholangiocytes (Strazzabosco et al. 1997). CA is also expressed in bile ducts, though compared to other HCO_3^- secretory organs, CA activity is rather low (Banales et al. 2006b).

Regarding the Na^+-coupled HCO_3^- import mechanisms across the basolateral membrane, there seem to be some interspecies variations (Fig. 5.5c). Na^+-dependent Cl^-/HCO_3^- exchanger, NDCBE1/*SLC4A8*, which has been cloned in humans (Grichtchenko et al. 2001) is assumed to import HCO_3^- across the basolateral membrane of cholangiocytes (Strazzabosco et al. 1997; Banales et al. 2006b). In rat cholangiocytes, several HCO_3^- transport mechanisms are functional as revealed by pH_i measurements (Strazzabosco et al. 1997). But apparently in the rat liver, only the NBC4c variant of NBCe2/NBC4/*SLC4A5* is expressed in cholangiocytes and hepatocytes, though in cholangiocytes, it is expressed apically (Abuladze et al. 2004). In murine cholangiocytes, NBCe1/*SLC4A4* is expressed on the luminal membrane (Uriarte et al. 2010), and in order to secrete HCO_3^- into the lumen, it would require coupling of several HCO_3^- to one Na^+ and highly hyperpolarized membrane potential.

K^+ channels are providing the driving force secretion, and at least two types of channels are expressed on the basolateral membrane –$K_{Ca}3.1$ and SK2/$K_{Ca}2.2$ (*KCNN2*) (Dutta et al. 2009; Feranchak et al. 2004). NKCC1 and Na^+/K^+ pump are also expressed on the basolateral membranes. A number of AQP are expressed in cholangiocytes, in particular AQP4 on the basolateral membrane and AQP1 on the luminal membrane, though there are species differences (Banales et al. 2006b; Tabibian et al. 2013).

There are additional transport systems in cholangiocytes in addition to those dealing with anion transport. Cholangiocytes take up bile salts via Na^+-dependent transporter ABAT or ASBT (*SLC10A2*) and released them across the basolateral membrane through truncated isoform of the same carrier (Tabibian et al. 2013; Banales et al. 2006b). Glucose is taken up by sodium-glucose transporter SGLT1 (*SLC5A1*) and released by GLUT1 (*SLC2A1*) (Tabibian et al. 2013). Glutathione, the tripeptide that is one of the principal driving forces for canalicular bile salt-independent secretion, is degraded by γ-glutamyl transpeptidase, expressed in cholangiocyte apical membranes, and resulting glutamate, cysteine and glycine are reabsorbed by Na^+-dependent transport in cholangiocytes (Tabibian et al. 2013).

5.4.4 Gallbladder Epithelium

The function of the gallbladder is to store bile and concentrate it during interdigestive phases by salt-dependent water reabsorption. Most studies on the rabbit and *Necturus* gallbladders showed that this transport was electroneutral (Na^+/H^+ and Cl^-/HCO_3^- exchangers in parallel) and these epithelia have low resistance and are regarded as a leaky type (Petersen and Reuss 1983; Petersen et al. 1990). However, in studies on human and primate gallbladders, it was discovered that these electrically silent absorptive organs can secrete, under the influence of secretin, HCO_3^--containing fluid after meals (Igimi et al. 1992; Svanvik et al. 1984). This secretion seems to depend on CFTR-like cAMP-activated channels that have relatively high permeability to HCO_3^- (Meyer et al. 2005). The importance of CFTR is highlighted in pig models of cystic fibrosis, where the disruption of CFTR leads to gallbladder and bile duct abnormalities (Rogers et al. 2008). Furthermore, prairie dog gallbladder shows forskolin-induced short-circuit current that was due to CFTR (Moser et al. 2007). This preparation is often used as an experimental model of human cholelithiasis, due to its unique propensity for developing gallstones on high-cholesterol chow. The basolateral HCO_3^- transport in gallbladder epithelium is carried out by pNBC1, and the driving force for secretion is maintained by cAMP-stimulated K^+ channels that are also sensitive to Ca^{2+} and pH, but their identity is not clear (Moser et al. 2007; Meyer et al. 2005).

5.4.5 *Regulation of Bile Formation*

The bile formation that occurs in at least three steps is also regulated at three levels. The first step is the canalicular bile fluid secretion by hepatocytes that occurs continuously and is relatively poorly regulated (Esteller 2008; Banales et al. 2006b). Interestingly, one regulator of canalicular HCO_3^- secretion is glucagon (Lenzen et al. 1997; Alvaro et al. 1995), which increases cAMP and stimulates insertion of vesicles containing Cl^-/HCO_3^- exchanger (AE2, *SLC4A2*) and AQP8 (Gradilone et al. 2003). In the last step, gallbladder secretion is regulated by secretin, and gallbladder contraction is stimulated by CCK.

The most extensive regulation of bile modification (secretion and absorption) occurs in bile ducts (Banales et al. 2006b; Tabibian et al. 2013). Firstly, there are regulatory factors modifying basal secretion of cholangiocytes. The most important hormone is secretin that is released as a physiological response to meals; the other two are bombesin (gastrin-releasing peptide) and vasoactive intestinal peptide (also a neurotransmitter). Secondly, acetylcholine and noradrenaline potentiate secretin-stimulated HCO_3^- and fluid secretion. Thirdly, many factors such as somatostatin, gastrin, insulin, dopaminergic agonists, α2 adrenergic receptor agonists, endothelin, GABA and cytokines (e.g. IL1, IL6, TNF-α) inhibit basal and secretin-stimulated cholangiocyte secretion. Lastly, there a number of bile-borne factors (flow and osmolality, amino acids, glucose, nucleotides, bile acids) that regulate cholangiocyte function. Here, we will focus on regulation by nucleotides and bile acids, similar to the section above dealing with pancreatic ducts.

5.4.5.1 Purinergic Signalling

Both hepatocytes and cholangiocytes release adenosine nucleotides, they express a number of ectonucleotidases, and both cell types express a number of P2 and P1 receptors (Schlosser et al. 1996; Chari et al. 1996; Fausther and Sevigny 2011). Cholangiocytes exhibit shear- and flow-sensitive and cell volume-sensitive ATP release, as well as release stimulated by forskolin and ionomycin, and it has been proposed that CFTR is regulating this ATP release (Schlosser et al. 1996; Chari et al. 1996; Fiorotto et al. 2007; Minagawa et al. 2007). Nevertheless, it seems that ATP release is larger in small upstream cholangiocytes compared with downstream ones, indicating that smaller ducts could signal to larger ones via paracrine ATP signal (Woo et al. 2008) and that other mechanisms than CFTR are important in regulating ATP secretion. In fact, recent studies indicate that ATP is also released by vesicular exocytosis similar to the pancreas (Feranchak et al. 2010; Woo et al. 2010).

Both P2 and adenosine receptors regulate cholangiocytes by stimulating K^+, Cl^- and HCO_3^- secretion. P2 receptors on the basolateral membrane of cholangiocytes, similar to acetylcholine evoked Ca^{2+} signalling, seem to have minimal effect on bile duct secretion (Nathanson et al. 1996), possibly due to rapid hydrolysis of nucleotides (Dranoff et al. 2001). The most prominent effects are relayed by the P2 receptors on the luminal membrane, many of which stimulate Ca^{2+} signalling or induce Ca^{2+} influx, which leads to the stimulation of CaCC, short-circuit current and ductal alkalization (Woo et al. 2010; Dranoff et al. 2001). Recently, it was shown that mechanosensitive ATP release stimulates TMEM16A/ANO1 in human cholangiocytes from small and large biliary ducts (Dutta et al. 2011, 2013). There a number of receptors expressed on cholangiocytes, e.g. mouse and rat express P2X4 and P2Y2, as well as P2Y1, P2Y4, P2Y6, P2Y12 and P2Y13 receptors (Woo et al. 2010; Dranoff et al. 2001; Doctor et al. 2005). Interestingly, the P2Y12 receptor is expressed on the primary cilium and induces cAMP signalling (Masyuk et al. 2008).

5.4.5.2 Bile Acids

Bile acids are synthetized from cholesterol in the liver and excreted into bile and subsequently into the small intestine, where they facilitate digestion and absorption of dietary fats and fat-soluble vitamins. Bile acids are also signalling hormone like molecules that act on many tissues/cells in our body and do so by activating several different receptors and sensing proteins including nuclear farnesoid X receptor (FXR), pregnane X receptor (PXR) and G-protein-coupled receptor TGR5 (Gpbar-1) (Keitel and Haussinger 2013; Pols et al. 2011). FXR and PXR are strongly expressed in hepatocytes, while TGR5 is strongly expressed in cholangiocytes.

Cholangiocytes are exposed to millimolar concentrations of bile acids and conjugated salts. Such concentrations would be toxic to cholangiocytes, and several defense mechanisms are at hand. One of these is the formation of micelles of phospholipids and bile salts. Another defense mechanism proposed is the biliary HCO_3^- secretion (Beuers et al. 2010). It has been suggested that BA profile/composition changes radially from the midstream to the apical membrane of cholangiocytes (Keitel and Haussinger 2013). Cholangiocytes express TGR5 in the primary cilia, the apical membrane and subcellular (Keitel et al. 2010; Masyuk et al. 2013). In particular, the primary cilia, which are mechano-, chemo- and osmosensors, could sense BA and via TGR5 receptors trigger cAMP signalling pathways that stimulate CFTR and AE2 and thus HCO_3^- secretion. Bile acids also stimulate CaCC directly (Shimokura et al. 1995) or through CFTR-regulated ATP release and P2 receptor-mediated stimulation of CaCC (Fiorotto et al. 2007). The outcome of these events would be the formation of unstirred HCO_3^- layer, so-called HCO_3^- umbrella, which together with dense glycocalyx creates a protective microenvironment at the apical membrane of cholangiocytes and prevents the protonation of glycine-conjugated bile acids dominating in human bile (Beuers et al. 2010; Keitel and Haussinger 2013). Therefore, cholangiocytes would be protected from diffusion of these polar and pro-apoptotic acids. Relevance of this system is demonstrated in patients with primary biliary cirrhosis, in which AE2 expression in bile ducts is reduced, leading to defective HCO_3^- secretion and BA-mediated cell injury (Keitel and Haussinger 2013).

5.5 Duodenum

The pH of luminal contents in the duodenum is more or less dictated by the acid chyme arising from the stomach until pH is neutralized by the pancreatic and hepatic outlets. Furthermore, in order to withstand the potentially damaging effect of low bulk pH, the duodenal epithelium creates a mucous layer, which is supplemented with HCO_3^- from the epithelial cells to create a protective barrier, and HCO_3^- secretion also contributes to pH neutralization of duodenal contents (Allen and Flemstrom 2005; Flemstrom and Kivilaakso 1983; Williams and Turnberg 1981; Ainsworth et al. 1990, 1992). In rabbit duodenum, HCO_3^- secretion amounts to 154 μEQ/cm/h, while other species have lower rates (Flemstrom et al. 1982). Duodenal ulcers result from an unbalance between protective factors such as mucus and HCO_3 and the aggressive factors gastric acid and pepsin and the infestation with pathogenic *Helicobacter pylori*. The long prevailing model for duodenal epithelial HCO_3^- secretion was inspired by studies of the exocrine pancreas given above (Novak and Greger 1988b). As mentioned, functional coupling of a luminal surface Cl^- channel and a Cl^-/HCO_3^- exchanger is fueled by intracellular conversion of CO_2 and H_2O to form HCO_3^- and H^+ catalysed by intracellular carbonic anhydrases. The intracellular pH is normalized by basolateral NHE-mediated extrusion of protons. The essence of this model is still standing, but a number of recent investigations have greatly expanded and refined the understanding of the molecular machinery for duodenal bicarbonate secretion (Fig. 5.6).

The involvement of interstitial CO_2, which enters the cells and CA and generates HCO_3^- for secretion, has been challenged. In the duodenum, the luminal pCO_2 is higher than anywhere else in the mammalian body (Rune and Henriksen 1969), due to the massive neutralization of gastric acid mainly by the pancreatic bicarbonate and hepatobiliary outlets. CO_2 crosses the luminal membrane to allow the enterocytes to sense and respond to luminal acid in a complex interplay with cytosolic carbonic anhydrase and a moderating effect of luminal carbonic

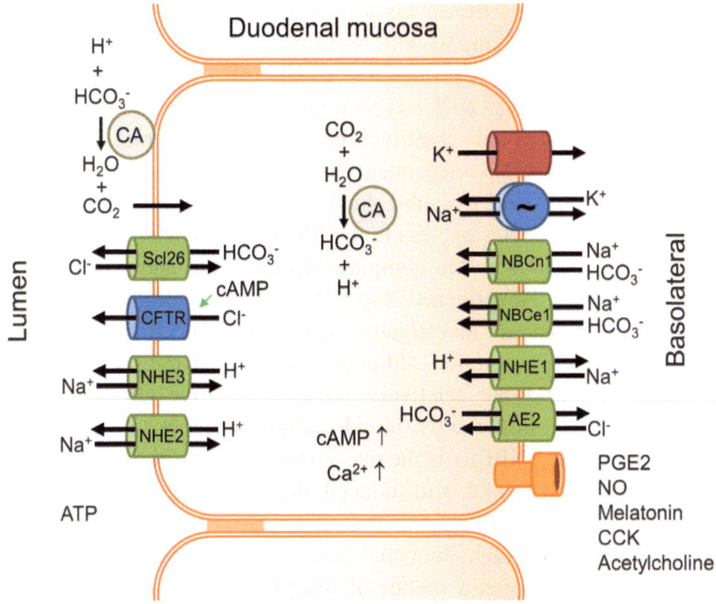

Fig. 5.6 Schematic diagram showing a current model for HCO_3^- secretion from duodenal enterocytes. In this model, secreted HCO_3^- does not arise from the blood side CO_2 and intracellular conversion to HCO_3^- and H^+. Instead CO_2 enters from the luminal side, which has high pCO_2, potentially counteracting the HCO_3^- secretion and challenging the cellular pH homeostasis. HCO_3^- for maintaining intracellular pH and for secretion may arise from the basolateral HCO_3^- import

anhydrases (Holm et al. 1998; Kaunitz and Akiba 2002, 2006a, b; Sjoblom et al. 2009). Besides the cytosolic CAII, the duodenal enterocytes expresses the membrane-bound luminal CAXIV and basolateral CAIX (Lonnerholm et al. 1989; Saarnio et al. 1998). As reviewed recently, it is most likely that the CA activity is centrally involved in acid sensing and signalling in the duodenal enterocytes rather than participating in HCO_3^- secretion directly (Sjoblom et al. 2009).

Isolated duodenal epithelial cells exhibit NHE activity (Ainsworth et al. 1996, 1998; Isenberg et al. 1993). The findings were first interpreted as basolateral NHE1 activity according to the general working model. However, a subsequent study of mouse duodenal enterocytes reported several distinct NHE activities and evidenced the molecular expression of NHE1 in the basolateral membrane and NHE2 and NHE3 in the luminal membrane (Praetorius et al. 2000; Praetorius 2010). NHE1 seems expressed at low abundance as compared with NHE2 and NHE3 and is an unlikely mechanism for eliminating intracellularly produced protons from duodenal enterocytes. Interestingly, Na^+-HCO_3^- cotransport may serve an important role in the duodenum as a supply for intracellular HCO_3^- as a defense against intracellular acid formed from CO_2 entering from the lumen (Akiba et al. 2001a, b). The robust expression of the HCO_3^- loaders NBCe1 and NBCn1 in the basolateral membrane of duodenal epithelial cells spurred the hypothesis that HCO_3^- is transported from the interstitial space to the lumen via a transcellular route, where the two transporters import HCO_3^- in symport with Na^+ from the interstitium to the cell.

Acidified rabbit enterocytes display potent Na^+-dependent recovery of pH_i in the presence of the CO_2/HCO_3^- buffer system (Ainsworth et al. 1996; Isenberg et al. 1993). DIDS and H_2DIDS only slightly affected the recovery rate from acid load. A similar DIDS-insensitive Na^+-dependent HCO_3^- import was observed in mouse (Praetorius et al. 2001) and in rabbit duodenal enterocytes (Jacob et al. 2000b). Human duodenal enterocytes express both NBCn1 and NBCe1, although immunoreactivity for the latter is subtle compared to the renal and pancreatic staining (Damkier et al. 2007). Nevertheless, the DIDS-sensitive electrogenic NBCe1 has been put forward by several investigators as the main HCO_3^- import mechanism for transcellular HCO_3^- secretion, although both NBCn1 transport and the duodenal Na^+-HCO_3^- import are relatively DIDS insensitive (Jacob et al. 2000a; Praetorius et al. 2001). Importantly, based on NBCn1 knockout studies in mice, it was recently shown that NBCn1 is the major contributor to Na^+-dependent HCO_3^- transport as well as basal and acid-induced HCO_3^- secretion (Chen et al. 2012; Singh et al. 2013b).

The mechanism by which duodenal enterocytes extrude HCO_3^- across the luminal membrane has been a matter of intense debate. Many transporters and channels have been investigated, among which CFTR and the *SLC26*-derived proteins DRA, PAT1 and *SLC26A9* have been the most prominent examples. First, it is firmly established that basal and agonist-induced duodenal HCO_3^- secretion strongly depends on CFTR expression (Clarke and Harline 1998; Hogan et al. 1997a, b; Seidler et al. 1997). It is still not clear how much CFTR conductance contributes to HCO_3^- secretion directly. First, HCO_3^- extrusion does not depend on CFTR, at least in isolated duodenal villus cells (Praetorius et al. 2002). Second, DRA encoded by *SLC26A3* mediates electroneutral Cl^-/HCO_3^- exchanger (Alper et al. 2011), is involved in NaCl absorption throughout the leaky intestinal segments and seems mainly expressed in the lower villus/crypt axis in duodenum (Walker et al. 2009). The study indicates that DRA contributes to bicarbonate secretion mainly in lower villus/crypt. The *SLC26A6* gene product PAT1 is mainly expressed in the proximal small intestine and is prevalent in the villus region where it also functions as the predominant anion exchanger (Simpson et al. 2007; Singh et al. 2013a; Walker et al. 2009). Knockout of PAT1 reduces basal duodenal HCO_3^- secretion by approximately 70 %, while knockout of DRA only induced a 30 % reduction (Simpson et al. 2007). However, only a minor role for PAT1 was found in acid-stimulated HCO_3^- secretion (Singh et al. 2013a). The *SLC26A9* gene product is mainly expressed in the proximal gastrointestinal tract and found in the crypt region of the duodenum (Liu et al. 2014). *SLC26A9* is associated with anion conductance, and the corresponding knockdown mice have reduced acid-induced duodenal HCO_3^- secretion (Singh et al. 2013a; Walker et al. 2009). An integrative model of duodenal bicarbonate secretion is shown in Fig. 5.6.

Among the many stimulators of duodenal bicarbonate secretion, luminal acid is a physiological and effective agonist (Flemstrom and Kivilaakso 1983; Isenberg et al. 1986; Allen and Flemstrom 2005). The neurohumoral control of duodenal

HCO_3^- transport is mediated either directly by cholinergic innervation acting on M3 receptors and indirectly via paracrine melatonin secretion from enterochromaffin cells, both using Ca^{2+} signalling to induce stimulation (Sjoblom and Flemstrom 2003). The same study shows that CCK also enhances HCO_3^- secretion through Ca^{2+} signalling. One of the first agonists to be discovered was PGE2, which acts via cAMP as well as Ca^{2+} signalling (Aoi et al. 2004), while the stimulatory effect of enterotoxin and guanylins occurs through cGMP increases (Guba et al. 1996; Joo et al. 1998). The intracellular signals are most likely to stimulate HCO_3^- secretion via the regulation of the luminal membrane transporters and channels, with CFTR being the best documented.

5.6 Renal Intercalated Cells

The overall function of the kidneys is broadly to secure homeostasis, and this is achieved by maintaining water and electrolyte balance, acid/base balance and blood pressure and rid the organism of waste products (Boron and Boulpaep 2012). The kidneys filter the blood plasma in the renal corpuscles and then modify the filtrate to conserve appropriate amounts of water, nutrients and salt by secretion and reabsorption processed in an intricate system of renal tubules. As an integral part of the process, the kidney helps to maintain normal acid/base balance by adjusting the secretion of acid/base equivalents. Among the many epithelial cell types in the renal tubular system, a single one has the capacity of secreting HCO_3^- which is the type B intercalated cell (IC). These cells are restricted to the late distal convoluted tubules, the connecting tubules and cortical collecting ducts and are functionally mirror images of the acid-secreting type A ICs (Kwon et al. 2012). In the mentioned renal cortical segments, the intercalated cells comprise almost 50 % of the tubular cells. Most of the cortical ICs are type B cells, whereas medullary collecting ducts exclusively contain type A ICs. Interestingly, the numbers of intercalated cell types can vary in a regulated and dynamic fashion. The number of type A ICs is increased in response to metabolic acidosis (Schwartz et al. 1985). It seems that the change in relative numbers relies on the ability of type B ICs to differentiate into type A ICs, given the proper stimulus (Schwartz et al. 1985). Hensin is a protein that is secreted by the epithelial cells and incorporated in the extracellular matrix and is necessary for cell differentiation (Gao et al. 2010; Al-Awqati 2013; Schwartz et al. 2002). It stimulated differentiation of type B ICs to type A ICs only in its polymeric forms induced by binding of certain activated integrins and the activity of extracellular galectin 3 (Hikita et al. 2000; Schwaderer et al. 2006; Vijayakumar et al. 2008). Figure 5.7a, b depict the two types of cells, and evidence for ion transporters is elaborated below.

The bicarbonate-secreting type B IC (or β-IC) was first described in the turtle bladder (Leslie et al. 1973; Stetson and Steinmetz 1985), which like similar cells in toad skin shares the functional significance for HCO_3^- secretion, as well as transport properties with mammalian renal type B ICs (Schwartz et al. 1985). The

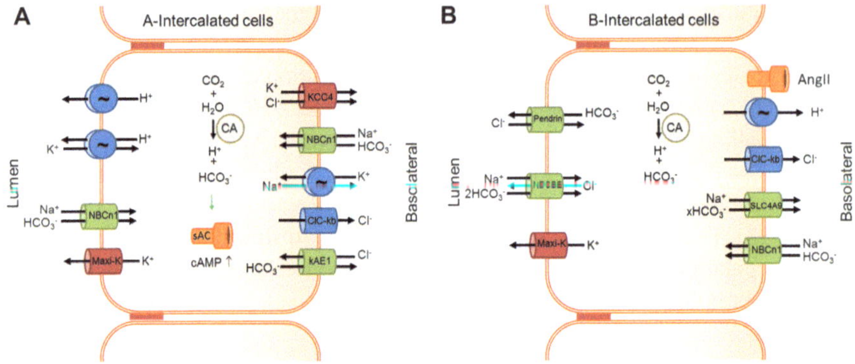

Fig. 5.7 (**a**) Schematic diagram showing the molecular machinery for acid/base transport in type A intercalated cells from the renal tubular system. (**b**) Schematic diagram showing the molecular machinery for acid/base transport in type B intercalated cells from the renal tubular system

studies showed that these cells were characterized by the luminal extrusion of HCO_3^- in exchange for Cl^-, thereby affecting pH homeostasis in the opposite direction as the already described type A ICs. The opposite function of the two cell types is accompanied by different polarization of V-ATPases and anion exchangers. The acid-secreting type A ICs expresses the V-ATPase in the luminal membrane, whereas type B ICs have basolateral V-ATPase (Brown et al. 1988a). It is now evident that the type B ICs use the *SLC26A4* gene product pendrin to extrude bicarbonate apically into the urine. Pendrin is an anion exchanger shown to function as a Cl^-/HCO_3^- exchanger in the type B ICs (Royaux et al. 2001). This was the first study to show that pendrin is expressed in the apical membrane of type B ICs and is involved in HCO_3^- secretion when animals were challenged by alkalosis, but not under baseline conditions. Importantly, the renal tubules lose apical Cl^-/HCO_3^- exchange and the capability to rid the body of excess base by the urine when the pendrin gene is disrupted (Amlal et al. 2010; Royaux et al. 2001). Type A ICs expresses a renal-specific variant of the anion exchanger AE1 (*SLC4A1*) in the basolateral membrane to extrude excess HCO_3^- (Alper et al. 1989). For both types of ICs, the source for acid/base extrusion is the intracellular conversion of CO_2 and H_2O to protons and HCO_3^-. The reaction rate is critically dependent on the intracellular carbonic anhydrase, CAII, and therefore inhibited by acetazolamide or CAII gene deletion (Breton et al. 1995; Sly et al. 1983). Pendrin function is sensitive to changes in local pH (Azroyan et al. 2011), whereas the expression level and function are reduced by metabolic acidosis and increased in alkalosis (Frische et al. 2003; Petrovic et al. 2003; Wagner et al. 2002). The key modulator of pendrin expression may not be systemic pH alone but include a concomitant change in circulating Cl^- (Hafner et al. 2008; Vallet et al. 2006; Verlander et al. 2006). At the hormone level, there is agreement that angiotensin II stimulates pendrin either directly or via the activation of the V-ATPase (Wagner et al. 2011).

Thus, the acid/base transport models for ICs are quite simple and well established but have nevertheless been challenged somewhat by more recent findings. First, antibodies against the human NBC3, also known as NBCn1 (*SLC4A7*), colocalized the protein with the V-ATPase in vesicles and the apical membrane of rat type A ICs and to the basolateral membrane of type B ICs (Kwon et al. 2000; Pushkin et al. 1999). The authors detected DIDS-sensitive NBC activity in the apical surface of type A ICs. These findings were surprising, as NBC would function as an electroneutral alternative transporter for type A ICs to acidify the urine. In the type B ICs, the function would be to acidify the interstitium in parallel to direct H^+ transport by V-ATPase. However, antibodies against the corresponding rat protein, NBCn1, localized the protein basolaterally in rat type A ICs (Vorum et al. 2000). This localization of the NBC is counterintuitive, as a main task for that membrane is in fact base extrusion. It should be mentioned that neither antibody produced labelling of human or mouse renal intercalated cells but performed as expected in the kidney in humans (Damkier et al. 2007). Thus, data on NBCn1 expression in the ICs are inconsistent, and the lack of renal phenotype of *SLC4A7*-depleted mice indicates that its function may be much less important than the H^+-ATPase function.

Second, recent discoveries has led to a change of the paradigm that intercalated cells are simply acid/base regulating cells but participate also in Cl^- transport. Previously, Cl^- was expected to take a paracellular route depending solely on a transepithelial potential difference and Cl^- permeable tight junctions such as the aldosterone regulated claudin-4 (Le et al. 2005). Pendrin was established as a transcellular route for connecting tubule and collecting duct Cl^- reabsorption (Wall et al. 2004). Genetic deletion of pendrin was shown to eliminate collecting duct Cl^- reabsorption in mineralocorticoid- and bicarbonate-treated mice. Thus, pendrin knockout mice are protected from mineralocorticoid-induced hypertension, while overexpression of pendrin leads to hypertension in mice on high-NaCl diet (Jacques et al. 2013; Verlander et al. 2003). The significance of HCO_3^- secretion by pendrin was also suggested to directly affect the function of the principal cell Na^+ loader ENaC (Pech et al. 2010). Whereas single gene knockout for the distal tubule NaCl reabsorption protein NCC and pendrin is surprisingly benign, mice with NCC-pendrin double knockout have a severe salt- and volume-wasting alkalotic phenotype, even under control conditions (Soleimani et al. 2012). Thus, according to the current model of collecting duct, Na^+ reabsorption and Cl^- reabsorption take place in separate cells—principal cells and type B ICs, respectively (Wall and Pech 2008). Pendrin now seems to take part in electroneutral NaCl reabsorption mechanism in addition to its role in alkalization in metabolic alkalosis. It is not known how these roles are separated and regulated.

The *SLC4A8* gene product NDCBE1 has been associated with electroneutral and thiazide-sensitive NaCl reabsorption in cortical collecting duct (Leviel et al. 2010). It may reside and function in the luminal membrane of the type B ICs, and in the proposed model (Fig. 5.7), NDCBE1 and pendrin work in concert with two turnovers of pendrin for each NDCBE1 turnover to yield electroneutral, pH-neutral NaCl reabsorption through type B ICs. The model is based on the observations that Na^+ reabsorption in collecting ducts is not completely inhibited by amiloride but contains a thiazide component under some experimental conditions. Also, the authors find luminal Cl^- and CO_2/HCO_3^- dependence of the thiazide-sensitive Na^+ transport in isolated cortical collecting ducts from Na^+-depleted animals. Furthermore, they detect NDCBE1 protein expression and mRNA in the relevant tubular segments and demonstrate the lack of such transport in tubules from *NDCBE1* knockout mice (Leviel et al. 2010). In a separate study, the *SLC4A9* gene product, usually named AE4, was suggested as the basolateral Na^+ exit pathway in these cells (Chambrey et al. 2013). It was shown that *AE4* gene deletion greatly reduced thiazide-sensitive Na^+ and Cl^- reabsorption. Interestingly, AE4 is suggested to work as an electrogenic Na^+-HCO_3^- extruder in their model. This would, however, render the total reabsorption process electrogenic, as extrusion of Na^+ and HCO_3^- (depending on the electrochemical gradients) would usually be in a 1:3 stoichiometry and in the model, only one Cl^- follows through ClC channels in the basolateral membrane. As all these discoveries have profound impact on our understanding of renal electrolyte handling, it is highly important to establish the electrochemical gradients of the type B ICs in order to validate the model, which also suggests the V-ATPase as the creating mechanism for transmembrane gradients (as opposed to the Na^+/K^+-ATPase). It is also necessary to localize the NDCBE1 protein in the apical membrane of type B ICs and directly evidence that thiazides target this protein e.g. by ligand binding assays. In *Xenopus laevis* oocytes, neither NDCBE1 nor pendrin seemed to be the site of action for thiazides, and carbonic anhydrase was only discharged as target in control experiments in untreated collecting ducts (Leviel et al. 2010). Furthermore, before NDCBE1 is recognized as important in renal Na^+ and Cl^- reabsorption and in blood pressure regulation, it is necessary to report normal urine electrolytes, pH, osmolarity and blood pressure values at baseline and in Na^+-depleted NDCBE1 KO mice. Whereas principal cell-specific ENaC subunit knockout has profound impact on baseline Na^+ and K^+ homeostasis, as well as urine output (Christensen et al. 2010), both pendrin KO and NDCBE1 KO phenotypes seem mild as far as they have been reported.

5.7 Choroid Plexus Epithelium

The choroid plexus epithelium (CPE) produces the vast majority of the intraventricular cerebrospinal fluid, CSF. Rougemont and colleague demonstrated directly the involvement of the choroid plexus in CSF formation in 1960 (Rougemont et al. 1960), although this was suggested much earlier by J. Faivre and H. Cushing (Cushing 1914; Faivre 1854). Rougemont's group established that the solute contents in the nascent CSF differed from a simple plasma ultrafiltrate. The reasons that CSF cannot be a simple ultrafiltrate are as follows: 1) CSF is approximately 5 mOsM hypertonic compared to plasma (Davson and Purvis 1954); 2) the $[Na^+]$ and $[HCO_3^-]$ are slightly higher, whereas $[K^+]$ and $[Cl^-]$ are lower than expected in the CSF at equilibrium (Ames, et al. 1964); and 3) there is a 5 mV lumen-positive electrical potential difference across the choroid plexus epithelium. Therefore, the choroid plexus must serve a secretory function, given that most intraventricular

CSF arises from this structure. The CSF has many important functions in the central nervous system. CSF fills the internal and external fluid spaces and thereby cushions the central nervous system and keeps the intracranial pressure at controlled levels, and it stabilizes the fluid around neurons, the brain extracellular fluid. Primarily, CPE-secreted fluid contains $NaHCO_3$ and, therefore, can neutralize acid produced by neuronal activity. The choroid plexus epithelium also secretes growth factors and nutrients into the CSF that are crucial for the brain development and function (Damkier et al. 2013). Importantly, CSF also helps to remove, i.e. absorb, K^+ produced by active neurons, as well as breakdown products of serotonin and dopamine. The composition of newly formed CSF is roughly 150 mM Na^+, 3 mM K^+, 130 mM Cl^- and 25 mM HCO_3^- (Davson and Segal 1996; Johanson and Murphy 1990). The choroid plexus produces up to 0.4 ml/g/min fluid, which gives approximately 60 μmol HCO_3^-/h/g tissue output.

While some mechanisms of CSF secretion are well established, other aspects of secretion are not fully understood. For example, the main transport mechanisms for Na^+ into the choroid plexus epithelium from the blood side are not defined. Another central question is how HCO_3^- is extruded from the cells into the CSF. Clues into the main transport pathways are reflected by the wide profile of drugs inhibiting CSF secretion: ouabain, acetazolamide, amiloride, DIDS, bumetanide and furosemide (Ames, et al. 1965; Davson and Segal 1970; Johanson and Murphy 1990; Melby et al. 1982; Wright 1972). The secretion rate of CSF is also influenced by the HCO_3^- concentration at the basolateral side (Mayer and Sanders-Bush 1993; Saito and Wright 1983). Thus, CO_2/HCO_3^- metabolism seems important for CSF secretion. Before describing the putative HCO_3^- extrusion mechanisms, we will give the general transport machinery that creates and sustains the gradients permitting HCO_3^- secretion. Lastly, other acid/base transporters that may assist or modify HCO_3^- secretion are considered.

5.7.1 *Basic Secretory Machinery*

As opposed to most other epithelia, the choroid plexus Na^+/K^+-ATPase is localized to the luminal membrane. Ventricular application of ouabain blocks the transepithelial net Na^+ flux as well as the CSF secretion (Ames et al. 1965; Davson and Segal 1970; Wright 1978). Luminal Na^+/K^+-ATPase is demonstrated by immunoreactivity and ouabain-binding studies (Ernst et al. 1986; Masuzawa et al. 1984; Praetorius and Nielsen 2006; Quinton et al. 1973; Siegel et al. 1984). The luminal Na^+/K^+-ATPase directly transports Na^+ into the CSF and also maintains the Na^+ and K^+ gradients that drive most other transport processes in CSF secretion (Fig. 5.8). Na^+/K^+-ATPase complexes consist of α1 and either β1 or β2 subunits, which are expressed in the choroid plexus together with the γ-subunit, phospholemman (FXYD1) (Feschenko et al. 2003; Klarr et al. 1997; Zlokovic et al. 1993).

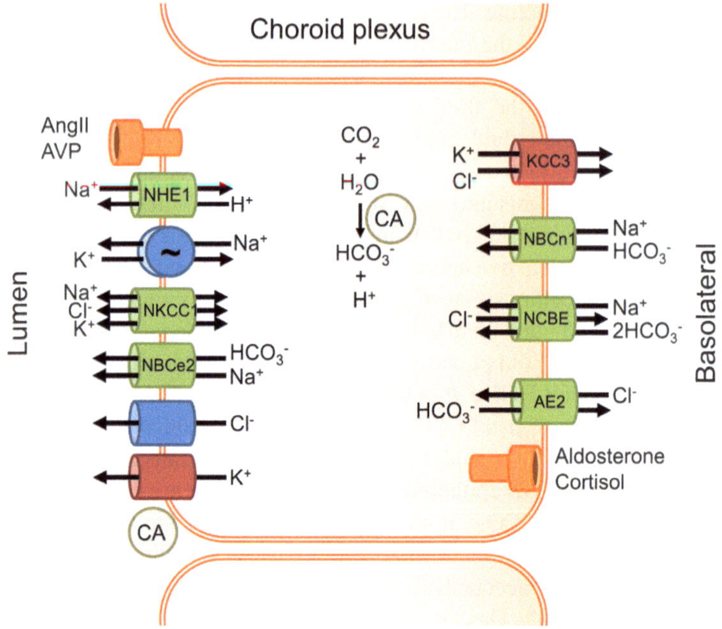

Fig. 5.8 Schematic diagram showing a simplified model of the CSF and thereby HCO_3^- secretion process by the choroid plexus epithelium. Although HCO_3^- secretion greatly depends on carbonic anhydrase activity, it also fully relies on a DIDS-sensitive basolateral supply of Na^+ and HCO_3^-. The luminal HCO_3^- extrusion is electrogenic, and CSF pH increases are followed by an increase in CSF Na^+ content, indicating a molecular or functional coupling of the extrusion of both ions. Interconversion between $CO_2 + H_2O$ and H^+ + HCO_3^- is catalysed by carbonic anhydrases luminal, i.e. CAXII and intracellular CAII. Cl^- conductances are characterized as a volume-regulated anion channel (VRAC) and an inward-rectifying Cl^- conductance (Clir), whereas the K^+ channels include Kv1.3, Kv1.1 and Kir7.1. KCC4 is also a luminal K^+ exit pathway but is left out for clarity

K^+ channels expressed on the luminal membrane have several functions. Firstly, they are important for K^+ recycling in connection with the Na^+/K^+-ATPase activity. Secondly, K^+ channels are the main determinant for the membrane potential. This is important for the luminal HCO_3^- secretion, which is electrogenic. Thirdly, certain K^+ channels (inward rectifiers) are also important for cellular uptake of K^+ in hyperpolarized voltages in order to provide a K^+ sink, e.g. removal of neuron-derived K^+. K^+ channels were first identified in the choroid plexus by patch clamping the luminal membrane of *Necturus maculosa* cells (Brown et al. 1988b; Christensen and Zeuthen 1987; Zeuthen and Wright 1981). Several K^+ conductances have been identified: an inward-rectifying conductance (Kir7.1 channels) and outward-rectifying conductances (Kv1.1 and Kv1.3) (Doring et al. 1998; Kotera and Brown 1994; Speake and Brown 2004). KCNQ1/KCNE2 channels may also contribute to the outward-rectifying conductance (Roepke et al. 2011). All these K^+ channel proteins are expressed in the luminal membrane of rat and mouse choroid plexus (Nakamura et al. 1999; Roepke et al. 2011; Speake and Brown 2004).

The CPE cells express abundant NKCC1 in the luminal membrane (Keep et al. 1994; Plotkin et al. 1997). With the typical ionic distribution across the luminal membrane in CPE, NKCC1 is close to equilibrium (Keep et al. 1994). Thus, both inward and outward NKCC1 transports have been observed (Bairamian et al. 1991; Wu et al. 1998). Additionally, luminal NKCC1 may also serve cell volume regulation or cation recycling. The K^+-Cl^- cotransporters (KCCs) are electroneutral, and transport of the ions out the cells is driven by the outward [K^+] gradient across cell membranes. KCCs are like NKCC1 inhibited by furosemide (Russell 2000). KCCs were described as a furosemide-sensitive K^+- and Cl^--dependent transport in the luminal membrane of bull frog choroid plexus (Zeuthen 1991). KCC4 (*SLC12A7*) most likely mediates the observed transport in the luminal membrane, whereas KCC3a (*SLC12A6*) is expressed in the basolateral membrane (Karadsheh et al. 2004; Pearson et al. 2001). KCC4 may, therefore, contribute to the recycling of K^+ across the luminal membrane helping to sustain the Na^+/K^+-ATPase activity (Fig. 5.8).

5.7.2 Luminal HCO_3^- Extrusion

A substantial Cl^- and HCO_3^- efflux across the luminal membrane of the choroid plexus occurs via electrogenic pathways, mainly anion channels for Cl^-. The mechanisms involved in HCO_3^- efflux are less well described but may overlap to some extent with the Cl^- extrusion. The majority of Cl^- moves via a transcellular route, because luminal DIDS minimizes Cl^- secretion (Deng and Johanson 1989). There are several potential candidates for Cl^- channels. Inward-rectifying anion conductances (Clir) have been shown by patch clamping in CPE cells from several species (Kajita et al. 2000; Kibble et al. 1996, 1997), and Clir channel activity is augmented by protein kinase A (Kibble et al. 1996) and inhibited by protein kinase C (Kajita et al. 2000). For the amphibian choroid plexus, Saito and Wright proposed that cAMP-regulated ion channels conducting Cl^- and HCO_3^- form the main efflux pathway for the anions in the luminal membrane (Saito and Wright 1984). Clir channels have high HCO_3^- permeability in mammals (Kibble et al. 1996). The molecular identity of the Clir channel remains to be determined.

The CFTR does not seem to contribute to cAMP-regulated Cl^- conductance in the choroid plexus (Kibble et al. 1996, 1997). ClC-2 channels are also unlikely to play a significant role in the choroid plexus transport, as Cl^- conductance is unaffected in the choroid plexus epithelium from ClC-2 knockout mice (Speake et al. 2002). The volume-regulated anion channels (VRACs) were also identified in choroid plexus epithelium (Kibble et al. 1996, 1997). VRACs are believed to be ubiquitously expressed but do not conduct HCO_3^- well (Duran et al. 2010). VRACs are most likely less important for CSF secretion, as their Cl^- conductance at normal cell volumes is low (Kibble et al. 1996; Millar and Brown 2008).

An alternative mechanism for luminal HCO_3^- extrusion to Cl^- channels is the electrogenic cotransporter NBCe2 (or NBC4), which mediates the transport of one Na^+ with two to three HCO_3^- ions (Sassani et al. 2002; Virkki et al. 2002). In rat choroid plexus, NBCe2 is situated in the luminal membrane (Bouzinova et al. 2005). In the mouse, it mediates the export of one Na^+ for three HCO_3^- across the luminal membrane (Millar and Brown 2008). A significant increase in the CSF Na^+ content without a reciprocal change in CSF K^+ occurs in rodents exposed to 11 % CO_2 at the CSF side (Nattie 1980). This finding indicates that Na^+-dependent acid/base transporters, such as NBCe2, are involved in compensating CSF pH for the increased pCO_2. Indeed, the composition of the CSF is changed in NBCe2 gene trap knockout mice with a significant decrease in [HCO_3^-] from 24 to 20 mM (Kao et al. 2011). These findings indicate that NBCe2 might participate in the regulation of CSF pH by extruding HCO_3^-.

5.7.3 *Other Acid/Base Transporters of Consequence for HCO_3^- Secretion*

The V-ATPase does not seem to be expressed in the luminal plasma membrane of the choroid plexus epithelium. Among the many V-ATPase subunits, only a few are associated with plasma membrane expression of the entire complex. The B1 subunit, which is involved in plasma membrane targeting, is not expressed at mRNA level in the choroid plexus, but B2 mRNA is detectable in these cells (Christensen, Damkier and Praetorius, unpublished data). Thus, the V-ATPase complexes seem restricted to the lysosomal system in these cells. The rabbit choroid plexus expresses mRNA encoding the non-gastric H^+/K^+-ATPase (Lindvall-Axelsson et al. 1992; Pestov et al. 1998). Interestingly, luminally applied omeprazole inhibited CSF secretion in another study indicating the gastric H^+/K^+-ATPase activity gastric (Lindvall-Axelsson et al. 1992). Both V-ATPase and H^+/K^+-ATPase are expressed at the luminal membrane together with HCO_3^- transporters, thus counteracting ongoing HCO_3^- secretion. Thus, the role of H^+ pumps is not clear in this epithelium.

NHE1 expressed in the choroid plexus epithelium. The mRNA encoding only NHE1 isoform was detected in the choroid plexus (Damkier et al. 2009; Kalaria et al. 1998) and Na^+/H^+ exchange activity of isolated epithelial cells is inhibited by EIPA (Bouzinova et al. 2005; Damkier et al. 2009). In most epithelia, NHE1 is localized to the basolateral membrane (Orlowski and Grinstein 2004), and this was also believed to be the case for the choroid plexus epithelium, as intravenous application of amiloride was reported to inhibit the rate of CSF secretion in vivo (Murphy and Johanson 1989). Nevertheless, NHE1 is located in the luminal membrane of both mouse and human choroid plexus (Damkier et al. 2009; Kao et al. 2011). Also functionally, NHE1 seems to be the only NHE in the normal choroid plexus plasma membrane, as no residual Na^+-dependent pH_i recovery was observed in *NHE1* knockout mouse (Damkier et al. 2009). The luminal membrane location of the NHE1 in CPE dictates that the protein mediates H^+ extrusion from the epithelium to the CSF. Thus, NHE1 may have a role in preventing increases in CSF pH, thereby counteracting efficient HCO_3^- secretion.

The *SLC4A10* gene product NCBE or NBCn2 has been described as a DIDS-sensitive, electroneutral Na^+-HCO_3^- cotransporter driven by the inward Na^+ gradient (Damkier et al. 2010; Giffard et al. 2003; Wang et al. 2000). These authors found that the Na^+-dependent HCO_3^- transport required intracellular Cl^- in mammalian cell lines, and the transporter was therefore called NCBE for Na^+-dependent Cl^-/HCO_3^- exchanger. However, Parker et al. showed that the human *SLC4A10* gene product did not extrude of Cl^- when expressed in *Xenopus laevis* oocytes and protein was renamed the protein NBCn2 by these scientists (Parker et al. 2008). This controversy is not resolved, and in this chapter, we use the rodent name NCBE for simplicity.

The NCBE is a basolateral plasma membrane protein in the choroid plexus (Praetorius and Nielsen 2006; Praetorius et al. 2004b), and it is believed to contribute to both Na^+ and HCO_3^- uptake from the blood side. The involvement in cellular Na^+ and HCO_3^- uptake is supported by a finding that reports 70 % decrease in the Na^+-dependent HCO_3^- import into choroid plexus cells in an NCBE knockout mouse model (Jacobs et al. 2008). In the murine choroid plexus, NCBE activity can account for almost all of the DIDS-sensitive Na^+-dependent HCO_3^- import (Damkier et al. 2009).

The electroneutral Na^+-HCO_3^- cotransporter 1, NBCn1, is also expressed in the choroid plexus epithelium (Praetorius et al. 2004b). It is normally found in the basolateral membrane of mammalian choroid plexus epithelial cells, but it is localized to the luminal membrane in certain mouse strains and regionally in human choroid plexus (Damkier et al. 2009; Praetorius and Nielsen 2006). The epithelial NBCn1 form seems to be DIDS insensitive. In rat choroid plexus, a large fraction of the pH_i recovery from acid load was mediated by DIDS-insensitive Na^+-HCO_3^- cotransport (Bouzinova et al. 2005). However, the NBC activity in choroid plexus from NBCn1 knockout mice is indistinguishable from that of wild-type littermates (Damkier, unpublished results). Thus, NBCn1 protein is not an obvious candidate to participate in transcellular HCO_3^- transport.

AE2 mediates electroneutral uptake of Cl^- in exchange for HCO_3^- and is the only anion exchanger from the *SLC4* family in the choroid plexus. AE2 was first localized in the basolateral membrane of mouse choroid plexus epithelium (Lindsey et al. 1990) and later found at the same site in choroid plexus tissue from other mammals (Alper et al. 1994; Praetorius and Nielsen 2006). AE2 is the only known basolateral entry pathway for Cl^- and may therefore be very important for CSF secretion. Two lines of evidence suggest the implication of basolateral AE in CSF formation. The inward chemical gradient for Cl^- would favour AE2-mediated Cl^- uptake across the basolateral membrane in the choroid plexus. Furthermore, DIDS applied from the blood side greatly reduce Cl^- transport into the CSF (Deng and Johanson 1989; Frankel and Kazemi 1983). The mRNA encoding the *SLC26A7*, *SLC26A10* and *SLC26A11* is expressed in rat and human choroid plexus

(Praetorius, unpublished observations), but the protein expression and a putative function in HCO_3^- transport have not been examined in the choroid plexus.

5.7.4 Model for Bicarbonate Secretion by the Choroid Plexus

A significant body of evidence shows that HCO_3^- transport is an essential element of the CSF secretion, and here, we summarize the most important transporters (Fig. 5.8). CSF secretion is greatly decreased by carbonic anhydrase inhibition (Vogh et al. 1987; Ames et al. 1965; Davson and Segal 1970; Welch 1963), and Na^+/K^+-ATPase activity is diminished in the absence of CO_2/HCO_3^- in amphibian CPE (Saito and Wright 1983). In humans, the intracranial pressure is also reduced by the carbonic anhydrase inhibition in humans (Cowan and Whitelaw 1991). It appears that the majority of the ions secreted to the CSF take a transcellular route. Na^+, Cl, HCO_3^- and Ca^{2+} are secreted into the CSF, while K^+ is reabsorbed. The transport of these ions is carefully regulated to provide a relatively constant CSF composition despite fluctuations in their plasma or BECF concentrations (Husted and Reed 1976; Jones and Keep 1988; Murphy et al. 1986). NCBE may be a good candidate for mediating HCO_3^- uptake from the blood side, while NBCn1 can only assist in cases where this protein is situated in the basolateral membrane. In the choroid plexus as elsewhere, cellular HCO_3^- is also generated from CO_2 and H_2O catalysed by carbonic anhydrases, and inhibition of the carbonic anhydrases by acetazolamide reduced CSF secretion by approximately 50 % (Vogh et al. 1987).

The choroid plexus expresses the cytosolic carbonic anhydrase CAII as well as the membrane-associated isoforms CAXII and CAIX (Kallio et al. 2006). For the actual extrusion of HCO_3^- to the lumen, NBCe2 is most likely key contributor. However, the luminal anion conductances and possibly *SLC26* proteins may also provide efficient HCO_3^- transport across the luminal membrane. Thus, the identification of the major HCO_3^- extrusion pathway in the choroid plexus is a major task for future research in the choroid plexus function.

5.7.5 Regulation of CP Bicarbonate Secretion

The control and regulation of choroid plexus secretion of HCO_3^- and CSF is a topic of major clinical importance. Several life-threatening conditions resulting in an increased intracranial pressure, such as head trauma, tumours, stroke and irradiation damage, can potentially be alleviated by acute reduction of CSF secretion. The CSF contains very little protein buffers, and appropriate pH level is most likely kept solely by the CO_2/HCO_3^- buffer system. Adjustments of the HCO_3^- extrusion from the choroid plexus can potentially control deviations of CSF pH from neutral values. However, virtually nothing is known about the potential acid/base sensing.

Arginine vasopressin (AVP or antidiuretic hormone ADH) decreases the CSF secretion and reduces the blood flow to the choroid plexus (Johanson et al. 1999). The choroid plexus expresses V1a receptors (Ostrowski et al. 1992; Phillips et al. 1988). AVP is actually produced by CPE stimulated by local angiotensin II through AT1 receptor-mediated activation. The activation of V1a receptor in the luminal membrane decreases CSF secretion by the CPE (Szmydynger-Chodobska et al. 2004). Accordingly, Ca^{2+} increases transiently in the choroid plexus epithelium upon vasopressin administration (Battle et al. 2000), without affecting cAMP levels (Crook et al. 1984). A local system of renin, angiotensinogen and angiotensin-converting enzyme (ACE) and AT1 receptors is expressed in the choroid plexus epithelium (Chai et al. 1987a, b; Gehlert et al. 1986; Imboden et al. 1987; Inagami et al. 1980). However, the significance of this local RAS system in the choroid plexus is unclear apart from increasing AVP production. The mineralocorticoid and glucocorticoid receptors are expressed in the choroid plexus (de Kloet et al. 2000; Weber et al. 2003). Plasma glucocorticoid levels usually greatly exceed aldosterone levels, and in aldosterone sensitive cells, the specificity of MR receptor activation is assured by intracellular conversion of cortisol to inactive metabolites by enzyme 11β-hydroxysteroid dehydrogenase type 2 (11βHSD2). Importantly, the choroid plexus expresses 11βHSD type 1 instead of 11βHSD2 (Sinclair et al. 2007), and it is therefore unlikely that aldosterone exerts specific MR-mediated actions on the choroid plexus epithelium.

Regarding intracellular signalling, it is unknown how Ca^{2+} inhibits CSF secretion by the choroid plexus. There are substantial numbers of the expressed transport mechanisms that are activated by cAMP and/or PKA. However, cAMP and/or PKA induction has only been shown for isoproterenol, prostaglandin, serotonin and histamine (Crook et al. 1984). Finally, it is noted that the choroid plexus also expresses receptors for atrial natriuretic factor, endothelin-1, serotonin, bradykinin and insulin, all of which could affect HCO_3^- secretion (Chodobski and Szmydynger-Chodobska 2001).

5.8 Conclusions and Perspectives

The importance of HCO_3^- secretion is clearly demonstrated by the large selection of ion transporters and channels involved in the process and the fact that mutations or dysregulation of these can lead to a large spectrum of diseases. In this chapter, we have covered a wide selection of, but not all, epithelia that secrete HCO_3^-. In leaky epithelia, HCO_3^- secretion is accompanied by fluid secretion, while in tight epithelia (such as salivary gland ducts and CCDs), HCO_3^- is usually exchanged for another anion. One of the analyses made in this chapter is that there is a great spectrum of secreted HCO_3^- concentrations, secretion rates and therefore HCO_3^- outputs, which range from 1 to 1000 μmol/h/g tissue or organ. These values could be more revealing if corrected for the mass of cells actually performing HCO_3^- secretion in the given organ. Nevertheless, it should be pointed out that HCO_3^-

secretion is usually only a part of a complex function any given epithelium performs, and it is regulated by specific neurotransmitters/hormones and locally generated agents or agents from neighbouring organs or tissues. Therefore, it is not unexpected to see that there are many molecularly different transporters and constellations for bicarbonate-secreting epithelia presented in this chapter. Nevertheless, we seek to deduce whether there are any common mechanisms for HCO_3^- secretion. It appears that the basolateral membranes of epithelia express one or more members of the NBC transporter family and/or NHEs and intracellular CA generating H^+ and HCO_3^- from CO_2. The HCO_3^- exit on the luminal membrane is little more problematic. It seems that Cl^- channels in combination with HCO_3^-/Cl^- exchangers (*SLC26* family), especially the electrogenic type of exchanger, are favoured and suffice for many epithelia. Also permeability of Cl^- channels to HCO_3^- has been proposed. In many epithelia, electrogenic NBCs have been proposed, but functional evidence for HCO_3^- secretion is missing. Nevertheless, transport mechanism producing fluid with more than 100 mM HCO_3^- concentrations is still elusive, and we are missing information about the electrochemical gradients (membrane voltages, intracellular anion concentrations) and CO_2 permeability, and the question still remains whether passive anion transporters are sufficient to explain HCO_3^- exit across the luminal membrane. Key components in the secretory models are also various Cl^- and K^+ channels that can be stimulated by agonists to open and provide accompanying ions and appropriate membrane voltages.

The most potent HCO_3^- secretors, such as the pancreas, some salivary glands and duodenum, could potentially affect body acid/base balance. Therefore, it is not surprising that these epithelia are very finely regulated and perform only periodically when stimulated, and at low secretion rates, HCO_3^- is reabsorbed or salvaged. This requires HCO_3^--reabsorbing mechanism, usually localized at the distal part of the complex structures of glands or gastrointestinal tract. These mechanisms could include certain isoforms of NBC transporters or NHEs and H^+ transporters.

The past 10–20 years of research has led to molecular identification of many transport proteins involved in bicarbonate secretion. Future challenges are to elucidate pH and CO_2 sensors and the effect of local regulating agents and to clarify which acid/base transporters are involved in pH_i regulation as opposed to pH_e regulation carried out by HCO_3^- secretion. Most importantly, studies of bicarbonate secretion and regulation on more integrated organ and whole-body settings could provide some important answers.

Acknowledgements Research projects founding basis for this chapter were supported by the Danish Council for Independent Research |Natural Sciences, the Danish Council for Independent Research |Medical Sciences, the Lundbeck Foundation, the Novo Nordisk Foundation and the Carlsberg Foundation.

References

Aalkjaer C, Cragoe EJ Jr (1988) Intracellular pH regulation in resting and contracting segments of rat mesenteric resistance vessels. J Physiol 402:391–410

Abuladze N, Lee I, Newman D, Hwang J, Boorer K, Pushkin A, Kurtz I (1998) Molecular cloning, chromosomal localization, tissue distribution, and functional expression of the human pancreatic sodium bicarbonate cotransporter. J Biol Chem 273:17689–95

Abuladze N, Pushkin A, Tatishchev S, Newman D, Sassani P, Kurtz I (2004) Expression and localization of rat NBC4c in liver and renal uroepithelium. Am J Physiol Cell Physiol 287: C781–9

Ahn W, Kim KH, Lee JA, Kim JY, Choi JY, Moe OW, Milgram SL, Muallem S, Lee MG (2001) Regulatory interaction between the cystic fibrosis transmembrane conductance regulator and HCO_3^- salvage mechanism in model systems and the mouse pancreatic duct. J Biol Chem 276:17236–43

Ahuja M, Jha A, Maleth J, Park S, Muallem S (2014) cAMP and Ca signaling in secretory epithelia: crosstalk and synergism. Cell Calcium 55:385–93

Ainsworth MA, Kjeldsen J, Olsen O, Christensen P, Schaffalitzky de Muckadell OB (1990) Duodenal disappearance rate of acid during inhibition of mucosal bicarbonate secretion. Digestion 47:121–9

Ainsworth MA, Svendsen P, Ladegaard L, Cantor P, Olsen O, Schaffalitzky de Muckadell OB (1992) Relative importance of pancreatic, hepatic, and mucosal bicarbonate in duodenal neutralization of acid in anaesthetized pigs. Scand J Gastroenterol 27:343–9

Ainsworth MA, Amelsberg M, Hogan DL, Isenberg JI (1996) Acid–base transport in isolated rabbit duodenal villus and crypt cells. Scand J Gastroenterol 31:1069–77

Ainsworth MA, Hogan DL, Rapier RC, Amelsberg M, Dreilinger AD, Isenberg JI (1998) Acid/base transporters in human duodenal enterocytes. Scand J Gastroenterol 33:1039–46

Akiba Y, Furukawa O, Guth PH, Engel E, Nastaskin I, Kaunitz JD (2001a) Acute adaptive cellular base uptake in rat duodenal epithelium. Am J Physiol Gastrointest Liver Physiol 280:G1083–92

Akiba Y, Furukawa O, Guth PH, Engel E, Nastaskin I, Sassani P, Dukkipatis R, Pushkin A, Kurtz I, Kaunitz JD (2001b) Cellular bicarbonate protects rat duodenal mucosa from acid-induced injury. J Clin Invest 108:1807–16

Al-Awqati Q (2013) Cell biology of the intercalated cell in the kidney. FEBS Lett 587:1911–4

Allen A, Flemstrom G (2005) Gastroduodenal mucus bicarbonate barrier: protection against acid and pepsin. Am J Physiol Cell Physiol 288:C1–19

Almassy J, Won JH, Begenisich TB, Yule DI (2012) Apical Ca^{2+}-activated potassium channels in mouse parotid acinar cells. J Gen Physiol 139:121–33

Alper SL (2006) Molecular physiology of SLC4 anion exchangers. Exp Physiol 91:153–61

Alper SL, Sharma AK (2013) The SLC26 gene family of anion transporters and channels. Mol Aspects Med 34:494–515

Alper SL, Natale J, Gluck S, Lodish HF, Brown D (1989) Subtypes of intercalated cells in rat kidney collecting duct defined by antibodies against erythroid band 3 and renal vacuolar H^+-ATPase. Proc Natl Acad Sci U S A 86:5429–33

Alper SL, Stuart-Tilley A, Simmons CF, Brown D, Drenckhahn D (1994) The fodrin-ankyrin cytoskeleton of choroid plexus preferentially colocalizes with apical Na^+K^+-ATPase rather than with basolateral anion exchanger AE2. J Clin Invest 93:1430–8

Alper SL, Stewart AK, Vandorpe DH, Clark JS, Horack RZ, Simpson JE, Walker NM, Clarke LL (2011) Native and recombinant *Slc26a3* (downregulated in adenoma, Dra) do not exhibit properties of 2Cl-/1HCO3- exchange. Am J Physiol Cell Physiol 300:C276–86

Alpini G, Glaser S, Robertson W, Rodgers RE, Phinizy JL, Lasater J, LeSage GD (1997) Large but not small intrahepatic bile ducts are involved in secretin-regulated ductal bile secretion. Am J Physiol Gastrointest Liver Physiol 272:G1064–74

Alvarez C, Fasano A, Bass BL (1998) Acute effects of bile acids on the pancreatic duct epithelium in vitro. J Surg Res 74:43–6

Alvarez L, Fanjul M, Carter N, Hollande E (2001) Carbonic anhydrase II associated with plasma membrane in a human pancreatic duct cell line (CAPAN-1). J Histochem Cytochem 49:1045–53

Alvarez BV, Loiselle FB, Supuran CT, Schwartz GJ, Casey JR (2003) Direct extracellular interaction between carbonic anhydrase IV and the human NBC1 sodium/bicarbonate co-transporter. Biochemistry 42:12321–9

Alvaro D, Della GP, Bini A, Gigliozzi A, Furfaro S, La RT, Piat C, Capocaccia L (1995) Effect of glucagon on intracellular pH regulation in isolated rat hepatocyte couplets. J Clin Invest 96:665–75

Ambudkar IS (2012) Polarization of calcium signaling and fluid secretion in salivary gland cells. Curr Med Chem 19:5774–81

Ames A III, Sakanoue M, Endo S (1964) Na, K, Ca, Mg, and Cl concentrations in choroid plexus fluid and cisternal fluid compared with plasma ultrafiltrate. J Neurophysiol 27:672–81

Ames A III, Higashi K, Nesbett FB (1965) Effects of P_{CO2} acetazolamide and ouabain on volume and composition of choroid-plexus fluid. J Physiol 181:516–24

Amlal H, Petrovic S, Xu J, Wang Z, Sun X, Barone S, Soleimani M (2010) Deletion of the anion exchanger *Slc26a4* (pendrin) decreases apical Cl^-/HCO_3^- exchanger activity and impairs bicarbonate secretion in kidney collecting duct. Am J Physiol Cell Physiol 299:C33–41

Andersen DH (1938) Cystic fibrosis of the pancreas and its relation to celiac disease: a clinical and pathological study. Am J Dis Child 56:344–99

Aoi M, Aihara E, Nakashima M, Takeuchi K (2004) Participation of prostaglandin E receptor EP4 subtype in duodenal bicarbonate secretion in rats. Am J Physiol Gastrointest Liver Physiol 287: G96–103

Aranda V, Martinez I, Melero S, Lecanda J, Banales JM, Prieto J, Medina JF (2004) Shared apical sorting of anion exchanger isoforms AE2a, AE2b1, and AE2b2 in primary hepatocytes. Biochem Biophys Res Commun 319:1040–6

Ashizawa N, Endoh H, Hidaka K, Watanabe M, Fukumoto S (1997) Three-dimensional structure of the rat pancreatic duct in normal and inflammated pancreas. Microsc Res Tech 37:543–56

Azroyan A, Laghmani K, Crambert G, Mordasini D, Doucet A, Edwards A (2011) Regulation of pendrin by pH: dependence on glycosylation. Biochem J 434:61–72

Bairamian D, Johanson CE, Parmelee JT, Epstein MH (1991) Potassium cotransport with sodium and chloride in the choroid plexus. J Neurochem 56:1623–9

Banales JM, Arenas F, Rodriguez-Ortigosa CM, Saez E, Uriarte I, Doctor RB, Prieto J, Medina JF (2006a) Bicarbonate-rich choleresis induced by secretin in normal rat is taurocholate-dependent and involves AE2 anion exchanger. Hepatology 43:266–75

Banales JM, Prieto J, Medina JF (2006b) Cholangiocyte anion exchange and biliary bicarbonate excretion. World J Gastroenterol 12:3496–511

Bardow A, Madsen J, Nauntofte B (2000) The bicarbonate concentration in human saliva does not exceed the plasma level under normal physiological conditions. Clin Oral Investig 4:245–53

Battle T, Preisser L, Marteau V, Meduri G, Lambert M, Nitschke R, Brown PD, Corman B (2000) Vasopressin V1a receptor signaling in a rat choroid plexus cell line. Biochem Biophys Res Commun 275:322–7

Beal AM (1984) Electrolyte composition of parotid saliva from sodium-replete red kangaroos (*Macropus rufus*). J Exp Biol 111:225–37

Begenisich T, Nakamoto T, Ovitt CE, Nehrke K, Brugnara C, Alper SL, Melvin JE (2004) Physiological roles of the intermediate conductance, Ca^{2+}-activated potassium channel Kcnn4. J Biol Chem 279:47681–7

Behrendorff N, Floetenmeyer M, Schwiening C, Thorn P (2010) Protons released during pancreatic acinar cell secretion acidify the lumen and contribute to pancreatitis in mice. Gastroenterology 139:1711–20

Bergmann F, Andrulis M, Hartwig W, Penzel R, Gaida MM, Herpel E, Schirmacher P, Mechtersheimer G (2011) Discovered on gastrointestinal stromal tumor 1 (DOG1) is expressed in pancreatic centroacinar cells and in solid-pseudopapillary neoplasms--novel evidence for a histogenetic relationship. Hum Pathol 42:817–23

Berridge MJ (2012) Calcium signalling remodelling and disease. Biochem Soc Trans 40:297–309

Beuers U, Hohenester S, de Buy Wenniger LJ, Kremer AE, Jansen PL, Elferink RP (2010) The biliary HCO_3^- umbrella: a unifying hypothesis on pathogenetic and therapeutic aspects of fibrosing cholangiopathies. Hepatology 52:1489–96

Beuers U, Maroni L, Elferink RO (2012) The biliary HCO_3^- umbrella: experimental evidence revisited. Curr Opin Gastroenterol 28:253–7

Bhattacharya S, Verrill DS, Carbone KM, Brown S, Yule DI, Giovannucci DR (2012) Distinct contributions by ionotropic purinoceptor subtypes to ATP-evoked calcium signals in mouse parotid acinar cells. J Physiol 590:2721–37

Bijman J, Cook DI, van Os CH (1981) Effect of amiloride on electrolyte transport parameters of the main duct of the rabbit mandibular gland. Pflugers Arch 398:96–102

Bilbao PS, Katz S, Boland R (2012) Interaction of purinergic receptors with GPCRs, ion channels, tyrosine kinase and steroid hormone receptors orchestrates cell function. Purinergic Signal 8:91–103

Billet A, Hanrahan JW (2013) The secret life of CFTR as a calcium-activated chloride channel. J Physiol 591:5273–8

Billet A, Luo Y, Balghi H, Hanrahan JW (2013) Role of tyrosine phosphorylation in the muscarinic activation of the cystic fibrosis transmembrane conductance regulator (CFTR). J Biol Chem 288:21815–23

Boedtkjer E, Praetorius J, Aalkjaer C (2006) NBCn1 (slc4a7) Mediates the Na^+-dependent bicarbonate transport important for regulation of intracellular pH in mouse vascular smooth muscle cells. Circ Res 98:515–23

Boron WF (2010) Evaluating the role of carbonic anhydrases in the transport of HCO_3^--related species. Biochim Biophys Acta 1804:410–21

Boron WF, Boulpaep EL (1983) Intracellular pH regulation in the renal proximal tubule of the salamander. Basolateral HCO_3^- transport. J Gen Physiol 81:53–94

Boron WF, Boulpaep EL (2012) Medical physiology. Saunders Elsevier, Philadelphia

Boron WF, Knakal RC (1992) Na^+-dependent Cl-HCO_3 exchange in the squid axon. Dependence on extracellular pH. J Gen Physiol 99:817–37

Bouwens L, Pipeleers DG (1998) Extra-insular beta cells associated with ductules are frequent in adult human pancreas. Diabetologia 41:629–33

Bouzinova EV, Praetorius J, Virkki LV, Nielsen S, Boron WF, Aalkjaer C (2005) Na^+-dependent HCO_3^- uptake into the rat choroid plexus epithelium is partially DIDS sensitive. Am J Physiol Cell Physiol 289:C1448–56

Boyer JL (2013) Bile formation and secretion. Compr Physiol 3:1035–78

Breton S, Brown D (2013) Regulation of luminal acidification by the V-ATPase. Physiology (Bethesda) 28:318–29

Breton S, Alper SL, Gluck SL, Sly WS, Barker JE, Brown D (1995) Depletion of intercalated cells from collecting ducts of carbonic anhydrase II-deficient (CAR2 null) mice. Am J Physiol 269: F761–74

Bro-Rasmussen F, Killmann SA, Thaysen JH (1956) The composition of pancreatic juice as compared to sweat, parotid saliva and tears. Acta Physiol Scand 37:97–113

Brown D, Hirsch S, Gluck S (1988a) An H^+-ATPase in opposite plasma membrane domains in kidney epithelial cell subpopulations. Nature 331:622–4

Brown PD, Loo DD, Wright EM (1988b) Ca^{2+}-activated K^+ channels in the apical membrane of Necturus choroid plexus. J Membr Biol 105:207–19

Brown D, Paunescu TG, Breton S, Marshansky V (2009) Regulation of the V-ATPase in kidney epithelial cells: dual role in acid–base homeostasis and vesicle trafficking. J Exp Biol 212:1762–72

Bruce JI, Shuttleworth TJ, Giovannucci DR, Yule DI (2002) Phosphorylation of inositol 1,4,5-trisphosphate receptors in parotid acinar cells. A mechanism for the synergistic effects of cAMP on Ca^{2+} signaling. J Biol Chem 277:1340–8
Burgen ASV (1956) The secretion of potassium in saliva. J Physiol Lond 132:20–39
Burghardt B, Elkaer ML, Kwon TH, Racz GZ, Varga G, Steward MC, Nielsen S (2003) Distribution of aquaporin water channels AQP1 and AQP5 in the ductal system of the human pancreas. Gut 52:1008–16
Burghardt B, Nielsen S, Steward MC (2006) The role of aquaporin water channels in fluid secretion by the exocrine pancreas. J Membr Biol 210:143–53
Burnstock G (2007) Physiology and pathophysiology of purinergic neurotransmission. Physiol Rev 87:659–797
Caflisch CR, Solomon S, Galey WR (1979) Exocrine ductal pCO_2 in the rabbit pancreas. Pflugers Arch 380:121–5
Caputo A, Caci E, Ferrera L, Pedemonte N, Barsanti C, Sondo E, Pfeffer U, Ravazzolo R, Zegarra-Moran O, Galietta LJ (2008) TMEM16A, a membrane protein associated with calcium-dependent chloride channel activity. Science 322:590–4
Case RM, Harper AA, Scratcherd T (1969) The secretion of electrolytes and enzymes by the pancreas of the anaesthetized cat. J Physiol Lond 201:335–48
Case RM, Scratcherd T, Wynne RDA (1970) The origin and secretion of pancreatic juice bicarbonate. J Physiol Lond 210:1–15
Case RM, Hotz J, Hutson D, Scratcherd T, Wynne RDA (1979) Electrolyte secretion by the isolated cat pancreas during replacement of extracellular bicarbonate by organic anions and chloride by inorganic anions. J Physiol Lond 286:563–76
Case RM, Conigrave AD, Novak I, Young JA (1980) Electrolyte and protein secretion by the perfused rabbit mandibular gland stimulated with acetylcholine or catecholamines. J Physiol Lond 300:467–87
Case RM, Conigrave AD, Favaloro EJ, Novak I, Thompson CH, Young JA (1982) The role of buffer anions and protons in secretion by the rabbit mandibular salivary gland. J Physiol Lond 322:273–86
Case RM, Hunter M, Novak I, Young JA (1984) The anionic basis of fluid secretion by the rabbit mandibular gland. J Physiol Lond 349:619–30
Catalan MA, Nakamoto T, Gonzalez-Begne M, Camden JM, Wall SM, Clarke LL, Melvin JE (2010) Cftr and ENaC ion channels mediate NaCl absorption in the mouse submandibular gland. J Physiol 588:713–24
Chai SY, McKinley MJ, Mendelsohn FA (1987a) Distribution of angiotensin converting enzyme in sheep hypothalamus and medulla oblongata visualized by in vitro autoradiography. Clin Exp Hypertens A 9:449–60
Chai SY, Mendelsohn FA, Paxinos G (1987b) Angiotensin converting enzyme in rat brain visualized by quantitative in vitro autoradiography. Neuroscience 20:615–27
Chambrey R, Kurth I, Peti-Peterdi J, Houillier P, Purkerson JM, Leviel F, Hentschke M, Zdebik AA, Schwartz GJ, Hubner CA, Eladari D (2013) Renal intercalated cells are rather energized by a proton than a sodium pump. Proc Natl Acad Sci U S A 110:7928–33
Chan HC, Cheung WT, Leung PY, Wu LJ, Chew SB, Ko WH, Wong PY (1996) Purinergic regulation of anion secretion by cystic fibrosis pancreatic duct cells. Am J Physiol Cell Physiol 271:C469–77
Chari RS, Schutz SM, Haebig JE, Shimokura GH, Cotton PB, Fitz JG, Meyers WC (1996) Adenosine nucleotides in bile. Am J Physiol Gastrointest Liver Physiol 270:G246–52
Chaturapanich G, Ishibashi H, Dinudom A, Young JA, Cook DI (1997) H^+ transporters in the main excretory duct of the mouse mandibular salivary gland. J Physiol 503:583–98
Chen M, Praetorius J, Zheng W, Xiao F, Riederer B, Singh AK, Stieger N, Wang J, Shull GE, Aalkjaer C, Seidler U (2012) The electroneutral Na^+:HCO_3^- cotransporter NBCn1 is a major pHi regulator in murine duodenum. J Physiol 590:3317–33

Chenevert J, Duvvuri U, Chiosea S, Dacic S, Cieply K, Kim J, Shiwarski D, Seethala RR (2012) DOG1: a novel marker of salivary acinar and intercalated duct differentiation. Mod Pathol 25:919–29

Cheng HS, Leung PY, Cheng Chew SB, Leung PS, Lam SY, Wong WS, Wang ZD, Chan HC (1998) Concurrent and independent HCO_3^- and Cl^- secretion in a human pancreatic duct cell line (CAPAN-1). J Membr Biol 164:155–67

Chey WY, Chang T (2001) Neural hormonal regulation of exocrine pancreatic secretion. Pancreatology 1:320–35

Choate KA, Kahle KT, Wilson FH, Nelson-Williams C, Lifton RP (2003) WNK1, a kinase mutated in inherited hypertension with hyperkalemia, localizes to diverse Cl^--transporting epithelia. Proc Natl Acad Sci U S A 100:663–8

Chodobski A, Szmydynger-Chodobska J (2001) Choroid plexus: target for polypeptides and site of their synthesis. Microsc Res Tech 52:65–82

Choi I, Romero MF, Khandoudi N, Bril A, Boron WF (1999) Cloning and characterization of a human electrogenic Na^+-HCO_3^- cotransporter isoform (hhNBC). Am J Physiol Cell Physiol 276:C576–84

Choi I, Aalkjaer C, Boulpaep EL, Boron WF (2000) An electroneutral sodium/bicarbonate cotransporter NBCn1 and associated sodium channel. Nature 405:571–5

Christensen O, Zeuthen T (1987) Maxi K^+ channels in leaky epithelia are regulated by intracellular Ca^{2+}, pH and membrane potential. Pflugers Arch 408:249–59

Christensen BM, Perrier R, Wang Q, Zuber AM, Maillard M, Mordasini D, Malsure S, Ronzaud C, Stehle JC, Rossier BC, Hummler E (2010) Sodium and potassium balance depends on αENaC expression in connecting tubule. J Am Soc Nephrol 21:1942–51

Clarke LL, Harline MC (1998) Dual role of CFTR in cAMP-stimulated HCO_3^- secretion across murine duodenum. Am J Physiol Gastrointest Liver Physiol 274:G718–26

Cohn JA, Strong TV, Picciotto MR, Nairn AC, Collins FS, Fitz JG (1993) Localization of the cystic fibrosis transmembrane conductance regulator in human bile duct epithelial cells. Gastroenterology 105:1857–64

Compton JS, Nelson J, Wright RD, Young JA (1980) A micropuncture investigation of electrolyte transport in the parotid glands of sodium-replete and sodium-deplete sheep. J Physiol Lond 309:429–46

Concepcion AR, Lopez M, Ardura-Fabregat A, Medina JF (2013) Role of AE2 for pH regulation in biliary epithelial cells. Front Physiol 4:413. doi:10.3389/fphys.2013.00413. eCollection 2013

Cook DI, Young JA (1989) Effect of K^+ channels in the apical plasma membrane on epithelial secretion based on secondary active Cl^- transport. J Membr Biol 110:139–46

Cook DI, Dinudom A, Komwatana P, Kumar S, Young JA (2002) Patch-clamp studies on epithelial sodium channels in salivary duct cells. Cell Biochem Biophys 36:105–13

Cowan F, Whitelaw A (1991) Acute effects of acetazolamide on cerebral blood flow velocity and pCO2 in the newborn infant. Acta Paediatr Scand 80:22–7

Crook RB, Farber MB, Prusiner SB (1984) Hormones and neurotransmitters control cyclic AMP metabolism in choroid plexus epithelial cells. J Neurochem 42:340–50

Cushing H (1914) Studies on the cerebrospinal fluid. J Med Res 26:1–19

Czako L, Yamamoto M, Otsuki M (1997) Exocrine pancreatic function in rats after acute pancreatitis. Pancreas 15:83–90

Damkier HH, Nielsen S, Praetorius J (2006) An anti-NH2-terminal antibody localizes NBCn1 to heart endothelia and skeletal and vascular smooth muscle cells. Am J Physiol Heart Circ Physiol 290:H172–80

Damkier HH, Nielsen S, Praetorius J (2007) Molecular expression of SLC4-derived Na^+-dependent anion transporters in selected human tissues. Am J Physiol Regul Integr Comp Physiol 293: R2136–46

Damkier HH, Prasad V, Hubner CA, Praetorius J (2009) Nhe1 is a luminal Na^+/H^+ exchanger in mouse choroid plexus and is targeted to the basolateral membrane in Ncbe/Nbcn2-null mice. Am J Physiol Cell Physiol 296:C1291–300
Damkier HH, Aalkjaer C, Praetorius J (2010) Na^+-dependent HCO_3^- import by the slc4a10 gene product involves Cl^- export. J Biol Chem 285:26998–7007
Damkier HH, Brown PD, Praetorius J (2013) Cerebrospinal fluid secretion by the choroid plexus. Physiol Rev 93:1847–92
Davson H, Purvis C (1954) Cryoscopic apparatus suitable for studies on aqueous humour and cerebro-spinal fluid. J Physiol 124:12–3P
Davson H, Segal MB (1970) The effects of some inhibitors and accelerators of sodium transport on the turnover of ^{22}Na in the cerebrospinal fluid and the brain. J Physiol 209:131–53
Davson H, Segal MB (1996) Physiology of the CSF and blood–brain barriers. CRC Press, Boca Raton
de Kloet ER, Van Acker SA, Sibug RM, Oitzl MS, Meijer OC, Rahmouni K, de Jong W (2000) Brain mineralocorticoid receptors and centrally regulated functions. Kidney Int 57:1329–36
de Ondarza J, Hootman SR (1997) Confocal microscopic analysis of intracellular pH regulation in isolated guinea pig pancreatic ducts. Am J Physiol Gastrointest Liver Physiol 272:G124–34
Demeter I, Hegyesi O, Nagy AK, Case MR, Steward MC, Varga G, Burghardt B (2009) Bicarbonate transport by the human pancreatic ductal cell line HPAF. Pancreas 38:913–20
Demitrack ES, Soleimani M, Montrose MH (2010) Damage to the gastric epithelium activates cellular bicarbonate secretion via SLC26A9 Cl^-/HCO_3^-. Am J Physiol Gastrointest Liver Physiol 299:G255–64
Deng QS, Johanson CE (1989) Stilbenes inhibit exchange of chloride between blood, choroid plexus and cerebrospinal fluid. Brain Res 501:183–7
Dinudom A, Komwatana P, Young JA, Cook DI (1995) A forskolin-activated Cl^- current in mouse mandibular duct cells. Am J Physiol Gastrointest Liver Physiol 268:G806–12
Dinudom A, Harvey KF, Komwatana P, Young JA, Kumar S, Cook DI (1998) Nedd4 mediates control of an epithelial Na^+ channel in salivary duct cells by cytosolic Na^+. Proc Natl Acad Sci U S A 95:7169–73
Dinudom A, Harvey KF, Komwatana P, Jolliffe CN, Young JA, Kumar S, Cook DI (2001) Roles of the C termini of α -, β -, and γ -subunits of epithelial Na^+ channels (ENaC) in regulating ENaC and mediating its inhibition by cytosolic Na^+. J Biol Chem 276:13744–9
Dissing S, Hansen HJ, Undén M, Nauntofte B (1990) Inhibitory effects of amitriptyline on the stimulation-induced Ca^{2+} increase in parotid acini. Eur J Pharmacol 177:43–54
Doctor RB, Matzakos T, McWilliams R, Johnson S, Feranchak AP, Fitz JG (2005) Purinergic regulation of cholangiocyte secretion: identification of a novel role for P2X receptors. Am J Physiol Gastrointest Liver Physiol 288:G779–86
Doring F, Derst C, Wischmeyer E, Karschin C, Schneggenburger R, Daut J, Karschin A (1998) The epithelial inward rectifier channel Kir7.1 displays unusual K^+ permeation properties. J Neurosci 18:8625–36
Dorwart MR, Shcheynikov N, Yang D, Muallem S (2008) The solute carrier 26 family of proteins in epithelial ion transport. Physiology (Bethesda) 23:104–14
Dranoff JA, Masyuk AI, Kruglov EA, LaRusso NF, Nathanson MH (2001) Polarized expression and function of P2Y ATP receptors in rat bile duct epithelia. Am J Physiol Gastrointest Liver Physiol 281:G1059–67
Duran C, Thompson CH, Xiao Q, Hartzell HC (2010) Chloride channels: often enigmatic, rarely predictable. Annu Rev Physiol 72:95–121
Dutta AK, Khimji AK, Sathe M, Kresge C, Parameswara V, Esser V, Rockey DC, Feranchak AP (2009) Identification and functional characterization of the intermediate-conductance Ca^{2+}-activated K^+ channel (IK-1) in biliary epithelium. Am J Physiol Gastrointest Liver Physiol 297: G1009–18
Dutta AK, Khimji AK, Kresge C, Bugde A, Dougherty M, Esser V, Ueno Y, Glaser SS, Alpini G, Rockey DC, Feranchak AP (2011) Identification and functional characterization of

TMEM16A, a Ca^{2+}-activated Cl^- channel activated by extracellular nucleotides, in biliary epithelium. J Biol Chem 286:766–76

Dutta AK, Woo K, Khimji AK, Kresge C, Feranchak AP (2013) Mechanosensitive Cl^- secretion in biliary epithelium mediated through TMEM16A. Am J Physiol Gastrointest Liver Physiol 304:G87–98

Ernst SA, Palacios JR, Siegel GJ (1986) Immunocytochemical localization of Na+, K + -ATPase catalytic polypeptide in mouse choroid plexus. J Histochem Cytochem 34:189–95

Esteller A (2008) Physiology of bile secretion. World J Gastroenterol 14:5641–9

Evans RL, Ashton N, Elliott AC, Green R, Argent BE (1996) Interaction between secretin and acetylcholine in the regulation of fluid secretion by isolated rat pancreatic ducts. J Physiol Lond 461:265–73

Evans RL, Park K, Turner RJ, Watson GE, Nguyen HV, Dennett MR, Hand AR, Flagella M, Shull GE, Melvin JE (2000) Severe impairment of salivation in $Na^+/K^+/2Cl^-$ cotransporter (NKCC1)-deficient mice. J Biol Chem 275:26720–6

Faivre J (1854) Structure du conarium et des plexus choro‹de chez l'hommes et des animaux. Gaz Med Paris 9:555–6

Fanjul M, Alvarez L, Salvador C, Gmyr V, Kerr-Conte J, Pattou F, Carter N, Hollande E (2004) Evidence for a membrane carbonic anhydrase IV anchored by its C-terminal peptide in normal human pancreatic ductal cells. Histochem Cell Biol 121:91–9

Fausther M, Sevigny J (2011) Extracellular nucleosides and nucleotides regulate liver functions via a complex system of membrane proteins. C R Biol 334:100–17

Fellenius E, Berglindh T, Sachs G, Olbe L, Elander B, Sjostrand SE, Wallmark B (1981) Substituted benzimidazoles inhibit gastric acid secretion by blocking (H^++K^+) ATPase. Nature 290:159–61

Feranchak AP, Doctor RB, Troetsch M, Brookman K, Johnson SM, Fitz JG (2004) Calcium-dependent regulation of secretion in biliary epithelial cells: the role of apamin-sensitive SK channels. Gastroenterology 127:903–13

Feranchak AP, Lewis MA, Kresge C, Sathe M, Bugde A, Luby-Phelps K, Antich PP, Fitz JG (2010) Initiation of purinergic signaling by exocytosis of ATP-containing vesicles in liver epithelium. J Biol Chem 285:8138–47

Fernandez-Salazar MP, Pascua P, Calvo JJ, Lopez MA, Case RM, Steward MC, San Roman JI (2004) Basolateral anion transport mechanisms underlying fluid secretion by mouse, rat and guinea-pig pancreatic ducts. J Physiol Lond 556:415–28

Feschenko MS, Donnet C, Wetzel RK, Asinovski NK, Jones LR, Sweadner KJ (2003) Phospholemman, a single-span membrane protein, is an accessory protein of Na, K-ATPase in cerebellum and choroid plexus. J Neurosci 23:2161–9

Fiorotto R, Spirli C, Fabris L, Cadamuro M, Okolicsanyi L, Strazzabosco M (2007) Ursodeoxycholic acid stimulates cholangiocyte fluid secretion in mice via CFTR-dependent ATP secretion. Gastroenterology 133:1603–13

Fitz JG, Persico M, Scharschmidt BF (1989) Electrophysiological evidence for Na^+-coupled bicarbonate transport in cultured rat hepatocytes. Am J Physiol Gastrointest Liver Physiol 256:G491–500

Fitz JG, Basavappa S, McGill J, Melhus O, Cohn JA (1993) Regulation of membrane chloride currents in rat bile duct epithelial cells. J Clin Invest 91:319–28

Flemstrom G, Kivilaakso E (1983) Demonstration of a pH gradient at the luminal surface of rat duodenum in vivo and its dependence on mucosal alkaline secretion. Gastroenterology 84:787–94

Flemstrom G, Garner A, Nylander O, Hurst BC, Heylings JR (1982) Surface epithelial HCO_3^- transport by mammalian duodenum in vivo. Am J Physiol Gastrointest Liver Physiol 243: G348–58

Fong P, Argent BE, Guggino WB, Gray MA (2003) Characterization of vectorial chloride transport pathways in the human pancreatic duct adenocarcinoma cell line, HPAF. Am J Physiol Cell Physiol 285:C433–45

Forte JG, Zhu L (2010) Apical recycling of the gastric parietal cell H, K-ATPase. Annu Rev Physiol 72:273–96

Foskett JK (1990) $[Ca^{2+}]_i$ modulation of Cl^- content controls cell volume in single salivary acinar cells during fluid secretion. Am J Physiol Cell Physiol 259:C998–1004

Frankel H, Kazemi H (1983) Regulation of CSF composition--blocking chloride-bicarbonate exchange. J Appl Physiol Respir Environ Exerc Physiol 55:177–82

Freedman SD, Scheele GA (1994) Acid–base interactions during exocrine pancreatic secretion. Primary role for ductal bicarbonate in acinar lumen function. Ann N Y Acad Sci 713:199–206

Freedman SD, Kern HF, Scheele GA (2001) Pancreatic acinar cell dysfunction in CFTR−/− mice is associated with impairments in luminal pH and endocytosis. Gastroenterology 121:950–7

Frische S, Kwon TH, Frokiaer J, Madsen KM, Nielsen S (2003) Regulated expression of pendrin in rat kidney in response to chronic NH_4Cl or $NaHCO_3$ loading. Am J Physiol Renal Physiol 284: F584–93

Frizzell RA, Hanrahan JW (2012) Physiology of epithelial chloride and fluid secretion. Cold Spring Harb Perspect Med 2:a009563

Gao X, Eladari D, Leviel F, Tew BY, Miro-Julia C, Cheema FH, Miller L, Nelson R, Paunescu TG, McKee M, Brown D, Al-Awqati Q (2010) Deletion of hensin/DMBT1 blocks conversion of beta- to alpha-intercalated cells and induces distal renal tubular acidosis. Proc Natl Acad Sci U S A 107:21872–7

Garrett JR (1987) The proper role of nerves in salivary secretion: a review. J Dent Res 66:387–97

Gawenis LR, Ledoussal C, Judd LM, Prasad V, Alper SL, Stuart-Tilley A, Woo AL, Grisham C, Sanford LP, Doetschman T, Miller ML, Shull GE (2004) Mice with a targeted disruption of the AE2 Cl^-/HCO_3^- exchanger are achlorhydric. J Biol Chem 279:30531–9

Gehlert DR, Speth RC, Wamsley JK (1986) Distribution of [125I]angiotensin II binding sites in the rat brain: a quantitative autoradiographic study. Neuroscience 18:837–56

Gerasimenko JV, Flowerdew SE, Voronina SG, Sukhomlin TK, Tepikin AV, Petersen OH, Gerasimenko OV (2006) Bile acids induce Ca^{2+} release from both the endoplasmic reticulum and acidic intracellular calcium stores through activation of inositol trisphosphate receptors and ryanodine receptors. J Biol Chem 281:40154–63

Giannuzzo A, Wang J, Hoffman EK, Pedersen SH, Novak I (2013) Role of H^+/K^+-ATPase and Na^+/Ca^{2+} exchangers in pancreatic ductal adenocarcinoma cells. FASEB J 27:953

Giffard RG, Lee YS, Ouyang YB, Murphy SL, Monyer H (2003) Two variants of the rat brain sodium-driven chloride bicarbonate exchanger (NCBE): developmental expression and addition of a PDZ motif. Eur J Neurosci 18:2935–45

Githens S (1988) The pancreas duct cell: proliferative capabilities, specific characteristics, metaplasia, isolation, and culture. J Pediatr Gastroenterol Nutr 7:486–506

Gmyr V, Belaich S, Muharram G, Lukowiak B, Vandewalle B, Pattou F, Kerr-Conte J (2004) Rapid purification of human ductal cells from human pancreatic fractions with surface antibody CA19-9. Biochem Biophys Res Commun 320:27–33

Gonzalez-Begne M, Nakamoto T, Nguyen HV, Stewart AK, Alper SL, Melvin JE (2007) Enhanced formation of a HCO_3^- transport metabolon in exocrine cells of Nhe1−/− mice. J Biol Chem 282:35125–32

Gradilone SA, Garcia F, Huebert RC, Tietz PS, Larocca MC, Kierbel A, Carreras FI, LaRusso NF, Marinelli RA (2003) Glucagon induces the plasma membrane insertion of functional aquaporin-8 water channels in isolated rat hepatocytes. Hepatology 37:1435–41

Gray MA, Greenwell JR, Argent BE (1988) Secretin-regulated chloride channel on the apical plasma membrane of pancreatic duct cells. J Membr Biol 105:131–42

Gray MA, Harris A, Coleman L, Greenwell JR, Argent BE (1989) Two types of chloride channel on duct cells cultured from human fetal pancreas. Am J Physiol Cell Physiol 257:C240–51

Gray MA, Greenwell JR, Garton AJ, Argent BE (1990) Regulation of maxi-K^+ channels on pancreatic duct cells by cyclic AMP-dependent phosphorylation. J Membr Biol 115:203–15

Gray MA, Plant S, Argent BE (1993) cAMP-regulated whole cell chloride currents in pancreatic duct cells. Am J Physiol Cell Physiol 264:C591–602

Gray MA, Winpenny JP, Porteous DJ, Dorin JR, Argent BE (1994) CFTR and calcium-activated chloride currents in pancreatic duct cells of a transgenic CF mouse. Am J Physiol Cell Physiol 266:C213–21

Greeley T, Shumaker H, Wang Z, Schweinfest CW, Soleimani M (2001) Downregulated in adenoma and putative anion transporter are regulated by CFTR in cultured pancreatic duct cells. Am J Physiol Gastrointest Liver Physiol 281:G1301–8

Gresz V, Kwon TH, Vorum H, Zelles T, Kurtz I, Steward MC, Aalkjaer C, Nielsen S (2002) Immunolocalization of electroneutral Na^+-HCO_{3-} cotransporters in human and rat salivary glands. Am J Physiol Gastrointest Liver Physiol 283:G473–80

Grichtchenko II, Choi I, Zhong X, Bray-Ward P, Russell JM, Boron WF (2001) Cloning, characterization, and chromosomal mapping of a human electroneutral Na^+-driven Cl-HCO_3 exchanger. J Biol Chem 276:8358–63

Grishin AV, Caplan MJ (1998) ATP1AL1, a member of the non-gastric H, K-ATPase family, functions as a sodium pump. J Biol Chem 273:27772–8

Grishin AV, Bevensee MO, Modyanov NN, Rajendran V, Boron WF, Caplan MJ (1996) Functional expression of the cDNA encoded by the human ATP1AL1 gene. Am J Physiol 271: F539–51

Grotmol T, Buanes T, Bros O, Raeder MG (1986) Lack of effect of amiloride, furosemide, bumetanide and triamterene on pancreatic $NaHCO_3$ secretion in pigs. Acta Physiol Scand 126:593–600

Guba M, Kuhn M, Forssmann WG, Classen M, Gregor M, Seidler U (1996) Guanylin strongly stimulates rat duodenal HCO_3^- secretion: proposed mechanism and comparison with other secretagogues. Gastroenterology 111:1558–68

Haanes KA, Novak I (2010) ATP storage and uptake by isolated pancreatic zymogen granules. Biochem J 429(2):303–11

Hafner P, Grimaldi R, Capuano P, Capasso G, Wagner CA (2008) Pendrin in the mouse kidney is primarily regulated by Cl^- excretion but also by systemic metabolic acidosis. Am J Physiol Cell Physiol 295:C1658–67

Hayashi M, Novak I (2013) Molecular basis of potassium channels in pancreatic duct epithelial cells. Channels (Austin) 7:432–41

Hayashi M, Wang J, Hede SE, Novak I (2012) An intermediate-conductance Ca^{2+}-activated K^+ channel is important for secretion in pancreatic duct cells. Am J Physiol Cell Physiol 303: C151–9

He X, Tse C-M, Danowitz M, Alper SL, Gabriel SE, Baum BJ (1997) Polarized distribution of key membrane transport proteins in the rat submandibular gland. Pflugers Arch 433:260–8

Hede SE, Amstrup J, Christoffersen BC, Novak I (1999) Purinoceptors evoke different electrophysiological responses in pancreatic ducts. P2Y inhibits K^+ conductance, and P2X stimulates cation conductance. J Biol Chem 274:31784–91

Hede SE, Amstrup J, Klaerke DA, Novak I (2005) P2Y2 and P2Y4 receptors regulate pancreatic Ca^{2+}-activated K^+ channels differently. Pflugers Arch 450:429–36

Hegyi P, Petersen OH (2013) The exocrine pancreas: the acinar-ductal tango in physiology and pathophysiology. Rev Physiol Biochem Pharmacol 165:1–30

Hegyi P, Maleth J, Venglovecz V, Rakonczay Z Jr (2011) Pancreatic ductal bicarbonate secretion: challenge of the acinar acid load. Front Physiol 2:36. doi:10.3389/fphys.2011.00036. eCollection 2011

Heitzmann D, Warth R (2008) Physiology and pathophysiology of potassium channels in gastrointestinal epithelia. Physiol Rev 88:1119–82

Hentschke M, Wiemann M, Hentschke S, Kurth I, Hermans-Borgmeyer I, Seidenbecher T, Jentsch TJ, Gal A, Hubner CA (2006) Mice with a targeted disruption of the Cl^-/HCO_3^- exchanger AE3 display a reduced seizure threshold. Mol Cell Biol 26:183–91

Hickson JC (1970) The secretion of pancreatic juice in response to stimulation of the vagus nerves in the pig. J Physiol 206:275–97

Hikita C, Vijayakumar S, Takito J, Erdjument-Bromage H, Tempst P, Al-Awqati Q (2000) Induction of terminal differentiation in epithelial cells requires polymerization of hensin by galectin 3. J Cell Biol 151:1235–46

Hogan DL, Crombie DL, Isenberg JI, Svendsen P, Schaffalitzky de Muckadell OB, Ainsworth MA (1997a) Acid-stimulated duodenal bicarbonate secretion involves a CFTR-mediated transport pathway in mice. Gastroenterology 113:533–41

Hogan DL, Crombie DL, Isenberg JI, Svendsen P, Schaffalitzky de Muckadell OB, Ainsworth MA (1997b) CFTR mediates cAMP- and Ca^{2+}-activated duodenal epithelial HCO_3^- secretion. Am J Physiol Gastrointest Liver Physiol 272:G872–8

Hoglund P, Haila S, Socha J, Tomaszewski L, Saarialho-Kere U, Karjalainen-Lindsberg ML, Airola K, Holmberg C, de la Chapelle A, Kere J (1996) Mutations of the down-regulated in adenoma (DRA) gene cause congenital chloride diarrhoea. Nat Genet 14:316–9

Hollander F, Birnbaum D (1952) The role of carbonic anhydrase in pancreatic secretion. Trans N Y Acad Sci 15:56–8

Holm M, Johansson B, Pettersson A, Fandriks L (1998) Carbon dioxide mediates duodenal mucosal alkaline secretion in response to luminal acidity in the anesthetized rat. Gastroenterology 115:680–5

Holst JJ (1993) Neural regulation of pancreatic exocrine function. In: Go VLW, DiMagno EP, Gardner JD, Lebenthal E, Reber HA, Scheele GA (eds) The pancreas. Biology, pathobiology, and disease. Raven, New York, pp 381–402

Hug M, Pahl C, Novak I (1994) Effect of ATP, carbachol and other agonists on intracellular calcium activity and membrane voltage of pancreatic ducts. Pflugers Arch 426:412–8

Humphrey SP, Williamson RT (2001) A review of saliva: normal composition, flow, and function. J Prosthet Dent 85:162–9

Huss M, Wieczorek H (2009) Inhibitors of V-ATPases: old and new players. J Exp Biol 212:341–6

Husted RF, Reed DJ (1976) Regulation of cerebrospinal fluid potassium by the cat choroid plexus. J Physiol 259:213–21

Hyde K, Reid CJ, Tebbutt SJ, Weide L, Hollingsworth MA, Harris A (1997) The cystic fibrosis transmembrane conductance regulator as a marker of human pancreatic duct development. Gastroenterology 113:914–9

Hyde K, Harrison D, Hollingsworth MA, Harris A (1999) Chloride-bicarbonate exchangers in the human fetal pancreas. Biochem Biophys Res Commun 263:315–21

Igimi H, Yamamoto F, Lee SP (1992) Gallbladder mucosal function: studies in absorption and secretion in humans and in dog gallbladder epithelium. Am J Physiol Gastrointest Liver Physiol 263:G69–74

Ignath I, Hegyi P, Venglovecz V, Szekely CA, Carr G, Hasegawa M, Inoue M, Takacs T, Argent BE, Gray MA, Rakonczay Z Jr (2009) CFTR expression but not Cl^- transport is involved in the stimulatory effect of bile acids on apical Cl^-/HCO_3^- exchange activity in human pancreatic duct cells. Pancreas 38:921–9

Imai Y (1965) Study of the secretion mechanism of the submaxillary gland of dog. Part 2. Effects of exchanging ions in the perfusate on salivary secretion and secretory potentials, with special reference to the ionic distribution in the gland tissue. Jpn J Physiol 27:313–24

Imboden H, Harding JW, Hilgenfeldt U, Celio MR, Felix D (1987) Localization of angiotensinogen in multiple cell types of rat brain. Brain Res 410:74–7

Inagami T, Celio MR, Clemens DL, Lau D, Takii Y, Kasselberg AG, Hirose S (1980) Renin in rat and mouse brain: immunohistochemical identification and localization. Clin Sci (Lond) 59 (Suppl 6):49 s–51s

Isenberg JI, Hogan DL, Koss MA, Selling JA (1986) Human duodenal mucosal bicarbonate secretion. Evidence for basal secretion and stimulation by hydrochloric acid and a synthetic prostaglandin E1 analogue. Gastroenterology 91:370–8

Isenberg JI, Ljungstrom M, Safsten B, Flemstrom G (1993) Proximal duodenal enterocyte transport: evidence for Na^+-H^+ and Cl^--HCO_3^- exchange and NaHCO3 cotransport. Am J Physiol Gastrointest Liver Physiol 265:G677–85

Ishibashi K, Okamura K, Yamazaki J (2008) Involvement of apical P2Y2 receptor-regulated CFTR activity in muscarinic stimulation of Cl^- reabsorption in rat submandibular gland. Am J Physiol Regul Integr Comp Physiol 294:R1729–36

Ishiguro H, Steward MC, Wilson RW, Case RM (1996) Bicarbonate secretion in interlobular ducts from guinea-pig pancreas. J Physiol Lond 495:179–91

Ishiguro H, Naruse S, Steward MC, Kitagawa M, Ko SB, Hayakawa T, Case RM (1998) Fluid secretion in interlobular ducts isolated from guinea-pig pancreas. J Physiol Lond 511:407–22

Ishiguro H, Naruse S, Kitagawa M, Hayakawa T, Case RM, Steward MC (1999) Luminal ATP stimulates fluid and HCO_3^- secretion in guinea-pig pancreatic duct. J Physiol Lond 519:551–8

Ishiguro H, Naruse S, Kitagawa M, Suzuki A, Yamamoto A, Hayakawa T, Case RM, Steward MC (2000) CO_2 permeability and bicarbonate transport in microperfused interlobular ducts isolated from guinea-pig pancreas. J Physiol 528:305–15

Ishiguro H, Namkung W, Yamamoto A, Wang Z, Worrell RT, Xu J, Lee MG, Soleimani M (2007) Effect of Slc26a6 deletion on apical Cl^-/HCO_3^- exchanger activity and cAMP-stimulated bicarbonate secretion in pancreatic duct. Am J Physiol Gastrointest Liver Physiol 292:G447–55

Ishiguro H, Steward MC, Naruse S, Ko SB, Goto H, Case RM, Kondo T, Yamamoto A (2009) CFTR functions as a bicarbonate channel in pancreatic duct cells. J Gen Physiol 133:315–26

Jacob P, Christiani S, Rossmann H, Lamprecht G, Vieillard-Baron D, Muller R, Gregor M, Seidler U (2000) Role of $Na^+HCO_3^-$ cotransporter NBC1, Na^+/H^+ exchanger NHE1, and carbonic anhydrase in rabbit duodenal bicarbonate secretion. Gastroenterology 119:406–19

Jacobs S, Ruusuvuori E, Sipila ST, Haapanen A, Damkier HH, Kurth I, Hentschke M, Schweizer M, Rudhard Y, Laatikainen LM, Tyynela J, Praetorius J, Voipio J, Hubner CA (2008) Mice with targeted Slc4a10 gene disruption have small brain ventricles and show reduced neuronal excitability. Proc Natl Acad Sci U S A 105:311–6

Jacques T, Picard N, Miller RL, Riemondy KA, Houillier P, Sohet F, Ramakrishnan SK, Busst CJ, Jayat M, Corniere N, Hassan H, Aronson PS, Hennings JC, Hubner CA, Nelson RD, Chambrey R, Eladari D (2013) Overexpression of pendrin in intercalated cells produces chloride-sensitive hypertension. J Am Soc Nephrol 24:1104–13

Johanson CE, Murphy VA (1990) Acetazolamide and insulin alter choroid plexus epithelial cell $[Na^+]$, pH, and volume. Am J Physiol 258:F1538–46

Johanson CE, Preston JE, Chodobski A, Stopa EG, Szmydynger-Chodobska J, McMillan PN (1999) AVP V1 receptor-mediated decrease in Cl^- efflux and increase in dark cell number in choroid plexus epithelium. Am J Physiol Cell Physiol 276:C82–90

Jones HC, Keep RF (1988) Brain fluid calcium concentration and response to acute hypercalcaemia during development in the rat. J Physiol 402:579–93

Joo NS, London RM, Kim HD, Forte LR, Clarke LL (1998) Regulation of intestinal Cl^- and HCO_3^-secretion by uroguanylin. Am J Physiol Gastrointest Liver Physiol 274:G633–44

Juhasz M, Chen J, Lendeckel U, Kellner U, Kasper HU, Tulassay Z, Pastorekova S, Malfertheiner P, Ebert MP (2003) Expression of carbonic anhydrase IX in human pancreatic cancer. Aliment Pharmacol Ther 18:837–46

Jung J, Lee MG (2014) Role of calcium signaling in epithelial bicarbonate secretion. Cell Calcium 55(6):376–84

Jung SR, Kim K, Hille B, Nguyen TD, Koh DS (2006) Pattern of Ca^{2+} increase determines the type of secretory mechanism activated in dog pancreatic duct epithelial cells. J Physiol 576:163–78

Jung SR, Hille B, Nguyen TD, Koh DS (2010) Cyclic AMP potentiates Ca^{2+}-dependent exocytosis in pancreatic duct epithelial cells. J Gen Physiol 135:527–43

Jung J, Nam JH, Park HW, Oh U, Yoon JH, Lee MG (2013) Dynamic modulation of ANO1/TMEM16A. Proc Natl Acad Sci U S A 110:360–5

Kahle KT, Gimenez I, Hassan H, Wilson FH, Wong RD, Forbush B, Aronson PS, Lifton RP (2004) WNK4 regulates apical and basolateral Cl^- flux in extrarenal epithelia. Proc Natl Acad Sci U S A 101:2064–9

Kahle KT, Rinehart J, Ring A, Gimenez I, Gamba G, Hebert SC, Lifton RP (2006) WNK protein kinases modulate cellular Cl^- flux by altering the phosphorylation state of the Na-K-Cl and K-Cl cotransporters. Physiology (Bethesda) 21:326–35
Kajita H, Whitwell C, Brown PD (2000) Properties of the inward-rectifying Cl^- channel in rat choroid plexus: regulation by intracellular messengers and inhibition by divalent cations. Pflugers Arch 440:933–40
Kalaria RN, Premkumar DR, Lin CW, Kroon SN, Bae JY, Sayre LM, LaManna JC (1998) Identification and expression of the Na^+/H^+ exchanger in mammalian cerebrovascular and choroidal tissues: characterization by amiloride-sensitive [3H]MIA binding and RT-PCR analysis. Brain Res Mol Brain Res 58:178–87
Kallio H, Pastorekova S, Pastorek J, Waheed A, Sly WS, Mannisto S, Heikinheimo M, Parkkila S (2006) Expression of carbonic anhydrases IX and XII during mouse embryonic development. BMC Dev Biol 6:22
Kao L, Kurtz LM, Shao X, Papadopoulos MC, Liu L, Bok D, Nusinowitz S, Chen B, Stella SL, Andre M, Weinreb J, Luong SS, Piri N, Kwong JM, Newman D, Kurtz I (2011) Severe neurologic impairment in mice with targeted disruption of the electrogenic sodium bicarbonate cotransporter NBCe2 (Slc4a5 gene). J Biol Chem 286:32563–74
Karadsheh MF, Byun N, Mount DB, Delpire E (2004) Localization of the KCC4 potassium-chloride cotransporter in the nervous system. Neuroscience 123:381–91
Kaunitz JD, Akiba Y (2002) Luminal acid elicits a protective duodenal mucosal response. Keio J Med 51:29–35
Kaunitz JD, Akiba Y (2006a) Duodenal carbonic anhydrase: mucosal protection, luminal chemosensing, and gastric acid disposal. Keio J Med 55:96–106
Kaunitz JD, Akiba Y (2006b) Review article: duodenal bicarbonate - mucosal protection, luminal chemosensing and acid–base balance. Aliment Pharmacol Ther 24(Suppl 4):169–76
Kawedia JD, Nieman ML, Boivin GP, Melvin JE, Kikuchi K, Hand AR, Lorenz JN, Menon AG (2007) Interaction between transcellular and paracellular water transport pathways through Aquaporin 5 and the tight junction complex. Proc Natl Acad Sci U S A 104:3621–6
Kay RNB (1960) The rate of flow and composition of various salivary secretions in sheep and calves. J Physiol Lond 150:515–37
Keep RF, Xiang J, Betz AL (1994) Potassium cotransport at the rat choroid plexus. Am J Physiol Cell Physiol 267:C1616–22
Keitel V, Haussinger D (2013) TGR5 in cholangiocytes. Curr Opin Gastroenterol 29:299–304
Keitel V, Ullmer C, Haussinger D (2010) The membrane-bound bile acid receptor TGR5 (Gpbar-1) is localized in the primary cilium of cholangiocytes. Biol Chem 391:785–9
Kerem B, Rommens JM, Buchanan JA, Markiewicz D, Cox TK, Chakravarti A, Buchwald M, Tsui LC (1989) Identification of the cystic fibrosis gene: genetic analysis. Science 245:1073–80
Kibble JD, Trezise AE, Brown PD (1996) Properties of the cAMP-activated Cl^- current in choroid plexus epithelial cells isolated from the rat. J Physiol 496:69–80
Kibble JD, Garner C, Colledge WH, Brown S, Kajita H, Evans M, Brown PD (1997) Whole cell Cl^- conductances in mouse choroid plexus epithelial cells do not require CFTR expression. Am J Physiol Cell Physiol 272:C1899–907
Kim JY, Kim KH, Lee JA, Namkung W, Sun AQ, Ananthanarayanan M, Suchy FJ, Shin DM, Muallem S, Lee MG (2002) Transporter-mediated bile acid uptake causes Ca^{2+}-dependent cell death in rat pancreatic acinar cells. Gastroenterology 122:1941–53
Kivela AJ, Parkkila S, Saarnio J, Karttunen TJ, Kivela J, Parkkila AK, Pastorekova S, Pastorek J, Waheed A, Sly WS, Rajaniemi H (2000) Expression of transmembrane carbonic anhydrase isoenzymes IX and XII in normal human pancreas and pancreatic tumours. Histochem Cell Biol 114:197–204
Klarr SA, Ulanski LJ, Stummer W, Xiang J, Betz AL, Keep RF (1997) The effects of hypo- and hyperkalemia on choroid plexus potassium transport. Brain Res 758:39–44
Knauf H, Lubcke R, Kreutz W, Sachs G (1982) Interrelationships of ion transport in rat submaxillary duct epithelium. Am J Physiol 242:F132–9

Ko SB, Luo X, Hager H, Rojek A, Choi JY, Licht C, Suzuki M, Muallem S, Nielsen S, Ishibashi K (2002a) AE4 is a DIDS-sensitive Cl^-/HCO_3^- exchanger in the basolateral membrane of the renal CCD and the SMG duct. Am J Physiol Cell Physiol 283:C1206–18

Ko SB, Shcheynikov N, Choi JY, Luo X, Ishibashi K, Thomas PJ, Kim JY, Kim KH, Lee MG, Naruse S, Muallem S (2002b) A molecular mechanism for aberrant CFTR-dependent HCO_3^- transport in cystic fibrosis. EMBO J 21:5662–72

Ko SB, Zeng W, Dorwart MR, Luo X, Kim KH, Millen L, Goto H, Naruse S, Soyombo A, Thomas PJ, Muallem S (2004) Gating of CFTR by the STAS domain of SLC26 transporters. Nat Cell Biol 6:343–50

Kodama T (1983) A light and electron microscopic study on the pancreatic ductal system. Acta Pathol Jpn 33:297–321

Kollert-Jons A, Wagner S, Hubner S, Appelhans H, Drenckhahn D (1993) Anion exchanger 1 in human kidney and oncocytoma differs from erythroid AE1 in its NH2 terminus. Am J Physiol 265:F813–21

Komwatana P, Dinudom A, Young JA, Cook DI (1996a) Control of the amiloride-sensitive Na^+ current in salivary duct cells by extracellular sodium. J Membr Biol 150:133–41

Komwatana P, Dinudom A, Young JA, Cook DI (1996b) Cytosolic Na^+ controls and epithelial Na^+ channel via the Go guanine nucleotide-binding regulatory protein. Proc Natl Acad Sci U S A 93:8107–11

Kopelman H, Durie P, Gaskin K, Weizman Z, Forstner G (1985) Pancreatic fluid secretion and protein hyperconcentration in cystic fibrosis. N Engl J Med 312:329–34

Kopelman H, Corey M, Gaskin K, Durie P, Weizman Z, Forstner G (1988) Impaired chloride secretion, as well as bicarbonate secretion, underlies the fluid secretory defect in the cystic fibrosis pancreas. Gastroenterology 95:349–55

Kordas KS, Sperlagh B, Tihanyi T, Topa L, Steward MC, Varga G, Kittel A (2004) ATP and ATPase secretion by exocrine pancreas in rat, Guinea pig, and human. Pancreas 29:53–60

Kotera T, Brown PD (1994) Evidence for two types of potassium current in rat choroid plexus epithelial cells. Pflugers Arch 427:317–24

Kowal JM, Haanes KA, CHristensen NM, Novak I (2015) Bile acid effects are mediated by ATP release and purinergic signalling in exocrine pancreatic cells. Cell Commun Signal 13:28. doi:10.1186/s12964-015-0107-9

Krane CM, Melvin JE, Nguyen HV, Richardson L, Towne JE, Doetschman T, Menon AG (2001) Salivary acinar cells from aquaporin 5-deficient mice have decreased membrane water permeability and altered cell volume regulation. J Biol Chem 276:23413–20

Kulaksiz H, Cetin Y (2002) The electrolyte/fluid secretion stimulatory peptides guanylin and uroguanylin and their common functional coupling proteins in the rat pancreas: a correlative study of expression and cell-specific localization. Pancreas 25:170–5

Kumpulainen T, Jalovaara P (1981) Immunohistochemical localization of carbonic anhydrase isoenzymes in the human pancreas. Gastroenterology 80:796–9

Kwon TH, Pushkin A, Abuladze N, Nielsen S, Kurtz I (2000) Immunoelectron microscopic localization of NBC3 sodium-bicarbonate cotransporter in rat kidney. Am J Physiol Renal Physiol 278:F327–36

Kwon TH, Fenton RA, Praetorius J, Nielsen S (2012) Section I: normal renal function: molecular, cellular, structural, and physiologic principles; chapter 2: anatomy and topography. In: Taal MW, Chertow GM, Marsden PA, Skoreckiand K, Yu ASL (eds) Brenner & Rector's the kidney. Saunders, Philadelphia

Larsen AM, Krogsgaard-Larsen N, Lauritzen G, Olesen CW, Honore HS, Boedtkjer E, Pedersen SF, Bunch L (2012) Gram-scale solution-phase synthesis of selective sodium bicarbonate co-transport inhibitor S0859: in vitro efficacy studies in breast cancer cells. ChemMedChem 7:1808–14

Le MC, Boulkroun S, Gonzalez-Nunez D, Dublineau I, Cluzeaud F, Fay M, Blot-Chabaud M, Farman N (2005) Aldosterone and tight junctions: modulation of claudin-4 phosphorylation in renal collecting duct cells. Am J Physiol Cell Physiol 289:C1513–21

Lee RJ, Foskett JK (2010) Mechanisms of Ca^{2+}-stimulated fluid secretion by porcine bronchial submucosal gland serous acinar cells. Am J Physiol Lung Cell Mol Physiol 298:L210–31

Lee MG, Schultheis PJ, Yan M, Shull GE, Bookstein C, Chang E, Tse M, Donowitz M, Park K, Muallem S (1998) Membrane-limited expression and regulation of Na^+-H^+ exchanger isoforms by P2 receptors in the rat submandibular gland duct. J Physiol 513:341–57

Lee MG, Ahn W, Choi JY, Luo X, Seo JT, Schultheis PJ, Shull GE, Kim KH, Muallem S (2000) Na^+-dependent transporters mediate HCO_3^- salvage across the luminal membrane of the main pancreatic duct. J Clin Invest 105:1651–8

Lee MG, Ohana E, Park HW, Yang D, Muallem S (2012) Molecular mechanism of pancreatic and salivary gland fluid and HCO_3^- secretion. Physiol Rev 92:39–74

Lenzen R, Elster J, Behrend C, Hampel KE, Bechstein WO, Neuhaus P (1997) Bile acid-independent bile flow is differently regulated by glucagon and secretin in humans after orthotopic liver transplantation. Hepatology 26:1272–81

Leslie BR, Schwartz JH, Steinmetz PR (1973) Coupling between Cl^- absorption and HCO_3^- secretion in turtle urinary bladder. Am J Physiol 225:610–7

Leviel F, Hubner CA, Houillier P, Morla L, El MS, Brideau G, Hassan H, Parker MD, Kurth I, Kougioumtzes A, Sinning A, Pech V, Riemondy KA, Miller RL, Hummler E, Shull GE, Aronson PS, Doucet A, Wall SM, Chambrey R, Eladari D (2010) The Na^+-dependent chloride-bicarbonate exchanger SLC4A8 mediates an electroneutral Na^+ reabsorption process in the renal cortical collecting ducts of mice. J Clin Invest 120:1627–35

Li X, Alvarez B, Casey JR, Reithmeier RA, Fliegel L (2002) Carbonic anhydrase II binds to and enhances activity of the Na^+/H^+ exchanger. J Biol Chem 277:36085–91

Li J, Koo NY, Cho IH, Kwon TH, Choi SY, Lee SJ, Oh SB, Kim JS, Park K (2006) Expression of the Na^+-HCO_3^- cotransporter and its role in pHi regulation in guinea pig salivary glands. Am J Physiol Gastrointest Liver Physiol 291:G1031–40

Lindsey AE, Schneider K, Simmons DM, Baron R, Lee BS, Kopito RR (1990) Functional expression and subcellular localization of an anion exchanger cloned from choroid plexus. Proc Natl Acad Sci U S A 87:5278–82

Lindvall-Axelsson M, Nilsson C, Owman C, Winbladh B (1992) Inhibition of cerebrospinal fluid formation by omeprazole. Exp Neurol 115:394–9

Liu S, Piwnica-Worms D, Lieberman M (1990) Intracellular pH regulation in cultured embryonic chick heart cells. Na^+-dependent Cl^-/HCO_3^- exchange. J Gen Physiol 96:1247–69

Liu X, Li T, Riederer B, Lenzen H, Ludolph L, Yeruva S, Tuo B, Soleimani M, Seidler U (2015) Loss of *Slc26a9* anion transporter alters intestinal electrolyte and HCO_3 transport and reduces survival in CFTR-deficient mice. Pflugers Arch 467(6):1261–75

Lohi H, Kujala M, Kerkela E, Saarialho-Kere U, Kestila M, Kere J (2000) Mapping of five new putative anion transporter genes in human and characterization of *SLC26A6*, a candidate gene for pancreatic anion exchanger. Genomics 70:102–12

Loiselle FB, Morgan PE, Alvarez BV (2003) Regulation of the human NBC3 Na^+/HCO_3^- cotransporter by carbonic anhydrase II and protein kinase A. Am J Physiol Cell Physiol 286: C1423–33

Lonnerholm G, Knutson L, Wistrand PJ, Flemstrom G (1989) Carbonic anhydrase in the normal rat stomach and duodenum and after treatment with omeprazole and ranitidine. Acta Physiol Scand 136:253–62

Lux SE, John KM, Kopito RR, Lodish HF (1989) Cloning and characterization of band 3, the human erythrocyte anion-exchange protein (AE1). Proc Natl Acad Sci U S A 86:9089–93

Ma T, Song Y, Gillespie A, Carlson EJ, Epstein CJ, Verkman AS (1999) Defective secretion of saliva in transgenic mice lacking aquaporin-5 water channels. J Biol Chem 274:20071–4

Mahieu I, Becq F, Wolfensberger T, Gola M, Carter N, Hollande E (1994) The expression of carbonic anhydrases II and IV in the human pancreatic cancer cell line (Capan 1) is associated with bicarbonate ion channels. Biol Cell 81:131–41

Maleth J, Venglovecz V, Razga Z, Tiszlavicz L, Rakonczay Z Jr, Hegyi P (2011) Non-conjugated chenodeoxycholate induces severe mitochondrial damage and inhibits bicarbonate transport in pancreatic duct cells. Gut 60:136–8

Mangos JA, Braun G, Hamann KF (1966) Micropuncture study of sodium and potassium excretion in the rat parotid saliva. Pflugers Arch 291:99–106

Mangos JA, McSherry NR, Nousia-Arvanitakis S, Irwin K (1973) Secretion and transductal fluxes of ions in exocrine glands of the mouse. Am J Physiol 225:18–24

Maren TH (1962) The binding of inhibitors to carbonic anhydrase in vivo: drugs as markers for enzyme. Biochem Pharmacol 9:39–48

Marino CR, Matovcik LM, Gorelick FS, Cohn JA (1991) Localization of the cystic fibrosis transmembrane conductance regulator in pancreas. J Clin Invest 88:712–6

Marino CR, Jeanes V, Boron WF, Schmitt BM (1999) Expression and distribution of the Na^+-HCO_3^- cotransporter in human pancreas. Am J Physiol Gastrointest Liver Physiol 277:G487–94

Marteau C, Silviani V, Ducroc R, Crotte C, Gerolami A (1995) Evidence for apical Na^+/H^+ exchanger in bovine main pancreatic duct. Dig Dis Sci 40:2336–40

Martin CJ, Young JA (1971) A microperfusion investigation of the effects of a sympathomimetic and a parasympathomimetic drug on water and electrolyte fluxes in the main duct of the rat submaxillary gland. Pflugers Arch 327:303–23

Martin AC, Willoughby D, Ciruela A, Ayling LJ, Pagano M, Wachten S, Tengholm A, Cooper DM (2009) Capacitative Ca^{2+} entry via Orai1 and stromal interacting molecule 1 (STIM1) regulates adenylyl cyclase type 8. Mol Pharmacol 75:830–42

Martinez JR, Cassity N (1983) Effects of transport inhibitors on secretion by perfused rat submandibular gland. Am J Physiol Gastrointest Liver Physiol 245:G711–6

Martinez JR, Holzgreve H, Frick A (1966) Micropuncture study of submaxillary glands of adult rats. Pflugers Arch 290:124–33

Martinez-Anso E, Castillo JE, Diez J, Medina JF, Prieto J (1994) Immunohistochemical detection of chloride/bicarbonate anion exchangers in human liver. Hepatology 19:1400–6

Maruyama Y, Gallacher DV, Petersen OH (1983) Voltage and Ca^{2+}-activated K^+ channel in basolateral acinar cell membranes of mammalian salivary glands. Nature 302:827–9

Masmoudi S, Charfedine I, Hmani M, Grati M, Ghorbel AM, Elgaied-Boulila A, Drira M, Hardelin JP, Ayadi H (2000) Pendred syndrome: phenotypic variability in two families carrying the same PDS missense mutation. Am J Med Genet 90:38–44

Masuzawa T, Ohta T, Kawamura M, Nakahara N, Sato F (1984) Immunohistochemical localization of Na^+, K^+-ATPase in the choroid plexus. Brain Res 302:357–62

Masyuk AI, Gradilone SA, Banales JM, Huang BQ, Masyuk TV, Lee SO, Splinter PL, Stroope AJ, LaRusso NF (2008) Cholangiocyte primary cilia are chemosensory organelles that detect biliary nucleotides via P2Y12 purinergic receptors. Am J Physiol Gastrointest Liver Physiol 295:G725–34

Masyuk AI, Huang BQ, Radtke BN, Gajdos GB, Splinter PL, Masyuk TV, Gradilone SA, LaRusso NF (2013) Ciliary subcellular localization of TGR5 determines the cholangiocyte functional response to bile acid signaling. Am J Physiol Gastrointest Liver Physiol 304:G1013–24

Matsuo R (2000) Role of saliva in the maintenance of taste sensitivity. Crit Rev Oral Biol Med 11:216–29

Mayer SE, Sanders-Bush E (1993) Sodium-dependent antiporters in choroid plexus epithelial cultures from rabbit. J Neurochem 60:1308–16

McCormick JA, Ellison DH (2011) The WNKs: atypical protein kinases with pleiotropic actions. Physiol Rev 91:177–219

McMurtrie HL, Cleary HJ, Alvarez BV, Loiselle FB, Sterling D, Morgan PE, Johnson DE, Casey JR (2004) The bicarbonate transport metabolon. J Enzyme Inhib Med Chem 19:231–6

Meier PJ, Knickelbein R, Moseley RH, Dobbins JW, Boyer JL (1985) Evidence for carrier-mediated chloride/bicarbonate exchange in canalicular rat liver plasma membrane vesicles. J Clin Invest 75:1256–63

Melby JM, Miner LC, Reed DJ (1982) Effect of acetazolamide and furosemide on the production and composition of cerebrospinal fluid from the cat choroid plexus. Can J Physiol Pharmacol 60:405–9

Melvin JE, Park K, Richardson L, Schultheis PJ, Shull GE (1999) Mouse down-regulated in adenoma (DRA) is an intestinal Cl^-/HCO_3^- exchanger and is up-regulated in colon of mice lacking the NHE3 Na^+/H^+ exchanger. J Biol Chem 274:22855–61

Mennone A, Biemesderfer D, Negoianu D, Yang CL, Abbiati T, Schultheis PJ, Shull GE, Aronson PS, Boyer JL (2001) Role of sodium/hydrogen exchanger isoform NHE3 in fluid secretion and absorption in mouse and rat cholangiocytes. Am J Physiol Gastrointest Liver Physiol 280: G247–54

Meyer G, Guizzardi F, Rodighiero S, Manfredi R, Saino S, Sironi C, Garavaglia ML, Bazzini C, Botta G, Portincasa P, Calamita G, Paulmichl M (2005) Ion transport across the gallbladder epithelium. Curr Drug Targets Immune Endocr Metabol Disord 5:143–51

Millar ID, Brown PD (2008) NBCe2 exhibits a 3 HCO_3^-:1 Na^+ stoichiometry in mouse choroid plexus epithelial cells. Biochem Biophys Res Commun 373:550–4

Minagawa N, Nagata J, Shibao K, Masyuk AI, Gomes DA, Rodrigues MA, Lesage G, Akiba Y, Kaunitz JD, Ehrlich BE, LaRusso NF, Nathanson MH (2007) Cyclic AMP regulates bicarbonate secretion in cholangiocytes through release of ATP into bile. Gastroenterology 133:1592–602

Modyanov NN, Petrukhin KE, Sverdlov VE, Grishin AV, Orlova MY, Kostina MB, Makarevich OI, Broude NE, Monastyrskaya GS, Sverdlov ED (1991) The family of human Na, K-ATPase genes. ATP1AL1 gene is transcriptionally competent and probably encodes the related ion transport ATPase. FEBS Lett 278:91–4

Modyanov NN, Mathews PM, Grishin AV, Beguin P, Beggah AT, Rossier BC, Horisberger JD, Geering K (1995) Human ATP1AL1 gene encodes a ouabain-sensitive H-K-ATPase. Am J Physiol Cell Physiol 269:C992–7

Moore-Hoon ML, Turner RJ (1998) Molecular and topological characterization of the rat parotid Na^+-K^+-$2Cl^-$ cotransporter1. Biochim Biophys Acta 1373:261–9

Moseley RH, Meier PJ, Aronson PS, Boyer JL (1986) Na-H exchange in rat liver basolateral but not canalicular plasma membrane vesicles. Am J Physiol Gastrointest Liver Physiol 250:G35–43

Moser AJ, Gangopadhyay A, Bradbury NA, Peters KW, Frizzell RA, Bridges RJ (2007) Electrogenic bicarbonate secretion by prairie dog gallbladder. Am J Physiol Gastrointest Liver Physiol 292:G1683–94

Mount DB, Romero MF (2004) The SLC26 gene family of multifunctional anion exchangers. Pflugers Arch 447:710–21

Murakami M, Murdiastuti K, Hosoi K, Hill AE (2006) AQP and the control of fluid transport in a salivary gland. J Membr Biol 210:91–103

Murphy VA, Johanson CE (1989) Alteration of sodium transport by the choroid plexus with amiloride. Biochim Biophys Acta 979:187–92

Murphy VA, Smith QR, Rapoport SI (1986) Homeostasis of brain and cerebrospinal fluid calcium concentrations during chronic hypo- and hypercalcemia. J Neurochem 47:1735–41

Nakahari T, Murakami M, Yoshida H, Miyamoto M, Sohma Y, Imai Y (1990) Decrease in rat submandibular acinar cell volume during ACh stimulation. Am J Physiol Gastrointest Liver Physiol 258:G878–86

Nakahari T, Murakami M, Sasaki Y, Kataoka T, Imai Y, Shiba Y, Kanno Y (1991) Dose effects of acetylcholine on the cell volume of rat mandibular salivary acini. Jpn J Physiol 41:153–68

Nakamoto T, Srivastava A, Romanenko VG, Ovitt CE, Perez-Cornejo P, Arreola J, Begenisich T, Melvin JE (2007) Functional and molecular characterization of the fluid secretion mechanism in human parotid acinar cells. Am J Physiol Regul Integr Comp Physiol 292:R2380–90

Nakamoto T, Romanenko VG, Takahashi A, Begenisich T, Melvin JE (2008) Apical maxi-K (KCa1.1) channels mediate K^+ secretion by the mouse submandibular exocrine gland. Am J Physiol Cell Physiol 294:C810–9

Nakamoto T, Brown DA, Catalan MA, Gonzalez-Begne M, Romanenko VG, Melvin JE (2009) Purinergic P2X7 receptors mediate ATP-induced saliva secretion by the mouse submandibular gland. J Biol Chem 284:4815–22

Nakamura N, Suzuki Y, Sakuta H, Ookata K, Kawahara K, Hirose S (1999) Inwardly rectifying K+ channel Kir7.1 is highly expressed in thyroid follicular cells, intestinal epithelial cells and choroid plexus epithelial cells: implication for a functional coupling with Na+, K+-ATPase. Biochem J 342:329–36

Namkung W, Lee JA, Ahn W, Han W, Kwon SW, Ahn DS, Kim KH, Lee MG (2003) Ca^{2+} activates cystic fibrosis transmembrane conductance regulator- and Cl^- -dependent HCO_3 transport in pancreatic duct cells. J Biol Chem 278:200–7

Namkung W, Finkbeiner WE, Verkman AS (2010) CFTR-adenylyl cyclase I association responsible for UTP activation of CFTR in well-differentiated primary human bronchial cell cultures. Mol Biol Cell 21:2639–48

Nathanson MH, Burgstahler AD, Mennone A, Boyer JL (1996) Characterization of cytosolic Ca^{2+} signaling in rat bile duct epithelia. Am J Physiol Gastrointest Liver Physiol 271:G86–96

Nattie EE (1980) Brain and cerebrospinal fluid ionic composition and ventilation in acute hypercapnia. Respir Physiol 40:309–22

Nauntofte B (1992) Regulation of electrolyte and fluid secretion in salivary acinar cells. Am J Physiol Gastrointest Liver Physiol 263:G823–37

Nauntofte B, Poulsen JH (1986) Effects of Ca^{2+} and furosemide on Cl^- transport on O_2 uptake in rat parotid acini. Am J Physiol Cell Physiol 251:C175–85

Nehrke K, Quinn CC, Begenisich T (2003) Molecular identification of Ca^{2+}-activated K^+ channels in parotid acinar cells. Am J Physiol Cell Physiol 284:C535–46

Nguyen TD, Moody MW, Savard CE, Lee SP (1998) Secretory effects of ATP on nontransformed dog pancreatic duct epithelial cells. Am J Physiol Gastrointest Liver Physiol 275:G104–13

Nguyen TD, Meichle S, Kim US, Wong T, Moody MW (2001) $P2Y_{11}$, a purinergic receptor acting via cAMP, mediates secretion by pancreatic duct epithelial cells. Am J Physiol Gastrointest Liver Physiol 280:G795–804

Nishimori I, Onishi S (2001) Carbonic anhydrase isozymes in the human pancreas. Dig Liver Dis 33:68–74

Nishimori I, FujikawaAdachi K, Onishi S, Hollingsworth MA (1999) Carbonic anhydrase in human pancreas: hypotheses for the pathophysiological roles of CA isozymes. Ann N Y Acad Sci 880:5–16

Novak I (1993) Cellular mechanism of salivary gland secretion. In: Gilles R (ed) Advances in comparative and environmental physiology, vol 15. Springer, Berlin, pp 1–43

Novak I (2008) Purinergic receptors in the endocrine and exocrine pancreas. Purinergic Signal 4:237–53

Novak I (2011) Purinergic signalling in epithelial ion transport: regulation of secretion and absorption. Acta Physiol (Oxf) 202:501–22

Novak I, Christoffersen BC (2001) Secretin stimulates HCO_3^- and acetate efflux but not Na^+/ HCO_3^- uptake in rat pancreatic ducts. Pflugers Arch 441:761–71

Novak I, Greger R (1988a) Electrophysiological study of transport systems in isolated perfused pancreatic ducts: properties of the basolateral membrane. Pflugers Arch 411:58–68

Novak I, Greger R (1988b) Properties of the luminal membrane of isolated perfused rat pancreatic ducts: effect of cyclic AMP and blockers of chloride transport. Pflugers Arch 411:546–53

Novak I, Greger R (1991) Effect of bicarbonate on potassium conductance of isolated perfused rat pancreatic ducts. Pflugers Arch 419:76–83

Novak I, Young JA (1986) Two independent anion transport systems in rabbit mandibular salivary glands. Pflugers Arch 407:649–56

Novak I, Hede SE, Hansen MR (2008) Adenosine receptors in rat and human pancreatic ducts stimulate chloride transport. Pflugers Arch 456:437–47

Novak I, Jans IM, Wohlfahrt L (2010) Effect of P2X7 receptor knockout on exocrine secretion of pancreas, salivary glands and lacrimal glands. J Physiol (Lond) 588(18):3615–27

Novak I, Wang J, Henriksen KL, Haanes KA, Krabbe S, Nitschke R, Hede SE (2011) Pancreatic bicarbonate secretion involves two proton pumps. J Biol Chem 286:280–9

Novak I, Haanes KA, Wang J (2013) Acid–base transport in pancreas-new challenges. Front Physiol 4:380. doi:10.3389/fphys.2013.00380

Odgaard E, Jakobsen JK, Frische S, Praetorius J, Nielsen S, Aalkjaer C, Leipziger J (2004) Basolateral Na^+-dependent HCO_3^- transporter NBCn1-mediated HCO_3^- influx in rat medullary thick ascending limb. J Physiol 555:205–18

Okolo C, Wong T, Moody MW, Nguyen TD (2002) Effects of bile acids on dog pancreatic duct epithelial cell secretion and monolayer resistance. Am J Physiol Gastrointest Liver Physiol 283:G1042–50

Olszewski U, Hlozek M, Hamilton G (2010) Activation of Na^+/H^+ exchanger 1 by neurotensin signaling in pancreatic cancer cell lines. Biochem Biophys Res Commun 393:414–9

Orlowski J, Grinstein S (2004) Diversity of the mammalian sodium/proton exchanger SLC9 gene family. Pflugers Arch 447:549–65

Ostrowski NL, Lolait SJ, Bradley DJ, O'Carroll AM, Brownstein MJ, Young WS III (1992) Distribution of V1a and V2 vasopressin receptor messenger ribonucleic acids in rat liver, kidney, pituitary and brain. Endocrinology 131:533–5

Pahl C, Novak I (1993) Effect of vasoactive intestinal peptide, carbachol and other agonists on cell membrane voltage of pancreatic duct cells. Pflugers Arch 424:315–20

Palk L, Sneyd J, Shuttleworth TJ, Yule DI, Crampin EJ (2010) A dynamic model of saliva secretion. J Theor Biol 266:625–40

Park HW, Lee MG (2012) Transepithelial bicarbonate secretion: lessons from the pancreas. Cold Spring Harb Perspect Med 2:155–70

Park HS, Lee YL, Kwon HY, Chey WY, Park HJ (1998) Significant cholinergic role in secretin-stimulated exocrine secretion in isolated rat pancreas. Am J Physiol Gastrointest Liver Physiol 274:G413–8

Park K, Olschowka JA, Richardson LA, Bookstein C, Chang EB, Melvin JE (1999) Expression of multiple Na^+/H^+ exchanger isoforms in rat parotid acinar and ductal cells. Am J Physiol Gastrointest Liver Physiol 276:G470–8

Park K, Case RM, Brown PD (2001a) Identification and regulation of K^+ and Cl^- channels in human parotid acinar cells. Arch Oral Biol 46:801–10

Park K, Evans RL, Watson GE, Nehrke K, Richardson L, Bell SM, Schultheis PJ, Hand AR, Shull GE, Melvin JE (2001b) Defective fluid secretion and NaCl absorption in the parotid glands of Na^+/H^+ exchanger-deficient mice. J Biol Chem 276:27042–50

Park K, Hurley PT, Roussa E, Cooper GJ, Smith CP, Thevenod F, Steward MC, Case RM (2002a) Expression of a sodium bicarbonate cotransporter in human parotid salivary glands. Arch Oral Biol 47:1–9

Park M, Ko SB, Choi JY, Muallem G, Thomas PJ, Pushkin A, Lee MS, Kim JY, Lee MG, Muallem S, Kurtz I (2002b) The cystic fibrosis transmembrane conductance regulator interacts with and regulates the activity of the HCO_3^- salvage transporter human Na^+-HCO_3^- cotransport isoform 3. J Biol Chem 277:50503–9

Park HW, Nam JH, Kim JY, Namkung W, Yoon JS, Lee JS, Kim KS, Venglovecz V, Gray MA, Kim KH, Lee MG (2010) Dynamic regulation of CFTR bicarbonate permeability by $[Cl^-]_i$ and its role in pancreatic bicarbonate secretion. Gastroenterology 139:620–31

Park S, Hong JH, Ohana E, Muallem S (2012) The WNK/SPAK and IRBIT/PP1 pathways in epithelial fluid and electrolyte transport. Physiology (Bethesda) 27:291–9

Park S, Shcheynikov N, Hong JH, Zheng C, Suh SH, Kawaai K, Ando H, Mizutani A, Abe T, Kiyonari H, Seki G, Yule D, Mikoshiba K, Muallem S (2013) Irbit mediates synergy between Ca^{2+} and cAMP signaling pathways during epithelial transport in mice. Gastroenterology 145:232–41

Parker MD, Musa-Aziz R, Rojas JD, Choi I, Daly CM, Boron WF (2008) Characterization of human SLC4A10 as an electroneutral Na/HCO_3 cotransporter (NBCn2) with Cl^- self-exchange activity. J Biol Chem 283:12777–88

Pascua P, Garcia M, Fernandez-Salazar MP, Hernandez-Lorenzo MP, Calvo JJ, Colledge WH, Case RM, Steward MC, San Roman JI (2009) Ducts isolated from the pancreas of CFTR-null mice secrete fluid. Pflugers Arch 459:203–14

Pastor-Soler NM, Hallows KR, Smolak C, Gong F, Brown D, Breton S (2008) Alkaline pH- and cAMP-induced V-ATPase membrane accumulation is mediated by protein kinase A in epididymal clear cells. Am J Physiol Cell Physiol 294:C488–94

Patterson K, Catalan MA, Melvin JE, Yule DI, Crampin EJ, Sneyd J (2012) A quantitative analysis of electrolyte exchange in the salivary duct. Am J Physiol Gastrointest Liver Physiol 303: G1153–63

Paulais M, Cragoe EJ Jr, Turner RJ (1994) Ion transport mechanisms in rat parotid intralobular striated ducts. Am J Physiol Cell Physiol 266:C1594–602

Pearson MM, Lu J, Mount DB, Delpire E (2001) Localization of the K^+-Cl^- cotransporter, KCC3, in the central and peripheral nervous systems: expression in the choroid plexus, large neurons and white matter tracts. Neuroscience 103:481–91

Pech V, Pham TD, Hong S, Weinstein AM, Spencer KB, Duke BJ, Walp E, Kim YH, Sutliff RL, Bao HF, Eaton DC, Wall SM (2010) Pendrin modulates ENaC function by changing luminal HCO_3^-. J Am Soc Nephrol 21:1928–41

Pedersen SF (2006) The Na^+/H^+ exchanger NHE1 in stress-induced signal transduction: implications for cell proliferation and cell death. Pflugers Arch 452:249–59

Pedersen AM, Bardow A, Jensen SB, Nauntofte B (2002) Saliva and gastrointestinal functions of taste, mastication, swallowing and digestion. Oral Dis 8:117–29

Pedersen AML, Sørensen CE, Dynesen AW, Jensen SB (2013a) Salivary glands structure and function and regulation of saliva secretion in health and disease. In: Braxton L, Quinn L (eds) Salivary glands: anatomy. Functions in digestion and role in disease. Nova Science, New York, pp 1–43

Pedersen SF, Hoffmann EK, Novak I (2013b) Cell volume regulation in epithelial physiology and cancer. Front Physiol 4:233. doi:10.3389/fphys.2013.00233

Perides G, Laukkarinen JM, Vassileva G, Steer ML (2010) Biliary acute pancreatitis in mice is mediated by the G-protein-coupled cell surface bile acid receptor Gpbar1. Gastroenterology 138:715–25

Pestov NB, Romanova LG, Korneenko TV, Egorov MV, Kostina MB, Sverdlov VE, Askari A, Shakhparonov MI, Modyanov NN (1998) Ouabain-sensitive H, K-ATPase: tissue-specific expression of the mammalian genes encoding the catalytic alpha subunit. FEBS Lett 440:320–4

Petersen OH, Gallacher DV (1988) Electrophysiology of pancreatic and salivary acinar cells. Annu Rev Physiol 50:65–80

Petersen OH, Poulsen JH (1967) The effects of varying the extracellular potassium concentration on the secretory rate and on resting and secretory potentials in the perfused cat submandibular gland. Acta Physiol Scand 70:293–8

Petersen KU, Reuss L (1983) Cyclic AMP-induced chloride permeability in the apical membrane of Necturus gallbladder epithelium. J Gen Physiol 81:705–29

Petersen OH, Tepikin AV (2008) Polarized calcium signaling in exocrine gland cells. Annu Rev Physiol 70:273–99

Petersen KU, Wehner F, Winterhager JM (1990) Transcellular bicarbonate transport in rabbit gallbladder epithelium: mechanisms and effects of cyclic AMP. Pflugers Arch 416:312–21

Petrovic S, Wang Z, Ma L, Soleimani M (2003) Regulation of the apical Cl^-/HCO_3^- exchanger pendrin in rat cortical collecting duct in metabolic acidosis. Am J Physiol Renal Physiol 284: F103–12

Phillips PA, Abrahams JM, Kelly J, Paxinos G, Grzonka Z, Mendelsohn FA, Johnston CI (1988) Localization of vasopressin binding sites in rat brain by in vitro autoradiography using a radioiodinated V1 receptor antagonist. Neuroscience 27:749–61

Pizzonia JH, Biemesderfer D, Abu-Alfa AK, Wu MS, Exner M, Isenring P, Igarashi P, Aronson PS (1998) Immunochemical characterization of Na^+/H^+ exchanger isoform NHE4. Am J Physiol 275:F510–7

Plotkin MD, Kaplan MR, Peterson LN, Gullans SR, Hebert SC, Delpire E (1997) Expression of the Na^+-K^+-$2Cl^-$ cotransporter BSC2 in the nervous system. Am J Physiol Cell Physiol 272:C173–83

Pols TW, Noriega LG, Nomura M, Auwerx J, Schoonjans K (2011) The bile acid membrane receptor TGR5: a valuable metabolic target. Dig Dis 29:37–44

Poronnik P, Schumann SY, Cook DI (1995) HCO_3^--dependent ACh-activated Na^+ influx in sheep parotid secretory endpieces. Pflugers Arch 429:852–8

Poulsen JH, Bundgaard M (1994) Quantitative estimation of the area of luminal and basolateral membranes of rat parotid acinar cells: some physiological applications. Pflugers Arch 429:240–4

Praetorius J (2010) Na^+-coupled bicarbonate transporters in duodenum, collecting ducts and choroid plexus. J Nephrol 23(Suppl 16):S35–42

Praetorius J, Nielsen S (2006) Distribution of sodium transporters and aquaporin-1 in the human choroid plexus. Am J Physiol Cell Physiol 291:C59–67

Praetorius J, Andreasen D, Jensen BL, Ainsworth MA, Friis UG, Johansen T (2000) NHE1, NHE2, and NHE3 contribute to regulation of intracellular pH in murine duodenal epithelial cells. Am J Physiol Gastrointest Liver Physiol 278:G197–206

Praetorius J, Hager H, Nielsen S, Aalkjaer C, Friis UG, Ainsworth MA, Johansen T (2001) Molecular and functional evidence for electrogenic and electroneutral Na^+-HCO_3^- cotransporters in murine duodenum. Am J Physiol Gastrointest Liver Physiol 280:G332–43

Praetorius J, Friis UG, Ainsworth MA, Schaffalitzky de Muckadell OB, Johansen T (2002) The cystic fibrosis transmembrane conductance regulator is not a base transporter in isolated duodenal epithelial cells. Acta Physiol Scand 174:327–36

Praetorius J, Kim YH, Bouzinova EV, Frische S, Rojek A, Aalkjaer C, Nielsen S (2004a) NBCn1 is a basolateral Na^+-HCO_3^- cotransporter in rat kidney inner medullary collecting ducts. Am J Physiol Renal Physiol 286:F903–12

Praetorius J, Nejsum LN, Nielsen S (2004b) A SLC4A10 gene product maps selectively to the basolateral plasma membrane of choroid plexus epithelial cells. Am J Physiol Cell Physiol 286: C601–10

Proctor GB, Carpenter GH (2014) Salivary secretion: mechanism and neural regulation. Monogr Oral Sci 24:14–29

Pushkin A, Abuladze N, Lee I, Newman D, Hwang J, Kurtz I (1999) Cloning, tissue distribution, genomic organization, and functional characterization of NBC3, a new member of the sodium bicarbonate cotransporter family. J Biol Chem 274:16569–75

Pushkin A, Abuladze N, Newman D, Lee I, Xu G, Kurtz I (2000) Cloning, characterization and chromosomal assignment of NBC4, a new member of the sodium bicarbonate cotransporter family. Biochim Biophys Acta 1493:215–8

Quinton PM (2008) Cystic fibrosis: impaired bicarbonate secretion and mucoviscidosis. Lancet 372:415–7

Quinton PM (2010) Role of epithelial HCO_3^- transport in mucin secretion: lessons from cystic fibrosis. Am J Physiol Cell Physiol 299:C1222–33

Quinton PM, Wright EM, Tormey JM (1973) Localization of sodium pumps in the choroid plexus epithelium. J Cell Biol 58:724–30

Rakonczay Z Jr, Fearn A, Hegyi P, Boros I, Gray MA, Argent BE (2006) Characterization of H^+ and HCO_3^- transporters in CFPAC-1 human pancreatic duct cells. World J Gastroenterol 12:885–95

Renner EL, Lake JR, Scharschmidt BF, Zimmerli B, Meier PJ (1989) Rat hepatocytes exhibit basolateral Na^+/HCO_3^- cotransport. J Clin Invest 83:1225–35

Riordan JR, Rommens JM, Kerem B, Alon N, Rozmahel R, Grzelczak Z, Zielenski J, Lok S, Plavsic N, Chou JL (1989) Identification of the cystic fibrosis gene: cloning and characterization of complementary DNA. Science 245:1066–73

Roepke TK, Kanda VA, Purtell K, King EC, Lerner DJ, Abbott GW (2011) KCNE2 forms potassium channels with KCNA3 and KCNQ1 in the choroid plexus epithelium. FASEB J 25:4264–73

Rogers CS, Stoltz DA, Meyerholz DK, Ostedgaard LS, Rokhlina T, Taft PJ, Rogan MP, Pezzulo AA, Karp PH, Itani OA, Kabel AC, Wohlford-Lenane CL, Davis GJ, Hanfland RA, Smith TL, Samuel M, Wax D, Murphy CN, Rieke A, Whitworth K, Uc A, Starner TD, Brogden KA, Shilyansky J, McCray PB Jr, Zabner J, Prather RS, Welsh MJ (2008) Disruption of the CFTR gene produces a model of cystic fibrosis in newborn pigs. Science 321:1837–41

Romanenko VG, Nakamoto T, Srivastava A, Begenisich T, Melvin JE (2007) Regulation of membrane potential and fluid secretion by Ca^{2+}-activated K^+ channels in mouse submandibular glands. J Physiol 581:801–17

Romanenko VG, Nakamoto T, Catalan MA, Gonzalez-Begne M, Schwartz GJ, Jaramillo Y, Sepulveda FV, Figueroa CD, Melvin JE (2008) Clcn2 encodes the hyperpolarization-activated chloride channel in the ducts of mouse salivary glands. Am J Physiol Gastrointest Liver Physiol 295:G1058–67

Romanenko VG, Catalan MA, Brown DA, Putzier I, Hartzell HC, Marmorstein AD, Gonzalez-Begne M, Rock JR, Harfe BD, Melvin JE (2010) Tmem16A encodes the Ca^{2+}-activated Cl^- channel in mouse submandibular salivary gland acinar cells. J Biol Chem 285:12990–3001

Romero MF, Hediger MA, Boulpaep EL, Boron WF (1997) Expression cloning and characterization of a renal electrogenic Na^+/HCO_3^- cotransporter. Nature 387:409–13

Rougemont JD, Ames A III, Nesbett FB, Hofmann HF (1960) Fluid formed by choroid plexus; a technique for its collection and a comparison of its electrolyte composition with serum and cisternal fluids. J Neurophysiol 23:485–95

Roussa E, Thevenod F (1998) Distribution of V-ATPase in rat salivary glands. Eur J Morphol 36 (Suppl):147–52

Roussa E, Thevenod F, Sabolic I, Herak-Kramberger CM, Nastainczyk W, Bock R, Schulz I (1998) Immunolocalization of vacuolar-type H^+-ATPase in rat submandibular gland and adaptive changes induced by acid–base disturbances. J Histochem Cytochem 46:91–100

Roussa E, Romero MF, Schmitt BM, Boron WF, Alper SL, Thevenod F (1999) Immunolocalization of anion exchanger AE2 and Na^+-HCO_3^- cotransporter in rat parotid and submandibular glands. Am J Physiol Gastrointest Liver Physiol 277:G1288–96

Roussa E, Alper SL, Thevenod F (2001) Immunolocalization of anion exchanger AE2, Na^+/H^+ exchangers NHE1 and NHE4, and vacuolar type H^+-ATPase in rat pancreas. J Histochem Cytochem 49:463–74

Roussa E, Shmukler BE, Wilhelm S, Casula S, Stuart-Tilley AK, Thevenod F, Alper SL (2002) Immunolocalization of potassium-chloride cotransporter polypeptides in rat exocrine glands. Histochem Cell Biol 117:335–44

Royaux IE, Suzuki K, Mori A, Katoh R, Everett LA, Kohn LD, Green ED (2000) Pendrin, the protein encoded by the Pendred syndrome gene (PDS), is an apical porter of iodide in the thyroid and is regulated by thyroglobulin in FRTL-5 cells. Endocrinology 141:839–45

Royaux IE, Wall SM, Karniski LP, Everett LA, Suzuki K, Knepper MA, Green ED (2001) Pendrin, encoded by the Pendred syndrome gene, resides in the apical region of renal intercalated cells and mediates bicarbonate secretion. Proc Natl Acad Sci U S A 98:4221–6

Rune SJ, Henriksen FW (1969) Carbon dioxide tensions in the proximal part of the canine gastrointestinal tract. Gastroenterology 56:758–62

Russell JM (2000) Sodium-potassium-chloride cotransport. Physiol Rev 80:211–76

Ryu SY, Peixoto PM, Won JH, Yule DI, Kinnally KW (2010) Extracellular ATP and P2Y2 receptors mediate intercellular Ca^{2+} waves induced by mechanical stimulation in submandibular gland cells: role of mitochondrial regulation of store operated Ca^{2+} entry. Cell Calcium 47:65–76

Saarnio J, Parkkila S, Parkkila AK, Waheed A, Casey MC, Zhou XY, Pastorekova S, Pastorek J, Karttunen T, Haukipuro K, Kairaluoma MI, Sly WS (1998) Immunohistochemistry of carbonic anhydrase isozyme IX (MN/CA IX) in human gut reveals polarized expression in the epithelial cells with the highest proliferative capacity. J Histochem Cytochem 46:497–504
Sachs G, Shin JM, Vagin O, Lambrecht N, Yakubov I, Munson K (2007) The gastric H, K ATPase as a drug target: past, present, and future. J Clin Gastroenterol 41(Suppl 2):S226–42
Saito Y, Wright EM (1983) Bicarbonate transport across the frog choroid plexus and its control by cyclic nucleotides. J Physiol 336:635–48
Saito Y, Wright EM (1984) Regulation of bicarbonate transport across the brush border membrane of the bull-frog choroid plexus. J Physiol 350:327–42
Sangan P, Thevananther S, Sangan S, Rajendran VM, Binder HJ (2000) Colonic H-K-ATPase alpha- and beta-subunits express ouabain-insensitive H-K-ATPase. Am J Physiol Cell Physiol 278:C182–9
Sassani P, Pushkin A, Gross E, Gomer A, Abuladze N, Dukkipati R, Carpenito G, Kurtz I (2002) Functional characterization of NBC4: a new electrogenic sodium-bicarbonate cotransporter. Am J Physiol Cell Physiol 282:C408–16
Satoh H, Moriyama N, Hara C, Yamada H, Horita S, Kunimi M, Tsukamoto K, Iso O, Inatomi J, Kawakami H, Kudo A, Endou H, Igarashi T, Goto A, Fujita T, Seki G (2003) Localization of Na^+-HCO_3^- cotransporter (NBC-1) variants in rat and human pancreas. Am J Physiol Cell Physiol 284:C729–37
Schlosser SF, Burgstahler AD, Nathason MH (1996) Isolated rat hepatocytes can signal to other hepatocytes and bile duct cells by release of nucleotides. Proc Natl Acad Sci U S A 93:9948–53
Schlue WR, Deitmer JW (1988) Ionic mechanisms of intracellular pH regulation in the nervous system. Ciba Found Symp 139:47–69
Schmitt BM, Biemesderfer D, Romero MF, Boulpaep EL, Boron WF (1999) Immunolocalization of the electrogenic Na^+-HCO_3^- cotransporter in mammalian and amphibian kidney. Am J Physiol 276:F27–38
Schneyer LH, Young JA, Schneyer CA (1972) Salivary secretion of electrolytes. Physiol Rev 52:720–77
Scholz W, Albus U, Counillon L, Gogelein H, Lang HJ, Linz W, Weichert A, Scholkens BA (1995) Protective effects of HOE642, a selective sodium-hydrogen exchange subtype 1 inhibitor, on cardiac ischaemia and reperfusion. Cardiovasc Res 29:260–8
Schroeder BC, Cheng T, Jan YN, Jan LY (2008) Expression cloning of TMEM16A as a calcium-activated chloride channel subunit. Cell 134:1019–29
Schulz I (1971) Influence of bicarbonate-CO_2 - and glycodiazine buffer on the secretion of the isolated cat's pancreas. Pflugers Arch 329:283–306
Schulz I, Ströver F, Ullrich KJ (1971) Lipid soluble weak organic acid buffers as "substrate" for pancreatic secretion. Pflugers Arch 323:121–40
Schwaderer AL, Vijayakumar S, Al-Awqati Q, Schwartz GJ (2006) Galectin-3 expression is induced in renal beta-intercalated cells during metabolic acidosis. Am J Physiol Renal Physiol 290:F148–58
Schwartz IL, Thaysen JH (1956) Excretion of sodium and potassium in human sweat. J Clin Invest 35:114–20
Schwartz GJ, Barasch J, Al-Awqati Q (1985) Plasticity of functional epithelial polarity. Nature 318:368–71
Schwartz GJ, Tsuruoka S, Vijayakumar S, Petrovic S, Mian A, Al-Awqati Q (2002) Acid incubation reverses the polarity of intercalated cell transporters, an effect mediated by hensin. J Clin Invest 109:89–99
Seidler U, Blumenstein I, Kretz A, Viellard-Baron D, Rossmann H, Colledge WH, Evans M, Ratcliff R, Gregor M (1997) A functional CFTR protein is required for mouse intestinal cAMP-, cGMP- and Ca^{2+}-dependent HCO_3^- secretion. J Physiol 505:411–23
Seow KTFP, Case RM, Young JA (1991) Pancreatic secretion by the anaesthetized rabbit in response to secretin, cholecystokinin, and carbachol. Pancreas 6:385–91

Sewell WA, Young JA (1975) Secretion of electrolytes by the pancreas of the anaesthetized rat. J Physiol Lond 252:379–96

Shcheynikov N, Wang Y, Park M, Ko SB, Dorwart M, Naruse S, Thomas PJ, Muallem S (2006) Coupling modes and stoichiometry of Cl^-/HCO_3^- exchange by slc26a3 and slc26a6. J Gen Physiol 127:511–24

Shcheynikov N, Yang D, Wang Y, Zeng W, Karniski LP, So I, Wall SM, Muallem S (2008) The Slc26a4 transporter functions as an electroneutral $Cl^-/I^-/HCO_3^-$ exchanger: role of Slc26a4 and Slc26a6 in I^- and HCO_3^- secretion and in regulation of CFTR in the parotid duct. J Physiol 586:3813–24

Shimokura GH, McGill JM, Schlenker T, Fitz JG (1995) Ursodeoxycholate increases cytosolic calcium concentration and activates Cl^- currents in a biliary cell line. Gastroenterology 109:965–72

Shitara A, Tanimura A, Sato A, Tojyo Y (2009) Spontaneous oscillations in intracellular Ca^{2+} concentration via purinergic receptors elicit transient cell swelling in rat parotid ducts. Am J Physiol Gastrointest Liver Physiol 297:G1198–205

Siegel GJ, Holm C, Schreiber JH, Desmond T, Ernst SA (1984) Purification of mouse brain (Na^++K^+)-ATPase catalytic unit, characterization of antiserum, and immunocytochemical localization in cerebellum, choroid plexus, and kidney. J Histochem Cytochem 32:1309–18

Simpson JE, Schweinfest CW, Shull GE, Gawenis LR, Walker NM, Boyle KT, Soleimani M, Clarke LL (2007) PAT-1 (Slc26a6) is the predominant apical membrane Cl^-/HCO_3^- exchanger in the upper villous epithelium of the murine duodenum. Am J Physiol Gastrointest Liver Physiol 292:G1079–88

Sinclair AJ, Onyimba CU, Khosla P, Vijapurapu N, Tomlinson JW, Burdon MA, Stewart PM, Murray PI, Walker EA, Rauz S (2007) Corticosteroids, 11beta-hydroxysteroid dehydrogenase isozymes and the rabbit choroid plexus. J Neuroendocrinol 19:614–20

Singh AK, Liu Y, Riederer B, Engelhardt R, Thakur BK, Soleimani M, Seidler U (2013a) Molecular transport machinery involved in orchestrating luminal acid-induced duodenal bicarbonate secretion in vivo. J Physiol 591:5377–91

Singh AK, Xia W, Riederer B, Juric M, Li J, Zheng W, Cinar A, Xiao F, Bachmann O, Song P, Praetorius J, Aalkjaer C, Seidler U (2013b) Essential role of the electroneutral Na^+-HCO_3^- cotransporter NBCn1 in murine duodenal acid–base balance and colonic mucus layer build-up in vivo. J Physiol 591:2189–204

Sjoblom M, Flemstrom G (2003) Melatonin in the duodenal lumen is a potent stimulant of mucosal bicarbonate secretion. J Pineal Res 34:288–93

Sjoblom M, Singh AK, Zheng W, Wang J, Tuo BG, Krabbenhoft A, Riederer B, Gros G, Seidler U (2009) Duodenal acidity "sensing" but not epithelial. Proc Natl Acad Sci U S A 106:13094–9

Sly WS, Hewett-Emmett D, Whyte MP, Yu YS, Tashian RE (1983) Carbonic anhydrase II deficiency identified as the primary defect in the autosomal recessive syndrome of osteopetrosis with renal tubular acidosis and cerebral calcification. Proc Natl Acad Sci U S A 80:2752–6

Soleimani M, Barone S, Xu J, Shull GE, Siddiqui F, Zahedi K, Amlal H (2012) Double knockout of pendrin and Na-Cl cotransporter (NCC) causes severe salt wasting, volume depletion, and renal failure. Proc Natl Acad Sci U S A 109:13368–73

Song Y, Yamamoto A, Steward MC, Ko SB, Stewart AK, Soleimani M, Liu BC, Kondo T, Jin CX, Ishiguro H (2012) Deletion of Slc26a6 alters the stoichiometry of apical Cl^-/HCO_3^- exchange in mouse pancreatic duct. Am J Physiol Cell Physiol 303:C815–24

Sorensen CE, Novak I (2001) Visualization of ATP release in pancreatic acini in response to cholinergic stimulus. Use of fluorescent probes and confocal microscopy. J Biol Chem 276 (35):32925–32

Speake T, Brown PD (2004) Ion channels in epithelial cells of the choroid plexus isolated from the lateral ventricle of rat brain. Brain Res 1005:60–6

Speake T, Kajita H, Smith CP, Brown PD (2002) Inward-rectifying anion channels are expressed in the epithelial cells of choroid plexus isolated from ClC-2 'knock-out' mice. J Physiol 539:385–90
Spirli C, Granato A, Zsembery K, Anglani F, Okolicsanyi L, LaRusso NF, Crepaldi G, Strazzabosco M (1998) Functional polarity of Na^+/H^+ and Cl^-/HCO_3^- exchangers in a rat cholangiocyte cell line. Am J Physiol Gastrointest Liver Physiol 275:G1236–45
Stetson DL, Steinmetz PR (1985) Alpha and beta types of carbonic anhydrase-rich cells in turtle bladder. Am J Physiol 249:F553–65
Steward MC, Ishiguro H (2009) Molecular and cellular regulation of pancreatic duct cell function. Curr Opin Gastroenterol 25:447–53
Steward MC, Poronnik P, Cook DI (1996) Bicarbonate transport in sheep parotid secretory cells. J Physiol 494:819–30
Steward MC, Ishiguro H, Case RM (2005) Mechanisms of bicarbonate secretion in the pancreatic duct. Annu Rev Physiol 67:377–409
Stewart AK, Yamamoto A, Nakakuki M, Kondo T, Alper SL, Ishiguro H (2009) Functional coupling of apical Cl^-/HCO_3^- exchange with CFTR in stimulated HCO_3^- secretion by guinea pig interlobular pancreatic duct. Am J Physiol Gastrointest Liver Physiol 296:G1307–17
Strazzabosco M, Joplin R, Zsembery A, Wallace L, Spirli C, Fabris L, Granato A, Rossanese A, Poci C, Neuberger JM, Okolicsanyi L, Crepaldi G (1997) Na^+-dependent and -independent Cl^-/HCO_3^- exchange mediate cellular HCO_3^- transport in cultured human intrahepatic bile duct cells. Hepatology 25:976–85
Stuenkel EL, Machen TE, Williams JA (1988) pH regulatory mechanism in rat pancreatic ductal cells. Am J Physiol Gastrointest Liver Physiol 254:G925–30
Supuran CT (2008) Carbonic anhydrases - an overview. Curr Pharm Des 14:603–14
Surprenant A, North RA (2009) Signaling at purinergic P2X receptors. Annu Rev Physiol 71:333–59
Svanvik J, Allen B, Pellegrini C, Bernhoft R, Way L (1984) Variations in concentrating function of the gallbladder in the conscious monkey. Gastroenterology 86:919–25
Swarts HG, Koenderink JB, Willems PH, De Pont JJ (2005) The non-gastric H, K-ATPase is oligomycin-sensitive and can function as an H+, NH4 + -ATPase. J Biol Chem 280:33115–22
Szalmay G, Varga G, Kajiyama F, Yang XS, Lang TF, Case RM, Steward MC (2001) Bicarbonate and fluid secretion evoked by cholecystokinin, bombesin and acetylcholine in isolated guinea-pig pancreatic ducts. J Physiol Lond 535:795–807
Szmydynger-Chodobska J, Chung I, Kozniewska E, Tran B, Harrington FJ, Duncan JA, Chodobski A (2004) Increased expression of vasopressin v1a receptors after traumatic brain injury. J Neurotrauma 21:1090–102
Szucs A, Demeter I, Burghardt B, Ovari G, Case RM, Steward MC, Varga G (2006) Vectorial bicarbonate transport by Capan-1 cells: a model for human pancreatic ductal secretion. Cell Physiol Biochem 18:253–64
Tabcharani JA, Chang X-B, Riordan JR, Hanrahan JW (1991) Phosphorylation-regulated Cl^- channel in CHO cells stably expressing the cystic fibrosis gene. Nature 352:628–31
Tabibian JH, Masyuk AI, Masyuk TV, O'Hara SP, LaRusso NF (2013) Physiology of cholangiocytes. Compr Physiol 3:541–65
Thaysen JH, Thorn NA, Schwartz IL (1954) Excretion of sodium, potassium, chloride, and carbon dioxide in human parotid saliva. Am J Physiol 178:155–9
Thompson J, Begenisich T (2006) Membrane-delimited inhibition of maxi-K channel activity by the intermediate conductance Ca^{2+}-activated K channel. J Gen Physiol 127:159–69
Thompson J, Begenisich T (2009) Mechanistic details of BK channel inhibition by the intermediate conductance, Ca^{2+}-activated K channel. Channels (Austin) 3:194–204
Turner RJ, George JN (1988) Cl^--HCO_3^- exchange is present with Na^+-K^+-Cl^- cotransport in rabbit parotid acinar basolateral membranes. Am J Physiol Cell Physiol 254:C391–6
Turner RJ, George JN, Baum BJ (1986) Evidence for a $Na^+/K^+/Cl^-$ cotransport system in basolateral membrane vesicles from the rabbit parotid. J Membr Biol 94:143–52

Uriarte I, Banales JM, Saez E, Arenas F, Oude Elferink RP, Prieto J, Medina JF (2010) Bicarbonate secretion of mouse cholangiocytes involves Na^+-HCO_3^- cotransport in addition to Na^+-independent Cl^-/HCO_3^- exchange. Hepatology 51:891–902

Vallet M, Picard N, Loffing-Cueni D, Fysekidis M, Bloch-Faure M, Deschenes G, Breton S, Meneton P, Loffing J, Aronson PS, Chambrey R, Eladari D (2006) Pendrin regulation in mouse kidney primarily is chloride-dependent. J Am Soc Nephrol 17:2153–63

Veel T, Villanger O, Holthe MR, Cragoe EJ, Raeder MG (1992) Na^+-H^+ exchange is not important for pancreatic HCO_3^- secretion in the pig. Acta Physiol Scand 144:239–46

Venglovecz V, Rakonczay Z Jr, Ozsvari B, Takacs T, Lonovics J, Varro A, Gray MA, Argent BE, Hegyi P (2008) Effects of bile acids on pancreatic ductal bicarbonate secretion in guinea pig. Gut 57:1102–12

Venglovecz V, Hegyi P, Rakonczay Z Jr, Tiszlavicz L, Nardi A, Grunnet M, Gray MA (2011) Pathophysiological relevance of apical large-conductance Ca^{2+}-activated potassium channels in pancreatic duct epithelial cells. Gut 60:361–9

Verlander JW, Hassell KA, Royaux IE, Glapion DM, Wang ME, Everett LA, Green ED, Wall SM (2003) Deoxycorticosterone upregulates PDS (Slc26a4) in mouse kidney: role of pendrin in mineralocorticoid-induced hypertension. Hypertension 42:356–62

Verlander JW, Kim YH, Shin W, Pham TD, Hassell KA, Beierwaltes WH, Green ED, Everett L, Matthews SW, Wall SM (2006) Dietary Cl^- restriction upregulates pendrin expression within the apical plasma membrane of type B intercalated cells. Am J Physiol Renal Physiol 291: F833–9

Vijayakumar S, Erdjument-Bromage H, Tempst P, Al-Awqati Q (2008) Role of integrins in the assembly and function of hensin in intercalated cells. J Am Soc Nephrol 19:1079–91

Villanger O, Veel T, Raeder MG (1993) Secretin causes H^+ secretion from intrahepatic bile ductules by vacuolar-type H^+-ATPase. Am J Physiol Gastrointest Liver Physiol 265:G719–24

Villanger O, Veel T, Ræder MG (1995) Secretin causes H^+/HCO_3^- secretion from pig pancreatic ductules by vacuolar-type H^+-adenosine triphosphatase. Gastroenterology 108:850–9

Virkki LV, Wilson DA, Vaughan-Jones RD, Boron WF (2002) Functional characterization of human NBC4 as an electrogenic Na^+-HCO_3^- cotransporter (NBCe2). Am J Physiol Cell Physiol 282:C1278–89

Virkki LV, Choi I, Davis BA, Boron WF (2003) Cloning of a Na^+-driven Cl/HCO_3 exchanger from squid giant fiber lobe. Am J Physiol Cell Physiol 285:C771–80

Vogh BP, Godman DR, Maren TH (1987) Effect of AlCl3 and other acids on cerebrospinal fluid production: a correction. J Pharmacol Exp Ther 243:35–9

Voronina S, Longbottom R, Sutton R, Petersen OH, Tepikin A (2002) Bile acids induce calcium signals in mouse pancreatic acinar cells: implications for bile-induced pancreatic pathology. J Physiol 540:49–55

Voronina SG, Barrow SL, Gerasimenko OV, Petersen OH, Tepikin AV (2004) Effects of secretagogues and bile acids on mitochondrial membrane potential of pancreatic acinar cells: comparison of different modes of evaluating DeltaPsim. J Biol Chem 279:27327–38

Vorum H, Kwon TH, Fulton C, Simonsen B, Choi I, Boron W, Maunsbach AB, Nielsen S, Aalkjaer C (2000) Immunolocalization of electroneutral Na-HCO_3^- cotransporter in rat kidney. Am J Physiol Renal Physiol 279:F901–9

Wagner CA, Finberg KE, Stehberger PA, Lifton RP, Giebisch GH, Aronson PS, Geibel JP (2002) Regulation of the expression of the Cl^-/anion exchanger pendrin in mouse kidney by acid–base status. Kidney Int 62:2109–17

Wagner CA, Mohebbi N, Capasso G, Geibel JP (2011) The anion exchanger pendrin (SLC26A4) and renal acid–base homeostasis. Cell Physiol Biochem 28:497–504

Walker NM, Simpson JE, Brazill JM, Gill RK, Dudeja PK, Schweinfest CW, Clarke LL (2009) Role of down-regulated in adenoma anion exchanger in HCO_3^- secretion across murine duodenum. Gastroenterology 136:893–901

Wall SM, Pech V (2008) The interaction of pendrin and the epithelial sodium channel in blood pressure regulation. Curr Opin Nephrol Hypertens 17:18–24

Wall SM, Kim YH, Stanley L, Glapion DM, Everett LA, Green ED, Verlander JW (2004) NaCl restriction upregulates renal Slc26a4 through subcellular redistribution: role in Cl^- conservation. Hypertension 44:982–7
Wan MH, Huang W, Latawiec D, Jiang K, Booth DM, Elliott V, Mukherjee R, Xia Q (2012) Review of experimental animal models of biliary acute pancreatitis and recent advances in basic research. HPB (Oxford) 14:73–81
Wang J, Novak I (2013) Ion transport in human pancreatic duct epithelium, Capan-1 cells, is regulated by secretin, VIP, acetylcholine, and purinergic receptors. Pancreas 42:452–60
Wang J, Barbuskaite D, Tozzi M, Giannuzzo A, Sørensen CE, Novak I (2015) Proton pump inhibitors inhibit pancreatic secretion: role of gastric and non-gastric H/K-ATPases. PLoS One 10:e0126432. doi:10.10371/journal.pone.0126432
Wang CZ, Yano H, Nagashima K, Seino S (2000) The Na^+-driven Cl^-/HCO_3^- exchanger. Cloning, tissue distribution, and functional characterization. J Biol Chem 275:35486–90
Wang Y, Soyombo AA, Shcheynikov N, Zeng W, Dorwart M, Marino CR, Thomas PJ, Muallem S (2006) Slc26a6 regulates CFTR activity in vivo to determine pancreatic duct HCO_3^- secretion: relevance to cystic fibrosis. EMBO J 25:5049–57
Wang J, Haanes KA, Novak I (2013) Purinergic regulation of CFTR and Ca^{2+}-activated Cl^- channels and K^+ channels in human pancreatic duct epithelium. Am J Physiol Cell Physiol 304: C673–84
Weber KT, Sun Y, Wodi LA, Munir A, Jahangir E, Ahokas RA, Gerling IC, Postlethwaite AE, Warrington KJ (2003) Toward a broader understanding of aldosterone in congestive heart failure. J Renin Angiotensin Aldosterone Syst 4:155–63
Wegman EA, Ishikawa T, Young JA, Cook DI (1992) Cation channels in basolateral membranes of sheep parotid secretory cells. Am J Physiol Gastrointest Liver Physiol 263:G786–94
Welch K (1963) Secretion of cerebrospinal fluid by the choroid plexus of the rabbit. Am J Physiol 205:617–24
Williams SE, Turnberg LA (1981) Demonstration of a pH gradient across mucus adherent to rabbit gastric mucosa: evidence for a 'mucus-bicarbonate' barrier. Gut 22:94–6
Wilschanski M, Novak I (2013) The cystic fibrosis of exocrine pancreas. Cold Spring Harb Perspect Med 3:a009746. doi:10.1101/cshperspect.a009746
Winpenny JP, Verdon B, McAlroy HL, Colledge WH, Ratcliff R, Evans MJ, Gray MA, Argent BE (1995) Calcium-activated chloride conductance is not increased in pancreatic duct cells of CF mice. Pflugers Arch 430:26–33
Winpenny JP, Harris A, Hollingsworth MA, Argent BE, Gray MA (1998) Calcium-activated chloride conductance in a pancreatic adenocarcinoma cell line of ductal origin (HPAF) and in freshly isolated human pancreatic duct cells. Pflugers Arch 435:796–803
Wizemann V, Schulz I (1973) Influence of amphotericin, amiloride, ionophores, and 2,4- dinitrophenol on the secretion of the isolated cats pancreas. Pflugers Arch 339:317–38
Woo K, Dutta AK, Patel V, Kresge C, Feranchak AP (2008) Fluid flow induces mechanosensitive ATP release, calcium signalling and Cl^- transport in biliary epithelial cells through a PKCzeta-dependent pathway. J Physiol 586:2779–98
Woo K, Sathe M, Kresge C, Esser V, Ueno Y, Venter J, Glaser SS, Alpini G, Feranchak AP (2010) Adenosine triphosphate release and purinergic (P2) receptor-mediated secretion in small and large mouse cholangiocytes. Hepatology 52:1819–28
Wright EM (1972) Mechanisms of ion transport across the choroid plexus. J Physiol 226:545–71
Wright EM (1978) Transport processes in the formation of the cerebrospinal fluid. Rev Physiol Biochem Pharmacol 83:3–34
Wright RD, Blair-West JR, Nelson JF (1986) Effects of ouabain, amiloride, monenesin, and other agents on ovine parotid secretion. Am J Physiol 250:F503–10
Wu Q, Delpire E, Hebert SC, Strange K (1998) Functional demonstration of Na^+-K^+-$2Cl^-$ cotransporter activity in isolated, polarized choroid plexus cells. Am J Physiol Cell Physiol 275:C1565–72

Yamaguchi S, Ishikawa T (2005) Electrophysiological characterization of native Na^+-HCO_3^- cotransporter current in bovine parotid acinar cells. J Physiol 568:181–97
Yamaguchi S, Ishikawa T (2012) IRBIT reduces the apparent affinity for intracellular Mg^{2+} in inhibition of the electrogenic Na^+-HCO_3^- cotransporter NBCe1-B. Biochem Biophys Res Commun 424:433–8
Yang CL, Liu X, Paliege A, Zhu X, Bachmann S, Dawson DC, Ellison DH (2007) WNK1 and WNK4 modulate CFTR activity. Biochem Biophys Res Commun 353:535–40
Yang YD, Cho H, Koo JY, Tak MH, Cho Y, Shim WS, Park SP, Lee J, Lee B, Kim BM, Raouf R, Shin YK, Oh U (2008) TMEM16A confers receptor-activated calcium-dependent chloride conductance. Nature 455:1210–5
Yang D, Shcheynikov N, Zeng W, Ohana E, So I, Ando H, Mizutani A, Mikoshiba K, Muallem S (2009) IRBIT coordinates epithelial fluid and HCO_3^- secretion by stimulating the transporters pNBC1 and CFTR in the murine pancreatic duct. J Clin Invest 119:193–202
Yang D, Li Q, So I, Huang CL, Ando H, Mizutani A, Seki G, Mikoshiba K, Thomas PJ, Muallem S (2011) IRBIT governs epithelial secretion in mice by antagonizing the WNK/SPAK kinase pathway. J Clin Invest 121:956–65
Yannoukakos D, Stuart-Tilley A, Fernandez HA, Fey P, Duyk G, Alper SL (1994) Molecular cloning, expression, and chromosomal localization of two isoforms of the AE3 anion exchanger from human heart. Circ Res 75:603–14
Yegutkin GG, Samburski SS, Jalkalen S, Novak I (2006) ATP-consuming and ATP-generating enzymes secreted by pancreas. J Biol Chem 281:29441–7
You CH, Rominger JM, Chey WY (1983) Potentiation effect of cholecystokinin-octapeptide on pancreatic bicarbonate secretion stimulated by a physiologic dose of secretin in humans. Gastroenterology 85:40–5
Young JA, Martin CJ (1971) The effect of a sympatho- and a parasympathomimetic drug on the electrolyte concentrations of primary and final saliva of the rat submaxillary gland. Pflugers Arch 327:285–302
Young JA, Schögel E (1966) Micropuncture investigation of sodium and potassium excretion in rat submaxillary saliva. Pflugers Arch 291:85–98
Young JA, van Lennep EW (1979) Transport in salivary and salt glands. In: Giebisch G, Tosteson DC, Ussing HH (eds) Membrane transport in biology, vol IVB. Springer, Berlin, pp 563–692
Young JA, Case RM, Conigrave AD, Novak I (1980) Transport of bicarbonate and other anions in salivary secretion. Ann N Y Acad Sci 341:172–90
Zeng W, Lee MG, Yan M, Diaz J, Benjamin I, Marino CR, Kopito R, Freedman S, Cotton C, Muallem S, Thomas P (1997) Immuno and functional characterization of CFTR in submandibular and pancreatic acinar and duct cells. Am J Physiol Cell Physiol 273:C442–55
Zeuthen T (1991) Secondary active transport of water across ventricular cell membrane of choroid plexus epithelium of Necturus maculosus. J Physiol 444:153–73
Zeuthen T, Wright EM (1981) Epithelial potassium transport: tracer and electrophysiological studies in choroid plexus. J Membr Biol 60:105–28
Zhao H, Star RA, Muallem S (1994) Membrane localization of H^+ and HCO_3^- transporters in the rat pancreatic ducts. J Gen Physiol 104:57–85
Zlokovic BV, Mackic JB, Wang L, McComb JG, McDonough A (1993) Differential expression of Na, K-ATPase alpha and beta subunit isoforms at the blood–brain barrier and the choroid plexus. J Biol Chem 268:8019–25

Chapter 6
Mathematical Modeling of Epithelial Ion Transport

David P. Nickerson, Kirk L. Hamilton, and Peter J. Hunter

Abstract In this chapter we provide a general introduction to mathematical modelling of epithelial ion transport. The basic mathematical concepts are introduced and the instantiation of these concepts into numerical simulation is demonstrated. Tools and technologies that aid scientists in the creation and use of epithelial ion transport models are also discussed.

Keywords Epithelial • Mathematical modelling • Ion transport physiology

6.1 Introduction

Epithelial cells are polarised cells in which different transport proteins reside within the apical and basolateral membranes (Fig. 6.1). It is this discrete localisation of transport proteins which allows the specific reabsorption of a solute species across the epithelial cell and, ultimately, its return to the blood. For instance, the reabsorption of sodium from the filtrate of the lumen of the proximal tubule in the renal nephron back to the blood requires the concerted action of both apical-specific and basolateral-specific transporters. The sodium-potassium pump (Na^+/K^+-ATPase) is a primary active transport protein located within the basolateral membrane of a cell and uses ATP to transport sodium against its electrochemical gradient, from the cell into the interstitium (from whence it can diffuse back into the blood, serosal solution in Fig. 6.1). The action of the Na^+/K^+ pump lowers the concentration of sodium inside the cell and thus establishes the electrochemical gradient favouring sodium entry from the lumen of the tubule (mucosal solution in Fig. 6.1) into the cell.

D.P. Nickerson (✉) • P.J. Hunter
Auckland Bioengineering Institute, University of Auckland, Private Bag 92019, Auckland Mail Centre, Auckland 1142, New Zealand
e-mail: d.nickerson@auckland.ac.nz; p.hunter@auckland.ac.nz

K.L. Hamilton
Department of Physiology, Otago School of Medical Sciences, University of Otago, Dunedin, New Zealand
e-mail: kirk.hamilton@otago.ac.nz

K.L. Hamilton, D.C. Devor (eds.), *Ion Channels and Transporters of Epithelia in Health and Disease*, Physiology in Health and Disease,
DOI 10.1007/978-1-4939-3366-2_6

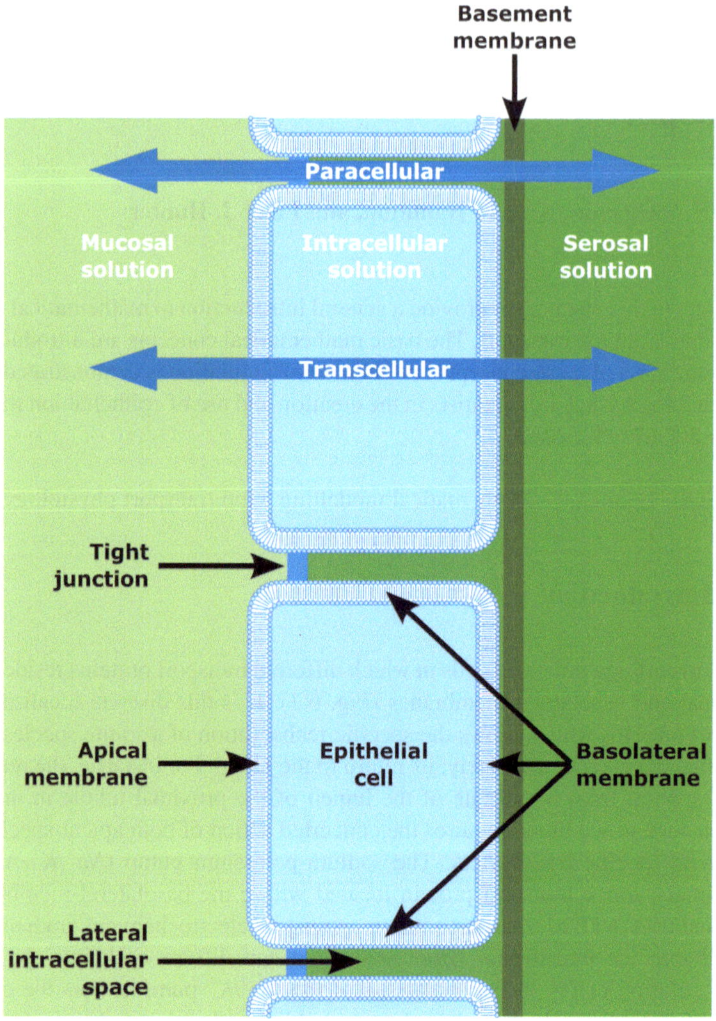

Fig. 6.1 A macroscopic schematic diagram of epithelial tissue illustrating the key anatomical features of the mathematical modelling framework discussed in this chapter. The *black text labels* indicate anatomical features and the *white text* indicates the functional compartments. The *white text with solid arrows* shows the two primary pathways for the transport of solutes between the mucosal and serosal solutions, either through the cell (transcellular) or via the tight junction (paracellular)

In this chapter, we present a general mathematical framework for modelling epithelial ion transport following the derivation of Latta et al. (1984). The mathematical modelling framework is based on the principles of mass conservation and maintenance of electroneutrality. Extending the work of Latta et al. (1984), we discuss recent advances in model description technologies which enable arbitrary extension of the mathematical model.

Figure 6.1 presents a schematic diagram of a macroscopic view of epithelial tissue. While this image clearly shows individual cells and their membranes from which the epithelium is constructed, the mathematical framework presented here takes a physiome-style multiscale view (see Hunter et al. 2010, 2013; de Bono and Hunter 2012, for examples). In this approach, a *CELL* represents the average behaviour of a region of tissue. Different modelling frameworks allow such cell models to be spatially distributed with varying properties to create tissue-scale models (e.g. Nickerson et al. 2011). Here, the cell-scale model is the upper level of the spatial hierarchy that we consider. The integration of subcelluar mechanisms is also discussed below.

Following the introduction of the mathematical modelling framework, we discuss the instantiation of the specific models into computational simulation experiments. As with the modelling framework, such simulation experiments build on the work of Latta et al. (1984). We also refer the reader to some open-source software developed by the authors which provide an example implementation of the described methods. The software and example epithelial transport models are available online at http://get.readthedocs.org/.

6.1.1 Model Exchange and Reproducible Science

The mathematical modelling framework presented here is able to represent models of epithelial transport of arbitrary complexity. In order for such models to be useful to the scientific community, it is important that we are able to share, distribute, and reuse the models. Several community-driven standards for encoding mathematical models have become established and are widely accepted (e.g. Cuellar et al. 2003; Hucka et al. 2003; Nickerson et al. 2006; Christie et al. 2009; Britten et al. 2013). CellML (Cuellar et al. 2003) is a standard that is focused on encoding the mathematical relationships in a modular and reusable manner (Nickerson and Buist 2008; Cooling et al. 2010). The modular hierarchical nature of CellML makes it an ideal format for encoding epithelial transport models as the actual transporters can be archived and managed independently and then assembled together to fit the epithelial model requirements. Section 6.4 discusses some of the benefits of this approach to model description.

In order to be able to reproduce a previous modelling study, one needs to know not only the mathematical model but also an unambiguous description of what tasks were performed with the model(s) to achieve the observed outcomes. In addition to the model encoding standards, guidelines have been established which define the minimal set of information required in order to enable a simulation experiment to be reproduced (Waltemath et al. 2011a). The simulation experiment description markup language (SED-ML) is an instantiation of these guidelines into an XML-based format for capturing descriptions of simulation experiments (Waltemath et al. 2011b). In developing the software tools discussed here, we endeavour to make use of standards such as CellML and SED-ML wherever possible to enable and support reproducible science.

6.2 Epithelial Cell Modelling

Many models can be found in the literature of mathematical models for specific types of epithelial cells (e.g. Weinstein 2010, 2011; Thomas and Dagher 1994; Thomas 2009). As with similar modelling efforts in other areas of computational physiology and systems biology (e.g. Nickerson and Hunter 2006; Land et al. 2014; Swainston et al. 2013), these models have evolved to include a wide range of biophysical details and are capable of investigating novel transport mechanisms. The adoption of model encoding standards in this area lags that of some other fields (e.g. Nickerson and Hunter 2006), and therefore, scientists are required to either implement the models themselves or make use of provided computer codes which conflate the mathematical model and numerical simulation methods.

Here, we describe a general epithelial transport mathematical modelling framework derived from the core conservation laws upon which all the existing models are based. This framework is based on the derivation of Latta et al. (1984) and readers are referred to that work for a comprehensive description of this approach. Understanding this general mathematical framework should help in the interpretation and reuse of existing models and also encourage the adoption of a modular and declarative approach to creating new models of epithelial transport. A graphical view of the mathematical modelling framework is presented in Fig. 6.2, and the common symbols and conventions are also defined in the figure. The following sections define the various components in this modelling framework.

6.2.1 *State Equations*

Conservation of cellular water yields the first state equation for the model, with the rate of change of cell volume, V, defined as

$$\frac{\mathrm{d}V}{\mathrm{d}t} = A^a J_w^a + A^b J_w^b, \tag{6.1}$$

where each of the total membrane water fluxes, J_w, is scaled by their respective membrane areas to take into account the averaged behaviour of the representative cell.

Assuming the mucosal and serosal solutions are sufficiently larger in size than the intracellular solution, we can simplify the model by treating the concentration of the ith solute species as constant in those solutions. Considering the concentration of the ith solute in the intracellular solution, the transport of solute i into the cell

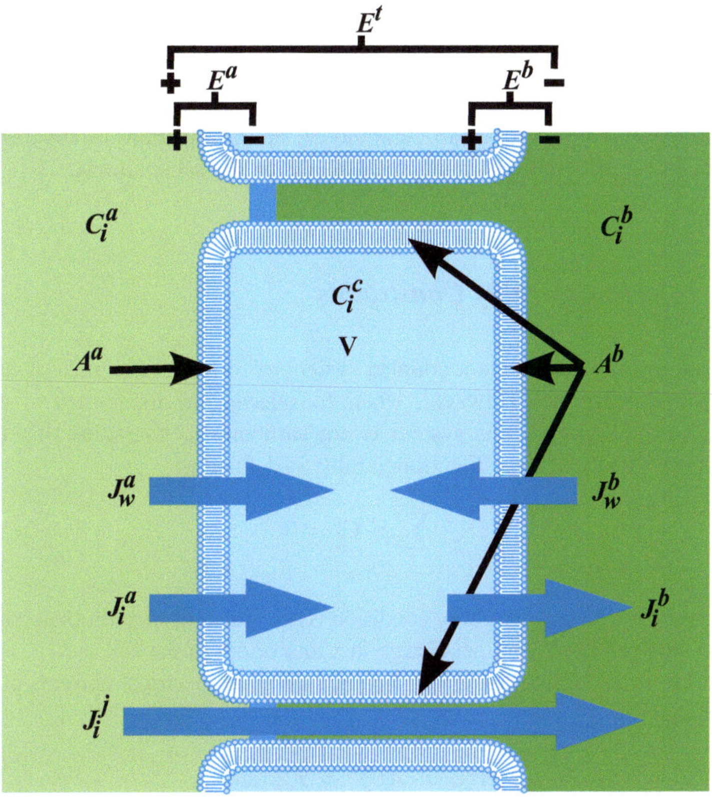

Fig. 6.2 A diagram of a general epithelial transport modelling framework, using the same anatomical schematic as presented in Fig. 6.1. The superscript *a* denotes quantities related to the apical membrane and mucosal solution; *b* denotes the basolateral membrane and serosal solution; and *j* denotes the paracellular pathway. The subscript *i* denotes the *i*th solute species included in the model. *E* represents the membrane potential, *C* the solute concentration, *V* the cell volume, and *A* the membrane area. J_w and J_i are the water and solute fluxes, respectively, and the *arrows* show the direction of positive flux. Based on the diagram in Fig. 1 from Latta et al. (1984)

through the apical membrane, A, is given by the total membrane flux for solute *i* scaled by the membrane area ($A^a J_i^a$), similarly for the transport of solute *i* out of the cell through the basolateral membrane ($A^b J_i^b$). In addition to the direct transport of solutes through the membranes, there is also a contribution from convective transport due to the flow of water through the membranes. This is modelled as $C_i^c\left(A^a J_w^a + A^b J_w^b\right)$, with the change in sign on the basolateral membrane flux due to positive water fluxes defined as influx into the cell (Fig. 6.2).

Combining these relations, mass conservation then defines the rate of change of the concentration of the *i*th solute species in the intracellular solution as

$$V\frac{\mathrm{d}C_i^c}{\mathrm{d}t} = A^a J_i^a - A^b J_i^b - C_i^c\left(A^a J_w^a + A^b J_w^b\right). \tag{6.2}$$

This equation holds for each solute being considered in a particular instantiation of the model. If required for a particular model, similar equations can be introduced for the solute concentrations in the mucosal and/or serosal solutions.

6.2.2 Electroneutrality Constraints

Equations 6.1 and 6.2 define a coupled system of differential equations ensuring that mass is conserved. However, when considering the movement of charged solutes, with valence z_i, this system is not sufficient to guarantee that the cell remains electrically neutral. Electroneutrality is defined as

$$\sum_i z_i\ C_i^c = 0. \tag{6.3}$$

To ensure the electroneutrality condition is not violated, an additional constraint must be imposed to prevent net charge flux into or out of the cell.

The membrane charge fluxes can be represented as electrical currents using the relationships

$$\begin{aligned} I^a &= F\ A^a \sum_i z_i\ J_i^a;\ \text{and} \\ I^b &= F\ A^b \sum_i z_i\ J_i^b, \end{aligned} \tag{6.4}$$

where F is Faraday's constant. Balancing the flow of charge into and out of the cell therefore results in the relationship

$$I^a = I^b, \tag{6.5}$$

which must hold true at all times. See Sect. 6.3 for a description of the numerical techniques that can be used to solve for the membrane potential across each membrane which satisfies Eq. 6.5.

6.2.3 Model Specialisation

The basic principles of mass conservation and maintenance of electroneutrality described above apply to epithelial transport in general. In order to instantiate the general model into a mathematical model for a specific epithelium, all that remains is to define the actual membrane solute and water fluxes of interest to create the specialised model.

Latta et al. (1984) present a minimal set of auxiliary flux equations that describe a specific model of the Koefoed-Johnsen and Ussing (1958) epithelial Na^+ transport. As an example of the specialisation of the general model, we reproduce the Latta et al. (1984) Na^+ model here and refer to the model as the Latta Na^+ model.

6.2.3.1 Water Fluxes

The Latta Na^+ model considers only osmotically induced water flux across the apical and basolateral membranes (Fig. 6.2) and assumes the paracellular pathway is impermeable to water. It is also assumed that the cell membranes are unable to support hydrostatic gradients. Thus the water fluxes are given by

$$\begin{aligned} J_w^a &= L_p^a RT \sum_i \sigma_i^a \left(C_i^c - C_i^a\right); \text{ and} \\ J_w^b &= L_p^b RT \sum_i \sigma_i^b \left(C_i^c - C_i^b\right). \end{aligned} \tag{6.6}$$

L_p^a and L_p^b are the hydraulic conductances of the apical and basolateral membranes, and σ_i^a and σ_i^b are the respective solute reflection coefficients.

6.2.3.2 Passive Solute Fluxes

In the Latta Na^+ model, passive solute fluxes across all membranes are assumed to occur by electrodiffusion and to conform to the Goldman-Hodgkin-Katz constant-field flux equation (Hodgkin and Katz 1949). Passive solute flux of the ith species across the apical membrane is therefore given by

$$J_{p,i}^a = P_i^a z_i \frac{E^a F}{RT} \frac{C_i^a - C_i^c \exp\left(-z_i \frac{E^a F}{RT}\right)}{1 - \exp\left(-z_i \frac{E^a F}{RT}\right)}. \tag{6.7}$$

P_i^a are the apical membrane permeabilities. Similarly, the passive solute flux across the basolateral membrane is given by

$$J_{p,i}^b = P_i^b z_i \frac{E^b F}{RT} \frac{C_i^c - C_i^b \exp\left(-z_i \frac{E^b F}{RT}\right)}{1 - \exp\left(-z_i \frac{E^b F}{RT}\right)}, \tag{6.8}$$

and P_i^b are the basolateral membrane permeabilities.

Finally, under conditions where the junctional permeability to the solutes is finite (i.e. leaky or moderately tight epithelium), the Latta Na^+ model assumes the paracellular flux for the *i*th species to also be passive and given by

$$J_{p,i}^{j} = P_i^{j} \, z_i \frac{E^t F}{RT} \frac{C_i^a - C_i^b \exp\left(-z_i \frac{E^t F}{RT}\right)}{1 - \exp\left(-z_i \frac{E^t F}{RT}\right)}. \tag{6.9}$$

P_i^j represents the solute permeabilities of the paracellular pathway determined by the combined properties of the tight junctions and lateral spaces.

6.2.3.3 Active Solute Fluxes

The mathematical modelling framework we describe here is able to represent arbitrary contributions to transport across the specialised epithelium. In Sect. 6.4, we discuss the use of modelling standards to enable the assembly of specialised models from a library of available transporter models. The Latta Na^+ model includes a sodium pump in the basolateral membrane defined by Lewis and Wills (1981) with the pump current given by

$$I^p = \frac{I_{\max}}{\left[1 + \left(K_{Na}/C_{Na}^c\right)^{n_{Na}}\right] \left[1 + \left(K_K/C_K^b\right)^{n_K}\right]}. \tag{6.10}$$

$I_{\max}$ is the maximum observable pump current, and κ_{NA} and κ_K are the concentrations of intracellular Na^+ and serosal K^+ at which half the maximal pump current is produced. The corresponding basolateral membrane Na^+ and K^+ fluxes can then be expressed in terms of the pump current as

$$J_{Na}^p = n_{Na} I^p / F; \quad J_K^p = -n_K I^p / F. \tag{6.11}$$

6.2.3.4 Total Membrane Solute Fluxes

The Latta Na^+ model assumes the solute flux of the *i*th species across the apical membrane is entirely passive. Similarly, the paracellular flux for the *i*th species is also assumed to be passive. This results in

$$J_i^a = J_{p,i}^a; \tag{6.12}$$

$$J_i^j = J_{p,i}^j. \tag{6.13}$$

For the basolateral membrane, the passive flux is supplemented by the active sodium pump, giving the solute flux of the *i*th species across the basolateral membrane as

$$J_i^b = J_{p,i}^b + J_i^p, \tag{6.14}$$

where J_i^p is given by Eq. 6.11 for Na^+ and K^+ and zero for all other species.

6.3 Computational Simulation

The equations of state described in Sect. 6.2.1 (Eqs. 6.1 and 6.2) can be numerically integrated as a function of time using any reasonable integration algorithm. We typically make use of the CVODE integrator from the SUNDIALS suite of solvers (Hindmarsh et al. 2005). The progression of the simulation follows the following steps, as per the algorithm of Latta et al. (1984):

1. Initialise the model to the desired values at the initial time.
2. Compute the membrane potentials (see Sect. 6.3.1).
3. Compute the solute and water fluxes using the potentials just computed and the flux equations described in Sect. 6.2.3.
4. Integrate Eqs. 6.1 and 6.2 to the next desired time point.
5. Increment the simulation time to the next time point. If the end of the desired simulation interval is reached, the computation ends; otherwise return to step 2 and continue the simulation.

6.3.1 Computation of Membrane Potentials

While integrating the system of differential equations for the complete specialised model, the electroneutrality constraint (Eq. 6.5) must be satisfied at each time point. Thus, the membrane potentials must be recomputed at each time point to ensure that the cell remains electroneutral. The method used to compute the potentials depends on the experimental preparation being simulated by model. Latta et al. (1984) define the numerical algorithms for two common preparations that suffice for the majority of simulations, namely, a preparation that is voltage clamped trans-epithelially, or one that is under current-clamp or open-circuit conditions.

6.3.1.1 Voltage-Clamp Case

Under voltage-clamp conditions, E^T is held at a constant value and the trans-epithelial current I^T is computed as the sum of the transcellular current, $I^A = I^B$, and a paracellular current which can be computed directly from Eq. 6.13 since

$$I^j = F\ A^a \sum_i z_i J_i^j. \tag{6.15}$$

Evaluating the value of the transcellular current requires either E^A or E^B to be defined, since if either is defined, the other can easily be determined given that $E^T = E^A + E^B$. Latta et al. (1984) describe an algorithm for refining an initial estimate of E^A using a Newton–Raphson method to minimise the difference between I^A and I^B. We update this algorithm here by introducing the use of KINSOL, also from the SUNDIALS suite (Hindmarsh et al. 2005) as a general-purpose nonlinear system solver. Given an initial estimate for E^A, usually the value from the previous time step, KINSOL is able to solve the system $I^A - I^B = 0$ to determine the value of E^A and E^B which satisfy the electroneutrality constraint under voltage-clamp conditions.

6.3.1.2 Open-Circuit and Current-Clamp Case

Under current-clamp conditions, I^T is held at a constant value and the resulting transepithelial potential E^T is computed. The open-circuit case corresponds to $I^T = 0$. For the current-clamp case, in addition to the cellular electroneutrality constraint of Eq. 6.5, the maintenance of mucosal and serosal solution electroneutrality requires that $I^A + I^J = I^T$. Therefore, in this case we must simultaneously determine two membrane potentials: E^T and either E^A or E^B. As above, we update the algorithm of Latta et al. (1984) to use KINSOL to solve the system $I^J + I^A - I^T = 0$. Internally to this solve, each estimate of E^T corresponds to the voltage-clamp case described above and that algorithm is used to solve for E^A and E^B for a given value of E^T. This nested solution process will result in membrane potentials E^T, E^A, and E^B which satisfy intracellular, serosal, and mucosal electroneutrality.

6.3.2 *Results*

We have implemented the algorithm described above as part of our freely available and open-source suite of tools available at http://get.readthedocs.org/. In Fig. 6.3, we present results obtained using this implementation with the Latta *Na^+* model to recreate the simulations described in Latta et al. (1984).

6.4 Transporter Modelling

The mathematical modelling framework described in the preceding sections is well suited to the creation of models encoded in a hierarchical and modular manner. The general model defined by the state equations in Sect. 6.2.1 and the electroneutrality

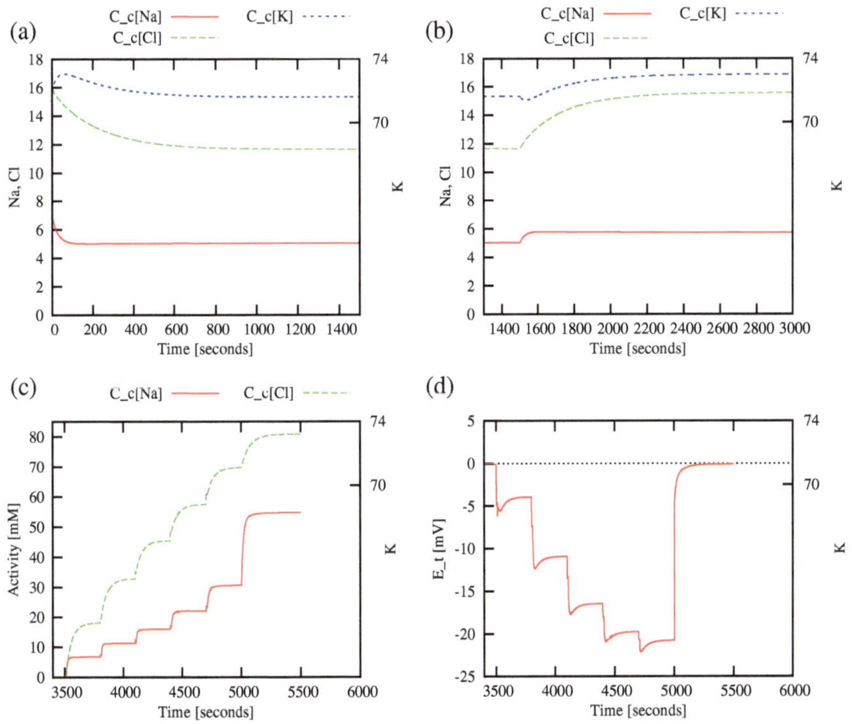

Fig. 6.3 Results from simulating the Latta Na^+ model with the computational algorithms described in Sect. 6.3. These results simply demonstrate the validity of the implementation of the computational algorithm and should be compared to the results discussed in Latta et al. (1984) as follows: (**a**) Fig. 5, (**b**) Fig. 7, (**c**) Fig. 8, and (**d**) Fig. 9

constraint defined in Sect. 6.2.2 provides the top-level model which varies only by the addition of different collections of solute species. The water and solute fluxes can then each be defined independently and integrated into the total flux equations, as demonstrated by the sodium pump in the Latta Na^+ model (Eq. 6.14).

This concept of a hierarchical model description is shown graphically in Fig. 6.4. In Fig. 6.4, we take it one step further and illustrate how the epithelial cellular transport models discussed here can be further integrated into an example tubule model, such as the renal nephron.

The modular and hierarchical nature of the CellML format (Cuellar et al. 2003) makes it an obvious candidate for encoding such multiscale epithelial cellular transport models. Indeed, the tools we develop do use CellML to encode the mathematical models (http://get.readthedocs.org/). As mentioned in Sect. 6.1.1, the use of such a standard format helps to ensure the models created using these tools are able to be widely disseminated and are able to be used by other scientists in their own work.

Using this approach, specialised models such as the one described in Sect. 6.2.3 can be assembled by extracting the required constituent transporter models from a

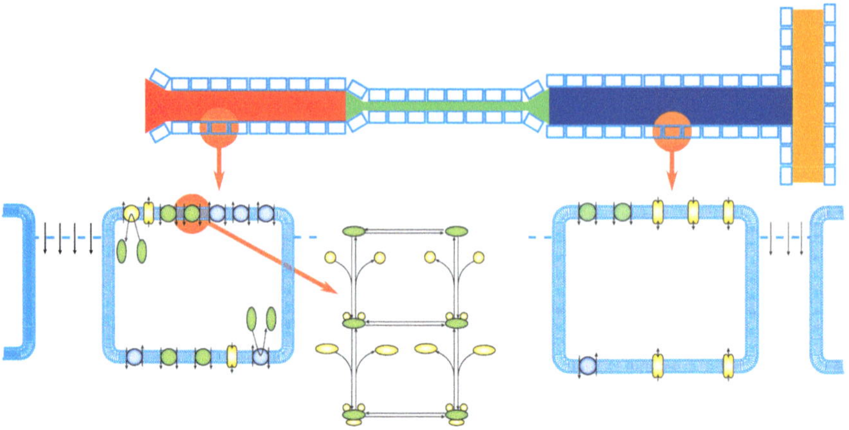

Fig. 6.4 A schematic diagram illustrating the spatial hierarchy and configuration of an example epithelial transport model. The wall of the tubule model displayed at the top of the figure consists of epithelial cells and the cells in different regions have different transport properties. The varying transport properties are due to the different transport mechanisms present in each of the cell variants—detail which can be captured in a hierarchical and modular description of the model

library or repository. The collection of transporter models can then be integrated into the cellular state equations (Eqs. 6.1 and 6.2) and simulations executed following the methods described in Sect. 6.3. Preliminary models are now freely available via the model library at http://get.readthedocs.org/ and current work is focused on extending this library and enhancing the automated model assembly tools. Presenting such model libraries, modelling tools, and simulation experiments via interactive web-based user interfaces is also an ongoing project, building on the work of Nickerson et al. (2011).

Acknowledgments DPN is supported by the Virtual Physiological Rat Project funded through NIH grant P50-GM094503 and the Maurice Wilkins Centre for Molecular Biodiscovery. KLH is supported by a University of Otago Research grant, a strategic grant from the Dean's Fund and the Department of Physiology, University of Otago.

References

Britten RD, Christie GR, Little C, Miller AK, Bradley C, Wu A, Yu T, Hunter P, Nielsen P (2013) FieldML, a proposed open standard for the physiome project for mathematical model representation. Med Biol Eng Comput 51(11):1191–1207. doi:10.1007/s11517-013-1097-7. http://www.ncbi.nlm.nih.gov/pmc/articles/PMC3825639/

Christie GR, Nielsen PM, Blackett SA, Bradley CP, Hunter PJ (2009) FieldML: concepts and implementation. Philos Trans A Math Phys Eng Sci 367(1895):1869–1884. doi:10.1098/rsta.2009.0025. http://www.ncbi.nlm.nih.gov/pmc/articles/PMC2665020/

Cooling MT, Rouilly V, Misirli G, Lawson J, Yu T, Hallinan J, Wipat A (2010) Standard virtual biological parts: a repository of modular modeling components for synthetic biology.

Bioinformatics 26(7):925–931. doi:10.1093/bioinformatics/btq063. http://bioinformatics.oxfordjournals.org/content/26/7/925

Cuellar AA, Lloyd CM, Nielsen PF, Bullivant DP, Nickerson DP, Hunter PJ (2003) An overview of CellML 1.1, a biological model description language. SIMULATION 79(12):740–747. doi:10.1177/0037549703040939. http://sim.sagepub.com/content/79/12/740

de Bono B, Hunter P (2012) Integrating knowledge representation and quantitative modelling in physiology. Biotechnol J 7(8):958–972. doi:10.1002/biot.201100304

Hindmarsh AC, Brown PN, Grant KE, Lee SL, Serban R, Shumaker DE, Woodward CS (2005) SUNDIALS: suite of nonlinear and differential/algebraic equation solvers. ACM Trans Math Softw 31(3):363–396. doi:10.1145/1089014.1089020. http://doi.acm.org/10.1145/1089014.1089020

Hodgkin AL, Katz B (1949) The effect of sodium ions on the electrical activity of the giant axon of the squid. J Physiol 108(1):37–77. http://www.ncbi.nlm.nih.gov/pmc/articles/PMC1392331/

Hucka M, Finney A, Sauro HM, Bolouri H, Doyle JC, Kitano H, Arkin AP, Bornstein BJ, Bray D, Cornish-Bowden A, Cuellar AA, Dronov S, Gilles ED, Ginkel M, Gor V, Goryanin II, Hedley WJ, Hodgman TC, Hofmeyr JH, Hunter PJ, Juty NS, Kasberger JL, Kremling A, Kummer U, Novre NL, Loew LM, Lucio D, Mendes P, Minch E, Mjolsness ED, Nakayama Y, Nelson MR, Nielsen PF, Sakurada T, Schaff JC, Shapiro BE, Shimizu TS, Spence HD, Stelling J, Takahashi K, Tomita M, Wagner J, Wang J (2003) The systems biology markup language (SBML): a medium for representation and exchange of biochemical network models. Bioinformatics 19(4):524–531. doi:10.1093/bioinformatics/btg015. http://bioinformatics.oxfordjournals.org/content/19/4/524

Hunter P, Coveney PV, de Bono B, Diaz V, Fenner J, Frangi AF, Harris P, Hose R, Kohl P, Lawford P, McCormack K, Mendes M, Omholt S, Quarteroni A, Skr J, Tegner J, Randall Thomas S, Tollis I, Tsamardinos I, van Beek JHGM, Viceconti M (2010) A vision and strategy for the virtual physiological human in 2010 and beyond. Philos Trans A Math Phys Eng Sci 368 (1920):2595–2614. doi:10.1098/rsta.2010.0048. http://www.ncbi.nlm.nih.gov/pmc/articles/PMC2944384/

Hunter P, Chapman T, Coveney PV, de Bono B, Diaz V, Fenner J, Frangi AF, Harris P, Hose R, Kohl P, Lawford P, McCormack K, Mendes M, Omholt S, Quarteroni A, Shublaq N, Skr J, Stroetmann K, Tegner J, Thomas SR, Tollis I, Tsamardinos I, van Beek JHGM, Viceconti M (2013) A vision and strategy for the virtual physiological human: 2012 update. Interface Focus 3(2):doi:10.1098/rsfs.2013.0004. http://www.ncbi.nlm.nih.gov/pmc/articles/PMC3638492/

Koefoed-Johnsen V, Ussing HH (1958) The nature of the frog skin potential. Acta Physiol Scand 42(3–4):298–308. doi:10.1111/j.1748-1716.1958.tb01563.x. http://onlinelibrary.wiley.com/doi/10.1111/j.1748-1716.1958.tb01563.x/abstract

Land S, Niederer SA, Louch WE, Re T, Aronsen JM, Stuckey DJ, Sikkel MB, Tranter MH, Lyon AR, Harding SE, Smith NP (2014) Computational modeling of takotsubo cardiomyopathy: effect of spatially varying -adrenergic stimulation in the rat left ventricle. Am J Physiol Heart Circ Physiol 307(10):H1487–H1496. doi:10.1152/ajpheart.00443.2014. http://www.ncbi.nlm.nih.gov/pmc/articles/PMC4233305/

Latta R, Clausen C, Moore LC (1984) General method for the derivation and numerical solution of epithelial transport models. J Membrane Biol 82(1):67–82. doi:10.1007/BF01870733. http://link.springer.com/article/10.1007/BF01870733

Lewis SA, Wills NK (1981) Interaction between apical and basolateral membranes during sodium transport across tight epithelia. Soc Gen Physiol Ser 36:93–107

Nickerson D, Buist M (2008) Practical application of CellML 1.1: the integration of new mechanisms into a human ventricular myocyte model. Prog Biophys Mol Biol 98(1):38–51. doi:10.1016/j.pbiomolbio.2008.05.006. http://www.sciencedirect.com/science/article/pii/S0079610708000461

Nickerson DP, Hunter PJ (2006) The Noble cardiac ventricular electrophysiology models in CellML. Prog Biophys Mol Biol 90(13):346–359. doi:10.1016/j.pbiomolbio.2005.05.007. http://www.sciencedirect.com/science/article/pii/S0079610705000313

Nickerson D, Nash M, Nielsen P, Smith N, Hunter P (2006) Computational multiscale modeling in the IUPS physiome project: modeling cardiac electromechanics. IBM J Res Dev 50(6): 617–630. doi:10.1147/rd.506.0617

Nickerson DP, Terkildsen JR, Hamilton KL, Hunter PJ (2011) A tool for multi-scale modelling of the renal nephron. Interface Focus 1(3):417–425. doi:10.1098/rsfs.2010.0032. http://rsfs.royalsocietypublishing.org/content/1/3/417

Swainston N, Mendes P, Kell DB (2013) An analysis of a community-driven reconstruction of the human metabolic network. Metabolomics 9(4):757–764. doi:10.1007/s11306-013-0564-3. http://www.ncbi.nlm.nih.gov/pmc/articles/PMC3715687/

Thomas SR (2009) Kidney modeling and systems physiology. WIREs Syst Biol Med 1(2): 172–190. doi:10.1002/wsbm.14. http://onlinelibrary.wiley.com/doi/10.1002/wsbm.14/abstract

Thomas SR, Dagher G (1994) A kinetic model of rat proximal tubule transport–load-dependent bicarbonate reabsorption along the tubule. Bull Math Biol 56(3):431–458

Waltemath D, Adams R, Beard DA, Bergmann FT, Bhalla US, Britten R, Chelliah V, Cooling MT, Cooper J, Crampin EJ, Garny A, Hoops S, Hucka M, Hunter P, Klipp E, Laibe C, Miller AK, Moraru I, Nickerson D, Nielsen P, Nikolski M, Sahle S, Sauro HM, Schmidt H, Snoep JL, Tolle D, Wolkenhauer O, Le Novre N (2011a) Minimum information about a simulation experiment (MIASE). PLoS Comput Biol 7(4), e1001,122. doi:10.1371/journal.pcbi.1001122

Waltemath D, Adams R, Bergmann FT, Hucka M, Kolpakov F, Miller AK, Moraru II, Nickerson D, Sahle S, Snoep JL, Novre NL (2011b) Reproducible computational biology experiments with SED-ML - the simulation experiment description markup language. BMC Syst Biol 5(1):198. doi:10.1186/1752-0509-5-198. http://www.biomedcentral.com/1752-0509/5/198/abstract

Weinstein AM (2010) A mathematical model of rat ascending Henle limb. I. Cotransporter function. Am J Physiol Renal Physiol 298(3):F512–F524. doi:10.1152/ajprenal.00230.2009. http://ajprenal.physiology.org/content/298/3/F512

Weinstein AM (2011) Potassium deprivation: a systems approach. Am J Physiol Renal Physiol 301(5): F967–F968. doi:10.1152/ajprenal.00430.2011. http://ajprenal.physiology.org/content/301/5/F967

Chapter 7
Molecular Mechanisms of Apical and Basolateral Sorting in Polarized Epithelial Cells

Ora A. Weisz and Heike Fölsch

Abstract The proper localization of ion channels and transporters to the apical or basolateral membrane domains is essential for correct ion transport through epithelia. Polarized sorting is achieved through elaborate intracellular sorting pathways encompassing the *trans*-Golgi network, early endosomes, and recycling endosomes that are fine-tuned for epithelial function and thus different from the secretory and endocytic systems in non-polarized cells. Polarized cells express epithelial cell-specific molecules, including sorting adaptors like the cytosolic clathrin adaptor AP-1B, that are required for efficient segregation and targeting of proteins to specific plasma membrane domains. In this book chapter, we will first discuss the various sorting stations in the cells followed by a review of apical and basolateral sorting signals and their interpretation at different compartments. Specific emphasis is placed on the trafficking of ion transporters.

Keywords Membrane traffic • Kidney • Biosynthetic • Endocytosis • Recycling • *Trans*-Golgi network • Endosome • AP-1B • Apical • Basolateral

Abbreviations

AEE	Apical early endosome
ARE	Apical RE
BEE	Basolateral early endosome
CRE	Common RE
RE	Recycling endosome

O.A. Weisz
Renal-Electrolyte Division, Department of Medicine, University of Pittsburgh, 3550 Terrace Street, Pittsburgh 15261, PA, USA

H. Fölsch (✉)
Department of Cell and Molecular Biology, Feinberg School of Medicine, Northwestern University, 303 East Chicago Avenue, Chicago, IL 60611, USA
e-mail: h-folsch@northwestern.edu

K.L. Hamilton, D.C. Devor (eds.), *Ion Channels and Transporters of Epithelia in Health and Disease*, Physiology in Health and Disease,
DOI 10.1007/978-1-4939-3366-2_7

TGN *trans*-Golgi network
TJ Tight junction

7.1 General Organization of Secretory and Endocytic Pathways

Typically, transmembrane proteins begin their journey to the cell surface in the endoplasmic reticulum (ER) during co-translational insertion into the ER membrane. In this compartment, N-glycans are added to consensus sequences within the protein ectodomains, and proteins are folded and in many cases assembled into oligomers. Properly folded proteins are transported via a coat protein complex II (COPII)-mediated pathway to the Golgi apparatus where they achieve their final glycosylation patterns. The most distal compartment of the Golgi complex, the *trans*-Golgi network (TGN), is the first major sorting station in the cells. From there, proteins may traffic to lysosomes, endosomes, and the cell surface. On their journey to the cell surface, they may traverse early or recycling endosomal populations. Often, the same sorting signals that are used during biosynthetic delivery are also used during endocytic recycling of transmembrane proteins (Matter et al. 1993; Potter et al. 2006b). In the case of polarized cells, an additional level of sorting exists to selectively direct proteins to and from the apical (luminal) and basolateral plasma membranes [reviewed in (Ang and Fölsch 2012; Fölsch et al. 2009; Rodriguez-Boulan et al. 2005)].

Much of the work on polarized sorting in epithelial cells has been performed in the Madin-Darby canine kidney (MDCK) cell line, derived in the 1950s from a female cocker spaniel. The type II subclone of these cells is easily and rapidly cultured as well-differentiated monolayers when plated on permeable support membranes. Moreover, these cells express highly regular microvilli similar to that observed in the brush border of the proximal tubule and elaborate a primary cilium. In MDCK and other cells of kidney origin, most newly synthesized apical and basolateral proteins are targeted vectorially (albeit not necessarily directly) to the plasma membrane. The fidelity of delivery to a given domain is never 100 %, and transcytotic retrieval pathways exist to reroute missorted proteins (Casanova et al. 1991). Other epithelial cell types make more use of this transcytotic pathway to deliver newly synthesized apical proteins. In hepatocytes, almost all transmembrane proteins are delivered first to the sinusoidal (basolateral) membrane, and bile canalicular (apical) targeting follows after internalization (Bartles et al. 1987). An exception to this rule is the class of polytopic membrane proteins, including the bile acid transporters, which are targeted vectorially from the TGN to the apical surface of liver cells (Sai et al. 1999). Intestinal cells, including the polarized Caco-2 cell line, also use the transcytotic pathway to a significant extent to deliver apical proteins, but a sizable fraction of some proteins is also targeted via vectorial pathways (Le Bivic et al. 1990; Matter et al. 1990). In the remainder of this book chapter, we will focus our attention on kidney cells and particularly on knowledge gained using MDCK cells.

After initial delivery to the apical or basolateral plasma membrane, transmembrane receptors may be internalized into early endosomes underlying the plasma membranes. Notably, epithelial cells maintain apical early endosomes (AEEs) and basolateral early endosomes (BEEs) that are biochemically distinct and physically segregated. Indeed, AEEs and BEEs did not fuse with one another in an in vitro fusion assay (Bomsel et al. 1990). Moreover, after treatment of MDCK cells with latrunculin B to disrupt the actin cytoskeleton, AEEs and BEEs were misplaced and intermixed in the cytoplasm, without undergoing fusion reactions (Sheff et al. 2002). Although it is currently unclear which proteins define AEEs and BEEs, their existence probably ensures correct delivery of internalized receptors back to their membrane of origin during rapid recycling from AEEs or BEEs.

In addition to rapid recycling, transmembrane receptors may be sorted further into the cells to reach lysosomes in case they are destined for degradation or recycling endosomes (REs) for sorting back to either their membrane of origin or the opposing membrane during apical-to-basolateral or basolateral-to-apical transcytosis. Unlike in non-polarized cells, REs in polarized cells constitute a second major sorting station of equal importance to the TGN. Indeed, REs in polarized cells are organized into subdomains to accommodate the various sorting needs of cargo in polarized epithelial cells. Two views of REs in MDCK cells have emerged based on conflicting data. In one model, REs are considered to be a single population of interconnected endosomal membranes with different subdomains specialized for targeting to diverse compartments including domains that cluster cargo destined for the apical or basolateral domains [Fig. 7.1b, reviewed in (Ang and Fölsch 2012)]. Notably, the basolateral targeting domain depends on and contains proteins that interact with the AP-1B clathrin adaptor (Fig. 7.1d; see also Sect. 7.3.4). The second model holds that two classes of REs exist: a transferrin receptor (TfnR)-positive common RE (CRE) that can receive internalized cargo from both apical and basolateral surface domains and a specialized apical RE (ARE), marked by Rab11a, that excludes basolaterally recycling cargo and communicates with the CRE and the apical domain [Fig. 7.1a, reviewed in (Mostov et al. 2003)]. These models may not be mutually exclusive, and it is possible that the structure of the RE system in MDCK cells reflects their differentiated state and the need for communication between apical and basolateral surface domains. From here on, we will use the terms ARE and CRE or simply just REs interchangeably. Furthermore, if not noted otherwise, reviewed studies were performed in MDCK cells.

7.2 Sorting to the Apical Membrane

A wide variety of transporters, enzymes, and receptors are present at the apical surface of kidney cells. The steady-state distribution of these proteins can be modulated by changes in the fidelity of biochemical and post-endocytic sorting as well as by alterations in the kinetics of delivery and removal from the surface. Dissecting the sorting signals on proteins that guide them to the apical or basolateral

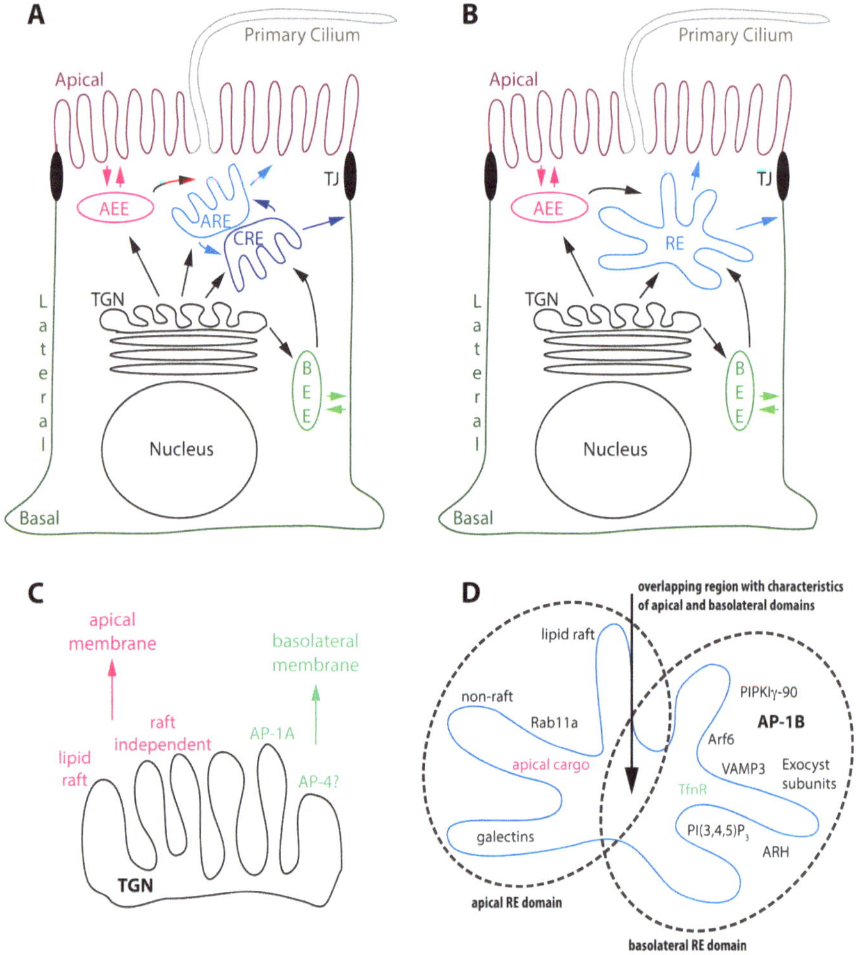

Fig. 7.1 This model figure depicts the intracellular sorting stations present in polarized epithelial cells. (**a**) illustrates the model of two separated populations of REs, whereas (**b**) illustrates the model championing only one RE. (**c**) depicts an expanded version of the TGN showing the different TGN domains responsible for apical and basolateral sorting. (**d**) depicts an expanded version of the RE in (**b**) showing its subdomains. Note that the overlapping region between the apical and the basolateral sorting domain may contain any of the components listed for each domain including, but not necessarily limited to, Rab11a, Arf6, and exocyst subunits. See Sect. 7.3.4 for a detailed description of the basolateral RE domain

surface is further complicated by the observation that many proteins contain multiple signals of varying strengths. Thus, mutation or ablation of a given signal on a protein does not always lead to its non-polarized distribution as one would expect, but rather can result in its profound retargeting to the opposing plasma membrane domain or to another compartment. These difficulties aside, we are slowly expanding our understanding of the specific signals that direct proteins apically and how these signals are interpreted by the cellular machinery.

7.2.1 Apical Sorting Signals

The apical sorting information on proteins described to date constitutes a remarkably diverse collection of signals that includes peptide sequences, posttranslational modifications, and structural information [reviewed in (Weisz and Rodriguez-Boulan 2009)].

7.2.2 GPI Anchors

The first known apical targeting signal to be identified was the presence of a glycosylphosphatidylinositol (GPI) lipid anchor on proteins expressed in MDCK cells (Lisanti et al. 1989). These anchors enable protein localization to glycolipid-enriched lipid microdomains (lipid rafts), and it was proposed that this in turn facilitates their segregation into apically destined sorting platforms at the TGN and possibly at distal sites along the biosynthetic pathway as well (Simons and Ikonen 1997). However, subsequent studies have shown that this is an overly simplistic model. There is now considerable evidence that additional clustering of GPI-anchored proteins (GPI-APs) into high-molecular-weight oligomers is required for apical sorting [reviewed in (Muniz and Zurzolo 2014)]. This clustering may be mediated by protein-protein, protein-lipid, and/or glycan-dependent interactions and may be cell type dependent (Imjeti et al. 2011; Paladino et al. 2004, 2008). Remodeling of GPI-AP acyl chains can also facilitate oligomer formation (Muniz and Zurzolo 2014).

7.2.3 Glycan-Dependent Sorting Signals

A second class of apical sorting signals is dependent on the addition or modification of N- or O-linked oligosaccharides [reviewed in (Potter et al. 2006a)]. Dissection of these signals can be difficult as N-glycans are often essential for the proper folding and ER export of proteins. That said, specific N-glycan structures or linkages have directly been implicated in apical sorting of several proteins. For example, sialylation of N-glycans on the sialomucin endolyn has been demonstrated to be essential for its efficient apical sorting (Mo et al. 2012).

The paucity of reagents and approaches to perturb O-linked glycans has made the role of this modification more challenging to address. Specifically, it has been difficult to determine whether the oligosaccharides themselves, as opposed to the structural consequences of O-glycosylation at a specific region of a protein, is essential for polarized sorting (Yeaman et al. 1997; Youker et al. 2013). For example, the O-glycosylated imperfect tandem repeats of MUC1 contain transferable apical sorting information, but no specific sugar residues have yet been shown to be essential for sorting (Kinlough et al. 2011).

How glycans function to direct sorting remains unclear, although there is increasing evidence for a role of galectins in mediating the apical sorting of a subset of proteins (Delacour et al. 2009). Galectins 3 and 9 have been shown to function in apical sorting of cargos that do not partition into lipid rafts in MDCK cells (Delacour et al. 2006; Mo et al. 2012). Similar to the clustering mechanism proposed for lipid raft-associated proteins, galectin 3-mediated oligomerization of apical cargos into high-molecular-weight complexes has been demonstrated (Delacour et al. 2007). Interestingly, galectins 4 and 9 also bind to glycolipids in lipid rafts in intestinal and kidney cells, respectively, and may also facilitate the clustering of raft-associated proteins (Delacour et al. 2005; Mishra et al. 2010; Stechly et al. 2009).

7.2.4 Peptide-Based Sorting Signals

Peptide sequences present in either the luminal, transmembrane, or cytoplasmically exposed regions of many proteins have been identified as important for their apical localization. Indeed, peptide-based signals are the most common type of sorting information identified for polytopic proteins, including many ion transporters and G-protein-coupled receptors. However, no unifying mechanism has emerged for how these sequences function to direct apical localization. Rather, these identified motifs have been variously implicated in polarized sorting, surface delivery, and cell surface retention. Sorting of proteins via selective retention at the plasma membrane will be discussed in more detail in Sect. 7.4.

The observation that some splice variants of ion transporters are targeted to opposing domains of epithelial cells has provided a unique opportunity to identify apical and basolateral sorting information encoded within these proteins. For example, in the presence of a 45-amino-acid insertion in the first cytosolic loop of the Ca^{2+}-ATPase, PMCA2 directs apical sorting, whereas splice variants lacking this sequence are targeted basolaterally in MDCK cells (Chicka and Strehler 2003). While this example might imply that the determination of peptide-based apical sorting information is straightforward, this is far from the case. Indeed, there are only a few cases where linear amino acid sequences that confer transferable apical localization have been demonstrated, and even then, the identification of key residues required for sorting has been fraught. For example, the sole difference between the basolaterally targeted human GLUT9a and the apically targeted splice variant GLUT9b is the absence of 29 amino acids within the amino terminus of GLUT9b. The shorter amino terminus of GLUT9b could redirect the normally basolaterally trafficked GLUT1 transporter to the apical surface, demonstrating that it contains transferable apical sorting information. However, no specific sequence unique to GLUT9b could be identified as essential for apical sorting (Bibee et al. 2013). One possible conclusion of these studies is that the amino terminal domain of GLUT9b interacts with or positions other regions of the transporter to create a conformation that somehow promotes apical sorting. These observations are consistent with a model in which apical sorting of some proteins is conferred by information dependent on a specific protein conformation as opposed to a specific amino acid sequence.

For other ion transporters, short stretches of amino acids have been identified most frequently within cytosolically oriented loops or tails, which direct their apical localization. A ten-amino-acid sequence (PIKPVFKGFS) that includes a putative β-turn within the C-terminus of the human sodium-dependent vitamin C transporter hSVCT1 is required for apical delivery in MDCK cells (Subramanian et al. 2004). Interestingly, similar motifs have also been shown to be important for apical targeting of the neuronal sodium-dependent glutamate transporter EAAT3 (Cheng et al. 2002) and in the rat sodium-dependent bile acid transporter (Sun et al. 2003). In addition, short and apparently unrelated peptide sequences direct the apical localization of the potassium MaxiK channel (Kwon and Guggino 2004), the mouse sodium-phosphate cotransporter NaPi-IIc [WLHSL; (Ito et al. 2010)], and the cystic fibrosis transmembrane conductance regulator [CFTR; (Milewski et al. 2000)]. It is not yet known how these nonterminal short sequences control apical polarity. One possibility is that they interact directly with apical targeting or retention machinery. In support of this, a five-amino-acid (SQDAL) sequence within a cytoplasmic loop of the ATP-binding cassette transporter isoform C2 (ABCC2; aka MRP2) is required for apical sorting in HepG2 cells via an apparently saturable mechanism (Bandler et al. 2008; Emi et al. 2012).

Despite the sizable literature on polarized sorting, we still know little about the mechanisms by which apical sorting determinants on polytopic proteins, including ion transporters, are interpreted. The difficulty in identifying specific sequences or transferable motifs that confer apical sorting of polytopic proteins may in part be due to the intrinsic challenges of generating mutations that do not disrupt their folding or interactions with accessory subunits. Alternatively, the sequences or domains required for apical sorting may influence targeting via steric or conformational effects rather than by binding to interacting proteins. Consistent with this idea, there are only a very few cases where apical sorting information has been demonstrated to be saturable or where similar peptide motifs have been identified in multiple proteins. Once proteins arrive at the apical surface, binding to the cytoskeleton via scaffolding proteins or PDZ-domain interactions can enhance their retention at this domain (see Sect. 7.4 for further details).

7.2.5 Apical Sorting Mechanisms

Clustering mediated by incorporation into glycolipid-enriched microdomains or by galectin-mediated cross-linking is implicated in the polarized sorting of lipid-raft-associated and glycan-dependent apical proteins. The possibility that peptide-based sorting determinants regulate the conformation of polytopic proteins suggests that clustering may be an underlying mechanism for the apical targeting of this class of proteins as well (Bigay and Antonny 2012; Dehmelt and Bastiaens 2010). This is not to say that other features of polytopic proteins do not contribute to their clustering and/or apical sorting. There is certainly precedence for incorporation of ion channels and transporters into lipid rafts (Ares and Ortiz 2012; Bravo-Zehnder et al. 2000; Chen et al. 2011; Hill et al. 2007; Riquier et al. 2009). Moreover, glycan processing may enhance the oligomerization of polytopic proteins (Ozaslan et al. 2003). Thus, various features of a given protein may guide its incorporation into homomeric or heteromeric arrays that facilitate apical delivery.

How might cargo clustering lead to apical sorting? Cytosolic domains of some apical proteins may bind directly to Rab proteins, molecular motors, or adaptors. Proteins that may facilitate clustering and/or trafficking to the apical membrane include but are not limited to VIP17/MAL, caveolin-1, annexins, and the four-phosphate adaptor protein 2 [FAPP2; reviewed in (Fölsch 2008)]. Clustering will increase the avidity of these interactions to create a sorting "platform" that may nucleate the assembly of an apically destined transport carrier. Many apically destined proteins have very short (<10 amino acid) cytoplasmic tails or, in the case of GPI-anchored proteins, no domains accessible to the cytosol. Co-clustering of these proteins with others that can recruit sorting machinery may be essential for their sorting. Moreover, incorporation of transmembrane VAMPs into these clusters may be necessary for eventual fusion of transport carriers with the apical surface. VAMPs implied in fusion of vesicles with the apical membrane are VAMP2, VAMP7 (aka TI-VAMP), and VAMP8 (Caceres et al. 2014; Galli et al. 1998; Pocard et al. 2007). It should be noted that the concept of spatially regulated protein self-assembly or scaffolding into higher-order clusters is not unique to apical sorting, and indeed models for self-organization have been demonstrated to drive numerous cellular functions [elegantly reviewed in (Bigay and Antonny 2012; Dehmelt and Bastiaens 2010)].

Another unresolved question is: where along the biosynthetic pathway does apical sorting occur? Newly synthesized apical and basolateral proteins are segregated at least by the time they leave the TGN. Indeed, there is also evidence that raft-associated and raft-independent apical proteins are segregated when leaving this compartment [Fig. 7.1c, (Guerriero et al. 2008; Jacob and Naim 2001)]. Moreover, it has become increasingly clear that these two classes of proteins take distinct routes to the apical surface and that these routes intersect different endocytic compartments (Cresawn et al. 2007; Fölsch et al. 2009). Lipid raft-associated proteins may pass through AEEs, whereas glycan-dependent and some other apical cargos transit Rab11a-positive AREs. In addition, intestinal and kidney cells also express the Rab11b isoform, which appears to be localized to a subdomain of the ARE (Lapierre et al. 2003). Apically recycling CFTR and the epithelial sodium channel ENaC traffic through Rab11b-positive endosomes, although it is not known whether newly synthesized proteins also access this compartment (Butterworth et al. 2012; Silvis et al. 2009). Complicating matters further, there are multiple exit routes for proteins that traffic through the ARE (Mattila et al. 2012).

How can the concept of cluster-dependent sorting be reconciled with the presence of multiple apical sorting stations? If there is continual segregation of apical proteins into different export pathways as they move from one compartment to another, there may be different subsets of clusters that form at each site. Indeed, clustering mediated by distinct mechanisms may be initiated in different compartments. It is believed that proteins are incorporated into lipid rafts during their transit through the Golgi (Guerriero et al. 2008; Jacob and Naim 2001), and galectin 9 that binds to the raft-enriched glycolipid Forssman antigen (as well as to non-raft-associated proteins) appears to access these structures at the TGN of MDCK cells (Mishra et al. 2010). In contrast, galectin 3 interaction with glycosylated proteins occurs after they leave the TGN (Delacour et al. 2007; Straube et al. 2013). Thus, non-raft-associated proteins that traffic together to the ARE (e.g., proteins with glycan-dependent and peptide-based sorting signals) might be segregated into distinct populations that differentially exit that compartment based on their ability to interact with galectin 3. Whether such hierarchical arrays of clustering actually occur remains to be determined.

7.3 Sorting to the Basolateral Membrane

Like apical targeting, there are multiple pathways that cargos may follow to reach the basolateral membrane during biosynthetic delivery and endocytic recycling (Arnspang et al. 2013; Farr et al. 2009; Fölsch et al. 2009). Many newly synthesized cargos travel from the TGN into REs during biosynthetic delivery, and cargos may be directed into the basolateral pathway from either of these sites (Ang et al. 2004; Fölsch et al. 2009; Nokes et al. 2008). Often, basolateral sorting signals are linear peptide motifs that are *cis*-dominant over apical sorting information such that mutation of the basolateral sorting signals results in a protein that is sorted to the apical membrane instead. In addition, basolateral sorting signals are often colinear with lysosomal sorting signals and endocytic motifs. Indeed, low amounts of the lysosomal-associated membrane glycoprotein Lamp1 and the cation-dependent mannose 6-phosphate receptor (CD-MPR) that cycles between the TGN and late endosomes exclusively cycle through the basolateral membrane in MDCK cells (Distel et al. 1998; Hunziker et al. 1991; Nabi et al. 1991).

7.3.1 Basolateral Sorting Signals and Adaptors

Linear peptide sorting motifs that target proteins to the basolateral membrane are frequently either tyrosine based [YxxØ, where Ø is a bulky hydrophobic amino acid (F, I, L, M, or V), and FxNPxY] or dileucine based (LL, LI, or LM) often in the form of [D/E]xxxL[L/I]. These signals are generally recognized by cytosolic adaptor complexes, most commonly heterotetrameric clathrin adaptor protein

(AP) complexes AP-1 (γ-β1-μ1-σ1), AP-2 (α-β2-μ2-σ2), AP-3 (δ-β3-μ3-σ3), and AP-4 (ε-β4-μ4-σ4) that does not associate with clathrin (Boehm and Bonifacino 2001; Hirst and Robinson 1998; Ohno et al. 1999). YxxØ signals are recognized by the AP medium subunits, whereas [D/E]xxxL[L/I]-signals are recognized simultaneously by σ1-γ – subunits of AP-1 as well as σ2-α of AP-2 and σ3-δ of AP-3 (Doray et al. 2007; Janvier et al. 2003; Mattera et al. 2011). Another form of dileucine signal (DxxLL) is recognized at the TGN by monomeric clathrin adaptor proteins of the GGA (Golgi-localized, γ-ear containing, Arf-binding) protein family [discussed in (Kirchhausen 2002)], perhaps leading to incorporation into AP-1A vesicles (Doray et al. 2002). Moreover, the cytoplasmic tail of the transforming growth factor α (TGFα) encodes a specialized LL signal that is aided by a cluster of negative charges and recognized by the cytosolic adaptor Naked2 for sorting to the basolateral membrane (Li et al. 2004, 2007). Finally, FxNPxY signals are bridged to clathrin adaptor complexes by connector proteins such as numb, Dab2, and autosomal recessive hypercholesterolemia protein (ARH) [reviewed in (Traub 2003)].

Other peptide-based sorting motifs have been described that seem to be far less abundant. For example, the basolateral sorting signal of CD147 consists of only a single leucine residue (Deora et al. 2004). A different, non-related mono-leucine signal for basolateral sorting was also discovered in amphiregulin (Gephart et al. 2011). No adaptor proteins have yet been described that recognize these mono-leucine sorting signals. In addition, proteins may be sorted to and selectively retained at the basolateral membrane by means of PDZ-binding motif interaction with PDZ domain-containing proteins [reviewed in (Brone and Eggermont 2005); see also Sect. 7.4]. PDZ-directed sorting from endosomes may be facilitated by the sorting nexin 27 [SNX27; (Lauffer et al. 2010)]. Furthermore, basolateral sorting from early endosomes may be facilitated by the "retromer" complex and SNX17 [reviewed in (Bonifacino 2014)]. SXN17 binds NPxY peptide motifs (Donoso et al. 2009; Stiegler et al. 2014).

Epithelial cells express two closely related AP-1 complexes, AP-1A and AP-1B, that differ only in the incorporation of their respective medium subunits μ1A or μ1B. Notably, μ1B expression is specific to columnar epithelial cells including MDCK and Caco-2 cell lines (Fölsch et al. 1999; Ohno et al. 1999). So far both AP-1A and AP-4 have been implicated in sorting of basolateral cargos at the TGN (Fig. 7.1c), and AP-1B was shown to facilitate cargo sorting in REs [reviewed in (Ang and Fölsch 2012; Rodriguez-Boulan et al. 2013)]. Interestingly, whereas the connector proteins numb and Dab2 specifically interact with AP-2, ARH also interacts with AP-1B in REs to facilitate basolateral sorting of cargos with FxNPxY signals (Kang and Fölsch 2011).

7.3.2 AP-1B Expression in the Kidneys

AP-1B is expressed in many regions of the kidney including the medullary and cortical thick ascending limbs (TALs) and the cortical collecting duct (CCD) (Schreiner et al. 2010). However, AP-1B is not expressed in the proximal tubules and cell lines that were derived from that region (e.g., LLC-PK1 cells) (Ohno et al. 1999; Schreiner et al. 2010). Therefore, LLC-PK1 cells are a useful model system for analyzing AP-1B function and to determine if a protein depends on AP-1B for steady-state localization at the basolateral membrane. To this end researchers use LLC-PK1 cells stably expressing exogenous copies of μ1B or μ1A as a control. Proteins that are apical in μ1A-expressing LLC-PK1 cells, but basolateral in μ1B-expressing cells, are then identified as AP-1B-dependent cargos (Fields et al. 2007; Fölsch et al. 1999). Because AP-1B is the only adaptor complex that localizes in REs to facilitate basolateral sorting during endocytic recycling (Fölsch et al. 1999, 2001; Gan et al. 2002), its cargo proteins will be missorted to the apical membrane in the absence of functional AP-1B (Fields et al. 2007; Fölsch et al. 1999; Gravotta et al. 2007; Sugimoto et al. 2002). Proteins dependent on AP-1B for basolateral localization include many nutrient receptors like low-density lipoprotein receptor (LDLR) and TfnR, as well as the adhesion proteins E-cadherin and β1 integrin, and the growth hormone receptor EGFR (Fölsch et al. 1999; Gravotta et al. 2007; Ling et al. 2007; Ryan et al. 2010).

7.3.3 ARH Expression in the Kidneys

ARH expression is found predominantly in the distal nephron tubules including the distal convoluted tubule (DCT), the connecting tubule (CNT), as well as the cortical collecting duct (CCD) and to a lesser extent in the regions of the thick ascending limbs (TALs) (Fang et al. 2009). Thus, ARH expression largely overlaps with AP-1B expression in the kidneys ensuring AP-1B-dependent cargo sorting of proteins with FxNPxY sorting motifs.

7.3.4 Mechanisms of AP-1B-Mediated Basolateral Sorting

What is known so far about AP-1B function on a molecular level? It is thought that recruitment of AP-1B onto REs depends on PI(3,4,5)P_3 and possibly Arf6 (Fields et al. 2010; Shteyn et al. 2011). Strikingly, PI(3,4,5)P_3 accumulates in REs only in epithelial cells that express AP-1B in a reaction that may involve the phosphatidylinositol 4-phosphate 5-kinase PIPKIγ-90 (Fields et al. 2010). Indeed PIPKIγ-90 was shown to directly interact with AP-1B, but not AP-1A, and to play a role in basolateral sorting of E-cadherin (Ling et al. 2007). After membrane recruitment,

AP-1B selects its cargos either directly or with the help of its co-adaptor ARH (Kang and Fölsch 2011). In addition, AP-1B recruits at least 2 subunits of the exocyst complex to REs for incorporation into AP-1B vesicles (Fölsch et al. 2003). AP-1B vesicles also contain the SNARE protein VAMP3 (aka cellubrevin) (Fields et al. 2007), and disruption of VAMP3 function in MDCK cells leads to dispersed REs to which AP-1B no longer localizes (Fields et al. 2007).

Tethering of AP-1B vesicles to the basolateral membrane is thought to depend on the exocyst complex (Grindstaff et al. 1998; Yeaman et al. 2001), and fusion of the vesicles may involve VAMP3 on the vesicles and syntaxin 4 at the basolateral membrane (Fields et al. 2007). Interestingly, basolateral sorting of syntaxin 4 itself depends on AP-1B expression, although how AP-1B influences basolateral sorting of syntaxin 4 is not yet entirely clear (Reales et al. 2011; Torres et al. 2011).

In addition to AP-1B-dependent sorting, syntaxin 4 is most likely also involved in fusion of vesicles that originated from BEEs. Indeed, syntaxin 4 was shown to directly interact with the early endosomal Rab4 protein (Li et al. 2001), and disruption of Rab4 function in MDCK cells resulted in apical missorting of basolateral membrane proteins (Mohrmann et al. 2002).

7.3.5 Basolateral Sorting of Transporters

Many transporters at the basolateral membrane contain tyrosine-based sorting motifs, but are independent of AP-1B. This is perhaps due to low internalization rates and/or stable retention at the basolateral membrane. For example, the kidney anion exchanger 1 (kAE1) contains a YDEV signal in its cytoplasmic tail that is needed for AP-1A-mediated basolateral sorting at the TGN (Almomani et al. 2012; Junking et al. 2014; Toye et al. 2004). However, although kAE1 is initially sorted to the basolateral membrane independent of AP-1B expression, AP-1B is probably needed during endocytic recycling (Almomani et al. 2012). Tyrosine-based motifs have also been described in the chicken kAE1-4 splice isoform that contains these signals in a 63-amino-acid N-terminal extension, which is missing in the apically sorted kAE1-3 splice isoform (Adair-Kirk et al. 1999). Interestingly, a disease-causing mutation in the C-terminal residue (M909T) of the human kAE1 results in the creation of a type I PDZ ligand and partial redirection of the protein to the apical surface (Fry et al. 2012). Because the wild-type protein is targeted directly to the basolateral domain, it was concluded that the presence of this PDZ-interacting motif causes apical misdirection of a subset of the mutant kAE1 within the biosynthetic pathway (Fry et al. 2012), although it is also possible that this mutation causes selective retention at the apical membrane.

Other transporters that do not rely on AP-1B for basolateral targeting are the homotetrameric water channels aquaporins 3 and 4 that are expressed in renal collecting ducts. Aquaporins have six transmembrane domains, and both the N- and C-termini are exposed to the cytosol. Aquaporin 3 contains a tyrosine-based sorting motif that overlaps with a dileucine-based signal (YRLL) in its N-terminus

(Rai et al. 2006). Mutation of either signal partially redirected aquaporin 3 to the apical membrane, and mutation of both motifs effectively disrupted surface delivery. Because aquaporin 3 is also localized to the basolateral membrane in LLC-PK1 cells, it was concluded that AP-1B is not needed for its basolateral sorting (Rai et al. 2006). Indeed, other receptors that contain dileucine-based motifs such as the Fc receptor FcRII-B2 likewise do not need AP-1B for basolateral targeting (Fields et al. 2007).

In contrast to aquaporin 3, aquaporin 4 contains basolateral sorting information in its C-terminus (Madrid et al. 2001). The first of two motifs is a tyrosine-based (S) YMEV(E) motif that also serves as an internalization motif and directly interacts with the medium subunits of AP-2 and AP-3, but not with AP-1A. The second motif is an acidic dileucine motif (ETEDLIL). Curiously, the YMEVE sequence may conform to the consensus sequence recognized by AP-4 (Burgos et al. 2010; Dell'Angelica et al. 1999), and it remains to be determined if it indeed interacts with this adaptor. Importantly, this motif overlaps with a casein kinase II (CKII) consensus site (SxxE/D), and phosphorylation by CKII introduces a negative charge directly upstream of the tyrosine leading to enhanced interaction with AP-3 and thus delivery of aquaporin 4 into lysosomes for degradation (Madrid et al. 2001). It is tempting to speculate that aquaporin 4 may be sorted at the TGN into AP-4 vesicles for basolateral targeting. After internalization, interaction with AP-3 may effectively remove the water channel from BEEs such that little if any aquaporin 4 may enter REs. Indeed, aquaporin 4 is also localized to the basolateral membrane in renal proximal tubules, indicating that AP-1B plays no role in its basolateral sorting (van Hoek et al. 2000).

7.4 Retention at the Cell Surface Through Interaction with PDZ Domains

Selective retention at the apical or basolateral membrane may be achieved through interaction with a specialized cytoskeleton. For example, ankyrin G and β2 spectrin specifically anchor proteins at the lateral membrane (Kizhatil et al. 2007). More common is retention through interaction of PDZ-binding motifs at the extreme C-terminus of transmembrane cargos with PDZ domains in scaffolding proteins. However, the presence of a PDZ-binding motif does not assure that it functions in retention, as disruption of the C-terminal PDZ-interacting sequences of hSMVT or ABCC2 has no effect on their localization (Nies et al. 2002; Subramanian et al. 2009). In other cases, retention at the apical membrane through a PDZ-binding motif may be overridden by a strong basolateral sorting signal as was the case for aquaporin 2 chimeras with aquaporin 3, in which the apical targeting signal in aquaporin 2, which is a PDZ-binding motif, was maintained, and the basolateral sorting information of aquaporin 3 was added to the N-terminus effectively directing the chimera to the basolateral surface (Rai et al. 2006).

Although PDZ-binding motifs are thought to primarily facilitate selective retention at the plasma membrane, they are often also important for surface delivery (Lauffer et al. 2010; Maday et al. 2008).

PDZ-binding motifs may be present in addition to other sorting signals. For example, the basolateral potassium channel Kir2.3 contains a type I PDZ-binding motif that partially overlaps with a basolateral sorting signal; both motifs are needed for robust basolateral localization of Kir2.3 (Le Maout et al. 2001). Importantly, the PDZ-binding motif is necessary for stable expression at the plasma membrane, and its deletion results in largely intracellular localization of the mutant channel. Retention at the basolateral membrane is most likely achieved through binding of Kir2.3 to the PDZ domain containing Lin-7/CASK protein complex (Olsen et al. 2002). Similarly, the epithelial γ-aminobutyric acid (GABA) transporter (BGT-1) is also retained at the basolateral membrane through interactions with Lin-7 (Perego et al. 1999). In MDCK cells, Lin-7 associates with adherens junctions through interaction with β-catenin (Perego et al. 2000).

Some transporters have tyrosine-based signals that are used for internalization, but not for basolateral targeting. For example, the renal outer medullary potassium secretory channel ROMK (aka Kir1.1) has an YDNPNFV motif that binds to ARH and serves as an endocytic signal (Fang et al. 2009). However, ROMK is localized to the apical membrane domain, although it is expressed in the kidney distal tubules that express both ARH and AP-1B (Fang et al. 2009; Schreiner et al. 2010). ROMK has a PDZ-binding motif at its C-terminus that is necessary for membrane expression of ROMK and interacts with the PDZ domain-containing proteins Na/H exchange regulatory factors NHERF1 and 2 (Yoo et al. 2004). Perhaps ROMK binding to NEHRF selectively stabilizes ROMK at the apical membrane effectively overriding basolateral sorting by AP-1B and ARH. PDZ-binding motifs have also been shown to contribute to the steady-state apical localization of other ion transporters (Brone and Eggermont 2005) including CFTR (Milewski et al. 2000; Moyer et al. 1999), TRPV4 and TRPV5 (van de Graaf et al. 2006), and PMCA2b (Antalffy et al. 2012).

Interestingly, stabilizing proteins at the apical membrane through binding to NHERF may also ensure their exclusion from the primary cilium that extends from the apical membrane. Primary cilia are important appendages needed for correct cell signaling and homeostasis, and their membrane composition is distinct from the apical (and basolateral) membranes. Most apical membrane proteins are excluded from the primary cilium, and it is thought that the base of the cilium contains a diffusion barrier that is formed by a member of the septin GTPase family, septin 2 (Hu et al. 2010). However, additional mechanisms exist. For example, it has been shown that interaction with NHERF1 prevents the apical membrane protein podocalyxin from entering the primary cilium (Francis et al. 2011). It is thought that NHERF1 immobilizes proteins in the apical membrane by cross-linking them to the apical actin network via ERM family members (Francis et al. 2011). It seems reasonable to assume that other apical proteins that interact with NHERF will be excluded from the primary cilium as well.

7.5 Sorting of Multi-subunit Transporters

Polarized sorting of heteromeric transporters may appear convoluted as each individual subunit may contain sorting information. Moreover, association with different subunit isoforms may drive the heteromers toward the apical or basolateral membrane. A well-studied example of this is the Na,K-ATPase, which in most tissues is localized basolaterally. The Na,K-ATPase consists of three subunits: the catalytic α-subunit, the β1-subunit that is thought to aid TGN exit of the complex, and the γ-subunit that modulates ion channel activity. Multiple sorting signals have been described for these subunits. For example, both the α-subunit and β1-subunit contain basolateral sorting information (Marrs et al. 1995; Muth et al. 1998; Shoshani et al. 2005). However, the signal in the α-subunit can be overturned by pairing with a different β-subunit (Vagin et al. 2005). Whereas expression of β1 directs the Na,K-ATPase α-subunit to the basolateral surface, β2 is responsible for apical localization of the Na,K-ATPase in gastric adenocarcinoma cells. Interestingly, this apical distribution of β2 is apparently mediated by its prevalence of N-glycans, although the mechanism is unknown. Moreover, addition of new N-glycosylation consensus sequences to the more sparsely glycosylated β1-subunit converts it to an apically localized protein (Vagin et al. 2005). Basolateral localization of the Na,K-ATPase has been further attributed to selective retention via interaction with the cytoskeleton (i.e., ankyrin G) (Kizhatil et al. 2007; Marrs et al. 1995) or through homotopic interaction of β1-subunits across adjacent cells (Shoshani et al. 2005). Interestingly though, basolateral localization of the β1-subunit in sensory hair cells in zebrafish depends on AP-1 expression (Clemens Grisham et al. 2013), indicating that at least the β1-subunit may have a tyrosine or dileucine-based motif that could be recognized by AP-1A at the TGN for basolateral sorting. Notably, AP-1B is not necessary for the basolateral localization of the Na,K-ATPase (Fölsch et al. 1999).

Both AP-1A and AP-1B have been implicated in the basolateral sorting of the H, K-ATPase β-subunit. This subunit has a reversed tyrosine-based signal, FRHY, that interacts with AP-1A and AP-1B in vitro (Duffield et al. 2004). The pump is localized to the basolateral membrane in MDCK cells and to the apical membrane in LLC-PK1 cells (Roush et al. 1998). Curiously though, the pump is also localized to the apical membrane in LLC-PK1 cells stably transfected to express AP-1B, which normally reinstates basolateral sorting of proteins with tyrosine-based sorting motifs. This suggests that the FRHY signal may be masked in LLC-PK1 cells by an unknown mechanism (Duffield et al. 2004).

Caplan and colleagues have employed the sequence homology and structural similarities between the apically targeted gastric H,K-ATPase and the basolateral Na,K-ATPase to dissect the apical targeting information in the former. Interestingly, they found that replacing the fourth transmembrane domain (TM4) of the Na, K-ATPase α-subunit with that of the closely related H,K-ATPase sequence was sufficient to redirect the Na,K-ATPase to the apical surface of polarized LLC-PK1 cells (Dunbar et al. 2000). In addition, apical targeting of the Na,K-ATPase was also observed when only the cytoplasmic and intracellular loop sequences surrounding this domain (but not TM4 itself) were replaced with the corresponding sequences from the H,K-ATPase. Similarly, a six-amino-acid sequence within the extracytoplasmic loop preceding TM4 of the related non-gastric H,K-ATPase was shown to be necessary for apical targeting in MDCK cells. In this case, a single point mutation in this region was sufficient to disrupt apical sorting (Lerner et al. 2006).

7.6 Challenges to the Field

Despite the many advances that were achieved over the years in determining polarized sorting mechanisms in epithelial cells, there remain many challenges to researchers in the field. For example, a confounding problem when working with ion transporters is that they are generally expressed at low levels in cells, and their folding, assembly, and surface delivery are often inefficient. Due to these technical challenges, polarized sorting determinants of polytopic proteins are often dissected by monitoring the steady-state distribution of heterologously expressed and sometimes epitope-tagged mutants. Unfortunately, by using this approach, it is often difficult to gather information about the mechanisms that enable polarized localization of a given protein (e.g., polarized biosynthetic targeting vs. surface retention).

Another challenge for the field is the lack of available well-differentiated cell culture models to study the trafficking of ion transporters in their native cell type. As noted above, polarized delivery routes differ significantly between cells of liver, intestinal, and kidney origin. Moreover, there can be differences in the regulation of membrane traffic between distinct cell types in a single organ. As one example, cells in the proximal tubule of the kidney are highly specialized for apical endocytosis as opposed to cells further downstream along the nephron, and the organization of apical endocytic compartments in these cells reflects the functional elaboration of this pathway (Mattila et al. 2014). Similarly, the absence of μ1B expression in the proximal tubules has profound implications for the localization of many proteins. However, few cell culture models are available that maintain both the expression of endogenous transporters and the polarization of distinct kidney nephron cell types. While many challenges remain, recent advances in protein expression and knockdown strategies as well as the development of live animal imaging modalities have facilitated the study of ion transporter trafficking in situ. These techniques, in combination with the use of MDCK cells as an in vitro model, should lead to a better understanding of surface delivery of ion transporters in the future.

Acknowledgments We apologize to colleagues whose work we could not cite due to space constraints. Work in the Weisz lab is supported by NIH R01 DK100357, and work in the Fölsch lab is supported by NIH R01 GM0707036.

References

Adair-Kirk TL, Cox KH, Cox JV (1999) Intracellular trafficking of variant chicken kidney AE1 anion exchangers: Role of alternative NH_2 termini in polarized sorting and Golgi recycling. J Cell Biol 147:1237–1248

Almomani EY, King JC, Netsawang J, Yenchitsomanus PT, Malasit P, Limjindaporn T, Alexander RT, Cordat E (2012) Adaptor protein 1 complexes regulate intracellular trafficking of the kidney anion exchanger 1 in epithelial cells. Am J Physiol Cell Physiol 303:C554–C566

Ang AL, Taguchi T, Francis S, Fölsch H, Murrells LJ, Pypaert M, Warren G, Mellman I (2004) Recycling endosomes can serve as intermediates during transport from the Golgi to the plasma membrane of MDCK cells. J Cell Biol 167:531–543

Ang SF, Fölsch H (2012) The role of secretory and endocytic pathways in the maintenance of cell polarity. Essays Biochem 53:29–39

Antalffy G, Mauer AS, Paszty K, Hegedus L, Padanyi R, Enyedi A, Strehler EE (2012) Plasma membrane calcium pump (PMCA) isoform 4 is targeted to the apical membrane by the w-splice insert from PMCA2. Cell Calcium 51:171–178

Ares GR, Ortiz PA (2012) Dynamin2, clathrin, and lipid rafts mediate endocytosis of the apical Na/K/2Cl cotransporter NKCC2 in thick ascending limbs. J Biol Chem 287:37824–37834

Arnspang EC, Sundbye S, Nelson WJ, Nejsum LN (2013) Aquaporin-3 and aquaporin-4 are sorted differently and separately in the trans-Golgi network. PLoS ONE 8, e73977

Bandler PE, Westlake CJ, Grant CE, Cole SP, Deeley RG (2008) Identification of regions required for apical membrane localization of human multidrug resistance protein 2. Mol Pharmacol 74:9–19

Bartles JR, Feracci HM, Stieger B, Hubbard AL (1987) Biogenesis of the rat hepatocyte plasma membrane in vivo: comparison of the pathways taken by apical and basolateral proteins using subcellular fractionation. J Cell Biol 105:1241–1251

Bibee KP, Augustin R, Gazit V, Moley KH (2013) The apical sorting signal for human GLUT9b resides in the N-terminus. Mol Cell Biochem 376:163–173

Bigay J, Antonny B (2012) Curvature, lipid packing, and electrostatics of membrane organelles: defining cellular territories in determining specificity. Dev Cell 23:886–895

Boehm M, Bonifacino JS (2001) Adaptins: the final recount. Mol Biol Cell 12:2907–2920

Bomsel M, Parton R, Kuznetsov SA, Schroer TA, Gruenberg J (1990) Microtubule- and motor-dependent fusion in vitro between apical and basolateral endocytic vesicles from MDCK cells. Cell 62:719–731

Bonifacino JS (2014) Adaptor proteins involved in polarized sorting. J Cell Biol 204:7–17

Bravo-Zehnder M, Orio P, Norambuena A, Wallner M, Meera P, Toro L, Latorre R, Gonzalez A (2000) Apical sorting of a voltage- and Ca^{2+} –activated K^{+} channel alpha -subunit in Madin-Darby canine kidney cells is independent of N-glycosylation. Proc Natl Acad Sci USA 97:13114–13119

Brone B, Eggermont J (2005) PDZ proteins retain and regulate membrane transporters in polarized epithelial cell membranes. Am J Physiol Cell Physiol 288:C20–C29

Burgos PV, Mardones GA, Rojas AL, daSilva LL, Prabhu Y, Hurley JH, Bonifacino JS (2010) Sorting of the Alzheimer's disease amyloid precursor protein mediated by the AP-4 complex. Dev Cell 18:425–436

Butterworth MB, Edinger RS, Silvis MR, Gallo LI, Liang X, Apodaca G, Frizzell RA, Johnson JP (2012) Rab11b regulates the trafficking and recycling of the epithelial sodium channel (ENaC). Am J Physiol Renal Physiol 302:F581–F590

Caceres PS, Mendez M, Ortiz PA (2014) Vesicle-associated Membrane Protein 2 (VAMP2) but Not VAMP3 Mediates cAMP-stimulated Trafficking of the Renal $Na^{+}–K^{+}–2Cl^{-}$ Co-transporter NKCC2 in Thick Ascending Limbs. J Biol Chem 289:23951–23962

Casanova JE, Mishumi Y, Ikehara Y, Hubbard AL, Mostov KE (1991) Direct apical sorting of rat liver dipeptidylpeptidase IV expressed in Madin-Darby canine kidney cells. J Biol Chem 266:24428–24432

Chen G, Howe AG, Xu G, Frohlich O, Klein JD, Sands JM (2011) Mature N-linked glycans facilitate UT-A1 urea transporter lipid raft compartmentalization. FASEB J 25:4531–4539
Cheng CL, Glover G, Banker G, Amara SG (2002) A novel sorting motif in the glutamate transporter excitatory amino acid transporter 3 directs its targeting in Madin-Darby canine kidney cells and hippocampal neurons. J Neurosci 22:10643–10652
Chicka MC, Strehler EE (2003) Alternative splicing of the first intracellular loop of plasma membrane Ca^{2+}-ATPase isoform 2 alters its membrane targeting. J Biol Chem 278:18464–18470
Clemens Grisham R, Kindt K, Finger-Baier K, Schmid B, Nicolson T (2013) Mutations in ap1b1 cause mistargeting of the Na^{+}/K^{+}-ATPase pump in sensory hair cells. PLoS ONE 8, e60866
Cresawn KO, Potter BA, Oztan A, Guerriero CJ, Ihrke G, Goldenring JR, Apodaca G, Weisz OA (2007) Differential involvement of endocytic compartments in the biosynthetic traffic of apical proteins. EMBO J 26:3737–3748
Dehmelt L, Bastiaens PI (2010) Spatial organization of intracellular communication: insights from imaging. Nat Rev 11:440–452
Delacour D, Cramm-Behrens CI, Drobecq H, Le Bivic A, Naim HY, Jacob R (2006) Requirement for galectin-3 in apical protein sorting. Curr Biol 16:408–414
Delacour D, Gouyer V, Zanetta JP, Drobecq H, Leteurtre E, Grard G, Moreau-Hannedouche O, Maes E, Pons A, Andre S, Le Bivic A, Gabius HJ, Manninen A, Simons K, Huet G (2005) Galectin-4 and sulfatides in apical membrane trafficking in enterocyte-like cells. J Cell Biol 169:491–501
Delacour D, Greb C, Koch A, Salomonsson E, Leffler H, Le Bivic A, Jacob R (2007) Apical sorting by galectin-3-dependent glycoprotein clustering. Traffic 8:379–388
Delacour D, Koch A, Jacob R (2009) The role of galectins in protein trafficking. Traffic 10:1405–1413
Dell'Angelica EC, Mullins C, Bonifacino JS (1999) AP-4, a novel protein complex related to clathrin adaptors. J Biol Chem 274:7278–7285
Deora AA, Gravotta D, Kreitzer G, Hu J, Bok D, Rodriguez-Boulan E (2004) The basolateral targeting signal of CD147 (EMMPRIN) consists of a single leucine and is not recognized by retinal pigment epithelium. Mol Biol Cell 15:4148–4165
Distel B, Bauer U, Le Borgne R, Hoflack B (1998) Basolateral sorting of the cation-dependent mannose 6-phosphate receptor in Madin-Darby canine kidney cells. Identification of a basolateral determinant unrelated to clathrin-coated pit localization signals. J Biol Chem 273:186–193
Donoso M, Cancino J, Lee J, van Kerkhof P, Retamal C, Bu G, Gonzalez A, Caceres A, Marzolo MP (2009) Polarized traffic of LRP1 involves AP1B and SNX17 operating on Y-dependent sorting motifs in different pathways. Mol Biol Cell 20:481–497
Doray B, Ghosh P, Griffith J, Geuze HJ, Kornfeld S (2002) Cooperation of GGAs and AP-1 in packaging MPRs at the trans-Golgi network. Science 297:1700–1703
Doray B, Lee I, Knisely J, Bu G, Kornfeld S (2007) The gamma/sigma1 and alpha/sigma2 hemicomplexes of clathrin adaptors AP-1 and AP-2 harbor the dileucine recognition site. Mol Biol Cell 18:1887–1896
Duffield A, Fölsch H, Mellman I, Caplan MJ (2004) Sorting of H,K-ATPase beta-subunit in MDCK and LLC-PK cells is independent of mu1B adaptin expression. Traffic 5:449–461
Dunbar LA, Aronson P, Caplan MJ (2000) A transmembrane segment determines the steady-state localization of an ion-transporting adenosine triphosphatase. J Cell Biol 148:769–778
Emi Y, Yasuda Y, Sakaguchi M (2012) A *cis*-acting five-amino-acid motif controls targeting of ABCC2 to the apical plasma membrane domain. J Cell Sci 125:3133–3143
Fang L, Garuti R, Kim BY, Wade JB, Welling PA (2009) The ARH adaptor protein regulates endocytosis of the ROMK potassium secretory channel in mouse kidney. J Clin Invest 119:3278–3289

Farr GA, Hull M, Mellman I, Caplan MJ (2009) Membrane proteins follow multiple pathways to the basolateral cell surface in polarized epithelial cells. J Cell Biol 186:269–282
Fields IC, King SM, Shteyn E, Kang RS, Fölsch H (2010) Phosphatidylinositol 3,4,5-trisphosphate localization in recycling endosomes is necessary for AP-1B-dependent sorting in polarized epithelial cells. Mol Biol Cell 21:95–105
Fields IC, Shteyn E, Pypaert M, Proux-Gillardeaux V, Kang RS, Galli T, Fölsch H (2007) v-SNARE cellubrevin is required for basolateral sorting of AP-1B-dependent cargo in polarized epithelial cells. J Cell Biol 177:477–488
Fölsch H (2008) Regulation of membrane trafficking in polarized epithelial cells. Curr Opin Cell Biol 20:208–213
Fölsch H, Mattila PE, Weisz OA (2009) Taking the scenic route: biosynthetic traffic to the plasma membrane in polarized epithelial cells. Traffic 10:972–981
Fölsch H, Ohno H, Bonifacino JS, Mellman I (1999) A novel clathrin adaptor complex mediates basolateral targeting in polarized epithelial cells. Cell 99:189–198
Fölsch H, Pypaert M, Maday S, Pelletier L, Mellman I (2003) The AP-1A and AP-1B clathrin adaptor complexes define biochemically and functionally distinct membrane domains. J Cell Biol 163:351–362
Fölsch H, Pypaert M, Schu P, Mellman I (2001) Distribution and function of AP-1 clathrin adaptor complexes in polarized epithelial cells. J Cell Biol 152:595–606
Francis SS, Sfakianos J, Lo B, Mellman I (2011) A hierarchy of signals regulates entry of membrane proteins into the ciliary membrane domain in epithelial cells. J Cell Biol 193:219–233
Fry AC, Su Y, Yiu V, Cuthbert AW, Trachtman H, Frankl FEK (2012) Mutation conferring apical-targeting motif on AE1 exchanger causes autosomal dominant distal RTA. J Am Soc Nephrol 23:1238–1249
Galli T, Zahraoui A, Vaidyanathan VV, Raposo G, Tian JM, Karin M, Niemann H, Louvard D (1998) A novel tetanus neurotoxin-insensitive vesicle-associated membrane protein in SNARE complexes of the apical plasma membrane of epithelial cells. Mol Biol Cell 9:1437–1448
Gan Y, McGraw TE, Rodriguez-Boulan E (2002) The epithelial-specific adaptor AP1B mediates post-endocytic recycling to the basolateral membrane. Nat Cell Biol 4:605–609
Gephart JD, Singh B, Higginbotham JN, Franklin JL, Gonzalez A, Fölsch H, Coffey RJ (2011) Identification of a novel mono-leucine basolateral sorting motif within the cytoplasmic domain of amphiregulin. Traffic 12:1793–1804
Gravotta D, Deora A, Perret E, Oyanadel C, Soza A, Schreiner R, Gonzalez A, Rodriguez-Boulan E (2007) AP1B sorts basolateral proteins in recycling and biosynthetic routes of MDCK cells. Proc Natl Acad Sci USA 104:1564–1569
Grindstaff KK, Yeaman C, Anandasabapathy N, Hsu SC, Rodriguez-Boulan E, Scheller RH, Nelson WJ (1998) Sec6/8 complex is recruited to cell-cell contacts and specifies transport vesicle delivery to the basal-lateral membrane in epithelial cells. Cell 93:731–740
Guerriero CJ, Lai Y, Weisz OA (2008) Differential sorting and Golgi export requirements for raft-associated and raft-independent apical proteins along the biosynthetic pathway. J Biol Chem 283:18040–18047
Hill WG, Butterworth MB, Wang H, Edinger RS, Lebowitz J, Peters KW, Frizzell RA, Johnson JP (2007) The epithelial sodium channel (ENaC) traffics to apical membrane in lipid rafts in mouse cortical collecting duct cells. J Biol Chem 282:37402–37411
Hirst J, Robinson MS (1998) Clathrin and adaptors. Biochim Biophys Acta 1404:173–193
Hu Q, Milenkovic L, Jin H, Scott MP, Nachury MV, Spiliotis ET, Nelson WJ (2010) A septin diffusion barrier at the base of the primary cilium maintains ciliary membrane protein distribution. Science 329:436–439
Hunziker W, Harter C, Matter K, Mellman I (1991) Basolateral sorting in MDCK cells requires a distinct cytoplasmic domain determinant. Cell 66:907–920

Imjeti NS, Lebreton S, Paladino S, de la Fuente E, Gonzalez A, Zurzolo C (2011) N-Glycosylation instead of cholesterol mediates oligomerization and apical sorting of GPI-APs in FRT cells. Mol Biol Cell 22:4621–4634

Ito M, Sakurai A, Hayashi K, Ohi A, Kangawa N, Nishiyama T, Sugino S, Uehata Y, Kamahara A, Sakata M, Tatsumi S, Kuwahata M, Taketani Y, Segawa H, Miyamoto K (2010) An apical expression signal of the renal type IIc Na^+–dependent phosphate cotransporter in renal epithelial cells. Am J Physiol Renal Physiol 299:F243–F254

Jacob R, Naim HY (2001) Apical membrane proteins are transported in distinct vesicular carriers. Curr Biol 11:1444–1450

Janvier K, Kato Y, Boehm M, Rose JR, Martina JA, Kim BY, Venkatesan S, Bonifacino JS (2003) Recognition of dileucine-based sorting signals from HIV-1 Nef and LIMP-II by the AP-1 gamma-sigma1 and AP-3 delta-sigma3 hemicomplexes. J Cell Biol 163:1281–1290

Junking M, Sawasdee N, Duangtum N, Cheunsuchon B, Limjindaporn T, Yenchitsomanus PT (2014) Role of adaptor proteins and clathrin in the trafficking of human kidney anion exchanger 1 (kAE1) to the cell surface. Traffic 15:788–802

Kang RS, Fölsch H (2011) ARH cooperates with AP-1B in the exocytosis of LDLR in polarized epithelial cells. J Cell Biol 193:51–60

Kinlough CL, Poland PA, Gendler SJ, Mattila PE, Mo D, Weisz OA, Hughey RP (2011) Core-glycosylated mucin-like repeats from MUC1 are an apical targeting signal. J Biol Chem 286:39072–39081

Kirchhausen T (2002) Single-handed recognition of a sorting traffic motif by the GGA proteins. Nat Struct Biol 9:241–244

Kizhatil K, Yoon W, Mohler PJ, Davis LH, Hoffman JA, Bennett V (2007) Ankyrin-G and beta2-spectrin collaborate in biogenesis of lateral membrane of human bronchial epithelial cells. J Biol Chem 282:2029–2037

Kwon SH, Guggino WB (2004) Multiple sequences in the C terminus of MaxiK channels are involved in expression, movement to the cell surface, and apical localization. Proc Natl Acad Sci USA 101:15237–15242

Lapierre LA, Dorn MC, Zimmerman CF, Navarre J, Burnette JO, Goldenring JR (2003) Rab11b resides in a vesicular compartment distinct from Rab11a in parietal cells and other epithelial cells. Exp Cell Res 290:322–331

Lauffer BE, Melero C, Temkin P, Lei C, Hong W, Kortemme T, von Zastrow M (2010) SNX27 mediates PDZ-directed sorting from endosomes to the plasma membrane. J Cell Biol 190:565–574

Le Bivic A, Quaroni A, Nichols B, Rodriguez-Boulan E (1990) Biogenetic pathways of plasma membrane proteins in Caco-2, a human intestinal epithelial cell line. J Cell Biol 111:1351–1361

Le Maout S, Welling PA, Brejon M, Olsen O, Merot J (2001) Basolateral membrane expression of a K^+ channel, Kir 2.3, is directed by a cytoplasmic COOH-terminal domain. Proc Natl Acad Sci USA 98:10475–10480

Lerner M, Lemke D, Bertram H, Schillers H, Oberleithner H, Caplan MJ, Reinhardt J (2006) An extracellular loop of the human non-gastric H, K-ATPase alpha-subunit is involved in apical plasma membrane polarization. Cell Physiol Biochem 18:75–84

Li C, Franklin JL, Graves-Deal R, Jerome WG, Cao Z, Coffey RJ (2004) Myristoylated Naked2 escorts transforming growth factor alpha to the basolateral plasma membrane of polarized epithelial cells. Proc Natl Acad Sci USA 101:5571–5576

Li C, Hao M, Cao Z, Ding W, Graves-Deal R, Hu J, Piston DW, Coffey RJ (2007) Naked2 acts as a cargo recognition and targeting protein to ensure proper delivery and fusion of TGF-alpha containing exocytic vesicles at the lower lateral membrane of polarized MDCK cells. Mol Biol Cell 18:3081–3093

Li L, Omata W, Kojima I, Shibata H (2001) Direct interaction of Rab4 with syntaxin 4. J Biol Chem 276:5265–5273

Ling K, Bairstow SF, Carbonara C, Turbin DA, Huntsman DG, Anderson RA (2007) Type Igamma phosphatidylinositol phosphate kinase modulates adherens junction and E-cadherin trafficking via a direct interaction with mu1B adaptin. J Cell Biol 176:343–353

Lisanti MP, Caras IW, Davitz MA, Rodriguez-Boulan E (1989) A glycophospholipid membrane anchor acts as an apical targeting signal in polarized epithelial cells. J Cell Biol 109:2145–2156

Maday S, Anderson E, Chang HC, Shorter J, Satoh A, Sfakianos J, Fölsch H, Anderson JM, Walther Z, Mellman I (2008) A PDZ-binding motif controls basolateral targeting of syndecan-1 along the biosynthetic pathway in polarized epithelial cells. Traffic 9:1915–1924

Madrid R, Le Maout S, Barrault MB, Janvier K, Benichou S, Merot J (2001) Polarized trafficking and surface expression of the AQP4 water channel are coordinated by serial and regulated interactions with different clathrin-adaptor complexes. EMBO J 20:7008–7021

Marrs JA, Andersson-Fisone C, Jeong MC, Cohen-Gould L, Zurzolo C, Nabi IR, Rodriguez-Boulan E, Nelson WJ (1995) Plasticity in epithelial cell phenotype: modulation by expression of different cadherin cell adhesion molecules. J Cell Biol 129:507–519

Matter K, Brauchbar M, Bucher K, Hauri HP (1990) Sorting of endogenous plasma membrane proteins occurs from two sites in cultured human intestinal epithelial cells (Caco-2). Cell 60:429–437

Matter K, Whitney JA, Yamamoto EM, Mellman I (1993) Common signals control low density lipoprotein receptor sorting in endosomes and the Golgi complex of MDCK cells. Cell 74:1053–1064

Mattera R, Boehm M, Chaudhuri R, Prabhu Y, Bonifacino JS (2011) Conservation and diversification of dileucine signal recognition by adaptor protein (AP) complex variants. J Biol Chem 286:2022–2030

Mattila PE, Raghavan V, Rbaibi Y, Baty CJ, Weisz OA (2014) Rab11a-positive compartments in proximal tubule cells sort fluid-phase and membrane cargo. Am J Physiol Cell Physiol 306: C441–C449

Mattila PE, Youker RT, Mo D, Bruns JR, Cresawn KO, Hughey RP, Ihrke G, Weisz OA (2012) Multiple biosynthetic trafficking routes for apically secreted proteins in MDCK cells. Traffic 13:433–442

Milewski MI, Forrest JK, Mickle JE, Stanton B, Cutting GR (2000) The role of PDZ-binding motif in polarized distribution of CFTR and other apical or basolateral membrane proteins. Am J Hum Genet 67:198–198

Mishra R, Grzybek M, Niki T, Hirashima M, Simons K (2010) Galectin-9 trafficking regulates apical-basal polarity in Madin-Darby canine kidney epithelial cells. Proc Natl Acad Sci USA 107:17633–17638

Mo D, Costa SA, Ihrke G, Youker RT, Pastor-Soler N, Hughey RP, Weisz OA (2012) Sialylation of N-linked glycans mediates apical delivery of endolyn in MDCK cells via a galectin-9-dependent mechanism. Mol Biol Cell 23:3636–3646

Mohrmann K, Leijendekker R, Gerez L, van Der Sluijs P (2002) rab4 regulates transport to the apical plasma membrane in Madin-Darby canine kidney cells. J Biol Chem 277:10474–10481

Mostov K, Su T, ter Beest M (2003) Polarized epithelial membrane traffic: conservation and plasticity. Nat Cell Biol 5:287–293

Moyer BD, Denton J, Karlson KH, Reynolds D, Wang S, Mickle JE, Milewski M, Cutting GR, Guggino WB, Li M, Stanton BA (1999) A PDZ-interacting domain in CFTR is an apical membrane polarization signal. J Clin Invest 104:1353–1361

Muniz M, Zurzolo C (2014) Sorting of GPI-anchored proteins from yeast to mammals - common pathways at different sites? J Cell Sci 127:2793–2801

Muth TR, Gottardi CJ, Roush DL, Caplan MJ (1998) A basolateral sorting signal is encoded in the alpha-subunit of Na-K-ATPase. Am J Physiol Cell Physiol 274:C688–C696

Nabi IR, Le Bivic A, Fambrough D, Rodriguez-Boulan E (1991) An endogenous MDCK lysosomal membrane glycoprotein is targeted basolaterally before delivery to lysosomes. J Cell Biol 115:1573–1584

Nies AT, Konig J, Cui YH, Brom M, Spring H, Keppler D (2002) Structural requirements for the apical sorting of human multidrug resistance protein 2 (ABCC2). Eur J Biochem 269:1866–1876

Nokes RL, Fields IC, Collins RN, Fölsch H (2008) Rab13 regulates membrane trafficking between TGN and recycling endosomes in polarized epithelial cells. J Cell Biol 182:845–853

Ohno H, Tomemori T, Nakatsu F, Okazaki Y, Aguilar RC, Fölsch H, Mellman I, Saito T, Shirasawa T, Bonifacino JS (1999) Mu1B, a novel adaptor medium chain expressed in polarized epithelial cells. FEBS Lett 449:215–220

Olsen O, Liu H, Wade JB, Merot J, Welling PA (2002) Basolateral membrane expression of the Kir 2.3 channel is coordinated by PDZ interaction with Lin-7/CASK complex. Am J Physiol Cell Physiol 282:C183–C195

Ozaslan D, Wang S, Ahmed BA, Kocabas AM, McCastlain JC, Bene A, Kilic F (2003) Glycosyl modification facilitates homo- and hetero-oligomerization of the serotonin transporter. A specific role for sialic acid residues. J Biol Chem 278:43991–44000

Paladino S, Lebreton S, Tivodar S, Campana V, Tempre R, Zurzolo C (2008) Different GPI-attachment signals affect the oligomerisation of GPI-anchored proteins and their apical sorting. J Cell Sci 121:4001–4007

Paladino S, Sarnataro D, Pillich R, Tivodar S, Nitsch L, Zurzolo C (2004) Protein oligomerization modulates raft partitioning and apical sorting of GPI-anchored proteins. J Cell Biol 167:699–709

Perego C, Vanoni C, Massari S, Longhi R, Pietrini G (2000) Mammalian LIN-7 PDZ proteins associate with beta-catenin at the cell-cell junctions of epithelia and neurons. EMBO J 19:3978–3989

Perego C, Vanoni C, Villa A, Longhi R, Kaech SM, Frohli E, Hajnal A, Kim SK, Pietrini G (1999) PDZ-mediated interactions retain the epithelial GABA transporter on the basolateral surface of polarized epithelial cells. EMBO J 18:2384–2393

Pocard T, Le Bivic A, Galli T, Zurzolo C (2007) Distinct v-SNAREs regulate direct and indirect apical delivery in polarized epithelial cells. J Cell Sci 120:3309–3320

Potter BA, Hughey RP, Weisz OA (2006a) Role of N- and O-glycans in polarized biosynthetic sorting. Am J Physiol Cell Physiol 290:C1–C10

Potter BA, Weixel KM, Bruns JR, Ihrke G, Weisz OA (2006b) N-glycans mediate apical recycling of the sialomucin endolyn in polarized MDCK cells. Traffic 7:146–154

Rai T, Sasaki S, Uchida S (2006) Polarized trafficking of the aquaporin-3 water channel is mediated by an NH2-terminal sorting signal. Am J Physiol Cell Physiol 290:C298–C304

Reales E, Sharma N, Low SH, Fölsch H, Weimbs T (2011) Basolateral sorting of syntaxin 4 is dependent on its N-terminal domain and the AP1B clathrin adaptor, and required for the epithelial cell polarity. PLoS ONE 6, e21181

Riquier AD, Lee DH, McDonough AA (2009) Renal NHE3 and NaPi2 partition into distinct membrane domains. Am J Physiol Cell Physiol 296:C900–C910

Rodriguez-Boulan E, Kreitzer G, Musch A (2005) Organization of vesicular trafficking in epithelia. Nat Rev 6:233–247

Rodriguez-Boulan E, Perez-Bay A, Schreiner R, Gravotta D (2013) Response: the "tail" of the twin adaptors. Dev Cell 27:247–248

Roush DL, Gottardi CJ, Naim HY, Roth MG, Caplan MJ (1998) Tyrosine-based membrane protein sorting signals are differentially interpreted by polarized Madin-Darby canine kidney and LLC-PK1 epithelial cells. J Biol Chem 273:26862–26869

Ryan S, Verghese S, Cianciola NL, Cotton CU, Carlin CR (2010) Autosomal recessive polycystic kidney disease epithelial cell model reveals multiple basolateral epidermal growth factor receptor sorting pathways. Mol Biol Cell 21:2732–2745

Sai Y, Nies AT, Arias IM (1999) Bile acid secretion and direct targeting of mdr1-green fluorescent protein from Golgi to the canalicular membrane in polarized WIF-B cells. J Cell Sci 112:4535–4545

Schreiner R, Frindt G, Diaz F, Carvajal-Gonzalez JM, Perez Bay AE, Palmer LG, Marshansky V, Brown D, Philp NJ, Rodriguez-Boulan E (2010) The absence of a clathrin adapter confers unique polarity essential to proximal tubule function. Kidney Int 78:382–388

Sheff DR, Kroschewski R, Mellman I (2002) Actin dependence of polarized receptor recycling in Madin-Darby canine kidney cell endosomes. Mol Biol Cell 13:262–275

Shoshani L, Contreras RG, Roldan ML, Moreno J, Lazaro A, Balda MS, Matter K, Cereijido M (2005) The polarized expression of Na^+, K^+ –ATPase in epithelia depends on the association between beta-subunits located in neighboring cells. Mol Biol Cell 16:1071–1081

Shteyn E, Pigati L, Fölsch H (2011) Arf6 regulates AP-1B-dependent sorting in polarized epithelial cells. J Cell Biol 194:873–887

Silvis MR, Bertrand CA, Ameen N, Golin-Bisello F, Butterworth MB, Frizzell RA, Bradbury NA (2009) Rab11b regulates the apical recycling of the cystic fibrosis transmembrane conductance regulator in polarized intestinal epithelial cells. Mol Biol Cell 20:2337–2350

Simons K, Ikonen E (1997) Functional rafts in cell membranes. Nature 387:569–572

Stechly L, Morelle W, Dessein AF, Andre S, Grard G, Trinel D, Dejonghe MJ, Leteurtre E, Drobecq H, Trugnan G, Gabius HJ, Huet G (2009) Galectin-4-regulated delivery of glycoproteins to the brush border membrane of enterocyte-like cells. Traffic 10:438–450

Stiegler AL, Zhang R, Liu W, Boggon TJ (2014) Structural determinants for binding of sorting nexin 17 (SNX17) to the cytoplasmic adaptor protein Krev interaction trapped 1 (KRIT1). J Biol Chem 289:25362–25373

Straube T, von Mach T, Honig E, Greb C, Schneider D, Jacob R (2013) pH-dependent recycling of galectin-3 at the apical membrane of epithelial cells. Traffic 14:1014–1027

Subramanian VS, Marchant JS, Boulware MJ, Ma TY, Said HM (2009) Membrane targeting and intracellular trafficking of the human sodium-dependent multivitamin transporter in polarized epithelial cells. Am J Physiol Cell Physiol 296:C663–C671

Subramanian VS, Marchant JS, Boulware MJ, Said HM (2004) A C-terminal region dictates the apical plasma membrane targeting of the human sodium-dependent vitamin C transporter-1 in polarized epithelia. J Biol Chem 279:27719–27728

Sugimoto H, Sugahara M, Fölsch H, Koide Y, Nakatsu F, Tanaka N, Nishimura T, Furukawa M, Mullins C, Nakamura N, Mellman I, Ohno H (2002) Differential recognition of tyrosine-based basolateral signals by AP-1B subunit mu1B in polarized epithelial cells. Mol Biol Cell 13:2374–2382

Sun AQ, Salkar R, Sachchidanand S, Xu L, Zeng MMZ, Suchy FJ (2003) A 14-amino acid sequence with a beta-turn structure is required for apical membrane sorting of the rat ileal bile acid transporter. J Biol Chem 278:4000–4009

Torres J, Funk HM, Zegers MM, ter Beest MB (2011) The syntaxin 4 N terminus regulates its basolateral targeting by munc18c-dependent and -independent mechanisms. J Biol Chem 286:10834–10846

Toye AM, Banting G, Tanner MJ (2004) Regions of human kidney anion exchanger 1 (kAE1) required for basolateral targeting of kAE1 in polarised kidney cells: mis-targeting explains dominant renal tubular acidosis (dRTA). J Cell Sci 117:1399–1410

Traub LM (2003) Sorting it out: AP-2 and alternate clathrin adaptors in endocytic cargo selection. J Cell Biol 163:203–208

Vagin O, Turdikulova S, Sachs G (2005) Recombinant addition of N-glycosylation sites to the basolateral Na, K-ATPase beta1 subunit results in its clustering in caveolae and apical sorting in HGT-1 cells. J Biol Chem 280:43159–43167

van de Graaf SF, Hoenderop JG, van der Kemp AW, Gisler SM, Bindels RJ (2006) Interaction of the epithelial Ca^{2+} channels TRPV5 and TRPV6 with the intestine- and kidney-enriched PDZ protein NHERF4. Pflugers Arch - Eur J Physiol 452:407–417

van Hoek AN, Ma T, Yang B, Verkman AS, Brown D (2000) Aquaporin-4 is expressed in basolateral membranes of proximal tubule S3 segments in mouse kidney. Am J Physiol Renal Physiol 278:F310–F316

Weisz OA, Rodriguez-Boulan E (2009) Apical trafficking in epithelial cells: signals, clusters and motors. J Cell Sci 122:4253–4266

Yeaman C, Grindstaff KK, Wright JR, Nelson WJ (2001) Sec6/8 complexes on trans-Golgi network and plasma membrane regulate late stages of exocytosis in mammalian cells. J Cell Biol 155:593–604

Yeaman C, Le Gall AH, Baldwin AN, Monlauzeur L, Le Bivic A, Rodriguez-Boulan E (1997) The O-glycosylated stalk domain is required for apical sorting of neurotrophin receptors in polarized MDCK cells. J Cell Biol 139:929–940

Yoo D, Flagg TP, Olsen O, Raghuram V, Foskett JK, Welling PA (2004) Assembly and trafficking of a multiprotein ROMK (Kir 1.1) channel complex by PDZ interactions. J Biol Chem 279:6863–6873

Youker RT, Bruns JR, Costa SA, Rbaibi Y, Lanni F, Kashlan OB, Teng H, Weisz OA (2013) Multiple motifs regulate apical sorting of p75 via a mechanism that involves dimerization and higher-order oligomerization. Mol Biol Cell 24:1996–2007

Chapter 8
Membrane Protein Folding and Structure

Liang Zhang and Patrick H. Thibodeau

Abstract Seminal structural and biochemical studies demonstrated that proteins are organized by specific structural elements and that the information necessary to form these ordered structures is encoded in the polypeptide chain. Questions related to the mechanisms of protein folding for both soluble and transmembrane proteins remain a large field of research utilizing a combination of biochemical, biophysical, genetic, and large-scale computational approaches. Understanding these processes has broad implications for the functional identification of unknown and putative proteins found in sequenced genomes, for the characterization of disease-causing mutations identified in genetic studies, and for rational design and protein engineering efforts to introduce novel biological function. This chapter will outline the basic principles related to transmembrane protein folding and provide an overview of transmembrane protein structure determination.

Keywords Protein folding • Protein biosynthesis • X-ray crystallography • Protein structure • ABC transporter • CFTR

8.1 Protein Folding and Biosynthesis

Seminal biochemical studies by Anfinsen demonstrated that the information for the specific structure and activity of a protein is contained in the polypeptide chain (Anfinsen and Redfield 1956). These observations, made initially on a soluble protein, RNaseA, established the field of protein folding and still underlie one of the great challenges in biology—understanding how the primary sequence of a protein facilitates folding into a compact and functional three-dimensional conformation. The fundamental observation was that a purified protein, when unfolded in chemical denaturants, could adopt its native conformation upon dilution from the denaturant in vitro. These results demonstrated, for the first time, that the process of

L. Zhang • P.H. Thibodeau (✉)
Department of Microbiology and Molecular Genetics, School of Medicine, University of Pittsburgh, 450 Technology Dr., 519 Bridgeside Point II, Pittsburgh, PA 15219, USA
e-mail: thibodea@pitt.edu

K.L. Hamilton, D.C. Devor (eds.), *Ion Channels and Transporters of Epithelia in Health and Disease*, Physiology in Health and Disease,
DOI 10.1007/978-1-4939-3366-2_8

protein folding, as well as the final protein structure, is encoded by the primary sequence of the polypeptide chain (Anfinsen 1973).

8.1.1 Physical Regulation of Protein Folding and Structure

A variety of parameters regulate the formation of these compact, folded structures through both thermodynamic and kinetic mechanisms (Anfinsen 1973; Schlesinger and Barrett 1965; Friedland and Hastings 1967; Baker et al. 1992). A quantitative, first-principle understanding of the physical processes has not been sufficiently developed to generally predict protein structure de novo (Baker 2014). However, the basic physical processes underlying protein folding have been described and include specific interactions within the polypeptide chain, both local and long distance, as well as interactions between the polypeptide chain and its environment. At the heart of the protein folding reaction is the thermodynamic minimization of the free energy of the polypeptide (Anfinsen 1973). The energetic interactions between residues in the polypeptide and interactions between the polypeptide and solvent facilitate protein folding by providing a driving force for polypeptide collapse and ordering.

The requirement for a driving force is illustrated by considering a random search in structure space, which was first formulated by Cyrus Levinthal (Levinthal 1968). A simple thought experiment considering only the polypeptide backbone and the two dihedral angles that describe its orientation, ϕ and ψ, illustrates the requirement for a protein folding "driving force." If one limits conformational space to three backbone positions for both dihedrals, the number of conformations is $3^{2(N-1)}$, where N is the number of amino acids in the polypeptide chain. If the polypeptide chain length is 100 amino acids, approximately 3^{198} (or 3×10^{94}) conformations would exist. Sampling each conformation at picosecond rates would require a timescale significantly longer than the lifespan of the known universe. Such sampling does not allow for the sub-millisecond folding seen in many experimental models. This formalization, while simplistic, illustrates the requirement for a driving force for protein folding and is known as the Levinthal paradox.

The initial and most prominent physical process that drives protein folding is likely the exclusion of water from the hydrophobic amino acids (Nozaki and Tanford 1971; Moelbert et al. 2004). For soluble proteins in aqueous environments, the exposure of hydrophobic amino acids results in an increase in solvent ordering (i.e., a reduction in solvent entropy) (Dill 1990; Sun et al. 1995). This ordering results from the loss of polar interactions between solvent and solute and results in the formation of an ordered cage around the hydrophobic solute. This is energetically unfavorable and facilitates hydrophobic collapse wherein the apolar residues are buried within the protein, driving the formation of structure and increasing solvent entropy (Matsumura et al. 1988; Karpusas et al. 1989). For membrane proteins, the integration of hydrophobic spans into the lipid bilayer similarly excludes polar solvent from the hydrophobic residues and provides a major constraint for the protein structures spanning and proximal to the membrane

(Buchegger et al. 1996; Trowbridge et al. 1991). The unique nature of the lipid bilayer places significant and specific constraints on the sequences of transmembrane proteins in comparison to soluble proteins, though both elements are generally enriched in densely packed hydrophobic amino acids.

Though these hydrophobic interactions provide a major energy source for the initial folding of the polypeptide chain, the native—or folded—state of a protein is stabilized by many small noncovalent interactions, including van der Waals interactions, hydrogen bonding, and electrostatic interactions (Dill 1990). The free energy associated with each of the individual interactions is small. However, the sum of thousands of these interactions is large for a folded protein. Electrostatic interactions are associated with formal charges in ionizable groups—the side chains of charged amino acids and the N- and C-termini of the polypeptide. In addition, dipoles found in polar groups and secondary structure elements provide an additional source of electrostatic forces within proteins. The electrostatic interactions, in sum, contribute significantly to the free energy of the folded state. In addition, these forces are unique in their ability to propagate long distances through space and in response to the local environment.

Hydrogen bonding and van der Waals interactions also contribute to the stabilization of the native protein structure, though to a lesser algebraic degree than electrostatic and hydrophobic interactions. In proteins, hydrogen bonds result from the differences in electronegativity between oxygen, nitrogen, and carbon atoms. The peptide backbone is regularly organized with hydrogen bond donors (N-H) and acceptors (C = O), in addition to the donors and acceptors found in the polar amino acids. Hydrogen bonding is responsible for the organization of secondary structure elements based on the specific orientation of the amine and carbonyl groups in the peptide backbone. Van der Waals interactions are weak, short-range interactions that facilitate atom-atom packing. These optimal energetic interactions are maximized as two atoms make contact without interpenetrating. These small forces underlie the energetic basis of steric hindrance.

8.1.2 Thermodynamic and Kinetic Regulation of Protein Folding

From a thermodynamic perspective, the main resistance to the proper folding and formation of a compact structure is the entropic penalty associated with the formation and constraint of the polypeptide chain (Fig. 8.1) (Dill 1990). In the absence of hydrophobic interactions, the unconstrained polypeptide chain would favor an extended, hydrated state with no regular structure. The entropic cost for structuring the polypeptide is the largest energetic barrier to protein folding. In an aqueous environment, the energies of solvent exclusion, electrostatics, hydrogen bonding, and van der Waals interactions offset the entropic cost and favor the folded state (Fig. 8.1). Though the total energy associated with protein folding is large, the difference in free energy states is relatively small, with the free energies

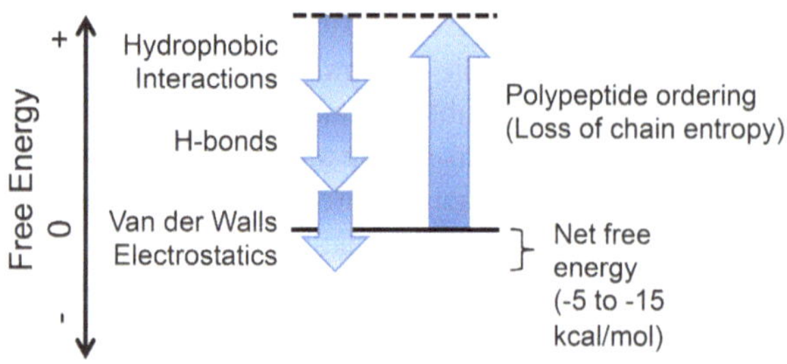

Fig. 8.1 A schematic of the free energies associated with protein folding is shown. The unfavorable entropic energy associated with polypeptide ordering is counterbalanced by hydrophobic interactions, hydrogen bonding, and electrostatic and van der Waals interactions. The net free energy for most folded proteins is relatively small, on the order of −1 to −15 kcal/mol

of folding often ranging between −5–15 kcal/mol (Pace and Shaw 2000). Native-state destabilization, resulting from changes to the primary sequence, is the molecular mechanism underlying many disease-causing protein mutations (Thomas et al. 1995).

The inherent complexity of the protein fold suggests that the kinetics of the folding reaction are similarly important. The formation of specific, individual structures occurs, in many cases, along a hierarchical pathway (Religa et al. 2005). For some proteins, folding pathways bifurcate into nonproductive branches wherein nonnative structures are formed and trap the nascent chain (Baker et al. 1992; King et al. 1996). This has been conceptualized by the transformation of the classic two-state reaction into a three-dimensional funnel with "conformational space" plotted against free energy (Fig. 8.2) (Dill and Chan 1997). The energetic and structural transitions through this landscape are complicated by nonproductive, local energetic minima.

The kinetic regulation of folding provides another mechanism to conceptually "steer" the folding along productive transitions to the native state (Dill et al. 1993). This work suggests that proteins can follow some specific "folding pathways" that facilitate folding in a nonrandom fashion. These may follow a vectorial or hierarchical pathway or be more global and nonvectorial in nature (Jansens et al. 2002; Zhang et al. 2012). The hierarchical or vectorial pathway has previously been assumed for multi-domain proteins; however, single-molecule studies and high-resolution solution studies provide evidence that protein folding can proceed through specific ensembles of protein conformations and intermediate structures (Dobson 2003; Vendruscolo et al. 2001). While these events are driven by the thermodynamics of solvent-solute interactions and contacts with the folding protein, the nascent chain is thought to provide some control of the kinetics for both on- and off-pathway events (Baker et al. 1992). These controls may include rate-limited conformational changes associated with specific amino acids (i.e., proline isomerization or disulfide formation/breakage) (Deber et al. 1986; Ho and Brasseur 2005). Alternatively, they may be related to the formation of specific structural elements and their interactions in the folded structure.

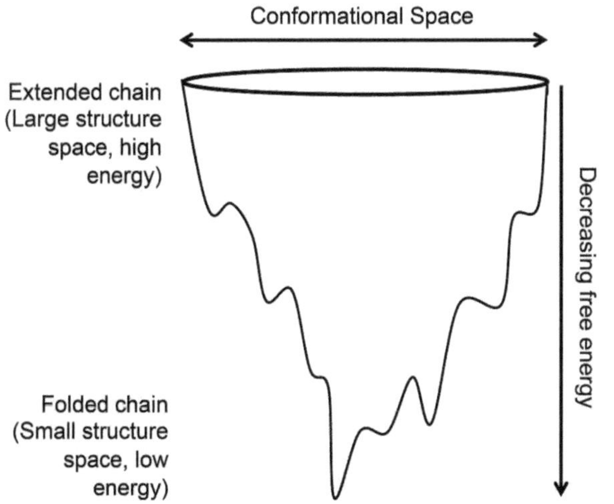

Fig. 8.2 A schematic of a protein landscape is shown with structure space plotted against free energy. The thermodynamics of the protein fold drive the conformational constraints of the polypeptide into smaller and smaller ensembles of protein structures and into the native state. Local energy minima, represented as dips or valleys in the energy profile, may trap folding intermediates and require energy from chaperones or other sources to facilitate proper folding

8.1.3 Protein Folding in the Cellular Environment

The evolution of protein sequence and function occurs within the context of specific biological systems. Thus, the native state function of a protein evolves to regulate specific cellular processes within the context of the cell or tissue in which it is expressed. Similarly, the primary sequence of the protein evolves in the context of this cellular environment and is selected for proper folding in this context. As a result, efficient folding of many proteins relies not just on the primary sequence but chaperones and other environmental factors associated with the cellular environment in which the protein evolved.

Soluble proteins rely on multiple chaperones systems to fold efficiently. These are too numerous to review in this context, but are well described in the primary and review literature. Chaperones play several key roles in facilitating protein folding. First, they hold the extended polypeptide chain in order to keep it soluble during protein translation and folding (Cha et al. 2013; Powis et al. 2013; Roy et al. 2014; Slepenkov and Witt 2002). This "holdase" function is critical during protein synthesis, where unpaired protein sequences must be kept soluble while waiting for their partner sequences to be translated. Chaperones may also serve to facilitate folding of the nascent chain mechanically (Cha et al. 2013; de Marco et al. 2005; Chan and Dill 1996; Schroder and Kaufman 2005; Wang and Tsou 1998). This "foldase" function is often coupled to nucleotide hydrolysis and alters the nascent chain conformation. Additionally, chaperones may utilize an "unfoldase" function to reverse the process of protein misfolding and aggregation (Burton and Baker

2005; Liu et al. 2013a; Sharma et al. 2009; Tamayo et al. 2011). Finally, chaperone systems have increasingly been recognized as playing a key role in protein degradation, providing the initial recognition and signaling for ubiquitin modification (Guerriero et al. 2013; Mehnert et al. 2014; Zacchi et al. 2014).

While the basic concepts of protein folding and chaperones are the same for soluble and transmembrane proteins, the unique nature of the transmembrane protein—that is, its insertion into the membrane bilayer—requires the evolution of machinery to facilitate its insertion and of sequences that preferentially interact with the membrane bilayer.

8.1.4 Protein Insertion in the Biosynthetic Pathway

Transmembrane protein folding is intrinsically more complicated than the folding of soluble, cytosolic proteins (Do et al. 1996). While many of the soluble chaperone systems contribute to the folding of ER-lumenal or cytosolic domains of a transmembrane protein, a specific set of cellular systems has evolved to handle the insertion and assembly of transmembrane protein sequences (Gorlich et al. 1990; Krieg et al. 1989; Sanders and Schekman 1992). These include proteins that recognize transmembrane sequences and those that facilitate the integration of the transmembrane protein into the lipid bilayer (Moller et al. 1998; Wang and Dobberstein 1999).

Translation of the mRNA sequence occurs within the ribosome and involves multiple chaperone and accessory proteins that regulate translational rate and pausing, delivery to and entry into the secretory pathway, insertion in the membrane, and cytosolic and ER-lumenal domain folding (Nilsson et al. 2014). The regulation of protein translation and insertion is mediated by the signal recognition particle (SRP), the Sec61 translocon, and a variety of accessory proteins that target the ribosome-nascent chain complex to the ER insertion site (Wang and Dobberstein 1999). The function of SRP is to identify and bind hydrophobic stretches in the nascent chain in order to facilitate their targeting to the ER membrane (Ataide et al. 2011; Jiang et al. 2008; Nilsson et al. 2014). The Sec61 translocon, a complex of proteins that forms a channel spanning the ER membrane, is responsible for membrane insertion of the nascent chain. Together, these proteins serve as the "gate keepers" to membrane insertion and entry into the secretory pathway (Moller et al. 1998).

Transmembrane protein identification begins with interactions of the extending, nascent chain with SRP (Gorlich et al. 1990). SRP complexes with the ribosome near the exit tunnel and binds hydrophobic polypeptide signal sequences (Flanagan et al. 2003; Johnson et al. 2001; Bornemann et al. 2008). This binding results in translational pausing, through allosteric interactions in the ribosome, and directs the ribosome-nascent chain complex to the ER via binding to the SRP receptor and the translocon (Noriega et al. 2014; Moller et al. 1998; Song et al. 2000; Lakkaraju et al. 2008; Mary et al. 2010). Once bound to the translocon, SRP dissociates and the transmembrane sequences are transferred to the translocon. Translation inhibition is then relieved, and the nascent chain continues its extension.

The translocon channel facilitates a complex series of events that includes the translocation of transmembrane sequence through the channel, assembly of these sequences into membrane-compatible structures, and integration of these structures into the ER membrane (Crowley et al. 1993, 1994; Devaraneni et al. 2011; Pitonzo et al. 2009; Sadlish et al. 2005). The first step involves interactions of the nascent chain with the aqueous translocon pore and residues lining the pore of the translocon (Crowley et al. 1994). Subsequent to this, the TM spanning sequences are transferred through structural rearrangement of the translocon to either docking sites that can temporarily hold the TM spans in the translocon complex or directly into the membrane (Do et al. 1996). Transfer directly to the membrane or transient holding by the translocon depends on the specific sequence of the TM spans and the topology of the protein (Devaraneni et al. 2011; Pitonzo et al. 2009; Sadlish et al. 2005). Single TM span proteins are often transferred directly into the membrane environment. These spans tend to be formed almost exclusively by hydrophobic residues, so the free energy transfer into the lipid bilayer is favorable (Chan and Dill 1997). Proteins with multiple TM spans can be held transiently by the translocon while waiting for interacting sequences to be translated and inserted (Sadlish et al. 2005). This is often required to pair charged residues and/or to stabilize secondary structures before transfer to the lipid bilayer. The ability to neutralize and mask charged and polar residues in multispan transmembrane proteins allows for additional sequence complexity in the transmembrane sequences but requires the translocon to serve as a temporary holding site for unpaired residues (Pitonzo et al. 2009; Sadlish et al. 2005).

Extracellular and cytosolic domains are often found in transmembrane proteins and are also recognized by the ribosome-translocon complex. Extracellular domains are inserted through the translocon pore and into the ER (McCormick et al. 2003). The specific interactions between the translocon and these sequences vary from those of the TM spanning sequences and facilitate their secretion into the lumen of the ER. In contrast, cytosolic domains require partial undocking of the ribosome from the translocon to allow those domains to be translated and fold in the cytosol. The mechanisms of this recognition are not fully elucidated, but are likely associated with changes in nascent chain hydrophobic properties and secondary structure.

8.1.5 Transmembrane Protein Sequences

The requirement that a protein span the lipid bilayer puts a unique and fundamental constraint on the evolution of the protein's primary sequence. Specifically, the favorable interaction of the polypeptide chain with the phospholipid bilayer generally restricts the amino acids that are found in transmembrane spanning structures (Bigelow and Rost 2009; Rost et al. 1995). The majority of mammalian transmembrane proteins are thought to span the bilayer using helical structures (Reeb et al. 2014). These helices are enriched with hydrophobic amino acids and are generally composed of between 16 and 25 amino acids (Reeb et al. 2014). The shorter helices likely adopt a conformation that is normal to the plane of the bilayer. Longer helices may adopt conformations that cross the membrane at obtuse angles, requiring additional helical turns to span the bilayer.

The hydrophobicity of these sequences is likely the key determinant for their identification as membrane spanning structures and their insertion into the lipid bilayer (Eisenberg et al. 1984). While any of the hydrophobic residues may be accommodated in the lipid bilayer, the distribution of amino acids in these sequences shows trends that provide additional specificity for the protein-membrane interaction. Two principle ideas emerge from statistical and energetic studies of model peptides and transmembrane proteins. First, the membrane bilayer is not uniform with respect to amino acid selection (Baeza-Delgado et al. 2013; Lehnert et al. 2004; Senes et al. 2001). The core of the bilayer, composed of the hydrocarbon tails of the lipids, preferentially interacts with small hydrophobic amino acids and the aromatics Phe and Trp. These interactions likely are the result of strong positive nonpolar interactions between the hydrocarbon and aromatic side chains with the lipid hydrocarbons. The interfacial regions of the membrane, those areas containing the phospho-head groups and bulk solvent, tend to be enriched in less apolar residues, though charged amino acids are generally disfavored. The subtle shift from hydrophobic to polar likely represents the need to interact favorably with both the hydrocarbon core and the electrostatics of the phospho-head groups. Second, charged residues that are found in the interfacial region of immediately flanking transmembrane spans show a propensity for positive residue to be located toward the cytoplasm (Wallin and von Heijne 1998; Engelman et al. 1986; Gafvelin et al. 1997; Hartmann et al. 1989). The enrichment of Lys and Arg residues on the cytoplasmic side of the membrane may facilitate proper topology of transmembrane helices when coupled with membrane potential. The membrane potential across the ER and plasma membranes is strongly negative, with increased negative charge distributed across the cytoplasmic leaves of these membranes. As a result, the propensity of transmembrane spans to include positive residues likely provides a means to complement the negative charge and provide orientation and anchoring for multiple transmembrane helices.

8.2 Transmembrane Protein Structure

High-resolution structural biology of ion channels and transporters begins with critical discoveries related to X-rays and materials science in the late nineteenth and early twentieth centuries. Today, the growing number of high-resolution membrane protein structures aids in our understanding of protein function, the molecular basis of disease pathophysiologies, and drug discovery and refinement.

8.2.1 The History of Crystallography

In the late nineteenth century, Wilhelm Röntgen began studying what would eventually be termed X-rays, a term that was established to describe their unknown ("X") qualities. For these observations, Röntgen won the Nobel Prize in 1901. The exact nature of this energy was not fully established until the twentieth century; however, experimentation with X-rays led to the imaging of multiple medical, biological, and chemical samples at the end of the nineteenth and early twentieth centuries (Glasser 1945; Keevil 1896). Though this technology was being explored in multiple applications, the real appreciation of its value—and its intrinsic danger—was not fully appreciated for decades after experimentation began.

In 1912, Max von Laue described the diffraction of X-rays by a solid matter crystal (Moffat et al. 1989). This seminal work established the basis for modern X-ray crystallography and is fundamentally unchanged today. Using salt crystals, Laue showed that an incoming light beam was diffracted, much like visible light through a diffraction grating, by a solid matter crystal. Moreover, when the diffracted X-rays were imaged using film, specific foci were generated with patterns that were dependent on the orientation and structure of the crystal itself. These observations led to the formulation of the Laue equations, which relate the diffraction pattern in reciprocal space to the real world structure of the object diffracting the X-rays. These equations, which are related to the Bragg equation discussed below, were central to solving the structures of small inorganic crystals in the early twentieth century and larger and more complicated macromolecules in the latter half of the century. In addition, this work demonstrated that X-rays behaved as waves with very short wavelengths, providing a partial description of the unknown properties of the X-rays described earlier by Röntgen.

In 1913, William Henry Bragg and William Lawrence Bragg, a father and son team, formalized a model that described the interactions between a solid matter crystal and X-rays (Bragg 1956). The physical interactions between the electrons with the incoming beam resulted in scattering and, depending on the wavelength of the incoming beam and the spacing between atoms in the crystal, produced constructive or destructive interference in the diffraction pattern (Fig. 8.3). With the right geometric spacing and wavelength, the incoming light beam was diffracted and effectively focused by outgoing constructive interference. Specific patterns, which related to the internal crystal lattice, could be observed and were dependent on the crystal's structure and the incoming light source. The Bragg's equation related the constructive interference seen to the spacing of the objects within the crystal. Thus, diffraction patterns collected from a sample could be used to back-calculate the foci of diffraction within the crystal.

Together, the efforts of the Braggs and Laue established the theoretical basis for the use of diffraction to determine spacial relationships in solid crystals. Laue was awarded the Nobel Prize in Physics in 1914 for his characterization of X-ray diffraction in solid matter crystals. For their theoretical contributions and the characterization of inorganic crystals and diamonds, the Braggs were awarded the

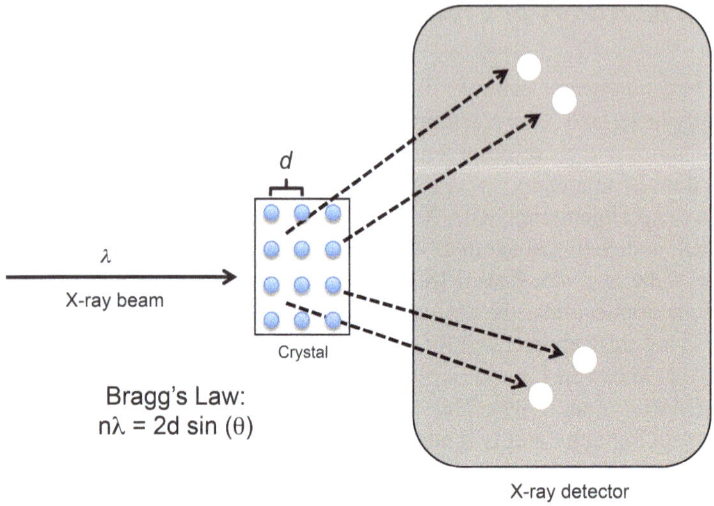

Fig. 8.3 A schematic showing the relationship between the light source and diffracting atoms in a crystal is shown. The constructive interference resulting from precisely spaced atoms in a crystal can be detected and utilized for structure determination. Bragg's law relates the incoming light beam and the space between atoms in an ordered crystal with the outgoing constructive interference as n is an integer scalar, λ is the wavelength of the incoming light, d is the distance between planes in the crystal, and Θ is the angle between the incident ray and the diffracting crystal planes. The constructive interference is detected in diffraction experiments, and the intensity and positions of the diffraction pattern are utilized to back-calculate the positions of atom in the crystal

Nobel Prize in Physics in 1915. They are the only father and son to share a Noble Prize, and William Lawrence Bragg, at the age of 25, remains the youngest winner of a scientific Nobel Prize to date.

8.2.2 *Biological Application of X-Ray Diffraction*

While these studies of the Braggs and Laue initially focused on small inorganic and organic compounds, the application of these approaches resulted in the first structures of biological molecules in the mid-twentieth century. The structure of DNA was solved using diffraction methodologies in 1953 by Watson and Crick and, in many ways, represents the beginning of modern molecular and structural biology (Watson and Crick 1953). In 1959 the first structure of a protein, whale myoglobin, was solved using X-ray crystallography (Kendrew et al. 1958). This structure provided the first detailed insights into the organization of secondary and tertiary structure in a folded protein and laid the groundwork for almost ~70 years of structural biology. John Kendrew shared the Nobel Prize in Chemistry in 1962 for this work on the structure of myoglobin. The fundamental processes of protein structure determination remains largely unchanged today, though X-ray crystallography is now complemented by nuclear magnetic resonance (NMR), electron microscopy, and other diffraction approaches (electron and neutron) for structure determination.

In 1985 the first structure of a membrane protein was solved. Prior work had focused on soluble proteins, in large part, because membrane proteins were technically challenging to extract from their cellular membranes and were not soluble in solution. In the early 1980s, Hartmut Michel demonstrated that the photosynthetic reaction complex from *Rhodopseudomonas viridis* could be purified and crystallized using a combination of detergents added during its extraction and purification (Michel 1982). Subsequent to this, Johann Deisenhofer, Robert Huber, and Hartmut Michel solved its structure—the first transmembrane crystal structure—at 3 Å resolution (Deisenhofer et al. 1985). This work showed the exquisite topology and orientation of the transmembrane and cytosolic components of the photosynthetic reaction center, providing insights into ATP generation and cellular energy homeostasis. This effort, which established the field of membrane protein crystallography, was acknowledged with the 1988 Nobel Prize in Chemistry awarded to Deisenhofer, Huber, and Michel.

8.2.3 Recent Advances in Structure Determination

Since 1959 the number of crystal structures has grown tremendously. Currently, there are over 105,000 protein structures representing approximately 35,000 proteins in the Protein Data Bank (PDB), the central repository for protein structures (pdb.org). These include proteins from all kingdoms of life, representing a variety of enzymatic and structural functions. The rapid expansion of the PDB was in part the result of large-scale structural biology programs designed to define "fold space," the range of protein folds that could be found naturally (Skolnick et al. 2000). To date, there are roughly 1300–1500 different protein folds identified in the PDB, depending on classification methods used in the definition of these folds. Since 1985, the number of transmembrane proteins has similarly grown (White 2009). There are approximately 515 unique transmembrane proteins in the PDB, with a total of ~1600 structure files (pdb.org; Structure).

Though technologies have improved, the fundamental challenges remain largely unchanged for X-ray crystallographers. The chief difficulties are the expression and purification of multiple milligrams of homogenous, pure protein and the identification of conditions that produce crystals that diffract to high resolution.

The first challenge—that of protein expression and purification—has been facilitated by the development of many recombinant technologies that allow for protein overexpression in prokaryotic and eukaryotic systems (Almo et al. 2013; Mus-Veteau 2010; Xiao et al. 2010). Recombinant expression strategies and efficient site-directed mutagenesis have been critical for screening protein variants with specific thermodynamic and/or functional properties. In addition, the recombinant incorporation of tags for purification facilitates the rapid capture of full-length proteins using efficient and economical purification technologies. Specifically, protein tags paired with specific and high-affinity column resins often provides a powerful single-step purification strategy.

Recombinant protein overexpression technologies have similarly advanced to facilitate purification. Prokaryotic systems have distinct advantages over eukaryotic systems with respect to time and cost. However, particularly with transmembrane proteins, bacterial systems tend to exhibit problems with poor cell growth and protein expression (Mus-Veteau 2010). These problems may relate to differences in the membrane protein biosynthesis pathways, chaperone systems, membrane lipid composition, and functional regulation of a heterologous membrane protein. Yeast and mammalian expression systems provide a more native-like environment for the production of mammalian membrane proteins. Multiple yeast models, including *Saccharomyces cerevisiae* and Pichia pastoris, have been utilized for transmembrane protein expression, as have mammalian cells grown in suspension. However, the added costs and time related to large-scale biomass generation in eukaryotic systems often limit their use.

The field of structural genomics has evolved as a result of difficulties in producing large quantities of transmembrane proteins (Skolnick et al. 2000). Here, multiple bacterial homologues of mammalian proteins are expressed in prokaryotic systems, and those that express and can be purified efficiently are selected for downstream structural studies. While these sequences may differ from those of their mammalian counterparts, if the evolutionary distance is sufficiently close, the structures of these homologues can serve as templates to facilitate our understanding of the mammalian counterparts. Though atomic details may differ slightly between evolutionarily related proteins, homologous structures provide the basis for our understanding of many transmembrane channels, exchangers, and transporters.

Recent advances in liquid handling, diffraction, and data acquisition technologies also continue to make structure determination more efficient. At its heart, the process of crystallization remains empirical (Skarina et al. 2014). Often this involves screening through thousands of conditions, altering pH, salt, detergents, ligands, additives, temperatures, and protein concentrations to identify buffer components that facilitate crystal formation and growth at high protein concentrations. This has traditionally required 10s of milligrams of purified protein. Liquid handling technologies have reduced the minimum quantities of protein required for crystal screening by 10–100 folds. Reducing the volumes of protein required from several microliters to 10–100 nanoliters per condition allows large-scale crystal trials to be attempted with smaller quantities of highly purified sample. Similarly, automated housing and handling robotics provide precise control of the conditions that support crystal growth and are capable of rapidly screening through thousands of crystallization conditions without human intervention.

The use of high-energy light sources and very-high-quality detectors enables the acquisition of the highest-resolution data. Using a combination of near-ideal protein crystals and high-energy synchrotron X-ray sources, the highest-resolution protein structures have been solved to roughly 1 Å, sufficient to clearly visualize the side-chain rings of an aromatic residue and the void in the ring itself (Rosenbaum et al. 2015). This is possible, in part, due to the high flux beam lines available from large particle accelerators. For pharmaceutical and academic applications, the data collection and acquisition systems are increasingly being automated with robotics controlling all aspects of crystal handling, diffraction events, and data collection. Combining the increased quality of the source X-rays with high-resolution and highly sensitive detection equipment allows for smaller and smaller crystals to be used for diffraction experiments, facilitating structure determination using less protein and for proteins that otherwise do not crystallize well (Rosenbaum et al. 2015).

Finally, new technologies are also emerging that allow for the collection of diffraction data from nanocrystals. Traditionally, protein crystals are grown such that the crystal dimensions are >10–50 μm in every given dimension. With higher energy sources and better detection systems, data from smaller crystals can be collected and used for structure determination. In addition, recent work with electron-free lasers has demonstrated that macroscopic crystals (>1–3 μm in any given dimension) may not be required for structure determination (Boutet et al. 2012; Hunter and Fromme 2011; Liu et al. 2013b). The first demonstration of this technology showed that microscopic crystals could be sprayed, in suspension, through a light source and used for structure determination. These crystals were destroyed by the energy of the laser beam, but only after they diffracted a small portion of the incident beam. Data could be collected on the stream of crystals and integrated to determine the diffraction patterns. This approach is still in its infancy, and only a small number of facilities exist with these light sources. However, the ability to collect sufficient data using these approaches with small crystals provides the possibility that structures of more membrane proteins and large macromolecular complexes will become increasingly available.

8.3 Biological Insights Derived from Transmembrane Protein Structures

The growing list of transmembrane proteins is too long to fully cover in the context of this work. Given the complexity and technical challenges associated with transmembrane protein structure determination, there are often only single representative structures for any given transmembrane protein. However, structures of a variety of ATP-binding cassette (ABC) transporters have been solved from multiple species and in multiple conformations. The first structures provided mechanistic insight into the mechano-chemistry associated with ATP binding and hydrolysis, while structures of the full-length protein now provide insight into the physical mechanisms by which solutes are pumped across membranes. This section will outline the structures that underlie our understanding of the ABC family of proteins.

8.3.1 ABC Transporters

The ABC transporter family of proteins is an evolutionarily conserved family of proteins which facilitate the import and export of solutes across the membranes of cells and organelles (Albers et al. 2004; Davidson and Chen 2004; Dean et al. 2001). These proteins are abundant in archaea, eubacteria, and eukaryotes. ABC transporters are composed of at least two core domains: a nucleotide-binding domain (NBD)—or ATP-binding cassette for which the family is named—and a transmembrane domain (TMD). These domains are expressed and function as dimers in the active protein, which is minimally composed of two NBDs and two TMDs. In addition, accessory domains are associated with a variety of ABC transporters and may include larger intracellular or extracellular globular domains or additional transmembrane domains.

While the basic protein architecture is well maintained in the ABC transport proteins, the functional complex may be formed by association of one or more individual polypeptide chains (Albers et al. 2004; Davidson and Chen 2004; Dean et al. 2001). In many mammalian transporters, the complete ABC transporter is encoded by a single gene, which encodes both the NBDs and TMDs. In contrast, multiple genes encode many bacterial transporters, each producing one or more of the individual domains that must assemble to form the functional tetrameric complex. Other transport systems are encoded as half transporters with a single NBD and TMD on a single polypeptide that then dimerizes to form the functional complex. These "half-transporter" molecules have been shown to homo- and hetero-dimerize, contributing to the evolutionary diversity of the family.

8.3.2 NBD Structure

The NBDs are functionally and structurally conserved domains across the ABC transporter protein family. Their function is highly conserved and couples the energy of ATP binding and hydrolysis to their dimerization and conformational movement within the TMDs (Carson and Welsh 1995; Bianchet et al. 1997; Senior and Gadsby 1997). The sequence and structural similarity are highly conserved within the NBDs, while the TMD components have evolved to recognize and transport a large number of solutes.

The core of the NBDs is composed of approximately 230 amino acids and contains the canonical nucleotide-binding sequence (Walker A and B) and the transport signature motif, sometimes referred to as the Walker C sequence (Schneider and Hunke 1998; Liu et al. 1999; Cui et al. 2001; Loo et al. 2002; Ambudkar et al. 2006; Buchaklian and Klug 2006). The Walker A (GxxGxKST) and Walker B ($Rx_{6\text{-}8}hydrophobic_4DE$) sequences are associated with ATP binding and hydrolysis. These sequences are more generally conserved among other nucleotide-binding proteins. The transport signature motif (LSGGQ) is more

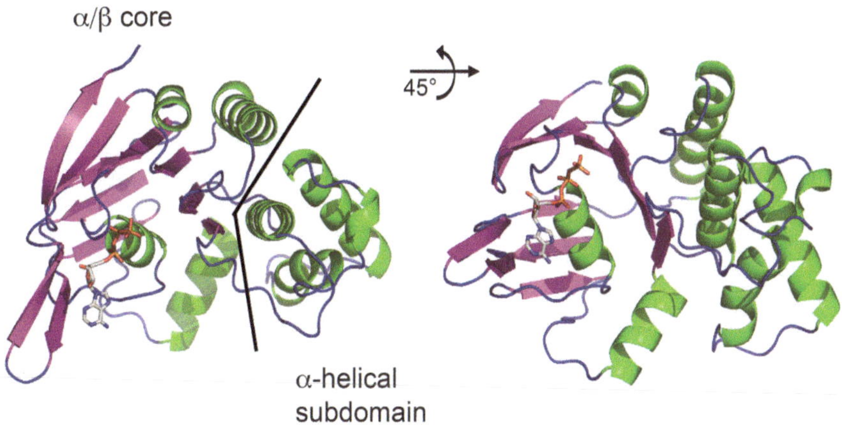

Fig. 8.4 A ribbon drawing of a bacterial nucleotide-binding domain from M. jannaschii is shown. The Mj0796 NBD fold is prototypical for the family of ABC transporters and is composed of two subdomains that undergo a small rigid body rotation upon nucleotide binding. Two views are shown, rotated by approximately 45°. The nucleotide is shown bound to the NBD monomer

specific to the ABC transporter family. These sequences represent the hallmark signatures of the NBDs, though other conserved elements (Q-loop, H-loop) are conserved throughout the family.

The structural organization of the NBDs is highly conserved (Fig. 8.4). Structures of multiple NBDs from a variety of bacterial and mammalian transporters have been solved and demonstrate that the core structural elements of these domains are highly similar (Hung et al. 1998; Yuan et al. 2001; Diederichs et al. 2000). The structures of the ABC transporter NBDs consist of two subdomains: a mixed α/β-core and an α-helical subdomain (Hung et al. 1998; Karpowich et al. 2001; Yuan et al. 2001). The α-/β-core is a Rossmann fold with high structural similarity to the RecA and F_1-core ATP-binding domains. The core contains both the Walker A and B sequences.

Nucleotides bind to the core of the domain via interactions between the Walker A and B sequences and the phosphate groups (Karpowich et al. 2001; Yuan et al. 2001). Additional ring-stacking interactions with an aromatic residue at the end of the first β strand and the adenine ring further stabilize the bound ATP. ATP binding at these sites occurs on the surface of the NBD and is not followed by nucleotide hydrolysis (Moody et al. 2002). The ATP bound at this "half-site" requires active site completion by complementation with the LSGGQ from the second NBD protomer in the dimer (Smith et al. 2002). Only after NBD dimerization are the necessary elements present and arranged to support nucleotide hydrolysis. Hydrolysis and release of the γ-phosphate alter the electrostatics of the nucleotides bound by the dimeric NBDs. This change in electrostatic potential is thought to cause dissociation of the dimer and allows for the release of the ADP.

The α-helical subdomain is divergent in both sequence and structural properties. This subdomain includes the LSGGQ sequence and is critical for the association of the NBDs with the TMDs in the functional transporter (Wilken et al. 1996; Locher et al. 2002; Pinkett et al. 2007). A flexible hinge connects the α-helical subdomain to the core domain and is termed the Q-loop for a highly conserved glutamine residue. ATP binding and hydrolysis induce a rigid body movement of the two subdomains via interactions with the Q-loop. This movement is thought to propagate into the TMDs and facilitate their structural rearrangement during the transport cycle (Locher 2004; Karpowich et al. 2001).

8.3.3 NBD-NBD Dimerization and Function

Initial studies of isolated NBDs suggested that ATPase activity required the dimerization of NBDs (Nikaido and Ames 1999; Hopfner et al. 2000; Moody et al. 2002; Smith et al. 2002; Chen et al. 2003; Zaitseva et al. 2005). Sequence analysis and mutagenesis suggested that both the Walker A and B sites, as well as the LSGGQ sequence, were required for ATP hydrolysis and were physically close during the ATPase reaction cycle. However, the initial structures of monomeric NBDs revealed that these sites were separated by 15–25 Å and likely could not coordinate ATP in a monomeric state. The structure of the Mj0796 dimer, solved in 2002, was the first to show the biologically relevant dimerization events that underlie ABC-protein function (Fig. 8.5) (Smith et al. 2002). This structure was enabled by the mutation of the Walker B catalytic glutamate to a glutamine, rendering the NBD inactive and trapped in an ATP-bound, dimeric form. This NBD dimer showed the physical juxtaposition of the Walker A/B half site from one monomer

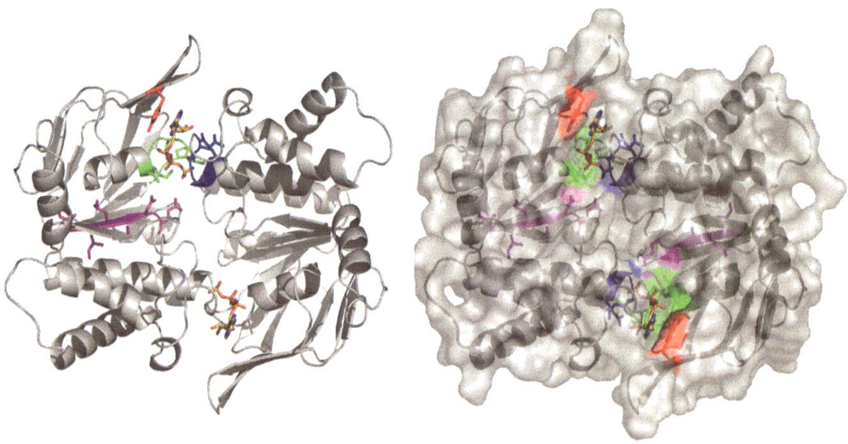

Fig. 8.5 A ribbon drawing of the ATP-driven NBD dimer is shown from Mj0796 in ribbon and surface renderings with the two bound nucleotides. The nucleotide-associated sequences are colored as aromatic loop, *red*; Walker A, *green*; Walker B, *magenta*; and LSGGQ/signature sequence, *blue*. Nucleotide binding drives association of the NBD protomers and results in the formation of two complete active ATPase sites

with the LSGGQ half-site of the opposing monomer. The ATP molecules were fully coordinated and buried within the NBD dimer, providing for the necessary chemistry for ATP hydrolysis.

Given the symmetry of the NBD dimers, two fully coordinated ATP molecules are bound by the juxtaposed Walker A and B sequences in the ATP-bound NBD dimer. Though ATP binding is required in at least one site to drive the formation of NBD dimers and transporter function, sequence analysis and biochemical studies show that the hydrolytic function is not strictly required (Nikaido and Ames 1999; Carson and Welsh 1995; Senior and Gadsby 1997; Lewis et al. 2004, 2005). Multiple studies have shown that transport function minimally requires a single active ATP site, and at least one family member, CFTR, has lost ATPase activity at one of the two NBD ATP-binding sites (Vergani et al. 2005).

8.3.4 TMD Structure

The structures of the TMDs are considerably more diverse than those of the NBDs. The TMD evolution likely results from the need to bind and transport a structurally diverse set of solutes. The TMDs are minimally composed of six transmembrane helices, though the numbers of helices increase in many bacterial transport systems (Dean et al. 2001). These domains dimerize in order to form the membrane-spanning structures that facilitate solute transport. Less is known about the structural and functional rearrangements in the TMDs than in the NBDs. Structures of the full-length ABC transporters provided evidence that the dimerization of the NBDs lead to the reorganization of the TMDs (Fig. 8.6) (Jin et al. 2012; Shintre et al. 2013; Dawson and Locher 2006, 2007). The TMDs are generally found in one of two orientations. In an ATP-bound form, the NBDs form a tight dimer, and the TMDs open into a vestibule on the outer leaflet of the membrane. This presumably would represent the low-affinity conformation associated with solute release in the exporters. In the second state, the NBDs are physically separated from one another, and the TMD have reoriented to face the inner leaflet or cytoplasm. These conformational changes are large and have not been established in a biological or in vivo system. Though this alternating access scheme likely represents the physical basis of the mammalian ABC transporters, the magnitude of this movement may vary from that seen in vitro structurally. Similarly, the activity of the bacterial ABC importers likely follows a similar general alternating access scheme, but has not been fully elucidated structurally. This model for alternating access is likely similar for multiple mammalian transporters and co-transporters as the conformational changes within the protein shield the substrate from interactions with the hydrophobic membrane (Zhou et al. 2014; Shimamura et al. 2010).

Little is known about the nature of solute binding in these domains, though the recent structures of Pgp and molecular modeling of ABCC6 provide some insight into solute recognition and selection (Aller et al. 2009; Hosen et al. 2014). In both the structures and molecular models, solute binding to the TMDs appears within the

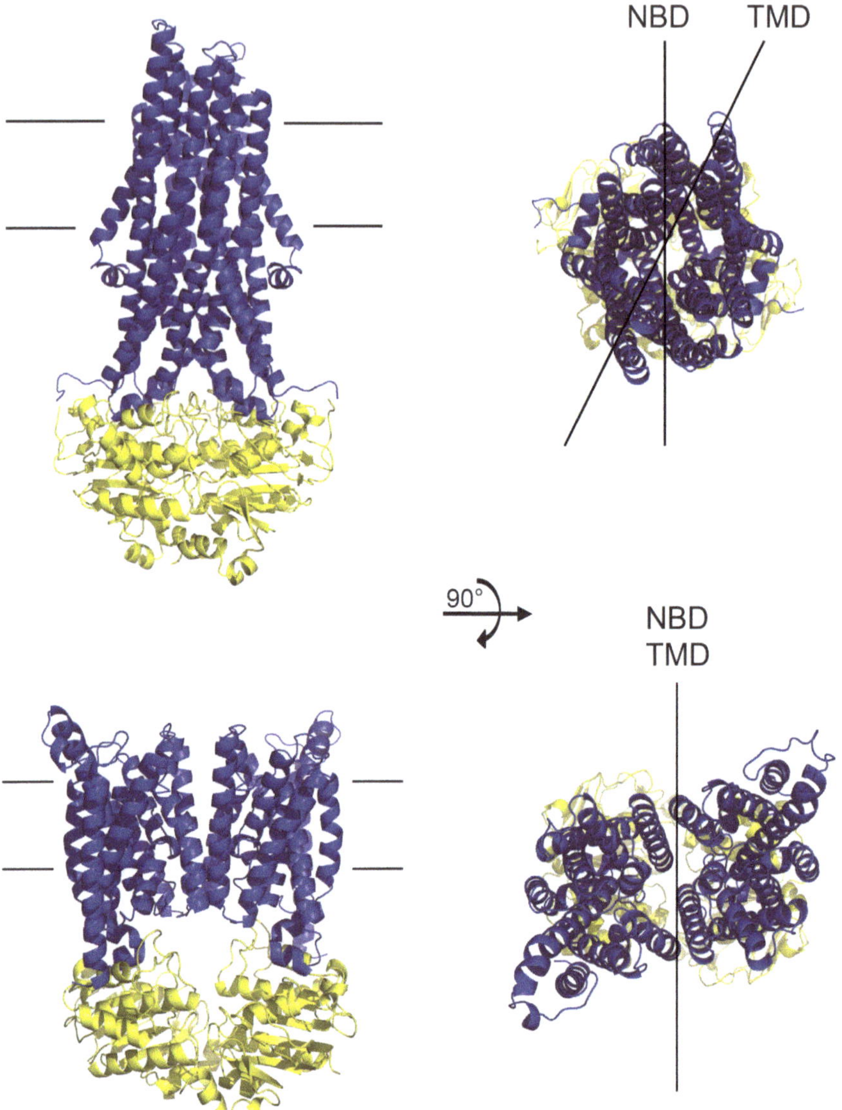

Fig. 8.6 A ribbon drawing showing the full-length structures of ABC transporters is shown. *Top*, the structure of Sav1866 is shown with the TMDs colored in *blue* and the NBDs colored in *yellow*. The TMDs are composed of 6 TM helices in each protomer. The axes of symmetry for both the TMD and NBD dimers are shown and are offset from one another. *Bottom*, the structure of BtuCD is shown and colored as for Sav1866. The TMDs of BtuC are composed of 10 TM helices in each protomer, and the axes of symmetry are roughly aligned in the BtuCD transporter. The independent evolution of the TMDs facilitates an expansion of function in prokaryotes. Mammalian ABC transporters putatively function exclusively as exporters and share the TMD organization seen in Sav1866

plane of the membrane. This could be readily accomplished by diffusion into the binding site by hydrophobic molecules found in the inner leaflet, as might be predicted for drug export, or through diffusion from the cytoplasm, as might be predicted for hydrophilic solutes. The exact mechanisms of solute binding are currently contentious and depend, in part, on the models used and the solutes chosen.

8.3.5 TMD-NBD Interactions

The interface between the TMDs and NBDs is a critical region for the structures of the ABC proteins (Lewis et al. 2004; Thibodeau et al. 2005; Wilken et al. 1996). The NBDs contact the TMDs via extended intracellular loops (Fig. 8.7). Biochemical studies of chimeric ABC-transporter molecules demonstrate that the α-helical subdomain in the NBD makes contacts with the loop sequences from the TMDs, providing specificity for the NBD-TMD interaction. These contacts are critical for both biosynthesis and function.

Disruption of this interface by mutation is common across multiple ABC-transporter family members (Fulop et al. 2009; Pomozi et al. 2014). The most studied mutant in this region is the F508del mutant in CFTR, which is found in >90 % of all patients with cystic fibrosis (Drumm et al. 1990; Cheng et al. 1990; Rommens et al. 1989). This mutation is located in the N-terminal NBD, and the side chain of the Phe508 residue makes direct contact with intracellular loops from the TMDs. Alterations at this site affect the local structure of the NBD by reducing its thermodynamic stability and folding efficiency and alter the interactions of the NBD with the TMD (He et al. 2010; Lukacs and Verkman 2012; Mendoza et al. 2012; Thibodeau et al. 2010). The combined effect of NBD misfolding and a loss of appropriate NBD-TMD interactions lead to the misfolding of the full-length protein in vivo. Full rescue of CFTR misfolding has been proposed to require the restoration of NBD structure and NBD-TMD assembly.

Though only partially characterized, this conserved NBD-TMD interface appears to be a structural "hotspot" for ABC-transporter mutations (Fulop et al. 2009). The large number of disease-causing mutations found within this interface suggests that disruption of the NBD-TMD interaction is deleterious across the family of ABC proteins. These alterations may be manifest as biosynthetic or functional defects, and the loss of functional protein likely explains the molecular pathophysiologies of these diseases.

8.3.6 ABC Transporter Folding

The majority of folding studies on ABC transporters have focused on mutations associated with CFTR, most notably F508del. However, studies of other ABC

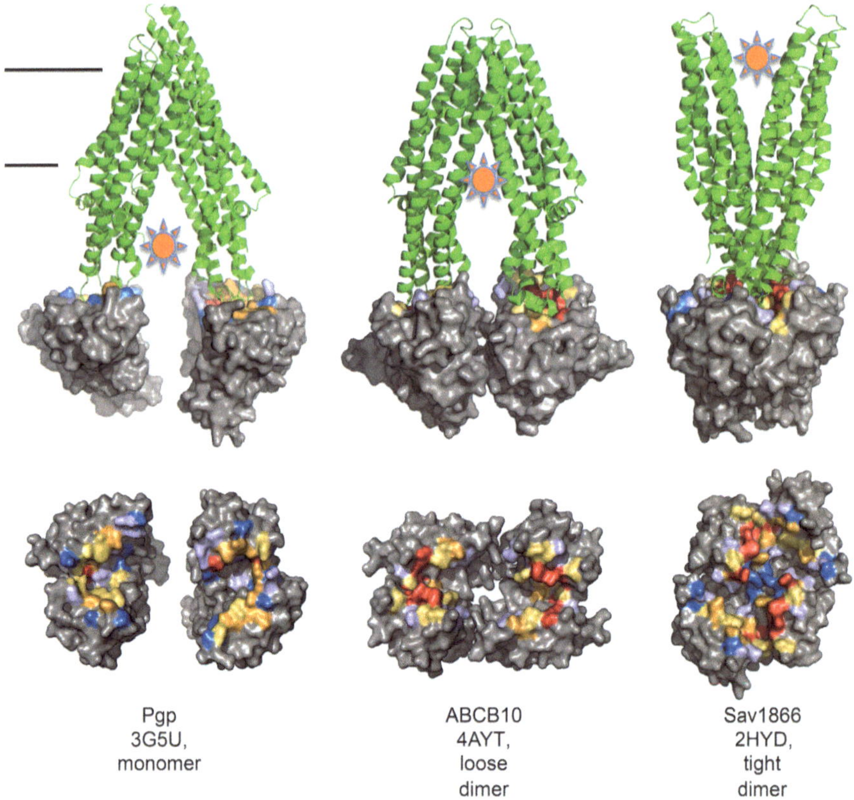

Fig. 8.7 A ribbon drawing of the ABC transporter reaction cycle is shown comprised of the ATP-free Pgp and ATP-bound ABCB10 and Sav1866 structures. The full-length transporters are shown with *green ribbons* for the transmembrane domains and surface representations of the nucleotide-binding domains. The postulated region of substrate binding is shown with an *orange star*. The Sav1866 transporter is rotated 90° about the *vertical axis* to show the extracellular vestibule. A heat map, corresponding to the tightness of the NBD-TMD interaction, is shown colored on the NBD surface and shows the dynamic structural changes associated with the catalytic and transport cycles

transporters suggest a common folding pathway (Pomozi et al. 2014). As might be expected for a multi-domain protein, folding appears to follow a hierarchical pathway wherein the individual domains must fold and acquire native or native-like structure before assembly into a quaternary complex (Du et al. 2005; Lukacs et al. 1994; Sharma et al. 2001; Thibodeau et al. 2010). The folding and assembly steps may be energetically coupled, but mutations appear to show discrete changes in either domain structure and/or assembly and thus seem to act at multiple steps in a linear pathway. These data suggest that each of the four core domains must first properly fold before assembly with their native-state binding partners. Failure of the TMDs to properly fold, integrate, and/or assemble leads to premature CFTR degradation. Similarly, disruption of local NBD structural stability or their association with the TMDs results in premature degradation. The NBD-NBD interface, which is critical for protein function, appears to be more tolerant of substitution, at least with respect to biosynthesis (DeCarvalho et al. 2002; Paddon et al. 1996; Teem et al. 1993, 1996). It is likely that mutations within this interface have deleterious effects on protein function.

8.4 Conclusions

Studies on protein folding and structure, though often very basic in nature, have provided key insights for understanding the molecular bases of disease and, perhaps more importantly, mechanisms by which defects in protein products can be corrected. Increasingly, pharmacological rescue of protein misfolding and instability is being recognized as a mechanism for therapeutic development (Gomes 2012). These ideas arise from the detailed structural and biophysical studies of protein folding and grow from decades of work to understand how protein structures are formed and function.

References

Albers SV, Koning SM, Konings WN, Driessen AJ (2004) Insights into ABC transport in archaea. J Bioenerg Biomembr 36(1):5–15

Aller SG, Yu J, Ward A, Weng Y, Chittaboina S, Zhuo R, Harrell PM, Trinh YT, Zhang Q, Urbatsch IL, Chang G (2009) Structure of P-glycoprotein reveals a molecular basis for poly-specific drug binding. Science 323(5922):1718–1722. doi:10.1126/science.1168750

Almo SC, Garforth SJ, Hillerich BS, Love JD, Seidel RD, Burley SK (2013) Protein production from the structural genomics perspective: achievements and future needs. Curr Opin Struct Biol 23(3):335–344. doi:10.1016/j.sbi.2013.02.014

Ambudkar SV, Kim IW, Xia D, Sauna ZE (2006) The A-loop, a novel conserved aromatic acid subdomain upstream of the Walker A motif in ABC transporters, is critical for ATP binding. FEBS Lett 580(4):1049–1055

Anfinsen CB (1973) Principles that govern the folding of protein chains. Science 181 (4096):223–230

Anfinsen CB, Redfield RR (1956) Protein structure in relation to function and biosynthesis. Adv Protein Chem 48(11):1–100

Ataide SF, Schmitz N, Shen K, Ke A, Shan SO, Doudna JA, Ban N (2011) The crystal structure of the signal recognition particle in complex with its receptor. Science 331(6019):881–886. doi:10.1126/science.1196473

Baeza-Delgado C, Marti-Renom MA, Mingarro I (2013) Structure-based statistical analysis of transmembrane helices. Eur Biophy J 42(2–3):199–207. doi:10.1007/s00249-012-0813-9

Baker D (2014) Centenary Award and Sir Frederick Gowland Hopkins Memorial Lecture. Protein folding, structure prediction and design. Biochem Soc Trans 42(2):225–229. doi:10.1042/BST20130055

Baker D, Sohl JL, Agard DA (1992) A protein-folding reaction under kinetic control. Nature 356 (6366):263–265

Bianchet MA, Ko YH, Amzel LM, Pedersen PL (1997) Modeling of nucleotide binding domains of ABC transporter proteins based on a F1-ATPase/recA topology: structural model of the nucleotide binding domains of the cystic fibrosis transmembrane conductance regulator (CFTR). J Bioenerg Biomembr 29(5):503–524

Bigelow H, Rost B (2009) Online tools for predicting integral membrane proteins. Methods Mol Biol 528:3–23. doi:10.1007/978-1-60327-310-7_1

Bornemann T, Jockel J, Rodnina MV, Wintermeyer W (2008) Signal sequence-independent membrane targeting of ribosomes containing short nascent peptides within the exit tunnel. Nat Struct Mol Biol 15(5):494–499. doi:10.1038/nsmb.1402

Boutet S, Lomb L, Williams GJ, Barends TR, Aquila A, Doak RB, Weierstall U, DePonte DP, Steinbrener J, Shoeman RL, Messerschmidt M, Barty A, White TA, Kassemeyer S, Kirian RA, Seibert MM, Montanez PA, Kenney C, Herbst R, Hart P, Pines J, Haller G, Gruner SM, Philipp HT, Tate MW, Hromalik M, Koerner LJ, van Bakel N, Morse J, Ghonsalves W, Arnlund D, Bogan MJ, Caleman C, Fromme R, Hampton CY, Hunter MS, Johansson LC, Katona G, Kupitz C, Liang M, Martin AV, Nass K, Redecke L, Stellato F, Timneanu N, Wang D, Zatsepin NA, Schafer D, Defever J, Neutze R, Fromme P, Spence JC, Chapman HN, Schlichting I (2012) High-resolution protein structure determination by serial femtosecond crystallography. Science 337(6092):362–364. doi:10.1126/science.1217737

Bragg L (1956) The diffraction of x-rays. Br J Radiol 29(339):121–126

Buchaklian AH, Klug CS (2006) Characterization of the LSGGQ and H motifs from the Escherichia coli lipid A transporter MsbA. Biochemistry 45(41):12539–12546

Buchegger F, Trowbridge IS, Liu LF, White S, Collawn JF (1996) Functional analysis of human/chicken transferrin receptor chimeras indicates that the carboxy-terminal region is important for ligand binding. Eur J Biochem 235(1–2):9–17

Burton BM, Baker TA (2005) Remodeling protein complexes: insights from the AAA+ unfoldase ClpX and Mu transposase. Protein Sci 14(8):1945–1954. doi:10.1110/ps.051417505

Carson MR, Welsh MJ (1995) Structural and functional similarities between the nucleotide-binding domains of CFTR and GTP-binding proteins. Biophys J 69(6):2443–2448

Cha JY, Ahn G, Kim JY, Kang SB, Kim MR, Su'udi M, Kim WY, Son D (2013) Structural and functional differences of cytosolic 90-kDa heat-shock proteins (Hsp90s) in Arabidopsis thaliana. Plant Physiol Biochem 70:368–373. doi:10.1016/j.plaphy.2013.05.039

Chan HS, Dill KA (1996) A simple model of chaperonin-mediated protein folding. Proteins 24 (3):345–351. doi:10.1002/(SICI)1097-0134(199603)24:3<345::AID-PROT7>3.0.CO;2-F

Chan HS, Dill KA (1997) Solvation: how to obtain microscopic energies from partitioning and solvation experiments. Annu Rev Biophys Biomol Struct 26:425–459. doi:10.1146/annurev.biophys.26.1.425

Chen J, Lu G, Lin J, Davidson AL, Quiocho FA (2003) A tweezers-like motion of the ATP-binding cassette dimer in an ABC transport cycle. Mol Cell 12(3):651–661

Cheng SH, Gregory RJ, Marshall J, Paul S, Souza DW, White GA, O'Riordan CR, Smith AE (1990) Defective intracellular transport and processing of CFTR is the molecular basis of most cystic fibrosis. Cell 63(4):827–834

Crowley KS, Reinhart GD, Johnson AE (1993) The signal sequence moves through a ribosomal tunnel into a noncytoplasmic aqueous environment at the ER membrane early in translocation. Cell 73(6):1101–1115

Crowley KS, Liao S, Worrell VE, Reinhart GD, Johnson AE (1994) Secretory proteins move through the endoplasmic reticulum membrane via an aqueous, gated pore. Cell 78(3):461–471

Cui L, Hou YX, Riordan JR, Chang XB (2001) Mutations of the Walker B motif in the first nucleotide binding domain of multidrug resistance protein MRP1 prevent conformational maturation. Arch Biochem Biophys 392(1):153–161

Davidson AL, Chen J (2004) ATP-binding cassette transporters in bacteria. Annu Rev Biochem 73:241–268

Dawson RJ, Locher KP (2006) Structure of a bacterial multidrug ABC transporter. Nature 443 (7108):180–185

Dawson RJ, Locher KP (2007) Structure of the multidrug ABC transporter Sav 1866 from Staphylococcus aureus in complex with AMP-PNP. FEBS Lett 581(5):935–938

de Marco A, Vigh L, Diamant S, Goloubinoff P (2005) Native folding of aggregation-prone recombinant proteins in Escherichia coli by osmolytes, plasmid- or benzyl alcohol-overexpressed molecular chaperones. Cell Stress Chaperones 10(4):329–339

Dean M, Rzhetsky A, Allikmets R (2001) The human ATP-binding cassette (ABC) transporter superfamily. Genome Res 11(7):1156–1166

Deber CM, Brandl CJ, Deber RB, Hsu LC, Young XK (1986) Amino acid composition of the membrane and aqueous domains of integral membrane proteins. Arch Biochem Biophys 251 (1):68–76

DeCarvalho AC, Gansheroff LJ, Teem JL (2002) Mutations in the nucleotide binding domain 1 signature motif region rescue processing and functional defects of cystic fibrosis transmembrane conductance regulator delta f508. J Biol Chem 277(39):35896–35905

Deisenhofer J, Epp O, Miki K, Huber R, Michel H (1985) Structure of the protein subunits in the photosynthetic reaction centre of Rhodopseudomonas viridis at 3A resolution. Nature 318 (6047):618–624

Devaraneni PK, Conti B, Matsumura Y, Yang Z, Johnson AE, Skach WR (2011) Stepwise insertion and inversion of a type II signal anchor sequence in the ribosome-Sec61 translocon complex. Cell 146(1):134–147. doi:10.1016/j.cell.2011.06.004

Diederichs K, Diez J, Greller G, Muller C, Breed J, Schnell C, Vonrhein C, Boos W, Welte W (2000) Crystal structure of MalK, the ATPase subunit of the trehalose/maltose ABC transporter of the archaeon Thermococcus litoralis. EMBO J 19(22):5951–5961

Dill KA (1990) Dominant forces in protein folding. Biochemistry 29(31):7133

Dill KA, Chan HS (1997) From Levinthal to pathways to funnels. Nat Struct Biol 4(1):10–19

Dill KA, Fiebig KM, Chan HS (1993) Cooperativity in protein-folding kinetics. Proc Natl Acad Sci USA 90(5):1942–1946

Do H, Falcone D, Lin J, Andrews DW, Johnson AE (1996) The cotranslational integration of membrane proteins into the phospholipid bilayer is a multistep process. Cell 85(3):369–378

Dobson CM (2003) Protein folding and misfolding. Nature 426(6968):884–890. doi:10.1038/nature02261

Drumm ML, Pope HA, Cliff WH, Rommens JM, Marvin SA, Tsui LC, Collins FS, Frizzell RA, Wilson JM (1990) Correction of the cystic fibrosis defect in vitro by retrovirus-mediated gene transfer. Cell 62(6):1227–1233

Du K, Sharma M, Lukacs GL (2005) The DeltaF508 cystic fibrosis mutation impairs domain-domain interactions and arrests post-translational folding of CFTR. Nat Struct Mol Biol 12 (1):17–25

Eisenberg D, Schwarz E, Komaromy M, Wall R (1984) Analysis of membrane and surface protein sequences with the hydrophobic moment plot. J Mol Biol 179(1):125–142

Engelman DM, Steitz TA, Goldman A (1986) Identifying nonpolar transbilayer helices in amino acid sequences of membrane proteins. Annu Rev Biophys Biophys Chem 15:321–353. doi:10.1146/annurev.bb.15.060186.001541

Flanagan JJ, Chen JC, Miao Y, Shao Y, Lin J, Bock PE, Johnson AE (2003) Signal recognition particle binds to ribosome-bound signal sequences with fluorescence-detected subnanomolar affinity that does not diminish as the nascent chain lengthens. J Biol Chem 278 (20):18628–18637. doi:10.1074/jbc.M300173200

Friedland J, Hastings JW (1967) The reversibility of the denaturation of bacterial luciferase. Biochemistry 6(9):2893–2900

Fulop K, Barna L, Symmons O, Zavodszky P, Varadi A (2009) Clustering of disease-causing mutations on the domain-domain interfaces of ABCC6. Biochem Biophys Res Commun 379 (3):706–709. doi:10.1016/j.bbrc.2008.12.142

Gafvelin G, Sakaguchi M, Andersson H, von Heijne G (1997) Topological rules for membrane protein assembly in eukaryotic cells. J Biol Chem 272(10):6119–6127

Glasser O (1945) Half a century of roentgen rays. Proc Rudolf Virchow Med Soc City N Y 4:96–102

Gomes CM (2012) Protein misfolding in disease and small molecule therapies. Curr Top Med Chem 12(22):2460–2469

Gorlich D, Prehn S, Hartmann E, Herz J, Otto A, Kraft R, Wiedmann M, Knespel S, Dobberstein B, Rapoport TA (1990) The signal sequence receptor has a second subunit and is part of a translocation complex in the endoplasmic reticulum as probed by bifunctional reagents. J Cell Biol 111(6 Pt 1):2283–2294

Guerriero CJ, Weiberth KF, Brodsky JL (2013) Hsp70 targets a cytoplasmic quality control substrate to the San1p ubiquitin ligase. J Biol Chem 288(25):18506–18520. doi:10.1074/jbc.M113.475905

Hartmann E, Rapoport TA, Lodish HF (1989) Predicting the orientation of eukaryotic membrane-spanning proteins. Proc Natl Acad Sci USA 86(15):5786–5790

He L, Aleksandrov LA, Cui L, Jensen TJ, Nesbitt KL, Riordan JR (2010) Restoration of domain folding and interdomain assembly by second-site suppressors of the DeltaF508 mutation in CFTR. FASEB J 24(8):3103–3112. doi:10.1096/fj.09-141788

Ho BK, Brasseur R (2005) The Ramachandran plots of glycine and pre-proline. BMC Struct Biol 5:14

Hopfner KP, Karcher A, Shin DS, Craig L, Arthur LM, Carney JP, Tainer JA (2000) Structural biology of Rad50 ATPase: ATP-driven conformational control in DNA double-strand break repair and the ABC-ATPase superfamily. Cell 101(7):789–800

Hosen MJ, Zubaer A, Thapa S, Khadka B, De Paepe A, Vanakker OM (2014) Molecular docking simulations provide insights in the substrate binding sites and possible substrates of the ABCC6 transporter. PLoS ONE 9(7), e102779. doi:10.1371/journal.pone.0102779

Hung LW, Wang IX, Nikaido K, Liu PQ, Ames GF, Kim SH (1998) Crystal structure of the ATP-binding subunit of an ABC transporter. Nature 396(6712):703–707

Hunter MS, Fromme P (2011) Toward structure determination using membrane-protein nanocrystals and microcrystals. Methods 55(4):387–404. doi:10.1016/j.ymeth.2011.12.006

Jansens A, van Duijn E, Braakman I (2002) Coordinated nonvectorial folding in a newly synthesized multidomain protein. Science 298(5602):2401–2403

Jiang Y, Cheng Z, Mandon EC, Gilmore R (2008) An interaction between the SRP receptor and the translocon is critical during cotranslational protein translocation. J Cell Biol 180 (6):1149–1161. doi:10.1083/jcb.200707196

Jin MS, Oldham ML, Zhang Q, Chen J (2012) Crystal structure of the multidrug transporter P-glycoprotein from Caenorhabditis elegans. Nature 490(7421):566–569. doi:10.1038/nature11448

Johnson AE, Chen JC, Flanagan JJ, Miao Y, Shao Y, Lin J, Bock PE (2001) Structure, function, and regulation of free and membrane-bound ribosomes: the view from their substrates and products. Cold Spring Harb Symp Quant Biol 66:531–541

Karpowich N, Martsinkevich O, Millen L, Yuan YR, Dai PL, MacVey K, Thomas PJ, Hunt JF (2001) Crystal structures of the MJ1267 ATP binding cassette reveal an induced-fit effect at the ATPase active site of an ABC transporter. Structure 9(7):571–586

Karpusas M, Baase WA, Matsumura M, Matthews BW (1989) Hydrophobic packing in T4 lysozyme probed by cavity-filling mutants. Proc Natl Acad Sci USA 86(21):8237–8241

Keevil GM (1896) The Roentgen rays. Br Med J 1(1833):433–434

Kendrew JC, Bodo G, Dintzis HM, Parrish RG, Wyckoff H, Phillips DC (1958) A three-dimensional model of the myoglobin molecule obtained by x-ray analysis. Nature 181 (4610):662–666

King J, Haase-Pettingell C, Robinson AS, Speed M, Mitraki A (1996) Thermolabile folding intermediates: inclusion body precursors and chaperonin substrates. FASEB J 10(1):57–66

Krieg UC, Johnson AE, Walter P (1989) Protein translocation across the endoplasmic reticulum membrane: identification by photocross-linking of a 39-kD integral membrane glycoprotein as part of a putative translocation tunnel. J Cell Biol 109(5):2033–2043

Lakkaraju AK, Mary C, Scherrer A, Johnson AE, Strub K (2008) SRP keeps polypeptides translocation-competent by slowing translation to match limiting ER-targeting sites. Cell 133 (3):440–451. doi:10.1016/j.cell.2008.02.049

Lehnert U, Xia Y, Royce TE, Goh CS, Liu Y, Senes A, Yu H, Zhang ZL, Engelman DM, Gerstein M (2004) Computational analysis of membrane proteins: genomic occurrence, structure prediction and helix interactions. Q Rev Biophys 37(2):121–146

Levinthal C (1968) Are there pathways for protein folding? J Chim Phys 65:44–45

Lewis HA, Buchanan SG, Burley SK, Conners K, Dickey M, Dorwart M, Fowler R, Gao X, Guggino WB, Hendrickson WA, Hunt JF, Kearins MC, Lorimer D, Maloney PC, Post KW, Rajashankar KR, Rutter ME, Sauder JM, Shriver S, Thibodeau PH, Thomas PJ, Zhang M, Zhao X, Emtage S (2004) Structure of nucleotide-binding domain 1 of the cystic fibrosis transmembrane conductance regulator. EMBO J 23(2):282–293. doi:10.1038/sj.emboj.7600040 [pii]

Lewis HA, Zhao X, Wang C, Sauder JM, Rooney I, Noland BW, Lorimer D, Kearins MC, Conners K, Condon B, Maloney PC, Guggino WB, Hunt JF, Emtage S (2005) Impact of the deltaF508 mutation in first nucleotide-binding domain of human cystic fibrosis transmembrane conductance regulator on domain folding and structure. J Biol Chem 280(2):1346–1353

Liu PQ, Liu CE, Ames GF (1999) Modulation of ATPase activity by physical disengagement of the ATP-binding domains of an ABC transporter, the histidine permease. J Biol Chem 274 (26):18310–18318

Liu J, Mei Z, Li N, Qi Y, Xu Y, Shi Y, Wang F, Lei J, Gao N (2013a) Structural dynamics of the MecA-ClpC complex: a type II AAA+ protein unfolding machine. J Biol Chem 288 (24):17597–17608. doi:10.1074/jbc.M113.458752

Liu W, Wacker D, Gati C, Han GW, James D, Wang D, Nelson G, Weierstall U, Katritch V, Barty A, Zatsepin NA, Li D, Messerschmidt M, Boutet S, Williams GJ, Koglin JE, Seibert MM, Wang C, Shah ST, Basu S, Fromme R, Kupitz C, Rendek KN, Grotjohann I, Fromme P, Kirian RA, Beyerlein KR, White TA, Chapman HN, Caffrey M, Spence JC, Stevens RC, Cherezov V (2013b) Serial femtosecond crystallography of G protein-coupled receptors. Science 342(6165):1521–1524. doi:10.1126/science.1244142

Locher KP (2004) Structure and mechanism of ABC transporters. Curr Opin Struct Biol 14 (4):426–431

Locher KP, Lee AT, Rees DC (2002) The E. coli BtuCD structure: a framework for ABC transporter architecture and mechanism. Science 296(5570):1091–1098

Loo TW, Bartlett MC, Clarke DM (2002) The "LSGGQ" motif in each nucleotide-binding domain of human P-glycoprotein is adjacent to the opposing walker A sequence. J Biol Chem 277 (44):41303–41306

Lukacs GL, Verkman AS (2012) CFTR: folding, misfolding and correcting the DeltaF508 conformational defect. Trends Mol Med 18(2):81–91. doi:10.1016/j.molmed.2011.10.003

Lukacs GL, Mohamed A, Kartner N, Chang XB, Riordan JR, Grinstein S (1994) Conformational maturation of CFTR but not its mutant counterpart (delta F508) occurs in the endoplasmic reticulum and requires ATP. EMBO J 13(24):6076–6086

Mary C, Scherrer A, Huck L, Lakkaraju AK, Thomas Y, Johnson AE, Strub K (2010) Residues in SRP9/14 essential for elongation arrest activity of the signal recognition particle define a positively charged functional domain on one side of the protein. RNA 16(5):969–979. doi:10.1261/rna.2040410

Matsumura M, Becktel WJ, Matthews BW (1988) Hydrophobic stabilization in T4 lysozyme determined directly by multiple substitutions of Ile 3. Nature 334(6181):406–410

McCormick PJ, Miao Y, Shao Y, Lin J, Johnson AE (2003) Cotranslational protein integration into the ER membrane is mediated by the binding of nascent chains to translocon proteins. Mol Cell 12(2):329–341

Mehnert M, Sommermeyer F, Berger M, Lakshmipathy SK, Gauss R, Aebi M, Jarosch E, Sommer T (2014) The interplay of Hrd3 and the molecular chaperone system ensures efficient degradation of malfolded secretory proteins. Mol Biol Cell. doi:10.1091/mbc.E14-07-1202

Mendoza JL, Schmidt A, Li Q, Nuvaga E, Barrett T, Bridges RJ, Feranchak AP, Brautigam CA, Thomas PJ (2012) Requirements for efficient correction of DeltaF508 CFTR revealed by analyses of evolved sequences. Cell 148(1–2):164–174. doi:10.1016/j.cell.2011.11.023

Michel H (1982) Three-dimensional crystals of a membrane protein complex. The photosynthetic reaction centre from Rhodopseudomonas viridis. J Mol Biol 158(3):567–572

Moelbert S, Emberly E, Tang C (2004) Correlation between sequence hydrophobicity and surface-exposure pattern of database proteins. Protein Sci 13(3):752–762

Moffat K, Bilderback D, Schildkamp W, Szebenyi D, Teng TY (1989) Laue photography from protein crystals. Basic Life Sci 51:325–330

Moller I, Jung M, Beatrix B, Levy R, Kreibich G, Zimmermann R, Wiedmann M, Lauring B (1998) A general mechanism for regulation of access to the translocon: competition for a membrane attachment site on ribosomes. Proc Natl Acad Sci USA 95(23):13425–13430

Moody JE, Millen L, Binns D, Hunt JF, Thomas PJ (2002) Cooperative, ATP-dependent association of the nucleotide binding cassettes during the catalytic cycle of ATP-binding cassette transporters. J Biol Chem 277(24):21111–21114

Mus-Veteau I (2010) Heterologous expression of membrane proteins for structural analysis. Methods Mol Biol 601:1–16. doi:10.1007/978-1-60761-344-2_1

Nikaido K, Ames GF (1999) One intact ATP-binding subunit is sufficient to support ATP hydrolysis and translocation in an ABC transporter, the histidine permease. J Biol Chem 274 (38):26727–26735

Nilsson I, Lara P, Hessa T, Johnson AE, von Heijne G, Karamyshev AL (2014) The code for directing proteins for translocation across ER membrane: SRP cotranslationally recognizes specific features of a signal sequence. J Mol Biol. doi:10.1016/j.jmb.2014.06.014

Noriega TR, Chen J, Walter P, Puglisi JD (2014) Real-time observation of signal recognition particle binding to actively translating ribosomes. eLife 3.doi:10.7554/eLife.04418

Nozaki Y, Tanford C (1971) The solubility of amino acids and two glycine peptides in aqueous ethanol and dioxane solutions. Establishment of a hydrophobicity scale. J Biol Chem 246 (7):2211–2217

Pace CN, Shaw KL (2000) Linear extrapolation method of analyzing solvent denaturation curves. Proteins Suppl 4:1–7

Paddon C, Loayza D, Vangelista L, Solari R, Michaelis S (1996) Analysis of the localization of STE6/CFTR chimeras in a Saccharomyces cerevisiae model for the cystic fibrosis defect CFTR delta F508. Mol Microbiol 19(5):1007–1017

pdb.org. http://www.pdb.org

Pinkett HW, Lee AT, Lum P, Locher KP, Rees DC (2007) An inward-facing conformation of a putative metal-chelate-type ABC transporter. Science 315(5810):373–377

Pitonzo D, Yang Z, Matsumura Y, Johnson AE, Skach WR (2009) Sequence-specific retention and regulated integration of a nascent membrane protein by the endoplasmic reticulum Sec61 translocon. Mol Biol Cell 20(2):685–698. doi:10.1091/mbc.E08-09-0902

Pomozi V, Brampton C, Fulop K, Chen LH, Apana A, Li Q, Uitto J, Le Saux O, Varadi A (2014) Analysis of pseudoxanthoma elasticum-causing missense mutants of ABCC6 in vivo; pharmacological correction of the mislocalized proteins. J Invest Dermatol 134(4):946–953. doi:10.1038/jid.2013.482

Powis K, Schrul B, Tienson H, Gostimskaya I, Breker M, High S, Schuldiner M, Jakob U, Schwappach B (2013) Get3 is a holdase chaperone and moves to deposition sites for aggregated proteins when membrane targeting is blocked. J Cell Sci 126(Pt 2):473–483. doi:10.1242/jcs.112151

Reeb J, Kloppmann E, Bernhofer M, Rost B (2014) Evaluation of transmembrane helix predictions in 2014. Proteins. doi:10.1002/prot.24749

Religa TL, Markson JS, Mayor U, Freund SM, Fersht AR (2005) Solution structure of a protein denatured state and folding intermediate. Nature 437(7061):1053–1056

Rommens JM, Iannuzzi MC, Kerem B, Drumm ML, Melmer G, Dean M, Rozmahel R, Cole JL, Kennedy D, Hidaka N et al (1989) Identification of the cystic fibrosis gene: chromosome walking and jumping. Science 245(4922):1059–1065

Rosenbaum G, Ginell SL, Chen JC (2015) Energy optimization of a regular macromolecular crystallography beamline for ultra-high-resolution crystallography. J Synchrotron Radiat 22 (Pt 1):172–174. doi:10.1107/S1600577514022619

Rost B, Casadio R, Fariselli P, Sander C (1995) Transmembrane helices predicted at 95% accuracy. Protein Sci 4(3):521–533. doi:10.1002/pro.5560040318

Roy SS, Patra M, Nandy SK, Banik M, Dasgupta R, Basu T (2014) In vitro holdase activity of E. coli small heat-shock proteins IbpA, IbpB and IbpAB: a biophysical study with some unconventional techniques. Protein Pept Lett 21(6):564–571

Sadlish H, Pitonzo D, Johnson AE, Skach WR (2005) Sequential triage of transmembrane segments by Sec61alpha during biogenesis of a native multispanning membrane protein. Nat Struct Mol Biol 12(10):870–878. doi:10.1038/nsmb994

Sanders SL, Schekman R (1992) Polypeptide translocation across the endoplasmic reticulum membrane. J Biol Chem 267(20):13791–13794

Schlesinger MJ, Barrett K (1965) The reversible dissociation of the alkaline phosphatase of Escherichia coli. I. Formation and reactivation of subunits. J Biol Chem 240(11):4284–4292

Schneider E, Hunke S (1998) ATP-binding-cassette (ABC) transport systems: functional and structural aspects of the ATP-hydrolyzing subunits/domains. FEMS Microbiol Rev 22(1):1–20

Schroder M, Kaufman RJ (2005) ER stress and the unfolded protein response. Mutat Res 569 (1–2):29–63. doi:10.1016/j.mrfmmm.2004.06.056

Senes A, Ubarretxena-Belandia I, Engelman DM (2001) The Calpha ---H...O hydrogen bond: a determinant of stability and specificity in transmembrane helix interactions. Proc Natl Acad Sci USA 98(16):9056–9061. doi:10.1073/pnas.161280798

Senior AE, Gadsby DC (1997) ATP hydrolysis cycles and mechanism in P-glycoprotein and CFTR. Semin Cancer Biol 8(3):143–150

Sharma M, Benharouga M, Hu W, Lukacs GL (2001) Conformational and temperature-sensitive stability defects of the delta F508 cystic fibrosis transmembrane conductance regulator in post-endoplasmic reticulum compartments. J Biol Chem 276(12):8942–8950

Sharma SK, Christen P, Goloubinoff P (2009) Disaggregating chaperones: an unfolding story. Curr Protein Pept Sci 10(5):432–446

Shimamura T, Weyand S, Beckstein O, Rutherford NG, Hadden JM, Sharples D, Sansom MS, Iwata S, Henderson PJ, Cameron AD (2010) Molecular basis of alternating access membrane transport by the sodium-hydantoin transporter Mhp1. Science 328(5977):470–473. doi:10.1126/science.1186303

Shintre CA, Pike AC, Li Q, Kim JI, Barr AJ, Goubin S, Shrestha L, Yang J, Berridge G, Ross J, Stansfeld PJ, Sansom MS, Edwards AM, Bountra C, Marsden BD, von Delft F, Bullock AN, Gileadi O, Burgess-Brown NA, Carpenter EP (2013) Structures of ABCB10, a human ATP-binding cassette transporter in apo- and nucleotide-bound states. Proc Natl Acad Sci USA 110(24):9710–9715. doi:10.1073/pnas.1217042110

Skarina T, Xu X, Evdokimova E, Savchenko A (2014) High-throughput crystallization screening. Methods Mol Biol 1140:159–168. doi:10.1007/978-1-4939-0354-2_12

Skolnick J, Fetrow JS, Kolinski A (2000) Structural genomics and its importance for gene function analysis. Nat Biotechnol 18(3):283–287. doi:10.1038/73723

Slepenkov SV, Witt SN (2002) The unfolding story of the Escherichia coli Hsp70 DnaK: is DnaK a holdase or an unfoldase? Mol Microbiol 45(5):1197–1206

Smith PC, Karpowich N, Millen L, Moody JE, Rosen J, Thomas PJ, Hunt JF (2002) ATP binding to the motor domain from an ABC transporter drives formation of a nucleotide sandwich dimer. Mol Cell 10(1):139–149

Song W, Raden D, Mandon EC, Gilmore R (2000) Role of Sec61alpha in the regulated transfer of the ribosome-nascent chain complex from the signal recognition particle to the translocation channel. Cell 100(3):333–343

Structure MPoKD. http://blanco.biomol.uci.edu/mpstruc/
Sun S, Brem R, Chan HS, Dill KA (1995) Designing amino acid sequences to fold with good hydrophobic cores. Protein Eng 8(12):1205–1213
Tamayo AG, Slater L, Taylor-Parker J, Bharti A, Harrison R, Hung DT, Murphy JR (2011) GRP78 (BiP) facilitates the cytosolic delivery of anthrax lethal factor (LF) in vivo and functions as an unfoldase in vitro. Mol Microbiol 81(5):1390–1401. doi:10.1111/j.1365-2958.2011.07770.x
Teem JL, Berger HA, Ostedgaard LS, Rich DP, Tsui LC, Welsh MJ (1993) Identification of revertants for the cystic fibrosis delta F508 mutation using STE6-CFTR chimeras in yeast. Cell 73(2):335–346
Teem JL, Carson MR, Welsh MJ (1996) Mutation of R555 in CFTR-delta F508 enhances function and partially corrects defective processing. Receptors Channels 4(1):63–72
Thibodeau PH, Brautigam CA, Machius M, Thomas PJ (2005) Side chain and backbone contributions of Phe508 to CFTR folding. Nat Struct Mol Biol 12(1):10–16
Thibodeau PH, Richardson JM 3rd, Wang W, Millen L, Watson J, Mendoza JL, Du K, Fischman S, Senderowitz H, Lukacs GL, Kirk K, Thomas PJ (2010) The cystic fibrosis-causing mutation deltaF508 affects multiple steps in cystic fibrosis transmembrane conductance regulator biogenesis. J Biol Chem 285(46):35825–35835. doi:10.1074/jbc.M110.131623
Thomas PJ, Qu BH, Pedersen PL (1995) Defective protein folding as a basis of human disease. Trends Biochem Sci 20(11):456–459
Trowbridge IS, Collawn J, Jing S, White S, Esekogwu V, Stangel M (1991) Structure-function analysis of the human transferrin receptor: effects of anti-receptor monoclonal antibodies on tumor growth. Curr Stud Hematol Blood Transfus 58:139–147
Vendruscolo M, Paci E, Dobson CM, Karplus M (2001) Three key residues form a critical contact network in a protein folding transition state. Nature 409(6820):641–645. doi:10.1038/35054591
Vergani P, Lockless SW, Nairn AC, Gadsby DC (2005) CFTR channel opening by ATP-driven tight dimerization of its nucleotide-binding domains. Nature 433(7028):876–880
Wallin E, von Heijne G (1998) Genome-wide analysis of integral membrane proteins from eubacterial, archaean, and eukaryotic organisms. Protein Sci 7(4):1029–1038. doi:10.1002/pro.5560070420
Wang L, Dobberstein B (1999) Oligomeric complexes involved in translocation of proteins across the membrane of the endoplasmic reticulum. FEBS Lett 457(3):316–322
Wang CC, Tsou CL (1998) Enzymes as chaperones and chaperones as enzymes. FEBS Lett 425 (3):382–384
Watson JD, Crick FH (1953) Molecular structure of nucleic acids; a structure for deoxyribose nucleic acid. Nature 171(4356):737–738
White SH (2009) Biophysical dissection of membrane proteins. Nature 459(7245):344–346. doi:10.1038/nature08142
Wilken S, Schmees G, Schneider E (1996) A putative helical domain in the MalK subunit of the ATP-binding-cassette transport system for maltose of Salmonella typhimurium (MalFGK2) is crucial for interaction with MalF and MalG. A study using the LacK protein of Agrobacterium radiobacter as a tool. Mol Microbiol 22(4):655–666
Xiao R, Anderson S, Aramini J, Belote R, Buchwald WA, Ciccosanti C, Conover K, Everett JK, Hamilton K, Huang YJ, Janjua H, Jiang M, Kornhaber GJ, Lee DY, Locke JY, Ma LC, Maglaqui M, Mao L, Mitra S, Patel D, Rossi P, Sahdev S, Sharma S, Shastry R, Swapna GV, Tong SN, Wang D, Wang H, Zhao L, Montelione GT, Acton TB (2010) The high-throughput protein sample production platform of the Northeast Structural Genomics Consortium. J Struct Biol 172(1):21–33. doi:10.1016/j.jsb.2010.07.011
Yuan YR, Blecker S, Martsinkevich O, Millen L, Thomas PJ, Hunt JF (2001) The crystal structure of the MJ0796 ATP-binding cassette. Implications for the structural consequences of ATP hydrolysis in the active site of an ABC transporter. J Biol Chem 276(34):32313–32321
Zacchi LF, Wu HC, Bell SL, Millen L, Paton AW, Paton JC, Thomas PJ, Zolkiewski M, Brodsky JL (2014) The BiP molecular chaperone plays multiple roles during the biogenesis of torsinA,

an AAA+ ATPase associated with the neurological disease early-onset torsion dystonia. J Biol Chem 289(18):12727–12747. doi:10.1074/jbc.M113.529123

Zaitseva J, Jenewein S, Jumpertz T, Holland IB, Schmitt L (2005) H662 is the linchpin of ATP hydrolysis in the nucleotide-binding domain of the ABC transporter HlyB. EMBO J 24 (11):1901–1910

Zhang L, Conway JF, Thibodeau PH (2012) Calcium-induced folding and stabilization of the Pseudomonas aeruginosa alkaline protease. J Biol Chem 287(6):4311–4322. doi:10.1074/jbc.M111.310300

Zhou X, Levin EJ, Pan Y, McCoy JG, Sharma R, Kloss B, Bruni R, Quick M, Zhou M (2014) Structural basis of the alternating-access mechanism in a bile acid transporter. Nature 505 (7484):569–573. doi:10.1038/nature12811

Part II
Epithelial Ion Channels and Transporters

Chapter 9
P2X Receptors in Epithelia

Jens Leipziger

Abstract P2X receptors are ubiquitously expressed in all epithelial tissues, but their functional roles are less well studied. Here we review the current state of knowledge by focusing on functional effects of P2X receptor in secretory and in absorptive tissues. In glandular tissue like the parotid gland, basolateral P2X receptors stimulate ion secretion via an increase of $[Ca^{2+}]_i$. In absorptive epithelia like the renal tubule, P2X receptor stimulation mediates the inhibition of NaCl, Mg^{2+}, and water transport in the thick ascending limb and the distal convoluted tubule, respectively. The underlying signaling pathways that inhibit epithelial absorption are currently not well understood. Epithelial P2X7 receptors show pronounced upregulation during various diseased states highlighting a role of purinergic signaling in epithelial pathophysiology. Importantly, functional effects of epithelial P2X receptors cover numerous other aspects ranging from modulation of sound transmission, activation of apoptosis, or production of oxygen radicals. Eventually, P2X receptors in epithelia are an understudied issue offering numerous novel and very attractive questions.

Keywords Epithelial transport • Exocrine secretion • Kidney • Renal tubule • Purinergic signaling • P2Y

9.1 Introduction

P2X receptors are nonselective ligand-gated cation ion channels that specifically use ATP as their agonist (North 2002). Each P2X receptor is composed of three subunits, which all have intracellular N- and C-termini, two transmembrane spanning α-helices, and a connecting acidic patch rich extracellular loop (Kawate et al. 2009). Three ATP-binding pockets are localized in between the subunits, and ATP binding likely conveys substantial spatial rearrangements of the receptor, eventually to permit its opening (Hattori and Gouaux 2012; Kawate et al. 2009). An

J. Leipziger (✉)
Department of Biomedicine – Physiology, Aarhus University, Ole Worms Allé 4, Building 1160, 8000 Aarhus C, Denmark
e-mail: leip@biomed.au.dk

K.L. Hamilton, D.C. Devor (eds.), *Ion Channels and Transporters of Epithelia in Health and Disease*, Physiology in Health and Disease,
DOI 10.1007/978-1-4939-3366-2_9

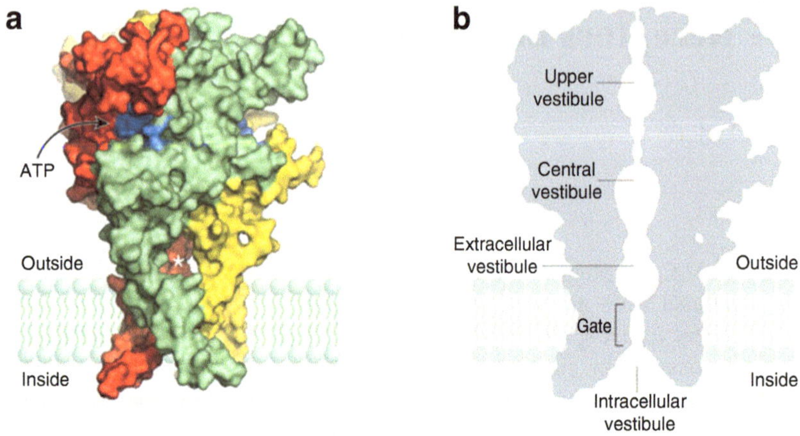

Fig. 9.1 Trimeric structure of the P2X receptor as taken from the original crystal structure description of the zebrafish P2X4 receptor (Kawate et al. 2009) [This figure is taken with permission from Silberberg and Swartz (2009)]. (**a**) Surface representation of the P2X4 receptor with the three identical subunits colored in *green, red*, and *yellow* and the ATP-binding sites colored in *blue*. The *white asterisk* marks one of three fenestrations between subunits where the extracellular vestibule opens to the exterior, providing a potential passage for ions to enter or exit from the extracellular side of the pore. (**b**) Representation of the P2X receptor ion channel showing the four vestibules and a gate formed by a plug of hydrophobic amino-acid residues within the pore

image derived from the crystal structure of a closed P2X4 receptor (no ATP bound) is shown in Fig. 9.1 (Silberberg and Swartz 2009). Seven different P2X receptor genes have been identified in the mammalian genome and cells commonly express several different P2X receptor transcripts and proteins. Except for the P2X6 receptors, all P2X receptor subunits can form functional homomeric ATP-gated channels, but in the native situation, heteromeric P2X receptors composed of two or three different subunits are often believed to mediate the physiological response (North 2002). ATP binding to P2X receptors causes a cellular increase of $[Ca^{2+}]_i$, a membrane depolarization via Na^+ entry, and an efflux of K^+, all caused by the movement of the respective cations through the channel. In excitable tissues like neurons or smooth muscle cells, P2X receptors stimulate the given cell by virtue of this membrane depolarizing effect moving the membrane voltage towards its activating threshold. Purinergic P2X receptors are ubiquitously expressed in nearly all mammalian and nonmammalian cells including non-excitable cells. They serve a large variety of different cell- and organ-specific functions. Moreover, also in epithelia, P2X receptors are widely expressed (Taylor et al. 1999). This review attempts to provide a comprehensive overview of P2X receptors in native mammalian epithelia and to discuss their possible physiological roles in these tissues. As the main purpose of epithelia is to govern vectorial transport of solutes and water, the overview will mostly approach this theme, with the renal epithelium as major area of interest.

9.2 P2X Receptors and Excitation-Secretion Coupling

Exocrine secretion is one of the prime functions of the gastrointestinal tract. The various glands serve to add the appropriate amount of digestive juices to the intraluminal intestinal milieu to permit mixing, breakdown, and eventually absorption of nutrients, salts, and water from the ingested food and beverages. Exocrine glands like the salivary, the gastric glands, or the pancreas are strongly innervated from both autonomous and enteric nerve fibers, and one of the prime agonists to activate secretion is the neurotransmitter acetylcholine (ACh). Commonly, the generation of inositol trisphosphate (IP_3) followed by an increase of $[Ca^{2+}]_i$ activates basolateral Ca^{2+}-activated K^+ channels and in parallel apical Ca^{2+}-activated Cl^- channels to lead to stimulated Cl^- secretion. The first functional description of an ATP-activated P2X receptor in epithelia was made in rat parotid acinar cells that displayed ATP-induced ionic current features and increases of $[Ca^{2+}]_i$ clearly independent of the generation of IP_3 (McMillian et al. 1988). These results were confirmed in lacrimal acinar cells and have led to a general revision of the understanding how agonists stimulate exocrine secretion in some glands (Sasaki and Gallacher 1990) (Fig. 9.2). Common for many ion secretion-stimulating agonists is their property to elevate $[Ca^{2+}]_i$ either by a G_q-protein-coupled mechanism in the case of ACh or by activation of a ligand-gated nonselective cation channel in

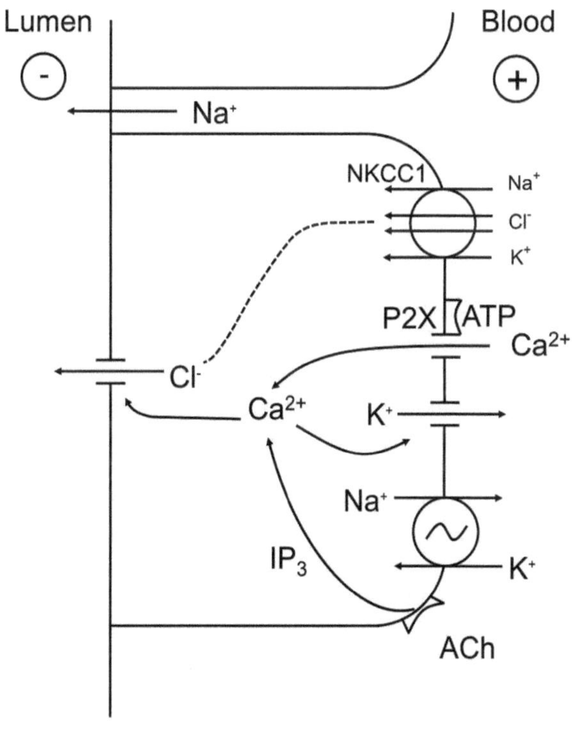

Fig. 9.2 Schematic cell model of the lacrimal and parotid exocrine acinar cell that highlights the central role of an increase of $[Ca^{2+}]_i$ for NaCl secretion activation. Note that a P2X receptor contributes to activate ion secretion via elevating $[Ca^{2+}]_i$

Table 9.1 P2X receptors in secretory epithelia

Epithelial tissue	Species	P2X receptor (subtype)	Intracellular signal	Regulated function	Reference
Parotid acinar cell	Rat	Undefined	$[Ca^{2+}]_i$	↑ ion secretion	McMillian et al. (1988)
Lacrimal acinar cell	Mouse	Undefined	$[Ca^{2+}]_i$	↑ ion secretion	Sasaki and Gallacher (1990)
Pancreatic duct	Rat	P2X1, P2X4, P2X7	$[Ca^{2+}]_i$ pH_i ↓	Suggested regulation of ductal ion transport	Hede et al. (1999), Henriksen and Novak (2003), Luo et al. (1999)

the case of extracellular ATP. Only a few reports have addressed this issue and describe P2X receptors in secretory epithelia as seen in Table 9.1. It is also a peculiar observation that not all exocrine acinar cells show robust functional expression of P2X receptors as shown in rat pancreatic acinar cells (Novak et al. 2002) as well as in colonic crypts (Larsen and Leipziger 2013). Furthermore, it is worth mentioning that ATP can also act via P2Y receptors to elevate $[Ca^{2+}]_i$ and ion secretion in some exocrine glands (Leipziger et al. 1997).

9.3 P2X Receptors in Absorptive Epithelia

Solute and water absorption is an essential function of the gastrointestinal, the respiratory, and especially the renal epithelium. Evidence is emerging that P2X receptors are abundantly expressed in most absorptive epithelia (Hede et al. 1999; Korngreen et al. 1998; Marques et al. 2012), and a summary of epithelial P2X receptors and their sites of expression is shown in Table 9.2. In the kidney, the first observation of a P2X receptor-like function was made in 1998. In the proximal tubule, epithelial cell line LLC-PK_1 ATP activated a nonselective cation conductance and an increase of $[Ca^{2+}]_i$, which was fully dependent on the presence of extracellular Ca^{2+} (Filipovic et al. 1998). In the meantime, further work by many groups has identified specific mRNAs for many P2X receptor subtypes along the different renal epithelial tissues often supported by positive immunohistochemistry results (IHC) (de Baaij et al. 2014; Marques et al. 2012; Turner et al. 2003; Wildman et al. 2008). Nonetheless, the understanding of a physiological role of P2X receptors in renal epithelia is still very limited. In 2001, a study using cultured mouse distal convoluted tubule (DCT) cells demonstrated that P2X receptor stimulation caused the inhibition of Mg^{2+} uptake into the cell's cytosol (Dai et al. 2001). Because this segment is responsible for transcellular Mg^{2+} reabsorption, it was speculated that ATP could inhibit Mg^{2+} reabsorption. At that time, the apical ion channel for Mg^{2+} reabsorption in the distal convoluted tubule (DCT) had not been

Table 9.2 P2X receptors in absorptive epithelia

Epithelial tissue	Species	P2X receptor (subtype)	Intracellular signal	Regulated function	Reference
LLC-PK_1 Proximal tubular origin	Pig	P2X1	$[Ca^{2+}]_i$	not known	Filipovic et al. (1998)
Distal convoluted tubule (MDCT cell line)	Mouse	P2X1-3 P2X4 P2X5	$[Ca^{2+}]_i$	↓ cytosolic Mg^{2+} uptake = ↓ Mg^{2+} reabsorption	Dai et al. (2001)
Distal convoluted tubule (DCT)	Mouse	P2X4 P2X6	not known	↓TRPM6	de Baaij et al. (2014)
Native medullary thick ascending limb (mTAL)	Mouse	P2X1 P2X4 P2X5	not known	↓ NaCl reabsorption	Marques et al. (2012)
Cortical collecting duct (CCD)	Rat	P2X1 P2X4 P2X6	PI3K	↓ and ↑ amiloride-sensitive currents	Wildman et al. (2008)
Entire renal tubule	Rat	P2X4 P2X6	not known	not known	Turner et al. (2003)
Medullary collecting duct and S3 segment of PT	Rat	P2X5	not known	not known	Turner et al. (2003)
A6, kidney epithelial cell line	*Xenopus laevis*	P2X4 like	not known	↑ ENaC-dependent Na^+ transport	Zhang et al. (2007)
IMCD-K2 Inner medullary collecting duct cell line	Mouse	P2X	not known	↓ NaCl reabsorption	McCoy et al. (1999)
Cochlear outer sulcus cells	Gerbil	P2X2?	not known	Stimulation of cation absorption	Lee et al. (2001)

discovered yet but is now identified as the TRPM6 channel (Schlingmann et al. 2002). Intriguingly, very recent results strongly support that P2X4 receptor stimulation indeed inhibits the TRPM6 channel (de Baaij et al. 2014). Furthermore, the P2X4 receptor is co-expressed with the TRPM6 channel in the DCT and in the distal colon, two important sites of epithelial Mg^{2+} reabsorption (de Baaij et al. 2014). Taken together, these results therefore strengthen the hypothesis that P2X receptor stimulation inhibits Mg^{2+} reabsorption. Direct evidence for this is currently pending, and P2X4$^{-/-}$ mice on a normal diet show unchanged urinary Mg^{2+} excretion (de Baaij et al. 2014). In isolated perfused mouse medullary thick ascending limbs (mTAL), it was recently shown that basolateral P2X4 receptor stimulation triggers a fast and fully reversible (~25 %) inhibition of NaCl reabsorption (Marques et al. 2012). Importantly, this study was the first to identify that P2X receptor stimulation in native renal epithelium modulates renal ion transport. It is also noteworthy that infusion of P2X receptor agonists into rats caused

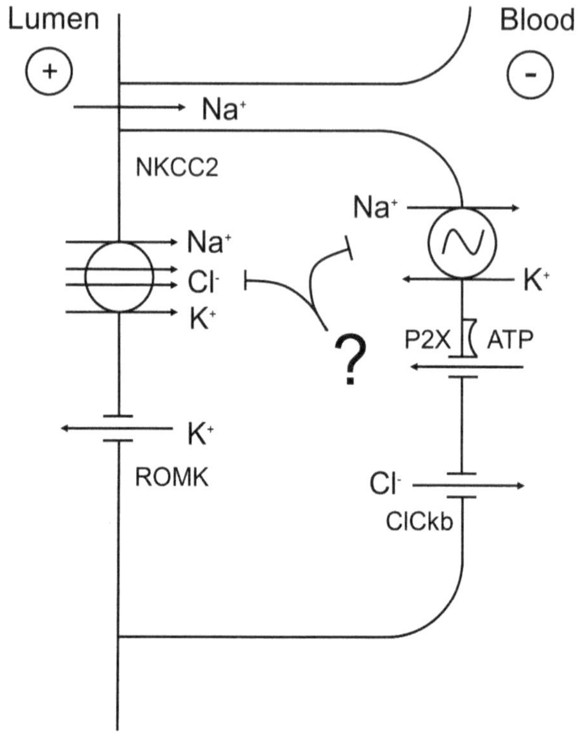

Fig. 9.3 P2X receptors are expressed in absorptive renal epithelia. Here the example of the mouse medullary thick ascending limb (mTAL) is shown. Basolateral ATP inhibits NaCl absorption via a P2X4 receptor, but the cellular mechanism of this inhibition is currently not understood

increased excretion of water and NaCl (Jankowski et al. 2011). In conjunction, these results suggest that P2X receptor stimulation in renal epithelia inhibits reabsorption of NaCl, Mg^{2+}, and water. Noteworthy, the field of P2Y receptor research in renal epithelia has a much longer and more extensive research history (Praetorius and Leipziger 2010; Vallon 2008). Unequivocal results from many groups have identified that P2Y receptor stimulation inhibits absorptive transport in most renal tubular segments. Taken together, local paracrine ATP signaling via both P2Y and P2X receptors inhibits renal tubular transport and thus opposes the transport activation effect of systemic hormones like arginine vasopressin or aldosterone (Fig. 9.3). The rather recent discoveries addressing functional effects of P2X receptors in absorptive renal epithelia are therefore in good agreement with the current understanding of intrarenal purinergic signaling: local extracellular ATP signals likely present an endogenous local diuretic signaling system (Praetorius and Leipziger 2010). It is interesting to note that the P2X4$^{-/-}$ mouse shows a hypertensive phenotype, which was suggested to be caused by endothelial dysfunction due to the absence of ATP/P2X4 receptor-dependent activation of the flow-induced endothelial nitric oxide (NO) system (Yamamoto et al. 2006). In the light of the functional data on P2X4 receptors in the renal tubule, one may speculate that tubular transport could be increased in the P2X4$^{-/-}$ mouse contributing to hypertension by an effect of plasma volume expansion. Eventually, the effect of ATP was tested on ENaC-mediated whole cell currents in rat split open cortical collecting duct cells (CCD) (Wildman et al. 2008). Much in agreement with numerous other studies, a marked inhibition of the epithelial Na^+ channel (ENaC) was measured, which was mediated via the P2Y2 receptor. In addition, P2X receptor-mediated effects were reported. Interestingly, under high extracellular Na^+ (145 mM), P2X receptor activation inhibited the amiloride-sensitive currents, but under low extracellular Na^+ (50 mM), P2X receptor activation apparently activated the amiloride-sensitive current in CCD cells. The authors therefore speculated that the P2X receptor could function as extracellular Na^+ sensor (Wildman et al. 2008). These results will require further attention to be fully understood. Nonetheless, the experimental evidence is interesting because an extracellular P2X receptor with Na^+ sensor features had previously been suggested in rabbit airway ciliated cells. In ciliated cells, the application of ATP stimulated a P2X receptor-dependent nonselective cation current (Korngreen et al. 1998). Interestingly, this ATP-induced inward current was strongly inhibited by raising the extracellular Na^+ concentration above 10 mM (Ma et al. 1999, 2006). Again, this finding remains to be fully understood and integrated into a physiological concept, and it is also uncertain whether the Na^+ concentrations in the airway surface liquid layer can reach these very low values. Nonetheless, the research question of the role of purinergic signaling in pulmonary mucociliary clearance and airway surface liquid layer regulation is of greatest interest and clinical relevance (Boucher 2003). Luminal ATP activates ciliary beat frequency and ion secretion (Korngreen and Priel 1996), but the specific role of epithelial P2X receptors in this physiology is not yet fully studied.

9.4 Identification of the Native P2X Receptor Responsible for a Physiological Response

It is a general feature of epithelia and most other cell types that they express multiple P2Y and P2X receptors (Praetorius and Leipziger 2010; Vallon 2008). By using specific P2Y receptor knockout mice together with reasonably specific receptor agonists and antagonists, it was in many cases possible to identify which of the eight different P2Y receptors is critically responsive for a certain nucleotide-induced effect. An example for this is the apical P2Y2 receptor in the collecting duct that markedly inhibits ENaC channels and thus Na^+ reabsorption (Pochynyuk et al. 2008).

Also, the identification of P2X receptor-mediated cellular effects has become reasonably straight forward. A number of arguments usually permit the safe conclusion that a P2X receptor was activated:

1. They are specifically activated by ATP (North 2002).
2. The ATP-stimulated $[Ca^{2+}]_i$ increase is fully dependent on the influx of Ca^{2+} from the extracellular space (Jensen et al. 2007). This argument can, however, be

confounded by co-expressed P2Y receptors in the same cell, leading to simultaneous release of Ca^{2+} from intracellular stores when ionotropic and metabotropic receptors are stimulated in parallel.
3. A nonselective cation current can be very rapidly activated and deactivated on addition and removal of ATP (North 2002).
4. P2X receptors are well inhibited by oxidized ATP (irreversible) leaving P2Y receptor-mediated effects apparently unaffected (Murgia et al. 1993).
5. Finally, other specific pharmacological probes help to study P2X receptors (Burnstock 2007).

The task to unequivocally identify the molecular nature of the P2X receptor in native tissue is for obvious reasons difficult. P2X receptors are built of three subunits, and theoretically three different ones could compose one receptor. Indeed, it was recently shown that heterologously expressed and epitope-tagged P2X2, P2X4, and P2X6 receptors form heterotrimeric channels (Antonio et al. 2014). It is difficult to answer this question using native tissue. We have identified a P2X receptor in the basolateral membrane of the mTAL, and on the level of mRNA, these mTAL renal tubules expressed the P2X1, P2X4, and P2X5 receptors (Marques et al. 2012). To unravel which P2X receptor subtype could mediate the physiological ATP effect, a knockout strategy was applied. In mTALs from $P2X4^{-/-}$ mice, the ATP effect was blunted but still present. These results clearly support the involvement of the P2X4 receptor subunit. This raises the question which other receptors could be activated in the $P2X4^{-/-}$. Apparently, P2X receptor subunits can interact promiscuously with numerous other P2X receptor subunits (North 2002). Most likely, the absence of one subunit in this example of the mTAL could be compensated by the remaining isoforms, possibly leading to homomeric P2X5 or heteromeric P2X1/P2X5 receptors. A double knockout of both receptor subunits, the P2X4 and the P2X5, could be attempted in this project, but this could be quite challenging. This describes the general problem of identifying the exact subunit composition of P2X receptors in native tissues. To the best of our knowledge, the unequivocal identification of the exact subunit composition of a native P2X receptor in multiple subunit expressing cells has not been successfully achieved in epithelia.

9.5 Intracellular Signaling Events Associated with Epithelial P2X Receptor Activation

In excitable cells, the activation of P2X receptors leads to a membrane voltage depolarization and thus has direct input to neuronal activation or inhibition by defining the voltage proximity to the electrical threshold. In excitation-secretion coupling in exocrine gland tissue, the direct influx of Ca^{2+} activates relevant Ca^{2+}-sensitive ion channels and thus ion secretion directly (Fig. 9.2). Thus, in exocrine secretion, the main P2X receptor-associated intracellular transduction signal is Ca^{2+} as shown in lacrimal and parotid acinar cells (McMillian et al. 1988; Sasaki and Gallacher 1990). In several renal absorptive epithelia, P2X receptors inhibit reabsorption, but the associated signaling mechanisms are not well understood. In all renal epithelial tissues, P2X receptor activation triggers as expected an increase of $[Ca^{2+}]_i$ (Filipovic et al. 1998; Jensen et al. 2007). In rat TAL, P2X receptor stimulation triggers an intracellular NO increase (Silva et al. 2006), and this in turn was suggested to inhibit NKCC2-dependent Cl^- reabsorption (Ortiz and Garvin 2002). Likely, the increase of $[Ca^{2+}]_i$ is upstream of the generation of NO. However, in perfused mTAL, we have shown that using BAPTA-AM loading to buffer intracellular Ca^{2+} or removal of extracellular Ca^{2+} to abrogate Ca^{2+} influx failed to block the ATP/P2X receptor-induced transport inhibition in mouse mTAL (unpublished). Currently, the role of an increase in $[Ca^{2+}]_i$ remains uncertain in the acute regulation of NaCl transport in the TAL. This is also highlighted by results showing that neither activation of the P2Y2 nor the Ca^{2+}-sensing receptor, two maneuvers well established to increase $[Ca^{2+}]_i$ in this tubule segment, has any acute effects on NaCl transport rate in mouse TAL (Desfleurs et al. 1998; Jensen et al. 2007; Marques et al. 2012). One may, therefore, consider other signaling events. One idea comes from in vivo studies, in which the systemic infusion of P2X receptor agonists triggers a marked diuresis and natriuresis (Jankowski et al. 2011). These authors also report a significant inhibition of the Na^+/K^+ ATPase activity after addition of P2X receptor agonists measured in freshly isolated proximal tubular cells. One may therefore speculate that P2X receptor stimulation in the mTAL, where both membrane proteins are in the same basolateral membrane domain, leads to an inhibition of the Na^+/K^+ ATPase and thus NaCl reabsorption. In the above discussed study that identified P2X4 receptor-stimulated inhibition of the TRPM6 channel, the authors also tested several protein kinase inhibitors (PKC, PKA, PI3K) but did not find any positive results (de Baaij et al. 2014). In summary, despite clear functional inhibitory effects on ion transport in renal epithelial cells, the underlying signaling events of P2X receptor stimulation remain currently poorly understood.

9.6 Apical and Basolateral P2X Receptors

Epithelial transport depends on the specific localization of ion pumps, transporters, and channels in either the apical or the basolateral membrane domain. Transcellular ion secretion follows the principal mechanism of "pump leak," where a primary or secondary active uptake of the ion over the basolateral membrane is followed by a passive leak over the apical membrane. In absorptive epithelial transport, this scheme can be thought "inverted" where initially an apical membrane channel or transporter permits cytosolic uptake followed by a secondary removal step over the basolateral membrane. Regulation of these processes is highly complex and most often involves hormone, neurotransmitter, and paracrine factor receptors in the basolateral membrane. It has also been recognized that most epithelia can be regulated by agonist signals via membrane receptors in their apical membranes.

Table 9.3 Apical and basolateral localization of P2X receptor in epithelia

Epithelial tissue	Species	P2X receptor Apical	P2X receptor Basolateral	Regulated function	Reference
Pancreatic duct	Rat	P2X7	P2X1 P2X4 P2X7	Ductal ion transport	Hede et al. (1999), Luo et al. (1999)
Medullary thick ascending limb	Mouse	None	P2X4 P2X5 P2X1	Inhibition of NaCl reabsorption	Marques et al. (2012)
Cortical collecting duct	Rat	P2X4 P2X6	P2X4 P2X6	Regulation of ENaC	Wildman et al. (2008)
Airway ciliated cells	Rabbit	P2X7? P2X4		? activation of mucociliary clearance	Ma et al. (2006)

This is especially well studied for the purinergic signaling system, where it was found that most epithelial cells express apical and basolateral P2Y receptors often in the same cell (Leipziger 2003). This has led to the appreciation that the luminal space, e.g., in the renal tubule, actually functions as an extracellular signaling compartment. The interested reader is directed to more comprehensive reviews on this topic (Leipziger 2003, 2011). Table 9.3 summarizes studies, which have addressed the apical and basolateral localization of epithelial P2X receptors. So far the number of studies is small, but these studies clearly indicate that P2X receptors in native tissue can be found in both the apical and basolateral membranes. One example is the P2X4 receptor, which can be found in the basolateral membrane of mouse mTAL and apparently also in the apical membrane of rat CCD and rabbit-ciliated respiratory epithelium. These results imply that some epithelia can pick up the information of luminal ATP concentrations via apical P2X receptors.

9.7 Epithelial P2X Receptors in Pathophysiology

In general, as discussed in this review, the studies that address the physiology of P2X receptors in epithelia still remain limited. Therefore, it is not surprising that only a few pathophysiological concepts involving epithelial P2X receptors have been developed and these often remain speculative. As mentioned above, the P2X4 receptor knockout is hypertensive (Yamamoto et al. 2006), and we wonder if an increased renal tubular reabsorption could be an underlying cause. For human hypertension, a single nucleotide polymorphism was found in the P2Y2 receptor, and current data support a role of the P2X7 but not for the P2X4 receptor in human diseases (Palomino-Doza et al. 2008; Wang et al. 2010). Much focus has been directed towards the P2X7 receptor as it is viewed to have multiple proinflammatory actions and is highly expressed in various immune cells. Interestingly, in normal renal epithelial cells including podocytes, P2X7 receptor expression apparently is low (Booth et al. 2012; Ilatovskaya et al. 2013; Marques et al. 2012; Vonend et al. 2004) but can be dramatically upregulated in diseased states. This has been shown in a rat experimental diabetes mellitus model and in a rat hypertensive model with a markedly increased P2X7 expression in podocytes (Vonend et al. 2004). P2X7 receptors also became significantly expressed in collecting ducts in a genetic model of recessive polycystic kidney disease and stimulation of in vitro cysts with BzATP, a potent P2X7 agonist that reduces cyst number (Hillman et al. 2004). Similarly, the P2X4 receptor is markedly upregulated in humans with diabetic nephropathy and is suggested to mediate a proinflammatory effect (Chen et al. 2013). Also in a zebrafish model of cystogenesis by knockdown of PKD2, a role of the P2X7 receptor was identified, but here, P2X7 receptor inhibitors caused a reduction of cystogenesis (Chang et al. 2014). These results suggest that normally no/low P2X7 expressing "healthy" epithelial cells can change their phenotype during damaging influences and become targetable for extracellular ATP. A deeper understanding of this phenotypical change remains to be developed. Similarly, also in salivary epithelial cells, the P2X7 receptor was recently described to trigger a proinflammatory response (Woods et al. 2012). These results highlight a general and critical role of the P2X7 receptor and possibly also the P2X4 receptor in epithelial pathology processes.

Unilateral ureter obstruction (UUO) is a powerful model to induce chronic renal inflammation and interstitial fibrosis. In P2X7 receptor knockout mice, the UUO-induced interstitial fibrosis was markedly reduced consistent with its proinflammatory role (Goncalves et al. 2006). Very recent results reinvestigated the same question in P2X4$^{-/-}$ mice and much to the authors' surprise found the contrary, namely, that P2X4$^{-/-}$ mice showed a significantly increased fibrosis (Kim et al. 2014). Apparently, one may speculate that the P2X4 receptor, known to be richly expressed in epithelia, could also mediate an anti-inflammatory effect in the kidney. In summary, epithelial P2X receptor physiology and pathophysiology is in many aspects still in its infancy, and much more basic work is necessary in this area. However, it has become clear that especially the P2X7 receptor could be a major disease causing element in a number of different epithelia, especially in pathophysiological conditions.

9.8 Other Functional Implications of Epithelial P2X Receptors

Beyond the effects of P2X receptors on the regulation of ion transport, other relevant epithelial functions have been identified, which are modulated by P2X receptors, and these are listed in Table 9.4. They cover a wide range of different biological phenomena, and many of the listed discoveries point to an important role of the P2X receptors in the control of an epithelial innate immune/inflammatory

Table 9.4 Other functions of epithelial P2X receptors

P2X receptor	Tissue	Function	Reference
P2X7	Human cervical epithelium Alveolar type I cells Intestinal epithelial cell	↑ apoptosis	Guo et al. (2014), Souza et al. (2012), Wang et al. (2004)
P2X4	Human cervical epithelium	↑ paracellular resistance	Gorodeski (2002)
P2X4/ P2X7	Alveolar type I cells	↑ AQP5 expression ? altered H_2O transport in the alveolus	Ebeling et al. (2014)
P2X7	Alveolar type I and II cells	↑ surfactant secretion from alveolar type II cells (paracrine)	Mishra et al. (2011)
P2X7	Retinal pigment epithelium	Lysosomal alkalinization	Guha et al. (2013)
P2X7/ P2X4	Gingival epithelial cell	↑ reactive oxygen species	Hung et al. (2013)
P2X7	Corneal epithelium	↑ cell migration, ↑ proliferation, ↑ cell death	Mankus et al. (2011)
P2X?	Epididymis	Regulation of H^+ secretion	Belleannee et al. (2010)
P2X2	Reissner membrane of the inner ear	? modulation of sound perception	King et al. (1998)
P2X4	Alveolar type II cells	FACE (= fusion-activated Ca^{2+} entry) ↑ surfactant secretion	Miklavc et al. (2013)
P2X4	HeLa cells	Inhibition of *Chlamydia trachomatis* infection via epithelial ATP release	Pettengill et al. (2012)

response and tissue regeneration and repair. One motif points to the epithelial barrier as a "sensory sheet" that decodes extracellular ATP concentrations to respond in a defense-oriented manner. This research area is very novel and underdeveloped and offers attractive perspectives to be pursued.

9.9 Final Concluding Remarks

The topic of P2X receptor research is emerging and is beginning to extend our understanding of how barrier forming epithelial sheets function. Local paracrine ATP signals have been identified to modify epithelial transport function via both P2Y and P2X receptors. The physiology of P2X receptors is currently best understood in renal absorptive epithelia, and here evidence from native mammalian tissue informs us that P2X receptor stimulation inhibits solute absorption thus producing a local diuretic event. Most intriguing is the rather dramatic upregulation of P2X receptors in processes that damage the epithelial tissue highlighting the possibility that P2X receptor biology is of substantial importance in tissue repair and regeneration.

Acknowledgments The author greatly thanks a very lively laboratory team, which is always willing to challenge the current working hypotheses and provides me with endless joy to go to work every day. These wonderful people are currently Helle A. Praetorius, Mads Vaarby Sørensen, Casper K. Larsen, Pauline I.A. de Bruijn, Iben Skov Jensen, Jeppe Olsen, Samuel Svendsen, Marianne Skals, Helle Jakobsen, and Karen Sjødt Sørensen. And I further thank the Danish Medical Research Council and the Lundbeck Foundation for many years of financial support for our research.

References

Antonio LS, Stewart AP, Varanda WA, Edwardson JM (2014) Identification of P2X2/P2X4/P2X6 heterotrimeric receptors using atomic force microscopy (AFM) imaging. FEBS Lett 588:2125–2128

Belleannee C, Da SN, Shum WW, Brown D, Breton S (2010) Role of purinergic signaling pathways in V-ATPase recruitment to apical membrane of acidifying epididymal clear cells. Am J Physiol Cell Physiol 298:C817–C830

Booth JW, Tam FW, Unwin RJ (2012) P2 purinoceptors: renal pathophysiology and therapeutic potential. Clin Nephrol 78:154–163

Boucher RC (2003) Regulation of airway surface liquid volume by human airway epithelia. Pflügers Arch 445:495–498

Burnstock G (2007) Physiology and pathophysiology of purinergic neurotransmission. Physiol Rev 87:659–797

Chang M-Y, Lu J-K, Tian Y-C, Chen Y-C, Hung C-C, Huang Y-H, Chen Y-H, Wu M-S, Yang C-W, Cheng Y-C (2014) Inhibition of the P2X7 receptor reduces cystogenesis in PKD. J Am Soc Nephrol 22:1696–1706

Chen K, Zhang J, Zhang W, Zhang J, Yang J, Li K, He Y (2013) ATP-P2X4 signaling mediates NLRP3 inflammasome activation: a novel pathway of diabetic nephropathy. Int J Biochem Cell Biol 45:932–943

Dai LJ, Kang HS, Kerstan D, Ritchie G, Quamme GA (2001) ATP inhibits Mg^{2+} uptake in mouse distal convoluted tubule cells via P2X purinoceptors. Am J Physiol 281:F833–F840

de Baaij JH, Blanchard MG, Lavrijsen M, Leipziger J, Bindels RJ, Hoenderop JG (2014) P2X4 receptor regulation of transient receptor potential melastatin type 6 (TRPM6) Mg channels. Pflugers Arch 466:1942–1952

Desfleurs E, Wittner M, Simeone S, Pajaud S, Moine G, Rajerison R, Di SA (1998) Calcium-sensing receptor: regulation of electrolyte transport in the thick ascending limb of Henle's loop. Kidney Blood Press Res 21:401–412

Ebeling G, Blasche R, Hofmann F, Augstein A, Kasper M, Barth K (2014) Effect of P2X7 receptor knockout on AQP-5 expression of type I alveolar epithelial cells. PLoS ONE 9, e100282

Filipovic DM, Adebanjo OA, Zaidi M, Reeves WB (1998) Functional and molecular evidence for P2X receptors in LLC-PK1 cells. Am J Physiol 274:F1070–F1077

Goncalves RG, Gabrich L, Rosario A Jr, Takiya CM, Ferreira ML, Chiarini LB, Persechini PM, Coutinho-Silva R, Leite M Jr (2006) The role of purinergic P2X7 receptors in the inflammation and fibrosis of unilateral ureteral obstruction in mice. Kidney Int 70:1599–1606

Gorodeski GI (2002) Expression, regulation, and function of P2X(4) purinergic receptor in human cervical epithelial cells. Am J Physiol Cell Physiol 282:C84–C93

Guha S, Baltazar GC, Coffey EE, Tu LA, Lim JC, Beckel JM, Patel S, Eysteinsson T, Lu W, O'Brien-Jenkins A, Laties AM, Mitchell CH (2013) Lysosomal alkalinization, lipid oxidation, and reduced phagosome clearance triggered by activation of the P2X7 receptor. FASEB J 27:4500–4509

Guo Y, Mishra A, Weng T, Chintagari NR, Wang Y, Zhao C, Huang C, Liu L (2014) Wnt3a mitigates acute lung injury by reducing P2X7 receptor-mediated alveolar epithelial type I cell death. Cell Death Dis 5, e1286

Hattori M, Gouaux E (2012) Molecular mechanism of ATP binding and ion channel activation in P2X receptors. Nature 485:207–212

Hede SE, Amstrup J, Christoffersen BC, Novak I (1999) Purinoceptors evoke different electrophysiological responses in pancreatic ducts. P2Y inhibits K^+ conductance, and P2X stimulates cation conductance. J Biol Chem 274:31784–31791

Henriksen KL, Novak I (2003) Effect of ATP on intracellular pH in pancreatic ducts involves P2X7 receptors. Cell Physiol Biochem 13:93–102

Hillman KA, Woolf AS, Johnson TM, Wade A, Unwin RJ, Winyard PJ (2004) The P2X7 ATP receptor modulates renal cyst development in vitro. Biochem Biophys Res Commun 322:434–439

Hung SC, Choi CH, Said-Sadier N, Johnson L, Atanasova KR, Sellami H, Yilmaz O, Ojcius DM (2013) P2X4 assembles with P2X7 and pannexin-1 in gingival epithelial cells and modulates ATP-induced reactive oxygen species production and inflammasome activation. PLoS ONE 8, e70210

Ilatovskaya DV, Palygin O, Levchenko V, Staruschenko A (2013) characterization of the P2 receptors profile in the podocytes of the freshly isolated rat glomeruli. Am J Physiol Cell Physiol 305:C1050–C1059

Jankowski M, Szamocka E, Kowalski R, Angielski S, Szczepanska-Konkel M (2011) The effects of P2X receptor agonists on renal sodium and water excretion in anaesthetized rats. Acta Physiol (Oxf) 202:193–201

Jensen ME, Odgaard E, Christensen MH, Praetorius HA, Leipziger J (2007) Flow-induced $[Ca^{2+}]_i$ increase depends on nucleotide release and subsequent purinergic signaling in the intact nephron. J Am Soc Nephrol 18:2062–2070

Kawate T, Michel JC, Birdsong WT, Gouaux E (2009) Crystal structure of the ATP-gated P2X (4) ion channel in the closed state. Nature 460:592–598

Kim MJ, Turner CM, Hewitt R, Smith J, Bhangal G, Pusey CD, Unwin RJ, Tam FW (2014) Exaggerated renal fibrosis in P2X4 receptor-deficient mice following unilateral ureteric obstruction. Nephrol Dial Transplant 29:1350–1361

King M, Housley GD, Raybould NP, Greenwood D, Salih SG (1998) Expression of ATP-gated ion channels by Reissner's membrane epithelial cells. Neuroreport 9:2467–2474

Korngreen A, Priel Z (1996) Purinergic stimulation of rabbit ciliated airway epithelia: control by multiple calcium sources. J Physiol 497(Pt 1):53–66

Korngreen A, Ma W, Priel Z, Silberberg SD (1998) Extracellular ATP directly gates a cation-selective channel in rabbit airway ciliated epithelial cells. J Physiol 508(Pt 3):703–720

Larsen CK, Leipziger J (2013) P2Y receptors in the gastrointestinal epithelium. WIREs Membr Transp Signal 2:27–36

Lee JH, Chiba T, Marcus DC (2001) P2X2 receptor mediates stimulation of parasensory cation absorption by cochlear outer sulcus cells and vestibular transitional cells. J Neurosci 21:9168–9174

Leipziger J (2003) Control of epithelial transport via luminal P2 receptors. Am J Physiol Renal Physiol 284:F419–F432

Leipziger J (2011) Luminal nucleotides are tonic inhibitors of renal tubular transport. Curr Opin Nephrol Hypertens 20:518–522

Leipziger J, Kerstan D, Nitschke R, Greger R (1997) ATP increases $[Ca^{2+}]_i$ and ion secretion via a basolateral P2Y receptor in rat distal colonic mucosa. Pflügers Arch 434:77–83

Luo X, Zheng W, Yan M, Lee MG, Muallem S (1999) Multiple functional P2X and P2Y receptors in the luminal and basolateral membranes of pancreatic duct cells. Am J Physiol 277:C205–C215

Ma W, Korngreen A, Uzlaner N, Priel Z, Silberberg SD (1999) Extracellular sodium regulates airway ciliary motility by inhibiting a P2X receptor. Nature 400:894–897

Ma W, Korngreen A, Weil S, Cohen EB, Priel A, Kuzin L, Silberberg SD (2006) Pore properties and pharmacological features of the P2X receptor channel in airway ciliated cells. J Physiol 571:503–517

Mankus C, Rich C, Minns M, Trinkaus-Randall V (2011) Corneal epithelium expresses a variant of P2X(7) receptor in health and disease. PLoS ONE 6, e28541

Marques RD, de Bruijn PI, Sorensen MV, Bleich M, Praetorius HA, Leipziger J (2012) Basolateral P2X receptors mediate inhibition of NaCl transport in mouse medullary thick ascending limb (mTAL). Am J Physiol Renal Physiol 302:F487–F494

McCoy DE, Taylor AL, Kudlow BA, Karlson K, Slattery MJ, Schwiebert LM, Schwiebert EM, Stanton BA (1999) Nucleotides regulate NaCl transport in mIMCD-K2 cells via P2X and P2Y purinergic receptors. Am J Physiol 277:F552–F559

McMillian MK, Soltoff SP, Lechleiter JD, Cantley LC, Talamo BR (1988) Extracellular ATP increases free cytosolic calcium in rat parotid acinar cells. Biochem J 255:291–300

Miklavc P, Thompson KE, Frick M (2013) A new role for P2X4 receptors as modulators of lung surfactant secretion. Front Cell Neurosci 7:171

Mishra A, Chintagari NR, Guo Y, Weng T, Su L, Liu L (2011) Purinergic P2X7 receptor regulates lung surfactant secretion in a paracrine manner. J Cell Sci 124:657–668

Murgia M, Hanau S, Pizzo P, Rippa M, Di VF (1993) Oxidized ATP. An irreversible inhibitor of the macrophage purinergic P2Z receptor. J Biol Chem 268:8199–8203

North RA (2002) Molecular physiology of P2X receptors. Physiol Rev 82:1013–1067

Novak I, Nitschke R, Amstrup J (2002) Purinergic receptors have different effects in rat exocrine pancreas. Calcium signals monitored by fura-2 using confocal microscopy. Cell Physiol Biochem 12:83–92

Ortiz PA, Garvin JL (2002) Role of nitric oxide in the regulation of nephron transport. Am J Physiol Renal Physiol 282:F777–F784

Palomino-Doza J, Rahman TJ, Avery PJ, Mayosi BM, Farrall M, Watkins H, Edwards CR, Keavney B (2008) Ambulatory blood pressure is associated with polymorphic variation in P2X receptor genes. Hypertension 52:980–985

Pettengill MA, Marques-da-Silva C, Avila ML, d'Arc dos Santos OS, Lam VW, Ollawa I, Abdul Sater AA, Coutinho-Silva R, Hacker G, Ojcius DM (2012) Reversible inhibition of Chlamydia trachomatis infection in epithelial cells due to stimulation of P2X(4) receptors. Infect Immun 80:4232–4238

Pochynyuk O, Bugaj V, Rieg T, Insel PA, Mironova E, Vallon V, Stockand JD (2008) Paracrine regulation of the epithelial Na + channel in the mammalian collecting duct by purinergic P2Y2 receptor tone. J Biol Chem 283:36599–36607

Praetorius HA, Leipziger J (2010) Intrarenal purinergic signaling in the control of renal tubular transport. Annu Rev Physiol 72:377–393

Sasaki T, Gallacher DV (1990) Extracellular ATP activates receptor-operated cation channels in mouse lacrimal acinar cells to promote calcium influx in the absence of phoshoinositide metabolism. FEBS Lett 264:130–134

Schlingmann KP, Weber S, Peters M, Niemann NL, Vitzthum H, Klingel K, Kratz M, Haddad E, Ristoff E, Dinour D, Syrrou M, Nielsen S, Sassen M, Waldegger S, Seyberth HW, Konrad M (2002) Hypomagnesemia with secondary hypocalcemia is caused by mutations in TRPM6, a new member of the TRPM gene family. Nat Genet 31:166–170

Silberberg SD, Swartz KJ (2009) Structural biology: trimeric ion-channel design. Nature 460:580–581

Silva G, Beierwaltes WH, Garvin JL (2006) Extracellular ATP stimulates NO production in rat thick ascending limb. Hypertension 47:563–567

Souza CO, Santoro GF, Figliuolo VR, Nanini HF, de Souza HS, Castelo-Branco MT, Abalo AA, Paiva MM, Coutinho CM, Coutinho-Silva R (2012) Extracellular ATP induces cell death in human intestinal epithelial cells. Biochim Biophys Acta 1820:1867–1878

Taylor AL, Schwiebert LM, Smith JJ, King C, Jones JR, Sorscher EJ, Schwiebert EM (1999) Epithelial P2X purinergic receptor channel expression and function. J Clin Invest 104:875–884

Turner CM, Vonend O, Chan C, Burnstock G, Unwin RJ (2003) The pattern of distribution of selected ATP-sensitive P2 receptor subtypes in normal rat kidney: an immunohistological study. Cells Tissues Organs 175:105–117

Vallon V (2008) P2 receptors in the regulation of renal transport mechanisms. Am J Physiol Renal Physiol 294:F10–F27

Vonend O, Turner CM, Chan CM, Loesch A, Dell'Anna GC, Srai KS, Burnstock G, Unwin RJ (2004) Glomerular expression of the ATP-sensitive P2X receptor in diabetic and hypertensive rat models. Kidney Int 66:157–166

Wang Q, Wang L, Feng YH, Li X, Zeng R, Gorodeski GI (2004) P2X7 receptor-mediated apoptosis of human cervical epithelial cells. Am J Physiol Cell Physiol 287:C1349–C1358

Wang Z, Nakayama T, Sato N, Izumi Y, Kasamaki Y, Ohta M, Soma M, Aoi N, Ozawa Y, Ma Y (2010) The purinergic receptor P2Y, G-protein coupled, 2 (P2RY2) gene associated with essential hypertension in Japanese men. J Hum Hypertens 24:327–335

Wildman SS, Marks J, Turner CM, Yew-Booth L, Peppiatt-Wildman CM, King BF, Shirley DG, Wang W, Unwin RJ (2008) Sodium-dependent regulation of renal amiloride-sensitive currents by apical P2 receptors. J Am Soc Nephrol 19:731–742

Woods LT, Camden JM, Batek JM, Petris MJ, Erb L, Weisman GA (2012) P2X7 receptor activation induces inflammatory responses in salivary gland epithelium. Am J Physiol Cell Physiol 303:C790–C801

Yamamoto K, Sokabe T, Matsumoto T, Yoshimura K, Shibata M, Ohura N, Fukuda T, Sato T, Sekine K, Kato S, Isshiki M, Fujita T, Kobayashi M, Kawamura K, Masuda H, Kamiya A, Ando J (2006) Impaired flow-dependent control of vascular tone and remodeling in $P2X_4$-deficient mice. Nat Med 12:133–137

Zhang Y, Sanchez D, Gorelik J, Klenerman D, Lab M, Edwards C, Korchev Y (2007) Basolateral P2X4-like receptors regulate the extracellular ATP-stimulated epithelial Na + channel activity in renal epithelia. Am J Physiol Renal Physiol 292:F1734–F1740

Chapter 10
Na^+/K^+-ATPase Drives Most Asymmetric Transports and Modulates the Phenotype of Epithelial Cells

Marcelino Cereijido, Isabel Larre, Omar Paez, Liora Shoshani, and Arturo Ponce

Abstract Usually the history of an enzyme is the narrative of the works to isolate and purify it, measure its molecular weight, determine its crystal configuration, measure its activity, and so on, along, say, 2 or 3 years. On the contrary, the history of Na^+/K^+-ATPase is a tortuous road full of pitfalls, skirmishes with physical chemistry, thermodynamics, and even philosophy. Fortunately it had a happy end, because it was the first known molecule to produce vectorial movement of ions using chemical energy; to cyclically modify its selectivity; to participate in the action potential; to act as a self-adhesion molecule at cell-cell contacts; to act as receptor of the hormone ouabain, whose main physiological role is to modulate cell contacts; and to generate a Na^+ gradient that enables co- and counter-transporters to transport net amount of ions, sugars, and amino acids, i.e., to act as secondary pumps. It is sufficient to say that one of its crucial properties, i.e., to be expressed polarizedly at the intercellular membrane of transporting epithelial cells, involves the β-subunit of the pump that happens to be an adhesion molecule which plays a crucial role in that polarization mechanism.

Keywords Active transport • Na^+/K^+-ATPase • Ouabain • Transporting epithelia • Cell contacts

10.1 Introduction

As recently as 1807 Humphry Davy was the first person to differentiate Na^+ from K^+. An ironic situation indeed, because if he was an average-sized person, he must have had 3.72×10^{13} cells (Bianconi et al. 2013), each one discriminating those cations along their whole life. Nevertheless, science would take another century and

M. Cereijido (✉) • I. Larre • O. Paez • L. Shoshani • A. Ponce
Department of Physiology, Biophysics and Neurosciences, Center for Research and Advanced Studies (Cinvestav), AP-14-740, México 07300, Mexico
e-mail: cereijido@fisio.cinvestav.mx

K.L. Hamilton, D.C. Devor (eds.), *Ion Channels and Transporters of Epithelia in Health and Disease*, Physiology in Health and Disease,
DOI 10.1007/978-1-4939-3366-2_10

a half to start learning how cells identify certain ion species and handle each one differently, but from the very beginning, it was clear that in order to explain ion distributions and movements in biological systems, the known laws of physical chemistry would not suffice. No wonder the history of Na^+/K^+-ATPase can be compared to a double-sided step ladder, one climbed by biologists that investigated its structure, subunits, and intrinsic mechanisms and the other by accomplished physicists who denounced (in a first step) but helped to solve (in a second step) formidable theoretical obstacles in the road toward understanding active transport. In a given moment, discrepancies were so sharp that led to fear that life would never be explained in terms of physical and chemical principles (Cereijido et al. 2003, 2004).

Thus, for a long while the peculiar composition of the cytoplasm was attributed to an alleged membrane impermeability to Na^+ because, if this ion was not able to penetrate to neutralize negative charges bound to macromolecules inside the cell, some other cations would have to do it to satisfy the principle of electroneutrality; hence, the high K^+ content in the cell was accounted for the claimed membrane impermeability to Na^+. This alternative was disproved right after the Second World War, when radioisotopes became available for biological research, and it was discovered that tracer Na^+ added to the bathing solution readily penetrates and distributes throughout the cytoplasm. This rekindled the question more emphatically: If Na^+ can readily penetrate into the cytoplasm, why does it remain at a concentration much lower (~5 mM) than in the extracellular water (~140 mM).

10.2 It All Started with Émil Du Bois Raymond

In the second half of the nineteenth century, Émil Du Bois Raymond observed that a frog skin (Fig. 10.1a, *green bar*) maintains an electrical potential difference between its inner and its outer sides (i.e., an asymmetry). Galeotti (1904, 1907) studied this electrical potential as a function of the ionic composition of the solutions and proposed that it would be accounted for if the flux of sodium from the outside toward the inside (*influx*) were larger than the *outflux* in the opposite direction (Fig. 10.1b, *black arrows*). His proposal was rejected on the basis that it would be in violation of the laws of thermodynamics. Thus, in a *Gedankenexperiment*, a frog skin mounted in a doughnut-shaped chamber containing saline solution, an asymmetric permeability would increase the concentration of Na^+ on the right-hand side of the epithelium (*small magenta dots*), and the accumulation of Na^+ will start the diffusion of this ion down to its gradient (*counterclockwise*), thus establishing a *perpetuum mobile* (*magenta arrow*) that would in fact perform a work without the corresponding dissipation of energy. Yet biologists observed that a frog skin dissected and mounted between two chambers dies in a few hours, and the electrical potential (Fig. 10.1c, *red curve*) (Ussing and Zerahn. 1951) vanishes; in other words, far from being *perpetuum*, the potential lasts as long as the skin is alive. "Perhaps, the energy is afforded by metabolic energy" they proposed. Yet, given that metabolism is the sum of chemical reactions, physicists claimed that physiologists were

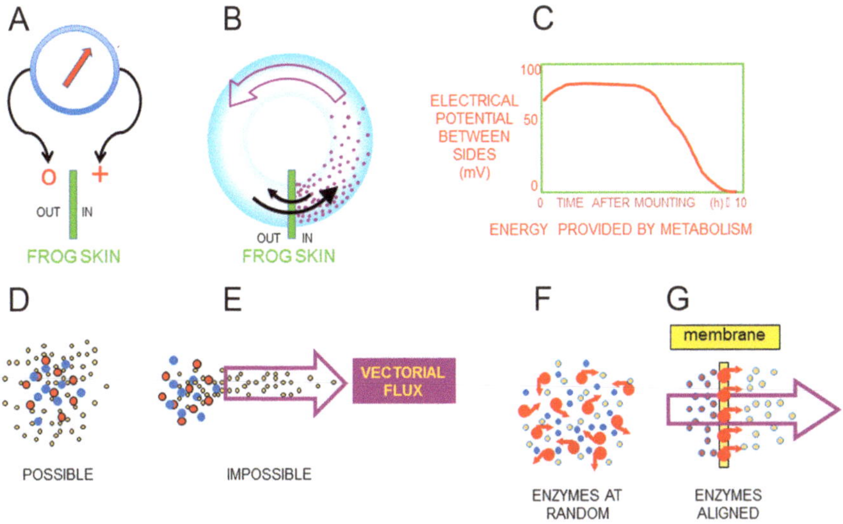

Fig. 10.1 Antecedents and roles of Na^+/K^+-ATPase. (**a**) Du Bois Raymond discovered that a frog skin can develop and maintain an electrical potential across. (**b**) In 1905 Galeotti proposed that such potential would be accounted for if the frog skin were more permeable to Na^+ in the a direction than in the opposite one. The proposal was easily refuted because it would originate a *perpetuum mobile*, i.e., Na^+ would rotate forever without a corresponding expenditure of free energy. (**c**) The electrical potential is not actually perpetual, but last as long as the skin is alive, and energy is provided by metabolism. (**d–e**) Yet, according to Curie principle, chemical reactions could not drive a unidirectional flux because this is a vectorial process. (**f**) Enzymes functioning as pumps are vectorial at microscopic level, but when studied macroscopically, it becomes unnoticeable. (**g**). Pumps ordered as in a plasma membrane exhibit macroscopic vectoriality

now violating Curie principle: “Chemical reactions are scalar phenomena” (again, a wrong assumption), they occur regardless of the orientation of the reacting molecules, and the resulting chemical products cannot diffuse in a particular direction (Fig. 10.1d); therefore, chemical reactions would never originate a vectorial flux across a frog skin as in Fig. 10.1e (another incorrect supposition).

Biologists argued that each molecule (Fig. 10.1f, *red*) may be asymmetric, but since there are millions of molecules oriented at random, asymmetry cannot be observed at a macroscopic scale. However, if they were ordered in a membrane as suggested in Fig. 10.1g, the asymmetry would be recovered, and a macroscopic flux would take place. This made theoreticians happy, yet where was “the pump,” i.e., the membrane molecule that would align and be responsible for the sided, asymmetrical movement of products? Around 1955–1959, Jens Christian Skou prepared an extract of crab tissue that contained an enzyme that splits molecules of ATP (hence deserving the name “ATPase”) into ADP + Pi, provided the medium contains K^+ and Na^+ ions at concentrations that compare with those in the cell and in the surrounding extracellular space. Therefore, the enzyme was aptly named Na^+/K^+-ATPase. Interestingly, Skou (1957) was able to inhibit the ATP splitting activity of his extract by adding *ouabain*, a substance of vegetal origin that, a few years earlier, was found to inhibit ion pumping when added from the outer, but not from

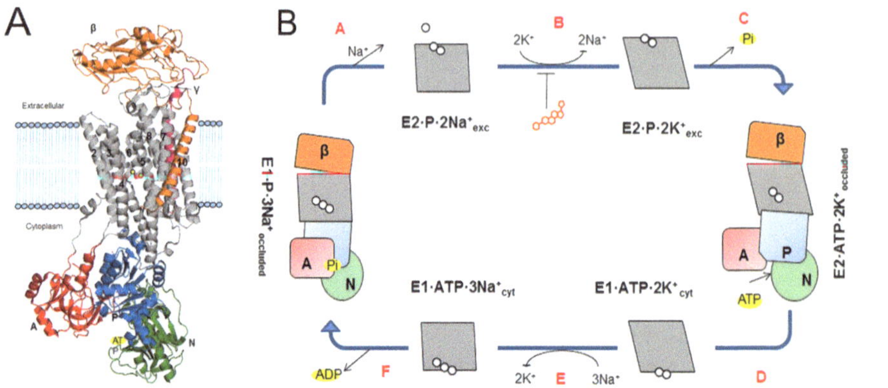

Fig. 10.2 (**a**) Ribbon model of the crystal structure of shark Na^+/K^+-ATPase [Shinoda et al. 2009 PDB code: 2ZXE] indicating alpha subunit transmembrane domains (TM1-9, *gray*) and cytoplasmic domains N, P, and A in *green*, *blue*, and *red*, respectively. β-subunit (*orange*). γ-subunit is colored in magenta. TM 1 and TM 9 are not visible from this projection. The two K^+ (*yellow*) are occluded in the crystal structure. In the N domain, the ATP binding site is also indicated. (**b**) Schematic depiction of the catalytic cycle of Na^+/K^+-ATPase. A structural rearrangement, especially in domain A (*red*), is suggested by the cartoons at both conformational states E1 and E2. For simplicity, the rest of the cycle stages are represented only with the TM region in gray. (1) Outward transport of three Na^+ ions (*white circles*) is coupled to the E1 to E1 transition. (2) Two K^+ (*red circles*) bind at binding sites oriented to the extracellular space. (3) Extracellularly bound K^+ activate dephosphorylation which in turn results in ion occlusion. Ouabain binds at this conformational state stopping the cycle. (4) ATP with a low affinity triggers acceleration of inward transport of K^+ through the pore as the pump enters the E1 conformational transition with low affinity for K^+. (5) Three Na^+ bind the intracellularly oriented sites. (6) Phosphorylation from ATP occurs and Na^+ are occluded again. Several transitional substates exist; nevertheless they are not depicted for simplification

the inner side of the cell membrane. By performing the K^+/Na^+ translocations cyclically, Na^+/K^+-ATPase transfers those ions in a net amount toward the extracellular medium and toward the cytoplasm respectively, so it was justified to call it "pump." Today, thanks to high-resolution crystallography (Shinoda et al. 2009), the molecular structure of Na^+/K^+-ATPase is known in great detail (Fig. 10.2a), and its cyclical reactions are described in Fig. 10.2b.

10.3 Polarized Distribution of Na^+/K^+-ATPase in Epithelial Cells

A unicellular organism in the ocean may consume nutrients and eliminate metabolic wastes without risking neither an exhaustion of nutrients nor pollution of the sea. In a metazoan, the ocean is represented by a narrow extracellular space that would be rapidly exhausted of nutrients and spoiled with wastes, were it not for a circulatory apparatus that moves fluids to and from tissues to transporting epithelia,

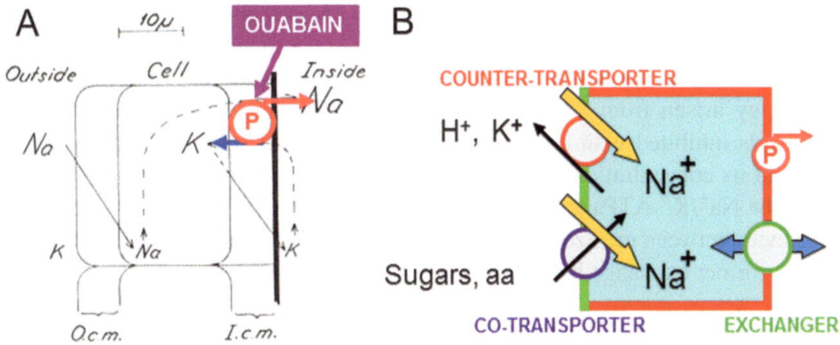

Fig. 10.3 Model proposed by Koefoed-Johnsen and Ussing (1958) to account for the net pond-to-blood transport of Na^+. (**a**) An outer cell membrane (*O.c.m.*) with a high selectivity for Na^+ and low selectivity for K^+ and an inner (*I.c.m.*) whose ion selectivity is high for K^+ and low for Na^+. A pump (P, *red circle*) located on the basal side takes Na^+ from the cytoplasm and transports it toward the blood (*inner*) side and takes K^+ from the inside and transports it toward the cytoplasm. This pump can be inhibited by ouabain added from the blood side. (**b**) The pump keeps the concentration of Na^+ in the cytoplasm at a level lower than in the extracellular fluid; this ion flows passively (*yellow arrow*) toward the cytoplasm, driving a cation in the opposite direction or an anion in the same direction. Furthermore, the penetration of Na^+ can also drive sugars and amino acids from the outer bathing solution to the cytoplasm

where the true exchange with the external milieu takes place. Thanks to the circulation of blood, in spite of being so small, this intercellular "ocean" behaves as constant and reliable reservoir (homeostasis). Figure 10.3a depicts the model put forward by Koefoed-Johnsen and Ussing (1958) in which the pump, represented by Na^+/K^+-ATPase, is assumed to be located on the basal side of the epithelial cell, and this asymmetric distribution, together with the specific Na permeability of the outer cell membrane (*Ocm*) and the specific K permeability of the inner facing membrane (*Icm*), is responsible for net movement of Na^+. This model served as a blueprint to understand the net transport of important biological substances across most epithelia.

Notice that the Na^+/K^+-ATPase is *primarily* responsible for the pumping of Na^+ and K^+, but the concentration gradients that it generates result in the *secondary* transport of amino acids, sugars, and ions other than Na^+ and K^+. Since transport occurs in a net amount and can be inhibited with ouabain, for a while this was taken as a proof that these glucose pumps exist, as well as other pumps for diverse amino acid species (Fig. 10.3b). Yet, eventually it was demonstrated that carriers for sugars and for amino acids are not pumps, as they are not *directly* coupled to metabolism, but their affinity for sugars and amino acids drastically increases when loaded with Na^+. Today, we are so used to talk about co- and counter-transport that we forget it caused another famous, but easy to understand, squirmish between biologists and physicists. Notice that co- and counter-transport assumed that the electrochemical potential difference for Na^+ can be used to move sugars and amino acids as depicted in Fig. 10.3b. Although we will not make a digression to explain how this important theoretical problem was solved, it is worth mentioning that for making a revolution in the Thermodynamics of Irreversible Processes, Lars Onsager was awarded the Nobel Prize in 1968. But since then, it was realized that, in principle, the flux of a substance (sugar, amino acid, chloride, etc.) can be moved by the energy arisen from any driving force present in the system. When Na^+/K^+-ATPase is inhibited with ouabain, the concentration of sodium in the cytoplasm rises and its concentration gradient across the apical membrane vanishes. In summary, the Na^+/K^+-ATPase is the *primum movens*, responsible for the exchange of substances between metazoan and the environment across transporting epithelia, as well as for net exchange between the internal milieu and the cytoplasm.

10.3.1 Why Do Epithelial Cells Express Na^+/K^+-ATPase in a Polarized Manner?

Single cells, e.g., a leukocyte, a bacterium, or an amoeba, have Na^+/K^+-ATPase distributed at random all over the plasma membrane. The cells of transporting epithelia instead have Na^+/K^+-ATPase distributed in a peculiar way; in the model of Koefoed-Johnsen and Ussing (1958), for instance (Fig. 10.3a), they are assumed to be expressed mainly on the basal membrane. For a while, the apical/basolateral polarity of epithelial cells was erroneously attributed to the tight junction (TJ). This mistaken idea was based on the demonstration that chelation of Ca^{2+} with EDTA or EGTA not only opens the TJs but destroys the apical/basolateral polarity, as well. However, TJs have no sorting mechanisms and, at most, maintain the apical/basolateral polarization that some molecular species have obtained by other mechanisms (e.g., lipids). This was clearly demonstrated by studies of Contreras and co-workers (1989), who found that MDCK (Madin-Darby canine kidney) cells adopt a spherical shape after harvesting and resuspension and a large fraction of plasma membrane is endocytosed along with a large amount of Na^+/K^+-ATPase units inserted on it contains, whereas those remaining at the surface lose its polarity and randomizes. When cells are seeded again, and subject to the "calcium switch protocol" (Gonzalez-Mariscal et al. 1990; Contreras et al. 1992), TJ forms so quickly that catches Na^+/K^+-ATPase still randomized, but in spite of the presence of the barrier constituted by the already sealed TJ, the enzyme bypasses this structure and polarizes correctly. Hence, to explain the nonrandom distribution of molecules in the plasma membrane, it was necessary to study the intracellular routes and sorting compartments involved, such as the endoplasmic reticulum (ER) and the trans-Golgi network (TGN), and different carrier vesicles for delivery to their corresponding plasma membrane domains (Weisz and Rodriguez-Boulan 2009; Rodriguez-Boulan and Macara 2014). For each of these routes, polarity depends critically on the existence of specific signals (motifs) encoded within the membrane proteins themselves. The basolateral sorting signals are short peptide sequences most often found within the cytoplasmic domain of the protein. Some basolateral sorting signals resemble endocytic signals, e.g., variations of the canonical endocytic dileucine, YXXΦ, and NPXY motifs. Other basolateral signals are unrelated to endocytic signals, e.g., the tyrosine motifs in the low-density lipoprotein receptor and the G-protein of the VSV. Early studies demonstrated that the Na^+/K^+-ATPase, comprised of α and β subunits, is sorted in the TGN and delivered directly to the basolateral membrane without significant appearance at the apical surface in certain strains of MDCK cells (Caplan et al. 1986; Mays et al. 1995). Therefore, a basolateral signal was assumed to exist in the α-subunit of the Na^+/K^+-ATPase. The Na^+/K^+-ATPase and H^+/K^+-ATPase are highly homologous ion pumps, yet in LLC-PK1 cells, they are polarized to the basolateral and apical domain, respectively. The polarized expression of chimeric constructs of the α-subunit of the H^+/K^+-ATPase and the Na^+/K^+-ATPase in LLC-PK1 cells has been studied (Blostein et al. 1993). An apical sorting motif was identified within the fourth transmembrane domain of the α-subunit of the H^+/K^+-ATPase that is sufficient to redirect the Na^+/K^+-ATPase from the basolateral to the apical surface of these cells (Dunbar et al. 2000). However, it remains unclear whether basolateral sorting information exists in the fourth transmembrane domain of the α-subunit of the Na^+/K^+-ATPase; thus, a still unknown, noncanonical molecular signal seems to be involved in the basolateral targeting of Na^+/K^+-ATPase.

Clathrin plays a fundamental role in basolateral sorting. It interacts with endocytic or basolateral proteins through a variety of clathrin adaptors. It has been shown that the adaptor involved in basolateral protein sorting is the epithelial cell-specific AP-1B (adaptor protein 1B). Nevertheless, the correct polarization of Na^+/K^+-ATPase is not significantly affected by knocking down clathrin expression. It also remains correctly polarized in both the μ1B-deficient cell line LLC-PK1 (Ohka et al. 2001) and in MDCK cells in which μ1B expression had been suppressed via RNAi (Duffield et al. 2004). By taking advantage of the SNAP tag system to reveal the trafficking itinerary of the newly synthesized Na^+/K^+-ATPase, it was shown that basolateral delivery of the Na^+/K^+-ATPase is very fast and does not involve passage through recycling endosomes en route to the plasma membrane. Moreover, Na^+/K^+-ATPase trafficking is not regulated by the same small GTPases as other basolateral proteins (Farr et al. 2009). Some membrane proteins may achieve polarity by selective retention at the apical or basolateral surface. Although less well understood, this polarity may reflect interactions with extracellular ligands or with intracellular scaffolds, such as cytoskeletal elements or arrays of PDZ domain-containing proteins (Zimmermann 2006). As described and discussed below, this is also the case of the epithelial Na^+/K^+-ATPase, which is retained at the lateral membrane domain due to the adhesion of its β_1-subunits with those of neighboring cells. Readers interested in apical and basolateral sorting of transport proteins are directed to Chap. 9 of this volume.

The β-subunit functions as a molecular chaperone of the catalytic α-subunit. It facilitates the correct membrane integration and packing of the newly synthesized catalytic α-subunit, which is necessary for its protection against cellular degradation, acquisition of functional properties, and routing to the plasma membrane (Geering 2008). In addition to its chaperone function, β-subunits influence the transport properties of mature Na^+/K^+-ATPase. α-Subunits associated with different β-isoforms exhibit different apparent potassium affinities, and the β-structure influences the apparent sodium affinity of Na^+/K^+-ATPase (Geering 2001). During the catalytic cycle, there is a conformational rearrangement between α- and β-subunits (Dempski et al. 2005).

10.3.2 Cues Leading to a Model of Na^+/K^+-ATPase Polarity

As with most transporting epithelia, the monolayer of MDCK expresses Na^+-K^+-ATPase polarizedly toward the basolateral side (Cereijido et al. 1980). Figure 10.4a shows a monolayer of MDCK cells with nuclei stained with propidium iodide (*red*) and Na^+/K^+-ATPase (*green*). Contrary to the assumption of Koefoed-Johnsen and Ussing (Koefoed-Johnsen and Ussing 1958), the pump is not located on the *basal* domain of the plasma membrane, but only in the *lateral* (*first cue*, Fig. 10.4a). Although green lines in Fig. 10.4a and b appear as a single, undivided green line, upon treating the monolayer with EGTA to sequester Ca^{2+}, the apparent single green line splits into two (Fig. 10.4c), indicating that in order to express Na^+/K^+-ATPase at a cell-cell contact, both neighboring cells have to contribute part of the enzyme (*second cue*). In order to express the Na^+/K^+-ATPase at a given lateral border, both contributing neighboring cells should be homotypic, for instance, MDCK/MDCK (dog/dog), but not heterotypic: MDCK/Ma104 (dog/monkey) (Shoshani et al. 2005), as depicted in Fig. 10.4d (*third cue*).

Figure 10.4e depicts the position and arrangement of Na^+/K^+-ATPase obtained by crystallography, showing the positioning of this trimmer: α-subunit (*violet*), β-subunit (*green*), and γ-subunit (*red*). The β-subunit has the typical structure of a cell-attachment protein (Geering 2008). Karlish's group has confirmed that the β-subunit ectodomain contains an immunoglobulin-like structure that would be responsible for its adhesive properties (Bab-Dinitz et al. 2009). Accordingly, the β-subunit has a short cytoplasmic tail, a single transmembrane segment, and a long extracellular segment heavily glycosylated. We also showed that β1-subunit immobilized on Ni beads could specifically bind to the soluble extracellular domain of β1-subunits of the same animal species (dog) (Padilla-Benavides et al. 2010). To test this property and to see if it works in MDCK cells, we transfected a gene coding for the β-subunit of the dog (remember that MDCK cells are also derived from a dog) into MDCK cells and demonstrated that this considerably increases cell-cell adhesion (Fig. 10.4f). This constituted our *fourth cue*. We also examined this property by transfecting other cell types with dog β-subunit (Fig. 10.4g), that is, the same experiment than the one in Fig. 10.4d, except that this time the "other" cell (CHO cell from Chinese hamster ovary) is transfected with dog β-subunit. Notice that a MDCK and a CHO cell transfected with a dog β-subunit offer the image of a single line, as in Fig. 10.4b. To complement the description of this cartoon, Fig. 10.4h and i shows two types of mixed monolayers, one with MDCK/NRK cells (NRK, from rat kidney) and the other with NRK transfected with dog β-subunit, showing, again, that MDCK cells only expose Na^+-K^+-ATPase, when

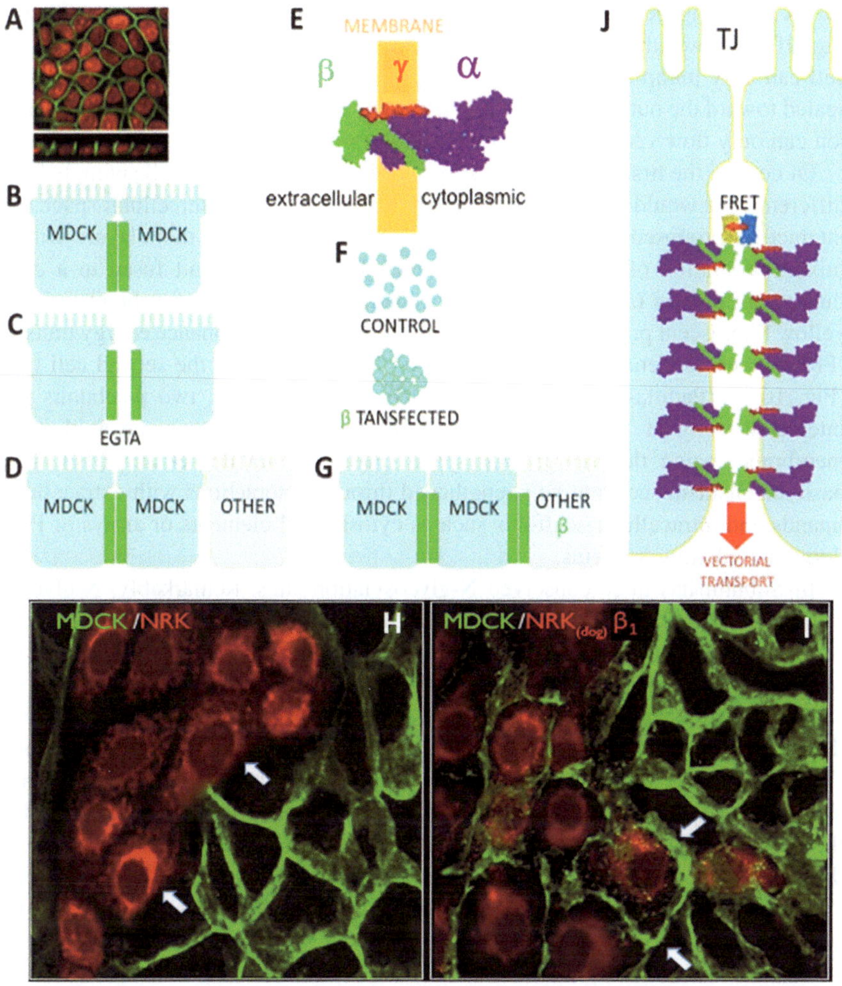

Fig. 10.4 Cues for the polarized distribution of Na^+/K^+-ATPase. (**a**) Monolayer of MDCK cells. Na^+/K^+-ATPase (*green*) and nuclei (*red*). The enzyme is expressed on the lateral membrane of the cells. (**b, c**). Although the image of the enzyme appears as a *single green line*, the use of EGTA splits the *line*, demonstrating that each cell contributes its own Na^+/K^+-ATPase. (**d**) In a monolayer prepared with a mixture of MDCK cells and epithelial cells of a different animal species (Chinese hamster ovary, CHO), the MDCK cell in the center only expresses its Na^+/K^+-ATPase on the side contacting another MDCK cell, but not on the side contacting the epithelial cell of a different animal species. Confocal image of a mixture of MDCK and NRK (normal rat kidney) cells is depicted in (**h**). *Arrows* indicate heterotypic borders lacking Na^+/K^+-ATPase. (**e**) The α-, β-, and γ-subunits of Na^+/K^+-ATPase as expressed in a cell membrane: most of the β-subunit (*green*) is exposed on the intercellular space. (**f**) (*above*). Suspension of CHO cells (*light blue*) with a poor attachment to each other. (*below*) CHO cells transfected with dog β-subunits tend to attach to each other and form aggregates, thereby demonstrating the self-adhesion tendency of this subunit. (**g**) A study essentially identical to the one described in (**d**), except that this time the third cell on the right is a CHO transfected with β-subunits of dog. MDCK cell at the center now expresses its β-subunits on both lateral sides. (**i**) A confocal image showing a mixture of MDCK and NRK cells transfected with dog β-subunit. *Arrows* indicate the presence of Na^+/K^+-ATPase at heterotypic borders. (**j**) Scheme showing the intercellular space of two neighbor MDCK cells in monolayer.

the neighboring cell exposes β-subunit from the same animal species. Finally, Fig. 10.4j shows that the Na^+/K^+-ATPase expressed at the lateral border of the cell can only pump Na^+ into the intercellular space, and, given that this space is sealed toward the outer side of the epithelium by the tight junctions (Fig. 10.4j), this ion can only flow vectorially toward the blood side.

Of course, the first question that arises is whether two matching β-subunits from different cells would get close enough to be able to span the intercellular space and interact as proposed. To answer this question, we prepared monolayers with a mixed population of MDCK cells transfected with a β-subunit fused to a cyan fluorescent protein (*blue*), with MDCK cells transfected with a β-subunit fused to yellow fluorescent protein, showing that in a fluorescence resonance energy transfer (FRET), analysis energy can be transferred from the first to the second cell type (Fig. 10.4j) (Padilla-Benavides et al. 2010). In other words, two β-subunits can interact directly at <10 nm, thereby anchoring the whole enzyme at the cell membrane facing the intercellular space (Padilla-Benavides et al. 2010). This basic mechanism seems to be strengthened through interactions with extracellular ligands and intracellular scaffolds such as cytoskeletal elements or arrays of PDZ domain-containing proteins.

β_1-Subunit has three conserved N-glycosylation sites. Remarkably, N-glycans stabilize the amino acid-mediated interactions between β_1_subunits (Tokhtaeva et al. 2010). Moreover, intercellular adhesion can be regulated via N-glycan-mediated modulation of β_1-β_1 interactions. This modulation occurs because of cell adhesion-dependent regulation of the level of the specific glycosyltransferases that control the degree of N-glycan branching of the Na^+/K^+-ATPase β_1 subunit (Tokhtaeva et al. 2011; Vagin et al. 2008). In this respect, several studies have shown that lateral targeting of the pump in transporting epithelial cells is related to the expression of the β_1 isoform, while apical targeting is related to expression of the β_2 isoform. Furthermore, this differential polarization is directly related to the variation of N-glycosylation between these two isoforms (Vagin et al. 2007a, b).

Fig. 10.4 (continued) Na^+/K^+-ATPase units are expressed at the lateral border, where they are anchored by the β-subunits. Notice that these pumps can only inject Na^+ into the intercellular space. Since this space is sealed toward the outer bathing solution by the tight junction (TJ), Na^+ can only diffuse toward the inner (*blood*) side. In summary, this cartoon depicts these two attributes whose mechanism of synthesis, assembly, and sealing, as well as the peculiar polarized distribution, are detailed elsewhere (Cereijido et al. 2008, 2012)

10.4 Structural Insights into the Na^+/K^+-ATPase Adhesion Mechanism

The crystal structure of the shark's Na^+/K^+-ATPase in the E2 state published by Shinoda and co-workers was the first resolving the atomic structure of the extracellular domain of the β-subunit (PDB: 2ZXE) (Shinoda et al. 2009). The extracellular C-terminal domain of the protein folds into an Ig-like β-sheet sandwich as predicted in silico (Bab-Dinitz et al. 2009); actually, deletion of this C-terminal domain abolishes the β_1 adhesion capacity (unpublished observations). However, a large number of adhesion and nonadhesion proteins contain domains with an immunoglobulin-like topology [CATH database]. Structural alignments of the β_1-subunit extracellular domain against other well-studied cell adhesion molecules reveal no structural homologue of β-subunits of any kind. Detailed inspection of the ectodomain structure uncovers several features distinctive to β-subunit family members. Namely, its Ig-like fold has a unique topology given that its β-sheet sandwich is interrupted by a long α-helix secondary structure and has an atypical β-sheet disposition in relation to classical Ig folds. Also, the β-subunit fold contains extensive loops, and therefore its length is twice as that of a typical Ig domain (unpublished observations). Furthermore, the β_1-subunit is structurally compromised with the catalytic α-subunit in such a way that the C-terminal fold must be more rigid than the typical flexibility of whole adhesion domains such as in cadherins. Altogether, these observations suggest that the β-subunit of the Na^+/K^+-ATPase must possess an adhesion mechanism that is particular to this family, as shown in Fig. 10.4f.

The first attempt to clarify this adhesion mechanism on a molecular base is related to the regions of the ectodomain involved in β_1-β_1 recognition. Given that the interaction between two β_1-subunits of the same species (dog-dog or rat-rat) is more effective than the interaction between rat and dog β_1-subunits (Padilla-Benavides et al. 2010), Tokhtaeva and colleagues (2012) looked for surface-exposed species-specific amino acids in the sequence of β_1-subunit and identified four residues which are different between both species and are contained in the 198–207 segment. Ratlike amino acid substitutions introduced in the dog β_1-subunit weaken its interaction with the endogenous dog β_1-subunit, whereas the insertion of the rat-specific Thr202 into the exogenous dog β_1-subunit impairs its interaction with the endogenous dog β_1-subunit to the level observed between dog and rat native subunits. The opposite effect is observed in the rat β_1-subunit upon the introduction of doglike residues and the deletion of Thr202. These results suggest that the amino acid residues important for β_1-β_1 binding are located upstream and downstream of the Thr insertion position. The insertion or removal of the Thr residues in one of the two interacting subunits probably misaligns these binding residues and thus causes the characteristic difference in affinity between the two species (Tokhtaeva et al. 2012).

How species-specific residues adjacent to Thr202 coordinate with residues residing at surrounding regions on the same β_1-subunit and with its interacting partner has yet to be elucidated. The segment 198–207 constitutes one of the characteristic protruding loops in the connecting β-strands B and C of the β_1-subunit extracellular fold. The majority of the ectodomain surface-exposed residues located most distal from the membrane reside within loops interconnecting β-strands, some of which must be involved in the dimer interface in conjunction with segment 198–207. Since the crystal structure of the Na^+/K^+-ATPase is now available (Nyblom et al. 2013; Toyoshima et al. 2011), we can model and predict interacting surfaces on beta subunit and localize amino acid candidates participating in β_1-β_1 interaction. This approach will soon lead us to uncover a detailed adhesion mechanism which is of great importance for epithelial physiology.

10.5 Cardiac Steroids

Cardiac steroids (CS) became a medical tool since William Withering (1741–1799) observed that patients with weak heart improved when drinking an herb tea of foxglove (*Digitalis purpurea*). He was emulated by European physicians using tisane of plants species from other regions, such as *Digitalis lanata* and West African plant (*Strophanthus gratus*) that led to coin the name Strophanthin (ouabain). In 1953, Hans J. Schatzmann observed that ouabain inhibits active potassium and sodium transport in erythrocyte membrane (Schatzmann 1953). Later on, Jens Christian Skou prepared an extract from crab nerves that contained an ATPase that hydrolyzed ATP into ADP and Pi. Since the activity of this extract was enhanced when exposed to a certain combination of Na^+, K^+, and Mg^{2+}, Skou hypothesized that the enzyme involved might be related to the active movement of sodium and potassium across the plasma membrane and properly named it Na^+/K^+-ATPase. Furthermore, this enzyme was inhibited with ouabain, an observation that reinforced the suspicion that there must be an endogenous analog (Skou 1957).

The remarkable affinity of cardiac steroids for Na^+/K^+-ATPase in mammalian organisms stressed the possibility that there was an endogenous substance analogous to the one extracted from plants. Hamlyn and co-workers (1991) and Mathews and co-workers (1991) demonstrated the presence of a substance in plasma that they could not distinguish from ouabain of vegetal origin, a finding confirmed with more advanced methods such as 1H-NMR and mass ionization spectrometry. There are other analog substances in human blood (Kawamura et al. 1999; Komiyama et al. 2005; Takahashi 2000; Tymiak et al. 1993) and in other animal species, such as digoxin-like factor (Goto et al. 1990; Qazzaz et al. 2004), marinobufagenin (Bagrov et al. 1998), telocinobufagenin (Komiyama et al. 2005), 19-nobufalin (Lichtstein et al. 1993), and proscillaridin A (Hilton et al. 1996; Li et al. 1997). Endogenous ouabain is synthesized and secreted by the hypothalamus (Ludens et al. 1992; el-Masri et al. 2002) and the adrenocortical gland (Doris et al. 1996; Laredo et al. 1994; Qazzaz et al. 2004; Schreiber and Stepan 1986). The status of the hormone ouabain was unambiguously recognized when it was demonstrated that it increases during exercise (Bauer et al. 2005), salty meals (Fedorova et al. 2000, 2005a, b), and pathological conditions such as arterial hypertension and myocardial infarction (Gottlieb et al. 1992; Manunta et al. 2009). Once ouabain was accepted as hormone, it became of interest to find its physiological role. Our laboratory showed that ouabain binding to the Na^+/K^+-ATPase modulates epithelial cell adhesion (Larre et al. 2011a).

10.5.1 The Search for the Physiological Role of Hormone Ouabain

Less than a decade ago, the evidence that ouabain was a hormone was gaining recognition; therefore, investigators began to wonder what the physiological role of ouabain might be. We observed that MDCK cells exposed to high concentrations of ouabain (≥ 1 μM) do not show sign of damage, yet retrieve from the plasma membrane molecules involved in cell-cell and cell-substrate attachment and, as a consequence, those cells detach from each other as well as from the substrate (Contreras et al. 1999). This suggests that there is a mechanism that relates occupancy of the pump (P) by ouabain to adhesion mechanisms (A). Accordingly, we called this mechanism $P \rightarrow A$ and put forward the working hypothesis that ouabain at nanomolar concentrations, i.e., within the hormonal range in mammalian plasma, may act on the same junctional structures without provoking irreversible damages (Larre et al. 2010). To explore the plausibility of this idea, we experimentally tested one cell contact at a time, as it will be described next.

10.5.2 Ouabain Modulates the Tight Junction

The tight junction is the most apical adhesion structure of cell-cell contact. It seals the intercellular space between epithelial cells, thereby transforming the layer of cells into an effective permeability barrier separating biological compartments such as the intestinal lumen from blood or the interior of a nephron and blood (Cereijido et al. 1978, 1980). Several observations indicate that the effects of β_1-subunit expression on tight junction formation may reflect the role of β-subunits in the maturation of Na^+/K^+-ATPase complexes and the consequent increase in Na^+/K^+-ATPase expression and activity (Rajasekaran et al. 2001; Cibrián-Uhalte et al. 2007). We have shown that high concentrations of ouabain ≥ 1 μM opens the TJ and causes a drastic drop of transepithelial resistance (TER), yet at 10–100 nM, it has the opposite effect, as it increases the degree of sealing of the TJ (Fig. 10.5a) and decreases the paracellular flux of neutral 3 kDa dextran (J_{DEX})

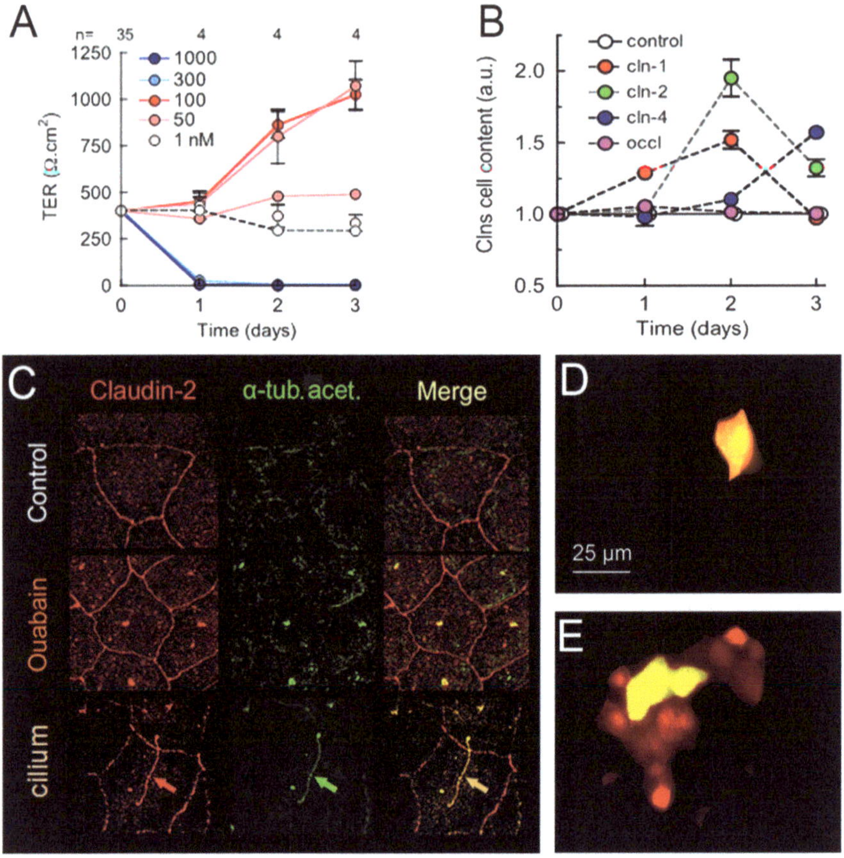

Fig. 10.5 Ouabain modulates several types of cell contacts. (**a**) Tight junctions (TJ): The hermeticity of TJs was gauged by measuring TER (transepithelial electrical resistance). At 1 μM (*blue circles*) ouabain is toxic and opens the TJ, and TER drops to zero. (**b**) Ouabain modulates the expression of claudins individually; each one varies with its own kinetics. (**c**) The cilium, located at the center of the apical domain, freely waves in the fluid bathing the epithelium. In the case of monolayers of MDCK claudin-2 adopts the pattern of a chicken fence (*red*), because it belongs to the TJ. When monolayers are treated with 10 nM ouabain, claudin-2 is also expressed at the procilium and cilium of each cell, as evidenced by its colocalization with α-tubulin (*green*) that is a specific protein of the cilium. (**d, e**) Gap junctional communication (GJC): Tracer injection assays made on MDCK cells in monolayers that were either left as control (**d**) or treated with 10 nM ouabain for 60 min (**e**). In both cases, a single cell was injected with a mix of neurobiotin as tracer and dextran-fluorescein as marker of injection. Neurobiotin diffusion to neighboring, GJ-coupled cells was revealed by streptavidin-rhodamine

(Larre et al. 2006, 2010; Rajasekaran et al. 2003; Rincon-Heredia et al. 2014). Figure 10.5b illustrates that ouabain modulates the permeability of the TJ by acting on the amount as well as the pattern of distribution of claudins of a given type independently.

10.5.3 Ouabain Modulates Adherens Junctions

Another prominent cell-cell contact is the adherens junction (AJ), and one of the scaffolding proteins of this junction is β-catenin, a key member of the Wnt signaling pathway, a family of highly conserved secreted signaling molecules that regulate cell-to-cell interactions during embryogenesis (Logan and Nusse 2004). During the activation of this pathway, a condition that can be achieved by changes in cell adhesion, β-catenin is translocated to the nucleus, where it modifies gene expression (Lien and Fuchs 2014). We have observed that 10 nM ouabain, as well as 1 μM, binding to Na^+/K^+-ATPase provokes the translocation of β-catenin to the nucleus at 8–10 h of ouabain challenge (Contreras et al. 2004). Liu and co-workers have recently found evidence that Na^+/K^+-ATPase and E-cadherin are closely associated, indicating that the pump might be a component of the adherens junction or that E-cadherin could be part of the signalosome of the Na^+-K^+-ATPase (Liu et al. 2011).

10.5.4 Ouabain Stimulates Ciliogenesis

We next tested ouabain at hormonal concentrations (10 nM) on the development of the cilium. The cilium is a slender protuberance polarizedly expressed at the center of the apical domain of most eukaryotic cells (Colegio et al. 2003; Hou et al. 2006). Its development (ciliogenesis) is the last step of the process of apical/basolateral polarization. Therefore, we gauged polarity through the expression of a cilium in MDCK cells. Ciliogenesis occurs in quiescent cells that have reached confluence and is drastically accelerated by ouabain binding to Na^+/K^+-ATPase (Larre et al. 2010). MDCK cells express claudin-2 at the TJ, where it accounts for specific cation permeation (Rosenthal et al. 2010; Overgaard et al. 2011) (Fig. 10.5c, *above*). Surprisingly, we observed that this claudin is also localized at the cilium that, contrary to TJs, does not separate into two different fluid compartments and therefore might not play a role in permeation (Larre et al. 2011b). Nevertheless, at the TJ, claudin-2 is responsible for the specific permeation of cations, in particular Na^+ (Günzel and Fromm 2012), indicating that it has a sequence of amino acids or adopts a configuration that is able to sense the presence of this ion. Therefore, the possibility exists that claudin-2 in the cilium may act as a sensor of Na^+ concentration in the fluid bathing the apical domain (Fig. 10.5c, *below*).

10.5.5 Ouabain Modulates Cell-Cell Communication Through Gap Junctions

Gap junctions (GJ) are structures where one cell can transfer water, ions, and small molecules like second messengers to its neighbors, bypassing the hydrophobic barriers constituted by their two plasma membranes arranged in series. In epithelia the GJ appears as an islet full of hemi-connexons expressed at the lateral membrane of the cells, where both neighboring cells can contact each other and arrange them in register, so the aqueous pore of one hemi-connexon may continue with that of the neighboring cell (hence the name *connexons*). These connexons are a parallel arrangement of proteins called connexins that surround an aqueous pore. We investigated whether 10 nM ouabain would affect the degree of communication through GJ using dye injection assays (Fig. 10.5d, e). We observed that 10 nM ouabain increases cell-cell communication by up to 510 % in 1 h (unpublished results). Therefore, ouabain should be listed among hormones that influence cell-cell communication, such as luteinizing hormones, which decrease the passage of cAMP from follicle cells to the oocyte (Dekel 2005); steroid hormones, which reduce cell communication in endometrial carcinoma (Saito et al. 2004); and fibroblast growth factor (FGF), which upregulates intercellular dye transfer in primary cultures of embryonic chick lens cells (Le and Musil 2001). Furthermore, silencing the expression of connexin 32 or 43 or both using targeted shRNAs significantly inhibited this effect of ouabain (unpublished observations).

10.5.6 Ouabain Modulates the Epithelial Transporting Phenotype

Apical/basolateral polarity and TJs are the two fundamental features of the so-called epithelial transporting phenotype (Cereijido et al. 2003, 2008). Given that hormone ouabain modulates the TJ as well as apical/basolateral polarity as gauged by ciliogenesis, we may conclude that the binding of ouabain to the sodium pump regulates this phenotype. It is pertinent to remember that up to 85 % of all human death are caused by the collapse of an epithelium (hepatic, renal, etc.) and most cancers in human adults start and occur in epithelia. Interestingly, corticoid ouabain helps develop transporting epithelia that are the site where other adrenal corticoids regulate the homeostasis of metazoan.

10.5.7 Na^+/K^+-ATPase Is a Receptor of Cardiac Steroids

The Na^+/K^+-ATPase and Src are enriched in caveolae of renal epithelial cells (Wang et al. 2004) and they directly interact with each other. More importantly, the Na^+/K^+-ATPase/c-Src forms a receptor complex that serves as a receptor for ouabain to increase protein tyrosine phosphorylation and consequently stimulate protein kinase cascades and phospholipase (Xie et al. 1999; Xie 2003; Tian et al. 2006).

The interaction between the Na^+/K^+-ATPase and Ins(1,4,5) P3R was first reported by Aperia's laboratory during the investigation of ouabain-induced calcium low-frequency oscillations in renal epithelial cells. This calcium oscillation activates and sends NF-κB to the nucleus (Aizman et al. 2001). Ouabain triggers the direct interaction between Na^+/K^+-ATPase and the Ins(1,4,5) P3R. Amino acid residues LKK in the N-terminus of the catalytic α-subunit of Na^+/K^+-ATPase are essential for binding with the Ins(1,4,5) P3R N-terminus. This interaction is also supported by the scaffolding protein Ankyrin-B. The ouabain/Na^+/K^+-ATPase/Ins (1,4,5) P3R complex triggers slow Ca^{2+} oscillations that subsequently activate the NF-κB p65 and protects from apoptosis (Fontana et al. 2013).

Wild MDCK (W-MDCK) is sensitive to extremely low concentrations of ouabain (Kd in the order of 10^{-8} M). We also studied a strand of MDCK resistant to high concentrations of ouabain (up to 10 μM) because, as mentioned above, their Na^+/K^+-ATPases have extremely low affinity to ouabain (R-MDCK cells) (Canessa et al. 1992; Soderberg et al. 1983; Cargnelli et al. 1983). 10 nM ouabain cannot elicit the hormonal effect on the TJ, adherens junction, and ciliogenesis nor gap junction in monolayers of R-MDCK cells, indicating that hormone ouabain elicits its effect when binding to the same site used by toxic concentrations of ouabain to block the pump (Larre and Cereijido 2010).

10.6 Signaling Pathways

Na^+-K^+-ATPase is able to interact with c-Src even in mutated pumps in which the pumping ability is suppressed (Tian et al. 2006). Because of this association, Na^+-K^+-ATPase also modulates different signaling routes involving ERK1/2 and oxygen-reactive species (Xie 2003; Xie et al. 1999). In keeping with these characteristics, we have found that the modulation of cell contacts by ouabain involves all these molecules and signaling routes. The function and composition of TJ is regulated by the receptor complex of Na^+/K^+-ATPase/c-Src that activates signaling pathways. For example, while TER depends on c-Src and partially on ERK1/2, J_{DEX} depends on ERK1/2 but not on c-Src, indicating that Na^+/K^+-ATPase can interact with other signal proteins and form a different receptor complex (Larre et al. 2010). The changes in the permeability of the TJ are accompanied by changes in the cell content and distribution patterns of specific claudins, each one regulated independently (Fig. 10.5b). Claudin-1 depends on c-Src and ERK1/2, but claudin-4 depends instead on an as-yet unidentified signal molecule *and* ERK1/2. Binding of ouabain to the Na^+/K^+-ATPase accelerates ciliogenesis through ERK1/2 pathway. At variance with the modulation of the TJ, that of adherens junctions and gap junctions completely depend on c-Src-ERK1/2 pathway. In turn, inhibitors c-Src as

PP2 and of ERK1/2 as PD98059 inhibit GJ communication, as well as the addressing of β-catenin to the nucleus.

Although all the effects attributed to hormone ouabain on the several cell contacts investigated are triggered by its binding to the Na^+/K^+-ATPase at the same moment, these effects are not simultaneous. Thus, the enhancement of GJC is observed in minutes, TJ effect is observed in hours, adherens junctions in 8–10 h, and ciliogenesis in 1–3 days.

10.7 Pathologies Related to Na^+/K^+-ATPase

In spite of the paramount importance of Na^+/K^+-ATPase and the number of fundamental cellular functions in which it participates, there are not many known pathologies directly associated to its malfunctioning. In neural tissue, where Na^+/K^+-ATPase is abundant, there has been an association between the Na^+/K^+-ATPase α-subunit isoforms and some neurological diseases. For instance, mutations in the gene encoding α2 isoform result in familial hemiplegic migraine type 2 (FHM2) (De Fusco et al. 2003), and mutations in the gene encoding the α3 subunit result in rapid-onset dystonia-parkinsonism (RDP) (Brashear et al. 2007). In epithelia, however, pathological conditions in which Na^+/K^+-ATPase is involved are more related to adhesion and polarity rather than transport function. Na^+/K^+-ATPase β_1-subunit is highly reduced in cells with deteriorated cell-cell adhesion, such as the cystic epithelia in polycystic kidney disease (Wilson et al. 1991), clear cell renal carcinoma cells (Rajasekaran et al. 1999), bladder cancer cells (Espineda et al. 2003), and poorly differentiated carcinoma cell lines derived from the colon, breast, kidney, and pancreas (Espineda et al. 2004) and the gastric carcinoma cell line, HGT-1 (Vagin et al. 2005).

As mentioned previously in this chapter, intercellular adhesion mediated in part by Na/K-ATPase β_1-subunits is essential for normal maturation of the epithelium, and so the loss of adhesion occurring in several forms of carcinoma may contribute to cancer metastasis. Alterations in the polarity of Na^+/K^+-ATPase have been described in autosomal-dominant polycystic kidney disease and were attributed to aberrant expression of β_2 isoforms that would target Na^+/K^+-ATPase to the apical membrane, thus contributing to cyst formation (Wilson et al. 1991, 2000). As for the γ-subunits of the Na^+/K^+-ATPase, the isoform FXYD3 is downregulated in lung cancer (Okudela et al. 2009) but overexpressed in some carcinomas such as FXYD3 in pancreatic ductal adenocarcinoma (Kayed et al. 2006), gastric adenocarcinoma, and esophageal squamous cell carcinoma, and although it has been suggested that FXYD3 may affect cellular adhesion and migration through interactions with Na^+/K^+-ATPase, this hypothesis is unclear (Zhu et al. 2010). In this regard, FXYD5 or dysadherin has been suggested as cell-cell contact molecule, with its unusually long extracellular domain, and is also overexpressed in various tumors (Lubarski et al. 2011). Lastly, an abnormal modulation of Na^+/K^+-ATPase transport by

FXYD1 resulting in atypical neuronal activity has been proposed as a direct contribution to Rett syndrome pathogenesis (Deng et al. 2007).

10.8 Horizons and Perspectives

Until a couple of decades ago, cell contacts were regarded as mere architectural structures to keep cells together, lest they would disaggregate on deformation. Yet the observation that they are constituted by a precise arrangement of dozens of specific protein species and the structure, physiological properties, and the supracellular architecture that they build up (e.g., a renal glomerulus, a hepatic lobule) are of paramount importance led to discover that cell contacts play fundamental roles such as gene expression, cell cycling, proliferation, differentiation, migration, tissue architecture, cancer, metastasis, and homing. Their importance can be hardly overestimated given that the brain, the most complex object in the universe, assembles itself depending on which cells contact which other cells, when, and how (i.e., the type of synapses established). Therefore, the finding that ouabain is a hormone that modulates cell contacts results promising from the physiological as well as the pathological views, although we have not yet systematically explored the physiological significance of the distribution in time of the different effects elicited by ouabain binding to the Na^+/K^+-ATPase complex receptor. Interestingly, hormone ouabain causes these effects acting on the same receptor site (the extracellular domain of the α-subunit of Na^+/K^+-ATPase) and activating similar signaling routes. Therefore, Na^+/K^+-ATPase is likely to play an important role in phenomena depending on cell contacts, such as proliferation, differentiation, migration, metastasis, and invasion.

Acknowledgments We wish to acknowledge the efficient and pleasant assistance of E. el Oso, A. Castillo, E. Méndez, and M. L. Roldan. Our experimental work was supported by the National Research Council of México (CONACYT) and the Research Council of Mexico City (ICyTDF). IL was recipient of a Postdoctoral Research Fellowship from ICyTDF. O Paez was recipient of a Doctoral Fellowship from CONACYT.

References

Aizman O, Uhlen P, Lal M, Brismar H, Aperia A (2001) Ouabain, a steroid hormone that signals with slow calcium oscillations. Proc Natl Acad Sci USA 98:13420–13424

Bab-Dinitz E, Albeck S, Peleg Y, Brumfeld V, Gottschalk KE, Karlish SJ (2009) A C-terminal lobe of the beta subunit of Na, K,-ATPase and H, K,-ATPase resembles cell adhesion molecules. Biochemistry 48:8684–8691

Bagrov AY, Fedorova OV, Dmitrieva RI, Howald WN, Hunter AP, Kuznetsova EA, Shpen VM (1998) Characterization of a urinary bufadienolide Na^+, K^+-ATPase inhibitor in patients after acute myocardial infarction. Hypertension 31:1097–1103

Bauer N, Müller-Ehmsen J, Krämer U, Hambarchian N, Zobel C, Schwinger RH, Neu H, Kirch U, Grünbaum EG, Schoner W (2005) Ouabain-like compound changes rapidly on physical exercise in humans and dogs: effects of beta-blockade and angiotensin-converting enzyme inhibition. Hypertension 45:1024–1028

Bianconi E, Piovesan A, Facchin F, Beraudi A, Casadei R, Frabetti F, Vitale L, Pelleri MC, Tassani S, Piva F, Perez-Amodio S, Strippoli P, Canaider S (2013) An estimation of the number of cells in the human body. Ann Hum Biol 40:463–471

Blostein R, Zhang R, Gottardi CJ, Caplan MJ (1993) Functional properties of an H/K-ATPase/Na/K-ATPase chimera. J Biol Chem 268:10654–10658

Brashear A, Dobyns WB, de Carvalho AP, Borg M, Frijns CJ, Gollamudi S, Green A, Guimaraes J, Haake BC, Klein C, Linazasoro G, Munchau A, Raymond D, Riley D, Saunders-Pullman R, Tijssen MA, Webb D, Zaremba J, Bressman SB, Ozelius LJ (2007) The phenotypic spectrum of rapid-onset dystonia-parkinsonism (RDP) and mutations in the ATP1A3 gene. Brain 130:828–835

Canessa CM, Horisberger JD, Louvard D, Rossier BC (1992) Mutation of a cysteine in the first transmembrane segment of Na, K-ATPase alpha subunit confers ouabain resistance. EMBO J 11:1681–1687

Caplan MJ, Anderson HC, Palade GE, Jamieson JD (1986) Intracellular sorting and polarized cell surface delivery of (Na^+/K^+)ATPase, an endogenous component of MDCK cell basolateral plasma membranes. Cell 46:623–631

Cargnelli G, Trevisi L, Soderberg K, Rossi B, Lazdunski M, Louvard D (1983) Characterization of ouabain resistant mutants of a canine kidney cell line, MDCK. J Biol Chem 258:12300–12307

Cereijido M, Robbins ES, Dolan WJ, Rotunno CA, Sabatini DD (1978) Polarized monolayers formed by epithelial cells on a permeable and translucent support. J Cell Biol 77:853–880

Cereijido M, Ehrenfeld J, Meza I, Martinez-Palomo A (1980) Structural and functional membrane polarity in cultured monolayers of MDCK cells. J Membr Biol 52:147–159

Cereijido M, Contreras RG, Shoshani L, García-Villegas MR (2003) Membrane targeting. Prog Biophys Mol Biol 81:81–115

Cereijido M, Contreras RG, Shoshani L (2004) Cell adhesion, polarity, and epithelia in the dawn of metazoans. Physiol Rev 84:1229–1262

Cereijido M, Contreras RG, Shoshani L, Flores-Benitez D, Larre I (2008) Tight junction and polarity interaction in the transporting epithelial phenotype. Biochim Biophys Acta 1778:770–793

Cereijido M, Contreras RG, Shoshani L, Larre I (2012) The Na^+-K^+-ATPase as self-adhesion molecule and hormone receptor. Am J Physiol Cell Physiol 302:C473–C481

Cibrián-Uhalte E, Langenbacher A, Shu X, Chen JN, Abdelilah-Seyfried S (2007) Involvement of zebrafish Na+, K+ ATPase in myocardial cell junction maintenance. J Cell Biol 176:223–230

Colegio OR, Van Itallie C, Rahner C, Anderson JM (2003) Claudin extracellular domains determine paracellular charge selectivity and resistance but not tight junction fibril architecture. Am J Physiol Cell Physiol 284:C1346–C1354

Contreras RG, Avila G, Gutierrez C, Bolívar JJ, González-Mariscal L, Darzon A, Beaty G, Rodriguez-Boulan E, Cereijido M (1989) Repolarization of Na^+-K^+ pumps during establishment of epithelial monolayers. Am J Physiol Cell Physiol 257:C896–C905

Contreras RG, Miller JH, Zamora M, González-Mariscal L, Cereijido M (1992) Interaction of calcium with plasma membrane of epithelial (MDCK) cells during junction formation. Am J Physiol Cell Physiol 263:C313–C318

Contreras RG, Shoshani L, Flores-Maldonado C, Lázaro A, Cereijido M (1999) Relationship between Na+, K+-ATPase and cell attachment. J Cell Sci 112:4223–4232

Contreras RG, Flores-Maldonado C, Lázaro A, Shoshani L, Flores-Benitez D, Larré I, Cereijido M (2004) Ouabain binding to Na^+/K^+-ATPase relaxes cell attachment and sends a specific signal (NACos) to the nucleus. J Membr Biol 198:147–158

De Fusco M, Marconi R, Silvestri L, Atorino L, Rampoldi L, Morgante L, Ballabio A, Aridon P, Casari G (2003) Haploinsufficiency of ATP1A2 encoding the Na^+/K^+ pump alpha2 subunit associated with familial hemiplegic migraine type 2. Nat Genet 33:192–196

Dekel N (2005) Cellular, biochemical and molecular mechanisms regulating oocyte maturation. Mol Cell Endocrinol 234:19–25

Dempski RE, Friedrich T, Bamberg E (2005) The beta subunit of the Na^+/K^+-ATPase follows the conformational state of the holoenzyme. J Gen Physiol 125:505–520

Deng V, Matagne V, Banine F, Frerking M, Ohliger P, Budden S, Pevsner J, Dissen GA, Sherman LS, Ojeda SR (2007) FXYD1 is an MeCP2 target gene overexpressed in the brains of Rett syndrome patients and Mecp2-null mice. Hum Mol Genet 16:640–650

Doris PA, Hayward-Lester A, Bourne D, Stocco DM (1996) Ouabain production by cultured adrenal cells. Endocrinology 137:533–539

Duffield A, Fölsch H, Mellman I, Caplan MJ (2004) Sorting of H/K-ATPase beta-subunit in MDCK and LLC-PK cells is independent of μ1B adaptin expression. Traffic 5:449–461

Dunbar LA, Aronson P, Caplan MJ (2000) A transmembrane segment determines the steady-state localization of an ion-transporting adenosine triphosphatase. J Cell Biol 148:769–778

el-Masri MA, Clark BJ, Qazzaz HM, Valdes R Jr (2002) Human adrenal cells in culture produce both ouabain-like and dihydroouabain-like factors. Clin Chem 48:1720–1730

Espineda C, Seligson DB, James Ball W Jr, Rao J, Palotie A, Horvath S, Huang Y, Shi T, Rajasekaran AK (2003) Analysis of the Na/K-ATPase α- and β-Subunit expression profiles of bladder cancer using tissue microarrays. Cancer 97:1859–1868

Espineda CE, Chang JH, Twiss J, Rajasekaran SA, Rajasekaran AK (2004) Repression of Na/K-ATPase beta1-subunit by the transcription factor snail in carcinoma. Mol Biol Cell 15:1364–1373

Farr GA, Hull M, Mellman I, Caplan MJ (2009) Membrane proteins follow multiple pathways to the basolateral cell surface in polarized epithelial cells. J Cell Biol 186:269–282

Fedorova OV, Lakatta EG, Bagrov AY (2000) Endogenous Na/K pump ligands are differentially regulated during acute NaCl loading of Dahl rats. Circulation 102:3009–3014

Fedorova OV, Agalakova NI, Talan MI, Lakatta EG, Bagrov AY (2005a) Brain ouabain stimulates peripheral marinobufagenin via angiotensin II signalling in NaCl-loaded Dahl-S rats. J Hypertens 23:1515–1523

Fedorova OV, Kolodkin NI, Agalakova NI, Namikas AR, Bzhelyansky A, St-Louis J, Lakatta EG, Bagrov AY (2005b) Antibody to marinobufagenin lowers blood pressure in pregnant rats on a high NaCl intake. J Hypertens 23:835–842

Fontana JM, Burlaka I, Khodus G, Brismar H, Aperia A (2013) Calcium oscillations triggered by cardiotonic steroids. FEBS J 280:5450–5455

Galeotti G (1904) Concerning the EMF which is generated at the surface of animal membranes on contact with different electrolytes. Z Phys Chem 49:542

Galeotti G (1907) Richerche di elettrofisiologia secondo i criteri dell'eletrochimica. Z Allg Physiol 6:99

Geering K (2001) The functional role of beta subunits in oligomeric P-type ATPases. J Bioenerg Biomembr 33:425–438

Geering K (2008) Functional roles of Na/K-ATPase subunits. Curr Opin Nephrol Hypertens 17:526–532

Gonzalez-Mariscal L, Contreras RG, Bolívar JJ, Ponce A, Chávez De Ramirez B, Cereijido M (1990) Role of calcium in tight junction formation between epithelial cells. Am J Physiol Cell Physiol 259:C978–C986

Goto A, Ishiguro T, Yamada K, Ishii M, Yoshioka M, Eguchi C, Shimora M, Sugimoto T (1990) Isolation of a urinary digitalis-like factor indistinguishable from digoxin. Biochem Biophys Res Commun 173:1093–1101

Gottlieb SS, Rogowski AC, Weinberg M, Krichten CM, Hamilton BP, Hamlyn JM (1992) Elevated concentrations of endogenous ouabain in patients with congestive heart failure. Circulation 86:420–425

Günzel D, Fromm M (2012) Claudins and other tight junction proteins. Compr Physiol 2:1819–1852
Hamlyn JM, Blaustein MP, Bova S, DuCharme DW, Harris DW, Mandel F, Mathews WR, Ludens JH (1991) Identification and characterization of a ouabain-like compound from human plasma. Proc Natl Acad Sci USA 88:6259–6263
Hilton PJ, White RW, Lord GA, Garner GV, Gordon DB, Hilton MJ, Forni LG, McKinnon W, Ismail FM, Keenan M, Jones K, Morden WE (1996) An inhibitor of the sodium pump obtained from human placenta. Lancet 348:303–305
Hou J, Gomes AS, Paul DL, Goodenough DA (2006) Study of claudin function by RNA interference. J Biol Chem 281:36117–36123
Kawamura A, Guo J, Itagaki Y, Bell C, Wang Y, Haupert GT Jr, Magil S, Gallagher RT, Berova N, Nakanishi K (1999) On the structure of endogenous ouabain. Proc Natl Acad Sci USA 96:6654–6659
Kayed H, Kleeff J, Kolb A, Ketterer K, Keleg S, Felix K, Giese T, Penzel R, Zentgraf H, Büchler MW, Korc M, Friess H (2006) FXYD3 is overexpressed in pancreatic ductal adenocarcinoma and influences pancreatic cancer cell growth. Int J Cancer 118:43–54
Koefoed-Johnsen V, Ussing HH (1958) On the nature of the frog skin potential. Acta Physiol Scand 42:298–308
Komiyama Y, Dong XH, Nishimura N, Masaki H, Yoshika M, Masuda M, Takahashi H (2005) A novel endogenous digitalis, telocinobufagin, exhibits elevated plasma levels in patients with terminal renal failure. Clin Biochem 38:36–45
Laredo J, Hamilton BP, Hamlyn JM (1994) Ouabain is secreted by bovine adrenocortical cells. Endocrinology 135:794–797
Larre I, Cereijido M (2010) Na+, K+-ATPase is the putative membrane receptor of hormone ouabain. Commun Integr Biol 3:625–628
Larre I, Ponce A, Fiorentino R, Shoshani L, Contreras RG, Cereijido M (2006) Contacts and cooperation between cells depend on the hormone ouabain. Proc Natl Acad Sci USA 103:10911–10916
Larre I, Lazaro A, Contreras RG, Balda MS, Matter K, Flores-Maldonado C, Ponce A, Flores-Benitez D, Rincon-Heredia R, Padilla-Benavides T, Castillo A, Shoshani L, Cereijido M (2010) Ouabain modulates epithelial cell tight junction. Proc Natl Acad Sci USA 107:11387–11392
Larre I, Castillo A, Flores-Maldonado C, Contreras RG, Galvan I, Muñoz-Estrada J, Cereijido M (2011a) Ouabain modulates ciliogenesis in epithelial cells. Proc Natl Acad Sci USA 108:20591–20596
Larre I, Contreras RG, Cereijido M (2011b) Ouabain modulates cell contacts as well as functions that depend on cell adhesion. Methods Mol Biol 763:155–168
Le AC, Musil LS (2001) A novel role for FGF and extracellular signal-regulated kinase in gap junction-mediated intercellular communication in the lens. J Cell Biol 154:197–216
Li SQ, Eim C, Schneider R, Sich B, Kirch U, Schoner W (1997) Proscillaridin A immunoreactivity: a new endogenous cardiac glycoside? Ann N Y Acad Sci 834:646–647
Lichtstein D, Gati I, Ovadia H (1993) Digitalis-like compounds in the toad Bufo viridis: interactions with plasma proteins. J Cardiovasc Pharmacol 22(Suppl 2):S102–S105
Lien WH, Fuchs E (2014) Wnt some lose some: transcriptional governance of stem cells by Wnt/β-catenin signaling. Genes Dev 28:1517–1532
Liu L, Ivanov AV, Gable ME, Jolivel F, Morrill GA, Askari A (2011) Comparative properties of caveolar and noncaveolar preparations of kidney Na^+/K^+-ATPase. Biochemistry 50:8664–8673
Logan CY, Nusse R (2004) The Wnt signaling pathway in development and disease. Annu Rev Cell Dev Biol 20:781–810
Lubarski I, Asher C, Garty H (2011) FXYD5 (dysadherin) regulates the paracellular permeability in cultured kidney collecting duct cells. Am J Physiol Renal Physiol 301:F1270–F1280
Ludens JH, Clark MA, Robinson FG, DuCharme DW (1992) Rat adrenal cortex is a source of a circulating ouabainlike compound. Hypertension 19:721–724

Manunta P, Ferrandi M, Bianchi G, Hamlyn JM (2009) Endogenous ouabain in cardiovascular function and disease. J Hypertens 27:9–18
Mathews WR, DuCharme DW, Hamlyn JM, Harris DW, Mandel F, Clark MA, Ludens JH (1991) Mass spectral characterization of an endogenous digitalislike factor from human plasma. Hypertension 930–935
Mays RW, Siemers KA, Fritz BA, Lowe AW, van Meer G, Nelson WJ (1995) Hierarchy of mechanisms involved in generating Na/K-ATPase polarity in MDCK epithelial cells. J Cell Biol 130:1105–1115
Nyblom M, Poulsen H, Gourdon P, Reinhard L, Andersson M, Lindahl E, Fedosova N, Nissen P (2013) Crystal structure of Na^+, K^+-ATPase in the Na^+-bound state. Science 342:123–127
Ohka S, Ohno H, Tohyama K, Nomoto A (2001) Basolateral sorting of human poliovirus receptor alpha involves an interaction with the μ1B subunit of the clathrin adaptor complex in polarized epithelial cells. Biochem Biophys Res Commun 287:941–948
Okudela K, Yazawa T, Ishii J, Woo T, Mitsui H, Bunai T, Sakaeda M, Shimoyamada H, Sato H, Tajiri M, Ogawa N, Masuda M, Sugimura H, Kitamura H (2009) Down-regulation of FXYD3 expression in human lung cancers: its mechanism and potential role in carcinogenesis. Am J Pathol 175:2646–2656
Overgaard CE, Daugherty BL, Mitchell LA, Koval M (2011) Claudins: control of barrier function and regulation in response to oxidant stress. Antioxid Redox Signal 15:1179–1193
Padilla-Benavides T, Roldán ML, Larre I, Flores-Benitez D, Villegas-Sepúlveda N, Contreras RG, Cereijido M, Shoshani L (2010) The polarized distribution of Na+, K+-ATPase: role of the interaction between β subunits. Mol Biol Cell 21:2217–2225
Qazzaz HM, Cao Z, Bolanowski DD, Clark BJ, Valdes R Jr (2004) De novo biosynthesis and radiolabeling of mammalian digitalis-like factors. Clin Chem 50:612–620
Rajasekaran SA, Ball WJ Jr, Bander NH, Liu H, Pardee JD, Rajasekaran AK (1999) Reduced expression of the beta-subunit of Na,/K-ATPase in human clear-cell renal cell carcinoma. J Urol 162:574–580
Rajasekaran SA, Palmer LG, Quan K, Harper JF, Ball WJ Jr, Bander NH, Peralta Soler A, Rajasekaran AK (2001) Na, K-ATPase beta-subunit is required for epithelial polarization, suppression of invasion, and cell motility. Mol Biol Cell 12:279–295
Rajasekaran SA, Hu J, Gopal J, Gallemore R, Ryazantsev S, Bok D, Rajasekaran AK (2003) Na, K-ATPase inhibition alters tight junction structure and permeability in human retinal pigment epithelial cells. Am J Physiol Cell Physiol 284:C1497–C1507
Rincon-Heredia R, Flores-Benitez D, Flores-Maldonado C, Bonilla-Delgado J, García-Hernández V, Verdejo-Torres O, Castillo AM, Larré I, Poot-Hernández AC, Franco M, Gariglio P, Reyes JL, Contreras RG (2014) Ouabain induces endocytosis and degradation of tight junction proteins through ERK1/2-dependent pathways. Exp Cell Res 320:108–118
Rodriguez-Boulan E, Macara IG (2014) Organization and execution of the epithelial polarity programme. Nat Rev Mol Cell Biol 15:225–242
Rosenthal R, Milatz S, Krug SM, Oelrich B, Schulzke JD, Amasheh S, Günzel D, Fromm M (2010) Claudin-2, a component of the tight junction, forms a paracellular water channel. J Cell Sci 123(Pt 11):1913–1921
Saito T, Tanaka R, Wataba K, Kudo R, Yamasaki H (2004) Overexpression of estrogen receptor-alpha gene suppresses gap junctional intercellular communication in endometrial carcinoma cells. Oncogene 23:1109–1116
Schatzmann HJ (1953) Cardiac glycosides as inhibitors of active potassium and sodium transport by erythrocyte membrane. Helv Physiol Pharmacol Acta 11:346–354
Schreiber V, Stepan J (1986) Digoxin-like immunoreactivity of 19-NOR and 19-OHandrost-4-ene-3,17-dione. Physiol Bohemoslov 35:190–192
Shinoda T, Ogawa H, Cornelius F, Toyoshima C (2009) Crystal structure of the sodium-potassium pump at 2.4 Å resolution. Nature 459:446–450
Shoshani L, Contreras RG, Roldán ML, Moreno J, Lázaro A, Balda MS, Matter K, Cereijido M (2005) The polarized expression of Na+, K+-ATPase in epithelia depends on the association between beta-subunits located in neighboring cells. Mol Biol Cell 16:1071–1081

Skou JC (1957) The influence of some cations on an adenosine triphosphatase from peripheral nerves. Biochim Biophys Acta 23:394–401

Soderberg K, Rossi B, Lazdunski M, Louvard D (1983) Characterization of ouabain-resistant mutants of a canine kidney cell line, MDCK. J Biol Chem 258:12300–12307

Takahashi H (2000) Endogenous digitalislike factor: an update. Hypertens Res 23(Suppl):S1–S5

Tian J, Cai T, Yuan Z, Wang H, Liu L, Haas M, Maksimova E, Huang XY, Xie ZJ (2006) Binding of Src to Na^+/K^+-ATPase forms a functional signaling complex. Mol Biol Cell 17:317–326

Tokhtaeva E, Munson K, Sachs G, Vagin O (2010) N-glycan-dependent quality control of the Na, K-ATPase β2 subunit. Biochemistry 49:3116–3128

Tokhtaeva E, Sachs G, Souda P, Bassilian S, Whitelegge JP, Shoshani L, Vagin O (2011) Epithelial junctions depend on intercellular trans-interactions between the Na/K-ATPase β_1 subunits. J Biol Chem 286:25801–25812

Tokhtaeva E, Sachs G, Sun H, Dada LA, Sznajder JI, Vagin O (2012) Identification of the amino acid region involved in the intercellular interaction between the β1 subunits of Na^+/K^+-ATPase. J Cell Sci 125:1605–1616

Toyoshima C, Kanai R, Cornelius F (2011) First crystal structures of Na+, K+-ATPase: new light on the oldest ion pump. Structure 19(12):1732–1738

Tymiak AA, Norman JA, Bolgar M, DiDonato GC, Lee H, Parker WL, Lo LC, Berova N, Nakanishi K, Haber E, Haupert GT Jr (1993) Physicochemical characterization of a ouabain isomer isolated from bovine hypothalamus. Proc Natl Acad Sci USA 90:8189–8193

Ussing HH, Zerahn K (1951) Active transport of sodium as the source of electric current in the short-circuited isolated frog skin. Acta Physiol Scand 23:110–127

Vagin O, Turdikulova S, Sachs G (2005) Recombinant addition of N-glycosylation sites to the basolateral Na/K-ATPase beta1 subunit results in its clustering in caveolae and apical sorting in HGT-1 cells. J Biol Chem 280:43159–43167

Vagin O, Sachs G, Tokhtaeva E (2007a) The roles of the Na/K-ATPase beta 1 subunit in pump sorting and epithelial integrity. J Bioenerg Biomembr 39:367–372

Vagin O, Turdikulova S, Tokhtaeva E (2007b) Polarized membrane distribution of potassium-dependent ion pumps in epithelial cells: different roles of the N-glycans of their beta subunits. Cell Biochem Biophys 47:376–391

Vagin O, Tokhtaeva E, Yakubov I, Shevchenko E, Sachs G (2008) Inverse correlation between the extent of N-glycan branching and intercellular adhesion in epithelia. Contribution of the Na, K-ATPase β1 subunit. J Biol Chem 283:2192–2202

Wang H, Haas M, Liang M, Cai T, Tian J, Li S, Xie Z (2004) Ouabain assembles signaling cascades through the caveolar Na^+/K^+-ATPase. J Biol Chem 279:17250–17259

Weisz OA, Rodriguez-Boulan E (2009) Apical trafficking in epithelial cells: signals, clusters and motors. J Cell Sci 122:4253–4266

Wilson PD, Sherwood AC, Palla K, Du J, Watson R, Norman JT (1991) Reversed polarity of Na^+-K^+-ATPase: mislocation to apical plasma membranes in polycystic kidney disease epithelia. Am J Physiol Renal Physiol 260:F420–F430

Wilson PD, Devuyst O, Li X, Gatti L, Falkenstein D, Robinson S, Fambrough D, Burrow CR (2000) Apical plasma membrane mispolarization of NaK-ATPase in polycystic kidney disease epithelia is associated with aberrant expression of the β_2 isoform. Am J Pathol 156:253–268

Xie Z (2003) Molecular mechanisms of Na/K-ATPase-mediated signal transduction. Ann N Y Acad Sci 986:497–503

Xie Z, Kometiani P, Liu J, Li J, Shapiro JI, Askari A (1999) Intracellular reactive oxygen species mediate the linkage of Na^+/K^+-ATPase to hypertrophy and its marker genes in cardiac myocytes. J Biol Chem 274:19323–19328

Zhu ZL, Zhao ZR, Zhang Y, Yang YH, Wang ZM, Cui DS, Wang MW, Kleeff J, Kayed H, Yan BY, Sun XF (2010) Expression and significance of FXYD-3 protein in gastric adenocarcinoma. Dis Markers 28:63–69

Zimmermann P (2006) PDZ domain-phosphoinositide interactions in cell-signaling. Verh K Acad Geneeskd Belg 68:271–286

Chapter 11
Na^+-K^+-$2Cl^-$ Cotransporter

Eric Delpire and Kenneth B. Gagnon

Abstract The conceptual breakthrough that the energy of the Na^+ gradient generated by the Na^+/K^+ pump could be used as the driving force for another membrane transport protein has led to the functional and molecular identification of multiple secondary active transporters. We have organized this chapter to address the expression, function, regulation, and evolutionary importance of the two isoforms of the electroneutral sodium-potassium-chloride cotransporter (NKCC). The combination of basolateral expression of the sodium-potassium pump and NKCC1 in various epithelia results in salt and water secretion, whereas basolateral expression of the pump with apical expression of NKCC2 in the thick ascending limb of Henle in the kidney nephron results in salt and water reabsorption. NKCCs are regulated by serine-threonine phosphorylation of specific residues in their amino-terminal domain, and the evolutionary conservation of these cotransporters from protists to humans confirms their vital role in cellular and whole-organism physiology.

Keywords Cl^- secretion • Na^+ reabsorption • Cell volume • Electroneutral transport • Loop diuretics

11.1 Introduction

Prokaryotic and eukaryotic cells have surrounded themselves with two leaflets of lipids forming a membrane or "oil" interface which isolates their biochemical reactions from the extracellular environment. However, complete isolation from the outside environment is incompatible with life, as uptake of nutrients and elimination of waste are necessary steps of cell metabolism. By insertion of transmembrane proteins, cells have acquired the capability to move water and

E. Delpire (✉)
Department of Anesthesiology, Vanderbilt University School of Medicine, T-4202 MCN, 1161 21st Avenue South, Nashville, TN 37232-2520, USA
e-mail: eric.delpire@vanderbilt.edu

K.B. Gagnon
Department of Medicine, Division of Nephrology and Hypertension, University of Louisville, 615 S. Preston Street, Louisville, KY 40202, USA

K.L. Hamilton, D.C. Devor (eds.), *Ion Channels and Transporters of Epithelia in Health and Disease*, Physiology in Health and Disease,
DOI 10.1007/978-1-4939-3366-2_11

solutes across this lipid barrier and gained the capacity to modulate the flow of these substrates. Sugars, amino acids, organic compounds, and inorganic ions such as Na^+, K^+, Cl^-, Ca^{2+}, Mg^{2+}, and HCO_3^- are able to cross the lipid membrane (in conjunction with water) in and out of cells. In a multicellular organism, depending on its specific function within the organism, a cell will express a certain array of channels, pumps, cotransporters, and exchangers at its plasma membrane, allowing for a defined and regulated movement of solutes.

11.2 Ouabain-Insensitive Cation Pump

In the early 1960s, physiologists believed that transport of Na^+ and K^+ across human red blood cell membranes was either energy dependent, through the ATP-dependent cation (Na^+/K^+) pump (see Chap. 10), or a result of passive diffusion in the direction of the ions electrochemical gradient. The Na^+/K^+ pump exchanges K^+ outside for Na^+ inside and is inhibited by cardiac glycosides (e.g., ouabain, digoxin, bufalin)). Identification of another transport pathway for Na^+ and K^+ which was "ouabain insensitive" but still dependent on external Na^+ prompted investigators to search for another type of cation pump (Hoffman and Kregenow 1966). In fact, what they considered to be an active transport mechanism dependent on external Na^+ was actually the first evidence of a secondary active transport mechanism (i.e., Na-K-2Cl cotransporter) dependent on the electrochemical gradient of Na^+ generated by the "ouabain-sensitive" cation pump. This "sodium-gradient hypothesis" was first proposed by RK Crane and represents a conceptual breakthrough in ion transport (Crane 1965).

In Fig. 11.1, the cation-exchange activity of the Na^+/K^+ pump maintains the intracellular Na^+ concentration significantly lower than the extracellular concentration while simultaneously maintaining the intracellular K^+ higher than the extracellular concentration. These concentration gradients provide the energy that secondary active transporters, such as the Na-K-2Cl cotransporter and the Na-glucose cotransporter, use to power the movement of other solutes against their own concentration gradients. In Box 11.1, the individual driving forces acting on Na^+, K^+, and Cl^- across the cell membrane can be added together to determine the net driving force and direction of the Na-K-2Cl cotransporter. Although accepted today, there was considerable resistance to the concept of a membrane transport protein using the downhill concentration gradient of one solute to move another solute uphill against its concentration gradient. Such a mechanism did not conform to the second law of thermodynamics. Today, there are multiple Na^+-coupled transporters which not only move inorganic ions but also amino acids, neurotransmitters, and even carbohydrates. In the remainder of this chapter, we will focus on a subset of the inorganic ion transporters, the Na^+-coupled cation-chloride cotransporters, and their role in epithelial health and disease.

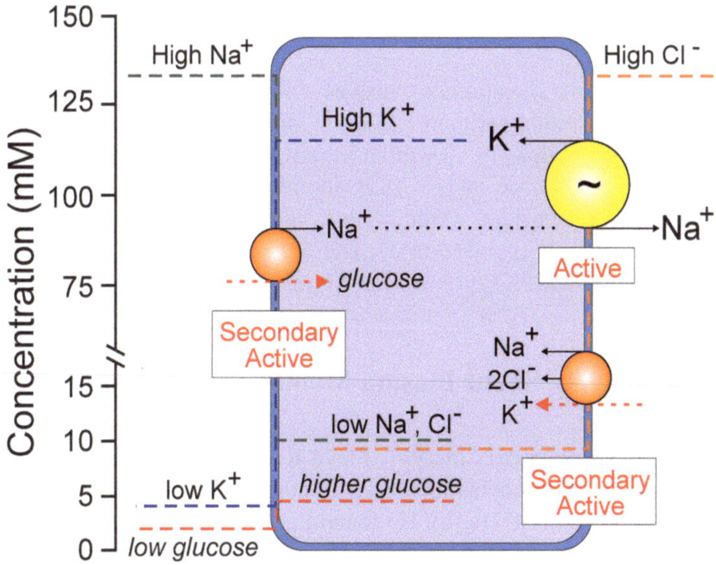

Fig. 11.1 The Na-K-2Cl cotransporter is a secondary active transport mechanism. Schematic representation of a cell with the Na^+/K^+-ATPase accumulating K^+ and extruding Na^+, thus creating energy in ionic gradients that can be used by other transport mechanisms. As the very low $[Na^+]$ inside the cell constitutes a driving force for the inward movement of glucose (even if higher glucose inside), the low $[Na^+]$ and $[Cl^-]$ inside the cell also facilitate the inward movement of ions through the Na-K-2Cl cotransporter

Box 11.1. Calculating the Electrochemical Driving Forces on the Na-K-2Cl Cotransporter

The electrochemical potential gradient for an ion, $\Delta\mu_{\text{ion}}$, is the sum of the chemical and electrical components: $\Delta\mu_{\text{ion}} = \text{RT ln }([\text{ion}]_i/[\text{ion}]_o) + \text{ZFE}_m$. For a transport system that couples the movement of multiple substrates, the net free energy or overall chemical potential gradient is the sum of the gradient for each substrate: $\Delta\mu_{\text{Na, K, Cl}} = \Delta\mu_{\text{Na}} + \Delta\mu_{\text{K}} + 2\Delta\mu_{\text{Cl}}$ (because of two Cl ions transported on NKCC). Note that there is no electrical component as the transport of two cations with two anions is electroneutral.

$$\begin{aligned}\text{Thus, } \Delta\mu_{\text{Na, K, Cl}} &= \text{RT ln}\left([\text{Na}]_i/[\text{Na}]_o\right) + \text{RT ln}\left([\text{K}]_i/[\text{K}]_o\right) \\ &\quad + 2\,\text{RT ln}\left([\text{Cl}]_i/[\text{Cl}]_o\right)\text{or} \\ &= \text{RT ln}\left([\text{Na}]_i[\text{K}]_i[\text{Cl}]_i^{\,2}\right)/\left([\text{Na}]_o[\text{K}]_o[\text{Cl}]_o^{\,2}\right).\end{aligned}$$

(continued)

Box 11.1 (continued)

We can see that the direction of transport will be dictated by the ratio of the products. For a ratio larger than 1, $\Delta\mu_{\mathrm{Na,\,K,\,Cl}}$ will be positive, whereas for a ratio lower than 1, $\Delta\mu_{\mathrm{Na,\,K,\,Cl}}$ will be negative. At 1, the chemical potential gradient will be 0 (since $\ln(1) = 0$), which represents thermodynamic equilibrium. If we consider external and internal ion concentrations of $[\mathrm{Na}]_i = 10$ mM, $[\mathrm{Cl}]_i = 30$ mM, $[\mathrm{K}]_i = 100$ mM, $[\mathrm{Na}]_o = 130$ mM, $[\mathrm{Cl}]_o = 110$ mM, and $[\mathrm{K}]_o = 3$, the ratio favors inward transport by 5:1.

11.3 Electrically Silent Plasma Membrane Transporter

In a set of companion papers in 1971, Floyd Kregenow described the two-phase response for duck erythrocytes incubated in nonhemolytic hypotonic and hypertonic media (Kregenow 1971a, b). He found that the instantaneous cell shrinkage induced by hypertonic shock was followed by a slower "volume regulatory phase" in which the cells swell back to their original size through a Na^+-dependent gain of K^+, Cl^-, and water (Kregenow 1971b). The accumulation of K^+ against its electrochemical gradient suggested an active transport mechanism; however, blockage of the Na^+/K^+ pump by ouabain had no effect on this volume-controlling mechanism. Interestingly, regardless of the severity of the hypertonic challenge, there was no water recovery when external K^+ concentration was maintained at normal levels (i.e., 2.8 mM). However, when the external K^+ concentration was increased to 15 mM, the rate of cell water recovery was faster the greater the hypertonic challenge (see Fig. 11.2a). Although at first counterintuitive, if we once again consider the individual driving forces acting on the Na-K-2Cl cotransporter (see Box 11.1), an increase in the external K^+ concentration actually reduces the resistance of the outwardly directed K^+ gradient on the net activity of the Na-K-2Cl cotransporter, allowing it to work more efficiently (ratio favors inward transport ~26:1) in transporting ions (and cell water) to promote cell volume recovery (see Fig. 11.2b).

Although we have been identifying the Na^+-dependent K^+ and Cl^- transport mechanism described by Kregenow as the Na-K-2Cl cotransporter, it actually took another 9 years before Geck and coworkers proposed the existence of a "secondary active" cotransport system for Na^+, K^+, and Cl^- in Ehrlich cells which exhibited the same "ouabain insensitivity" and "volume control" identified by Kregenow. Thermodynamic experiments revealed a tight stoichiometric coupling of 1 Na^+, 1 K^+, and 2 Cl^- ions in every transport cycle (Geck et al. 1980).

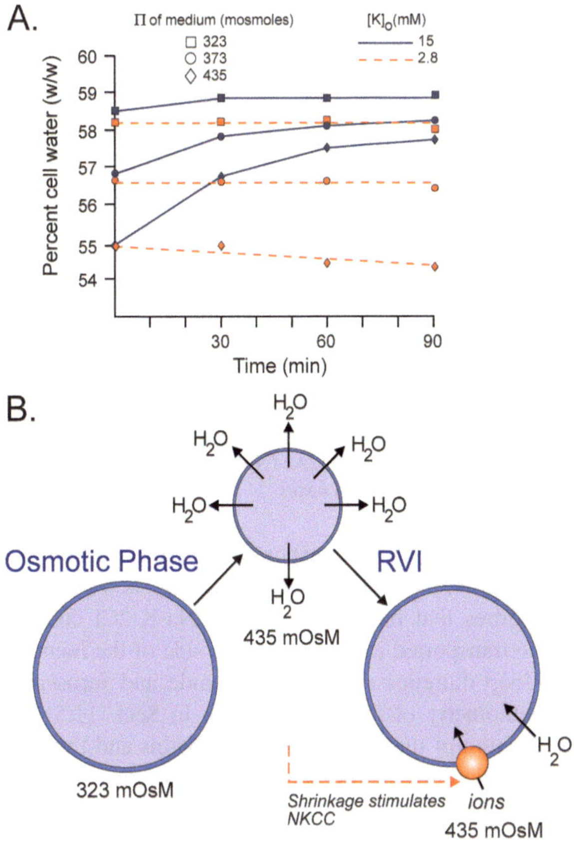

Fig. 11.2 The Na-K-2Cl cotransporter participates in cell volume regulation. (**a**) Data redrawn from Kregenow showing cell water in duck red blood cells recovering after an initial water loss due to osmotic phase during a hypertonic stimulus (Kregenow 1971b). Note that the recovery is larger in cells exposed to larger osmotic shock and it occurs only when the driving force for NKCC transport is increased by raising external K^+. (**b**) Schematic representation of a cell responding to a hypertonic stimulus. The cell loses water during the osmotic phase, leading to NKCC1 activation, uptake of ions, and obligatory water

11.4 More Than One Isoform of the Na-K-2Cl Cotransporter

In 1994, independent research teams identified cDNA sequences for two Na-K-2Cl cotransporter isoforms. The first isoform, termed NKCC1, was isolated from fish (Xu et al. 1994) and mammals (Delpire et al. 1994), appears to be ubiquitously expressed, and has been mapped to chromosome 5q23 in humans and chromosome 18 in mice. The second isoform, termed NKCC2, was isolated from mammalian kidney (Gamba et al. 1994; Payne and Forbush 1994), is expressed exclusively in the apical membrane of the thick ascending limb of Henle, and has been mapped to chromosome 15 in humans and chromosome 2 in mice. In Fig. 11.3, amino acid sequences of the two mouse isoforms have been aligned to illustrate identical (yellow), conserved (blue), and non-conserved (white) regions of the two proteins. The predominance of yellow color throughout the protein illustrates the very high (56 % overall and 76 % in the transmembrane core) degree of amino acid identity between the two cotransporters. Alternating light-pink and light-blue boxes identify

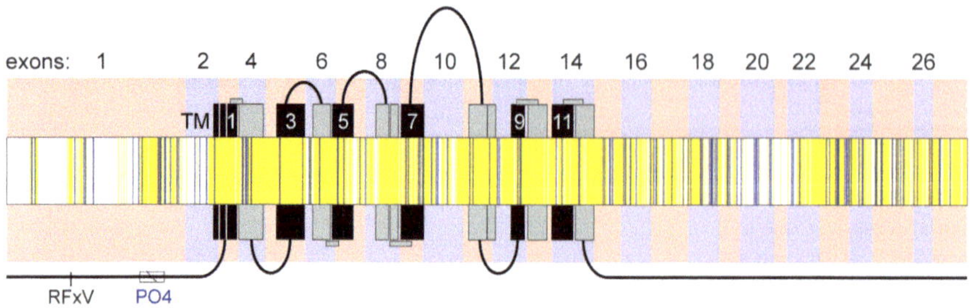

Fig. 11.3 Schematic representation of NKCC1 and NKCC2 proteins. Alignment of the full-length mouse NKCC1 and NKCC2 amino acid sequences illustrating identical (*yellow*), conserved (*blue*), and non-conserved (*white*) regions of the two proteins are superimposed on the topology of the transporter with odd-numbered transmembrane domains in black and even numbered domains in gray. Also shown in the cytosolic amino-terminal domain are conserved SPAK/OSR1 binding sites (RFxV sequences) and a region of multiple key phosphorylation sites (PO4). The exons of the *Slc12A1* (NKCC2) and *Slc12A2* (NKCC1) genes are represented in the background by alternate *pink* (*odd*) and *blue* (*even*) *boxes*

the exons of the cotransporters and the portion of the protein (NH_2-terminal, transmembrane, or COOH-terminal) encoded by each exon. There are three unique properties that functionally define Na-K-2Cl cotransport: 1) the presence of all three transported ions on the same side of the membrane, 2) transporter inhibition by loop diuretics such as bumetanide and furosemide, and 3) electrically silent stoichiometry of ion translocation. In Sect. 11.5, we will discuss the functional regulation of these cotransporter proteins and the involvement of these conserved kinase binding sites and phosphorylation sites present on the NH_2-terminal domain.

11.4.1 NKCC1

The *Slc12A2* gene encodes for a 1205 amino acid protein with 58 % similarity with the second kidney-specific isoform encoded by the *Slc12A1* gene (see Sect. 14.4.2). NKCC1 has 12 alpha-helical membrane-spanning regions flanked by large intracellular cytosolic NH_2- and COOH-terminal domains. There are also several *N*-linked glycosylation sites on a large extracellular loop between transmembrane domain 7 and 8. RNA and protein expression studies have demonstrated that NKCC1 is expressed in various tissues including the eye, stomach, heart, lung, brain, thymus, smooth and skeletal muscle, neurons, testis, colon, and red blood cells (Delpire et al. 1994; Payne et al. 1995). This ubiquitous distribution suggests multiple physiological roles beyond salt reabsorption including Cl^- and fluid secretion (Cook and Young 1989; Liedkte 1992), acid secretion (Soybel et al. 1995), cell volume homeostasis (Palfrey and O'Donnell 1992), and possibly even cell division and proliferation (Panet et al. 1994, 2000).

Contrary to the salt reabsorption observed with NKCC2 (see Sect. 14.4.2), the basolateral expression of NKCC1 in most epithelial cells results in salt secretion.

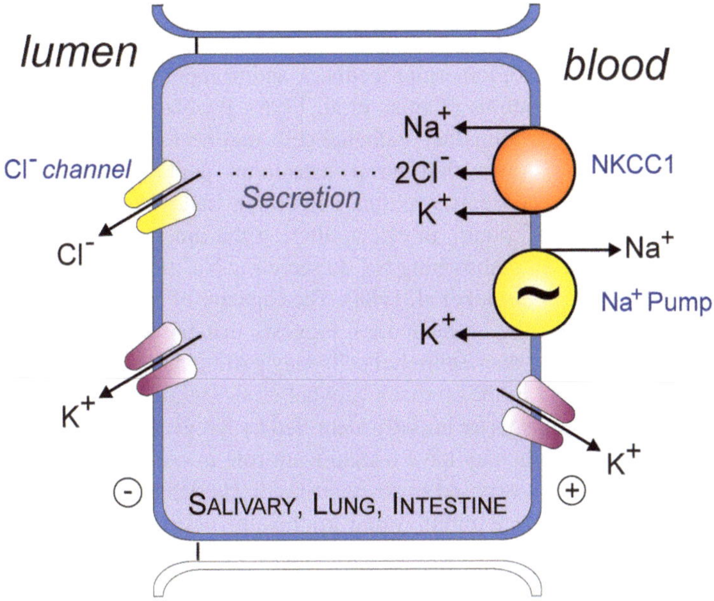

Fig. 11.4 Schematic representation of a prototypical Cl^--secreting epithelial cell. In this model, the Na^+/K^+ pump provides the energy for Cl^- transport from the basolateral (blood) side to the apical (luminal) side of the epithelium. Cl^- is secreted to the lumen at the apical membrane through a Cl^- channel (typically CFTR), and the Na-K-2Cl cotransporter on the basolateral membrane constitutes the entry pathway for Cl^- into the epithelial cell. Activation of the Cl^- channel (e.g., through cAMP) results in a drop in intracellular Cl^- which stimulates NKCC1 function. Na^+ ions entering through the cotransporter are recycled by the pump, and K^+ ions entering through the cotransporter and the pump are recycled mostly by basolateral K^+ channels

In Fig. 11.4, a model of a prototypical epithelial cell illustrates how the Na^+ gradient generated by the basolateral expression of the Na^+/K^+ pump is utilized by NKCC1 to import Na^+, K^+, and Cl^- into the cell. The increase in intracellular Cl^- above electrochemical equilibrium provides the driving force for Cl^- secretion into the lumen through apical Cl^- channels. In many epithelia, the apical Cl^- channel is a cAMP/PKA-regulated Cl^- channel, the Cystic Fibrosis Transmembrane Regulator or CFTR (see Chap. 16). Upon cAMP/PKA stimulation, Cl^- secretion is activated depleting the cells of Cl^-, which is replenished through NKCC1. This generic model is applicable to salivary, sweat, and lacrimal glands, as well as airway, gastric, and intestinal epithelial cells. NKCC1 has also been localized to several structures within the mammalian kidney. In the mouse, the cotransporter was detected in smooth muscle cells of the afferent arteriole of the glomerulus, the extraglomerular mesangium, and the inner medullary collecting duct (Kaplan et al. 1996). In the rat, expression was also found in alpha-intercalated cells of outer medullary collecting duct (Ginns et al. 1996). These cells are acid-secreting cells, and the cotransporter on the basolateral membrane might provide a pathway for Cl^- entry from the blood side, thus helping H^+ and Cl^- secretion on the apical side. Note that the species difference in expression of the cotransporter in the medulla may reflect significant physiological differences between the two species. Expression of NKCC1 in renin-producing smooth muscle cells suggests a role for the cotransporter in tubuloglomerular feedback and/or renin release, both sensitive to luminal Cl^- concentrations (Kaplan et al. 1996). Basolateral expression of the cotransporter in the collecting duct epithelial cells may serve several roles. Physiological studies have shown that atrial natriuretic peptide (ANP) stimulates upregulation of Na-K-2Cl cotransporter expression and downregulation of the Na^+ channel and Na^+/K^+ pump, in effect, altering the inner medullary collecting duct epithelial cells from reabsorbing Na^+ to secreting Na^+ (Rocha and Kudo 1990; Sonnenberg et al. 1990; Zeidel et al. 1988). The capacity of NH_4^+ to displace K^+ at the binding site within the cotransporter suggests that basolateral expression of NKCC1 in the collecting duct epithelial cells may participate in H^+/NH_4^+ secretion similar to that observed in the stomach (Soybel et al. 1995). Finally, the "volume sensitivity" of the cotransporter initially identified by Kregenow (Kregenow 1971a, b) in duck red blood cells may have a significant role in collecting duct epithelial cells which are routinely exposed to severe osmotic challenges.

The stria vascularis is a multilayered epithelium of the inner ear extending from Reissner's membrane to the spiral prominence. This unusual tissue has three primary cell types (marginal, intermediate, and basal cells) as well as intraepithelial capillaries. Basolateral expression of NKCC1 in the marginal cells, which directly face the endolymphatic compartment, does not participate in Cl^- secretion like in most other epithelia but in K^+ secretion which generates a low-sodium, high-potassium cochlear endolymph (Tasaki and Spiropoulos 1959). On the apical membrane of these cells, a K^+ channel (KCNQ1) associated with a beta subunit (KCNE1) participates in the apical secretion of the cation. Mutations in either of the subunits result in Jervell and Lange-Nielsen syndrome, a type of long QT syndrome, associated with severe, bilateral hearing loss (Tyson et al. 1997). Knockout in mice of the KCNQ1 channel (Casimiro et al. 2001) or of its beta subunit (Vetter et al. 1996) results in sensorineural deafness. Knockout of the cotransporter in mice similarly results in sensorineural deafness (Delpire et al. 1999; Flagella et al. 1999). This precise localization of the cotransporter in the stria vascularis also explains the ototoxicity caused by the use of large doses of loop diuretics (Ikeda et al. 1997; Rybak 1993).

One tissue where NKCC1 is actually localized on the apical membrane is the choroid plexus. As illustrated in Fig. 11.5, choroid plexus epithelial cells participate in the secretion of cerebrospinal fluid (CSF). Fluid movement is generated by Na^+ movement through the Na^+/K^+ pump. What is actually secreted in the ventricle is Na^+ bicarbonate. On the basolateral side, Na^+ enters through Na^+-coupled bicarbonate transporters, such as the Na^+-dependent Cl^-/HCO_3^- exchanger (NCBE) or the Na^+-HCO_3^- cotransporter (NBCn1, see Chap. 13) (Christensen et al. 2013). Location of NKCC1 on the apical membrane is counter to the movement of Na^+ as Na-K-2Cl cotransport is inward. It is also known that choroid plexus epithelial cells can participate in K^+ reabsorption. This is an important function, as the CSF K^+ content, and consequently the K^+ concentration around neurons needs to be tightly regulated. The Na^+/K^+ pump can also serve as the primary mechanism for K^+ reabsorption. However, placing the cotransporter alongside the pump has the

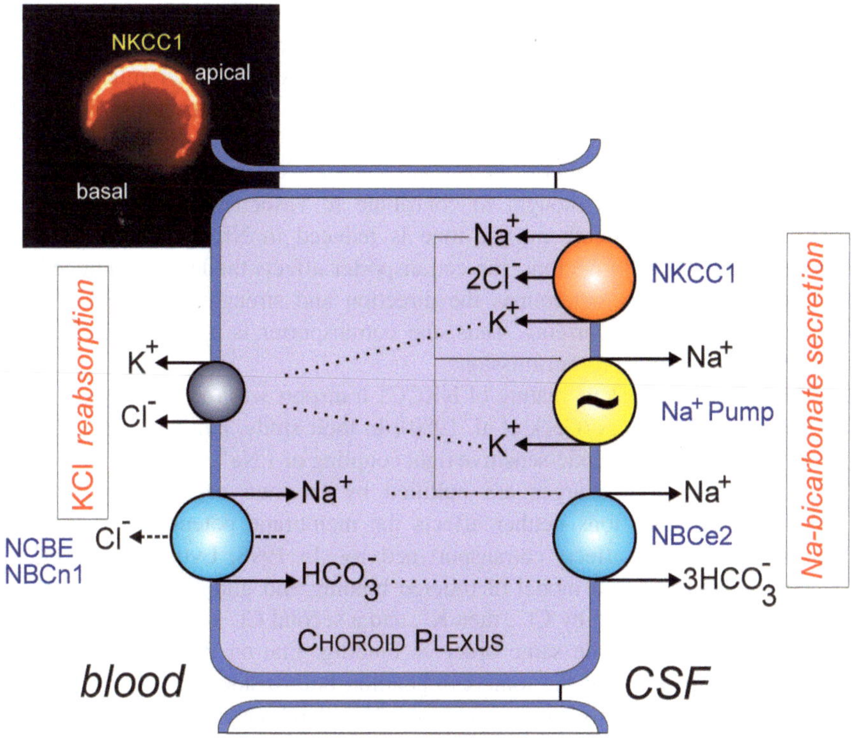

Fig. 11.5 Na-K-2Cl cotransport in choroid plexus. *Picture*: Single choroid plexus cell isolated with collagenase to keep apical versus basolateral polarity. After fixation, the cell was permeabilized and exposed to rabbit anti-NKCC1 polyclonal antibody followed by Cy3-conjugated anti-rabbit secondary antibody. *Model*: In these cuboidal epithelial cells, NKCC1 is located on the apical membrane alongside the Na$^+$/K$^+$ pump. The pump provides the energy for the secretion of Na bicarbonate from the blood side to the luminal (cerebrospinal fluid (CSF)) side. Bicarbonate enters and exits the cell through distinct Na$^+$-coupled bicarbonate transporters. The pump also provides the driving force for K$^+$ reabsorption or movement of K$^+$ from CSF to the blood. Expression of NKCC1 on the apical membrane may allow for uncoupling of K$^+$ reabsorption from Na$^+$ secretion

advantage of uncoupling the pump's obligatory exchange of Na$^+$ with K$^+$. Indeed, the Na$^+$ that exits through the pump can now be reclaimed by the cotransporter (see Fig. 11.5).

Aside from being expressed in epithelial cells, NKCC1 is also found in many other cell types, such as myocytes, vascular smooth muscle cells, and neurons. In the muscle, the cotransporter participates in conjunction with the Na$^+$/K$^+$-ATPase in the accumulation of K$^+$. This is a critical function as skeletal muscle cells store 70–75 % of the body K$^+$ (Gosmanov et al. 2003). Furthermore, a high K$^+$ concentration is needed for proper contraction, as a significant decrease in plasma K$^+$ (and therefore muscle K$^+$) caused by high doses of bumetanide is known to produce musculoskeletal pain, cramping, and/or muscle weakness (Howard and Dunn 1997; Vaduganathan et al. 2013). As chloride conductances in muscle cells are critical in stabilizing the membrane potential at the resting level, NKCC1 might also be involved in this stabilization by accumulating Cl^- above electrochemical potential equilibrium (Aickin et al. 1989). In addition, NKCC1 activity is also critical in helping muscle cells maintain their water content (Gosmanov et al. 2003). In vasculature, NKCC1 is thought to contribute to vasoconstriction (Akar et al. 1999), as vascular smooth muscle tone is reduced in NKCC1 knockout mice (Meyer et al. 2002). In neurons, the cotransporter affects the level of intracellular Cl^- which ultimately determines the direction and strength of GABA- and/or glycine-mediated Cl^- currents. Thus, the cotransporter is a key determinant of inhibitory synaptic neurotransmission.

The "electrically silent" nature of NKCC1 transport was recognized very early by Geck and coworkers (Geck et al. 1980). In their study, pulse-response experiments revealed a furosemide-sensitive tight coupling of 1 Na^+, 1 K^+, and 2 Cl^- ions. As the two positive charges are nullified by the two negative charges, the cotransport of these ions neither affects the membrane potential nor does the membrane potential affects cotransport activity. In 1998, Lytle and coworkers (Lytle 1998) proposed a model of ordered binding and gliding symmetry where Na^+ binds first, followed by Cl^-, then K^+, and a second Cl^- and on the inside, the ions were released in the same order as binding: first-on first-off (Fig. 11.6a). Through manipulation of the ionic composition both inside and outside the cell, they found that the tightly coupled ratio of 1 Na^+, 1 K^+, and 2 Cl^- was maintained regardless of the internal and external ion composition. They also observed two partial reaction cycles: Na^+/Na^+ exchange in cells with high $[Na^+]_i$ and K^+/K^+ exchange in cells with high $[K^+]_i$ (Lytle 1998). Although their model adequately explained the 1 Na^+, 1 K^+, and 2 Cl^- stoichiometry of the cotransporter, their first-on first-off gliding symmetry model does not explain why Na^+/Na^+ exchange would be dependent on external K^+ or why K^+/K^+ exchange would require internal Na^+ but not external Na^+.

We and others have observed atypical NKCC transport stoichiometries in many cell types which cannot be resolved by first-on first-off gliding symmetry (Canessa et al. 1986; Gagnon and Delpire 2010; Hall and Ellory 1985; Lytle 1998; Orlov et al. 1996; Russell 1983). In 2011, we proposed a different model for ion binding based on velocity equations for K^+ influx under both rapid-equilibrium assumptions and combined equilibrium and steady-state assumptions. Our model also has ions binding in a preferred order but with Cl^- ion binding first, followed by a Na^+, the second Cl^-, and finally a K^+ ion (Fig. 11.6b). Upon binding, the fully loaded cotransporter undergoes a conformational change ("translocates") and releases the ions inside the cell in the reverse order to which they bound (i.e., K^+, Cl^-, Na^+, and Cl^-). In rapid-equilibrium kinetics, the rate-limiting steps are the translocation steps, with the "fully loaded" cotransporter translocating much faster than the "empty" cotransporter. In this model, some partial reactions (i.e., slippage) are also permitted, while preserving transport electroneutrality. Rapid translocation of fully loaded transporter and slippage steps can explain why under certain

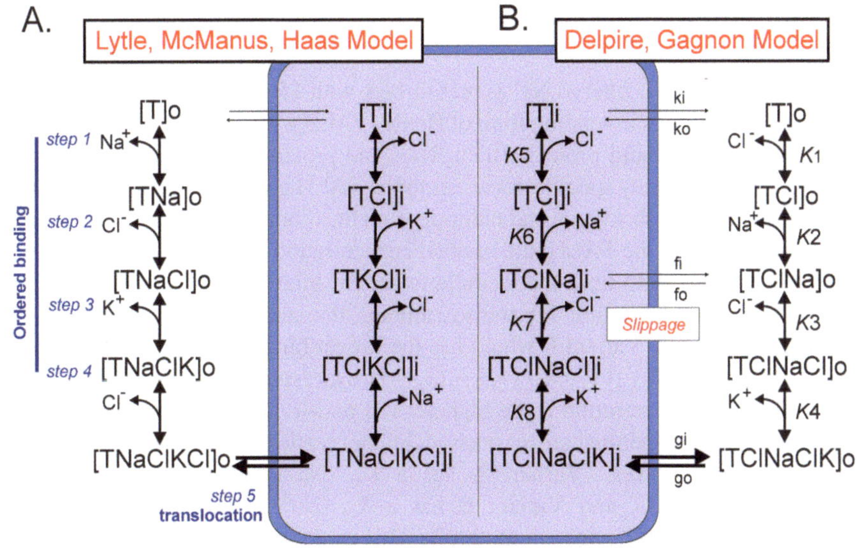

Fig. 11.6 Kinetic models of NKCC1 transport. (**a**) Model proposed by Lytle et al. (1998). The transporter outside loads ions in an *ordered* fashion with Na^+ binding first, followed by Cl^-, K^+, and then the second Cl^- (steps 1–4). The fully loaded transporter then translocates (step 5) and releases ions on the other side on a first-on first-off basis. (**b**) Model proposed by Delpire and Gagnon (2011) where the transporter outside also loads ions in an *ordered* fashion but with Cl^- binding first, followed by Na^+, a second Cl^-, and then K^+. After translocation, the release occurs in a more traditional fashion. Binding rate constants for each reaction are indicated by labels above and below the directional *arrows*. Computer simulations performed in Delpire and Gagnon (2011) provided several sets of rate constants, depending upon the number of transporters inserted in the membrane. Note that the translocation of the loaded transporter is much faster (*thicker arrows*) than the translocation of the empty transporter (*thinner arrows*). Kinetics allow for some partial reactions to occur (slippage step)

conditions, hypertonicity, for instance, the transporter unidirectionally moves more K^+ than Cl^- and more Cl^- than Na^+. In these conditions, the preferred mode of transport is K^+/K^+ exchange, followed by K-Cl cotransport (Na^+ is not released), and finally Na-K-2Cl cotransport (Delpire and Gagnon 2011).

In the Lytle study (Lytle 1998), they measured endogenous transport activity in duck red blood cells, whereas in our study (Delpire and Gagnon 2011), we measured transport activity of mouse NKCC1 heterologously overexpressed in *Xenopus laevis* oocytes. In our study, we altered the tonicity of the extracellular solutions, whereas the Haas study replaced transportable cations (Na^+ and K^+) with less-transportable cations (Li^+ and Rb^+), respectively. Our study used radiolabeled $^{22}Na^+$, $^{86}Rb^+$, and $^{36}Cl^-$ as tracers, whereas, the Haas study used a combination of radiolabeled tracers ($^{22}Na^+$, $^{86}Rb^+$, and $^{36}Cl^-$) and ion measurements from dilute perchloric acid extracts via air-acetylene flame spectrophotometry. Any combination of these differences could explain the two divergent models, and more experiments will be necessary to distinguish which better characterizes actual transport activity.

11.4.2 NKCC2

In man, 99.8 % of the filtered Na^+ is reabsorbed, with 15–25 % of this reabsorption occurring in the thick ascending limb of Henle (TALH). The *Slc12A1* gene encodes for an 1100 amino acid protein which RNA and protein expression studies have demonstrated as kidney specific, localizes to the TALH epithelial apical membrane, and serves to reabsorb sodium and chloride from the tubule lumen. However, as salt is reabsorbed along the TALH, the luminal concentration of Na^+ and Cl^- decreases approximately fivefold from the medulla to cortex. Interestingly, alternative splicing of the exons encoding for transmembrane domain 2 has resulted in three variants of NKCC2. Variant F which has the lowest binding affinity (K_m (Na^+) = 66.72 ± 5.8 and a K_m (Cl^-) = 111.3 ± 13.4) is expressed in the medullar TALH where the Na^+ concentration is the highest. The mid- (variant A) and high-affinity (variant B) cotransporters are expressed in the cortical TALH where the Na^+ concentration is lowest. Variant A has a K_m (Na^+) = 16.45 ± 1.9 and a K_m (Cl^-) = 44.65 ± 3.87), and Variant B has a K_m (Na^+) = 20.65 ± 2.4 and a K_m (Cl^-) = 8.95 ± 1.3) (Giménez et al. 2002). This expression pattern results in maximal reabsorption of luminal salt and water (see Fig. 11.7a). Similar to NKCC1, the

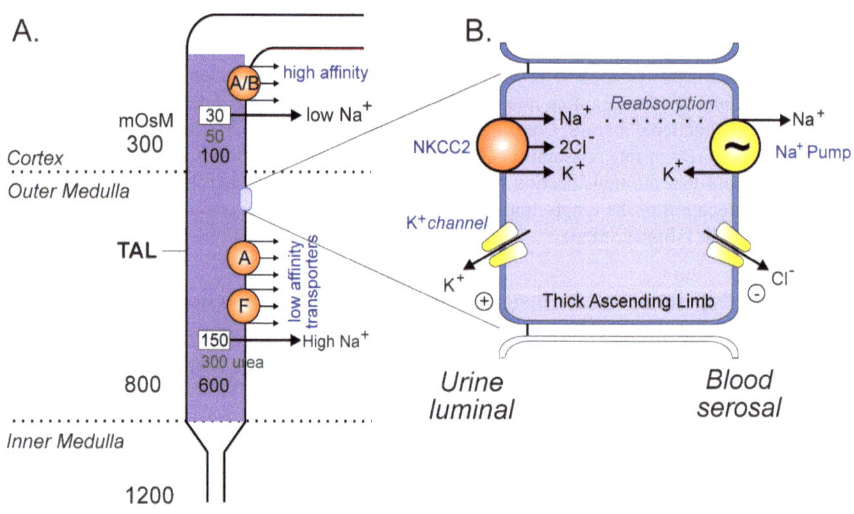

Fig. 11.7 NKCC2 function in the kidney. (**a**) Schematic representation of the thick ascending limb of Henle (TALH) showing its location in the cortex and medulla. In inner medulla, the osmolarity is elevated, whereas in superficial cortex, the osmolarity is equivalent to the blood. Therefore, there is a gradient of Na^+ in the TALH from inner medulla to the cortex. Represented are three variants of the Na-K-2Cl cotransporter, NKCC2. In the medulla where the Na^+ concentration is high, the low (variant F)- and mid (variant A)-affinity cotransporters are expressed, whereas in the cortex where the Na^+ is low, the high-affinity cotransporter (variant B) is expressed. (**b**) Schematic representation of a TALH epithelial cell with expression of NKCC2 and K^+ channel (ROMK) at the urine or luminal membrane and the Na^+/K^+ pump and Cl channels (CLCKA & B) at the blood or serosal membrane

driving force for apical NKCC2 cotransport activity is provided by the basolateral Na^+/K^+ pump. However, now combined with Cl^- channels on the basolateral membrane, apical NKCC2 serves to absorb Na^+ and Cl^- into the blood. The K^+ ions transported by NKCC2 are recycled back to the lumen through ROMK, the renal outer medullary K^+ channel (see Fig. 11.7b and Chap. 19). Human mutations in NKCC2 which disrupt cotransporter function are responsible for type I Bartter syndrome, a rare inherited disorder characterized by prenatal onset, low potassium levels (hypokalemic), low blood pressure (hypotension), increased blood pH (alkalosis), excessive urination (polyuria), and excessive amounts of calcium in the urine (hypercalciuria) (Bartter et al. 1962). More than 70 mutations in the *Slc12a1* gene have been found in patients suffering from type I Bartter syndrome (Simon et al. 1996). Both homozygous mutations (same mutation on both alleles) and compound heterozygous mutations (each allele has a different mutation) will cause this salt-wasting disorder to become clinically relevant. Symptoms of type I Bartter syndrome often appear before birth; however, patients can also manifest the disorder later in life as a result of residual NKCC2 function (Pressler et al. 2006).

Along with NKCC2 in the TALH, Na^+ reabsorption in the distal kidney tubule is accomplished by the thiazide-sensitive Na-Cl cotransporter (NCC, see Chap. 12) and the amiloride-sensitive epithelial Na^+ channel (ENaC, see Chap. 18). Each of these transport mechanisms is a major target for pharmacological modification. Proper water and salt retention involves these transport mechanisms along with activation of the renin-angiotensin-aldosterone system, a secondary target for antihypertensive drugs. In Fig. 11.8, the inhibitory actions of furosemide, thiazide, and amiloride determine the amount of fluid reabsorbed into the blood and ultimately blood pressure. The juxtaglomerular cells in the macula densa (MD) match glomerular filtration rate (GFR) with tubular reabsorption through a tubuloglomerular feedback (TGF) mechanism. Renin release into the blood from the MD converts angiotensinogen to angiotensin I (Ang I) which is then converted to angiotensin II (Ang II) by angiotensin-converting enzyme (ACE). Ang II increases the blood pressure through several mechanisms: 1) increase in sympathetic activity of the heart; 2) stimulation of aldosterone release from the adrenal gland which increases tubular Na^+, Cl^- reabsorption, K^+ excretion, and water retention; 3) arteriolar vasoconstriction; and 4) stimulation of antidiuretic hormone release from the pituitary which increases water absorption in the collecting duct. Beta blockers acting on adrenergic receptors, ACE inhibitors preventing the conversion of Ang I to Ang II, spironolactone acting on mineralocorticoid (aldosterone) receptor, and the loop diuretics are used to regulate salt and water reabsorption in the kidney tubule (Fig. 11.8).

11.5 NKCC Activity Is Regulated by Phosphorylation

Activity of the Na-K-2Cl cotransporter is increased by phosphorylation and decreased by dephosphorylation. Consensus exists as to the major phosphorylation event affecting the intrinsic transport turnover of carriers already inserted in the

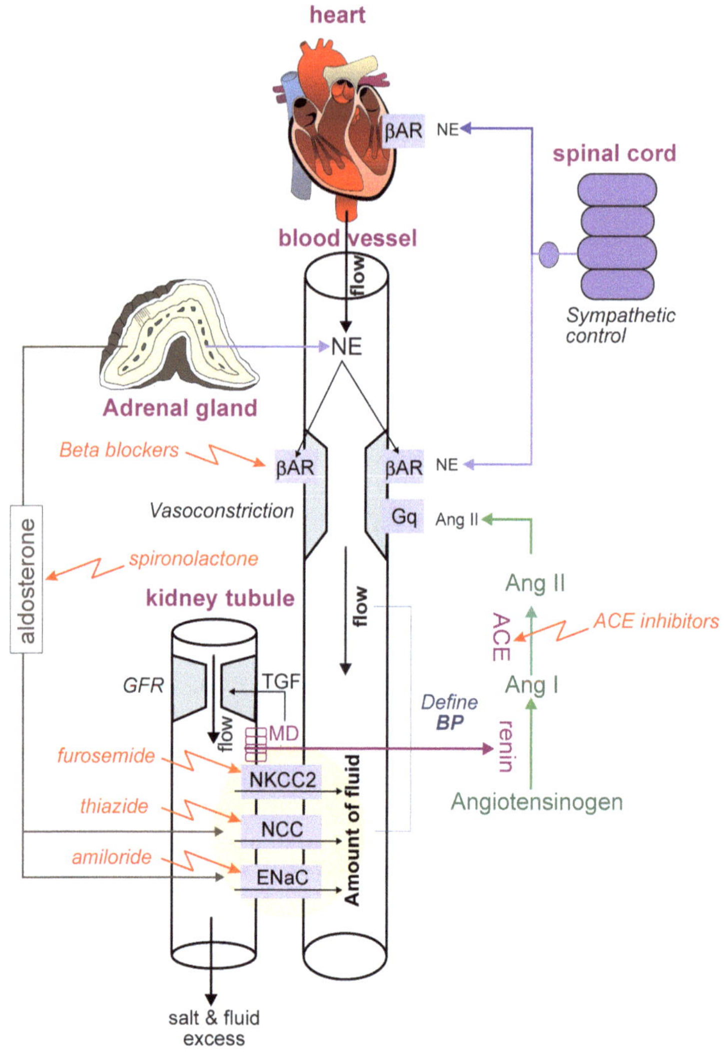

Fig. 11.8 Schematic representation of major targets of antihypertensive drugs. The Na^+ reabsorptive transport mechanisms (NKCC2, NCC, and ENaC) in kidney tubules constitute major drug targets. The renin-angiotensin-aldosterone system, which affects both blood vessel contraction and Na^+ transport, constitutes a second site of drug targets. Both angiotensin-converting enzyme (ACE) and aldosterone receptor are targets of antihypertensive drugs. By acting on adrenergic receptors of the heart and blood vessels, beta blockers decrease blood flow and contractility of vessels. There are two major sources of catecholamines: the adrenal gland and postganglionic sympathetic fibers. The flow of the urine, which is regulated by glomerular filtration rate (GFR) and renin-angiotensin-mediated tubuloglomerular feedback (TGF) mechanism, also affects the rate of Na^+ reabsorption. *BP* blood pressure, *MD* macula densa, *GFR* glomerular filtration rate, *TGF* tubuloglomerular feedback, *ACE* angiotensin-converting enzyme, *Ang* angiotensin, *NE* norepinephrine, and ***β***AR, beta-adrenergic receptors

plasma membrane instead of affecting trafficking. This consensus does not mean that other phosphorylation events cannot affect the trafficking or recycling of transporters to or from the plasma membrane. This will be briefly discussed at the end of this section.

Original evidence that NKCC is activated by phosphorylation comes from studies performed by John Russell in 1988 in squid giant axon (Altamirano et al. 1988). The study showed that ATP depletion causes a marked decrease in Na-K-Cl transport activity. While this decrease could have been due to the collapse of the Na^+ gradient generated by the ATP-driven Na^+/K^+ pump, the fact that it was diminished by the addition of protein phosphatase inhibitors (such as orthovanadate or fluoride) indicated that instead, NKCC phosphorylation was likely affected by ATP depletion. This was later confirmed in studies performed in shark rectal gland by Chris Lytle and Biff Forbush (Lytle and Forbush 1992a, b). They demonstrated that a specific residue in the cytosolic NH_2-terminal tail of the transporter acquired a phosphate residue upon cotransporter activation. In fact, we know now that multiple neighboring threonine and serine residues are phosphorylated upon cotransporter activation (Fig. 11.9). How phosphorylation of the NH_2-terminus leads to conformational changes in the protein core, consequently increasing transport activity is currently not known. Moreover, whether specificity exists in the multiple phosphorylation sites that were identified is also unknown.

Identification of the kinases involved in the activation of the cotransporter was made 10 years later (Piechotta et al. 2002, 2003; Dowd and Forbush 2003). There are still many details that are not resolved, mostly concerning the precise role of each kinase identified as regulators of the Na-K-2Cl cotransporter. We will start by addressing the current consensus. Upstream of the phosphorylation sites in both NKCC1 and NKCC2 reside a short sequence that constitutes a binding site for serine-threonine kinases. The minimal amino acid sequence is an arginine residue, followed by a phenylalanine, any residue, and then a valine or isoleucine (RFxV/I). This site is found once in the cytosolic amino-terminal domain of NKCC2 and twice in the amino-terminal domain of NKCC1. The kinases that bind to the RFxV/I site are members of a relatively large family of serine-threonine kinases: the mammalian Ste20p-like kinase family (Delpire 2009). SPAK (STE20/SPS1-related proline-/alanine-rich kinase, also known as PASK and STK39) and OSR1 (oxidative stress response 1, also known as OxSR1) are the two members regulating the cotransporters. OSR1 is the original kinase, as it is found from protist to human, whereas SPAK originated from gene duplication late during vertebrate evolution (Gagnon and Delpire 2012). The two kinases possess an NH_2-terminal catalytic domain, followed by a COOH-terminal regulatory domain, whose last 90 residues form a protein fold that accommodates or binds RFxV/I peptides (Austin et al. 2014; Villa et al. 2007). Thus, constitutively active SPAK and OSR1 when overexpressed in heterologous expression systems, like *Xenopus laevis* oocytes, significantly activate both Na-K-2Cl cotransporters. Interestingly, when overexpressed as native kinases in oocytes, they fail to activate the cotransporters and require activation by an upstream With No lysine (K) or WNK kinase. The C-terminal domain of WNK kinases also contains RFxV/I sequences which make them binding partners of SPAK/OSR1, thus facilitating their activation by these WNK kinases (Fig. 11.10). A recent study (Ponce-Coria et al. 2014) has indicated

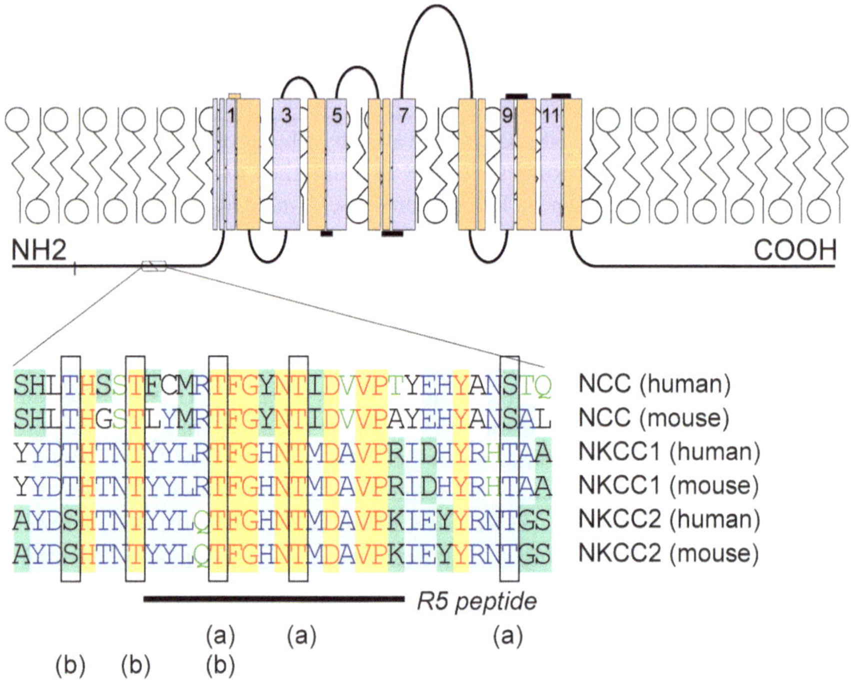

Fig. 11.9 Phosphorylation "hot spot" in Na^+-dependent cation-chloride cotransporters. Depicted is an amino acid alignment of human and mouse sequences for NCC, NKCC1, and NKCC2 of a small portion of the cytosolic NH_2-terminal tail of the cotransporters. Note the presence of five conserved threonine/serine residues. *Highlighted by boxes* are phosphorylation sites identified by mass spectrometry in two studies: (**a**) Darman and Forbush (2002) and (**b**) Vitari et al. (2006). *Horizontal bar* delineates the peptide epitope of a phospho-specific antibody (anti-phospho NKCC1 R5 antibody) created by the Forbush Laboratory (Flemmer et al. 2002)

that WNK kinases also possess a portion of the binding domain of SPAK, which means that this upstream kinase can also bind directly to the cotransporters (Fig. 11.10). Figure 11.10 also highlights an adaptor protein (Cab39) which has been shown to facilitate Ste20 and WNK kinase activation, as well as the serum kinase SGK1 as a possible upstream regulator of WNK kinases. Finally, recent structural studies have uncovered the presence of a Cl^- ion within the crystal structure of the WNK1 catalytic domain, indicating a possible role for the anion in modulating kinase function (Piala et al. 2014).

As kinases that directly bind and phosphorylate NKCC1 and NKCC2, SPAK and OSR1 are likely to be essential components of cotransporter function. Thus, disruption of their expression should have significant physiological consequences. Global knockout mice were generated for both kinases. While complete absence of OSR1 results in embryonic lethality, the absence of SPAK results in a viable mouse with no overt phenotype (Delpire and Gagnon 2008). However, upon closer

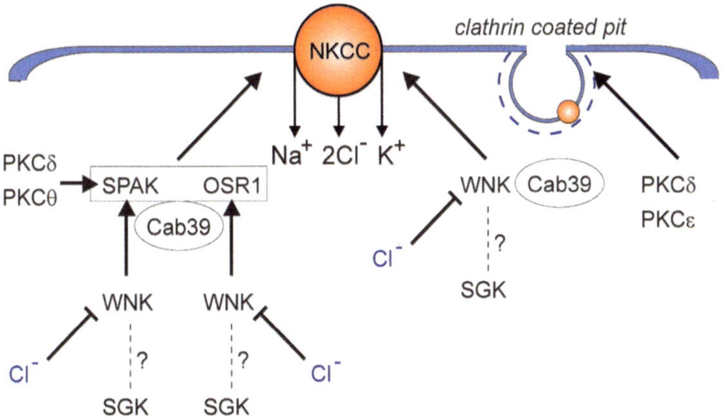

Fig. 11.10 Illustration of phosphoregulation of NKCC by WNK-SPAK pathway. Depiction of the signaling cascade leading to NKCC activation with two members of the Ste20p-like family of protein kinases (SPAK and OSR1) anchoring to the cytosolic NH_2-terminal tail and phosphorylating the cotransporter. Upstream of the Ste20p kinases are WNK kinases that can, through a similar binding mechanism (RFxV/I motif), anchor and activate SPAK and OSR1. In addition, an adaptor protein (Cab39) is depicted as a facilitator of kinase activation. Upstream regulatory mechanisms of WNK kinases, such as the serum glucocorticoid-regulated kinase (SGK) and intracellular Cl^-, are also shown. WNK kinase also possesses a portion of the binding domain of SPAK, which means that this upstream kinase can bind directly to the cotransporter. Also depicted is the PKC-induced clathrin-dependent internalization of the cotransporter

analysis, SPAK knockout mice demonstrate a renal phenotype that resembles the loss of NCC function in the distal convoluted tubule: salt-sensitive hypotension accompanied by electrolyte imbalance, such as hypokalemia, hypomagnesemia, and hypercalciuria (Grimm et al. 2012; McCormick et al. 2011; Yang et al. 2010). In contrast, targeted deletion of OSR1 in the kidney led to a viable mouse with a phenotype resembling, in this case, the loss of NKCC2 function. Indeed, the mice displayed impaired Na^+ reabsorption in the TALH on a low-Na^+ diet with significant reduction in NKCC2 phosphorylation and blunted response to furosemide (Lin et al. 2011). Thus, whereas these two kinases fulfill the same function in heterologous expression systems, they seem to have distinct functions in the kidney: OSR1 mainly regulating NKCC2 and SPAK mostly regulating NCC. In agreement with SPAK mostly regulating NCC, genetic variations in the human SPAK gene have been linked to hypertension (Wang et al. 2009). In other tissues, such as primary afferent sensory neurons, the two kinases seem to be complementing each other, as disruption of one kinase reduces by half NKCC1 function (Geng et al. 2009).

Mutations in WNK1 and WNK4 have been identified in the human population causing a disorder called Gordon syndrome or pseudohypoaldosteronism type II (PHAII). Features of the disorder are opposite to Gitelman's, indicating an increase instead of a decrease in NCC function. It is now recognized that these mutations prevent degradation of WNK1 or WNK4 by the cullin-3/Kelch-like 3 ubiquitinylation-mediated protein degradation pathway. In fact, mutations in both proteins also result in Gordon syndrome (Boyden et al. 2012). The consequence of all these mutations is an increased abundance of WNK proteins, leading to increased function of the Na-Cl cotransporter, and increased distal Na^+ reabsorption. This phenotype has been recapitulated in the mouse (Yang et al. 2007). Furthermore, disruption of the WNK4 gene in mice results in a mild Gitelman phenotype, consistent with WNK4 being a positive modulator of NCC function (Castañeda-Bueno et al. 2012; Ohta et al. 2008). Much less effort has been placed on determining whether manipulation of the WNK kinases also affects the function of the Na-K-2Cl cotransporters.

Aside from the major phosphorylation event that leads to stimulation of the cotransporters, there is evidence that PKC mediates internalization of NKCC1 (Mykoniatis et al. 2010; Tang et al. 2010) Fig. 11.10). Using a GFP-tagged NKCC1 cotransporter expressed in MDCK cells, it was shown that activation of PKC with phorbol ester leads to internalization of NKCC1 through a clathrin-dependent endocytic pathway (Mykoniatis et al. 2010). Using ShRNA delivered with adenovirus, PKCδ and PKCε were identified as the isotopes modulating cell surface expression of the cotransporter (Tang et al. 2010).

As discussed, in Sect. 11.4.1, activation of PKA through cAMP also leads to an increase in NKCC1 stimulation. However, the effect is indirect and mediated by the drop in the intracellular Cl^- concentration that follows the cAMP-opening of CFTR (Fig. 11.4). This effect not only pertains to the cAMP stimulation of colonic NKCC1 in mammals but also to the cAMP/PKA-mediated stimulation of NKCC1 in the shark rectal gland (Lytle and Forbush 1992a, b).

11.6 Gene Structure, Transporter Family, and Superfamily

We mentioned in Sect. 11.4 that human NKCC1 and NKCC2 are the products of two distinct genes located on chromosomes 5 and 15, respectively. Analysis of animal genomes revealed that the *Slc12A2* gene originated first since NKCC1 is found in bacteria and protists (e.g., NCBI accession numbers: WP_008693040 and EFW46279), whereas *Slc12A1* originated from gene duplication 400–500 million years ago in early vertebrate evolution, as NKCC2 is found in cartilaginous fish (NCBI accession number: AAM749868). Examination of their genetic exons structure revealed that despite this very significant length of time since duplication, the exons have been well conserved (Fig. 11.11). Indeed, 21 out of 27 exons are conserved (including two exons which have a single codon added). Therefore, there are only six exons which are somewhat different between the two genes: exons 1 and 2 for *Slc12A1* (NKCC2) are different from exon 1 in *Slc12A2* (NKCC1). Despite exon 3 of *Slc12A1* and exon 2 of *Slc12A2* starting differently, they end up with conserved sequences. Another difference can be seen in exons 21

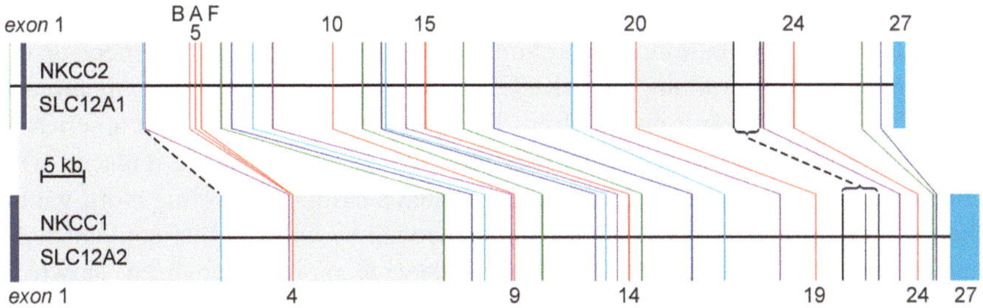

Fig. 11.11 Alignment of the human Slc12A1 and Slc12A2 genes. The gene structure with all 27 exons spanning 100 kb genomic DNA is displayed. The vast majority of exons are conserved with 21 out of 27 exons having identical sizes and coding the same protein fragments. Exons 1–3 which encode the NH_2-terminal tails of both cotransporters show the lowest degree of conservation. The 3′-untranslated region of the transcript, encoded by exon 27, is typically gene specific and differs significantly between the two genes. Exon 5 in the *Slc12A1* gene is triplicated to give rise to the three splice variants: NKCC2F, NKCC2A, and NKCC2B. A short 48 bp alternatively spliced cassette in exon 21 is unique to the *Slc12A2* gene. Large genomic regions present between exons are depicted by a gray background

and 22 for *Slc12A1* which correspond to exons 20, 21, and 22 in *Slc12A2* with very little conservation, if we exclude the terminal part of exon 22. Finally, exon 27 encodes the end of the open reading frame of both cotransporters, followed by gene-specific 3'-untranslated region. This high degree of overall exon (Fig. 11.11) and protein (Fig. 11.3) conservation reflects the evolutionary pressure that was imposed to keep intact the vital function of the two cotransporters. At the end, there are very few functional differences between the two cotransporters; their strict stoichiometry with three (even four if we include the second Cl^-) distinctions require tertiary, secondary, and primary structure integrity. To this note, we can add the surprising finding that while sequences of the 5' end of the N-terminal tails are highly divergent, the transporters have managed to retain a SPAK/OSR1 binding motif.

Very little is known about transcriptional promoters for the two genes. While the promoter of *Slc12A2*/NKCC1 contains features of housekeeping genes with a TATA-less promoter and several activator protein-2 (AP-2) and specificity protein 1 (SPS1) putative binding sites, the promoter of *Slc12A1*/NKCC2 contains a TATA box and seems to be under the control of hepatocyte nuclear factor (Igarashi et al. 1996). There is no indication that either transcript can be generated from alternative promoters. In contrast, as already mentioned, NKCC2 exists in three variants, due to alternative splicing of exon 5, whereas NKCC1 exists in two variants, one with and one without exon 21 (Randall et al. 1997). Exon 5, as shown in Fig. 11.3, encodes for the second transmembrane domain and the sequence variations that affect the cotransporters affinity for ions (Giménez et al. 2002).

As mentioned earlier, the membrane targeting of the two NKCC isoforms is directly related to their different physiological roles. In an elegant set of experiments, Biff Forbush and coworkers generated chimeras of the two isoforms to identify what motifs within the two proteins mediate their basolateral (NKCC1) versus apical (NKCC2) membrane expression (Carmosino et al. 2008).

Transposition of the cytosolic amino-terminus of NKCC1 onto NKCC2 did not alter the apical distribution of the chimeric protein, whereas replacement of the cytosolic carboxy-terminus of NKCC2 onto NKCC1 altered the membrane targeting of the chimeric protein from basolateral to apical. Through a series of chimeras where regions of the carboxy-terminus of NKCC1 were replaced with corresponding regions of NKCC2, they identified a basolateral sorting motif within the carboxy-terminus of NKCC1 which is encoded by exon 21 that contains a di-leucine motif (Carmosino et al. 2008). Di-leucine motifs have been shown to mediate endocytosis and basolateral sorting of membrane receptors in epithelial cells (Hunziker and Fumey 1994). Both the alternative NKCC1 isoform, which lacks exon 21, and the apically targeted NKCC2 lack this di-leucine motif. Whether the alternatively spliced NKCC1 isoform, which is abundantly expressed in the brain, is localized to the apical membrane in choroid plexus epithelial cells due to the absence of this di-leucine motif is still unknown. However, the absence of this motif does explain the localization of NKCC2 on the apical membrane of the thick ascending limb of Henle (Carmosino et al. 2008).

The two Na-K-2Cl cotransporters are also closely related to the thiazide-sensitive Na-Cl cotransporter (NCC). Together, these transporters constitute the Na^+-transporting branch of the cation-chloride cotransporters and share high homology to four Na^+-independent K-Cl cotransporters (KCC1-KCC4; (Gagnon and Delpire 2013). All these transporters belong to the family of *Slc12A* transporters which itself shares ancestry with many other families of transport proteins which share the basic structure of an inverted repeat of five transmembrane domains (TM1-5 and TM6-10), followed by two additional transmembrane α-helical segments located either at their N- and/or C-termini (Wong et al. 2012). Because of high conservation at the structure level, the three-dimensional resolution of related transporter structures (e.g., a bacterium sodium-galactose symporter PBD#:3dh4 and a bacterium glutamate/GABA antiporter PDB#:4dji) allows for modeling of NKCC proteins. Using the Phyre2 protein fold recognition server (www.sbg.bio.ic.ac.uk/phyre2/), we drew a model of the first ten transmembrane domains of NKCC1 (Fig. 11.12). We can see that specific helices run parallel to each other (TM1 and 6 in blue, TM2 and 7 in red, TM3 and 8 in green, TM4 and 9 in white). Considering the distance that separates LeuTAa, a bacterial homologue of a neurotransmitter sodium symporter, it is extraordinary that four out of six residues in TM9 and that seven out of ten residues in TM10, shown to promote LeuTAa dimerization (Yamashita et al. 2005), are conserved in NKCC1. In a 2012 study, Biff Forbush and coworkers identified several pore-lining residues within TM3 (green helix in Fig. 11.12) as functionally important in ion coordination (Tyr-383, Ala-379, Ala-375, Asn-376, Ile-368, and Gly-369), loop diuretic binding (Phe-372 and Ile-371), and complete bumetanide insensitivity (Met-382) (Somasekharan et al. 2012). Conserved identity within the transmembrane domains of the cation-chloride cotransporter family suggests that future homology studies may help understand which pore-lining residues define the specific characteristics of ion transport in both the Na^+-dependent and Na^+-independent cotransporters.

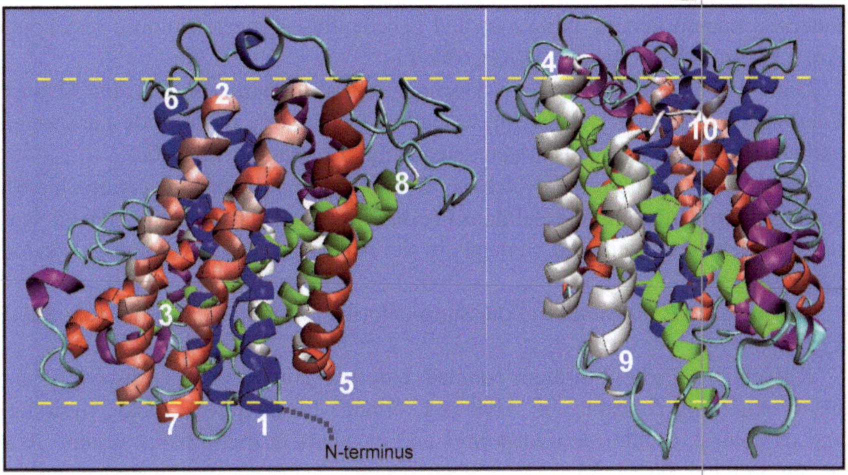

Fig. 11.12 Three-dimensional model of the transmembrane core of NKCC1. Three-dimensional resolution of two related transporter structures: a bacterium sodium-galactose symporter (PDB#:3dh4) and a bacterium glutamate/GABA antiporter (PDB#:4dji) were used to model the transmembrane core of NKCC1. Antiparallel transmembrane helices from the first and second halves of the transmembrane core are identically colored. The structure is rotated 180° (*right*) to visualize the helices visible from the back. The model was created using the Phyre2 software (see text)

11.7 Summary

Over the past 50 years, physiologists have used functional, biochemical, and molecular biological experiments to go from the initial Na^+/K^+ pump II hypothesis to the conceptual breakthrough of secondary active transport where the energy of the Na^+ gradient generated by the Na^+/K^+ pump is used by the Na-K-2Cl cotransporter in multiple organ systems. Expression of the older of the two isoforms, NKCC1, in various tissues suggests multiple physiological roles including Cl^- and fluid secretion, acid secretion, cell volume homeostasis, and possibly cell division and proliferation. The kidney-specific expression of three variants of NKCC2 to optimize salt reabsorption demonstrates its vital role in renal homeostasis and regulation of systemic blood pressure. The evolutionary conservation of these cotransporters from protists to humans confirms their vital role in cellular and whole-organism physiology.

11.8 Selected Readings

Gagnon KB, Delpire E (2013) Physiology of Slc12 transporters: lessons from inherited human genetic mutations and genetically-engineered mouse knockouts. Am J Physiol Cell Physiol 304(8):C693–714

Gamba G (2009) The sodium-dependent chloride cotransporters. In: Alvarez-Leefmans FJ, Delpire E (eds) Physiology and pathology of chloride transporters and channels in the nervous system. Academic, London, pp 307–332

Gamba G, Garbarini N, Delpire E (2009) Regulation of cation-chloride cotransporters. In: Alvarez-Leefmans FJ, Delpire E (eds) Physiology and pathology of chloride transporters and channels in the nervous system. Academic, London, pp 357–382

Russell JM (2000) Sodium-potassium-chloride cotransport. Physiol Rev 80(1): 212–276

Russell JM (2009) Sodium-coupled chloride cotransporters: discovery and newly emerging concepts. In: Alvarez-Leefmans FJ, Delpire E (eds) Physiology and pathology of chloride transporters and channels in the nervous system. Academic, London, pp 17–26

References

Aickin CC, Betz WJ, Harris GL (1989) Intracellular chloride and the mechanism for its accumulation in rat lumbrical muscle. J Physiol 411:437–455

Akar F, Skinner E, Klein JD, Jena M, Paul RJ, O'Neill WC (1999) Vasoconstrictors and nitrovasodilators reciprocally regulate the Na^{+}-K^{+}-$2Cl^{-}$ cotransporter in rat aorta. Am J Physiol 276:C1383–C1390

Altamirano AA, Breitwieser GE, Russel JM (1988) Vanadate and fluoride effects on Na-K-Cl cotransport in squid giant axon. Am J Physiol 254:C582–C586

Austin T, Nannemann DP, Deluca SL, Meiler J, Delpire E (2014) In silico analysis and experimental verification of OSR1 kinase - peptide interaction. J Struct Biol 187:58–65

Bartter FC, Pronove P, Gill JRJ, Maccardle RC (1962) Hyperplasia of the juxtaglomerular complex with hyperaldosteronism and hypokalemic alkalosis. A new syndrome. Am J Med 33:811–828

Boyden LM et al (2012) Mutations in kelch-like 3 and cullin 3 cause hypertension and electrolyte abnormalities. Nature 482:98–102

Canessa M, Brugnara C, Cusi D, Tosteson DC (1986) Modes of operation and variable stoichiometry of the furosemide- sensitive Na and K fluxes in human red cells. J Gen Physiol 87: 113–142

Carmosino M, Giménez I, Caplan M, Forbush B (2008) Exon loss accounts for differential sorting of Na-K-Cl cotransporters in polarized epithelial cells. Mol Biol Cell 19:4341–4351

Casimiro MC, Knollmann BC, Ebert SN, Vary JC Jr, Greene AE, Franz MR, Grinberg A, Huang SP, Pfeifer K (2001) Targeted disruption of the Kcnq1 gene produces a mouse model of Jervell and Lange-Nielsen syndrome. Proc Natl Acad Sci USA 98:2526–2531

Castañeda-Bueno M, Cervantes-Pérez LG, Vázquez N, Uribe N, Kantesaria S, Morla L, Bobadilla NA, Doucet A, Alessi DR, Gamba G (2012) Activation of the renal Na^{+}:Cl^{-} cotransporter by angiotensin II is a WNK4-dependent process. Proc Natl Acad Sci USA 109:7929–7934

Christensen HL, Nguyen AT, Pedersen FD, Damkier HH (2013) Na(+) dependent acid–base transporters in the choroid plexus; insights from slc4 and slc9 gene deletion studies. Front Physiol 4:1–10

Cook DI, Young JA (1989) Fluid and electrolyte secretion by salivary glands. In: Handbook of physiology. The gastrointestinal system. Salivary, pancreatic, gastric and hepatobiliary secretion. American Physiological Society, Bethesda, MD, pp 1–23.

Crane RK (1965) Na^+ -dependent transport in the intestine and other animal tissues. Fed Proc 24: 1000–1006

Darman RB, Forbush B (2002) A regulatory locus of phosphorylation in the N terminus of the Na-K-Cl cotransporter, NKCC1. J Biol Chem 277:37542–37550

Delpire E (2009) The Mammalian family of Sterile20p-like protein kinases. Pflugers Arch 458: 953–967

Delpire E, Gagnon KB (2008) SPAK and OSR1: STE20 kinases involved in the regulation of ion homoeostasis and volume control in mammalian cells. Biochem J 409:321–331

Delpire E, Gagnon KB (2011) Kinetics of hyperosmotically-stimulated Na-K-2Cl cotransporter in Xenopus laevis oocytes. Am J Physiol Cell Physiol 301:C1074–C1085

Delpire E, Rauchman MI, Beier DR, Hebert SC, Gullans SR (1994) Molecular cloning and chromosome localization of a putative basolateral Na-K-2Cl cotransporter from mouse inner medullary collecting duct (mIMCD-3) cells. J Biol Chem 269:25677–25683

Delpire E, Lu J, England R, Dull C, Thorne T (1999) Deafness and imbalance associated with inactivation of the secretory Na-K-2Cl co-transporter. Nat Genet 22:192–195

Dowd BF, Forbush B (2003) PASK (proline-alanine-rich STE20-related kinase), a regulatory kinase of the Na-K-Cl cotransporter (NKCC1). J Biol Chem 278:27347–27353

Flagella M, Clarke LL, Miller ML, Erway LC, Giannella RA, Andringa A, Gawenis LR, Kramer J, Duffy JJ, Doetschman T, Lorenz JN, Yamoah EN, Cardell EL, Shull GE (1999) Mice lacking the basolateral Na-K-2Cl cotransporter have impaired epithelial chloride secretion and are profoundly deaf. J Biol Chem 274:26946–26955

Flemmer AW, Gimenez I, Dowd BF, Darman RB, Forbush B (2002) Activation of the Na-K-Cl cotransporter NKCC1 detected with a phospho-specific antibody. J Biol Chem 277: 37551–37558

Gagnon KB, Delpire E (2010) Molecular determinants of hyperosmotically activated NKCC1-mediated K^+/K^+ exchange. J Physiol Lond 588:3385–3396

Gagnon KB, Delpire E (2012) Molecular physiology of SPAK and OSR1: two Ste20-related protein kinases regulating ion transport. Physiol Rev 92:1577–1617

Gagnon KB, Delpire E (2013) Physiology of SLC12 transporters: lessons from inherited human genetic mutations and genetically-engineered mouse knockouts. Am J Physiol Cell Physiol 304:C693–C714

Gamba G, Miyanoshita A, Lombardi M, Lytton J, Lee W-S, Hediger M, Hebert SC (1994) Molecular cloning, primary structure, and characterization of two members of the mammalian electroneutral sodium-(potassium)-chloride cotransporter family expressed in kidney. J Biol Chem 269: 17713–17722

Geck P, Pietrzyk C, Burckhardt B-C, Pfeiffer B, Heinz E (1980) Electrically silent cotransport of Na+, K+ and Cl – in Ehrlich cells. Biochim Biophys Acta 600:432–447

Geng Y, Hoke A, Delpire E (2009) The Ste20 kinases SPAK and OSR1 regulate NKCC1 function in sensory neurons. J Biol Chem 284:14020–14028

Giménez I, Isenring P, Forbush B (2002) Spatially distributed alternative splice variants of the renal Na-K-Cl cotransporter exhibit dramatically different affinities for the transported ions. J Biol Chem 277:8767–8770

Ginns SM, Knepper MA, Ecelbarger CA, Terris J, He X, Coleman RA, Wade JB (1996) Immunolocalization of the secretory isoform of Na-K-Cl cotransporter in rat renal intercalated cells. J Am Soc Nephrol 7:2533–2542

Gosmanov AR, Lindinger MI, Thomason DB (2003) Riding the tides: K^+ concentration and volume regulation by muscle $Na^+-K^+-2Cl^-$ cotransport activity. News Physiol Sci 18: 196–200

Grimm PR, Taneja TK, Liu J, Coleman R, Chen YY, Delpire E, Wade JB, Welling PA (2012) SPAK isoforms and OSR1 regulate sodium-chloride co-transporters in a nephron-specific manner. J Biol Chem 287:37673–37690

Hall AC, Ellory JC (1985) Measurement and stoichiometry of bumetanide-sensitive (2Na:1K:3Cl) cotransport in ferret red cells. J Membr Biol 85:205–213

Hoffman JF, Kregenow FM (1966) The characterization of new energy dependent cation transport processes in red blood cells. Ann N Y Acad Sci 137:566–576

Howard PA, Dunn MI (1997) Severe musculoskeletal symptoms during continuous infusion of bumetanide. Chest 111:359–364

Hunziker W, Fumey C (1994) A di-leucine motif mediates endocytosis and basolateral sorting of macrophage IgG Fc receptors in MDCK cells. EMBO J 13:2963–2969

Igarashi P, Whyte DA, Li K, Nagami GT (1996) Cloning and kidney cell-specific activity of the promoter of the murine renal Na-K-Cl cotransporter gene. J Biol Chem 271:9666–9674

Ikeda K, Oshima T, Hidaka H, Takasaka T (1997) Molecular and clinical implications of loop diuretic ototoxicity. Hear Res 107:1–8

Kaplan MR, Plotkin MD, Brown D, Hebert SC, Delpire E (1996) Expression of the mouse Na-K-2Cl cotransporter, mBSC2, in the terminal IMCD, the glomerular and extraglomerular mesangium and the glomerular afferent arteriole. J Clin Invest 98:723–730

Kregenow FM (1971a) The response of duck erythrocytes to nonhemolytic hypotonic media. Evidence for a volume-controlling mechanism. J Gen Physiol 58:372–395

Kregenow FM (1971b) The response of duck erythrocytes to hypertonic media. Further evidence for a volume-controlling mechanism. J Gen Physiol 58:396–411

Liedkte CM (1992) Electrolyte transport in the epitheliun of pulmonary segments of normal and cystic fibrosis lung. FASEB J 6:3076–3084

Lin SH, Yu IS, Jiang ST, Lin SW, Chu P, Chen A, Sytwu HK, Sohara E, Uchida S, Sasaki S, Yang SS (2011) Impaired phosphorylation of Na^{+}-K^{+}-$2Cl^{-}$ cotransporter by oxidative stress-responsive kinase-1 deficiency manifests hypotension and Bartter-like syndrome. Proc Natl Acad Sci USA 108:17538–17543

Lytle C (1998) A volume-sensitive protein kinase regulates the Na-K-2Cl cotransporter in duck red blood cells. Am J Physiol Cell Physiol 274:C1002–C1010

Lytle C, Forbush BI (1992a) Na-K-Cl cotransport in the shark rectal gland. II. Regulation in isolated tubules. Am J Physiol Cell Physiol 262:C1009–C10117

Lytle C, Forbush BI (1992b) The Na-K-Cl cotransport protein of shark rectal gland. II. Regulation by direct phosphorylation. J Biol Chem 267:25438–25443

Lytle C, McManus TJ, Haas M (1998) A model of Na-K-2Cl cotransport based on ordered ion binding and glide symmetry. Am J Physiol 274:C299–C309

McCormick JA, Mutig K, Nelson JH, Saritas T, Hoorn EJ, Yang C-L, Rogers S, Curry J, Delpire E, Bachmann S, Ellison DH (2011) A SPAK isoform switch modulates renal salt transport and blood pressure. Cell Metab 14:352–364

Meyer JW, Flagella M, Sutliff RL, Lorenz JN, Nieman ML, Weber CS, Paul RJ, Shull GE (2002) Decreased blood pressure and vascular smooth muscle tone in mice lacking basolateral Na(+)-K(+)-2Cl(−) cotransporter. Am J Physiol Heart Circ Physiol 283:H1846–H1855

Mykoniatis A, Shen L, Fedor-Chaiken M, Tang J, Tang X, Worrell RT, Delpire E, Turner JR, Matlin KS, Bouyer P, Matthews JB (2010) Phorbol 12-myristate 13-acetate-induced endocytosis of the Na-K-2Cl cotransporter in MDCK cells is associated with a clathrin-dependent pathway. Am J Physiol Cell Physiol 298:C85–C97

Ohta A, Rai T, Yui N, Chiga M, Yang SS, Lin SH, Sohara E, Sasaki S, Uchida S (2008) Targeted disruption of the Wnk4 gene decreases phosphorylation of Na-Cl cotransporter, increases Na excretion and lowers blood pressure. Hum Mol Genet 18:3978–3986

Orlov SN, Tremblay J, Hamet P (1996) Bumetanide-sensitive ion fluxes in vascular smooth muscle cells: lack of functional Na^{+}, K^{+}, 2 Cl^{-} cotransport. J Membr Biol 153:125–135

Palfrey HC, O'Donnell ME (1992) Characteristics and regulation of the Na/K/2Cl cotransporter. Cell Physiol Biochem 2:293–307

Panet R, Markus M, Atlan H (1994) Bumetanide and furosemide inhibited vascular endothelial cell proliferation. J Cell Physiol 158:121–127

Panet R, Marcus M, Atlan H (2000) Overexpression of the Na$^+$/K$^+$/Cl$^-$ cotransporter gene induces cell proliferation and phenotypic transformation in mouse fibroblasts. J Cell Physiol 182: 109–118

Payne JA, Forbush BI (1994) Alternatively spliced isoforms of the putative renal Na-K-Cl cotransporter are differentially distributed within the rabbit kidney. Proc Natl Acad Sci USA 91:4544–4548

Payne JA, Xu J-C, Haas M, Lytle CY, Ward D, Forbush BI (1995) Primary structure, functional expression, and chromosome localization of the bumetanide sensitive Na-K-Cl cotransporter in human colon. J Biol Chem 270:17977–17985

Piala AT, Moon TM, Akella R, He H, Cobb MH, Goldsmith EJ (2014) Chloride sensing by WNK1 involves inhibition of autophosphorylation. Sci Signal 7:ra41

Piechotta K, Lu J, Delpire E (2002) Cation-chloride cotransporters interact with the stress-related kinases SPAK and OSR1. J Biol Chem 277:50812–50819

Piechotta K, Garbarini NJ, England R, Delpire E (2003) Characterization of the interaction of the stress kinase SPAK with the Na$^+$-K$^+$-2Cl$^-$ cotransporter in the nervous system: evidence for a scaffolding role of the kinase. J Biol Chem 278:52848–52856

Ponce-Coria J, Markadieu N, Austin T, Flammang L, Rios K, Welling PA, Delpire E (2014) A novel SPAK-independent pathway involving Cab39 and WNK4 in the activation of Na-K-Cl cotransporters. J Biol Chem 289:17680–17688

Pressler CA, Heinzinger J, Jeck N, Waldegger P, Pechmann U, Reinalter S, Konrad M, Beetz R, Seyberth HW, Waldegger S (2006) Late-onset manifestation of antenatal Bartter syndrome as a result of residual function of the mutated renal Na$^+$-K$^+$-2Cl$^-$ co-transporter. J Am Soc Nephrol 17:2136–2142

Randall J, Thorne T, Delpire E (1997) Partial cloning and characterization of *Slc12a2*: the gene encoding the secretory Na$^+$-K$^+$-2Cl$^-$ cotransporter. Am J Physiol Cell Physiol 273: C1267–C1277

Rocha AS, Kudo LH (1990) Atrial peptide and cGMP effects on NaCl transport in inner medullary collecting duct. Am J Physiol 259:F258–F268

Russell JM (1983) Cation-coupled chloride influx in squid axon. Role of potassium and stoichiometry of the transport process. J Gen Physiol 81:909–925

Rybak LP (1993) Ototoxicity of loop diuretics. Otolaryngol Clin North Am 26:829–844

Simon DB, Karet FE, Rodriquez-Soriano J, Hamdan JH, DiPietro A, Trachtman H, Sanjad SA, Lifton RP (1996) Genetic heterogeneity of Bartter's syndrome revealed by mutations in the K$^+$ channel, ROMK. Nat Genet 14:152–156

Somasekharan S, Tanis J, Forbush B (2012) Loop diuretic and ion-binding residues revealed by scanning mutagenesis of transmembrane helix 3 (TM3) of Na-K-Cl cotransporter (NKCC1). J Biol Chem 287:17308–17317

Sonnenberg H, Honrath U, Wilson DR (1990) In vivo microperfusion of inner medullary collecting duct in rats: effect of amiloride and ANF. Am J Physiol 259:F222–F226

Soybel DI, Gullans SR, Maxwell F, Delpire E (1995) Role of basolateral Na-K-Cl cotransport in HCl secretion by Amphibian gastric mucosa. Am J Physiol Cell Physiol 269:C242–C249

Tang J, Bouyer P, Mykoniatis A, Buschmann M, Matlin KS, Matthews JB (2010) Activated PKC {delta} and PKC{epsilon} inhibit epithelial chloride secretion response to cAMP via inducing internalization of the Na$^+$-K$^+$-2Cl$^-$ cotransporter NKCC1. J Biol Chem 285:34072–34085

Tasaki I, Spiropoulos CS (1959) Stria vascularis as source of endocochlear potential. J Neurophysiol 22:149–155

Tyson J, Tranebjaerg L, Bellman S, Wren C, Taylor JFN, Bathen J, Aslaksen B, Sorland SJ, Lund O, Malcolm S, Pembrey M, Bhattacharya S, Bitner-Glindzicz M (1997) IsK and KvLQT1: mutation in either of the two subunits of the slow component of the delayed rectifier potassium channel can cause Jervell and Lange-Nielsen syndrome. Hum Mol Genet 61: 2179–2185

Vaduganathan M, Allegretti AS, Manchette AM, Patel SS, Olson KR, Bazari H (2013) Intravenous moderate-dose bumetanide continuous infusion and severe musculoskeletal pain. Int J Cardiol 168:e29–e31

Vetter DE, Mann JR, Wangemann P, Liu J, McLaughlin KJ, Lesage F, Marcus DC, Lazdunski M, Heinemann SF, Barhanin J (1996) Inner ear defects induced by null mutation of the *isk* gene. Neuron 17:1251–1264

Villa F, Goebel J, Rafiqi FH, Deak M, Thastrup J, Alessi DR, van Aalten DMF (2007) Structural insights into the recognition of substrates and activators by the OSR1 kinase. EMBO Rep 8: 839–845

Vitari AC, Thastrup J, Rafiqi FH, Deak M, Morrice NA, Karlsson HK, Alessi DR (2006) Functional interactions of the SPAK/OSR1 kinases with their upstream activator WNK1 and downstream substrate NKCC1. Biochem J 397:223–231

Wang Y et al (2009) Whole-genome association study identifies STK39 as a hypertension susceptibility gene. Proc Natl Acad Sci USA 106:226–231

Wong FH, Chen JS, Reddy V, Day JL, Shlykov MA, Wakabayashi ST, Saier MHJ (2012) The amino acid-polyamine-organocation superfamily. J Mol Microbiol Biotechnol 22:105–113

Xu J-C, Lytle C, Zhu TT, Payne JA, Benz EJ, Forbush BI (1994) Molecular cloning and functional expression of the bumetanide-sensitive Na-K-2Cl cotransporter. Proc Natl Acad Sci USA 91: 2201–2205

Yamashita A, Singh SK, Kawate T, Jin Y, Gouaux E (2005) Crystal structure of a bacterial homologue of Na^+/Cl^--dependent neurotransmitter transporters. Nature 437:215–223

Yang SS, Morimoto T, Rai T, Chiga M, Sohara E, Ohno M, Uchida K, Lin SH, Moriguchi T, Shibuya H, Kondo Y, Sasaki S, Uchida S (2007) Molecular pathogenesis of pseudohypoaldosteronism type II: generation and analysis of a Wnk4(D561A/+) knockin mouse model. Cell Metab 5:331–344

Yang SS, Lo YF, Wu CC, Lin SW, Yeh CJ, Chu P, Sytwu HK, Uchida S, Sasaki S, Lin SH (2010) SPAK-knockout mice manifest Gitelman syndrome and impaired vasoconstriction. J Am Soc Nephrol 21:1868–1877

Zeidel ML, Kikeri D, Silva P, Burrowes M, Brenner BM (1988) Atrial natriuretic peptides inhibit conductive sodium uptake by rabbit inner medullary collecting duct cells. J Clin Invest 82: 1067–1074

Chapter 12
Thiazide-Sensitive NaCl Cotransporter

Arohan R. Subramanya

Abstract The thiazide-sensitive NaCl cotransporter (NCC, SLC12A3) is a member of the solute carrier 12 (SLC12) family of electroneutral cotransporters. NCC is expressed in the distal convoluted tubule (DCT) of the kidney, where it mediates the transport of sodium with chloride in a 1:1 stoichiometry. NCC is responsible for reabsorbing approximately 5–10 % of sodium from the glomerular filtrate. Because NCC is expressed in the distal portion of the nephron, its activity helps dictate the final sodium chloride content in the urine. Thus, it plays a key role in determining the blood pressure set point through the regulation of extracellular fluid volume. Consistent with its critical role in blood pressure and volume control, NCC is one of the major molecular targets of thiazide-type diuretics, drugs commonly used as first-line agents in the treatment of essential hypertension and edema. Since its cloning over two decades ago, much has been learned about the physiological transport properties of NCC, its regulation, and its connections to human disease. In this chapter, I provide an overview of the biology of this important cotransporter.

Keywords NCC • Thiazide • WNK • SPAK • OSR1 • Distal convoluted tubule • Gitelman syndrome • Pseudohypoaldosteronism type 2

12.1 Early Studies and Cloning of NCC

The first studies of thiazide-sensitive salt transport in the kidney were conducted in the 1950s. These studies were motivated by the search for new diuretic drugs that lacked the toxic effects of organic mercurial diuretics such as mersalyl (Freeman et al. 1962). Proximal tubule carbonic anhydrase inhibitors, such as the sulfonamide acetazolamide, were among the first nontoxic diuretics identified, although it was immediately recognized that their ability to stimulate Na and water excretion waned over repeated uses. In addition, the diuretic effects of acetazolamide are

A.R. Subramanya (✉)
Renal Section, VA Pittsburgh Healthcare System, Renal-Electrolyte Division, Department of Medicine, University of Pittsburgh School of Medicine, S828A Scaife Hall, 3550 Terrace Street, Pittsburgh, PA 15261, USA
e-mail: ars129@pitt.edu

K.L. Hamilton, D.C. Devor (eds.), *Ion Channels and Transporters of Epithelia in Health and Disease*, Physiology in Health and Disease,
DOI 10.1007/978-1-4939-3366-2_12

almost always accompanied by clinically significant metabolic acidosis (Novello and Sprague 1957). In an effort to develop new sulfonamides with enhanced effects on sodium chloride excretion, investigators at Merck synthesized the first thiazide diuretic, chlorothiazide (Diuril), in 1957 (Novello and Sprague 1957). Although chlorothiazide inhibited carbonic anhydrase to some degree, it also stimulated chloride excretion in a manner similar to mercurial compounds, which act more distally. In the 1960s, experiments by Orloff and others suggested that chlorothiazide blocks NaCl reabsorption in the distal nephron (Earley and Orloff 1964). In the mid-1970s, Kunau et al. narrowed the site of action to the distal convolution (Kunau 1974), studies that were later verified by Constanzo et al. (Costanzo 1988; Costanzo and Windhager 1978). Ellison et al. established that thiazide diuretics block a coupled sodium chloride cotransport pathway in the DCT (Ellison et al. 1987). These critically important papers established that the natriuretic effects of thiazides were predominantly due to the inhibition of a distal nephron salt cotransporter, not proximal tubule carbonic anhydrase.

Initial efforts to identify the molecular identity of the "thiazide receptor" from the kidney using an expression cloning approach were unsuccessful, possibly because the DCT is the shortest tubular segment of the nephron and NCC therefore comprises only a tiny fraction of total mRNA from a kidney extract. Thus, a radically different strategy had to be employed to clone NCC. In the 1980s, studies by Stokes (1984) and Renfro (1977) identified a salt cotransporter in the bladder of the winter flounder (*Pseudopleuronectes americanus*) whose activity was sensitive to thiazides. The transport properties were thought to be identical to the NCC in human DCT. Thus, in 1993, Gamba and Hebert astutely used mRNA from flounder bladder as starting material to derive the first NCC cDNA (Gamba et al. 1993). The flounder sequence was then used as a template to isolate mammalian NCC cDNAs via homology-based strategies (Gamba et al. 1994).

12.2 Primary Structure and Molecular Architecture

Analysis of the primary structure of NCC suggested a complex topology consisting of 12 transmembrane domains (TMs) flanked by large intracellular amino- and carboxy-termini (Gamba et al. 1993). It is believed that the N-terminal intracellular domain is unstructured, while the C-terminus adopts a complex and compact fold. A large extracellular loop containing N-glycosylation sites was predicted to reside between TMs 7 and 8 (Fig. 12.1). In mammals, this extracellular loop contains two N-glycosylation sites, whereas in teleosts, three sites are present (Gamba 2005). Only one of these asparagines, Asn-403 in flounder, is conserved among all NCCs. This general topology is consistent with that of the two other electroneutral sodium-coupled cotransporters within the SLC12 family, the bumetanide-sensitive Na^+–K^+–$2Cl^-$ cotransporters NKCC1 (SLC12A2) and NKCC2 (SLC12A1).

A more detailed version of the transmembrane topology of NCC can be extrapolated from the recently solved crystal structures of other sodium-coupled

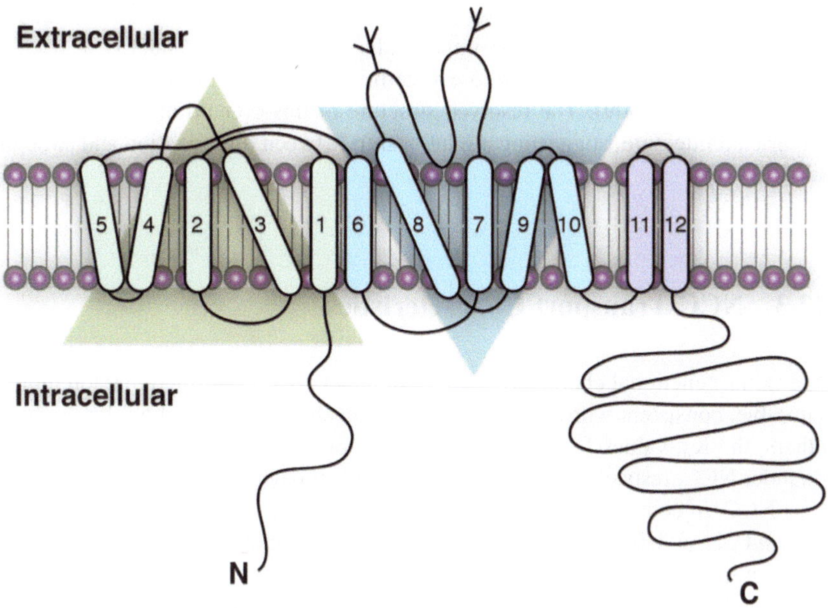

Fig. 12.1 Putative two-dimensional topology diagram of the thiazide-sensitive NaCl cotransporter (NCC), based on the 5 + 5 transmembrane inverted repeat common to members of the APC superfamily. The inverted "pseudosymmetric" folds are indicated with triangles. NCC has two large intracellular termini and a sizeable extracellular loop that contains N-glycosylation sites. The N-terminus is a regulatory domain that undergoes phosphorylation, while the structured C-terminus is important for homodimerization

secondary transporters, such as the bacterial sodium-coupled leucine transporter LeuT (Krishnamurthy and Gouaux 2012). Based on the fold of this transporter and other members of the amino acid–polyamine–cation (APC) transporter superfamily (of which the SLC12 cotransporters are a part of), the first ten TMs probably organize into two internally folded domains (Krishnamurthy et al. 2009). The two domains each consist of five helices that are oriented perpendicular to the membrane in opposite directions to each other in a "pseudosymmetric" manner (Fig. 12.1). The LeuT fold suggests that the site for sodium binding may be located near TMs 1 and 6. Discussed in further detail below, phosphorylation of the intracellular amino-terminus increases NCC activity. Since this intracellular domain is immediately contiguous with TM1, it is conceivable that phosphorylation of the NCC N-terminus may provoke structural alterations in the positioning of TM1 relative to TM6, thereby enhancing the efficiency of sodium transport.

Previous studies by de Jong et al. indicate that NCC assembles into homodimers prior to plasma membrane delivery (de Jong et al. 2003). Like many proteins (Zerangue et al. 1999), NCC assembly into macromolecular structures likely takes places in the ER, since high mannose immature ER-localized glycoforms of NCC readily form dimers (unpublished observations). Biochemical protein interaction studies suggest that the large and highly folded carboxy-terminus is essential for homodimer formation. These studies have been supported by the recently reported crystal structure of the intracellular C-terminus of MaCCC, a bacterial cation chloride cotransporter whose C-terminus shares some homology to NCC (Warmuth et al. 2009). The resolved structure of this domain naturally adopted a dimeric architecture, suggesting a relationship between dimerization and structural stability.

12.3 NCC Transport Characteristics

The initial functional characterization of flounder NCC confirmed that ion influx is saturable, consistent with a carrier-mediated mechanism. According this study and others, the K_m values for Na^+ and Cl^- are approximately 58 mM and 22 mM in flounder NCC, respectively (Gamba et al. 1993; Vazquez et al. 2002). Mammalian (rat) NCC, in contrast, exhibits a much higher affinity for ions with K_m values for Na^+ and Cl^- of <10 mM (Vazquez et al. 2002). The mammalian NCC appears to be much more thiazide-sensitive, with an IC_{50} for metolazone of 0.3 μM, compared to the flounder NCC IC_{50} of 4.0 μM (Moreno et al. 2006; Vazquez et al. 2002).

To date, little empirical information is available about the structural requirements for ion translocation. Challenges that interfere with the study of NCC structure–function relationships include the complex topology of the cotransporter and its tendency to misfold easily with single-nucleotide mutations. A decade ago, studies carried out with NCC–NKCC2 chimeras revealed that, as expected, the hydrophobic 12 transmembrane domain region of NCC confers ion translocation and diuretic sensitivity (Tovar-Palacio et al. 2004). This is consistent with more recent structural information obtained from the solved crystal structures of other APC superfamily members (Krishnamurthy and Gouaux 2012; Krishnamurthy et al. 2009; Singh et al. 2008). In 2006, Moreno et al. (2006) took advantage of the functional differences between rat and flounder NCC to define sites of ion translocation and diuretic binding. Using several rat–flounder chimeric constructs, they concluded that transmembrane domains 1–7 contained residues that alter chloride affinity. In contrast, swapping transmembrane domains 8–12 between species altered the thiazide binding affinity. This indicates that the binding sites for chloride and thiazide diuretics are different. Further analysis of these transmembrane domains revealed that substituting residues within TM11 was sufficient to confer altered sensitivity to metolazone (Castaneda-Bueno et al. 2010). Mutating serine 575 in the rat cotransporter to the corresponding residue in flounder (a cysteine) was sufficient to mediate this effect. Because only cysteine substitution at this site was capable of altering thiazide sensitivity, it is possible that the flounder cotransporter contains an extra intramembranous disulfide bond that alters the NCC fold. Although it is unclear if rat serine 575 participates in a local binding site for metolazone, these findings implicate TM11 as an important structural component of NCC that defines its thiazide sensitivity.

Mammalian NCC undergoes N-linked glycosylation on two asparagines harbored within the third extracellular loop, located between TM7 and TM8 (Hoover et al. 2003). N-glycosylation at these sites may play an important role in thiazide and chloride binding, since mutation of the glycosylation site asparagines to glutamine decreased the rat NCC K_m for chloride binding and IC_{50} for thiazide inhibition substantially, without affecting its affinity for sodium (Hoover et al. 2003). As mentioned above, the binding sites of chloride and thiazide diuretics are different (Moreno et al. 2006). Therefore, these findings suggest that N-linked glycosylation alters NCC conformation at multiple sites. One interpretation of the data is that glycosylation may introduce broad changes into the NCC tertiary structure, allowing it to achieve a mature conformation that can mediate thiazide-sensitive salt transport. This would be consistent with the well-appreciated role of N-glycosylation in the folding of other polytopic membrane proteins, such as the cystic fibrosis transmembrane conductance regulator (CFTR) (Glozman et al. 2009).

12.4 NCC Biogenesis

12.4.1 NCC Processing in the ER

Given the complex topology of NCC, it must undergo considerable processing within the biosynthetic pathway before it can achieve a proper fold. Current data suggest that, similar to other polytopic membrane proteins (Vembar and Brodsky 2008), this process is slow and/or inefficient. This is sensed by protein quality control mechanisms, which target suboptimally folded NCC conformers for ER-associated degradation (ERAD). Once the NCC polypeptide becomes integrated into the plasma membrane, molecular chaperones facilitate its proper folding. Heterologous expression studies in yeast and in mammalian cells have identified a host of cytoplasmic chaperone proteins that are responsible for this process (Fig. 12.2). These molecular chaperones are exclusively located within the cytoplasm. ER luminal chaperones, in contrast, appear to be less important (Needham et al. 2011). The cytoplasmic Hsp70/Hsp40 chaperone complex participates in one of the earliest quality control steps during NCC processing in the ER (Needham et al. 2011). Within this complex, Hsp70 is critically important, while Hsp40 is bound to Hsp70 and NCC but is dispensable for NCC quality control. If Hsp70 is able to fold NCC into a stable intermediate conformation, it will then be transferred from Hsp70 to Hsp90 via the Hsp70/Hsp90 organizer cochaperone (HOP/Stip1) (Donnelly et al. 2013) (Fig. 12.2). Hsp90 will then engage NCC and attempt to fold it further. If the cotransporter is interacting with either of these chaperones and becomes terminally misfolded, it will be targeted for ubiquitylation and ERAD (Donnelly et al. 2013). Thus far, three E3 ubiquitin ligases have been identified that can mediate this process. Based on studies of NCC ERAD in yeast, suboptimal

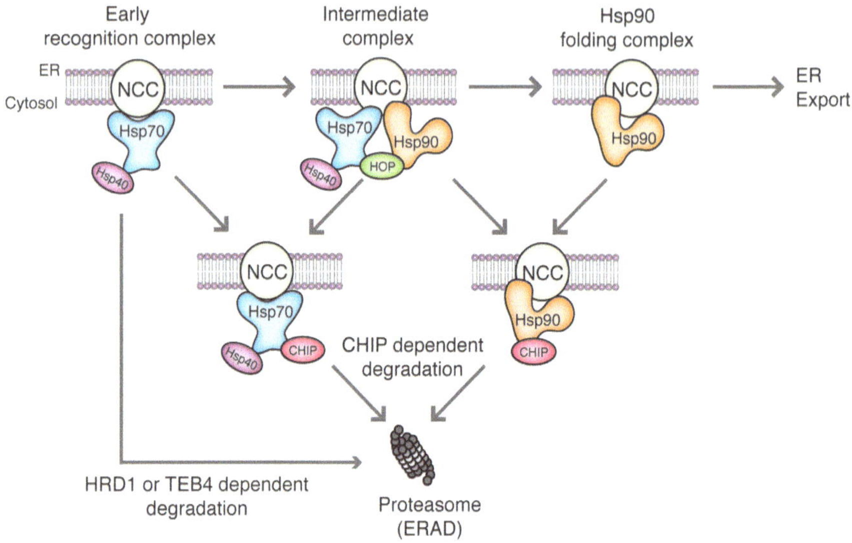

Fig. 12.2 Early steps in NCC biogenesis. NCC is recognized at an early stage of its biogenesis by an Hsp70/Hsp40 complex. Hsp70 recognizes misfolded regions of the cotransporter and attempts to fold it into a stable conformation. NCC will then be transferred from Hsp70 to Hsp90 via HOP. Cotransporters that achieve a stable Hsp90-bound conformational intermediate will then advance further along the biosynthetic pathway and out of the ER. At various stages of this pathway, the E3 ubiquitin ligases HRD, TEB4, and CHIP can ubiquitylate misfolded NCC conformers, targeting them for ER-associated proteasomal degradation (ERAD)

folding and aggregation of transmembrane domains are likely a common mishap, resulting in targeted degradation by the E3 ligase HRD1 (Needham et al. 2011). Misfolded cytoplasmic regions of NCC, on the other hand, can be recognized by the E3 ligase Doa10/TEB4 (Needham et al. 2011) or by the C-terminus of Hsp70-interacting protein (CHIP/Stub1) (Donnelly et al. 2013) (Fig. 12.2). Following ubiquitylation, NCC is then extracted from the ER membrane and extruded into the cytosol by the AAA+ ATPase Cdc48 and degraded by the 26S proteasome (Needham et al. 2011). In each case, interaction with Hsp70 or Hsp90 appears to be a prerequisite prior to ubiquitylation and ERAD (Needham et al. 2011; Donnelly et al. 2013). As will be discussed in detail below, altered interaction between NCC and this system of chaperones, cochaperones, and E3 ubiquitin ligases often underlies the pathogenesis of Gitelman syndrome, a hereditary salt-wasting nephropathy caused by loss-of-function mutations of NCC (Donnelly et al. 2013; Kunchaparty et al. 1999).

As NCC is being folded in the ER, it undergoes glycosylation on two asparagines harbored within its third extracellular loop (Hoover et al. 2003). Like other transmembrane glycoproteins, such as CFTR (Ward and Kopito 1994), the N-linked sugars attached to NCC are processed in the ER into a high-mannosylated form. Upon transit to the Golgi, the mannoses are cleaved into mature shorter glycan chains. Proper glycosylation is essential for NCC to exit the biosynthetic pathway and reach the plasma membrane (Hoover et al. 2003).

12.4.2 Post-ER Processing and Endosomal Storage of NCC

Little is known about how later steps of NCC maturation are carried out from within the Golgi. Presumably, most NCC will be processed in the Golgi cisternae when it is already in its homodimeric form, as oligomerization events commonly take place in the ER (Geva and Schuldiner 2014). As it matures, the dimeric NCC passes through the Golgi cisternae, is packaged into vesicles at the trans-Golgi network (TGN), and is sent to the apical plasma membrane via an endosomal system. In some cases, however, a subpopulation of cotransporters can be sorted directly from the TGN to the lysosome for degradation. Protein interaction studies suggest that this process may require the adaptor protein complex AP-3 (Subramanya et al. 2009) or the lysosomal sorting receptor sortilin (Zhou et al. 2009). In heterologous expression systems, this alternative TGN-to-lysosome transport pathway appears to be greatly enhanced in cells overexpressing with-no-lysine kinase 4 (WNK4, discussed in detail below).

Ultrastructural studies performed in rodent kidneys commonly demonstrate that NCC exists in apical and subapical populations in DCT cells (Lee et al. 2009; Sandberg et al. 2006). The subapical population resides within endosomes, which can be rapidly mobilized to the surface by hormones such as vasopressin and angiotensin II (Pedersen et al. 2010; Mutig et al. 2010; Sandberg et al. 2007). Conversely, physiological states such as acute hypertension can cause the NCC to quickly retract from the plasma membrane into endosomes and lysosomes (Lee et al. 2009).

12.5 NCC Activity at the Plasma Membrane

12.5.1 NCC Phosphorylation

In order for NCC to be active, it must first be "switched on" by phosphorylation (Pacheco-Alvarez et al. 2006). These phosphoactivation events specifically occur on serines and threonines present in the intracellular NCC amino-terminus. Figure 12.3a illustrates sites of phosphorylation that have been identified, either through phosphoproteomic studies or through analysis of homologous phosphorylation sites experimentally identified on other sodium-coupled cotransporters that are members of the SLC12 family, such as NKCC1 (Darman and Forbush 2002). To date, seven phosphoacceptor residues have been identified. According to the human NCC numeration (which I will use throughout this chapter), these amino acids

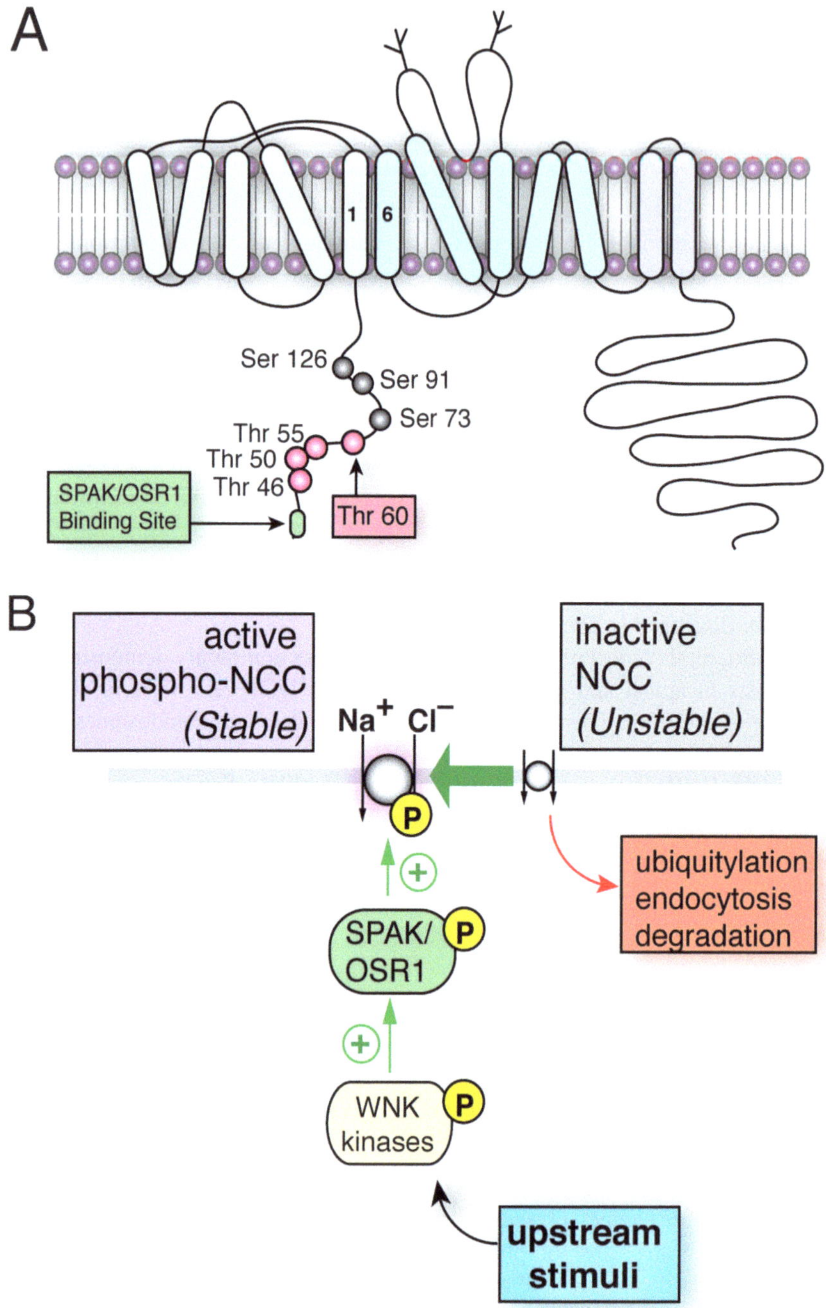

Fig. 12.3 Control of NCC activity by phosphorylation. (**a**) A regulatory locus of phosphorylation consisting of a cluster of serines and threonines is contained within the cytoplasmic amino-terminus of NCC. The amino-terminus is contiguous with TM1, which likely forms a sodium-binding site with TM6. A binding motif for the serine/threonine kinases SPAK and OSR1 is

correspond to threonines 46, 50, 55, and 60 and serines 73, 91, and 126 (Richardson and Alessi 2008; Richardson et al. 2008; Rosenbaek et al. 2012). Phosphorylation at each of these sites enhances NCC activity, although some evidence suggests that threonine 60 may have a unique and dominant role in dictating NCC-mediated sodium transport (Pacheco-Alvarez et al. 2006) (Fig. 12.3a). Since all of these phosphosites are harbored within the NCC amino-terminus, they are located immediately proximal to the first transmembrane domain. Thus, phosphorylation of the NCC N-terminus may somehow alter the positioning or structure of residues related to TM 1. Since structural data from other members of the APC superfamily indicate that TMs 1 and 6 help form a sodium-binding site (Krishnamurthy et al. 2009), phosphorylation may enhance the efficiency of sodium transport by altering the positioning of TM1 relative to TM6.

Recent data also indicate a relationship between NCC phosphorylation status and expression at the plasma membrane. Immunolocalization and ultrastructural studies using phosphospecific antibodies to one or more of the aforementioned NCC phosphorylation sites indicate that phosphorylated NCC is tightly restricted to the DCT apical membrane, whereas the unphosphorylated form appears to be more diffusely spread in both plasma membrane and intracellular compartments (Yang et al. 2007b; Mutig et al. 2010; Sandberg et al. 2006; Lee et al. 2009). This suggests that phosphorylation enhances NCC accumulation at the plasma membrane. Consistent with this, phosphorylated forms of NCC are not internalized via clathrin-mediated endocytosis as rapidly as unphosphorylated forms of the cotransporter and exhibit decreased ubiquitylation (Rosenbaek et al. 2014). Although the mechanism by which the phosphorylated form of NCC is relatively spared from endocytosis is unclear, it seems to be related to a change in its membrane complex properties. Blue native PAGE electrophoresis (BN-PAGE) studies suggest that upon phosphorylation, plasma membrane NCC shifts from a 400 kDa complex to heavier complexes of ~700 kDa in molecular mass (Lee et al. 2013). Included in the 700 kDa complex is γ-adducin, a membrane cytoskeletal protein that stimulates NCC (Lee et al. 2013; Dimke et al. 2011). The incorporation of phosphorylated NCC into high-molecular-weight complexes at the plasma membrane may prevent the cotransporter from being incorporated into clathrin-coated vesicles.

Fig. 12.3 (continued) located at the extreme end of the N-terminus. SPAK/OSR1 phosphorylation sites are shown in pink. Threonine 60 is critical for NCC activity. Other known phosphoacceptor residues are shown in *gray*. The kinases that phosphorylate these residues are unknown, but some of them (including serine 73) are phosphorylated in concert with SPAK/OSR1 activation. This suggests that other kinases may cooperate with SPAK/OSR1 to activate NCC. (**b**) Current model for SPAK/OSR1 activation. SPAK and OSR1 are activated by WNK kinases and phosphorylate NCC at the plasma membrane. The WNKs are a convergent regulatory point for upstream stimuli, including low intracellular chloride levels and hormones that stimulate renal salt reabsorption. Phosphorylation increases NCC stability at the surface. Inactive nonphosphorylated forms of NCC do not transport NaCl and are ubiquitylated and endocytosed more rapidly from the plasma membrane

12.5.2 The WNK–SPAK/OSR1 Signaling Pathway

12.5.2.1 SPAK and OSR1

Two structurally homologous serine threonine kinases directly phosphorylate the NCC amino-terminus. These kinases are the Ste20-like proline–alanine-rich kinase (SPAK, encoded by the gene *STK39*) and oxidative stress-responsive kinase 1 (OSR1, encoded by *OXSR1*) (Richardson et al. 2008). The relationship between SPAK, OSR1, and sodium-coupled cation chloride cotransporters was discovered initially in yeast two-hybrid studies of the secretory Na^+–K^+–$2Cl^-$ cotransporter NKCC1 (*SLC12A2*) (Piechotta et al. 2003). These studies identified SPAK as a binding partner with the NKCC1 amino-terminus. The findings were then extended to OSR1, whose amino acid sequence is 67 % identical to SPAK (Delpire and Gagnon 2006). Both SPAK and OSR1 bind to the NKCC1 N-terminus via an [R/K] FX[V/I] motif (where X can be any amino acid residue) (Piechotta et al. 2003). Later, these motifs were identified on NCC and the renal thick ascending limb Na^+–K^+–$2Cl^-$ cotransporter NKCC2 (*SLC12A1*) and were confirmed to be SPAK and OSR1 binding sites on those cotransporters as well (Gagnon et al. 2006; Richardson et al. 2008; Richardson and Alessi 2008; Villa et al. 2007). Consistent with a broad role for SPAK and OSR1 in stimulating the activity of SLC12 cotransporters in the kidney, SPAK and OSR1 are highly expressed in the TAL and DCT (McCormick et al. 2011). Moreover, SPAK mutant knock-in mice that express a catalytically inactive form of SPAK that cannot be phosphorylated by upstream signals in its kinase activation loop (i.e., "T-loop") exhibit decreased salt reabsorption and NKCC2 and NCC phosphorylation in both the TAL and DCT (Rafiqi et al. 2010). Presumably, this inactive form of SPAK acts as a dominant negative, preventing association of native SPAK and/or OSR1 with the N-terminal binding sites in NKCC2 and NCC. Many of the identified phosphorylation sites on NCC are SPAK/OSR1 phosphoacceptor residues. These sites correspond to amino acids 46, 50, 55, and 60 of the human NCC N-terminus (Richardson et al. 2008; Richardson and Alessi 2008). Serine 73 may also be directly phosphorylated by SPAK; as SPAK KO mice exhibit decreased phosphorylation at this residue (Yang et al. 2010), this site was not modified in in vitro studies. Likewise, phosphorylation of serine 91 is diminished in cells lacking SPAK activity, although this residue was not a direct SPAK/OSR1 phosphorylation site. Thus, some of the phosphoacceptor residues in the NCC N-terminus may be targeted by other kinases that cooperate with SPAK and OSR1 to stimulate salt transport in the DCT.

Studies in SPAK and OSR1 knockout animals illustrate the importance of these kinases in renal salt transport. Although both SPAK and OSR1 can activate NCC, their genetic ablation in the kidney results in disparate renal salt-wasting phenotypes. Mice deficient in SPAK exhibit salt wasting similar to a mild form of Gitelman syndrome (Yang et al. 2010; McCormick et al. 2011), suggesting that this kinase stimulates salt reabsorption in the DCT. These mice exhibit decreased NCC phosphorylation, but their mild salt wasting is due to the fact that they exhibit a dramatic *increase* in NKCC2 activity. In contrast, OSR1 knockout mice exhibit a more Bartter-like phenotype, due to a strong decrease in NKCC2 activity in the thick ascending limb of the loop of Henle (TAL) (Lin et al. 2011). NCC activity is relatively unaffected.

Current evidence suggests that these different phenotypes are due to discrepancies in the expression and activity of SPAK and OSR1 in the TAL and DCT. In the TAL, SPAK activity appears to be low, due to the expression of short forms of the kinase (McCormick et al. 2011; Grimm et al. 2012). These short forms of SPAK are exclusively expressed in the TAL and are either truncated transcripts or proteolytic fragments that lack kinase activity (McCormick et al. 2008; Markadieu et al. 2014). These kinase-deficient species can suppress the kinase activity of full-length SPAK and OSR1, functioning in a dominant-interfering manner. Genetic deletion of SPAK eliminates the expression of both its long and short forms in the TAL; this effectively releases OSR1 from inhibition in the TAL, causing it to hyperphosphorylate NKCC2. Since only the full-length form of SPAK is expressed in the DCT, SPAK KO mice exhibit a net decrease in NCC phosphorylation, resulting in the Gitelman-like phenotype. In the absence of SPAK, OSR1 redistributes into an intracellular location, collecting in large puncta (McCormick et al. 2011; Grimm et al. 2012; Terker et al. 2014). Although the reason for this finding is unclear, it seems that this aberrant mislocalization may explain why OSR1 is unable to compensate for SPAK ablation. More work is needed to further elucidate the relationship between SPAK, OSR1, and cation chloride cotransporters in the kidney, but the current body of work suggests that SPAK is the dominant activator of NCC in the DCT, while OSR1 serves as the main activator of NKCC2 in the TAL.

12.5.2.2 WNK Kinases

In order for SPAK and OSR1 to phosphorylate and activate the sodium-coupled cation chloride cotransporters, they must first be in an active state. SPAK and OSR1 activation occurs through phosphorylation of key residues in their kinase domains (Vitari et al. 2005, 2006; Thastrup et al. 2012). A family of proteins called with-no-lysine (amino acid = [K]; WNK) kinases mediates this process. The mammalian WNK family consists of four genes, which encode large serine threonine kinases found in many eukaryotes. All four WNK kinases share a similar domain organization, consisting of an N-terminal kinase domain, an autoinhibitory domain that suppresses intrinsic kinase activity and can cross-inhibit other WNKs, 2–3 coiled-coil domains which facilitate oligomerization, and multiple SPAK/OSR1 binding motifs (Richardson and Alessi 2008). They were named "with-no-lysine" kinases after bioinformatic analysis of the kinase domain revealed that the catalytic lysine required for ATP binding and phosphoryl transfer is missing from its usual location in β strand 3 of the active site; instead, this lysine is positioned in β strand 2 (Xu et al. 2000; Min et al. 2004). Due to this unusual arrangement of residues within the kinase domain, chloride can bind directly to the catalytic site and block its ability to trigger autophosphorylation and kinase domain activation (Piala et al. 2014). Thus, the unique structure of WNK kinases allows them to act as intracellular chloride sensors.

The WNKs were identified as NCC regulators over a decade ago, through studies of a rare chloride-dependent thiazide-sensitive Mendelian blood pressure disorder called familial hyperkalemic hypertension (FHHt, also known as pseudohypoaldosteronism type 2 or Gordon syndrome). In 2001, mutations in two WNK kinases, WNK1 and WNK4, were linked to FHHt (Wilson et al. 2001). Both of these kinases were expressed with NCC in the DCT, along with another member of the family, WNK3 (Wilson et al. 2001; Rinehart et al. 2005). WNK2, on the other hand, is a brain, heart, and intestine-expressed WNK kinase that is relatively excluded from the kidney (Rinehart et al. 2011).

WNK1 is unique from WNK4 and WNK3 in that it contains multiple promoters and undergoes extensive alternative splicing in the kidney (O'Reilly et al. 2003; Delaloy et al. 2003). Thus, its renal expression pattern consists of multiple isoforms with disparate functions. These isoforms can be classified into two major functional groups: long kinase active isoforms ("L-WNK1") with intact serine–threonine kinase activity or short kidney-specific isoforms that are kinase defective ("KS-WNK1"). KS-WNK1 isoforms are generated by an alternative promoter located between exons 4 and 5. This promoter is only active in the kidney and drives the expression of KS-WNK1 in the TAL, DCT, and collecting duct (O'Reilly et al. 2006; Vidal-Petiot et al. 2012; Liu et al. 2011).

Although mutations in WNK1 and WNK4 were the first to be implicated in the pathogenesis of FHHt, most FHHt affected harbor mutations in two other genes, the E3 ubiquitin ligase *Cullin 3* (CUL3) and its adaptor *Kelch-like 3* (KLHL3) (Boyden et al. 2012; Louis-Dit-Picard et al. 2012). The CUL3–KLHL3 complex binds to WNK1 and WNK4 and facilitates their ubiquitylation and degradation (Shibata et al. 2013; Ohta et al. 2013).

WNK4

The first studies defining relationships between the WNKs and NCC were carried out in the *Xenopus laevis* oocyte expression system. This in vitro system had been used for the initial expression cloning studies that permitted the isolation of the cDNAs encoding NCC and other members of the cation chloride cotransporter family. Thus, it was a natural extension to utilize the system to test the effects of WNK kinases on NCC function. The general approach for these experiments was to coinject in vitro transcribed RNA encoding a WNK kinase along with NCC and measure NCC activity by performing 22Na^+ radioisotopic fluxes in the presence or absence of thiazide. These initial experiments indicated that overexpression of WNK4 suppresses NCC activity (Yang et al. 2003; Wilson et al. 2003). Later, this finding was also seen in mammalian overexpression systems and mouse distal convoluted tubule cell lines (Subramanya et al. 2009; Ko et al. 2012; Cai et al. 2006; Zhou et al. 2009). Follow-up studies revealed that this overexpression phenotype is due to the suppression of NCC trafficking from the TGN to the plasma membrane (Golbang et al. 2006; Subramanya et al. 2009). Instead of being trafficked from the TGN to the surface, NCC is routed directly to lysosomes for degradation via an intracellular route. This event was associated with enhanced interaction between NCC and the AP-3 adaptor protein complex, which sorts cargo to the lysosome from the TGN (Subramanya et al. 2009; Simpson et al. 1997; Dell'Angelica et al. 1997). In addition, WNK4 overexpression promotes NCC packaging into vesicles that contain the lysosomal sorting receptor sortilin (Zhou et al. 2009). Although some studies suggested that intact WNK4 kinase activity is required for this inhibitory effect on NCC forward trafficking (Wilson et al. 2003; Golbang et al. 2006; Cai et al. 2006), other studies have argued that the effect is mediated by the WNK4 C-terminus and therefore is a kinase-independent process (Yang et al. 2005). The MAP kinase ERK1/2 may mediate the inhibitory effect of WNK4 on NCC, as ERK1/2 activation is associated with decreased NCC plasma membrane expression, and WNK4 overexpression can activate the ERK1/2 signaling pathway (Zhou et al. 2012; Ko et al. 2007). However, these studies remain largely correlative to date, and conflicting experiments have implicated ERK1/2 as a stimulator of NCC endocytosis, a trafficking operation that is not affected by WNK4 overexpression (Golbang et al. 2006; Subramanya et al. 2009).

Although numerous laboratories have independently corroborated this "inhibitory" effect of WNK4 (Subramanya et al. 2009; Yang et al. 2003, 2005; Golbang et al. 2006; Wilson et al. 2003; Cai et al. 2006; Ko et al. 2012), caution should be exercised when interpreting these experiments. Importantly, nearly all of the in vitro studies looking at this aspect of WNK4-NCC regulation have been performed in heterologous expression systems. Overexpression of a WNK kinase can cause aberrant sequestration and mislocalization of other components of the signaling pathway, including other WNK kinases or SPAK and OSR1. Thus, the inhibitory phenotype seen in in vitro systems may not have an in vivo correlate. Indeed, transgenic mice overexpressing wild-type WNK4 have provided conflicting results; one of the mouse models exhibited strong inhibition of NCC expression and DCT atrophy, while the other mouse actually exhibited *increased* NCC activity due to overstimulation of SPAK and OSR1 (Wakabayashi et al. 2013; Lalioti et al. 2006).

Consistent with this second mouse model, a number of studies have shown that WNK4 can activate NCC via SPAK/OSR1 (Vitari et al. 2005; Yang et al. 2007b). In contrast to the inhibitory effect of WNK4 described above, there is a general consensus that the stimulatory effect requires direct phosphorylation of SPAK/OSR1 by WNK4. This, in turn, results in downstream enhanced SPAK-/OSR1-mediated phosphorylation of NCC. Moreover, several hormones that activate salt transport via NCC (discussed in detail below) appear to trigger NCC phosphorylation by stimulating WNK4-mediated activation of SPAK/OSR1. Thus, many experts in the field view the stimulatory effect of WNK4 on NCC phosphorylation

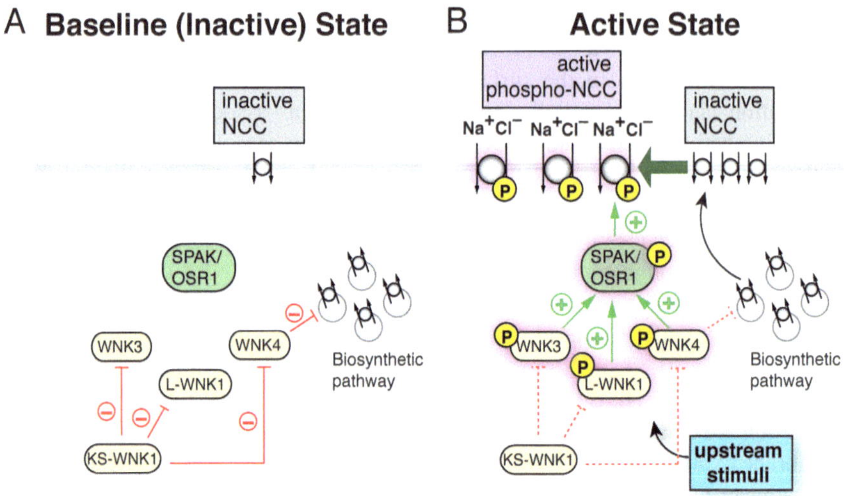

Fig. 12.4 WNK kinases. A model for WNK-mediated regulation of SPAK/OSR1 and NCC. (**a**) In the baseline unstimulated state, WNK4 may act as an inhibitor, suppressing NCC forward trafficking from endosomes derived from the biosynthetic pathway. WNK4, WNK3, and full-length WNK1 (L-WNK1) are all expressed in the DCT. The kinase-defective short isoform of WNK1, KS-WNK1, suppresses the kinase activity of WNK4, WNK3, and L-WNK1 at baseline. (**b**) When the DCT is challenged with a physiological stimulus, the kinase activity of WNK4, WNK3, and L-WNK1 increases relative to KS-WNK1. This converts these kinases into activators of SPAK and OSR1, which triggers NCC phosphorylation

status as its primary and more physiologically relevant mode of operation (Takahashi et al. 2014; Richardson and Alessi 2008).

In summary, current data on the role of WNK4 in NCC regulation provides evidence on both sides of the fence: some studies support its role as an inhibitor of NCC trafficking to the plasma membrane, while others have shown that it can stimulate NCC activity via SPAK- and OSR1-mediated phosphorylation. So, how does one reconcile these two radically different effects into a unifying model? One view is that the kinase inactive form of WNK4 may act as an inhibitor at baseline, holding NCC in an intracellular compartment via a mechanism that does not require intact kinase activity (Subramanya and Ellison 2014; McCormick et al. 2008) (Fig. 12.4a). Upon exposure to physiological factors that stimulate salt reabsorption in the DCT, WNK4 is converted into an active state. This results in WNK4 kinase domain activation and downstream SPAK and OSR1 phosphorylation and loss of the inhibitory effect (Fig. 12.4b). This would increase NCC phosphorylation and the number of active NCC cotransporters expressed at the plasma membrane due to enhanced forward trafficking. Recent studies utilizing RNA interference in DCT-derived cell lines that endogenously express SPAK, WNK4, and NCC provide evidence in support of such a model (Ko et al. 2010, 2012).

WNK1

Unlike WNK4, early studies with L-WNK1 in the *Xenopus* oocyte expression system were carried out with the original L-WNK1 cDNA, which was derived from rat brain (Xu et al. 2000). Studies with this cDNA revealed no effect of L-WNK1 on NCC activity or surface expression when it was coexpressed with the cotransporter in oocytes. However, when WNK4, L-WNK1, and NCC were coexpressed together, L-WNK1 associated with WNK4 and reversed its inhibitory effect on NCC transport back to baseline levels (Yang et al. 2003). This effect was not seen when either the kinase-defective KS-WNK1 isoform or a kinase-dead mutant of L-WNK1 was coexpressed with WNK4 and NCC, suggesting that intact L-WNK1 kinase activity is required to suppress WNK4-mediated inhibition (Yang et al. 2005; Subramanya et al. 2006b). As expected, like the other WNKs, L-WNK1 can also stimulate SPAK/OSR1 activation by phosphorylating residues within the SPAK and OSR1 kinase domains (Vitari et al. 2005). A more recent study that used human L-WNK1 found a stimulatory effect on NCC in *Xenopus* oocytes, via SPAK/OSR1 activation (Chavez-Canales et al. 2014). Collectively, these studies indicate that L-WNK1 can activate SPAK and OSR1 and can also override the inhibitory effect of WNK4 commonly seen in heterologous expression systems (Fig. 12.4b).

As mentioned above, KS-WNK1 isoforms do not contain a functional kinase domain. They do, however, contain "autoinhibitory" regions that can suppress L-WNK1 kinase activity (Subramanya et al. 2006b). Upon binding of KS-WNK1 to L-WNK1, these regions block the ability of L-WNK1 and other WNKs to phosphorylate and activate SPAK and OSR1 and can restore the inhibitory effect of WNK4 on NCC. Thus, KS-WNK1 functions as a dominant negative inhibitor of NCC (Fig. 12.4a). Consistent with this, KS-WNK1 knockout mice exhibit increased NCC total protein abundance and phosphorylation (Hadchouel et al. 2010).

WNK3

Comparative studies of the WNK kinases indicate that WNK3 is a potent NCC activator (Yang et al. 2007a; Rinehart et al. 2005) (Fig. 12.4b). In vitro, WNK3 stimulates NCC transport activity by enhancing its phosphorylation and surface expression (Rinehart et al. 2005; Yang et al. 2007a; Chavez-Canales et al. 2014). This effect can be blocked by KS-WNK1 (Yang et al. 2007a). Immunolocalization studies in the kidney have identified WNK3 in several segments of the renal tubule, including the proximal tubule, loop of Henle, DCT, and collecting duct (Rinehart et al. 2005). Based on the strong stimulatory effects of WNK3 on NCC activity, and its colocalization with NCC in the DCT, some have proposed that WNK3 may be a critically important regulator of thiazide-sensitive NaCl reabsorption. This hypothesis predicts that deleting WNK3 expression from the kidney should cause substantial salt wasting. Surprisingly, however, WNK3 knockout mice do not have an appreciable renal phenotype (Oi et al. 2012; Mederle et al. 2013). Although the reason for this is not entirely clear, compensation by L-WNK1 may play a role, since WNK1 mRNA is increased in WNK3 knockout mice relative to controls (Mederle et al. 2013).

12.5.2.3 NCC Dephosphorylation by Protein Phosphatases

Although much of the work on NCC phosphorylation has been centered on understanding how protein kinases stimulate NCC activity, a growing body of evidence indicates that protein phosphatases can inhibit NCC. Some of these phosphatases likely directly remove phosphate groups from the NCC amino-terminus, while others may influence NCC phosphorylation status by inhibiting the WNK–SPAK/OSR1 pathway.

Early studies of the sodium-coupled SLC12 cotransporters indicated that inhibitors of the ubiquitously expressed protein phosphatase 1 (PP1) could suppress NKCC1 activity through dephosphorylation (Darman and Forbush 2002; Lytle and Forbush 1996). Indeed, PP1 also appears to play an important role in NCC dephosphorylation; as one of its canonical inhibitors, PP1 inhibitor 1 (Ppp1r1a) is highly coexpressed with the cotransporter in the DCT (Picard et al. 2014). Deficiency of Ppp1r1a leads to a mild thiazide-sensitive salt-wasting phenotype. This suggests that the absence of Ppp1r1a leads to enhanced PP1 activity and NCC dephosphorylation. In 2010, Glover et al. identified PP4 as another phosphatase that is expressed in the DCT (Glover et al. 2010). Coexpression of PP4 with NCC reduces its phosphorylation at threonine 60 and therefore decreases its transport activity.

Protein phosphatase 3 (also known as calcineurin) was also recently identified as a potential negative regulator of NCC activity. The calcineurin inhibitor tacrolimus (FK506, "Prograf") commonly causes hypertension, hyperkalemia, hypercalciuria, and metabolic acidosis, despite normal glomerular filtration rate (GFR) (Hoorn et al. 2011). This constellation of findings is similar to the hypertension and hyperkalemia seen in FHHt (Hadchouel et al. 2006; Mayan et al. 2002). In 2012, Hoorn and colleagues showed that calcineurin inhibitors cause hypertension by activating thiazide-sensitive NaCl cotransport in the kidney. The mechanism appears to be directly related to an increase in NCC phosphorylation status. WNK4 abundance and SPAK/OSR1 activation were also increased. Although additional work will be necessary to determine the mechanism of calcineurin action, these findings suggest that tacrolimus may stimulate NCC activity through an indirect effect, possibly by preventing the dephosphorylation and inactivation of WNK kinases.

12.6 NCC Endocytosis

NCC is removed from the plasma membrane by dynamin- and clathrin-dependent endocytosis (Ko et al. 2010; Rosenbaek et al. 2014). Studies of overexpressed NCC in MDCK renal epithelial cells indicate that the cotransporter can be constitutively endocytosed, with one third of the total NCC surface pool undergoing internalization within 10 min; this endocytic rate is compatible with other kidney tubule-expressed membrane proteins such as the epithelial sodium channel (ENaC) (Butterworth et al. 2005). Clathrin-dependent endocytic retrieval of NCC is associated with increased cotransporter ubiquitylation and routing to the lysosomal pathway (Rosenbaek et al. 2014). While some groups have reported that endocytosed NCC undergoes polyubiquitylation (Ko et al. 2010), others have reported that the endocytosed ubiquitylated cotransporter migrates at molecular mass similar to monomeric NCC, suggesting oligo- or monoubiquitylation (Rosenbaek et al. 2014). Endocytosis of NCC is dependent on the phosphorylation status of the protein, since phosphorylation at threonines 55, 60, and 73 stabilizes NCC surface expression by slowing the rate of endocytosis (Rosenbaek et al. 2014; Yang et al. 2013). This, in turn, is associated with decreased NCC ubiquitylation. Ubiquitylation of the surface NCC pool is mediated by the aldosterone-regulated E3 ubiquitin ligase Nedd4-2 (Ronzaud et al. 2013; Arroyo et al. 2011), though other currently unidentified E3 ubiquitin ligases also probably participate in this process.

Treatment of cultured mDCT cells with the diacylglycerol analog 12-*O*-tetradecanoylphorbol-13-acetate (TPA) inhibits NCC by stimulating its endocytosis and ubiquitylation (Ko et al. 2010). Although TPA is classically used in the laboratory to activate protein kinase C (PKC), the effect appeared to be unrelated to this kinase, since PKC blockers did not reverse the inhibitory effect of TPA. Instead, the alternative TPA target Ras guanyl-releasing protein 1 (RasGRP1) was the primary mediator of NCC inhibition (Ko et al. 2007). The MAP kinase ERK1/2 may act downstream of RasGRP1 to trigger NCC endocytosis (Ko et al. 2010).

12.7 Mendelian Disorders of NCC Dysfunction

12.7.1 Gitelman Syndrome

In 1996, Simon et al. reported that mutations in the *SLC12A3* gene encoding NCC cause Gitelman syndrome, an autosomal recessive salt-wasting disorder (Simon et al. 1996). Patients may be hypotensive, but often have normal blood pressures, since the urinary salt wasting is offset by hyperreninemia and upregulation of the renin–angiotensin system. Hypokalemia and hypomagnesemia, however, are always seen in the disorder. Other manifestations include hypocalciuria and metabolic alkalosis. Thus, the overall phenotype of Gitelman syndrome is similar to the constellation of findings seen in patients taking thiazide diuretics.

Since the initial linkage of NCC to Gitelman syndrome, hundreds of disease-causing mutations have been identified. These mutations are generally scattered throughout the NCC protein, with a slight clustering of lesions in the intracellular C-terminus. In vitro analyses indicate that mutations that cause the disorder generally promote NCC misfolding (Kunchaparty et al. 1999). These so-called "Class 2" mutations cause the NCC to be retained in the endoplasmic reticulum and targeted for ERAD (Ellison 2003). Consistent with this finding, a recent study found that Gitelman mutations localized to the NCC C-terminus cause the cotransporter to interact strongly with a molecular chaperone complex containing Hsp70, Hsp40, and the cochaperone/E3 ubiquitin ligase CHIP (Donnelly et al. 2013). On the other hand, some mutations that cause Gitelman syndrome do not affect the biosynthetic processing of NCC; cotransporters with these "Class 1" mutations appear to be processed normally and traffic to the plasma membrane without difficulty (Sabath et al. 2004; Riveira-Munoz et al. 2007; Ellison 2003).

In 2007, a study by Ji et al. suggested that mutations of NCC that are associated with Gitelman syndrome play an important role in determining susceptibility to essential hypertension (Ji et al. 2008). Based on the prevalence of Gitelman syndrome, about 1 % of the general population should be heterozygous for the disorder. On average, individuals in a large study population who were carriers of a loss of function NCC variant had a systolic blood pressure that was 6.3 mmHg lower throughout life, compared with individuals who had two wild-type copies of NCC. Thus, the carrier state for Gitelman syndrome conferred lifelong protection from the development of hypertension (Acuna et al. 2011; Subramanya and Welling 2011). As the cost of next-generation sequencing technology continues to drop, one could imagine that screening for these mutations in normotensive individuals could provide information about the risk of developing hypertension over time.

12.7.2 Familial Hyperkalemic Hypertension

Familial hyperkalemic hypertension (FHHt) is a rare Mendelian syndrome featuring the unusual triad of chloride-dependent hypertension, severe hyperkalemia, and normal glomerular filtration rate. The disease was initially described by Paver and Pauline in the 1960s (Paver and Pauline 1964), and later by Gordon (1986). For years, the disease was known as "Gordon syndrome" or "pseudohypoaldosteronism type 2," although more recently, some have elected to rename the syndrome "FHHt" (Yang et al. 2005; Hadchouel et al. 2006). The chloride-dependent hypertension and hyperkalemia can be cured by low doses of thiazide diuretics (Mayan et al. 2002). Those familiar with the syndrome immediately recognized that its constellation of findings are essentially a "mirror image" of Gitelman syndrome (Hadchouel et al. 2006). A natural conclusion, then, was that gain-of-function mutations of NCC should be linked to the disease. Candidate gene approaches, however, indicated that this was not the case, as targeted sequencing of the *SLC12A3* gene revealed no mutations (Simon et al. 1995).

In 2001, Wilson et al. linked mutations in WNK1 and WNK4 to FHHt (Wilson et al. 2001). Although mutations in WNK kinases were the first to be implicated in the pathogenesis of FHHt, they only account for a minority of families affected by the disorder. As mentioned above, most FHHt affected harbor mutations in the genes encoding the RING E3 ubiquitin ligase scaffold CUL3 and its adaptor, KLHL3 (Boyden et al. 2012). These two proteins participate in a protein complex that degrades WNK1 and WNK4. Recent studies from several laboratories show that CUL3 assembles a multisubunit E3 ubiquitin ligase that covalently attaches ubiquitin molecules to WNK1 and WNK4, marking them for disposal (Shibata et al. 2013; Ohta et al. 2013). In order to do so, KLHL3 must be present, since it connects CUL3 to the WNKs.

An elaborate interrelationship between WNK1, WNK4, and the KLHL3/CUL3 complex underlies the pathogenesis of FHHt. FHHt-causing mutations in WNK4 are missense mutations that reduce WNK4 binding to KLHL3 (Shibata et al. 2013). Total WNK4 expression is therefore increased in patients who have these mutations. Disease-causing WNK4 mutants can phosphorylate SPAK and OSR1 and enhance NCC trafficking to the plasma membrane relative to the wild-type protein (Yang et al. 2007b). Thus, increased mutant WNK4 protein expression causes NCC overactivation. FHHt mutations in WNK1 are intron deletions that do not alter the kinase's protein structure. Instead, these mutations increase total WNK1 mRNA and protein expression (Wilson et al. 2001). The increased protein expression probably overrides ubiquitylation and degradation by the KLHL3/CUL3 complex, and the net effect of the mutation is to increase total L-WNK1 kinase activity (Vidal-Petiot et al. 2013). Thus, in FHHt caused by mutations in WNK1, increased L-WNK1 protein expression should suppress the inhibitory effect of WNK4 on NCC traffic while stimulating SPAK and OSR1, causing NCC to be overactive.

FHHt-associated mutations in KLHL3 reduce binding of the CUL3–KLHL3 complex to WNK1 and WNK4 (Shibata et al. 2013; Ohta et al. 2013; Wakabayashi et al. 2013). Thus, mutations in KLHL3 diminish the amount of CUL3-mediated WNK1 and WNK4 degradation. This increases WNK1 and WNK4 abundance, resulting in enhanced WNK-dependent signaling and NCC activation. Mutations in cullin-3 always cause skipping of an exon, resulting in an in-frame deletion (Boyden et al. 2012; Glover et al. 2014). It seems likely that this specific change in the cullin-3 protein sequence reduces WNK1 and WNK4 ubiquitylation, but this remains to be tested.

12.8 Physiological Regulation of NCC

The discovery of the WNK–SPAK/OSR1 signaling pathway and the advent of new tools to study NCC function, including cell lines, genetically modified mice, and phosphospecific antibodies, have led to a number of fascinating insights into the underlying mechanisms that regulate NCC activity. This section summarizes current understanding of the molecular bases of physiologic NCC regulation.

12.8.1 Regulation by Intracellular Chloride

SLC12 cation chloride cotransporters are important regulators of intracellular chloride concentration (Gamba 2005). Studies examining the effect of chloride on NKCC1 activity found that intracellular Cl^- depletion increases NKCC1 activity (Flatman 2002). The stimulatory effect of chloride depletion on NKCC1 appears to be more than a consequence of generating a simple chemical gradient. Intracellular chloride depletion stimulates phosphorylation of the sodium-coupled SLC12 cotransporters at the same residues that are phosphorylated by SPAK and OSR1 (Richardson and Alessi 2008). Moreover, as described above, WNK kinases are chloride sensors that are activated by chloride depletion (Piala et al. 2014). Thus, the WNK–SPAK/OSR1 pathway represents an important intracellular signaling mechanism that responds to changes in intracellular Cl^- levels to activate NaCl influx via these electroneutral cotransporters. NCC is activated by intracellular chloride depletion in a manner identical to NKCC1 (Pacheco-Alvarez et al. 2006). In *Xenopus* oocytes and mammalian cells, preincubating cells expressing NCC with chloride-free solutions strongly activate NCC transport activity by stimulating amino-terminal phosphorylation at canonical SPAK/OSR1 target sites (Pacheco-Alvarez et al. 2006; Richardson et al. 2008).

12.8.2 Regulation by Extracellular Potassium

NCC activity is exquisitely sensitive to changes in the concentration of extracellular K^+. Hyperkalemia has been shown to be a potent NCC inhibitor, while hypokalemia activates NCC (Vallon et al. 2009; McDonough and Youn 2013; Rengarajan et al. 2014). These effects can occur within the physiologic range; a plasma $[K^+]$ of 7 mmol/L is associated with NCC inhibition (Sorensen et al. 2013), while a plasma $[K^+]$ of 3 mmol/L is sufficient to trigger NCC activation (Vallon et al. 2009). Potassium predominantly affects NCC phosphorylation status. Within minutes of ingesting a high potassium load, rodents develop a natriuresis that is accompanied by rapid NCC dephosphorylation at the canonical SPAK/OSR1 phosphorylation sites (Sorensen et al. 2013). Conversely, hypokalemia appears to strongly trigger NCC phosphorylation at those same sites (Vallon et al. 2009). In both cases, the effects of potassium can occur independently of aldosterone action.

Several groups have proposed that extracellular potassium may regulate SPAK/OSR1 activity by adjusting the balance of WNK1 isoforms. In this regard, a high-K^+ diet has been shown by multiple groups to stimulate the mRNA expression of KS-WNK1 relative to L-WNK1 (Wade et al. 2006; Lazrak et al. 2006). Thus, by increasing the abundance of the kinase-defective kidney-specific isoform, hyperkalemia reduces the L-WNK1/KS-WNK1 ratio. Since KS-WNK1 contains inhibitory regions that block L-WNK1 kinase activity, this would be an effective way to shut off WNK-mediated signaling cascades in the DCT, resulting in diminished downstream SPAK/OSR1 and NCC phosphorylation (Liu et al. 2011; Subramanya et al. 2006a; Hadchouel et al. 2010).

Extracellular potassium may also regulate NCC by altering the basolateral conductance of the DCT. In theory, this should affect chloride-dependent activity of the WNK–SPAK/OSR1 signaling pathway. Consider, for example, the physiological alterations seen during hyperkalemia. A high extracellular K^+ concentration in the blood should cause the basolateral membrane of the DCT to become depolarized (Brown 1991). This decreased negativity of the cell membrane potential would diminish Cl^- exit via the basolateral DCT chloride channel ClC-Kb, raising the intracellular chloride concentration (Zhang et al. 2014). As described above, the corresponding increase in intracellular Cl^- would suppress WNK kinase autoactivation (Piala et al. 2014), resulting in reduced downstream SPAK/OSR1 and NCC activity. Supporting this concept, a recent study showed that knockout mice lacking the major K^+ channel mediating basolateral K^+ conductance in the DCT, *KCNJ10* (Kir4.1), exhibit reduced NCC activity caused by depolarization of the DCT membrane potential and a corresponding increase in intracellular chloride, which inhibited SPAK and OSR1 (Zhang et al. 2014).

12.8.3 Luminal NaCl Delivery

NCC-mediated NaCl reabsorption in the DCT is dependent on the delivered load of NaCl (Khuri et al. 1975). The DCT responds to chronic increases in the delivery of NaCl with an increase in the capacity for NaCl cotransport, as well as marked ultrastructural changes in the DCT cell. These morphologic changes include an increase in the size of DCT cells, increased basolateral membrane surface area, and an increase in DCT mitochondrial mass (Ellison et al. 1989). Accompanying the functional and morphologic changes are an increase in basolateral Na–K ATPase activity and an increase in thiazide-binding sites (Chen et al. 1990). It is not clear if these changes are due to transcriptional or posttranslational effects on NCC expression, although they do appear to be a consequence of increased sodium entry into the DCT. Inhibition of NaCl entry into DCT cells with chronic thiazide treatment resulted in a loss of cell height and polarity and triggered apoptosis (Morsing et al. 1991). The cellular mechanisms whereby NaCl entry affects NCC transport function and DCT morphology are currently not known.

12.8.4 Hormonal Regulation of NCC

12.8.4.1 Angiotensin II

The peptide hormone angiotensin II (AngII) plays a critically important role in blood pressure homeostasis and tubular salt handling. AngII has dual effects on

NCC trafficking and phosphorylation. Infusion of AngII into animals causes the cotransporter to traffic from intracellular vesicles to the plasma membrane, while treatment of rodents with angiotensin converting enzyme (ACE) inhibitors redistributes NCC to intracellular vesicles (Sandberg et al. 2007). The surface expressed cotransporter appears to be highly phosphorylated, based on immunolocalization studies and biochemical inquiry of the aforementioned canonical SPAK/OSR1 phosphorylation sites. The WNK–SPAK/OSR1 pathway mediates these effects, as the stimulatory effects of angiotensin II are associated with SPAK and OSR1 activation (Talati et al. 2010). Moreover, in vitro and in vivo experiments implicate WNK4 as being critically important (Castaneda-Bueno et al. 2012; San-Cristobal et al. 2009). Studies in the *Xenopus* oocyte system suggest that AngII converts WNK4 from its inhibitory state into a SPAK/OSR1 dependent NCC activator, and NCC expression and phosphorylation are unaffected by AngII infusion in WNK4 knockout mice. NCC stimulation could be seen in adrenalectomized animals, suggesting that the AngII effect occurred independently of aldosterone (van der Lubbe et al. 2011).

The classic view of the renin–angiotensin system is that it functions as a whole body blood pressure and volume control mechanism that acts through the concerted effects of multiple organs (Guyton 1991). Renin release from the kidney triggers conversion of liver-derived angiotensinogen to angiotensin I, which in turn is converted to AngII by lung-expressed ACE (Atlas 2007). However, it is clear that an intrarenal renin–angiotensin system also exists within the kidney and that this system is critically important for the effects of AngII on renal hemodynamics (Navar and Rosivall 1984). This intrinsic system has important effects on the activity of NCC, since the kidney-specific ablation of intrarenal ACE decreases NCC activity and phosphorylation (Gonzalez-Villalobos et al. 2013).

12.8.4.2 Adrenal Steroids

Current evidence indicates that adrenal steroid hormones stimulate NCC-mediated NaCl cotransport. The presence of both mineralocorticoid and glucocorticoid receptors in the DCT has been demonstrated by immunohistochemistry and by hormone-binding experiments (Farman and Bonvalet 1983; Farman et al. 1991). Although mineralocorticoid receptors are expressed throughout the entire DCT, only the "late" portion of this segment (also known as the DCT2) expresses the steroid metabolizing enzyme 11-β-hydroxysteroid dehydrogenase 2 (11-βHSD2) (Velazquez et al. 1998) (Bostanjoglo et al. 1998). This enzyme metabolizes cortisol to the inactive metabolite cortisone, thereby preventing circulating glucocorticoids from binding to mineralocorticoid receptors expressed in the DCT2. Because the mineralocorticoid receptors in the DCT2 can only be occupied by aldosterone, the DCT2 is highly sensitive to changes in circulating aldosterone levels.

Early studies by Chen et al. (1994a), Velazquez et al. (1995), and Kim et al. (1998) collectively demonstrated that aldosterone increases total NCC protein abundance without affecting its mRNA expression, indicating that mineralocorticoids stimulate NCC activity through posttranslational mechanisms. More recently, McDonough (Sandberg et al. 2006) showed that mineralocorticoid receptor blockade decreases NCC trafficking to the plasma membrane while simultaneously increasing its accumulation in lysosomes. Thus, one way in which aldosterone can increase NCC activity is by enhancing its expression at the plasma membrane. The molecular mechanism likely involves the serum- and glucocorticoid-regulated kinase SGK1. The transcription of this serine–threonine kinase is upregulated in response to aldosterone binding to the mineralocorticoid receptor (Pearce 2001). The active form of SGK1 has been reported to regulate NCC trafficking via two mechanisms. First, in in vitro systems, SGK1 can reverse the inhibitory effect of overexpressed WNK4 on NCC delivery to the plasma membrane (Rozansky et al. 2009a). Second, SGK1 phosphorylates and inactivates Nedd4-2, an E3 ubiquitin ligase, which ubiquitylates and negatively regulates NCC surface expression (Arroyo et al. 2011). Although the first of these two findings has only been shown in the *Xenopus* oocyte expression system (Rozansky et al. 2009b), the in vitro inhibitory effect of Nedd4-2 on NCC has been verified in mouse models (Arroyo et al. 2011; Ronzaud et al. 2013).

Aldosterone also appears stimulate NCC transport activity by promoting NCC phosphorylation. In 2008, Chiga et al. (2008) showed that NCC phosphorylation is dynamically regulated by adjustments in dietary NaCl intake and that this effect is at least partially dependent on the mineralocorticoid receptor (MR). Vallon et al. echoed these findings 1 year later (Vallon et al. 2009). In both reports, careful measurements of NCC abundance demonstrated enhanced NCC phosphorylation at all of its canonical SPAK/OSR1 phosphorylation residues. Consistent with this, aldosterone increased SPAK and OSR1 activation in a mineralocorticoid receptor-dependent manner (Chiga et al. 2008). A study by van der Lubbe et al. (2012) implicated WNK4 as a mediator of the stimulatory effect of aldosterone on SPAK/OSR1 and NCC phosphorylation, though other WNK kinases expressed in the DCT may also be involved.

Glucocorticoids such as dexamethasone increase thiazide-sensitive NaCl transport and increase the number of metolazone-binding sites in the kidneys of adrenalectomized rodents (Chen et al. 1994a). This strongly suggests that glucocorticoids stimulate NCC-mediated salt transport in the DCT. The DCT1 may be the more relevant site of action of glucocorticoids, since, as mentioned above, the levels of active cortisol in the DCT2 should be relatively low due to its metabolism by 11β-HSD (Bostanjoglo et al. 1998). The WNK–SPAK/OSR1 pathway is expressed in both DCT1 and DCT2; thus, many of the same players involved in mineralocorticoid receptor-dependent regulation of NCC are likely involved. Although NCC overactivity may participate in the salt-sensitive hypertension seen in disorders of cortisol excess such as Cushing's syndrome, the role of glucocorticoids in the physiological regulation of Na^+ transport in the DCT remains understudied.

12.8.4.3 Gonadal Steroids and NCC

Gonadal steroid hormones may also influence NaCl cotransport in the DCT. Chen et al. (1994b) reported gender differences in the density of thiazide-binding sites and in the natriuretic response of thiazides in rodents. Specifically, female rats had higher levels of "thiazide receptors" in the renal cortex than males. The number of thiazide-binding sites decreased following ovariectomy, while levels in males rose following orchiectomy. Consistent with these results, Verlander et al. (1998) found that estradiol treatments increased NCC expression in the DCT. Collectively, these results are consistent with the view that male sex hormones (e.g., testosterone) may downregulate NCC expression and salt transport, while estrogens increase NCC expression and salt transport in the DCT. The physiological relevance of these findings with respect to human hypertension is currently unclear, as no clear gender difference in the response of humans to thiazide diuretics has been established.

12.8.4.4 Vasopressin

Recent work indicates that vasopressin is a potent NCC activator (Gamba 2010). The vasopressin analog deamino-Cys-1, d-Arg-8 vasopressin (dDAVP) triggers an increase in NCC abundance, trafficking to the plasma membrane, and activation through SPAK/OSR1-mediated phosphorylation of its amino-terminus (Saritas et al. 2013; Mutig et al. 2010; Pedersen et al. 2010). The effect probably requires cyclic AMP and protein kinase A (PKA), well-established intermediaries of vasopressin-dependent signaling (Nedvetsky et al. 2009). The effect on NCC abundance may be a Nedd4-2-dependent process, since PKA can phosphorylate and inactivate Nedd4-2 through mechanisms similar to SGK1 (Snyder et al. 2004); however, this hypothesis is yet to be tested.

12.8.4.5 Insulin

One of the first studies suggesting a potential link between insulin and NCC was performed by Bickel et al., who noted that insulin-resistant obese Zucker rats retain salt and exhibit increased NCC abundance (Bickel et al. 2001). Subsequent studies confirmed that NCC is hyperphosphorylated in these animals, caused by an increase in total SPAK/OSR1 activity (Komers et al. 2012; Chavez-Canales et al. 2013). Similar results were noted in another model of type 2 diabetes, the leptin receptor-deficient db/db mouse. Crossing db/db mice with SPAK/OSR1 catalytically inactive "T-loop" knock-in mice that cannot be activated by WNKs resulted in the stimulatory effect of insulin on NCC being completely abrogated (Nishida et al. 2012). This indicates that insulin activates NCC via the WNK–SPAK/OSR1 pathway. Although it seems clear that SPAK/OSR1 are the key downstream mediators of NCC stimulation, there have been conflicting data with regard to which of the upstream WNK kinases mediate the effect. Data from Chavez-Canales et al. (2013) suggest that WNK3 is required for insulin to stimulate NCC activity, while studies by Sohara et al. (2011) and Takahashi et al. (2014) provide evidence that WNK4 is the dominant WNK kinase that mediates the effect. The phosphoinositol 3-kinase/-mTORC2-Akt pathway transduces the upstream signal from insulin receptors to the WNK–SPAK/OSR1 network (Nishida et al. 2012; Chavez-Canales et al. 2013).

12.9 Concluding Remarks

Although interest in the thiazide-sensitive cotransporter had only been modest for years, exciting genetic studies, the discovery of key NCC regulators, and the advent of tools to biochemically monitor NCC activation status have yielded an explosion of new insights into its role in physiology. Indeed, many of the papers cited in this chapter have been published only within the last 5 years. Furthermore, the recent linkage of KLHL3 and CUL3 to FHHt, as well as the generation of new genetic models to study the WNK–SPAK/OSR1 pathway, has opened up new and important questions and avenues of investigation. As the underlying molecular wiring that connects important blood pressure, potassium, and volume control systems to NCC in the DCT is elucidated further, perhaps new targets for the treatment of clinical disorders such as salt-sensitive hypertension, hyperkalemia, edema, and nephrolithiasis will be established.

Acknowledgments I would like to thank David Ellison and Gerardo Gamba for helpful discussions. This work was supported by grants from the NIH (R01-DK098145 and R01-HL088120), by the Pittsburgh Center for Kidney Research (P30-DK079307), and by the VA Pittsburgh Healthcare System Research Service.

References

Acuna R, Martinez-de-la-Maza L, Ponce-Coria J, Vazquez N, Ortal-Vite P, Pacheco-Alvarez D, Bobadilla NA, Gamba G (2011) Rare mutations in SLC12A1 and SLC12A3 protect against hypertension by reducing the activity of renal salt cotransporters. J Hypertens 29(3):475–483. doi:10.1097/HJH.0b013e328341d0fd

Arroyo JP, Lagnaz D, Ronzaud C, Vazquez N, Ko BS, Moddes L, Ruffieux-Daidie D, Hausel P, Koesters R, Yang B, Stokes JB, Hoover RS, Gamba G, Staub O (2011) Nedd4-2 modulates renal Na^+-Cl^- cotransporter via the aldosterone-SGK1-Nedd4-2 pathway. J Am Soc Nephrol 22(9):1707–1719. doi:10.1681/ASN.2011020132

Atlas SA (2007) The renin-angiotensin aldosterone system: pathophysiological role and pharmacologic inhibition. J Manag Care Pharm 13(8 Suppl B):9–20

Bickel CA, Verbalis JG, Knepper MA, Ecelbarger CA (2001) Increased renal Na-K-ATPase, NCC, and beta-ENaC abundance in obese Zucker rats. Am J Physiol Renal Physiol 281(4): F639–648

Bostanjoglo M, Reeves WB, Reilly RF, Velazquez H, Robertson N, Litwack G, Morsing P, Dorup J, Bachmann S, Ellison DH (1998) 11Beta-hydroxysteroid dehydrogenase, mineralocorticoid receptor, and thiazide-sensitive Na-Cl cotransporter expression by distal tubules. J Am Soc Nephrol 9(8):1347–1358

Boyden LM, Choi M, Choate KA, Nelson-Williams CJ, Farhi A, Toka HR, Tikhonova IR, Bjornson R, Mane SM, Colussi G, Lebel M, Gordon RD, Semmekrot BA, Poujol A, Valimaki MJ, De Ferrari ME, Sanjad SA, Gutkin M, Karet FE, Tucci JR, Stockigt JR, Keppler-Noreuil KM, Porter CC, Anand SK, Whiteford ML, Davis ID, Dewar SB, Bettinelli A, Fadrowski JJ, Belsha CW, Hunley TE, Nelson RD, Trachtman H, Cole TR, Pinsk M, Bockenhauer D, Shenoy M, Vaidyanathan P, Foreman JW, Rasoulpour M, Thameem F, Al-Shahrouri HZ, Radhakrishnan J, Gharavi AG, Goilav B, Lifton RP (2012) Mutations in kelch-like 3 and cullin 3 cause hypertension and electrolyte abnormalities. Nature 482(7383):98–102. doi:10.1038/nature10814

Brown RH Jr (1991) Molecular basis for hyperkalemic periodic paralysis. Int J Neurol 25–26:89–96

Butterworth MB, Edinger RS, Johnson JP, Frizzell RA (2005) Acute ENaC stimulation by cAMP in a kidney cell line is mediated by exocytic insertion from a recycling channel pool. J Gen Physiol 125(1):81–101. doi:10.1085/jgp.200409124

Cai H, Cebotaru V, Wang YH, Zhang XM, Cebotaru L, Guggino SE, Guggino WB (2006) WNK4 kinase regulates surface expression of the human sodium chloride cotransporter in mammalian cells. Kidney Int 69(12):2162–2170

Castaneda-Bueno M, Vazquez N, Bustos-Jaimes I, Hernandez D, Rodriguez-Lobato E, Pacheco-Alvarez D, Carino-Cortes R, Moreno E, Bobadilla NA, Gamba G (2010) A single residue in transmembrane domain 11 defines the different affinity for thiazides between the mammalian and flounder NaCl transporters. Am J Physiol Renal Physiol 299(5):F1111–1119. doi:10.1152/ajprenal.00412.2010

Castaneda-Bueno M, Cervantes-Perez LG, Vazquez N, Uribe N, Kantesaria S, Morla L, Bobadilla NA, Doucet A, Alessi DR, Gamba G (2012) Activation of the renal Na^{+}:Cl^{-} cotransporter by angiotensin II is a WNK4-dependent process. Proc Natl Acad Sci USA 109(20):7929–7934. doi:10.1073/pnas.1200947109

Chavez-Canales M, Arroyo JP, Ko B, Vazquez N, Bautista R, Castaneda-Bueno M, Bobadilla NA, Hoover RS, Gamba G (2013) Insulin increases the functional activity of the renal NaCl cotransporter. J Hypertens 31(2):303–311. doi:10.1097/HJH.0b013e32835bbb83

Chavez-Canales M, Zhang C, Soukaseum C, Moreno E, Pacheco-Alvarez D, Vidal-Petiot E, Castaneda-Bueno M, Vazquez N, Rojas-Vega L, Meermeier NP, Rogers S, Jeunemaitre X, Yang CL, Ellison DH, Gamba G, Hadchouel J (2014) WNK-SPAK-NCC cascade revisited: WNK1 stimulates the activity of the Na-Cl cotransporter via SPAK, an effect antagonized by WNK4. Hypertension. doi:10.1161/HYPERTENSIONAHA.114.04036

Chen ZF, Vaughn DA, Beaumont K, Fanestil DD (1990) Effects of diuretic treatment and of dietary sodium on renal binding of 3H-metolazone. J Am Soc Nephrol 1(1):91–98

Chen Z, Vaughn DA, Blakely P, Fanestil DD (1994a) Adrenocortical steroids increase renal thiazide diuretic receptor density and response. J Am Soc Nephrol 5(6):1361–1368

Chen Z, Vaughn DA, Fanestil DD (1994b) Influence of gender on renal thiazide diuretic receptor density and response. J Am Soc Nephrol 5(4):1112–1119

Chiga M, Rai T, Yang SS, Ohta A, Takizawa T, Sasaki S, Uchida S (2008) Dietary salt regulates the phosphorylation of OSR1/SPAK kinases and the sodium chloride cotransporter through aldosterone. Kidney Int 74(11):1403–1409. doi:10.1038/ki.2008.451

Costanzo LS (1988) Mechanism of action of thiazide diuretics. Semin Nephrol 8(3):234–241

Costanzo LS, Windhager EE (1978) Calcium and sodium transport by the distal convoluted tubule of the rat. Am J Physiol 235(5):F492–506

Darman RB, Forbush B (2002) A regulatory locus of phosphorylation in the N terminus of the Na-K-Cl cotransporter, NKCC1. J Biol Chem 277(40):37542–37550. doi:10.1074/jbc.M206293200 de Jong JC, Willems PH, Mooren FJ, van den Heuvel LP, Knoers NV, Bindels RJ (2003) The structural unit of the thiazide-sensitive NaCl cotransporter is a homodimer. J Biol Chem 278 (27):24302–24307. doi:10.1074/jbc.M303101200

Delaloy C, Lu J, Houot AM, Disse-Nicodeme S, Gasc JM, Corvol P, Jeunemaitre X (2003) Multiple promoters in the WNK1 gene: one controls expression of a kidney-specific kinase-defective isoform. Mol Cell Biol 23(24):9208–9221

Dell'Angelica EC, Ohno H, Ooi CE, Rabinovich E, Roche KW, Bonifacino JS (1997) AP-3: an adaptor-like protein complex with ubiquitous expression. EMBO J 16(5):917–928

Delpire E, Gagnon KB (2006) SPAK and OSR1, key kinases involved in the regulation of chloride transport. Acta Physiol (Oxf) 187(1–2):103–113

Dimke H, San-Cristobal P, de Graaf M, Lenders JW, Deinum J, Hoenderop JG, Bindels RJ (2011) gamma-Adducin stimulates the thiazide-sensitive NaCl cotransporter. J Am Soc Nephrol 22 (3):508–517. doi:10.1681/ASN.2010060606

Donnelly BF, Needham PG, Snyder AC, Roy A, Khadem S, Brodsky JL, Subramanya AR (2013) Hsp70 and Hsp90 multichaperone complexes sequentially regulate thiazide-sensitive cotransporter endoplasmic reticulum-associated degradation and biogenesis. J Biol Chem 288(18):13124–13135. doi:10.1074/jbc.M113.455394

Earley LE, Orloff J (1964) Thiazide diuretics. Annu Rev Med 15:149. doi:10.1146/Annurev.Me.15.020164.001053

Ellison DH (2003) The thiazide-sensitive Na-Cl cotransporter and human disease: reemergence of an old player. J Am Soc Nephrol 14(2):538–540

Ellison DH, Velazquez H, Wright FS (1987) Thiazide-sensitive sodium chloride cotransport in early distal tubule. Am J Physiol 253(3 Pt 2):F546–554

Ellison DH, Velazquez H, Wright FS (1989) Adaptation of the distal convoluted tubule of the rat. Structural and functional effects of dietary salt intake and chronic diuretic infusion. J Clin Invest 83(1):113–126. doi:10.1172/JCI113847

Farman N, Bonvalet JP (1983) Aldosterone binding in isolated tubules. III. Autoradiography along the rat nephron. Am J Physiol 245(5 Pt 1):F606–614

Farman N, Oblin ME, Lombes M, Delahaye F, Westphal HM, Bonvalet JP, Gasc JM (1991) Immunolocalization of gluco- and mineralocorticoid receptors in rabbit kidney. Am J Physiol 260(2 Pt 1):C226–233

Flatman PW (2002) Regulation of Na-K-2Cl cotransport by phosphorylation and protein-protein interactions. Biochim Biophys Acta 1566(1–2):140–151

Freeman RB, Maher JF, Schreiner GE, Mostofi FK (1962) Renal tubular necrosis due to nephrotoxicity of organic mercurial diuretics. Ann Intern Med 57:34–43

Gagnon KB, England R, Delpire E (2006) Characterization of SPAK and OSR1, regulatory kinases of the Na-K-2Cl cotransporter. Mol Cell Biol 26(2):689–698

Gamba G (2005) Molecular physiology and pathophysiology of electroneutral cation-chloride cotransporters. Physiol Rev 85(2):423–493. doi:10.1152/physrev.00011.2004

Gamba G (2010) The nanopeptide hormone vasopressin is a new player in the modulation of renal Na^{+}-Cl^{-} cotransporter activity. Kidney Int 78(2):127–129. doi:10.1038/ki.2010.147

Gamba G, Saltzberg SN, Lombardi M, Miyanoshita A, Lytton J, Hediger MA, Brenner BM, Hebert SC (1993) Primary structure and functional expression of a cDNA encoding the thiazide-sensitive, electroneutral sodium-chloride cotransporter. Proc Natl Acad Sci USA 90 (7):2749–2753

Gamba G, Miyanoshita A, Lombardi M, Lytton J, Lee WS, Hediger MA, Hebert SC (1994) Molecular cloning, primary structure, and characterization of two members of the mammalian electroneutral sodium-(potassium)-chloride cotransporter family expressed in kidney. J Biol Chem 269(26):17713–17722

Geva Y, Schuldiner M (2014) The back and forth of cargo exit from the endoplasmic reticulum. Curr Biol 24(3):R130–136. doi:10.1016/j.cub.2013.12.008

Glover M, Mercier Zuber A, Figg N, O'Shaughnessy KM (2010) The activity of the thiazide-sensitive Na^+-Cl^- cotransporter is regulated by protein phosphatase PP4. Can J Physiol Pharmacol 88(10):986–995. doi:10.1139/y10-080

Glover M, Ware JS, Henry A, Wolley M, Walsh R, Wain LV, Xu S, Van't Hoff WG, Tobin MD, Hall IP, Cook S, Gordon RD, Stowasser M, O'Shaughnessy KM (2014) Detection of mutations in KLHL3 and CUL3 in families with FHHt (familial hyperkalaemic hypertension or Gordon's syndrome). Clin Sci (Lond) 126(10):721–726. doi:10.1042/CS20130326

Glozman R, Okiyoneda T, Mulvihill CM, Rini JM, Barriere H, Lukacs GL (2009) N-glycans are direct determinants of CFTR folding and stability in secretory and endocytic membrane traffic. J Cell Biol 184(6):847–862. doi:10.1083/jcb.200808124

Golbang AP, Cope G, Hamad A, Murthy M, Liu CH, Cuthbert AW, O'Shaughnessy KM (2006) Regulation of the expression of the Na/Cl cotransporter (NCCT) by WNK4 and WNK1: evidence that accelerated dynamin-dependent endocytosis is not involved. Am J Physiol Renal Physiol 291(6):F1369–76

Gonzalez-Villalobos RA, Janjoulia T, Fletcher NK, Giani JF, Nguyen MT, Riquier-Brison AD, Seth DM, Fuchs S, Eladari D, Picard N, Bachmann S, Delpire E, Peti-Peterdi J, Navar LG, Bernstein KE, McDonough AA (2013) The absence of intrarenal ACE protects against hypertension. J Clin Invest 123(5):2011–2023. doi:10.1172/JCI65460

Gordon RD (1986) Syndrome of hypertension and hyperkalemia with normal glomerular filtration rate. Hypertension 8(2):93–102

Grimm PR, Taneja TK, Liu J, Coleman R, Chen YY, Delpire E, Wade JB, Welling PA (2012) SPAK isoforms and OSR1 regulate sodium-chloride co-transporters in a nephron-specific manner. J Biol Chem 287(45):37673–37690. doi:10.1074/jbc.M112.402800

Guyton AC (1991) Blood pressure control--special role of the kidneys and body fluids. Science 252(5014):1813–1816

Hadchouel J, Delaloy C, Faure S, Achard JM, Jeunemaitre X (2006) Familial hyperkalemic hypertension. J Am Soc Nephrol 17(1):208–217. doi:10.1681/ASN.2005030314

Hadchouel J, Soukaseum C, Busst C, Zhou XO, Baudrie V, Zurrer T, Cambillau M, Elghozi JL, Lifton RP, Loffing J, Jeunemaitre X (2010) Decreased ENaC expression compensates the increased NCC activity following inactivation of the kidney-specific isoform of WNK1 and prevents hypertension. Proc Natl Acad Sci USA 107(42):18109–18114. doi:10.1073/pnas.1006128107

Hoorn EJ, Walsh SB, McCormick JA, Furstenberg A, Yang CL, Roeschel T, Paliege A, Howie AJ, Conley J, Bachmann S, Unwin RJ, Ellison DH (2011) The calcineurin inhibitor tacrolimus activates the renal sodium chloride cotransporter to cause hypertension. Nat Med 17 (10):1304–1309. doi:10.1038/nm.2497

Hoover RS, Poch E, Monroy A, Vazquez N, Nishio T, Gamba G, Hebert SC (2003) N-Glycosylation at two sites critically alters thiazide binding and activity of the rat thiazide-sensitive Na^+:Cl^- cotransporter. J Am Soc Nephrol 14(2):271–282

Ji W, Foo JN, O'Roak BJ, Zhao H, Larson MG, Simon DB, Newton-Cheh C, State MW, Levy D, Lifton RP (2008) Rare independent mutations in renal salt handling genes contribute to blood pressure variation. Nat Genet 40(5):592–599. doi:10.1038/ng.118

Khuri RN, Strieder N, Wiederholt M, Giebisch G (1975) Effects of graded solute diuresis on renal tubular sodium transport in the rat. Am J Physiol 228(4):1262–1268

Kim GH, Masilamani S, Turner R, Mitchell C, Wade JB, Knepper MA (1998) The thiazide-sensitive Na-Cl cotransporter is an aldosterone-induced protein. Proc Natl Acad Sci USA 95 (24):14552–14557

Ko B, Joshi LM, Cooke LL, Vazquez N, Musch MW, Hebert SC, Gamba G, Hoover RS (2007) Phorbol ester stimulation of RasGRP1 regulates the sodium-chloride cotransporter by a PKC-independent pathway. Proc Natl Acad Sci USA 104(50):20120–20125. doi:10.1073/pnas.0709506104

Ko B, Kamsteeg EJ, Cooke LL, Moddes LN, Deen PM, Hoover RS (2010) RasGRP1 stimulation enhances ubiquitination and endocytosis of the sodium-chloride cotransporter. Am J Physiol Renal Physiol 299(2):F300–309. doi:10.1152/ajprenal.00441.2009

Ko B, Mistry AC, Hanson L, Mallick R, Cooke LL, Hack BK, Cunningham P, Hoover RS (2012) A new model of the distal convoluted tubule. Am J Physiol Renal Physiol 303(5):F700–710. doi:10.1152/ajprenal.00139.2012

Komers R, Rogers S, Oyama TT, Xu B, Yang CL, McCormick J, Ellison DH (2012) Enhanced phosphorylation of Na^+-Cl^- co-transporter in experimental metabolic syndrome: role of insulin. Clin Sci (Lond) 123(11):635–647. doi:10.1042/CS20120003

Krishnamurthy H, Gouaux E (2012) X-ray structures of LeuT in substrate-free outward-open and apo inward-open states. Nature 481(7382):469–474. doi:10.1038/nature10737

Krishnamurthy H, Piscitelli CL, Gouaux E (2009) Unlocking the molecular secrets of sodium-coupled transporters. Nature 459(7245):347–355. doi:10.1038/nature08143

Kunau R (1974) Nature of chloruretic effect of thiazide diuretics. Kidney Int 6(6):A62–A63

Kunchaparty S, Palcso M, Berkman J, Velazquez H, Desir GV, Bernstein P, Reilly RF, Ellison DH (1999) Defective processing and expression of thiazide-sensitive Na-Cl cotransporter as a cause of Gitelman's syndrome. Am J Physiol 277(4 Pt 2):F643–649

Lalioti MD, Zhang J, Volkman HM, Kahle KT, Hoffmann KE, Toka HR, Nelson-Williams C, Ellison DH, Flavell R, Booth CJ, Lu Y, Geller DS, Lifton RP (2006) Wnk4 controls blood pressure and potassium homeostasis via regulation of mass and activity of the distal convoluted tubule. Nat Genet 38(10):1124–1132. doi:10.1038/ng1877

Lazrak A, Liu Z, Huang CL (2006) Antagonistic regulation of ROMK by long and kidney-specific WNK1 isoforms. Proc Natl Acad Sci USA 103(5):1615–1620. doi:10.1073/pnas.0510609103

Lee DH, Riquier AD, Yang LE, Leong PK, Maunsbach AB, McDonough AA (2009) Acute hypertension provokes acute trafficking of distal tubule Na-Cl cotransporter (NCC) to subapical cytoplasmic vesicles. Am J Physiol Renal Physiol 296(4):F810–818. doi:10.1152/ajprenal.90606.2008

Lee DH, Maunsbach AB, Riquier-Brison AD, Nguyen MT, Fenton RA, Bachmann S, Yu AS, McDonough AA (2013) Effects of ACE inhibition and ANG II stimulation on renal Na-Cl cotransporter distribution, phosphorylation, and membrane complex properties. Am J Physiol Cell Physiol 304(2):C147–163. doi:10.1152/ajpcell.00287.2012

Lin SH, Yu IS, Jiang ST, Lin SW, Chu P, Chen A, Sytwu HK, Sohara E, Uchida S, Sasaki S, Yang SS (2011) Impaired phosphorylation of Na^+-K^+-$2Cl^-$ cotransporter by oxidative stress-responsive kinase-1 deficiency manifests hypotension and Bartter-like syndrome. Proc Natl Acad Sci USA 108(42):17538–17543. doi:10.1073/pnas.1107452108

Liu Z, Xie J, Wu T, Truong T, Auchus RJ, Huang CL (2011) Downregulation of NCC and NKCC2 cotransporters by kidney-specific WNK1 revealed by gene disruption and transgenic mouse models. Hum Mol Genet 20(5):855–866. doi:10.1093/hmg/ddq525

Louis-Dit-Picard H, Barc J, Trujillano D, Miserey-Lenkei S, Bouatia-Naji N, Pylypenko O, Beaurain G, Bonnefond A, Sand O, Simian C, Vidal-Petiot E, Soukaseum C, Mandet C, Broux F, Chabre O, Delahousse M, Esnault V, Fiquet B, Houillier P, Bagnis CI, Koenig J, Konrad M, Landais P, Mourani C, Niaudet P, Probst V, Thauvin C, Unwin RJ, Soroka SD, Ehret G, Ossowski S, Caulfield M, International Consortium for Blood Pressure, Bruneval P, Estivill X, Froguel P, Hadchouel J, Schott JJ, Jeunemaitre X (2012) KLHL3 mutations cause familial hyperkalemic hypertension by impairing ion transport in the distal nephron. Nat Genet 44(4):456–460. doi:10.1038/ng.2218, S451–453

Lytle C, Forbush B 3rd (1996) Regulatory phosphorylation of the secretory Na-K-Cl cotransporter: modulation by cytoplasmic Cl. Am J Physiol 270(2 Pt 1):C437–448

Markadieu N, Rios K, Spiller BW, McDonald WH, Welling PA, Delpire E (2014) Short forms of SPAK in the kidney are created by Dnpep-mediated proteolytic cleavage. J Biol Chem. doi:10.1074/jbc.M114.604009

Mayan H, Vered I, Mouallem M, Tzadok-Witkon M, Pauzner R, Farfel Z (2002) Pseudohypoaldosteronism type II: marked sensitivity to thiazides, hypercalciuria, normomagnesemia, and low bone mineral density. J Clin Endocrinol Metab 87(7):3248–3254
McCormick JA, Yang CL, Ellison DH (2008) WNK kinases and renal sodium transport in health and disease: an integrated view. Hypertension 51(3):588–596
McCormick JA, Mutig K, Nelson JH, Saritas T, Hoorn EJ, Yang CL, Rogers S, Curry J, Delpire E, Bachmann S, Ellison DH (2011) A SPAK isoform switch modulates renal salt transport and blood pressure. Cell Metab 14(3):352–364. doi:10.1016/j.cmet.2011.07.009
McDonough AA, Youn JH (2013) Need to quickly excrete K^+? Turn off NCC. Kidney Int 83 (5):779–782. doi:10.1038/ki.2012.468
Mederle K, Mutig K, Paliege A, Carota I, Bachmann S, Castrop H, Oppermann M (2013) Loss of WNK3 is compensated for by the WNK1/SPAK axis in the kidney of the mouse. Am J Physiol Renal Physiol 304(9):F1198–1209. doi:10.1152/ajprenal.00288.2012
Min X, Lee BH, Cobb MH, Goldsmith EJ (2004) Crystal structure of the kinase domain of WNK1, a kinase that causes a hereditary form of hypertension. Structure (Camb) 12(7):1303–1311
Moreno E, Cristobal PS, Rivera M, Vazquez N, Bobadilla NA, Gamba G (2006) Affinity-defining domains in the Na-Cl cotransporter: a different location for Cl^- and thiazide binding. J Biol Chem 281(25):17266–17275. doi:10.1074/jbc.M602614200
Morsing P, Velazquez H, Wright FS, Ellison DH (1991) Adaptation of distal convoluted tubule of rats. II. Effects of chronic thiazide infusion. Am J Physiol 261(1 Pt 2):F137–143
Mutig K, Saritas T, Uchida S, Kahl T, Borowski T, Paliege A, Bohlick A, Bleich M, Shan Q, Bachmann S (2010) Short-term stimulation of the thiazide-sensitive Na^+-Cl^- cotransporter by vasopressin involves phosphorylation and membrane translocation. Am J Physiol Renal Physiol 298(3):F502–509. doi:10.1152/ajprenal.00476.2009
Navar LG, Rosivall L (1984) Contribution of the renin-angiotensin system to the control of intrarenal hemodynamics. Kidney Int 25(6):857–868
Nedvetsky PI, Tamma G, Beulshausen S, Valenti G, Rosenthal W, Klussmann E (2009) Regulation of aquaporin-2 trafficking. Handb Exp Pharmacol 190:133–157. doi:10.1007/978-3-540-79885-9_6
Needham PG, Mikoluk K, Dhakarwal P, Khadem S, Snyder AC, Subramanya AR, Brodsky JL (2011) The thiazide-sensitive NaCl cotransporter is targeted for chaperone-dependent endoplasmic reticulum-associated degradation. J Biol Chem 286(51):43611–43621. doi:10.1074/jbc.M111.288928
Nishida H, Sohara E, Nomura N, Chiga M, Alessi DR, Rai T, Sasaki S, Uchida S (2012) Phosphatidylinositol 3-kinase/Akt signaling pathway activates the WNK-OSR1/SPAK-NCC phosphorylation cascade in hyperinsulinemic db/db mice. Hypertension 60(4):981–990. doi:10.1161/HYPERTENSIONAHA.112.201509
Novello FC, Sprague JM (1957) Benzothiadiazine dioxides as novel diuretics. J Am Chem Soc 79 (8):2028–2029. doi:10.1021/Ja01565a079
O'Reilly M, Marshall E, Speirs HJ, Brown RW (2003) WNK1, a gene within a novel blood pressure control pathway, tissue-specifically generates radically different isoforms with and without a kinase domain. J Am Soc Nephrol 14(10):2447–2456
O'Reilly M, Marshall E, Macgillivray T, Mittal M, Xue W, Kenyon CJ, Brown RW (2006) Dietary electrolyte-driven responses in the renal WNK kinase pathway in vivo. J Am Soc Nephrol 17 (9):2402–2413
Ohta A, Schumacher FR, Mehellou Y, Johnson C, Knebel A, Macartney TJ, Wood NT, Alessi DR, Kurz T (2013) The CUL3-KLHL3 E3 ligase complex mutated in Gordon's hypertension syndrome interacts with and ubiquitylates WNK isoforms: disease-causing mutations in KLHL3 and WNK4 disrupt interaction. Biochem J 451(1):111–122. doi:10.1042/BJ20121903
Oi K, Sohara E, Rai T, Misawa M, Chiga M, Alessi DR, Sasaki S, Uchida S (2012) A minor role of WNK3 in regulating phosphorylation of renal NKCC2 and NCC co-transporters in vivo. Biol Open 1(2):120–127. doi:10.1242/bio.2011048

Pacheco-Alvarez D, Cristobal PS, Meade P, Moreno E, Vazquez N, Munoz E, Diaz A, Juarez ME, Gimenez I, Gamba G (2006) The Na^{+}:Cl^{-} cotransporter is activated and phosphorylated at the amino-terminal domain upon intracellular chloride depletion. J Biol Chem 281 (39):28755–28763. doi:10.1074/jbc.M603773200

Paver WK, Pauline GJ (1964) Hypertension and hyperpotassaemia without renal disease in a young male. Med J Aust 2:305–306

Pearce D (2001) The role of SGK1 in hormone-regulated sodium transport. Trends Endocrinol Metab 12(8):341–347, doi:S1043-2760(01)00439-8 [pii]

Pedersen NB, Hofmeister MV, Rosenbaek LL, Nielsen J, Fenton RA (2010) Vasopressin induces phosphorylation of the thiazide-sensitive sodium chloride cotransporter in the distal convoluted tubule. Kidney Int 78(2):160–169. doi:10.1038/ki.2010.130

Piala AT, Moon TM, Akella R, He H, Cobb MH, Goldsmith EJ (2014) Chloride sensing by WNK1 involves inhibition of autophosphorylation. Sci Signal 7(324):ra41. doi:10.1126/scisignal.2005050

Picard N, Trompf K, Yang CL, Miller RL, Carrel M, Loffing-Cueni D, Fenton RA, Ellison DH, Loffing J (2014) Protein phosphatase 1 inhibitor-1 deficiency reduces phosphorylation of renal NaCl cotransporter and causes arterial hypotension. J Am Soc Nephrol 25(3):511–522. doi:10.1681/ASN.2012121202

Piechotta K, Garbarini N, England R, Delpire E (2003) Characterization of the interaction of the stress kinase SPAK with the Na^{+}-K^{+}-$2Cl^{-}$ cotransporter in the nervous system: evidence for a scaffolding role of the kinase. J Biol Chem 278(52):52848–52856

Rafiqi FH, Zuber AM, Glover M, Richardson C, Fleming S, Jovanovic S, Jovanovic A, O'Shaughnessy KM, Alessi DR (2010) Role of the WNK-activated SPAK kinase in regulating blood pressure. EMBO Mol Med 2(2):63–75. doi:10.1002/emmm.200900058

Renfro JL (1977) Interdependence of Active Na^{+} and Cl^{-} transport by the isolated urinary bladder of the teleost, Pseudopleuronectes americanus. J Exp Zool 199(3):383–390. doi:10.1002/jez.1401990311

Rengarajan S, Lee DH, Oh YT, Delpire E, Youn JH, McDonough AA (2014) Increasing plasma [K+] by intravenous potassium infusion reduces NCC phosphorylation and drives kaliuresis and natriuresis. Am J Physiol Renal Physiol 306(9):F1059–1068. doi:10.1152/ajprenal.00015.2014

Richardson C, Alessi DR (2008) The regulation of salt transport and blood pressure by the WNK-SPAK/OSR1 signalling pathway. J Cell Sci 121(Pt 20):3293–3304. doi:10.1242/jcs.029223

Richardson C, Rafiqi FH, Karlsson HK, Moleleki N, Vandewalle A, Campbell DG, Morrice NA, Alessi DR (2008) Activation of the thiazide-sensitive Na^{+}-Cl^{-} cotransporter by the WNK-regulated kinases SPAK and OSR1. J Cell Sci 121(Pt 5):675–684

Rinehart J, Kahle KT, de Los HP, Vazquez N, Meade P, Wilson FH, Hebert SC, Gimenez I, Gamba G, Lifton RP (2005) WNK3 kinase is a positive regulator of NKCC2 and NCC, renal cation-Cl^{-} cotransporters required for normal blood pressure homeostasis. Proc Natl Acad Sci USA 102(46):16777–16782. doi:10.1073/pnas.0508303102

Rinehart J, Vazquez N, Kahle KT, Hodson CA, Ring AM, Gulcicek EE, Louvi A, Bobadilla NA, Gamba G, Lifton RP (2011) WNK2 kinase is a novel regulator of essential neuronal cation-chloride cotransporters. J Biol Chem 286(34):30171–30180. doi:10.1074/jbc.M111.222893

Riveira-Munoz E, Chang Q, Godefroid N, Hoenderop JG, Bindels RJ, Dahan K, Devuyst O, Belgian Network for Study of Gitelman Syndrome (2007) Transcriptional and functional analyses of SLC12A3 mutations: new clues for the pathogenesis of Gitelman syndrome. J Am Soc Nephrol 18(4):1271–1283. doi:10.1681/ASN.2006101095

Ronzaud C, Loffing-Cueni D, Hausel P, Debonneville A, Malsure SR, Fowler-Jaeger N, Boase NA, Perrier R, Maillard M, Yang B, Stokes JB, Koesters R, Kumar S, Hummler E, Loffing J, Staub O (2013) Renal tubular NEDD4-2 deficiency causes NCC-mediated salt-dependent hypertension. J Clin Invest 123(2):657–665. doi:10.1172/JCI61110

Rosenbaek LL, Assentoft M, Pedersen NB, MacAulay N, Fenton RA (2012) Characterization of a novel phosphorylation site in the sodium-chloride cotransporter, NCC. J Physiol 590 (Pt 23):6121–6139. doi:10.1113/jphysiol.2012.240986

Rosenbaek LL, Kortenoeven ML, Aroankins TS, Fenton RA (2014) Phosphorylation decreases ubiquitylation of the thiazide-sensitive cotransporter NCC and subsequent clathrin-mediated endocytosis. J Biol Chem 289(19):13347–13361. doi:10.1074/jbc.M113.543710

Rozansky DJ, Cornwall T, Subramanya AR, Rogers S, Yang YF, David LL, Zhu X, Yang CL, Ellison DH (2009a) Aldosterone mediates activation of the thiazide-sensitive Na-Cl cotransporter through an SGK1 and WNK4 signaling pathway. J Clin Invest 119 (9):2601–2612. doi:10.1172/JCI38323

Rozansky DJ, Cornwall T, Subramanya AR, Rogers S, Yang YF, David LL, Zhu XM, Yang CL, Ellison DH (2009b) Aldosterone mediates activation of the thiazide-sensitive Na-Cl cotransporter through an SGK1 and WNK4 signaling pathway. J Clin Investig 119 (9):2601–2612. doi:10.1172/Jci38323

Sabath E, Meade P, Berkman J, de los Heros P, Moreno E, Bobadilla NA, Vazquez N, Ellison DH, Gamba G (2004) Pathophysiology of functional mutations of the thiazide-sensitive Na-Cl cotransporter in Gitelman disease. Am J Physiol-Renal 287(2):F195–F203. doi:10.1152/Ajprenal.00044.2004

San-Cristobal P, Pacheco-Alvarez D, Richardson C, Ring AM, Vazquez N, Rafiqi FH, Chari D, Kahle KT, Leng Q, Bobadilla NA, Hebert SC, Alessi DR, Lifton RP, Gamba G (2009) Angiotensin II signaling increases activity of the renal Na-Cl cotransporter through a WNK4-SPAK-dependent pathway. Proc Natl Acad Sci USA 106(11):4384–4389. doi:10.1073/pnas.0813238106

Sandberg MB, Maunsbach AB, McDonough AA (2006) Redistribution of distal tubule Na^+-Cl^- cotransporter (NCC) in response to a high-salt diet. Am J Physiol Renal Physiol 291(2): F503–508. doi:10.1152/ajprenal.00482.2005

Sandberg MB, Riquier AD, Pihakaski-Maunsbach K, McDonough AA, Maunsbach AB (2007) ANG II provokes acute trafficking of distal tubule Na^+-Cl^- cotransporter to apical membrane. Am J Physiol Renal Physiol 293(3):F662–669. doi:10.1152/ajprenal.00064.2007

Saritas T, Borschewski A, McCormick JA, Paliege A, Dathe C, Uchida S, Terker A, Himmerkus N, Bleich M, Demaretz S, Laghmani K, Delpire E, Ellison DH, Bachmann S, Mutig K (2013) SPAK differentially mediates vasopressin effects on sodium cotransporters. J Am Soc Nephrol 24(3):407–418. doi:10.1681/ASN.2012040404

Shibata S, Zhang J, Puthumana J, Stone KL, Lifton RP (2013) Kelch-like 3 and Cullin 3 regulate electrolyte homeostasis via ubiquitination and degradation of WNK4. Proc Natl Acad Sci USA 110(19):7838–7843. doi:10.1073/pnas.1304592110

Simon DBFZ, Ellison D, Bia M, Tucci J, Lifton RP (1995) Examination of the thiazide-sensitive NaCl cotransporter as a candidate gene in Gordon's syndrome. J Am Soc Nephrol 6:632

Simon DB, Nelson-Williams C, Bia MJ, Ellison D, Karet FE, Molina AM, Vaara I, Iwata F, Cushner HM, Koolen M, Gainza FJ, Gitleman HJ, Lifton RP (1996) Gitelman's variant of Bartter's syndrome, inherited hypokalaemic alkalosis, is caused by mutations in the thiazide-sensitive Na-Cl cotransporter. Nat Genet 12(1):24–30

Simpson F, Peden AA, Christopoulou L, Robinson MS (1997) Characterization of the adaptor-related protein complex, AP-3. J Cell Biol 137(4):835–845

Singh SK, Piscitelli CL, Yamashita A, Gouaux E (2008) A competitive inhibitor traps LeuT in an open-to-out conformation. Science 322(5908):1655–1661. doi:10.1126/science.1166777

Snyder PM, Olson DR, Kabra R, Zhou R, Steines JC (2004) cAMP and serum and glucocorticoid-inducible kinase (SGK) regulate the epithelial Na^+ channel through convergent phosphorylation of Nedd4-2. J Biol Chem 279(44):45753–45758. doi:10.1074/jbc.M407858200

Sohara E, Rai T, Yang SS, Ohta A, Naito S, Chiga M, Nomura N, Lin SH, Vandewalle A, Ohta E, Sasaki S, Uchida S (2011) Acute insulin stimulation induces phosphorylation of the Na-Cl cotransporter in cultured distal mpkDCT cells and mouse kidney. PLoS ONE 6(8), e24277. doi:10.1371/journal.pone.0024277

Sorensen MV, Grossmann S, Roesinger M, Gresko N, Todkar AP, Barmettler G, Ziegler U, Odermatt A, Loffing-Cueni D, Loffing J (2013) Rapid dephosphorylation of the renal sodium chloride cotransporter in response to oral potassium intake in mice. Kidney Int 83(5):811–824. doi:10.1038/ki.2013.14

Stokes JB (1984) Sodium chloride absorption by the urinary bladder of the winter flounder. A thiazide-sensitive, electrically neutral transport system. J Clin Invest 74(1):7–16. doi:10.1172/JCI111420

Subramanya AR, Ellison DH (2014) Distal convoluted tubule. Clin J Am Soc Nephrol. doi:10.2215/CJN.05920613

Subramanya AR, Welling PA (2011) Toward an understanding of hypertension resistance. Am J Physiol Renal Physiol 300(4):F838–839. doi:10.1152/ajprenal.00078.2011

Subramanya AR, Yang CL, McCormick JA, Ellison DH (2006a) WNK kinases regulate sodium chloride and potassium transport by the aldosterone-sensitive distal nephron. Kidney Int 70 (4):630–634. doi:10.1038/sj.ki.5001634

Subramanya AR, Yang CL, Zhu X, Ellison DH (2006b) Dominant-negative regulation of WNK1 by its kidney-specific kinase-defective isoform. Am J Physiol Renal Physiol 290(3):F619–624. doi:10.1152/ajprenal.00280.2005

Subramanya AR, Liu J, Ellison DH, Wade JB, Welling PA (2009) WNK4 diverts the thiazide-sensitive NaCl cotransporter to the lysosome and stimulates AP-3 interaction. J Biol Chem 284 (27):18471–18480. doi:10.1074/jbc.M109.008185

Takahashi D, Mori T, Nomura N, Khan MZ, Araki Y, Zeniya M, Sohara E, Rai T, Sasaki S, Uchida S (2014) WNK4 is the major WNK positively regulating NCC in the mouse kidney. Biosci Rep 34(3):e00107. doi:10.1042/BSR20140047

Talati G, Ohta A, Rai T, Sohara E, Naito S, Vandewalle A, Sasaki S, Uchida S (2010) Effect of angiotensin II on the WNK-OSR1/SPAK-NCC phosphorylation cascade in cultured mpkDCT cells and in vivo mouse kidney. Biochem Biophys Res Commun 393(4):844–848. doi:10.1016/j.bbrc.2010.02.096

Terker AS, Yang CL, McCormick JA, Meermeier NP, Rogers SL, Grossmann S, Trompf K, Delpire E, Loffing J, Ellison DH (2014) Sympathetic stimulation of thiazide-sensitive sodium chloride cotransport in the generation of salt-sensitive hypertension. Hypertension 64 (1):178–184. doi:10.1161/Hypertensionaha.114.03335

Thastrup JO, Rafiqi FH, Vitari AC, Pozo-Guisado E, Deak M, Mehellou Y, Alessi DR (2012) SPAK/OSR1 regulate NKCC1 and WNK activity: analysis of WNK isoform interactions and activation by T-loop trans-autophosphorylation. Biochem J 441(1):325–337. doi:10.1042/BJ20111879

Tovar-Palacio C, Bobadilla NA, Cortes P, Plata C, de los Heros P, Vazquez N, Gamba G (2004) Ion and diuretic specificity of chimeric proteins between apical Na + -K + -2Cl − and Na +-Cl − cotransporters. Am J Physiol Renal Physiol 287(3):F570–577. doi:10.1152/ajprenal.00124.2004

Vallon V, Schroth J, Lang F, Kuhl D, Uchida S (2009) Expression and phosphorylation of the Na^+-Cl^- cotransporter NCC in vivo is regulated by dietary salt, potassium, and SGK1. Am J Physiol Renal Physiol 297(3):F704–712. doi:10.1152/ajprenal.00030.2009

van der Lubbe N, Lim CH, Fenton RA, Meima ME, Jan Danser AH, Zietse R, Hoorn EJ (2011) Angiotensin II induces phosphorylation of the thiazide-sensitive sodium chloride cotransporter independent of aldosterone. Kidney Int 79(1):66–76. doi:10.1038/ki.2010.290

van der Lubbe N, Lim CH, Meima ME, van Veghel R, Rosenbaek LL, Mutig K, Danser AH, Fenton RA, Zietse R, Hoorn EJ (2012) Aldosterone does not require angiotensin II to activate NCC through a WNK4-SPAK-dependent pathway. Pflugers Arch 463(6):853–863. doi:10.1007/s00424-012-1104-0

Vazquez N, Monroy A, Dorantes E, Munoz-Clares RA, Gamba G (2002) Functional differences between flounder and rat thiazide-sensitive Na-Cl cotransporter. Am J Physiol Renal Physiol 282(4):F599–607. doi:10.1152/ajprenal.00284.2001

Velazquez H, Bernstein PL, Bartiss A, Kunchaparty S, Ellison DH (1995) Thiazide-sensitive NaCl transport capacity of rat renal distal tubule is enhanced by adrenal-steroids. FASEB J 9(3): A586–A586

Velazquez H, Naray-Fejes-Toth A, Silva T, Andujar E, Reilly RF, Desir GV, Ellison DH (1998) Rabbit distal convoluted tubule coexpresses NaCl cotransporter and 11 beta-hydroxysteroid dehydrogenase II mRNA. Kidney Int 54(2):464–472. doi:10.1046/j.1523-1755.1998.00036.x

Vembar SS, Brodsky JL (2008) One step at a time: endoplasmic reticulum-associated degradation. Nat Rev Mol Cell Biol 9(12):944–957. doi:10.1038/nrm2546

Verlander JW, Tran TM, Zhang L, Kaplan MR, Hebert SC (1998) Estradiol enhances thiazide-sensitive NaCl cotransporter density in the apical plasma membrane of the distal convoluted tubule in ovariectomized rats. J Clin Invest 101(8):1661–1669. doi:10.1172/JCI601

Vidal-Petiot E, Cheval L, Faugeroux J, Malard T, Doucet A, Jeunemaitre X, Hadchouel J (2012) A new methodology for quantification of alternatively spliced exons reveals a highly tissue-specific expression pattern of WNK1 isoforms. PLoS ONE 7(5), e37751. doi:10.1371/journal.pone.0037751

Vidal-Petiot E, Elvira-Matelot E, Mutig K, Soukaseum C, Baudrie V, Wu S, Cheval L, Huc E, Cambillau M, Bachmann S, Doucet A, Jeunemaitre X, Hadchouel J (2013) WNK1-related familial hyperkalemic hypertension results from an increased expression of l-wnk1 specifically in the distal nephron. Proc Natl Acad Sci USA 110(35):14366–14371. doi:10.1073/pnas.1304230110

Villa F, Goebel J, Rafiqi FH, Deak M, Thastrup J, Alessi DR, van Aalten DM (2007) Structural insights into the recognition of substrates and activators by the OSR1 kinase. EMBO Rep 8 (9):839–845. doi:10.1038/sj.embor.7401048

Vitari AC, Deak M, Morrice NA, Alessi DR (2005) The WNK1 and WNK4 protein kinases that are mutated in Gordon's hypertension syndrome phosphorylate and activate SPAK and OSR1 protein kinases. Biochem J 391(Pt 1):17–24

Vitari AC, Thastrup J, Rafiqi FH, Deak M, Morrice NA, Karlsson HK, Alessi DR (2006) Functional interactions of the SPAK/OSR1 kinases with their upstream activator WNK1 and downstream substrate NKCC1. Biochem J 397(1):223–231. doi:10.1042/BJ20060220

Wade JB, Fang L, Liu J, Li D, Yang CL, Subramanya AR, Maouyo D, Mason A, Ellison DH, Welling PA (2006) WNK1 kinase isoform switch regulates renal potassium excretion. Proc Natl Acad Sci USA 103(22):8558–8563. doi:10.1073/pnas.0603109103

Wakabayashi M, Mori T, Isobe K, Sohara E, Susa K, Araki Y, Chiga M, Kikuchi E, Nomura N, Mori Y, Matsuo H, Murata T, Nomura S, Asano T, Kawaguchi H, Nonoyama S, Rai T, Sasaki S, Uchida S (2013) Impaired KLHL3-mediated ubiquitination of WNK4 causes human hypertension. Cell Rep 3(3):858–868. doi:10.1016/j.celrep.2013.02.024

Ward CL, Kopito RR (1994) Intracellular turnover of cystic fibrosis transmembrane conductance regulator. Inefficient processing and rapid degradation of wild-type and mutant proteins. J Biol Chem 269(41):25710–25718

Warmuth S, Zimmermann I, Dutzler R (2009) X-ray structure of the C-terminal domain of a prokaryotic cation-chloride cotransporter. Structure 17(4):538–546, doi:S0969-2126(09)00090-2 [pii]10.1016/j.str.2009.02.009

Wilson FH, Disse-Nicodeme S, Choate KA, Ishikawa K, Nelson-Williams C, Desitter I, Gunel M, Milford DV, Lipkin GW, Achard JM, Feely MP, Dussol B, Berland Y, Unwin RJ, Mayan H, Simon DB, Farfel Z, Jeunemaitre X, Lifton RP (2001) Human hypertension caused by mutations in WNK kinases. Science 293(5532):1107–1112. doi:10.1126/science.1062844

Wilson FH, Kahle KT, Sabath E, Lalioti MD, Rapson AK, Hoover RS, Hebert SC, Gamba G, Lifton RP (2003) Molecular pathogenesis of inherited hypertension with hyperkalemia: the Na-Cl cotransporter is inhibited by wild-type but not mutant WNK4. Proc Natl Acad Sci USA 100(2):680–684. doi:10.1073/pnas.242735399

Xu B, English JM, Wilsbacher JL, Stippec S, Goldsmith EJ, Cobb MH (2000) WNK1, a novel mammalian serine/threonine protein kinase lacking the catalytic lysine in subdomain II. J Biol Chem 275(22):16795–16801

Yang CL, Angell J, Mitchell R, Ellison DH (2003) WNK kinases regulate thiazide-sensitive Na-Cl cotransport. J Clin Invest 111(7):1039–1045. doi:10.1172/JCI17443
Yang CL, Zhu X, Wang Z, Subramanya AR, Ellison DH (2005) Mechanisms of WNK1 and WNK4 interaction in the regulation of thiazide-sensitive NaCl cotransport. J Clin Invest 115 (5):1379–1387. doi:10.1172/JCI22452
Yang CL, Zhu X, Ellison DH (2007a) The thiazide-sensitive Na-Cl cotransporter is regulated by a WNK kinase signaling complex. J Clin Invest 117(11):3403–3411. doi:10.1172/JCI32033
Yang SS, Morimoto T, Rai T, Chiga M, Sohara E, Ohno M, Uchida K, Lin SH, Moriguchi T, Shibuya H, Kondo Y, Sasaki S, Uchida S (2007b) Molecular pathogenesis of pseudohypoaldosteronism type II: generation and analysis of a Wnk4(D561A/+) knockin mouse model. Cell Metab 5(5):331–344
Yang SS, Lo YF, Wu CC, Lin SW, Yeh CJ, Chu P, Sytwu HK, Uchida S, Sasaki S, Lin SH (2010) SPAK-knockout mice manifest Gitelman syndrome and impaired vasoconstriction. J Am Soc Nephrol 21(11):1868–1877. doi:10.1681/ASN.2009121295
Yang SS, Fang YW, Tseng MH, Chu PY, Yu IS, Wu HC, Lin SW, Chau T, Uchida S, Sasaki S, Lin YF, Sytwu HK, Lin SH (2013) Phosphorylation regulates NCC stability and transporter activity in vivo. J Am Soc Nephrol 24(10):1587–1597. doi:10.1681/ASN.2012070742
Zerangue N, Schwappach B, Jan YN, Jan LY (1999) A new ER trafficking signal regulates the subunit stoichiometry of plasma membrane K(ATP) channels. Neuron 22(3):537–548
Zhang C, Wang L, Zhang J, Su XT, Lin DH, Scholl UI, Giebisch G, Lifton RP, Wang WH (2014) KCNJ10 determines the expression of the apical Na-Cl cotransporter (NCC) in the early distal convoluted tubule (DCT1). Proc Natl Acad Sci USA 111(32):11864–11869. doi:10.1073/pnas.1411705111
Zhou B, Zhuang J, Gu D, Wang H, Cebotaru L, Guggino WB, Cai H (2009) WNK4 enhances the degradation of NCC through a sortilin-mediated lysosomal pathway. J Am Soc Nephrol, doi: ASN.2008121275 [pii]10.1681/ASN.2008121275
Zhou B, Wang D, Feng X, Zhang Y, Wang Y, Zhuang J, Zhang X, Chen G, Delpire E, Gu D, Cai H (2012) WNK4 inhibits NCC protein expression through MAPK ERK1/2 signaling pathway. Am J Physiol Renal Physiol 302(5):F533–539. doi:10.1152/ajprenal.00032.2011

Chapter 13
NBCe1, an Electrogenic Na^+ Bicarbonate Cotransporter, in Epithelia

Aleksandra Sinđić and Michael F. Romero

Abstract Na^+/bicarbonate (HCO_3^-) cotransporters (NBCs) are involved in maintaining both cellular and systemic pH homeostasis. Thus far, NBCs are found in the SLC4 gene family, which also includes Na^+-independent anion exchangers. Of the ten SLC4 genes, five genes are Na^+ bicarbonate cotransport systems: 2 electrogenic transporters, NBCe1 and 2 (SLC4A4 and A5); 2 electroneutral, NBCn1 and 2 (SLC4A7 and A8); and a Na^+-driven Cl^-/HCO_3^- exchanger (NDCBE; SLC4A10). There are five NBCe1 isoforms: NBCe1-A (kidney isoform), NBCe1-B (pancreatic and general isoform), NBCe1-C (brain isoform), NBCe1-D, and NBCe1-E (unknown localization). Malfunction of NBCe1 transporters in epithelial cells leads to inappropriate bicarbonate transport and cellular buffering. In humans, recessive NBCe1 (SLC4A4; OMIM #603345) mutations cause development of proximal renal tubular acidosis (pRTA), cataracts, glaucoma, and defective tooth enamel with occasional occurrence of intestinal obstruction. This chapter will focus on physiological and pathological NBCe1 protein function in epithelial tissues, especially those resulting in human disease.

Keywords Bicarbonate • Sodium • Epithelial transport • SLC4 • Electrogenic • Acid–base homeostasis • Proximal renal tubular acidosis • Glaucoma • Kidney • Eye • Pancreas

13.1 Introduction

For maintaining of pH in physiological range, transport of bicarbonate (HCO_3^-) is of crucial importance. There are numerous transporters involved in this regulation, Cl^-/HCO_3^- exchangers and Na^+/HCO_3^- cotransporters (NBC). Many of the

A. Sinđić
Physiology and Immunology, School of Medicine, University of Zagreb, Šalata 3, 10000 Zagreb, Croatia

M.F. Romero (✉)
Physiology and Biomedical Engineering, Nephrology and Hypertension, Mayo Clinic College of Medicine, 200 First St SW, Rochester, MN 55905, USA
e-mail: romero.michael@mayo.edu

K.L. Hamilton, D.C. Devor (eds.), *Ion Channels and Transporters of Epithelia in Health and Disease*, Physiology in Health and Disease,
DOI 10.1007/978-1-4939-3366-2_13

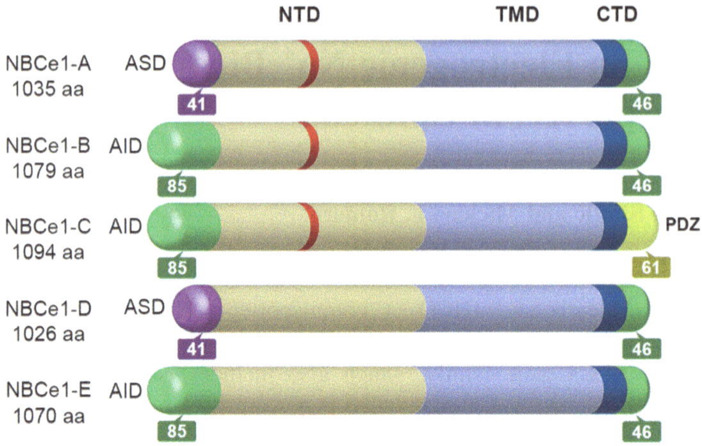

Fig. 13.1 Sequence differences of NBCe1 isoforms. NBCe1-A and NBCe1-D include an auto-stimulatory domain (ASD). All other variants include an auto-inhibitory domain (AID). NBCe1-C terminates with a PDZ-domain binding sequence. *NTD* N-terminal domain; *TMD* transmembrane domains; *CTD* C-terminal domain

proteins with these functions are in the SLC4 gene family (Romero et al. 2004, 2013a; Parker and Boron 2013). NBCs are divided into 2 electrogenic transporters, NBCe1 and 2 (SLC4A4 and A5); 2 electroneutral, NBCn1 and 2 (SLC4A7 and A8); and Na^+-driven Cl^-/HCO_3^- exchanger (NDCBE; SLC4A10). In contrast to electroneutral NBCs, which transport 1 Na^+ and 1 HCO_3^-, electrogenic NBCs have a stoichiometry 1:2 when transporting HCO_3^- into the cell and 1:3 when bicarbonate is transported out of the cell (see Sect. 13.3.1 below).

Electrogenic Na^+/HCO_3^- function is found in many tissues such as the kidney, pancreas, brain, small intestine, colon, eye, heart, liver, salivary glands, stomach, testis, epididymis, and uterus (Chinigarzadeh et al. 2014). Five isoforms of NBCe1 (Fig. 13.1) are encoded by the SLC4A4 gene which has 26 exons (Abuladze et al. 2000) located at human chromosome 4q21 (Romero et al. 1998a; Romero and Boron 1999; Abuladze et al. 2000). NBCe1-A (rNBC, aNBC, kNBC, kNBC1, hkNBCe1) was cloned initially using functional expression by Romero and colleagues (Romero et al. 1997), and subsequently human (Burnham et al. 1997) and rodent (Romero et al. 1998b) sequences were identified. A year later, NBCe1-B (pNBC, pNBC1, rpNBC, hh-NBC, hcNBC, rb1NBC) was cloned (Abuladze et al. 1998; Choi et al. 1999). In 2000, a C-terminal variation from the brain was reported (NBCe1-C, bNBC1, rb2NBC) (Bevensee et al. 2000). NBCe1-D and NBCe1-E transcripts have the identical sequence as NBCe1-A and NBCe1-B except they are missing a 9 amino-acid segment at the N-terminus (Fig. 13.1), but their protein expression has not yet been shown (Liu et al. 2011). NBCe1 proteins have N- and C-terminal ends located in the cytoplasm of the cell, 14 transmembrane domains (TM), and a large extracellular glycosylated loop between the 5th and 6th TM and a smaller one between TM 7 and 8 (Fig. 13.2). All NBCe1 isoforms share the same TM domain architecture but differ in the N- and C-termini (Fig. 13.2).

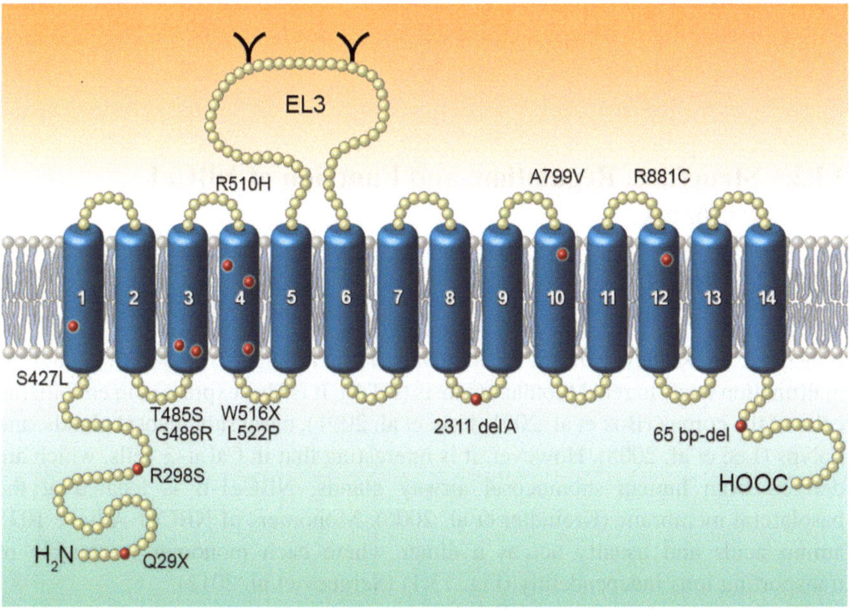

Fig. 13.2 Predicted membrane structure of NBCe1 transporter proteins. All isoforms have 14 transmembrane domains and C- and N-terminal ends which are located in the cell cytoplasm (1–14). Between TM5 and TM6 is a large extracellular glycosylated loop and smaller one between TM7 and 8. Mutations in human NBCe1-A are illustrated

To date, there are 12 reported recessive mutations and one compound, heterozygous mutation of NBCe1 (SLC4A4; OMIM #603345) (Fig. 13.2):

(a) Missense mutations: R298S, S427L, T485S, G486R, R510H, L522P, A799V, R881C
(b) Nonsense mutations: Q29X, W516X, with a frame shift at Δ2311A
(c) C-terminal 65-base-pair deletion (Kurtz and Zhu 2013; Romero et al. 2013a)
(d) R510H / Q913R compound, heterozygote (Myers, et al. 2015)

In addition to the systemic acidosis, several patients have extrarenal manifestations, e.g., short stature, cataracts, band-keratopathy, glaucoma, mental retardation, basal ganglia calcifications, migraine headaches, and tooth defects. Currently, there is some controversy regarding the nonrenal symptoms and their etiology, i.e., do these symptoms develop due to NBCe1-A mutations, systemic acidosis due to renal bicarbonate loss, or mutations in one of the other NBCe1 isoforms altering organ-specific physiology.

This chapter will focus on our current understanding of NBCe1-isoform localization, function, and regulation in epithelial tissues. The reader should keep in mind that several areas are understudied and that some epithelia and tissues make use of more than one NBCe1 isoform for normal tissue function. Likewise, the second messenger regulation of NBCe1 isoforms has not been exhaustively studied, but emerging themes will be noted. Finally, several expression systems, biophysical approaches, cell biology techniques, animal models, and human clinical phenotypes have been used by investigators to determine cell, tissue, and systemic physiology and pathophysiology associated with NBCe1 expression in epithelia.

13.2 Structure, Regulation, and Function of NBCe1 Isoforms

13.2.1 NBCe1-A

NBCe1-A or kidney isoform is involved in bicarbonate reabsorption, and its malfunction leads to renal tubular acidosis (RTA). It is also expressed in endothelial cells of the cornea (Bok et al. 2001; Usui et al. 2001), nasal submucosal glands, and polyps (Lee et al. 2005). However, it is interesting that in CaLu–3 cells, which are derived from human submucosal airway glands, NBCe1-B is located at the basolateral membrane (Kreindler et al. 2006). Monomers of NBCe1-A have 1035 amino acids and usually act as a dimer where each monomer is capable of transporting ions independently (Fig. 13.1) (Sergeev et al. 2012).

The N-terminal region of NBCe1-A has an auto-stimulatory domain (ASD) (Fig. 13.1) (McAlear et al. 2006). At the same region, Arg298 may participate in forming a "HCO_3^- tunnel" (Chang et al. 2008), which is disrupted in the R298S mutation (Igarashi et al. 1999; Horita et al. 2005; Chang et al. 2008; Shnitsar et al. 2013). The C-terminal region is important for membrane processing, and partial deletion of the C-terminal tail leads to intracellular localization of the transporter (Suzuki et al. 2010). Also, the sequence for basolateral targeting in kidney cells is located at the C-terminal end of the protein (Fig. 13.1) (Li et al. 2004). The most investigated mutations of NBCe1 are shown in Fig. 13.1, and the majority of the known mutations are located in TM domains (Fig. 13.2) (Zhu et al. 2010a, b). NBCe1-A shares the same 14 TM domains with other isoform: Each of the 14 TM domains has a different function in the protein. TM1 has 31 amino acids and it is important for ion permeation which is disrupted by the S427L mutation (Dinour et al. 2004; Zhu et al. 2013a). Ion interaction sites are located at TM3, and they are responsible for stoichiometry of the ion transport. Mutation in the TM3, T485S, reduces its function and converts an electrogenic transporter to electroneutral (Horita et al. 2005; Zhu et al. 2010b, 2013b). Two mutations (R510H and L522P) are known at TM4 which result in abnormal folding of the protein which does not reach the cellular membrane (Li et al. 2004; Demirci et al. 2006; Suzuki et al. 2008; Yamazaki et al. 2013). Specificity of TM5 is Asp555 which was identified as an important site of the transport inhibition by DIDS and selectivity to bicarbonate (Lu and Boron 2007; Yang et al. 2009b). The A799V mutation in TM10 causes, in addition to RTA, extrarenal symptoms like hypokalemia, diarrhea, vomiting, and muscle weakness (Horita et al. 2005). TM11 to 14 are important for maintaining protein stability. The R881C mutation in this region does not change protein function; rather, the protein is dysfunctional since it is not transported to the membrane (Horita et al. 2005; Toye et al. 2006; Zhu et al. 2010a, b).

Experimentally, NBCe1-A can be controlled by a variety of signaling pathways. Different steps in the Ca^{2+} signaling pathway are involved, starting with phosphatidylinositol 4,5-bisphosphate (PIP_2) which stimulates NBCe1-A (Wu et al. 2009). By activation of phospholipase C, PIP_2 is degraded to inositol 1,4,5-triphosphate (IP_3) and diacyl-glycerol (DAG). IP_3 releases Ca^{2+} which together with DAG activates protein kinase C (PKC) and Ca^{2+}/calmodulin-dependent protein kinase II (CaMKII). Both kinases regulate function of NBCe1-A probably by changing the expression of the protein on the cell membrane (Perry et al. 2007, 2009). Angiotensin II (AngII) inhibits this transporter by activation of Ca^{2+}-independent PKC (Perry et al. 2006). Another signaling pathway is involved in regulation, cAMP, and protein kinase A (PKA). Phosphorylation of Ser982 by PKA leads to a shift in transport stoichiometry from 1:3 to 1:2 (Gross et al. 2001b). Regulation of NBCe1-A by carbonic anhydrase (CA) is controversial. However, regulation by intracellular isoforms of CA increases transport activity (Schueler et al. 2011). This transporter is also regulated by Mg^{2+} and ATP (Heyer et al. 1999; Wu et al. 2009).

13.2.2 NBCe1-B

NBCe1-B begins with 85 different amino acids compared to NBCe1-A (45 unique amino acids at the N-terminus) for a total of 1079 amino acids (Fig. 13.1). NBCe1-B was originally cloned from the pancreas where it plays an important role in bicarbonate secretion (Abuladze et al. 1998) by the exocrine pancreas. NBCe1-B has been identified in the ameloblasts (Lacruz et al. 2010; Zheng et al. 2011), salivary gland (Roussa et al. 1999; Park et al. 2002; Koo et al. 2006), stomach (Rossmann et al. 1999; Konturek et al. 2004; Damkier et al. 2007), small intestine (Rossmann et al. 1999; Bartolo et al. 2009; Charoenphandhu et al. 2011; Jakab et al. 2011), colon (Bachmann et al. 2003, 2006; Zhao et al. 2005; Gawenis et al. 2007; Xue et al. 2007; Yu et al. 2009), thyroid (Galeza-Kulik et al. 2006), lungs (Kreindler et al. 2006), eye (Igarashi et al. 1999; Bok et al. 2001; Turner et al. 2001; Sun and Bonanno 2003), heart (Choi et al. 1999; Khandoudi et al. 2001; Ten Hove et al. 2004; Yamamoto et al. 2007; Orlowski et al. 2014; Wang et al. 2014), epididymis, testis, prostate, and uterus (Liu et al. 2011). In the brain, this NBCe1-B transporter is expressed in astrocytes, while NBCe1-C is expressed in neurons (Bevensee et al. 2000).

The N-terminus of NBCe1-B and NBCe1-C contains an auto-inhibitory domain (AID) shown to decrease NBCe1 transport (McAlear et al. 2006; Lee et al. 2012b). When this AID-domain interacts with IRBIT (inositol 1,4,5-*tris*-phosphate (IP_3) receptor binding protein), NBCe1-B/NBCe1-C are activated by relief of inhibition (Shirakabe et al. 2006). IRBIT activates protein phosphatase 1 (PP1) which attenuates the inhibitory effects of WNK/SPAK and AID (Yang et al. 2011; Lee et al. 2012b) (Fig. 13.4). In some cases, cAMP has been shown to increase NBCe1-B activity directly (Gross et al. 2003) and by increase in membrane expression (Yu et al. 2009).

13.2.3 NBCe1-C

A third, C-terminal variant of NBCe1 was cloned from the brain (Bevensee et al. 2000). NBCe1-C has 1094 amino acids and is similar to NBCe1-B except the C-end terminus of the protein which posses with PDZ-domain binding sequence (Bevensee et al. 2000) (Fig. 13.1). The NBCe1-C mRNA and protein were subsequently found in the hippocampus, cerebellum, olfactory bulb, and piriform cortex (Douglas et al. 1998, 2001; Schmitt et al. 2000; Majumdar et al. 2008). Messenger RNA for NBCe1-C has also been found in epithelia of the epididymis and testis (Liu et al. 2011).

13.3 Biophysics of NBCe1

13.3.1 Theoretical Considerations

Membranes in general present a barrier to ion and solute movements, but this is especially true for epithelia through which vectorial transport occurs. Cell membranes are mosaics of phospholipids, cholesterol, and proteins. These phospholipids compose the majority of the membrane and biophysically make a bilayer. The lipid bilayer is composed of two discrete zones: (a) the hydrophilic (water-liking) phosphate and sugar head-groups facing both the outside world and the inside world and (b) the middle hydrophobic (water repelling) core contains the C_{14}–C_{22} acyl chains. Physically, the cell bilayer functions as an electrical capacitor, separating charge, thereby creating a membrane potential (electrical gradient, V_m). However, an intact bilayer without proteins, unlike a capacitor, does not allow ions or solutes (uncharged molecules) passage. This separation of ions or solutes creates an additional solute concentration gradient (chemical gradient).

There are four major ways to move ions and solutes across this otherwise non-permeable membrane, all of which require a membrane protein. Mode 1 is active transport, i.e., hydrolysis of ATP (cellular energy) to move solutes against their gradients. Modes 2,3,4 (exchange, cotransport, and uniport) use the "potential energy" (Gibbs free energy, ΔG) stored as the electrochemical gradient or electrochemical potential difference, to move solutes across a membrane. The names imply their respective roles. The electrochemical potential difference ($\Delta\mu_{ion}$)[1] has an electrical part ($zF\Delta\Psi$)[2] and a chemical part ($RT\bullet\ln\{[ion]_{inside}/[ion]_{outside}\}$).[3] The resulting thermodynamic equation can be used to predict ions and solute directional movements, i.e., into or out of a cell (or vesicle). Which solutes contribute to the $\Delta\mu$

---

[1] $\Delta\mu_{ion} = RT\bullet\ln\{[ion]_{inside}/[ion]_{outside}\} + zF\Delta\Psi$.

[2] "z" is the net charge on a molecule, F is the Faraday constant, and $\Delta\Psi = \Psi(\text{inside}) - \Psi(\text{outside})$, which is also equal to the membrane potential (V_m).

[3] "R" is the gas constant, "T" is temperature in Kelvin degrees, and "ln" is the natural log.

equation are indicated the specific protein performing the transport activity. There are two important considerations. First, if no membrane protein is present, then there can be no transmembrane movement of ions/solutes. Second, the $\boldsymbol{\Delta\mu}$ calculation is a thermodynamic expression of potential energy and does not provide any information about rate of transport (kinetics, kinetic energy).

From cloning and expression experiments, there are two electrogenic Na^+/HCO_3^- cotransporters in mammals (Romero et al. 2013a): NBCe1 (SLC4A4) and NBCe2 (NBC4/ SLC4A5). A third NBCe is found in the invertebrate nervous system (Piermarini et al. 2007). Typically, these NBCe proteins are thought to cotransport Na^+ and 2–3 HCO_3^- across epithelial membranes or neuronal membranes (below "2NBCe" designates $1Na^+ + \underline{2}HCO_3^-$ (Eqs. 13.1 and 13.2) and "3NBCe" designates $1Na^+ + \underline{3}HCO_3^-$ (Eqs. 13.3 and 13.4)). There is some indication of regulation of the stoichiometry "property" (Heyer et al. 1999; Ducoudret et al. 2001; Gross et al. 2001a, b; Muller-Berger et al. 2001). This stoichiometry dictates that both membrane potential (V_m) and the ion gradients (Na^+ and HCO_3^-) determine the electrochemical gradient and thus dictate the direction of ion transport (see below).

Transport models

2:1	$\Delta\mu_{2NBCe} = 2\Delta\mu_{HCO3} + \Delta\mu_{Na}$	Eq. (13.1)
or	$\Delta\mu_{2NBCe} = RT \bullet \ln\left(\left\{[HCO_3^-]_i^2[Na^+]_i\right\} / \left\{[HCO_3^-]_o^2[Na^+]_i\right\}\right) + 2(-1)FV_m + (+1)FV_m$	
or	$\Delta\mu_{2NBCe} = 2.3RT \bullet \log_{10}\left(\left\{[HCO_3^-]_i^2[Na^+]_o\right\} / \left\{[HCO_3^-]_o^2[Na^+]_i\right\}\right) - FV_m$	Eq. (13.2)
3:1	$\Delta\mu_{3NBCe} = 3\Delta\mu_{HCO3} + \Delta\mu_{Na}$	Eq. (13.3)
or	$\Delta\mu_{3NBCe} = RT \bullet \ln\left(\left\{[HCO_3^-]_i^3[Na^+]_i\right\} / \left\{[HCO_3^-]_o^3[Na^+]_i\right\}\right) + 3(-1)FV_m + (+1)FV_m$	
or	$\Delta\mu_{3NBCe} = 2.3RT \bullet \log_{10}\left(\left\{[HCO_3^-]_i^3[Na^+]_o\right\} / \left\{[HCO_3^-]_o^3[Na^+]_i\right\}\right) - 2FV_m$	Eq. (13.4)

From these thermodynamic considerations, it is widely thought that the 3:1 stoichiometry is needed for HCO_3^- absorption by the renal proximal tubule (movement from cell ***out to*** the blood). Similarly, these same equations indicate that a 2:1 stoichiometry would mean Na^+ and HCO_3^- cotransport from the blood ***into*** a cell. While these models are helpful to understand transport processes in various tissues, the models ignore the amount of NBCe1 protein expression, second messenger regulation, interacting proteins, and other possible ion combinations (e.g., $NaCO_3^-$ ion pair). Not surprisingly, likely due to these missing factors, there are active controversies about NBCe1-mediated transport. Thus, continued biophysical, cell, tissue, and animal experiments are needed to document "real physiology."

13.3.2 *Transported Ions*

13.3.2.1 Na^+ Dependence

Before NBCe1 was cloned, early experiments with basolateral membrane vesicles of rabbit kidney cortex and ^{22}Na uptake indicated that the Na/HCO_3 cotransporter activity was supported by Li^+, K^+, and choline$^+$ (Soleimani et al. 1991). Using BSC-1 cells (corneal cells), Jentsch and coworkers (Jentsch et al. 1985) showed electrogenic, DIDS-inhibitable, Na/HCO_3 cotransporter activity with apparent K_m for Na^+ of 20–40 mM at 28 mM HCO_3^-. However, these investigators were unable to support this transport activity with Li^+ or K^+.

Since NBCe1 cloning, the biophysics of ion substitution has been revisited. Amlal and colleagues reported that after transfecting human NBCe1-A in HEK-293 cells and monitoring pH_i using BCECF, NBCe1-A has a low affinity for Li^+ and lesser affinity for K^+ (Amlal et al. 1998). Nevertheless, when NBCe1-A from various species are evaluated by electrophysiology, little to no substitution of Na^+ by Li^+ or K^+ was measured (Romero et al. 1998b; Sciortino and Romero 1999; Lee et al. 2013). Regardless, all of these NBCe1 clones for both influx (outward current) and efflux (inward current) are dependent on extracellular Na^+ and voltage (Sciortino and Romero 1999; Becker and Deitmer 2004; Lee et al. 2013). That is, for V_m more positive than −40 mV, extracellular [Na^+] influx (outward I increasing with depolarization) is always measured. Likewise, for V_m more negative than −100 mV, efflux (inward I increasing with hyperpolarization) is always measured for NBCe1 clones regardless of expression system. These experiments combined indicate that (a) the extracellular Na^+ apparent affinity is ~30 mM (for $NaHCO_3$ efflux or influx), (b) for physiological voltages (−40 to −100 mV) both influx and efflux modes depend on the [Na^+] gradient and V_m, and (c) reduced $[Na^+]_o$ in this physiological V_m range elicits NBCe1-mediated $NaHCO_3$ efflux.

13.3.2.2 HCO_3^- Dependence

The renal proximal tubule, NBCe1 protein (NBCe1-A), seems to be the major, perhaps exclusive, mode of "HCO_3^-" exit from the cell into the blood. This renal NBCe1 mediates absorption of up to 90 % of renal HCO_3^- (Boron et al. 1997). This notion is supported by patients with NBCe1 (SLC4A4; OMIM #603345) mutations who have a very severe metabolic acidosis ($pH_{blood} < 7.05$; [HCO_3^-] = 3–12 mM).

However, the chemical form of "HCO_3^-" (i.e., HCO_3^-, CO_3^{2-}, or the $NaCO_3^-$ ion pair) transported by the NBCe1 protein is still disputed.

HCO_3^- dependence of NBCe1-A clones has been evaluated directly. Grichtchenko and colleagues examined extracellular [$HCO^-{}_3$] dependence of NBCe1-A (salamander and rat) in *Xenopus* oocytes (Grichtchenko et al. 2000). For these experiments, they used a brief pulse protocol: varying [HCO_3^-] by varying [CO_2] at pH-7.5. These experiments showed that the apparent K_m for external HCO_3^- was ~12 mM for NBCe1-A (Grichtchenko et al. 2000).

Early BLMV (basolateral membrane vesicles) from rabbit renal cortex vesicle experiments indicated that raising $[CO_3^{2-}]_o$ at constant $[HCO_3^-]_o$ increased ^{22}Na influx, consistent with CO_3^{2-} as a substrate of NBCe1 (Soleimani and Aronson 1989). However, elevating $[CO_3^{2-}]_o$ at constant $[HCO_3^-]_o$ requires raising pH_o at equilibrium. Other experiments indicated that sulfite (SO_3^{2-}) also increases ^{22}Na uptake in vesicles, supporting the hypothesis that SO_3^{2-} and CO_3^{2-} are transported by NBCe1 (Soleimani and Aronson 1989). SO_3^{2-} in solution at pH 7.4 also contains significant HSO_3^-, which could imply that HSO_3^- is an NBCe1 substrate. Unfortunately, experiments with rat NBCe1-A in *Xenopus* oocytes illustrated that neither SO_3^{2-} nor HSO_3^- were transported (Grichtchenko et al. 2000). Using nonequilibrium conditions, Moss and coworkers have recently provided evidence that CO_3^{2-} is transported by NBCe1A (Moss et al. 2014). These experiments indicate an apparent K_m (CO_3^{2-}) = 4 μM (constant pH_o 7.5 and 5 % CO_2) (Moss et al. 2014).

13.3.3 Biophysical Control of Transport

13.3.3.1 pH_i Dependence

Acid–base transporters move acid–base equivalents (e.g., H^+, HCO_3^-) but can additionally be controlled by pH *per se* (Boron 1986, 1992; Alper 1991). Early BBMV (brush border membrane vesicles) studies, using ^{22}Na uptake to assess Na/HCO_3 cotransporter activity, indicated that ^{22}Na uptake increased as a function of the membrane pH gradient (constant pH_i with varying pH_o) (Soleimani et al. 1992). Likewise, cellular voltage clamp experiments showed similar results (Gross and Hopfer 1999). Later experiments of the same group using human NBCe1-A in HEK-293 cells showed that pH_i increases after raising pH_o from 7.4 to 7.8 (Amlal et al. 1998). Electrophysiology experiments with NBCe1-A also indicate that changing pH_o will dictate NBCe1-A function (Becker and Deitmer 2004; Moss et al. 2014).

13.3.3.2 Voltage Dependence

For channels and electrogenic transporters, one would expect, especially from thermodynamic considerations (see above), transport across a membrane should be sensitive to the voltage difference across a membrane, i.e., membrane potential (V_m). Before NBCe1 cloning, this voltage dependence was studied in whole cell recordings of retinal glial cells, indicating a stoichiometry of 1 Na^+:3 HCO_3^- (Newman and Astion 1991). There is some indication of regulation of the stoichiometry "property" (Heyer et al. 1999; Ducoudret et al. 2001; Gross et al. 2001a, b; Muller-Berger et al. 2001). All experiments with expressed NBCe1 isoforms show voltage dependence (Romero et al. 2004, 2013a).

13.4 Role of NBCe1 in Epithelial Tissues

Currently there are 12 patients reported with recessive mutations in NBCe1 (SLC4A4; OMIM #603345). The common pathophysiologies of this rare disease are severe proximal renal tubular acidosis (pRTA), glaucoma, cataracts, and defective tooth enamel. Similarly, there are three reported mouse knockout models of the *nbce1* gene (Gawenis et al. 2007; Lo et al. 2011; Romero et al. 2014). The unifying defects found in these animals were pRTA, defective teeth, and ocular disorders. Chen and Romero have initial reports of an *nbce1A*-isoform knockout (Romero et al. 2014), which has pRTA, some ocular abnormalities, but normal teeth. Comparing the human disease to the mouse phenotypes has highlighted NBCe1's role in various epithelial tissues.

13.4.1 Brain

Dysfunction of NBCe1 was reported to lead to numerous other less consistent disorders (for reviews, see Romero et al. 2004, 2013a; Parker and Boron 2013). The primary manifestation in central nervous system is characterized by calcification of the basal ganglia, epilepsy, mental retardation, and migraine (Igarashi et al. 1999, 2001, 2002, 2003; Deda et al. 2001; Horita et al. 2005; Demirci et al. 2006; Suzuki et al. 2008, 2010; Lo et al. 2011; Ruminot et al. 2011; Banales et al. 2012). The exact cellular etiology of these pathophysiologies has not been elaborated. However, most of the CNS pathophysiologies seem to be associated with NBCe1 activity to regulate either intracellular pH or extracellular pH. While other Slc4 bicarbonate transporters have been shown to control secretion of the CSF by the choroid plexus (i.e., NDBCE (Damkier et al. 2007; Chen et al. 2008a, b), NBCe2 (Bouzinova et al. 2005; Damkier et al. 2007; Kao et al. 2011; Fukuda et al. 2013)), it is still possible that NBCe1 has a role as mRNA is found in the choroid plexus (Bevensee et al. 2000; Schmitt et al. 2000; Douglas et al. 2001).

13.4.2 Eye

Expression of NBCe1 in the eye is complex, and the eye seems to be the only organ which expresses NBCe1-A, NBCe1-B, and NBCe1-C. The complex expression pattern is likely due to the eye being an extension of the CNS which embryologically induces surrounding mesenchyme to form both lens epithelia and the cornea. Loss of NBCe1-B function causes band-keratopathy (Ca^{2+} and Mg^{2+} deposits in corneal endothelial cells), cataracts (opacity of lens epithelia), and glaucoma (neuropathy of retina and optic nerve) (Bok et al. 2001). The function of NBCe1 in lens physiology and production of aqueous humor is still not clear. For example, NBCe1 is highly expressed in the ciliary body (secretion of ions, solutes, and water to form aqueous humor), but human mutations and mouse knockouts all have displayed decreased function. Thus, decreased NBCe1 function should lead to decreased aqueous fluid production and consequently decrease intraocular pressure (IOP). However, both patient clinical finds and mouse data have indicated that decreased NBCe1 function *elevates* IOP (see reviews Romero et al. 2004, 2013a).

13.4.3 Teeth: Ameloblasts

Patients with NBCe1 mutations have poor dentition (see reviews Romero et al. 2004, 2013a). This pathophysiology was originally attributed to either the chronic pRTA or acid (less basic) secretions of the mature salivary glands (Dinour et al. 2004; Demirci et al. 2006). More recent studies in *nbce1* knockout mice revealed this phenotype of poorly formed enamel (Paine et al. 2008; Lacruz et al. 2010; Zheng et al. 2011; Jalali et al. 2014; Wen et al. 2014). These studies showed that enamel was hypomineralized and has abnormal architecture directly resulting from altered ameloblasts (Paine et al. 2008; Lacruz et al. 2010). Enamel mineralization, a function of polarized NBCe1 expression (Paine et al. 2008; Lacruz et al. 2010), seems to occur by NBCe1-B/C/D/E in mice (Zheng et al. 2011; Jalali et al. 2014) but NBCe1-A in humans (Zheng et al. 2011). Remarkably, providing an appropriate environment (pH and HCO_3^-) using a transplant technique, *nbce1*$^{(-/-)}$ teeth develop normally (Jalali et al. 2014).

13.4.4 Kidney

One of the kidney's major jobs is to filter the blood, and in doing this, 4320 mmol of bicarbonate is filtered daily by the glomerulus. This massive filtration occurs because the kidney receives 20 % of the cardiac output (1 L blood/min) and 10 % of this volume (~180 L/d) is filtered by the glomerulus to become ultrafiltrate, i.e., plasma without blood cells. The epithelia of the nephron and collecting duct system then modify this fluid by absorbing components which are needed by the body and secreting waste products.

This daily-filtered HCO_3^- is then transported to the interstitial space and blood. 85 % of filtrated bicarbonate is reabsorbed in proximal tubule (PT), 10 % in thick ascending limb, and 5 % in collecting ducts (Boron and Boulpaep 2003). The reabsorption of bicarbonate by the proximal tubule involves H^+ secretion at the apical cell membrane and transport of the bicarbonate into the interstitial space (Fig. 13.3). Specifically, filtered bicarbonate combines with H^+ secreted in PT lumen by Na^+/H^+ exchanger type 3 (NHE3), which in turn is converted by CA

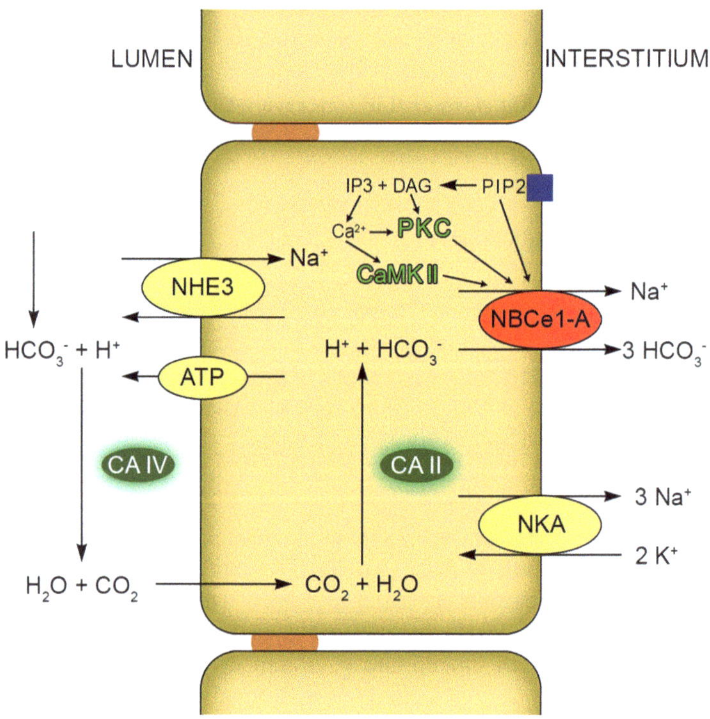

Fig. 13.3 Bicarbonate transport by proximal tubule cells. Filtered bicarbonate is turned to CO_2 and H_2O by carbonic anhydrase (CA) IV. CAII in cells turns CO_2 into H^+ and bicarbonate. The bicarbonate is transported into the interstitium by NBCe1-A, while H^+ is transported into the tubular lumen by the Na^+/H^+ exchanger type 3 (NHE_3) and proton pump (H^+ - ATPase). NKA = Na^+/K^+ ATPase

type IV to CO_2 and H_2O. Soluble CO_2 then moves into PT cells, in part via aquaporin 1 (Nakhoul et al. 1998; Prasad et al. 1998; Cooper et al. 2002; Musa-Aziz et al. 2009), and then CO_2 is rehydrated to regenerate HCO_3^- and H^+ by action of an intracellular CAII (Boron and Boulpaep 2003). Bicarbonate leaves the PT cell at the basolateral membrane via NBCe1-A (Romero et al. 1997; Abuladze et al. 1998; Maunsbach et al. 2000; Skelton et al. 2010). This PT HCO_3^- absorption can be increased to ~90 % of all renal HCO_3^- absorption through activation of PT ammoniagenesis (Giebisch and Windhager 2003; Hirata et al. 2003). Ammoniagenesis (usually glutamine metabolism) generates both H^+ for apical secretion and intracellular HCO_3^-, which is transported to the blood via NBCe1-A as above.

If the HCO_3^- reabsorption by NBCe1-A is abolished, proximal RTA occurs (Igarashi et al. 1999). To date, there are 12 reported recessive mutations of NBCe1-A:

(1) Missense mutations: R298S, S427L, T485S, G486R, R510H, L522P, A799V, R881C
(2) Nonsense mutations: Q29X, W516X with a frame shift at 2311A
(3) C-terminal 65-base-pair deletion (Kurtz and Zhu 2013; Romero et al. 2013a) Furthermore, patients have extrarenal manifestations (e.g., short stature, cataracts, band-keratopathy, glaucoma, mental retardation, basal ganglia calcifications, migraine headaches, and tooth defects) for which there is controversy regarding if these nonrenal symptoms are due to the NBCe1-A mutation, systemic acidosis, or alteration in one of the other NBCe1 isoforms. Interestingly, it has been suggested that NBCe1-A mutations that are retained in the endoplasmatic reticulum lead also to less expression of NBCe1-B in the cell membranes especially of the brain (Suzuki et al. 2010; Yamazaki et al. 2013).

13.4.5 Pancreas

Exocrine pancreas produces and secretes pro-digestive enzymes with HCO_3^- to neutralize the low pH content released from the stomach. Pancreatic acinar cells are responsible for enzyme secretion, while pancreatic ductal cells are responsible for the secretion of ions and water. Of course, enzyme secretion via zymogen granules also involves ion and water transport to solubilize and "flush out" contents via compound exocytosis (De Lisle and Hopfer 1986; Gasser et al. 1988a, b; Thevenod et al. 1990, 1992). Since HCO_3^- secretion by duodenal epithelia is activated when luminal pH in the duodenum is low (<4), the pancreas secretes a high pH solution. To prevent premature activation of the pancreatic digestive enzymes, the pancreatic juice is virtually isotonic $NaHCO_3$ and thus alkalotic (pH 8.2–8.5) and highly buffered (Romero 2001; Sinđić et al. 2010).

NBCe1-B is detected in the B cells of Langerhans islands, while NBCe1-A is diffusely expressed in islet cells. In the exocrine pancreas, NBCe1-B is present in acinar and duct cells. Dysfunction of NBCe1-B might therefore result in pancreatitis (Soyfoo et al. 2009).

Ductal cells have bicarbonate transporters on the apical and basolateral membrane in order to secret bicarbonate. The cystic fibrosis transmembrane conductance regulator (CFTR) and Slc26 proteins (a3, a6, and a9) transport bicarbonate at the apical membrane (Sinđić et al. 2010; Lee et al. 2012a). NBCe1-B and Cl^-/HCO_3^- anion exchanger (AE3) are responsible for basolateral transport of bicarbonate (Abuladze et al. 1998; Ishiguro et al. 1998; Marino et al. 1999; Thévenod et al. 1999); for reviews see (Steward et al. 2005; Sinđić et al. 2010; Lee et al. 2012a). NBCe1-B transports bicarbonate into the ductal cells using the electrochemical gradients maintained by the Na^+/K^+ ATPase, and NBCe1-B is activated by IRBIT, particularly after carbachol action (Shirakabe et al. 2006; Yang et al. 2009a, 2011; Lee et al. 2012b; Park et al. 2012; Hong et al. 2013). Secretin regulates bicarbonate secretion by pancreatic ducts by increasing intracellular cAMP, which increase the activity of this transporter directly, and by an increase in membrane expression (Gross et al. 2003; Yu et al. 2009). In addition to transported HCO_3^- entering from the basolateral side, intracellular CAs can also produce HCO_3^- for secretion. To maintain this CA activity, H^+ are transported out of the cell by basolateral Na^+/H^+ exchanger, NHE1 (Fig. 13.4).

13.4.6 Gastrointestinal Tract

The digestive tract absorbs nutrients but also ions such as Na^+, K^+, and Cl^-. While absorbing these solutes, the small intestine and colon secrete bicarbonate. As in pancreatic duct epithelia, HCO_3^- transport at the apical membrane is due to the combined action of CFTR and members of the Slc26 family (a3, a6). To secrete this HCO_3^-, the epithelia produces cellular HCO_3^-, and HCO_3^- is transported from blood via NBCe1-B and NBCn1 in the basolateral membrane (Fig. 13.5) (Rossmann et al. 1999; Jacob et al. 2000; Seidler et al. 2001; Gawenis et al. 2007; Kurita et al. 2008a; Charoenphandhu et al. 2011; Jakab et al. 2011). This same system in the euryhaline fish intestine is responsible for $CaCO_3$ precipitation (Kurita et al. 2008b; Gregorio et al. 2014).

NBCe1-B is widely expressed in the lower digestive tract (Table 13.1) (Jacob et al. 2000; Praetorius et al. 2001; Bartolo et al. 2009; Yu et al. 2009; Charoenphandhu et al. 2011; Jakab et al. 2011). A small amount of NBCe1-A has also been detected in the rabbit digestive tract (Jacob et al. 2000) and mouse gut (Khandoudi et al. 2001; Praetorius et al. 2001). NBCe1 is strongly expressed in villus enterocytes, yet expression decreases closer to the crypts (Jakab et al. 2011). From these localizations and regulation, NBCe1 can be hypothesized to contribute to the severity of ileal obstruction in newborns with cystic fibrosis (dysfunction of CFTR) (Dorfman et al. 2009; Lee et al. 2012a). This hypothesis is supported by the severe intestinal phenotype found in the NBCe1 knockout mice (Gawenis et al. 2007).

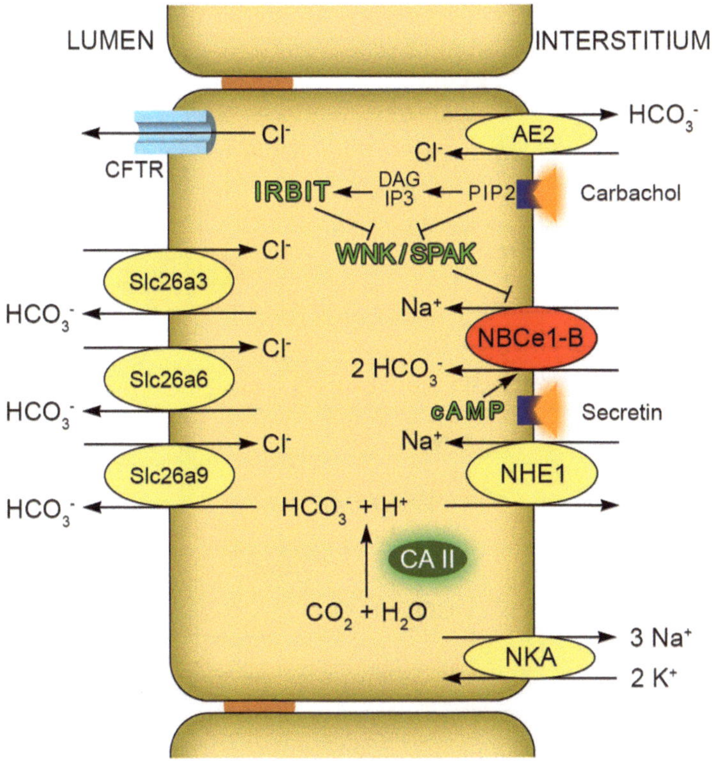

Fig. 13.4 Bicarbonate transport by pancreatic duct epithelia. Apical players are Slc26a3, Slc26a6, and Slc26a9 which can interact with CFTR. NBCe1-B is the major basolateral bicarbonate transporter, which is activated by IP_3 receptor-binding protein (IRBIT), thereby increasing bicarbonate secretion by pancreatic duct cells. Secretin increases intracellular cAMP concentration which activates NBCe1-B and therefore enhances bicarbonate secretion by pancreatic ducts. *AE2* anion exchanger 2; *IP_3* inositol 1,4,5-triphosphate; *DAG* diacyl-glycerol; *NKA* Na^+/K^+ ATPase; *NHE1* Na^+/H^+ exchanger

13.4.7 Male Reproductive Tract: Epididymis and Prostate

The male reproductive tract must maintain an acidic environment for proper sperm maturation (Rodriguez-Martinez et al. 1990; Pastor-Soler et al. 2005; Joseph et al. 2010). To accomplish this luminal acidification, the proximal epididymis absorbs HCO_3^- and secretes H^+ using the same complement of transporters as the renal proximal tubule (see above) (Pastor-Soler et al. 2005; Joseph et al. 2010). In support of this model, NBCe1 has been localized to the basolateral membrane (Jensen et al. 1999). HCO_3^- activates sperm (Tajima et al. 1987).

The role, if any, of the testis, seminal vesicles, and prostate is unclear from the HCO_3^- transport standpoint. Activation of soluble adenylate cyclase (sAC) is HCO_3^- dependent, resulting in sAC being called a metabolic and bicarbonate sensor (Chen et al. 2000; Sinclair et al. 2000; Zippin et al. 2001; Kleinboelting

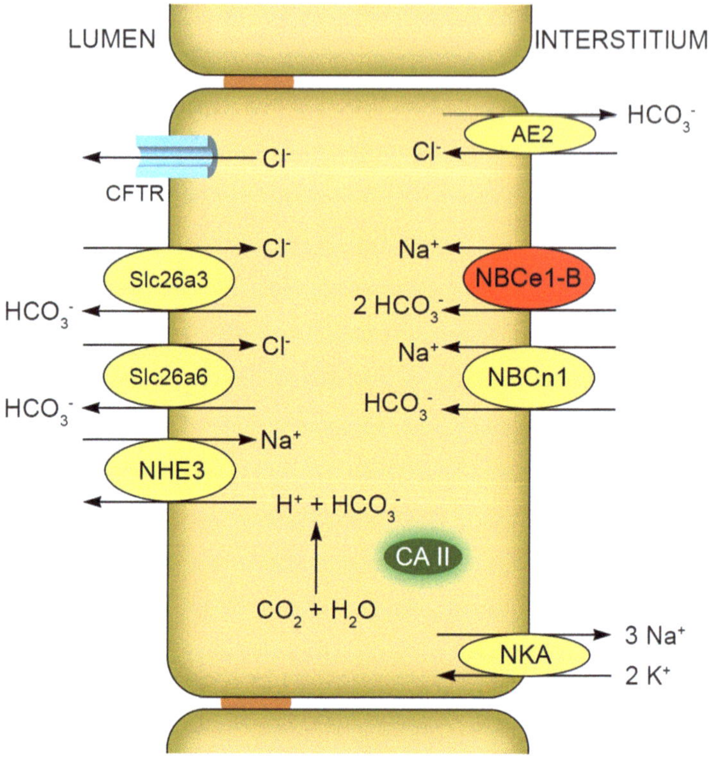

Fig. 13.5 Bicarbonate transport by intestinal epithelia. Bicarbonate produced by carbonic anhydrase (CA) and transported by basolateral bicarbonate transporters (NBCe1-B, NBCn1) is secreted from the cells via apical proteins (Slc26a3 and Slc26a6 which can interact with CFTR; Ko et al. 2002, 2004). *AE2* anion exchanger 2; *NKA* Na^+/K^+ ATPase; *NHE₃* Na^+/H^+ exchanger

et al. 2014). Action of sAC is known to control the trafficking of the V-ATPase in clear cells of the male reproductive tract (Pastor-Soler et al. 2003). Since HCO_3^- is needed to activate sperm as well as to buffer the acidity of the female reproductive tract, HCO_3^- secretion must occur. Evidence from congenital chloride diarrhea (CLD) patients indicates that part of the mechanism is similar to epithelial HCO_3^- secretion by the pancreas involving CFTR and Slc26a3 at the apical membrane (Hihnala et al. 2006). By analogy, NBCe1-B should be at these basolateral membranes to facilitate HCO_3^- secretion. NBCe1 mRNA in the prostate is extremely high, i.e., greater than renal proximal tubule and the pancreas (Romero and Boron 1999; Thévenod et al. 1999). However, whether the seminal vesicles or the prostate gland is the major HCO_3^- secretory epithelium of the male reproductive tract has not been investigated.

Table 13.1 NBCe1 isoform mutations linked to kidney, eye, and GI disorders

NBCe1 isoform	Localization	Pathology
NBCe1-A	Kidney	Proximal renal tubular acidosis
	Ciliary body, trabecular meshwork, Schlemm's canal	Glaucoma
	Nasal glands	
NBCe1-B	**GI**	
	Ameloblasts	Defective enamel deposition
	Salivary gland	
	Pancreas	Possible pancreatitis
	Liver	
	Stomach	
	Intestine	Possible obstruction
	Colon	
	Eye	
	Cornea	Band-keratopathy Cataracts Glaucoma
	Lens epithelium	
	Reproductive system	
	Epididymis	
	Testis	
	Prostate	
	Uterus	
	Other tissues	
	Lungs	
	Heart	
	Thyroid	
	Spinal cord	
	Astrocytes	
NBCe1-C	Brain	
	Epididymis	
	Testis	
NBCe1-D	Unknown	Unknown
NBCe1-E	Unknown	Unknown

13.4.8 Female Reproductive Tract: Ovarian Duct and Uterus

Similar to the male reproductive tract, molecular mechanisms of HCO_3^- secretion by the female reproductive tract are understudied. NBCe1 is expressed in the uterus with mRNA and protein expression increasing during estrous (Gholami et al. 2014). Chinigarzadeh and coworkers have also shown that both NBCe1-A and NBCe1-B and Slc26a6 are present in the uterus (Chinigarzadeh et al. 2014). Further investigations are needed to determine if NBCe1 isoforms play critical physiological roles in these tissues.

13.4.9 Salivary Glands

Similar to glandular, HCO_3^- secretory tissues (e.g., pancreas, prostate), salivary glands (i.e., parotid, submandibular), and related glands (i.e., lacrimal glands and sweat ducts) express NBCe1. The common features of these epithelial organs are glandular and ductal epithelial tubes. While somewhat understudied for bicarbonate transport, these tissues secrete NaCl, $NaHCO_3$, and water. The general model is similar to that of the intestine and pancreas (Figs. 13.4 and 13.5). With stimulation, saliva of guinea pigs and humans is ~16 mM HCO_3^- (Koo et al. 2006). Koo and colleagues were able to clone and show NBCe1 mRNA in both the parotid and submandibular glands. Luo and coworkers also demonstrated NBCe1-B mRNA in submandibular glands, as well as some isoforms of NBCn1 (Luo et al. 2001). Nevertheless, the role of NBCe1 isoforms in the physiology of these epithelia remains to be determined.

13.5 Conclusion

It is evident that the electrogenic Na^+ bicarbonate cotransporters, NBCe1 (aka, Slc4a4), have major physiological and pathophysiological roles in the kidney, eye, pancreas, and brain. Knockout animals and humans with recessive NBCe1 (SLC4A4) mutations have greatly facilitated our understanding. However, there are currently 12 patients and two published knockout models (Gawenis et al. 2007; Lo et al. 2011). For example, an NBCe1A-isoform knockout has been reported (Chen et al. 2014) which should allow a better understanding of this isoform. Each of these is a valuable resource; however, further investigation is needed to understand NBCe1's role in acid–base homeostasis as well as epithelial cell function. Given that NBCe1 forms dimers, it is attractive to speculate that single nucleotide polymorphisms (SNPs) of NBCe1 could affect epithelial physiology under certain circumstances such as aging (Romero et al. 2013b). Future experiments and additional molecular tools will be needed to address some of the unanswered or undocumented physiology of NBCe1.

Acknowledgments We thank our collaborators and past members of the Romero lab who contributed to some of the basic physiology. This work was supported by NIH EY017732 (MFR), the Cystic Fibrosis Foundation (Sindic-06F0, Romero-06G0), the Arlene and Robert Kogod Aging Center, and the Mayo Foundation.

References

Abuladze N, Lee I, Newman D, Hwang J, Boorer K, Pushkin A, Kurtz I (1998) Molecular cloning, chromosomal localization, tissue distribution, and functional expression of the human pancreatic sodium bicarbonate cotransporter. J Biol Chem 273(28):17689–17695

Abuladze N, Song M, Pushkin A, Newman D, Lee I, Nicholas S, Kurtz I (2000) Structural organization of the human NBC1 gene: kNBC1 is transcribed from an alternative promoter in intron 3. Gene 251(2):109–122

Alper SL (1991) The band 3-related anion exchanger (AE) gene family. Annu Rev Physiol 53:549–564

Amlal H, Wang Z, Burnham C, Soleimani M (1998) Functional characterization of a cloned human kidney Na^+:HCO_3^- cotransporter. J Biol Chem 273(27):16810–16815

Bachmann O, Rossmann H, Berger UV, Colledge WH, Ratcliff R, Evans MJ, Gregor M, Seidler U (2003) cAMP-mediated regulation of murine intestinal/pancreatic Na^+/HCO_3^- cotransporter subtype pNBC1. Am J Physiol Gastrointest Liver Physiol 284(1):G37–G45

Bachmann O, Reichelt D, Tuo B, Manns MP, Seidler U (2006) Carbachol increases Na^+/HCO_3^- cotransport activity in murine colonic crypts in a M3-, Ca^{2+}/calmodulin-, and PKC-dependent manner. Am J Physiol Gastrointest Liver Physiol 291(4):G650–G657

Banales JM, Saez E, Uriz M, Sarvide S, Urribarri AD, Splinter P, Tietz Bogert PS, Bujanda L, Prieto J, Medina JF et al (2012) Upregulation of mir-506 leads to decreased AE2 expression in biliary epithelium of patients with primary biliary cirrhosis. Hepatology 56(2):687–697

Bartolo RC, Harfoot N, Gill M, McLeod BJ, Butt AG (2009) Secretagogues stimulate electrogenic HCO_3^- secretion in the ileum of the brushtail possum, *Trichosurus vulpecula*: evidence for the role of a Na^+/HCO_3^- cotransporter. J Exp Biol 212(Pt 16):2645–2655

Becker HM, Deitmer JW (2004) Voltage dependence of H^+ buffering mediated by sodium bicarbonate cotransport expressed in *Xenopus* oocytes. J Biol Chem 279(27):28057–28062

Bevensee MO, Schmitt BM, Choi I, Romero MF, Boron WF (2000) An electrogenic Na/HCO_3 cotransporter (NBC) with a novel C terminus, cloned from rat brain. Am J Physiol Cell Physiol 278(6):C1200–C1211

Bok D, Schibler MJ, Pushkin A, Sassani P, Abuladze N, Naser Z, Kurtz I (2001) Immunolocalization of electrogenic sodium-bicarbonate cotransporters pNBC1 and kNBC1 in the rat eye. Am J Physiol Renal Physiol 281(5):F920–F935

Boron WF (1986) Intracellular pH regulation in epithelial cells. Annu Rev Physiol 48:377–388

Boron WF (1992) Control of intracellular pH. In: Seldin DW, Giebisch G (eds) The kidney: physiology and pathophysiology, 2nd edn. Raven, New York

Boron WF, Boulpaep EL (2003) Medical physiology. Saunders, Philadelphia, PA

Boron WF, Fong P, Hediger MA, Boulpaep EL, Romero MF (1997) The electrogenic Na/HCO_3 cotransporter. Wien Klin Wochenschr 109(12–13):445–456

Bouzinova EV, Praetorius J, Virkki LV, Nielsen S, Boron WF, Aalkjaer C (2005) Na^+– dependent HCO_3^- uptake into the rat choroid plexus epithelium is partially DIDS sensitive. Am J Physiol Cell Physiol 289(6):C1448–C1456

Burnham CE, Amlal H, Wang Z, Shull GE, Soleimani M (1997) Cloning and functional expression of a human kidney Na^+:HCO_3^- cotransporter. J Biol Chem 272(31):19111–19114

Chang M-H, DiPiero JM, Sönnichsen FD, Romero MF (2008) Entry to "HCO_3 tunnel" revealed by human mutation and structural model. J Biol Chem 283(26):18402–18410

Charoenphandhu N, Laohapitakworn S, Kraidith K, Nakkrasae LI, Jongwattanapisan P, Tharabenjasin P, Krishnamra N (2011) Electrogenic Na/HCO_3 co-transporter-1 is essential for the parathyroid hormone-stimulated intestinal HCO secretion. Biochem Biophys Res Commun 409(4):775–779

Chen Y, Cann MJ, Litvin TN, Iourgenko V, Sinclair ML, Levin LR, Buck J (2000) Soluble adenylyl cyclase as an evolutionarily conserved bicarbonate sensor. Science 289 (5479):625–628

Chen LM, Kelly ML, Parker MD, Bouyer P, Gill HS, Felie JM, Davis BA, Boron WF (2008a) Expression and localization of Na-driven Cl^--HCO_3^- exchanger (SLC4A8) in rodent CNS. Neuroscience 153(1):162–174

Chen LM, Kelly ML, Rojas JD, Parker MD, Gill HS, Davis BA, Boron WF (2008b) Use of a new polyclonal antibody to study the distribution and glycosylation of the sodium-coupled bicarbonate transporter NCBE in rodent brain. Neuroscience 151(2):374–385

Chen A-P, Holmes H, Chang M-H, Carleton L, Romero MF (2014) Targeted NBCe1A deletion causes proximal RTA: *whole nbce1 (sh_nbce1)* KO vs. *nbce1A* KO mice. J Am Soc Nephrol 25:71A

Chinigarzadeh A, Kasim NF, Muniandy S, Kassim NM, Salleh N (2014) Genistein induces increase in fluid pH, Na^+ and HCO_3^- concentration, SLC26A6 and SLC4A4 (NBCe1)-B expression in the uteri of ovariectomized rats. Int J Mol Sci 15(1):958–976

Choi I, Romero MF, Khandoudi N, Bril A, Boron WF (1999) Cloning and characterization of an electrogenic Na/HCO_3 cotransporter from human heart. Am J Physiol 274(3 Pt 1):C576–C584

Cooper GJ, Zhou Y, Bouyer P, Grichtchenko II, Boron WF (2002) Transport of volatile solutes through AQP1. J Physiol 542(Pt 1):17–29

Damkier HH, Nielsen S, Praetorius J (2007) Molecular expression of SLC4 derived Na^+ dependent anion transporters in selected human tissues. Am J Physiol Regul Integr Comp Physiol 293(5): R2136–R2146

De Lisle RC, Hopfer U (1986) Electrolyte permeabilities of pancreatic zymogen granules: implications for pancreatic secretion. Am J Physiol 250(4 Pt 1):G489–G496

Deda G, Ekim M, Guven A, Karagol U, Tumer N (2001) Hypopotassemic paralysis: a rare presentation of proximal renal tubular acidosis. J Child Neurol 16(10):770–771

Demirci FY, Chang M-H, Mah TS, Romero MF, Gorin MB (2006) Proximal renal tubular acidosis and ocular pathology: L522P, a novel missense mutation in the gene (SLC4A4) for Sodium Bicarbonate Cotransporter protein (NBCe1). Mol Vis 12:324–330

Dinour D, Chang M-H, Satoh J-I, Smith BL, Angle N, Knecht A, Serban I, Holtzman EJ, Romero MF (2004) A novel missense mutation in the sodium bicarbonate cotransporter (NBCe1/SLC4A4) causes proximal tubular acidosis and glaucoma through ion transport defects. J Biol Chem 279(50):52238–52246

Dorfman R, Li W, Sun L, Lin F, Wang Y, Sandford A, Pare PD, McKay K, Kayserova H, Piskackova T et al (2009) Modifier gene study of meconium ileus in cystic fibrosis: statistical considerations and gene mapping results. Hum Genet 126(6):763–778

Douglas R, Xia Y, Schmitt BM, Bevensee MO, Biemesderfer D, Boron WF, Haddad GG (1998) Developmental and stress-induced profile of two CNS pH regulatory proteins: the Na/H exchanger and the Na/HCO_3 cotransporter. Soc Neurosci Absrt 24:1605

Douglas RM, Schmitt BM, Xia Y, Bevensee MO, Biemesderfer D, Boron WF, Haddad GG (2001) Sodium-hydrogen exchangers and sodium-bicarbonate co-transporters: ontogeny of protein expression in the rat brain. Neuroscience 102(1):217–228

Ducoudret O, Diakov A, Mueller-Berger S, Romero MF, Boron WF, Frömter E (2001) The renal Na-HCO_3-cotransporter expressed in Xenopus laevis oocytes: inhibition by tenidap and benzamil and effect of temperature on transport rate and stoichiometry. Pflugers Arch 442 (5):709–717

Fukuda H, Hirata T, Nakamura N, Kawahara K, Wakabayashi S, Romero MF, Hirose S (2013) Na^+/HCO_3 cotransporter 4 (NBC4) is predominantly expressed in choroid plexus and involved in cerebrospinal fluid production. Biochem J 450:179–187

Galeza-Kulik M, Zebracka J, Szpak-Ulczok S, Czarniecka AK, Kukulska A, Gubala E, Stojcev Z, Wiench M (2006) Expression of selected genes involved in transport of ions in papillary thyroid carcinoma. Endokrynologia Polska 57(Suppl A):26–31

Gasser KW, DiDomenico J, Hopfer U (1988a) Potassium transport by pancreatic and parotid zymogen granule membranes. Am J Physiol 255(6 Pt 1):C705–C711

Gasser KW, DiDomenico J, Hopfer U (1988b) Secretagogues activate chloride transport pathways in pancreatic zymogen granules. Am J Physiol 254(1 Pt 1):G93–G99

Gawenis LR, Bradford EM, Prasad V, Lorenz JN, Simpson JE, Clarke LL, Woo AL, Grisham C, Sanford LP, Doetschman T et al (2007) Colonic anion secretory defects and metabolic acidosis in mice lacking the NBC1 Na^+/HCO_3^- cotransporter. J Biol Chem 282(12):9042–9052

Gholami K, Muniandy S, Salleh N (2014) Modulation of sodium-bicarbonate co-transporter (SLC4A4/NBCe1) protein and mRNA expression in rat's uteri by sex-steroids and at different phases of the oestrous cycle. Res Vet Sci 96(1):164–170

Giebisch G, Windhager E (2003) Transport of acids and bases. In: Boron WF, Boulpaep EL (eds) Medical physiology. Saunders, Philadelphia, PA

Gregorio SF, Carvalho ES, Campinho MA, Power DM, Canario AV, Fuentes J (2014) Endocrine regulation of carbonate precipitate formation in marine fish intestine by stanniocalcin and PTHrP. J Exp Biol 217(Pt 9):1555–1562

Grichtchenko II, Romero MF, Boron WF (2000) Extracellular HCO_3 dependence of electrogenic Na/HCO_3 cotransporters (NBC) cloned from salamander and rat kidney. J Gen Physiol 115 (5):533–546

Gross E, Hopfer U (1999) Effects of pH on kinetic parameters of the Na-HCO3 cotransporter in renal proximal tubule. Biophys J 76(6):3066–3075

Gross E, Hawkins K, Abuladze N, Pushkin A, Cotton CU, Hopfer U, Kurtz I (2001a) The stoichiometry of the electrogenic sodium bicarbonate cotransporter NBC1 is cell-type dependent. J Physiol 531(3):597–603

Gross E, Hawkins K, Pushkin A, Sassani P, Dukkipati R, Abuladze N, Hopfer U, Kurtz I (2001b) Phosphorylation of Ser(982) in the sodium bicarbonate cotransporter kNBC1 shifts the HCO (3)(−): Na(+) stoichiometry from 3: 1 to 2: 1 in murine proximal tubule cells. J Physiol 537 (Pt 3):659–665

Gross E, Fedotoff O, Pushkin A, Abuladze N, Newman D, Kurtz I (2003) Phosphorylation-induced modulation of pNBC1 function: distinct roles for the amino- and carboxy-termini. J Physiol 549(Pt 3):673–682

Heyer M, Müller-Berger S, Romero MF, Boron WF, Frömter E (1999) Stoichiometry of rat kidney $Na\text{-}HCO_3$ cotransporter (rkNBC) expressed in Xenopus laevis oocytes. Pflugers Arch 438 (3):322–329

Hihnala S, Kujala M, Toppari J, Kere J, Holmberg C, Hoglund P (2006) Expression of SLC26A3, CFTR and NHE3 in the human male reproductive tract: role in male subfertility caused by congenital chloride diarrhoea. Mol Hum Reprod 12(2):107–111

Hirata T, Kaneko T, Ono T, Nakazato T, Furukawa N, Hasegawa S, Wakabayashi S, Shigekawa M, Chang M-H, Romero MF et al (2003) Mechanism of acid adaptation of a fish living in a pH 3.5 lake. Am J Physiol Regul Integr Comp Physiol 284(5):R1199–R1212

Hong JH, Yang D, Shcheynikov N, Ohana E, Shin DM, Muallem S (2013) Convergence of IRBIT, phosphatidylinositol (4,5) bisphosphate, and WNK/SPAK kinases in regulation of the $Na^+–HCO_3^-$ cotransporters family. Proc Natl Acad Sci USA 110(10):4105–4110

Horita S, Yamada H, Inatomi J, Moriyama N, Sekine T, Igarashi T, Endo Y, Dasouki M, Ekim M, Al-Gazali L et al (2005) Functional analysis of NBC1 mutants associated with proximal renal tubular acidosis and ocular abnormalities. J Am Soc Nephrol 16(8):2270–8

Igarashi T, Inatomi J, Sekine T, Cha SH, Kanai Y, Kunimi M, Tsukamoto K, Satoh H, Shimadzu M, Tozawa F et al (1999) Mutations in SLC4A4 cause permanent isolated proximal renal tubular acidosis with ocular abnormalities. Nat Genet 23(3):264–266

Igarashi T, Inatomi J, Sekine T, Seki G, Shimadzu M, Tozawa F, Takeshima Y, Takumi T, Takahashi T, Yoshikawa N et al (2001) Novel nonsense mutation in the Na^+/HCO_3^- cotransporter gene (SLC4A4) in a patient with permanent isolated proximal renal tubular acidosis and bilateral glaucoma. J Am Soc Nephrol 12(4):713–718

Igarashi T, Sekine T, Inatomi J, Seki G (2002) Unraveling the molecular pathogenesis of isolated proximal renal tubular acidosis. J Am Soc Nephrol 13(8):2171–2177

Igarashi T, Inatomi J, Sekine T, Seki G, Yamada H, Horita S, Fujita T (2003) Mutational and functional analysis of the Na^+/HCO_3^- cotransporter gene (SLC4A4) in patients with

permanent isolated proximal renal tubular acidosis and ocular abnormalities. J Am Soc Nephrol 14(Suppl S):302A–302A
Ishiguro H, Naruse S, Steward MC, Kitagawa M, Ko SB, Hayakawa T, Case RM (1998) Fluid secretion in interlobular ducts isolated from guinea-pig pancreas. J Physiol Lond 511 (Pt 2):407–422
Jacob P, Christiani S, Rossmann H, Lamprecht G, Vieillard-Baron D, Muller R, Gregor M, Seidler U (2000) Role of Na^+/HCO_3^- cotransporter NBC1, Na^+/H^+ exchanger NHE1, and carbonic anhydrase in rabbit duodenal bicarbonate secretion. Gastroenterology 119(2):406–419
Jakab RL, Collaco AM, Ameen NA (2011) Physiological relevance of cell-specific distribution patterns of CFTR, NKCC1, NBCe1, and NHE3 along the crypt-villus axis in the intestine. Am J Physiol Gastrointest Liver Physiol 300(1):G82–G98
Jalali R, Guo J, Zandieh-Doulabi B, Bervoets TJ, Paine ML, Boron WF, Parker MD, Bijvelds MJ, Medina JF, DenBesten PK et al (2014) NBCe1 (SLC4A4) a potential pH regulator in enamel organ cells during enamel development in the mouse. Cell Tissue Res 358(2):433–442
Jensen LJ, Schmitt BM, Berger UV, Nsumu NN, Boron WF, Hediger MA, Brown D, Breton S (1999) Localization of sodium bicarbonate cotransporter (NBC) protein and messenger ribonucleic acid in rat epididymis. Biol Reprod 60(3):573–579
Jentsch TJ, Schill BS, Schwartz P, Matthes H, Keller SK, Wiederholt M (1985) Kidney epithelial cells of monkey origin (BSC-1) express a sodium bicarbonate cotransport. Characterization by $^{22}Na^+$ flux measurements. J Biol Chem 260(29):15554–15560
Joseph A, Hess RA, Schaeffer DJ, Ko C, Hudgin-Spivey S, Chambon P, Shur BD (2010) Absence of estrogen receptor alpha leads to physiological alterations in the mouse epididymis and consequent defects in sperm function. Biol Reprod 82(5):948–957
Kao L, Kurtz LM, Shao X, Papadopoulos MC, Liu L, Bok D, Nusinowitz S, Chen B, Stella SL, Andre M et al (2011) Severe neurologic impairment in mice with targeted disruption of the electrogenic sodium bicarbonate cotransporter NBCe2 (Slc4a5 gene). J Biol Chem 286 (37):32563–32574
Khandoudi N, Albadine J, Robert P, Krief S, Berrebi-Bertrand I, Martin X, Bevensee MO, Boron WF, Bril A (2001) Inhibition of the cardiac electrogenic sodium bicarbonate cotransporter reduces ischemic injury. Cardiovasc Res 52(3):387–396
Kleinboelting S, Diaz A, Moniot S, van den Heuvel J, Weyand M, Levin LR, Buck J, Steegborn C (2014) Crystal structures of human soluble adenylyl cyclase reveal mechanisms of catalysis and of its activation through bicarbonate. Proc Natl Acad Sci USA 111(10):3727–3732
Ko SB, Shcheynikov N, Choi JY, Luo X, Ishibashi K, Thomas PJ, Kim JY, Kim KH, Lee MG, Naruse S et al (2002) A molecular mechanism for aberrant CFTR-dependent HCO_3^- transport in cystic fibrosis. EMBO J 21(21):5662–5672
Ko SB, Zeng W, Dorwart MR, Luo X, Kim KH, Millen L, Goto H, Naruse S, Soyombo A, Thomas PJ et al (2004) Gating of CFTR by the STAS domain of SLC26 transporters. Nat Cell Biol 6 (4):343–350
Konturek SJ, Konturek PC, Pawlik T, Sliwowski Z, Ochmanski W, Hahn EG (2004) Duodenal mucosal protection by bicarbonate secretion and its mechanisms. J Physiol Pharmacol 55 (Suppl 2):5–17
Koo NY, Li J, Hwang SM, Choi SY, Lee SJ, Oh SB, Kim JS, Lee JH, Park K (2006) Molecular cloning and functional expression of a sodium bicarbonate cotransporter from guinea-pig parotid glands. Biochem Biophys Res Commun 342(4):1114–1122
Kreindler JL, Peters KW, Frizzell RA, Bridges RJ (2006) Identification and membrane localization of electrogenic sodium bicarbonate cotransporters in Calu-3 cells. Biochim Biophys Acta 1762 (7):704–710
Kurita Y, Nakada T, Kato A, Doi H, Mistry AC, Chang M-H, Romero MF, Hirose S (2008a) Identification of intestinal bicarbonate transporters involved in formation of carbonate precipitates to stimulate water absorption in marine teleost fish. FASEB J 22 (PMID EB-abstract)
Kurita Y, Nakada T, Kato A, Mistry AC, Chang M-H, Romero MF, Hirose S (2008b) Identification of intestinal bicarbonate transporters involved in formation of carbonate precipitates to

stimulate water absorption in marine teleost fish. Am J Physiol Regul Integr Comp Physiol 284 (4):R1402–R1412

Kurtz I, Zhu Q (2013) Proximal renal tubular acidosis mediated by mutations in NBCe1-A: unraveling the transporter's structure-functional properties. Front Physiol 4:350

Lacruz RS, Nanci A, White SN, Wen X, Wang H, Zalzal SF, Luong VQ, Schuetter VL, Conti PS, Kurtz I et al (2010) The sodium bicarbonate cotransporter (NBCe1) is essential for normal development of mouse dentition. J Biol Chem 285(32):24432–24438

Lee SH, Park JH, Jung HH, Oh JW, Lee HM, Jun HS, Cho WJ, Lee JY (2005) Expression and distribution of ion transport mRNAs in human nasal mucosa and nasal polyps. Acta Otolaryngol 125(7):745–752

Lee MG, Ohana E, Park HW, Yang D, Muallem S (2012a) Molecular mechanism of pancreatic and salivary gland fluid and HCO_3 secretion. Physiol Rev 92(1):39–74

Lee SK, Boron WF, Parker MD (2012b) Relief of autoinhibition of the electrogenic Na-HCO_3 cotransporter NBCe1-B: role of IRBIT vs.amino-terminal truncation. Am J Physiol Cell Physiol 302(3):C518–C526

Lee SK, Boron WF, Parker MD (2013) Substrate specificity of the electrogenic sodium/bicarbonate cotransporter NBCe1-A (SLC4A4, variant A) from humans and rabbits. Am J Physiol Renal Physiol 304(7):F883–F899

Li HC, Worrell RT, Matthews JB, Husseinzadeh H, Neumeier L, Petrovic S, Conforti L, Soleimani M (2004) Identification of a carboxyl-terminal motif essential for the targeting of Na^+/HCO_3^- cotransporter NBC1 to the basolateral membrane. J Biol Chem 279(41):43190–43197

Liu Y, Xu JY, Wang DK, Wang L, Chen LM (2011) Cloning and identification of two novel NBCe1 splice variants from mouse reproductive tract tissues: a comparative study of NCBT genes. Genomics 98(2):112–119

Lo YF, Yang SS, Seki G, Yamada H, Horita S, Yamazaki O, Fujita T, Usui T, Tsai JD, Yu IS et al (2011) Severe metabolic acidosis causes early lethality in NBC1 W516X knock-in mice as a model of human isolated proximal renal tubular acidosis. Kidney Int Suppl 79(7):730–741

Lu J, Boron WF (2007) Reversible and irreversible interactions of DIDS with the human electrogenic Na/HCO_3 cotransporter NBCe1-A: role of lysines in the KKMIK motif of TM5. Am J Physiol Cell Physiol 292(5):C1787–C1798

Luo X, Choi JY, Ko SB, Pushkin A, Kurtz I, Ahn W, Lee MG, Muallem S (2001) HCO_3^- salvage mechanisms in the submandibular gland acinar and duct cells. J Biol Chem 3:3

Majumdar D, Maunsbach AB, Shacka JJ, Williams JB, Berger UV, Schultz KP, Harkins LE, Boron WF, Roth KA, Bevensee MO (2008) Localization of electrogenic Na/bicarbonate cotransporter NBCe1 variants in rat brain. Neuroscience 155(3):818–832

Marino CR, Jeanes V, Boron WF, Schmitt BM (1999) Expression and distribution of the Na^+-HCO_3^- cotransporter in human pancreas. Am J Physiol 277(2 Pt 1):G487–G494

Maunsbach AB, Vorum H, Kwon TH, Nielsen S, Simonsen B, Choi I, Schmitt BM, Boron WF, Aalkjaer C (2000) Immunoelectron microscopic localization of the electrogenic Na/HCO_3 cotransporter in rat and ambystoma kidney. J Am Soc Nephrol 11(12):2179–2189

McAlear SD, Liu X, Williams JB, McNicholas-Bevensee CM, Bevensee MO (2006) Electrogenic Na/HCO_3 cotransporter (NBCe1) variants expressed in Xenopus oocytes: functional comparison and roles of the amino and carboxy termini. J Gen Physiol 127(6):639–658

Moss F, Lee S, Parker M, Boron W (2014) Distinguishing HCO_3^- from $CO_3^=$ transport by the electrogenic Na/HCO_3 cotransporter NBCe1 (SLC4A4). FASEB J 28(1):a1098.7

Muller-Berger S, Ducoudret O, Diakov A, Fromter E (2001) The renal Na-HCO_3-cotransporter expressed in Xenopus laevis oocytes: change in stoichiometry in response to elevation of cytosolic Ca^{2+} concentration. Pflugers Arch 442(5):718–728

Musa-Aziz R, Chen LM, Pelletier MF, Boron WF (2009) Relative CO_2/NH_3 selectivities of AQP1, AQP4, AQP5, AmtB, and RhAG. Proc Natl Acad Sci USA 106(13):5406–5411

Myers E, Yuan L, Felmlee M, Lin YY, Jiang Y, Pei Y, Wang O, Li M, Xing XP, Marshall A et al (2015) A novel missense mutation in the NBCe1 gene (SLC4A4) and a novel mode of inheritance for proximal renal tubular acidosis (pRTA). FASEB J 29(1 Supp):809.24, abstract

Nakhoul NL, Davis BA, Romero MF, Boron WF (1998) Effect of expressing the water channel aquaporin-1 on the CO_2 permeability of Xenopus oocytes [see comments]. Am J Physiol 274 (2 Pt 1):C543–C548

Newman EA, Astion ML (1991) Localization and stoichiometry of electrogenic sodium bicarbonate cotransport in retinal glial cells. Glia 4(4):424–428

Orlowski A, Ciancio MC, Caldiz CI, De Giusti VC, Aiello EA (2014) Reduced sarcolemmal expression and function of the NBCe1 isoform of the Na_+-HCO_3^- cotransporter in hypertrophied cardiomyocytes of spontaneously hypertensive rats: role of the renin-angiotensin system. Cardiovasc Res 101(2):211–219

Paine ML, Snead ML, Wang HJ, Abuladze N, Pushkin A, Liu W, Kao LY, Wall SM, Kim YH, Kurtz I (2008) Role of NBCe1 and AE2 in secretory ameloblasts. J Dent Res 87(4):391–395

Park K, Hurley PT, Roussa E, Cooper GJ, Smith CP, Thevenod F, Steward MC, Case RM (2002) Expression of a sodium bicarbonate cotransporter in human parotid salivary glands. Arch Oral Biol 47(1):1–9

Park S, Hong JH, Ohana E, Muallem S (2012) The WNK/SPAK and IRBIT/PP1 pathways in epithelial fluid and electrolyte transport. Physiology 27(5):291–299

Parker MD, Boron WF (2013) The divergence, actions, roles, and relatives of sodium-coupled bicarbonate transporters. Physiol Rev 93(2):803–959

Pastor-Soler N, Beaulieu V, Litvin TN, Da Silva N, Chen Y, Brown D, Buck J, Levin LR, Breton S (2003) Bicarbonate-regulated adenylyl cyclase (sAC) is a sensor that regulates pH-dependent V-ATPase recycling. J Biol Chem 278(49):49523–49529

Pastor-Soler N, Pietrement C, Breton S (2005) Role of acid/base transporters in the male reproductive tract and potential consequences of their malfunction. Physiology 20:417–428

Perry C, Blaine J, Le H, Grichtchenko II (2006) PMA- and ANG II-induced PKC regulation of the renal Na^+-HCO_3^- cotransporter (hkNBCe1). Am J Physiol Renal Physiol 290(2):F417–F427

Perry C, Le H, Grichtchenko II (2007) ANG II and calmodulin/CaMKII regulate surface expression and functional activity of NBCe1 via separate means. Am J Physiol Renal Physiol 293(1): F68–F77

Perry C, Baker OJ, Reyland ME, Grichtchenko II (2009) PKCαβγ- and PKCδ dependent endocytosis of NBCe1-A and NBCe1-B in salivary parotid acinar cells. Am J Physiol Cell Physiol 297 (6):C1409–C1423

Piermarini PM, Choi I, Boron WF (2007) Cloning and characterization of an electrogenic Na/HCO_3^- cotransporter from the squid giant fiber lobe. Am J Physiol Cell Physiol 292(6): C2032–C2045

Praetorius J, Hager H, Nielsen S, Aalkjaer C, Friis UG, Ainsworth MA, Johansen T (2001) Molecular and functional evidence for electrogenic and electroneutral Na^+-HCO_3^- cotransporters in murine duodenum. Am J Physiol Gastrointest Liver Physiol 280(3):G332–G343

Prasad GVR, Coury LA, Finn F, Zeidel ML (1998) Reconstituted aquaporin 1 water channels transport CO_2 across membranes. J Biol Chem 273(50):33123–33126

Rodriguez-Martinez H, Ekstedt E, Einarsson S (1990) Acidification of epididymal fluid in the boar. Int J Androl 13(3):238–243

Romero MF (2001) The electrogenic Na^+/HCO_3^- cotransporter, NBC. JOP 2(4 Suppl):182–191

Romero MF, Boron WF (1999) Electrogenic Na/HCO_3 cotransporters: expression cloning and physiology. Ann Rev Physiol 61:699–723

Romero MF, Hediger MA, Boulpaep EL, Boron WF (1997) Expression cloning and characterization of a renal electrogenic Na^+/HCO_3^- cotransporter. Nature 387(6631):409–413

Romero MF, Davis BA, Sussman CR, Bray-Ward P, Ward D, Boron WF (1998a) Identification of multiple genes for human electrogenic Na/HCO_3 cotransporters (NBC) on 4q. J Am Soc Nephrol 9:11A

Romero MF, Fong P, Berger UV, Hediger MA, Boron WF (1998b) Cloning and functional expression of rNBC, an electrogenic Na^+-HCO_3^- cotransporter from rat kidney. Am J Physiol 274(2 Pt 2):F425–F432

Romero MF, Fulton CM, Boron WF (2004) The SLC4 gene family of HCO_3^- transporters. Pflugers Arch 447(5):495–509

Romero MF, Chen AP, Parker MD, Boron WF (2013a) The SLC4 family of bicarbonate (HCO_3^-) transporters. Mol Asp Med 34(2–3):159–182

Romero MF, Holmes HL, Chowdhury UR, Hann CR, Chang M-H, Fautsch MP, Chen A-P (2013b) Ocular indications of partial penetrance in mice heterozygous for NBCe1 (Slc4a4). Invest Ophthalmol Vis Sci 54(15):5646, abstract

Romero MF, Holmes HL, Chowdhury UR, Hann CR, Chang M-H, Fautsch MP, Chen A-P (2014) Mutants in NBCe1A isoform have elevated intraocular pressure. Invest Ophthalmol Vis Sci 55: 2415, abstract

Rossmann H, Bachmann O, Vieillard-Baron D, Gregor M, Seidler U (1999) Na^+/HCO_3^- cotransport and expression of NBC1 and NBC2 in rabbit gastric parietal and mucous cells. Gastroenterology 116(6):1389–1398

Roussa E, Romero MF, Schmitt BM, Boron WF, Alper SL, Thévenod F (1999) Immunolocalization of AE2 anion exchanger and Na^+-HCO_3^- cotransporter in rat parotid and submandibular glands. Am J Physiol 277(6 Pt 1):G1288–G1296

Ruminot I, Gutierrez R, Pena-Munzenmayer G, Anazco C, Sotelo-Hitschfeld T, Lerchundi R, Niemeyer MI, Shull GE, Barros LF (2011) NBCe1 mediates the acute stimulation of astrocytic glycolysis by extracellular K^+. J Neurosci 31(40):14264–14271

Schmitt BM, Berger UV, Douglas RM, Bevensee MO, Hediger MA, Haddad GG, Boron WF (2000) Na/HCO_3 cotransporters in rat brain: expression in glia, neurons, and choroid plexus. J Neurosci 20(18):6839–6848

Schueler C, Becker HM, McKenna R, Deitmer JW (2011) Transport activity of the sodium bicarbonate cotransporter NBCe1 Is enhanced by different isoforms of carbonic anhydrase. PLoS ONE 6(11), e27167

Sciortino CM, Romero MF (1999) Cation and voltage dependence of rat kidney, electrogenic Na/HCO cotransporter, rkNBC, expressed in oocytes. Am J Physiol 277(4 Pt 2):F611–F623

Seidler U, Bachmann O, Jacob P, Christiani S, Blumenstein I, Rossmann H (2001) Na^+/HCO_3^- cotransport in normal and cystic fibrosis intestine. JOP 2(4 Suppl):247–256

Sergeev M, Godin AG, Kao L, Abuladze N, Wiseman PW, Kurtz I (2012) Determination of membrane protein transporter oligomerization in native tissue using spatial fluorescence intensity fluctuation analysis. PLoS ONE 7(4), e36215

Shirakabe K, Priori G, Yamada H, Ando H, Horita S, Fujita T, Fujimoto I, Mizutani A, Seki G, Mikoshiba K (2006) IRBIT, an inositol 1,4,5-trisphosphate receptor-binding protein, specifically binds to and activates pancreas-type Na^+/HCO_3^- cotransporter 1 (pNBC1). Proc Natl Acad Sci USA 103(25):9542–9547

Shnitsar V, Li J, Li X, Calmettes C, Basu A, Casey JR, Moraes TF, Reithmeier RA (2013) A substrate access tunnel in the cytosolic domain is not an essential feature of the solute carrier 4 (SLC4) family of bicarbonate transporters. J Biol Chem 288(47):33848–33860

Sinclair ML, Wang XY, Mattia M, Conti M, Buck J, Wolgemuth DJ, Levin LR (2000) Specific expression of soluble adenylyl cyclase in male germ cells. Mol Reprod Dev 56(1):6–11

Sinđić A, Sussman CR, Romero MF (2010) Primers on molecular pathways - bicarbonate transport by the pancreas. Pancreatology 10(6):660–663

Skelton LA, Boron WF, Zhou Y (2010) Acid–base transport by the renal proximal tubule. J Nephrol 23(Suppl 16):S4–S18

Soleimani M, Aronson PS (1989) Ionic mechanism of Na^+-HCO_3^- cotransport in rabbit renal basolateral membrane vesicles. J Biol Chem 264(31):18302–18308

Soleimani M, Lesoine GA, Bergman JA, Aronson PS (1991) Cation specificity and modes of the Na^+:CO_3^{2-}:HCO_3^- cotransporter in renal basolateral membrane vesicles. J Biol Chem 266 (14):8706–8710

Soleimani M, Hattabaugh YJ, Bizal GL (1992) pH sensitivity of the Na^+:HCO_3^- cotransporter in basolateral membrane vesicles isolated from rabbit kidney cortex. J Biol Chem 267 (26):18349–18355

Soyfoo MS, Bulur N, Virreira M, Louchami K, Lybaert P, Crutzen R, Perret J, Delporte C, Roussa E, Thevenod F et al (2009) Expression of the electrogenic Na^+-HCO_3^--cotransporters NBCe1-A and NBCe1-B in rat pancreatic islet cells. Endocrine 35(3):449–458

Steward MC, Ishiguro H, Case RM (2005) Mechanisms of bicarbonate secretion in the pancreatic duct. Annu Rev Physiol 67:377–409

Sun XC, Bonanno JA (2003) Identification and cloning of the Na/HCO_3^- cotransporter (NBC) in human corneal endothelium. Exp Eye Res 77(3):287–295

Suzuki M, Vaisbich MH, Yamada H, Horita S, Li Y, Sekine T, Moriyama N, Igarashi T, Endo Y, Cardoso TP et al (2008) Functional analysis of a novel missense NBC1 mutation and of other mutations causing proximal renal tubular acidosis. Pflugers Arch 455(4):583–593

Suzuki M, Van Paesschen W, Stalmans I, Horita S, Yamada H, Bergmans BA, Legius E, Riant F, De Jonghe P, Li Y et al (2010) Defective membrane expression of the Na^+-HCO_3^- cotransporter NBCe1 is associated with familial migraine. Proc Natl Acad Sci USA 107 (36):15963–15968

Tajima Y, Okamura N, Sugita Y (1987) The activating effects of bicarbonate on sperm motility and respiration at ejaculation. Biochim Biophys Acta 924(3):519–529

Ten Hove M, Nederhoff MG, Van Echteld CJ (2004) Relative contributions of Na^+/H^+ exchange and Na^+/HCO_3^- co-transport to ischemic Na^+_i overload in isolated rat hearts. Am J Physiol Heart Circ Physiol 288(1):H287–H292

Thevenod F, Gasser KW, Hopfer U (1990) Dual modulation of chloride conductance by nucleotides in pancreatic and parotid zymogen granules. Biochem J 272(1):119–126

Thevenod F, Chathadi KV, Jiang B, Hopfer U (1992) ATP-sensitive K^+ conductance in pancreatic zymogen granules: block by glyburide and activation by diazoxide. J Membr Biol 129 (3):253–266

Thévenod F, Roussa E, Schmitt BM, Romero MF (1999) Cloning and immunolocalization of a rat pancreatic Na^+ bicarbonate cotransporter. Biochem Biophys Res Commun 264(1):291–298

Toye AM, Parker MD, Daly CM, Lu J, Virkki LV, Pelletier MF, Boron WF (2006) The human NBCe1-A mutant R881C, associated with proximal renal tubular acidosis, retains function but is mistargeted in polarized renal epithelia. Am J Physiol Cell Physiol 291(4):C788–C801

Turner HC, Alvarez LJ, Candia OA (2001) Identification and localization of acid–base transporters in the conjunctival epithelium. Exp Eye Res 72(5):519–531

Usui T, Hara M, Satoh H, Moriyama N, Kagaya H, Amano S, Oshika T, Ishii Y, Ibaraki N, Hara C et al (2001) Molecular basis of ocular abnormalities associated with proximal renal tubular acidosis. J Clin Invest 108(1):107–115

Wang HS, Chen Y, Vairamani K, Shull GE (2014) Critical role of bicarbonate and bicarbonate transporters in cardiac function. World J Biol Chem 5(3):334–345

Wen X, Kurtz I, Paine ML (2014) Prevention of the disrupted enamel phenotype in *Slc4a4*-null mice using explant organ culture maintained in a living host kidney capsule. PLoS ONE 9(5), e97318

Wu J, McNicholas CM, Bevensee MO (2009) Phosphatidylinositol 4,5-bisphosphate (PIP_2) stimulates the electrogenic Na/HCO_3 cotransporter NBCe1-A expressed in *Xenopus* oocytes. Proc Natl Acad Sci USA 106(33):14150–14155

Xue H, Tian YM, Yan M, Yang N, Chen X, Xing Y, Zhu JX (2007) Appearance of segmental discrepancy of anion transport in rat distal colon. Biol Pharm Bull 30(8):1407–1411

Yamamoto T, Shirayama T, Sakatani T, Takahashi T, Tanaka H, Takamatsu T, Spitzer KW, Matsubara H (2007) Enhanced activity of ventricular Na^+-HCO_3^- cotransport in pressure overload hypertrophy. Am J Physiol Heart Circ Physiol 293(2):H1254–H1264

Yamazaki O, Yamada H, Suzuki M, Horita S, Shirai A, Nakamura M, Satoh N, Fujita T, Seki G (2013) Identification of dominant negative effect of L522P mutation in the electrogenic Na_+-HCO_3^- cotransporter NBCe1. Pflugers Arch 465(9):1281–1291

Yang D, Shcheynikov N, Zeng W, Ohana E, So I, Ando H, Mizutani A, Mikoshiba K, Muallem S (2009a) IRBIT coordinates epithelial fluid and HCO_3^- secretion by stimulating the transporters pNBC1 and CFTR in the murine pancreatic duct. J Clin Invest 119(1):193–202

Yang HS, Kim E, Lee S, Park HJ, Cooper DS, Rajbhandari I, Choi I (2009b) Mutation of aspartate 555 of the sodium/bicarbonate transporter SLC4A4/NBCe1 induces chloride transport. J Biol Chem 284(23):15970–15979

Yang D, Li Q, So I, Huang CL, Ando H, Mizutani A, Seki G, Mikoshiba K, Thomas PJ, Muallem S (2011) IRBIT governs epithelial secretion in mice by antagonizing the WNK/SPAK kinase pathway. J Clin Invest 121(3):956–965

Yu H, Riederer B, Stieger N, Boron WF, Shull GE, Manns MP, Seidler UE, Bachmann O (2009) Secretagogue stimulation enhances NBCe1 (electrogenic Na^+/HCO_3^- cotransporter) surface expression in murine colonic crypts. Am J Physiol Gastrointest Liver Physiol 297(6):G1223–G1231

Zhao WC, Duan DX, Wang ZJ, Tang N, Yan M, Zhang GH, Xing Y (2005) The underlying cellular mechanism in the effect of tetramethylpyrazine on the anion secretion of colonic mucosa. Jpn J Physiol 55(6):325–329

Zheng L, Zhang Y, He P, Kim J, Schneider R, Bronckers AL, Lyaruu DM, DenBesten PK (2011) NBCe1 in mouse and human ameloblasts may be indirectly regulated by fluoride. J Dent Res 90(6):782–787

Zhu Q, Kao L, Azimov R, Abuladze N, Newman D, Pushkin A, Liu W, Chang C, Kurtz I (2010a) Structural and functional characterization of the C-terminal transmembrane region of NBCe1-A. J Biol Chem 285(48):37178–37187

Zhu Q, Kao L, Azimov R, Newman D, Liu W, Pushkin A, Abuladze N, Kurtz I (2010b) Topological location and structural importance of the NBCe1-A residues mutated in proximal renal tubular acidosis. J Biol Chem 285(18):13416–13426

Zhu Q, Liu W, Kao L, Azimov R, Newman D, Abuladze N, Kurtz I (2013a) Topology of NBCe1 protein transmembrane segment 1 and structural effect of proximal renal tubular acidosis (pRTA) S427L mutation. J Biol Chem 288(11):7894–7906

Zhu Q, Shao XM, Kao L, Azimov R, Weinstein AM, Newman D, Liu W, Kurtz I (2013b) Missense mutation T485S alters NBCe1-A electrogenicity causing proximal renal tubular acidosis. Am J Physiol Cell Physiol 305(4):C392–C405

Zippin JH, Levin LR, Buck J (2001) CO_2/HCO_3^--responsive soluble adenylyl cyclase as a putative metabolic sensor. Trends Endocrinol Metab 12(8):366–370

Chapter 14
Properties and Function of the Solute Carrier 26 Family of Anion Transporters

Nikolay Shcheynikov, Ehud Ohana, and Shmuel Muallem

Abstract The SLC26 family of anion transporters consists of 10 members that display remarkable functional and substrate diversity. Mutations in several members of the family have been identified as causing a variety of human diseases and mouse phenotype when deleted. The family drew the attention and strong interest of epithelial biologist with the identification of the first elusive luminal Cl^-/HCO_3^- exchange, which turned to be the third member of the family SLC26A3. Fairly quick, progress since revealed that members of the family transport all halides, NO_3^-, SO_4^{2-}, oxalate, and formate, among others. Members of the family can be grouped into three subgroups based on substrate selectivity and transport mode: the SO_4^{2-} transporters SLC26A1 and SLC26A2; the anion exchangers $2Cl^-/1HCO_3^-$ SLC26A3, $1Cl^-/1HCO_3^-$ SLC26A4, and $1Cl^-/2HCO_3^-$ SLC26A6; and the Cl^- channels SLC26A7, SLC26A9, and SLC26A11. This chapter discusses structural features, transport properties, and regulation of the transporters that are essential to understand their functions and roles in human diseases.

Keywords SLC26 transporters • Anions • Exchangers • Channels • Function, regulation, disease

14.1 General Features of the SLC26 Transporters

This chapter discusses the properties and function of the anion transporting SLC26 family. Several SLC26 transporters are associated with human disease or mouse phenotype with implication to human disease. Members of the family transport diverse monovalent and polyvalent anions including Cl^-, HCO_3^-, I^-, SO_4^{2-},

N. Shcheynikov • S. Muallem (✉)
Epithelial Signaling and Transport Section, Molecular Physiology and Therapeutics Branch, National Institute of Dental and Craniofacial Research, National Institute of Health, NIH, NIDCR, Building 10, Room 1N-112, Bethesda, MD 20892, USA
e-mail: shmuel.muallem@nih.gov

E. Ohana
Department of Clinical Biochemistry and Pharmacology, Ben Gurion University of the Negev, Beer Sheva, Israel

K.L. Hamilton, D.C. Devor (eds.), *Ion Channels and Transporters of Epithelia in Health and Disease*, Physiology in Health and Disease,
DOI 10.1007/978-1-4939-3366-2_14

oxalate, and formate. They display equally diverse modes of transport, with several functioning as electroneutral exchangers, electrogenic exchangers, or bona fide ion channels and even transporters that can function as both electrogenic exchangers and ion channels, depending on the substrate. As such, the SLC26 transporters are involved in various cellular functions in many cell types. In epithelia, their best documented and understood function is fluid and electrolyte secretion (Dorwart et al. 2008; Lee et al. 2012b; Park et al. 2012).

Fluid and electrolyte secretion is the most crucial function of most epithelia and is driven by Cl^- absorption (or secretion). Another critical function of Cl^- transport is HCO_3^- secretion (or absorption). Together, the two functions control the volume, electrolyte composition, and pH of the secreted fluids in a variety of epithelia discussed in this book. A special example in which HCO_3^- secretion is of paramount importance is secretory glands, as discussed in detail by Blouquit-Laye and Chinet (2007) and Lee et al. (2012b). In general, vectorial Cl^- absorption and HCO_3^- secretion are mediated by HCO_3^- entry at the basolateral membrane, which is principally mediated by the ubiquitous Na^+-HCO_3^- co-transporter NBCe1-B. Basolateral HCO_3^- influx by the Cl^-/HCO_3^- exchanger, AE2, coupled with the activity of the Na^+-H^+ exchanger, NHE1, mediates a small portion of transcellular HCO_3^- transport in secretory glands (Bachmann et al. 2006; Melvin et al. 2005; Steward et al. 2005; Zhao et al. 1994). However, HCO_3^- loading by AE2 appears to be equally important as NBCe1-B in other epithelia, like the airway serous glands (Huang et al. 2012; Shan et al. 2012). HCO_3^- then exits the luminal membrane by the combined action of CFTR and members of the SLC26 transporters, most commonly the $1Cl^-/2HCO_3^-$ exchanger SLC26A6 (Dorwart et al. 2008; Ko et al. 2004; Lee et al. 2012b). Accordingly, many SLC26 transporters are expressed in the luminal membrane and co-localize with CFTR, as illustrated for SLC26A6 and SLC26A4 in Fig. 14.1b.

The SLC26 transporter family is coded by 11 genes with SLC26A10 being a pseudogene in humans (Dorwart et al. 2008; Park et al. 2012). The relatedness between members of the family is illustrated in Fig. 14.1a. Further analysis of the SulP clad can be found in manuscripts by Compton et al. (2014) and Shelden et al. (2010). The SLC26 transporter family was established with the identification of SLC26A1 as a liver SO_4^{2-} transporter (Bissig et al. 1994). The second member to be identified was the SO_4^{2-} transporter SlC26A2 as the protein mutated in the human disease dystrophic dysplasia (DTDST; Hastbacka et al. 1994). The family grabbed the interest of epithelial biologists with the discovery that congenital Cl^- diarrhea is caused by mutations in SLC26A3 (Hoglund et al. 1996) and the subsequent demonstration that SLC26A3 is expressed in the luminal membrane of colonic epithelium and functions as a Cl^- and HCO_3^- transporter (Moseley et al. 1999). This was the first identification of the elusive luminal membrane transporter that mediates Cl^-/HCO_3^- exchange. The function of SLC26A3 as a coupled Cl^-/HCO_3^- exchanger was then described by expressing the protein in HEK cells (Melvin et al. 1999). SLC26A4 was then identified as the transporter mutated in Pendred syndrome (Everett et al. 1997) and was later shown to function as a Cl^-/

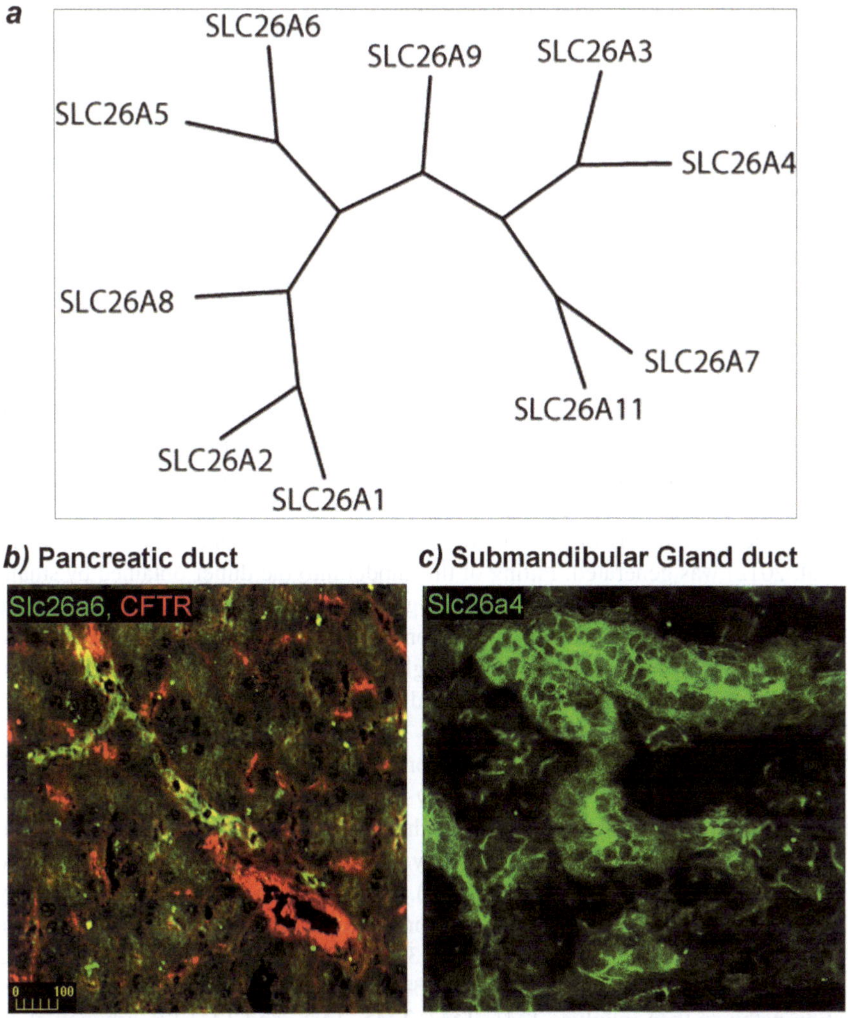

Fig. 14.1 The SLC26 transporter family. Panel (**a**) shows a dendrogram of the SLC26 transporter family. Panel (**b**) shows co-localization of CFTR (*red*) and SLC26A6 (*green*) in the mouse pancreatic duct [reproduced from Ko et al. (2004)], and panel (**c**) shows localization of SLC26A4 (*green*) in the submandibular duct

HCO_3^-/I^- exchanger (Shcheynikov et al. 2008). SLC26A5 (prestin) was identified as the protein mediating electromotility of outer hair cells (Zheng et al. 2000). Other members of the family were identified by database searches (Dorwart et al. 2008).

14.1.1 Structure of the SLC26 Transporters

SLC26 transporters are found in many species, from human to bacteria (Compton et al. 2014; Dorwart et al. 2008). There is limited structural information on these transporters. Domain analysis of bacterial SLC26 homologues established 12 transmembrane domains and a cytoplasmic C-terminus STAS domain for the SLC26 transporter superfamily (Compton et al. 2014; Ohana et al. 2011; Shelden et al. 2010). Analysis of a bacterial homologue of SLC26A2 by small-angle neutron scattering indicated that SLC26 transporters are dimers (Compton et al. 2011) and provided a general shape for the transporters with the dimerization mediated by a central lobe formed by the two transmembrane sectors of the two monomers, which are flanked by peripheral domains formed by the STAS domains (Compton et al. 2014). Interestingly, the monomers within the dimer appear to influence the activity of each other, as concluded from assembly of a dimer with novel features by co-expressing wild-type and mutant SLC26A5 monomers (Detro-Dassen et al. 2008). Based on potential structural homology, and functional similarity to the CLC transporters, a model for SLC26A6 (Ohana et al. 2011) and SLC26A2 (Ohana et al. 2012) was generated. Fitting of this model into the dimer obtained by small-angle neutron scattering is shown in Fig. 14.2a. In this model, the permeation pathway has a similar orientation of a conserved glutamate and tyrosine as found in the CLC transporters. The conserved glutamate appears to determine the transport mode of the transporters as coupled or as channel-like (Ohana et al. 2011, 2012).

Recent modeling of SLC26A5, based on the structure of a related bacterial uracil transporter, UraA, came to a significantly different conclusion, suggesting that the SLC26 transporter transmembrane sector has 14 transmembrane spans grouped into two lobes of structural repeats related by a twofold inversion axis parallel to the membrane bilayer (Gorbunov et al. 2014). The latter was verified for SLC26A5 by cysteine accessibility scan. The model predicts an ion binding site formed by a cavity that is bordered by parts of TMs 1, 3, and 5 of sector 1 and TMs 8, 10, and 12 of sector 2 and includes the antiparallel β-strands of TM3 and TM10. The permeation pathway then involves TM1, TM3, TM5, TM8, and TM12 (Gorbunov et al. 2014). The model further suggests that the glutamate conserved in all SLC26 transporters and affecting ion permeation interacts with conserved SLC26 transporters tyrosine within the SulP motif in TM2 to mediate TM2–TM9 association between the two sectors (Gorbunov et al. 2014). The model on SLC26A2 developed based on the model of SLC26A5 fitted into the small-angle neutron scattering image is shown in Fig. 14.2b. However, it is not obvious how this function of the glutamate residue can explain its role in controlling ion selectivity and transport mode (coupled transport and channel-like; Ohana et al. 2011). Whether the model developed for SLC26A5 holds for all SLC26 transporters needs to be tested experimentally. Further, it is important to note that models are only predictors, and the actual structure of bacterial and mammalian SLC26 transporters needs to be resolved to understand the structure and function of the transmembrane sector of the transporters.

a

b

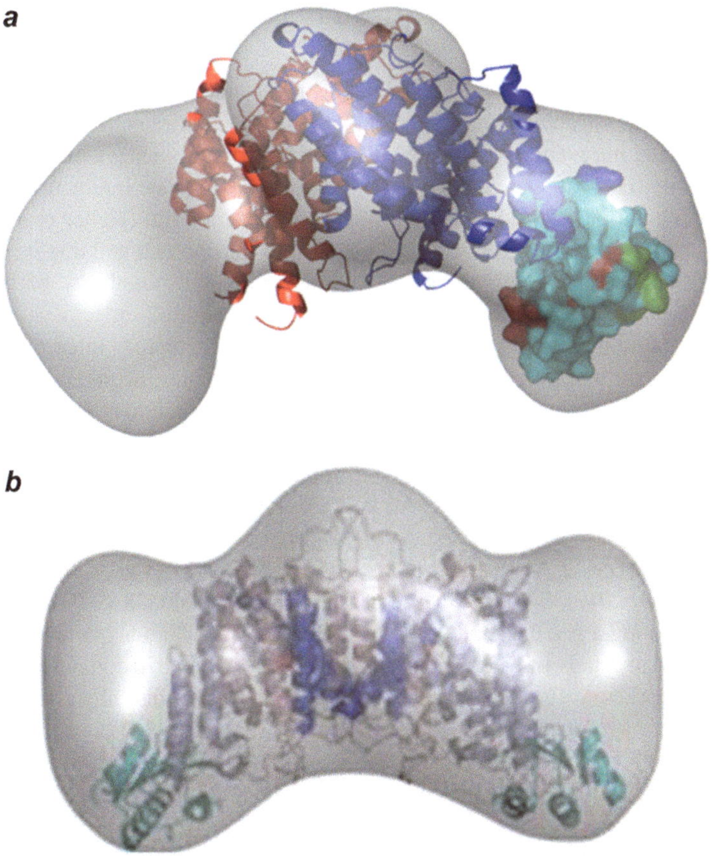

Fig. 14.2 Structural features of the SLC26 transporters. Panel (**a**) shows fitting of a model of SLC26A6 transmembrane domains (*red* and *blue*) and the STAS domain (*turquoise*) (see Ohana et al. 2011) into the small-angle neutron scattering image of bacterial SLC26A2 (reproduced from Compton et al. 2011). Panel (**b**) shows fitting of a model of SLC26A5 transmembrane domains (core domain in *pink*, β strands in *red*, and gate domain in *blue*) and the STAS domain (*turquoise*) (see Gorbunov et al. 2014) into the small-angle neutron scattering image of bacterial SLC26A2 (reproduced from Compton et al. 2014)

The only established SLC26 transporters structure is the C-terminus protein-protein interacting STAS domain. The STAS domain was predicted based on similarity to the anti-sigma factor antagonist (Aravind and Koonin 2000), and the crystal structure of the SLC26A5 (prestin) STAS domain was then solved (Pasqualetto et al. 2010). The structure of the *E. coli* YchM STAS domain in complex with acyl carrier protein (Babu et al. 2010) and of Rv1739c from *Mycobacterium tuberculosis* (Sharma et al. 2011) are also available.

14.1.2 Regulation of the SLC26 Transporters

An important feature of the STAS domain in the mammalian transporter is the presence of an unstructured and divergent variable loop between the $\alpha 1$ helix and $\beta 3$ strand (named intervening sequence, IVS) that is different in size among members of the family (Dorwart et al. 2008). The IVS appears to mediate interaction with and possible regulation of the SLC26 transporters by calmodulin (Keller et al. 2014). Of note is the variable length of the IVS in SLC26 transporters and that it is almost missing in SLC26A2 (Dorwart et al. 2008), raising the possibility of isoform-specific regulation of the SLC26 transporters by calmodulin.

A major role of the STAS domain is mediating the regulatory interaction of the SLC26 transporters with neighboring proteins within plasma membrane microdomains. The best established and understood form of regulation is the mutual regulation of SLC26 transporters and CFTR that is mediated by interaction of the STAS domain with the CFTR R domain (Fig. 14.3). This was first demonstrated for interaction of SLC26A3 and SLC26A6 STAS domains with the R domain (Ko et al. 2004). Since its discovery, the interaction, and in some cases the mutual regulation, was shown for SLC26A4 (Garnett et al. 2011; Shcheynikov et al. 2008), SLC26A5 (Homma et al. 2010), SLC26A8 (Rode et al. 2012), and SLC26A9 (Bertrand et al. 2009), indicating a general mode of regulation of the SLC26 transporters by CFTR. The molecular details of interaction between the STAS and R domains have been examined biochemically and structurally. The R domain is intrinsically disordered and interacts with several partners (Bozoky et al. 2013a). The non-phosphorylated R domain avidly interacts with the nucleotide-binding domains (NBDs) of CFTR to inhibit channel activity, while phosphorylation of the R domain reduces its interaction with the NBDs and activates the channel (Baker et al. 2007; Bozoky et al. 2013b). A reciprocal situation is observed with the STAS domain, in which phosphorylation of the R domain increases its interaction with the STAS domain (Bozoky et al. 2013b; Ko et al. 2004) to activate the SLC26 transporters (Ko et al. 2004). Hence, the two transporters are mutually regulated by interaction of the R and STAS domains (Fig. 14.3).

The SLC26 transporters are regulated by the IP_3 receptor-binding protein released with IP_3 (IRBIT) (Ando et al. 2003, 2006) and the WNK/SPAK kinases (Park et al. 2013; Yang et al. 2009, 2011a), although the molecular determinants mediating the interactions are not known at present. The WNK/SPAK pathway reduces surface expression of the transporters, while IRBIT recruits protein phosphatase 1 (PP1) to dephosphorylate the transporters, increase their surface expression, and activate them (Park et al. 2013; Yang et al. 2009, 2011a), probably by removal of autoinhibition (Hong et al. 2013). Moreover, IRBIT mediates the synergistic activation of the transporters by the cAMP and Ca^{2+} signaling pathways (Ahuja et al. 2014; Park et al. 2013).

Other interactions of the STAS domain include interaction with the dicarboxylate transporter, NaDC-1 (Ohana et al. 2013), and with the microtubule-associated protein MAP1S (Bai et al. 2010). The SLC26A6 STAS domain interacts with

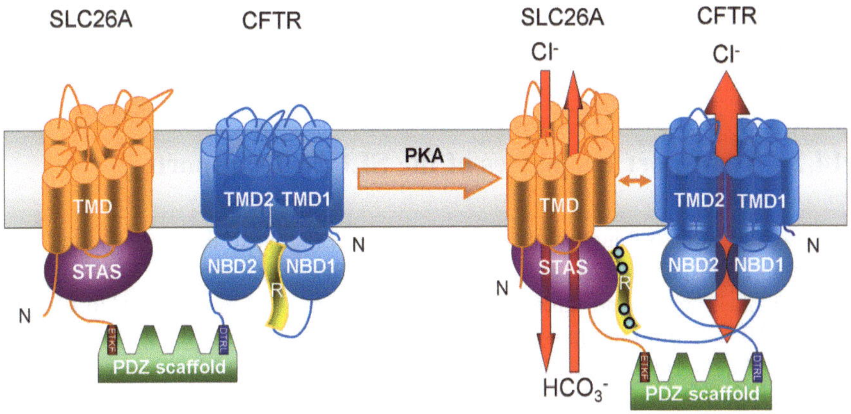

Fig. 14.3 Mutual regulation of CFTR and the SLC26 transporters. The SLC26 transporter and CFTR are assembled into complexes by binding of their PDZ motif to PDZ domain-containing scaffolds like NHERF4. In the resting state, the unphosphorylated R domain interacts with CFTR NBD1 to prevent CFTR and SLC26 transporter activation. Phosphorylation of the R domain by PKA reduces its binding to NBD1 and at the same time increases its binding to the SLC26 transporter STAS domain, resulting in the mutual activation of CFTR and the SLC26 transporter (see Dorwart et al. 2008; Ko et al. 2004 for further details)

an NaDC-1 region that encompasses residues 45–118. Based on the recently resolved crystal structure of a bacterial homologue of NaDC-1, vcINDY, this segment is expected to include TMDs 3 and 4. The STAS domain interacts with this region to stimulate SLC26A6 activity and at the same time strongly inhibit NaDC-1 activity. This interaction operates in the kidney, intestine, and likely other tissues expressing the two proteins to regulate urinary or luminal space citrate levels to prevent formation of kidney stones (Ohana et al. 2013). Interaction of MAP1S with the STAS domain of SLC26A5 increases SLC26A5 surface expression and thus activity at the plasma membrane (Bai et al. 2010). Since MAP1S is ubiquitously expressed and interacts with the conserved STAS domain, it likely affects the activity of other SLC26 transporters. Interestingly, MAP1S appears to coordinate several components of the autophagic pathways to control autophagosome biogenesis and degradation (Xie et al. 2011). It should be of interest to determine if there is any relationship between the roles of MAP1S in the regulation of SLC26 transporters and autophagy and whether the two MAP1S roles are related.

The interactions of the STAS domain with other proteins described above are only the few known. It is certain that the SLC26 transporters STAS domain interacts with many other proteins to mediate additional modes of transporter regulation. In addition, most SLC26 transporters have C-terminus PDZ ligands that mediate their recruitment and assembly into complexes. The role of the PDZ ligands of several transporters in interaction with PDZ domains possessing scaffolds have been described, including their role in CFTR-SLC26A3 and CFTR-SLC26A6 interactions (Ko et al. 2004; Lamprecht and Seidler 2006; Lee et al.

2012a). Again, this increases the potential for interaction of the SLC26 transporters with other proteins that may involve domains other than the STAS domain.

14.2 Transport Properties of the SLC26 Transporters

Below, the SLC26 transporters are discussed in relation to their transported substrate and transport modes. The first group is the SO_4^{2-} transporters, SLC26A1 and SLC26A2. The second group is the exchangers, SLC26A3, SLC26A4, and SLC26A6. The third group is the ion channels, SLC26A7, SLC26A9, and possibly SLC26A11 (Dorwart et al. 2008). The mammalian SLC26A5 functions as a motor protein that does not transport ions, although the invertebrate SLC26A5 functions as an exchanger and their function is described elsewhere (Detro-Dassen et al. 2007; Gorbunov et al. 2014; Schaechinger and Oliver 2007). The transport mode and substrate specificity of SLC26A8 are not known at present.

14.2.1 *The SO_4^{2-} Transporters*

SLC26A1 is expressed in the basolateral membrane and transports SO_4^{2-} (Karniski et al. 1998; Regeer et al. 2003) and oxalate (Xie et al. 2002), but not Cl^-, OH^-, or HCO_3^- (Karniski et al. 1998; Regeer et al. 2003) by an unknown transport mode. SO_4^{2-} transport by SLC26A1 is activated by extracellular halides and acidic pH (Xie et al. 2002), further indicating that SLC26A1 does not transport Cl^- and does not function as a SO_4^{2-}/Cl^- exchanger. The molecular feature of the interaction of SLC26A1with halides has not been examined, but it may be mediated by a mechanism similar to that reported for activation of SLC26A5 by extracellular Cl^- (Ohana et al. 2012). The physiological role for SLC26A1 is not well understood, although some information became available from the SLC26A1$^{-/-}$ mice. That is, when fed with high oxalate, the SLC26A1$^{-/-}$ mice exhibit hyposulfatemia, hypersulfaturia, urolithiasis, and nephrocalcinosis (Dawson et al. 2010). These findings are consistent with a role of SLC26A1 in maintaining systemic SO_4^{2-} homeostasis and a role in oxalate clearance either by the intestine or the kidney.

SLC26A2 was first identified as the gene mutated in diastrophic dysplasia (Hastbacka et al. 1994). Subsequently, additional mutations all along the protein associated with the disease were identified (Jackson et al. 2012). SLC26A2 is ubiquitous, but is expressed at high levels in cartilage, connective tissues, the adrenal cortex, and the apical membrane of several epithelia (Chapman and Karniski 2010; Haila et al. 2001; Hastbacka et al. 1994; Spyroglou et al. 2014). Expression in *Xenopus* oocytes showed that SLC26A2 functions as an electroneutral Cl^--regulated $SO_4^{2-}/Cl^-/OH^-$ exchanger (Ohana et al. 2012) and can also exchange Cl^- for oxalate (Heneghan et al. 2010; Ohana et al. 2012). The $SO_4^{2-}/Cl^-/OH^-$ exchange mode is illustrated in Fig. 14.4a. The coupling of SO_4^{2-} transport to Cl^- or OH^- depends on the cellular Cl^- and pH gradients. For example,

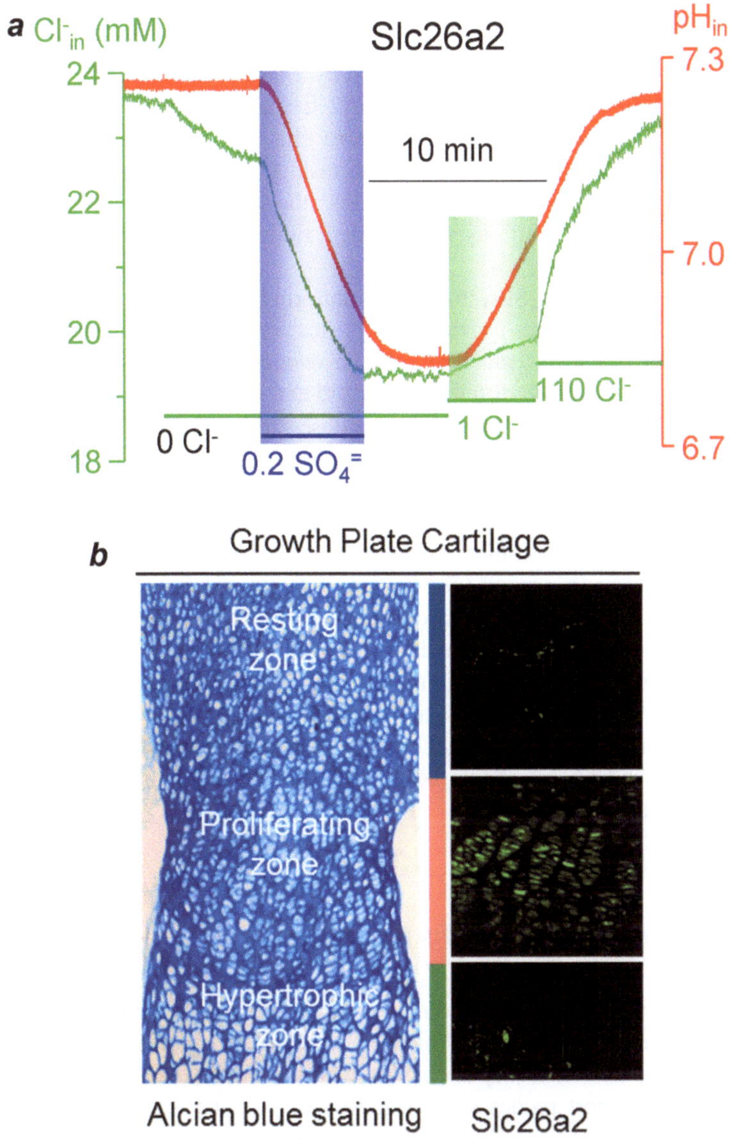

Fig. 14.4 SLC26A2-mediated SO_4^{2-} transport and expression in growth plate cartilage cells. Panel (**a**) shows simultaneous measurement of pH_{in} and Cl^-_{in} in *Xenopus* oocytes expressing SLC26A2. Note that SO_4^{2-} transport is coupled to both Cl^- and OH^- (or H^+) transport. Regulation by Cl^-_{out} is shown in the *green square* in which 1 mM Cl^-_{out} potently increases $SO_4^{2-}{}_{in}/OH^-$ exchange (reproduced from Ohana et al. 2012). Panel (**b**) shows expression levels of SLC26A2 in chondrocytes at different zones of the growth plate cartilage (reproduced from Park et al. 2014)

at acidic extracellular pH, SO_4^{2-} influx is mostly coupled to OH^- efflux (or H^+ cotransport), while at high extracellular pH, the opposite holds. Interestingly, SLC26A2 activity is exquisitely regulated by extracellular Cl^- (the green bar in Fig. 14.4a) which indicates that SLC26A2 has two distinct Cl^- binding sites, regulatory and catalytic. While the regulatory site remains elusive, at the catalytic site, Cl^- most likely interacts with a Phe residue found within a GFXXP motif (Ohana et al. 2012). When mutated to alanine, the affinity of SLC26A2 to Cl^- was attenuated, while SO_4^{2-} affinity was elevated, further indicating that Cl^- competes with SO_4^{2-} for binding.

In bone, expression of SLC26A2 is not uniform but varies between growth plate zones (Fig. 14.4b). Besides providing SO_4^{2-} for proteoglycan sulfation (Gualeni et al. 2013), SLC26A2 has multiple roles in chondrocyte biology, including cell proliferation, chondrocyte differentiation, and cell mass (Park et al. 2014). This is reflected in the multiple skeletal alterations observed in diastrophic dysplasia and in the mouse model of the disease (Forlino et al. 2005). The function of SLC26A2 may even be broader. That is, a recent study showed that SLC26A2 is expressed in the adrenal cortex and regulates aldosterone secretion (Spyroglou et al. 2014). A genome-wide association study found association between the aldosterone/renin ratio and a locus coding for SLC26A2. SLC26A2 suppresses aldosterone secretion as knockdown of SLC26A2 enhanced aldosterone secretion. These effects may not be direct, but rather through remodeling, steroidogenic, and Ca^{2+} signaling pathways in the knockdown cells (Spyroglou et al. 2014). Regulation of aldosterone levels and the potential effect on Ca^{2+} signaling can account for the multiple roles of SLC26A2 in the biology of chondrocytes and other cell types that express high levels of SLC26A2. This should be a very attractive topic to pursue in the near future.

14.2.2 The Anion Exchangers

The SLC26 transporters that function as anion exchangers are SLC26A3, SLC26A4, and SLC26A6. The functions of these transporters were determined largely by expression in *Xenopus* oocytes, which is an ideal experimental system to characterize transport modes since it lacks many of the regulatory functions of mammalian cells, and the large oocyte allows simultaneous monitoring of multiple ion concentrations and the membrane potential with the aid of ion selective microelectrodes (Shcheynikov et al. 2006b). SLC26A2, SLC26A4, and SLC26A6 function as obligatory Cl^-/HCO_3^- exchangers, but with isoform-specific stoichiometry (Shcheynikov et al. 2006a, b, 2008). Examples of some of the transport modes mediated by each transporter are shown in Fig. 14.5.

Simultaneous measurement of Cl^-_{in} and pH_{in} in oocytes expressing SLC26A3 showed that SLC26A3 functions as a $2Cl^-/1HCO_3^-$ coupled exchanger (Ko et al. 2002; Ohana et al. 2009; Shcheynikov et al. 2006b). However, whether SLC26A3

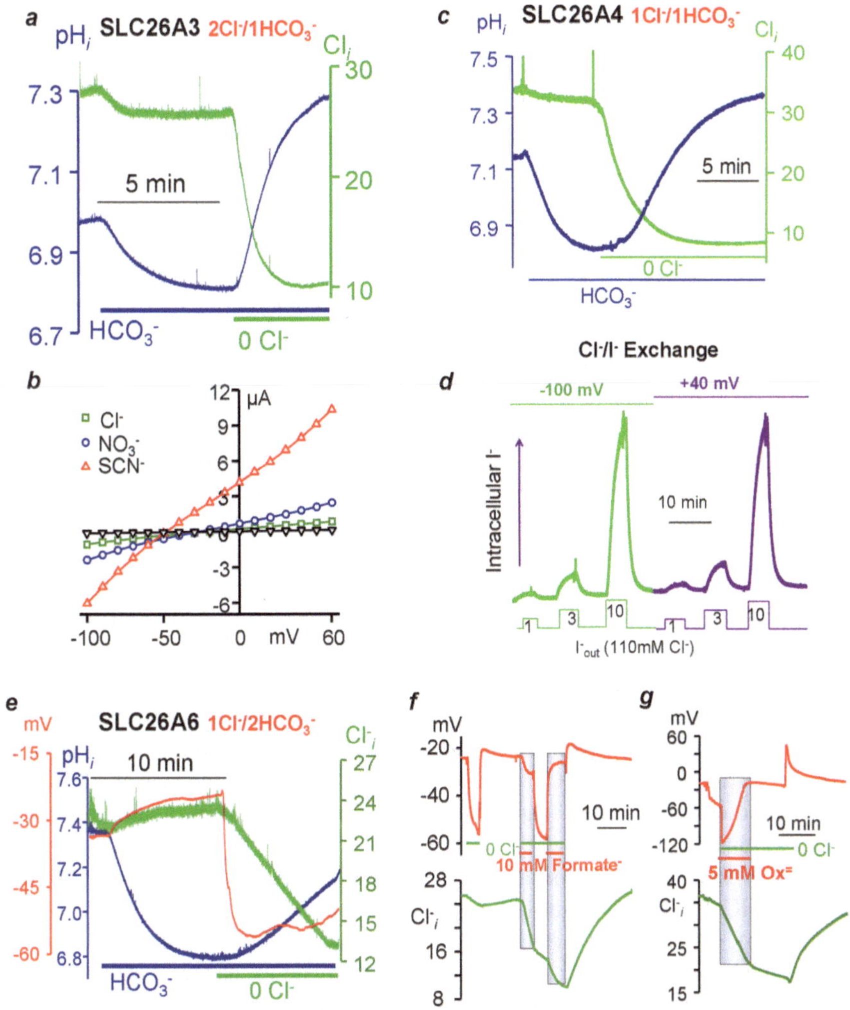

Fig. 14.5 Cl^-/HCO_3^- exchange and channel function of SLC26A3, SLC26A4, and SLC26A6. The results in panels (**a**, **b**), and (**e–g**) are reproduced from Ohana et al. (2011). Panels (**a**) and (**b**) show the SLC26A3-mediated $2Cl^-/1HCO_3^-$ exchange (**a**) and the NO_3^- and SCN^- current (**b**). Panels (**c**) and (**d**) show Cl^-/HCO_3^- exchange (**c**) and I^-/Cl^- exchange at two membrane potentials (**d**) by SLC26A4 (reproduced from Shcheynikov et al. 2008). Panel (**e**) shows the SLC26A6-mediated $1Cl^-/2HCO_3^-$ exchange and the associated change in membrane potential, and panels (**f**) and (**g**) show the SLC26A6-mediated formate (**f**) and oxalate (**g**) exchange with Cl^-

functions as a $2Cl^-/1HCO_3^-$ exchanger in vivo remains to be determined. Measurement of colonic SLC26A3-mediated Cl^- and HCO_3^- fluxes suggested that SLC26A3 may mediate electroneutral Cl^-/HCO_3^- exchange (Alper et al. 2011). Notably, the electrogenic properties of SLC26A3 are strongly supported by its function as a channel-like transporter when conducting NO_3^- and SCN^-

(Ko et al. 2002; Ohana et al. 2009; Shcheynikov et al. 2006b). Figure 14.5b shows that SLC26A3 conducts several anions in the sequence and selectivity $SCN^- > NO_3^- > Cl^-$. A recent study identified a conserved glutamate in all SLC26 transporters that determines the properties of the gate by determining whether SLC26A3 (and SLC26A3, see below) functions as a coupled or as an uncoupled conductive transporter (Ohana et al. 2012).

SLC26A4 functions as electroneutral $1Cl^-/1HCO_3^-$ exchanger (Fig. 14.5c, d) and Cl^-/I^- and I^-/HCO_3^- exchanger (Shcheynikov et al. 2008). SLC26A4 has particularly high affinity for I^-, and even in the presence of physiological intracellular and extracellular Cl^-, it transports I^- by mediating largely Cl^-/I^- exchange (Fig. 14.5d) and if HCO_3^- is present also I^-/HCO_3^- exchange (Pesce et al. 2012; Shcheynikov et al. 2008).

SLC26A6 functions as a $1Cl^-/2HCO_3^-$ exchanger, both in expression systems (Ko et al. 2002; Shcheynikov et al. 2006a, b, 2008; Xie et al. 2002) and in vivo (Kim et al. 2013; Song et al. 2012; Stewart et al. 2009). This is also illustrated in Fig. 14.5e. Measurement of the actual Cl^- and HCO_3^- fluxes resulted in an $1Cl^-/2HCO_3^-$ stoichiometry. Accordingly, removal of external Cl^- caused rapid and marked membrane hyperpolarization due to charge redistribution. All fluxes and the change in membrane potential were blocked by mutation of the conserved glutamate essential for the coupled transport (Ohana et al. 2009) and by DIDS (Shcheynikov et al. 2006b). The channel-like function was spared by the Glu/Ala mutation (Ohana et al. 2009). In this respect, the activity of SLC26A3 is largely resistant to DIDS, which can be used as a tool to distinguish the contribution of the two exchangers when expressed in the same cells/tissues. Another study failed to see an electrogenic Cl^-/HCO_3^- exchange by SLC26A6 (Chernova et al. 2005), but these studies suffer from significant inconsistencies and difficult to explain observations.

SLC26A6 is the most versatile SLC26 transporter described so far. In addition to Cl^- and HCO_3^-, SLC26A6 also transports formate and oxalate in exchange for Cl^- (Fig. 14.5f, g, respectively; Jiang et al. 2002; Knauf et al. 2001; Ohana et al. 2009; Shcheynikov et al. 2006b). Interestingly, Cl^-/formate exchange is electroneutral, and that may reflect fitting of only a single formate in the permeation pathway. Cl^-/oxalate exchange is electrogenic due to the oxalate charge, and thus, SLC26A6 likely mediates an exchange of 1oxalate for $1Cl^-$. Accordingly, SLC26A6 has a major role in intestinal and renal oxalate transport (Jiang et al. 2006) and in regulating dicarboxylate transport through the regulation of NaDC-1 (Ohana et al. 2013). The combination of reduced SLC26A6-mediated intestinal oxalate excretion that increases urine oxalate load (Jiang et al. 2006) and inhibition of NaDC-1 by SLC26A6 that reduces urine citrate (Ohana et al. 2013) result in Ca^{2+}-oxalate stone formation in the kidney.

SLC26A3 and SLC26A6 are present in the luminal membrane of several epithelia, although they do not always overlap (Haila et al. 2000; Hoglund et al. 1996, 2001; Lohi et al. 2002, 2003; Xia et al. 2013). A good example is the intestine, where SLC26A6 is more abundant in the proximal intestine, while SLC26A3 is expressed at high levels in the colon (Xia et al. 2013). Mutations in the *hSLC26A3* gene result in the disease congenital chloride diarrhea (Hoglund et al. 1996) with more than 60 known mutations and several founders (Wedenoja et al. 2011). Similarly, the SLC26A3$^{-/-}$ mouse exhibits Cl^--losing diarrhea and volume depletion (Schweinfest et al. 2006). As mentioned above, SLC26A3 interacts with and is activated by CFTR (Ko et al. 2004). SLC26A3 may also be regulated by PI3 kinase through increased surface expression (Singla et al. 2010). In this respect, the surface expression of SLC26A3 (Park et al. 2010) and SLC26A6 (Park et al. 2013) is regulated by the WNK/SPAK pathway. It should be of interest to examine whether the regulation of the transporters surface expression by the two pathways is related or not.

An interesting alteration in SLC26A3 function is in the inflammatory disease ulcerative colitis, where mRNA (Yang et al. 1998) and protein (Farkas et al. 2011; Xiao et al. 2012) are reduced. This may be related to the reduced levels of SLC26A3 by inflammatory mediators (Malakooti et al. 2011). Another link of SLC26A3 to inflammatory diseases is the association of polymorphism in SLC26A3 with ulcerative colitis in Japanese patients (Asano et al. 2009) and with Crohn's disease in Korean patients (Yang et al. 2011b). A potential functional link between SLC26A3 and SLC26A6 function and inflammatory diseases is their central role in Cl^- absorption and HCO_3^- secretion in several epithelia, including the intestine (Jacob et al. 2002; Simpson et al. 2007; Xia et al. 2013), pancreas (Song et al. 2012; Wang et al. 2006), and salivary glands (Shcheynikov et al. 2008). Hence, recent work showed that intestinal HCO_3^- secretion is essential for mucosal integrity. Deletion of SLC26A3 eliminated mid-distal colonic HCO_3^- secretion and the associated fluid absorption, which led to a damaged, thin inner mucus layer and absence of stratified structures and consequently increased susceptibility to colitis (Xiao et al. 2014).

SLC26A4 is expressed in the luminal membrane of several tissues and at high levels in thyroid follicular cells, inner ear (Everett et al. 1997; Royaux et al. 2000), renal cortical collecting duct (Royaux et al. 2001; Soleimani et al. 2001), airway serous cells (Pedemonte et al. 2007), and the salivary gland ducts (Shcheynikov et al. 2008). The major role of SLC26A4 is transcellular I^- transport and Cl^-/HCO_3^- exchange. Mutations in SLC26A4 are associated with Pendred syndrome, which is characterized by impaired I^- organification in the thyroid and goiter (Campbell et al. 2001; Everett et al. 1997). This is likely due to impaired HCO_3^-/I^- exchange and limited I^- secretion into the follicular space.

Pendred syndrome is associated with hearing loss (Campbell et al. 2001; Coyle et al. 1996; Everett et al. 1997). This is likely due to impaired SLC26A4-mediated Cl^-/HCO_3^- exchange and altered endolymph pH. Indeed, SLC26A4 mediates HCO_3^- secretion by epithelial cells of the inner ear to alkalinize the pH of the endolymph. This function is absent in the SLC26A4$^{-/-}$ mouse with resultant reduced endocochlear potential due to reduced stria vascularis K^+ channel activity and accumulation of Na^+ (Li et al. 2013b; Wangemann et al. 2007). Interestingly, SLC26A4 is obligatory for development of normal hearing and balance only during early development. Hence, deletion of SLC26A4 at any time before P2 resulted in hearing and balance loss, but deletion of SLC26A4 at P2 or later did not result in deafness or loss of balance, which remained normal (Choi et al. 2011). Moreover, normal HCO_3^- secretion, endolymph pH, hearing, and balance can be rescued in the SLC26A4$^{-/-}$ mouse by expression of SLC26A4 exclusively in the endolymphatic sac (Li et al. 2013a).

Functions of SLC26A4 in other tissues include expression of SLC26A4 in the β-intercalated cells of the cortical collecting duct where it has a significant role in renal acid–base homeostasis and the control of blood pressure (Wall 2005; Wall et al. 2004). Although basal blood pressure is normal in the SLC26A4$^{-/-}$ mice, their hypertensive response to treatment with angiotensin II (Verlander et al. 2011) or DOCA-salt is reduced (Verlander et al. 2003). In the absence of SLC26A4, the activity of ENaC is reduced (Pech et al. 2010), perhaps accounting for part of the reduced hypertension. SLC26A4 has a prominent role in the airway, in particular during inflammation such as in asthma and perhaps CF. SLC26A4 is induced by IL-13 stimulation, allergy, and rhinovirus infection, leading to increased thickness of airway surface liquid (Nakagami et al. 2008). Regulation of SLC26A4 expression by cytokines appears to be mediated by interferon-γ. SLC26A4 is upregulated in a mouse model of asthma and chronic lung disease to increase the levels of MUC5AC production and recruitment of neutrophils (Nakao et al. 2008). Whether the increased SLC26A4 is in response to the increased MUC5AC or is the cause remains to be determined using the SLC26A4$^{-/-}$ mice. If SLC26A4 is the cause of the MUC5AC increase, it can be a very promising target in asthma.

14.2.3 The Anion Channels

The two established SLC26 transporter ion channels are SLC26A7 and SLC26A9, with the possibility that SLC26A11 also functions as a channel. Examples of the currents mediated by these channels and their HCO_3^- permeability are shown in Fig. 14.6. Figure 14.6a shows measurements of SLC26A7 currents expressed in *Xenopus* oocytes and Fig. 14.6b the low HCO_3^- permeability of the channel. SLC26A7 functions as a Cl^- channel when expressed in *Xenopus* oocytes and in HEK cells and probably in parietal cells (Kim et al. 2005; Kosiek et al. 2007). In addition, SLC26A7 is expressed in Reissner's membrane epithelial cells where it functions as a Cl^- channel to participate in Cl^- transport by these cells (Kim et al. 2014). SLC26A7 shows low HCO_3^- permeability, and HCO_3^- does not appreciably affect channel function or compete with Cl^- (Fig. 14.6b). However, HCO_3^- increases the selectivity of SLC26A7 for Cl^- due to regulation of SLC26A7 by intracellular H^+ (Fig. 14.6b and Kim et al. 2005).

SLC26A7 was reported to be expressed in the basolateral membrane of parietal cells (Petrovic et al. 2003) where it may participate in Cl^- loading during acid secretion (Kosiek et al. 2007). On the other hand, SLC26A7 was localized to the subapical region in proximal tubule cells and on the basolateral surface of thick ascending limb cells in the kidney (Dudas et al. 2006), raising the possibility that

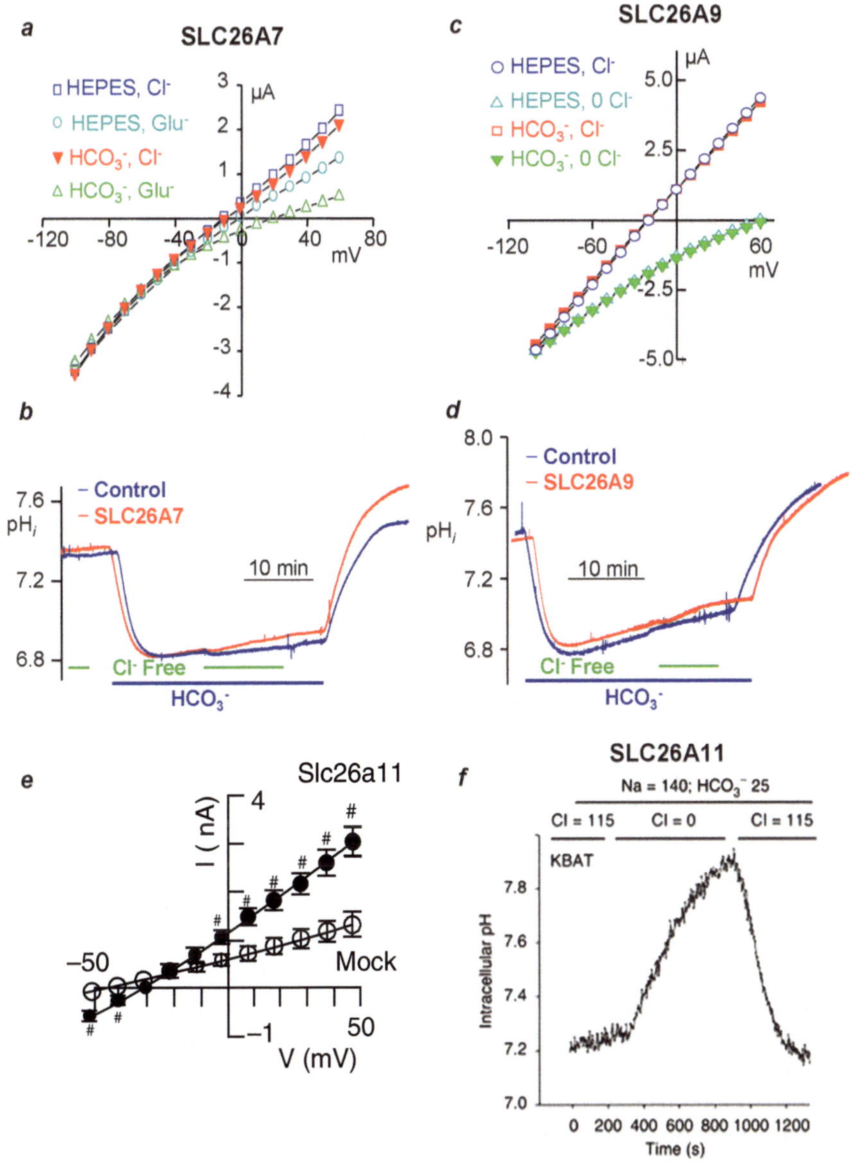

Fig. 14.6 Cl^- channel activity of SLC26A7 and SLC26A9 and SLC26A11. Panel (**a**) shows the SLC26A7-mediated Cl^- current expressed in *Xenopus* oocytes, and panel (**b**) shows the minimal Cl^-/HCO_3^- exchange mediated by SLC26A7 [reproduced from Kim et al. (2005)]. Panel (**c**) shows the SLC26A9-mediated Cl^- current expressed in *Xenopus* oocytes, and panel (**d**) shows the minimal Cl^-/HCO_3^- exchange mediated by SLC26A9 [reproduced from Dorwart et al. (2007)]. Panel (**e**) shows the SLC26A11-mediated Cl^- current expressed in *Xenopus* oocytes [reproduced from Rahmati et al.2013)], and panel (**f**) shows the Cl^-/HCO_3^- exchange activity of SLC26A11 expressed in HEK cells [reproduced from Xu et al. (2011)]

SLC26A7 may have different roles in different cell types. The SLC26A7$^{-/-}$ mouse revealed several unexpected functions of SLC26A7. The mice developed distal renal tubular acidosis and gastric hypochlorhydria (Xu et al. 2009), suggesting a role in acid secretion by parietal cells and outer medullary collecting duct cells. The specific role of a basolateral Cl^- channel in acid secretion by the two cell type is not clear at present.

Following the initial demonstration that SLC26A9 expressed in *Xenopus* oocytes functions as a Cl^- channel that is regulated by the WNK kinases (Dorwart et al. 2007), it is now firmly established that SLC26A9 is a Cl^- channel with minimal or no OH^-/HCO_3^- permeability (Bertrand et al. 2009; Chen et al. 2012; Loriol et al. 2008). Figure 14.6c, d illustrate the Cl^- channel function of SLC26A9 and its low HCO_3^- permeability. However, further analysis reported that HCO_3^- slightly stimulates Cl^- channel activity by SLC26A9 (Loriol et al. 2008). SLC26A9 is expressed at high level in gastric gland surface epithelial cells (Xu et al. 2005) and in the parietal cell tubulovesicles (Xu et al. 2008). Based on the loss of acid secretion in older SLC26A9$^{-/-}$ mice, it was suggested that SLC26A9 may be the elusive Cl^- channel mediating acid secretion. However, this needs to be further established as stimulated acid secretion in young Slc26a9$^{-/-}$ mice was normal, and the achlorhydria developed with time as is damage to other cells in the gland, which may explain the reduction in the number of parietal cells and acid secretion with age (Xu et al. 2008). Another function of SLC26A9 is a potential role in the control of Cl^- secretion in the medullary collecting duct and consequently arterial blood pressure. The SLC26A9$^{-/-}$ mice are hypertensive and the hypertension is exacerbated by high NaCl diet and water deprivation (Amlal et al. 2013).

Interest in SLC26A9 dramatically increased when a link between CF and SLC26A9 was established by a genome-wide association study with marked upregulation of SLC26A9 in cystic fibrosis (Sun et al. 2012). SLC26A9 was found to be pleiotropic for meconium ileus and pancreatic damage (Li et al. 2014). SLC26A9 is also associated with CF-related diabetes (Blackman et al. 2013; Soave et al. 2014). Interestingly, two missense variants in *SLC26A9* were identified in heterozygote patients presenting with diffuse idiopathic bronchiectasis with one patient being Δ508F/+. The SLC26A9 mutations link to bronchiectasis eliminated Cl^- channel activity and, significantly, activation of CFTR by SLC26A9 (Bakouh et al. 2013). Most interestingly, in a mouse model of CF, deletion of CFTR resulted in mild mortality. However, when SLC26A9 was deleted in these mice, mortality increased dramatically due to inhibition of intestinal secretion and likely the function of other organs affected in CF (Liu et al. 2014). Further evidence for a central role of SLC26A9 in the airway was obtained in SLC26A9$^{-/-}$ mice treated with IL-13 as an asthma model. IL-13-stimulated Cl^- secretion was absent in SLC26A9$^{-/-}$ mice that exhibited airway mucus obstruction, and a SLC26A9 polymorphism causing reduction in protein expression is associated with asthma (Anagnostopoulou et al. 2012). These findings are certain to draw further attention to SLC26A9 both in terms of its function and regulation and as a potential therapeutic target.

The function and physiological roles of SLC26A11 are not well understood. Only recently, some functional data became available. As shown in Fig. 14.6e, f, it was reported that when expressed in the plasma membrane, SLC26A11 can function as a Cl^- channel (Rahmati et al. 2013) and as HCO_3^- transporter (Xu et al. 2011). Yet, another study concluded that SLC26A11 has minimal HCO_3^- transport capacity (Stewart et al. 2011). The site(s) of SLC26A11 expression and its function are not certain. Immunolocalization in the kidney intercalated cells and in Purkinje neurons concluded plasma membrane localization (Rahmati et al. 2013; Xu et al. 2011), although the resolution of the images was not definitive to exclude mostly intracellular localization of the native SLC26A11. On the other hand, endogenous expression of SLC26A11 in the liver shows lysosomal localization similar to what is observed with transfected SLC26A11 (Della Valle et al. 2011; Palmieri et al. 2011; Schroder et al. 2010). Moreover, SLC26A11 expression is controlled by the master lysosomal biogenesis transcription factor TFEB (Palmieri et al. 2011; Sardiello et al. 2009). This raises the intriguing possibility that SLC26A11 is a lysosomal Cl^- channel that participates in the function of the lysosomes. It will be interesting to pursue this possibility.

14.3 Conclusion

This chapter illustrates the remarkable diversity of the SLC26 transporters and their multiple physiological roles. The structural basis for their functional diversity and wide substrate specificity are not understood and await structural information of the transmembrane domain, especially that of SLC26A6. Understanding regulation of the transporters is another fertile field of research. Finally, we are at the very beginning in appreciating the diverse physiological roles of the SLC26 transporters, let alone their tissue-specific function. The availability of mice knockout models for most of the SLC26 transporters should facilitate the latter.

References

Ahuja M, Jha A, Maleth J, Park S, Muallem S (2014) cAMP and Ca(2)(+) signaling in secretory epithelia: crosstalk and synergism. Cell Calcium 55:385–93

Alper SL, Stewart AK, Vandorpe DH, Clark JS, Horack RZ, Simpson JE, Walker NM, Clarke LL (2011) Native and recombinant Slc26a3 (downregulated in adenoma, Dra) do not exhibit properties of 2Cl-/1HCO3- exchange. Am J Physiol Cell Physiol 300:C276–86

Amlal H, Xu J, Barone S, Zahedi K, Soleimani M (2013) The chloride channel/transporter Slc26a9 regulates the systemic arterial pressure and renal chloride excretion. J Mol Med 91:561–72

Anagnostopoulou P, Riederer B, Duerr J, Michel S, Binia A, Agrawal R, Liu X, Kalitzki K, Xiao F, Chen M, Schatterny J, Hartmann D, Thum T, Kabesch M, Soleimani M, Seidler U, Mall MA (2012) SLC26A9-mediated chloride secretion prevents mucus obstruction in airway inflammation. J Clin Invest 122:3629–34

Ando H, Mizutani A, Matsu-ura T, Mikoshiba K (2003) IRBIT, a novel inositol 1,4,5-trisphosphate (IP3) receptor-binding protein, is released from the IP3 receptor upon IP3 binding to the receptor. J Biol Chem 278:10602–12

Ando H, Mizutani A, Kiefer H, Tsuzurugi D, Michikawa T, Mikoshiba K (2006) IRBIT suppresses IP3 receptor activity by competing with IP3 for the common binding site on the IP3 receptor. Mol Cell 22:795–806

Aravind L, Koonin EV (2000) The STAS domain - a link between anion transporters and antisigma-factor antagonists. Curr Biol 10:R53–5

Asano K, Matsushita T, Umeno J, Hosono N, Takahashi A, Kawaguchi T, Matsumoto T, Matsui T, Kakuta Y, Kinouchi Y, Shimosegawa T, Hosokawa M, Arimura Y, Shinomura Y, Kiyohara Y, Tsunoda T, Kamatani N, Iida M, Nakamura Y, Kubo M (2009) A genome-wide association study identifies three new susceptibility loci for ulcerative colitis in the Japanese population. Nat Genet 41:1325–9

Babu M, Greenblatt JF, Emili A, Strynadka NC, Reithmeier RA, Moraes TF (2010) Structure of a SLC26 anion transporter STAS domain in complex with acyl carrier protein: implications for E. coli YchM in fatty acid metabolism. Structure 18:1450–62

Bachmann O, Reichelt D, Tuo B, Manns MP, Seidler U (2006) Carbachol increases Na+−HCO3-cotransport activity in murine colonic crypts in a M3-, Ca2+/calmodulin-, and PKC-dependent manner. Am J Physiol Gastrointest Liver Physiol 291:G650–7

Bai JP, Surguchev A, Ogando Y, Song L, Bian S, Santos-Sacchi J, Navaratnam D (2010) Prestin surface expression and activity are augmented by interaction with MAP1S, a microtubule-associated protein. J Biol Chem 285:20834–43

Baker JM, Hudson RP, Kanelis V, Choy WY, Thibodeau PH, Thomas PJ, Forman-Kay JD (2007) CFTR regulatory region interacts with NBD1 predominantly via multiple transient helices. Nat Struct Mol Biol 14:738–45

Bakouh N, Bienvenu T, Thomas A, Ehrenfeld J, Liote H, Roussel D, Duquesnoy P, Farman N, Viel M, Cherif-Zahar B, Amselem S, Taam RA, Edelman A, Planelles G, Sermet-Gaudelus I (2013) Characterization of SLC26A9 in patients with CF-like lung disease. Hum Mutat 34:1404–14

Bertrand CA, Zhang R, Pilewski JM, Frizzell RA (2009) SLC26A9 is a constitutively active, CFTR-regulated anion conductance in human bronchial epithelia. J Gen Physiol 133:421–38

Bissig M, Hagenbuch B, Stieger B, Koller T, Meier PJ (1994) Functional expression cloning of the canalicular sulfate transport system of rat hepatocytes. J Biol Chem 269:3017–21

Blackman SM, Commander CW, Watson C, Arcara KM, Strug LJ, Stonebraker JR, Wright FA, Rommens JM, Sun L, Pace RG, Norris SA, Durie PR, Drumm ML, Knowles MR, Cutting GR (2013) Genetic modifiers of cystic fibrosis-related diabetes. Diabetes 62:3627–35

Blouquit-Laye S, Chinet T (2007) Ion and liquid transport across the bronchiolar epithelium. Respir Physiol Neurobiol 159:278–82

Bozoky Z, Krzeminski M, Chong PA, Forman-Kay JD (2013a) Structural changes of CFTR R region upon phosphorylation: a plastic platform for intramolecular and intermolecular interactions. FEBS J 280:4407–16

Bozoky Z, Krzeminski M, Muhandiram R, Birtley JR, Al-Zahrani A, Thomas PJ, Frizzell RA, Ford RC, Forman-Kay JD (2013b) Regulatory R region of the CFTR chloride channel is a dynamic integrator of phospho-dependent intra- and intermolecular interactions. Proc Natl Acad Sci USA 110:E4427–36

Campbell C, Cucci RA, Prasad S, Green GE, Edeal JB, Galer CE, Karniski LP, Sheffield VC, Smith RJ (2001) Pendred syndrome, DFNB4, and PDS/SLC26A4 identification of eight novel mutations and possible genotype-phenotype correlations. Hum Mutat 17:403–11

Chapman JM, Karniski LP (2010) Protein localization of SLC26A2 (DTDST) in rat kidney. Histochem Cell Biol 133:541–7

Chen AP, Chang MH, Romero MF (2012) Functional analysis of nonsynonymous single nucleotide polymorphisms in human SLC26A9. Hum Mutat 33:1275–84

Chernova MN, Jiang L, Friedman DJ, Darman RB, Lohi H, Kere J, Vandorpe DH, Alper SL (2005) Functional comparison of mouse slc26a6 anion exchanger with human SLC26A6 polypeptide variants: differences in anion selectivity, regulation, and electrogenicity. J Biol Chem 280: 8564–80

Choi BY, Kim HM, Ito T, Lee KY, Li X, Monahan K, Wen Y, Wilson E, Kurima K, Saunders TL, Petralia RS, Wangemann P, Friedman TB, Griffith AJ (2011) Mouse model of enlarged vestibular aqueducts defines temporal requirement of Slc26a4 expression for hearing acquisition. J Clin Invest 121:4516–25

Compton EL, Karinou E, Naismith JH, Gabel F, Javelle A (2011) Low resolution structure of a bacterial SLC26 transporter reveals dimeric stoichiometry and mobile intracellular domains. J Biol Chem 286:27058–67

Compton EL, Page K, Findlay HE, Haertlein M, Moulin M, Zachariae U, Norman DG, Gabel F, Javelle A (2014) Conserved structure and domain organisation amongst bacterial Slc26 transporters. Biochem J 463(2):297–307

Coyle B, Coffey R, Armour JA, Gausden E, Hochberg Z, Grossman A, Britton K, Pembrey M, Reardon W, Trembath R (1996) Pendred syndrome (goitre and sensorineural hearing loss) maps to chromosome 7 in the region containing the nonsyndromic deafness gene DFNB4. Nat Genet 12:421–3

Dawson PA, Russell CS, Lee S, McLeay SC, van Dongen JM, Cowley DM, Clarke LA, Markovich D (2010) Urolithiasis and hepatotoxicity are linked to the anion transporter Sat1 in mice. J Clin Invest 120:706–12

Della Valle MC, Sleat DE, Zheng H, Moore DF, Jadot M, Lobel P (2011) Classification of subcellular location by comparative proteomic analysis of native and density-shifted lysosomes. Mol Cell Proteomics 10(M110):006403

Detro-Dassen S, Schanzler M, Lauks H, Martin I, Meyer Zu Berstenhorst S, Nothmann D, Torres-Salazar D, Hidalgo P, Schmalzing G, Fahlke C (2007) Conserved dimeric subunit stoichiometry of SLC26 multifunctional anion exchangers. J Biol Chem 283(7):4177–88

Detro-Dassen S, Schanzler M, Lauks H, Martin I, zu Berstenhorst SM, Nothmann D, Torres-Salazar D, Hidalgo P, Schmalzing G, Fahlke C (2008) Conserved dimeric subunit stoichiometry of SLC26 multifunctional anion exchangers. J Biol Chem 283:4177–88

Dorwart MR, Shcheynikov N, Wang Y, Stippec S, Muallem S (2007) SLC26A9 is a Cl(−) channel regulated by the WNK kinases. J Physiol 584:333–45

Dorwart MR, Shcheynikov N, Yang D, Muallem S (2008) The solute carrier 26 family of proteins in epithelial ion transport. Physiology (Bethesda) 23:104–14

Dudas PL, Mentone S, Greineder CF, Biemesderfer D, Aronson PS (2006) Immunolocalization of anion transporter Slc26a7 in mouse kidney. Am J Physiol Renal Physiol 290:F937–45

Everett LA, Glaser B, Beck JC, Idol JR, Buchs A, Heyman M, Adawi F, Hazani E, Nassir E, Baxevanis AD, Sheffield VC, Green ED (1997) Pendred syndrome is caused by mutations in a putative sulphate transporter gene (PDS). Nat Genet 17:411–22

Farkas K, Yeruva S, Rakonczay Z Jr, Ludolph L, Molnar T, Nagy F, Szepes Z, Schnur A, Wittmann T, Hubricht J, Riederer B, Venglovecz V, Lazar G, Kiraly M, Zsembery A, Varga G, Seidler U, Hegyi P (2011) New therapeutic targets in ulcerative colitis: the importance of ion transporters in the human colon. Inflamm Bowel Dis 17:884–98

Forlino A, Piazza R, Tiveron C, Della Torre S, Tatangelo L, Bonafe L, Gualeni B, Romano A, Pecora F, Superti-Furga A, Cetta G, Rossi A (2005) A diastrophic dysplasia sulfate transporter (SLC26A2) mutant mouse: morphological and biochemical characterization of the resulting chondrodysplasia phenotype. Hum Mol Genet 14:859–71

Garnett JP, Hickman E, Burrows R, Hegyi P, Tiszlavicz L, Cuthbert AW, Fong P, Gray MA (2011) Novel role for pendrin in orchestrating bicarbonate secretion in cystic fibrosis transmembrane conductance regulator (CFTR)-expressing airway serous cells. J Biol Chem 286:41069–82

Gorbunov D, Sturlese M, Nies F, Kluge M, Bellanda M, Battistutta R, Oliver D (2014) Molecular architecture and the structural basis for anion interaction in prestin and SLC26 transporters. Nat Commun 5:3622

Gualeni B, de Vernejoul MC, Marty-Morieux C, De Leonardis F, Franchi M, Monti L, Forlino A, Houillier P, Rossi A, Geoffroy V (2013) Alteration of proteoglycan sulfation affects bone growth and remodeling. Bone 54:83–91

Haila S, Saarialho-Kere U, Karjalainen-Lindsberg ML, Lohi H, Airola K, Holmberg C, Hastbacka J, Kere J, Hoglund P (2000) The congenital chloride diarrhea gene is expressed in seminal vesicle, sweat gland, inflammatory colon epithelium, and in some dysplastic colon cells. Histochem Cell Biol 113:279–86

Haila S, Hastbacka J, Bohling T, Karjalainen-Lindsberg ML, Kere J, Saarialho-Kere U (2001) SLC26A2 (diastrophic dysplasia sulfate transporter) is expressed in developing and mature cartilage but also in other tissues and cell types. J Histochem Cytochem 49:973–82

Hastbacka J, de la Chapelle A, Mahtani MM, Clines G, Reeve-Daly MP, Daly M, Hamilton BA, Kusumi K, Trivedi B, Weaver A et al (1994) The diastrophic dysplasia gene encodes a novel sulfate transporter: positional cloning by fine-structure linkage disequilibrium mapping. Cell 78:1073–87

Heneghan JF, Akhavein A, Salas MJ, Shmukler BE, Karniski LP, Vandorpe DH, Alper SL (2010) Regulated transport of sulfate and oxalate by SLC26A2/DTDST. Am J Physiol Cell Physiol 298:C1363–75

Hoglund P, Haila S, Socha J, Tomaszewski L, Saarialho-Kere U, Karjalainen-Lindsberg ML, Airola K, Holmberg C, de la Chapelle A, Kere J (1996) Mutations of the Down-regulated in adenoma (DRA) gene cause congenital chloride diarrhoea. Nat Genet 14:316–9

Hoglund P, Sormaala M, Haila S, Socha J, Rajaram U, Scheurlen W, Sinaasappel M, de Jonge H, Holmberg C, Yoshikawa H, Kere J (2001) Identification of seven novel mutations including the first two genomic rearrangements in SLC26A3 mutated in congenital chloride diarrhea. Hum Mutat 18:233–42

Homma K, Miller KK, Anderson CT, Sengupta S, Du GG, Aguinaga S, Cheatham M, Dallos P, Zheng J (2010) Interaction between CFTR and prestin (SLC26A5). Biochim Biophys Acta 1798:1029–40

Hong JH, Yang D, Shcheynikov N, Ohana E, Shin DM, Muallem S (2013) Convergence of IRBIT, phosphatidylinositol (4,5) bisphosphate, and WNK/SPAK kinases in regulation of the Na +–HCO3- cotransporters family. Proc Natl Acad Sci USA 110:4105–10

Huang J, Shan J, Kim D, Liao J, Evagelidis A, Alper SL, Hanrahan JW (2012) Basolateral chloride loading by the anion exchanger type 2: role in fluid secretion by the human airway epithelial cell line Calu-3. J Physiol 590:5299–316

Jackson GC, Mittaz-Crettol L, Taylor JA, Mortier GR, Spranger J, Zabel B, Le Merrer M, Cormier-Daire V, Hall CM, Offiah A, Wright MJ, Savarirayan R, Nishimura G, Ramsden SC, Elles R, Bonafe L, Superti-Furga A, Unger S, Zankl A, Briggs MD (2012) Pseudoachondroplasia and multiple epiphyseal dysplasia: a 7-year comprehensive analysis of the known disease genes identify novel and recurrent mutations and provides an accurate assessment of their relative contribution. Hum Mutat 33:144–57

Jacob P, Rossmann H, Lamprecht G, Kretz A, Neff C, Lin-Wu E, Gregor M, Groneberg DA, Kere J, Seidler U (2002) Down-regulated in adenoma mediates apical Cl-/HCO3- exchange in rabbit, rat, and human duodenum. Gastroenterology 122:709–24

Jiang Z, Grichtchenko II, Boron WF, Aronson PS (2002) Specificity of anion exchange mediated by mouse Slc26a6. J Biol Chem 277:33963–7

Jiang Z, Asplin JR, Evan AP, Rajendran VM, Velazquez H, Nottoli TP, Binder HJ, Aronson PS (2006) Calcium oxalate urolithiasis in mice lacking anion transporter Slc26a6. Nat Genet 38: 474–8

Karniski LP, Lotscher M, Fucentese M, Hilfiker H, Biber J, Murer H (1998) Immunolocalization of sat-1 sulfate/oxalate/bicarbonate anion exchanger in the rat kidney. Am J Physiol 275: F79–87

Keller JP, Homma K, Duan C, Zheng J, Cheatham MA, Dallos P (2014) Functional regulation of the SLC26-family protein prestin by calcium/calmodulin. J Neurosci 34:1325–32

Kim KH, Shcheynikov N, Wang Y, Muallem S (2005) SLC26A7 is a Cl- channel regulated by intracellular pH. J Biol Chem 280:6463–70

Kim HJ, Myers R, Sihn CR, Rafizadeh S, Zhang XD (2013) Slc26a6 functions as an electrogenic Cl-/HCO3- exchanger in cardiac myocytes. Cardiovasc Res 100:383–91

Kim KX, Sanneman JD, Kim HM, Harbidge DG, Xu J, Soleimani M, Wangemann P, Marcus DC (2014) Slc26a7 chloride channel activity and localization in mouse Reissner's membrane epithelium. PLoS ONE 9, e97191

Knauf F, Yang CL, Thomson RB, Mentone SA, Giebisch G, Aronson PS (2001) Identification of a chloride-formate exchanger expressed on the brush border membrane of renal proximal tubule cells. Proc Natl Acad Sci USA 98:9425–30

Ko SB, Shcheynikov N, Choi JY, Luo X, Ishibashi K, Thomas PJ, Kim JY, Kim KH, Lee MG, Naruse S, Muallem S (2002) A molecular mechanism for aberrant CFTR-dependent HCO(3) (−) transport in cystic fibrosis. EMBO J 21:5662–72

Ko SB, Zeng W, Dorwart MR, Luo X, Kim KH, Millen L, Goto H, Naruse S, Soyombo A, Thomas PJ, Muallem S (2004) Gating of CFTR by the STAS domain of SLC26 transporters. Nat Cell Biol 6: 343–50

Kosiek O, Busque SM, Foller M, Shcheynikov N, Kirchhoff P, Bleich M, Muallem S, Geibel JP (2007) SLC26A7 can function as a chloride-loading mechanism in parietal cells. Pflugers Arch 454:989–98

Lamprecht G, Seidler U (2006) The emerging role of PDZ adapter proteins for regulation of intestinal ion transport. Am J Physiol Gastrointest Liver Physiol 291:G766–77

Lee JH, Nam JH, Park J, Kang DW, Kim JY, Lee MG, Yoon JS (2012a) Regulation of SLC26A3 activity by NHERF4 PDZ-mediated interaction. Cell Signal 24:1821–30

Lee MG, Ohana E, Park HW, Yang D, Muallem S (2012b) Molecular mechanism of pancreatic and salivary gland fluid and HCO3 secretion. Physiol Rev 92:39–74

Li X, Sanneman JD, Harbidge DG, Zhou F, Ito T, Nelson R, Picard N, Chambrey R, Eladari D, Miesner T, Griffith AJ, Marcus DC, Wangemann P (2013a) SLC26A4 targeted to the endolymphatic sac rescues hearing and balance in Slc26a4 mutant mice. PLoS Genet 9, e1003641

Li X, Zhou F, Marcus DC, Wangemann P (2013b) Endolymphatic Na(+) and K(+) concentrations during cochlear growth and enlargement in mice lacking Slc26a4/pendrin. PLoS ONE 8, e65977

Li W, Soave D, Miller MR, Keenan K, Lin F, Gong J, Chiang T, Stephenson AL, Durie P, Rommens J, Sun L, Strug LJ (2014) Unraveling the complex genetic model for cystic fibrosis: pleiotropic effects of modifier genes on early cystic fibrosis-related morbidities. Hum Genet 133:151–61

Liu X, Li T, Riederer B, Lenzen H, Ludolph L, Yeruva S, Tuo B, Soleimani M, Seidler U (2014) Loss of Slc26a9 anion transporter alters intestinal electrolyte and HCO transport and reduces survival in CFTR-deficient mice. Pflugers Arch 467(6):1261–75

Lohi H, Kujala M, Makela S, Lehtonen E, Kestila M, Saarialho-Kere U, Markovich D, Kere J (2002) Functional characterization of three novel tissue-specific anion exchangers SLC26A7, −A8, and -A9. J Biol Chem 277:14246–54

Lohi H, Lamprecht G, Markovich D, Heil A, Kujala M, Seidler U, Kere J (2003) Isoforms of SLC26A6 mediate anion transport and have functional PDZ interaction domains. Am J Physiol 284:C769–79

Loriol C, Dulong S, Avella M, Gabillat N, Boulukos K, Borgese F, Ehrenfeld J (2008) Characterization of SLC26A9, facilitation of Cl(−) transport by bicarbonate. Cell Physiol Biochem 22: 15–30

Malakooti J, Saksena S, Gill RK, Dudeja PK (2011) Transcriptional regulation of the intestinal luminal Na(+) and Cl(−) transporters. Biochem J 435:313–25

Melvin JE, Park K, Richardson L, Schultheis PJ, Shull GE (1999) Mouse down-regulated in adenoma (DRA) is an intestinal Cl(−)/HCO(3)(−) exchanger and is up-regulated in colon of mice lacking the NHE3 Na(+)/H(+) exchanger. J Biol Chem 274:22855–61

Melvin JE, Yule D, Shuttleworth T, Begenisich T (2005) Regulation of fluid and electrolyte secretion in salivary gland acinar cells. Annu Rev Physiol 67:445–69
Moseley RH, Hoglund P, Wu GD, Silberg DG, Haila S, de la Chapelle A, Holmberg C, Kere J (1999) Downregulated in adenoma gene encodes a chloride transporter defective in congenital chloride diarrhea. Am J Physiol 276:G185–92
Nakagami Y, Favoreto S Jr, Zhen G, Park SW, Nguyenvu LT, Kuperman DA, Dolganov GM, Huang X, Boushey HA, Avila PC, Erle DJ (2008) The epithelial anion transporter pendrin is induced by allergy and rhinovirus infection, regulates airway surface liquid, and increases airway reactivity and inflammation in an asthma model. J Immunol 181:2203–10
Nakao I, Kanaji S, Ohta S, Matsushita H, Arima K, Yuyama N, Yamaya M, Nakayama K, Kubo H, Watanabe M, Sagara H, Sugiyama K, Tanaka H, Toda S, Hayashi H, Inoue H, Hoshino T, Shiraki A, Inoue M, Suzuki K, Aizawa H, Okinami S, Nagai H, Hasegawa M, Fukuda T, Green ED, Izuhara K (2008) Identification of pendrin as a common mediator for mucus production in bronchial asthma and chronic obstructive pulmonary disease. J Immunol 180:6262–9
Ohana E, Yang D, Shcheynikov N, Muallem S (2009) Diverse transport modes by the solute carrier 26 family of anion transporters. J Physiol 587:2179–85
Ohana E, Shcheynikov N, Yang D, So I, Muallem S (2011) Determinants of coupled transport and uncoupled current by the electrogenic SLC26 transporters. J Gen Physiol 137:239–51
Ohana E, Shcheynikov N, Park M, Muallem S (2012) Solute carrier family 26 member a2 (Slc26a2) protein functions as an electroneutral SOFormula/OH-/Cl- exchanger regulated by extracellular Cl. J Biol Chem 287:5122–32
Ohana E, Shcheynikov N, Moe OW, Muallem S (2013) SLC26A6 and NaDC-1 transporters interact to regulate oxalate and citrate homeostasis. J Am Soc Nephrol 24(10):1617–26
Palmieri M, Impey S, Kang H, di Ronza A, Pelz C, Sardiello M, Ballabio A (2011) Characterization of the CLEAR network reveals an integrated control of cellular clearance pathways. Hum Mol Genet 20:3852–66
Park HW, Nam JH, Kim JY, Namkung W, Yoon JS, Lee JS, Kim KS, Venglovecz V, Gray MA, Kim KH, Lee MG (2010) Dynamic regulation of CFTR bicarbonate permeability by [Cl-]i and its role in pancreatic bicarbonate secretion. Gastroenterology 139:620–31
Park S, Hong JH, Ohana E, Muallem S (2012) The WNK/SPAK and IRBIT/PP1 pathways in epithelial fluid and electrolyte transport. Physiology (Bethesda) 27:291–9
Park S, Shcheynikov N, Hong JH, Zheng C, Suh SH, Kawaai K, Ando H, Mizutani A, Abe T, Kiyonari H, Seki G, Yule D, Mikoshiba K, Muallem S (2013) Irbit mediates synergy between ca(2+) and cAMP signaling pathways during epithelial transport in mice. Gastroenterology 145:232–41
Park M, Ohana E, Choi SY, Lee MS, Park JH, Muallem S (2014) Multiple roles of the SO4(2-)/Cl-/OH- exchanger protein Slc26a2 in chondrocyte functions. J Biol Chem 289:1993–2001
Pasqualetto E, Aiello R, Gesiot L, Bonetto G, Bellanda M, Battistutta R (2010) Structure of the cytosolic portion of the motor protein prestin and functional role of the STAS domain in SLC26/SulP anion transporters. J Mol Biol 400:448–62
Pech V, Pham TD, Hong S, Weinstein AM, Spencer KB, Duke BJ, Walp E, Kim YH, Sutliff RL, Bao HF, Eaton DC, Wall SM (2010) Pendrin modulates ENaC function by changing luminal HCO3. J Am Soc Nephrol 21:1928–41
Pedemonte N, Caci E, Sondo E, Caputo A, Rhoden K, Pfeffer U, Di Candia M, Bandettini R, Ravazzolo R, Zegarra-Moran O, Galietta LJ (2007) Thiocyanate transport in resting and IL-4-stimulated human bronchial epithelial cells: role of pendrin and anion channels. J Immunol 178:5144–53
Pesce L, Bizhanova A, Caraballo JC, Westphal W, Butti ML, Comellas A, Kopp P (2012) TSH regulates pendrin membrane abundance and enhances iodide efflux in thyroid cells. Endocrinology 153:512–21
Petrovic S, Ju X, Barone S, Seidler U, Alper SL, Lohi H, Kere J, Soleimani M (2003) Identification of a basolateral Cl-/HCO3- exchanger specific to gastric parietal cells. Am J Physiol Gastrointest Liver Physiol 284:G1093–103

Rahmati N, Kunzelmann K, Xu J, Barone S, Sirianant L, De Zeeuw CI, Soleimani M (2013) Slc26a11 is prominently expressed in the brain and functions as a chloride channel: expression in Purkinje cells and stimulation of V H(+)-ATPase. Pflugers Arch 465:1583–97

Regeer RR, Lee A, Markovich D (2003) Characterization of the human sulfate anion transporter (hsat-1) protein and gene (SAT1; SLC26A1). DNA Cell Biol 22:107–17

Rode B, Dirami T, Bakouh N, Rizk-Rabin M, Norez C, Lhuillier P, Lores P, Jollivet M, Melin P, Zvetkova I, Bienvenu T, Becq F, Planelles G, Edelman A, Gacon G, Toure A (2012) The testis anion transporter TAT1 (SLC26A8) physically and functionally interacts with the cystic fibrosis transmembrane conductance regulator channel: a potential role during sperm capacitation. Hum Mol Genet 21:1287–98

Royaux IE, Suzuki K, Mori A, Katoh R, Everett LA, Kohn LD, Green ED (2000) Pendrin, the protein encoded by the Pendred syndrome gene (PDS), is an apical porter of iodide in the thyroid and is regulated by thyroglobulin in FRTL-5 cells. Endocrinology 141:839–45

Royaux IE, Wall SM, Karniski LP, Everett LA, Suzuki K, Knepper MA, Green ED (2001) Pendrin, encoded by the Pendred syndrome gene, resides in the apical region of renal intercalated cells and mediates bicarbonate secretion. Proc Natl Acad Sci U S A 98:4221–6

Sardiello M, Palmieri M, di Ronza A, Medina DL, Valenza M, Gennarino VA, Di Malta C, Donaudy F, Embrione V, Polishchuk RS, Banfi S, Parenti G, Cattaneo E, Ballabio A (2009) A gene network regulating lysosomal biogenesis and function. Science 325:473–7

Schaechinger TJ, Oliver D (2007) Nonmammalian orthologs of prestin (SLC26A5) are electrogenic divalent/chloride anion exchangers. Proc Natl Acad Sci USA 104:7693–8

Schroder BA, Wrocklage C, Hasilik A, Saftig P (2010) The proteome of lysosomes. Proteomics 10:4053–76

Schweinfest CW, Spyropoulos DD, Henderson KW, Kim JH, Chapman JM, Barone S, Worrell RT, Wang Z, Soleimani M (2006) slc26a3 (dra)-deficient mice display chloride-losing diarrhea, enhanced colonic proliferation, and distinct up-regulation of ion transporters in the colon. J Biol Chem 281:37962–71

Shan J, Liao J, Huang J, Robert R, Palmer ML, Fahrenkrug SC, O'Grady SM, Hanrahan JW (2012) Bicarbonate-dependent chloride transport drives fluid secretion by the human airway epithelial cell line Calu-3. J Physiol 590:5273–97

Sharma AK, Ye L, Baer CE, Shanmugasundaram K, Alber T, Alper SL, Rigby AC (2011) Solution structure of the guanine nucleotide-binding STAS domain of SLC26-related SulP protein Rv1739c from Mycobacterium tuberculosis. J Biol Chem 286:8534–44

Shcheynikov N, Ko SB, Zeng W, Choi JY, Dorwart MR, Thomas PJ, Muallem S (2006a) Regulatory interaction between CFTR and the SLC26 transporters. Novartis Found Symp 273:177–86; discussion 186–92, 261–74

Shcheynikov N, Wang Y, Park M, Ko SB, Dorwart M, Naruse S, Thomas PJ, Muallem S (2006b) Coupling modes and stoichiometry of Cl-/HCO3- exchange by slc26a3 and slc26a6. J Gen Physiol 127:511–24

Shcheynikov N, Yang D, Wang Y, Zeng W, Karniski LP, So I, Wall SM, Muallem S (2008) The Slc26a4 transporter functions as an electroneutral Cl-/I-/HCO3- exchanger: role of Slc26a4 and Slc26a6 in I- and HCO3- secretion and in regulation of CFTR in the parotid duct. J Physiol 586:3813–24

Shelden MC, Howitt SM, Price GD (2010) Membrane topology of the cyanobacterial bicarbonate transporter, BicA, a member of the SulP (SLC26A) family. Mol Membr Biol 27:12–22

Simpson JE, Schweinfest CW, Shull GE, Gawenis LR, Walker NM, Boyle KT, Soleimani M, Clarke LL (2007) PAT-1 (Slc26a6) is the predominant apical membrane Cl-/HCO3- exchanger in the upper villous epithelium of the murine duodenum. Am J Physiol Gastrointest Liver Physiol 292:G1079–88

Singla A, Dwivedi A, Saksena S, Gill RK, Alrefai WA, Ramaswamy K, Dudeja PK (2010) Mechanisms of lysophosphatidic acid (LPA) mediated stimulation of intestinal apical Cl-/OH- exchange. Am J Physiol Gastrointest Liver Physiol 298:G182–9

Soave D, Miller MR, Keenan K, Li W, Gong J, Ip W, Accurso F, Sun L, Rommens JM, Sontag M, Durie PR, Strug LJ (2014) Evidence for a causal relationship between early exocrine pancreatic disease and cystic fibrosis-related diabetes: a Mendelian randomization study. Diabetes 63: 2114–9

Soleimani M, Greeley T, Petrovic S, Wang Z, Amlal H, Kopp P, Burnham CE (2001) Pendrin: an apical Cl-/OH-/HCO3- exchanger in the kidney cortex. Am J Physiol Renal Physiol 280: F356–64

Song Y, Yamamoto A, Steward MC, Ko SB, Stewart AK, Soleimani M, Liu BC, Kondo T, Jin CX, Ishiguro H (2012) Deletion of Slc26a6 alters the stoichiometry of apical Cl-/HCO-3 exchange in mouse pancreatic duct. Am J Physiol Cell Physiol 303:C815–24

Spyroglou A, Bozoglu T, Rawal R, De Leonardis F, Sterner C, Boulkroun S, Benecke AG, Monti L, Zennaro MC, Petersen AK, Doring A, Rossi A, Bidlingmaier M, Warth R, Gieger C, Reincke M, Beuschlein F (2014) Diastrophic dysplasia sulfate transporter (SLC26A2) is expressed in the adrenal cortex and regulates aldosterone secretion. Hypertension 63:1102–9

Steward MC, Ishiguro H, Case RM (2005) Mechanisms of bicarbonate secretion in the pancreatic duct. Annu Rev Physiol 67:377–409

Stewart AK, Yamamoto A, Nakakuki M, Kondo T, Alper SL, Ishiguro H (2009) Functional coupling of apical Cl-/HCO3- exchange with CFTR in stimulated HCO3- secretion by guinea pig interlobular pancreatic duct. Am J Physiol Gastrointest Liver Physiol 296:G1307–17

Stewart AK, Shmukler BE, Vandorpe DH, Reimold F, Heneghan JF, Nakakuki M, Akhavein A, Ko S, Ishiguro H, Alper SL (2011) SLC26 anion exchangers of guinea pig pancreatic duct: molecular cloning and functional characterization. Am J Physiol Cell Physiol 301:C289–303

Sun L, Rommens JM, Corvol H, Li W, Li X, Chiang TA, Lin F, Dorfman R, Busson PF, Parekh RV, Zelenika D, Blackman SM, Corey M, Doshi VK, Henderson L, Naughton KM, O'Neal WK, Pace RG, Stonebraker JR, Wood SD, Wright FA, Zielenski J, Clement A, Drumm ML, Boelle PY, Cutting GR, Knowles MR, Durie PR, Strug LJ (2012) Multiple apical plasma membrane constituents are associated with susceptibility to meconium ileus in individuals with cystic fibrosis. Nat Genet 44:562–9

Verlander JW, Hassell KA, Royaux IE, Glapion DM, Wang ME, Everett LA, Green ED, Wall SM (2003) Deoxycorticosterone upregulates PDS (Slc26a4) in mouse kidney: role of pendrin in mineralocorticoid-induced hypertension. Hypertension 42:356–62

Verlander JW, Hong S, Pech V, Bailey JL, Agazatian D, Matthews SW, Coffman TM, Le T, Inagami T, Whitehill FM, Weiner ID, Farley DB, Kim YH, Wall SM (2011) Angiotensin II acts through the angiotensin 1a receptor to upregulate pendrin. Am J Physiol Renal Physiol 301:F1314–25

Wall SM (2005) Recent advances in our understanding of intercalated cells. Curr Opin Nephrol Hypertens 14:480–4

Wall SM, Kim YH, Stanley L, Glapion DM, Everett LA, Green ED, Verlander JW (2004) NaCl restriction upregulates renal Slc26a4 through subcellular redistribution: role in Cl- conservation. Hypertension 44:982–7

Wang Y, Soyombo AA, Shcheynikov N, Zeng W, Dorwart M, Marino CR, Thomas PJ, Muallem S (2006) Slc26a6 regulates CFTR activity in vivo to determine pancreatic duct HCO3- secretion: relevance to cystic fibrosis. EMBO J 25:5049–57

Wangemann P, Nakaya K, Wu T, Maganti RJ, Itza EM, Sanneman JD, Harbidge DG, Billings S, Marcus DC (2007) Loss of cochlear HCO3- secretion causes deafness via endolymphatic acidification and inhibition of Ca2+ reabsorption in a Pendred syndrome mouse model. Am J Physiol Renal Physiol 292:F1345–53

Wedenoja S, Pekansaari E, Hoglund P, Makela S, Holmberg C, Kere J (2011) Update on SLC26A3 mutations in congenital chloride diarrhea. Hum Mutat 32:715–22

Xia W, Yu Q, Riederer B, Singh AK, Engelhardt R, Yeruva S, Song P, Tian DA, Soleimani M, Seidler U (2013) The distinct roles of anion transporters Slc26a3 (DRA) and Slc26a6 (PAT-1) in fluid and electrolyte absorption in the murine small intestine. Pflugers Arch 466(8):1541–56

Xiao F, Juric M, Li J, Riederer B, Yeruva S, Singh AK, Zheng L, Glage S, Kollias G, Dudeja P, Tian DA, Xu G, Zhu J, Bachmann O, Seidler U (2012) Loss of downregulated in adenoma (DRA) impairs mucosal HCO3(−) secretion in murine ileocolonic inflammation. Inflamm Bowel Dis 18:101–11

Xiao F, Yu Q, Li J, Johansson ME, Singh AK, Xia W, Riederer B, Engelhardt R, Montrose M, Soleimani M, Tian DA, Xu G, Hansson GC, Seidler U (2014) Slc26a3 deficiency is associated with loss of colonic HCO3 (−) secretion, absence of a firm mucus layer and barrier impairment in mice. Acta Physiol 211:161–75

Xie Q, Welch R, Mercado A, Romero MF, Mount DB (2002) Molecular characterization of the murine Slc26a6 anion exchanger: functional comparison with Slc26a1. Am J Physiol Renal Physiol 283:F826–38

Xie R, Nguyen S, McKeehan K, Wang F, McKeehan WL, Liu L (2011) Microtubule-associated protein 1S (MAP1S) bridges autophagic components with microtubules and mitochondria to affect autophagosomal biogenesis and degradation. J Biol Chem 286:10367–77

Xu J, Henriksnas J, Barone S, Witte D, Shull GE, Forte JG, Holm L, Soleimani M (2005) SLC26A9 is expressed in gastric surface epithelial cells, mediates Cl-/HCO3- exchange, and is inhibited by NH4+. Am J Physiol Cell Physiol 289:C493–505

Xu J, Song P, Miller ML, Borgese F, Barone S, Riederer B, Wang Z, Alper SL, Forte JG, Shull GE, Ehrenfeld J, Seidler U, Soleimani M (2008) Deletion of the chloride transporter Slc26a9 causes loss of tubulovesicles in parietal cells and impairs acid secretion in the stomach. Proc Natl Acad Sci USA 105:17955–60

Xu J, Song P, Nakamura S, Miller M, Barone S, Alper SL, Riederer B, Bonhagen J, Arend LJ, Amlal H, Seidler U, Soleimani M (2009) Deletion of the chloride transporter slc26a7 causes distal renal tubular acidosis and impairs gastric acid secretion. J Biol Chem 284:29470–9

Xu J, Barone S, Li H, Holiday S, Zahedi K, Soleimani M (2011) Slc26a11, a chloride transporter, localizes with the vacuolar H(+)-ATPase of A-intercalated cells of the kidney. Kidney Int 80: 926–37

Yang H, Jiang W, Furth EE, Wen X, Katz JP, Sellon RK, Silberg DG, Antalis TM, Schweinfest CW, Wu GD (1998) Intestinal inflammation reduces expression of DRA, a transporter responsible for congenital chloride diarrhea. Am J Physiol 275:G1445–53

Yang D, Shcheynikov N, Zeng W, Ohana E, So I, Ando H, Mizutani A, Mikoshiba K, Muallem S (2009) IRBIT coordinates epithelial fluid and HCO3- secretion by stimulating the transporters pNBC1 and CFTR in the murine pancreatic duct. J Clin Invest 119:193–202

Yang D, Li Q, So I, Huang CL, Ando H, Mizutani A, Seki G, Mikoshiba K, Thomas PJ, Muallem S (2011a) IRBIT governs epithelial secretion in mice by antagonizing the WNK/SPAK kinase pathway. J Clin Invest 121:956–65

Yang SK, Jung Y, Kim H, Hong M, Ye BD, Song K (2011b) Association of FCGR2A, JAK2 or HNF4A variants with ulcerative colitis in Koreans. Dig Liver Dis 43:856–61

Zhao H, Star RA, Muallem S (1994) Membrane localization of H+ and HCO3- transporters in the rat pancreatic duct. J Gen Physiol 104:57–85

Zheng J, Shen W, He DZ, Long KB, Madison LD, Dallos P (2000) Prestin is the motor protein of cochlear outer hair cells. Nature 405:149–55

Chapter 15
ClC-2 Chloride Channels

John Cuppoletti, Danuta H. Malinowska, and Ryuji Ueno

Abstract This article focuses on activators and inhibitors of ClC-2. ClC-2 Cl^- channels from mouse, rabbit, rat, and human have been studied extensively in recombinant form by whole-cell patch-clamp and in single-channel studies. There are similarities in responses of these channels from different species to protons, ATP, lipids, and lubiprostone for the recombinant channels. However, there are differences in channel properties, notably in the activation of human and rabbit (but not rat or mouse) forms by PKA. Two PKA phosphorylation sites of human ClC-2 have been identified by mutagenesis. One is required for lipid and lubiprostone activation. Structural studies suggest that these sites interact with cystathionine-β-synthase (CBS) domains to regulate ClC-2. Single-channel studies have been carried out on recombinant ClC-2, cortical astrocytes, and A6 cells which have both ClC-2 and CFTR in the apical membrane. Methadone and GaTx2 differentially inhibit ClC-2, while DASU-02 differentially inhibits CFTR. CFTRinh172 and glibenclamide inhibit both. Knockdown of ion channels combined with pharmacology is more powerful than using pharmacological agents alone. Studies using human cells and tissues such as intestine and lung have used various activators and inhibitors, and these studies are reviewed in detail. This review indicates that it is essential to verify the specificity of all agents experimentally to correctly interpret experimental results. Some studies provide insight into the possible roles of ClC-2 in physiological processes.

Keywords ClC-2 • ClC-2 phosphorylation • ClC-2 single channels • Lubiprostone • Intestinal chloride transport • Cystic fibrosis • CFTR • I_{sc} • Intestinal chloride • Airway chloride • Nasal potential difference

J. Cuppoletti (✉) • D.H. Malinowska
Department of Molecular and Cellular Physiology, University of Cincinnati, 231 Albert Sabin Way, Cincinnati, OH 45267-0576, USA
e-mail: John.Cuppoletti@uc.edu; Danuta.Malinowska@uc.edu

R. Ueno
VLP Therapeutics, 19 Firstfield Road, Suite 100B, Gaithersburg, MD 20878, USA
e-mail: ktolliver@vlptherapeutics.com

K.L. Hamilton, D.C. Devor (eds.), *Ion Channels and Transporters of Epithelia in Health and Disease*, Physiology in Health and Disease,
DOI 10.1007/978-1-4939-3366-2_15

15.1 Introduction

This chapter will cover only ClC-2 Cl^- channels (except to provide a background of the ClC field based on the work of Miller and coworkers). ClC-2 is found in a variety of epithelial and non-epithelial tissues including the retina, testes, lung, stomach, intestine, and salivary ducts. This review of ClC-2 addresses the basic properties of recombinant ClC-2 from different species, ClC-2 functional studies in whole-cell current studies and in single-channel studies, ClC-2 in physiological processes, and activators of ClC-2 and inhibitors of ClC-2, which have been shown to distinguish between ClC-2 and other channels such as CFTR. This review outlines some controversies in the ClC-2 literature, some of which appear to derive from differences in amino acid sequence of the proteins from different species, use of inhibitors or other chemical agents which have not been validated, and apparent differences in localization of ClC-2 in various tissues and species. Lubiprostone, a bicyclic fatty acid, is used to treat constipation by acting in the small intestine to cause fluid secretion. A number of studies using lubiprostone and forskolin to identify either ClC-2 or CFTR in Cl^- secretory function have appeared which are directed toward the repair of barrier properties and increasing fluid secretion in the intestine and in lung fluid. A number of articles have suggested that lubiprostone is an EP_4 receptor agonist and have suggested that the mechanism of action of lubiprostone is to activate CFTR instead of ClC-2. The lack of EP_4 receptor agonist activity and lack of effect of lubiprostone on human CFTR will be discussed. Thus, the present review will focus on the function and regulation of ClC-2 where it has been studied from various species and in epithelia from various tissues.

15.2 Review Articles

A comprehensive review of ClC-2 with a focus on Cl^- channelopathies of ClC-2 has recently appeared (Bi et al. 2013). The various classes of Cl^- channels have been reviewed (Duran et al. 2010). The biophysical properties of ClC Cl^- channels have been reviewed (Zifarelli and Pusch 2007; Fahlke 2004). The structures of ClC Cl^- channels have been reviewed (Dutzler 2004a, b, 2006, 2007; Chen 2005). Miller (2006) has reviewed ClC channel structure in light of findings that some of the ClC channels are proton-coupled Cl^- transporters. The molecular physiology, pathophysiology, and proposed channelopathies of ClC-type Cl^- channels have also been reviewed (Stauber et al. 2012; Jentsch et al. 2005). Unique ways to study various Cl^- channels against a background of numerous other Cl^- channels appear to be a promising approach in the future (Duan 2013). An excellent older review of the pharmacology of CFTR channel activity is recommended (Schultz et al. 1999). A very recent review by Linsdell (2014) has appeared which reviews the newest generation of CFTR inhibitors.

15.3 The Beginning: ClC-0

The *Torpedo electroplax* Cl^- channel, in impressive and pioneering studies, was purified and studied in bilayers as well as in patch clamp of the protein reconstituted into liposomes. Voltage-dependent macroscopic currents were measured, ion selectivities were P_x/P_{Cl} Cl (1.0) Br (0.58), and iodide and thiocyanate blocked the channel (Miller and White 1980). It was blocked by SITS and DIDS (White and Miller 1979). It was shown to be a dimeric channel of 10 and 20 pS with a closed state in 200 mM Cl^-, and the probability of forming an open channel was voltage dependent and increased with reduced pH (Hanke and Miller 1983). The channel was extensively characterized in a series of further studies (Miller 1982). The cloned *Torpedo* clectroplax Cl^- channel protein was expressed in oocytes and had essentially identical properties as that of the protein purified from electroplax (Bauer et al. 1991). It was later named ClC-0, and the properties of ClC-0 were a guide for studies of other ClC channels.

15.4 Properties of Rat ClC-2

15.4.1 Cloned Rat ClC-2

ClC-2 was initially cloned from rat (Thiemann et al. 1992). A construct expressed in *Xenopus* oocytes gave voltage-activated currents with ion selectivity of Cl > Br > I. ClC-2 was shown to be widely expressed in a variety of rat tissues including the skeletal muscle, heart, brain, lung, kidney pancreas, stomach, intestine, and liver by northern analysis. ClC-2 was also found in a large variety of cell lines from human sources. As was determined with ClC-0 (Hanke and Miller 1983), acidification altered gating and activated the channel (Jordt and Jentsch 1997).

15.4.2 ClC-2 in the Rat Airway

ClC-2 was found in the apical membranes of airway tissues of fetal rat lung (Murray et al. 1995) and it was downregulated after birth (Murray et al. 1996). It was postulated that ClC-2 might be involved in lung development. However, $ClC\text{-}2^{-/-}$ mice have normal lungs (Chu et al. 1999; Vij and Zeitlin 2006), suggesting a different role for rat lung ClC-2. The downregulation after birth in the rat may be mediated by the transcription factors Sp1 and Sp3 (Chu et al. 1999; Vij and Zeitlin 2006; Holmes et al. 2003). Growth factors can stimulate the expression of ClC-2 in rat lung (Blaisdell et al. 1999). Cl^- secretion, thought to involve ClC-2, can be stimulated by reduced pH in the fetal rat lung (Blaisdell et al. 2000). The fetal rat lung appears to be a good model for studies of ClC-2 in intact tissue.

Transcriptional regulation by Sp1 and Sp3 and low pH activation might allow higher rates of Cl^- secretion needed to play a role as a rescue channel for cystic fibrosis (CF). The adult rat lung has low levels of ClC-2 (Murray et al. 1996) making it a poor model for functional studies. MacDonald et al. (2008), Schiffhauer et al. (2013), Joo et al. (2009), and Cuthbert (2011) have studied Cl^- transport in the adult airway. These and other model systems will be discussed in Sect. 9.

15.4.3 Cloned Human ClC-2

Human ClC-2 was first cloned from T84 cells by Cid et al. (1995). By northern analysis, it was found to be present in a large variety of human tissues including the pancreas, brain, heart, placenta, skeletal muscle, and kidney. Weak signals were also found in the lung and liver, in addition to being present in T84 cells (Cid et al. 1995). ClC-2 mRNA was also present in IB3-1 cells, an airway epithelial cell line from a CF patient. The investigators envisaged that this channel might be a potential rescue channel when CFTR is defective, using such techniques as gene replacement therapy or augmentation using ClC-2 as the alternative Cl^- secretory pathway. The authors noted that although human ClC-2 had 94 % identity with the rat form, there were sequence differences between rat and human ClC-2 that might confer PKA activation and regulation by other kinases, including ten potential sites for phosphorylation by PKC in the human form. Human ClC-2 was characterized in IB3-1 cells and then overexpressed in those cells (Schwiebert et al. 1998). In IB3-1 cells which contained a low level of ClC-2, the channel gave a linear *I*/*V* curve with $Cl > I >$ gluconate (Glu) anion selectivity. The channel was activated by low pH, with pH 3.79 the lowest pH tested. Time- and voltage-activated currents were minimal. However, upon overexpression of human ClC-2 in IB3-1 cells, ClC-2 gave inward rectification and time- and voltage-activated currents with pH activation and $Cl > I^- >$ Glu anion selectivity. Cadmium (Cd^{2+}) gave a concentration-dependent block of the currents in both IB3-1 cells and IB3-1 cells overexpressing human ClC-2. The authors suggested that pharmacological agents that might increase the acid sensitivity of the channel might lead to it being used as a rescue channel in CF.

The properties of the cloned human channel were similar to those observed in T84 cells in earlier studies by Fritsch and Edelman (1996). In this study, they showed that ClC-2 in T84 cells could be inhibited by treatment with the protein kinase C activator phorbol 12-myristate 13-acetate. Staurosporine and calphostin C increased activation of the channel. Neither forskolin nor myristoylated protein kinase A inhibitor fragment 5–24 amide (mPKI) was shown to have significant consistent effects in that study. They concluded that human ClC-2 could be modulated by phosphorylation, as later indicated from the consensus sequences identified by Cid et al. (1995). This group also showed that ClC-2 was affected by TNF-α (Bali et al. 2001) and membrane cholesterol (Hinzpeter et al. 2007). The distribution of ClC-2 in both human and rat tissues (Lipecka et al. 2002) was also studied. It was concluded that there were differences in expression of the channel between rat and human tissues, particularly in the intestine. Thus, findings in rat intestine might not extrapolate to human intestine. However, the human and rat airways appeared to have similar locations of ClC-2 expression.

15.5 Single-Channel Studies of ClC-2

Single-channel recordings of human ClC-2 were used in studies of channel gating (Stölting et al. 2013). The authors found one closed state and two open states in the presence of cytosolic ATP. The single-channel current was found to be approximately 0.23 pA at −100 mV. The same authors also carried out single-channel measurements of ClC-2 as part of a study of chimeras of human ClC-2 with ClC-1. ClC-2 showed similar single-channel current magnitudes and again showed double-barreled behavior (Stölting et al. 2014).

The cloned human channel was studied in lipid bilayers using 800 mM tetraethylammonium chloride (TEACl) to increase the small-channel currents (Sherry et al. 1997). At −80 mV, the channel gave currents of approximately 1.5 pA, which would be correspondingly smaller if measured at the lower concentration of salt. PKA and acid activation of the human channel were demonstrated.

Single-channel recordings of rabbit ClC-2 were carried out in studies of a peptide toxin isolated from *Leiurus quinquestriatus hebraeus* venom, GaTx2, that inhibits ClC-2 channels in a voltage-dependent manner with an apparent K_D of ~20 pm (Thompson et al. 2009). Toxin inhibition is highly complex, being protocol dependent, showing poor inhibition of the open channel, but affecting the slow gating of the channel. The single-channel currents from rabbit ClC-2 at −100 mV appear to be approximately 0.3 pA, slightly larger than the currents reported for human ClC-2.

Rabbit ClC-2 was cloned and studied at the single-channel level in lipid bilayers, using high concentrations of TEACl (800 mM) to increase channel currents (Malinowska et al. 1995). Channel records at −100 mV gave single-channel currents of approximately 2.2–3 pA at 800 mM TEACl. Given that the high concentration of Cl^- in this study was approximately 5.3 times the concentration used in the later study of rabbit ClC-2 by Thompson et al. (2009), these currents would be approximately similar in magnitude at the lower concentration of salt. PKA and low pH also activated rabbit ClC-2 (Malinowska et al. 1995).

Rat ClC-2 from astrocytes was studied at the single-channel level by Nobile et al. (2000), and conductance levels of 0.3 and 0.6 pA at −100 mV were observed, similar to the channel currents seen in human and rabbit ClC-2.

All of the above studies are somewhat limited in terms of detailed characterization of the single-channel characteristics of ClC-2 from any species, perhaps reflecting in part the difficulty of such measurements.

15.6 Single-Channel Studies of ClC-2 in the Presence of CFTR and Other Cl⁻ Channels

Single-channel studies provide a way to study ClC-2 in the presence of CFTR, since ClC-2 and CFTR have distinctly different biophysical properties at the single-channel level. A very complete study of the single-channel characteristics of A6 cell apical membrane ClC-2 and CFTR with an included comparison with human ClC-2 has appeared using lubiprostone to stimulate *Xenopus* A6 cell ClC-2 (Bao et al. 2008). The characteristics of A6 cell ClC-2 studied at the single-channel level were a unit conductance of 3–4 pS and a high degree of rectification, with voltage dependence of mean open time and open probability, both of which decreased with depolarization. The anion selectivity of A6 ClC-2 was $Cl > Br > NO_3 > I > SCN$. ClC-2 was potently activated by lubiprostone ($EC_{50} = 69$ nM). CFTR was also present in A6 apical membranes, but was activated by lubiprostone in a cAMP-independent manner with an EC_{50} about ten times higher than the EC_{50} for activation of ClC-2. During these studies, the authors determined that both diphenylcarboxylic acid and glibenclamide were blockers of both ClC-2 and CFTR. They found that glibenclamide was an open-channel blocker. When Cd^{2+} was added to the bath, it eliminated not only ClC-2 currents but also all other channel currents, suggesting a generally toxic effect of Cd^{2+}.

Human ClC-2 expressed in HEK293 cells as used previously (Tewari et al. 2000; Cuppoletti et al. 2001, 2004a, b, 2013) yielded single-channel currents activated by lubiprostone which were "indistinguishable from" those observed in *Xenopus* A6 cells (Bao et al. 2008).

Interestingly this was the first report and characterization of ClC-2 in the apical membrane of the widely studied *Xenopus* A6 cell line. Previous studies had studied apical membrane Cl^- transport by short-circuit current (I_{sc}) in Ussing chambers and I_{sc} fluctuation by noise analysis (Atia et al. 1999). The authors found evidence only for CFTR and a Ca^{2+}-activated Cl^- channel with no evidence of a Cl^- channel with the characteristics of ClC-2 in *Xenopus* A6 cells.

Single-channel studies of *Xenopus* A6 cell apical membranes (Price et al. 1996) showed that *Xenopus* CFTR when activated by forskolin had a slightly larger single-channel conductance than the human form and that the anion selectivity sequence varied from human. The authors did not mention finding ClC-2-like single channels suggesting that *Xenopus* ClC-2 is not activated by PKA and consistent with the lack of ClC-2 (or CFTR) channels in A6 cells observed by Bao et al. (2008) before the addition of lubiprostone. Supporting this view, CFTR and other Cl^- channels were not seen in A6 cells unless treated by forskolin in another single-channel study (Ling et al. 1997).

Thus, lubiprostone, used in the Bao et al. (2008) study, was the first to unmask A6 cell ClC-2. Previous studies showed that prostaglandins (which are metabolized to prostones by NAD^+PGDH) activated I_{sc} and Cl^- fluxes in *Xenopus* A6 cells (Keeler and Wong 1986). Studies using impedance measurements suggested that PGE_2 only activated Na^+ fluxes in *Xenopus* A6 cells (Păunescu and Helman 2001).

15.7 PKA Phosphorylation of ClC-2

Rabbit ClC-2 contains consensus phosphorylation sites for PKA and can be activated by PKA (Malinowska et al. 1995). Porcine choroid plexus contains ClC-2 activated by extracellular VIP and PKA catalytic subunit (Kajita et al. 2000). Cid et al. (1995) were the first to identify a PKA consensus sequence when the human ClC-2 was cloned. They noted that the site was not present in the rat sequence.

Recombinant human ClC-2 was studied in lipid bilayers and shown to be activated by the catalytic subunit of PKA plus ATP (Sherry et al. 1997). When expressed in HEK cells and studied by patch clamp, human ClC-2 was activated by treatment of the cells with a combination of forskolin and IBMX. In another study, human ClC-2 was activated by arachidonic acid in a PKA-independent manner (Tewari et al. 2000). Forskolin-IBMX was also shown to activate human ClC-2 in an mPKI-sensitive manner, and both methadone and Cd^{2+} inhibited forskolin-IBMX-stimulated human ClC-2 (Cuppoletti et al. 2013).

A detailed study of two unique PKA consensus sites $RRAT_{655}$ and $RGET_{691}$ on human ClC-2 was undertaken by Cuppoletti et al. (2004a). In these studies, the effects of phosphorylation at one of the sites were studied in channels mutated at the other site at neutral and low extracellular pH. For each of the channels with mutated consensus phosphorylation sites, the effect of forskolin-IBMX at neutral and low pH was studied. In each case, activity of the mutant channels was confirmed by arachidonic acid activation, which is not dependent upon phosphorylation (Tewari et al. 2000). The effect of forskolin-IBMX plus phosphatase inhibitors was also tested to help ensure that the phosphodiesterase was inhibited and that phosphorylation of the sites was not differentially susceptible to phosphatase action. The double phosphorylation site mutant $RRAT(A)_{655}$ plus $RGET(A)_{691}$ lost PKA activation and also lost activation by low pH. Either $RRAT_{655}$ or $RGET_{691}$ was sufficient for activation, as long as phosphatase inhibitors were present. It was demonstrated that $RRAT_{655}(A)$ and $RGET_{691}(A)$ retained PKA activation at pH 7.4. $RGET_{691}$, but not $RRAT_{655}$, was required for PKA activation at pH 6. An $RGED_{691}$ mutant was constitutively active and could be further activated by PKA at low pH. These studies demonstrated that the consensus phosphorylation sites, $RRAT_{655}$ and $RGET_{691}$, were responsible for the observed PKA activation of human ClC-2 and showed that there was interplay between activation by phosphorylation and by low pH.

Rabbit ClC-2 also contains two consensus PKA phosphorylation sites RRQS (for PKA, CaMKII and PKC) and KRKS (for PKA and PKC) in the same general region of the channel shown to be responsible for PKA activation of the human channel. Activation of the native and recombinant rabbit ClC-2 channel was shown to be activated by ATP plus PKA catalytic subunit in planar lipid bilayers (Malinowska et al. 1995).

Alioth et al. (2007) showed that this region of the channel (where PKA phosphorylation sites are situated) contains organized structures, which appear to interact with the CBS domains, and they suggested that this mechanism is responsible for regulation of the channel. This same region is also phosphorylated in a cell cycle-dependent ClC anion channel (Denton et al. 2005). It has been recently suggested that phosphorylation in this region may be a general mechanism for regulation of ClC channels (Yamada et al. 2013; Miyazaki et al. 2012). Proteins involved in regulation of phosphorylation have been identified (Denton et al. 2005; Rutledge et al. 2002; Furukawa et al. 2002). GCK-3 is a kinase that has been shown to alter fast gating in *C. elegans* CLH-3b, an ortholog of ClC-2. GSK-3 binds to a specific site on CLH-3b and mutation of the binding site abolishes regulation by GCK-3 (Denton et al. 2005). Activation of CLH-3b during meiosis and oocyte swelling is also regulated by *C. elegans* PP1-type phosphatases CeGlC-7α and β (Rutledge et al. 2002). Furukawa et al. (2002) showed phosphorylation and functional regulation of rabbit ClC-2 by M cyclin-dependent protein kinase and that PP1α and PP1β physically interact with the COOH terminus of rabbit ClC-2.

Park et al. (2001) showed that rat ClC-2 could be phosphorylated by PKA, but without a change in the extent or rate of activation. These contrasting results are presumably because rat ClC-2 does not contain the same type of phosphorylation sites as those in human and rabbit in these regions (Cid et al. 1995; Sherry et al. 1997; Cuppoletti et al. 2004a). Mouse ClC-2 (which has the same sequence as the rat channel in this region) expressed in HEK293EBNA cells was not activated by forskolin-IBMX using the same conditions where human ClC-2 is activated (Cuppoletti et al. unpublished).

Porcine choroid plexus ClC-2 was activated by extracellular VIP as well as intracellular PKA catalytic subunit (Kajita et al. 2000), although the sites of PKA phosphorylation in porcine ClC-2 are not known.

15.8 Small-Molecule Activators of ClC-2

15.8.1 Protons

Protons were initially shown to activate ClC-0 (Hanke and Miller 1983). Protons also activate human ClC-2 currents (Sherry et al. 1997; Schwiebert et al. 1998; Tewari et al. 2000; Cuppoletti et al. 2001, 2004a; Stölting et al. 2014). Rabbit ClC-2 was activated by acid (Malinowska et al. 1995; Stroffekova et al. 1998). Chemical modification (amidation) resulted in activation and loss of further pH activation. Site-directed mutagenesis of the channel resulted in identification of residues (EELE) responsible for the pH effect (Stroffekova et al. 1998). Rat ClC-2 is also activated by low pH (Blaisdell et al. 2000). Jordt and Jentsch (1997) studied rat ClC-2 and showed that the cytoplasmic loop between D7 and D8 was an important region for acid activation. Protonation of a glutamate residue in the pore of guinea pig ClC-2 has been found to be important for gating (Niemeyer et al. 2003, 2009).

15.8.2 ATP

ATP has been shown to be important for endocytosis (Dhani et al. 2008). ATP depletion inhibits endocytosis of the channel (Dhani et al. 2008), but these effects appear to be through effects on proteins involved in intracellular trafficking.

Fritsch and Edelman (1996) showed that T84 cell ClC-2 could be measured with ATP in the pipette, but that omission of ATP or use of the non-hydrolyzable ATP analogue 5′-adenylimidodiphosphate accelerated activation kinetics and increased amplitude of the currents. Stölting et al. (2013) further investigated the ATP effect, showing that there was concentration-dependent slowing of the activation kinetics over the range of 1–5 mM ATP or ADP and the non-hydrolyzable 5′(β-γ-imido) triphosphate (but not AMP). ATP is known to bind to the CBS domains of the ClC channels (Meyer et al. 2007). Thus, ATP binding to the CBS domains slows gating of the channel (Stölting et al. 2013). A variety of natural mutants in the region of the CBS domain were also studied by Stölting et al. (2013), and these have altered gating expected for disruption of the ATP-/ADP-binding site. Studies on ATP/ADP and non-hydrolyzable analogues were carried out without lipid activators or phosphorylation to increase the activity of the human channel.

15.8.3 Activation by Lipids and by Lubiprostone

Lipids including arachidonic acid activate human ClC-2 (Tewari et al. 2000; Cuppoletti et al. 2001, 2004a, 2013, 2014b). Tewari et al. (2000) showed that recombinant human ClC-2 was activated by a variety of lipids including oleic acid (C:18 *cis*D9), elaidic acid (C:18 *trans*D9), arachidonic acid (AA; C:20 *cis*D5,8,11,14), and 5,8,11,14-eicosatetraynoic acid (ETYA; C:20 *trans*D5,8,11,14). Lipid activation of the human ClC-2 channel was not inhibited by mPKI or staurosporine and was therefore independent of PKA or protein kinase C activation. Amino acids essential for lipid activation of human ClC-2 have been identified. Site-directed mutagenesis of one of the two PKA phosphorylation sites leads to ablation of lipid activation of ClC-2 (Cuppoletti et al. 2014a) (see Sect. 8.3.3).

15.8.3.1 Lubiprostone Stimulation of Recombinant ClC-2

Lubiprostone, a bicyclic fatty acid, also activates human ClC-2 at low concentrations (EC_{50} approximately 20–30 nM; Cuppoletti et al. 2004b, 2013, 2014b). These studies were carried out using recombinant human ClC-2 expressed in HEK293 cells (Cuppoletti et al. 2004b) and in a HEK293EBNA cells (Cuppoletti et al. 2013, 2014b), which express a higher level of ClC-2 than HEK293 cells (Cuppoletti et al. 2013). ClC-2 currents in the latter cells give time- and voltage-dependent current recordings more readily than ClC-2 in HEK293 cells, and the currents were fourfold higher in control and lubiprostone-stimulated cells.

15.8.3.2 Lubiprostone Activation in Cell Lines Containing Both ClC-2 and CFTR

In T84 cells in I_{sc} studies (Cuppoletti et al. 2004b), lubiprostone gave a similar EC_{50} as measured for recombinant human ClC-2 in HEK293 cells, but the basis for lubiprostone stimulation of I_{sc} of T84 cells was not determined in those studies. However, recent patch-clamp studies using various more selective inhibitors of ClC-2 and CFTR and knockdowns of ClC-2 and CFTR in T84 cells have allowed assignment of lubiprostone stimulation to ClC-2 (Cuppoletti et al. 2008b, 2014b). Since lubiprostone at concentrations up to 1 μM did not activate recombinant human CFTR in HEK293 cells (Cuppoletti et al. 2004b, 2013), it was concluded that T84 cell CFTR was not activated by lubiprostone. In later studies, Cuppoletti et al. (2008b) showed that in native T84 cells ($n = 6$), 20 nM lubiprostone significantly stimulated Cl^- currents ($P < 0.0025$) (control, -34.2 ± 3.4 pA/pF; with lubiprostone -78.2 ± 9.0 pA/pF). Cd^{2+} (500 μM) significantly ($P < 0.001$) inhibited the stimulated currents to -25.8 ± 3.5 pA/pF. It was further shown that knockdown of T84 ClC-2 with siRNA caused a loss of lubiprostone-stimulated Cl^- currents. Similar findings were demonstrated recently (Cuppoletti et al. 2014b), where shRNA was used to knock down ClC-2 in T84 cells. When ClC-2 protein was reduced to very low levels, lubiprostone-stimulated currents were abolished compared to wild-type T84 cells. CFTR function was retained, stimulated by forskolin-IBMX, not affected by methadone, inhibited by DASU-02, and inhibited to very low levels by CFTRinh172. In contrast, when shRNA to CFTR was used to knock down CFTR in T84 cells, CFTR protein was reduced to very low levels, and lubiprostone-stimulated current was retained at levels comparable to those of the wild-type cells. The current was not inhibited by DASU-02 but reduced by methadone and further reduced by Cd^{2+}. Since human ClC-2 is also activated by PKA (Cuppoletti et al. 2004a, b, 2013, 2014b), mPKI inhibited the response to forskolin-IBMX but did not affect lubiprostone stimulation of ClC-2 (Cuppoletti et al. 2013). Interestingly, treatment with mPKI reduced the control, basal (unstimulated) currents slightly without affecting lubiprostone stimulation suggesting that a part of the basal activity may be due to partially phosphorylated ClC-2. Similar effects were seen with treatment with H-89. H-89 results are unpublished, since H-89 has numerous nonspecific effects which cannot be controlled for without H89's inactive form, H-85, which is no longer available (see Sect. 12 on H-89). Thus, in T84 cells, CFTR does not contribute to lubiprostone-stimulated Cl^- currents, and ClC-2 activation is independent of the presence of CFTR, as demonstrated in HEK293 cells transfected separately with ClC-2 or CFTR (Cuppoletti et al. 2004b), in HEK293EBNA cells (Cuppoletti et al. 2013, 2014b) transfected with ClC-2 alone, and in T84 cells lacking CFTR (Cuppoletti et al. 2014b). Knockdown of ClC-2 in T84 cells causes loss of lubiprostone stimulation (Cuppoletti et al. 2014b).

In studies of T84 cells, Bijvelds et al. (2009) found no inhibition by Cd^{2+}, which however has been shown to inhibit T84 lubiprostone-stimulated currents in patch-clamp studies (Cuppoletti et al. 2008b, 2013, 2014b). It is not clear why Cd^{2+} responses were not seen in this study, especially since it is generally toxic to epithelial cells (Bao et al. 2008; Moeser et al. 2004, 2007). Bijvelds et al. (2009) also used CFTRinh172 assuming it is CFTR specific and found block of all Cl^- currents and ascribed all of the current activated by lubiprostone to CFTR. CFTRinh172 has now been shown to be equipotent in inhibiting both CFTR and ClC-2 (Cuppoletti et al. 2014b). When lubiprostone and forskolin were used together, the experimental design did not take into account the fact that lubiprostone and forskolin-IBMX both stimulate human ClC-2 (Cuppoletti et al. 2004b, 2013, 2014b). Yet they ascribed all of the current increases only to CFTR. Bijvelds et al. (2009) found cAMP was raised by lubiprostone, but to levels only about 30 % of that of forskolin. Despite this, lubiprostone and forskolin gave similar activation of I_{sc}. This is not expected from the studies of Cartwright et al. (1985) showing that anion transport in T84 cells is proportional to cAMP levels over the ranges reported by Bijvelds et al. (2009). Neither Cuppoletti et al. (2008a) nor Bao et al. (2008) found increased cAMP levels with lubiprostone.

Lubiprostone similarly activated I_{sc} in Calu-3 cells, and knockdown of >95 % of CFTR had no effect on lubiprostone stimulation with or without the addition of GlyH-101 (MacVinish et al. 2007). Thus, also in Calu-3 cells, lubiprostone does not stimulate CFTR, but rather it stimulated ClC-2. CFTR is not required for stimulation of ClC-2 by lubiprostone.

15.8.3.3 Site of Action of Lubiprostone in Stimulation of ClC-2

When one of two phosphorylation sites in human ClC-2 is mutated, $RGET_{691} > RGEA$, the EC_{50} for lubiprostone shifted from 27.6 ± 4.5 nM ($n = 4$) to 445.8 ± 2.1 nM (Cuppoletti et al. 2004a). In the triple mutant $RGET_{691} > AGAA$, complete loss of lubiprostone stimulation occurred (Cuppoletti et al. 2014a), but forskolin-IBMX acting at $RRAT_{655}$ (Cuppoletti et al. 2004a) still activated human ClC-2 via PKA. Therefore, the $RGET_{691}$ site appears to be essential for lubiprostone stimulation.

15.8.3.4 Lubiprostone Stimulation of Recombinant ClC-2 from Other Species

Lubiprostone also activated ClC-2 channels from other species, as determined by whole-cell patch clamp of recombinant ClC-2 from various species expressed in HEK293EBNA cells (Cuppoletti et al. 2014a). Rabbit ClC-2 gave an EC_{50} of 350 ± 37.9 nM ($n = 3$) and mouse ClC-2 gave an EC_{50} of 64.3 ± 3.7 nM ($n = 4$). By comparison, the EC_{50} for human ClC-2 expressed in these cells was 28.2 ± 2.2 nM ($n = 4$) (Cuppoletti et al. 2013). Norimatsu et al. (2012) were unable to activate rabbit ClC-2 with lubiprostone expressed in *Xenopus* oocytes. In contrast, rabbit ClC-2 expressed in HEK293 cells was stimulated by lubiprostone with an EC_{50} of approximately 350 nM (Cuppoletti et al. 2014a). It is unclear why Norimatsu et al. (2012) were unable to observe activation of rabbit ClC-2 in the *Xenopus* oocyte expression system. It is possible that it was pre-activated by PKA before the addition of lubiprostone.

15.8.3.5 Lack of Involvement of PKA Phosphorylation in Lubiprostone Stimulation

Lubiprostone stimulation of human ClC-2 is independent of PKA, as determined from experiments where mPKI inhibited activation by forskolin-IBMX, but not activation by lubiprostone (Cuppoletti et al. 2004b, 2013). It has also been shown that a double phosphorylation knockout mutant of ClC-2 cannot be activated by forskolin-IBMX (Cuppoletti et al. 2004a) but is activated by 20 nM lubiprostone (Cuppoletti et al. 2008a). Wild-type ClC-2 control currents at −140 mV ($n = 6$) of -27.2 ± 3.5 pA/pF significantly ($P < 0.0005$) increased to -100.9 ± 6.5 pA/pF with lubiprostone, and 500 μM Cd^{2+} reduced the current to control levels. The mutant channel currents were also significantly ($P < 0.005$) increased by lubiprostone and inhibited by $CdCl_2$. Control currents were -27.1 ± 6.9 pA/pF which increased to -81.9 ± 5.1 pA/pF with lubiprostone and were reduced with $CdCl_2$ to -29.4 ± 3.7 pA/pF ($P < 0.005$) ($n = 6$). Thus, PKA phosphorylation is not involved in lubiprostone stimulation of human ClC-2.

It is clear from the above that human ClC-2 and not human CFTR was activated by lubiprostone and that PKA is not involved in lubiprostone stimulation. Bao et al. (2008) also carried out single-channel studies of the apical membrane of A6 epithelial cells and human ClC-2. The EC_{50} for A6 cell ClC-2 single channels was 69 ± 18 nM. However, in this study, there was also a non-cAMP-dependent activation of CFTR with an EC_{50} of 791 ± 273 nM approximately tenfold higher than the EC_{50} for ClC-2. The mechanism whereby lubiprostone stimulates *Xenopus* A6 cell CFTR is not known.

15.9 ClC-2 Function in Epithelia

Bösl et al. (2001) showed that ClC-$2^{-/-}$ mice showed only retinal and testes degeneration. Mice with a double CFTR and ClC-2 knockout (Zdebik et al. 2004) did not have increased mortality over $CFTR^{-/-}$ mice, and there were no anomalies in electrocardiograms or the morphology of the intestine, lung, or pancreas. ClC-$2^{-/-}$ mice crossed into a background of CFTR ΔF508/ΔF508 lived longer than $CFTR^{\Delta F508/\Delta F508}$ mice. Increased Cl^- secretion was seen in the colon of ClC-$2^{-/-}$ mice. An important question then is whether ClC-2 has any additional functions, as might be expected with it being ubiquitously expressed.

Bösl et al. (2001) also reported that histamine-stimulated gastric acid secretion was normal in ClC-$2^{-/-}$ mice. This was somewhat unexpected, since the gastric mucosa of ClC-$2^{-/-}$mice has abnormal histology, reduced numbers of parietal cells, and reduced H/K ATPase as well as other changes which would be expected to result in less HCl secretion in the ClC-$2^{-/-}$ mouse, regardless of the Cl^- ion channel involved in acid secretion (Nighot et al. 2011, 2014). Indeed, Ranker et al. (2004) found that ClC-$2^{-/-}$ mice secreted significantly less acid than wild-type (ClC-$2^{+/+}$) mice when stimulated with histamine plus diphenhydramine and carbachol. Acid secretion in ClC-$2^{-/-}$ mice (4.0 ± 1.6 μEq H^+/15 min, $n = 4$) was significantly lower ($P < 0.02$) at 30 min (peak secretion) compared to ClC-$2^{+/+}$ mice (11.1 ± 1.2 μEq H^+/15 min, $n = 3$), while pepsinogen secretion was similar in ClC-$2^{+/+}$ and ClC-$2^{-/-}$ mice. Recently, it has been shown that ClC-$2^{-/-}$ mice also have significant loss of electroneutral absorption in the distal colon (Catalán et al. 2012). In addition, ClC-$2^{-/-}$ mice are more susceptible to damage of the intestine (Nighot et al. 2009). Thus, it appears that under certain conditions, ClC-2 may play a role in tissues other than the retina and testes. Additional examples of the possible importance of ClC-2 in epithelia will be discussed.

It has been shown by Gyömörey et al. (2000) that in the small intestine of mice, there exists a basal Cl^- current and a hyperpolarization-stimulated current which can be activated by hypertonic shock. These currents were sensitive to glibenclamide, an open-channel blocker of ClC-2 (Bao et al. 2008) as well as NPPB. These currents were present in both wild-type CFTR$^{+/+}$ and CFTR$^{-/-}$ mice. The channel had a distribution in the small intestine at the cellular level similar to that seen in Caco-2 cells (Mohammad-Panah et al. 2001), namely, predominantly at the tight junction complex between cells with diffuse labeling of the apical brush border membrane. Gyömörey et al. (2000) concluded that ClC-2 contributes to Cl^- secretion in both the wild-type and CFTR$^{-/-}$ mouse, as was concluded in studies of Caco-2 cells (Mohammad-Panah et al. 2001). Catalán et al. (2002, 2004) demonstrated that ClC-2 is expressed in the basolateral membrane of the mouse distal colon and indeed showed that the ClC-$2^{-/-}$ mouse has an absorptive defect in the distal colon (Catalán et al. 2012). It must be noted however that the basolateral location of ClC-2 in the colon does not mean that the work of others showing apical distribution in other parts of the intestine and lung are incorrect.

As described earlier (Sect. 8.3.2), Bijvelds et al. (2009) found no evidence for ClC-2 function in T84 cells based on the lack of effect of Cd^{2+}, the use of CFTRinh172 assumed to be CFTR specific, and assuming that only CFTR is activated by forskolin. Using the same approaches, Bijvelds et al. (2009) also did not find evidence for ClC-2 in any experiments in healthy or CF rectum or ileum, CFTR$^{-/-}$ mice, or CFTR$^{\Delta F508/\Delta F508}$ mice. These findings are in conflict with those of Gyömörey et al. (2000), Catalán et al. (2012), Fei et al. (2009), and Moeser et al. (2004, 2007) who have readily demonstrated ClC-2 function in the intestine.

In another study in guinea pig intestine and colon, Fei et al. (2009) found that lubiprostone increased I_{sc} in the intestine, with an EC_{50} of 42.5 nM, and in the colon, with an EC_{50} of 31.7 nM. Lubiprostone responses were blocked by glibenclamide, which has been shown to be an open-channel blocker of ClC-2 (Bao et al. 2008) in addition to inhibiting CFTR. Antagonists of EP_1, EP_2, EP_3, or EP_4 receptors did not suppress I_{sc}, although they blocked PGE_2 responses. The finding by Fei et al. (2009) that an EP_4 antagonist GW627368X did not block I_{sc} responses conflict strongly with the results and conclusions of Bijvelds et al. (2009) where L161,982, also an EP_4 antagonist, inhibited all I_{sc} responses. This suggests that L161,982 at high concentrations will affect other prostaglandin receptors and may have unknown or nonspecific effects (see Sect. 11). In contrast, the guinea pig appears to be well suited for studies of ClC-2 in intestinal function.

The possible involvement of ClC-2 in Cl^- transport function and barrier function has also been studied in the porcine intestine (Moeser et al. 2004, 2007). Ischemia, NSAIDS, and pathogens disrupt the intestinal epithelial barrier. PGE_2 has been shown to repair damaged porcine intestine and to increase Cl^- secretion as measured by I_{sc} in the small intestine (Moeser et al. 2004). Three Cl^- channels are present in the porcine intestine: Ca^{2+}-activated Cl^- channels, CFTR, and ClC-2. DNDS and DASU-02, inhibitors of Ca^{2+}-activated Cl^- channels (outwardly rectified), and CFTR, respectively, were used. Neither of these inhibitors prevented PGE_2-stimulated improvement in barrier function, as measured by transepithelial resistance (TER) or PGE_2-stimulated I_{sc}. However, Zn^{2+}, an inhibitor of ClC-2, inhibited both PGE_2-stimulated Cl^- transport and restitution of the barrier properties. Cd^{2+} was not used because histological examination showed that it caused damage to the tissue. ClC-2 channel protein was upregulated upon ischemic injury, while CFTR channel protein was downregulated. Occludin was found to immunoprecipitate with ClC-2 suggesting an association of ClC-2 with the tight junction. Immunogold electron microscopy showed ClC-2 localization at the tight junction during recovery. Immunofluorescence showed that ClC-2 was localized to the apical aspect of mucosal villi as well as in mucosal crypts, as indicated by others. Thus, this porcine model of ischemic injury and recovery has demonstrated physiological roles for ClC-2 in barrier repair and associated Cl^- secretion. Careful selection of Cl^- channel inhibitors and activators was a strength of this study and will be instructive to the field.

In the same ischemia-injured porcine tissue, lubiprostone was also shown to give a concentration-dependent increase in TER and reduction of ^{3}H-mannitol flux. It also promoted recovery of ClC-2 to tight junctions. I_{sc} was stimulated by lubiprostone in a concentration-dependent manner, and the recovery of TER and I_{sc} increases were significantly correlated with lubiprostone treatment, whereas there was a poor correlation with PGE_2 treatment (Moeser et al. 2007). TER recovery and I_{sc} were blocked by Zn^{2+}, whereas DASU-02 was without effect, suggesting that ClC-2, and not CFTR, was responsible for the barrier repair and associated Cl^- secretion. The porcine intestine appears to be well suited for studies of the physiological functions of ClC-2.

In addition, it has been shown in a ClC-$2^{-/-}$ mouse model of ulcerative colitis, that ClC-2 plays an important role in maintaining and improving recovery of loss of barrier function and maintaining tight junctions (Nighot et al. 2013). The model is used to mimic ulcerative colitis patient intestinal problems, and this result again suggests that mice might be useful for studies of ClC-2 function under some circumstances (Nighot et al. 2009). In the mouse intestine, ClC-2 appears to be essential for recovery of barrier function. In uninjured animals ClC-2 knockout appears to improve intestinal barrier function (Nighot et al. 2009), but in the ischemic mouse intestine, ClC-2 appears to be essential for recovery of barrier function (Nighot and Blikslager 2010).

A comprehensive study by Joo et al. (2009) on the effects of lubiprostone on airway submucosal fluid secretion and surface epithelium I_{sc} in pigs, sheep, and man has appeared. According to the authors, "...a major reason for these studies was to examine ClC-2 channels as possible surrogates for CFTR using the putative ClC-2 activator lubiprostone in the airways...," and the authors cite the findings of Schwiebert et al. (1998) showing that ClC-2 currents can be obtained in human CF airway cells. Humans, pigs, and sheep all responded to lubiprostone. The authors found that lubiprostone also stimulated anion currents at the apical surface epithelium with a K_d of 10.5 nM, with a single site of action, similar to that seen by others using recombinant human ClC-2 and T84 cells (Cuppoletti et al. 2004a, 2013). I_{sc} responses were sensitive to removal of Cl^- and were sensitive to 300 μM Cd^{2+}. I_{sc} responses to lubiprostone were reversible, and they were sensitive to bumetanide, to furosemide, and to the removal of Cl^-, thus likely emanating from Cl^- secretion. GlyH-101 caused a rightward shift in the concentration response curve, and the authors noted a personal communication from Dr. Verkman indicating that GlyH-101 may have some nonspecific actions, in addition to its action on CFTR. GlyH-101 was previously shown by MacVinish et al. (2007) to not inhibit Cl^- secretion in Calu-3 cells when CFTR was ablated. Fluid secretion in airway glands of human, pigs, and sheep was also increased by lubiprostone. Removal of Cl^- and treatment with furosemide prevented lubiprostone stimulation of gland fluid secretion. In the presence of a maximal concentration of forskolin in pig, human, and sheep submucosal glands, lubiprostone caused a further increase in secretion, suggesting that forskolin and lubiprostone affect different targets. Gland fluid secretion was stimulated, but the increases were not significant in samples from two out of four CF patients and in two of three chronic obstructive pulmonary disease and idiopathic pulmonary fibrosis patient samples. No bronchoconstriction was noted with lubiprostone. Together, these results support the view that airway surface epithelium and glandular fluid secretion can be modulated by activators of ClC-2. These studies suggest that the tracheal surface epithelium and submucosal glands are a good model system for studies of activators of ClC-2 in the presence of normal or defective CFTR.

A contrasting conclusion was made by Cuthbert (2011). Lubiprostone was shown to stimulate sheep surface epithelium I_{sc} and submucosal gland secretion. Following the various protocols and approaches of Bijvelds et al. (2009), Cuthbert (2011) found that 10 μM L161,982 (an EP_4 antagonist), when added to the apical bathing solution, but not when added to the basolateral bathing solution of the surface epithelium, inhibited the lubiprostone response. 1 μM L161,982 shifted the concentration of lubiprostone required to achieve 50 % activation by more than tenfold. Moreover, again following the approaches of Bijvelds et al. (2009),

Cuthbert (2011) used 40 μM H-89, which blocked the lubiprostone response by only 50 %. H-89 is discussed in Sect. 12, and experiments using H89 must be interpreted with great caution because of possible off-target or nonspecific effects. Similar considerations may apply to L161,982. Cuthbert (2011) further showed that the EP_4 agonist L-90288 inhibited the lubiprostone response, but not in a predictable manner. It is not known whether ovine ClC-2 is activated by PKA as is human ClC-2 (Cuppoletti et al. 2004a, 2014b), so it is difficult to evaluate if L161,982 should prevent activation of ovine ClC-2 by lowering $[cAMP]_i$. It has been clearly shown that ClC-2 (non-CFTR Cl^- channel) is activated by lubiprostone in Calu-3 cells lacking CFTR (MacVinish et al. 2007). Others have shown that Calu-3 cells have EP_4 receptors which are coupled to anion transport (Joy and Cowley 2008). However, no test was done by Cuthbert (2011) to determine ClC-2 function in the airway. The interpretation of all of his experiments was that no functional ClC-2 was observed.

Acidic pH hyperpolarizes nasal potential difference (Uwaifo et al. 2006), suggesting the presence of ClC-2 in the nasal epithelia. A recent paper studied the effects of treatment of murine nasal epithelia with lubiprostone. Schiffhauer et al. (2013) suggest that CFTR and ClC-2 are activated by lubiprostone in mouse nasal epithelia. The authors treated mouse nasal epithelia with 20 μM lubiprostone. They used mice which conditionally (doxycycline regulated) overexpressed hClC-2 in the airway, $CFTR^{-/-}$ mice and $ClC\text{-}2^{-/-}$ mice. Human ClC-2 was found in the apical region of the doxycycline-treated mice and at lower levels in untreated mice and in the $CFTR^{-/-}$ mice. CFTR expression was similar in untreated, doxycycline-treated, and $ClC\text{-}2^{-/-}$ mice. Lubiprostone (20 μM) stimulated nasal Cl^- potential difference (PD) in doxycycline-untreated mice and stimulated nasal Cl^- PD to higher levels in the doxycycline-treated mice. Nasal Cl^- PD measurements were carried out in the presence of CFTRinh172, which may have decreased the response since it is now known that CFTRinh172 inhibits ClC-2 as well as CFTR (Cuppoletti et al. 2014b). 200 nM GaTx2 inhibited the response in doxycycline-treated mice. Lubiprostone stimulated nasal Cl^- PD in $CFTR^{-/-}$ mice and GaTx2 completely blocked the response. CFTRinh172 was not present in this experiment.

In addition, Schiffhauer et al. (2013) showed that lubiprostone also stimulated nasal Cl^- PD in $ClC\text{-}2^{-/-}$ mice. CFTRinh172 dose dependently inhibited nasal Cl^- PD in $ClC\text{-}2^{-/-}$ mice. The results of the study were interpreted to show that CFTR was activated by lubiprostone in the $ClC\text{-}2^{-/-}$ mice. This result was in conflict with findings in a previous study (MacDonald et al. 2008) where $ClC\text{-}2^{-/-}$ mice did not give a response to lubiprostone. This difference was not discussed and remains to be explained.

In the MacDonald et al. (2008) study, there was no effect of lubiprostone on $ClC\text{-}2^{-/-}$ mice, but wild-type and $CFTR^{-/-}$ mice showed activation consistent with stimulation of ClC-2. CFTRinh172 did not prevent the response, and it was concluded that CFTR was not required for the response to lubiprostone. The lack of effect of lubiprostone in $ClC\text{-}2^{-/-}$ mice suggests that CFTR is not activated by 20 μM lubiprostone, in contrast to the effect of lubiprostone seen in the $ClC\text{-}2^{-/-}$ mouse in the Schiffhauer et al. (2013) study. It should be considered that at high concentrations (20 μM) of lubiprostone, there may be nonspecific effects. Further experiments using lower concentrations of lubiprostone might help clarify these conflicting results.

To reiterate, in the case of the doxycycline-treated mice overexpressing human ClC-2, 20 μM lubiprostone was used, which is 1000 times the EC_{50} of lubiprostone for human ClC-2 as determined in patch clamp of human ClC-2 expressed in HEK293 cells (Cuppoletti et al. 2004a, 2013). Nevertheless, the results taken together suggest that ClC-2 is activated by lubiprostone, and the differences in the findings on CFTR in $ClC2^{-/-}$ mice between the two studies suggest that additional studies are needed to clarify whether indeed lubiprostone at these very high concentrations (20 μM) can have an effect on CFTR in addition to ClC-2. Also, future studies must take into account the equipotent inhibitory effect of CFTRinh172 on ClC-2, as well as CFTR.

Lubiprostone has been shown to cause a dose-dependent increase in Cl^--rich fluid in the small intestine of rats, without affecting the serum electrolyte levels (Ueno et al. 2004). Lubiprostone, given orally, is only known to reach the small intestine (Ueno R, personal communication).

15.10 Lubiprostone in Human Treatments

15.10.1 Chronic Idiopathic Constipation, Irritable Bowel Syndrome, and Opioid-Induced Constipation

Lubiprostone is approved for use in the treatment of chronic idiopathic constipation (Lembo et al. 2011; Barish et al. 2010; Johanson et al. 2008a). It is also used for treatment of irritable bowel syndrome (Chey et al. 2012; Carter and Scott 2009; Drossman et al. 2009; Johanson et al. 2008b) and opioid-induced constipation (Wong and Camilleri 2011; Cryer et al. 2014).

15.10.2 Effect of Lubiprostone in the Intestine of CF Patients

Pilot studies of the effect of lubiprostone on treatment of constipation in CF patients have been carried out (O'Brien et al. 2011). Lubiprostone improved overall symptoms of constipation in these CF patients.

15.11 Lubiprostone Is Not an EP_4 Receptor Agonist; Rather, It Is a Weak EP_4 Receptor Antagonist

Bijvelds et al. (2009) and Cuthbert (2011) used L161,191 at high concentrations and found that it inhibited Cl^- transport as measured by I_{sc}. The K_i for blocking EP_4 receptors by L161,982 is 0.032 nM (Weinreb et al. 2006). Its specificity is 200 times greater for EP_4 than that for EP_{1-3} ($K_i = 6.4$ μM) (Weinreb et al. 2006). Bijvelds et al. (2009) used concentrations of L161,982 as high as 20 μM, which is 625 times the IC_{50} for EP_4 receptor antagonism, and Cuthbert (2011) used L161,982 at concentrations as high as 10 μM. At these concentrations, L161,982 will also act upon EP_{1-3} receptors and could have other nonspecific effects. Thus, it is not possible to interpret the effects of L161,982 as being due to the binding of lubiprostone to EP_4 receptors alone. Other EP_4 receptor antagonists had no effect on lubiprostone stimulation of I_{sc} (Fei et al. 2009).

Lubiprostone is not an agonist of cloned human EP_4 receptors, but rather a weak antagonist of EP_4 receptors with an IC_{50} of 143 nM (Cuppoletti et al. 2008c). Also, lubiprostone activated ClC-2 whether expressed in HEK293 cells (Cuppoletti et al. 2004a, 2008a) or overexpressed in HEK293EBNA cells (Cuppoletti et al. 2013, 2014a, b). These cells do not contain EP receptors, and for this reason, they are used for expression of EP receptors (Boie et al. 1997). Thus, lubiprostone binding to EP_4 receptors is not involved in lubiprostone stimulation of ClC-2 mediated Cl^- transport. Studies of the effects of 10–20 μM L161,982 on ClC-2 and CFTR in native cells and on recombinant ClC-2 and CFTR channel function and cell physiology in general would be required before it is possible to interpret L161,982 effects. In view of these reports, we examined the effects of L161,982 on the lubiprostone and forskolin/IBMX responses of recombinant ClC-2 and CFTR and knockdown T84 cells where only CFTR or ClC-2 is present. Lubiprostone stimulated ClC-2, but not CFTR in these systems, and forskolin/IBMX stimulated both ClC-2 and CFTR. In all cases there was no effect of L161,982 (unpublished data). This suggests that L161,982 might have other effects such as upon T84 cell culture barrier properties. The high concentrations used also mean that EP_{1-4} receptors would be occupied by L161,982. L161,982 effects clearly cannot be interpreted as antagonizing the binding of lubiprostone at only EP_4 receptors. In addition, other EP_4 antagonists do not inhibit lubiprostone-stimulated I_{sc} (Fei et al. 2009). To reiterate, lubiprostone at concentrations as high as 1 μM is not an agonist of cloned human EP_4 receptors, and it is only a weak antagonist (Cuppoletti et al. 2008a).

15.12 H-89 Effects

Bijvelds et al. (2009) studied the effect of treatment with 20 μM H-89 for 30 min on lubiprostone-stimulated I_{sc} in T84 cells and found partial (70 %) inhibition, as we found in studies with mPKI (Cuppoletti et al. 2013) and H89 (see Sect. 8.3.2). Cuthbert (2011) studied the effect of 40 μM H-89 for 30 min on lubiprostone stimulation of ovine lung I_{sc} and also found partial (50 %) inhibition. Both investigators assigned the effect to inhibition of CFTR. Strohmeier et al. (1995) and Hoque et al. (2010) both showed approximately 80–90 % inhibition of forskolin-stimulated anion transport in T84 cells using 60 μM and 50 μM H-89, respectively. H-89 was much less effective in inhibition of anion transport in the Cuthbert (2011) studies. The inability of H-89 to more fully inhibit is consistent with lubiprostone stimulation of ClC-2 rather than CFTR.

We have examined the effect of a more specific PKA inhibitor, mPKI (Lochner and Moolman 2006), on lubiprostone-stimulated currents (Cuppoletti et al. 2004a, 2013). It does not inhibit lubiprostone-stimulated currents. mPKI readily inhibited forskolin-stimulated currents of HEK293EBNA cells containing only ClC-2 and also (not significantly) inhibited basal (non-stimulated) currents (Cuppoletti et al. 2013). Cuppoletti et al. (unpublished) found that H-89 behaves similarly and inhibits basal currents of recombinant ClC-2, but it did not inhibit lubiprostone-stimulated currents in HEK293EBNA cells containing only ClC-2. It is not possible to determine the specificity of H-89 without the use of H-85 (an inactive analogue) as explained below.

H-89 has been widely used to study ion channels. Park et al. (2007) showed that H-89 and its inactive analogue H-85 had similar effects, suggesting direct modulation of Ca^{2+}-activated K^+ currents (BK) with an EC_{50} of 0.5 μM H-89. Bijvelds et al. (2009) used 40 times this concentration (20 μM), and Cuthbert (2011) used 80 times this concentration (40 μM). Similarly, Son et al. (2006) showed that H89 and H85 both inhibited an arterial Kv channel with an K_d of 1.2 μM. Sun Park et al. (2006) also found direct inhibition of K_{ATP} and K_{ir} channels with IC_{50}'s of 1.2 and 3.78 μM, respectively. Niisato et al. (1999) showed that epithelial Na^+ channels were instead activated by treatment with H89, H85, H7, and H8, whereas mPKI and KT-5720 decreased the currents. Marunaka and Niisato (2003) showed that H-89 stimulated Na^+ transport by transporting ENaC to the apical membrane. Thus, in the absence of testing whether there are effects of H-85 as well as with H-89 on ClC-2, it is not formally possible to interpret the effect of H-89. A review of the various effects of H-89 and the precautions that must be taken when interpreting the effects of this compound has appeared (Lochner and Moolman 2006).

15.13 Small-Molecule Effectors

15.13.1 Compounds That May Be Useful for Studies of ClC-2 in the Presence of CFTR and Other Cl$^-$ Channels

GlyH-101 was discovered by high-throughput screening. It is water soluble up to 1 mM and is more active at positive voltages and has an IC_{50} of 5 μM. It inhibits CFTR-dependent nasal PDs in mice and was effective in blocking cholera toxin-induced intestinal fluid secretion (Muanprasat et al. 2004). It has also been used to differentiate between ClC-2 and CFTR in model systems. For example, GlyH-101 appears to have a more focused inhibitory effect on CFTR than ClC-2 based on the finding that GlyH-101 did not inhibit lubiprostone-stimulated currents thought to be ClC-2 currents before or after knockdown of CFTR (MacVinish et al. 2007). Similar findings were reported using GlyH-101 by Joo et al. (2009) in studies of airway surface I_{sc} and airway serous gland secretion. In that study, concentration-dependent stimulation by lubiprostone of I_{sc} persisted, but the concentration curves were shifted to the right, suggesting that GlyH-101 might also affect ClC-2 or another Cl$^-$ channel and may not be CFTR specific. GlyH-101 may also affect mitochondrial function (Kelly et al. 2010). No studies have yet appeared to determine whether ClC-2 is affected by GlyH-101.

DASU-02 (N-(4-methylphenylsulfonyl)-N′-4-trifluromethyl-phenyl)urea) is a CFTR inhibitor (Schultz et al. 1996, 1999). DASU-02 and several other related disulfonylurea analogues were tested on enterotoxin-induced porcine colonic secretion (O'Donnell et al. 2000). DASU-02 also inhibited *E. coli* EAST 1 toxin-induced I_{sc} in T84 cell cultures (Veilleux et al. 2008). DASU-02 was also used to study ClC-2 present with Ca^{2+}-activated Cl$^-$ channels and CFTR (Moeser et al. 2004, 2007). Schultz et al. (1999) refer to unpublished data that showed that DASU-02 did not inhibit recombinant rabbit ClC-2 activated by forskolin-IBMX. We have recently shown that DASU-02 will inhibit recombinant human CFTR expressed in HEK293 cells. However, DASU-02 did not inhibit lubiprostone-stimulated Cl$^-$ currents in HEK293EBNA cells expressing recombinant ClC-2 or in T84 cells where CFTR was knocked down with CFTR siRNA, but inhibited Cl$^-$ currents in T84 cells where ClC-2 was knocked down with ClC-2 siRNA (Cuppoletti et al. 2014b). Thus, DASU-02 might be useful in studies which require separation of ClC-2 from CFTR in complex systems that might contain both channels.

Methadone has been shown to inhibit recombinant human ClC-2 without inhibition of recombinant human CFTR in HEK cell lines and Cl$^-$ transport in T84 cell cultures (Cuppoletti et al. 2013). In that study, methadone inhibited lubiprostone-stimulated and PKA-activated recombinant human ClC-2 expressed in HEK293EBNA cells. Methadone also inhibited lubiprostone-stimulated I_{sc} in T84 cultures, as well as Cl$^-$ currents measured by whole-cell patch clamp of T84 cells in which CFTR was knocked down by siRNA, but not in T84 cells in which ClC-2 was knocked down by siRNA. Methadone was most effective when added before patch formation and before stimulation by drugs. To emphasize, it was much less effective when added after ClC-2 was activated by voltage or drugs. Methadone also prevented lubiprostone-induced protection against inflammatory cytokines (Cuppoletti et al. 2012). Methadone might also be useful in separating ClC-2 and CFTR effects in complex systems.

The *Leiurus quinqueststriatus hebraeus* (scorpion) toxin GaTx2 has been shown to inhibit rabbit ClC-2 but not ClC-1, ClC-3, and ClC-4, CFTR, GABAc, Shaker B-IR, or Kv1.2 (Thompson et al. 2009). GaTx2 inhibits ClC-2 only from the extracytoplasmic face of the channel with an IC_{50} of 20 pM. The inhibition is partial, and it does not inhibit the activated channel (as is the case with methadone). Despite these limitations, the specificity of this drug for ClC-2 versus other specific channels is defined, and it has a very high affinity for ClC-2. GaTx2 has been shown to inhibit ClC-2 in nasal PD measurements in lubiprostone-stimulated $CFTR^{-/-}$ mice as well as in mice overexpressing human ClC-2 in the nasal epithelia (Schiffhauer et al. 2013).

15.13.2 Compounds That May Not Be Useful for Studies of ClC-2 in the Presence of CFTR and Other Cl^- Channels

Cd^{2+} has been shown to inhibit ClC-2 in a large number of studies, but it appears to have an effect on other channels, including CFTR in A6 cells, and is considered generally toxic to cells (Bao et al. 2008). Other inhibitors, for example, glibenclamide, an inhibitor of CFTR, also inhibit A6 cell ClC-2 (Bao et al. 2008). CFTRinh172 has now been shown to also be a potent inhibitor of human ClC-2 (Cuppoletti et al. 2014b). Effects of a large number of other inhibitors of CFTR have been described, but the specificity of these inhibitors has not been examined and will not be discussed here. While methadone does not inhibit CFTR at concentrations up to 1 μM, it is known to inhibit numerous other channels (Cuppoletti et al. 2013). Only methadone and GaTx2 have been shown to be more effective on ClC-2 than CFTR and DASU-02 on CFTR and not ClC-2.

15.14 Summary

In summary, ClC-2 appears to have numerous physiological functions other than only involvement in the testes and retina as described initially in $ClC\text{-}2^{-/-}$ mice (Bösl et al. 2001). Other roles of ClC-2 are being discovered at a rapid pace. Inhibitors have now been identified and documented which might be useful for determination of ClC-2 function in the presence of CFTR. Knockouts and conditional knock-ins of ClC-2 in cells and tissues have been described. Combinations of pharmacological and genetic approaches will be useful to further delineate the functions of ClC-2 in different tissues and in different physiological contexts.

Disclosure JC and DHM are supported by a research grant from Sucampo AG, Zug, Switzerland. JC, DHM, and RU have financial interests in Sucampo Pharmaceuticals, Inc.

References

Alioth S, Meyer S, Dutzler R, Pervushin K (2007) The cytoplasmic domain of the chloride channel ClC-0: structural and dynamic characterization of flexible regions. J Mol Biol 369:1163–1169

Atia F, Zeiske W, van Driessche W (1999) Secretory apical Cl^- channels in A6 cells: possible control by cell Ca^{2+} and cAMP. Pflugers Arch 438:344–353

Bali M, Lipecka J, Edelman A, Fritsch J (2001) Regulation of ClC-2 chloride channels in T84 cells by TGF-alpha. Am J Physiol Cell Physiol 280:C1588–C1598

Bao HF, Liu L, Self J, Duke BJ, Ueno R, Eaton DC (2008) A synthetic prostone activates apical chloride channels in A6 epithelial cells. Am J Physiol Gastrointest Liver Physiol 295:G234–G251

Barish CF, Drossman D, Johanson JF, Ueno R (2010) Efficacy and safety of lubiprostone in patients with chronic constipation. Dig Dis Sci 55:1090–1097

Bauer CK, Steinmeyer K, Schwarz JR, Jentsch TJ (1991) Completely functional double-barreled chloride channel expressed from a single Torpedo cDNA. Proc Natl Acad Sci U S A 88:11052–11056

Bi MM, Hong S, Zhou HY, Wang HW, Wang LN, Zheng YJ (2013) Chloride channelopathies of ClC-2. Int J Mol Sci 15:218–249

Bijvelds MJ, Bot AG, Escher JC, De Jonge HR (2009) Activation of intestinal Cl^- secretion by lubiprostone requires the cystic fibrosis transmembrane conductance regulator. Gastroenterology 137:976–985

Blaisdell CJ, Pellettieri JP, Loughlin CE, Chu S, Zeitlin PL (1999) Keratinocyte growth factor stimulates CLC-2 expression in primary fetal rat distal lung epithelial cells. Am J Respir Cell Mol Biol 20:842–847

Blaisdell CJ, Edmonds RD, Wang XT, Guggino S, Zeitlin PL (2000) pH-regulated chloride secretion in fetal lung epithelia. Am J Physiol Lung Cell Mol Physiol 278:L1248–L1255

Boie Y, Stocco R, Sawyer N, Slipetz DM, Ungrin MD, Neuschäfer-Rube F, Püschel GP, Metters KM, Abramovitz M (1997) Molecular cloning and characterization of the four rat prostaglandin E_2 prostanoid receptor subtypes. Eur J Pharmacol 340:227–241

Bösl MR, Stein V, Hübner C, Zdebik AA, Jordt SE, Mukhopadhyay AK, Davidoff MS, Holstein AF, Jentsch TJ (2001) Male germ cells and photoreceptors, both dependent on close cell-cell interactions, degenerate upon ClC-2 Cl^- channel disruption. EMBO J 20:1289–1299

Carter NJ, Scott LJ (2009) Lubiprostone: in constipation-predominant irritable bowel syndrome. Drugs 69:1229–1237

Cartwright CA, McRoberts JA, Mandel KG, Dharmsathaphorn K (1985) Synergistic action of cyclic adenosine monophosphate- and calcium-mediated chloride secretion in a colonic epithelial cell line. J Clin Invest 76:1837–1842

Catalán M, Cornejo I, Figueroa CD, Niemeyer MI, Sepúlveda FV, Cid LP (2002) ClC-2 in guinea pig colon: mRNA, immunolabeling, and functional evidence for surface epithelium localization. Am J Physiol Gastrointest Liver Physiol 283:G1004–G1013

Catalán M, Niemeyer MI, Cid LP, Sepúlveda FV (2004) Basolateral ClC-2 chloride channels in surface colon epithelium: regulation by a direct effect of intracellular chloride. Gastroenterology 126:1104–1114

Catalán MA, Flores CA, González-Begne M, Zhang Y, Sepúlveda FV, Melvin JE (2012) Severe defects in absorptive ion transport in distal colons of mice that lack ClC-2 channels. Gastroenterology 142:346–354

Chen TY (2005) Structure and function of ClC channels. Annu Rev Physiol 67:809–839

Chey WD, Drossman DA, Johanson JF, Scott C, Panas RM, Ueno R (2012) Safety and patient outcomes with lubiprostone for up to 52 weeks in patients with irritable bowel syndrome with constipation. Aliment Pharmacol Ther 35:587–599

Chu S, Blaisdell CJ, Liu MZ, Zeitlin PL (1999) Perinatal regulation of the ClC-2 chloride channel in lung is mediated by Sp1 and Sp3. Am J Physiol 276:L614–L624

Cid LP, Montrose-Rafizadeh C, Smith DI, Guggino WB, Cutting GR (1995) Cloning of a putative human voltage-gated chloride channel (CIC-2) cDNA widely expressed in human tissues. Hum Mol Genet 4:407–413

Cryer B, Katz S, Vallejo R, Popescu A, Ueno R (2014) A randomized study of lubiprostone for opioid-induced constipation in patients with chronic noncancer pain. Pain Med. doi:10.1111/pme.12437

Cuppoletti J, Tewari KP, Sherry AM, Kupert EY, Malinowska DH (2001) ClC-2 Cl^- channels in human lung epithelia: activation by arachidonic acid, amidation, and acid-activated omeprazole. Am J Physiol Cell Physiol 281:C46–C54

Cuppoletti J, Tewari KP, Sherry AM, Ferrante CJ, Malinowska DH (2004a) Sites of protein kinase A activation of the human ClC-2 Cl^- channel. J Biol Chem 279:21849–21856

Cuppoletti J, Malinowska DH, Tewari KP, Li QJ, Sherry AM, Patchen ML, Ueno R (2004b) SPI-0211 activates T84 cell chloride transport and recombinant human ClC-2 chloride currents. Am J Physiol Cell Physiol 287:C1173–C1183

Cuppoletti J, Malinowska DH, Tewari K, Chakrabarti J, Mende K, Ueno R (2008a) Lubiprostone activation of Cl^- currents does not involve Ca^{2+}, cAMP, or PKA. Gastroenterology 134(Suppl 1):A581–A582

Cuppoletti J, Malinowska DH, Tewari KP, Chakrabarti J, Ueno R (2008b) ClC-2-targeted siRNA eliminates lubiprostone activation of Cl^- currents in T84 cells. Gastroenterology 134(Suppl 1): A-582

Cuppoletti J, Malinowska DH, Ueno R (2008c) Prostaglandin receptor activation properties of lubiprostone. Am J Gastroenterol 103(Suppl 1):S95–S118

Cuppoletti J, Blikslager AT, Chakrabarti J, Nighot PK, Malinowska DH (2012) Contrasting effects of linaclotide and lubiprostone on restitution of epithelial cell barrier properties and cellular homeostasis after exposure to cell stressors. BMC Pharmacol 12:3

Cuppoletti J, Chakrabarti J, Tewari K, Malinowska DH (2013) Methadone but not morphine inhibits lubiprostone-stimulated Cl^- currents in T84 intestinal cells and recombinant human ClC-2, but not CFTR Cl^- currents. Cell Biochem Biophys 66:53–63

Cuppoletti J, Danuta Malinowska D, Chakrabarti J, Tewari K, Ueno R (2014a) Amino acids involved in lubiprostone activation of human ClC-2. FASEB J 28(Suppl):896.11

Cuppoletti J, Chakrabarti J, Tewari KP, Malinowska DH (2014b) Differentiation between human ClC-2 and CFTR channels with pharmacological agents. Am J Physiol Cell Physiol 307:C479-C492

Cuthbert AW (2011) Lubiprostone targets prostanoid EP_4 receptors in ovine airways. Br J Pharmacol 162:508–520

Denton J, Nehrke K, Yin X, Morrison R, Strange K (2005) GCK-3, a newly identified Ste20 kinase, binds to and regulates the activity of a cell cycle-dependent ClC anion channel. J Gen Physiol 125:113–125

Dhani SV, Kim Chiaw P, Huan LJ, Bear CE (2008) ATP depletion inhibits endocytosis of ClC-2. J Cell Physiol 214:273–280

Drossman DA, Chey WD, Johanson JF, Fass R, Scott C, Panas R, Ueno R (2009) Clinical trial: lubiprostone in patients with constipation-associated irritable bowel syndrome–results of two randomized, placebo-controlled studies. Aliment Pharmacol Ther 29:329–341

Duan DD (2013) Phenomics of cardiac chloride channels. Compr Physiol 3:667–692

Duran C, Thompson CH, Xiao Q, Hartzell HC (2010) Chloride channels: often enigmatic, rarely predictable. Annu Rev Physiol 72:95–121
Dutzler R (2004a) Structural basis for ion conduction and gating in ClC chloride channels. FEBS Lett 564:229–233
Dutzler R (2004b) The structural basis of ClC chloride channel function. Trends Neurosci 27:15–20
Dutzler R (2006) The ClC family of chloride channels and transporters. Curr Opin Struct Biol 16:439–446
Dutzler R (2007) A structural perspective on ClC channel and transporter function. FEBS Lett 581:2839–2844
Fahlke C (2004) Molecular physiology and pathophysiology of ClC-type chloride channels. Adv Mol Cell Physiol 32:189–217
Fei G, Wang YZ, Liu S, Hu HZ, Wang GD, Qu MH, Wang XY, Xia Y, Sun X, Bohn LM, Cooke HJ, Wood JD (2009) Stimulation of mucosal secretion by lubiprostone (SPI-0211) in guinea pig small intestine and colon. Am J Physiol Gastrointest Liver Physiol 296:G823–G832
Fritsch J, Edelman A (1996) Modulation of the hyperpolarization-activated Cl^- current in human intestinal T84 epithelial cells by phosphorylation. J Physiol 490:115–128
Furukawa T, Ogura T, Zheng YJ, Tsuchiya H, Nakaya H, Katayama Y, Inagaki N (2002) Phosphorylation and functional regulation of ClC-2 chloride channels expressed in Xenopus oocytes by M cyclin-dependent protein kinase. J Physiol 540:883–893
Gyömörey K, Yeger H, Ackerley C, Garami E, Bear CE (2000) Expression of the chloride channel ClC-2 in the murine small intestine epithelium. Am J Physiol Cell Physiol 279:C1787–C1794
Hanke W, Miller C (1983) Single chloride channels from Torpedo electroplax. Activation by protons. J Gen Physiol 82:25–45
Hinzpeter A, Fritsch J, Borot F, Trudel S, Vieu DL, Brouillard F, Baudouin-Legros M, Clain J, Edelman A, Ollero M (2007) Membrane cholesterol content modulates ClC-2 gating and sensitivity to oxidative stress. J Biol Chem 282:2423–2432
Holmes KW, Hales R, Chu S, Maxwell MJ, Mogayzel PJ Jr, Zeitlin PL (2003) Modulation of Sp1 and Sp3 in lung epithelial cells regulates ClC-2 chloride channel expression. Am J Respir Cell Mol Biol 29:499–505
Hoque KM, Woodward OM, van Rossum DB, Zachos NC, Chen L, Leung GP, Guggino WB, Guggino SE, Tse CM (2010) Epac1 mediates protein kinase A-independent mechanism of forskolin-activated intestinal chloride secretion. J Gen Physiol 135:43–58
Jentsch TJ, Poët M, Fuhrmann JC, Zdebik AA (2005) Physiological functions of CLC Cl^- channels gleaned from human genetic disease and mouse models. Annu Rev Physiol 67:779–807
Johanson JF, Morton D, Geenen J, Ueno R (2008a) Multicenter, 4-week, double-blind, randomized, placebo-controlled trial of lubiprostone, a locally-acting type-2 chloride channel activator, in patients with chronic constipation. Am J Gastroenterol 103:170–177
Johanson JF, Drossman DA, Panas R, Wahle A, Ueno R (2008b) Clinical trial: phase 2 study of lubiprostone for irritable bowel syndrome with constipation. Aliment Pharmacol Ther 27:685–696
Joo NS, Wine JJ, Cuthbert AW (2009) Lubiprostone stimulates secretion from tracheal submucosal glands of sheep, pigs, and humans. Am J Physiol Lung Cell Mol Physiol 296:L811–L824
Jordt SE, Jentsch TJ (1997) Molecular dissection of gating in the ClC-2 chloride channel. EMBO J 16:1582–1592
Joy AP, Cowley EA (2008) 8-iso-PGE_2 stimulates anion efflux from airway epithelial cells via the EP_4 prostanoid receptor. Am J Respir Cell Mol Biol 38:143–152
Kajita H, Omori K, Matsuda H (2000) The chloride channel ClC-2 contributes to the inwardly rectifying Cl^- conductance in cultured porcine choroid plexus epithelial cells. J Physiol 523:313–324
Keeler R, Wong NL (1986) Evidence that prostaglandin E_2 stimulates chloride secretion in cultured A6 renal epithelial cells. Am J Physiol 250:F511–F515

Kelly M, Trudel S, Brouillard F, Bouillaud F, Colas J, Nguyen-Khoa T, Ollero M, Edelman A, Fritsch J (2010) Cystic fibrosis transmembrane regulator inhibitors CFTRinh-172 and GlyH-101 target mitochondrial functions, independently of chloride channel inhibition. J Pharmacol Exp Ther 333:60–69

Lembo AJ, Johanson JF, Parkman HP, Rao SS, Miner PB Jr, Ueno R (2011) Long-term safety and effectiveness of lubiprostone, a chloride channel (ClC-2) activator, in patients with chronic idiopathic constipation. Dig Dis Sci 56:2639–2645

Ling BN, Zuckerman JB, Lin C, Harte BJ, McNulty KA, Smith PR, Gomez LM, Worrell RT, Eaton DC, Kleyman TR (1997) Expression of the cystic fibrosis phenotype in a renal amphibian epithelial cell line. J Biol Chem 272:594–600

Linsdell P (2014) Cystic fibrosis transmembrane conductance regulator chloride channel blockers: pharmacological, biophysical and physiological relevance. World J Biol Chem 5:26–39

Lipecka J, Bali M, Thomas A, Fanen P, Edelman A, Fritsch J (2002) Distribution of ClC-2 chloride channel in rat and human epithelial tissues. Am J Physiol Cell Physiol 282:C805–C816

Lochner A, Moolman JA (2006) The many faces of H89: a review. Cardiovasc Drug Rev 24:261–274

MacDonald KD, McKenzie KR, Henderson MJ, Hawkins CE, Vij N, Zeitlin PL (2008) Lubiprostone activates non-CFTR-dependent respiratory epithelial chloride secretion in cystic fibrosis mice. Am J Physiol Lung Cell Mol Physiol 295:L933–L940

MacVinish LJ, Cope G, Ropenga A, Cuthbert AW (2007) Chloride transporting capability of Calu-3 epithelia following persistent knockdown of the cystic fibrosis transmembrane conductance regulator, CFTR. Br J Pharmacol 150:1055–1065

Malinowska DH, Kupert EY, Bahinski A, Sherry AM, Cuppoletti J (1995) Cloning, functional expression, and characterization of a PKA-activated gastric Cl^- channel. Am J Physiol 268: C191–C200

Marunaka Y, Niisato N (2003) H89, an inhibitor of protein kinase A (PKA), stimulates Na^+ transport by translocating an epithelial Na^+ channel (ENaC) in fetal rat alveolar type II epithelium. Biochem Pharmacol 66:1083–1089

Meyer S, Savaresi S, Forster IC, Dutzler R (2007) Nucleotide recognition by the cytoplasmic domain of the human chloride transporter ClC-5. Nat Struct Mol Biol 14:60–67

Miller C (1982) Open-state substructure of single chloride channels from Torpedo electroplax. Philos Trans R Soc Lond B Biol Sci 299:401–411

Miller C (2006) ClC chloride channels viewed through a transporter lens. Nature 440:484–489

Miller C, White MM (1980) A voltage-dependent chloride conductance channel from Torpedo electroplax membrane. Ann NY Acad Sci 341:534–551

Miyazaki H, Yamada T, Parton A, Morrison R, Kim S, Beth AH, Strange K (2012) CLC anion channel regulatory phosphorylation and conserved signal transduction domains. Biophys J 103:1706–1718

Moeser AJ, Haskell MM, Shifflett DE, Little D, Schultz BD, Blikslager AT (2004) ClC-2 chloride secretion mediates prostaglandin-induced recovery of barrier function in ischemia-injured porcine ileum. Gastroenterology 127:802–815

Moeser AJ, Nighot PK, Engelke KJ, Ueno R, Blikslager AT (2007) Recovery of mucosal barrier function in ischemic porcine ileum and colon is stimulated by a novel agonist of the ClC-2 chloride channel, lubiprostone. Am J Physiol Gastrointest Liver Physiol 292:G647–G656

Mohammad-Panah R, Gyomorey K, Rommens J, Choudhury M, Li C, Wang Y, Bear CE (2001) ClC-2 contributes to native chloride secretion by a human intestinal cell line, Caco-2. J Biol Chem 276:8306–8313

Muanprasat C, Sonawane ND, Salinas D, Taddei A, Galietta LJ, Verkman AS (2004) Discovery of glycine hydrazide pore-occluding CFTR inhibitors: mechanism, structure-activity analysis, and in vivo efficacy. J Gen Physiol 124:125–137

Murray CB, Morales MM, Flotte TR, McGrath-Morrow SA, Guggino WB, Zeitlin PL (1995) CIC-2: a developmentally dependent chloride channel expressed in the fetal lung and downregulated after birth. Am J Respir Cell Mol Biol 12:597–604

Murray CB, Chu S, Zeitlin PL (1996) Gestational and tissue-specific regulation of C1C-2 chloride channel expression. Am J Physiol 271:L829–L837

Niemeyer MI, Cid LP, Zúñiga L, Catalán M, Sepúlveda FV (2003) A conserved pore-lining glutamate as a voltage- and chloride-dependent gate in the ClC-2 chloride channel. J Physiol 553:873–879

Niemeyer MI, Cid LP, Yusef YR, Briones R, Sepúlveda FV (2009) Voltage-dependent and -independent titration of specific residues accounts for complex gating of a ClC chloride channel by extracellular protons. J Physiol 587:1387–1400

Nighot PK, Blikslager AT (2010) ClC-2 regulates mucosal barrier function associated with structural changes to the villus and epithelial tight junction. Am J Physiol Gastrointest Liver Physiol 299:G449–G456

Nighot PK, Moeser AJ, Ryan KA, Ghashghaei T, Blikslager AT (2009) ClC-2 is required for rapid restoration of epithelial tight junctions in ischemic-injured murine jejunum. Exp Cell Res 315:110–118

Nighot MP, Blikslager A, Nighot PK (2011) The chloride channel ClC-2 is involved in organization of murine gastric glands. Gastroenterology 140(Suppl 1):S93

Nighot P, Young K, Nighot M, Rawat M, Sung EJ, Maharshak N, Plevy SE, Ma T, Blikslager A (2013) Chloride channel ClC-2 is a key factor in the development of DSS-induced murine colitis. Inflamm Bowel Dis 13:2867–2877

Nighot MP, Nighot PK, Ma TY, Blikslager A (2014) Genetic absence of chloride channel ClC-2 results in disruption of organization and function of murine gastric glands. Gastroenterology 146:S505

Niisato N, Ito Y, Marunaka Y (1999) Effects of PKA inhibitors, H-compounds, on epithelial Na^+ channels via PKA-independent mechanisms. Life Sci 65:109–114

Nobile M, Pusch M, Rapisarda C, Ferroni S (2000) Single-channel analysis of a ClC-2-like chloride conductance in cultured rat cortical astrocytes. FEBS Lett 479:10–14

Norimatsu Y, Moran AR, MacDonald KD (2012) Lubiprostone activates CFTR, but not ClC-2, via the prostaglandin receptor EP_4. Biochem Biophys Res Commun 426:374–379

O'Brien CE, Anderson PJ, Stowe CD (2011) Lubiprostone for constipation in adults with cystic fibrosis: a pilot study. Ann Pharmacother 45:1061–1066

O'Donnell EK, Sedlacek RL, Singh AK, Schultz BD (2000) Inhibition of enterotoxin-induced porcine colonic secretion by diarylsulfonylureas *in vitro*. Am J Physiol 279:G1104–G1112

Park K, Begenisich T, Melvin JE (2001) Protein kinase A activation phosphorylates the rat ClC-2 Cl^- channel but does not change activity. J Membr Biol 182:31–37

Park WS, Son YK, Kim N, Youm JB, Warda M, Ko JH, Ko EA, Kang SH, Kim E, Earm YE, Han J (2007) Direct modulation of Ca^{2+}-activated K^+ current by H-89 in rabbit coronary arterial smooth muscle cells. Vascul Pharmacol 46:105–113

Păunescu TG, Helman SI (2001) PGE_2 activation of apical membrane Cl^- channels in A6 epithelia: impedance analysis. Biophys J 81:852–866

Price MP, Ishihara H, Sheppard DN, Welsh MJ (1996) Function of Xenopus cystic fibrosis transmembrane conductance regulator (CFTR) Cl channels and use of human-Xenopus chimeras to investigate the pore properties of CFTR. J Biol Chem 271:25184–25191

Ranker J, Cuppoletti J, Melvin JE, Shull GE, Malinowska DH (2004) ClC-2 knockout mice exhibit reduced stimulated HCl secretion. FASEB J 18:456.2 (Abstract)

Rutledge E, Denton J, Strange K (2002) Cell cycle- and swelling-induced activation of a Caenorhabditis elegans ClC channel is mediated by CeGLC-7alpha/beta phosphatases. J Cell Biol 158:435–444

Schiffhauer ES, Vij N, Kovbasnjuk O, Kang PW, Walker D, Lee S, Zeitlin PL (2013) Dual activation of CFTR and CLCN2 by lubiprostone in murine nasal epithelia. Am J Physiol Lung Cell Mol Physiol 304:L324–L331

Schultz BD, Singh AK, Frizzell RA, Bridges RJ (1996) Developing potent modulators of CFTR channel gating. Pediatr Pulmonol 13(Suppl):258

Schultz BD, Singh AK, Devor DC, Bridges RJ (1999) Pharmacology of CFTR chloride channel activity. Physiol Rev 79:S109–S144
Schwiebert EM, Cid-Soto LP, Stafford D, Carter M, Blaisdell CJ, Zeitlin PL, Guggino WB, Cutting GR (1998) Analysis of ClC-2 channels as an alternative pathway for chloride conduction in cystic fibrosis airway cells. Proc Natl Acad Sci U S A 95:3879–3884
Sherry AM, Stroffekova K, Knapp LM, Kupert EY, Cuppoletti J, Malinowska DH (1997) Characterization of the human pH- and PKA-activated ClC-2G(2α) Cl^- channel. Am J Physiol 273:C384–C393
Son YK, Park WS, Kim SJ, Earm YE, Kim N, Youm JB, Warda M, Kim E, Han J (2006) Direct inhibition of a PKA inhibitor, H-89 on Kv channels in rabbit coronary arterial smooth muscle cells. Biochem Biophys Res Commun 341:931–937
Stauber T, Weinert S, Jentsch TJ (2012) Cell biology and physiology of CLC chloride channels and transporters. Compr Physiol 2:1701–1744
Stölting G, Teodorescu G, Begemann B, Schubert J, Nabbout R, Toliat MR, Sander T, Nürnberg P, Lerche H, Fahlke C (2013) Regulation of ClC-2 gating by intracellular ATP. Pflugers Arch 465:1423–1437
Stölting G, Fischer M, Fahlke C (2014) ClC-1 and ClC-2 form hetero-dimeric channels with novel protopore functions. Pflugers Arch 466(12):2191–2204
Stroffekova K, Kupert EY, Malinowska DH, Cuppoletti J (1998) Identification of the pH sensor and activation by chemical modification of the ClC-2G Cl^- channel. Am J Physiol 275:C1113–C1123
Strohmeier GR, Reppert SM, Lencer WI, Madara JL (1995) The A2b adenosine receptor mediates cAMP responses to adenosine receptor agonists in human intestinal epithelia. J Biol Chem 270:2387–2394
Sun Park W, Kyoung Son Y, Kim N, Boum Youm J, Joo H, Warda M, Ko JH, Earm YE, Han J (2006) The protein kinase A inhibitor, H-89, directly inhibits K_{ATP} and Kir channels in rabbit coronary arterial smooth muscle cells. Biochem Biophys Res Commun 340:1104–1110
Tewari KP, Malinowska DH, Sherry AM, Cuppoletti J (2000) PKA and arachidonic acid activation of human recombinant ClC-2 chloride channels. Am J Physiol Cell Physiol 279:C40–C50
Thiemann A, Gründer S, Pusch M, Jentsch TJ (1992) A chloride channel widely expressed in epithelial and non-epithelial cells. Nature 356:57–60
Thompson CH, Olivetti PR, Fuller MD, Freeman CS, McMaster D, French RJ, Pohl J, Kubanek J, McCarty NA (2009) Isolation and characterization of a high affinity peptide inhibitor of ClC-2 chloride channels. J Biol Chem 284:26051–26062
Ueno R, Osama H, Habe T, Engelke K, Patchen M (2004) Oral SPI-0211 increases intestinal fluid secretion without altering serum electrolyte levels. Gastroenterology 126:A276
Uwaifo O, Bamford P, Zeitlin PL, Blaisdell CJ (2006) Acidic pH hyperpolarizes nasal potential difference. Pediatr Pulmonol 41:151–157
Veilleux S, Holt N, Schultz BD, Dubreuil JD (2008) Escherichia coli EAST1 toxin toxicity of variants 17-2 and O 42. Comp Immunol Microbiol Infect Dis 31:567–578
Vij N, Zeitlin PL (2006) Regulation of the ClC-2 lung epithelial chloride channel by glycosylation of SP1. Am J Respir Cell Mol Biol 34:754–759
Weinreb M, Shamir D, Machwate M, Rodan GA, Harada S, Keila S (2006) Prostaglandin E_2 (PGE_2) increases the number of rat bone marrow osteogenic stromal cells (BMSC) via binding the EP_4 receptor, activating sphingosine kinase and inhibiting caspase activity. Prostaglandins Leukot Essent Fatty Acids 75:81–90
White MM, Miller C (1979) A voltage-gated anion channel from the electric organ of Torpedo californica. J Biol Chem 254:10161–10166
Wong BS, Camilleri M (2011) Lubiprostone for the treatment of opioid-induced bowel dysfunction. Expert Opin Pharmacother 12:983–990
Yamada T, Bhate MP, Strange K (2013) Regulatory phosphorylation induces extracellular conformational changes in a CLC anion channel. Biophys J 104:1893–1904

Zdebik AA, Cuffe JE, Bertog M, Korbmacher C, Jentsch TJ (2004) Additional disruption of the ClC-2 Cl^- channel does not exacerbate the cystic fibrosis phenotype of cystic fibrosis transmembrane conductance regulator mouse models. J Biol Chem 279:22276–22283

Zifarelli G, Pusch M (2007) CLC chloride channels and transporters: a biophysical and physiological perspective. Rev Physiol Biochem Pharmacol 158:23–76

Chapter 16
CFTR and Cystic Fibrosis

William T. Harris and Kevin L. Kirk

Abstract The cystic fibrosis transmembrane conductance regulator (CFTR) is an essential anion channel that regulates fluid and electrolyte transport across a variety of epithelia. Mutations that reduce CFTR protein expression or channel activity cause cystic fibrosis, the most common lethal recessive genetic disorder among Caucasians. CFTR occupies a prominent position among ion channels, as physiologic investigation of genotype/phenotype correlation has led to significant biomedical advancement with direct consequence on quality of life and survival. Here, we review our current understanding of how CFTR functions as an ATP-gated and phosphorylation-regulated anion channel that is the only identified ion channel in the large superfamily of ATP-binding cassette (ABC) transporters. The CFTR gating mechanism is also compared to the gating mechanisms of other more conventional ligand-gated channels with an interpretation of how prospective drugs may affect channel function. The various mechanisms by which different classes of CFTR mutations disrupt the function of this anion channel to cause cystic fibrosis are also discussed. Finally, we review the clinical manifestations of cystic fibrosis disease and current treatment options with an emphasis on the recent development of drugs that directly target the basic anion channel defect.

Keywords Cystic fibrosis • Ion channel • ATP-binding cassette transporter • Ligand • Allosteric • Phosphorylation • Airways • Mucus • CFTR modulation • Ivacaftor

W.T. Harris (✉)
Department of Pediatrics, University of Alabama at Birmingham, Birmingham, AL 35294, USA

Children's of Alabama, Suite ACC 620, Birmingham, AL 35233, USA
e-mail: tharris@peds.uab.edu

K.L. Kirk
Department of Cell, Developmental and Integrative Biology, University of Alabama at Birmingham, Birmingham, AL 35294, USA

K.L. Hamilton, D.C. Devor (eds.), *Ion Channels and Transporters of Epithelia in Health and Disease*, Physiology in Health and Disease,
DOI 10.1007/978-1-4939-3366-2_16

16.1 Introduction

The CFTR anion channel is an essential regulator of chloride, bicarbonate, and fluid transport across multiple epithelial tissues. Loss-of-function mutations in the CFTR polypeptide cause cystic fibrosis (CF), the most common lethal recessive genetic disorder among individuals of Northern European descent (Tsui 1991). CF is characterized by thick mucus in the airways and gut, poor mucociliary clearance by the airways, and consequent lung infection and inflammation (Davis 2006). CFTR hyperactivation in the intestine also causes disease; in particular, the secretory diarrheas that are induced by bacterial toxins such as cholera toxin that stimulate the overproduction of cyclic nucleotides and subsequent CFTR phosphorylation (Thiagarajah and Verkman 2003).

This chapter addresses the following topics: (1) our current view of how CFTR functions as an anion channel that is gated by ATP binding and regulated by phosphorylation, (2) the various mechanisms by which CF-causing mutations disrupt the expression or function of this channel, (3) the clinical manifestations of CF, and (4) the standard of care of CF patients including a discussion of newer therapies. The specific physiologic roles that CFTR channels play in epithelial chloride transport (Chap. 1), bicarbonate secretion (Chap. 5), pancreatic function (Chap. 28), and secretory diarrheas (Chap. 29) are discussed in other chapters. Other interesting cell biologic issues that relate to CFTR biosynthesis (Chap. 8) or the coupling between CFTR channels and other transporters such as the SLC transporters also are addressed in detail elsewhere in this volume (e.g., Chap. 14).

16.2 CFTR: An Anion Channel with a Unique Evolutionary History

16.2.1 CFTR as an ATP-Binding Cassette Transporter

CFTR has the distinction of being the only known ion channel in the very large superfamily of ATP-binding cassette (ABC) transporters. ABC transporters are found in all organisms where they typically function as active transport ATPases (Moitra and Dean 2011). Other biomedically important ABC transporters include P-glycoprotein (MDR1 or ABCB1; Amin 2013) and multidrug resistance proteins (MRPs), which are the closest homologues of CFTR (Keppler 2011). ABC transporters generally possess two nucleotide-binding domains (NBDs) that dimerize in response to binding two ATP molecules at the NBD dimer interface (Locher 2009). ATP-induced NBD dimerization promotes a conformational shift that is propagated to the transmembrane domains (TMs) that form the substrate translocation pathway. The crystal structures of several ABC exporters in the presence and absence of nucleotide have been solved (Dawson and Locher 2006, 2007; Ward et al. 2007; Li et al. 2014; e.g., see Fig. 16.1). Inspection of these various structures reveals two

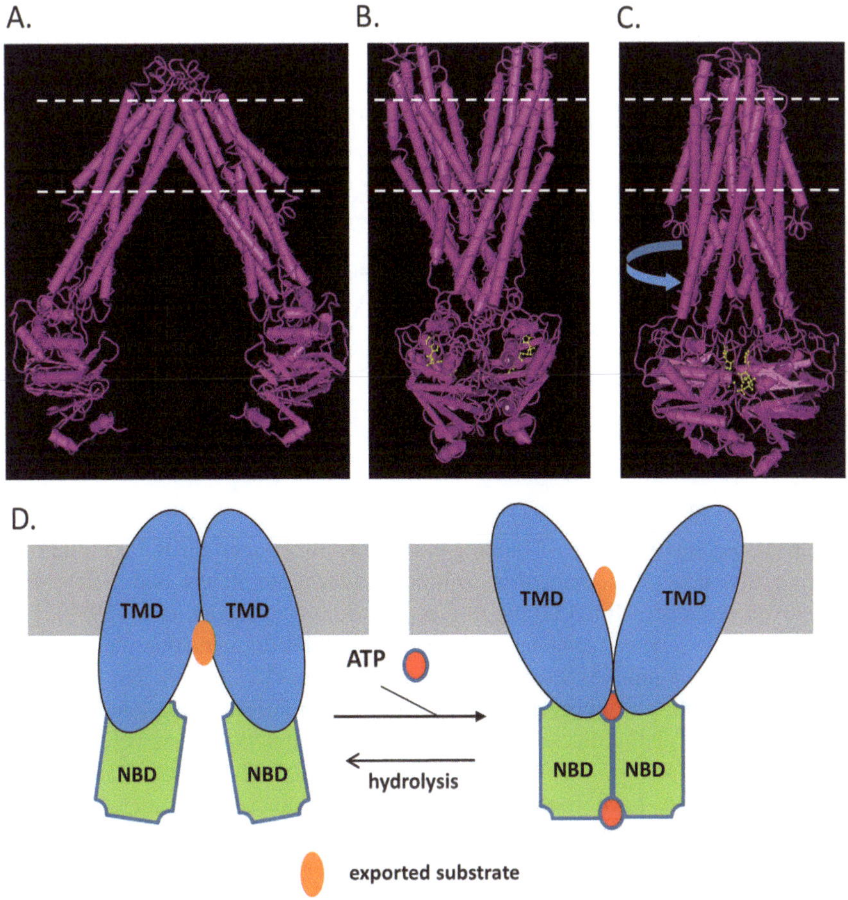

Fig. 16.1 Structures and functions of ABC exporters. (**a**) Inward-facing structure of the MsbA exporter from *E. coli* in the absence of nucleotide (Adapted from PDB ID: 3B5W and Ward et al. 2007). (**b**, **c**) Two views of outward-facing MsbA structure from *S. typhimurium* with bound AMP-PNP in yellow (Adapted from PDB ID 3B5Z and Ward et al. 2007). (**d**) Alternating access model of a generic ABC exporter

extreme conformations: (1) a nucleotide-free conformation in which the NBDs are separated and the TMs are oriented in an inward-facing arrangement and (2) a nucleotide-bound conformation in which the NBDs are tightly dimerized and the TMs are oriented in an outward-facing conformation. These conformational differences support an alternating access transport mechanism in which an ABC exporter flips between an inward-facing conformation with high substrate binding affinity and an outward-facing conformation with low substrate affinity in response to ATP binding and NBD dimerization (Ward et al. 2007; Locher 2009). The proposed role of subsequent ATP hydrolysis is to destabilize the NBD dimer and reset the exporter to the inward-facing conformation (see Fig. 16.1d).

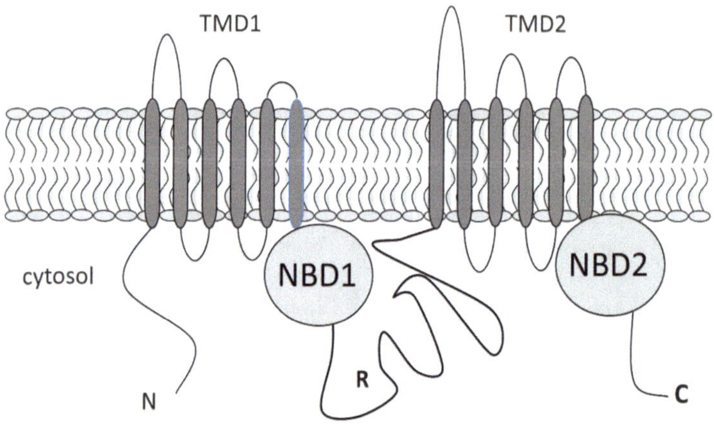

Fig. 16.2 Linear schematic of the major CFTR protein domains

CFTR belongs to the ATP-binding cassette (ABC) superfamily by virtue of it sharing this general domain organization and because it has two NBDs with conserved sequences for ATP binding and hydrolysis (Riordan et al. 1989). Figure 16.2 shows a linear schematic of the major CFTR domains that include these two NBDs plus a unique, centrally located regulatory (R) domain. The R domain contains multiple phosphorylation sites for cyclic nucleotide-dependent kinases, and its phosphorylation is essential for CFTR channel activity (see below). High-resolution structures of the intact CFTR polypeptide are not yet available, but homology models have been produced using the crystal structures of several bacterial ABC exporters as guides (Mornon et al. 2009; Norimatsu et al. 2012; e.g., Fig. 16.3). As noted above, CFTR has the thermodynamic distinction of functioning as an anion channel through which ion transport is driven solely by the prevailing electrochemical gradients of the permeant anions (primarily Cl^- and HCO_3^-). Like other ABC transporters, CFTR's two NBDs apparently dimerize in response to ATP binding based on the results of cysteine cross-linking studies (Mense et al. 2006). CFTR also possesses an ATPase activity, but this activity is used to control ligand (ATP) occupancy rather than to energize substrate translocation (more on this below). Another distinction is that the CFTR translocation pathway or pore must remain open to both sides of the membrane for a period of time (typically 0.1–1 s). This is unlike an ABC exporter or importer that presumably flips between two conformations each of which is accessible to the substrate from one side only. This difference between the CFTR channel and an ABC pump is conceptualized as the degraded gate hypothesis, which posits that CFTR lacks one of the gates that for a typical ABC transporter would prevent substrate entry into the translocation pathway from one side of the membrane (Gadsby 2009).

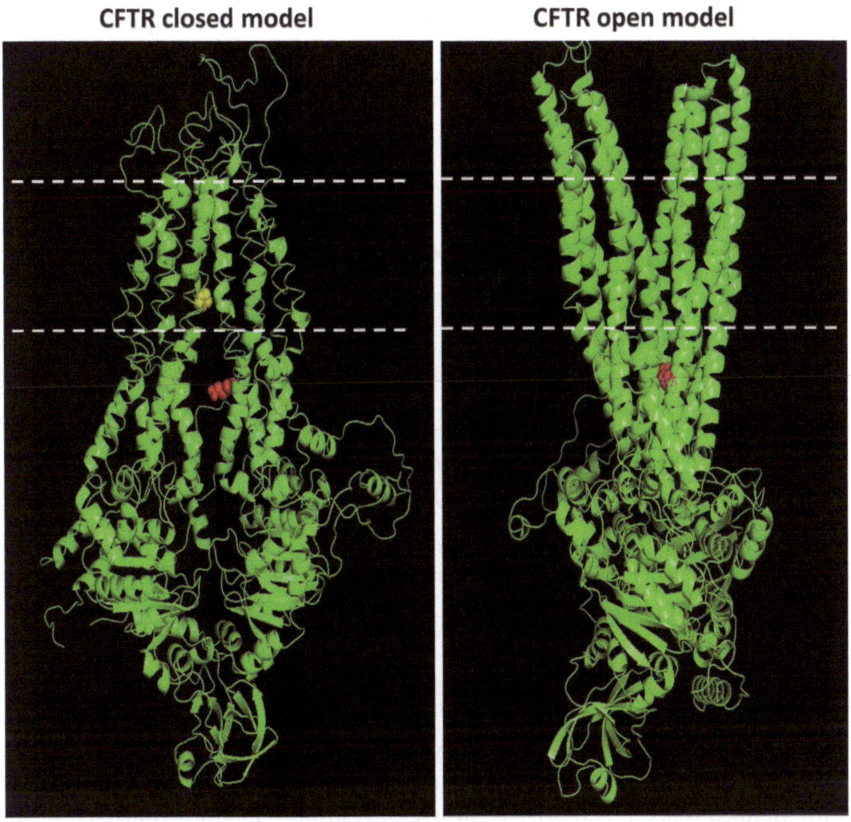

Fig. 16.3 Homology models of the CFTR closed and open states adapted from Mornon et al. (2009). Closed model is based on the nucleotide-free MsbA structure from *V. cholerae* (PDB ID: 3B5X and Ward et al. 2007). The open model is based on the nucleotide-bound structure of the Sav1866 exporter from *S. aureus* (PDB ID: 2ONJ and Dawson and Locher 2007). Two residues that are known to be sites for producing CFTR gain-of-function mutations are shown (K978 in *red* (Wang et al. 2010) and P355 in *yellow*) (Adapted from Wei et al. 2014)

16.2.2 How CFTR Channels Open and Close in Response to ATP Binding and Hydrolysis

A commonly held view of the CFTR gating mechanism is that channel opening and closing are strictly coupled to ATP binding and subsequent hydrolysis (e.g., Csanády et al. 2010 and Fig. 16.4). This view is a good first approximation for highly phosphorylated channels with ATP as the ligand (see below for discussion of the role of phosphorylation in channel activation). But, as will be discussed below, CFTR channels can open and close without ATP binding or NBD dimerization, and gain-of-function (GOF) mutations that increase ATP-free channel activity have been discovered. These observations are better explained by an allosteric activation

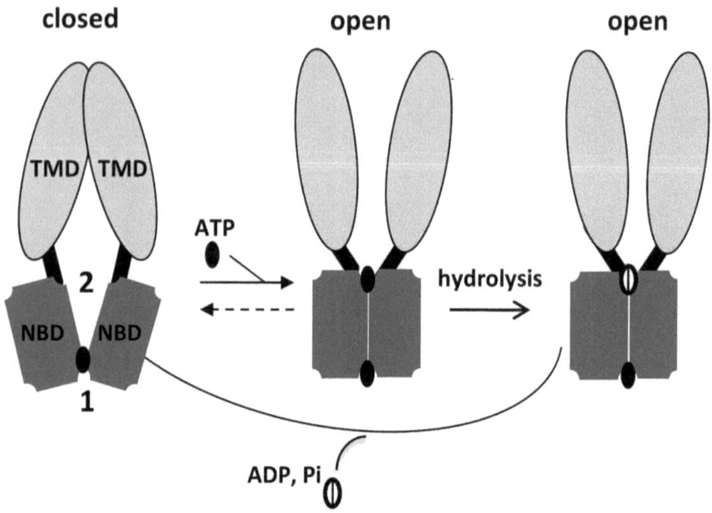

Fig. 16.4 Strict coupling model of CFTR gating that assumes that all channel openings follow ATP binding to both sites and that channel closings follow ATP hydrolysis at site 2 (Adapted from Kirk and Wang 2011)

model like those that explain the gating of conventional ligand-gated channels, as will be discussed below.

The strict coupling model is based on the findings that for highly phosphorylated CFTR channels, nearly all openings follow ATP binding to both nucleotide-binding sites (termed sites 1 and 2) and nearly all closings follow hydrolysis at site 2 (Csanády et al. 2010). Site 1 binds ATP but has negligible catalytic activity due to its lacking residues that are known to be important for ATP hydrolysis in other ABC transporters (e.g., a glutamate in the Walker B sequence of NBD1 that is essential for ATP hydrolysis in other ABC transporters; Orelle et al. 2003; Aleksandrov et al. 2002; Basso et al. 2003). The NBD dimer is so tight following ATP binding that the nonhydrolytic unbinding of ATP is very slow (ATP occupancy >5 s without hydrolysis; Csanády et al. 2010). ATP hydrolysis at site 2 speeds ligand dissociation from this site and destabilizes the dimer by converting ATP into its hydrolysis products (ADP and Pi) which have a much lower affinity for the channel (in part, due to the electrostatic repulsion between ADP and Pi). Mutations that inhibit ATP hydrolysis at site 2 greatly lengthen channel openings (more correctly, open channel bursts) as would be expected for such a gating mechanism (Gunderson and Kopito 1995; Vergani et al. 2003). Channels are nearly always open during these long bursts although brief closings can be detected at high sampling frequencies (Vergani et al. 2003). This means that ATP has a very high efficacy to gate open phosphorylated CFTR channels, with efficacy defined as the open-closed equilibrium constant for ligand-bound channels (# of ATP-bound open channels/# of ATP-bound closed channels; Colquhoun 1998).

Recent evidence supports an allosteric activation mechanism for CFTR that shares features with the gating mechanisms of conventional ligand-gated channels

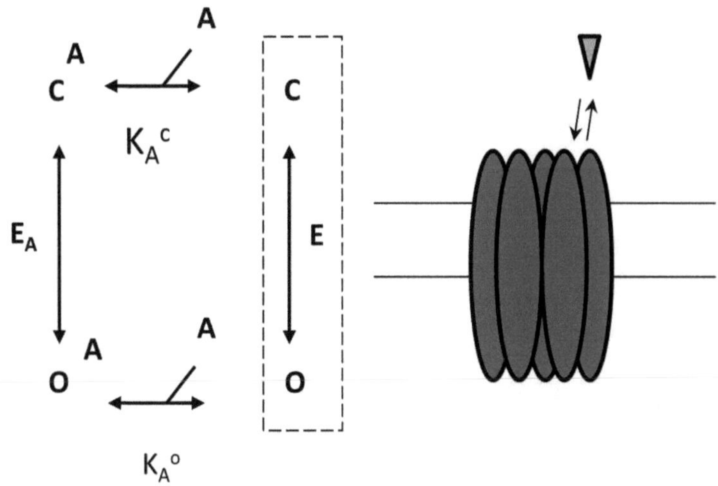

Fig. 16.5 Cyclic allosteric gating scheme for a generic ligand-gated channel with one ligand- or agonist (A)-binding site. C and O represent closed and open channels, respectively. E and E_A represent the open-closed equilibrium constants for unbound and agonist-bound channels, respectively. $K_A{}^c$ and $K_A{}^o$ represent the agonist equilibrium binding constants for closed and open channels, respectively, where $K_A{}^c > K_A{}^o$. The unliganded closed-open transition, or ligand-free isomerization, is indicated within the *dotted box* (Adapted from Kirk and Wang 2011)

that reversibly bind rather than hydrolyze their ligands (e.g., the neurotransmitter-gated channels). The gating of a conventional ligand-gated channel can be described by a classical allosteric activation scheme of the type originally proposed by Changeux and colleagues to describe cooperative oxygen binding to hemoglobin and later used to describe the activation of hormone receptors and neurotransmitter-gated channels by the binding of an activating ligand (agonist; see Fig. 16.5) (Monod et al. 1965; Changeux and Edelstein 2005; Auerbach 2012). Such an allosteric scheme predicts that channels can open in the absence of ligand (usually infrequently) and close when ligand is bound (a function of ligand efficacy; see above) unlike a strict coupling mechanism. Ligand binds to both closed and open channels but with a higher affinity to the open conformation. For such a cyclic activation scheme, the affinities of the closed and open conformations for agonist differ in proportion to the open-closed equilibrium constants of agonist-free (E) and agonist-bound channels (E_A, where $E_A > E$) (Changeux and Edelstein 2005; Auerbach 2012). Because channels can open in the absence of agonist, GOF mutations that increase ligand-free activity have been discovered for many hormone receptors and ligand-gated channels (Chang and Weiss 1999; Changeux and Edelstein 2005; Auerbach 2012). These are termed constitutive or isomerization mutations because they increase the open-closed equilibrium constant (E), or channel isomerization, in the absence of ligand binding. An important consequence of an isomerization or constitutive mutation is that such a mutation also increases the sensitivity to agonist because it biases the equilibrium to the conformation with the higher agonist affinity (open state; Chang and Weiss 1999; Changeux and

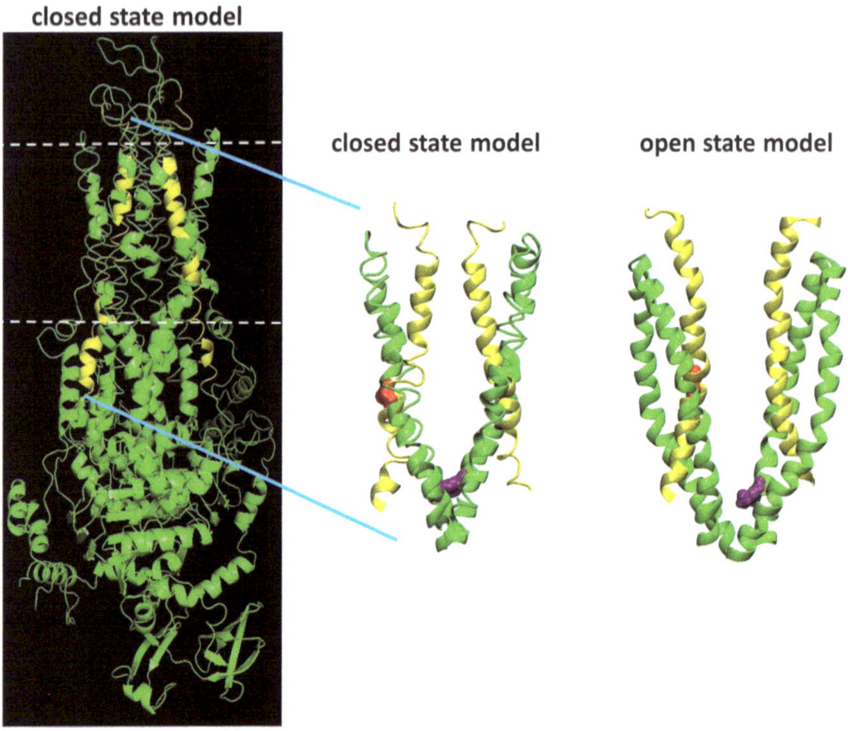

Fig. 16.6 Cutout views of the major pore-lining transmembrane domains (TMs 6 and 12 in *yellow*) and the putative outer TM collar (TMs 3/4 and 9/10 in *green*) from the CFTR closed and open homology models shown in Fig. 18.3. The sites of two classes of GOF mutations are indicated in space fill as K978 in *purple* and P355 in *red* (Adapted from Wei et al. 2014)

Edelstein 2005; Auerbach 2012). This is the concept of reciprocity that is a feature of all allosteric activation schemes.

CFTR channel gating by ATP exhibits many of the same properties. Wild-type channels open infrequently but detectably in the absence of ATP (Hennager et al. 2001; Wang et al. 2010; Szollosi et al. 2010). Several groups have reported GOF mutations in the CFTR polypeptide that promote channel activity in the absence of ATP or NBD dimerization. These mutations map to multiple regions of the CFTR protein including the NBDs (F1296S and N1303Q; Szollosi et al. 2010), the cytosolic loops that couple the NBDs to the TMs (K190C in loop 1 and several K978 mutations in loop 3; Wang et al. 2010), and the TMs themselves (e.g., several P355 mutations at the base of TM6; Wei et al. 2014). The majority of these GOF mutations locate at or near the putative symmetry axis that extends from the center of the NBD dimer to the presumed pore in CFTR homology models (see Fig. 16.6). Most GOF mutations in conventional ligand-gated channels similarly lie along the axis between the ligand-binding domains and the pore (Chang and Weiss 1999; Changeux and Edelstein 2005; Cadugan and Auerbach 2010; Auerbach 2012). Several CFTR GOF mutations increase activities of channel constructs that are unresponsive to ATP either because they have CF-causing mutations in the NBDs (e.g., G551D-CFTR, the most common CF regulation mutant) or because they cannot dimerize the NBDs (e.g., Δ1198-CFTR, a truncation mutant lacking NBD2) (Wang et al. 2010; Wei et al. 2014). Two classes of CFTR GOF mutation (P355 mutations in TM6 and K978 mutations in loop 3) also have been shown to increase the ATP sensitivity of channel activation as predicted for allosteric or reciprocal coupling between the ATP occupancies of the NBDs and the open channel conformation (Wang et al. 2010; Wei et al. 2014).

The discovery of CFTR GOF mutations that promote ATP-independent channel activity and reciprocally enhance the ATP sensitivity of channel gating has two implications. The first (and obvious) implication is that CFTR channel gating is not strictly coupled to ATP occupancy. Instead, an allosteric activation scheme of the type proposed for conventional ligand-gated channels better describes the data. The opening and closing of phosphorylated CFTR channels appear to be tightly coupled to ATP binding and hydrolysis because ATP has a very high efficacy to open phosphorylated channels. This high efficacy of ATP would not make CFTR especially different from conventional ligand-gated channels. In this regard, the efficacy of acetylcholine to open the mouse nicotinic acetylcholine receptor (nAchR) at saturating ligand concentrations has been reported to be about 20 (Salamone et al. 1999), which converts to an intraburst P_o of about 0.95 (meaning that ligand-saturated nAchR channels are open 95 % of the time). What makes CFTR unique is its use of a catalytic activity to speed the removal of its ligand ATP from one of the binding sites. ATP binds too tightly otherwise to permit reversible (nonhydrolytic) unbinding on a physiologically relevant (sub-second) time scale.

The second implication is that the locations and physicochemical properties of different classes of GOF mutations reveal distinct structural elements that function together to control the energy barrier to channel opening. For example, GOF mutations that map to loop 3 at the base of TM9 (i.e., K978 mutations) may distort a collar of TMs that surrounds the major pore-lining TMs (TMs 6 and 12) and that normally restricts the conformational movements of these TMs in the absence of ATP binding and NBD dimerization (Wei et al. 2014 and Fig. 16.6). These outer TM collar mutations have additive effects when combined with GOF mutations of a conserved TM6 proline (P355) that appears to function as a switch to control TM6 movements (e.g., rotation) during channel opening (Wei et al. 2014). Other GOF mutations, as they are discovered, likely will reveal additional structural elements that contribute to the CFTR gating process.

16.2.3 The CFTR Pore

TMs 6 and 12 have been argued to be the principal pore-lining TMs as noted above. This argument is supported by the results of a number of cysteine scanning/ accessibility experiments and pore blocker studies (McDonough et al. 1994; Linsdell et al. 2000; Bai et al. 2010, 2011; Norimatsu et al. 2012). Figure 16.6

shows a view of these pore-lining TMs (in yellow) surrounded by the outer TM collar noted above (in green) adapted from one of several published CFTR homology models. Portions of other TMs likely also contribute to forming the CFTR pore (e.g., TMs 1 and 9; Wang et al. 2011; Norimatsu et al. 2012), but TMs 6 and 12 probably line nearly its entire length. The CFTR pore has a large cytoplasmic vestibule that permits the entry of large organic anions that block the channel in a voltage-dependent fashion such as the sulfonylureas, like glibenclamide (e.g., see Cui et al. 2012). This vestibule is lined by multiple positively charged residues that increase anion throughput by attracting anions and increasing their local concentrations in the vicinity of the cytosolic entrance to the pore (Aubin and Linsdell 2006).

Anion selectivity studies of the CFTR pore have revealed that, like most other anion channels, CFTR does not select between anions with exceptionally high fidelity. Many anions can permeate the CFTR pore with chloride and bicarbonate being two of the most physiologically relevant. CFTR selects between anions almost entirely on the basis of their hydration/dehydration energies (termed a lyotropic sequence; e.g., Linsdell et al. 1997). Dawson and colleagues have modeled the CFTR pore as a dielectric tunnel that selects between anions solely on the basis of their different interaction energies with water and the lower dielectric pore wall (Smith et al. 1999; Liu et al. 2003). This heuristic model explains much of the CFTR permeability data without the need to invoke a specific selectivity filter. On the other hand, there is good functional evidence for a narrow region of the pore near the extracellular surface that is formed by opposing residues from TMs 6 and 12 (see Fig. 16.6 and Linsdell et al. 2000; Cui et al. 2012). This narrowing of the pore probably influences the permselectivity of the CFTR channel and likely contributes to the binding sites for extracellular and intracellular pore blockers. Regarding the latter point, Norimatsu et al. (2012) used a combination of molecular modeling, molecular dynamics simulations, and cysteine scanning results to predict how this narrow region binds GlyH101, a prospective antidiarrheal drug that was identified by the Verkman group as a potent and selective extracellular CFTR blocker (Muanprasat et al. 2004). In summary, CFTR lacks a high-fidelity selectivity filter like those used by certain cation channels to discriminate between the equally abundant sodium and potassium ions. Instead, it selects against cations in part by possessing intracellular and extracellular vestibules that are lined by positively charged residues and selects between anions primarily on the basis of their different hydration/dehydration energies.

16.2.4 CFTR Regulation by Phosphorylation of Its Unique R Domain

Unphosphorylated CFTR channels open rarely. The frequency of channel openings increases in response to CFTR phosphorylation by the cyclic nucleotide-dependent protein kinases (PKA and PKG) (Cheng et al. 1991; French et al. 1995; Mathews et al. 1998). A number of other kinases also regulate CFTR traffic and function, but we will limit our discussion to the regulation of CFTR gating by PKA, which is best understood. Another reason for focusing our discussion on PKA or PKG is that the phosphorylation of CFTR by these kinases is the main physiologic stimulus of CFTR-dependent salt and fluid secretion in secretory glands and intestinal crypts. This is the mechanism by which bacterial toxins that stimulate the intestinal overproduction of cyclic AMP (cholera) or cyclic GMP (heat-stable enterotoxin) cause secretory diarrhea (Thiagarajah and Verkman 2003). The CFTR polypeptide has at least ten different consensus sites for PKA or PKG phosphorylation (Hegedüs et al. 2009). Most of these sites reside within the unique R domain, an approximately 150 residue domain that links NBD1 to TM7 in the middle of the polypeptide (Fig. 16.2). The unphosphorylated R domain limits CFTR activity, but how this occurs, and how PKA phosphorylation relieves this inhibition, is unclear. Below, we describe two plausible mechanisms that are not mutually exclusive.

The first possible mechanism is the regulation of NBD dimerization by R domain phosphorylation. The central location of the R domain between NBD1 and TM7 makes this an attractive idea. Two lines of evidence support this mechanism. First, Mense et al. (2006) showed that the efficiency of cross-linking cysteine residues that were introduced across the presumed NBD dimer interface into co-expressed halves of the CFTR protein was enhanced substantially by PKA phosphorylation. Second, Baker et al. (2007) showed in NMR studies of recombinant NBD1 and an R domain fragment (residues 654–838) that the unphosphorylated R domain bound to regions of NBD1 that would be predicted to be important for interacting with NBD2. Phosphorylating the R domain fragment with PKA inhibited its binding to recombinant NBD1. Together, these results support a simple model in which NBD dimerization is prevented, or at least inhibited, by the unphosphorylated R domain, an inhibitory effect that is relieved by PKA phosphorylation. This model is consistent with the well-accepted role of ATP-induced NBD dimerization in promoting CFTR channel opening.

The second possible mechanism is that the R domain regulates the packing or flexibility of the cytosolic loops and TMs. This mechanism is consistent with the finding that the activity of a CFTR truncation construct that lacks NBD2 but possesses a GOF mutation in loop 3 near the base of TM9 (K978C) is strongly dependent on PKA phosphorylation of its R domain (Wang et al. 2010). This finding clearly argues for an NBD dimerization-independent mechanism. Hegedüs et al. (2008) proposed that the R domain functions as an entropic spring that regulates TM packing on the basis of the results of a computational analysis of the structures of the unphosphorylated and phosphorylated R domain. A related possibility is that portions of the unphosphorylated R domain physically interact with the cytosolic loops and/or TMs to restrict the movements that they must undergo to open the channel. Such frictional or occlusive interactions presumably would be reduced by phosphorylation of this domain.

These two plausible mechanisms are not mutually exclusive, as was noted above. The R domain is a large domain with multiple PKA sites that likely interacts with several other regions of the CFTR polypeptide including the NBDs, cytosolic loops, and TMs. Consistent with this point of view, Forman-Kay and colleagues have proposed that the bulk of the R domain is an intrinsically disordered structure that engages in dynamic and multiple intramolecular and intermolecular interactions (Bozoky et al. 2013). Which of these various interactions dominate presumably would be determined by which and/or how many sites are phosphorylated.

It should be noted that CFTR has several other PKA sites outside the R domain that appear to be functionally significant as well. Two of these map to unique sequences in NBD1 termed the regulatory insertion (RI) and regulatory extension (RE). In NMR studies of recombinant NBD1 and a peptide mimic of cytosolic loop 1, Kanelis et al. (2010) reported a binding interaction between these CFTR fragments that was enhanced by PKA phosphorylation. Thus, PKA phosphorylation may also strengthen the coupling between NBD1 and the translocation pathway (loops/TMs) as an additional mechanism to increase channel activity.

16.3 How CFTR Channel Function Is Disrupted by the Common CF-Causing Mutations: Prospects for Drugs That Directly Target These Channel Defects

Cystic fibrosis (CF) is a recessive genetic disorder; i.e., affected individuals have two defective copies of the *CFTR* gene. The cystic fibrosis mutation database lists more than 1800 variations in the gene (CFTR1; www.genet.sickkids.on.ca/cftr/app). Although most of these variations have not been determined to be disease causing, at least 80 *CFTR* gene mutations have been confirmed to result in CF (www.cftr2.org). CF-causing mutations are found throughout the *CFTR* gene and coding region and can be classified by how they primarily reduce CFTR channel expression or function, as will be discussed in more detail later in this chapter.

16.3.1 F508del

Because one mutation (F508del) is by far the most abundant in the CF patient population, it is instructive at this point to discuss how this particular mutation impacts CFTR function. Unfortunately, this single amino acid deletion in NBD1 has multiple effects on CFTR biosynthesis and channel gating. The first effect is to cause misfolding of the nascent CFTR protein in the ER and its consequent ER-associated degradation (ERAD) (Cheng et al. 1990; Younger et al. 2006). Few F508del-CFTR polypeptides exit the ER and subsequently make it to the cell surface under physiologic conditions. Those mutant proteins that do make it to the cell surface likely have low channel activity and reduced residence times at the cell surface as well. In this regard, F508del-CFTR is a temperature-sensitive processing mutant for which greater numbers of channels can exit the ER and reach the surface of cultured cells if those cells are incubated at reduced temperature (26–28 °C; Denning et al. 1992). Detailed analyses of the channel activities and cell surface retention of the "low-temperature rescued" F508del-CFTR polypeptide have revealed much lower single-channel open probabilities (P_os) of this mutant (Wang et al. 2005a; see also Dalemans et al. 1991) and far shorter residence times at the plasma membrane as compared to wild-type CFTR (Lukacs et al. 1993). Thus, the F508del mutation probably disrupts multiple aspects of CFTR synthesis and function.

Perhaps not surprisingly, the multiple effects of the F508del mutation on CFTR expression and function are mirrored by this mutation having multiple biochemical effects on CFTR domain folding and domain interactions. In CFTR homology models, F508 maps to an interface between NBD1 and the cytosolic loops that link the NBDs to the TMs and translocation pathway. The F508del mutation reportedly disrupts the coupling between NBD1 and TMs, in particular, the interface between NBD1 and the fourth cytosolic loop between TMs 10 and 11 (Serohijos et al. 2008). This mutation also disrupts the folding of NBD1 as deduced from the reduced solubility of recombinant NBD1 constructs with the F508del mutation (Qu et al. 1997). NBD2 folding also appears to be affected by this mutation by a mechanism that is perhaps less intuitively obvious but probably also functionally significant (Du et al. 2005). The effect of the F508del mutation in NBD1 on NBD2 folding supports the sort of cooperative domain folding model that recently has been proposed to explain the folding of complex polytopic membrane proteins such as CFTR (e.g., see Chap. 8).

From the standpoint of F508del-CFTR pharmacology, the obvious hope is that a single drug that corrects all of the biochemical and functional defects in CFTR can be found. In lieu of discovering such a drug, combination therapies will have to be considered and may offer promise in the near term. Lukacs' group has characterized three different drug classes each with a different biochemical effect on the F508del-CFTR polypeptide (improved NBD1 or NBD2 solubility or stabilization of the NBD1-TM interfaces). By combining first-generation "drugs" from all three classes, the authors observed robust rescue of the expression, surface localization, and channel activity of the F508del mutant to near wild-type levels in cell culture and ex vivo models (Okiyoneda et al. 2013). Their findings support the view that combination therapies should be considered for CF patients with the F508del mutation as safe (FDA-approved) drugs within these different classes become available.

16.3.2 G551D and Vertex-770

The most exciting recent development in CF therapy is the discovery of a drug that benefits CF patients with the most common regulation mutation, G551D (ca, 5 % of all CF patients; www.cftr2.org). This mutation in NBD1 disrupts a conserved sequence that lines the ATP-binding sites in ABC transporters termed the ABC signature sequence (Locher 2009). G551D-CFTR channels have almost no channel activity and are completely unresponsive to ATP (Bompadre et al. 2007). But this mutant is synthesized, processed, and trafficked to the cell surface much like wild-type CFTR. Van Goor et al. (2009) discovered a potentiator of G551D-CFTR channel activity in a high-throughput screen of a compound library searching for molecules that increase its activity in cultured cells. This approach led to the identification of an orally bioavailable drug, vertex (VX)-770 or ivacaftor, that has been shown to be safe and effective both for adult and juvenile G551D patients (Ramsey et al. 2011; Davies et al. 2013).

VX-770 also increases the channel activity of the F508del mutant after its surface expression has been "temperature rescued" (Van Goor et al. 2009). These findings have motivated multiple clinical trials of VX-770 in combination with a compound that partially rescues F508del-CFTR processing and cell surface expression in F508del patients (Van Goor et al. 2011; clinicaltrials.gov). Although appealing in theory, the results of two recent in vitro studies indicate that chronic VX-770 treatment leads to decreased F508del-CFTR protein amount due to increased protein turnover which may reduce the efficacies of compounds that correct the F508del folding defect (Cholon et al. 2014; Veit et al. 2014). The mechanism of action for VX-770 is unclear, but it probably directly binds to the CFTR polypeptide given that it activates CFTR constructs that have been purified and reconstituted in lipid bilayers (Eckford et al. 2012). Because this compound stimulates CFTR channel activity in the absence of ATP, VX-770 may bypass rather than rescue the primary G551D defect, i.e., the inability to be activated by ATP (Jih and Hwang 2013; Eckford et al. 2012). Returning to our earlier discussion of an allosteric gating mechanism for CFTR, VX-770 can be classified as an allosteric activator that increases ligand-free channel activity probably by a mechanism that is similar to the effect of a constitutive or isomerization mutation to promote ATP-independent channel activity. Identifying the binding sites for this and other allosteric modulators of CFTR channel gating (Wang et al. 2005a) would facilitate the rational design of next-generation CF drugs.

16.4 Clinical Manifestations of CFTR Dysfunction

Despite its monogenic etiology, phenotypic manifestations of CFTR dysfunction are diffuse, with distinct consequences in separate organs. Therapeutic strategies that address downstream consequences of CFTR defects have significantly increased survival expectations among individuals with cystic fibrosis (Fig. 16.7). In 2012, median life expectancy for CF patients was estimated to be 41.1 years (Cystic Fibrosis Foundation 2013a). Improvements in nutritional care, aggressive antimicrobial therapy including chronic use of inhaled therapies, advancements in mucolytic therapy, and a commitment to quality improvement have all contributed to this progress in clinical outcome.

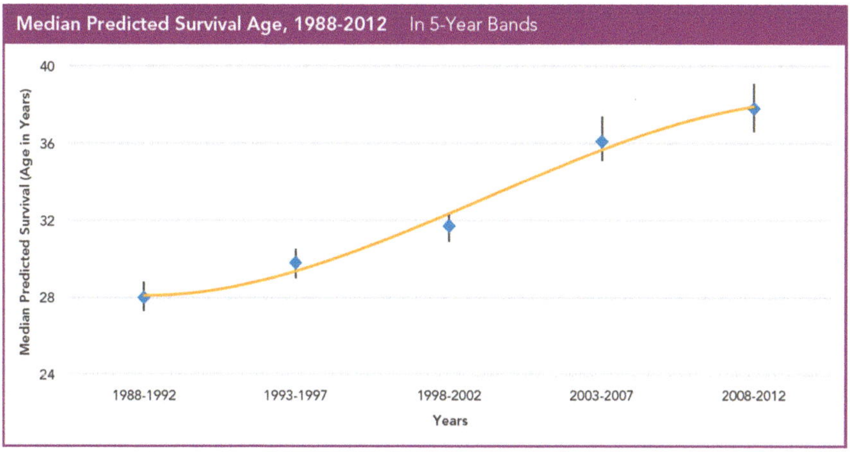

Fig. 16.7 Increased life expectancy in CF patients over the last 20 years. Further improvements are expected with the development of Newborn screening and CFTR modulator therapy (*Source*: CF 2012 Registry Data; with permission from the Cystic Fibrosis Foundation, www.CFF.org)

"Cystic fibrosis of the pancreas" was initially described as a nutritional, malabsorptive disease marked by fibrocystic changes of the pancreas (Andersen 1938). Later, a summer heat wave in New York City exposed defects in sweat composition and susceptibility to dehydration (Kessler and Andersen 1951), prompting the development of sweat chloride analysis (Di Sant'Agnese et al. 1953) that continues to be the pathognomic test for the disease (Farrell et al. 2008). As CFTR regulates chloride resorption from sweat, analyzing sweat electrolytes is the most clinically available measure of CFTR function. While nutritional limitations and gastrointestinal problems continue to plague CF individuals, respiratory disease is now the major cause of mortality. Questions that continue to plague CF clinicians, such as defining the primary defect that generates such a harsh disease and overcoming limitations of available antimicrobials, are as pertinent now as in the early descriptions of CF disease (Andersen 1949).

16.4.1 Gastrointestinal Involvement

Approximately 90 % of CF patients will have pancreatic insufficiency (malabsorptive symptoms secondary to loss of exocrine pancreatic function). Typically, this is considered a consequence of blocked pancreatic enzymatic secretions resulting in self-degradation of the pancreatic tissue. The recent development of a porcine model of cystic fibrosis successfully models pancreatic disease (Meyerholz et al. 2010b), with genomic analysis indicating increased inflammatory and remodeling pathways contributing to destruction (Abu-El-Haija et al. 2012). The loss of pancreatic function is manifest by fat malabsorption presenting as frequent, large malodorous stools. In utero, loss of pancreatic function may lead to meconium plugging of the gastrointestinal tract. This obstruction occurs most frequently at the terminal ileum. This meconium ileus occurs in about 15 % of CF gestations (Kerem et al. 1989; van der Doef et al. 2011). Intestinal obstruction frequently continues to plague children and adults with CF, particularly at the ileocecal valve, ranging from constipation to complete obstruction (Distal Intestinal Obstruction Syndrome, DIOS). Loss of pancreatic secretions also alters intestinal pH (Uc et al. 2012), further contributing to malabsorption.

Liver dysfunction is also present in cystic fibrosis but more difficult to quantify. Inspissated biliary secretions may lead to focal biliary cirrhosis. The most common manifestation of CF-associated liver disease manifests as increased transaminases (increased AST/ALT) with hepatic ultrasound often showing hepatic steatosis. Less frequently (approximately 5 %), frank hepatic failure will develop with portal hypertension, esophageal varices, and coagulopathy that may ultimately require liver transplantation. The onset of severe CF liver disease typically occurs during adolescence and rarely develops de novo in adulthood. Genetic modifier studies indicate that CF patients with the Z allele for alpha-1 antitrypsin (SERPINA1) have a nearly fivefold greater risk of developing more severe CF liver disease (Bartlett et al. 2009).

16.4.2 Pulmonary Disease

Respiratory disease is the major cause of mortality in CF. Previously, clinicians described CF lungs to be developmentally normal at birth and then have respiratory decline as a consequence of airway infection and inflammation. Recent careful analysis of both the CF pig (and mouse) indicates abnormal tracheal development that may lead to alterations in physiology and airway clearance (Bonvin et al. 2008; Fischer et al. 2013; Meyerholz et al. 2010a). Controversy continues regarding the primary consequence of CFTR dysfunction that initiates the cycle of impaired mucociliary clearance, chronic infection, and neutrophilic inflammation. Dysfunctional CFTR impedes chloride (and bicarbonate) transport that decreases hydration of the airway surface layer (ASL) (Matsui et al. 1998). Additionally, this ASL has defective ability to impede bacterial growth, perhaps through alterations in pH (Pezzulo et al. 2012). CFTR dysfunction itself may be pro-inflammatory (Bonfield et al. 2012). Recent porcine studies suggest that impaired bacterial clearance generates inflammation that results in purulent bronchitis (Stoltz et al. 2010). Neutrophilic inflammation is predominant. Externalized DNA contributes to the tenacity of airway secretions. The chronic inflammation and plugged airways result in bronchiectasis and airway remodeling. The accumulation of these destructive processes ultimately leads to respiratory failure.

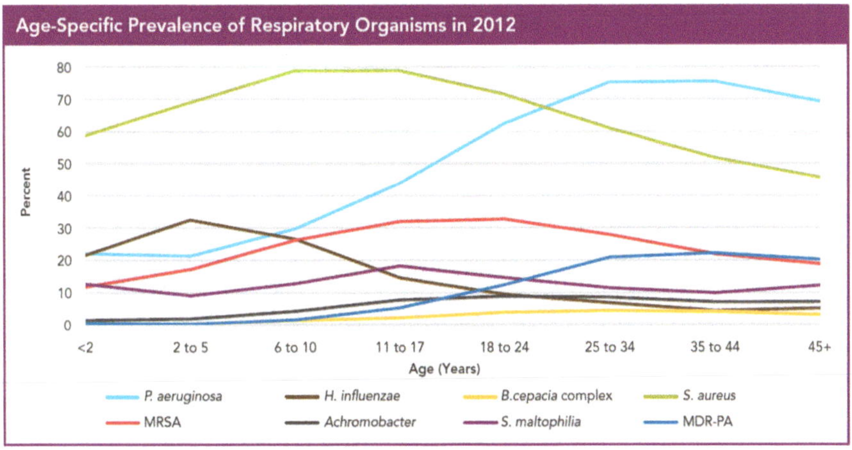

Fig. 16.8 Microbiology of CF airways. *Staphylococcus aureus* and *Pseudomonas aeruginosa* are two common organisms encountered in CF, with *Pseudomonas* becoming more predominant with age (*Source*: CF 2012 Registry Data; with permission from the Cystic Fibrosis Foundation, www.CFF.org)

16.4.3 Microbiology

The chief bacteria that infect CF lungs are *Staphylococcus aureus* and *Pseudomonas aeruginosa* (Fig. 16.8). Typically in toddlers, *S. aureus* and *Haemophilus influenzae* are most commonly encountered (CF 2012 Registry Data; www.cff.org). With time, *P. aeruginosa* becomes more prevalent, likely becoming more dominant as it adapts to the airway environment. Drug resistance also becomes a therapeutic concern. Methicillin-resistant *S. aureus* (MRSA) is now present in more than 25 % of CF children, and multidrug-resistant *P. aeruginosa* (MDR-PA) becomes more present in adulthood.

16.4.4 Other Organ Systems

Although respiratory and nutritional diseases are the most overt clinical manifestations of CFTR dysfunction, multiple other organ systems are involved (Fig. 16.9). Most patients will have refractory sinusitis and/or nasal polyposis that often requires surgical intervention. The congenital absence of the vas deferens makes nearly all CF males infertile. CF females can conceive, but with a reduced fertility rate. Bone density is also low: a consequence of chronic inflammation, Vitamin D deficiency, and CFTR dysfunction in the osteoblast. Digits are clubbed, and bone/joint pain is not unusual. Puberty is often delayed, and height is often shortened.

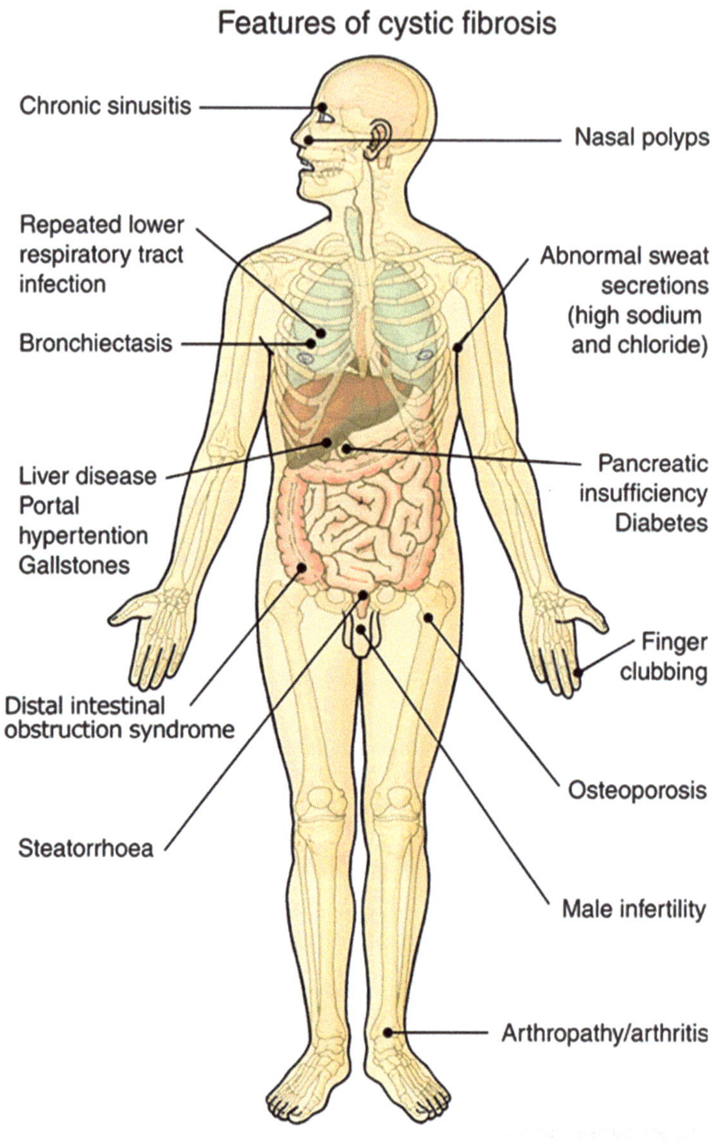

Fig. 16.9 Multiorgan involvement of CFTR dysfunction (with permission from cfmedicine.com)

With disease progression, cystic fibrosis-related diabetes frequently develops, caused by both insulin deficiency and resistance. Other endocrine abnormalities may be present, such as short stature thought to be partially due to IGF-1 deficiency (Rogan et al. 2010).

16.5 Diagnosis of Cystic Fibrosis

Though significant advancements in detection of CFTR mutations have been made, the diagnosis of cystic fibrosis continues to be based largely upon phenotypic confirmation. Current consensus guidelines (Farrell et al. 2008; Rosenstein and Cutting 1998) recommend that a CF diagnosis be made on the basis of (1) characteristic phenotypic features (sinopulmonary disease, gastrointestinal/nutritional disease, salt loss syndromes, or congenital absence of the vas deferens), (2) the presence of CF in a sibling, or (3) positive newborn screen *plus* laboratory evidence of CFTR abnormality by (1) elevated sweat chloride and (2) abnormalities in the measurement of nasal potential difference (PD) or identification of CFTR genetic abnormalitics known to cause disease.

As the name implies, sweat chloride testing involves stimulation and collection of sweat on filter paper and measurement of chloride concentration (Gibson and Cooke 1959; LeGrys et al. 2007). CFTR dysfunction leads to impaired resorption of chloride from sweat, with consequent reduction in sodium absorption due to electrical coupling of their transepithelial fluxes (Quinton 1983). Often, parents will report the sweat of affected infants as tasting salty. A minimal sweat quantity (e.g., 75 mg or 15 μL depending on technique) is required to insure an adequate sample. Sweat chlorides above 60 mmol/L are considered positive, values between 40 and 60 mmol/L are considered indeterminate, and values less than 40 mmol/L are considered negative. In infants less than 6 months, these cutoffs are lessened, such that a sweat chloride value >30 should be considered abnormal.

When CF is suspected, but sweat chloride is indeterminate, measurement of the PD across the nasal mucosa can be helpful (Rowe et al. 2011). Nasal PD may also better reflect CFTR function in the lower airways, so it perhaps is a better measure of the clinical impact of CFTR restoration. A nasal probe is placed on the ciliated nasal epithelium and a reference probe in a separate interstitial compartment (typically subcutaneous forearm). A characteristic readout is shown in Fig. 16.10. In CF, the nasal potential difference is increased at baseline due to heightened ENaC activity. Addition of amiloride blocks ENaC activity and depolarizes the potential difference. Addition of a chloride-free solution and later of cyclic AMP (through isoproterenol) is expected to stimulate CFTR activity that repolarizes the nasal epithelium in non-CF subjects but does not have a significant effect in subjects with defective CFTR. Final perfusion of the nares with ATP increases the activity of calcium-activated chloride channels (CaCCs), to repolarize the CF mucosa (Rowe et al. 2011). Nasal PD is available in CF specialty centers and is often used as a research tool to evaluate alterations in CFTR function in the sinopulmonary tract.

In the last decade, more children are being diagnosed with cystic fibrosis through newborn screening (NBS) programs. Newborn screening for CF began in the 1980s with programs in Colorado (1982) and Wisconsin (1985) pioneering the effort. Data from these programs indicated long-term nutritional benefits that prompted the US Centers for Disease Control (CDC) to recommend newborn screening in 2004

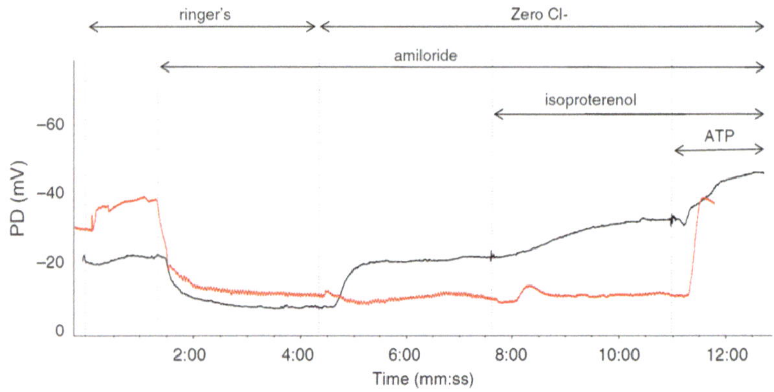

This figure will be printed in b/w

Fig. 16.10 Characteristic nasal potential difference readout in CF (*red*) vs. non-CF (*black*) patients. CF patients have an increased PD at baseline that is reduced with amiloride (ENaC blockade), is refractory to chloride-free solution and isoproterenol (CFTR stimulation), and is responsive to ATP stimulation of CaCC channels (From Rowe et al. 2011 with permission)

(Grosse et al. 2004). Now, all states perform newborn screening, and >60 % of new CF diagnoses are made by NBS (Cystic Fibrosis Foundation 2013b). Screening programs rely on elevated immunoreactive trypsinogen (IRT) measured in dried blood spots (Guthrie cards) obtained from newborns. IRT in CF infants is higher than non-CF subjects, secondary to leakage of the pancreatic enzyme precursor into the circulation. States have distinct protocols for identification of a positive screen. Some states use an IRT/IRT screen, in which two positive IRT tests prompt referral for sweat chloride analysis. Others utilize an IRT/DNA protocol where an initially elevated IRT prompts DNA analysis that identifies the most common disease-causing mutations (Watson et al. 2004). Importantly, both IRT/IRT and IRT/DNA are considered screening algorithms utilized to identify infants in need of sweat chloride testing. As with all newborn screening regimens, the vast majority of positive screens will not have disease. In the case of CF NBS, infants with a positive screen and negative sweat chloride often will be carriers of one CFTR mutation and are expected to have normal life expectancy. Although avoiding the nutritional and pulmonary phenotype of CF, carriers of one mutation of CFTR may have increased the prevalence of sinus disease, pancreatic disease, and male infertility (Anguiano et al. 1992; Chillon et al. 1995; Griesenbach et al. 1999; Sharer et al. 1998; Wang et al. 2000, 2005b; Yu et al. 2012).

16.5.1 Effect of CFTR Genotype on Disease Phenotype

Although significant disease heterogeneity exists among CF patients sharing the same genetic mutation (e.g., F508del/F508del), an increasing appreciation has developed that some mutation classes have a more favorable outcome than others.

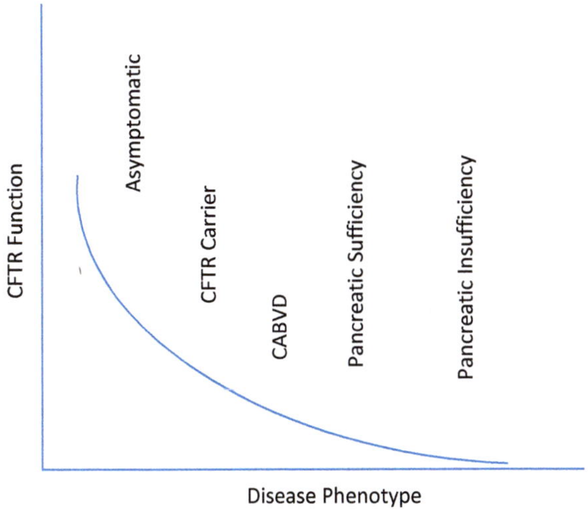

Fig. 16.11 Effect of CFTR mutation on disease manifestation. CF disease manifestations become more prominent with severity of CFTR dysfunction (Adapted from Zielenski 2000; Amaral 2005)

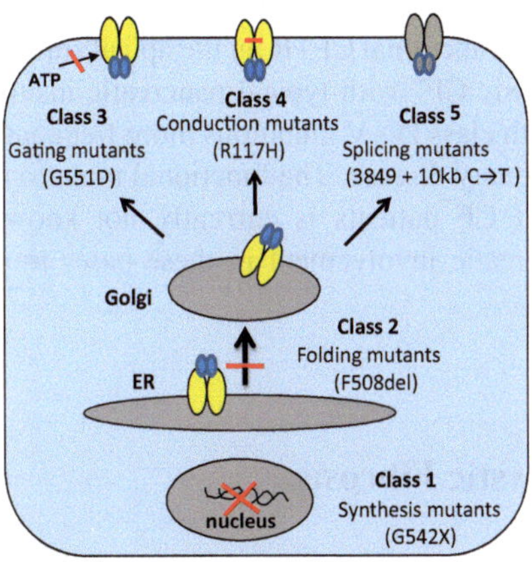

Diagram of CFTR functional mutational classes. (Adapted from Rowe NEJM 2005 and Rowntree Ann Hum Genet 2003)

Mutational Class	Summary Description of Defect	Functional Consequence	Example (% of US patients with mutation)
Class 1	Premature Stop Mutation	Interrupted transcription	G542X (4.6%), W1282X (2.3%)
Class 2	Folding Defect	Degradation in Golgi	F508del (86.7%)
Class 3	Gating Defect	Closed Channel	G551D (4.3%)
Class 4	Conductance Defect	Misshapen pore	R117H (2.7%)
Class 5	Splicing Defect	Decreased protein levels	3849 + 10kb C→T (1.5%)

Summary Table of CFTR functional classes with examples of specific mutations and prevalence. (Data from 2012 CFF Registry data, with adaptation from Karem Peds Pulm 2005, www. cfmedicine.com, and www.cftr.info

Fig. 16.12 CFTR functional mutational classes (*top*) with examples (*bottom*) (Adapted from Rowe et al. 2005; Rowntree and Harris 2003; Zielenski and Tsui 1995)

Previously, this distinction was made between those who were pancreatic sufficient (15 % of patients) and pancreatic insufficient (85 % of CF patients). Now, it is recognized that the clinical manifestation of pancreatic sufficiency may be indicative of increased CFTR function (Fig. 16.11; Amaral 2005; Zielenski 2000). CFTR mutations can be divided into functional classes (Fig. 16.12), with these categories influencing phenotype (Rowe et al. 2005; Rowntree and Harris 2003; Zielenski 2000). *Class I* mutations are secondary to premature stop mutations that inhibit protein biosynthesis. Examples are G542X (5 % of CF patients) and W1282X (2 % of CF patients). *Class II* mutations are secondary to defective protein folding leading to misprocessing, with degradation in the endoplasmic reticulum (ER). The major example of a class II mutation is F508del, the most common CFTR mutation, present in nearly 90 % of CF patients. The mutant polypeptide is degraded and does not reach the apical surface, as indicated earlier. Polypeptides with *class III* mutations reach the apical epithelial surface but do not function properly. These mutations prevent the necessary conformational change that allows ion transport. G551D, present in 5 % of the CF population, is the best known of these mutations. G551D is of significance as this is the first CFTR mutation to be addressed by an FDA-approved modulator compound that improves both lung function and nutritional status (e.g., the Vertex-770 compound whose mechanism of action to promote G551D-CFTR channel opening was discussed earlier). Proteins with *class IV* mutations similarly reach the apical surface, but chloride conductance is suboptimal. These CFTR channels respond to stimuli, but with decreased function. *Class V* mutations are typically the result of splice mutations such as $3849 + 10$ kb $C \rightarrow T$ that result in a decreased amount of RNA transcript. Additionally, mutations that decrease protein stability (sometimes called *class VI* mutations) decrease but do not eliminate the functional CFTR on the apical surface. Class I–III mutations typically cause "classic CF" with typical pancreatic insufficiency and respiratory disease. Patients with class IV–V mutations more frequently are pancreatic sufficient with milder pulmonary disease. The functional mutational classification for approximately 20 % of CF patients is currently not known, although lung function and rates of pancreatic involvement in these cases more closely resemble "classic" disease.

16.6 Therapeutic Options for Cystic Fibrosis

16.6.1 Nutritional Support

One of the first and primary therapies to treat cystic fibrosis is nutritional support (Borowitz et al. 2002; Stallings et al. 2008). At the time of diagnosis, a patient suspected of having pancreatic insufficiency (either by history or mutational analysis) is immediately placed on pancreatic enzymes and supplemental fat-soluble vitamins. The development of pancreatic enzyme replacement has significantly improved nutritional outcomes and allows CF patients to ingest a high-fat, high-calorie diet. Pancreatic enzymes are dosed on the units of lipase. Out of convenience, these are typically dosed by unit weight, assuming that body weight correlates with fat ingestion. Pancreatic enzyme dose is 1000–2000 U/kg/meal and often half of this dose with snacks. Daily maximums are typically 10,000 U/

kg/day. Infants that ingest a relatively high-fat diet with frequent meals often push these doses to higher limits. Most CF physicians will not prescribe in excess of 2500 U/kg/meal or 12,500 U/kg/day for fear of inducing fibrosing colonopathy. Enzymes are taken before meals and are active for about 30 min. Acid blockade can help potentiate enzyme efficacy. Diminished pancreatic bicarbonate secretion in CF limits neutralization of gastric pH. This is corrected with acid-reducing therapy. Even with pancreatic enzymes supplementation, caloric supplementation is often needed to achieve normal growth. A strong association exists between weight and lung function, so current guidelines aim for a body mass index percentile (BMI%) of >50 %. Fat-soluble vitamin supplementation (vitamins A, D, E, and K) is also initiated at the time of diagnosis. Vitamin levels are typically checked annually, with Vitamin D often requiring additional supplementation.

16.6.2 Airway Clearance

Enhancing mucociliary clearance is another fundamental therapy in CF respiratory care (Flume et al. 2009; Lester and Flume 2009) and initiated at the time of diagnosis. Multiple modalities of chest physiotherapy (CPT) exist, with few comparative studies showing a benefit of one technique over the other. Rather, patient preference combined with financial considerations defines which technique is chosen. Therapies can be divided into three categories: external vibration, breathing techniques, and exercise. Manual chest percussion is the standard form of CPT and often the one started in infancy. As patients age and seek more independence, high-frequency chest compression (such with the Vest®) becomes preferred. Typical CPT frequency is twice daily. Breathing techniques are used less commonly in the USA but are more encouraged in Europe. These techniques offer increased independence and mobility. Examples include positive expiratory pressure, in which the patient breathes against resistance (such as bubbles or valves). Vibratory PEP is popular with some, in which the exhaled breath causes a ball to vibrate. Active cycle breathing and autogenic drainage require more training to learn variation in the depth of inspiration/expiration, but mastery of these techniques is very appealing as they have no cost, offer independence, and can be inconspicuous. In addition to standard CPT, exercise is an important adjunctive therapy to improve airway clearance. Exercise results in variations in respiratory depth as well as vibration in the chest wall and has positive effects on quality of life.

16.6.3 Mucolytic Therapy

Inhaled therapies are also available to help mobilize airway secretions. Two of the most commonly used are DNase (Pulmozyme®) and hypertonic saline (7 %). Both have been tested in phase III clinical trials with improvement in forced expiratory volume in 1 s (FEV1) (Fuchs et al. 1994). As spirometry is a less reliable endpoint in children less than 6 years, recommendations for use in the younger population are less established. Some of the thick tenacity of CF airway secretions is secondary to extracellular DNA extruded from neutrophils. DNase degrades this sticky DNA, with a 6 % improvement in FEV1 (Fuchs et al. 1994). Hypertonic saline theoretically helps correct a primary defect of CF (diminished intraluminal salt), with a potential to stimulate hydration of respiratory secretions. Hypertonic saline also induces a more vigorous cough, with bronchodilator pretreatment (e.g., with albuterol) necessary before administration. Similar to Pulmozyme®, hypertonic saline results in a modest improvement in FEV1 and a decrease in respiratory exacerbations.

16.6.4 Antibiotic Therapy

Chronic endobronchitis is of primary concern in CF. Antibiotics are frequently used to control respiratory flares and to suppress chronic pathogens. On average, CF patients are admitted once annually for intravenous antibiotic therapy, with this frequency increasing with age and respiratory decline. Early in disease, the most common respiratory pathogens are *Staphylococcus aureus* and *Haemophilus influenzae*. The physician response to the presence of these organisms varies depending upon region. In the USA, antibiotics are provided if recovery of these organisms is associated with respiratory symptoms. Physicians in the USA do not routinely prescribe antistaphylococcal prophylaxis, due to concern for antibiotic resistance and engendering an environment more favorable to *Pseudomonas* acquisition. In an influential study, children randomized to cephalexin had decreased prevalence of *S. aureus* but increased colonization with *Pseudomonas* (Stutman et al. 2002). The higher prevalence of methicillin-resistant *S. aureus* (MRSA) is another reason US physicians hesitate to prescribe prophylactic antibiotics. Physicians in the UK, where MRSA is less prevalent, prescribe a narrower spectrum antistaphylococcal agent (Flucloxacillin) for prophylaxis and do not report increased pseudomonal colonization. The use of prophylactic azithromycin has increased in the USA, but this is justified by anti-inflammatory considerations rather than antimicrobial effect.

When *Pseudomonas* is recovered from respiratory cultures, attempts are made to eradicate it, even in the absence of respiratory symptoms. Eradication regimens often include an inhaled antipseudomonal (such as tobramycin, e.g., Tobi®) for at least a month, with an oral agent such as ciprofloxacin added if patients fail to respond. In patients who continue to harbor *Pseudomonas*, chronic inhaled antipseudomonal antibiotics (Tobi in the USA; colistin in the UK) are recommended.

When a patient has a respiratory flare (demarcated as increased cough, congestion, or decline in lung function), an initial attempt is often made to treat as an outpatient with oral antibiotics. For those without *Pseudomonas*, a broad-spectrum antibiotic with beta-lactamase resistance (e.g., amoxicillin-clavulanic acid or cefdinir) is a frequent choice to treat *H. influenzae* and sensitive *S. aureus*. For MRSA, co-trimoxazole (brand name Bactrim) is a common choice. For patients with *Pseudomonas*, an initial attempt at treatment with a combination of inhaled antipseudomonal (e.g., Tobi) and oral fluoroquinolone (e.g., ciprofloxacin) is prescribed. If the respiratory exacerbation does not respond to outpatient management, patients will be hospitalized for intravenous (IV) antibiotics and respiratory care. Typical duration of IV antibiotics is 14 days, although some of this may be completed at home. Usually, *Pseudomonas* is treated with two IV antibiotics. Previous response to antibiotic regimen and in vitro bacterial culture sensitivities guide antibiotic selection. Typically, one is a beta-lactam (e.g., ceftazidime) and the other an aminoglycoside (e.g., IV tobramycin) that are believed to act synergistically. If MRSA is present, it is often addressed with an additional IV antibiotic (vancomycin in the USA; teicoplanin in the UK).

16.6.5 Other Microbes

Although *S. Aureus* and *P. aeruginosa* are the most common pathogens in CF, multiple other microbes grow in CF respiratory secretions. Of these, *Burkholderia cenocepacia* is of particular importance as it can herald a rapid decline in lung function (Lipuma 2005; Lynch 2009). Intrinsically resistant to aminoglycosides, *Burkholderia* sp. often develops a multidrug-resistant profile. *B. cenocepacia* can be quite immunogenic, triggering a "cepacia syndrome" demarcated by fever, sepsis, and rapid respiratory decline. The impact of other gram-negative bacteria such as *Stenotrophomonas* and *Acinetobacter* is less clear, but these can complicate antibiotic regimens and response to therapy. Nontuberculous mycobacteria (NTM) also may lead to CF respiratory decline. *Mycobacteria avium complex* (MAC) is more frequent in adults and perhaps of less virulence (Olivier et al. 2003). *Mycobacterium abscessus* is more prevalent in children and associated with lung function decline. Recovery of fungi from CF respiratory secretions is not uncommon (Esther et al. 2005; Leung and Olivier 2013). Currently, recovery of fungal elements in respiratory culture is not standardly treated. An exception is *Aspergillus* when it induces allergic bronchopulmonary aspergillosis (ABPA) that induces lung function decline (Thia and Balfour Lynn 2009). Traditional culture techniques likely underestimate the prevalence of anaerobic bacteria. Culture-independent techniques that recognize bacteria DNA suggest the importance of anaerobes and suggest that antibiotic therapy directed at anaerobes may be helpful (Zemanick et al. 2011).

16.6.6 Anti-inflammatory Therapy

Addressing the hyper-inflammatory state of CF has been an appealing but elusive target in CF care. Chronic oral corticosteroids have been investigated with improvement in lung function but have intolerable adverse effects on growth and glucose tolerance (Eigen et al. 1995; Lai et al. 2000). Inhaled corticosteroids are not of significant benefit in the absence of an asthma phenotype. High-dose ibuprofen has been among the most successful anti-inflammatory agents in clinical trials (Konstan et al. 1995, 2007), but utilization by CF clinicians has been low. Reasons for this poor utilization have been concern about GI side effects, nephrotoxicity, and necessity of monitoring drug levels. In contrast, utilization of thrice-weekly azithromycin (every Monday, Wednesday, Friday) is one of the most used chronic respiratory medications. In patients with *Pseudomonas* infection, chronic azithromycin increased FEV1 by 6 % (Saiman et al. 2003). In younger patients without *Pseudomonas*, chronic azithromycin halved the exacerbation rate, decreased inflammatory markers, and increased weight gain (Ratjen et al. 2012; Saiman et al. 2010).

16.6.7 Other Common Therapies

Insulin is required to treat CF-related diabetes (CFRD). Often, a combination of basal long-acting insulin once daily with rapid-acting insulin administered at meal times is sufficient. In addition to benefits on hyperglycemia, insulin therapy is associated with weight gain, improved lung function, and decreased respiratory exacerbations. Oral agents to control hyperglycemia are currently not recommended in CFRD (Moran et al. 2010). Ursodiol is frequently used to treat CF-related liver disease (CFLD), theoretically improving bile flow. Ursodiol is most effective at decreasing transaminases in the early stages of CF lung disease, although no data are available on its survival benefit (Colombo 2007). Finally, cyproheptadine (Periactin), an antihistamine, is often used for its side effect of appetite stimulation in patients with nutritional failure and poor intake. Often, the benefit is brief but can sometimes jump-start an improvement in nutritional status.

16.6.8 Transplantation

Ultimately, the consequences of "classic" cystic fibrosis will lead to respiratory failure. At that point, the CF patient (and family) will decide if they want to attempt lung transplantation. A multifaceted lung allocation score prioritizes recipients, but typically transplant is a pertinent treatment option when FEV1 declines below 30 %. At this point, life expectancy from CF disease intersects with expected posttransplant survival. In many ways, transplantation is trading one disease for another. Survival from lung transplantation lags behind other organs, typically reported to be 50–70 % at 5 years (Christie et al. 2012; Hirche et al. 2014; Liou et al. 2005; Valapour et al. 2014).

16.6.9 Clinical Benefit of Direct CFTR Channel Modulators

Most previous CF therapies have focused on the downstream consequences of CFTR dysfunction such as inflammation, infection, or malabsorption. Recently, direct intervention on CFTR has been evaluated with promising results. High-throughput screening has identified compounds that modify CFTR channel function in vitro. These compounds are mutation (or mutational class) specific, although combination therapy is being explored. The greatest success is with the aforementioned ivacaftor (Vertex-770) for the G551D mutation. Administration of ivacaftor to patients with G551D (and other class III) mutations results in a 10 % improvement in FEV1, significant weight gain, and near normalization of sweat chloride, with results sustained over 48 weeks (Davies et al. 2013; Ramsey et al. 2011). These results led to ivacaftor being the first FDA-approved CFTR modulator and prompting significant hope that CFTR modulation can halt or even partially reverse clinical disease. Currently approved for patients 6 years and older, studies are ongoing to demonstrate efficacy in the youngest populations. Other CFTR modulators have shown promise but not demonstrated as much clinical impact. Ataluren (PTC124) is a compound that fosters ribosomal read-through to overcome class I (stop) mutations such as G542X and W1282X. While ataluren administration does improve nasal potential difference measurements (Sermet-Gaudelus et al. 2010; Wilschanski et al. 2011), this impact has not manifest a meaningful improvement in lung function (Kerem et al. 2014). For the most common CFTR defect, F508del, a class II mutation, lumacaftor (VX-809) is in clinical development. In vitro studies indicate that lumacaftor improves CFTR processing within the endoplasmic reticulum (34 % normal) with increased apical CFTR activity in primary human epithelial cells (14 % normal; Van Goor et al. 2011). A phase II clinical study showed a decrease in sweat chloride of 8 mmol/L (from a baseline value 98 mmol/L) but no significant difference in lung function or other clinical outcomes (Clancy et al. 2012). Combination therapy with lumacaftor and ivacaftor (addressing ER processing and apical surface function) has recently been evaluated in a complicated, multi-arm phase II study. In subjects on higher-dose therapy, sweat chloride improved 8–10 mmol/L with a mild improvement in FEV1 (3.6 %). These findings will be tested further in upcoming phase III clinical studies. In addition to lumacaftor (VX-809), Vertex Pharmaceuticals has two other CFTR correctors in drug development (VX-661 and VX-983) to address the F508del mutation. Additionally, expansion of ivacaftor to other mutations such as R117H is currently under investigation.

16.7 Conclusions

Cystic fibrosis is a remarkable disease in that mutations in a single ion channel manifest as a life-shortening disease with multiorgan involvement. Research efforts to prolong the lives of CF patients have led to an extensive understanding of CFTR mutations and function. CFTR mutational class is important to both disease presentation and potential for intervention. Improved therapeutics addressing downstream nutritional and respiratory consequences of CFTR dysfunction have significantly improved life expectancy of CF patients to over 40 years of age. Newborn screening for CF allows the opportunity of therapeutic intervention before irreversible disease progresses. Recent advancements now offer the potential to intervene directly on the CFTR defect. For the 5 % of CF patients with class III mutations such as G551D, successful potentiation of CFTR channel activity is within grasp. For the remainder of CF patients, such as those homozygous for F508del-CFTR, therapeutic possibilities are in development. The study of CFTR and cystic fibrosis continues to be a fascinating interplay of science and human disease, with a promising future.

References

Abu-El-Haija M, Ramachandran S, Meyerholz DK et al (2012) Pancreatic damage in fetal and newborn cystic fibrosis pigs involves the activation of inflammatory and remodeling pathways. Am J Pathol 181(2):499–507

Aleksandrov L, Aleksandrov AA, Chang XB et al (2002) The first nucleotide binding domain of cystic fibrosis conductance regulator is a site of stable nucleotide interaction, whereas the second is a site of rapid turnover. J Biol Chem 277:15419–15425

Amaral MD (2005) Processing of CFTR: traversing the cellular maze–how much CFTR needs to go through to avoid cystic fibrosis? Pediatr Pulmonol 39(6):479–491

Amin ML (2013) P-glycoprotein inhibition for optimal drug delivery. Drug Target Insights 7:27–34

Andersen DH (1938) Cystic fibrosis of the pancreas and its relationship to celiac disease: a clinical and pathologic study. Am J Dis Child 56(2):344–399

Andersen DH (1949) The present diagnosis and therapy of cystic fibrosis of the pancreas. Proc R Soc Med 42(1):25–32

Anguiano A, Oates RD, Amos JA et al (1992) Congenital bilateral absence of the vas deferens. A primarily genital form of cystic fibrosis. JAMA 267(13):1794–1797

Aubin CN, Linsdell P (2006) Positive charges at the intracellular mouth of the pore regulate anion conduction in the CFTR chloride channel. J Gen Physiol 128:535–545

Auerbach A (2012) Thinking in cycles: MWC is a good model for acetylcholine receptor-channels. J Physiol 590:93–98

Bai Y, Li M, Hwang TC (2010) Dual roles of the sixth transmembrane segment of the CFTR chloride channel in gating and permeation. J Gen Physiol 136:293–309

Bai Y, Li M, Hwang TC (2011) Structural basis for the channel function of a degraded ABC transporter, CFTR (ABCC7). J Gen Physiol 138:495–507

Baker JMR, Hudson RP, Kanelis V et al (2007) CFTR regulatory region interacts with NBD1 predominantly via multiple transient helices. Nat Struct Mol Biol 14:738–745

Bartlett JR, Friedman KJ, Ling SC et al (2009) Genetic modifiers of liver disease in cystic fibrosis. JAMA 302(10):1076–1083

Basso C, Vergani P, Nairn AC et al (2003) Prolonged nonhydrolytic interaction of nucleotide with CFTR's NH_2 terminal nucleotide binding domain and its role in channel gating. J Gen Physiol 122:333–348

Bompadre SG, Sohma Y, Li M et al (2007) G551D and G1349D, two CF-associated mutations in the signature sequences of CFTR, exhibit distinct gating defects. J Gen Physiol 129:285–298

Bonfield TL, Hodges CA, Cotton CU et al (2012) Absence of the cystic fibrosis transmembrane regulator (Cftr) from myeloid-derived cells slows resolution of inflammation and infection. J Leukoc Biol 92(5):1111–1122

Bonvin E, Le Rouzic P, Bernaudin JF et al (2008) Congenital tracheal malformation in cystic fibrosis transmembrane conductance regulator-deficient mice. J Physiol 586(13):3231–3243

Borowitz D, Baker RD, Stallings V (2002) Consensus report on nutrition for pediatric patients with cystic fibrosis. J Pediatr Gastroenterol Nutr 35(3):246–259

Bozoky Z, Krzeminski M, Muhandiram R et al (2013) Regulatory R region of the CFTR chloride channel is a dynamic integrator of phospho-dependent intra- and intermolecular interactions. Proc Natl Acad Sci U S A 110:4427–4436

Cadugan DJ, Auerbach A (2010) Linking the acetylcholine receptor-channel agonist-binding sites with the gate. Biophys J 99:798–807

CFTR2, Clinical and functional translation of CFTR website (2014) http://www.cftr2.org

Chang Y, Weiss DS (1999) Allosteric activation mechanism of the $\alpha 1\beta 2\gamma 2$ γ-Aminobutyric acid type A receptor revealed by mutation of the conserved M2 leucine. Biophys J 77:2542–2551

Changeux JP, Edelstein SJ (2005) Allosteric mechanisms of signal transduction. Science 308:1424–1428

Cheng SH, Gregory RJ, Marshall J et al (1990) Defective intracellular transport and processing of CFTR is the molecular basis of most cystic fibrosis. Cell 63:827–834

Cheng SH, Rich DP, Marshall J et al (1991) Phosphorylation of the R domain by cAMP-dependent protein kinase regulates the CFTR chloride channel. Cell 66:1027–1036

Chillon M, Casals T, Mercier B et al (1995) Mutations in the cystic fibrosis gene in patients with congenital absence of the vas deferens. N Engl J Med 332(22):1475–1480

Cholon DM, Quinney NL, Fulcher ML et al (2014) Potentiator ivacaftor abrogates pharmacological correction of ΔF508 CFTR in cystic fibrosis. Sci Transl Med 6:246ra96

Christie JD, Edwards LB, Kucheryavaya AY et al (2012) The Registry of the International Society for Heart and Lung Transplantation: 29th adult lung and heart-lung transplant report. J Heart Lung Transplant 31(10):1073–1086

Clancy JP, Rowe SM, Accurso FJ et al (2012) Results of a phase IIa study of VX-809, an investigational CFTR corrector compound, in subjects with cystic fibrosis homozygous for the F508del-CFTR mutation. Thorax 67(1):12–18

Colombo C (2007) Liver disease in cystic fibrosis. Curr Opin Pulm Med 13(6):529–536

Colquhoun D (1998) Binding, gating, affinity and efficacy: the interpretation of structure-activity relationships for agonists and of the effects of mutating receptors. Br J Pharmacol 125:924–947

Csanády L, Vergani P, Gadsby DC (2010) Strict coupling between CFTR's catalytic cycle and gating of its Cl^- ion pore revealed by distributions of open channel burst durations. Proc Natl Acad Sci U S A 107:1241–1246

Cui G, Song B, Turki HW et al (2012) Differential contribution of TM6 and TM12 to the pore of CFTR identified by three sulfonylurea-based blockers. Pflugers Arch 463:405–418

Cystic Fibrosis Foundation (2013a) Cystic Fibrosis Foundation Patient Registry: 2012 annual data report. http://www.cff.org

Cystic Fibrosis Foundation (2013b) Cystic Fibrosis Foundation Patient Registry: 2012 annual data report to the Center Directors. http://www.cff.org

Cystic Fibrosis Mutation Database (2014) http://www.genet.sickkids.on.ca/cftr/app

Dalemans W, Barbry P, Champigny G et al (1991) Altered chloride ion channel kinetics associated with the ΔF508 cystic fibrosis mutation. Nature 354:526–528

Davies JC, Wainwright CE, Canny GJ et al (2013) Efficacy and safety of ivacaftor in patients aged 6 to 11 years with cystic fibrosis with a G551D mutation. Am J Respir Crit Care Med 187:1219–1225
Davis PB (2006) Cystic fibrosis since 1938. Am J Crit Care Med 173:475–482
Dawson RJ, Locher KP (2006) Structure of a bacterial multidrug ABC transporter. Nature 443:180–185
Dawson RJ, Locher KP (2007) Structure of the multidrug ABC transporter Sav 1866 from Staphylococcus aureus in complex with AMP-PNP. FEBS Lett 581:935–938
Denning GM, Anderson MP, Amara JF et al (1992) Processing of mutant cystic fibrosis transmembrane conductance regulator is temperature-sensitive. Nature 358:761–764
Di Sant'Agnese PA, Darling RC, Perera GA et al (1953) Abnormal electrolyte composition of sweat in cystic fibrosis of the pancreas; clinical significance and relationship to the disease. Pediatrics 12(5):549–563
Du K, Sharma M, Lukacs GL (2005) The ΔF508 mutation impairs domain-domain interactions and arrests post-translational folding of CFTR. Nat Struct Mol Biol 12:17–25
Eckford PD, Li C, Ramjeesingh M et al (2012) Cystic fibrosis transmembrane conductance regulator (CFTR) potentiator VX-770 (ivacaftor) opens the defective gate of mutant CFTR in a phosphorylation-dependent but ATP-independent manner. J Biol Chem 287:36639–36649
Eigen H, Rosenstein BJ, FitzSimmons S et al (1995) A multicenter study of alternate-day prednisone therapy in patients with cystic fibrosis. Cystic Fibrosis Foundation Prednisone Trial Group. J Pediatr 126(4):515–523
Esther CR Jr, Henry MM, Molina PL et al (2005) Nontuberculous mycobacterial infection in young children with cystic fibrosis. Pediatr Pulmonol 40(1):39–44
Farrell PM, Rosenstein BJ, White TB et al (2008) Guidelines for diagnosis of cystic fibrosis in newborns through older adults: Cystic Fibrosis Foundation consensus report. J Pediatr 153(2): S4–S14
Fischer AJ, Singh SB, Adam RJ et al (2013) Tracheomalacia is associated with lower FEV and Pseudomonas acquisition in children with CF. Pediatr Pulmonol. doi:10.1002/ppul.22922
Flume PA, Robinson KA, O'Sullivan BP et al (2009) Cystic fibrosis pulmonary guidelines: airway clearance therapies. Respir Care 54(4):522–537
French PJ, Bijman J, Edixhoven M et al (1995) Isotype-specific activation of cystic fibrosis transmembrane conductance regulator-chloride channels by cGMP-dependent protein kinase II. J Biol Chem 270:26626–26631
Fuchs HJ, Borowitz DS, Christiansen DH et al (1994) Effect of aerosolized recombinant human DNase on exacerbations of respiratory symptoms and on pulmonary function in patients with cystic fibrosis. The Pulmozyme Study Group. N Engl J Med 331(10):637–642
Gadsby DC (2009) Ion channels versus ion pumps: the principal difference, in principle. Nat Rev Mol Cell Biol 10:344–352
Gibson LE, Cooke RE (1959) A test for concentration of electrolytes in sweat in cystic fibrosis of the pancreas utilizing pilocarpine by iontophoresis. Pediatrics 23(3):545–549
Griesenbach U, Geddes DM, Alton EW (1999) The pathogenic consequences of a single mutated CFTR gene. Thorax 54(Suppl 2):S19–S23
Grosse SD, Boyle CA, Botkin JR et al (2004) Newborn screening for cystic fibrosis: evaluation of benefits and risks and recommendations for state newborn screening programs. MMWR Recomm Rep 53(RR-13):1–36
Gunderson KL, Kopito RR (1995) Conformational states of CFTR associated with channel gating: the role of ATP binding and hydrolysis. Cell 82:231–239
Hegedüs T, Serohijos AW, Dokholyan NV et al (2008) Computational studies reveal phosphorylation-dependent changes in the unstructured R domain of CFTR. J Mol Biol 378:1052–1063
Hegedüs T, Aleksandrov A, Mengos A et al (2009) Role of individual phosphorylation sites in CFTR regulation by protein kinase A. Biochim Biophys Acta 1788:1341–1349

Hennager DJ, Ikuma M, Hoshi T et al (2001) A conditional probability analysis of cystic fibrosis transmembrane conductance regulator gating indicates that ATP has multiple effects during the gating cycle. Proc Natl Acad Sci U S A 98:3594–3599

Hirche TO, Knoop C, Hebestreit H et al (2014) Practical guidelines: lung transplantation in patients with cystic fibrosis. Pulm Med 621342. doi:10.1155/2014/621342

Jih KY, Hwang TC (2013) Vx-770 potentiates CFTR function by promoting decoupling between the gating cycle and ATP hydrolysis cycle. Proc Natl Acad Sci U S A 110:4404–4409

Kanelis V, Hudson RP, Thibodeau PH et al (2010) NMR evidence for differential phosphorylation-dependent interactions in WT and ΔF508 CFTR. EMBO J 29:263–277

Keppler D (2011) Multidrug resistance proteins (MRPs, ABCCs): importance for pathophysiology and drug therapy. Handb Exp Pharmacol 201:299–323

Kerem E, Corey M, Kerem B et al (1989) Clinical and genetic comparisons of patients with cystic fibrosis, with or without meconium ileus. J Pediatr 114(5):767–773

Kerem E, Konstan MW, De Boeck K et al (2014) Ataluren for the treatment of nonsense-mutation cystic fibrosis: a randomised, double-blind, placebo-controlled phase 3 trial. Lancet Respir Med 2(7):539–547

Kessler WR, Andersen DH (1951) Heat prostration in fibrocystic disease of the pancreas and other conditions. Pediatrics 8(5):648–656

Kirk KL, Wang W (2011) A unified view of Cystic Fibrosis Transmembrane Conductance Regulator (CFTR) gating: combining the allosterism of a ligand-gated channel with the enzymatic activity of an ATP-binding cassette (ABC) transporter. J Biol Chem 286:12813–12819

Konstan MW, Byard PJ, Hoppel CL et al (1995) Effect of high-dose ibuprofen in patients with cystic fibrosis. N Engl J Med 332(13):848–854

Konstan MW, Schluchter MD, Xue W et al (2007) Clinical use of Ibuprofen is associated with slower FEV1 decline in children with cystic fibrosis. Am J Respir Crit Care Med 176 (11):1084–1089

Lai HC, FitzSimmons SC, Allen DB et al (2000) Risk of persistent growth impairment after alternate-day prednisone treatment in children with cystic fibrosis. N Engl J Med 342 (12):851–859

LeGrys VA, Yankaskas JR, Quittell LM et al (2007) Diagnostic sweat testing: the Cystic Fibrosis Foundation guidelines. J Pediatr 151(1):85–89

Lester MK, Flume PA (2009) Airway-clearance therapy guidelines and implementation. Respir Care 54(6):733–750

Leung JM, Olivier KN (2013) Nontuberculous mycobacteria: the changing epidemiology and treatment challenges in cystic fibrosis. Curr Opin Pulm Med 19(6):662–669

Li J, Jaimes KF, Aller SG (2014) Refined structures of mouse P-glycoprotein. Protein Sci 23:34–46

Linsdell P, Tabcharani JA, Rommens JM et al (1997) Permeability of wild-type and mutant cystic fibrosis transmembrane conductance regulator chloride channels to polyatomic anions. J Gen Physiol 110:355–364

Linsdell P, Evagelidis A, Hanrahan JW (2000) Molecular determinants of anion selectivity in the cystic fibrosis transmembrane conductance regulator chloride channel pore. Biophys J 78:2973–2982

Liou TG, Adler FR, Huang D (2005) Use of lung transplantation survival models to refine patient selection in cystic fibrosis. Am J Respir Crit Care Med 171(9):1053–1059

Lipuma JJ (2005) Update on the Burkholderia cepacia complex. Curr Opin Pulm Med 11 (6):528–533

Liu X, Smith SS, Dawson DC (2003) CFTR: what's it like inside the pore? J Exp Zool A Comp Exp Biol 300:69–75

Locher KP (2009) Review: Structure and mechanism of ATP-binding cassette transporters. Philos Trans R Soc Lond B Biol Sci 364:239–245

Lukacs GL, Chang XB, Bear C et al (1993) The ΔF508 mutation decreases the stability of cystic fibrosis transmembrane conductance regulator in the plasma membrane. Determination of functional half-lives on transfected cells. J Biol Chem 268:21592–21598

Lynch JP III (2009) Burkholderia cepacia complex: impact on the cystic fibrosis lung lesion. Semin Respir Crit Care Med 30(5):596–610

Mathews CJ, Tabcharani JA, Chang X-B et al (1998) Dibasic protein kinase A sites regulate bursting rate and nucleotide sensitivity of the cystic fibrosis transmembrane conductance regulator chloride channel. J Physiol 508:365–377

Matsui H, Grubb BR, Tarran R et al (1998) Evidence for periciliary liquid layer depletion, not abnormal ion composition, in the pathogenesis of cystic fibrosis airways disease. Cell 95 (7):1005–1015

McDonough S, Davidson N, Lester HA et al (1994) Novel pore-lining residues in CFTR that govern permeation and open-channel block. Neuron 13:623–634

Mense M, Vergani P, White DM et al (2006) In vivo phosphorylation of CFTR promotes formation of a nucleotide binding domain heterodimer. EMBO J 25:4728–4739

Meyerholz DK, Stoltz DA, Namati E et al (2010a) Loss of cystic fibrosis transmembrane conductance regulator function produces abnormalities in tracheal development in neonatal pigs and young children. Am J Respir Crit Care Med 182(10):1251–1261

Meyerholz DK, Stoltz DA, Pezzulo AA et al (2010b) Pathology of gastrointestinal organs in a porcine model of cystic fibrosis. Am J Pathol 176(3):1377–1389

Moitra K, Dean M (2011) Evolution of ABC transporters by gene duplication and their role in human disease. Biol Chem 392:29–37

Monod J, Wyman J, Changeux JP (1965) On the nature of allosteric transitions: a plausible model. J Mol Biol 12:88–118

Moran A, Brunzell C, Cohen RC et al (2010) Clinical care guidelines for cystic fibrosis-related diabetes: a position statement of the American Diabetes Association and a clinical practice guideline of the Cystic Fibrosis Foundation, endorsed by the Pediatric Endocrine Society. Diabetes Care 33(12):2697–2708

Mornon JP, Lehn P, Callebaut I (2009) Molecular models of the open and closed states of the whole human CFTR protein. Cell Mol Life Sci 66:3469–3486

Muanprasat C, Sonawane ND, Salinas D et al (2004) Discovery of glycine hydrazide pore-occluding CFTR inhibitors: mechanism, structure-activity analysis, and in vivo efficacy. J Gen Physiol 124:125–137

Norimatsu Y, Ivetac A, Alexander C et al (2012) Locating a plausible binding site for an open-channel blocker, GlyH-101, in the pore of the cystic fibrosis transmembrane conductance regulator. Mol Pharmacol 82:1042–1055

Okiyoneda T, Veit G, Dekkers JF et al (2013) Mechanism-based corrector combination restores ΔF508-CFTR folding and function. Nat Chem Biol 9:444–454

Olivier KN, Weber DJ, Lee JH et al (2003) Nontuberculous mycobacteria. II: Nested-cohort study of impact on cystic fibrosis lung disease. Am J Respir Crit Care Med 167(6):835–840

Orelle C, Dalmas O, Gros P et al (2003) The conserved glutamate residue adjacent to the Walker B motif is the catalytic base for ATP hydrolysis in the ATP-binding cassette transporter BmrA. J Biol Chem 278:47002–47008

Pezzulo AA, Tang XX, Hoegger MJ et al (2012) Reduced airway surface pH impairs bacterial killing in the porcine cystic fibrosis lung. Nature 487(7405):109–113

Qu BH, Strickland E, Thomas PJ (1997) Cystic fibrosis: a disease of altered protein folding. J Bioenerg Biomembr 29:483–490

Quinton PM (1983) Chloride impermeability in cystic fibrosis. Nature 301(5899):421–422

Ramsey BW, Davies J, McElvaney NG et al (2011) A CFTR potentiator in patients with cystic fibrosis and the G551D mutation. N Engl J Med 365:1663–1672

Ratjen F, Saiman L, Mayer-Hamblett N et al (2012) Effect of azithromycin on systemic markers of inflammation in patients with cystic fibrosis uninfected with Pseudomonas aeruginosa. Chest 142(5):1259–1266

Riordan JR, Rommens JM, Kerem B et al (1989) Identification of the cystic fibrosis gene: cloning and characterization of complementary DNA. Science 245:1066–1073

Rogan MP, Reznikov LR, Pezzulo AA et al (2010) Pigs and humans with cystic fibrosis have reduced insulin-like growth factor 1 (IGF1) levels at birth. Proc Natl Acad Sci U S A 107 (47):20571–20575

Rosenstein BJ, Cutting GR (1998) The diagnosis of cystic fibrosis: a consensus statement. Cystic Fibrosis Foundation Consensus Panel. J Pediatr 132(4):589–595

Rowe SM, Miller S, Sorscher EJ (2005) Cystic fibrosis. N Engl J Med 352(19):1992–2001

Rowe SM, Clancy JP, Wilschanski M (2011) Nasal potential difference measurements to assess CFTR ion channel activity. Methods Mol Biol 741:69–86

Rowntree RK, Harris A (2003) The phenotypic consequences of CFTR mutations. Ann Hum Genet 67(Pt 5):471–485

Saiman L, Marshall BC, Mayer-Hamblett N et al (2003) Azithromycin in patients with cystic fibrosis chronically infected with Pseudomonas aeruginosa: a randomized controlled trial. JAMA 290(13):1749–1756

Saiman L, Anstead M, Mayer-Hamblett N et al (2010) Effect of azithromycin on pulmonary function in patients with cystic fibrosis uninfected with Pseudomonas aeruginosa: a randomized controlled trial. JAMA 303(17):1707–1715

Salamone FN, Zhou M, Auerbach A (1999) A re-examination of adult mouse nicotinic acetylcholine receptor activation kinetics. J Physiol 516:315–330

Sermet-Gaudelus I, Boeck KD, Casimir GJ et al (2010) Ataluren (PTC124) induces cystic fibrosis transmembrane conductance regulator protein expression and activity in children with nonsense mutation cystic fibrosis. Am J Respir Crit Care Med 182(10):1262–1272

Serohijos AW, Hegedüs T, Aleksandrov AA et al (2008) Phenylalanine-508 mediates a cytoplasmic-membrane domain contact in the CFTR 3D structure crucial to assembly and channel function. Proc Natl Acad Sci U S A 105:3256–3261

Sharer N, Schwarz M, Malone G et al (1998) Mutations of the cystic fibrosis gene in patients with chronic pancreatitis. N Engl J Med 339(10):645–652

Smith SS, Steinle ED, Meyerhoff ME et al (1999) Cystic fibrosis transmembrane conductance regulator. Physical basis for lyotropic selectivity patterns. J Gen Physiol 114:799–818

Stallings VA, Stark LJ, Robinson KA et al (2008) Evidence-based practice recommendations for nutrition-related management of children and adults with cystic fibrosis and pancreatic insufficiency: results of a systematic review. J Am Diet Assoc 108(5):832–839

Stoltz DA, Meyerholz DK, Pezzulo AA et al (2010) Cystic fibrosis pigs develop lung disease and exhibit defective bacterial eradication at birth. Sci Transl Med 2(29):29ra31. doi:10.1126/scitranslmed.3000928

Stutman HR, Lieberman JM, Nussbaum E et al (2002) Antibiotic prophylaxis in infants and young children with cystic fibrosis: a randomized controlled trial. J Pediatr 140(3):299–305

Szollosi A, Vergani P, Csanády L (2010) Involvement of F1296 and N1303 of CFTR in induced-fit conformational change in response to ATP binding at NBD2. J Gen Physiol 136:407–423

Thia LP, Balfour Lynn IM (2009) Diagnosing allergic bronchopulmonary aspergillosis in children with cystic fibrosis. Paediatr Respir Rev 10(1):37–42

Thiagarajah JR, Verkman AS (2003) CFTR pharmacology and its role in intestinal fluid secretion. Curr Opin Pharmacol 3:594–599

Tsui LC (1991) Probing the basic defect in cystic fibrosis. Curr Opin Genet Dev 1:4–10

Uc A, Giriyappa R, Meyerholz DK et al (2012) Pancreatic and biliary secretion are both altered in cystic fibrosis pigs. Am J Physiol Gastrointest Liver Physiol 303(8):G961–G968

Valapour M, Skeans MA, Heubner BM et al (2014) OPTN/SRTR 2012 Annual Data Report: lung. Am J Transplant 14(Suppl 1):139–165

van der Doef HP, Kokke FT, van der Ent CK et al (2011) Intestinal obstruction syndromes in cystic fibrosis: meconium ileus, distal intestinal obstruction syndrome, and constipation. Curr Gastroenterol Rep 13(3):265–270

Van Goor F, Hadida S, Grootenhuis PD et al (2009) Rescue of CF airway epithelial cell function in vitro by a CFTR potentiator, VX-770. Proc Natl Acad Sci U S A 106:18825–18830
Van Goor F, Hadida S, Grootenhuis PD et al (2011) Correction of the F508del-CFTR protein processing defect in vitro by the investigational drug VX-809. Proc Natl Acad Sci U S A 108:18843–18848
Veit PM, Avramescu RG, Perdomo D et al (2014) Some gating potentiators, including VX-770, diminish ΔF508-CFTR functional expression. Sci Transl Med 6:246ra97
Vergani P, Nairn AC, Gadsby DC (2003) On the mechanism of MgATP-dependent gating of CFTR Cl^- channels. J Gen Physiol 121:17–36
Wang X, Moylan B, Leopold DA et al (2000) Mutation in the gene responsible for cystic fibrosis and predisposition to chronic rhinosinusitis in the general population. JAMA 284 (14):1814–1819
Wang W, Li G, Clancy JP et al (2005a) Activating cystic fibrosis transmembrane conductance regulator channels with pore blocker analogs. J Biol Chem 280:23622–23630
Wang X, Kim J, McWilliams R et al (2005b) Increased prevalence of chronic rhinosinusitis in carriers of a cystic fibrosis mutation. Arch Otolaryngol Head Neck Surg 131(3):237–240
Wang W, Wu J, Bernard K et al (2010) ATP-independent CFTR channel gating and allosteric modulation by phosphorylation. Proc Natl Acad Sci U S A 107:3888–3893
Wang W, El Hiani Y, Linsdell P (2011) Alignment of transmembrane regions in the cystic fibrosis transmembrane conductance regulator chloride channel pore. J Gen Physiol 138:165–178
Ward A, Reyes CL, Yu J et al (2007) Flexibility in the ABC transporter MsbA: alternating access with a twist. Proc Natl Acad Sci U S A 104:19005–19010
Watson MS, Cutting GR, Desnick RJ et al (2004) Cystic fibrosis population carrier screening: 2004 revision of American College of Medical Genetics mutation panel. Genet Med 6 (5):387–391
Wei S, Roessler BC, Chauvet S et al (2014) Conserved allosteric hot spots in the transmembrane domains of cystic fibrosis transmembrane conductance regulator (CFTR) channels and multidrug resistance protein (MRP) pumps. J Biol Chem 289:19942–19957
Wilschanski M, Miller LL, Shoseyov D et al (2011) Chronic ataluren (PTC124) treatment of nonsense mutation cystic fibrosis. Eur Respir J 38(1):59–69
Younger JM, Chen L, Ren HY et al (2006) Sequential quality-control checkpoints triage misfolded cystic fibrosis transmembrane conductance regulator. Cell 126:571–582
Yu J, Chen Z, Ni Y et al (2012) CFTR mutations in men with congenital bilateral absence of the vas deferens (CBAVD): a systemic review and meta-analysis. Hum Reprod 27(1):25–35
Zemanick ET, Sagel SD, Harris JK (2011) The airway microbiome in cystic fibrosis and implications for treatment. Curr Opin Pediatr 23(3):319–324
Zielenski J (2000) Genotype and phenotype in cystic fibrosis. Respiration 67(2):117–133
Zielenski J, Tsui LC (1995) Cystic fibrosis: genotypic and phenotypic variations. Annu Rev Genet 29:777–807

Chapter 17
TMEM16 Proteins (Anoctamins) in Epithelia

Paolo Scudieri and Luis J.V. Galietta

Abstract The members of transmembrane 16 (TMEM16) family, also known as anoctamins, are membrane proteins characterized by eight putative transmembrane segments and regulated by cytosolic calcium. TMEM16A (ANO1) and TMEM16B (ANO2) are calcium-activated chloride channels, whereas other anoctamins, such as TMEM16F (ANO6), are associated with calcium-activated scrambling of membrane glycerophospholipids and sphingolipids. TMEM16F also shows ion channel activity although at very high calcium concentrations. In airway, salivary, biliary, and pancreatic epithelial cells, TMEM16A represents a route for chloride and bicarbonate secretion across the apical membrane. In airway epithelial cells, TMEM16A expression is strongly stimulated by Th2 cytokines IL-4 and IL-13. This upregulation occurs in mucus-producing goblet cells thus indicating a particular role of TMEM16A in mucus release and expansion. In intestinal epithelial cells, TMEM16A shows instead a basolateral localization and a different mechanism to modulate transepithelial ion transport.

Keywords Chloride channel • Chloride secretion • Airway epithelium • Goblet cell • Mucociliary clearance • Lipid scramblase • Intracellular calcium • Intestinal epithelium

17.1 General Characteristics of TMEM16 Proteins

In the last years, TMEM16 proteins, also known as anoctamins, have appeared as an important class of macromolecules residing in cell membranes and involved in a variety of physiological functions. In 2008, TMEM16A (ANO1) and TMEM16B (ANO2) were found to act as calcium-activated chloride channels (CaCCs) (Caputo et al. 2008; Schroeder et al. 2008; Yang et al. 2008). Therefore, it was presumed that

P. Scudieri • L.J.V. Galietta (✉)
Istituto Giannina Gaslini, Genova, Italy
e-mail: galietta@unige.it

K.L. Hamilton, D.C. Devor (eds.), *Ion Channels and Transporters of Epithelia in Health and Disease*, Physiology in Health and Disease,
DOI 10.1007/978-1-4939-3366-2_17

the other eight members of the family were also involved in ion transport. However, a few years later, another study revealed that TMEM16F (ANO6) expression is associated with scramblase activity, i.e., translocation of phosphatidylserine from one leaflet to the other of plasma membrane (Suzuki et al. 2010). More recently, it has been reported that other members of the family (TMEM16C, TMEM16D, TMEM16G, and TMEM16J) are also involved in glycerophospholipid and sphingolipid scrambling (Suzuki et al. 2013). Furthermore, it has been shown that a fungal TMEM16 protein has both ion channel and lipid scramblase function (Malvezzi et al. 2013). Hence, anoctamin family may include proteins with diversified and multiple functions.

The precise role and characteristics of each member of the family are not clear yet. Anoctamins seem to share a common structural organization (Fig. 17.1a) with multiple transmembrane segments (probably eight) and the amino- and carboxy-termini protruding into the cytosol (Pedemonte and Galietta 2014). Most, if not all, anoctamins are also regulated by cytosolic calcium concentrations. TMEM16A is activated when calcium reaches concentrations in the submicromolar range (Caputo et al. 2008; Schroeder et al. 2008; Yang et al. 2008) (Fig. 17.1b). TMEM16B instead requires values above 1 μM (Pifferi et al. 2009). TMEM16F is even less sensitive, with activation of membrane currents occurring at 10 μM or higher (Grubb et al. 2013; Shimizu et al. 2013; Yang et al. 2012). Despite the different calcium sensitivity, the mechanism of activation may involve the binding of calcium to two highly conserved acidic amino acid residues localized in the third intracellular loop. In TMEM16A this putative binding site is constituted by two glutamic acid residues at positions 702 and 705 (Yu et al. 2012). Mutagenesis of these residues in TMEM16A and TMEM16F (Yang et al. 2012; Yu et al. 2012), as well as in the fungal TMEM16 (Malvezzi et al. 2013), markedly decreases the apparent affinity for calcium. The presence of the acidic residues in all anoctamins suggests that these channels are all regulated by calcium. Calmodulin has also been postulated to mediate activation of TMEM16A or to modulate its ion selectivity (Tian et al. 2011; Jung et al. 2013). However, the involvement of calmodulin in TMEM16A function has been questioned in a set of recent studies (Yu et al. 2014a, b).

Anoctamins have different patterns of expression among organs and tissues. TMEM16B, TMEM16C, and TMEM16D have a predominant expression in neuronal cells. In this respect, TMEM16B appears to code for the calcium-dependent chloride channel of olfactory neurons and photoreceptors (Stephan et al. 2009; Stöhr et al. 2009). Instead, TMEM16C is involved in the regulation of a sodium-activated potassium channel in neurons (Huang et al. 2013). TMEM16E has a selective expression in skeletal muscle and bone. Indeed, mutations in TMEM16E cause two different human genetic diseases, a form of limb-girdle muscular dystrophy (Bolduc et al. 2010) and gnathodiaphyseal dysplasia (Tsutsumi et al. 2004). Other anoctamins have a more ubiquitous distribution. This is the case of TMEM16F and, to a lower extent, of TMEM16K (Schreiber et al. 2010). According to various studies and to gene expression databases, TMEM16F is the anoctamin with the highest expression in all tissues (Schreiber et al. 2010).

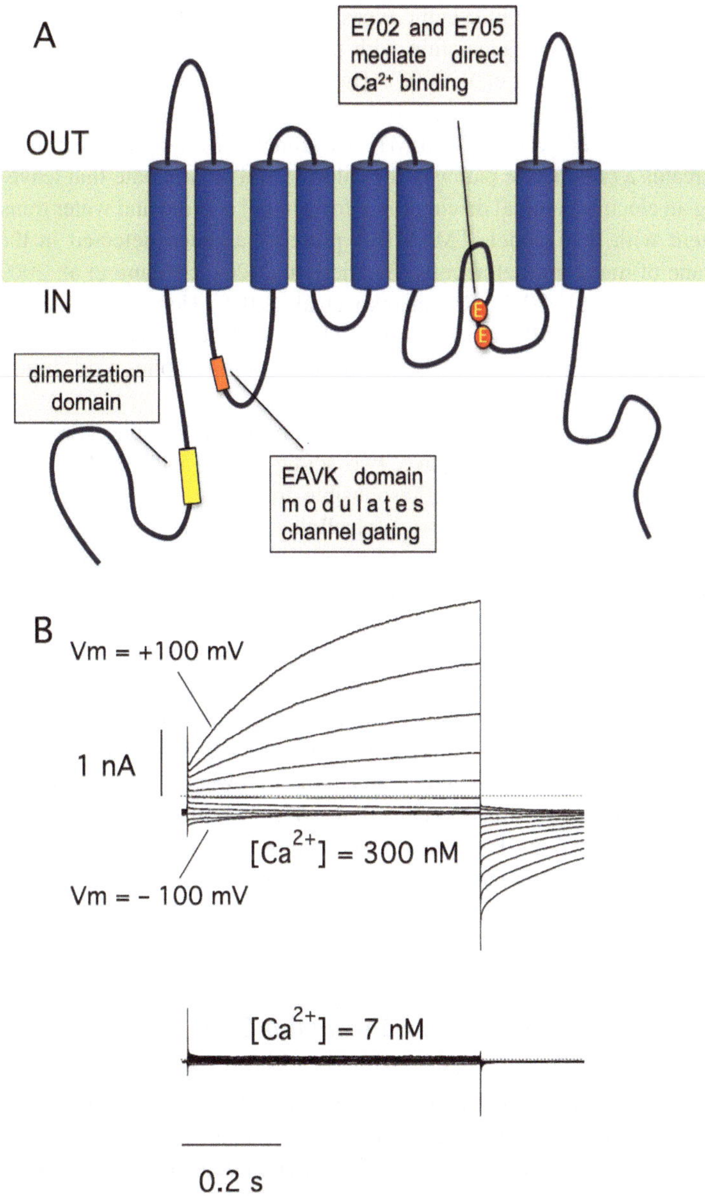

Fig. 17.1 TMEM16A/ANO1 as a calcium-activated chloride channel. (**a**) Predicted structure of TMEM16A. The protein has eight transmembrane segments with both amino- and carboxy-termini exposed to the intracellular environment. A region between the sixth and seventh transmembrane segment, containing two critical glutamic acid residues (E702 and E705), is directly involved in the activation of TMEM16A by intracellular calcium (Yu et al. 2012). A Glu-Ala-Val-Lys (EAVK) sequence (positions 448–451) in the first intracellular loop is also important for channel opening (Xiao et al. 2011). A short sequence in the amino-terminus (positions 161–179) is requested for the formation of TMEM16A homodimers (Tien et al. 2013). (**b**) The figure shows superimposed membrane currents elicited in cells expressing TMEM16A by application of membrane voltages in the range +100 to −100 mV. Currents were recorded with two intracellular

Although present also in specific types of neurons and in some smooth muscle cells, TMEM16A is the anoctamin with particular localization in epithelial cells (Huang et al. 2009; Yang et al. 2008). This suggests a direct role in transepithelial anion secretion. Elevation of cytosolic calcium concentrations by physiological agonists is expected to open TMEM16A channels in the apical membrane. This event creates a conductive pathway for chloride and bicarbonate that leave the cell creating an electrochemical driving force for parallel sodium and water transport. In agreement with this model, TMEM16A protein has been detected in the apical membrane of many epithelial cells (Scudieri et al. 2012; Huang et al. 2009, 2012; Perez-Cornejo et al. 2012; Romanenko et al. 2010). However, as discussed in following sections, TMEM16A has been also found in the basolateral membrane (He et al. 2011; Schreiber et al. 2015). This localization implies a different and still to be clarified function.

Among all anoctamins, TMEM16A is the most studied one (Pedemonte and Galietta 2014). As discussed above, two acidic amino acid residues in the third intracellular loop appear directly responsible for channel activation by cytosolic calcium (Yu et al. 2012). The first intracellular loop, a region whose amino acid sequence is affected by alternative splicing, is also involved in the modulation of channel activity by calcium and membrane voltage (Ferrera et al. 2009; Xiao et al. 2011). Another region of 19 amino acid residues in the amino-terminus (Tien et al. 2013) is instead involved in the formation of TMEM16A homodimers (Fallah et al. 2011; Sheridan et al. 2011). It is not clear yet if the dimer forms a single or a double conductive pathway. In one study, several proteins were found to interact with TMEM16A by using a proteomics approach. In particular, the ezrin-radixin-moesin complex appeared to bind TMEM16A in the plasma membrane and to regulate its activity (Perez-Cornejo et al. 2012). Another functionally important interacting protein is the IP3 receptor. Interaction between TMEM16A and IP3R1 tethers the plasma membrane to endoplasmic reticulum thus generating microdomains important for calcium signaling (Jin et al. 2013).

17.2 TMEM16A in Airway Epithelial Cells

One of the three studies that first reported TMEM16A as a chloride channel was specifically focused on airway epithelial cells (Caputo et al. 2008). The starting point in this study was the observation that calcium-activated chloride secretion in human bronchial epithelial cells is strongly upregulated by chronic treatment with interleukin-4 (IL-4) (Danahay et al. 2002; Galietta et al. 2002). This observation suggested a possible mechanism based on increased expression of the gene coding

---

Fig. 17.1 (continued) calcium concentrations: 7 nM and 300 nM. With low calcium, no currents are activated at any membrane voltage. At high calcium, large currents are detected, with increasing activation as the membrane potential is made more positive

for the calcium-activated chloride channel. Based on this hypothesis, TMEM16A was identified with a strategy based on global gene expression profiling of cells treated with IL-4 followed by silencing of the candidate genes with siRNA (Caputo et al. 2008). Other studies subsequently confirmed that TMEM16A is expressed in airway epithelial cells and involved in electrolyte/fluid secretion. In particular, with the generation of a knockout mouse, it was possible to demonstrate that Tmem16a loss of function causes a reduced anion secretion and impaired mucociliary clearance in the airways (Ousingsawat et al. 2009; Rock et al. 2009). Interestingly, the knockout mouse showed also a malformation of trachea at birth, i.e., tracheomalacia (Rock et al. 2008). It is not clear if this defect results from defective TMEM16A function in the epithelium or in the smooth muscle/connective tissue.

IL-4 and IL-13 are cytokines particularly involved in the Th2 immune response that underlies asthma. In the airway epithelium, Th2 cytokines promote mucous metaplasia, i.e., an increase in the number of mucus-secreting goblet cells and a parallel decrease in ciliated cells (Curran and Cohn 2010; Park et al. 2007) (Fig. 17.2a). The induction of TMEM16A by Th2 cytokines suggested a relationship with mucus production. Indeed, two studies revealed that TMEM16A is particularly expressed in goblet cells (Huang et al. 2012; Scudieri et al. 2012). This finding was obtained in cultured cells treated with Th2 cytokines, in mice with experimentally induced Th2 airway inflammation, and in asthmatic patients. Interestingly, the release of mucus from goblet cells was reduced using a pharmacological inhibitor of TMEM16A (Huang et al. 2012). These findings suggest that TMEM16A may be a target for lung diseases characterized by mucus hypersecretion.

The specific localization of TMEM16A in goblet cells suggests a role in the mucin hypersecretion that characterizes asthmatic airways. In this respect, it has been shown that secretion of anions, particularly of bicarbonate, is required to support mucin release and expansion from surface goblet cells and submucosal glands and that TMEM16A is permeable to bicarbonate (Garcia et al. 2009; Gustafsson et al. 2012; Hoegger et al. 2014). Therefore, upregulation of TMEM16A could be a mechanism to adjust anion secretion levels to increased production of mucus.

It is important to note that airway epithelia express the cystic fibrosis transmembrane conductance regulator (CFTR) protein as another type of chloride channel (Riordan 2008). CFTR, which is regulated by cAMP-dependent phosphorylation, is absolutely important to maintain the human airway surface clean and microbial-free. In patients with cystic fibrosis (CF), loss of CFTR function in the airways causes impaired mucociliary clearance, accumulation of dense mucus, and loss of bactericidal activity (Hoegger et al. 2014; Matsui et al. 1998; Pezzulo et al. 2012). Because of their similar activity as anion channels, it may be argued that CFTR and TMEM16A have a complementary or overlapping function in the respiratory system. However, it is not clear yet if TMEM16A has a role in CF. The presence of a severe lung disease in CF patients could be taken as a demonstration that TMEM16A cannot compensate for CFTR deficit. On the other hand, the large differences in lung phenotype observed in CF patients having the same set of CFTR mutations indicate that other genes, acting as

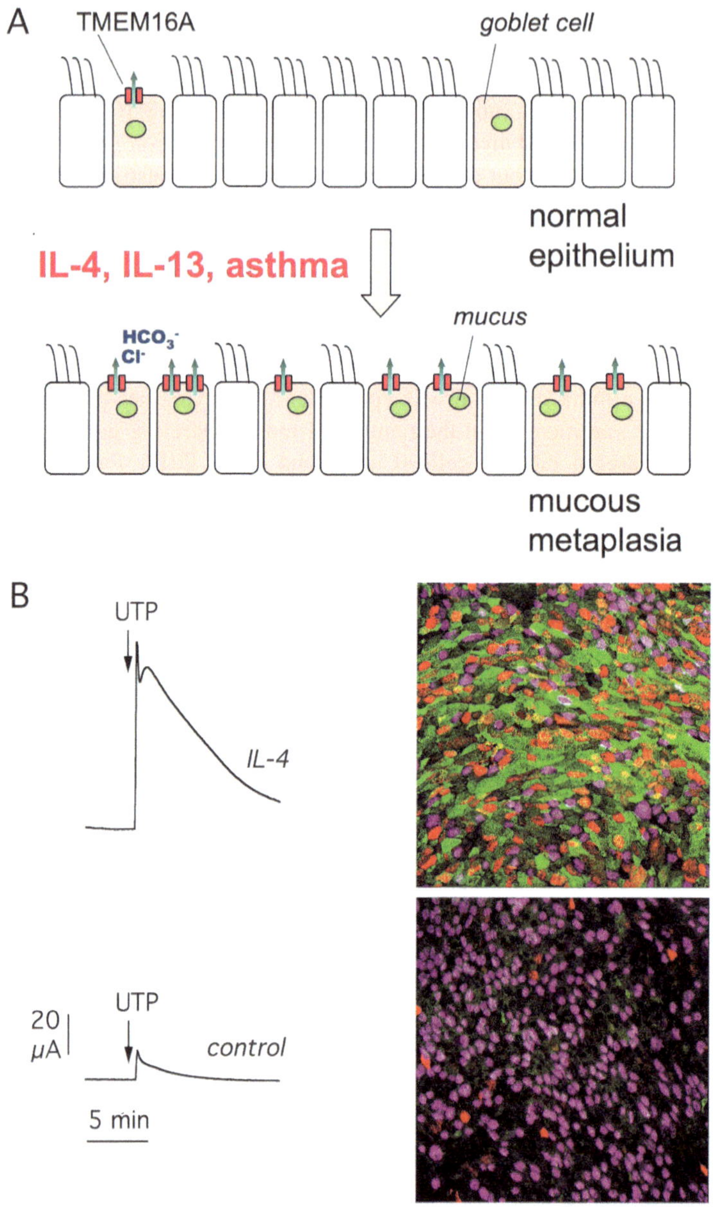

Fig. 17.2 TMEM16A/ANO1 in airway epithelium. (**a**) Mucous metaplasia induced by Th2 cytokines in the airway epithelium. IL-4 and IL-13 increase the abundance of mucus-producing goblet cells and decrease the number of ciliated cells. (**b**) Increased expression and function of TMEM16A in cells treated with Th2 cytokines. Treatment of cultured human bronchial epithelial cells with IL-4 increases calcium-dependent chloride secretion as demonstrated by the larger peak of short-circuit current induced by a purinergic agonist, UTP (*left panel*). Treatment with IL-4 also increases TMEM16A protein expression, with a particular co-localization with MUC5AC mucin, a marker of goblet cells (*right panel*). The figure shows confocal microscope images reporting detection by immunofluorescence of TMEM16A (*green*), MUC5AC (*red*), and acetylated tubulin (*magenta*)

modifiers, are also involved. It could be hypothesized that differences in TMEM16A expression and function among CF patients modulate the severity of the disease. In fact, the increased calcium-activated chloride secretion is suspected to be the reason why CF mice do not show a lung disease (Grubb et al. 1994). It would be interesting to study Tmem16a knockout mice to determine if they develop a CF-like lung disease. However, the severity of the phenotype shown by the TMEM16A−/− animal has precluded further studies to understand the role of the protein in adult airways. Indeed, general knockout mice mostly die a few days after birth (Ousingsawat et al. 2009; Rock et al. 2009). More useful information will be obtained by generating a conditional knockout animal with selective destruction of TMEM16A in the airway epithelium. This animal model would be important for the development of therapeutic strategies aiming at TMEM16A as a target. In fact, it is not clear if activators or inhibitors are needed to treat patients with CF or asthma. On one hand, stimulation of TMEM16A function is theoretically useful to hydrate airway surface and to support mucociliary clearance. On the other hand, inhibition of TMEM16A was found to reduce mucin release (Huang et al. 2012), an effect that could prevent accumulation of mucus in airway lumen.

As discussed above, it is quite established that TMEM16A protein underlies the calcium-dependent anion secretion in airway epithelial cells under inflammatory conditions (Caputo et al. 2008; Huang et al. 2012; Scudieri et al. 2012). It is less clear if another type of calcium-activated chloride channel is prevalent under resting conditions. Actually, in a study based on a pharmacological inhibitor, it was postulated that TMEM16A has a minor role in the airway epithelium (Namkung et al. 2011a). This contrasts with studies in which silencing or genetic knockdown of TMEM16A expression resulted in marked decrease in CaCC activity (Caputo et al. 2008; Ousingsawat et al. 2009; Rock et al. 2009). Furthermore, the expression of TMEM16A in airway epithelia is also supported by results obtained with a TMEM16A pharmacological activator. This compound significantly stimulated calcium-dependent chloride secretion (Namkung et al. 2011b). In conclusion, the participation of TMEM16A to transepithelial anion transport in the airways is demonstrated by several lines of evidence. However, the possible existence of another type of CaCC cannot be ruled out. Regarding the relative contribution of TMEM16A to total anion transport in airway epithelia, we have to consider that the expression of this protein appears to change widely by different stimuli. As seen before, TMEM16A is markedly upregulated by Th2 cytokines (Huang et al. 2012; Scudieri et al. 2012). Recently, downregulation of TMEM16A expression and function has been found when airway cells are treated with TGF-beta (Sun et al. 2014). Therefore, the expression of TMEM16A in the airway epithelium is a highly dynamic phenomenon probably affected in vivo by a variety of physiological and pathological stimuli.

In addition to airway surface epithelial cells, submucosal glands also show expression of TMEM16A protein (Fischer et al. 2010; Lee and Foskett 2010). In particular, TMEM16A was found more abundant in mucous than in serous cells, a finding consistent with a relationship between TMEM16A-dependent anion secretion and mucus secretion (Fischer et al. 2010).

17.3 TMEM16A Expression in the Gastrointestinal Tract

TMEM16A is present in different cells of the gastrointestinal tract. In salivary glands, TMEM16A has a strong expression responsible for fluid secretion. In vivo silencing of TMEM16A in mice by injection of siRNA resulted in decreased salivary fluid secretion (Yang et al. 2008). Also, salivary gland cells from Tmem16a knockout mice were devoid of calcium-activated chloride currents and of carbachol-induced chloride efflux (Romanenko et al. 2010).

TMEM16A protein was also found in human, rat, and mouse biliary epithelial cells (Dutta et al. 2011, 2013). TMEM16A channel activity underlies the calcium-dependent chloride and fluid secretion elicited by purinergic agonists. This function probably contributes to the volume and composition of bile. Interestingly, treatment of biliary epithelial cells with IL-4 increased TMEM16A expression and function (Dutta et al. 2011). This result indicates that upregulation by Th2 cytokines is not restricted to the airway epithelium but is a more general phenomenon.

The physiological role of TMEM16A in the intestinal epithelium seems different from that of other organs. Indeed, two separate studies found evidence for localization of TMEM16A protein in the basolateral and not in the apical membrane (He et al. 2011; Schreiber et al. 2015). However, the two studies arrived to different conclusions about the contribution of TMEM16A to chloride secretion. In one study, the use of a TMEM16A inhibitor, $CaCC_{inh}$-A01, resulted in increased chloride secretion (He et al. 2011). The authors concluded that opening of basolateral TMEM16A channels favors potassium secretion and limits the rate of chloride secretion. In the other study, TMEM16A contribution to intestinal ion transport was investigated by generating a conditional knockout mouse, with selective ablation of TMEM16A gene in intestinal epithelial cells (Schreiber et al. 2015). Calcium-dependent chloride secretion was significantly diminished in the ileum and large intestine of TMEM16A−/− animals. Importantly, this effect was due to a reduction of calcium mobilization by cholinergic agents (Schreiber et al. 2015). Therefore, based on these results, TMEM16A would be important to support chloride secretion not as an apical membrane conductive pathway but as a facilitator of the calcium increase that is needed to activate basolateral potassium channels (Schreiber et al. 2015). The link between TMEM16A and calcium increase is not clear, but Schreiber et al. (2015) postulated a possible role in the release of calcium from the endoplasmic reticulum (ER). According to this model, TMEM16A in intestinal cells would be an ER channel that provides the counterions that need to match the exit of calcium. Another possibility is that TMEM16A tethers the ER to the plasma membrane thus facilitating the release of calcium from intracellular stores and the activation of potassium channels (Schreiber et al. 2015). In this respect, Ist2 protein, a TMEM16A yeast homologue, is involved in the formation of contact sites between ER and plasma membrane (Wolf et al. 2012).

TMEM16A is presumably expressed also in exocrine pancreas. This assumption is based on the finding that TMEM16A is highly expressed in several cell lines derived from the pancreatic ducts such as CFPAC-1, AsPC-1, BxPC-3, and Capan-1 (Caputo et al. 2008; Wang et al. 2013; Sauter et al. 2015). In the duct, TMEM16A could represent a route for the calcium-dependent regulation of bicarbonate secretion. However, the pancreatic duct cell lines used in the different studies are derived from cancers of the exocrine pancreas, and therefore it is not clear if part of the high expression is due to the tumoral transformation. In fact, TMEM16A expression has been linked to cancer since it seems to affect cell proliferation and/or migration (Britschgi et al. 2013; Qu et al. 2014; Sauter et al. 2015). More studies are needed to assess the expression and role of TMEM16A protein in the normal pancreas in vivo.

17.4 TMEM16A in the Kidney

TMEM16A was detected in human and mouse kidney, with a predominantly apical membrane localization in proximal tubular epithelial cells and additional expression in podocytes and the collecting duct (Faria et al. 2014; Svenningsen et al. 2014). TMEM16A was also found expressed in the renal cyst-lining epithelium of individuals affected by autosomal dominant polycystic kidney disease (ADPKD) and in vitro cyst models (Buchholz et al. 2014). Interestingly, the in vitro models revealed a dynamic cellular localization of TMEM16A controlled by cAMP. Indeed, non-stimulated cysts derived from principal-like MDCK cells (PLCs) showed a diffuse cytosolic and plasma membrane TMEM16A expression, whereas stimulation with forskolin led to luminal membrane localization of TMEM16A protein.

Polycystic disease is characterized by multiple renal cysts that gradually enlarge leading to compression of the adjacent normal tissue and decline of the renal function. Enlargement of kidney cysts is due to both proliferation of the cyst-lining cells and fluid secretion into the cyst lumen. Besides a cAMP-dependent chloride secretion through CFTR that represents the major driving force, it was thought that an additional ion channel might also contribute to electrolyte secretion that leads to cyst enlargement. Buchholz and colleague showed indeed that MDCK cells display calcium-dependent chloride secretion in response to purinergic stimulation and that knockdown of TMEM16A protein expression or pharmacological inhibition of TMEM16A function abolishes this process (Buchholz et al. 2014). Importantly, knockdown or inhibition of TMEM16A also resulted in reduced cyst growth in PLCs and in the metanephric kidney cyst model. Furthermore, knockdown of TMEM16A also decreased the number of cysts proliferating cells, a finding consistent with the involvement of TMEM16A in promoting cell proliferation (Qu et al. 2014).

Important information about TMEM16A function in kidney was obtained through the generation of kidney-specific knockout mice. Selective ablation of TMEM16A in tubular epithelial cells, but not in podocytes, affects renal function (Faria et al. 2014). Tubular knockout of TMEM16A seems associated with a defect in proximal tubular protein reabsorption that leads to accumulation of large intracellular vesicles in proximal tubular epithelial cells, enhanced urine protein excretion with macroalbuminuria, and reduced urine electrolyte concentrations (Faria et al. 2014). Proximal tubular epithelial cells from wild-type animals exhibit chloride currents activated by purinergic stimulation and inhibited by CaCC inhibitor tannic acid. Chloride currents in these cells, but not in cells derived from the TMEM16A−/− animal, were also activated by extracellular acidification, thus suggesting a role of TMEM16A during proton secretion (Faria et al. 2014).

17.5 TMEM16F and Other Anoctamins in Epithelial Cells

TMEM16F expression is associated with the calcium-activated scrambling of phosphatidylserine in the plasma membrane (Suzuki et al. 2010). It has been shown that TMEM16F has also ion channel activity. However, the characteristics of the ion currents generated by TMEM16F are a matter of controversy. In one study, TMEM16F was reported to form a cation selective channel (Yang et al. 2012). In other studies, TMEM16F was instead found to carry anions (Grubb et al. 2013; Shimizu et al. 2013). The reason for these discrepancies is not clear. Another intriguing characteristic is the activation at nonphysiological high calcium concentrations and the rectification of current-voltage relationship. Regarding this latter aspect, TMEM16F currents were essentially seen only at very positive membrane potentials.

Because of its high expression in all cell types, TMEM16F is expected to have a physiological function also in epithelial cells (Schreiber et al. 2010). However, patients affected by Scott syndrome do not show any abnormality besides a defect in blood coagulation (Weiss 1994). This problem is due to reduced externalization of phosphatidylserine on the surface of platelets. The coagulation defect was also confirmed in a TMEM16F knockout mouse (Yang et al. 2012). Besides a prolonged bleeding time, this animal did not show other defects thus indicating that absence of TMEM16F in epithelial cells is without clear consequences. In a second type of knockout mouse, TMEM16F loss resulted in a significant defect in osteoblast function with severe skeletal abnormalities (Ehlen et al. 2013) but without a clear dysfunction in epithelial cells. As a scramblase, TMEM16F may be involved in a variety of cell functions. To explain the absence of an epithelial phenotype in TMEM16F−/− animals, it is possible to hypothesize that other anoctamins compensate for the absence of TMEM16F.

Very little information is available on the role of other anoctamins in epithelial cells. However, in a recent study, a mouse with selective deletion of Tmem16K (Ano10) in intestinal epithelial cells has been generated (Schreiber et al. 2015).

This animal showed defective calcium-dependent chloride secretion in the jejunum in contrast to the conditional TMEM16Aa−/− mouse in which the defect was observed in the ileum and large intestine. Therefore, TMEM16A and TMEM16K have a similar role in supporting chloride secretion but in separate regions of the intestine. It is not clear if TMEM16K has a basolateral localization similar to that of TMEM16A in intestinal epithelial cells (Schreiber et al. 2015).

17.6 Perspectives

The recent discovery of anoctamins has provided the molecular basis for calcium-activated chloride transport and phospholipid scrambling. However, future studies, also based on animal models and human genetic diseases, may reveal novel and unsuspected functions for this family of membrane proteins. Particularly interesting is the possibility that anoctamins represent novel pharmacological targets for the treatment of a series of diseases. In epithelial cells, pharmacological stimulation or inhibition of anoctamins may be useful to modulate the volume and composition of secreted fluids. The association of TMEM16A expression with mucus hypersecretion requires further studies to understand if pharmacological modulation of TMEM16A is beneficial for cystic fibrosis, asthma, and other respiratory diseases.

Acknowledgments Our work was supported by Ministero della Salute (Ricerca Finalizzata: GR-2009-1596824; Ricerca Corrente: Cinque per mille), Telethon Foundation (GGP10026), and Fondazione Italiana Fibrosi Cistica (FFC#2/2012).

References

Bolduc V, Marlow G, Boycott KM, Saleki K, Inoue H, Kroon J, Itakura M, Robitaille Y, Parent L, Baas F, Mizuta K, Kamata N, Richard I, Linssen WH, Mahjneh I, de Visser M, Bashir R, Brais B (2010) Recessive mutations in the putative calcium-activated chloride channel Anoctamin 5 cause proximal LGMD2L and distal MMD3 muscular dystrophies. Am J Hum Genet 86:213–221

Britschgi A, Bill A, Brinkhaus H, Rothwell C, Clay I, Duss S, Rebhan M, Raman P, Guy CT, Wetzel K, George E, Popa MO, Lilley S, Choudhury H, Gosling M, Wang L, Fitzgerald S, Borawski J, Baffoe J, Labow M, Gaither LA, Bentires-Alj M (2013) Calcium-activated chloride channel ANO1 promotes breast cancer progression by activating EGFR and CAMK signaling. Proc Natl Acad Sci U S A 110:E1026–E1034

Buchholz B, Faria D, Schley G, Schreiber R, Eckardt KU, Kunzelmann K (2014) Anoctamin 1 induces calcium-activated chloride secretion and proliferation of renal cyst–forming epithelial cells. Kidney Int 85:1058–1067

Caputo A, Caci E, Ferrera L, Pedemonte N, Barsanti C, Sondo E, Pfeffer U, Ravazzolo R, Zegarra-Moran O, Galietta LJ (2008) TMEM16A, a membrane protein associated with calcium-dependent chloride channel activity. Science 322:590–594

Curran DR, Cohn L (2010) Advances in mucous cell metaplasia: a plug for mucus as a therapeutic focus in chronic airway disease. Am J Respir Cell Mol Biol 42:268–275

Danahay H, Atherton H, Jones G, Bridges RJ, Poll CT (2002) Interleukin-13 induces a hypersecretory ion transport phenotype in human bronchial epithelial cells. Am J Physiol Lung Cell Mol Physiol 282:L226–L236

Dutta AK, Khimji AK, Kresge C, Bugde A, Dougherty M, Esser V, Ueno Y, Glaser SS, Alpini G, Rockey DC, Feranchak AP (2011) Identification and functional characterization of TMEM16A, a Ca^{2+}-activated Cl^- channel activated by extracellular nucleotides, in biliary epithelium. J Biol Chem 286:766–776

Dutta AK, Woo K, Khimji AK, Kresge C, Feranchak AP (2013) Mechanosensitive Cl^- secretion in biliary epithelium mediated through TMEM16A. Am J Physiol Gastrointest Liver Physiol 304:G87–G98

Ehlen HW, Chinenkova M, Moser M, Munter HM, Krause Y, Gross S, Brachvogel B, Wuelling M, Kornak U, Vortkamp A (2013) Inactivation of anoctamin-6/Tmem16f, a regulator of phosphatidylserine scrambling in osteoblasts, leads to decreased mineral deposition in skeletal tissues. J Bone Miner Res 28:246–259

Fallah G, Römer T, Detro-Dassen S, Braam U, Markwardt F, Schmalzing G (2011) TMEM16A(a)/anoctamin-1 shares a homodimeric architecture with CLC chloride channels. Mol Cell Proteomics 10:1–10

Faria D, Rock JR, Romao AM, Schweda F, Bandulik S, Witzgall R, Schlatter E, Heitzmann D, Pavenstädt H, Herrmann E, Kunzelmann K, Schreiber R (2014) The calcium-activated chloride channel Anoctamin 1 contributes to the regulation of renal function. Kidney Int 85:1369–1381

Ferrera L, Caputo A, Ubby I, Bussani E, Zegarra-Moran O, Ravazzolo R, Pagani F, Galietta LJ (2009) Regulation of TMEM16A chloride channel properties by alternative splicing. J Biol Chem 284:33360–33368

Fischer H, Illek B, Sachs L, Finkbeiner WE, Widdicombe JH (2010) CFTR and calcium-activated chloride channels in primary cultures of human airway gland cells of serous or mucous phenotype. Am J Physiol Lung Cell Mol Physiol 299:L585–L594

Galietta LJ, Pagesy P, Folli C, Caci E, Romio L, Costes B, Nicolis E, Cabrini G, Goossens M, Ravazzolo R, Zegarra-Moran O (2002) IL-4 is a potent modulator of ion transport in the human bronchial epithelium in vitro. J Immunol 168:839–845

Garcia MA, Yang N, Quinton PM (2009) Normal mouse intestinal mucus release requires cystic fibrosis transmembrane regulator-dependent bicarbonate secretion. J Clin Invest 119:2613–2622

Grubb BR, Vick RN, Boucher RC (1994) Hyperabsorption of Na^+ and raised Ca^{2+}-mediated Cl^- secretion in nasal epithelia of CF mice. Am J Physiol Cell Physiol 266:C1478–C1483

Grubb S, Poulsen KA, Juul CA, Kyed T, Klausen TK, Larsen EH, Hoffmann EK (2013) TMEM16F (Anoctamin 6), an anion channel of delayed Ca^{2+} activation. J Gen Physiol 141:585–600

Gustafsson JK, Ermund A, Ambort D, Johansson ME, Nilsson HE, Thorell K, Hebert H, Sjövall H, Hansson GC (2012) Bicarbonate and functional CFTR channel are required for proper mucin secretion and link cystic fibrosis with its mucus phenotype. J Exp Med 209:1263–1272

He Q, Halm ST, Zhang J, Halm DR (2011) Activation of the basolateral membrane Cl^- conductance essential for electrogenic K^+ secretion suppresses electrogenic Cl^- secretion. Exp Physiol 96:305–316

Hoegger MJ, Fischer AJ, McMenimen JD, Ostedgaard LS, Tucker AJ, Awadalla MA, Moninger TO, Michalski AS, Hoffman EA, Zabner J, Stoltz DA, Welsh MJ (2014) Cystic fibrosis. Impaired mucus detachment disrupts mucociliary transport in a piglet model of cystic fibrosis. Science 345:818–822

Huang F, Rock JR, Harfe BD, Cheng T, Huang X, Jan YN, Jan LY (2009) Studies on expression and function of the TMEM16A calcium-activated chloride channel. Proc Natl Acad Sci U S A 106:21413–21418

Huang F, Zhang H, Wu M, Yang H, Kudo M, Peters CJ, Woodruff PG, Solberg OD, Donne ML, Huang X, Sheppard D, Fahy JV, Wolters PJ, Hogan BL, Finkbeiner WE, Li M, Jan YN, Jan LY, Rock JR (2012) Calcium-activated chloride channel TMEM16A modulates mucin secretion and airway smooth muscle contraction. Proc Natl Acad Sci U S A 109:16354–16359

Huang F, Wang X, Ostertag EM, Nuwal T, Huang B, Jan YN, Basbaum AI, Jan LY (2013) TMEM16C facilitates Na^+-activated K^+ currents in rat sensory neurons and regulates pain processing. Nat Neurosci 16:1284–1290

Jin X, Shah S, Liu Y, Zhang H, Lees M, Fu Z, Lippiat JD, Beech DJ, Sivaprasadarao A, Baldwin SA, Zhang H, Gamper N (2013) Activation of the Cl^- channel ANO1 by localized calcium signals in nociceptive sensory neurons requires coupling with the IP3 receptor. Sci Signal 6: ra73

Jung J, Nam JH, Park HW, Oh U, Yoon JH, Lee MG (2013) Dynamic modulation of ANO1/ TMEM16A HCO_3^- permeability by Ca^{2+}/calmodulin. Proc Natl Acad Sci U S A 110:360–365

Lee RJ, Foskett JK (2010) Mechanisms of Ca^{2+}-stimulated fluid secretion by porcine bronchial submucosal gland serous acinar cells. Am J Physiol Lung Cell Mol Physiol 298:L210–L231

Malvezzi M, Chalat M, Janjusevic R, Picollo A, Terashima H, Menon AK, Accardi A (2013) Ca^{2+}-dependent phospholipid scrambling by a reconstituted TMEM16 ion channel. Nat Commun 4:2367

Matsui H, Randell SH, Peretti SW, Davis CW, Boucher RC (1998) Coordinated clearance of periciliary liquid and mucus from airway surfaces. J Clin Invest 102:1125–1131

Namkung W, Phuan PW, Verkman AS (2011a) TMEM16A inhibitors reveal TMEM16A as a minor component of calcium-activated chloride channel conductance in airway and intestinal epithelial cells. J Biol Chem 286:2365–2374

Namkung W, Yao Z, Finkbeiner WE, Verkman AS (2011b) Small-molecule activators of TMEM16A, a calcium-activated chloride channel, stimulate epithelial chloride secretion and intestinal contraction. FASEB J 25:4048–4062

Ousingsawat J, Martins JR, Schreiber R, Rock JR, Harfe BD, Kunzelmann K (2009) Loss of TMEM16A causes a defect in epithelial Ca^{2+}-dependent chloride transport. J Biol Chem 284:28698–28703

Park KS, Korfhagen TR, Bruno MD, Kitzmiller JA, Wan H, Wert SE, Khurana Hershey GK, Chen G, Whitsett JA (2007) SPDEF regulates goblet cell hyperplasia in the airway epithelium. J Clin Invest 117:978–988

Pedemonte N, Galietta LJ (2014) Structure and function of TMEM16 proteins (anoctamins). Physiol Rev 94:419–459

Perez-Cornejo P, Gokhale A, Duran C, Cui Y, Xiao Q, Hartzell HC, Faundez V (2012) Anoctamin 1 (Tmem16A) Ca^{2+}-activated chloride channel stoichiometrically interacts with an ezrin-radixin-moesin network. Proc Natl Acad Sci U S A 109:10376–10381

Pezzulo AA, Tang XX, Hoegger MJ, Alaiwa MH, Ramachandran S, Moninger TO, Karp PH, Wohlford-Lenane CL, Haagsman HP, van Eijk M, Bánfi B, Horswill AR, Stoltz DA, McCray PB Jr, Welsh MJ, Zabner J (2012) Reduced airway surface pH impairs bacterial killing in the porcine cystic fibrosis lung. Nature 487:109–113

Pifferi S, Dibattista M, Menini A (2009) TMEM16B induces chloride currents activated by calcium in mammalian cells. Pflugers Arch 458:1023–1038

Qu Z, Yao W, Yao R, Liu X, Yu K, Hartzell C (2014) The Ca^{2+}-activated Cl^- channel, ANO1 (TMEM16A), is a double-edged sword in cell proliferation and tumorigenesis. Cancer Med 3:453–461

Riordan JR (2008) CFTR function and prospects for therapy. Annu Rev Biochem 77:701–726

Rock JR, Futtner CR, Harfe BD (2008) The transmembrane protein TMEM16A is required for normal development of the murine trachea. Dev Biol 321:141–149

Rock JR, O'Neal WK, Gabriel SE, Randell SH, Harfe BD, Boucher RC, Grubb BR (2009) Transmembrane protein 16A (TMEM16A) is a Ca^{2+}-regulated Cl^- secretory channel in mouse airways. J Biol Chem 284:14875–14880

Romanenko VG, Catalán MA, Brown DA, Putzier I, Hartzell HC, Marmorstein AD, Gonzalez-Begne M, Rock JR, Harfe BD, Melvin JE (2010) Tmem16A encodes the Ca^{2+}-activated Cl^{-} channel in mouse submandibular salivary gland acinar cells. J Biol Chem 285:12990–13001
Sauter DR, Novak I, Pedersen SF, Larsen EH, Hoffmann EK (2015) ANO1 (TMEM16A) in pancreatic ductal adenocarcinoma (PDAC). Pflugers Arch 467(7):1495–1508
Schreiber R, Uliyakina I, Kongsuphol P, Warth R, Mirza M, Martins JR, Kunzelmann K (2010) Expression and function of epithelial anoctamins. J Biol Chem 285:7838–7845
Schreiber R, Faria D, Skryabin BV, Wanitchakool P, Rock JR, Kunzelmann K (2015) Anoctamins support calcium-dependent chloride secretion by facilitating calcium signaling in adult mouse intestine. Pflugers Arch 467(6):1203–1213
Schroeder BC, Cheng T, Jan YN, Jan LY (2008) Expression cloning of TMEM16A as a calcium-activated chloride channel subunit. Cell 134:1019–1029
Scudieri P, Caci E, Bruno S, Ferrera L, Schiavon M, Sondo E, Tomati V, Gianotti A, Zegarra-Moran O, Pedemonte N, Rea F, Ravazzolo R, Galietta LJ (2012) Association of TMEM16A chloride channel overexpression with airway goblet cell metaplasia. J Physiol 590:6141–6155
Sheridan JT, Worthington EN, Yu K, Gabriel SE, Hartzell HC, Tarran R (2011) Characterization of the oligomeric structure of the Ca^{2+}-activated Cl^{-} channel Ano1/TMEM16A. J Biol Chem 286:1381–1388
Shimizu T, Iehara T, Sato K, Fujii T, Sakai H, Okada Y (2013) TMEM16F is a component of a Ca^{2+}-activated Cl^{-} channel but not a volume-sensitive outwardly rectifying Cl^{-} channel. Am J Physiol Cell Physiol 304:C748–C759
Stephan AB, Shum EY, Hirsh S, Cygnar KD, Reisert J, Zhao H (2009) ANO2 is the cilial calcium-activated chloride channel that may mediate olfactory amplification. Proc Natl Acad Sci U S A 106:11776–11781
Stöhr H, Heisig JB, Benz PM, Schöberl S, Milenkovic VM, Strauss O, Aartsen WM, Wijnholds J, Weber BH, Schulz HL (2009) TMEM16B, a novel protein with calcium-dependent chloride channel activity, associates with a presynaptic protein complex in photoreceptor terminals. J Neurosci 29:6809–6818
Sun H, Harris WT, Kortyka S, Kotha K, Ostmann AJ, Rezayat A, Sridharan A, Sanders Y, Naren AP, Clancy JP (2014) TGF-beta downregulation of distinct chloride channels in cystic fibrosis-affected epithelia. PLoS One 9, e106842
Suzuki J, Umeda M, Sims PJ, Nagata S (2010) Calcium-dependent phospholipid scrambling by TMEM16F. Nature 468:834–838
Suzuki J, Fujii T, Imao T, Ishihara K, Kuba H, Nagata S (2013) Calcium-dependent phospholipid scramblase activity of TMEM16 protein family members. J Biol Chem 288:13305–13316
Svenningsen P, Nielsen MR, Marcussen N, Walter S, Jensen BL (2014) TMEM16A is a Ca^{2+}-activated Cl^{-} channel expressed in the renal collecting duct. Acta Physiol 212:166–174
Tian Y, Kongsuphol P, Hug M, Ousingsawat J, Witzgall R, Schreiber R, Kunzelmann K (2011) Calmodulin-dependent activation of the epithelial calcium-dependent chloride channel TMEM16A. FASEB J 25:1–11
Tien J, Lee HY, Minor DL Jr, Jan YN, Jan LY (2013) Identification of a dimerization domain in the TMEM16A calcium-activated chloride channel (CaCC). Proc Natl Acad Sci U S A 110:6352–6357
Tsutsumi S, Kamata N, Vokes TJ, Maruoka Y, Nakakuki K, Enomoto S, Omura K, Amagasa T, Nagayama M, Saito-Ohara F, Inazawa J, Moritani M, Yamaoka T, Inoue H, Itakura M (2004) The novel gene encoding a putative transmembrane protein is mutated in gnathodiaphyseal dysplasia (GDD). Am J Hum Genet 74:1255–1261
Wang J, Haanes KA, Novak I (2013) Purinergic regulation of CFTR and Ca^{2+}-activated Cl^{-} channels and K^{+} channels in human pancreatic duct epithelium. Am J Physiol Cell Physiol 304:C673–C684
Weiss HJ (1994) Scott syndrome: a disorder of platelet coagulant activity. Semin Hematol 31:312–319

Wolf W, Kilic A, Schrul B, Lorenz H, Schwappach B, Seedorf M (2012) Yeast Ist2 recruits the endoplasmic reticulum to the plasma membrane and creates a ribosome-free membrane microcompartment. PLoS One 7, e39703

Xiao Q, Yu K, Perez-Cornejo P, Cui Y, Arreola J, Hartzell HC (2011) Voltage- and calcium-dependent gating of TMEM16A/Ano1 chloride channels are physically coupled by the first intracellular loop. Proc Natl Acad Sci U S A 108:8891–8896

Yang YD, Cho H, Koo JY, Tak MH, Cho Y, Shim WS, Park SP, Lee J, Lee B, Kim BM, Raouf R, Shin YK, Oh U (2008) TMEM16A confers receptor-activated calcium-dependent chloride conductance. Nature 455:1210–1215

Yang H, Kim A, David T, Palmer D, Jin T, Tien J, Huang F, Cheng T, Coughlin SR, Jan YN, Jan LY (2012) TMEM16F forms a Ca^{2+}-activated cation channel required for lipid scrambling in platelets during blood coagulation. Cell 151:111–122

Yu K, Duran C, Qu Z, Cui YY, Hartzell HC (2012) Explaining calcium-dependent gating of anoctamin-1 chloride channels requires a revised topology. Circ Res 110:990–999

Yu K, Zhu J, Qu Z, Cui YY, Hartzell HC (2014a) Activation of the Ano1 (TMEM16A) chloride channel by calcium is not mediated by calmodulin. J Gen Physiol 143:253–267

Yu Y, Kuan AS, Chen TY (2014b) Calcium-calmodulin does not alter the anion permeability of the mouse TMEM16A calcium-activated chloride channel. J Gen Physiol 144:115–124

Chapter 18
Epithelial Sodium Channels (ENaCs)

Tiffany L. Tai, He-Ping Ma, and Douglas C. Eaton

Abstract Amiloride-sensitive epithelial sodium (Na^+) channels (ENaCs) play a crucial role in Na^+ transport and fluid reabsorption in the kidney, lung, and colon. ENaC is made up of three homologous subunits, α, β, and γ. The magnitude of ENaC-mediated Na^+ transport in epithelial cells depends on the average open probability of the channels and the number of channels on the apical surface of epithelial cells. The open probability of ENaC depends upon the association of ENaCs with inositol lipid phosphates and association with a collection of chaperone proteins including calmodulin and MARCKS (myristoylated alanine-rich C-kinase substrate). MARCKS when associated with the membrane sequesters PIP_2 and increases open probability and channel stability. The number of channels in the apical membrane, in turn, depends upon a balance between the rate of ENaC insertion and the rate of removal from the apical membrane. To this end, ENaC is associated with an additional collection of proteins, the ENaC regulatory complex (ERC) that contains proteins including SGK1, GILZ1, ERK, and Nedd4-2 that regulate channel stability in the membrane. The ERC under basal conditions contains several molecules including ERK and Nedd4-2 that inhibit ENaC activity. The C-terminal domain of all three subunits is intracellular and contains a proline-rich motif (PPxY) to which one isoform of the ubiquitin ligase, Nedd4-2 (neural precursor cell-expressed developmentally downregulated protein), binds, thereby, promoting ubiquitin conjugation of the channel subunits followed by internalization and degradation or recycling. In the presence of aldosterone, several proteins including SGK1, GILZ1, and MARCKS are recruited to the complex and stimulate ENaC activity. In addition, aldosterone promotes insertion of channels from intracellular pools and transcription/translation of new ENaC. Other agents that alter ENaC activity work by modulating PIP_2 presentation to the channel or by altering the rate of retrieval, recycling, insertion, or transcription/translation of ENaC.

T.L. Tai • H.-P. Ma • D.C. Eaton (✉)
Department of Physiology, Emory University School of Medicine, 615 Michael Street, Atlanta, GA 30322, USA

The Center for Cell and Molecular Signaling, Emory University School of Medicine, 615 Michael Street, Atlanta, GA 30322, USA
e-mail: deaton@emory.edu

K.L. Hamilton, D.C. Devor (eds.), *Ion Channels and Transporters of Epithelia in Health and Disease*, Physiology in Health and Disease,
DOI 10.1007/978-1-4939-3366-2_18

Keywords ENaC • Epithelial sodium channels • Aldosterone • ENaC structure • ENaC regulation • ENaC regulatory complex • MARCKS • ENaC trafficking • ENaC pathology • Hypertension • Kidney • Lung

18.1 Introduction

Epithelial tissues, in general, are designed to form a barrier between two body compartments and to facilitate transport from one compartment to the other, thereby regulating the composition of the compartments. Many epithelial tissues are responsible for the movement of salt and water to maintain total body salt and water balance. This regulation of salt and water balance involves primary active transport of sodium (Na^+) and secondary active transport of chloride (Cl^-) followed by, in eukaryotes, osmotically driven water movement. That is, water movement depends upon prior transport of osmotic equivalents, usually salt. This implies that the proximate event in controlling blood osmolarity, blood, pressure, and blood volume is sodium transport. The major hormonally controlled mechanism for sodium transport is epithelial sodium channels (ENaCs). Therefore, regulation of water and salt balance requires the ability to regulate both salt reabsorption and salt secretion (Garty and Benos 1988; Garty and Palmer 1997; Rossier 1997; Rossier et al. 2002). Besides control of blood pressure, ENaC-mediated sodium transport in the lungs is responsible for normal fluid clearance from the alveolar space (Hummler et al. 1997; Matalon and O'Brodovich 1999) and consequently for normal gas exchange in the lungs.

Not surprisingly then, abnormalities of ENaC function or ENaC regulation have been linked to disorders of total body Na^+ homeostasis, blood volume, blood pressure, and lung fluid balance (Boucher 2003; Garty and Palmer 1997; Hummler et al. 1996; Mall et al. 2004). ENaCs are also expressed in other tissues: distal colon, sweat ducts, salivary ducts, inner ear, lingual epithelium, keratinocytes, lymphocytes, vascular smooth muscle, endothelium, and in the eye (epithelia within retina, lens, and pigmented ciliary body and iris) (Brouard et al. 1999; Bubien et al. 2001; Couloigner et al. 2001; Drummond et al. 2004; Grunder et al. 2001; Kretz et al. 1999; Mirshahi et al. 1999). The role of ENaC within some of these tissues is, at present, unclear. We will describe the properties of ENaC and its regulation.

18.1.1 Mechanism of Salt Transport Across Epithelial Tissues

This concept of epithelial salt and water transport is generally associated with both salt concentration gradients and electrical gradients across epithelial tissue. The ability of one epithelial tissue, the living frog skin, to generate an electrical potential has been known since the time of Emil du Bois-Reymond in 1857 (Du Bois-Reymond et al. 1857). The origin of the potential was unknown at the time but was described as a "vital force" since it was not present in dead frog skin. There were many additional studies of the transepithelial potential (Dean and Gatty 1937; Francis and Pumphryey 1933; Orbeli 1910), but the origin of the potential was unclear until the work of Ussing and Zerahn (1951) later codified by Koefoed-Johnsen and Ussing (1958). They unambiguously showed that the frog skin bioelectric potential was due to active sodium transport mediated by a two-step process in which "...sodium movement through the outer [apical] border is passive, but highly selective..." and sodium movement across the basolateral membrane is via active transport of sodium.

Koefoed-Johnsen and Ussing's cellular model of epithelial sodium absorption has been repeatedly confirmed in all sodium-transporting epithelial tissues. These tissues include amphibian and mammalian urinary bladder, distal colon, various segments of the mammalian distal nephron, respiratory epithelia, sweat ducts, and numerous cultured cell lines derived from various tight absorptive epithelial tissues.

18.1.2 The Sodium-Selective Entry Pathway Is Blocked by the Diuretic, Amiloride, and Is an Ion Channel

In Ussing's work, the nature of the entry step remained unclear, but in 1967 a potassium-sparing diuretic, amiloride, was approved. Amiloride, after apical application, could reversibly reduce the frog skin potential close to zero (Bentley 1968; Dorge and Nagel 1970; Nagel and Dorge 1970). Many of these results strongly suggested that the sodium entry pathway was an ion-conducting channel protein. Initial microelectrode recordings from several epithelia lent weight to the idea that sodium moved down its electrochemical gradient through an ion channel (Fromter and Gebler 1977; Lewis et al. 1976; Wills et al. 1979). This idea was further strengthened by fluctuation analysis by Lindemann and Van Driessche (1977) which strongly suggested sodium-specific membrane channels blocked by amiloride in frog skin. The experimental evidence of amiloride-inhibitable epithelial Na^+ channels was finally confirmed by patch-clamp studies by Hamilton and Eaton (Hamilton et al. 1986; Hamilton and Eaton 1985) and Palmer and Frindt (1986). Over 35 years after Koefoed-Johnsen and Ussing published their model, molecular studies by Canessa et al. (1993, 1994b) identified the genes for the epithelial Na^+ channel, now simply known as ENaC.

18.2 Molecular Properties of ENaC

ENaC apparently consists of three homologous subunits, α, β, and γ, presumably arranged in a trimeric stoichiometry (Kashlan and Kleyman 2011; Stockand et al. 2008) (Fig. 18.1). cDNAs for epithelial Na channels were originally described by two groups of investigators who both used expression cloning methods to isolate αENaC cDNA from rat distal colon (Canessa et al. 1993; Lingueglia et al. 1993). Canessa (Canessa et al. 1994b) subsequently identified by complementation the cDNAs for β- and γ-rENaC (rat epithelial Na channel). Co-injection into *Xenopus* oocytes of cRNA prepared from all three cDNAs induces expression of amiloride-sensitive Na channels (Ahn et al. 1999; Canessa et al. 1994b; George et al. 1989; Hinton and Eaton 1989; Rossier et al. 1994) with characteristics virtually identical to those observed in renal cells in culture (Eaton and Hamilton 1988) or renal cortical collecting tubule principal cells (Palmer and Frindt 1988). The three subunits show limited sequence homology to each other although, across species, the same subunit has significant amino acid sequence homology (between 70 % and 90 %), even though there are some regions which show only limited homology. Other subunits, the so-called δ- and ε-subunits, have been described which, when expressed in heterologous expression systems, can by themselves or when associated with β- and γ- (and possibly α)-subunits produce cationic channels (Babini et al. 2003; Biasio et al. 2004; Chang and McDonald 2007; Ji et al. 2004, 2006). The ε-subunit has only been reported in *Xenopus* renal cells once (Babini et al. 2003) and bears no significant homology to any proteins other than those in *Xenopus* (where the homology is less than 30 %). It has not been cloned from any other genus than

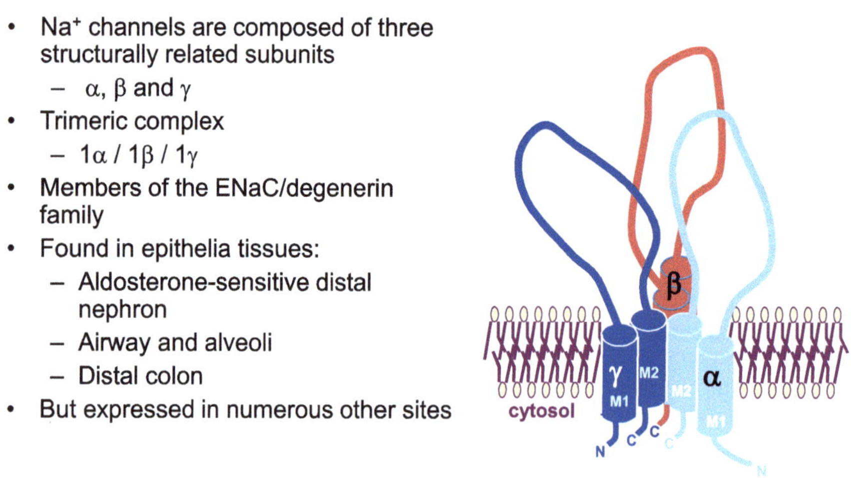

Fig. 18.1 ENaC structure. ENaC is composed of three structurally related subunits: α, β, and γ in a trimeric complex 1α/1β/1γ. ENaC is a member of the ENaC/degenerin family and is found in epithelia tissues including aldosterone-sensitive distal nephron, airway and alveoli, and distal colon but also expressed in numerous other sites

Xenopus. Its properties are poorly described; however recently, based upon homology considerations, this subunit has been reclassified as *Xenopus*-δ (Althaus et al. 2007). The δ-subunit originally appeared to be primate specific (human and chimp) although there are apparent predicted homologs in mammals and a few non-mammals, only a few of which have been successfully cloned. Its channel-forming properties are similar to the α-subunit when expressed in heterologous expression systems (Chang et al. 2011; Giraldez et al. 2012; Haerteis et al. 2009; Ji et al. 2012; Schwagerus et al. 2015). The highest expression levels of δ-ENaC mRNA are in the brain, testis, ovary, and pancreas, suggesting that the primary function of δ-ENaC may not involve sodium transport in epithelial tissue. This surmise is reinforced by the observation that δ-ENaC cannot compensate for the loss of α-ENaC in the lung or kidney in α-ENaC knockout mice which die shortly after birth from a failure to clear fluid from their lungs (Hummler et al. 1996; Schwagerus et al. 2012).

Although other combinations may exist (Fyfe and Canessa 1998; McNicholas and Canessa 1997), channels composed of the three subunits, α, β, and γ, have the following characteristic features: the channel is highly selective for sodium over potassium; has a small single-channel conductance of 4–6 pS; is inhibited by the diuretic, amiloride, at sub-micromolar concentrations; and is typically found in the apical membranes of sodium-transporting epithelial cells. Each subunit has only two predicted membrane spanning domains, an unusual topology for an ion channel. Topologic analyses suggest that the amino and carboxyl terminal regions of the ENaCs are cytoplasmic and that each subunit has a large extracellular loop present between the membrane spanning regions (Canessa 1996; Fyfe et al. 1998; Schild and Kellenberger 2001). The α-, β-, and γ-ENaCs have cysteine-rich regions within putative extracellular loops which are postulated to form intersubunit disulfide bonds which are important in oligomerization of Na channels (Collier et al. 2014; Sheng et al. 2005). The three ENaC subunits also possess proline-rich regions within their intracellular C-terminal domains that are important for interaction with accessory proteins that regulate the stability of functional channels in the apical membrane of renal cells (Canessa et al. 1994b; Debonneville et al. 2001; Rotin et al. 2000; Staub et al. 1997a, 2000).

18.3 Structure of ENaC

ENaC subunits are members of the ENaC/degenerin gene family (Corey and Garcia-Anoveros 1996; Goodman et al. 2002; Huang and Chalfie 1994). Functional ENaC containing α-, β-, and γ-subunits likely forms a heterotrimer (Kashlan and Kleyman 2011; Stewart et al. 2011; Stockand et al. 2008) since the closely related gene family member, acid-sensing ion channel 1 (ASIC1), forms a homotrimer (Jasti et al. 2007).

The three subunits share modest (~30–40 %) sequence identity. Each subunit has two membrane spanning helices (TM1 and TM2) resulting in cytoplasmic amino and carboxyl termini (Canessa et al. 1994a; Renard et al. 1994). The cytoplasmic domains have several sites of posttranslational modification and protein–protein and protein–lipid interactions that affect channel gating or trafficking. The structure and function of ENaC have been interpreted based on similarities to the crystal structure of the homologous acid-sensing ion channel 1 (ASIC1). A large extracellular region lies between the transmembrane spanning domains. Based on homology to ASIC1, this region is organized in distinct domains with well-conserved β-sheet domains at the center of the fold and poorly conserved α-helical domains surrounding the protein core (Fig. 18.2) (Kashlan et al. 2011; Kashlan and Kleyman 2011). This extracellular domain contains sites for ligand-dependent gating, proteolytic activation, and mechanosensitivity. The proposed structure of the extracellular and pore regions allows some understanding of channel assembly, processing, and transduction of external environment. Unfortunately, the homology modeling does not provide information about the structure of the cytosolic domains so that the mechanisms by which posttranslational events modify channel structure remain unclear.

The extracellular crystal structure of ASIC1 looks like an outstretched hand containing a ball with so-called finger, thumb, palm, knuckle, and β-ball domains (Jasti et al. 2007). There appears to be structural homology between ENaC subunits

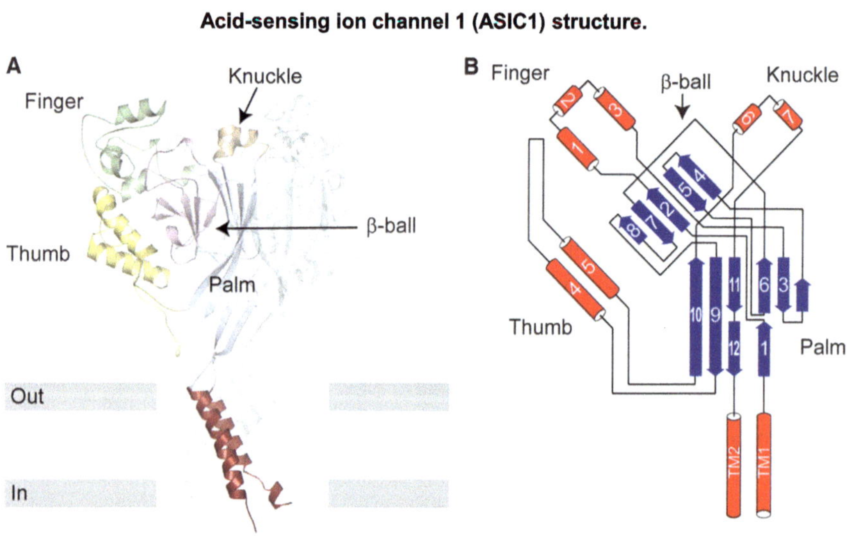

Fig. 18.2 Acid-sensing ion channel 1 (ASIC1) structure. (**a**) Ribbon diagram of ASIC1 trimer (pdb code: 3HGC). One subunit is shown as a solid ribbon, with domains in the extracellular region indicated. The other two subunits are shown as transparent ribbons. The approximate positions of the outer (Out) and inner (In) borders of the cell membrane are indicated. (**b**) Topology diagram of ASIC1. α-Helices and β-strands are represented by *red cylinders* and *blue arrows*, respectively. Labeling of secondary structure elements as in Jasti et al. (2007). This figure was adapted from a figure originally published in Ossama B. Kashlan and Thomas R. Kleyman Am J Physiol Renal Physiol 2011;301:F684–F696

and ASIC1 in most of these extracellular domains (Kashlan et al. 2011; Kashlan and Kleyman 2011). ENaC subunits have a predicted secondary structure that matches the observed ASIC1a structures in most regions of the extracellular region (for a thorough discussion, see Kashlan et al. 2011; Kashlan and Kleyman 2011) (Fig. 18.3).

Despite different suggestions for ENaC subunit stoichiometry, the similarity between the structures of ENaC and ASIC1a strongly implied that a heterotrimeric subunit stoichiometry was most likely. Subsequent resolution of ENaC channels with atomic force microscopy appears to confirm the heterotrimeric structure (Stewart et al. 2011). If ENaC is a heterotrimer, then the question arises of the radial organization. Is ENaC an α-β-γ-trimer or an α-γ-β-trimer? Recent work examining Cl^--dependent inhibition of ENaC (Collier and Snyder 2009) shows that ENaC is likely an α-γ-β-trimer (clockwise, viewed from above). Based on Cu^{2+} binding at subunit interfaces, Chen et al. (2011) suggested a similar arrangement.

The TM1 and TM2 helices that form the ENaC pore are homologous to their ASIC1 counterparts; the sequence identity for the TM2 domain is considerably

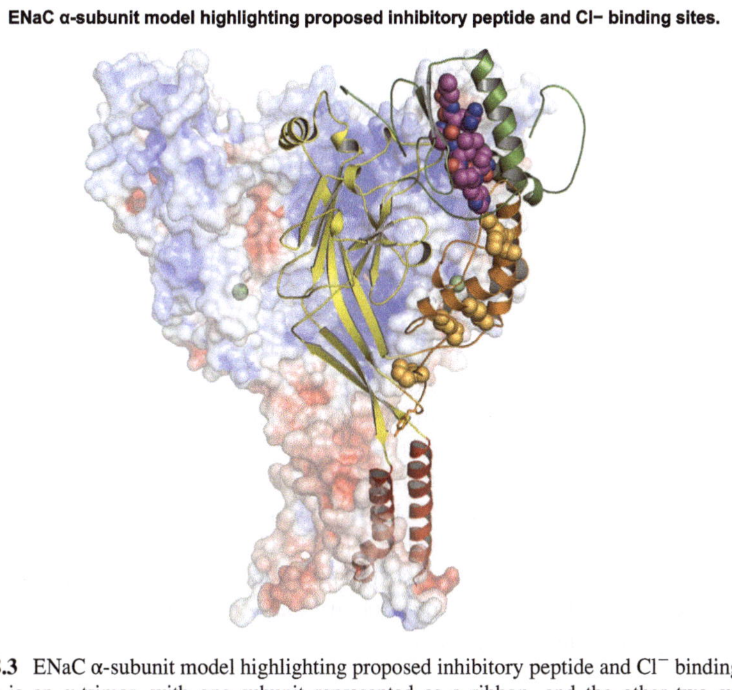

Fig. 18.3 ENaC α-subunit model highlighting proposed inhibitory peptide and Cl^- binding sites. Shown is an α-trimer, with one subunit represented as a ribbon, and the other two subunits represented as a surface with relatively basic (*blue*) and acidic (*red*) areas indicated. Positions of bound Cl^- (*green spheres*) were derived from the ASIC1 structure (pdb code: 3HGC). Ribbon is colored by domain: *red*, TM helices; *yellow*, palm, β-ball, and knuckle; *green*, finger; *orange*, thumb. Five disulfide bonds in the thumb are highlighted (*orange spheres*). The eight inhibitory tract residues essential for peptide inhibition are highlighted (*purple spheres*), while the remaining residues of the furin-excised inhibitory tract are removed. From Ossama B. Kashlan, and Thomas R. Kleyman Am J Physiol Renal Physiol 2011;301:F684–F696

higher than TM1 (Stockand et al. 2008). This apparent similarity and structure of the pore regions is supported by the relative accessibility of TM2 residues (Kellenberger et al. 1999a, b, 2001; Li et al. 2003; Schild et al. 1997; Sheng et al. 2001, 2005).

One characteristic of the ENaC pore is its remarkably high selectivity for Li^+ and Na^+ over K^+. The exact selectivity for Na^+ over K^+ is difficult to determine, but it is at least 100:1 and may be substantially higher. While ENaC does transport some K^+, ENaC is absolutely impermeable to larger ions (Palmer 1982). ENaC's high selectivity is unique among the members of the ENaC/degenerin family of proteins (Favre et al. 1996; Kellenberger et al. 1999b; Sheng et al. 2005). Examination of point mutations in the pore region determined that a Gly/Ser-X-Ser tract in TM2 of each ENaC subunit was responsible for determining ENaC selectivity (Kellenberger et al. 1999b, 2001; Li et al. 2003; Sheng et al. 2000; Snyder et al. 1999).

The proposed model of the ENaC extracellular domain also provides some insight into the site for amiloride binding. In particular, mutations within the Gly/Ser-X-Ser tract affected amiloride affinity (Kellenberger et al. 1999a, b, 2001; Li et al. 2003; Sheng et al. 2000; Waldmann et al. 1995). Sheng et al. and Kashlan et al. suggest that the pyrazine ring of amiloride binds at αSer583/βGly525/γGly537 and the guanidinium group is positioned at the outer mouth of the Gly/Ser-X-Ser selectivity tract (Kashlan et al. 2006; Kashlan and Kleyman 2011; Sheng et al. 2005).

18.4 Cellular Regulation of ENaC

Sodium transport across the tight epithelia of Na^+ reabsorbing tissues, such as the distal nephron and colon, is the major factor determining total body Na^+ levels, and sodium transport in lung epithelium is the major determinant of lung fluid balance. Sodium reabsorption in these epithelia is a two-step process. First, sodium enters renal cells from the tubular lumen through ENaC in the apical membrane before it is actively transported out of the cell by the basolateral Na/K-ATPase. Normally, there is excess capacity of the basolateral Na/K-ATPase so that the activity of apical ENaC determines the overall rate of sodium transport. Therefore, sodium transport is described by the flux of sodium through ENaC (measured as sodium current) across the apical membrane given as

$$I_{Na} = iNP_o \qquad (18.1)$$

where I_{Na} is total sodium flux, i is the flux through a single open ENaC, N is the number of ENaC in the surface membrane, and P_o is the probability that a channel is open. In theory, total sodium transport could be regulated by altering any of the factors in the equation above; but, in practice, sodium transport is almost always regulated by altering the probability that a channel is open or by regulating the number of functional channels in the apical membrane.

The total number of ENaC subunits within the cytosol of epithelial cells is relatively large compared with the number in the apical membrane (which can functionally transport sodium into the cell). Therefore, in polarized epithelia, it appears that assembly and trafficking of ENaC subunits out of the endoplasmic reticulum is an inefficient process. Estimates of the relative amounts of ENaC in the cell interior vs. the plasma membrane of epithelial cells vary enormously probably at least partially because of the type of the cells being studied. In transfected systems, only 1 or 2 % of ENaC is outside the ER; in native cells, it appears to be much higher (10–30 %) (Cheng et al. 1998; Hughey et al. 2002, 2004c; Prince and Welsh 1998, 1999). The large variability in the native cells appears to depend upon the hormonal state of the cell; e.g., EGFR ligands can dramatically reduce ENaC outside the ER and also total cellular ENaC (Liu et al. 2009; Tong and Stockand 2005). The exact point in the trafficking pathway at which ENaC subunits are assembled into a functional channel is unclear. Nonetheless, one of the mechanisms by which ENaC functional activity can be regulated is by altering the rate of delivery of assembled ENaC to the surface membrane. This mechanism has been suggested as the mechanism by which the hormones, vasopressin, and, possibly, aldosterone alter sodium transport (Eaton et al. 2001).

18.4.1 Small Molecule Agents That Modify ENaC Activity

18.4.1.1 Amiloride

As mentioned above (Sect. 18.1.2), amiloride has played a central role in our understanding of ENaC. Until the advent of single-channel measurements, ENaC current was defined as that component of transepithelial or whole-cell current that could be blocked by amiloride. In the 1980s, Palmer found that amiloride and other large cations blocked apical Na^+ channels in a voltage-dependent manner, suggesting that amiloride entered the channel pore and coulombically blocked sodium movement (Palmer 1984). The cloning of ENaC subunits allowed a further characterization of the amiloride-binding site. Schild et al. (1997) described mutations in all three subunits that reduced the efficacy of amiloride block, and the mutation sites were only accessible from the extracellular space. When Cys residues were introduced at these sites, they were susceptible to chemical modification by methanethiosulfonate compounds from the extracellular space (Carattino et al. 2005b; Kashlan et al. 2005; Li et al. 2003; Sheng et al. 2001, 2005; Snyder et al. 1999) implying that amiloride entered a relatively wide channel mouth before blocking the channel at its narrowest point.

18.4.1.2 Ion Activity

Sodium Ion Activity

ENaC activity is regulated by changes in both extracellular and intracellular sodium. This was first suggested by MacRobbie and Ussing (1961) to explain why, when the basolateral Na^{+},K^{+}-ATPase was inhibited, that frog skin epithelial cells did not swell as expected. They suggested that there was a mechanism by which intracellular sodium could limit the apical entry of sodium and thereby prevent sodium loading of cells. Subsequently, it was recognized that there were two categories of effect. A rapid change in sodium permeability or channel activity when extracellular sodium is changed has been called sodium self-inhibition (Fuchs et al. 1977). A slower change (over hours) in channel activity caused by changes in intracellular sodium concentration has been called feedback inhibition.

Both self-inhibition and feedback inhibition make strong evolutionary and physiological sense. ENaC plays a key role in the regulation of extracellular fluid volume and blood pressure. Therefore, from a regulatory standpoint, sodium transport through ENaC should respond to signals related to blood pressure and ECF volume. If net sodium transport changed dramatically in response to changes in lumenal sodium or intracellular sodium, the system would be unstable and subject to oscillatory changes unrelated to blood pressure or ECF volume. Self-inhibition and feedback inhibition provide a short-term and a long-term damping effect on overall transport that tend to keep transport relatively constant and to match apical entry with basolateral exit, thereby preventing large changes in intracellular sodium activity and epithelial cell volume (Palmer et al. 1998; Turnheim 1991).

Sodium Self-Inhibition

Self-inhibition is the ability of extracellular sodium to alter the activity of ENaC independent of changes in intracellular sodium activity (Lindemann 1984). Self-inhibition will reduce the permeability of the apical membrane when lumenal sodium concentration increases. In frog skin, when sodium concentration on the pond side of the frog skin is rapidly increased, short circuit current rapidly increases and then decreases in a matter of seconds (Fuchs et al. 1977). The phenomenon was suggested to be due to the presence of an allosteric receptor for sodium separate from the pore that could reduce channel activity. Treatment of ENaC with trypsin eliminated self-inhibition suggesting that self-inhibition is reduced in proteolytically cleaved channels. Noncleaved channels have low activity that appears to result from enhanced channel inhibition by external sodium (Sheng et al. 2006).

The mechanism for self-inhibition has been illuminated by considerations of the proposed structure of the channel (see Sect. 18.3). The structural model of the extracellular region of ENaC (Fig. 18.3) has clefts with acidic residues. On the basis of this model, Sheng et al. (2004) identified histidine residues in these acidic clefts that modulate sodium self-inhibition.

Sodium Feedback Inhibition

Several investigators initially found little evidence for a direct effect of intracellular sodium on channel activity (Eaton 1981; Palmer 1985). Recently, however, in mouse mandibular duct cells, a G protein-dependent decrease in Na^+ permeability was observed when intracellular sodium was increased during whole-cell patch-clamp recording (Komwatana et al. 1996). This is consistent with prior work in excised patches from renal cells in culture, in which it had been shown that G protein activation could inhibit ENaC activity (Ohara et al. 1993). More recently, some of the same investigators, again using mouse mandibular duct cells, showed that increases in intracellular sodium could activate the ubiquitin ligase, Nedd4-2, and Nedd4-2 in turn could reduce ENaC activity by promoting channel internalization (see below) (Dinudom et al. 2001).

However, the original situation described by MacRobbie and Ussing was more definitively examined by Silver et al. (1993) in cell-attached patches on tubules that were split open to expose the luminal surface; ouabain was added to the bath to increase cell Na. Channel activity decreased dramatically, and the mechanism of inhibition did not appear to be a direct effect of intracellular sodium, but rather appeared to involve increased intracellular calcium since calcium measured with Fura-2 increased and removal of calcium eliminated the effects of ouabain and increased sodium. Increasing intracellular calcium by apical addition of the calcium ionophore, ionomycin, produces a similar inhibition of ENaC (Palmer and Frindt 1987; Thai et al. 2014).

Chloride Ion Activity

Collier and Snyder (Collier et al. 2014; Collier and Snyder 2011) recently showed that Cl^- inhibits ENaC currents with an apparent K_i of 30 mM, suggesting that urinary Cl^- has a physiological role in modulating ENaC activity in the collecting duct. External Cl^- inhibits ENaC by reducing P_o with an apparent time course similar to the other allosteric inhibitors of ENaC. This inhibitory effect has a degree of anion selectivity; Cl^-, I^-, and Br^- inhibition of ENaC is approximately the same; and F^- and larger sulfates and phosphates have little effect.

ASIC1 crystal structure revealed a bound Cl^- at each of the three subunit interfaces, with residues from the thumb and palm domains of different subunits defining each binding site (Kashlan and Kleyman 2011). Collier and Snyder (Collier et al. 2014; Collier and Snyder 2011) examined whether homologous sites within ENaC are involved in Cl^- binding (Fig. 18.3). Mutations of key residues at these sites had differing effects on Cl^- inhibition as well as Na^+ self-inhibition. Mutants at the αγ-interface had little effect on Cl^- inhibition, mutants at the βγ-interface reduced Cl^- inhibition, and mutants at the αβ-interface reduced both Cl^- inhibition and Na^+ self-inhibition. These mutations suggested a clockwise α-γ-β-subunit arrangement when viewed from the extracellular side.

Hydrogen Ion

Protons regulate electrogenic sodium absorption in a variety of epithelia, including the cortical collecting duct, frog skin, and urinary bladder. When various combinations of α-, β-, and γ-subunits of ENaC are expressed in *Xenopus*, ENaC activity was reduced by intracellular acidification and increased by intracellular. Changes in extracellular pH had no effect on ENaC (Chalfant et al. 1999). A more direct examination of the effects of pH used patches excised from the apical membrane of rat principal cells and varied the cytosolic pH. ENaC open probability increased with alkalinization of the cytosolic side of the patch increasing approximately eightfold between pH 6.4 and 7.4 (Palmer and Frindt 1987).

The effects of pH may be physiologically relevant. Although there is little direct effect of extracellular pH on oocytes or on excised membrane patches, because there are other pathways for the movement of hydrogen ion, intracellular pH often tracks extracellular pH particularly in the presence of CO_2 or bicarbonate (Clark and Eaton 1983; Ifshin et al. 1983). In a specific example, pendrin in collecting duct intercalated cells secretes bicarbonate into the lumen which increases ENaC activity in adjacent principal cells, presumably by alkalinization of the principal cell interiors (Pech et al. 2010).

Calcium Ion Activity

As already mentioned above, there is ample evidence that increasing intracellular calcium reduces ENaC activity (Alli et al. 2012a; Ishikawa et al. 1998; Palmer and Frindt 1987; Poulsen et al. 2005; Taruno et al. 2008; Thai et al. 2014; Wang and Chan 2000). Most of these studies probably involved an indirect effect of calcium on ENaC. The question remains of whether there is a direct effect. In excised, inside-out patches, varying the free Ca^{2+} concentration of the solution bathing the cytoplasmic side of patches had no measurable effect on channel activity even at nonphysiologically high calcium activities (Palmer and Frindt 1987).

Heavy Metal Ions

For many years heavy metals have been known to alter sodium transport. In fact, prior to the availability of amiloride, mercuric chloride was used as an inhibitor of apical sodium entry in frog skin (Linderholm 1952; Lodin et al. 1963). But the effects of heavy metal poisoning on the kidney are profound and often involve failure to reabsorb sodium (Barbier et al. 2005). Yu et al. (2007b) examined the effects of Zn^{2+}, Ni^{2+}, Cu^{2+}, Pb^{2+}, Cd^{2+}, and Hg^{2+} on the activity of single ENaC in a renal epithelial cell line. None of the cations changed the single-channel conductance. However, ENaC activity was decreased by Cd^{2+} and Hg^{2+} and increased by Cu^{2+}, Zn^{2+}, and Ni^{2+} but was not changed by Pb^{2+}. Cu^{2+} increased channel density and stimulated open probability by reducing sodium self-inhibition as others have also shown (Chen et al. 2011; Sheng et al. 2002). Ni^{2+} binds histidine residues and the effect of the histidine-modifying reagent, diethyl pyrocarbonate, mimicked that of Ni^{2+}. This result is contrary to the result of Sheng (Sheng et al. 2002) who found that Ni^{2+} inhibited by binding to specific extracellular histidines. On the other hand several of the heavy metals that inhibit ENaC (especially Hg^{2+}) strongly bind cysteines and cysteines are known to be important for subunit assembly (Sheng et al. 2002, 2007).

18.4.2 Posttranslational Modifications of ENaC

Posttranslational modifications are a common mechanism to regulate proteins especially membrane proteins. ENaC is no exception: it is modified by at least six separate modifications.

18.4.2.1 Acetylation

Recently, Butler et al. (2015) showed that α-, β-, and γ-ENaC are substrates for lysine acetylation. Trichostatin A (TSA), a histone deacetylase inhibitor, enhanced ENaC acetylation and increased ENaC abundance in the total cell lysate and at the cell surface and increased transepithelial current in tissue culture and native renal epithelial cells. ENaC abundance and activity were reduced by overexpression of histone deacetylase 7 (HDAC7) and was increased by silencing HDAC7. There was also a reciprocal relationship between acetylation and ubiquitination; TSA reduced ENaC ubiquitination, whereas HDAC7 increased ubiquitination. By reducing ENaC ubiquitination, TSA decreased the rate of ENaC degradation. Thus, acetylation increases epithelial Na^{+} absorption by antagonizing ENaC ubiquitination. This stabilizes ENaC and, hence, increases its abundance at the cell surface.

18.4.2.2 Methylation

Methylation is a modification that in some senses is analogous to phosphorylation except that the rate of demethylation reactions is much slower, making methylation more persistent than phosphorylation; there are several enzymes that promote addition of methyl groups to isoprenylcysteine, aspartate, leucine, or lysine (Lanouette et al. 2014; Mato et al. 2002; Morgunkova and Barlev 2006; Wei et al. 2014; Zakrzewicz et al. 2012). Direct methylation increases the activity of ENaC reconstituted into lipid bilayers (Ismailov et al. 1994). In a tissue model of renal principal cells, inhibiting all protein methylation with 3-deazaadenosine dramatically reduced ENaC single-channel activity (Becchetti et al. 2000). 3-Deazaadenosine also blocked the increase in channel open probability associated with the addition of aldosterone. A methyl donor, *S*-adenosyl-L-methionine, activated ENaC in excised, inside-out patches. The combination of the methyl donor with intracellular GTP stabilized ENaC channel activity in excised patches for relatively long periods of time.

Isoprenylcysteine-O-carboxyl methyltransferase (pcMTase) is responsible for methylation of terminal cysteines modified by isoprenyl groups, and the enzyme can be detected immunologically in renal principal cells. Aldosterone increased pcMTase activity without increasing pcMTase protein levels. These results suggest that a critical transmethylation is necessary for aldosterone-mediated increases in Na^+ transport (Stockand et al. 1999b, 2000b). Aldosterone via pcMTase does induce methylation of the small G protein, K-Ras, and membrane-associated K-Ras does activate ENaC. However, methylation also has a more direct role in ENaC activation. Rokaw et al. (1998) showed that, in cultured mouse cortical collecting duct epithelial cells, the target for aldosterone-induced methylation was β-ENaC and that β-ENaC is a substrate for methylation by an enzyme other than pcMTase contained in the mouse CCD cells. Subsequently, Edinger et al. (2006) identified a protein, the product of gene ENSMUSG00000021773, that was expressed in CCD cells and was responsible for β-ENaC methylation. This protein product is catechol O-methyltransferase domain-containing protein 1 (Accession: Q8BIG7) (Carninci et al. 2005).

18.4.2.3 Ubiquitination

Ubiquitin is a highly conserved 76-amino-acid polypeptide with a role in a variety of cellular functions. Ubiquitin plays a major role in protein degradation because it serves as a tag for internalization and degradation of membrane proteins (Hershko and Ciechanover 1998; Hicke 1997, 2001). Ubiquitin is covalently linked to substrate proteins via the ε-amino group of lysine residues. Ubiquitination of proteins generally requires three different enzymatic activities mediated by either two or three enzymes. First, a ubiquitin-activating enzyme (E1) activates ubiquitin by forming a thioester bond in an ATP-dependent reaction. Ubiquitin is then transferred to a ubiquitin-conjugating enzyme (E2), followed by the addition of ubiquitin to target proteins by a ubiquitin ligase (E3). Single ubiquitin molecules can be added to different lysine residues on the target protein so there will be one ubiquitin per lysine on multiple lysines; this modification is known as monoubiquitination or multiubiquitination. Alternatively, additional ubiquitins can be added to ubiquitins already on lysines resulting in a chain of multiple ubiquitins, a process known as polyubiquitination (Hershko and Ciechanover 1998).

Ubiquitination of membrane proteins promotes internalization; but once internalized, the degradative pathway that proteins follow appears to depend upon whether the proteins are monoubiquitinated or polyubiquitinated.

Polyubiquitination directs proteins to the proteasome for degradation (Coux et al. 1996; Hershko and Ciechanover 1998; Mitch and Goldberg 1996); monoubiquitination is a signal for endocytosis and lysosomal degradation (Bonifacino and Weissman 1998; Mimnaugh et al. 1999; Pickart 2000a, b; Pickart and VanDemark 2000; Rotin et al. 2000; Thrower et al. 2000; van Kerkhof et al. 2001; van Kerkhof and Strous 2001; Yu and Kopito 1999). Proteins degraded by the proteasome tend to be associated with lipid rafts and internalized via caveolae (Felley-Bosco et al. 2002; Minshall et al. 2003; Waterman and Yarden 2001). Once internalized, the targeted protein is unfolded, ubiquitin is removed from the proteins by deubiquitinating enzymes, and the proteins are cleaved to small peptides by proteasomal proteolytic activity (Hershko and Ciechanover 1998; Hicke 2001; Weissman 2001).

In contrast, monoubiquitinated membrane proteins are internalized and degraded by proteases in lysosomes (Hicke 1997, 2001; Jeffers et al. 1997). Ubiquitination is not a prerequisite for degradation of all proteins by the lysosome; however, several receptors and ENaC must undergo ubiquitination before degradation (Hicke 2001; Jeffers et al. 1997; van Kerkhof et al. 2001). In the process of internalization via clathrin-mediated endocytosis, ubiquitinated proteins pass through several endosomal compartments to lysosomes. The first compartment is an early endosome (also referred to as recycling endosome). Proteins from this compartment can deubiquitinated and recycled back to the plasma membrane. If early endosomes are not recycled back to the plasma membrane, then they become late endosomes by acquiring rab7-GTP protein which finally fuse with lysosomes that contain proteolytic enzymes which degrade internalized plasma membrane proteins (Van IJzendoorn et al. 2000; Van IJzendoorn and Mostov 2000).

A Specific Ubiquitin Ligase Isoform, Nedd4-2, Ubiquitinates ENaC

Ubiquitin conjugation is required for ENaC internalization and subsequent degradation. Ubiquitin conjugation of ENaC subunits requires a specific ubiquitin ligase, Nedd4 (neural precursor cell-expressed developmentally downregulated protein). Nedd4 contains an E6-AP carboxyl terminus (Hect) domain that is homologous to other ubiquitin ligases, three WW domains in the rat and mouse protein or four WW domains in human and *Xenopus* protein, and a calcium-/lipid-binding domain (CaLB/C2) (Staub and Rotin 1996). In humans, Nedd4 has nine known isoforms (for reviews, see Ingham et al. 2004; Rotin and Kumar 2009; Yang and Kumar 2010). Nedd4-2 appears to be the isoform responsible for physiologically relevant ubiquitination of ENaC in all mammalian epithelial cells (Snyder et al. 2004b), but other E3 ligases can modify ENaC. WWP2 is expressed in human adult and embryonic epithelial tissues, and transepithelial current is inhibited when co-expressed with ENaC subunits in Fischer rat thyroid epithelial cells (McDonald et al. 2002). Nonetheless, there is direct evidence that Nedd4-2 is the ligase responsible for ENaC ubiquitination and that Nedd4-2 is associated with membrane ENaC molecules (Malik et al. 2005b; Snyder 2005; Zhou et al. 2007). Nedd4-2 co-localizes with ENaC in renal cortical and outer medulla collecting duct principal cells and in airway epithelia (Staub et al. 1997b). Nedd4-2 apparently requires direct association with ENaC to allow ubiquitin conjugation as suggested by the direct binding of Nedd4 to all three ENaC subunits (Debonneville et al. 2001; Dinudom et al. 1998; Staub et al. 1996). Interaction of the Nedd4-2 WW domain with the PY motif of ENaC β-subunits involves contact between tyrosine 618 and leucine 621 and WW domain of Nedd4-2 (Kanelis et al. 2001). Nedd4-2 WW domain binding to ENaC β- and γ-subunits is practically irreversible because of a very slow dissociation rate when determined by surface plasmon resonance (Asher et al. 2001). Co-expression of ENaC and Nedd4-2 reduces whole-cell current and the amount of ENaC subunit protein at the cell surface. ENaC mutants lacking the PPxY motif are unaffected by co-expression of Nedd4-2 (Abriel et al. 1999; Goulet et al. 1998). In renal cells in culture, Nedd4-2 co-immunoprecipitates with α- and β-ENaC and is associated with apical membrane proteins (Malik et al. 2005b).

One reason for alternative degradative pathways, lysosomal vs. proteasomal, could be different levels of ENaC protein expression: in native cells the surface expression of ENaC channels ranges from about 40 channels per cell in mouse cortical collecting duct cells (M1) to as low as four per cell in the *Xenopus* renal cell line (A6) (with alveolar type 2 cells somewhere in between), whereas in transfected cells, surface expression levels may exceed several thousand per cell. The large number in the membrane overwhelms the normal internalization and degradative machinery of the proteasome and promotes clathrin-mediated endocytosis and degradation via a lysosomal pathway at a rate that is independent of proteasome activity. In addition, the large excess of ENaC protein may limit the extent of ubiquitination, thus promoting monoubiquitination (and lysosomal degradation) rather than polyubiquitination (and proteasomal degradation).

Although it has yet to be demonstrated for α-, β-, or γ-ENaC, ubiquitination of the δ-subunit can be promoted by interactions of the COMMD1 (formerly MURR1) protein with Nedd4-2 (Biasio et al. 2004; Chang and McDonald 2007). COMMD1 plays numerous other roles in cellular regulation, so the significance of its regulation of the δ-subunit remains to be determined (as does the role of the δ-subunit itself in epithelial function).

Deubiquitination of ENaC

All of the mechanisms mentioned above involve changes in the affinity of Nedd4-2 for ENaC with the consequent change in the rate of ubiquitination and the rate of internalization. If ENaC subunits remain ubiquitinated, they will be degraded by the proteasome or the lysosome, but ubiquitins can be removed from ENaC. USP8 removes ubiquitin from ENaC before ENaC leaves the membrane (Zhou et al. 2013). After internalization, in early endosomes, the deubiquitinating enzyme, UCH-L3, appears to remove approximately half of the ubiquitin from internalized subunits (Butterworth et al. 2007). At some later point in the endosomal progression, another deubiquitinating enzyme, USP2-45, can be induced by steroids and promote removal of ubiquitin to allow functional ENaC to return to the surface membrane (Fakitsas et al. 2007).

18.4.2.4 Phosphorylation

Phosphorylation of ENaC was first described by Shimkets et al. (1998) as a possible mechanism for the regulation of ENaC. Phosphorylation is central to regulating ENaC ubiquitination and degradation (Bhalla and Hallows 2008; Snyder 2009; Soundararajan et al. 2010a). Phosphorylation of Nedd4-2 controls ENaC ubiquitination (Debonneville et al. 2001).

G protein receptor kinase 2 (GRK2) phosphorylates Nedd4-2 to reduce interaction with ENaC and prevent ubiquitination and degradation (Sanchez-Perez et al. 2007). Nedd4-2 function is also inhibited by cAMP-dependent protein kinase A (PKA) phosphorylation (Snyder et al. 2004a). Transmitter agents which increase cAMP (vasopressin in the kidney, β-adrenergic agents in the lung) stimulate ENaC activity. The stimulation has been widely attributed to PKA-induced ENaC insertion, but clearly increases of cAMP might also control ENaC function by regulating Nedd4-2 function. This implies that there is again a concerted response with one action of the transmitter agents promoting addition of channels to the membrane and the other action preventing their removal.

Phosphorylation of ENaC itself plays a role in altering Nedd4 ubiquitination. MAPK phosphorylation of ENaC promotes Nedd4-2 binding. Adenosine monophosphate-activated protein kinase (AMPK) promotes Nedd4-2 retrieval of ENaC and AMPK knockout mice have excess ENaC activity (Almaca et al. 2009). Whether AMPK has its effect by phosphorylating ENaC or Nedd4 remains to be determined. In contrast, casein kinase 2 phosphorylates ENaC and prevents interaction with Nedd4 (Bachhuber et al. 2008).

Protein kinase C (PKC), which may be stimulated through extracellular purinergic receptor activation, inhibits both ENaC P_o and surface expression (Booth and Stockand 2003; Stockand et al. 2000a). The change in surface expression is due to signaling through the mitogen-activated protein kinase/ERK pathway. The change in open probability is due to PKC phosphorylation of MARCKS (myristoylated alanine-rich C-kinase substrate). The phosphorylation inactivates MARCKS and reduces MARCKS ability to present PIP_2 to ENaC. Reduced PIP_2 reduces ENaC open probability (Alli et al. 2012a) (and see below).

ERK2 and casein kinase 2 (CK2) directly phosphorylate the COOH-termini of β- and γ-ENaC and promote Nedd4-2-mediated ENaC internalization (Shi et al. 2002a, b). A recently discovered functional polymorphism in the COOH-terminal tail of human α-ENaC also prompted investigation of its regulation by kinase-mediated changes in ENaC trafficking (Samaha et al. 2004). The COOH-terminal tail of α-ENaC is regulated by CK1 or PKC-δ to increase insertion of channels in the plasma membrane (Yan et al. 2006, 2007). SGK1 also directly phosphorylates and stimulates α-ENaC (Diakov and Korbmacher 2004).

Some kinases can indirectly regulate ENaC trafficking. IκB kinase-β interacts with ENaC to increase cell surface expression and Na^+ current in CCD cells (Edinger et al. 2009; Lebowitz et al. 2004). WNK1 (with no lysine [K] kinase 1) was recently demonstrated to alter SGK1 activity and hence stimulate ENaC (Xu et al. 2005). In contrast, WNK4 inhibits ENaC and the inhibition is not mediated through an action of SGK1 or Nedd4-2 (Yu et al. 2013). Akt (similar to SGK1) also disrupts ENaC–Nedd4-2 binding and hence stimulates ENaC-mediated Na^+ current in response to insulin (Holzman et al. 2007; Lee et al. 2007a). Protein kinase A was shown previously to inhibit Nedd4-2 (Snyder et al. 2004a) (similar to SGK1) but has also now been suggested to modulate tyrosine kinase activity to increase apical expression of ENaC (Woollhead and Baines 2006). Protein kinase D also rapidly promotes insertion of ENaC into the plasma membrane (McEneaney et al. 2007). The metabolic sensor AMP-activated protein kinase (AMPK), which is sensitive to a host of external stimuli and intracellular energy stores, indirectly inhibits ENaC surface expression by promoting an interaction between Nedd4-2 and ENaC (Bhalla et al. 2006a; Carattino et al. 2005a).

18.4.2.5 Proteolysis

Initially, ENaC activation by endogenous serine proteases was shown in cultured renal cells, in which it was found that the general protease inhibitor, aprotinin, partially blocks amiloride-sensitive Na^+ uptake (Vallet et al. 1997). Expression cloning identified channel-activating peptidase 1, also called prostasin, as a factor that increases ENaC currents when co-expressed in oocytes. Similar effects were observed with trypsin in mouse collecting duct M-1 cells (Nakhoul et al. 1998). Masilamani et al. (1999) showed direct biochemical evidence for cleavage by showing the apparent molecular weight of the γ-subunit shifts from approximately 85 to approximately 70 kDa in rats treated with aldosterone or a low-salt diet.

Patch-clamp studies showed that proteolytic cleavage substantially increased ENaC single-channel open probability from "near silent" to frequently open (Caldwell et al. 2004). It was also reported that furin, a member of the pro-protein convertase family thought to reside in the trans-Golgi network, was important in cleaving the α-subunit at two sites and the γ-subunit at one site in the extracellular loops (ECL) (Hughey et al. 2003, 2004a, b) (Fig. 18.4). Furin-mediated cleavage of α-ENaC activates ENaC by releasing an inhibitory peptide from the ECL (Carattino et al. 2006). Furin-dependent proteolysis of α-ENaC activates the channel by relieving Na^+ self-inhibition, thereby increasing Na^+ transport at higher extracellular Na^+ concentrations (Carattino et al. 2006; Hughey et al. 2007; Sheng et al. 2006). Recently, it was also shown that prostasin, likely present in the extracellular apical compartment, also cleaves γ-ENaC at a site in the ECL near furin cleavage sites, which, as with the dual furin cleavage sites in α-ENaC, releases an inhibitory peptide that further activates ENaC (Bruns et al. 2007; Carattino et al. 2006). Proteolysis of the channel per se is not strictly necessary for activation, because deletion of these inhibitory tracts in α- and γ-subunits is sufficient to enhance

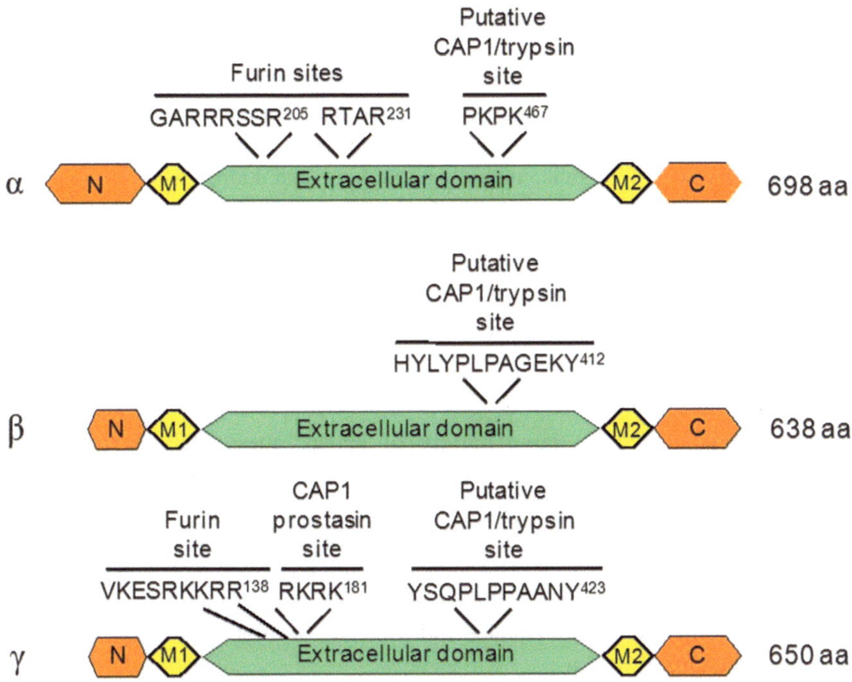

Fig. 18.4 Schematic representation of the rat ENaC subunits and their identified and putative sites for furin, CAP1, trypsin, and prostasin cleavage. M1, M2, transmembrane domains; N, C, intracellular amino and carboxyl termini, respectively. From Ahmed Chraibi and Stéphane Renauld (2013). New Insights into the Epithelial Sodium Channel Using Directed Mutagenesis, Genetic Manipulation of DNA and Protein—Examples from Current Research, Dr David Figurski (Ed.), ISBN: 978-953-51-0994-5, InTech, DOI: 10.5772/50309. Available from: http://www.intechopen.com/books/genetic-manipulation-of-dna-and-protein-examples-from-current-research/new-insights-into-the-epithelial-sodium-channel-using-directed-mutagenesis. Open access

channel activity without any cleavage. Moreover, addition of the inhibitory peptides to wild-type ENaC confers inhibition of the channel (Carattino et al. 2006, 2008).

Both processed (complex glycosylated with cleaved subunits, active) and nonprocessed (noncleaved, inactive) forms of ENaC subunits exist in cells and tissues expressing endogenous ENaC (Ergonul et al. 2006; Hughey et al. 2004a; Masilamani et al. 1999). Inactive, near-silent channels at the cell surface are probably uncleaved and may serve as a channel pool that can be activated by extracellular proteases. Recent evidence suggests the extent of ENaC proteolysis is dependent on channel residency time in the plasma membrane, because channels with mutations that block ubiquitin-dependent endocytosis remain on the surface longer, have an increased proportion of mature (cleaved) subunits, and have a correspondingly higher activity (Carattino et al. 2008).

There is at least one additional level of regulation since there are endogenous protease inhibitors, such as protease nexin 1 (Wakida et al. 2006). Indeed, recent studies suggest the proteolytic state of ENaC subunits reflects a balance between the expression of activating proteases and protease inhibitors (Myerburg et al. 2006). ENaC regulation by proteases and protease inhibitors is likely to be physiologically significant in several organs, including the lung and kidney (for reviews, see Hughey et al. 2007; Kleyman et al. 2006; Soundararajan et al. 2010a).

18.4.3 Transmitter and Humoral Agents That Modify ENaC Activity

ENaC is regulated by a large variety of agents; these include transmitters interacting with G protein-coupled receptors (e.g., purinergic, adrenergic, and dopaminergic agents), circulating hormones (e.g., glucocorticoids, angiotensin, and eicosanoids), chemokines (e.g., TNF-α, TGF-β, interleukins), and reactive oxygen and nitrogen species (e.g., superoxide and nitric oxide). The large number of regulatory agents underscores the fact that regulation of lung Na^+ absorption is critical for normal lung fluid balance and function. Several agents have a particularly profound effect on Na^+.

18.4.3.1 Regulation of ENaC by Adrenergic Agents

Lung epithelial cells are a well-known target for adrenergic transmitters, but the kidney contains a more diverse set of adrenergic receptors.

Regulation of ENaC by α-Adrenergic Agents

There are few reports of alpha adrenergic regulation of ENaC. However, recently Mansley et al. (Mansley et al. 2015) described an increase in ENaC-mediated transepithelial current in response to norepinephrine applied to confluent mCCDcl1 murine cortical collecting duct cells. This effect was concentration dependent and mediated via basolateral α2-adrenoceptors.

Regulation of ENaC by β-Adrenergic Agents

Sympathetic innervation of the kidney tends to be mostly on the proximal nephron and juxtaglomerular apparatus (Badve et al. 2011) with, at most, modest effects on ENaC in the distal nephron. On the other hand, β-adrenergic stimulation of ENaC in the lungs is one of the primary mechanisms controlling ENaC-mediated lung fluid clearance (Davis and Matalon 2007; Fronius 2013). Activation of β_2 receptors on alveolar cells stimulates adenylyl cyclase that, in turn, increases intracellular cyclic AMP levels and increases the density of ENaC in the apical membrane. The effects of increased cAMP are totally blocked by the β-antagonist, propranolol, and by the protein kinase A (PKA) blocker, H89 (Chen et al. 2002; Downs et al. 2012; Eaton et al. 2004).

18.4.3.2 Regulation of ENaC by Vasopressin

Excessive sodium reabsorption by the kidney has long been known to participate in the pathogenesis of some forms of hypertension. In the kidney, the final control of NaCl reabsorption takes place in the distal nephron through ENaC. Although aldosterone is usually thought to be the main hormone regulating ENaC activity, several studies in animal models and in humans highlight the important effect of vasopressin on ENaC regulation and sodium transport. The selective V2 receptor agonist, dDAVP, not only increases urine osmolality and reduces urine flow rate but also reduces sodium excretion in rats and humans. Chronic V2 receptor stimulation increases blood pressure in rats, and blood pressure and urine concentration are significantly correlated in healthy humans. Thus, excessive vasopressin-dependent ENaC stimulation could be a risk factor for sodium retention and high blood pressure (Bankir et al. 2005, 2010; Nicco et al. 2001).

Marunaka and Eaton (1991) examined single amiloride-blockable Na channels in membrane patches from cultured distal nephron cells before and after treatment with arginine vasopressin. Vasopressin increases ENaC activity by increasing the number of ENaC with only a modest increase in the open probability of individual channels. Cells pretreated with cholera toxin or dibutyryl-cAMP to increase intracellular cAMP appeared similar to cells treated with vasopressin; that is, the number of Na channels per patch increased with a small effect on the open probability of individual Na channels. These observations suggest that vasopressin works by increasing cAMP and promoting insertion of clusters of new sodium channels. Subsequent work by Snyder (2000) on cells transfected with labeled channels confirmed this idea. He examined the appearance of modified channels at the surface membrane and found that the number increased in response to cAMP. Also, inhibition of vesicle trafficking by incubating epithelia at 15 °C prevented the cAMP-mediated stimulation of ENaC. Morris and Schafer (2002) used more quantitative methods to examine the effects of vasopressin. They concluded that the increased density of ENaC subunits in the apical membrane can account completely for the transepithelial current increase produced by cAMP. However, their work made several assumptions and a direct measure of ENaC open probability using patch-clamp methods in split-open mouse collecting duct showed that there was a rapid increase in ENaC open probability immediately after addition of vasopressin followed by an increase in channel density (Bugaj et al. 2009). In addition, stimulated production of cAMP with forskolin and phosphodiesterase inhibitors also enhanced ENaC open probability. The stimulatory effect of cAMP was dependent on the presence of intact PY motifs in the C termini of the channel (Yang et al. 2006). The change in open probability was not due to direct phosphorylation of ENaC by protein kinase A, but rather involved ERK1/2 phosphorylation of two sites, T613 in β-ENaC and T623 in γ-ENaC. Phosphorylation of these sites enhances the interaction of ENaC and Nedd4 (Shi et al. 2002a). Thus, an increase in intracellular cAMP appears to promote dephosphorylation of the two ERK sites, reducing channel retrieval and increasing open probability by reducing ENaC–Nedd4 interaction.

Vasopressin stimulates ENaC activity in adrenalectomized mice (Mironova et al. 2012). This implies that the ability of vasopressin to stimulate ENaC is independent of aldosterone. Vasopressin modulates sodium reabsorption in the collecting duct through adenylyl cyclase-stimulated cyclic AMP. The specific isoform involved in vasopressin-stimulated sodium transport appears to be adenylyl cyclase type VI since mice deficient in this isoform in the principal cells of the collecting duct are no longer responsive to vasopressin (Roos et al. 2013).

The observations on vasopressin-stimulated sodium reabsorption have important physiological consequences. Ordinarily, vasopressin is viewed as only affecting water reabsorption through aquaporins, but concomitant vasopressin-induced increases in salt reabsorption will enhance the driving force for water reabsorption. Thus, while vasopressin decreases free water excretion, it does this, in part, by promoting ENaC-mediated sodium reabsorption (Stockand 2010).

18.4.3.3 Regulation of ENaC by Purinergic Agonists

Epithelial tissues expressing ENaC often also contain an intrinsic (autocrine/paracrine) mechanism of ENaC regulation. In the best example of this intrinsic regulation, circulating aldosterone can be excessively high, but ENaC activity in the kidney remains stagnant after an initial increase (Knox et al. 1980). This phenomenon is referred to as aldosterone escape. While many factors are thought to mediate this response, one is an intrinsic regulatory mechanism that occurs with an increase in renal perfusion pressure. The source of this intrinsic ENaC inhibition in the presence of high aldosterone remained mysterious until Stockand et al. found that mice lacking P2Y2 purinergic receptors that bind locally produced ATP have impaired aldosterone escape (Stockand et al. 2010). While other mechanisms such as atrial natriuretic peptide, arginine vasopressin, and even other channels such as the sodium chloride cotransporter (NCC) have been suggested to mediate aldosterone escape (Kelly and Nelson 1987; Stockand et al. 2010; Wang et al. 2001), it is clear that locally produced ATP participates in this response. While the best studied example is that of renal epithelia, ATP is produced intrinsically in many tissues that express ENaC and has varying actions on the channel as listed below.

Mechanisms of Apical ATP Release in Epithelia

In the kidney, apical ATP release is well studied. ATP is released from the apical surface of CCD in response to an increase in tubular flow or renal perfusion pressure (Hovater et al. 2008). While ENaC is expressed on the apical surface of principal cells, connexin 30 hemichannels are expressed on the apical surface of intercalated cells (McCulloch et al. 2005). Placing an ATP sensor cell next to a principal or intercalated cell from an isolated mouse cortical collecting duct (CCD) showed that ATP is produced by the intercalated and not the principal cells (Sipos et al. 2009). Furthermore, when this sensor was placed close to an intercalated cell on a CCD isolated from a connexin 30 knockout mouse, no ATP release was sensed. CCD cells may also release ATP packaged into vesicles that are stimulated to release their contents upon increases in intracellular calcium or hypotonicity (Bjaelde et al. 2013).

In pulmonary epithelia, the generally accepted primary mechanism of ATP release from the apical surface of the cell involves pannexin 1 (Panx1), a channel expressed on the apical pole of airway epithelia. In human airway epithelia, inhibiting Panx1 through knockdown or pharmacology impaired ATP secretion both under basal and hypotonic (ATP stimulating) conditions (Ransford et al. 2009). Panx1-induced ATP release is thought to be a response to mechanical and hypotonic stress (Ohbuchi et al. 2014). Panx1-mediated ATP only accounts for 60 % of ATP release both under basal and stimulated conditions. The remainder of ATP release is likely mediated by vesicular stores of ATP maintained in goblet cells that are released by a Ca^{2+}-dependent mechanism (Lazarowski et al. 2003). This process is thought to mediate increased ATP release under pathological conditions such as inflammation (Okada et al. 2013). Many other mediators of airway ATP release have been suggested over the years including CFTR, VDAC1, and connexins, but all of these have been proven not to directly facilitate ATP production.

Mechanisms of Basal ATP Release in Epithelia

The mechanisms underlying serosal ATP release in ENaC expressing epithelia are not as thoroughly characterized as apical pathways. In the renal CCD, ATP release is stimulated by a decrease in serum osmolality (Schwiebert and Zsembery 2003), and in the airway, ATP release occurs downstream of mechanotransduction pathways (Homolya et al. 2000). The molecular mechanisms underlying this ATP release and the channels that the ATP travels through on the basolateral side of these epithelia have yet to be identified.

Purinergic Receptor Families

Purinergic (nucleotide binding) receptor classes include P1 receptors which bind adenosine and P2 receptors which bind ATP (Burnstock 2006). The P2 receptors are divided into several families which include P2X, Y, Z, U, and T. P2X and Y regulate ENaC as detailed below. P2Y receptors are metabotropic, signaling through G protein-coupled receptors of the Gq class, activating phospholipase C (PLC), and generating IP3 and DAG. P2X receptors are ionotropic, meaning they are ligand-gated (ATP being the ligand) ion channels capable of transporting a variety of cations including Na^+ and Ca^{2+}. P2X receptors can function as monomers or in homo- or heterodimeric complexes.

Effects of Apical ATP on ENaC in the Kidney

In the distal nephron where ENaC is expressed, P2Y and P2X receptors are expressed on the apical surface and basolateral surface. On the apical surface, P2Y2, P2Y4, P2X4, and P2X6 have been shown to play some sort of role in ENaC regulation. The vast majority of the work detailing the effects of purinergic signaling on renal ENaC function has focused on P2Y2 receptor-mediated ENaC inhibition. Early evidence for an effect of ATP on renal ENaC came from *Xenopus* distal nephron cells (A6 line) (Ma et al. 2002). A non-hydrolysable ATP analog, ATPγS, applied to the apical surface of these cells, decreased the probability of opening ENaC. Furthermore, ENaC activity only increased in response to stretch when the P2 receptor inhibitor suramin was added to the patch pipet, indicating that tonic inhibition by P2 receptors prevents a stretch response in this cell line. The effects of ATP on ENaC in this model are dependent on phospholipase C (PLC) activity. P2Y2 receptors activate PLC which cleaves PIP2 to produce IP3 and DAG. These molecules can work together to activate protein kinase C (PKC) and decrease available PIP2 in the process. Since PIP2 typically functions to hold ENaC into the membrane and Ca^{2+} can inhibit ENaC by a variety of mechanisms, it was important to determine whether decreasing PIP2 or increasing intracellular Ca^{2+} and activating PKC is inhibiting ENaC. Pochynyuk et al. showed that inhibiting PLC attenuated ENaC's response to apical ATP, whereas inhibiting PKC had no effect, suggesting that the decrease in PIP2 mediated the ATP response (Pochynyuk et al. 2008b). Contrary to the generally accepted view that P2Y2 inhibits renal ENaC, however, mice lacking P2Y2 receptors have decreased ENaC expression (but increased P_o of the channels that are there) and salt-resistant hypertension (but ENaC did not appropriately downregulate in response to high-salt diet) (Pochynyuk et al. 2008a, 2010). This response is explained by suppressed renin angiotensin aldosterone system activation. These mice had an enhanced response to increasing plasma aldosterone, however, suggesting that they could not "escape" from aldosterone and that $P2Y_2$ receptors participate in aldosterone escape. Stockand et al. showed that $P2Y_2$ KO mice did in fact have impaired ability to escape from high aldosterone levels (Stockand et al. 2010). The compensatory changes in RAAS in

$P2Y_2$ knockout mice make it difficult to assess purinergic effects on renal ENaC function in this model. To address the role of luminal ATP in countering aldosterone-induced ENaC stimulation, transgenic mice that globally overexpress the ectonucleotidase NTPDase1 were created. These mice have decreased extracellular ATP throughout all tissues. When aldosterone is clamped in these animals, they show impaired natriuresis on a high-salt diet which correlated with an inability to downregulate ENaC suggesting that ATP-induced ENaC inhibition allows the body to maintain sodium balance during periods of high-salt intake (Zhang et al. 2015).

In addition to P2Y2, P2Y4, P2X4, and P2X6 are all expressed on the apical membrane of distal nephron principal cells and have been reported to have inhibitory effects on ENaC activity. Interestingly, however, in the presence of low extracellular Na^+, P2X channels on the apical membrane may stimulate ENaC via a PI3 kinase-dependent mechanism (Wildman et al. 2009).

Effects of Basolateral ATP on ENaC in the Kidney

While luminal renal ATP production is thought to help the kidney escape from the effects of aldosterone, serosal ATP has been postulated to mediate aldosterone's effects. In 1963, Edelman et al. showed that the effects of aldosterone on ENaC were dependent on ATP production (Edelman et al. 1963). It was not until 50 years later that this phenomenon was revisited. Gorelik et al. found that aldosterone stimulates ATP release on the basolateral side of *Xenopus* cortical collecting duct A6 cells and that this basolateral ATP causes cell contraction, cytoskeletal reorganization, and ENaC activation (Gorelik et al. 2005). The same group used a complicated pharmacological approach to determine that this response was likely mediated by basolateral $P2X_4$ receptor-mediated increases in PI3 kinase (Zhang et al. 2007b). We used cell-attached patch-clamp techniques and a dominant negative $P2X_4$ receptor construct to confirm that basolaterally expressed $P2X_4$ channels enhance ENaC activity (Thai et al. 2014).

P2Y2 receptors are also expressed on the basolateral surface of ENaC expressing principal cells and have an inhibitory effect on amiloride-sensitive current in M1 mouse collecting duct principal cells (Cuffe et al. 2000). This effect is independent of increases in intracellular Ca^{2+}, but is interestingly dependent, in part, on changes in intracellular pH (Thomas et al. 2001).

Effects of ATP on ENaC in Airway

While the effect of serosal ATP on ENaC in the airway has not been studied, apical ATP can be stimulatory or inhibitory. Cellular products of *Pseudomonas* inhibit ENaC in type II alveolar cells by stimulating ATP release and activation of P2Y receptors (Kunzelmann et al. 2006). Lipopolysaccharide was identified as the trigger for ATP release, and, unlike in the kidney where $P2Y_2$ receptors inhibit

ENaC by PLC but not Ca^{2+} or PKC, both a decrease in PIP_2 availability and an increase in PKC activation mediate the effects of apical ATP on ENaC in type II alveolar cells (Boncoeur et al. 2010). In type I alveolar cells, on the other hand, apical P2Y stimulation increases short circuit current, and this effect is mediated by ENaC activation rather than inhibition (Yang et al. 2009). In intact rabbit airway epithelium, ENaC is inhibited by ATP via a P2Y receptor-mediated pathway that involves increases in intracellular Ca^{2+} (Poulsen et al. 2005). Interestingly, this same study found no effect of increasing basolateral ATP on short circuit current suggesting that serosal ATP may not affect ENaC activity in airway epithelium.

ENaC Regulation by Purinergic Receptors in Other Tissues

The general consensus in most tissues where ENaC is expressed is that apical stimulation of purinergic receptors inhibits ENaC. Experiments looking at the effects of basolateral ATP on ENaC have not been done. In the distal colon, ENaC is inhibited by apical addition of ATP (Matos et al. 2007). This effect is mediated by $P2Y_2$ receptors and mice lacking these receptors show no change in colonic sodium uptake in response to ATP. Apical $P2Y_4$ receptors inhibit ENaC in Reissner's membrane of the chochlea by a mechanism that is PLC-dependent (Kim et al. 2010). ENaC in mouse endometrial epithelium is also inhibited by apical ATP via a P2Y receptor-mediated mechanism, but in this tissue the inhibition is Ca^{2+} dependent (Wang and Chan 2000). In primary cultures of mammary epithelia, basolateral P2Y receptors stimulate ENaC by a mechanism that is dependent on intracellular Ca^{2+} increases (Lee et al. 2007b).

18.4.3.4 Regulation of ENaC by Dopamine

Dopamine increases lung liquid clearance under basal conditions and in situations where edema accompanies lung injury. Alveolar epithelial cells appear to contain both type 1 and 2 dopamine receptors (D1 and D2). Stimulation of D1 or D2 receptors on the basolateral surface of T2 cells activates Na,K-ATPase, but some of the increase in lung fluid clearance is mediated by D1 activation of apical ENaC. Apical D1 receptors stimulate the production of cyclic AMP but do not activate protein kinase A. Rather the cAMP, through a complicated signaling pathway that involves activation of the small G protein, rap-1, by cAMP activation of a GDP exchange protein, EPAC, finally increases the activity of individual ENaC proteins. The effect of dopamine on both ENaC and (Na,K)-ATPase suggests a coordinated response of both transporters to promote maximal increases in transport and alveolar fluid clearance. The action of dopamine also demonstrates that signaling within alveolar cells is compartmentalized. Basolateral production of cAMP by β-adrenergic receptors produces a protein kinase A-dependent increase in the number of ENaC in the apical membrane with little or no change in the activity of individual channels. On the other hand, the apical production of cAMP by D1 receptors produces a protein kinase A-independent increase in the activity of individual channels with little or no change in the number of channels.

18.4.3.5 Regulation of ENaC by Cholinergic Agonists

There are few reports about the action of cholinergic agonists on ENaC. Acetylcholine receptors (AChRs) are present in the lung in the same cells as ENaC (Racke et al. 2006) and airway surface liquid contains acetylcholine (Yang et al. 2009). Takemura et al. (2013) using cell-attached patch-clamp recordings found that, in alveolar type 2 cells, the muscarinic agonists carbachol and oxotremorine activated ENaC in a dose-dependent manner but that nicotine did not. Carbachol-induced activation of ENaC was blocked by atropine. Western blotting, PCR, and immunohistochemistry showed that muscarinic M2 and M3 receptors but not nicotinic receptors were present in alveolar type 2 cells (Takemura et al. 2013) and airway epithelial cells (Yang et al. 2009). The small G protein, RhoA, and its activated form, GTP-RhoA, increased in response to CCh and the increase was reduced by pretreatment with atropine. An inhibitor of Rho-associated protein kinase, Y-27632, blocked carbachol-induced ENaC activity. Interestingly, while carbachol activates ENaC in alveolar cells, it does not appear to activate ENaC in tracheal epithelial cells but does activate other ion channels (Chen et al. 2006).

18.4.3.6 Regulation of ENaC by Angiotensin II

Angiotensin II contributes to the regulation of total body salt and water balance; therefore, it would seem evolutionarily sensible for it to also regulate ENaC. In fact, in biochemical studies, angiotensin II treatment increased the amount of ENaC subunit proteins and angiotensin II receptor antagonists like candesartan decreased ENaC protein. Immunocytochemistry confirmed the increase in β- and γ-ENaC protein abundance and demonstrated candesartan-induced ENaC internalization in collecting duct cells (Beutler et al. 2003). When intracellular Na^+ concentration was used as a measure of Na^+ transport across the apical membrane, angiotensin II added to the lumen of isolated cortical collecting ducts increased intracellular Na^+ and apical sodium permeability (Peti-Peterdi et al. 2002). Angiotensin II added to the basolateral surface also increased intracellular Na^+; however, luminal addition was much more effective. An amiloride analog, benzamil, almost completely inhibited the elevations in intracellular Na^+. These results suggest that angiotensin II directly stimulates ENaC activity in the CCD.

A more direct examination of the effects of angiotensin II on ENaC combined patch-clamp electrophysiology and immunohistochemistry in freshly isolated split-opened distal nephrons of mice to determine the mechanism and molecular signaling pathway of angiotensin II regulation of ENaC (Mamenko et al. 2012). Angiotensin II acutely increases ENaC open probability and prolonged exposure induces translocation of α-ENaC to the apical membrane (cp. Beutler et al. 2003). The actions of angiotensin II are independent of the effects of aldosterone on ENaC. Although angiotensin II produces large increases in intracellular calcium in smooth muscle cells, it does not in the renal epithelial cells. It does, however, produce an increase in reactive oxygen species sufficient to explain the increase in ENaC open probability (see below). Inhibition of NADPH oxidase with apocynin abolished the angiotensin II-mediated increases in ENaC open probability.

The results show that agonist binding to the angiotensin II receptor regulates ENaC abundance and functional activity, implying a physiological role for angiotensin II to increase sodium reabsorption in the face of reduced blood pressure or blood volume.

18.4.3.7 Regulation of ENaC by Reactive Oxygen Species

All cells produce reactive oxygen species as part of their normal mitochondrial metabolism. Although there is growing evidence indicating the involvement of NADPH oxidase in renal pathology, there is a paucity of information on the role of NADPH oxidase in the regulation of normal renal function, but some of the cells containing ENaC may have especially high levels of reactive oxygen species. For example, besides the endogenously derived reactive oxygen species (ROS), lung cells are also exposed to relatively high levels of free oxygen in inspired air that can lead to production of additional ROS. In the past, ROS have generally been considered to damage cells, and exposure of cells to high concentrations of ROS under pathological circumstances for extended periods of time does result in inflammation and fibrosis and finally programmed cell death. However, low concentrations of ROS, rather than producing adverse events, appear to act as signaling molecules that detect the oxidative state of cells and play a significant role in regulating ENaC activity (Downs and Helms 2013).

There are several sources of ROS within cells, some endogenous and some exogenous. NADPH oxidases are one of these sources. There are seven different isoforms of NADPH oxidases (Nox1–5, Duox1, and Duox2). Nox4 is the predominant form in the kidney, although Nox2 is also expressed. Nox4 has been implicated in the basal production of ROS in the kidney and in pathological conditions such as diabetic nephropathy and CKD; upregulation of Nox4 may be important in renal oxidative stress and kidney injury (Sedeek et al. 2013).

The potential importance of ROS-dependent signaling can be appreciated by remembering that shortly after birth, the fetal lung must change from a fluid-secreting to fluid-absorbing organ for independent breathing to occur. Perinatal exposure of lung cells to oxygen could be one signal for turning on the active transport of Na ions necessary to promote fluid clearance from the newborn lung. Cell culture models of the alveolar epithelium do show that increases in oxygen tension enhance ENaC activity. ENaC's oxygen sensitivity could be due to a direct effect of oxygen; however, changes in oxygen tension also produce changes in ROS, such as superoxide anion (O_2^-), and there does appear to be ROS regulation of ENaC since addition of a cell-permeable, O_2^- scavenger [2,2,6,6-

tetramethylpiperidine 1-oxyl (TEMPOL)] significantly decreases ENaC activity (Rafii et al. 1998; Yu et al. 2007a). In fetal distal lung epithelial (FDLE) cells, Na^+ transport increases 6 h after an increase in oxygen tension (up to 100 mmHg) (Baines et al. 2001). Besides the direct effect on Na^+ transport, ROS can influence transport properties of the alveolar epithelium in the long term by altering ENaC expression. Long-term exposure (48 h) to oxygen or ROS increases ENaC subunit promoter activity (Baines et al. 2001), and following extended exposure to high O_2 tension treatment, total ENaC protein increases (Baines et al. 2001; Rafii et al. 1998; Thome et al. 2003). The promoter regions of ENaC subunits have NF-κB and AP-1 response elements (Bremner et al. 2002; Rafii et al. 1998) and both transcription factors respond to the redox state of the cell. Transcription factor regulation of ENaC gene expression in response to ROS could account for ROS-induced increases in ENaC protein.

ROS can apparently be the final step in signaling mechanisms that modify ENaC activity. For example, ENaC activity is acutely increased by EGF, insulin, and IGF-1 and is associated with a comparable increase in ROS (H_2O_2) production. Pretreatment with the generic NADPH oxidase inhibitor, apocyanin, reduced both H_2O_2 production and the increase in ENaC activity. Knocking down a key NADPH oxidase subunit (Rac1) eliminated EGF activation of ENaC (Ilatovskaya et al. 2013). These observations imply that EGF, insulin, and IGF-1 signaling pathways converge at a common ENaC stimulatory effect mediated by ROS production.

Exogenous agents that modify ENaC activity also converge on ROS production. Acute exposure of a renal cell tissue culture model to ethanol activates ENaC by significantly increasing both ENaC open probability and the active ENaC density in the membrane. The effects of ethanol were mimicked by acetaldehyde, the first metabolic product of ethanol. The effects of ethanol on ENaC were abolished by a superoxide scavenger, 4-hydroxy-2,2,6,6-tetramethylpiperidinyloxy (TEMPOL). Consistent with an effect of ethanol-induced reactive oxygen species (ROS) on ENaC, ethanol and acetaldehyde elevated intracellular ROS (Bao et al. 2012). The same effects are observed in intact alveolar cells. ROS production increased in rat lung slices and in vivo mouse lung in response to acute EtOH exposure. ROS production was Rac1 dependent (Downs and Helms 2013).

18.4.3.8 Regulation of ENaC by Nitric Oxide

Nitric oxide (NO) is a highly diffusible, short-lived free radical that is synthesized in cells from L-arginine in a reaction catalyzed by NO synthase (NOS). NO is involved in the regulation of sodium balance. NO and, to a lesser extent, peroxynitrite ($ONOO^-$ formed from the reaction of O_2^- and NO) are produced by many different cell types. There is a large literature on the role of NO in chronic kidney disease and pulmonary vascular disease (Baylis 2008, 2012; Yamagishi and Matsui 2011; Yousefipour et al. 2010). In terms of ENaC-mediated lung fluid reabsorption, nitric oxide is known for its deleterious effects: elevated levels of NO metabolites have been detected in pulmonary edema fluid (Zhu et al. 2001), and several studies report nitric oxide inhibition of ENaC activity (Guo et al. 1998; Helms et al. 2005b; Nielsen et al. 2000; Ruckes-Nilges et al. 2000; Yu et al. 2007a). Direct nitric oxide inhibition of ENaC is certainly possible, given that there are several tyrosine and cysteine residues on the transmembrane domains and extracellular loops of ENaC subunits, respectively, that could be nitrated or nitrosylated directly by NO; but there is currently no strong evidence for NO-mediated modification of ENaC. On the other hand, the more traditional NO signaling pathway involving soluble guanylate cyclase (sGC) activation and production of cyclic guanosine 3′,5′-monophosphate (cGMP) does inhibit ENaC function (Jain et al. 1998). This pathway also inhibits ENaC in a tissue culture model of renal principal cells (Guo et al. 2013). All of these studies examined the effects of relatively high concentrations of NO in which NO reduced ENaC open probability. On the other hand, if NO is completely removed by treatment with an iNOS inhibitor from airway (by IV infusion) or lung cells in culture for several hours, ENaC channel density is reduced (Hardiman et al. 2004). This implies that NO derived from iNOS under basal conditions is necessary for amiloride-sensitive Na^+ transport across lung epithelial cells and modulates the amount of α- and γ-ENaC via posttranscriptional, cGMP-independent mechanisms or gene regulation.

18.4.3.9 Interaction of NO and ROS Signaling in ENaC Regulation

Since nitric oxide is an inhibitor of ENaC activity, it might be necessary for epithelial cells to decrease the production or activity of NO under edematous conditions. In fact, the bioavailability of nitric oxide can be limited directly by ROS. Increased O_2^- prevents the immediate NO inhibition of ENaC that is normally observed in Na-transporting epithelia (Yu et al. 2007a). This effect is presumably due to the reaction of O_2^- with NO to form peroxynitrite. This may be an important mechanism regulating lung fluid balance, since AT1 cells appear to generate very high levels of O_2^- compared to AT2 cells (Helms et al. 2008; Yu et al. 2007a). This explains why exogenous NO given to reduce pulmonary vascular resistance and increase lung perfusion does not cause edema: NO presumably reduces ENaC activity in AT2 cells, but ENaC activity is unaffected in AT1 cells since NO is scavenged by the high levels of superoxide.

18.4.3.10 Regulation of ENaC by Hydrogen Sulfide

Hydrogen sulfide decreases β-adrenergic agonist-stimulated lung liquid clearance by inhibiting ENaC-mediated transepithelial sodium absorption (Agne et al. 2015). Application of an H_2S donor reduced both baseline and adrenergic stimulated amiloride-sensitive alveolar fluid clearance. It also blocked amiloride-sensitive transepithelial current in lung cells in tissue culture. H_2S prevents the stimulation of ENaC by cAMP/PKA and, thereby, inhibits the proabsorptive effect of β-adrenergic agonists on lung liquid clearance.

18.4.3.11 Regulation of ENaC by Sex Hormones

Female sex predisposes individuals to poorer outcomes from respiratory disorders like cystic fibrosis and influenza-associated pneumonia (Carey et al. 2007; Klein et al. 2010; Olesen et al. 2010; Serfling et al. 1967; Stephenson et al. 2011; World Health Organization 2008; Zeitlin 2008). Both pathologies have abnormal alveolar fluid clearance and reduced ENaC activity. Recently, two sex hormones, estradiol and prolactin, were shown to directly regulate expression and activity of alveolar ENaC. Overnight exposure to estradiol increased ENaC channel activity (NP_o) through an increase in channel open probability (P_o) and an increased number of patches with observable channel activity. Apical plasma membrane abundance of the ENaC α-subunit more than doubled in response to estradiol as determined by cell surface biotinylation. α-ENaC in the apical membrane was approximately threefold greater in the lungs from female rats in proestrus, when serum estradiol is greatest, compared with diestrus, when it is lowest. These effects were mediated by the G protein-coupled estrogen receptor (Gper). In contrast to estradiol, there was no effect of progesterone (Greenlee et al. 2013).

Overnight exposure to the peptide, prolactin, also increased ENaC channel NP_o through an increase in channel open probability and increased the number of patches with observable channel activity. Inhibition of protein kinase A with H-89 abolished the effect of prolactin on ENaC, and prolactin also increased cyclic AMP (cAMP) in a renal principal cell line, consistent with signaling through the cAMP-dependent PKA pathway. These observations show that alveolar ENaC is at least partially responsible for the sex differences observed in the pathogenesis of several pulmonary diseases (Greenlee et al. 2015).

18.5 Regulation of ENaC Density in the Apical Membrane

As mentioned in Sect. 18.4, sodium reabsorption through ENaC can be altered in two ways: by changing the open probability of channels already in the membrane and by increasing the number of functional channels in the membrane. These two mechanisms are not mutually exclusive and some combination of the two does occur in response to several agents which alter ENaC activity (e.g., aldosterone, vasopressin, sex hormones). Nonetheless, because altering membrane channel density is a major means to alter sodium transport, it is important to understand how ENaC is inserted into the surface membrane and the mechanisms for ENaC trafficking. This section will deal with trafficking of ENaC to the membrane, ENaC recycling, and, to a lesser extent, ENaC retrieval (since it was dealt with in Sect. 18.4 above where we discussed the role of ubiquitination in retrieval of

ENaC from the surface membrane). There have been several recent reviews on ENaC trafficking (Butterworth et al. 2009; Butterworth 2010; Myerburg et al. 2010; Snyder 2005).

18.5.1 ENaC Trafficking to the Apical Membrane

As a starting point, Fig. 18.5 (Butterworth et al. 2009) shows a schematic diagram depicting common, well-characterized trafficking routes found in many epithelial cells. In many ways ENaC trafficking recapitulates these generic pathways.

Experiments in heterologous expression systems show that individual ENaC subunits can traffic from the ER to the cell surface, but this process is inefficient (Firsov et al. 1996; Prince and Welsh 1998; Valentijn et al. 1998; Weisz et al. 2000). Efficient trafficking of ENaC out of the ER requires assembly of the heterotrimeric complex. Any excess subunits not part of a heterotrimer are targeted for degradation in the proteasome (Staub et al. 1997a; Valentijn et al. 1998). The mechanism by which subunits are restricted to the ER in the absence of assembly may be that all three subunits are required for proper protein folding and assembly (Duncan and Mata 2011). Alternatively, ER retention signals might prevent release of individual subunits from the ER until a complex is formed between all three subunits, similar to trafficking of K^+(ATP) channel complexes (Zerangue et al. 1999).

As with many other proteins, after heterotrimeric assembly, ENaC is core glycosylated in the endoplasmic reticulum with more complex modifications and additions to the glycans beyond the ER in the cis- and trans-Golgi (Ergonul et al. 2006; Prince and Welsh 1998; Rotin et al. 2001). There has been a suggestion that there is an alternative population of ENaC that appears to bypass Golgi processing and channels with "immature" n-linked glycans have been observed in cultured renal cells (Hughey et al. 2004c; Yu et al. 2008). The alternative interpretation of these results is that, like the ribophorin proteins (Rosenfeld et al. 1984), not all the ER glycosylation sites are modified in the Golgi. The α-mannosidases that are the initial step in Golgi glycan modification and addition are extremely sensitive to steric restrictions (Stanley 2011), and in the absence of mannose trimming, no other glycan processing can take place at these sites in the Golgi. If enough are modified, the proteins will move out of the cis-Golgi to the trans-Golgi and subsequently to the apical membrane, but an examination of their sensitivity to endoglycosidase H will suggest there are "immature" channels in the membrane because some sites will still have high mannose glycosylation and be cleaved by endoglycosidase H.

There is some controversy about the role of lipid rafts in the delivery and stability of ENaC at the apical membrane. Lipid rafts are domains that contain high concentrations of inositol phospholipids and the enzymes which produce the lipids (Li et al. 2004; Matsuura et al. 2007; Tong et al. 2008). There seems little question that some fraction of ENaC traffics to and is present in the apical

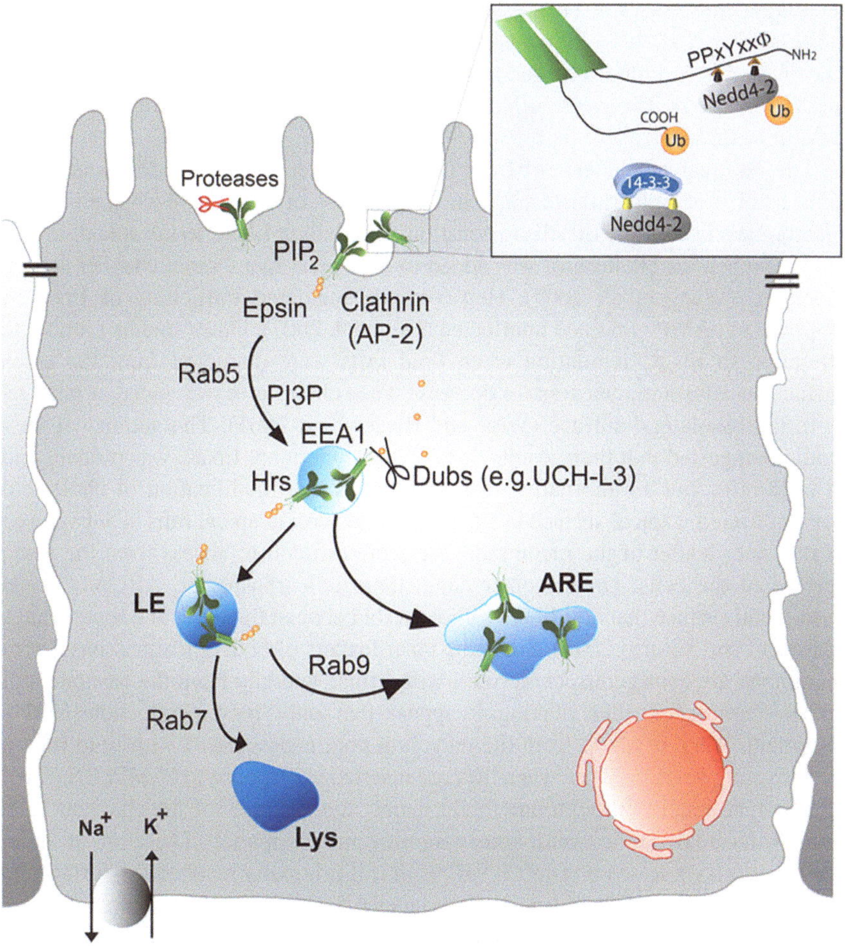

Fig. 18.5 Schematic diagram outlining steps in epithelial sodium channel (ENaC) endocytosis and progress through early endosomes as it is recycled. ENaC internalization is initiated by the binding of Nedd4-2 to the intracellular NH2 terminus (NH2) of each subunit and the addition of ubiquitin (Ub) moieties to the COOH tail (COOH) of the ENaC subunits (enlarged inset). Binding of 14-3-3 proteins to phosphorylated Nedd4-2 (*yellow* symbols on Nedd4-2) prevents Nedd4-2 interaction with the PY motifs on ENaC. Following ubiquitination, the channel is internalized by clathrin-dependent endocytosis through an interaction with epsin and passes through an early endosomal compartment (EEA1 and Hrs positive). If ENaC is deubiquitinated by deubiquitinating enzymes (Dubs), it will be transferred to the apical recycling endosome (ARE) for recycling to the apical membrane. If it remains ubiquitinated, it is likely trafficked via the late endosomes (LE) to the lysosomes (Lys) for degradation. From studies on other transporters, it is likely that Rab proteins may be involved in rescuing ENaC from the LE back to the ARE or facilitating its movement to Lys for degradation. From Michael B. Butterworth et al. Am J Physiol Renal Physiol 2009;296:F10–F24

membrane in rafts (Alli et al. 2012b; Balut et al. 2006; Hill et al. 2002, 2007; Reifenberger et al. 2014; Shlyonsky et al. 2003) that contain high levels of inositol phospholipids and other charged lipids (Bao et al. 2007; Ma et al. 2002, 2004; Ma and Eaton 2005). The work of Hill et al. (2002, 2007) has implied a role for lipid rafts in trafficking ENaC.

The removal of cholesterol from the surface membranes of kidney epithelial cells to disrupt lipid rafts resulted in the absence of any active channels in the membrane and a failure of cells to constitutively deliver ENaC to the apical surface. When exogenous cholesterol was added to the apical membrane, channel activity increased (Wang et al. 2009). However, the regulated trafficking of ENaC in response to cAMP remained unaffected (Hill et al. 2007). Other studies found little alteration in ENaC regulation when lipid rafts were disrupted from the apical surface but substantial increase or decrease when cholesterol was added or removed from the basolateral surface (West and Blazer-Yost 2005). The authors of these studies suggested that their results implied that functional ENaC was probably not in lipid rafts; but an alternative interpretation is that modification of cholesterol content from the apical surface is difficult. Cholesterol in apical rafts is only present in the inner leaflet of the membrane, therefore difficult to access from the apical surface of the cells. On the other hand, there is a transporter, ABCA1, on the basolateral surface that equilibrates cholesterol between the outer and inner leaflets (Albrecht and Viturro 2007). Since the inner leaflets of both apical and basolateral membrane are contiguous, cholesterol will diffuse laterally from the basolateral to apical membranes, thus making it appear that only basolateral cholesterol is important. Also, if ENaC from the recycling pool is associated with lipid rafts in the recycling vesicles, then, when they are inserted in response to cAMP, ENaC will still be functional (although one might expect the lifetimes of the channels to be less). Rafts are dynamic, continuously forming and dissipating. Therefore, it should not be surprising to see non-raft ENaC even if it originally reached the surface in a raft. Artificial disruption of lipid rafts reduces ENaC activity, but the prevalence of lipid rafts also depends upon hormonal factors (e.g., tumor necrosis factor-α (Bao et al. 2007), activators of protein kinase C, or PTEN tumor suppressor). Inositol phospholipids stabilize functional ENaC and are necessary for channel trafficking and channel gating as well (Ma et al. 2007; Pochynyuk et al. 2005, 2007b; Staruschenko et al. 2007). Overexpression of caveolin-1, a signature protein of caveolae, which promotes formation of caveolae and internalization, reduces the activity and membrane surface expression of ENaC (Lee et al. 2009). One interesting possibility consistent with these results is that ENaC is more stable in rafts. Only when it leaves the raft environment can it be ubiquitinated, retrieved, and degraded. The fact that the ENaC regulatory complex (see below) is associated with rafts lends some credence to this idea.

A potential mechanism linking lipid rafts to apical membrane traffic may involve the phosphatidyl-4-phosphate adaptor protein 2 (FAPP2) (D'Angelo et al. 2012; Godi et al. 2004; Vieira et al. 2005, 2006). FAPP2 is involved in glycolipid production and targeting in the trans-Golgi that could allow the formation of lipid rafts. FAPP2 links 4-PIP and activated Arf1 in the trans-Golgi (D'Angelo et al.

2007) and may facilitate the clustering of apical membrane proteins in lipid rafts (Balla et al. 2005; Godi et al. 2004) Disruption of FAPP2 function leads to a failure of cells to deliver membrane proteins to the apical membrane.

As with many other proteins, ENaC-containing vesicles likely use the cytoskeleton and molecular motors to move from the trans-Golgi to the apical membrane (Weisz and Rodriguez-Boulan 2009). When either actin or tubulin is disrupted using pharmacological agents, there is a generalized failure to deliver ENaC to the apical membrane (Butterworth et al. 2005), but ENaC density in the apical membrane does not change rapidly, even though the transport activity is reduced (Reifenberger et al. 2014). It therefore appears likely that actin may play an important role in ENaC trafficking (Mazzochi et al. 2006; Smith et al. 1991; Zuckerman et al. 1999).

A more specific schematic of ENaC trafficking is shown in Fig. 18.6 (Butterworth et al. 2009). This figure shows a role of SNARE proteins in the final vesicle fusion events that would deliver ENaC to the apical membrane. Disruption of SNAREs or SNARE binding proteins that facilitate SNARE complex formation prevents the fusion of ENaC vesicles with the apical membrane. SNARE proteins also likely play a role in the delivery of ENaC to the apical membrane. The delivery of transmembrane proteins to the apical surface of epithelial cells following synthesis involves sorting and packaging in the Golgi and forward traffic from the trans-Golgi network to the apical membrane (Folsch et al. 2009; Stoops and Caplan 2014; Weisz and Rodriguez-Boulan 2009). In addition, a direct association between syntaxin and the cytoplasmic domains of ENaC has been reported (Berdiev et al. 2004; Condliffe et al. 2003, 2004; Peters et al. 2001; Qi et al. 1999; Saxena et al. 1999, 2006, 2007). Several accessory proteins are essential to ENaC trafficking. These include small GTPases, Rab and Rho proteins (Butterworth 2010; Butterworth et al. 2012), and motor proteins to move membrane vesicles to the desired compartment or apical surface (Weisz and Rodriguez-Boulan 2009). The role of Rab proteins is complicated by the interplay between Rab-GEFS and Rab-GAPs. A specific example of this is the Rab-GAP, AS160. In renal epithelia, Liang et al. (2010) showed that aldosterone increased AS160 protein expression threefold. In the absence of aldosterone, AS160 overexpression increased total ENaC expression but did not increase apical membrane ENaC or amiloride-sensitive Na current, but when AS160 was overexpressed, aldosterone was more effective at increasing apical ENaC and transepithelial ENaC current. Aldosterone promoted AS160 binding to the steroid-induced 14-3-3 isoforms, β and ε. These findings suggest that AS160 stabilizes ENaC in an intracellular compartment under basal conditions and that aldosterone-dependent AS160 phosphorylation permits ENaC forward trafficking to the apical membrane to augment Na absorption.

When the vesicles do arrive at the apical membrane, SNARE protein family members promote docking and fusion of vesicles (Duman and Forte 2003; Szule and Coorssen 2003). ENaC trafficking may be unique in that it appears to involve a direct interaction between cargo (ENaC) and syntaxins (Condliffe et al. 2003; Saxena et al. 2007). Besides altering ENaC trafficking, syntaxin 1A inhibits ENaC by altering its gating (Berdiev et al. 2004; Qi et al. 1999).

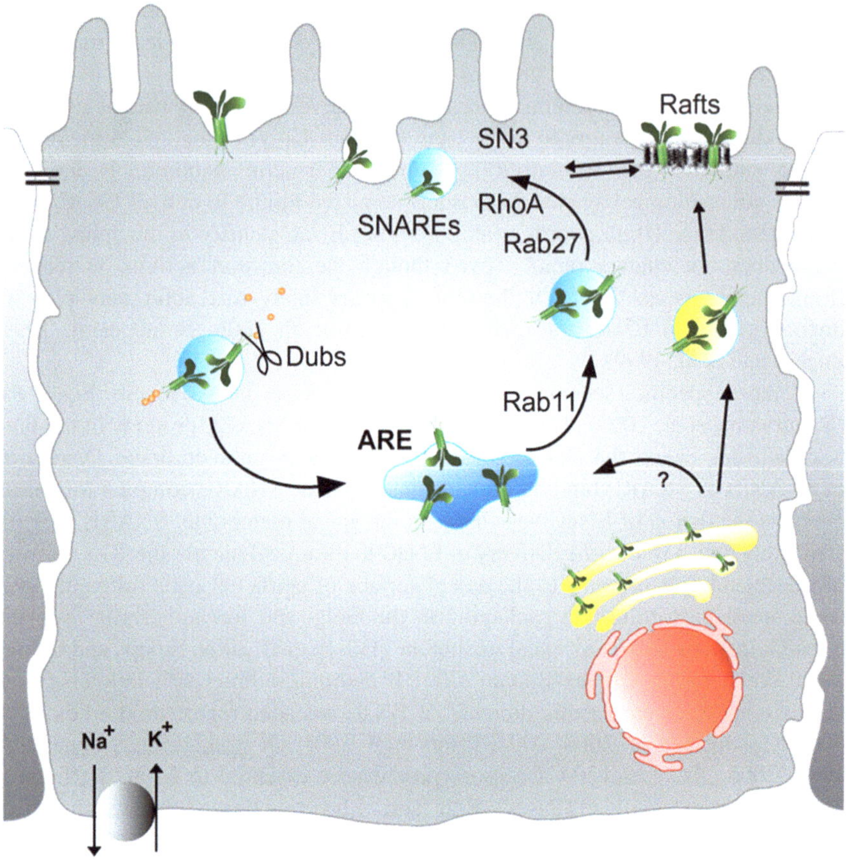

Fig. 18.6 Schematic diagram outlining the steps in ENaC recycling and delivery to the apical membrane. Following synthesis and assembly (*yellow* Golgi apparatus), channels are transported to the apical surface (*yellow* vesicle) and can be localized in lipid raft membrane domains (rafts). It is also possible that ENaC enters the regulated recycling pathway directly (*question mark* and *arrow*), but no support for this idea has yet emerged. Those channels retrieved from the apical surface (see Fig. 18.1) and delivered to the ARE are trafficked back to the apical membrane through the involvement of Rab proteins (most likely Rab11, 27). Delivery of ENaC to the apical surface is regulated by RhoA and sorting nexin 3 (SN3) and vesicle fusion is mediated by SNARE proteins (SNAREs). Details of the mechanisms of ENaC delivery are provided in the text. From Michael B. Butterworth et al. Am J Physiol Renal Physiol 2009;296:F10–F24

18.5.2 ENaC Leaves the Apical Membrane

As we have discussed previously, the first step in ENaC internalization is ubiquitination by Nedd4-2 (Eaton et al. 2010; Malik et al. 2001, 2005a, 2006). The next step in ENaC (and other membrane proteins) endocytosis is the association of ENaC with specialized structures on the membrane surface, clathrin-coated pits or caveolae: the two types of structures tend to promote endocytosis from different regions of the surface membrane. Chlathrin-coated pits form from areas of the membrane that include normal regions of the membrane; caveolae tend to form from lipid rafts. In general, membrane proteins in lipid rafts that are internalized via caveolae are likely to be degraded by the proteasome, while membrane proteins in non-raft areas of the membrane tend to be internalized via chlathrin-coated pits and degraded in lysosomes. The mechanism for ENaC retrieval remains ambiguous; however, in oocytes at least, ENaC appears to be internalized by a dynamin-2-dependent process into either clathrin-coated pits or caveolae (Shimkets et al. 1997). At this point, the addition of ubiquitin targets ENaC for internalization. However, ENaC can be deubiquitinated by the deubiquitinating enzyme, USP2-45 (ubiquitin-specific protease 2) (Fakitsas et al. 2007; Ruffieux-Daidie et al. 2008; Verrey et al. 2008). USP2-45 associates with Nedd4-2 to reverse its effects. If ENaC is deubiquitinated at this point, it remains in the membrane (Oberfeld et al. 2011).

If not deubiquitinated, ENaC is internalized into an EEA1-positive early endosomal compartment; ENaC has several alternative fates. In the endosome, ubiquitinated ENaC is detected and sorted to lysosomes by Hrs (hepatocyte growth factor-regulated tyrosine kinase substrate) (Zhou et al. 2010). Together with STAM (signal-transducing adaptor molecule), Hrs forms a complex called ESCRT-0 (Piper and Luzio 2007; Raiborg and Stenmark 2009). Disruption of this complex reduces ENaC degradation and increases its recycling to the cell surface (Zhou et al. 2010). If it remains ubiquitinated, it will be recognized by the ESCRT machinery and proceed to late endosomes and then into lysosome or proteasomes for degradation. However, several deubiquitinating enzymes can remove ubiquitin in the early endosome. The deubiquitinating enzyme, USP8 (ubiquitin-specific peptidase 8), binds to the ESCRT-0 complex through an interaction with STAM (Kato et al. 2000). USP8 regulates ENaC trafficking selectively at the endocytic sorting step and prevents it from progressing to degradative compartments.

The evidence for a recycling pool is relatively clear-cut. Many investigators have shown that cyclic AMP stimulation promotes ENaC trafficking to the apical surface to increase channel number and transcellular ENaC current (Butterworth et al. 2005; Marunaka and Eaton 1991; Morris and Schafer 2002; Robins et al. 2013; Yang et al. 2006). Edinger et al. (2012) subsequently showed that ENaC plays a role in establishing and maintaining the pool of vesicles that respond to cAMP stimulation and allow ENaC exocytosis. The ENaC channels responsive to cAMP are derived in part from a recycling pool of channels (Fig. 18.6), although it is likely that a pool of newly synthesized channels also contributes (Butterworth et al. 2001).

18.6 An Extended ENaC Regulatory Complex

Membrane proteins have an inherent problem: if they are free to move in the plane of the membrane, interactions with regulatory molecules will be hit-or-miss in a true Markovian statistical sense. The evolutionary solution to this problem is conceptually simple: put the effector molecule and all the relevant signaling molecules together into a stable complex tethered together through direct interaction of chaperone proteins or localized to sub-domains of the membrane (for reviews, see Calder and Yaqoob 2007; Cho and Stahelin 2005; Kapus and Janmey 2013). ENaC is part of a surprisingly large complex of signaling molecules that are intrinsic to its normal function and regulation. In a previous section, we have described the interaction between ENaC and Nedd4-2, the initial interacting ENaC partner. The initial description of a large complex of proteins associated with ENaC was based on biochemical isolation and purification (Benos et al. 1987). Under nonreducing conditions, these authors described a protein band of approximately 730 kDa. After reduction, they resolved the complex into five major polypeptide bands with apparent average molecular weights of 315, 149, 95, 71, and 55 kDa and suggested that the 150 kDa band contained the amiloride binding domain, i.e., that it was ENaC. They did not formally report on the identity of the other bands. Subsequently, the concept of a large ENaC regulatory complex (ERC) in which the components are identifiable is due to Pearce and his colleagues (Bhalla et al. 2006b; Pearce et al. 2015; Soundararajan et al. 2009, 2010a, 2012a, b) who have described many ENaC-interacting proteins including SGK1 (Chen et al. 1999; Pearce 2001). Not surprisingly, some of the same regulatory molecules we have already discussed are part of or can be recruited to the ERC. Both the small G protein Ras and its downstream cascade of Raf, MEK, and ERK are part of the complex (Soundararajan et al. 2010a). ENaC is inhibited by the Raf–MEK–ERK1/2 MAPK pathway (Falin et al. 2005; Grossmann et al. 2004; Nicod et al. 2002), but another aldosterone-induced protein, glucocorticoid-induced leucine zipper protein 1 (GILZ1), binds to and inhibits the initial member of the cascade, Raf-1, leading to the relief of the ERK1/2 inhibition of ENaC. GILZ1 associates with other members of the ERC, in particular with SGK1 and Nedd4-2 (Soundararajan et al. 2009) (Fig. 18.7). The ERC is dynamic in the sense that, in the absence of aldosterone with little SGK1 or GILZ1 in cells, the complex likely consists of only Nedd4-2 and Raf-1, along with other components of the Raf–MEK–ERK signaling pathway that are tightly associated with Raf-1. In the presence of aldosterone, GILZ1 associates with SGK1 and this association prevents SGK1 from entering endoplasmic reticulum-associated protein degradation pathway where misfolded proteins of the endoplasmic reticulum (i.e., ones not associated with a chaperone like SGK1 and GILZ1) are ubiquitinated and degraded. Together, GILZ1 and SGK1 move into the complex. Once there, they cooperatively inhibit Nedd4-2 and Raf-1 and stimulate ENaC surface expression and activity (Soundararajan et al. 2009).

However, there are other components of the ERC. One protein connector enhancer of kinase suppressor of Ras isoform 3 (CNK3) acts as a scaffold in

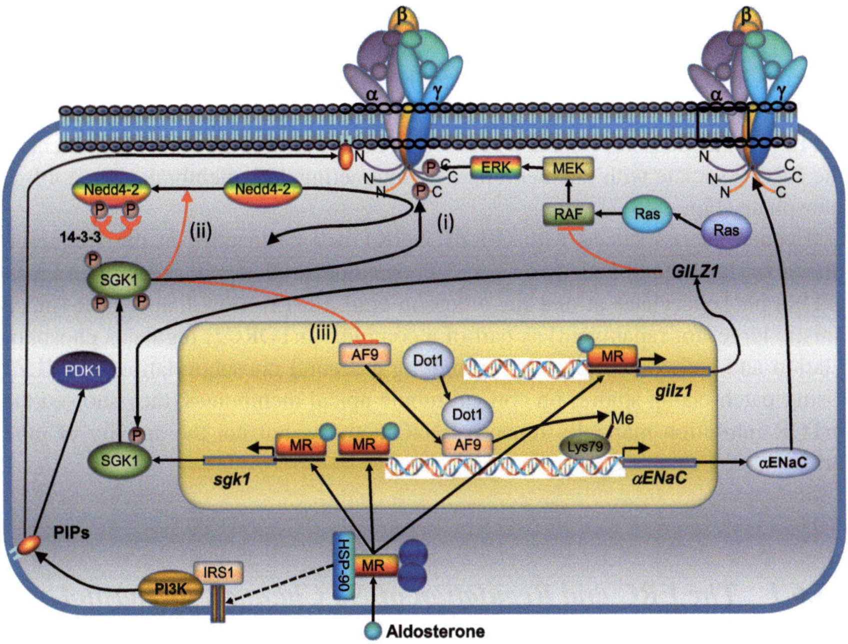

Fig. 18.7 SGK1 acts at multiple levels to regulate ENaC-mediated Na^+ transport in renal collecting duct cells. At low aldosterone (Aldo) levels, α-ENaC gene transcription is low because of tonic inhibition by a complex of Dot1a and AF9, which methylates Lys79 on histone H3. At the same time, Nedd4-2 is active and maintains plasma membrane ENaCs at low levels through ubiquitinylation, which results in channel internalization and degradation. With an elevation in aldosterone levels, activated MR translocates to the nucleus, binds to MREs in the SGK1 5′ regulatory region, and rapidly stimulates SGK1 transcription (*solid arrow*). α-ENaC is not initially responsive to aldosterone-activated MR because the Dot1a-AF9 complex maintains surrounding chromatin in a hypermethylated, stable state, which blocks entry of transcription factors, including MR itself, to their DNA-binding sites (*dashed arrow*). SGK1 catalytic activity is stimulated by PI3K-dependent generation of 3-phosphorylated phosphoinositides (PtdIns), which activate phosphoinositide-dependent protein kinase 1 (PDK1) and ultimately SGK1 itself. SGK1 activates ENaC through phosphorylation of several key substrates: the channel itself (i), resulting in increased open probability; Nedd4-2 (ii), resulting in inhibition of Nedd4-2 through recruitment of inhibitory 14-3-3 chaperone proteins and subsequent decreased channel internalization and degradation; and AF9 (iii), resulting in inhibition of the Dot1a-AF9 complex and diminished H3 hypermethylation in the vicinity of the α-ENaC promoter. This latter effect is predicted to enhance MR access to chromatin, binding to MREs, and stimulation of transcription. *IRS1* insulin receptor substrate 1; *Me* lysine 79 methylation; *α, β, γ*, label ENaCα, -β, and -γ subunits in the surface membrane (*top* of figure). Modified from Bhalla V et al. Am J Physiol Renal Physiol 2006;291: F714–F721

which the other members of the ERC assemble and are organized (Soundararajan et al. 2012c). CNK family members function as protein scaffolds, regulating the activity and the subcellular localization of RAS-activated RAF. Mammals have 3 CNK proteins (CNK1, CNK2, and CNK3). All of the CNK members contain a PSD-95/DLG-1/ZO-1 (PDZ) domain, and, with the exception of CNK3, a PH domain (http://www.ncbi.nlm.nih.gov/Structure/cdd/cddsrv.cgi?uid=cd01260).

Soundararajan et al. (2012c) show that aldosterone induces the formation of a 1.0–1.2-MDa plasma membrane complex which contains CNK3. CNK3 physically interacts with ENaC, Nedd4-2, and SGK1 and stimulates ENaC function in a PDZ domain-dependent manner. It is interesting that CNK3 is the one member of the family that does not contain a pH domain since many of the other components of the ERC associate with the membrane by interacting with membrane inositol lipid phosphate (Fig. 18.8).

Even when associated with GILZ1, SGK1 will be inactive in the absence of phosphorylation at at least two sites. 3-Phosphoinositide-dependent protein kinase (PDK) phosphorylates at one site. Gleason et al. (2015) show that mTOR (mechanistic target for rapamycin) as part of a complex (mTORC2) regulates phosphorylation and activation SGK1 by promoting a second phosphorylation by GILZ1. Using patch-clamp studies on cortical tubule apical membranes, they showed that mTOR inhibition markedly reduces ENaC activity, but not the activity of other principal cell channels.

18.6.1 The ERC and Regulation of Na Channels by Inositol Lipids

Besides the elements described by Pearce and his co-workers, there may well be other elements of the ERC associated with membrane lipids. Phosphatidylinositol phosphates (PIPs) are known to regulate epithelial sodium channels (ENaC) (Blazer-Yost et al. 1999; Pochynyuk et al. 2008b; Yue et al. 2002). Phosphatidylinositol 4,5-bisphosphate (PIP_2) and phosphatidylinositol 3,4,5-trisphosphate (PIP_3) are trace, but ubiquitous, components of eukaryotic membrane phospholipids. The traditional view of phosphoinositide metabolism suggests that phosphatidylinositol is sequentially phosphorylated by phosphatidylinositol 4-kinase and phosphatidylinositol 5-kinase to produce 4,5-PIP_2. Once formed, PIP_2 can be further phosphorylated to produce phosphatidylinositol 3,4,5-PIP_3 by phosphatidylinositol 3-kinase, or PIP_2 can be hydrolyzed by phospholipase C to produce 1,4,5-inositol trisphosphate (IP3) and diacylglycerol (DAG). Until recently, it was assumed that the sole purpose of PIP_2 production was to act as a precursor for IP3 and DAG which activate protein kinase C. It now seems clear that PIP_2 is a signaling molecule in and of itself that can act directly to activate other effector and signaling proteins and localize them at their sites of action. In particular, PIP_2 modulates the activity of ENaC apparently through interactions with hydrophobic and positively charged residues in the cytosolic tails of the β- and γ-subunits. Lipid binding assays and coimmunoprecipitation showed that the amino-terminal domain of the β- and γ-subunits of *Xenopus* ENaC can directly bind to phosphatidylinositol 4,5-bisphosphate (PIP_2), phosphatidylinositol 3,4,5-trisphosphate (PIP_3), and phosphatidic acid (PA) (Alli et al. 2012a). Other experiments defined the regions in the ENaC subunits that interacted with PIP_2 and PIP_3

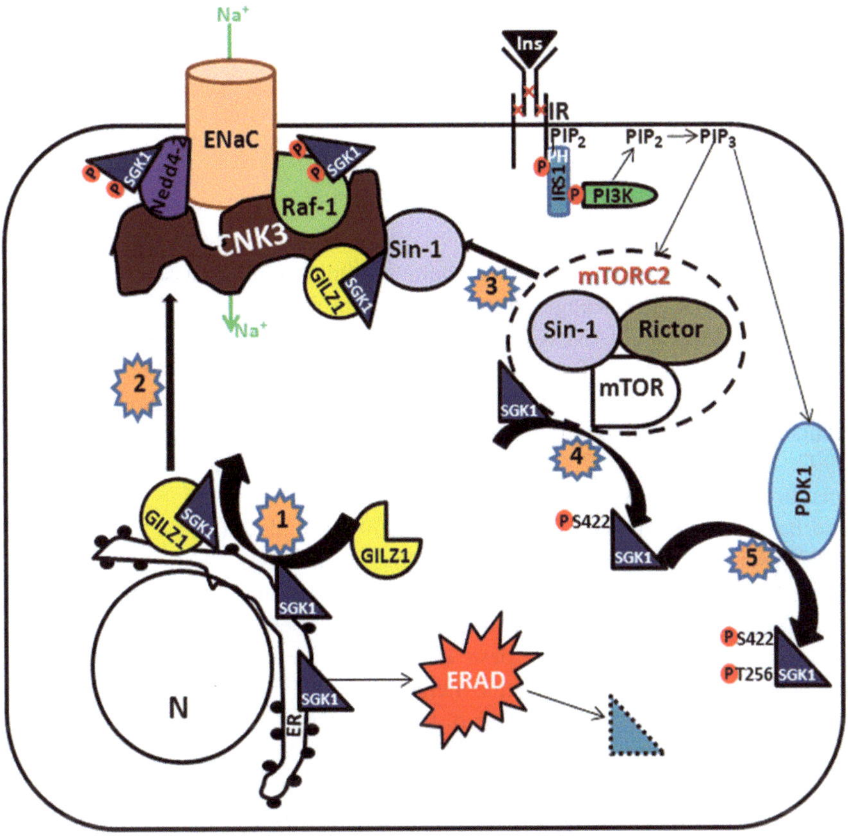

Fig. 18.8 Hypothetical model for SGK1 activation and regulation of ENaC by the ENaC-regulatory complex (or ERC). ENaC regulation occurs within a unique multi-protein complex called the ERC. The various components of the complex are regulated by diverse signaling networks, including steroid receptor-, PI3K-, mTOR-, and Raf–MEK–ERK-dependent pathways. Aldosterone coordinately induces the expression of SGK1, GILZ1, as well as the scaffold protein CNK3, among many others. Highlights 1–5 depict sequential steps in the selective recruitment and activation of SGK1 (not meant to depict spatial/subcellular localization). GILZ1 protects SGK1 from rapid endoplasmic reticulum-associated degradation and redirects it away from the endoplasmic reticulum and into the ERC that possibly includes CNK3. CNK3 provides a platform for organizing various components of the ENaC regulatory machinery by virtue of its ability to interact with multiple proteins including mSin1. mTOR couples with Rictor and mSin1 to form the functional TORC2 complex, which phosphorylates and activates SGK1 at S422. Subsequent phosphorylation of SGK1 at T256 by PDK1 is required for its full activation. It is possible that activation of SGK1 occurs at the ERC and that mTORC2 is recruited to the ERC via mSin1. It is also possible that SGK1 is brought into the ERC after it has been activated by mTORC2 and PDK1 elsewhere. *ENaC* epithelial Na^+ channel, *ERAD* endoplasmic reticulum-associated degradation, *mTOR* mammalian target of rapamycin, *PDK1* phosphoinositide-dependent kinase-1, *SGK1* serum- and glucocorticoid-induced kinase 1. With permission, from Soundararajan R1, Lu M, Pearce D. Organization of the ENaC-regulatory machinery. Crit Rev Biochem Mol Biol. 2012 Jul-Aug;47(4):349–59. doi:10.3109/10409238.2012.678285

(Helms et al. 2005a; Ma et al. 2002, 2007; Ma and Eaton 2005; Pochynyuk et al. 2005, 2006, 2007a, b, 2008b; Staruschenko et al. 2007; Tong et al. 2004; Yue et al. 2002; Zhang et al. 2010).

18.6.2 The Role of MARCKS Protein

Besides binding to ENaC subunits, additional direct binding assays showed that various PIPs can bind strongly to another native protein, myristoylated alanine-rich C-kinase substrate (MARCKS), but weakly or not at all to a mutant form of MARCKS (Alli et al. 2012a). MARCKS consists of a myristoylated amino-terminal domain and a basic effector domain, which contains sites for PKC phosphorylation, filamentous (F)-actin binding, and calcium/calmodulin binding (Hartwig et al. 1992; Yamauchi et al. 2003). MARCKS reversibly associates with the plasma membrane through hydrophobic and electrostatic interactions of its myristoylation and basic effector domain, respectively (Gambhir et al. 2004; McLaughlin et al. 2002; McLaughlin and Aderem 1995; Wang et al. 2004). MARCKS cross-links actin, but binding is regulated by PKC phosphorylation and calmodulin binding (Arbuzova et al. 1998; Kim et al. 1994; McLaughlin et al. 2005). PKC phosphorylates serine residues within the basic effector domain of MARCKS, leading to the translocation of MARCKS from the membrane to the cytoplasm. Phosphorylation prevents the electrostatic interaction of the effector region MARCKS with plasma membrane PIP_2 after which MARCKS enters the cytosol (George and Blackshear 1992; Kim et al. 1994; Taniguchi and Manenti 1993). In epithelial cells, MARCKS also cross-links actin (Yarmola et al. 2001) and acts as a source of PIP_2 for other membrane proteins (including ENaC, SGK1, and PDK) (Fig. 18.9).

Confocal microscopy demonstrated co-localization between MARCKS and PIP_2 (Alli et al. 2012a). Confocal microscopy also showed that in renal principal cells in culture, MARCKS redistributes from the apical membrane to the cytoplasm after PMA-induced MARCKS phosphorylation or ionomycin-induced calmodulin activation. When MARCKS is dephosphorylated (i.e., at the membrane), it can present PIP_2 to ENaC and activate ENaC. In PKC-α knockout mice with no constitutive phosphorylation of MARCKS, ENaC is persistently activated (Bao et al. 2014). Thus, another component of the ERC is MARCKS and calmodulin which move from the complex to the cytosol if MARCKS is phosphorylated (Fig. 18.10).

18.6.3 Role of the Cytoskeleton

MARCKS acts as an organizing center for the actin cytoskeleton, and CNK3 has a PSD domain that could interact with the MARCKS-organized cytoskeletal proteins. Numerous reports have linked cytoskeleton-associated proteins with the

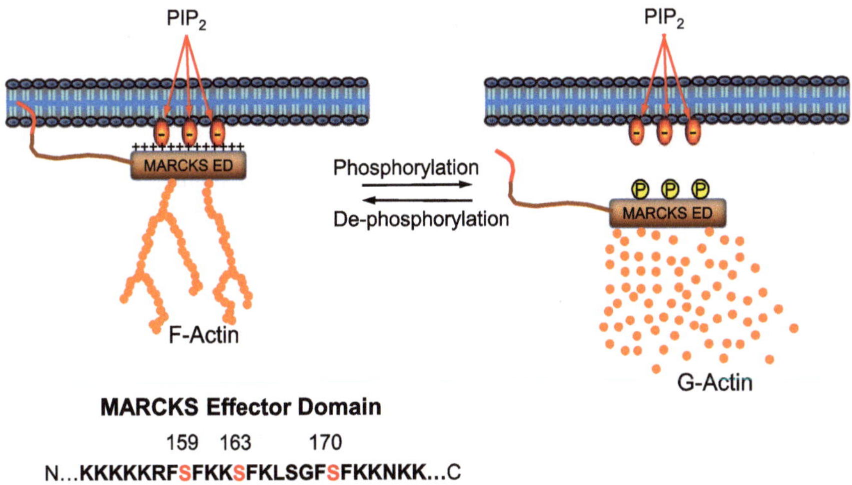

Fig. 18.9 MARCKS function. Dephospho-MARCKS binds to the plasma membrane via a myristoyl group and electrostatic interaction between the effector basic residues and membrane PIP2. It sequesters three PIP2 per MARCKS and cross-links actin. Phospho-MARCKS detaches from the plasma membrane to enter the cytosol and disrupts actin filaments and leaves PIP2 subject to hydrolysis. *Below*: MARCKS effector domain sequence with phosphorylation sites in *red*. It mainly consists of basic amino acids (K and R) and has serine residues which can be phosphorylated by protein kinase C and by Rho kinase (ROCK)

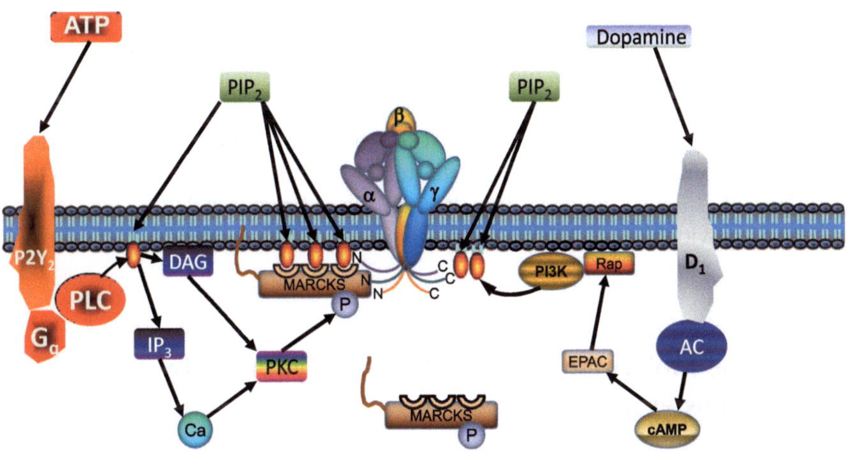

Fig. 18.10 Role of G protein-linked receptors and MARCKS in regulating ENaC activity. In the absence of purinergic ligands, MARCKS sequesters PIP2 near ENaC to increase ENaC open probability. Two receptors, purinergic p2Y2 on the left and dopaminergic D1 on the right, couple to phospholipase C or adenylyl cyclase. In the absence of purinergic ligands, MARCKS sequesters PIP2 near ENaC to increase ENaC open probability. In the presence of P2Y2 ligand, the receptor activates protein kinase C that phosphorylates MARCKS. MARCKS leaves the membrane and ENaC is no longer stimulated by PIP2. In the presence of a dopaminergic ligand, cAMP is produced that activates EPAC and finally PI-3-K that produces inositol lipid phosphates that stimulate ENaC

regulation of ENaC activity (Awayda and Subramanyam 1998; Ilatovskaya et al. 2011; Karpushev et al. 2010a, b; Wang et al. 2013; Zuckerman et al. 1999). Recently Reifenberger et al. (2014) showed that cytochalasin E or latrunculin treatment for 60 min can disrupt the integrity of the actin cytoskeleton in cultured renal principal cells. In single-channel patch-clamp experiments, there were few active channels in treated vs. untreated cells. However, the number of ENaC subunits on the surface of the cells measured by in situ biotinylation was unchanged by disruption of the cytoskeleton. The analysis of those patches that did contain ENaC activity after treatment had significantly decreased channel open probability. The subcellular expression of fodrin changed significantly, and several protein elements of the cytoskeleton decreased at least twofold after 60 min of cytochalasin E treatment. Cytochalasin E treatment also disrupted the association between ENaC and myristoylated alanine-rich C-kinase substrate. These results suggest that disruption of the actin cytoskeleton can reduce ENaC activity by destabilizing the ENaC-MARCKS complex and possibly the entire ERC.

18.7 ENaC Mechanosensitivity

ENaC mechanosensitivity is not surprising given that the ENaC/deg family originally evolved in the first metazoans as mechanosensitive channels such as Mecs and Degs (Studer et al. 2011). Whether ENaC is activated by membrane stretch caused by hydrostatic pressure or cell swelling remains controversial (Fronius and Clauss 2008); however, fluid flow under physiologically relevant conditions increases ENaC open probability in perfused rabbit collecting ducts and in ENaC-expressing *Xenopus* oocytes (Althaus et al. 2007; Carattino et al. 2004, 2005b; Satlin et al. 2001). As pointed out above, the cytoskeleton regulates ENaC, as the COOH-terminus of α-ENaC interacts directly with the actin cytoskeleton (Mazzochi et al. 2006) providing a mechanism to transduce flow shear to intracellular events.

18.8 The Molecular Basis for ENaC Regulation by Aldosterone

Aldosterone is the primary hormonal mechanism for regulating ENaC. In an evolutionary sense, ENaC and aldosterone are likely to have coevolved. Aldosterone dates from approximately 400 million years ago at approximately the time of divergence of teleosts and tetrapods (Colombo et al. 2006; Fournier et al. 2012). Aldosterone synthase, encoded by a member of the P450 family, CYP11B2, arose by duplication of the CYP11B gene in the sarcopterygian or lobe-finned fish/tetrapod line after its divergence from the actinopterygian or ray-finned fish line

420 Mya. This was before the beginning of the colonization of land by tetrapods in the late Devonian period, around 370 Mya. The fact that aldosterone was already present in Dipnoi, which occupy an evolutionary transition between water and air breathing but are fully aquatic, suggests that the role of this steroid was to potentiate the corticoid response to hypoxia, rather than to prevent dehydration out of the water (Colombo et al. 2006). But it would also be consistent with the requirements of sarcopterygii and early tetrapods to regulate salt balance separately from water balance in an estuarine or partially aquatic environment. Members of the ENaC/degenerin family are much older dating from the very first metazoans (Studer et al. 2011); however, the advent of recognizable ENaC function probably also dates from the late teleosts (Wilson et al. 2000) and sequences for all three subunits are detectable in dipnoid osteicthys. On the other hand, a well-developed ability to link blood pressure and blood volume to sodium reabsorption was only finally developed in mammals (Studer et al. 2011). In states of volume depletion, decreased renal perfusion results in release of renin from the juxtaglomerular apparatus, activating the renin–angiotensin–aldosterone pathway. Most of the actions of aldosterone require gene transcription (Garty 1986), although some actions only require receptor binding (Holzman et al. 2007) and a few are non-genomic (Christ et al. 1995; Falkenstein and Wehling 2000; Funder 2001; Losel et al. 2002). Binding of aldosterone to the mineralocorticoid receptor activates transcription of a variety of genes, which increases ENaC current in two phases: an initial phase which increases transport four- to sixfold in the first 0–6 h and a late phase which requires 12–48 h and increases transport another three- to fourfold (Garty 1986). The mechanisms for the early and late phases appear to be different. Aldosterone, like other steroid hormones, enters target cells (in the case of the kidney, one major target is principal cells of the collecting duct) and binds to cytosolic, mineralocorticoid receptor complexes. In mammalian principal cells, mineralocorticoid receptors are protected from activation by the high levels of circulating glucocorticoids, by the enzyme, 11-β-hydroxysteroid dehydrogenase that alters glucocorticoids to a form that is unable to bind aldosterone receptors (Monder 1991; Naray-Fejes-Toth et al. 1991). After some rearrangement, the aldosterone-bound receptor acts as a DNA binding protein which targets mineralocorticoid response elements (MREs) on genetic DNA (Funder 1997; Horisberger and Rossier 1992). Binding to the response elements alters gene expression (Horisberger and Rossier 1992). Increases in Na^+ transport can be measured within the first half hour of exposure to aldosterone, and this increase is dependent on gene transcription and translation with the gene products generically referred to as aldosterone-induced proteins (AIPs). Aldosterone and increased intracellular sodium also promote increased sodium pump expression in the long term.

Many attempts using a variety of differential screening methods have been made to identify the aldosterone-induced gene products (Chen et al. 1999; Mastroberardino et al. 1998; Pearce et al. 2003; Spindler et al. 1997; Spindler and Verrey 1999; Verrey 1998). These attempts have met with varying levels of success, but the relationship between the genes identified and the mechanism by which aldosterone increases Na^+ transport was often unclear. Originally, because of the observation that protein synthesis was required for aldosterone to increase sodium transport, it was hypothesized that aldosterone induced ENaC synthesis and insertion. While the evidence from electrophysiological measurements remains somewhat controversial, all of the available biochemical evidence suggests that the number of sodium channel proteins in the apical membrane does not change in the early period after aldosterone treatment (at least in the first 2–4 h when the increase in sodium transport is most dramatic) (Helman et al. 1996; Kleyman et al. 1989, 1992; Tousson et al. 1989). More evidence also supports the view that the early effects of aldosterone are indirect and that the proteins which are synthesized are modulatory proteins which convert poorly transporting ENaC in the apical membrane into ones which readily transport sodium (Chen et al. 1999; Loffing et al. 2001; Verrey et al. 2000).

The mechanism for the long-term effect of aldosterone to increase sodium transport is also not completely clear. There is evidence in a variety of tissues that the total cellular amount of ENaC subunits may increase; however, whether this is due to an increase in translation is controversial since changes in message levels are quite variable in different tissues. Therefore, the long-term changes could be due to an increase in translation, an increase in trafficking of subunits to the membrane, posttranslational modifications of channels in the membrane, or a reduction in degradation of membrane channels.

On cursory examination, the magnitude of the changes induced by aldosterone is surprising; the expression of so many genes and the activity of so many proteins are somewhat overwhelming. However, the overall effects of aldosterone are exactly what one would expect for a graded physiological response. The actions can be summarized as four successive molecular events. (1) Activate existing channels by increasing their open probability; (2) increase channel density by reducing the rate at which channels are removed from the membrane; (3) increase the rate at which channels are inserted into the membrane from cellular pools; and (4) increase transcription and translation to add to the total amount of ENaC in the membrane and cell.

18.8.1 Activate Existing Channels by Increasing Their Open Probability

Likely the most important early event is activation of PI-3-kinase to produce more inositol lipid phosphates (Blazer-Yost et al. 1999; Pochynyuk et al. 2006, 2007b, 2008c; Staruschenko et al. 2007; Tong et al. 2004; Wang et al. 2008). This will increase ENaC open probability (Frindt and Palmer 2015; Kemendy et al. 1992; Yue et al. 2002). Another important early event is expression and activation of K-Ras (Mastroberardino et al. 1998; Verrey et al. 2000). K-Ras is an early gene, but K-Ras requires posttranslational modification to activate ENaC. Aldosterone-induced methylation may directly activate ENaC (Edinger et al. 2006; Rokaw et al. 1998), but it also promotes methylation of K-Ras, an essential step in its association with the membrane and activation (Al-Baldawi et al. 2000; Becchetti et al. 2000; Stockand et al. 1999b, 2000b). The isoprenyl-cysteine-O-carboxy-methyl-transferase responsible for K-Ras2A methylation is not an aldosterone-induced protein, but the intrinsic activity of the enzyme does increase in response to aldosterone (by protein kinase C phosphorylation). In addition, the activity of the other enzyme intimately involved in protein methylation, SAH hydrolase, is also increased by aldosterone (Stockand et al. 1999a, 2001). Once active, K-Ras couples PI-3-kinase to EnaC (Staruschenko et al. 2004, 2005) and increases open probability. Thus, at least part of the aldosterone-induced short-term increase in ENaC activity is due to an increase in anionic lipids that directly activate existing ENaC channels.

18.8.2 Keep Existing (and Any New) ENaC in the Membrane

ENaC internalization and degradation involve Nedd4-2 ubiquitination (Staub et al. 1997a; Staub and Rotin 1997). Aldosterone controls Nedd4-2 activity by promoting expression of SGK1 (Bhargava et al. 2001; Chen et al. 1999; Debonneville et al. 2001; Loffing et al. 2001; Pearce 2001; Verrey et al. 2000). SGK1 is a kinase that phosphorylates Nedd4-2 at two locations after which Nedd4-2 is inactivated by interaction with the scaffold protein, 14-3-3σ (Bhalla et al. 2005). SGK1 is normally inactive after translation and is activated by phosphorylation at three or four sites. The first phosphorylation is by the aldosterone-induced chaperone protein, GILZ1. GILZ1 also protects SGK1 from ERAD proteasomal degradation (Soundararajan et al. 2010b, 2012a). The second phosphorylation is by PDK1 (3-phosphoinositide dependent kinase) (Faletti et al. 2002). PDK1 is activated by PI-3-kinase production of phosphatidyl inositol phosphates (produced in step 1) (Hemmings and Restuccia 2012). Subsequently, mTORC2 (mTOR complex 2) phosphorylates the hydrophobic motif of SGK1 (Garcia-Martinez and Alessi 2008; Gleason et al. 2015; Lu et al. 2010). mTORC2 activation may be part of an aldosterone-induced long-term proliferative response (Alessi et al. 2009). Finally, for full activation, SGK1 may be additionally phosphorylated by another kinase, WNK1 (Chen et al. 2009).

Nedd4-2 interaction with ENaC is normally promoted by ERK phosphorylation of ENaC (Soundararajan et al. 2009, 2012a). It is unclear whether SGK1-mediated inactivation of Nedd4-2 is complete, but, besides altering SGK1 activity, GILZ1 also inhibits ERK activation (Soundararajan et al. 2009, 2012a). There may be additional kinases that also affect Nedd4-2 activity, but their relationship to aldosterone activation of ENaC is unclear (Hallows et al. 2010).

18.8.3 Recruit More Active Channels to the Membrane

There are two processes by which aldosterone recruits active channels to the membrane: recruitment from cytosolic recycling pools and promoting proteolytic cleavage to convert inactive to active channels.

Volume depletion and aldosterone administration are associated with an increase in the proteolytic processing of ENaC subunits in rodent kidney (Ergonul et al. 2006; Kleyman et al. 2006; Masilamani et al. 1999). An increase in the residency time of channels at the plasma membrane in response to aldosterone may contribute to the increase in ENaC subunit proteolysis, since noncleaved channels at the plasma membrane will have more time to be processed by proteases (Knight et al. 2006). Altering the protease/protease inhibitor balance may also contribute to the increase in ENaC subunit proteolysis observed with volume depletion and aldosterone administration (Narikiyo et al. 2002).

Aldosterone does promote incorporation of ENaC from subapical pools of channels (Loffing et al. 2000; O'Neill et al. 2008). The mechanisms by which this occurs are often unclear; however, a few cases are clearer. AS160 phosphorylation permits ENaC forward trafficking to the apical membrane to augment Na absorption (Liang et al. 2010). Also, there is more trafficking from a subapical pool of channels (Edinger et al. 2012).

18.8.4 Transcribe and Translate New ENaC Subunits

Aldosterone controls the transcription of ENaC subunits (Auerbach et al. 2000; Mick et al. 2001; Sayegh et al. 1999; Thomas et al. 1996, 1999, 2002), and this transcription leads to increased numbers of functional channels in the apical membrane. Only recently have some of the specifics of the transcriptional control been described. Zhang et al. (2013) have shown that aldosterone activates α-ENaC transcription by reducing expression of Dot1a (disruption of telomeric silencing 1) and Af9 (ALL1-fused gene from chromosome 9 protein) and by impairing Dot1a–Af9 interaction. Members of the Dot1 family including mouse Dot1l specifically methylate histone H3 K79 (van Leeuwen et al. 2002; Zhang et al. 2004). Dot1a is highly expressed in the kidney (Zhang et al. 2004). Under basal conditions, Dot1a and Af9 form a repression complex, which binds to specific sites of the α-ENaC promoter, leading to targeted histone H3 K79 hypermethylation and repression of α-ENaC. Aldosterone attenuates the effects of the Dot1a–Af9 complex by reducing expression of Dot1a and Af9, and Sgk1 impairs Dot1a–Af9 interaction by phosphorylating Af9 (Zhang et al. 2007a), leading to histone H3 K79 hypomethylation of the α-ENaC promoter and release of α-ENaC repression. In addition, the mineralocorticoid receptor competes with Dot1a to bind Af9.

As a separate mechanism for controlling ENaC transcription/translation, Edinger et al. (2014) have recently described regulation of microRNA expression by aldosterone in both cultured mCCD and isolated primary distal nephron principal cells. Reducing the expression of these miRs increased epithelial sodium channel (ENaC)-mediated sodium transport in mCCD cells. For a review, see (Butterworth 2015).

18.8.5 Summary

Thus, aldosterone, the hormone primarily responsible for altering epithelial Na transport, apparently accomplishes the increase in many ways. Initially, as a rapid response, aldosterone promotes the production of PIP_2 (and PIP_3) which rapidly increases the activity of ENaCs that are already in the surface membrane. Later, aldosterone apparently increases the number of functional channels in the surface membrane, at least in part, by reducing the degradation rate of ENaC and allowing an increase in the surface membrane pool of functional ENaC. Finally, aldosterone promotes an increase in ENaC transcription and translation.

Acknowledgments This work was supported by DHHS R37 DK037963 to DCE, R01-DK100582 to HPM, and AHA POST16820072 to TLT.

References

Abriel H, Loffing J, Rebhun JF, Pratt JH, Schild L, Horisberger JD, Rotin D, Staub O (1999) Defective regulation of the epithelial Na+ channel by Nedd4 in Liddle's syndrome. J Clin Invest 103(5):667–673

Agne AM, Baldin JP, Benjamin AR, Orogo-Wenn MC, Wichmann L, Olson KR, Walters DV, Althaus M (2015) Hydrogen sulfide decreases beta-adrenergic agonist-stimulated lung liquid clearance by inhibiting ENaC-mediated transepithelial sodium absorption. Am J Physiol Regul Integr Comp Physiol 308(7):R636–R649

Ahn YJ, Brooker DR, Kosari F, Harte BJ, Li J, Mackler SA, Kleyman TR (1999) Cloning and functional expression of the mouse epithelial sodium channel. Am J Physiol 277(1 Pt 2):F121–F129

Al-Baldawi NF, Stockand JD, Al Khalili OK, Yue G, Eaton DC (2000) Aldosterone induces *Ras* methylation in A6 epithelia. Am J Physiol (Cell Physiol) 279(2):C429–C439

Albrecht C, Viturro E (2007) The ABCA subfamily–gene and protein structures, functions and associated hereditary diseases. Pflugers Arch 453(5):581–589

Alessi DR, Pearce LR, Garcia-Martinez JM (2009) New insights into mTOR signaling: mTORC2 and beyond. Sci Signal 2(67), e27

Alli AA, Bao HF, Alli AA, Aldrugh Y, Song JZ, Ma HP, Yu L, Al-Khalili O, Eaton DC (2012a) Phosphatidylinositol phosphate-dependent regulation of Xenopus ENaC by MARCKS protein. Am J Physiol Renal Physiol 303(6):F800–F811

Alli AA, Bao HF, Alli AA, Aldrugh Y, Zhou Y, Yu L, Eaton DC (2012b) Calmodulin and CaM kinase II govern MARCKS-mediated PIP2-dependent regulation of ENaC. FASEB J 26:867.15

Almaca J, Kongsuphol P, Hieke B, Ousingsawat J, Viollet B, Schreiber R, Amaral MD, Kunzelmann K (2009) AMPK controls epithelial Na(+) channels through Nedd4-2 and causes an epithelial phenotype when mutated. Pflugers Arch 458(4):713–721

Althaus M, Bogdan R, Clauss WG, Fronius M (2007) Mechano-sensitivity of epithelial sodium channels (ENaCs): laminar shear stress increases ion channel open probability. FASEB J 21 (10):2389–2399

Arbuzova A, Murray D, McLaughlin SG (1998) MARCKS, membranes, and calmodulin: kinetics of their interaction. Biochim Biophys Acta 1376(3):369–379

Asher C, Chigaev A, Garty H (2001) Characterization of interactions between Nedd4 and β and γENaC using surface plasmon resonance. Biochem Biophys Res Commun 286(5):1228–1231

Auerbach SD, Loftus RW, Itani OA, Thomas CP (2000) Human amiloride-sensitive epithelial Na+ channel gamma subunit promoter: functional analysis and identification of a polypurine-polypyrimidine tract with the potential for triplex DNA formation. Biochem J 347(Pt 1):105–114

Awayda MS, Subramanyam M (1998) Regulation of the epithelial Na^+ channel by membrane tension. J Gen Physiol 112(2):97–111

Babini E, Geisler HS, Siba M, Grunder S (2003) A new subunit of the epithelial Na+ channel identifies regions involved in Na+ self-inhibition. J Biol Chem 278(31):28418–28426

Bachhuber T, Almaca J, Aldehni F, Mehta A, Amaral MD, Schreiber R, Kunzelmann K (2008) Regulation of the epithelial Na+ channel by the protein kinase CK2. J Biol Chem 283 (19):13225–13232

Badve SV, Roberts MA, Hawley CM, Cass A, Garg AX, Krum H, Tonkin A, Perkovic V (2011) Effects of beta-adrenergic antagonists in patients with chronic kidney disease. A systematic review and meta-analysis. J Am Coll Cardiol 58(11):1152–1161

Baines DL, Ramminger SJ, Collett A, Haddad JJ, Best OG, Land SC, Olver RE, Wilson SM (2001) Oxygen-evoked Na+ transport in rat fetal distal lung epithelial cells. J Physiol 532(Pt 1):105–113

Balla A, Tuymetova G, Tsiomenko A, Varnai P, Balla T (2005) A plasma membrane pool of phosphatidylinositol 4-phosphate is generated by phosphatidylinositol 4-kinase type-III alpha: studies with the PH domains of the oxysterol binding protein and FAPP1. Mol Biol Cell 16 (3):1282–1295

Balut C, Steels P, Radu M, Ameloot M, Driessche WV, Jans D (2006) Membrane cholesterol extraction decreases Na+ transport in A6 renal epithelia. Am J Physiol Cell Physiol 290(1):87–94

Bankir L, Fernandes S, Bardoux P, Bouby N, Bichet DG (2005) Vasopressin-V2 receptor stimulation reduces sodium excretion in healthy humans. J Am Soc Nephrol 16(7):1920–1928

Bankir L, Bichet DG, Bouby N (2010) Vasopressin V2 receptors, ENaC, and sodium reabsorption: a risk factor for hypertension? Am J Physiol Renal Physiol 299(5):F917–F928

Bao HF, Zhang ZR, Liang YY, Ma JJ, Eaton DC, Ma HP (2007) Ceramide mediates inhibition of the renal epithelial sodium channel by tumor necrosis factor-alpha through protein kinase C. Am J Physiol Renal Physiol 293(4):F1178–F1186

Bao HF, Song JZ, Duke BJ, Ma HP, Denson DD, Eaton DC (2012) Ethanol stimulates epithelial sodium channels by elevating reactive oxygen species. Am J Physiol Cell Physiol 303(11): C1129–C1138

Bao HF, Thai TL, Yue Q, Ma HP, Eaton AF, Cai H, Klein JD, Sands JM, Eaton DC (2014) ENaC activity is increased in isolated, split-open cortical collecting ducts from protein kinase Calpha knockout mice. Am J Physiol Renal Physiol 306(3):F309–F320

Barbier O, Jacquillet G, Tauc M, Cougnon M, Poujeol P (2005) Effect of heavy metals on, and handling by, the kidney. Nephron Physiol 99(4):105–110

Baylis C (2008) Nitric oxide deficiency in chronic kidney disease. Am J Physiol Renal Physiol 294 (1):F1–F9

Baylis C (2012) Nitric oxide synthase derangements and hypertension in kidney disease. Curr Opin Nephrol Hypertens 21(1):1–6

Becchetti A, Kemendy AE, Stockand JD, Sariban-Sohraby S, Eaton DC (2000) Methylation increases the open probability of the epithelial sodium channel in A6 epithelia. J Biol Chem 275(22):16550–16559

Benos DJ, Saccomani G, Sariban-Sohraby S (1987) The epithelial sodium channel. Subunit number and location of the amiloride binding site. J Biol Chem 262(22):10613–10618

Bentley PJ (1968) Amiloride: a potent inhibitor of sodium transport across the toad bladder. J Physiol 195(2):317–330

Berdiev BK, Jovov B, Tucker WC, Naren AP, Fuller CM, Chapman ER, Benos DJ (2004) ENaC subunit-subunit interactions and inhibition by syntaxin 1A. Am J Physiol Renal Physiol 286(6): F1100–F1106

Beutler KT, Masilamani S, Turban S, Nielsen J, Brooks HL, Ageloff S, Fenton RA, Packer RK, Knepper MA (2003) Long-term regulation of ENaC expression in kidney by angiotensin II. Hypertension 41(5):1143–1150

Bhalla V, Hallows KR (2008) Mechanisms of ENaC regulation and clinical implications. J Am Soc Nephrol 19(10):1845–1854

Bhalla V, Daidie D, Li H, Pao AC, LaGrange LP, Wang J, Vandewalle A, Stockand JD, Staub O, Pearce D (2005) Serum- and glucocorticoid-regulated kinase 1 regulates ubiquitin ligase neural precursor cell-expressed, developmentally down-regulated protein 4-2 by inducing interaction with 14-3-3. Mol Endocrinol 19(12):3073–3084

Bhalla V, Oyster NM, Fitch AC, Wijngaarden MA, Neumann D, Schlattner U, Pearce D, Hallows KR (2006a) AMP-activated kinase inhibits the epithelial Na+ channel through functional regulation of the ubiquitin ligase Nedd4-2. J Biol Chem 281(36):26159–26169

Bhalla V, Soundararajan R, Pao AC, Li H, Pearce D (2006b) Disinhibitory pathways for control of sodium transport: regulation of ENaC by SGK1 and GILZ. Am J Physiol Renal Physiol 291(4): F714–F721

Bhargava A, Fullerton MJ, Myles K, Purdy TM, Funder JW, Pearce D, Cole TJ (2001) The serum- and glucocorticoid-induced kinase is a physiological mediator of aldosterone action. Endocrinology 142(4):1587–1594

Biasio W, Chang T, McIntosh CJ, McDonald FJ (2004) Identification of Murr1 as a regulator of the human delta epithelial sodium channel. J Biol Chem 279(7):5429–5434

Bjaelde RG, Arnadottir SS, Overgaard MT, Leipziger J, Praetorius HA (2013) Renal epithelial cells can release ATP by vesicular fusion. Front Physiol 4:238

Blazer-Yost BL, Punescu TG, Helman SI, Lee KD, Vlahos CJ (1999) Phosphoinositide 3-kinase is required for aldosterone-regulated sodium reabsorption. Am J Physiol 277(3 Pt 1):C531–C536

Boncoeur E, Tardif V, Tessier MC, Morneau F, Lavoie J, Gendreau-Berthiaume E, Grygorczyk R, Dagenais A, Berthiaume Y (2010) Modulation of epithelial sodium channel activity by lipopolysaccharide in alveolar type II cells: involvement of purinergic signaling. Am J Physiol Lung Cell Mol Physiol 298(3):L417–L426

Bonifacino JS, Weissman AM (1998) Ubiquitin and the control of protein fate in the secretory and endocytic pathways. Annu Rev Cell Dev Biol 14:19–57

Booth RE, Stockand JD (2003) Targeted degradation of ENaC in response to PKC activation of the ERK1/2 cascade. Am J Physiol Renal Physiol 284(5):F938–F947

Boucher RC (2003) Regulation of airway surface liquid volume by human airway epithelia. Pflugers Arch 445(4):495–498

Bremner HR, Freywald T, O'Brodovich HM, Otulakowski G (2002) Promoter analysis of the gene encoding the beta-subunit of the rat amiloride-sensitive epithelial sodium channel. Am J Physiol (Lung Cell Mol Physiol) 282(1):L124–L134

Brouard M, Casado M, Djelidi S, Barrandon Y, Farman N (1999) Epithelial sodium channel in human epidermal keratinocytes: expression of its subunits and relation to sodium transport and differentiation. J Cell Sci 112(Pt 19):3343–3352

Bruns JB, Carattino MD, Sheng S, Maarouf AB, Weisz OA, Pilewski JM, Hughey RP, Kleyman TR (2007) Epithelial Na+ channels are fully activated by furin- and prostasin-dependent release of an inhibitory peptide from the gamma-subunit. J Biol Chem 282(9):6153–6160

Bubien JK, Watson B, Khan MA, Langloh AL, Fuller CM, Berdiev B, Tousson A, Benos DJ (2001) Expression and regulation of normal and polymorphic epithelial sodium channel by human lymphocytes. J Biol Chem 276(11):8557–8566

Bugaj V, Pochynyuk O, Stockand JD (2009) Activation of the epithelial Na+ channel in the collecting duct by vasopressin contributes to water reabsorption. Am J Physiol Renal Physiol 297(5):F1411–F1418

Burnstock G (2006) Purinergic signalling--an overview. Novartis Found Symp 276:26–48

Butler PL, Staruschenko A, Snyder PM (2015) Acetylation stimulates the epithelial sodium channel by reducing its ubiquitination and degradation. J Biol Chem 290(20):12497–12503

Butterworth MB (2010) Regulation of the epithelial sodium channel (ENaC) by membrane trafficking. Biochim Biophys Acta 1802(12):1166–1177

Butterworth MB (2015) MicroRNAs and the regulation of aldosterone signaling in the kidney. Am J Physiol Cell Physiol 308(7):C521–C527

Butterworth MB, Helman SI, Els WJ (2001) cAMP-sensitive endocytic trafficking in A6 epithelia. Am J Physiol Cell Physiol 280(4):C752–C762

Butterworth MB, Edinger RS, Johnson JP, Frizzell RA (2005) Acute ENaC stimulation by cAMP in a kidney cell line is mediated by exocytic insertion from a recycling channel pool. J Gen Physiol 125(1):81–101

Butterworth MB, Edinger RS, Ovaa H, Burg D, Johnson JP, Frizzell RA (2007) The deubiquitinating enzyme UCH-L3 regulates the apical membrane recycling of the epithelial sodium channel. J Biol Chem 282(52):37885–37893

Butterworth MB, Edinger RS, Frizzell RA, Johnson JP (2009) Regulation of the epithelial sodium channel by membrane trafficking. Am J Physiol Renal Physiol 296(1):F10–F24

Butterworth MB, Edinger RS, Silvis MR, Gallo LI, Liang X, Apodaca G, Frizzell RA, Johnson JP (2012) Rab11b regulates the trafficking and recycling of the epithelial sodium channel (ENaC). Am J Physiol Renal Physiol 302(5):F581–F590

Calder PC, Yaqoob P (2007) Lipid rafts–composition, characterization, and controversies. J Nutr 137(3):545–547

Caldwell RA, Boucher RC, Stutts MJ (2004) Serine protease activation of near-silent epithelial Na + channels. Am J Physiol Cell Physiol 286(1):C190–C194

Canessa CM (1996) What is new about the structure of the epithelial Na^+ channel? News Physiol Sci 11(5):195–201

Canessa CM, Horisberger JD, Rossier BC (1993) Epithelial sodium channel related to proteins involved in neurodegeneration. Nature 361:467–470

Canessa CM, Merillat AM, Rossier BC (1994a) Membrane topology of the epithelial sodium channel in intact cells. Am J Physiol 267(6 Pt 1):C1682–C1690

Canessa CM, Schild L, Buell G, Thorens B, Gautschi I, Horisberger JD, Rossier BC (1994b) Amiloride-sensitive epithelial Na+ channel is made of three homologous subunits. Nature 367 (6462):463–467

Carattino MD, Sheng S, Kleyman TR (2004) Epithelial Na+ channels are activated by laminar shear stress. J Biol Chem 279(6):4120–4126

Carattino MD, Edinger RS, Grieser HJ, Wise R, Neumann D, Schlattner U, Johnson JP, Kleyman TR, Hallows KR (2005a) Epithelial sodium channel inhibition by AMP-activated protein kinase in oocytes and polarized renal epithelial cells. J Biol Chem 280(18):17608–17616

Carattino MD, Sheng S, Kleyman TR (2005b) Mutations in the pore region modify epithelial sodium channel gating by shear stress. J Biol Chem 280(6):4393–4401

Carattino MD, Sheng S, Bruns JB, Pilewski JM, Hughey RP, Kleyman TR (2006) The epithelial Na+ channel is inhibited by a peptide derived from proteolytic processing of its alpha subunit. J Biol Chem 281(27):18901–18907

Carattino MD, Passero CJ, Steren CA, Maarouf AB, Pilewski JM, Myerburg MM, Hughey RP, Kleyman TR (2008) Defining an inhibitory domain in the alpha-subunit of the epithelial sodium channel. Am J Physiol Renal Physiol 294(1):F47–F52

Carey MA, Card JW, Voltz JW, Germolec DR, Korach KS, Zeldin DC (2007) The impact of sex and sex hormones on lung physiology and disease: lessons from animal studies. Am J Physiol Lung Cell Mol Physiol 293:L272–L278

Carninci P, Kasukawa T, Katayama S, Gough J, Frith MC, Maeda N, Oyama R, Ravasi T, Lenhard B, Wells C, Kodzius R, Shimokawa K, Bajic VB, Brenner SE, Batalov S, Forrest AR, Zavolan M, Davis MJ, Wilming LG, Aidinis V, Allen JE, Ambesi-Impiombato A, Apweiler R, Aturaliya RN, Bailey TL, Bansal M, Baxter L, Beisel KW, Bersano T, Bono H, Chalk AM, Chiu KP, Choudhary V, Christoffels A, Clutterbuck DR, Crowe ML, Dalla E, Dalrymple BP, de Bono B, Della GG, di Bernardo D, Down T, Engstrom P, Fagiolini M, Faulkner G, Fletcher CF, Fukushima T, Furuno M, Futaki S, Gariboldi M, Georgii-Hemming P, Gingeras TR, Gojobori T, Green RE, Gustincich S, Harbers M, Hayashi Y, Hensch TK, Hirokawa N, Hill D, Huminiecki L, Iacono M, Ikeo K, Iwama A, Ishikawa T, Jakt M, Kanapin A, Katoh M, Kawasawa Y, Kelso J, Kitamura H, Kitano H, Kollias G, Krishnan SP, Kruger A, Kummerfeld SK, Kurochkin IV, Lareau LF, Lazarevic D, Lipovich L, Liu J, Liuni S, McWilliam S, Madan BM, Madera M, Marchionni L, Matsuda H, Matsuzawa S, Miki H, Mignone F, Miyake S, Morris K, Mottagui-Tabar S, Mulder N, Nakano N, Nakauchi H, Ng P, Nilsson R, Nishiguchi S, Nishikawa S, Nori F, Ohara O, Okazaki Y, Orlando V, Pang KC, Pavan WJ, Pavesi G, Pesole G, Petrovsky N, Piazza S, Reed J, Reid JF, Ring BZ, Ringwald M, Rost B, Ruan Y, Salzberg SL, Sandelin A, Schneider C, Schonbach C, Sekiguchi K, Semple CA, Seno S, Sessa L, Sheng Y, Shibata Y, Shimada H, Shimada K, Silva D, Sinclair B, Sperling S, Stupka E, Sugiura K, Sultana R, Takenaka Y, Taki K, Tammoja K, Tan SL, Tang S, Taylor MS, Tegner J, Teichmann SA, Ueda HR, van Nimwegen E, Verardo R, Wei CL, Yagi K, Yamanishi H, Zabarovsky E, Zhu S, Zimmer A, Hide W, Bult C, Grimmond SM, Teasdale RD, Liu ET, Brusic V, Quackenbush J, Wahlestedt C, Mattick JS, Hume DA, Kai C, Sasaki D, Tomaru Y, Fukuda S, Kanamori-Katayama M, Suzuki M, Aoki J, Arakawa T, Iida J, Imamura K, Itoh M, Kato T, Kawaji H, Kawagashira N, Kawashima T, Kojima M, Kondo S, Konno H, Nakano K, Ninomiya N, Nishio T, Okada M, Plessy C, Shibata K, Shiraki T, Suzuki S, Tagami M, Waki K, Watahiki A, Okamura-Oho Y, Suzuki H, Kawai J, Hayashizaki Y (2005) The transcriptional landscape of the mammalian genome. Science 309(5740):1559–1563

Chalfant ML, Denton JS, Berdiev BK, Ismailov II, Benos DJ, Stanton BA (1999) Intracellular H+ regulates the alpha-subunit of ENaC, the epithelial Na+ channel. Am J Physiol 276(2 Pt 1): C477–C486

Chang T, McDonald FJ (2007) COMMD1 downregulates δ epithelial sodium channel (δENaC) through trafficking and ubiquitination. FASEB J 21:Ib465

Chang T, Ke Y, Ly K, McDonald FJ (2011) COMMD1 regulates the delta epithelial sodium channel (deltaENaC) through trafficking and ubiquitination. Biochem Biophys Res Commun 411(3):506–511

Chen SY, Bhargava A, Mastroberardino L, Meijer OC, Wang J, Buse P, Firestone GL, Verrey F, Pearce D (1999) Epithelial sodium channel regulated by aldosterone-induced protein sgk. Proc Natl Acad Sci U S A 96(5):2514–2519

Chen XJ, Eaton DC, Jain L (2002) Beta-adrenergic regulation of amiloride-sensitive lung sodium channels. Am J Physiol (Lung Cell Mol Physiol) 282(4):609–620

Chen J, Chen Z, Chintagari NR, Bhaskaran M, Jin N, Narasaraju T, Liu L (2006) Alveolar type I cells protect rat lung epithelium from oxidative injury. J Physiol 572(Pt 3):625–638

Chen W, Chen Y, Xu BE, Juang YC, Stippec S, Zhao Y, Cobb MH (2009) Regulation of a third conserved phosphorylation site in SGK1. J Biol Chem 284(6):3453–3460

Chen J, Myerburg MM, Passero CJ, Winarski KL, Sheng S (2011) External Cu2+ inhibits human epithelial Na+ channels by binding at a subunit interface of extracellular domains. J Biol Chem 286(31):27436–27446

Cheng C, Prince LS, Snyder PM, Welsh MJ (1998) Assembly of the epithelial Na^+ channel evaluated using sucrose gradient sedimentation analysis. J Biol Chem 273(35):22693–22700

Cho W, Stahelin RV (2005) Membrane-protein interactions in cell signaling and membrane trafficking. Annu Rev Biophys Biomol Struct 34:119–151
Christ M, Meyer C, Sippel K, Wehling M (1995) Rapid aldosterone signaling in vascular smooth muscle cells: involvement of phospholipase C, diacylglycerol and protein kinase C. Biochem Biophys Res Commun 213(1):123–129
Clark MA, Eaton DC (1983) Effect of CO2 on neurons of the house cricket, Acheta domestica. J Neurobiol 14(3):237–250
Collier DM, Snyder PM (2009) Extracellular chloride regulates the epithelial sodium channel. J Biol Chem 284(43):29320–29325
Collier DM, Snyder PM (2011) Identification of epithelial Na+ channel (ENaC) intersubunit Cl− inhibitory residues suggests a trimeric alpha gamma beta channel architecture. J Biol Chem 286(8):6027–6032
Collier DM, Tomkovicz VR, Peterson ZJ, Benson CJ, Snyder PM (2014) Intersubunit conformational changes mediate epithelial sodium channel gating. J Gen Physiol 144(4):337–348
Colombo L, Dalla VL, Fiore C, Armanini D, Belvedere P (2006) Aldosterone and the conquest of land. J Endocrinol Invest 29(4):373–379
Condliffe SB, Carattino MD, Frizzell RA, Zhang H (2003) Syntaxin 1A regulates ENaC via domain-specific interactions. J Biol Chem 278(15):12796–12804
Condliffe SB, Zhang H, Frizzell RA (2004) Syntaxin 1A regulates ENaC channel activity. J Biol Chem 279(11):10085–10092
Corey DP, Garcia-Anoveros J (1996) Mechanosensation and the DEG/ENaC ion channels. Science 273(5273):323–324
Couloigner V, Fay M, Djelidi S, Farman N, Escoubet B, Runembert I, Sterkers O, Friedlander G, Ferrary E (2001) Location and function of the epithelial Na channel in the cochlea. Am J Physiol Renal Physiol 280(2):F214–F222
Coux O, Tanaka K, Goldberg AL (1996) Structure and functions of the 20S and 26S proteasomes. Annu Rev Biochem 65:801–847
Cuffe JE, Bielfeld-Ackermann A, Thomas J, Leipziger J, Korbmacher C (2000) ATP stimulates Cl− secretion and reduces amiloride-sensitive Na+ absorption in M-1 mouse cortical collecting duct cells. J Physiol (London) 524(Pt 1):77–90
D'Angelo G, Polishchuk E, Di TG, Santoro M, Di CA, Godi A, West G, Bielawski J, Chuang CC, van der Spoel AC, Platt FM, Hannun YA, Polishchuk R, Mattjus P, De Matteis MA (2007) Glycosphingolipid synthesis requires FAPP2 transfer of glucosylceramide. Nature 449 (7158):62–67
D'Angelo G, Rega LR, De Matteis MA (2012) Connecting vesicular transport with lipid synthesis: FAPP2(). Biochim Biophys Acta 1821(8):1089–1095
Davis IC, Matalon S (2007) Epithelial sodium channels in the adult lung–important modulators of pulmonary health and disease. Adv Exp Med Biol 618:127–140
Dean RB, Gatty O (1937) The bioelectrical properties of frog skin. Trans Faraday Soc 33:1040–1046
Debonneville C, Flores SY, Kamynina E, Plant PJ, Tauxe C, Thomas MA, Munster C, Chraibi A, Pratt JH, Horisberger JD, Pearce D, Loffing J, Staub O (2001) Phosphorylation of Nedd4-2 by Sgk1 regulates epithelial Na(+) channel cell surface expression. EMBO J 20(24):7052–7059
Diakov A, Korbmacher C (2004) A novel pathway of epithelial sodium channel activation involves a serum- and glucocorticoid-inducible kinase consensus motif in the C terminus of the channel's alpha-subunit. J Biol Chem 279(37):38134–38142
Dinudom A, Harvey KF, Komwatana P, Young JA, Kumar S, Cook DI (1998) Nedd4 mediates control of an epithelial Na+ channel in salivary duct cells by cytosolic Na+. Proc Natl Acad Sci U S A 95(12):7169–7173
Dinudom A, Harvey KF, Komwatana P, Jolliffe CN, Young JA, Kumar S, Cook DI (2001) Roles of the C termini of α-, β-, and γ-subunits of epithelial Na^+ channels (ENaC) in regulating ENaC and mediating its inhibition by cytosolic Na^+. J Biol Chem 276(17):13744–13749

Dorge A, Nagel W (1970) Effect of amiloride on sodium transport in frog skin. II. Sodium transport pool and unidirectional fluxes. Pflugers Arch 321(2):91–101
Downs CA, Helms MN (2013) Regulation of ion transport by oxidants. Am J Physiol Lung Cell Mol Physiol 305(9):L595–L603
Downs CA, Kriener LH, Yu L, Eaton DC, Jain L, Helms MN (2012) beta-Adrenergic agonists differentially regulate highly selective and nonselective epithelial sodium channels to promote alveolar fluid clearance in vivo. Am J Physiol Lung Cell Mol Physiol 302(11):L1167–L1178
Drummond HA, Gebremedhin D, Harder DR (2004) Degenerin/epithelial Na+ channel proteins: components of a vascular mechanosensor. Hypertension 44(5):643–648
Du Bois-Reymond E, Churchill J, Jones B (1857) On animal electricity being an abstract of the discoveries of Emil du Bois-Reymond. Member of the Academy of Sciences of Berlin, etc. John Churchill, London
Duman JG, Forte JG (2003) What is the role of SNARE proteins in membrane fusion? Am J Physiol Cell Physiol 285(2):C237–C249
Duncan CDS, Mata J (2011) Widespread cotranslational formation of protein complexes. PLoS Genet 7(12), e1002398
Eaton DC (1981) Intracellular sodium ion activity and sodium transport in rabbit urinary bladder. J Physiol 316:527–544
Eaton DC, Hamilton KL (1988) The amiloride-blockable sodium channel of epithelial tissue. Ion Channels 1:251–282
Eaton DC, Malik B, Saxena NC, Al Khalili OK, Yue G (2001) Mechanisms of aldosterone's action on epithelial Na^+ transport. J Membr Biol 184(3):313–319
Eaton DC, Chen XJ, Ramosevac S, Matalon S, Jain L (2004) Regulation of Na+ channels in lung alveolar type II epithelial cells. Proc Am Thorac Soc 1(1):10–16
Eaton DC, Malik B, Bao HF, Yu L, Jain L (2010) Regulation of epithelial sodium channel trafficking by ubiquitination. Proc Am Thorac Soc 7(1):54–64
Edelman IS, Bogoroch R, Porter GA (1963) On the mechanism of action of aldosterone on sodium transport: the role of protein synthesis. Proc Natl Acad Sci U S A 50:1169–1177
Edinger RS, Yospin J, Perry C, Kleyman TR, Johnson JP (2006) Regulation of epithelial Na+ channels (ENaC) by methylation: a novel methyltransferase stimulates ENaC activity. J Biol Chem 281(14):9110–9117
Edinger RS, Lebowitz J, Li H, Alzamora R, Wang H, Johnson JP, Hallows KR (2009) Functional regulation of the epithelial Na+ channel by I{kappa}B kinase-{beta} occurs via phosphorylation of the ubiquitin ligase Nedd4-2. J Biol Chem 284(1):150–157
Edinger RS, Bertrand CA, Rondandino C, Apodaca GA, Johnson JP, Butterworth MB (2012) The epithelial sodium channel (ENaC) establishes a trafficking vesicle pool responsible for its regulation. PLoS One 7(9), e46593
Edinger RS, Coronnello C, Bodnar AJ, LaFramboise WA, Benos PV, Ho J, Johnson JP, Butterworth MB (2014) Aldosterone regulates microRNAs in the cortical collecting duct to alter sodium transport. J Am Soc Nephrol 25(11):2445–2457
Ergonul Z, Frindt G, Palmer LG (2006) Regulation of maturation and processing of ENaC subunits in the rat kidney. Am J Physiol Renal Physiol 291(3):F683–F693
Fakitsas P, Adam G, Daidie D, van Bemmelen MX, Fouladkou F, Patrignani A, Wagner U, Warth R, Camargo SM, Staub O, Verrey F (2007) Early aldosterone-induced gene product regulates the epithelial sodium channel by deubiquitylation. J Am Soc Nephrol 18(4):1084–1092
Faletti CJ, Perrotti N, Taylor SI, Blazer-Yost BL (2002) sgk: an essential convergence point for peptide and steroid hormone regulation of ENaC-mediated Na+ transport. Am J Physiol (Cell Physiol) 282(3):C494–C500
Falin R, Veizis IE, Cotton CU (2005) A role for ERK1/2 in EGF- and ATP-dependent regulation of amiloride-sensitive sodium absorption. Am J Physiol Cell Physiol 288(5):C1003–C1011
Falkenstein E, Wehling M (2000) Nongenomically initiated steroid actions. Eur J Clin Invest 30 (Suppl 3):51–54

Favre I, Moczydlowski E, Schild L (1996) On the structural basis for ionic selectivity among Na+, K+, and Ca2+ in the voltage-gated sodium channel. Biophys J 71(6):3110–3125
Felley-Bosco E, Bender F, Quest AF (2002) Caveolin-1-mediated post-transcriptional regulation of inducible nitric oxide synthase in human colon carcinoma cells. Biol Res 35(2):169–176
Firsov D, Schild L, Gautschi I, Merillat AM, Schneeberger E, Rossier BC (1996) Cell surface expression of the epithelial Na channel and a mutant causing Liddle syndrome: a quantitative approach. Proc Natl Acad Sci U S A 93(26):15370–15375
Folsch H, Mattila PE, Weisz OA (2009) Taking the scenic route: biosynthetic traffic to the plasma membrane in polarized epithelial cells. Traffic 10(8):972–981
Fournier D, Luft FC, Bader M, Ganten D, Andrade-Navarro MA (2012) Emergence and evolution of the renin-angiotensin-aldosterone system. J Mol Med (Berl) 90(5):495–508
Francis WL, Pumphryey RJ (1933) The electrical properties of frog skin: Part I. Introductory. J Exp Biol 10:379–385
Frindt G, Palmer LG (2015) Acute effects of aldosterone on the epithelial Na channel in rat kidney. Am J Physiol Renal Physiol 308(6):F572–F578
Fromter E, Gebler B (1977) Electrical properties of amphibian urinary bladder epithelia. III. The cell membrane resistances and the effect of amiloride. Pflugers Arch 371(1–2):99–108
Fronius M (2013) Treatment of pulmonary edema by ENaC activators/stimulators. Curr Mol Pharmacol 6(1):13–27
Fronius M, Clauss W (2008) Mechano-sensitivity of ENaC: may the (shear) force be with you. Pflugers Arch Eur J Physiol 455(5):775–785
Fuchs W, Larsen EH, Lindemann B (1977) Current-voltage curve of sodium channels and concentration dependence of sodium permeability in frog skin. J Physiol 267(1):137–166
Funder JW (1997) Glucocorticoid and mineralocorticoid receptors: biology and clinical relevance. Annu Rev Med 48:231–240
Funder JW (2001) Non-genomic actions of aldosterone: role in hypertension. Curr Opin Nephrol Hypertens 10(2):227–230
Fyfe GK, Canessa CM (1998) Subunit composition determines the single channel kinetics of the epithelial sodium channel. J Gen Physiol 112:423–432
Fyfe GK, Quinn A, Canessa CM (1998) Structure and function of the Mec-ENaC family of ion channels. Semin Nephrol 18(2):138–151
Gambhir A, Hangyas-Mihalyne G, Zaitseva I, Cafiso DS, Wang J, Murray D, Pentyala SN, Smith SO, McLaughlin SG (2004) Electrostatic sequestration of PIP2 on phospholipid membranes by basic/aromatic regions of proteins. Biophys J 86(4):2188–2207
Garcia-Martinez JM, Alessi DR (2008) mTOR complex 2 (mTORC2) controls hydrophobic motif phosphorylation and activation of serum- and glucocorticoid-induced protein kinase 1 (SGK1). Biochem J 416(3):375–385
Garty H (1986) Mechanisms of aldosterone action in tight epithelia. J Membr Biol 90:193–205
Garty H, Benos DJ (1988) Characteristics and regulatory mechanisms of the amiloride-blockable Na+ channel. Physiol Rev 68(2):309–373
Garty H, Palmer LG (1997) Epithelial sodium channels: function, structure, and regulation. Physiol Rev 77(2):359–396
George DJ, Blackshear PJ (1992) Membrane association of the myristoylated alanine-rich C kinase substrate (MARCKS) protein appears to involve myristate-dependent binding in the absence of a myristoyl protein receptor. J Biol Chem 267(34):24879–24885
George AL Jr, Staub O, Geering K, Rossier BC, Kleyman TR, Kraehenbuhl JP (1989) Functional expression of the amiloride-sensitive sodium channel in Xenopus oocytes. Proc Natl Acad Sci U S A 86(18):7295–7298
Giraldez T, Rojas P, Jou J, Flores C, Alvarez de la Rosa D (2012) The epithelial sodium channel delta-subunit: new notes for an old song. Am J Physiol Renal Physiol 303(3):F328–F338
Gleason CE, Frindt G, Cheng CJ, Ng M, Kidwai A, Rashmi P, Lang F, Baum M, Palmer LG, Pearce D (2015) mTORC2 regulates renal tubule sodium uptake by promoting ENaC activity. J Clin Invest 125(1):117–128

Godi A, Campli AD, Konstantakopoulos A, Tullio GD, Alessi DR, Kular GS, Daniele T, Marra P, Lucocq JM, Matteis MAD (2004) FAPPs control Golgi-to-cell-surface membrane traffic by binding to ARF and PtdIns(4)P. Nat Cell Biol 6(5):393–404

Goodman MB, Ernstrom GG, Chelur DS, O'Hagan R, Yao CA, Chalfie M (2002) MEC-2 regulates C. elegans DEG/ENaC channels needed for mechanosensation. Nature 415 (6875):1039–1042

Gorelik J, Zhang Y, Sanchez D, Shevchuk A, Frolenkov G, Lab M, Klenerman D, Edwards C, Korchev Y (2005) Aldosterone acts via an ATP autocrine/paracrine system: the Edelman ATP hypothesis revisited. Proc Natl Acad Sci U S A 102:15000–15005

Goulet CC, Volk KA, Adams CM, Prince LS, Stokes JB, Snyder PM (1998) Inhibition of the epithelial Na^+ channel by interaction of Nedd4 with a PY motif deleted in Liddle's syndrome. J Biol Chem 273(45):30012–30017

Greenlee MM, Mitzelfelt JD, Yu L, Yue Q, Duke BJ, Harrell CS, Neigh GN, Eaton DC (2013) Estradiol activates epithelial sodium channels in rat alveolar cells through the G protein-coupled estrogen receptor. Am J Physiol Lung Cell Mol Physiol 305(11):L878–L889

Greenlee MM, Mitzelfelt JD, Duke BJ, Al-Khalili OK, Bao HF, Eaton DC (2015) Prolactin stimulates sodium and chloride ion channels in A6 renal epithelial cells. Am J Physiol Renal Physiol 308(7):F697–F705

Grossmann C, Freudinger R, Mildenberger S, Krug AW, Gekle M (2004) Evidence for epidermal growth factor receptor as negative-feedback control in aldosterone-induced Na+ reabsorption. Am J Physiol Renal Physiol 286(6):F1226–F1231

Grunder S, Muller A, Ruppersberg JP (2001) Developmental and cellular expression pattern of epithelial sodium channel alpha, beta and gamma subunits in the inner ear of the rat. Eur J Neurosci 13(4):641–648

Guo Y, Duvall MD, Crow JP, Matalon S (1998) Nitric oxide inhibits Na^+ absorption across cultured alveolar type II monolayers. Am J Physiol 274(3 Pt 1):L369–L377

Guo LJ, Alli AA, Eaton DC, Bao HF (2013) ENaC is regulated by natriuretic peptide receptor-dependent cGMP signaling. Am J Physiol Renal Physiol 304(7):F930–F937

Haerteis S, Krueger B, Korbmacher C, Rauh R (2009) The delta-subunit of the epithelial sodium channel (ENaC) enhances channel activity and alters proteolytic ENaC activation. J Biol Chem 284(42):29024–29040

Hallows KR, Bhalla V, Oyster NM, Wijngaarden MA, Lee JK, Li H, Chandran S, Xia X, Huang Z, Chalkley RJ, Burlingame AL, Pearce D (2010) Phosphopeptide screen uncovers novel phosphorylation sites of Nedd4-2 that potentiate its inhibition of the epithelial Na+ channel. J Biol Chem 285(28):21671–21678

Hamilton KL, Eaton DC (1985) Single-channel recordings from amiloride-sensitive epithelial sodium channel. Am J Physiol 249(3 Pt 1):200–207

Hamilton KL, Good DW, Eaton DC (1986) Single sodium channel events in apical membrane of rat cortical duct (CCD). Kidney Int 29:397

Hardiman KM, McNicholas-Bevensee CM, Fortenberry J, Myles CT, Malik B, Eaton DC, Matalon S (2004) Regulation of amiloride-sensitive Na(+) transport by basal nitric oxide. Am J Respir Cell Mol Biol 30(5):720–728

Hartwig JH, Thelen M, Resen A, Janmey PA, Nairn AC, Aderem A (1992) MARCKS is an actin filament crosslinking protein regulated by protein kinase C and calcium-calmodulin. Nature 356(6370):618–622

Helman SI, Liu X, Blazer-Yost BL (1996) Early response of A6 epithelia to aldosterone is mediated by vesicle trafficking of apical Na^+ channels and not open probability. FASEB J 10:452

Helms MN, Liu L, Liang YY, Al-Khalili O, Vandewalle A, Saxena S, Eaton DC, Ma HP (2005a) Phosphatidylinositol 3,4,5-trisphosphate mediates aldosterone stimulation of epithelial sodium channel (ENaC) and interacts with gamma-ENaC. J Biol Chem 280(49):40885–40891

Helms MN, Yu L, Malik B, Kleinhenz DJ, Hart CM, Eaton DC (2005b) Role of SGK1 in nitric oxide inhibition of ENaC in Na^+-transporting epithelia. Am J Physiol Cell Physiol 289(3): C717–C726
Helms MN, Jain L, Self JL, Eaton DC (2008) Redox regulation of epithelial sodium channels examined in alveolar type 1 and 2 cells patch-clamped in lung slice tissue. J Biol Chem 283 (33):22875–22883
Hemmings BA, Restuccia DF (2012) PI3K-PKB/Akt pathway. Cold Spring Harb Perspect Biol 4 (9):a011189
Hershko A, Ciechanover A (1998) The ubiquitin system. Annu Rev Biochem 67:425–479
Hicke L (1997) Ubiquitin-dependent internalization and down-regulation of plasma membrane proteins. FASEB J 11(14):1215–1226
Hicke L (2001) Protein regulation by monoubiquitin. Nat Rev Mol Cell Biol 2(3):195–201
Hill WG, An B, Johnson JP (2002) Endogenously expressed epithelial sodium channel is present in lipid rafts in A6 cells. J Biol Chem 277(37):33541–33544
Hill WG, Butterworth MB, Wang H, Edinger RS, Lebowitz J, Peters KW, Frizzell RA, Johnson JP (2007) The epithelial sodium channel (ENaC) traffics to apical membrane in lipid rafts in mouse cortical collecting duct cells. J Biol Chem 282(52):37402–37411
Hinton CF, Eaton DC (1989) Expression of amiloride-blockable Na channels in *Xenopus* oocytes. Am J Physiol 257:C825–C829
Holzman JL, Liu L, Duke BJ, Kemendy AE, Eaton DC (2007) Transactivation of the IGF-1R by aldosterone. Am J Physiol Renal Physiol 292(4):F1219–F1228
Homolya L, Steinberg TH, Boucher RC (2000) Cell to cell communication in response to mechanical stress via bilateral release of ATP and UTP in polarized epithelia. J Cell Biol 150(6):1349–1360
Horisberger JD, Rossier BC (1992) Aldosterone regulation of gene transcription leading to control of ion transport. Hypertension 19(3):221–227
Hovater MB, Olteanu D, Hanson EL, Cheng NL, Siroky B, Fintha A, Komlosi P, Liu W, Satlin LM, Bell PD, Yoder BK, Schwiebert EM (2008) Loss of apical monocilia on collecting duct principal cells impairs ATP secretion across the apical cell surface and ATP-dependent and flow-induced calcium signals. Purinergic signal 4:155–170
Huang M, Chalfie M (1994) Gene interactions affecting mechanosensory transduction in Caenorhabditis elegans. Nature 367(6462):467–470
Hughey RP, Kinlough CL, Bruns JB, Mueller GM, Poland PA, Kleyman TR (2002) Expression of functional epithelial sodium channels (ENaC) is dependent on a lectin-specific quality control pathway in the endoplasmic reticulum. Mol Biol Cell 13:235A–236A
Hughey RP, Mueller GM, Bruns JB, Kinlough CL, Poland PA, Harkleroad KL, Carattino MD, Kleyman TR (2003) Maturation of the epithelial Na+ channel involves proteolytic processing of the alpha- and gamma-subunits. J Biol Chem 278(39):37073–37082
Hughey RP, Bruns JB, Kinlough CL, Harkleroad KL, Poland PA, Carattino MD, Mueller GM, Kleyman TR (2004a) Furin-dependent processing of the epithelial sodium channel (ENaC). FASEB J 18(5):A1040
Hughey RP, Bruns JB, Kinlough CL, Harkleroad KL, Tong Q, Carattino MD, Johnson JP, Stockand JD, Kleyman TR (2004b) Epithelial sodium channels are activated by furin-dependent proteolysis. J Biol Chem 279(18):18111–18114
Hughey RP, Bruns JB, Kinlough CL, Kleyman TR (2004c) Distinct pools of epithelial sodium channels are expressed at the plasma membrane. J Biol Chem 279(47):48491–48494
Hughey RP, Carattino MD, Kleyman TR (2007) Role of proteolysis in the activation of epithelial sodium channels. Curr Opin Nephrol Hypertens 16(5):444–450
Hummler E, Barker PM, Gatzy J, Beermann F, Verdumo C, Schmidt A, Boucher RC, Rossier BC (1996) Early death due to defective neonatal lung liquid clearance in alpha-ENaC-deficient mice. Nat Genet 12(3):325–328
Hummler E, Barker PM, Beermann F, Gatzy J, Verdumo C, Boucher RC, Rossier BC (1997) Role of the epithelial sodium channel in lung liquid clearance. Chest 111(6 Suppl):113S

Ifshin MS, Johnson KE, Eaton DC (1983) Acid pH and weak acids induce Na–Cl cotransport in the rabbit urinary bladder. J Membr Biol 76:151–164

Ilatovskaya DV, Pavlov TS, Levchenko V, Negulyaev YA, Staruschenko A (2011) Cortical actin binding protein cortactin mediates ENaC activity via Arp2/3 complex. FASEB J 25(8):2688–2699

Ilatovskaya DV, Pavlov TS, Levchenko V, Staruschenko A (2013) ROS production as a common mechanism of ENaC regulation by EGF, insulin, and IGF-1. Am J Physiol Cell Physiol 304(1): C102–C111

Ingham RJ, Gish G, Pawson T (2004) The Nedd4 family of E3 ubiquitin ligases: functional diversity within a common modular architecture. Oncogene 23(11):1972–1984

Ishikawa T, Marunaka Y, Rotin D (1998) Electrophysiological characterization of the rat epithelial Na+ channel (rENaC) expressed in MDCK cells. Effects of Na+ and Ca2+. J Gen Physiol 111:825–846

Ismailov II, McDuffie JH, Sariban-Sohraby S, Johnson JP, Benos DJ (1994) Carboxyl methylation activates purified renal amiloride-sensitive Na^+ channels in planar lipid bilayers. J Biol Chem 269(35):22193–22197

Jain L, Chen XJ, Brown LA, Eaton DC (1998) Nitric oxide inhibits lung sodium transport through a cGMP-mediated inhibition of epithelial cation channels. Am J Physiol 274(4 Pt 1):L475–L484

Jasti J, Furukawa H, Gonzales EB, Gouaux E (2007) Structure of acid-sensing ion channel 1 at 1.9 A resolution and low pH. Nature 449(7160):316–323

Jeffers M, Taylor GA, Weidner KM, Omura S, VandeWoude GF (1997) Degradation of the Met tyrosine kinase receptor by the ubiquitin-proteasome pathway. Mol Cell Biol 17(2):799–808

Ji HL, Bishop LR, Anderson SJ, Fuller CM, Benos DJ (2004) The role of Pre-H2 domains of alpha- and delta-epithelial Na+ channels in ion permeation, conductance, and amiloride sensitivity. J Biol Chem 279(9):8428–8440

Ji HL, Su XF, Kedar S, Li J, Barbry P, Smith PR, Matalon S, Benos DJ (2006) delta-subunit confers novel biophysical features to alpha beta gamma-human ENaC via a physical interaction. J Biol Chem 281(12):8233–8241

Ji HL, Zhao RZ, Chen ZX, Shetty S, Idell S, Matalon S (2012) delta ENaC: a novel divergent amiloride-inhibitable sodium channel. Am J Physiol Lung Cell Mol Physiol 303(12):L1013–L1026

Kanelis V, Rotin D, Forman-Kay JD (2001) Solution structure of a Nedd4 WW domain-ENaC peptide complex. Nat Struct Biol 8(5):407–412

Kapus A, Janmey P (2013) Plasma membrane–cortical cytoskeleton interactions: a cell biology approach with biophysical considerations. Compr Physiol 3(3):1231–1281

Karpushev AV, Ilatovskaya DV, Pavlov TS, Negulyaev YA, Staruschenko A (2010a) Intact cytoskeleton is required for small G protein dependent activation of the epithelial Na+ channel. PLoS One 5(1), e8827

Karpushev AV, Ilatovskaya DV, Staruschenko A (2010b) The actin cytoskeleton and small G protein RhoA are not involved in flow-dependent activation of ENaC. BMC Res Notes 3:210

Kashlan OB, Kleyman TR (2011) ENaC structure and function in the wake of a resolved structure of a family member. Am J Physiol Renal Physiol 301(4):F684–F696

Kashlan OB, Sheng S, Kleyman TR (2005) On the interaction between amiloride and its putative alpha-subunit epithelial Na+ channel binding site. J Biol Chem 280(28):26206–26215

Kashlan OB, Maarouf AB, Kussius C, Denshaw RM, Blumenthal KM, Kleyman TR (2006) Distinct structural elements in the first membrane-spanning segment of the epithelial sodium channel. J Biol Chem 281(41):30455–30462

Kashlan OB, Adelman JL, Okumura S, Blobner BM, Zuzek Z, Hughey RP, Kleyman TR, Grabe M (2011) Constraint-based, homology model of the extracellular domain of the epithelial Na+ channel alpha subunit reveals a mechanism of channel activation by proteases. J Biol Chem 286(1):649–660

Kato M, Miyazawa K, Kitamura N (2000) A deubiquitinating enzyme UBPY interacts with the Src homology 3 domain of Hrs-binding protein via a novel binding motif PX(V/I)(D/N)RXXKP. J Biol Chem 275(48):37481–37487

Kellenberger S, Gautschi I, Schild L (1999a) A single point mutation in the pore region of the epithelial Na+ channel changes ion selectivity by modifying molecular sieving. Proc Natl Acad Sci U S A 96(7):4170–4175

Kellenberger S, Hoffmann-Pochon N, Gautschi I, Schneeberger E, Schild L (1999b) On the molecular basis of ion permeation in the epithelial Na+ channel. J Gen Physiol 114(1):13–30

Kellenberger S, Auberson M, Gautschi I, Schneeberger E, Schild L (2001) Permeability properties of ENaC selectivity filter mutants. J Gen Physiol 118(6):679–692

Kelly TM, Nelson DH (1987) Sodium excretion and atrial natriuretic peptide levels during mineralocorticoid administration. A mechanism for the escape from hyperaldosteronism. Endocr Res 13:363–383

Kemendy AE, Kleyman TR, Eaton DC (1992) Aldosterone alters the open probability of amiloride-blockable sodium channels in A6 epithelia. Am J Physiol 263(4 Pt 1):825–837

Kim J, Shishido T, Jiang X, Aderem A, McLaughlin S (1994) Phosphorylation, high ionic strength, and calmodulin reverse the binding of MARCKS to phospholipid vesicles. J Biol Chem 269 (45):28214–28219

Kim CH, Kim HY, Lee HS, Chang SO, Oh SH, Lee JH (2010) P2Y4-mediated regulation of Na+ absorption in the Reissner's membrane of the cochlea. J Neurosci 30:3762–3769

Klein SL, Passaretti C, Anker M, Olukoya P, Pekosz A (2010) The impact of sex, gender and pregnancy on 2009 H1N1 disease. Biol Sex Differ 1(1):5

Kleyman TR, Cragoe EJ Jr, Kraehenbuhl JP (1989) The cellular pool of Na+ channels in the amphibian cell line A6 is not altered by mineralocorticoids. Analysis using a new photoactive amiloride analog in combination with anti-amiloride antibodies. J Biol Chem 264(20):11995–12000

Kleyman TR, Coupaye-Gerard B, Ernst SA (1992) Aldosterone does not alter apical cell-surface expression of epithelial Na+ channels in the amphibian cell line A6. J Biol Chem 267 (14):9622–9628

Kleyman TR, Myerburg MM, Hughey RP (2006) Regulation of ENaCs by proteases: an increasingly complex story. Kidney Int 70(8):1391–1392

Knight KK, Olson DR, Zhou R, Snyder PM (2006) Liddle's syndrome mutations increase Na^+ transport through dual effects on epithelial Na^+ channel surface expression and proteolytic cleavage. Proc Natl Acad Sci U S A 103(8):2805–2808

Knox FG, Burnett JC Jr, Kohan DE, Spielman WS, Strand JC (1980) Escape from the sodium-retaining effects of mineralocorticoids. Kidney Int 17:263–276

Koefoed-Johnsen V, Ussing HH (1958) The nature of the frog skin potential. Acta Physiol Scand 42(3–4):298–308

Komwatana P, Dinudom A, Young JA, Cook DI (1996) Cytosolic Na^+ controls an epithelial Na^+ channel via the G_o guanine nucleotide-binding regulatory protein. Proc Natl Acad Sci U S A 93 (15):8107–8111

Kretz O, Barbry P, Bock R, Lindemann B (1999) Differential expression of RNA and protein of the three pore-forming subunits of the amiloride-sensitive epithelial sodium channel in taste buds of the rat. J Histochem Cytochem 47(1):51–64

Kunzelmann K, Scheidt K, Scharf B, Ousingsawat J, Schreiber R, Wainwright B, McMorran B (2006) Flagellin of Pseudomonas aeruginosa inhibits Na+ transport in airway epithelia. FASEB J 20:545–546

Lanouette S, Mongeon V, Figeys D, Couture JF (2014) The functional diversity of protein lysine methylation. Mol Syst Biol 10:724

Lazarowski ER, Boucher RC, Harden TK (2003) Mechanisms of release of nucleotides and integration of their action as P2X- and P2Y-receptor activating molecules. Mol Pharmacol 64(4):785–795

Lebowitz J, Edinger RS, An B, Perry CJ, Onate S, Kleyman TR, Johnson JP (2004) Ikappab kinase-beta (ikkbeta) modulation of epithelial sodium channel activity. J Biol Chem 279 (40):41985–41990
Lee IH, Dinudom A, Sanchez-Perez A, Kumar S, Cook DI (2007a) Akt mediates the effect of insulin on epithelial sodium channels by inhibiting Nedd4-2. J Biol Chem 282(41):29866–29873
Lee SY, Palmer ML, Maniak PJ, Jang SH, Ryu PD, O'Grady SM (2007b) P2Y receptor regulation of sodium transport in human mammary epithelial cells. Am J Physiol Cell Physiol 293: C1472–C1480
Lee IH, Campbell CR, Song SH, Day ML, Kumar S, Cook DI, Dinudom A (2009) The activity of the epithelial sodium channels is regulated by caveolin-1 via a Nedd4-2-dependent mechanism. J Biol Chem 284(19):12663–12669
Lewis SA, Eaton DC, Diamond JM (1976) The mechanism of Na^+ transport by rabbit urinary bladder. J Membr Biol 28:41–70
Li J, Sheng S, Perry CJ, Kleyman TR (2003) Asymmetric organization of the pore region of the epithelial sodium channel. J Biol Chem 278(16):13867–13874
Li X, Leu S, Cheong A, Zhang H, Baibakov B, Shih C, Birnbaum MJ, Donowitz M (2004) Akt2, phosphatidylinositol 3-kinase, and PTEN are in lipid rafts of intestinal cells: role in absorption and differentiation. Gastroenterology 126(1):122–135
Liang X, Butterworth MB, Peters KW, Frizzell RA (2010) AS160 modulates aldosterone-stimulated epithelial sodium channel forward trafficking. Mol Biol Cell 21(12):2024–2033
Lindemann B (1984) Fluctuation analysis of sodium channels in epithelia. Annu Rev Physiol 46:497–515
Lindemann B, Van Driessche W (1977) Sodium-specific membrane channels of frog skin are pores: current fluctuations reveal high turnover. Science 195:292–294
Linderholm H (1952) Active transport of ions through frog skin with special reference to the action of certain diuretics; a study of the relation between electrical properties, the flux of labelled ions, and respiration. Acta Physiol Scand Suppl 27(97):1–144
Lingueglia E, Voilley N, Waldmann R, Lazdunski M, Barbry P (1993) Expression cloning of an epithelial amiloride-sensitive Na+ channel. A new channel type with homologies to Caenorhabditis elegans degenerins. FEBS Lett 318(1):95–99
Liu L, Duke BJ, Malik B, Yue Q, Eaton DC (2009) Biphasic regulation of ENaC by TGF-{alpha} and EGF in renal epithelial cells. Am J Physiol Renal Physiol 296:F1417–F1427
Lodin Z, Janacek K, Mueller JF (1963) The localization and effects of mercury bound by the surviving frog skin. J Cell Physiol 62:215–230
Loffing J, Pietri L, Aregger F, Bloch-Faure M, Ziegler U, Meneton P, Rossier BC, Kaissling B (2000) Differential subcellular localization of ENaC subunits in mouse kidney in response to high- and low-Na diets. Am J Physiol Renal Physiol 279(2):F252–F258
Loffing J, Zecevic M, Feraille E, Kaissling B, Asher C, Rossier BC, Firestone GL, Pearce D, Verrey F (2001) Aldosterone induces rapid apical translocation of ENaC in early portion of renal collecting system: possible role of SGK. Am J Physiol Renal Physiol 280(4):F675–F682
Losel RM, Feuring M, Falkenstein E, Wehling M (2002) Nongenomic effects of aldosterone: cellular aspects and clinical implications. Steroids 67(6):493–498
Lu M, Wang J, Jones KT, Ives HE, Feldman ME, Yao LJ, Shokat KM, Ashrafi K, Pearce D (2010) mTOR complex-2 activates ENaC by phosphorylating SGK1. J Am Soc Nephrol 21(5):811–818
Ma HP, Eaton DC (2005) Acute regulation of epithelial sodium channel by anionic phospholipids. J Am Soc Nephrol 16(11):3182–3187
Ma HP, Saxena S, Warnock DG (2002) Anionic phospholipids regulate native and expressed ENaC. J Biol Chem 277:7641–7644
Ma HP, Al-Khalili O, Ramosevac S, Saxena S, Liang YY, Warnock DG, Eaton DC (2004) Steroids and exogenous gamma-ENaC subunit modulate cation channels formed by alpha-ENaC in human B lymphocytes. J Biol Chem 279(32):33206–33212

Ma HP, Chou CF, Wei SP, Eaton DC (2007) Regulation of the epithelial sodium channel by phosphatidylinositides: experiments, implications, and speculations. Pflugers Arch 455 (1):169–180
MacRobbie EAC, Ussing HH (1961) Osmotic behavior of the epithelial cells of frog skin. Acta Physiol Scand 53:348–365
Malik B, Schlanger L, Al-Khalili O, Bao HF, Yue G, Price SR, Mitch WE, Eaton DC (2001) ENaC degradation in A6 cells by the ubiquitin-proteosome proteolytic pathway. J Biol Chem 276:12903–12910
Malik B, Yue Q, Yue G, Chen XJ, Price SR, Mitch WE, Eaton DC (2005a) Role of Nedd4-2 and polyubiquitination in epithelial sodium channel degradation in untransfected renal A6 cells expressing endogenous ENaC subunits. Am J Physiol Renal Physiol 2005(03/17):F107–F116
Malik B, Yue Q, Yue GC, Ma HP, Bao HF, Eaton DC (2005b) Growth conditions determine ENaC channel type. FASEB J 19(5):A1179
Malik B, Price SR, Mitch WE, Yue Q, Eaton DC (2006) Regulation of epithelial sodium channels by the ubiquitin-proteasome proteolytic pathway. Am J Physiol Renal Physiol 290(6):F1285–F1294
Mall M, Grubb BR, Harkema JR, O'Neal WK, Boucher RC (2004) Increased airway epithelial Na + absorption produces cystic fibrosis-like lung disease in mice. Nat Med 10(5):487–493
Mamenko M, Zaika O, Ilatovskaya DV, Staruschenko A, Pochynyuk O (2012) Angiotensin II increases activity of the epithelial Na+ channel (ENaC) in distal nephron additively to aldosterone. J Biol Chem 287(1):660–671
Mansley MK, Neuhuber W, Korbmacher C, Bertog M (2015) Norepinephrine stimulates the epithelial Na+ channel in cortical collecting duct cells via alpha2-adrenoceptors. Am J Physiol Renal Physiol 308(5):F450–F458
Marunaka Y, Eaton DC (1991) Effects of vasopressin and cAMP on single amiloride-blockable Na channels. Am J Physiol 260(5 Pt 1):C1071–C1084
Masilamani S, Kim GH, Mitchell C, Wade JB, Knepper MA (1999) Aldosterone-mediated regulation of ENaC alpha, beta, and gamma subunit proteins in rat kidney. J Clin Invest 104 (7):R19–R23
Mastroberardino L, Spindler B, Forster I, Loffing J, Assandri R, May A, Verrey F (1998) Ras pathway activates epithelial Na+ channel and decreases its surface expression in Xenopus oocytes. Mol Biol Cell 9(12):3417–3427
Matalon S, O'Brodovich HM (1999) Sodium channels in alveolar epithelial cells: molecular characterization, biophysical properties, and physiological significance. Annu Rev Physiol 61:627–661
Mato JM, Corrales FJ, Lu SC, Avila MA (2002) S-Adenosylmethionine: a control switch that regulates liver function. FASEB J 16(1):15–26
Matos JE, Sorensen MV, Geyti CS, Robaye B, Boeynaems JM, Leipziger J (2007) Distal colonic Na(+) absorption inhibited by luminal P2Y(2) receptors. Pflugers Arch 454:977–987
Matsuura D, Taguchi K, Yagisawa H, Maekawa S (2007) Lipid components in the detergent-resistant membrane microdomain (DRM) obtained from the synaptic plasma membrane of rat brain. Neurosci Lett 423(2):158–161
Mazzochi C, Benos DJ, Smith PR (2006) Interaction of epithelial ion channels with the actin-based cytoskeleton. Am J Physiol Renal Physiol 291(6):F1113–F1122
McCulloch F, Chambrey R, Eladari D, Peti-Peterdi J (2005) Localization of connexin 30 in the luminal membrane of cells in the distal nephron. Am J Physiol Renal Physiol 289:F1304–F1312
McDonald FJ, Western AH, McNeil JD, Thomas BC, Olson DR, Snyder PM (2002) Ubiquitin-protein ligase WWP2 binds to and downregulates the epithelial Na^+ channel. Am J Physiol 283 (3):F431–F436
McEneaney V, Harvey BJ, Thomas W (2007) Aldosterone rapidly activates protein kinase D via a mineralocorticoid receptor/EGFR trans-activation pathway in the M1 kidney CCD cell line. J Steroid Biochem Mol Biol 107(3–5):180–190

McLaughlin SG, Aderem A (1995) The myristoyl-electrostatic switch: a modulator of reversible protein-membrane interactions. Trends Biochem Sci 20(7):272–276

McLaughlin SG, Wang J, Gambhir A, Murray D (2002) PIP(2) and proteins: interactions, organization, and information flow. Annu Rev Biophys Biomol Struct 31:151–175

McLaughlin S, Hangyas-Mihalyne G, Zaitseva I, Golebiewska U (2005) Reversible – through calmodulin – electrostatic interactions between basic residues on proteins and acidic lipids in the plasma membrane. Biochem Soc Symp 72:189–198

McNicholas CM, Canessa CM (1997) Diversity of channels generated by different combinations of epithelial sodium channel subunits. J Gen Physiol 109(6):681–692

Mick VE, Itani OA, Loftus RW, Husted RF, Schmidt TJ, Thomas CP (2001) The alpha-subunit of the epithelial sodium channel is an aldosterone-induced transcript in mammalian collecting ducts, and this transcriptional response is mediated via distinct cis-elements in the 5′-flanking region of the gene. Mol Endocrinol 15(4):575–588

Mimnaugh EG, Bonvini P, Neckers L (1999) The measurement of ubiquitin and ubiquitinated proteins. Electrophoresis 20(2):418–428

Minshall RD, Sessa WC, Stan RV, Anderson RG, Malik AB (2003) Caveolin regulation of endothelial function. Am J Physiol Lung Cell Mol Physiol 285(6):L1179–L1183

Mironova E, Bugaj V, Roos KP, Kohan DE, Stockand JD (2012) Aldosterone-independent regulation of the epithelial Na+ channel (ENaC) by vasopressin in adrenalectomized mice. Proc Natl Acad Sci U S A 109(25):10095–10100

Mirshahi M, Nicolas C, Mirshahi S, Golestaneh N, d'Hermies F, Agarwal MK (1999) Immunochemical analysis of the sodium channel in rodent and human eye. Exp Eye Res 69(1):21–32

Mitch WE, Goldberg AL (1996) Mechanisms of muscle wasting. The role of the ubiquitin-proteasome pathway. N Engl J Med 335(25):1897–1905

Monder C (1991) Heterogeneity of 11 beta-hydroxysteroid dehydrogenase in rat tissues. J Steroid Biochem Mol Biol 40:533–536

Morgunkova A, Barlev NA (2006) Lysine methylation goes global. Cell Cycle 5(12):1308–1312

Morris RG, Schafer JA (2002) cAMP increases density of ENaC subunits in the apical membrane of MDCK cells in direct proportion to amiloride-sensitive Na(+) transport. J Gen Physiol 120 (1):71–85

Myerburg MM, McKenna EE, McKenna EE, Peters KW, Frizzell RA, Kleyman TR, Pilewski JM (2006) Airway surface liquid volume regulates ENaC by altering the serine protease-protease inhibitor balance: a mechanism for sodium hyperabsorption in cystic fibrosis. J Biol Chem 281 (38):27942–27949

Myerburg MM, Harvey PR, Heidrich EM, Pilewski JM, Butterworth MB (2010) Acute regulation of the epithelial sodium channel in airway epithelia by proteases and trafficking. Am J Respir Cell Mol Biol 43(6):712–719

Nagel W, Dorge A (1970) Effect of Amiloride on sodium transport of frog skin. I. Action on intracellular sodium content. Pflugers Arch 317(1):84–92

Nakhoul NL, Hering-Smith KS, Gambala CT, Hamm LL (1998) Regulation of sodium transport in M-1 cells. Am J Physiol 275(6 Pt 2):F998–F1007

Naray-Fejes-Toth A, Watlington CO, Fejes-Toth G (1991) 11β-Hydroxysteroid dehydrogenase activity in the renal target cells of aldosterone. Endocrinology 129:17–21

Narikiyo T, Kitamura K, Adachi M, Miyoshi T, Iwashita K, Shiraishi N, Nonoguchi H, Chen LM, Chai KX, Chao J, Tomita K (2002) Regulation of prostasin by aldosterone in the kidney. J Clin Invest 109(3):401–408

Nicco C, Wittner M, DiStefano A, Jounier S, Bankir L, Bouby N (2001) Chronic exposure to vasopressin upregulates ENaC and sodium transport in the rat renal collecting duct and lung. Hypertension 38(5):1143–1149

Nicod M, Michlig S, Flahaut M, Salinas M, Fowler JN, Horisberger JD, Rossier BC, Firsov D (2002) A novel vasopressin-induced transcript promotes MAP kinase activation and ENaC downregulation. EMBO J 21(19):5109–5117

Nielsen VG, Baird MS, Chen L, Matalon S (2000) DETANONOate, a nitric oxide donor, decreases amiloride-sensitive alveolar fluid clearance in rabbits. Am J Respir Crit Care Med 161(4 Pt 1):1154–1160

O'Neill H, Lebeck J, Collins PB, Kwon TH, Frokiaer J, Nielsen S (2008) Aldosterone-mediated apical targeting of ENaC subunits is blunted in rats with streptozotocin-induced diabetes mellitus. Nephrol Dial Transplant 23(5):1546–1555

Oberfeld B, Ruffieux-Daidie D, Vitagliano JJ, Pos KM, Verrey F, Staub O (2011) Ubiquitin-specific protease 2-45 (Usp2-45) binds to epithelial Na+ channel (ENaC)-ubiquitylating enzyme Nedd4-2. Am J Physiol Renal Physiol 301(1):F189–F196

Ohara A, Matsunaga H, Eaton DC (1993) G protein activation inhibits amiloride-blockable highly selective sodium channels in A6 cells. Am J Physiol 264:C352–C360

Ohbuchi T, Takenaga F, Hohchi N, Wakasugi T, Ueta Y, Suzuki H (2014) Possible contribution of pannexin-1 to ATP release in human upper airway epithelia. Physiol Rep 2, e00227

Okada SF, Ribeiro CM, Sesma JI, Seminario-Vidal L, Abdullah LH, van Heusden C, Lazarowski ER, Boucher RC (2013) Inflammation promotes airway epithelial ATP release via calcium-dependent vesicular pathways. Am J Respir Cell Mol Biol 49:814–820

Olesen HV, Pressler T, Hjelte L, Mared L, Lindblad A, Knudsen PK, Laerum BN, Johannesson M (2010) Gender differences in the Scandinavian cystic fibrosis population. Pediatr Pulmonol 45:959–965

Orbeli AL (1910) The electrical properties of frog skin. Zeitschrift fuer Biologie 59:329–339

Palmer LG (1982) Ion selectivity of the apical membrane Na channel in the toad urinary bladder. J Membr Biol 67:91–98

Palmer LG (1984) Voltage-dependent block by amiloride and other monovalent cations of apical Na channels in the toad urinary bladder. J Membr Biol 80:153–165

Palmer LG (1985) Modulation of apical Na permeability of the toad urinary bladder by intracellular Na, Ca, and H. J Membr Biol 83:57–69

Palmer LG, Frindt G (1986) Amiloride-sensitive Na channels from the apical membrane of the rat cortical collecting tubule. Proc Natl Acad Sci U S A 83(8):2767–2770

Palmer LG, Frindt G (1987) Effects of cell Ca and pH on Na channels from rat cortical collecting tubule. Am J Physiol 253(2 Pt 2):F333–F339

Palmer LG, Frindt G (1988) Conductance and gating of epithelial Na channels from rat cortical collecting tubule. Effects of luminal Na and Li. J Gen Physiol 92(1):121–138

Palmer LG, Sackin H, Frindt G (1998) Regulation of Na^+ channels by luminal Na^+ in rat cortical collecting tubule. J Physiol 509(Pt 1):151–162

Pearce D (2001) The role of SGK1 in hormone-regulated sodium transport. Trends Endocrinol Metab 12(8):341–347

Pearce D, Bhargava A, Cole TJ (2003) Aldosterone: its receptor, target genes, and actions. Vitam Horm 66:29–76

Pearce D, Soundararajan R, Trimpert C, Kashlan OB, Deen PM, Kohan DE (2015) Collecting duct principal cell transport processes and their regulation. Clin J Am Soc Nephrol 10(1):135–146

Pech V, Pham TD, Hong S, Weinstein AM, Spencer KB, Duke BJ, Walp E, Kim YH, Sutliff RL, Bao HF, Eaton DC, Wall SM (2010) Pendrin modulates ENaC function by changing luminal HCO3-. J Am Soc Nephrol 21(11):1928–1941

Peters KW, Qi J, Johnson JP, Watkins SC, Frizzell RA (2001) Role of snare proteins in CFTR and ENaC trafficking. Pflugers Arch 443(Suppl 1):S65–S69

Peti-Peterdi J, Warnock DG, Bell PD (2002) Angiotensin II directly stimulates ENaC activity in the cortical collecting duct via AT(1) receptors. J Am Soc Nephrol 13(5):1131–1135

Pickart CM (2000a) Ubiquitin biology: an old dog learns an old trick. Nat Cell Biol 2(8):E139–E141

Pickart CM (2000b) Ubiquitin in chains. Trends Biochem Sci 25(11):544–548

Pickart CM, VanDemark AP (2000) Opening doors into the proteasome. Nat Struct Biol 7 (11):999–1001

Piper RC, Luzio JP (2007) Ubiquitin-dependent sorting of integral membrane proteins for degradation in lysosomes. Curr Opin Cell Biol 19(4):459–465

Pochynyuk O, Staruschenko A, Tong Q, Medina J, Stockand JD (2005) Identification of a functional phosphatidylinositol 3,4,5-trisphosphate binding site in the epithelial Na+ channel. J Biol Chem 280(45):37565–37571

Pochynyuk O, Tong Q, Staruschenko A, Ma HP, Stockand JD (2006) Regulation of the epithelial Na^+ channel (ENaC) by phosphatidylinositides. Am J Physiol Renal Physiol 290(5):F949–F957

Pochynyuk O, Tong Q, Medina J, Vandewalle A, Staruschenko A, Bugaj V, Stockand JD (2007a) Molecular determinants of PI(4,5)P2 and PI(3,4,5)P3 regulation of the epithelial Na+ channel. J Gen Physiol 130(4):399–413

Pochynyuk O, Tong Q, Staruschenko A, Stockand JD (2007b) Binding and direct activation of the epithelial Na+ channel (ENaC) by phosphatidylinositides. J Physiol 580(Pt 2):365–372

Pochynyuk O, Bugaj V, Rieg T, Insel PA, Mironova E, Vallon V, Stockand JD (2008a) Paracrine regulation of the epithelial Na^+ channel in the mammalian collecting duct by purinergic P2Y2 receptor tone. J Biol Chem 283(52):36599–36607

Pochynyuk O, Bugaj V, Stockand JD (2008b) Physiologic regulation of the epithelial sodium channel by phosphatidylinositides. Curr Opin Nephrol Hypertens 17(5):533–540

Pochynyuk O, Bugaj V, Vandewalle A, Stockand JD (2008c) Purinergic control of apical plasma membrane PI(4,5)P2 levels sets ENaC activity in principal cells. Am J Physiol Renal Physiol 294(1):F38–F46

Pochynyuk O, Rieg T, Bugaj V, Schroth J, Fridman A, Boss GR, Insel PA, Stockand JD, Vallon V (2010) Dietary Na+ inhibits the open probability of the epithelial sodium channel in the kidney by enhancing apical P2Y2-receptor tone. FASEB J 24(6):2056–2065

Poulsen AN, Klausen TL, Pedersen PS, Willumsen NJ, Frederiksen O (2005) Regulation of ion transport via apical purinergic receptors in intact rabbit airway epithelium. Pflugers Arch 450:227–235

Prince LS, Welsh MJ (1998) Cell surface expression and biosynthesis of epithelial Na+ channels. Biochem J 336(Pt 3):705–710

Prince LS, Welsh MJ (1999) Effect of subunit composition and Liddle's syndrome mutations on biosynthesis of ENaC. Am J Physiol 276(6 Pt 1):C1346–C1351

Qi J, Peters KW, Liu C, Wang JM, Edinger RS, Johnson JP, Watkins SC, Frizzell RA (1999) Regulation of the amiloride-sensitive epithelial sodium channel by syntaxin 1A. J Biol Chem 274(43):30345–30348

Racke K, Juergens UR, Matthiesen S (2006) Control by cholinergic mechanisms. Eur J Pharmacol 533(1–3):57–68

Rafii B, Tanswell AK, Otulakowski G, Pitkanen O, Belcastro-Taylor R, O'Brodovich H (1998) O2-induced ENaC expression is associated with NF-kappaB activation and blocked by superoxide scavenger. Am J Physiol 275(4 Pt 1):L764–L770

Raiborg C, Stenmark H (2009) The ESCRT machinery in endosomal sorting of ubiquitylated membrane proteins. Nature 458(7237):445–452

Ransford GA, Fregien N, Qiu F, Dahl G, Conner GE, Salathe M (2009) Pannexin 1 contributes to ATP release in airway epithelia. Am J Respir Cell Mol Biol 41:525–534

Reifenberger MS, Yu L, Bao HF, Duke BJ, Liu BC, Ma HP, Alli AA, Eaton DC, Alli AA (2014) Cytochalasin E alters the cytoskeleton and decreases ENaC activity in Xenopus 2F3 cells. Am J Physiol Renal Physiol 307(1):F86–F95

Renard S, Lingueglia E, Voilley N, Lazdunski M, Barbry P (1994) Biochemical analysis of the membrane topology of the amiloride-sensitive Na+ channel. J Biol Chem 269(17):12981–12986

Robins GG, Maclennan KA, Boot-Handford RP, Sandle GI (2013) Rapid stimulation of human renal ENaC by cAMP in Xenopus laevis oocytes. J Physiol Biochem 69(3):419–427

Rokaw MD, Wang JM, Edinger RS, Weisz OA, Hui D, Middleton P, Shlyonsky V, Berdiev BK, Ismailov II, Eaton DC, Benos DJ, Johnson JP (1998) Carboxylmethylation of the β subunit of xENaC regulates channel activity. J Biol Chem 273:28746–28784

Roos KP, Bugaj V, Mironova E, Stockand JD, Ramkumar N, Rees S, Kohan DE (2013) Adenylyl cyclase VI mediates vasopressin-stimulated ENaC activity. J Am Soc Nephrol 24(2):218–227

Rosenfeld MG, Marcantonio EE, Hakimi J, Ort VM, Atkinson PH, Sabatini D, Kreibich G (1984) Biosynthesis and processing of ribophorins in the endoplasmic reticulum. J Cell Biol 99 (3):1076–1082

Rossier BC (1997) Cum Grano Salis: the epithelial sodium channel and the control of blood pressure. J Am Soc Nephrol 8(6):980–992

Rossier BC, Canessa CM, Schild L, Horisberger JD (1994) Epithelial sodium channels. Curr Opin Nephrol Hypertens 3(5):487–496

Rossier BC, Pradervand S, Schild L, Hummler E (2002) Epithelial sodium channel and the control of sodium balance: interaction between genetic and environmental factors. Annu Rev Physiol 64:877–897

Rotin D, Kumar S (2009) Physiological functions of the HECT family of ubiquitin ligases. Nat Rev Mol Cell Biol 10(6):398–409

Rotin D, Staub O, Haguenauer-Tsapis R (2000) Ubiquitination and endocytosis of plasma membrane proteins: role of Nedd4/Rsp5p family of ubiquitin-protein ligases. J Membr Biol 176 (1):1–17

Rotin D, Kanelis V, Schild L (2001) Trafficking and cell surface stability of ENaC. Am J Physiol 281(3):F391–F399

Ruckes-Nilges C, Lindemann H, Klimek T, Glanz H, Weber WM (2000) Nitric oxide has no beneficial effects on ion transport defects in cystic fibrosis human nasal epithelium. Pflugers Arch 441(1):133–137

Ruffieux-Daidie D, Poirot O, Boulkroun S, Verrey F, Kellenberger S, Staub O (2008) Deubiquitylation regulates activation and proteolytic cleavage of ENaC. J Am Soc Nephrol 19 (11):2170–2180

Samaha FF, Rubenstein RC, Yan W, Ramkumar M, Levy DI, Ahn YJ, Sheng S, Kleyman TR (2004) Functional polymorphism in the carboxyl terminus of the alpha-subunit of the human epithelial sodium channel. J Biol Chem 279(23):23900–23907

Sanchez-Perez A, Kumar S, Cook DI (2007) GRK2 interacts with and phosphorylates Nedd4 and Nedd4-2. Biochem Biophys Res Commun 359(3):611–615

Satlin LM, Sheng S, Woda CB, Kleyman TR (2001) Epithelial Na(+) channels are regulated by flow. Am J Physiol Renal Physiol 280(6):F1010–F1018

Saxena S, Quick MW, Tousson A, Oh Y, Warnock DG (1999) Interaction of syntaxins with the amiloride-sensitive epithelial sodium channel. J Biol Chem 274(30):20812–20817

Saxena SK, George CM, Pinskiy V, McConnell B (2006) Epithelial sodium channel is regulated by SNAP-23/syntaxin 1A interplay. Biochem Biophys Res Commun 343(4):1279–1285

Saxena SK, Singh M, Kaur S, George C (2007) Distinct domain-dependent effect of syntaxin1A on amiloride-sensitive sodium channel (ENaC) currents in HT-29 colonic epithelial cells. Int J Biol Sci 3(1):47–56

Sayegh R, Auerbach SD, Li X, Loftus RW, Husted RF, Stokes JB, Thomas CP (1999) Glucocorticoid induction of epithelial sodium channel expression in lung and renal epithelia occurs via trans-activation of a hormone response element in the 5′-flanking region of the human epithelial sodium channel alpha subunit gene. J Biol Chem 274:12431–12437

Schild L, Kellenberger S (2001) Structure function relationships of ENaC and its role in sodium handling. Adv Exp Med Biol 502:305–314

Schild L, Schneeberger E, Gautschi I, Firsov D (1997) Identification of amino acid residues in the alpha, beta, and gamma subunits of the epithelial sodium channel (ENaC) involved in amiloride block and ion permeation. J Gen Physiol 109(1):15–26

Schwagerus E, Sladek S, Buckley ST, Armas-Capote N, de la Rosa DA, Harvey BJ, Fischer H, Illek B, Huwer H, Schneider-Daum N, Lehr CM, Ehrhardt C (2015) Expression and function of the epithelial sodium channel delta-subunit in human respiratory epithelial cells in vitro. Pflugers Arch 467(11):2257–73

Schwagerus E, Sladek S, Buckley ST, Armas-Capote N, de la Rosa DA, Harvey BJ, Fischer H, Illek B, Huwer H, Schneider-Daum N, Lehr CM, Ehrhardt C (2015) Expression and function of the epithelial sodium channel delta-subunit in human respiratory epithelial cells in vitro. Pflugers Arch

Schwiebert EM, Zsembery A (2003) Extracellular ATP as a signaling molecule for epithelial cells. Biochim Biophys Acta 1615(1–2):7–32

Sedeek M, Nasrallah R, Touyz RM, Hébert RL (2013) NADPH oxidases, reactive oxygen species, and the kidney: friend and foe. J Am Soc Nephrol 24(10):1512–1518

Serfling RE, Sherman IL, Houseworth WJ (1967) Excess pneumonia-influenza mortality by age and sex in three major influenza A2 epidemics, United States, 1957–58, 1960 and 1963. Am J Epidemiol 86:433–441

Sheng S, Li J, Mcnulty KA, Avery D, Kleyman TR (2000) Characterization of the selectivity filter of the epithelial sodium channel. J Biol Chem 275(12):8572–8581

Sheng S, Li J, Mcnulty KA, Kieber-Emmons T, Kleyman TR (2001) Epithelial sodium channel pore region. structure and role in gating. J Biol Chem 276(2):1326–1334

Sheng S, Perry CJ, Kleyman TR (2002) External nickel inhibits epithelial sodium channel by binding to histidine residues within the extracellular domains of alpha and gamma subunits and reducing channel open probability. J Biol Chem 277(51):50098–50111

Sheng S, Bruns JB, Kleyman TR (2004) Extracellular histidine residues crucial for Na^+ self-inhibition of epithelial Na^+ channels. J Biol Chem 279(11):9743–9749

Sheng S, Perry CJ, Kashlan OB, Kleyman TR (2005) Side chain orientation of residues lining the selectivity filter of epithelial Na+ channels. J Biol Chem 280(9):8513–8522

Sheng S, Carattino MD, Bruns JB, Hughey RP, Kleyman TR (2006) Furin cleavage activates the epithelial Na+ channel by relieving Na+ self-inhibition. Am J Physiol Renal Physiol 290(6): F1488–F1496

Sheng S, Maarouf AB, Bruns JB, Hughey RP, Kleyman TR (2007) Functional role of extracellular loop cysteine residues of the epithelial Na+ channel in Na+ self-inhibition. J Biol Chem 282 (28):20180–20190

Shi H, Asher C, Chigaev A, Yung Y, Reuveny E, Seger R, Garty H (2002a) Interactions of beta and gamma ENaC with Nedd4 can be facilitated by an ERK-mediated phosphorylation. J Biol Chem 277(16):13539–13547

Shi H, Asher C, Yung Y, Kligman L, Reuveny E, Seger R, Garty H (2002b) Casein kinase 2 specifically binds to and phosphorylates the carboxy termini of ENaC subunits. Eur J Biochem 269(18):4551–4558

Shimkets RA, Lifton RP, Canessa CM (1997) The activity of the epithelial sodium channel is regulated by clathrin-mediated endocytosis. J Biol Chem 272(41):25537–25541

Shimkets RA, Lifton R, Canessa CM (1998) In vivo phosphorylation of the epithelial sodium channel. Proc Natl Acad Sci U S A 95:3301–3305

Shlyonsky VG, Mies F, Sariban-Sohraby S (2003) Epithelial sodium channel activity in detergent-resistant membrane microdomains. Am J Physiol Renal Physiol 284(1):F182–F188

Silver RB, Frindt G, Windhager EE, Palmer LG (1993) Feedback regulation of Na channels in rat CCT. I. Effects of inhibition of Na pump. Am J Physiol 264:F557–F564

Sipos A, Vargas SL, Toma I, Hanner F, Willecke K, Peti-Peterdi J (2009) Connexin 30 deficiency impairs renal tubular ATP release and pressure natriuresis. J Am Soc Nephrol 20:1724–1732

Smith PR, Saccomani G, Joe EH, Angelides KJ, Benos DJ (1991) Amiloride-sensitive sodium channel is linked to the cytoskeleton in renal epithelial cells. Proc Natl Acad Sci U S A 88 (16):6971–6975

Snyder PM (2000) Liddle's syndrome mutations disrupt cAMP-mediated translocation of the epithelial Na^+ channel to the cell surface. J Clin Invest 105(1):45–53

Snyder PM (2005) Minireview: regulation of epithelial Na+ channel trafficking. Endocrinology 146(12):5079–5085

Snyder PM (2009) Down-regulating destruction: phosphorylation regulates the E3 ubiquitin ligase Nedd4-2. Sci Signal 2(79), e41

Snyder PM, Olson DR, Bucher DB (1999) A pore segment in DEG/ENaC Na(+) channels. J Biol Chem 274(40):28484–28490

Snyder PM, Olson DR, Kabra R, Zhou R, Steines JC (2004a) cAMP and serum and glucocorticoid-inducible kinase (SGK) regulate the epithelial Na(+) channel through convergent phosphorylation of Nedd4-2. J Biol Chem 279(44):45753–45758

Snyder PM, Steines JC, Olson DR (2004b) Relative contribution of Nedd4 and Nedd4-2 to ENaC regulation in epithelia determined by RNA interference. J Biol Chem 279(6):5042–5046

Soundararajan R, Melters D, Shih IC, Wang J, Pearce D (2009) Epithelial sodium channel regulated by differential composition of a signaling complex. Proc Natl Acad Sci U S A 106 (19):7804–7809

Soundararajan R, Pearce D, Hughey RP, Kleyman TR (2010a) Role of epithelial sodium channels and their regulators in hypertension. J Biol Chem 285(40):30363–30369

Soundararajan R, Wang J, Melters D, Pearce D (2010b) Glucocorticoid-induced Leucine zipper 1 stimulates the epithelial sodium channel by regulating serum- and glucocorticoid-induced kinase 1 stability and subcellular localization. J Biol Chem 285(51):39905–39913

Soundararajan R, Lu M, Pearce D (2012a) Organization of the ENaC-regulatory machinery. Crit Rev Biochem Mol Biol 47(4):349–359

Soundararajan R, Pearce D, Ziera T (2012b) The role of the ENaC-regulatory complex in aldosterone-mediated sodium transport. Mol Cell Endocrinol 350(2):242–247

Soundararajan R, Ziera T, Koo E, Ling K, Wang J, Borden SA, Pearce D (2012c) Scaffold protein connector enhancer of kinase suppressor of Ras isoform 3 (CNK3) coordinates assembly of a multiprotein epithelial sodium channel (ENaC)-regulatory complex. J Biol Chem 287 (39):33014–33025

Spindler B, Verrey F (1999) Aldosterone action: induction of p21(ras) and fra-2 and transcription-independent decrease in myc, jun, and fos. Am J Physiol 276(5 Pt 1):C1154–C1161

Spindler B, Mastroberardino L, Custer M, Verrey F (1997) Characterization of early aldosterone-induced RNAs identified in A6 kidney epithelia. Pflugers Arch 434(3):323–331

Stanley P (2011) Golgi glycosylation. Cold Spring Harb Perspect Biol 3(4). doi:10.1101/cshperspect.a005199

Staruschenko A, Patel P, Tong Q, Medina JL, Stockand JD (2004) Ras activates the epithelial Na (+) channel through phosphoinositide 3-OH kinase signaling. J Biol Chem 279(36):37771–37778

Staruschenko A, Pochynyuk OM, Tong Q, Stockand JD (2005) Ras couples phosphoinositide 3-OH kinase to the epithelial Na^+ channel. Biochim Biophys Acta 1669(2):108–115

Staruschenko A, Pochynyuk O, Vandewalle A, Bugaj V, Stockand JD (2007) Acute regulation of the epithelial Na+ channel by phosphatidylinositide 3-OH kinase signaling in native collecting duct principal cells. J Am Soc Nephrol 18(6):1652–1661

Staub O, Rotin D (1996) WW domains. Structure 4(5):495–499

Staub O, Rotin D (1997) Regulation of ion transport by protein-protein interaction domains. Curr Opin Nephrol Hypertens 6(5):447–454

Staub O, Dho S, Henry P, Correa J, Ishikawa T, McGlade J, Rotin D (1996) WW domains of Nedd4 bind to the proline-rich PY motifs in the epithelial Na+ channel deleted in Liddle's syndrome. EMBO J 15(10):2371–2380

Staub O, Gautschi I, Ishikawa T, Breitschopf K, Ciechanover A, Schild L, Rotin D (1997a) Regulation of stability and function of the epithelial Na^+ channel (ENaC) by ubiquitination. EMBO J 16(21):6325–6336

Staub O, Yeger H, Plant PJ, Kim H, Ernst SA, Rotin D (1997b) Immunolocalization of the ubiquitin-protein ligase Nedd4 in tissues expressing the epithelial Na+ channel (ENaC). Am J Physiol 272(6 Pt 1):C1871–C1880

Staub O, Abriel H, Plant P, Ishikawa T, Kanelis V, Saleki R, Horisberger JD, Schild L, Rotin D (2000) Regulation of the epithelial Na^+ channel by Nedd4 and ubiquitination. Kidney Int 57 (3):809–815

Stephenson A, Hux J, Tullis E, Austin PC, Corey M, Ray J (2011) Higher risk of hospitalization among females with cystic fibrosis. J Cyst Fibros 10:93–99

Stewart AP, Haerteis S, Diakov A, Korbmacher C, Edwardson JM (2011) Atomic force microscopy reveals the architecture of the epithelial sodium channel (ENaC). J Biol Chem 286 (37):31944–31952

Stockand JD (2010) Vasopressin regulation of renal sodium excretion. Kidney Int 78(9):849–856

Stockand JD, Al-Baldawi NF, Al-Khalili OK, Worrell RT, Eaton DC (1999a) S-adenosyl-L-homocysteine hydrolase regulates aldosterone-induced Na+ transport. J Biol Chem 274 (6):3842–3850

Stockand JD, Edinger RS, Al-Baldawi NF, Sariban-Sohraby S, Al-Khalili OK, Eaton DC, Johnson JP (1999b) Isoprenylcysteine-O-carboxyl methyltransferase regulates aldosterone-sensitive Na' reabsorption. J Biol Chem 274(38):26912–26916

Stockand JD, Bao HF, Schenck J, Malik B, Middleton P, Schlanger LE, Eaton DC (2000a) Differential effects of protein kinase C on the levels of epithelial Na+ channel subunit proteins. J Biol Chem 275(33):25760–25765

Stockand JD, Edinger RS, Eaton DC, Johnson JP (2000b) Toward understanding the role of methylation in aldosterone-sensitive $Na(^+)$ transport. News Physiol Sci 15:161–165

Stockand JD, Zeltwanger S, Bao HF, Becchetti A, Worrell RT, Eaton DC (2001) S-adenosyl-L-homocysteine hydrolase is necessary for aldosterone-induced activity of epithelial $Na(^+)$ channels. Am J Physiol Cell Physiol 281(3):C773–C785

Stockand JD, Staruschenko A, Pochynyuk O, Booth RE, Silverthorn DU (2008) Insight toward epithelial Na^+ channel mechanism revealed by the acid-sensing ion channel 1 structure. IUBMB Life 60(9):620–628

Stockand JD, Mironova E, Bugaj V, Rieg T, Insel PA, Vallon V, Peti-Peterdi J, Pochynyuk O (2010) Purinergic inhibition of ENaC produces aldosterone escape. J Am Soc Nephrol 21:1903–1911

Stoops EH, Caplan MJ (2014) Trafficking to the apical and basolateral membranes in polarized epithelial cells. J Am Soc Nephrol 25(7):1375–1386

Studer RA, Person E, Robinson-Rechavi M, Rossier BC (2011) Evolution of the epithelial sodium channel and the sodium pump as limiting factors of aldosterone action on sodium transport. Physiol Genomics 43(13):844–854

Szule JA, Coorssen JR (2003) Revisiting the role of SNAREs in exocytosis and membrane fusion. Biochim Biophys Acta Mol Cell Res 1641(2–3):121–135

Takemura Y, Helms MN, Eaton AF, Self J, Ramosevac S, Jain L, Bao HF, Eaton DC (2013) Cholinergic regulation of epithelial sodium channels in rat alveolar type 2 epithelial cells. Am J Physiol Lung Cell Mol Physiol 304(6):L428–L437

Taniguchi H, Manenti S (1993) Interaction of myristoylated alanine-rich protein kinase C substrate (MARCKS) with membrane phospholipids. J Biol Chem 268(14):9960–9963

Taruno A, Niisato N, Marunaka Y (2008) Intracellular calcium plays a role as the second messenger of hypotonic stress in gene regulation of SGK1 and ENaC in renal epithelial A6 cells. Am J Physiol Renal Physiol 294(1):F177–F186

Thai TL, Yu L, Eaton DC, Duke BJ, Al-Khalili O, Lam HY, Ma H, Bao HF (2014) Basolateral P2X(4)channels stimulate ENaC activity in Xenopus cortical collecting duct A6 cells. Am J Physiol Renal Physiol 307(7):F806–F813

Thomas CP, Doggett NA, Fisher R, Stokes JB (1996) Genomic organization and the 5′ flanking region of the gamma subunit of the human amiloride-sensitive epithelial sodium channel. J Biol Chem 271(42):26062–26066

Thomas CP, Auerbach SD, Zhang C, Stokes JB (1999) The structure of the rat amiloride-sensitive epithelial sodium channel gamma subunit gene and functional analysis of its promoter. Gene 228(1–2):111–122

Thomas J, Deetjen P, Ko WH, Jacobi C, Leipziger J (2001) P2Y(2) receptor-mediated inhibition of amiloride-sensitive short circuit current in M-1 mouse cortical collecting duct cells. J Membr Biol 183:115–124

Thomas CP, Zhou J, Liu KZ, Mick VE, MacLaughlin E, Knowles M (2002) Systemic pseudohypoaldosteronism from deletion of the promoter region of the human Beta epithelial na(+) channel subunit. Am J Respir Cell Mol Biol 27(3):314–319

Thome UH, Davis IC, Nguyen SV, Shelton BJ, Matalon S (2003) Modulation of sodium transport in fetal alveolar epithelial cells by oxygen and corticosterone. Am J Physiol Lung Cell Mol Physiol 284(2):L376–L385

Thrower JS, Hoffman L, Rechsteiner M, Pickart CM (2000) Recognition of the polyubiquitin proteolytic signal. EMBO J 19(1):94–102

Tong Q, Stockand JD (2005) Receptor tyrosine kinases mediate epithelial Na(+) channel inhibition by epidermal growth factor. Am J Physiol Renal Physiol 288(1):F150–F161

Tong Q, Gamper N, Medina JL, Shapiro MS, Stockand JD (2004) Direct activation of the epithelial Na(+) channel by phosphatidylinositol 3,4,5-trisphosphate and phosphatidylinositol 3,4-bisphosphate produced by phosphoinositide 3-OH kinase. J Biol Chem 279(21):22654–22663

Tong J, Nguyen L, Vidal A, Simon SA, Skene JH, McIntosh TJ (2008) Role of GAP-43 in sequestering phosphatidylinositol 4,5-bisphosphate to Raft bilayers. Biophys J 94(1):125–133

Tousson A, Alley CD, Sorscher EJ, Brinkley BR, Benos DJ (1989) Immunochemical localization of amiloride-sensitive sodium channels in sodium-transporting epithelia. J Cell Sci 93(Pt 2):349–362

Turnheim K (1991) Intrinsic regulation of apical sodium entry in epithelia. Physiol Rev 71:429–445

Ussing HH, Zerahn K (1951) Active transport of sodium as the source of electric current in the short-circuited isolated frog skin. Acta Physiol Scand 23(2–3):110–127

Valentijn JA, Fyfe GK, Canessa CM (1998) Biosynthesis and processing of epithelial sodium channels in Xenopus oocytes. J Biol Chem 273(46):30344–30351

Vallet V, Chraibi A, Gaeggeler HP, Horisberger JD, Rossier BC (1997) An epithelial serine protease activates the amiloride-sensitive sodium channel. Nature 389(6651):607–610

Van IJzendoorn SC, Mostov KE (2000) Connecting apical endocytosis to the intracellular traffic infrastructure in polarized hepatocytes. Gastroenterology 119(6):1791–1794

Van IJzendoorn SC, Maier O, Van Der Wouden JM, Hoekstra D (2000) The subapical compartment and its role in intracellular trafficking and cell polarity. J Cell Physiol 184(2):151–160

van Kerkhof P, Strous GJ (2001) The ubiquitin-proteasome pathway regulates lysosomal degradation of the growth hormone receptor and its ligand. Biochem Soc Trans 29(Pt 4):488–493

van Kerkhof P, Alves dos Santos CM, Sachse M, Klumperman J, Bu G, Strous GJ (2001) Proteasome inhibitors block a late step in lysosomal transport of selected membrane but not soluble proteins. Mol Biol Cell 12(8):2556–2566

van Leeuwen F, Gafken PR, Gottschling DE (2002) Dot1p modulates silencing in yeast by methylation of the nucleosome core. Cell 109(6):745–756

Verrey F (1998) Early aldosterone effects. Exp Nephrol 6(4):294–301

Verrey F, Pearce D, Pfeiffer R, Spindler B, Mastroberardino L, Summa V, Zecevic M (2000) Pleiotropic action of aldosterone in epithelia mediated by transcription and post-transcription mechanisms. Kidney Int 57(4):1277–1282

Verrey F, Fakitsas P, Adam G, Staub O (2008) Early transcriptional control of ENaC (de) ubiquitylation by aldosterone. Kidney Int 73(6):691–696

Vieira OV, Verkade P, Manninen A, Simons K (2005) FAPP2 is involved in the transport of apical cargo in polarized MDCK cells. J Cell Biol 170(4):521–526

Vieira OV, Gaus K, Verkade P, Fullekrug J, Vaz WLC, Simons K (2006) FAPP2, cilium formation, and compartmentalization of the apical membrane in polarized Madinâ€"Darby canine kidney (MDCK) cells. Proc Natl Acad Sci 103(49):18556–18561

Wakida N, Kitamura K, Tuyen DG, Maekawa A, Miyoshi T, Adachi M, Shiraishi N, Ko T, Ha V, Nonoguchi H, Tomita K (2006) Inhibition of prostasin-induced ENaC activities by PN-1 and regulation of PN-1 expression by TGF-beta1 and aldosterone. Kidney Int 70(8):1432–1438

Waldmann R, Champigny G, Lazdunski M (1995) Functional degenerin-containing chimeras identify residues essential for amiloride-sensitive Na+ channel function. J Biol Chem 270 (20):11735–11737

Wang XF, Chan HC (2000) Adenosine triphosphate induces inhibition of Na(+) absorption in mouse endometrial epithelium: a Ca(2+)-dependent mechanism. Biol Reprod 63:1918–1924

Wang XY, Masilamani S, Nielsen J, Kwon TH, Brooks HL, Nielsen S, Knepper MA (2001) The renal thiazide-sensitive Na-Cl cotransporter as mediator of the aldosterone-escape phenomenon. J Clin Invest 108:215–222

Wang J, Gambhir A, McLaughlin SG, Murray D (2004) A computational model for the electrostatic sequestration of PI(4,5)P2 by membrane-adsorbed basic peptides. Biophys J 86(4):1969–1986

Wang J, Knight ZA, Fiedler D, Williams O, Shokat KM, Pearce D (2008) Activity of the p110-alpha subunit of phosphatidylinositol-3-kinase is required for activation of epithelial sodium transport. Am J Physiol Renal Physiol 295(3):F843–F850

Wang J, Zhang ZR, Chou CF, Liang YY, Gu Y, Ma HP (2009) Cyclosporine stimulates the renal epithelial sodium channel by elevating cholesterol. Am J Physiol Renal Physiol 296(2):F284–F290

Wang Q, Dai XQ, Li Q, Tuli J, Liang G, Li SS, Chen XZ (2013) Filamin interacts with epithelial sodium channel and inhibits its channel function. J Biol Chem 288(1):264–273

Waterman H, Yarden Y (2001) Molecular mechanisms underlying endocytosis and sorting of ErbB receptor tyrosine kinases. FEBS Lett 490(3):142–152

Wei H, Mundade R, Lange KC, Lu T (2014) Protein arginine methylation of non-histone proteins and its role in diseases. Cell Cycle 13(1):32–41

Weissman AM (2001) Themes and variations on ubiquitylation. Nat Rev Mol Cell Biol 2(3):169–178

Weisz OA, Rodriguez-Boulan E (2009) Apical trafficking in epithelial cells: signals, clusters and motors. J Cell Sci 122(Pt 23):4253–4266

Weisz OA, Wang JM, Edinger RS, Johnson JP (2000) Non-coordinate regulation of endogenous epithelial sodium channel (ENaC) subunit expression at the apical membrane of A6 cells in response to various transporting conditions. J Biol Chem 275(51):39886–39893

West TA, Blazer-Yost BL (2005) Modulation of basal and peptide hormone-stimulated Na^+ transport by membrane cholesterol content in the A6 epithelial cell line. Cell Physiol Biochem 16(4–6):263–270

Wildman SS, Kang ES, King BF (2009) ENaC, renal sodium excretion and extracellular ATP. Purinergic Signal 5:481–489

Wills NK, Lewis SA, Eaton DC (1979) Active and passive properties of rabbit descending colon: a microelectrode and nystatin study. J Membr Biol 45:81–108

Wilson JM, Laurent P, Tufts BL, Benos DJ, Donowitz M, Vogl AW, Randall DJ (2000) NaCl uptake by the branchial epithelium in freshwater teleost fish: an immunological approach to ion-transport protein localization. J Exp Biol 203(Pt 15):2279–2296

Woollhead AM, Baines DL (2006) Forskolin-induced cell shrinkage and apical translocation of functional enhanced green fluorescent protein-human alphaENaC in H441 lung epithelial cell monolayers. J Biol Chem 281(8):5158–5168

World Health Organization (2008) Update: WHO-confirmed human cases of avian influenza A (H5N1) infection, November 2003–May 2008. Releve epidemiologique hebdomadaire/Section d'hygiene du Secretariat de la Societe des Nations = Wkly Epidemiol Rec/Health Section of the Secretariat of the League of Nations 83:415–420

Xu B, Stippec S, Chu PY, Lazrak A, Li XJ, Lee BH, English JM, Ortega B, Huang CL, Cobb MH (2005) WNK1 activates SGK1 to regulate the epithelial sodium channel. Proc Natl Acad Sci U S A 102(29):10315–10320

Yamagishi S, Matsui T (2011) Nitric oxide, a janus-faced therapeutic target for diabetic microangiopathy–friend or foe? Pharmacol Res 64(3):187–194

Yamauchi E, Nakatsu T, Matsubara M, Kato H, Taniguchi H (2003) Crystal structure of a MARCKS peptide containing the calmodulin-binding domain in complex with Ca2+-calmodulin. Nat Struct Mol Biol 10(3):226–231

Yan W, Suaud L, Kleyman TR, Rubenstein RC (2006) Differential modulation of a polymorphism in the COOH terminus of the alpha-subunit of the human epithelial sodium channel by protein kinase Cdelta. Am J Physiol Renal Physiol 290(2):F279–F288

Yan W, Spruce L, Rosenblatt MM, Kleyman TR, Rubenstein RC (2007) Intracellular trafficking of a polymorphism in the COOH terminus of the alpha-subunit of the human epithelial sodium channel is modulated by casein kinase 1. Am J Physiol Renal Physiol 293(3):F868–F876

Yang B, Kumar S (2010) Nedd4 and Nedd4-2: closely related ubiquitin-protein ligases with distinct physiological functions. Cell Death Differ 17(1):68–77

Yang LM, Rinke R, Korbmacher C (2006) Stimulation of the epithelial sodium channel (ENaC) by cAMP involves putative ERK phosphorylation sites in the C termini of the channel's beta- and gamma-subunit. J Biol Chem 281(15):9859–9868

Yang C, Su L, Wang Y, Liu L (2009) UTP regulation of ion transport in alveolar epithelial cells involves distinct mechanisms. Am J Physiol Lung Cell Mol Physiol 297:L439–L454

Yarmola EG, Edison AS, Lenox RH, Bubb MR (2001) Actin filament cross-linking by MARCKS: characterization of two actin-binding sites within the phosphorylation site domain. J Biol Chem 276(25):22351–22358

Yousefipour Z, Oyekan A, Newaz M (2010) Interaction of oxidative stress, nitric oxide and peroxisome proliferator activated receptor gamma in acute renal failure. Pharmacol Ther 125 (3):436–445

Yu H, Kopito RR (1999) The role of multiubiquitination in dislocation and degradation of the alpha subunit of the T cell antigen receptor. J Biol Chem 274(52):36852–36858

Yu L, Bao HF, Self JL, Eaton DC, Helms MN (2007a) Aldosterone-induced increases in superoxide production counters nitric oxide inhibition of epithelial Na channel activity in A6 distal nephron cells. Am J Physiol Renal Physiol 293(5):F1666–F1677

Yu L, Eaton DC, Helms MN (2007b) Effect of divalent heavy metals on epithelial Na+ channels in A6 cells. Am J Physiol Renal Physiol 293(1):F236–F244

Yu L, Helms MN, Yue Q, Eaton DC (2008) Single-channel analysis of functional epithelial sodium channel (ENaC) stability at the apical membrane of A6 distal kidney cells. Am J Physiol Renal Physiol 295(5):F1519–F1527

Yu L, Cai H, Yue Q, Alli AA, Wang D, Al-Khalili O, Bao HF, Eaton DC (2013) WNK4 inhibition of ENaC is independent of Nedd4-2-mediated ENaC ubiquitination. Am J Physiol Renal Physiol 305(1):F31–F41

Yue G, Malik B, Yue G, Eaton DC (2002) Phosphatidylinositol 4,5-bisphosphate (PIP2) stimulates epithelial sodium channel activity in A6 cells. J Biol Chem 277(14):11965–11969

Zakrzewicz D, Zakrzewicz A, Preissner KT, Markart P, Wygrecka M (2012) Protein Arginine Methyltransferases (PRMTs): promising targets for the treatment of pulmonary disorders. Int J Mol Sci 13(10):12383–12400

Zeitlin PL (2008) Cystic fibrosis and estrogens: a perfect storm. J Clin Invest 118:3841–3844

Zerangue N, Schwappach B, Jan YN, Jan LY (1999) A new ER trafficking signal regulates the subunit stoichiometry of plasma membrane K(ATP) channels. Neuron 22(3):537–548

Zhang W, Hayashizaki Y, Kone BC (2004) Structure and regulation of the mDot1 gene, a mouse histone H3 methyltransferase. Biochem J 377(Pt 3):641–651

Zhang W, Xia X, Reisenauer MR, Rieg T, Lang F, Kuhl D, Vallon V, Kone BC (2007a) Aldosterone-induced Sgk1 relieves Dot1a-Af9-mediated transcriptional repression of epithelial Na+ channel alpha. J Clin Invest 117(3):773–783

Zhang Y, Sanchez D, Gorelik J, Klenerman D, Lab M, Edwards C, Korchev Y (2007b) Basolateral P2X4-like receptors regulate the extracellular ATP-stimulated epithelial Na+ channel activity in renal epithelia. Am J Physiol Renal Physiol 292:F1734–F1740

Zhang ZR, Chou CF, Wang J, Liang YY, Ma HP (2010) Anionic phospholipids differentially regulate the epithelial sodium channel (ENaC) by interacting with alpha, beta, and gamma ENaC subunits. Pflugers Arch 459(3):377–387

Zhang X, Zhou Q, Chen L, Berger S, Wu H, Xiao Z, Pearce D, Zhou X, Zhang W (2013) Mineralocorticoid receptor antagonizes Dot1a-Af9 complex to increase alphaENaC transcription. Am J Physiol Renal Physiol 305(10):F1436–F1444

Zhang Y, Robson SC, Morris KL, Heiney KM, Dwyer KM, Kishore BK, Ecelbarger CM (2015) Impaired natriuretic response to high-NaCl diet plus aldosterone infusion in mice overexpressing human CD39, an ectonucleotidase (NTPDase1). Am J Physiol Renal Physiol 308 (12):F1398–F1408

Zhou R, Patel SV, Snyder PM (2007) Nedd4-2 catalyzes ubiquitination and degradation of cell surface ENaC. J Biol Chem 282(28):20207–20212

Zhou R, Kabra R, Olson DR, Piper RC, Snyder PM (2010) Hrs controls sorting of the epithelial Na + channel between endosomal degradation and recycling pathways. J Biol Chem 285 (40):30523–30530

Zhou R, Tomkovicz VR, Butler PL, Ochoa LA, Peterson ZJ, Snyder PM (2013) Ubiquitin-specific peptidase 8 (USP8) regulates endosomal trafficking of the epithelial Na+ channel. J Biol Chem 288(8):5389–5397

Zhu S, Ware LB, Geiser T, Matthay MA, Matalon S (2001) Increased levels of nitrate and surfactant protein a nitration in the pulmonary edema fluid of patients with acute lung injury. Am J Respir Crit Care Med 163(1):166–172

Zuckerman JB, Chen X, Jacobs JD, Hu B, Kleyman TR, Smith PR (1999) Association of the epithelial sodium channel with Apx and alpha-spectrin in A6 renal epithelial cells. J Biol Chem 274(33):23286–23295

Chapter 19
ROMK and Bartter Syndrome Type 2

Paul A. Welling

Abstract The renal outer medullary potassium (K) channel, ROMK (aka $K_{ir}1.1$, encoded by the KCNJ1 gene), is the founding member of the inwardly rectifying potassium channel family. It is primarily expressed in the kidney where it transports potassium into the pro-urine, important for maintaining salt reabsorption in the thick ascending limb, and potassium-homeostasis. Loss-of-function mutations in ROMK cause Bartter syndrome (type 2), a familial salt-losing nephropathy. This review provides a bedside-to-bench understanding of the disease, tracking the clinical phenotype to defects in tubule transport to the mutations in the channel.

Keywords ROMK • KCNJ1 • Potassium • Hypokalemia • Hyperkalemia • Blood pressure • Salt • Hypertension • Hypotension • Thick ascending limb of Henle's loop • Kidney • Channel

19.1 Introduction

Bartter syndrome (Bartter et al. 1962) is a rare kidney disorder caused by impaired salt reabsorption in the thick ascending limb of Henle's loop (Fig. 19.1). Typically, the disease is manifested in urinary salt wasting, hypokalemic alkalosis, hypotension, hyperrenin-hyperaldosteronism, and a constellation of secondary sequelae, similar to those observed with loop diuretic toxicity. The antenatal or infantile form (ABS) presents with maternal polyhydramnios and severe, life-threatening intravascular volume contraction, hyperprostaglandinuria, and failure to thrive in the postnatal period (Seyberth and Schlingmann 2011).

ABS is inherited through autosomal recessive transmission and has heterogenous genetic underpinnings. Remarkably, loss-of-function mutations in any one of the four major components of the NaCl transport machinery in the TAL can produce ABS (Fig. 19.1). These include the $Na^+/K^+/2Cl^-$ co-transporter, NKCC2

P.A. Welling (✉)
Department of Physiology, University of Maryland School of Medicine, 655 West Baltimore St., BRB 5-029, Baltimore, MD 21201, USA
e-mail: pwelling@umaryland.edu

K.L. Hamilton, D.C. Devor (eds.), *Ion Channels and Transporters of Epithelia in Health and Disease*, Physiology in Health and Disease,
DOI 10.1007/978-1-4939-3366-2_19

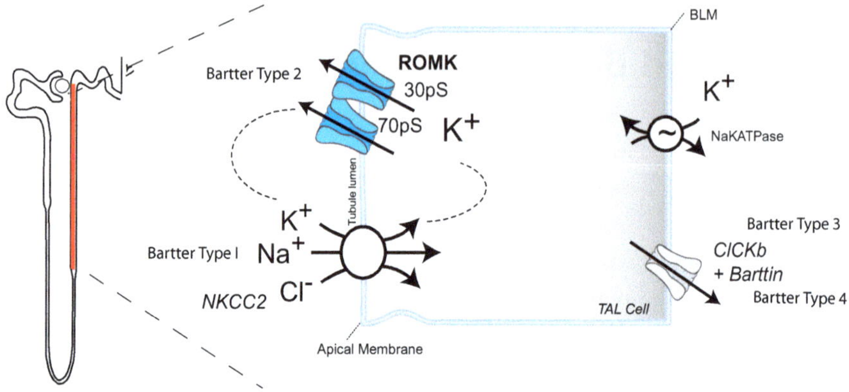

Fig. 19.1 The thick ascending limb of Henle's loop reabsorbs ~20 % of the filtered load of sodium chloride through a transcellular transport pathway. The major elements of the sodium chloride reabsorptive machine are shown. Loss-of-function mutations in each of them cause Bartter syndrome

(SCL12a1) (Simon et al. 1996), the renal outer medullary potassium channel, ROMK ($K_{ir}1.1$ of the KCJ1 gene) (Simon et al. 1997a; Karolyi et al. 1997), the basolateral Cl^- channel, CLCNKB (Simon et al. 1997b; Konrad et al. 2000), and the accessory subunit of CLCNKB and BSND (Birkenhager et al. 2001). Gain-of-function mutations in the calcium-sensing G-protein-coupled receptor, CASR, also lead to a Bartter phenotype in addition to hypocalcemia and hypoparathyroidism (Watanabe et al. 2002; Vargas-Poussou et al. 2002). Subtypes of ABS are classified by the gene that is affected (Table 19.1).

This review focuses on type 2 Bartter syndrome and the ROMK channel.

19.2 The ROMK Channel

ROMK [aka $K_{ir}1.1$ or KCNJ1 (Ho et al. 1993)] is the founding member of the inward-rectifier "K_{ir}" type potassium channel family (Welling and Ho 2009), which are encoded by KCN "J" type genes. Each ROMK subunit has two transmembrane spanning regions and a large cytoplasmic domain, typical of K_{ir} channels. The transmembrane domains of four ROMK subunits assemble to create a canonical potassium conduction pathway, which includes the potassium selectivity filter and a regulated gate structure. Cytoplasmic N- and C-termini of each subunit fold to create a large regulatory structure with a centrally located pore (see Fig. 19.2). Missense mutations that are associated with Bartter syndrome are located in each of these major domains (see Sects. 19.6–19.9).

Like other inward rectifiers, ROMK exhibits a nonlinear current-voltage relationship, distinguished by a larger inward current than outward current. Nevertheless, ROMK exhibits the weakest inward rectification of all members of the K_{ir}

Table 19.1 Bartter syndrome subtypes

Type	Gene	Chromosome	OMIM	Unique features
1	SLC12a1	15q21.1	#601678	
2	KCNJ1	11q24.3	#241200	Transient hyperkalemia followed by mild hypokalemia
3	CLCNKB	1p36.13	#607364	Hypocalciuria
4a	BSND	1p32.3	#602522	Sensorineural deafness
4b	CLCNKA/B	1p36.13	#613090	Sensorineural deafness
5	CASR	3q21.1	#601198	Autosomal dominant hypocalcemia1/ hypoparathyroidism

Table lists the subtype by affected gene. Chromosome location of gene, OMIM catalog number, and distinguishing features of each subtype are provided

family. In fact, ROMK channels carry significant outward potassium flux under physiological conditions. Together with their high open probability ($P_o = 0.9$), ROMK channels support robust potassium efflux at physiological membrane potentials. These properties make ROMK channels ideally suited for their physiological role in secreting potassium into the pro-urine.

ROMK channels have three major functions in the kidney. In the thick ascending limb (TAL), ROMK channels are expressed on the apical membrane and mediate the K^+ efflux into the tubule lumen to help sustain the turnover of NKCC2 for NaCl transport (see Sect. 19.3). They also contribute to the transepithelial current flow and potential difference in the TAL, important for paracellular Na^+ and divalent cation reabsorption. In the late distal tubule, connecting tubule, and collecting duct, ROMK channels function in the final step of urinary potassium excretion, a highly regulated process that ensures potassium is precisely excreted into the urine to match variations in dietary potassium intake for K^+ homeostasis (Hebert et al. 2005; Welling 2013).

Patch clamp studies of the apical membrane in split-open TAL cells and connecting tubule (CNT) and cortical collecting duct (CCD) principal cells revealed that the predominate apical K_{ir} channels in the TAL and CCD have biophysical and pharmacological properties closely matching recombinant ROMK (~single channel conductance 35 pS, high open probability) (Palmer and Frindt 1999; Wang et al. 1990; Bleich et al. 1990). Ablation of the ROMK gene (Lorenz et al. 2002) eliminates the 35-pS channel in the mouse TAL and CCD (Lu et al. 2002), providing strong evidence that ROMK encodes the renal secretory channel. Interestingly, another potassium channel on the TAL apical membrane has been described. It exhibits ROMK-like gating characteristics except for its higher single channel conductance (70 pS) (Wang 1994). This channel is also lost in the ROMK null mouse, strongly suggesting that ROMK is a critical subunit of the higher conductance channel (Lu et al. 2002). The identity of the conductance modifier is presently unknown.

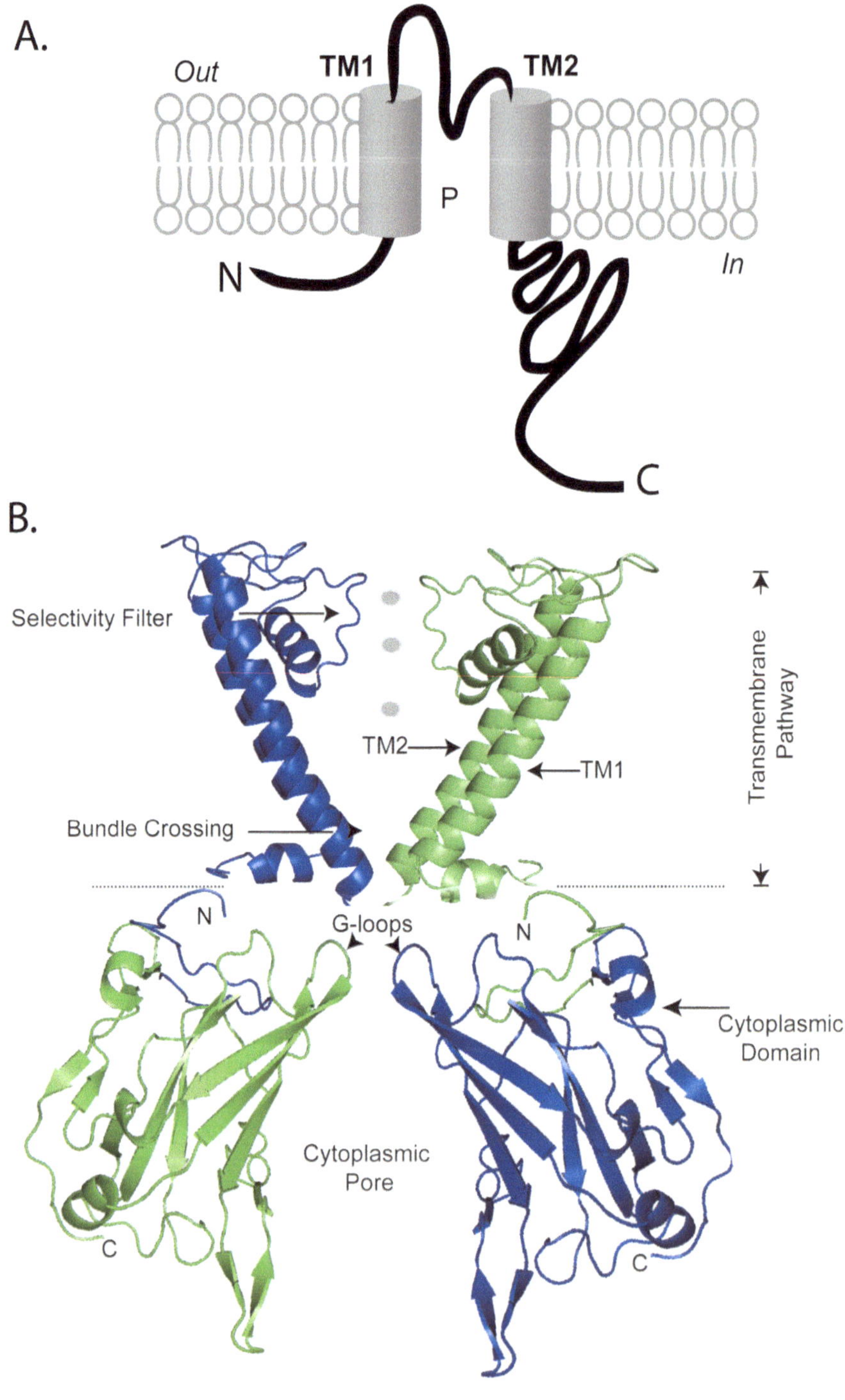

Fig. 19.2 ROMK structure. (**a**) Cartoon of a ROMK channel subunit, with a two-transmembrane body plan, typical of inwardly rectifying potassium channels. (**b**) Atomic resolution model of

19.3 ROMK Function in the TAL

Elegant physiologic studies in the isolated perfused tubule provide a context to understand why ROMK channels are required for salt reabsorption in the TAL (Greger and Schlatter 1981) and how their dysfunction leads to Bartter syndrome. Normally, the thick ascending limb reabsorbs ~20 % of the filtered load of sodium chloride through the multistep process shown in Fig. 19.1. In this scheme NaCl, along with potassium, is transported from the tubule lumen into the cell by the apically localized $Na^+/K^+/2Cl^-$ co-transporter, NKCC2 (product of the SLC12A1 gene); sodium is then actively transported across the basolateral membrane through the Na^+/K^+-ATPase, while Cl^- diffuses down a favorable electrochemical gradient into the interstitium through the basolaterally located CLCNKB channels. Apical potassium channels, formed by ROMK (Lorenz et al. 2002; Lu et al. 2002), allow the obligatory uptake of potassium through NKCC2, to recycle across the apical membrane. In fact, it has been believed that the ROMK-mediated recycling process is necessary to supply a sufficient level of potassium to drive the turnover of NKCC2 (Greger and Schlatter 1981; Hebert et al. 1984; Hebert and Andreoli 1984; Hebert 2003). Consistent with this, in the isolated perfused tubule preparation, sodium reabsorption is critically dependent on the secretory movement of potassium (Greger and Schlatter 1981).

The discovery that loss-of-function mutations in ROMK cause Bartter syndrome, together with subsequent studies in ROMK knockout mice (Lu et al. 2002; Lorenz et al. 2002; Bailey et al. 2006), which recapitulate the disease, provided solid evidence of the transport mechanism. Nevertheless, mathematical modeling of salt transport in the thick ascending limb in vivo reveals a wrinkle in the transport scheme (Weinstein 2010). Remarkably, the model fails to capture the inhibition of salt transport in type 2 Bartter syndrome when NH_4^+ is included. In this case, the model reveals that there is little impact of luminal membrane K^+ permeability on overall Na^+ reabsorption because NH_4^+ can take the place of K^+ in maintaining NKCC2 turnover. By contrast, sodium reabsorption in the isolated perfused TAL preparation requires potassium secretion (Greger and Schlatter 1981) because NH_4^+ is not typically included.

Viewed in this light, it seems likely that ROMK does more to sustain sodium transport than provide potassium to the tubule lumen. Two mechanisms should be considered. The ROMK channel contributes to the generation of a lumen positive transepithelial potential, which drives the passive reabsorptive movement of cations through the paracellular pathway. In the absence of the ROMK, the passive reabsorptive process, which accounts for about a third of the total reabsorptive flux of sodium, will be impaired. The expression of NKCC2 is compromised in

Fig. 19.2 (continued) ROMK based on the structure of $K_{ir}2.2$ (pdb 3JYC) and $K_{ir}3.1$ (pdb 2QKS). Two subunits are shown. Front and back subunits have been removed to see the potassium permeation pathway

ROMK knockout mice (Wagner et al. 2008), suggesting that ROMK and NKCC2 might be more directly linked together. The basis for ROMK-NKCC2 coupling is not known, but it may relate to the mutual dependence of ROMK and NKCC2 for a glycosylphosphatidylinositol-anchored uromodulin (aka Tamm-Horsfall protein) to be efficiently expressed on the apical membrane (Mutig et al. 2011; Renigunta et al. 2011).

19.4 Transient Hyperkalemia: A Unique Phenotype of Bartter Type 2

Characteristically, profound hypokalemia occurs in Bartter syndrome as a consequence of urinary potassium wasting. This transpires because factors that normally stimulate potassium secretion in the distal nephron [increased delivery of sodium, increased tubular fluid flow rate and elevation of aldosterone (Welling 2013)] are pathologically activated when sodium reabsorption in the TAL is impaired. Additionally, reduced potassium absorption in the TAL can account for significant potassium losses in Bartter syndrome (Bailey et al. 2006).

In remarkable contrast, newborns with Bartter type 2 present with hyperkalemia (Brochard et al. 2009; Finer et al. 2003; Peters et al. 2003). This most likely occurs because loss of ROMK function in the distal nephron leaves the neonatal kidney without a means to efficiently excrete potassium. The hyperkalemia is transient as potassium returns to normal levels within a few weeks after birth, and then persistent mild hypokalemia can follow throughout life. Free-flow micropuncture and stationary microperfusion studies in ROMK knockout mice revealed that the renal potassium excretion is restored in type 2 Bartter by compensatory overexpression of BK channels (Bailey et al. 2006), similar to the compensatory upregulation of ROMK channels that preserves potassium secretion in BK channel knockout animals (Rieg et al. 2007). BK channels are especially suited to stimulate potassium excretion in the high-flow, high-sodium delivery setting of Bartter syndrome (Bailey et al. 2006) because they are activated by flow and membrane depolarization (Liu et al. 2007; Pluznick et al. 2003). In the mouse model of type 2 Bartter syndrome, urinary potassium wasting is further exaggerated by reduced potassium reabsorption in the TAL (Bailey et al. 2006).

19.5 Carriers of Bartter Mutations Are Protected from Hypertension

Recent genetic and physiological studies indicate that rare mutations in ROMK also contribute to blood pressure variation in the general population (Ji et al. 2008; Fang et al. 2010). In a large-scale exome sequencing effort in the Framingham heart study cohort, Ji et al. determined that carriers of known or inferred loss-of-function mutations in ROMK (as well as NKCC2 and Gitelman-associated mutations in the thiazide-sensitive sodium chloride co-transporter (NCC); see Chap. 12) exhibited a significantly lower blood pressure throughout their life than the rest of the cohort. The trait conferred protection from the development of hypertension (Ji et al. 2008). Subsequent physiological studies revealed that these rare mutations do, in fact, cause a loss of ROMK (Fang et al. 2010), NKCC2, or NCC function (Acuna et al. 2011; Monette et al. 2011), verifying and reinforcing the role of functional variants in kidney salt handling genes in blood pressure. In the case of ROMK, one defective allele appears sufficient to reduce the sodium reabsorption in the thick ascending limb enough to shift the blood pressure "set point" for sodium excretion and confer a protective trait. These observations have inspired the development of ROMK inhibitors as a new class of antikaliuretic diuretics (Garcia et al. 2014; Tang et al. 2012; Bhave et al. 2011).

The discovery that rare independent mutations in renal salt handling genes, such as ROMK, contribute to blood pressure variation in the general population has important implications for the genetic bases of hypertension (Subramanya and Welling 2011). Because the ROMK, NKCC2, and NCC mutants represent only a small fraction of variant genes that alter blood pressure and many more remain to be discovered, Ji et al. reasonably argued that the combined effects of many rare independent mutations will likely account for a substantial fraction of blood pressure variation in the general population (Ji et al. 2008). It will be interesting to learn how many of these genes impinge on the regulation of ROMK.

19.6 Bartter Mutations in ROMK: Overview

Over 35 different Bartter disease mutations have been identified in ROMK. Several introduce nonsense codons or frameshifts that produce truncated proteins and obvious loss of function. Over half of the missense mutations reduce or even eliminate surface expression of ROMK (Peters et al. 2003), presumably as a result of global misfolding or mistrafficking. Others alter channel opening by disrupting the potassium permeation pathway or by upsetting the way the channel is regulated by phosphorylation, PIP2, or intracellular pH. The underlying mechanisms are discussed below.

19.7 Bartter Mutations in the Potassium Permeation Structure

Like all known potassium channels, the transmembrane linkers of each ROMK subunit project into the membrane to form the narrowest part of the open pore and create the potassium selectivity filter around the conserved sequence, "T[V/I]GYG" (Heginbotham et al. 1994). The sequence adopts a strand conformation where backbone carbonyl oxygens point into the pore, effectively mimicking the inner hydration shell of potassium but not of sodium (MacKinnon 2004). In this way, the selectivity filter creates energetically favorable sites for potassium to selectively diffuse from the aqueous environment into the pore. A Bartter mutation, $I_{142}T$, in the second residue of the selectivity sequence blocks channel function, presumably by collapsing the potassium permeation pathway (Peters et al. 2003). Other mutations in the vicinity do the same (Fig. 19.3a). In some cases, mutations in this region change ion permeation. In the related $K_{ir}3.4$ channel (KCNJ5), for example, mutations that are in or near the conserved sequence cause the channel to lose potassium selectivity and gain sodium permeation (Choi et al. 2011). These gain-of-function mutations underlie the pathogenesis of aldosterone-producing adrenal adenoma formation in some patients.

19.8 Bartter Mutations That Alter Regulated Gates

PKA phosphorylation (Xu et al. 1996; McNicholas et al. 1994) and PIP2 binding (Huang et al. 1998; Huang 2007) are required to maintain ROMK in a high open probability state, while intracellular acidification (Zhou et al. 1994) and ATP binding close the channel (McNicholas et al. 1996; Ruknudin et al. 1998). According to current understanding, the effects of phosphorylation and ligand binding within the cytoplasmic domain are transmitted to gating structures, located in the transmembrane helices, and the apex of the cytoplasmic domain. Bartter mutations affect these gates (Fig. 19.3).

Regulated channel opening of the transmembrane gate is governed by movement of the end of second transmembrane domain (TM2), which pulls the putative gate residue, L_{179}, (Sackin et al. 2005; Nanazashvili et al. 2007) away from the pore (Fig. 19.3). This is influenced by interactions between the two transmembrane helices, involving A_{177} on TM2 and K_{80} on TM1 (Rapedius et al. 2006, 2007). A Bartter mutation that replaces A_{177} with threonine (Peters et al. 2003) disrupts the interaction and causes a profound alkaline shift in the pH sensitivity of ROMK (Rapedius et al. 2006), effectively turning off channel function at physiological pH_i. Another Bartter mutation, involving a nearby lipid-facing residue, Y_{79}, at the base of TM1, disrupts regulated channel opening in a similar way (Peters et al. 2003). Interestingly, mutations in the comparable residues of the $K_{ir}6.2$ K_{ATP} channel alter ATP-dependent gating and produce neonatal diabetes (Hattersley and Ashcroft

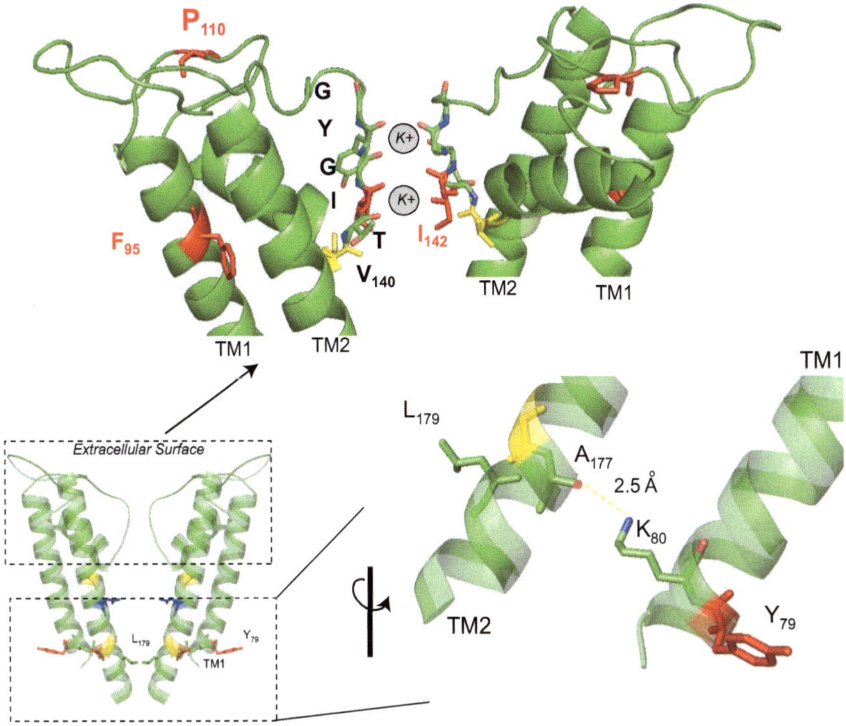

Fig. 19.3 Bartter mutations affect the transmembrane potassium permeation pathway at the selectivity filter (*top*) and the regulated gate (*bottom*). Residues that are mutated in Bartter syndrome are shown in *red*

2005). It would seem that residues at the base of TM1 and TM2 strongly influence the energetics of regulated K^+ channel opening, making these structures especially vulnerable to disease-causing mutations. Furthermore, inhibitory effects of Bartter mutations in other regions of the channel, especially those that influence pH and PIP2-dependent gating, can be rescued by a second-site suppressor mutation, $K_{80}M$, in the transmembrane gate (Fig. 19.3) (Flagg et al. 2002; Schulte et al. 1999), highlighting closed-TM gate stabilization as a broad mechanism for loss of ROMK channel function in Bartter disease.

An adjacent gating structure involves a narrow opening at the apex of the cytoplasmic pore, formed by the coalescence of four identical loop structures from each subunit (Fig. 19.2), called the G-loop. The structure is believed to create a flexible diffusion barrier between the cytoplasmic and transmembrane pore. G-loop residues, which face into the central axis of the cytoplasmic pore, have been implicated in channel gating (Pegan et al. 2005). One of these residues is affected by a Bartter mutation ($A_{306}T$), which disrupts potassium conduction (Peters et al. 2003). Disease-causing mutations in other K_{ir} channels also affect the G-loop, underscoring the general importance of this structure in the inward rectifiers (Ma et al. 2007; Pegan et al. 2006).

19.9 Bartter Mutations That Disrupt Channel Modulation

The large cytoplasmic domain structure has an enormous solvent-exposed surface area, important for docking modulators (PIP2, pH) and serving as a substrate for PKA phosphorylation. A number of Bartter mutations affect residues in this domain, and these mutations alter the way the modulators control channel gating.

19.9.1 PIP2

The X-ray structure of the $K_{ir}2.2$ channel in complex with a derivative of PIP2 revealed that PIP2 binds to residues at the top of the cytoplasmic domain, and upon binding, an adjacent linker contracts to move the entire cytoplasmic domain toward the membrane and pull the inner helix gate open (Hansen et al. 2011). Systematic mutagenesis studies in ROMK and other K_{ir} channels (Lopes et al. 2002; Zeng et al. 2002) support a common mechanism. Indeed, these studies revealed that PIP2-dependent gating is determined by a group of shared residues near or within the PIP2 binding surface and gating transducer that was identified in $K_{ir}2.2$. Many of the Bartter mutations ($C_{49}Y$, $I_{51}T$, $P_{166}S$, $R_{169}H$, $A_{214}V$ and $L_{220}F$) are found in the vicinity of this region and disrupt PIP2-dependent gating (Lopes et al. 2002; Fang et al. 2010). Similar observations with disease-causing mutations in $K_{ir}2.1$ [Andersen-Tawil, Plaster et al. 2001)] have been interpreted to indicate that a decrease in channel PIP2 interactions underlies many K_{ir} channelopathies (Lopes et al. 2002).

The effects of several Bartter type 2 mutations [$P_{166}S$, $R_{169}H$, $L_{220}F$ (Fang et al. 2010; Srivastava et al. 2013)] are not observed until PIP_2 hydrolysis is stimulated (Fig. 19.4). Wild-type ROMK channels are relatively resistant to physiological changes in PIP_2 levels because of their high apparent binding affinity for PIP_2 (Lopes et al. 2002; Huang 2007). By contrast, the P166S and R169H mutations destabilize PIP2 binding, and this causes channels bearing these Bartter mutations to rapidly close in response to activation of GPCR-Gq-PLC signaling, which stimulates PIP2 hydrolysis (Fang et al. 2010). Astonishingly, P_{166} and R_{169} reside in the PIP2 binding site, flanking the two residues that interact with the phosphates on the phosphatidylinositol moiety of PIP2 (Fig. 19.4). Although $L_{220}F$ also causes a loss of function that is conditional on PIP2 hydrolysis, this Bartter mutation is located near the G-loop and presumably destabilizes the PIP2-dependent gating mechanism rather than the PIP2 binding site.

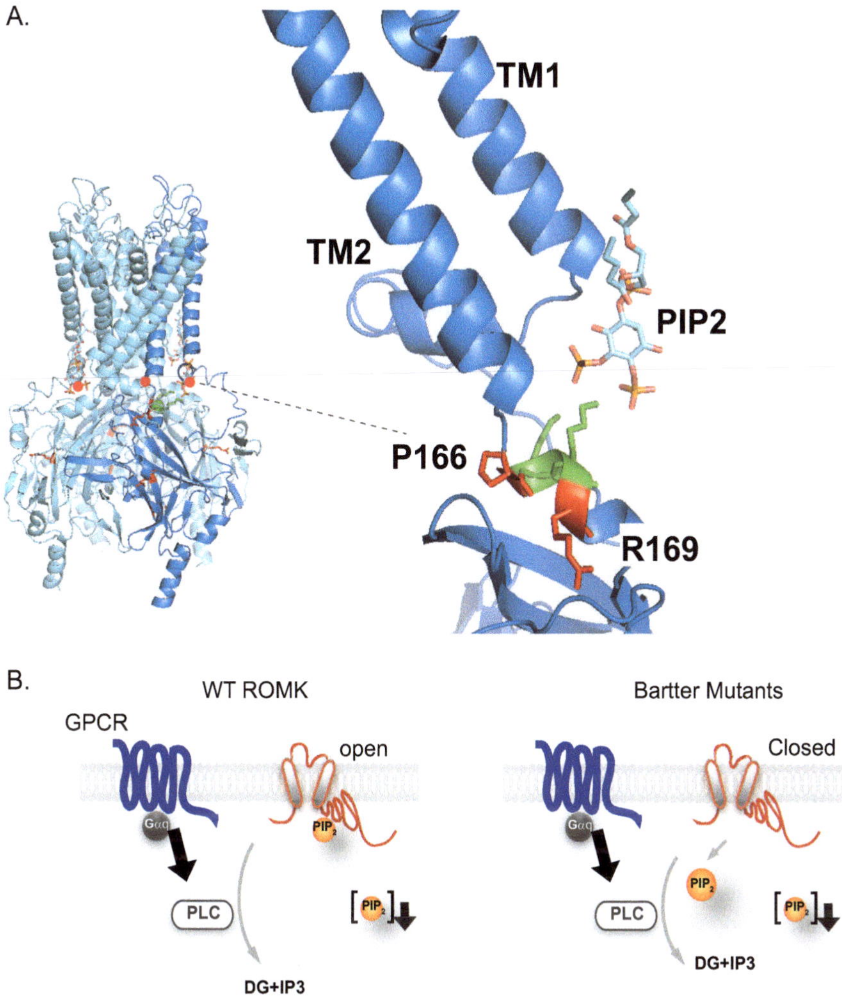

Fig. 19.4 Bartter mutations near the PIP2 binding site can be conditionally manifested. (**a**) Atomic resolution model of ROMK with PIP2 based on structure of $K_{ir}2.2$ in complex with PIP2 (pdb 3SPG), showing positions of P166 and R169, flanking the basic residues that bind PIP2. Bartter mutations at these positions destabilize PIP2 binding. (**b**) Channels bearing P166S and R169H mutations are open until activation of GPCR-Gq-PLC signaling simulates PIP2 hydrolysis, while wild-type (WT) ROMK stay open because they have a tight binding affinity for PIP2

19.9.2 PKA Phosphorylation Sites

ROMK is maintained in a high open probability state by phosphorylation of two key PKA sites in the cytoplasmic domain (Xu et al. 1996; McNicholas et al. 1994). Two Bartter disease-causing mutations replace these phosphorylatable residues [S_{219}R

(Simon et al. 1997a) and $S_{313}C$ (Starremans et al. 2002)] and thereby alter the regulated gating process.

19.9.3 pH-Dependent Gating

ROMK channels close in response to cytoplasmic acidification with an apparent p*K* a near neutral pH (Welling and Ho 2009). Many Bartter disease-associated mutations in the cytoplasmic domain shift the p*K* a in the alkaline direction and cause channel closure at physiological pH (Flagg et al. 1999, 2002; Schulte et al. 1998, 1999; Peters et al. 2003). Several of the affected residues were once believed to form the pH sensor (Schulte et al. 1999), but the atomic structures of related channels indicate that this is not true (Rapedius et al. 2006). The precise location of the pH sensor remains elusive but likely involves titratable residues in the cytoplasmic domain that stabilize the TM gate in an open configuration. Recent mutagenesis studies reveal the pH sensing is coupled to gating by a large network of residues that link the cytoplasmic domain to the TM gate (Bollepalli et al. 2014). Several Bartter mutants (e.g., Asp74, Ala177, and Ala306) likely increase pH sensitivity by destabilizing contacts in the network.

19.10 Summary

The discovery that loss-of-function mutations in ROMK cause Bartter disease type 2, together with studies in ROMK knockout mice, provided airtight evidence that ROMK is necessary for sodium chloride reabsorption in the thick ascending limb of Henle's loop and the control of potassium balance. It has been suggested that ROMK mediates a potassium recycling process in the TAL that is necessary to sustain the turnover of NKCC2, but further studies will be required to more precisely establish the physiological mechanism underlying the requirement for ROMK in TAL salt reabsorption. Nevertheless, efforts to understand the underlying molecular mechanisms for loss-of-channel function have revealed and reinforced new ideas about the structural basis for channel function and regulation.

References

Acuna R, Martinez-De-La-Maza L, Ponce-Coria J, Vazquez N, Ortal-Vite P, Pacheco-Alvarez D, Bobadilla NA, Gamba G (2011) Rare mutations in SLC12A1 and SLC12A3 protect against hypertension by reducing the activity of renal salt cotransporters. J Hypertens 29:475–483

Bailey MA, Cantone A, Yan Q, Macgregor GG, Leng Q, Amorim JB, Wang T, Hebert SC, Giebisch G, Malnic G (2006) Maxi-K channels contribute to urinary potassium excretion in the ROMK-deficient mouse model of Type II Bartter's syndrome and in adaptation to a high-K diet. Kidney Int 70:51–59

Bartter FC, Pronove P, Gill JR Jr, Maccardle RC (1962) Hyperplasia of the juxtaglomerular complex with hyperaldosteronism and hypokalemic alkalosis: a new syndrome. Am J Med 33:811–828

Bhave G, Chauder BA, Liu W, Dawson ES, Kadakia R, Nguyen TT, Lewis LM, Meiler J, Weaver CD, Satlin LM, Lindsley CW, Denton JS (2011) Development of a selective small-molecule inhibitor of Kir1.1, the renal outer medullary potassium channel. Mol Pharmacol 79:42–50

Birkenhager R, Otto E, Schurmann MJ, Vollmer M, Ruf EM, Maier-Lutz I, Beekmann F, Fekete A, Omran H, Feldmann D, Milford DV, Jeck N, Konrad M, Landau D, Knoers NV, Antignac C, Sudbrak R, Kispert A, Hildebrandt F (2001) Mutation of BSND causes Bartter syndrome with sensorineural deafness and kidney failure. Nat Genet 29:310–314

Bleich M, Schlatter E, Greger R (1990) The luminal K^+ channel of the thick ascending limb of Henle's loop. Pflugers Arch 415:449–460

Bollepalli MK, Fowler PW, Rapedius M, Shang L, Sansom MS, Tucker SJ, Baukrowitz T (2014) State-dependent network connectivity determines gating in a K^+ channel. Structure 22: 1037–1046

Brochard K, Boyer O, Blanchard A, Loirat C, Niaudet P, Macher MA, Deschenes G, Bensman A, Decramer S, Cochat P, Morin D, Broux F, Caillez M, Guyot C, Novo R, Jeunemaitre X, Vargas-Poussou R (2009) Phenotype-genotype correlation in antenatal and neonatal variants of Bartter syndrome. Nephrol Dial Transplant 24:1455–1464

Choi M, Scholl UI, Yue P, Bjorklund P, Zhao B, Nelson-Williams C, Ji W, Cho Y, Patel A, Men CJ, Lolis E, Wisgerhof MV, Geller DS, Mane S, Hellman P, Westin G, Akerstrom G, Wang W, Carling T, Lifton RP (2011) K^+ channel mutations in adrenal aldosterone-producing adenomas and hereditary hypertension. Science 331:768–772

Fang L, Li D, Welling PA (2010) Hypertension resistance polymorphisms in ROMK (Kir1.1) alter channel function by different mechanisms. Am J Physiol Renal Physiol 299:F1359–F1364

Finer G, Shalev H, Birk O, Galron D, Jeck N, Sinai-Treiman L, Landau D (2003) Transient neonatal hyperkalemia in the antenatal (ROMK defective) Bartter syndrome. J Pediatr 142: 318–323

Flagg TP, Tate M, Merot J, Welling PA (1999) A mutation linked with Bartter's syndrome locks Kir 1.1a (ROMK1) channels in a closed state. J Gen Physiol 114:685–700

Flagg TP, Yoo D, Sciortino CM, Tate M, Romero MF, Welling PA (2002) Molecular mechanism of a COOH-terminal gating determinant in the ROMK channel revealed by a Bartter's disease mutation. J Physiol 544:351–362

Garcia ML, Priest BT, Alonso-Galicia M, Zhou X, Felix JP, Brochu RM, Bailey T, Thomas-Fowlkes B, Liu J, Swensen A, Pai LY, Xiao J, Hernandez M, Hoagland K, Owens K, Tang H, De Jesus RK, Roy S, Kaczorowski GJ, Pasternak A (2014) Pharmacologic inhibition of the renal outer medullary potassium channel causes diuresis and natriuresis in the absence of kaliuresis. J Pharmacol Exp Ther 348:153–164

Greger R, Schlatter E (1981) Presence of luminal K^+, a prerequisite for active NaCl transport in the cortical thick ascending limb of Henle's loop of rabbit kidney. Pflugers Arch 392:92–94

Hansen SB, Tao X, Mackinnon R (2011) Structural basis of PIP2 activation of the classical inward rectifier K^+ channel Kir2.2. Nature 477:495–498

Hattersley AT, Ashcroft FM (2005) Activating mutations in Kir6.2 and neonatal diabetes: new clinical syndromes, new scientific insights, and new therapy. Diabetes 54:2503–2513

Hebert SC (2003) Bartter syndrome. Curr Opin Nephrol Hypertens 12:527–532

Hebert SC, Andreoli TE (1984) Effects of antidiuretic hormone on cellular conductive pathways in mouse medullary thick ascending limbs of Henle: II. Determinants of the ADH-mediated increases in transepithelial voltage and in net Cl-absorption. J Membr Biol 80:221–233

Hebert SC, Friedman PA, Andreoli TE (1984) Effects of antidiuretic hormone on cellular conductive pathways in mouse medullary thick ascending limbs of Henle: I. ADH increases transcellular conductance pathways. J Membr Biol 80:201–219

Hebert SC, Desir G, Giebisch G, Wang W (2005) Molecular diversity and regulation of renal potassium channels. Physiol Rev 85:319–371

Heginbotham L, Lu Z, Abramson T, Mackinnon R (1994) Mutations in the K^+ channel signature sequence. Biophys J 66:1061–1067

Ho K, Nichols CG, Lederer WJ, Lytton J, Vassilev PM, Kanazirska MV, Hebert SC (1993) Cloning and expression of an inwardly rectifying ATP-regulated potassium channel. Nature 362:31–38

Huang CL (2007) Complex roles of PIP2 in the regulation of ion channels and transporters. Am J Physiol Renal Physiol 293:F1761–F1765

Huang CL, Feng S, Hilgemann DW (1998) Direct activation of inward rectifier potassium channels by PIP2 and its stabilization by Gbetagamma. Nature 391:803–806

Ji W, Foo JN, O'roak BJ, Zhao H, Larson MG, Simon DB, Newton-Cheh C, State MW, Levy D, Lifton RP (2008) Rare independent mutations in renal salt handling genes contribute to blood pressure variation. Nat Genet 40:592–599

Karolyi L, Konrad M, Köckerling A, Ziegler A, Zimmermann DK, Roth B, Wieg C, Grzeschik K-H, Koch MC, Seyberth HW, Vargas R, Forestier L, Jean G, Deschaux M, Rizzoni GF, Niaudet P, Antignac C, Feldmann D, Lorridon F, Cougoureux E, Laroze F, Alessandri J-L, David L, Saunier P, Deschenes G, Hildebrandt F, Vollmer M, Proesmans W, Brandis M, van den Heuvel LPWJ, Lemmink HH, Nillesen N, Monnens LAH, Knoers NVAM, Guay-Woodford LM, Wright CJ, Madrigal G, Hebert SC (1997) Mutations in the gene encoding the inwardly-rectifying renal potassium channel, ROMK, cause the antenatal variant of Bartter syndrome: evidence for genetic heterogeneity. International Collaborative Study Group for Bartter-like Syndromes. Hum Mol Genet 6:17–26

Konrad M, Vollmer M, Lemmink HH, Van Den Heuvel LP, Jeck N, Vargas-Poussou R, Lakings A, Ruf R, Deschenes G, Antignac C, Guay-Woodford L, Knoers NV, Seyberth HW, Feldmann D, Hildebrandt F (2000) Mutations in the chloride channel gene CLCNKB as a cause of classic Bartter syndrome. J Am Soc Nephrol 11:1449–1459

Liu W, Morimoto T, Woda C, Kleyman TR, Satlin LM (2007) Ca^{2+} dependence of flow-stimulated K secretion in the mammalian cortical collecting duct. Am J Physiol Renal Physiol 293:F227–F235

Lopes CM, Zhang H, Rohacs T, Jin T, Yang J, Logothetis DE (2002) Alterations in conserved Kir channel-PIP2 interactions underlie channelopathies. Neuron 34:933–944

Lorenz JN, Baird NR, Judd LM, Noonan WT, Andringa A, Doetschman T, Manning PA, Liu LH, Miller ML, Shull GE (2002) Impaired renal NaCl absorption in mice lacking the ROMK potassium channel, a model for type II Bartter's syndrome. J Biol Chem 277:37871–37880

Lu M, Wang T, Yan Q, Yang X, Dong K, Knepper MA, Wang W, Giebisch G, Shull GE, Hebert SC (2002) Absence of small conductance K^+ channel (SK) activity in apical membranes of thick ascending limb and cortical collecting duct in ROMK (Bartter's) knockout mice. J Biol Chem 277:37881–37887

Ma D, Tang XD, Rogers TB, Welling PA (2007) An Andersen-Tawil syndrome mutation in Kir2.1 (V302M) alters the G-loop cytoplasmic K^+ conduction pathway. J Biol Chem 282:5781–5789

Mackinnon R (2004) Nobel Lecture. Potassium channels and the atomic basis of selective ion conduction. Biosci Rep 24:75–100

McNicholas CM, Wang W, Ho K, Hebert SC, Giebisch G (1994) Regulation of ROMK1 K^+ channel activity involves phosphorylation processes. Proc Natl Acad Sci U S A 91:8077–8081

McNicholas CM, Guggino WB, Schwiebert EM, Hebert SC, Giebisch G, Egan ME (1996) Sensitivity of a renal K^+ channel (ROMK2) to the inhibitory sulfonylurea compound glibenclamide is enhanced by coexpression with the ATP-binding cassette transporter cystic fibrosis transmembrane regulator. Proc Natl Acad Sci U S A 93:8083–8088

Monette MY, Rinehart J, Lifton RP, Forbush B (2011) Rare mutations in the human Na-K-Cl cotransporter (NKCC2) associated with lower blood pressure exhibit impaired processing and transport function. Am J Physiol Renal Physiol 300:F840–F847

Mutig K, Kahl T, Saritas T, Godes M, Persson P, Bates J, Raffi H, Rampoldi L, Uchida S, Hille C, Dosche C, Kumar S, Castaneda-Bueno M, Gamba G, Bachmann S (2011) Activation of the bumetanide-sensitive Na+, K+,2Cl− cotransporter (NKCC2) is facilitated by Tamm-Horsfall protein in a chloride-sensitive manner. J Biol Chem 286:30200–30210

Nanazashvili M, Li H, Palmer LG, Walters DE, Sackin H (2007) Moving the pH gate of the Kir1.1 inward rectifier channel. Channels (Austin) 1:21–28

Palmer LG, Frindt G (1999) Regulation of apical K channels in rat cortical collecting tubule during changes in dietary K intake. Am J Physiol Renal Physiol 277:F805–F812

Pegan S, Arrabit C, Zhou W, Kwiatkowski W, Collins A, Slesinger PA, Choe S (2005) Cytoplasmic domain structures of Kir2.1 and Kir3.1 show sites for modulating gating and rectification. Nat Neurosci 8:279–287

Pegan S, Arrabit C, Slesinger PA, Choe S (2006) Andersen's syndrome mutation effects on the structure and assembly of the cytoplasmic domains of Kir2.1. Biochemistry 45:8599–8606

Peters M, Ermert S, Jeck N, Derst C, Pechmann U, Weber S, Schlingmann KP, Seyberth HW, Waldegger S, Konrad M (2003) Classification and rescue of ROMK mutations underlying hyperprostaglandin E syndrome/antenatal Bartter syndrome. Kidney Int 64:923–932

Plaster NM, Tawil R, Tristani-Firouzi M, Canun S, Bendahhou S, Tsunoda A, Donaldson MR, Iannaccone ST, Brunt E, Barohn R, Clark J, Deymeer F, George AL Jr, Fish FA, Hahn A, Nitu A, Ozdemir C, Serdaroglu P, Subramony SH, Wolfe G, Fu YH, Ptacek LJ (2001) Mutations in Kir2.1 cause the developmental and episodic electrical phenotypes of Andersen's syndrome. Cell 105:511–519

Pluznick JL, Wei P, Carmines PK, Sansom SC (2003) Renal fluid and electrolyte handling in BK_{Ca}-b1$^{-/-}$ mice. Am J Physiol Renal Physiol 284:F1274–F1279

Rapedius M, Haider S, Browne KF, Shang L, Sansom MS, Baukrowitz T, Tucker SJ (2006) Structural and functional analysis of the putative pH sensor in the Kir1.1 (ROMK) potassium channel. EMBO Rep 7:611–616

Rapedius M, Fowler PW, Shang L, Sansom MS, Tucker SJ, Baukrowitz T (2007) H bonding at the helix-bundle crossing controls gating in Kir potassium channels. Neuron 55:602–614

Renigunta A, Renigunta V, Saritas T, Decher N, Mutig K, Waldegger S (2011) Tamm-Horsfall glycoprotein interacts with renal outer medullary potassium channel ROMK2 and regulates its function. J Biol Chem 286:2224–2235

Rieg T, Vallon V, Sausbier M, Sausbier U, Kaissling B, Ruth P, Osswald H (2007) The role of the BK channel in potassium homeostasis and flow-induced renal potassium excretion. Kidney Int 72:566–573

Ruknudin A, Schulze DH, Sullivan SK, Lederer WJ, Welling PA (1998) Novel subunit composition in a renal K_{ATP} channel. J Biol Chem 273:14165–14171

Sackin H, Nanazashvili M, Palmer LG, Krambis M, Walters DE (2005) Structural locus of the pH gate in the Kir1.1 inward rectifier channel. Biophys J 88:2597–2606

Schulte U, Hahn H, Wiesinger H, Ruppersberg JP, Fakler B (1998) pH-dependent gating of ROMK (Kir1.1) channels involves conformational changes in both N and C termini. J Biol Chem 273:34575–34579

Schulte U, Hahn H, Konrad M, Jeck N, Derst C, Wild K, Weidemann S, Ruppersberg JP, Fakler B, Ludwig J (1999) pH gating of ROMK (K_{ir}1.1) channels: control by an Arg-Lys-Arg triad disrupted in antenatal Bartter syndrome. Proc Natl Acad Sci U S A 96:15298–15303

Seyberth HW, Schlingmann KP (2011) Bartter- and Gitelman-like syndromes: salt-losing tubulopathies with loop or DCT defects. Pediatr Nephrol 26:1789–1802

Simon DB, Karet FE, Hamdan JM, Dipietro A, Sanjad S, Lifton RP (1996) Bartter's syndrome, hypokalaemic alkalosis with hypercalciuria, is caused by mutations in the Na-K-2Cl cotransporter NKCC2. Nat Genet 13:183–188

Simon D, Karet F, Rodriguez-Soriano J, Hamdan J, Dipietro A, Sanjad S, Lifton R (1997a) Genetic heterogeneity of Bartter's syndrome revealed by mutations in the K^+ channel, ROMK. Nat Genet 14:152–156

Simon DB, Bindra RS, Mansfield TA, Nelson-Williams C, Mendonca E, Stone R, Schurman S, Nayir A, Alpay H, Bakkaloglu A, Rodriguez-Soriano J, Morales JM, Sanjad SA, Taylor CM, Pilz D, Brem A, Trachtman H, Griswold W, Richard GA, John E, Lifton RP (1997b) Mutations in the chloride channel gene, CLCNKB, cause Bartter's syndrome type III. Nat Genet 17: 171–178

Srivastava S, Li D, Edwards N, Hynes AM, Wood K, Al-Hamed M, Wroe AC, Reaich D, Moochhala SH, Welling PA, Sayer JA (2013) Identification of compound heterozygous KCNJ1 mutations (encoding ROMK) in a kindred with Bartter's syndrome and a functional analysis of their pathogenicity. Physiol Rep 1, e00160

Starremans PG, Van Der Kemp AW, Knoers NV, Van Den Heuvel LP, Bindels RJ (2002) Functional implications of mutations in the human renal outer medullary potassium channel (ROMK2) identified in Bartter syndrome. Pflugers Arch 443:466–472

Subramanya AR, Welling PA (2011) Toward an understanding of hypertension resistance. Am J Physiol Renal Physiol 300:F838–F839

Tang H, Walsh SP, Yan Y, De Jesus RK, Shahripour A, Teumelsan N, Zhu Y, Ha S, Owens KA, Thomas-Fowlkes BS, Felixv JP, Liu J, Kohler M, Priest BT, Bailey T, Brochu R, Alonso-Galicia M, Kaczorowski GJ, Roy S, Yang L, Mills SG, Garcia ML, Pasternak A (2012) Discovery of selective small molecule ROMK inhibitors as potential new mechanism diuretics. ACS Med Chem Lett 3:367–672

Vargas-Poussou R, Huang C, Hulin P, Houillier P, Jeunemaitre X, Paillard M, Planelles G, Dechaux M, Miller RT, Antignac C (2002) Functional characterization of a calcium-sensing receptor mutation in severe autosomal dominant hypocalcemia with a Bartter-like syndrome. J Am Soc Nephrol 13:2259–2266

Wagner CA, Loffing-Cueni D, Yan Q, Schulz N, Fakitsas P, Carrel M, Wang T, Verrey F, Geibel JP, Giebisch G, Hebert SC, Loffing J (2008) Mouse model of type II Bartter's syndrome. II. Altered expression of renal sodium- and water-transporting proteins. Am J Physiol Renal Physiol 294:F1373–F1380

Wang WH (1994) Two types of K^+ channel in thick ascending limb of rat kidney. Am J Physiol Renal Physiol 267:F599–F605

Wang WH, Schwab A, Giebisch G (1990) Regulation of small-conductance K^+ channel in apical membrane of rat cortical collecting tubule. Am J Physiol Renal Physiol 259:F494–F502

Watanabe S, Fukumoto S, Chang H, Takeuchi Y, Hasegawa Y, Okazaki R, Chikatsu N, Fujita T (2002) Association between activating mutations of calcium-sensing receptor and Bartter's syndrome. Lancet 360:692–694

Weinstein AM (2010) A mathematical model of rat ascending Henle limb. III. Tubular function. Am J Physiol Renal Physiol 298:F543–F556

Welling PA (2013) Regulation of renal potassium secretion: molecular mechanisms. Semin Nephrol 33:215–228

Welling PA, Ho K (2009) A comprehensive guide to the ROMK potassium channel: form and function in health and disease. Am J Physiol Renal Physiol 297:F849–F863

Xu ZC, Yang Y, Hebert SC (1996) Phosphorylation of the ATP-sensitive, inwardly rectifying K^+ channel, ROMK, by cyclic AMP-dependent protein kinase. J Biol Chem 271:9313–9319

Zeng WZ, Liou HH, Krishna UM, Falck JR, Huang CL (2002) Structural determinants and specificities for ROMK1-phosphoinositide interaction. Am J Physiol Renal Physiol 282: F826–F834

Zhou H, Tate SS, Palmer LG (1994) Primary structure and functional properties of an epithelial K channel. Am J Physiol Cell Physiol 266:C809–C824

Chapter 20
KCa3.1 in Epithelia

Daniel C. Devor, Claudia A. Bertuccio, and Kirk L. Hamilton

Abstract Calcium-mediated agonists have long been known to stimulate transepithelial ion and fluid transport across a wide array of epithelial tissues. A key component of this response are the Ca^{2+}-activated K^+ channels, whose activation results in a hyperpolarization of the basolateral and apical membranes, thereby maintaining the electrochemical driving force for ion transport. In 1997, the intermediate conductance, Ca^{2+}-activated K^+ channel, KCa3.1, was cloned and subsequently shown to be localized to both the basolateral and apical membranes of secretory epithelia where it is activated by Ca^{2+}-mediated agonists. Herein, we review the data confirming the critical role that KCa3.1 plays in transepithelial ion transport as well as the regulation, gating, trafficking of this channel, and this channel's role in cell proliferation. Finally, we summarize the evidence for the role of KCa3.1 in epithelial diseases.

Keywords KCa3.1 • IK1 • SK4 • KCNN4 • Epithelia

20.1 Introduction

20.1.1 Early Evidence for Ca^{2+}-Mediated Regulation of Transepithelial Transport

The evidence that parasympathetic stimulation is important in intestinal salt and water secretion was first obtained by several investigators in the 1880s when they demonstrated that injection of the muscarinic agonist pilocarpine, in large doses, could induce colonic secretion (Florey et al. 1941). While this early work suggested

D.C. Devor (✉) • C.A. Bertuccio
Department of Cell Biology, School of Medicine, University of Pittsburgh, S331 BST, 3500 Terrace St., Pittsburgh, PA 15261, USA
e-mail: dd2@pitt.edu; claudia.bertuccio@yale.edu

K.L. Hamilton
Department of Physiology, Otago School of Medical Sciences, University of Otago, Dunedin, New Zealand
e-mail: kirk.hamilton@otago.ac.nz

K.L. Hamilton, D.C. Devor (eds.), *Ion Channels and Transporters of Epithelia in Health and Disease*, Physiology in Health and Disease,
DOI 10.1007/978-1-4939-3366-2_20

an important role for parasympathetic innervation of the intestine, very little additional work was done in this area until the 1940s. In a classic paper, Wright et al. (1940) demonstrated that direct vagal stimulation resulted in the flow of intestinal juices that continued only as long as the stimulation was maintained. It was further shown that this vagally induced secretion was blocked by the classic muscarinic antagonist, atropine (Wright et al. 1940). This potentially fruitful field of investigation again lays dormant until the 1960s when Tidball (1961) demonstrated, in the anesthetized dog, that administration of the cholinergic agonist, bethanechol, resulted in a change in the direction of intestinal water and Cl^- movement from absorption to secretion. It was further demonstrated that Cl^- was being secreted against both electrical and chemical gradients, demonstrating that secretion must be an active process (Tidball 1961). Similar increases in the transepithelial potential difference (PD_{te}) and Cl^- transport were reported for both in vivo (Hardcastle and Eggenton 1973; Hubel 1976, 1977) and in vitro (Hardcastle and Eggenton 1973) rat intestinal preparations. Isaacs et al. (1976) confirmed the above results in a stripped ileal mucosa preparation from human intestine and further demonstrated, using isotopic flux measurements, that the increase in PD_{te} was due to an increase in the unidirectional flux of Cl^- from serosa to mucosa. Although the above results clearly demonstrated that parasympathetic stimulation was involved in secretion, all of these results were from small intestinal preparations. It was not until 1977 that acetylcholine (Ach) was shown to cause fluid and electrolyte secretion from the colon, both in vivo and in vitro (Browning et al. 1977).

Initial evidence that sympathetic stimulation of ion and water secretion was due to direct innervation of epithelia came from staining for acetylcholinesterase, thereby localizing the nerve endings at the mucosa (Jacobowitz 1965; Browning et al. 1977; Isaacs et al. 1976). The functional release of neurotransmitter in the proximity of colonic epithelial cells was demonstrated by showing the uptake of radiolabeled choline and the subsequent synthesis and release of labeled Ach (Wu et al. 1982). This localization of functional nerve endings at the mucosa suggested that the physiological response associated with parasympathetic stimulation was due to the actions of Ach directly on the epithelial cells. This was confirmed by Rimele et al. (1981) when they demonstrated the existence of muscarinic receptors on intestinal epithelial cells. Zimmerman and Binder (1982) further demonstrated that the ability of muscarinic agonists to increase the short-circuit current (I_{sc}) was directly related to agonist binding to specific muscarinic receptors on colonic epithelial cells.

While these early studies confirmed that parasympathetic stimulation of the intestine resulted in the direct stimulation of Cl^- and fluid secretion from the epithelia, a great many additional studies were required to elucidate the mechanisms by which this fluid secretion occurred, as detailed in earlier chapters in this volume. Drawing upon a wealth of electrophysiological data, a general Cl^- secretory model has been elucidated (Welsh et al. 1982; Reuss et al. 1983; Greger et al. 1984; Suzuki and Petersen 1985) as shown in Fig. 20.1. This model includes (a) the uptake of Cl^- from the interstitial fluid by an electroneutral Na^+-K^+-$2Cl^-$ cotransporter; (b) the recycling of Na^+ and K^+ across the basolateral (serosal)

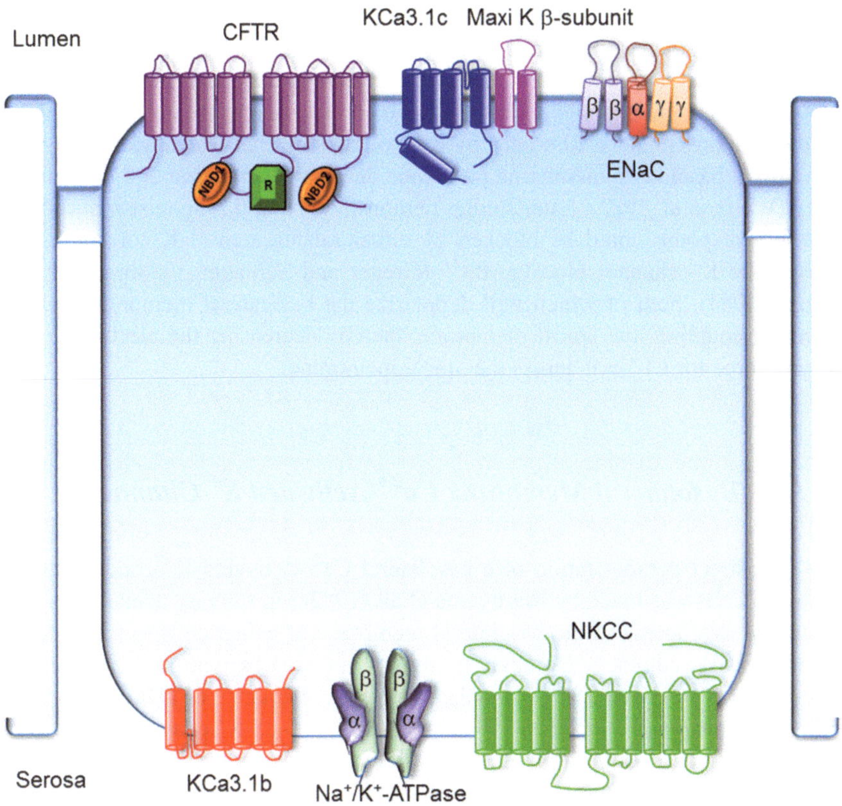

Fig. 20.1 Generalized epithelial cell model that includes the transport proteins for both Cl^- secretion and Na^+ absorption of epithelial cells. For Cl^- secretion, the uptake of Cl^- across the basolateral membrane is via the Na^+-K^+-$2Cl^-$ cotransporter (NKCC). Once stimulated, Cl^- moves passively across the apical membrane through Cl^- channels. The recycling of Na^+ and K^+ across the basolateral membrane (BM) is performed by Na^+/K^+-ATPase and K^+ channels, respectively. For Na^+ absorption, Na^+ is transported down its electrochemical gradient via ENaC and Na^+ exits the BM by the Na^+/K^+-ATPase. Activation of KCa3.1 in the apical and basolateral membranes results in hyperpolarization that maintains the electrochemical driving force for both Cl^- secretion and Na^+ absorption. *CFTR* cystic fibrosis transmembrane conductance regulator, *R* regulatory domain, *NBD* nuclear binding domain, *ENaC* epithelial Na^+ channel, *KCa3.1c* apical KCa3.1 channel, *KCa3.1b* basolateral KCa3.1; Maxi-K β-subunit. See text for details

membrane by the Na^+/K^+-ATPase and K^+ channel, respectively; and (c) the movement of Cl^- across the apical (mucosal) membrane through a Cl^- channel by passive diffusion. Each of these transporters and channels is covered in detail in additional chapters in this volume. In the present chapter, we focus our attention on the basolateral, Ca^{2+}-activated K^+ channel activated by Ca^{2+}-mediated agonists, including Ach.

Early electrophysiological evidence for the role of an increased basolateral K^+ conductance during stimulation came from cAMP-mediated agonists, rather than

Ca^{2+}-mediated agonists, using canine trachea (Welsh et al. 1982) and later spiny dogfish rectal gland (Greger and Schlatter 1984a). In canine trachea, intracellular microelectrode measurements demonstrated that epinephrine caused an initial depolarization of membrane potential, due to Cl^- exit across the apical membrane, followed approximately 20 s later by a repolarization that was associated with a decrease in basolateral membrane resistance due to the activation of a K^+ conductance (Welsh et al. 1982). Later studies demonstrated that this agonist-induced Cl^- secretory response could be blocked by either raising serosal K^+ or adding the nonspecific K^+ channel blocker, Ba^{2+} (Greger and Schlatter 1984b; Smith and Frizzell 1984), both of which will depolarize the basolateral membrane and, by electrical coupling, the apical membrane, thereby decreasing the electrochemical driving force for Cl^- exit across the apical membrane.

20.1.2 Basolateral Membrane Ca^{2+}-Activated K^+ Channels

The first direct demonstration of a basolateral Ca^{2+}-activated K^+ conductance in epithelial cells was made by Maruyama et al. (1983a) in salivary acinar cells. The relatively easy access of the basolateral membrane of acinar cells to patch-clamp electrodes allowed others to quickly follow up on these observations by identifying additional Ca^{2+}-activated K^+ channels. Petersen and colleagues (Maruyama et al. 1983b; Maruyama and Petersen 1984) identified a 200 pS K^+ channel activated by Ach and cholecystokinin in pig pancreatic acinar cells and later identified a 50 pS channel in the basolateral membrane. Early on, Ca^{2+}-activated K^+ channels were also identified in lacrimal (Trautmann and Marty 1984) and parotid (Foskett et al. 1989) acinar cells. In contrast to acinar cells, basolateral membrane Ca^{2+}-activated K^+ channels were only identified later in intestinal cells. Chang and Dawson (1988) identified a Ca^{2+}-activated K^+ conductance in permeabilized turtle colon, whereas Sepulveda and Mason (1985) characterized 36 pS and 90 pS Ca^{2+}-activated K^+ conductances in rabbit enterocytes, while Morris et al. (1986) identified a 250 pS K^+ channel in rat enterocytes. However, these initial observations were all made on absorptive cells of the intestine rather than the secretory crypts. Subsequently, Loo and Kaunitz (1989) patch-clamped the basolateral membrane of crypt cells from rabbit distal colon and identified a 130 pS K^+ channel that was activated by both Ca^{2+} and cAMP. Further, Burckhardt and Gogelein (1992) identified a 12 pS K^+ channel in rat distal colonic crypts. In 1993, we identified a Ca^{2+}-activated K^+ channel in the T84 colonic cell line that was activated by Ach and taurodeoxycholate (Devor and Frizzell 1993; Devor et al. 1993). We had previously shown that bile acids increase intracellular Ca^{2+} via an IP_3-mediated mechanism, such that this activation of KCa3.1 was identical to the Ach-mediated activation of the channel (Devor et al. 1993). Further, this channel was inwardly rectified, with chord conductances of 10 pS at +100 mV and 35 pS at −100 mV in symmetric K^+.

This rectification was independent of voltage and the channel was blocked by extracellular charybdotoxin (CTX) in a voltage-dependent manner (Devor and Frizzell 1993). This colonic Ca^{2+}-activated K^+ channel appeared to be identical to that described by Welsh and colleagues in tracheal epithelia (Welsh and McCann 1985; McCann et al. 1990) in which they characterized a channel with a slope conductance of 19 pS at 0 mV that was activated by intracellular Ca^{2+} and blocked by extracellular CTX. To the best of our knowledge, these studies represented the first identification of the Ca^{2+}-activated K^+ channel in epithelia that would eventually be realized to be the KCNN4 gene product, KCa3.1 (syn, IK1, SK4, KCa4).

20.2 Cloning of KCa3.1

As is apparent from the above, there are multiple forms of Ca^{2+}-activated K^+ channels. These have historically been grouped into (a) small conductance, SK channels, with single channel conductances of ~5–20 pS, many of which are sensitive to block by the bee venom, apamin; (b) the intermediate conductance, IK channels, having single channel conductances of ~20–80 pS and which are blocked by clotrimazole and charybdotoxin; and (c) the large conductance, maxi-K channels (also called BK), having single channel conductances of ~100–250 pS that are blocked by charybdotoxin, iberiotoxin, and paxilline. The maxi-K channels (KCa1.1) have been discussed in Chap. 21 and will not be discussed herein. In 1996, Kohler et al. described the cloning of the SK family of channels, including SK1, SK2, and SK3, now referred to as KCa2.1, KCa2.2, and KCa2.3, respectively. These channels were predicted to have a classic 6-transmembrane domain architecture with both the N- and C-termini being cytosolic and the pore region being between the fifth and sixth transmembrane domains (S5 and S6), similar to the voltage-gated K^+ channels (K_v), although these channels reside on distinct evolutionary branches. Kohler et al. (1996) demonstrated that these SK channels were Ca^{2+} dependent, being half-maximally activated at ~600–700 nM intracellular Ca^{2+} with a very steep dependence on Ca^{2+}, having Hill coefficients of between 4 and 5. These authors further demonstrated the SK channels were voltage independent and that SK2 and SK3 were apamin sensitive, whereas SK1 was insensitive to apamin block. In total, these characteristics were completely consistent with these channels being responsible for the wide array of physiological events attributed to SK channels previously. As KCa2.x (SK) channels have rarely been reported to be expressed in epithelia as of the writing of this chapter, these channels will not be discussed in any detail, but rather we will contrast some of their attributes with KCa3.1 where appropriate. Interested readers are referred to Adelman et al. (2012) for further details about KCa2.x channels.

Based on the SK channel clones, Joiner et al. (1997) and Ishii et al. (1997) screened the expressed sequence tag database and identified a fourth family member.

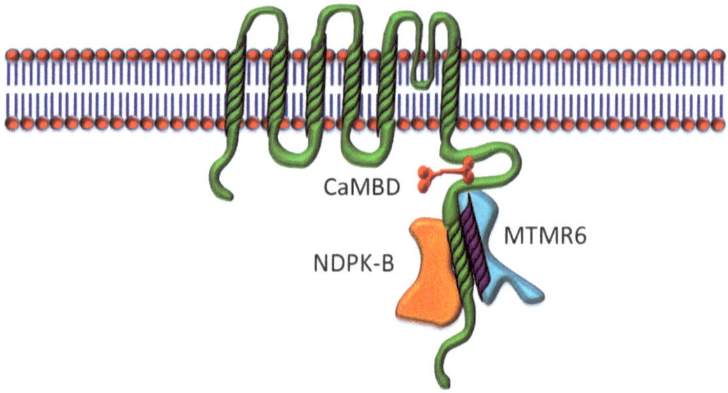

Fig. 20.2 Schematic of KCa3.1 consisting of six transmembrane domains (TMD) with both the N- and C-termini within the intracellular compartment. The pore loop region of the channel lies between TMDs S5 and S6. Additionally, regulatory subunits which have been demonstrated to have a role in the modulation of KCa3.1 include the calmodulin-binding domain (CaMBD), myotubularin 6 (MTMR6, a lipid (PI(3)P) phosphatase), and nucleoside diphosphate kinase B (NDPK-B, a histidine kinase). See text for further details

Joiner et al. (1997) referred to this clone as SK4, based on its homology to the rest of the SK family members, whereas Ishii et al. (1997) dubbed this channel IK1 based on this clone having both conductance and blocker characteristics historically associated with IK channels. Logsdon et al. (1997) also cloned this channel from T lymphocytes and referred to it as KCa4. Based on the standard nomenclature derived from the International Union of Pharmacology (IUPHAR; Gutman et al. 2003), this channel is now referred to as KCa3.1. Full-length KCa3.1 is 427 amino acids in length with an architecture identical to that for the SK and K_v channels (Fig. 20.2). Indeed, KCa3.1 is ~40 % identical to the KCa2.x channels, with the highest degree of homology being in the transmembrane domains, pore region, and the proximal C-terminus (Joiner et al. 1997; Ishii et al. 1997). In contrast, the cytosolic N-terminus and distal C-terminus exhibit virtually no homology within the gene family, and numerous unique regulatory and trafficking motifs have been identified in this region of KCa3.1, as detailed below. KCa3.1 was shown to activate in response to increasing intracellular Ca^{2+}, with a half-maximal concentration reported anywhere between 100 and 600 nM and a Hill coefficient of ~2 (Joiner et al. 1997; Ishii et al. 1997; Gerlach et al. 2000; Bailey et al. 2010). Note that these rather disparate $K_{0.5}$ values for Ca^{2+} are likely associated with differences in posttranslational modifications (see below). At the single channel level, KCa3.1 was shown to exhibit slight inward rectification in symmetric K^+, having chord conductances of ~10 pS at +100 mV and ~35 pS at −100 mV (Fig. 20.3), similar to what had previously been reported for IK channels in epithelia (McCann et al. 1990; Devor and Frizzell 1993). Further, in these initial reports, it was demonstrated that KCa3.1 was blocked

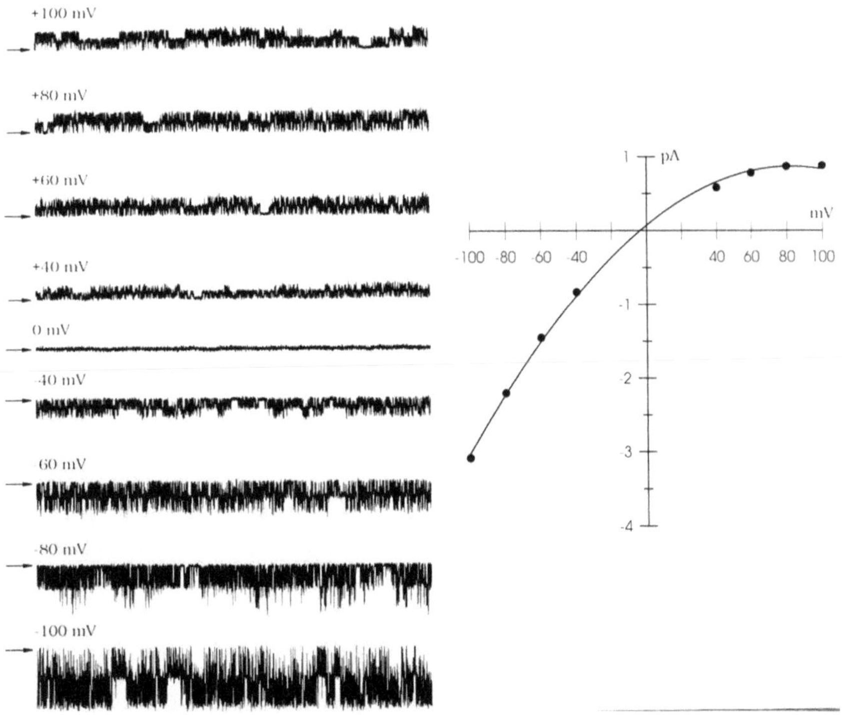

Fig. 20.3 Single channel recording and current-voltage relationship of KCa3.1. *Left panel*: single channel recordings of KCa3.1 from an inside-out patch recording made at the indicated voltages (referenced to the inside surface of the membrane patch) in symmetric 150 mM K^+-gluconate solutions. The bath Ca^{2+} concentration was 400 nM free Ca^{2+}. *Arrows* indicate the closed state of the channel. *Right panel*: average current–voltage relationship of KCa3.1 recorded in symmetric 150 mM K^+-gluconate ($n = 4$). From Devor et al. (1999)

by clotrimazole and charybdotoxin (Fig. 20.4), known blockers of IK channels in epithelia (McCann et al. 1990; Devor and Frizzell 1993; Rufo et al. 1996; Devor et al. 1997). Importantly, transcripts for KCa3.1 were found in numerous epithelial containing tissues, including the placenta, lung, trachea, salivary gland, kidney, pancreas, colon, bladder, stomach, and prostate (Joiner et al. 1997; Ishii et al. 1997; Logsdon et al. 1997; Jensen et al. 1998). Warth et al. (1999) and Gerlach et al. (2000) subsequently identified KCa3.1 by RT-PCR and Northern blot, respectively, in the T84 cell line, a human colonic crypt cell model. Gerlach et al. (2000) also identified KCa3.1 in Calu-3 cells, which are a model for serous cells from human airway. These results clearly implicated KCa3.1 as being the previously characterized IK1 channel in airway and colonic epithelia, although additional studies would be required to confirm this speculation.

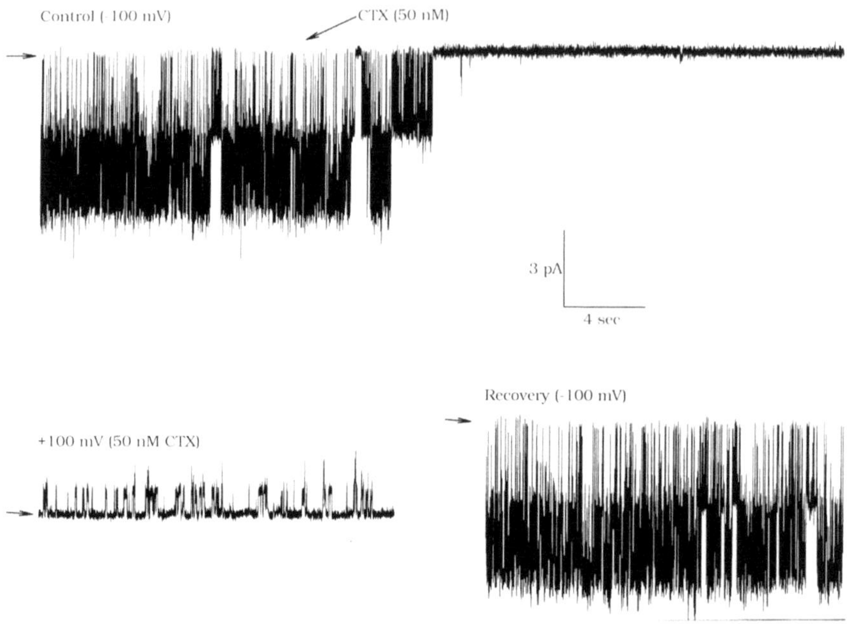

Fig. 20.4 Inhibition of KCa3.1 by charybdotoxin (CTX). *Upper panel*: addition of CTX (50 nM) to the extracellular surface of KCa3.1 in an excised outside-out patch held at −100 mV resulted in complete inhibition of the channel within 5 s. *Lower left panel*: the inhibitory effect of CTX was reversed at +100 mV due to knockoff of the CTX by K+ flowing from the inside of the channel to the outside. *Lower right panel*: the inhibition of KCa3.1 was totally reversed upon washout of the CTX (−100 mV). Modified from Devor et al. (1999)

20.3 Role of Basolateral KCa3.1 in Transepithelial Ion Transport

As is clear from the above, Ca^{2+}-dependent K^+ channels are required in the basolateral membrane to maintain the electrochemical driving force for transepithelial Cl^- secretion during Ca^{2+}-mediated agonist secretion. Below, we will detail the experimental results that have led to our understanding of the role of KCa3.1 in the basolateral membrane of colonic, airway, salivary acinar, and pancreatic duct epithelia.

20.3.1 KCa3.1 in the Basolateral Membrane of Intestinal Epithelium

An important breakthrough in the study of Ca^{2+}-mediated transepithelial Cl^- secretion came from the work of Dharmsathaphorn and colleagues when they characterized the T84 cell line as a colonic crypt model system which formed high-resistance monolayers capable of vectorial electrolyte transport (Dharmsathaphorn et al. 1984, 1985, 1989; Weymer et al. 1985; Dharmsathaphorn and Pandol 1986; Mandell et al. 1986; Madara et al. 1987; Wasserman et al. 1988). Using $^{86}Rb^+$ efflux measurements (note that Rb^+ is often used as a surrogate for K^+ in these studies), Dharmsathaphorn and Pandol (1986) demonstrated that both the cAMP-mediated agonists, vasoactive intestinal peptide (VIP) and prostaglandin E_1, as well as the Ca^{2+}-mediated agonist, carbachol, increased K^+ efflux across the basolateral membrane, while having no effect on K^+ efflux across the apical membrane, in T84 cells. These authors further demonstrated that Ba^{2+} blocked the VIP-stimulated basolateral K^+ efflux pathway, whereas the carbachol-dependent pathway was insensitive to Ba^{2+} block. Finally, these two K^+ efflux pathways were additive in nature, indicative of them being unique conductances. Duffey and colleagues (Devor et al. 1990, 1991; Devor and Duffey 1992) characterized this carbachol-induced K^+ conductance using the whole-cell patch-clamp technique, demonstrating that it was activated by the increase in intracellular Ca^{2+} induced by carbachol and insensitive to block by Ba^{2+} as predicted based on the $^{86}Rb^+$ efflux experiments of Dharmsathaphorn and Pandol (1986). Further studies by Devor and Frizzell (1993), using on-cell and excised patch-clamp techniques, defined the basolateral membrane, carbachol-activated K^+ channel, in T84 cells as being Ca^{2+} dependent, inwardly rectifying and blocked by extracellular charybdotoxin.

Further evidence for the role of this channel in transepithelial Cl^- secretion came when Devor et al. (1996a, b) characterized the first known pharmacological opener of this channel, 1-ethyl-2-benzimidazolinone (1-EBIO). 1-EBIO was subsequently shown to activate KCa3.1 by numerous labs (Devor et al. 1996a, b; Pedersen et al. 1999; Syme et al. 2000; Jensen et al. 2001), as detailed in Chap. 26; thus the pharmacology of these modulators will not be discussed herein. Critically, 1-EBIO induced a sustained transepithelial Cl^- secretory response in T84 cells that was blocked by the basolateral addition of charybdotoxin, with an inhibitory constant of 3.6 nM, indicative of IK (KCa3.1) being the basolateral K^+ channel activated. In contrast, the chromanol, 293B (Lohrmann et al. 1995), an inhibitor of the cAMP-dependent K^+ channel (now known to be KCNQ1), had no effect on this 1-EBIO-induced current. These results, as illustrated in Fig. 20.5, clearly demonstrated, using selective K^+ channel blockers, that the Ca^{2+}- and cAMP-mediated basolateral K^+ conductances were distinct, as initially proposed based upon their Ba^{2+} sensitivity. Similar results were observed in rat colonic epithelium, consistent with this IK channel playing a key role in native ex vivo tissue (Devor et al. 1996a, b). Another critical piece of evidence for these IK channels being the basolateral K^+ channel activated by Ca^{2+}-dependent agonists came when Rufo et al. (1996) demonstrated that clotrimazole inhibited transepithelial Cl^- secretion across T84 cells via an inhibition of $^{86}Rb^+$ efflux across the basolateral membrane with no effect on other transporters critical to this Cl^- secretory process. These authors further demonstrated that clotrimazole inhibited the basolateral K^+ conductance in T84 cells using nystatin-permeabilized monolayers (Rufo et al. 1997). Finally, Devor et al. (1997) demonstrated a direct

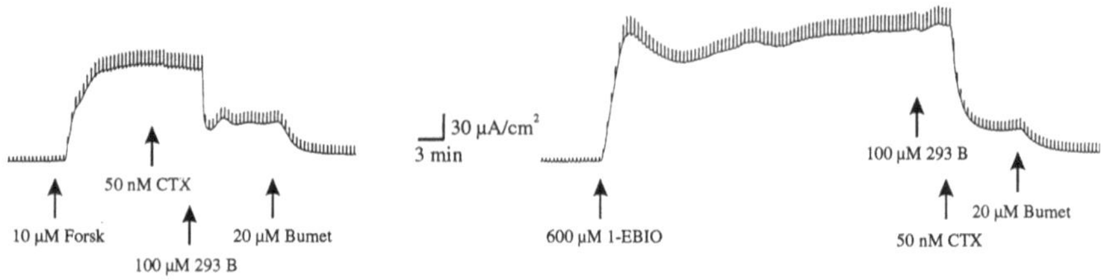

Fig. 20.5 The effects of activation and inhibition of endogenous KCa3.1 in T84 cell monolayers as measured with the short-circuit current (I_{sc}) technique using Ussing chambers. *Left panel*: Forskolin (10 μM, a cAMP-dependent agonist) stimulated a sustained I_{sc} that was not inhibited by CTX (50 nM); however, the stimulated I_{sc} was partially blocked by 293B (100 μM, serosal addition, an inhibitor of cAMP-dependent K^+ channels). Bumetanide (20 μM, serosal addition, an inhibitor of the Na^+-K^+-$2Cl^-$ cotransporter) blocked the remainder of the stimulated I_{sc}. *Right panel*: 1-EBIO (600 μM, a KCa3.1 agonist) activated a prolonged I_{sc} that was not inhibited by 293B, but was partially inhibited by CTX and further blocked by bumetanide. Taken together, these results demonstrate that cAMP- and Ca^{2+}-mediated agonists activate basolateral K^+ channels with unique pharmacologies. See text for additional details. From Devor et al. (1996a)

effect of clotrimazole on IK channels in excised patch-clamp experiments from T84 cells and further demonstrated that clotrimazole inhibited the 1-EBIO-induced Cl^- secretory response. Subsequently, Hamilton et al. (1999) demonstrated that 1-EBIO stimulated transepithelial Cl^- secretion across mouse jejunum and further demonstrated that 1-EBIO directly activated IK channels in the basolateral membrane of isolated jejunal crypts using the on-cell patch-clamp technique. Similarly, Warth et al. (1999) demonstrated in rat and rabbit colonic mucosa that 1-EBIO stimulated clotrimazole-sensitive Cl^- secretion as well as activated IK channels in excised patches from the basolateral membrane of rat colonic crypts. These authors also demonstrated that antisense probes to KCa3.1 attenuated the carbachol-induced Cl^- secretory current across T84 cells, confirming the critical role of this channel in Ca^{2+}-mediated Cl^- secretion in these cells (Warth et al. 1999). In an additional series of studies, Cuthbert and colleagues (Cuthbert et al. 1999; Cuthbert 2001; MacVinish et al. 2001) demonstrated that 1-EBIO stimulated Cl^- secretion across mouse colon and that this was dependent upon a charybdotoxin-sensitive K^+ conductance in the basolateral membrane. As noted above, subsequent to the cloning of KCa3.1 and the realization that this was the IK channel previously reported, Warth et al. (1999) and Gerlach et al. (2000) confirmed by RT-PCR and Northern blot, respectively, that this was the channel expressed in T84 cells. Finally, Flores et al. (2007) carried out studies on KCa3.1 knockout mice and demonstrated that Ca^{2+}-mediated Cl^- secretion was completely eliminated in both distal colon and small intestinal epithelium which resulted in a marked reduction in water content in the stools. In total, these studies unequivocally demonstrate that KCa3.1 is the basolateral membrane K^+ channel activated by Ca^{2+}-mediated agonists as a means of maintaining transepithelial Cl^- secretion across intestinal epithelia.

20.3.2 KCa3.1 in the Basolateral Membrane of Airway Epithelium

Similar to colonic epithelia, early microelectrode data demonstrated that cAMP-mediated agonists activate a basolateral K^+ conductance necessary for the maintenance of transepithelial Cl^- secretion (Welsh et al. 1982) and that Ba^{2+} applied to the basolateral side blocked this conductance, resulting in a diminution of Cl^- secretion (Greger and Schlatter 1984b; Smith and Frizzell 1984). Welsh and McCann (1985) first characterized the Ca^{2+}-activated K^+ channel in primary cultures of canine tracheal epithelial cells, demonstrating that it was activated by epinephrine, which increases Ca^{2+} in tracheal epithelia. These authors further demonstrated that increasing Ca^{2+} induced $^{86}Rb^+$ efflux from subconfluent canine tracheal epithelial cells that was blocked by charybdotoxin. Additional patch-clamp studies showed that the K^+ channel was blocked by charybdotoxin (McCann et al. 1990). Finally, these authors (McCann and Welsh 1990) confirmed these studies on primary cultures of canine tracheal epithelial cells grown on semipermeable supports, thereby demonstrating that this IK channel (called KCLIC by these authors) was the basolateral membrane Ca^{2+}-activated K^+ channel in airway epithelia.

With the identification of 1-EBIO as a direct activator of IK channels, Devor et al. (1996a, b) also demonstrated that this compound stimulated transepithelial Cl^- secretion across murine tracheal epithelium. Subsequent studies in both the serous cell model cell line, Calu-3 (Devor et al. 1999), and primary cultures of human bronchial epithelia (Devor et al. 2000) demonstrated that 1-EBIO and the higher affinity analogue, DCEBIO (5,6-dichloro-1-ethyl-1,3-dihydro-2H-benzimidazol-2-one) (Singh et al. 2001), stimulated transepithelial Cl^- secretion that was blocked by basolateral addition of either charybdotoxin or clotrimazole. Additional patch-clamp (Devor et al. 1999) and Northern blot analysis (Gerlach et al. 2000) confirmed expression of KCa3.1 in Calu-3 cells. Subsequent studies by Cowley and Linsdell (2002) and Mall et al. (2003) confirmed expression of KCa3.1 in Calu-3 cells and native human nasal epithelia from normal and cystic fibrosis (CF) patient samples by RT-PCR. Similar studies in the 16HBE14o- (Bernard et al. 2003) and H441 (Wilson et al. 2006) human airway cell lines confirmed expression of KCa3.1 by RT-PCR and that this channel was activated by 1-EBIO resulting in sustained transepithelial Cl^- secretion that was blocked by clotrimazole. In total, these studies clearly demonstrate expression of KCa3.1 in the basolateral membrane of airway epithelia and that activation of this channel is required for Ca^{2+}-mediated Cl^- secretion.

Interestingly, Smith and Welsh (1992) demonstrated that increasing cAMP stimulated HCO_3^- secretion across normal airway epithelia but not in airway from CF airway epithelia, while Ashton et al. (1991) also suggested that pancreatic duct epithelial cells could be differentially stimulated to secrete either Cl^- or HCO_3^-. In this regard, it has been shown that the cystic fibrosis transmembrane conductance regulator (CFTR) conducts both Cl^- and HCO_3^- (Gray et al. 1990; Linsdell et al. 1997; Ishiguro et al. 2009). Further, Devor et al. (1999) found that activation of KCa3.1 acted as an “ionic switch,” dictating whether the human airway serous cell line, Calu-3, secreted HCO_3^- or Cl^-. That is, stimulation of Calu-3 cells with forskolin, to increase cAMP, resulted in a HCO_3^- secretory response as assessed by both the lack of bumetanide sensitivity and isotopic flux experiments. In contrast, stimulating Calu-3 cells with the KCa3.1 opener, 1-EBIO, resulted in Cl^- secretion. Finally, stimulation with both forskolin and 1-EBIO resulted in a primarily Cl^- secretory response with a decrease in HCO_3^- secretion relative to forskolin alone. Thus, these results suggest that activation of basolateral KCa3.1 and the concomitant basolateral hyperpolarization dictate the anion transport properties of the epithelium.

Finally, Devor et al. (2000) showed that activation of basolateral KCa3.1 also increased the amiloride-sensitive transepithelial Na^+ absorption (Fig. 1) across primary cultures of human bronchial epithelia (HBE) expressing ΔF508 CFTR. These results indicate that the KCa3.1-induced hyperpolarization was sufficient to drive Na^+ entry though the epithelial Na^+ channel (ENaC) in the apical membrane of HBEs. Gao et al. (2001) confirmed that 1-EBIO increased transepithelial Na^+ transport in monolayers of the ΔF508 CFTR-expressing human airway cell line, CFT1. These authors further demonstrated that 1-EBIO did not activate apical ENaC in basolateral membrane amphotericin B-permeabilized CFT1 monolayers or in *Xenopus* oocytes expressing ENaC, confirming that activation of basolateral KCa3.1 results in an increased transepithelial Na^+ absorption. In contrast to these studies, Devor et al. (1996a) demonstrated that 1-EBIO did not influence Na^+ transport in either rat colonic or murine tracheal epithelia. These apparently disparate results are likely the result of the unique physiologies of human airway versus mouse airway and rat colon such that the rate-limiting conductance for Na^+ absorption is different in these cell types.

20.3.3 KCa3.1 in the Basolateral Membrane of Salivary Acinar and Pancreatic Duct Epithelium

Salivary acinar cells secrete electrolytes and fluid by mechanisms similar to those described above and recently reviewed by Catalan et al. (2014). Importantly, both maxi-K (KCa1.1) and KCa3.1 channels have been molecularly described in human and rodent salivary gland acinar cells (Nehrke et al. 2003; Begenisich et al. 2004). Surprisingly, genetic knockout of either KCa1.1 or KCa3.1 alone in mice had no effect on salivary gland fluid secretion, indicating that expression of either of these channels is sufficient to maintain fluid secretion (Begenisich et al. 2004; Romanenko et al. 2006). This result was explained by the observation that knockout of either of these channels did not affect the ability of the acinar cells to hyperpolarize toward the K^+ reversal potential (E_K). In contrast, mice lacking both KCa1.1 and KCa3.1 channels exhibited a severely reduced fluid secretion rate in response to Ca^{2+}-mediated agonists, and this was paralleled by an inhibition of the membrane hyperpolarization toward E_K (Romanenko et al. 2006). These results clearly demonstrate a role for KCa3.1 in salivary acinar cell physiology but also demonstrate that these cells can bypass deficits in KCa3.1 function by recruiting KCa1.1 channels.

Finally, Nguyen and Moody (1998) demonstrated that an 1-EBIO activated and charybdotoxin and clotrimazole inhibited IK channel in primary cultures of dog pancreatic duct epithelial cells that were localized to the basolateral membrane based on both $^{86}Rb^+$ efflux studies and blocker sidedness. Similar to the above, this channel has now been positively identified as KCa3.1 in rodent and human pancreatic duct cell lines (Capan-1, PANC-1, CFPAC-1) as well as in ducts isolated from rodent pancreas (Hede et al. 2005; Hayashi et al. 2012; Wang et al. 2013). In these cells, DCEBIO stimulated, and clotrimazole inhibited, anion secretion, although it was not determined whether this was HCO_3^- or Cl^- secretion. As Ashton et al. (1991) have suggested that pancreatic duct epithelial cells can be differentially stimulated to secrete either Cl^- or HCO_3^-, it will be important to determine whether activation of KCa3.1 regulates this process as was described in human airway (Devor et al. 2000). The study of the regulation and function of KCa3.1 in the pancreatic ducts is still in its early stages and the next few years are likely to shed additional light on this important subject.

20.4 Role of KCa3.1 in the Apical Membrane

In addition to K^+ channels playing a pivotal role in the basolateral membrane to modulate transepithelial Cl^- secretion, it is also well known that epithelia of the airway, pancreas, kidney, parotid gland, and intestine secrete K^+ across the apical membrane, as detailed in Chap. 3. While the K^+ channels that have been historically shown to be involved in this K^+ secretory process include the maxi-K (KCa1.1, KCNMA1, Chap. 21) and ROMK (Kir1.1, KCNJ1, Chap. 19) channels, evidence has accumulated to implicate KCa3.1 in this process as well. Indeed, Bernard et al. (1993) demonstrated that the KCa3.1 blocker, clotrimazole (Rufo et al. 1996; Devor et al. 1997; Wulff and Castle 2010), inhibited $^{86}Rb^+$ efflux across both the basolateral and apical membranes of the human bronchial cell line, 16HBE14o-. Further, 1-EBIO stimulated $^{86}Rb^+$ efflux across both basolateral and apical membranes in a clotrimazole-dependent manner, implicating a role for KCa3.1 in the apical membrane of airway epithelia (Bernard et al. 1993). Similarly, Joiner et al. (2003) demonstrated that the apical addition of clotrimazole completely inhibited the carbachol-induced $^{86}Rb^+$ secretion across rat proximal colon. These authors also isolated apical versus basolateral membrane and demonstrated that KCa3.1 was expressed in both membrane fractions by immunoblot. Further, it was demonstrated that rats deprived of K^+ in their diet exhibited a loss of KCa3.1 mRNA that correlated with $^{86}Rb^+$ secretion being abolished (Joiner et al. 2003). At the same time, Neylon and colleagues (Furness et al. 2003) demonstrated, by immunofluorescent localization, that KCa3.1 was expressed in both apical and basolateral membranes of rat intestine. Finally, Almassy et al. (2012) demonstrated expression of KCa3.1 in the apical membrane of parotid acinar cells. In these experiments, caged Ca^{2+} was released in close proximity to the apical membrane while simultaneously recording whole-cell K^+ currents. These authors demonstrated that the release of Ca^{2+} exclusively adjacent to the apical membrane resulted in the activation of K^+ currents that were partially blocked by both paxilline and TRAM-34, inhibitors of maxi-K and KCa3.1 channels, respectively. Further, in parotid acinar cells from maxi-K null mice, apical TRAM-34-sensitive KCa3.1 currents were still observed following release of Ca^{2+} at the apical membrane, providing compelling evidence for KCa3.1 in the apical membrane of parotic acinar cells. Taken together, these results implicate KCa3.1 in K^+ secretion across the apical membrane of various epithelial tissues.

While there is clear data supporting a role for KCa3.1 in both the basolateral and apical membranes of secretory epithelia, these data beg the question of how this channel is targeted to both apical and basolateral membranes. The answer to this question began to come into focus in 2010 when Rajendran and colleagues (Barmeyer et al. 2010) identified a splice variant of KCa3.1, KCNN4c from rat colon that lacked exon 2 which encodes 29 amino acids, including 5 amino acids in the S1–S2 extracellular linker as well as all of S2 and 2 amino acids in the proposed intracellular S2–S3 linker. As expected, deletion of S2 resulted in a channel that failed to traffic to the plasma membrane when heterologously expressed. However, when KCNN4c was co-expressed with the maxi-K β1-subunit, the channel was expressed at the membrane in *Xenopus* oocytes (see Fig. 20.2). KCNN4c was confirmed to form a functional channel by measuring $^{86}Rb^+$ efflux from oocytes that was activated by the KCa3.1 channel opener, DCEBIO (Singh et al. 2001), and inhibited by the KCa3.1 blocker, TRAM-34 (Wulff et al. 2000), although the sensitivity of TRAM-34 block was significantly reduced in the KCNN4c channel, being 0.6 μM for full-length KCa3.1 and 7.8 μM for KCNN4c plus β1-subunit (Barmeyer et al. 2010). Further studies demonstrated that the KCNN4c/β1 channels had single channel properties that are similar to full-length KCa3.1, i.e., they display inward rectification and have a chord conductance of 31 pS in symmetric K^+ (Basalingappa et al. 2011). Interestingly, these authors also demonstrated that KCNN4c did not co-assemble and, therefore, inhibit expression of the full-length KCa3.1 channel (Barmeyer et al. 2010). Importantly, immunoblots of apical versus basolateral membranes demonstrated an ~37 kDa band in the apical membrane lysates, while an ~40 kDa band was detected in the basolateral membrane lysates, suggesting that this lower molecular mass product corresponds to KCNN4c being expressed in the apical membrane of rat colonic epithelia (Barmeyer et al. 2010).

Further studies by Singh et al. (2012) demonstrated that a Na^+-free diet, which increases circulating aldosterone levels, increases KCNN4c mRNA and protein expression in rat distal colon. These authors also demonstrated that in the presence of aldosterone, a TRAM-34-dependent K^+ secretory pathway was revealed, consistent with an increase in apical membrane expression of KCNN4c under these conditions. More recent studies have identified mineralocorticoid response elements in KCNN4 introns that respond to aldosterone (O'Hara et al. 2014). In an exciting series of studies, Hoque and colleagues demonstrated that binding of cAMP to exchange protein directly activated by cAMP (Epac) activates a PKA-independent, Ca^{2+}-dependent Cl^- secretory process in T84 cells that involves stimulation of Rap2-phospholipase Cε (Hoque et al. 2010). More recently, these authors demonstrated that Epac1 regulates the apical surface expression of KCNN4c in T84 cells (Sheikh et al. 2013). That is, knockdown of Epac1 both significantly decreased the TRAM-34-dependent apical potassium conductance (G_K) and also resulted in the redistribution of KCNN4c into subapical vesicles. Furthermore, inhibitors of both RhoA (GGT1298) and Rho-associated kinase (ROCK; H1152) reduced apical G_K, and ROCK inhibition also caused KCNN4c to redistribute out of the apical membrane (Sheikh et al. 2013). To our knowledge, these studies represent the first demonstration of a dynamic redistribution of KCNN4c into and out of the apical membrane of polarized epithelia.

20.5 Gating of KCa3.1

The gating of an ion channel can be simply thought of as the process by which the pore transitions between nonconducting (closed) and conducting (open) states. As outlined above, KCa3.1 gating is dictated by changes in the intracellular Ca^{2+} concentration. Calmodulin is constitutively bound to the calmodulin-binding domain (CaMBD) of KCa3.1 (see Fig. 20.2), located in the C-terminus just distal to S6 (Fanger et al. 1999; Schumacher et al. 2001, 2004). When Ca^{2+} binds to calmodulin, this results in a conformational change in calmodulin and, therefore, the CaMBD, allowing the channel to transition from the closed to the open conformation (Keen et al. 1999; Li et al. 2009). The mechanism by which Ca^{2+} binding is translated to KCa3.1 gating has received considerable attention, as detailed below.

20.5.1 A Gating Model for KCa3.1

The first attempt to define the open-closed gating kinetics of KCa3.1 was carried out by Grygorczyk and Schwarz (1985) using the single channel patch-clamp technique on human red blood cells. These authors identified a single open state, as assessed by fitting an open-time (τ_o) distribution plot to a single exponential, as well as two closed states. However, the authors speculated an additional long closed time (τ_c) that was >60 ms in duration that could not be resolved during their recordings to fully account for their entire data set. The τ_o was shown to increase with increasing Ca^{2+}, whereas the two τ_c did not significantly depend on Ca^{2+}, although the mean closed time was decreased by increasing Ca^{2+}, indicative of the long, unresolved τ_c being reduced in duration. In 1992, Leinders et al. (1992) again evaluated the gating kinetics of KCa3.1 in human erythrocytes. These authors found that both the τ_o and τ_c distributions were fit to two exponentials, indicating two open as well as two closed states. However, these authors similarly proposed a long-lived third τ_c to account for all of their data. While Leinders et al. (1992) found no dependence of τ_o on Ca^{2+}, distinct from Grygorczyk and Schwarz (1985), they proposed that the long-lived third τ_c was dependent on Ca^{2+}, similar to Grygorczyk and Schwarz (1985). Dunn (1998) conducted a gating study on KCa3.1 in human erythrocytes and identified two τ_o and three τ_c, indicative of two open and three closed gating steps. Indeed, this third τ_c averaged 155 ms in agreement with previous reports. However, Dunn did not evaluate the effect of Ca^{2+} on these gating kinetics. Syme et al. (2000) studied KCa3.1 heterologously expressed in HEK cells and also identified two τ_o and three τ_c. However, based on an analysis of the relative amplitudes of each closed component, the data they could resolve would have accounted for a P_o of 0.3, when the single channel patches used to obtain the data had an actual P_o of 0.18. Thus, similar to the authors above, Syme et al. (2000) concluded that a longer fourth τ_c of ~200 ms, which could not be resolved, must be present. Again, these authors did not evaluate the effect of changing Ca^{2+} on these gating parameters.

A significant advancement in our understanding of the gating of the KCNN gene family members came in 1998 when Hirschberg et al. (1998) studied the single channel gating kinetics of KCa2.2 heterologously expressed in *Xenopus* oocytes. An analysis of the single channel behavior revealed two τ_o and three τ_c. Further, it was demonstrated that neither of the τ_o nor the shorter τ_c were Ca^{2+} dependent. Rather, it was the long τ_c that was shown to decrease with increasing concentrations of Ca^{2+}, similar to what was described by Grygorczyk and Schwarz (1985) for KCa3.1. When Hirschberg et al. (1998) attempted to mathematically model KCa2.2 channel gating kinetics, they found that a model with two τ_o and three τ_c would only yield a reasonable fit to their experimental data if they assumed that the derived rate constants had a nonlinear dependence on Ca^{2+} concentration. To obtain the desired linear dependence on Ca^{2+} concentration, these authors required a fourth τ_c that was not experimentally resolved, similar to what had been described for KCa3.1, to accurately model their data. The six-state open-closed gating model for KCa2.2 is shown in Fig. 20.6a with the forward transitions between closed states being Ca^{2+} dependent.

Bailey et al. (2010) undertook a study to model the Ca^{2+} dependence of KCa3.1 gating behavior based upon the model of Hirschberg et al. (1998) described above. In contrast to previous studies on single channel patches, Bailey et al. (2010) utilized macro-patches containing hundreds or thousands of channels heterologously expressed in HEK cells and carried out rapid step changes in Ca^{2+} (~5 ms) from 0.4 to 10 μM, after which the macroscopic currents were fit to a series of exponentials and gating kinetics modeled. Consistent with other reports, the activation kinetics of KCa3.1 were found to be Ca^{2+} dependent, whereas the deactivation kinetics were Ca^{2+} independent. Using the six-state model of Hirschberg et al. (1998), Bailey et al. (2010) obtained an accurate description of their entire data set except for the activation rate at saturating Ca^{2+} (Fig. 20.6c). Based on this, these authors assumed that the Ca^{2+}-dependent steps in KCa3.1

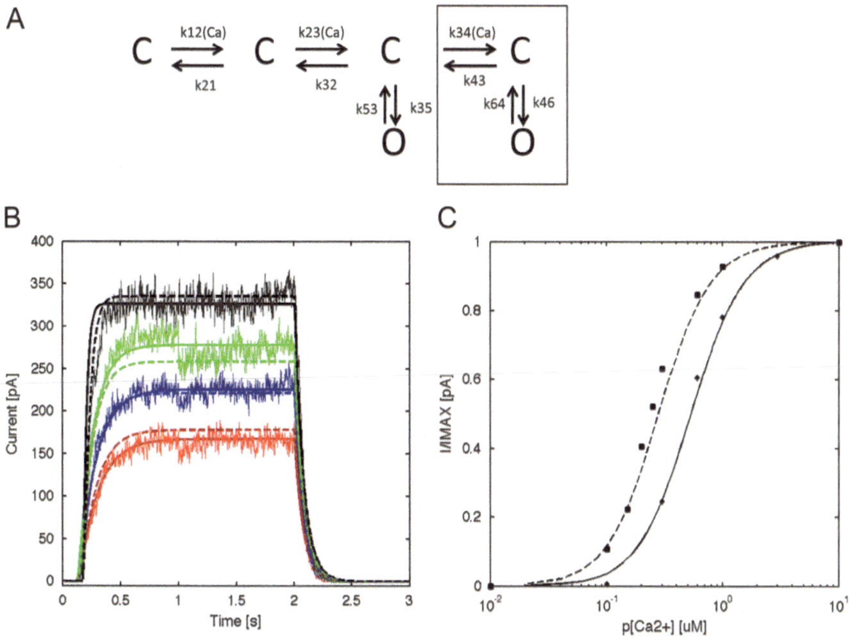

Fig. 20.6 Kinetic model for KCa3.1 gating. (**a**) The gating scheme to describe the activation and deactivation kinetics of KCa2.3 required a six-state kinetic model with four closed states and two open states, with the forward transitions between closed states dependent upon Ca^{2+} and the other states Ca^{2+} independent (Hirschberg et al. 1998). The *box* indicates kinetic states that are not required to fit the gating kinetics of KCa3.1 as determined by sensitivity analysis (not shown), such that the gating of KCa3.1 was described by three closed and one open state. (**b**) Representative records for activation and deactivation with two variations of the model shown in the *upper panel*. The *colors* refer to varying concentrations of Ca^{2+}; *red* = 0.5, *blue* = 0.7, *green* = 1.0, and *black* = 10 μM Ca^{2+}. The *dashed line* indicates the fitting assuming Ca^{2+}-dependent rate constants have a nonlinear dependence on Ca^{2+} concentration, $k = A \cdot [Ca]/(B + [Ca])$. The *solid line* represents the fitting assuming Ca^{2+}-dependent rate constants have a linear dependence on Ca^{2+} concentration, $k = A \cdot [Ca]$. (**c**) The model can be used to predict the shift in the apparent Ca^{2+} affinity of PCMBS. Plot of normalized <*I*> current against the corresponding Ca^{2+}_i for KCa3.1 (*diamonds*) and KCa3.1 + PCMBS (*squares*). These data demonstrate that a four-state kinetic model accurately explains the Ca^{2+}-dependent gating of KCa3.1. See text for additional details. From Bailey et al. (2010)

gating depended nonlinearly on Ca^{2+} concentration such that the Ca^{2+}-binding rate constants could saturate at high Ca^{2+} concentrations. With this assumption, these authors were able to fit their entire data set (Fig. 20.6b). However, a further sensitivity analysis to evaluate the importance of each parameter in the model revealed that the data could be fit with a simpler four-state kinetic model that excluded the final C-C and C-O transitions shown in the box in Fig. 20.6a such that the data were fit with a single τ_o and three τ_c with the Ca^{2+}-dependent rate constants saturating at high Ca^{2+}. Finally, these authors reported unpublished observations suggesting that the gating behavior of KCa2.2 could also be described by a simpler four-state gating scheme with the Ca^{2+}-dependent transitions being nonlinear. The nonlinear dependence of the opening transitions on Ca^{2+} observed by Bailey et al. (2010) compared to the linear dependence proposed by Hirschberg et al. (1998) may be based on the fact that Bailey et al. (2010) utilized 10 μM Ca^{2+} as a saturating level and this may begin to produce channel block (Ledoux et al. 2008). Whether this becomes important in physiological or pathophysiological situations is unclear.

20.5.2 KCa3.1 Pore Architecture and Allostery

The work of Simoes et al. (2002) was the first to examine the pore architecture of KCa3.1. These authors used the substituted cysteine accessibility method (SCAM) that combines site-directed mutagenesis and chemical modification with thiol-specific reagents to analyze the access of introduced cysteines (Falke et al. 1988; Karlin and Akabas 1998; Roberts et al. 1986; Akabas et al. 1992, 1994). Thus, by combining SCAM with the patch-clamp technique, gating-sensitive amino acids can be identified by alterations in channel gating (Holmgren et al. 1996). In this regard, Simoes et al. (2002) identified two regions (Val275–Val282 and Ala283–Ala286) along S6 of KCa3.1 that display functional differences to MTSET ([2-(trimethylammonium) ethyl] methanethiosulfonate bromide) binding. That is, in the open state, the addition of MTSET caused inhibition at positions Val275, Thr278, and Val282, implying that these residues line the lumen of the pore. In both channel open and closed configurations, the Ala283–Ala286 region is accessible to MTSET, suggesting this region is not embedded in the membrane and participates in the conformational changes in CaM/CaMBD to open the KCa3.1 channel's pore.

Additionally, Klein et al. (2007) studied the KCa3.1 channel pore structure in the closed configuration and found that the modification rates of MTSEA (2-aminoethyl methanethiosulfonate hydrobromide), with a diameter 4.6 Å, for a residue located in the central cavity (Val275Cys) compared to a residue located at the C-terminal end of S6 (Ala286Cys) were found to differ by less than sevenfold, whereas experiments performed with MTSET, with a diameter of 5.8 Å, resulted in a modification rate 10^3–10^4 faster for the cysteine located in the C-terminal end compared to the cysteine in the central cavity. Modification rates of Val275Cys using $Et\text{-}Hg^+$ (4.1 Å diameter) and Ag^+ (2.55 Å diameter), which are smaller than MTSET, were found to be closed/open state independent, while modification rates for MTSET were 10^3 times faster for the open compared to the closed state. Based on these results, the authors proposed that the closed structure of KCa3.1 can be represented by a narrow passage centered at Val282, connecting the channel inner cavity to the cytosolic medium, rather than to the inverted teepee-like structure as described for KcsA channel (Doyle et al. 1998) (Fig. 20.7). As passage through the pore would not be restrictive to the diffusion of small reagents such as MTSEA, $Et\text{-}Hg^+$, and Ag^+, similar in size to the K^+ ion, these data suggest that in

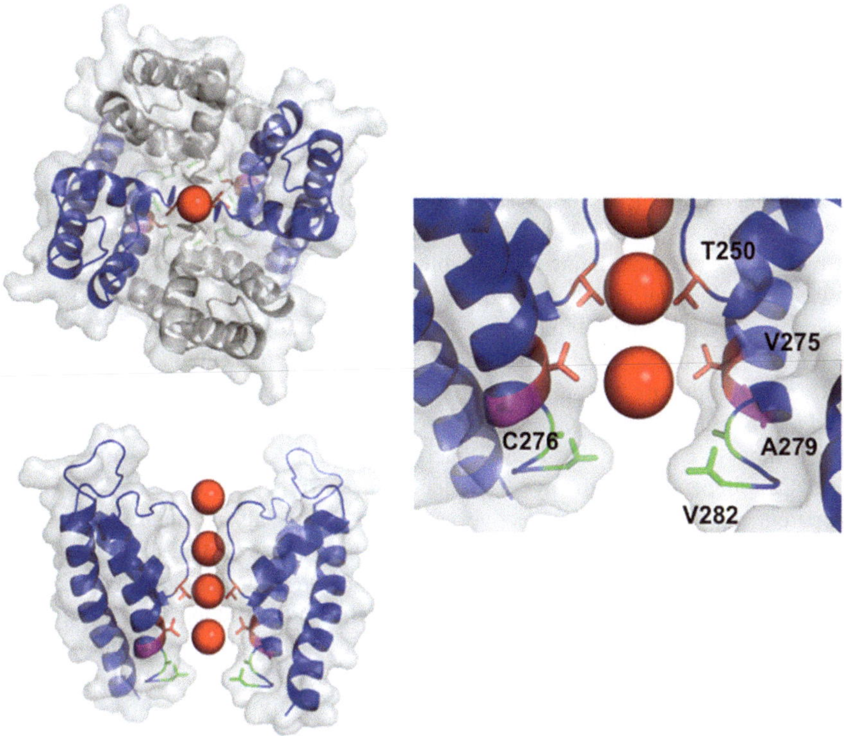

Fig. 20.7 Schematic and surface rendering of human KCa3.1 is shown in two views. *Left*: a view from the extracellular space into the channel tetramer is shown, *top*. The K^+ ions visualized in the pore are colored *red*. The surface is rendered in *gray* with the secondary structures colored in *blue* or *gray*. *Bottom*, the channel pore is viewed along the plane of the membrane with two KCa3.1 subunits removed for clarity. *Right*: close up view of the inner pore and selectivity filter is shown. The residues associated with pore blocking, T250 and V275, are colored in *red*. Residues defining the inner pore, V282 and A279, are colored in *green*. Cysteine 276, which allosterically regulates channel gating, is colored in *magenta* and labeled on the opposing subunit. See text for details

the closed conformation, the cytoplasmic ends of S6 do not function as an obstructive barrier to the movement of K^+ ions (Klein et al. 2007).

The substitution of Val188 (Val282 in KCa3.1) by less hydrophobic residues in the GIRK2 channel (Yi et al. 2001) or Pro475 (Ala283 in KCa3.1) by a more hydrophilic residue in the *Shaker* channel (Sukhareva et al. 2003) resulted in GIRK2 channels that were constitutively active or in *Shaker* channels that were unstable in the closed configuration. Using a glycine (Gly) scan analysis, Garneau et al. (2009) observed that the substitutions Ala279Gly and Val282Gly in S6 of KCa3.1 resulted in the channel being constitutively active in zero Ca^{2+}. In contrast, when residues between Cys276 and Ala286, as well as Ala279 to Val282, were substituted with Gly, constitutive activation of KCa3.1 was not observed. These results demonstrate that hydrophobic interactions involving Val282 are key determinants of KCa3.1 gating (Fig. 20.7). The authors further showed that Ag^+ enjoyed free access to the channel cavity in both an ion-conducting mutant, Val275Cys/Val282Gly, and the closed channel mutant, Val275Cys, further arguing against the activation gate of KCa3.1 being localized to the distal end of S6 (Garneau et al. 2009).

Although KCa3.1 has four cysteines in S6 (Cys267, 269, 276, 277), Simoes et al. (2002) demonstrated that MTSET only transiently inhibited KCa3.1 current, indicative of these cysteines being inaccessible for modification, i.e., these cysteines face away from the pore lumen. Subsequently, Bailey et al. (2010) evaluated the effect of another cysteine-modifying agent, parachloromercuribenze sulfonate (PCMBS) on KCa3.1 channel activity. In contrast to the MTS reagent, PCMBS activated KCa3.1 by inducing both an increase in current at saturating Ca^{2+} concentrations and a leftward shift in the Ca^{2+} activation curve to higher affinity. These results led the authors to conclude that PCMBS activated KCa3.1 via both Ca^{2+}-independent and Ca^{2+}-dependent pathways as reflected in the changes in both $P_{o(max)}$ and apparent Ca^{2+} affinity, respectively. Mutational analysis demonstrated that Cys276 was required for the PCMBS-mediated activation of KCa3.1. These authors further carried out a partial tryptophan scan of S6 and showed that both Leu281Trp and Val282Trp substitutions resulted in dramatic shifts in apparent Ca^{2+} affinity, similar to what was observed with PCMBS (Bailey et al. 2010). As crystallographic data have shown that PCMBS modification of cysteines can result in downstream amino acid side-chain perturbations (Ackerman et al. 2002), it was posited that the binding of PCMBS to Cys276 altered the neighboring Leu281 and Val282 residues (Bailey et al. 2010) (Fig. 20.7). In total, the results from Bailey et al. (2010) and Sauvé and colleagues (Simoes et al. 2002; Klein et al. 2007; Garneau et al. 2009) demonstrate that both luminal and non-luminal amino acids along S6 participate in the activation mechanism of KCa3.1 and that the C-terminal region of S6 must be allosterically coupled to the activation gate (Fig. 20.7).

An interesting characteristic of KCa3.1 gating is that the $P_{o(max)}$, at saturating Ca^{2+} concentrations, is ~0.1–0.2 compared with a $P_{o(max)}$ of ~0.8 for the KCa2.2 channel (Hirschberg et al. 1998). As noted above, binding of Ca^{2+} to the CaM N-lobe results in CaM associating with the CaMBD of adjacent KCa2.2 subunits, within the tetramer (Schumacher et al. 2001, 2004), and it is presumed by homology that this holds true for KCa3.1. Thus, Morales et al. (2013) undertook a series of mutagenesis studies designed to identify amino acids in the CaMBD that are critical to determining KCa3.1 $P_{o(max)}$. In this regard, these authors demonstrated that hydrophobic effects at Ser367 contribute to $P_{o(max)}$, whereas electrostatic interactions involving Arg362 and Glu363 regulate channel activation rate (τ_{on}). That is, the Ser367Cys mutation resulted in an increase in $P_{o(max)}$ from 0.22 to 0.62 coupled with an increase in channel open time and a decrease in channel closed time. Additional substitutions with nonpolar amino acids of greater surface area (Thr, Leu, Trp) also increased $P_{o(max)}$, whereas substitution with an amino acid of similar surface area (Ala) had no effect. It was also demonstrated that neutralizing the charged amino acids at positions Arg362 and Glu363 resulted in dramatic shifts in

τ_{on}, indicative of amino acids not involved in KCa3.1-CaM binding but important in channel activation.

20.5.3 Calmodulin and KCa3.1 Gating

Following the cloning of KCa2.x, and subsequently KCa3.1, channels that exhibited Ca^{2+}-dependent gating characteristics similar to what had been historically noted for the small and intermediate conductance, Ca^{2+}-activated K^+ channels, it raised the question of how Ca^{2+} mechanistically regulated these channels. It was noted by Kohler et al. (1996) for KCa2.x channels, as well as by both Joiner et al. (1997) and Ishii et al. (1997) for KCa3.1 channels, that these channels did not share homology with the maxi-K channels and did not possess motifs similar to an E-F hand capable of binding Ca^{2+} in order to induce gating. Shortly after the cloning of these channels, in a series of elegant studies, Adelman and colleagues (Xia et al. 1998; Keen et al. 1999) demonstrated that calmodulin bound constitutively to the proximal C-terminus of the KCa2.x channels at the CaMBD just distal to the sixth transmembrane domain. These authors further demonstrated that one calmodulin was bound to each of the four subunits of the tetrameric channel and that it was specifically the binding of Ca^{2+} to the N-lobe E-F hands of calmodulin that induced channel gating via a conformational change in calmodulin that was transmitted to the KCa2.x channel α-subunit. In contrast, the C-lobe E-F hands of calmodulin did not bind Ca^{2+} when it was associated with KCa2.x channels. Fanger et al. (1999) similarly demonstrated that KCa3.1 channels were constitutively bound to calmodulin that resulted in their Ca^{2+}-dependent gating.

A significant breakthrough in understanding how the binding of Ca^{2+} to calmodulin resulted in channel gating came from the work of Schumacher et al. (2001, 2004). These authors solved the crystal structure for calmodulin bound to the CaMBD of KCa2.2 in both the apo and calcified states, thereby providing a structural mechanism for Ca^{2+} binding and the subsequent conformational changes leading to KCa2.2 gating. Based on these studies, it was proposed that the binding of Ca^{2+} to the N-lobes of CaM in KCa2.2 causes the formation of a dimeric complex between two adjacent CaMBD, while initiating a >90° rotation of the S6 helices leading to the opening of the channel pore. Because there is a high degree of homology in the calmodulin/CaMBD between KCa3.1 and KCa2.2, the gating mechanism proposed in KCa2.2 likely applies to KCa3.1. Therefore, the C-terminal portion of the S6 in KCa3.1 is thought to be involved in the transformation of energy from the calmodulin/CMBD into a mechanical force, modifying the conformational state of the pore (Schumacher et al. 2001, 2004; Simoes et al. 2002; Klein et al. 2007; Garneau et al. 2009). Unfortunately, the linker between S6 and the CaMBD could not be resolved in these crystal structures and so the exact mechanism by which this force is exerted remains to be determined.

20.5.4 Role of KCa3.1 in Gating of Maxi-K (KCa1.1)

Interestingly, the gating of KCa3.1 has also been shown to directly regulate the gating of maxi-K (KCa1.1) channels. That is, Thompson and Begenisich (2006) showed that activation of KCa3.1 channels inhibits maxi-K channels in acinar cells, where the two channels are endogenously expressed, and in heterologous expression systems. Similar reports were described in human parotid (Nakamoto et al. 2007) and in mouse submandibular glands (Romanenko et al. 2007). The inhibition of maxi-K channels occurred whether the KCa3.1 channels were activated by direct perfusion of the cell with elevated Ca^{2+}, by muscarinic receptor stimulation, or by the exogenous chemical activator DCEBIO at low Ca^{2+} concentration. In addition, the inhibition of maxi-K currents was blocked by TRAM-34, an inhibitor of KCa3.1 (Thompson and Begenisich 2006; Romanenko et al. 2009). Quantitative analysis led these authors to postulate that each maxi-K channel may be surrounded by four KCa3.1 channels and it would be inhibited if any one of these KCa3.1 channels opens (Thompson and Begenisich 2006). Further, since KCa3.1-induced inhibition of maxi-K channels takes place in excised membrane patches, the authors proposed a close interaction between the two channels without any intermediaries such that the block of the ion flow through the pore in the maxi-K channel may occur by inserting the cytoplasmic N-terminal domains of KCa3.1 (Thompson and Begenisich 2006), as was described for two maxi-K channel β-subunits (β2 and β3) (Wallner et al. 1999; Xia et al. 1999; Zhang et al. 2006; Li et al. 2007). In this regard, Thompson and Begenisich (2009) observed that the N-terminus of KCa3.1 shares general similarity with the N-terminal regions of these two maxi-K β-subunits, the N-terminal region peptide of KCa3.1 blocked maxi-K channels, and the activation of KCa3.1 competed with the N-terminus peptide for blocking of maxi-K channels. In total, these data resulted in these authors proposing a model where KCa3.1 channels are located sufficiently close to the maxi-K channels such that the activation of KCa3.1 results in a conformational change in their N-termini, which inserts through a cytoplasmic side portal in the BK channel, thereby inhibiting permeation through the maxi-K pore.

20.6 Regulation of KCa3.1

As detailed above, the main regulator of KCa3.1 gating is Ca^{2+} binding to calmodulin constitutively associated with the CaMBD. However, shortly after the mechanism by which Ca^{2+}-dependent gating was unraveled, it was shown that phosphorylation modified this Ca^{2+}-dependent gating process. That is, Gerlach et al. (2000) first demonstrated a role for kinases in the regulation of KCa3.1 by showing that hydrolyzable ATP analogues, but not other nucleoside triphosphates, activated the channel via an increase in $P_{o(max)}$ without changing the apparent Ca^{2+} sensitivity. Addition of alkaline phosphatase eliminated this ATP-dependent activation of KCa3.1, indicative of a phosphorylation event. Indeed, this ATP-dependent activation was shown to be mediated through PKA when KCa3.1 was heterologously expressed in *Xenopus* oocytes or when endogenously expressed in T84 cells, but was independent of PKA when heterologously expressed in HEK cells, suggesting additional kinases may also regulate KCa3.1. In contrast to KCa3.1, KCa2.3 was not activated by ATP under similar conditions. Thus, using a series of KCa3.1/KCa2.3 chimeras, these authors further demonstrated that the kinase-dependent activation of KCa3.1 could be localized to a 14 amino acid domain within the C-terminus (Arg^{355}–Met^{368}), distal to the CaMBD (Gerlach et al. 2001). Additional studies by Jones et al. (2007) indicated that the ATP-dependent activation of KCa3.1 also depended upon an N-terminal RKR motif. That is, mutation of $^{15}RKR^{17}$ to alanines resulted in a channel that was no longer sensitive to ATP and alkaline phosphatase no longer reduced channel activity. Finally, the open probability (P_o) of these mutated channels was reduced fourfold compared to WT KCa3.1, suggesting that the ATP-/kinase-dependent regulation of KCa3.1 observed by these authors is dependent upon a close N-/C-terminal association. In this regard, it has similarly been demonstrated that KCa2.2 associates with both protein kinase CK2 and protein phosphatase 2A (PP2A), in a manner that brings the N- and C-termini into close proximity, such that they phosphorylate and dephosphorylate the associated calmodulin, respectively, resulting in a shift in the apparent Ca^{2+}-sensitivity of the channel (Bildl et al. 2004; Allen et al. 2007).

Additional studies have shown that PKC regulates KCa3.1 function (Wulf and Schwab 2002), although this appears to be an indirect effect as mutation of the four conserved PKC phosphorylation sites has no effect on activation (Gerlach et al. 2000). Also, AMP-activated protein kinase has been shown to directly interact with the distal C-terminus of KCa3.1 between Asp^{380} and Ala^{400}, where activation of AMP-activated protein kinase results in an inhibition of KCa3.1 function and subsequent inhibition of transepithelial Cl^- secretion (Klein et al. 2009). In additional studies, Skolnik and colleagues (Srivastava et al. 2005) utilized a yeast two-hybrid approach to identify the myotubularin, MTMR6, a lipid (PI(3)P) phosphatase, as an interacting protein with a *C. elegans* KCa2.x (KCNL) family member. This interaction occurred via the channel coiled-coil domain, a region Syme et al. (2003) had previously shown to be required for proper channel assembly and trafficking (see below). Srivastava et al. (2006) further demonstrated that this MTMR6-dependent regulation required the same 14 amino acid region of KCa3.1 previously shown to be required for ATP-/kinase-dependent activation (Gerlach et al. 2001) (Fig. 20.2). In an exciting next step, Skolnik and colleagues demonstrated that nucleoside diphosphate kinase B (NDPK-B) (Fig. 20.2), a histidine kinase, directly binds and activates KCa3.1 by phosphorylating His^{358} and that this is reversed by protein histidine phosphatase (Srivastava et al. 2008). Further, NDPK-B knockout mice exhibit a reduced KCa3.1 activity (Di et al. 2010), indicative of this pathway being functional in vivo, as well. Finally, it has also been demonstrated that the nitric oxide (NO) donor, sodium nitroprusside, activated KCa3.1 channels in interstitial cells of Cajal (Zhu et al. 2007). It was subsequently shown that NO activates KCa3.1 via the PKG pathway in human dermal fibroblasts (Bae et al. 2014). In total, these results suggest that a kinase/phosphatase protein regulatory complex exists for the careful regulation of KCa3.1.

Devor and Frizzell (1998) demonstrated that IK channels in the T84 epithelial cell line were inhibited by numerous fatty acids, including arachidonic acid with an apparent K_i of 425 nM. Indeed, in intact epithelial monolayers, inhibition of cytosolic phospholipase A2 resulted in a significant increase in Ca^{2+}-mediated agonist transepithelial Cl^- secretion, suggesting that agonist-mediated generation of arachidonic acid may blunt the response to these agonists. Subsequently, Hamilton et al. (2003) identified two amino acids in the pore of KCa3.1 that were responsible for the arachidonic-mediated inhibition of the channel and further demonstrated that introduction of these amino acids into KCa2.2 resulted in this channel being sensitive to arachidonic acid block. Interestingly, these two amino acids, Thr^{250} and Val^{275}, had previously been shown to be responsible for the clotrimazole-dependent inhibition of KCa3.1 (Wulff et al. 2001) (Fig. 20.7).

20.7 Trafficking of KCa3.1

The physiological response of an ion channel is dependent upon both its gating, i.e., the likelihood that the channel is in open or closed state (P_o), and the number of channels (N) at the cell surface. The N of channels is in turn determined by the relative rates of trafficking along the anterograde and retrograde pathways. Studies over the past 15 years on the trafficking of KCa3.1 channels have revealed key molecular components in the assembly, delivery to the plasma membrane, and subsequent internalization and degradation of these channels. A schematic representation of the known intracellular trafficking pathways for KCa3.1 is presented in Fig. 20.8.

20.7.1 Anterograde Trafficking of KCa3.1

Prior to being delivered to the plasma membrane via the biosynthetic route, KCa3.1 must be correctly folded and assembled into tetramers in the endoplasmic reticulum (ER; Ellgaard et al. 1999; Papazian 1999; Deutsch 2002). Since the cloning of the KCa2.x/KCa3.1 channels, numerous investigators have identified motifs or regions within the N- and C-termini of KCa3.1 that are required for proper subunit folding and association. In an initial series of studies, Kaczmarek and colleagues (Joiner et al. 2001) and Adelman and colleagues (Lee et al. 2003) demonstrated that, apart from its role in channel gating, constitutive association of calmodulin with the proximal C-terminal CaMBD of KCa3.1 and KCa2.x channels regulates the surface expression of these channels by regulating their multimerization. In addition to the CaMBD, the distal C-termini of KCa3.1 and KCa2.x channels contain a leucine

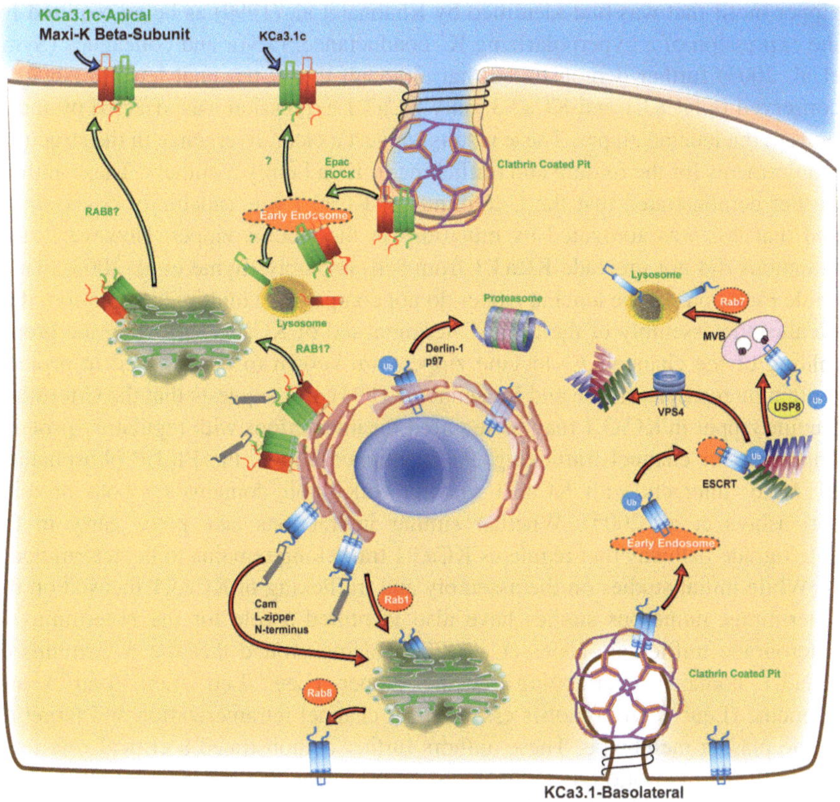

Fig. 20.8 Schematic of anterograde and retrograde trafficking pathways for KCa3.1 channels (KCa3.1 basolateral, and KCa3.1c apical). The important motifs and modulatory proteins in the channel assembly are shown. The molecular players of the anterograde trafficking of KCa channels are still emerging. KCa3.1 basolateral (KCa3.1b, *blue channel*) channels are trafficked to the basolateral membrane via Rab1- and Rab8-dependent pathways with additional trafficking motifs within the N-terminus and leucine zipper (L-zipper). The forward trafficking of KCa3.1c apical (*green channel*) and maxi-K β-subunit (*red channel*) to the apical membrane is not known. Hence, the involvement of Rab1 and Rab8 in apical membrane targeting is not known. The retrograde trafficking of both KCa3.1 basolateral and KCa3.1c apical is via clathrin-coated pit mechanisms. Endocytosed KCa3.1 is quickly targeted for lysosomal degradation through an endosomal sorting complexes required for transport (ESCRT)- and Rab7-dependent pathways. Ubiquitylation (Ub) facilitates the targeting of KCa3.1b to the lysosomes, and USP8 (ubiquitin-specific protease 8) modulates the rate of KCa3.1b degradation by deubiquitylating KCa3.1b prior to delivery to the lysosome. In contrast, KCa3.1c apical uses an Epac- and ROCK-dependent pathway on route to early endosome. Whether KCa3.1c apical is recycled is not known and the mechanism(s) for lysosomal degradation is also currently not known. *Epac* exchange protein directly activated by cAMP, *ROCK* Rho-associated kinase, *MSV* multivesicular body, *Cam* calmodulin, *VPS4* vacuolar protein sorting-associated protein 4. See text for details

zipper motif that was first identified by Khanna et al. (1999) as being required for the expression of a hyperpolarizing K^+ conductance. Devor and colleagues (Syme et al. 2003) further demonstrated that although this C-terminal leucine zipper is conserved in KCa3.1 and KCa2.3, only KCa3.1 expression was affected by mutations in the leucine zipper. These results suggest a clear divergence in the structural requirements for the folding and trafficking of KCa family members. These authors further demonstrated that the C-terminus of KCa3.1 was capable of co-assembly and that this was abrogated by mutations in the leucine zipper. However, these mutations did not preclude KCa3.1 from self-assembly (Syme et al. 2003). Thus, while mutations of the leucine zipper do not compromise multimer formation, they do alter the assembly of the distal C-terminal tail of KCa3.1, affecting the proper folding of the channel. As leucine zippers are known to be involved in protein-protein interactions (Kobe and Deisenhofer 1994), this suggests that the C-terminal leucine zipper in KCa3.1 may be required for interactions with regulatory proteins important for channel trafficking. As mentioned above, the PI(3)P phosphatase, MTMR6, interacts with KCa3.1 via the coiled-coil domains on both proteins (Srivastava et al. 2005). Whether similar interactions take place early in the anterograde pathway that regulates KCa3.1 trafficking remains to be determined.

While initial studies on the assembly and trafficking of KCa3.1 focused on the C-terminus, numerous studies have also identified a role for the N-terminus in anterograde trafficking. Jones et al. (2004) demonstrated that the N-terminus of KCa3.1 contains overlapping leucine zipper ($Leu^{18}/Leu^{25}/Leu^{32}/Leu^{39}$) and dileucine ($Leu^{18}/Leu^{19}$) motifs essential for channel tetramerization and targeting to the plasma membrane. These authors further demonstrated a critical role for a lysine in the cytosolic S4–S5 linker of KCa3.1 (Lys^{197}) and KCa2.3 (Lys^{453}) in protein trafficking (Jones et al. 2005). Mutation of these lysines resulted in these KCa channels failing to exit the ER, suggesting that critical associations required for channel trafficking are disrupted (Fig. 20.8). Additional studies demonstrated that misfolded KCa3.1 and KCa2.3 channels are polyubiquitylated prior to degradation (Gao et al. 2008). These authors further demonstrated a role for the ER integral membrane protein, Derlin-1, and the cytosolic chaperone AAA-ATPase p97 in the ER translocation, protein dislocation, and targeting of misfolded and polyubiquitylated KCa3.1 and KCa2.3 channels for proteasomal degradation (Fig. 20.8).

The anterograde trafficking of KCa channels requires that they be properly assembled into tetramers. It is generally believed that the 6-TMD KCNN gene family members (KCa2.x, KCa3.1) are evolutionarily older than the voltage-gated K^+ channels (K_v) (Anderson and Greenberg 2001), and from the crystal structure data of K_v channels, it is clear that the additional four TM domains (S1–S4), which were appended to the central pore (S5-pore-S6) to form the 6-TMD channels, form a unique structural domain (Lee et al. 2005). A fundamental difference between the K_v and KCa3.1/KCa2.x channels is that the KCa channels exhibit no voltage dependence to their gating. The voltage dependence of K_v channels is conferred by a series of 4–7 positively charged amino acids in S4 as well as critical negatively charged amino acids in S2 and S3 (Aggarwal and MacKinnon 1996; Seoh et al.

1996). In addition to conferring voltage dependence, Papazian and colleagues (Papazian et al. 1995; Tiwari-Woodruff et al. 1997, 2000; Myers et al. 2004) demonstrated that these charged amino acids are essential for channel biogenesis. That is, key salt bridges are required to maintain the fundamental fold of this S1–S4 domain. Indeed, it has been shown that these electrostatic interactions are required at the initial folding process as the nascent polypeptide chain is being formed prior to entering the ER (Sato et al. 2002, 2003). Although KCa3.1 and KCa2.x channels are voltage independent, these channels conserve two critical arginines in S4 as well as a glutamic acid in S3. Devor and colleagues (Gao et al. 2008) demonstrated that mutation of these charged amino acids in S3 or S4 resulted in the rapid degradation of KCa3.1, KCa2.3, and KCa2.2 suggesting that, similar to K_V channels, key electrostatic interactions required for folding of the S1–S4 domain are abrogated, resulting in channel degradation. The observation that S4 arginines, as well as an S3 glutamic acid, share a conserved function between K_V channels and the ligand-gated KCa channels led the authors to speculate that the role of these charged amino acids in establishing the proper folding of the S1–S4 domain evolved prior to the establishment of voltage dependence and that the basic structure of the S1–S4 domain is conserved between the K_V and KCa channels.

As noted, the splice variant KCNN4c is missing 29 amino acids, including the entire second transmembrane domain (S2) and is only expressed at the cell surface in the presence of the maxi-K β1-subunit (Barmeyer et al. 2010) (Fig. 20.8). Thus, a key remaining question is what is the structural topology of KCNN4c? That is, does the S1 segment of KCNN4c still act as a TMD or is it perhaps cytosolic in the final folded KCNN4c structure? In this way, the two TMDs of the β1-subunit may substitute for S1–S2 in KCNN4c. In the case of the KAT1, Shaker-type K^+ channel, S1 and S2 exhibited typical topogenic properties in which S1 is inserted into the ER membrane followed by the insertion of S2 (Sato et al. 2002). In contrast, in $K_V1.3$, the S2 segment appears to function as the signal sequence to establish topology such that S1 remains cytosolic and translocation into the Sec61 translocon and the ER membrane is initiated by S2 (Tu et al. 2000). Similarly, in the cystic fibrosis transmembrane conductance regulator (CFTR) Cl^- channel, the second TMD initiates translocation rather than TMD1 as a means of establishing proper channel topology (Lu et al. 1998; Sadlish and Skach 2004). It has also been demonstrated in the KAT1 channel that S3 and S4 are posttranslationally inserted into the ER membrane following release of the positively charged S4 from the ribosome (Sato et al. 2002). Indeed, this was shown to be dependent upon the known S3–S4 salt-bridge interactions in this channel (Sato et al. 2002). As noted, we have shown that these S3–S4 interactions exist in KCa3.1, although this channel is voltage independent (Gao et al. 2008). Thus, interactions between S3 and S4 may similarly be required for the insertion of this domain into the membrane in KCa3.1. Finally, it has been shown that the S5-P-S6 domain can be integrated independent of the S1–S4 TMDs of KAT1 and $K_V1.3$, consistent with these regions of the channels being unique structural domains (Tu et al. 2000; Sato et al. 2002). Based on these previous studies, it is interesting to speculate that S1 of KCNN4c remains in the cytosol as it is typically inserted into the translocon and ER membrane together with S2. The subsequent translation of S3 and S4 results in salt-bridge pairings such that these TMDs are inserted into the membrane, thereby establishing the proper topology of the channel, with the S5-P-S6 region being inserted as in the full-length channel. Future studies will undoubtedly answer these interesting questions concerning KCNN4c biogenesis in addition to defining how and where the β1-subunit associates with KCNN4c to divert the channel from the degradative pathway along the biosynthetic route.

20.7.2 Retrograde Trafficking of KCa3.1

Compared to the extensive studies on the KCa channels and anterograde trafficking, the mechanisms of endocytosis and downstream sorting of these channels have only recently begun to be unraveled. To begin to elucidate these mechanisms, Devor and colleagues (Gao et al. 2010) developed a strategy for monitoring the endocytic routes of KCa3.1 based on the work of Ting and colleagues (Howarth et al. 2005). For these studies, the biotin ligase acceptor peptide (BLAP) sequence was inserted into the second extracellular loop of KCa3.1, followed by a fast and specific biotinylation of the channels at the cell surface using recombinant biotin ligase (BirA) (Gao et al. 2010; Balut et al. 2010a). Addition of fluorophore-conjugated streptavidin allows the fate of the endocytosed channels to be addressed using a combination of biochemical and imaging techniques. Using these approaches, Balut et al. (2010a) demonstrated that KCa3.1 is endocytosed from the plasma membrane within 60–90 min. This endocytic step of KCa3.1 has been proposed to be clathrin dependent by Schwab et al. (2012) in migrating MDCK cells. In addition, a dileucine motif present in the C-terminus of KCa3.1 (Leu^{344}/Leu^{345}) appears to be important for channel internalization (Schwab et al. 2012), consistent with adaptor protein binding and clathrin-dependent endocytosis. Subsequent to endocytosis, KCa3.1 is targeted for lysosomal degradation via a Rab7- and MVB (multivesicular body)/ESCRT (endosomal sorting complex required for transport)-dependent pathway (Balut et al. 2010a) (Fig. 20.8). By combining BLAP-tagged KCa3.1 with TUBEs (tandem ubiquitin-binding entities) and deubiquitylase Chip methodologies, Balut et al. (2011) further demonstrated that polyubiquitylation mediates the targeting of internalized KCa3.1 to the lysosomes and also that USP8 (ubiquitin-specific protease) regulates the rate of KCa3.1 degradation by deubiquitylating KCa3.1 prior to lysosomal delivery (Fig. 20.8). Importantly, it has been shown that KCa3.1 endocytosis can be reduced by blocking the ubiquitin-activating enzyme E1 (Balut et al. 2010b, 2011), suggesting that ubiquitylation of either the channel itself or associated adaptors functions as a signal for KCa3.1 endocytosis. In contrast to our data on the KCNN family member KCa2.3, which rapidly recycles back to the plasma membrane following endocytosis (Gao et al. 2010), we did not observe recycling of KCa3.1 following endocytosis suggesting this channel is targeted for degradation following endocytosis. The plausibility of a ubiquitin-dependent pathway is supported by data showing that in CD4 T cells the tripartite motif containing protein 27 (TRIM27)-dependent ubiquitylation of the class II PI3K-C2β kinase acts as a negative regulator of KCa3.1 activation (Cai et al. 2011). Thus, TRIM27 functions as an E3 ligase and mediates lysine 48 polyubiquitylation of PI3K-C2β kinase, leading to an inhibition of the PI3K enzyme activity. This leads to decreased levels of PI(3)P which, as discussed above, results in a decreased histidine phosphorylation and activation of KCa3.1 by NDPK-B (Cai et al. 2011).

20.7.3 Plasma Membrane Targeting of KCa3.1 in Polarized Cells

In contrast to the studies above, the trafficking of KCa3.1 in polarized cells has only just begun to be investigated. In T lymphocytes it has been shown that KCa3.1 is evenly distributed on the plasma membrane but rapidly translocates to the immunological synapse upon antigen stimulation as part of the signaling complex that facilitates T cell activation (Nicolaou et al. 2007). It has also been shown that a KCa3.1 isoform lacking the N-terminus is retained within the cytosol and acts as a dominant negative on the full-length channel in human lymphoid tissues (Ohya et al. 2011). This is consistent with the previously published role of the N-terminus in KCa3.1 trafficking (Jones et al. 2004). Additionally, it was demonstrated that KCa3.1 localizes with F-actin to the immunological synapse, emphasizing the role of cytoskeleton in providing directionality for channel movement (Nicolaou et al. 2007). An interesting observation is that signaling molecules such as PKA and PKC, which are known regulators of KCa3.1 function, as discussed above, accumulate in the immunological synapse on T cell activation, as well (Bi et al. 2001; Skalhegg et al. 1994; Zhou et al. 2004), and this close proximity could provide a regulatory mechanism for KCa3.1 function at the synapse.

To date, very few studies have explored the trafficking of KCa3.1 in polarized epithelia. While it has been shown that in migrating MDCK cells KCa3.1 displays a polarized distribution by accumulating at the leading edge of the lamellipodium (Schwab et al. 2006), these cells do not demonstrate classic apical-to-basolateral polarization. We recently utilized polarized MDCK, Caco-2, and FRT cells, grown on semipermeable Transwell® supports, to investigate the anterograde and retrograde trafficking of full-length KCa3.1 in polarized epithelia (Bertuccio et al. 2014). Given that Rabs 1, 2, and 6 have been shown to be involved in the trafficking of proteins from the ER to Golgi, while Rabs 8 and 10 are known to be involved in the trafficking of proteins from the Golgi to the plasma membrane (Martinez and Goud 1998; Babbey et al. 2006; Zhang et al. 2009; Dong et al. 2010), we determined whether any of these Rabs played a crucial role in the anterograde trafficking of KCa3.1 in polarized epithelia. In this regard, we demonstrated that Rab1 and Rab8 are required for the ER/Golgi exit of KCa3.1 and subsequent trafficking to the basolateral membrane (Fig. 20.8), whereas Rabs 2, 6, and 10 did not play a role in this trafficking step (Bertuccio et al. 2014). In contrast, trafficking of KCa2.3 was unaffected by siRNA-mediated knockdown or overexpression of dominant negatives of these Rabs, thereby confirming the specificity of these Rabs within the gene family. It has also been shown that the sorting of numerous proteins to the basolateral domain of polarized epithelia is the result of an association with the μ1 subunit of the adaptor protein-1 complex (AP-1) (Ohno et al. 1998; Muth and Caplan 2003; Cancino et al. 2007). The AP-1 complex is the only clathrin-associated adaptor complex implicated in basolateral sorting in polarized epithelia (Futter et al. 1998). Importantly, while μ1B is expressed in MDCK cells, it is not expressed in LLC-PK_1 cells, and the lack of the μ1B subunit in LLC-PK_1 cells has been shown to result in the mistargeting of some basolateral proteins to the apical surface, while stable expression of μ1B restores proper basolateral targeting (Fölsch et al. 1999). However, we demonstrated correct basolateral targeting of KCa3.1 in LLC-PK1 cells, indicating μ1B is not required for basolateral sorting of KCa3.1. Finally, it has been shown that proteins destined for the basolateral membrane can traffic either (1) from the Golgi to the basolateral membrane directly or (2) from the Golgi through recycling endosomes to the basolateral membrane (Fuller et al. 1985; Orzech et al. 2000). To determine whether KCa3.1 traffics via recycling endosomes, we carried out a series of studies to block trafficking through the recycling endosomes. We demonstrated that KCa3.1 did not traffic to the basolateral membrane via transferrin- and RME-1-positive recycling endosomes, suggesting this channel traffics directly to the basolateral membrane from the Golgi (Bertuccio et al. 2014). Finally, we demonstrated that, similar to our studies in non-polarized cells, KCa3.1 is endocytosed from the basolateral membrane of polarized MDCK, Caco-2, and FRT cells in a ubiquitin-dependent manner and that the channels are then targeted to the lysosome and proteasome for degradation (Bertuccio et al. 2014).

Whether the correct apical targeting of KCNN4c similarly requires Rabs 1 and 8 and whether this splice variant of KCa3.1 is directly targeted to the apical membrane or via the apical recycling endosomal compartment have yet to be addressed (Fig. 20.8). As detailed above, apical KCNN4c expression is regulated by knockdown of Epac1 with KCNN4c being redistributed to subapical vesicles. Similarly, inhibition of RhoA and ROCK induced a redistribution of KCNN4c out of the apical membrane (Sheikh et al. 2013). Whether KCNN4c is capable of recycling back to the apical membrane from this vesicular pool and whether this involves ubiquitylation and deubiquitylation as well as the ultimate degradative pathway for KCNN4c from the apical membrane remain open questions yet to be addressed.

20.8 Role of KCa3.1 in the Cell Cycle and Cell Proliferation

During cell proliferation, a cell progresses through various phases of the cell cycle (G0, G1, S, G2, and M). It has been known, for some time, that intracellular Ca^{2+}, $[Ca^{2+}]_i$, levels alter cell proliferation (Hazelton et al. 1979). Additionally, Nilius and Wohlrab (1992), using human melanoma cells, provided the first evidence linking changes in the electrochemical gradient for Ca^{2+} entry into the cell due to membrane hyperpolarization and K^+ channels. Eventually, Wonderlin and colleagues (Strobl et al. 1995; Wonderlin et al. 1995; Wonderlin and Strobl 1996) suggested that Ca^{2+}-activated K^+ channels might play a significant role in the regulation of the cell through the cell cycle. The coupled action of the cellular hyperpolarization via K^+ channels and increased $[Ca^{2+}]_i$ has been demonstrated to enhance passing of cells through G0/G1 and into S phase, resulting in cell proliferation (Schreiber 2005; Wonderlin and Strobl 1996; Wonderlin et al. 1995).

The basis of our appreciation of the possible role of KCa3.1 in cell proliferation in epithelial cells has arisen from cell differentiation and blocker pharmacology studies of lymphocytes. Cahalan and co-workers (Decoursey et al. 1984) first provided evidence for the relationship between K^+ channels and the proliferative processes of T lymphocytes. Originally, the characterization of KCa3.1 in cell activation/proliferation was demonstrated in erythroid cells, T-lymphoblasts, and T-lymphocytes. Alper and colleagues (Vandorpe et al. 1998) reported, in mouse embryonic stem cells (ES), that mRNA for mIK1 (mouse KCa3.1) was maintained at low levels in uninduced ES cells; however, it exhibited sustained high mRNA levels during erythroid differentiation of ES cells. Further, they demonstrated that increasing concentrations of clotrimazole (even as low as 10 nM) inhibited cell proliferation of human peripheral blood stem cells, and thus, these studies provide strong evidence that KCa3.1 participates in cell differentiation and proliferation (Vandorpe et al. 1998).

The information available about the role of KCa3.1 and the cell cycle and cell proliferation in epithelial cells is limited. Strobl and colleagues (Wang et al. 1998), using quinidine (a generic K^+ channel blocker), provided early evidence that K^+ channels might participate in the G phase of the cell cycle of MCF-7 breast cancer epithelial cells. Afterward, Ouadid-Ahidouch and colleagues demonstrated that the current density of KCa3.1 increased in MCF-7 cells that had been synchronized at the end of G1 or S phase when compared to those cells in the early G1 phase of the cell cycle (Ouadid-Ahidouch et al. 2004). Additionally, they reported that the increase in current density was in congruence with increased mRNA levels of KCa3.1 and the enhanced negative membrane potential was observed. To examine whether KCa3.1 influenced cell progression by altering the $[Ca^{2+}]_i$, Ouadid-Ahidouch et al. (2004) examined whether there was a correlation between effects of $[Ca^{2+}]_i$ on K^+ channel activity. They reported that $[Ca^{2+}]_i$ was lower in the early G1 phase (49 ± 6 nM) when the activity of KCa3.1 was low and the membrane potential was depolarized. Additionally, the $[Ca^{2+}]_i$ level was higher (240 ± 23 nM) in cells arrested at the end of G1 and $[Ca^{2+}]_i$ reached high levels as cells progress into the S phase (323 ± 20 nM) when the activity of KCa3.1 was high and the membrane was hyperpolarized. Further, Ouadid-Ahidouch et al. (2004) demonstrated that clotrimazole caused a depolarization of membrane potential of cells arrested at the end of G1 and during the S phase, providing evidence of active KCa3.1, while clotrimazole had no effect on the membrane potential of cells arrested in early G1 phase. Finally, the treatment of MCF-7 cells with clotrimazole and econazole (another inhibitor of KCa3.1) caused a dose-dependent reduction in cell proliferation, and these inhibitors resulted in an increase of cells in the G1 phase and decrease in the number of cells moving into the S phase (Ouadid-Ahidouch et al. 2004). In agreement, Parihar et al. (2003), using LNCaP and PC-3 (prostate cancer epithelial cells), reported that both 1-EBIO and riluzole (activators of KCa3.1) increased cell proliferation of both types of prostate cells and resulted in decreased cell proliferation when administered in the presence of charybdotoxin and clotrimazole.

There have been advances in our understanding of the "link" between KCa3.1 and the Ca^{2+} entry step into cells in the cell proliferation process. Lallet-Daher et al. (2009) examined aspects of the control of Ca^{2+} entry in human prostate cancer cell proliferation using LNCaP and PC-3 prostate cells that has furthered our understanding of the role of KCa3.1 in cell proliferation. They examined the mechanism by which the inhibition of KCa3.1 prevented cell proliferation by examining the changes in expression of $p21^{Cip1}$ (cyclin-dependent kinase inhibitor 1) that is known to participate in the G1 phase (Ghiani et al. 1999). Indeed, when KCa3.1 was inhibited by TRAM-34, the expression (mRNA) of $p21^{Cip1}$ of both LNCaP and PC-3 cells was increased by two- to fourfold, and they also reported a similar fold increase in $p21^{Cip1}$ when they knocked down KCa3.1 with siRNA. Further, Lallet-Daher et al. (2009) provided a role of the Ca^{2+} channel, TRPV6, in the signaling pathway, when they demonstrated an association of TRPV6 and KCa3.1 by co-immunoprecipitation. Further, they also reported a novel role of TRPV6 in maintaining the hyperpolarization-activated Ca^{2+} entry by using an siRNA approach to knockdown TRPV6 and Ca^{2+} entry was not stimulated by activation of KCa3.1. These data certainly suggested a "close" association of KCa3.1 and TRPV6 in the regulation of cell proliferation. Finally, Ouadid-Ahdouch and colleagues (Dhennin-Duthille et al. 2011) and others (Bolanz et al. 2008) have now identified TRPV6 in MCF-7 cells, while Dhennin-Duthille et al. (2011) reported highly expressed levels of TRPV6 in human breast ductal adenocarcinoma cells compared to adjacent non-tumoral tissue. Therefore, the link between KCa3.1, TRPV6, and cell proliferation is gaining strength.

There is an exciting caveat to the KCa3.1 and cell proliferation story as Bruce and co-workers (Millership et al. 2011) reported that both a nonfunctional pore mutant of KCa3.1 ($KCa3.1^{GYG/AAA}$) and a non-trafficking mutant of KCa3.1 ($KCa3.1^{L18A/L25A}$; Jones et al. 2004) resulted in enhanced cell proliferation using a heterologous expression system (human embryonic kidney cells, HEK). These results indicated that KCa3.1 induced cell proliferation, but did not require membrane localized KCa3.1.

It appears that the role of KCa3.1 in cell proliferation in epithelial cells is just at its infancy. No doubt, it will be very interesting to see how this story develops over the next 5 years.

20.9 Role of KCa3.1 in Epithelial Diseases

Although KCa3.1 is expressed in a wide variety of epithelial tissues, no human disease has been directly associated with mutations in this channel. However, given the critical role of KCa3.1 in maintaining transepithelial ion and fluid transport, it is not surprising that this channel has been suggested to play a role in the etiology of various diseases (Table 20.1). In the kidney, autosomal-dominant polycystic kidney disease (ADPKD) is characterized by the progressive development and enlargement of multiple bilateral fluid-filled cysts. Evidence indicates that cyst formation in ADPKD is produced by tubular cell proliferation, anomalies in the extracellular matrix, and a net transepithelial Cl^- secretion toward the cyst lumen (Ye and Grantham 1993; Li et al. 2004; Wallace et al. 1996; Mangos et al. 2010). Thus, a clear objective in ADPKD treatment is to slow or stop the growth of the cysts (Grantham et al. 2006a, b). Based on the known role of KCa3.1 in maintaining transepithelial Cl^- secretion, it was demonstrated that inhibition of KCa3.1 by TRAM-34 blocks Cl^- secretion and ameliorates cyst formation and enlargement in vitro (Albaqumi et al. 2008). A second kidney disease in which epithelial KCa3.1 has been implicated is diabetic nephropathy, which is a chronic complication that affects ~30 % of patients with diabetes mellitus (Reddy et al. 2008). In this case, Huang et al. (2013) demonstrated that KCa3.1 protein expression was upregulated in proximal tubular cells from patients with diabetic nephropathy (Huang et al. 2013). These authors similarly observed an increase in KCa3.1 expression in mouse models of diabetes and found that administration of TRAM-34 significantly reversed the increased KCa3.1 expression in the proximal tubule. Thus, inhibition of KCa3.1 may provide a novel approach in the treatment of diabetic nephropathy.

Table 20.1 KCa3.1 and its role in epithelial diseases

Organ	Disease	Function	References
Kidney	ADPKD	TRAM-34 blocks Cl^- secretion and ameliorates cyst formation and enlargement in vitro	Albaqumi et al. (2008)
Kidney	Diabetic nephropathy	TRAM-34 inhibited fibroblast proliferation	Huang et al. (2013)
Pancreas	Pancreas ductal adenocarcinoma	TRAM-34 blocks the proliferation of these pancreatic cells	Jager et al. (2004)
Colon	Ulcerative colitis	Decreased basolateral expression and activity of KCa3.1 channel	Al-Hazza et al. (2012)
Distal colon and rectum	Inflammatory bowel diseases	Blockade of KCa3.1 in T cells decreased inflammatory cell infiltration and tissue destruction	Al-Hazza et al. (2012); Di et al. (2010)
Respiratory tract	Cystic fibrosis and chronic obstructive pulmonary disease (COPD)	Activation of basolateral KCa3.1 with 1-EBIO or DCEBIO stimulates transepithelial Cl^- secretion	Devor et al. (1999, 2000); Singh et al. (2001); Roth et al. (2011)

As noted above, Ca^{2+}-activated K^+ channels are known to mediate the secretion of fluid and electrolytes in epithelial cells of pancreatic ducts (Nguyen and Moody 1998) and this channel has been identified as KCa3.1 (Jung et al. 2006; Hayashi et al. 2012). Whether KCa3.1 channels are essential in the dysfunction of pancreatic epithelial transport in cystic fibrosis remains unknown, however. In additional studies, it has been demonstrated that the mRNA levels of KCa3.1 are upregulated in patients with pancreas ductal adenocarcinoma and inhibitors of KCa3.1 block the proliferation of these pancreatic cells (Jager et al. 2004).

As detailed above, and in other chapters in this volume, the human colon absorbs Na^+, Cl^-, and water and simultaneously secretes K^+ and HCO_3^- (Sandle 1998). In the distal colon, the electrogenic absorption of Na^+ generates a lumen-negative electrical potential difference (PD_{te}), and it is known that the downregulation of Na^+ absorption promotes the pathogenesis of diarrhea in patients with ulcerative colitis (UC) (Sandle et al. 1986; Hawker et al. 1980; Amasheh et al. 2004; Greig et al. 2004). As indicated above, basolateral KCa3.1 plays a critical role in maintaining transepithelial ion transport by maintaining this transepithelial PD_{te}. Thus, Sandle et al. (1990) have shown that the changes in transepithelial PD_{te} associated with the inflamed human colon are at least partially the result of a depolarized basolateral membrane. Indeed, a variant of the KCa3.1 gene, KCNN4, has been reported to be associated with ileal Crohn's disease (Simms et al. 2010). In this regard, Al-Hazza et al. (2012) demonstrated that the reduced Na^+ absorption in UC was associated with an ~75 % decrease in basolateral KCa3.1 channel expression and activity in patients with UC, whereas crypts from patients with Crohn's collagenous and infective colitis showed identical levels of KCa3.1 channel protein to those of control patients.

Bridges and colleagues (Devor et al. 1999, 2000; Singh et al. 2001) demonstrated that pharmacological activation of basolateral KCa3.1 with 1-EBIO or DCEBIO in airway epithelia from both wild-type CFTR-expressing and CF airway stimulates transepithelial Cl^- secretion. These results led the authors to propose pharmacological modulation of basolateral KCa3.1 as a potential therapeutic strategy in airway diseases, including CF and chronic obstructive pulmonary disease (COPD). Indeed, Roth et al. (2011) demonstrated that activation of basolateral KCa3.1 with 1-EBIO potentiated the Cl^- secretory response to cAMP in rectal biopsy tissues from CF patients compared to age-matched controls, thereby supporting this proposal.

While KCa3.1 has also been shown to play a critical role in cells of the immune system and the targeting of this channel in T cells has shown promise for the treatment of various diseases associated with the lung, intestine, and kidney, these studies have not been discussed in this chapter due to our emphasis on epithelial function. However, we would direct the reader to several excellent reviews on this subject that have been recently written (Wulff and Zhorov 2008; Wulff and Castle 2010; Huang et al. 2014).

20.10 Conclusions

Based on the above, it is clear that basolateral membrane localized KCa3.1 plays a critical role in maintaining the electrochemical driving force for transepithelial ion and fluid transport across a wide array of epithelial tissues. More recent evidence also points to the crucial role of this channel in maintaining a hyperpolarized apical membrane potential, although additional studies are required to fully understand its regulation, trafficking, and function in this membrane. The study of any ion channel is dramatically augmented by having both activators and inhibitors such that its physiology can be explored, and it is clear that the identification of both activators (1-EBIO, DCEBIO) and inhibitors (clotrimazole) prior to the cloning of KCa3.1 undoubtedly propelled our rapid understanding of the role of these channels in epithelial physiology. The identification of both higher specificity and affinity activators and inhibitors, as reviewed in Chap. 26, has further fostered this advance. Clearly, there is still much to be learned about how KCa3.1 can be targeted for therapeutic gain in a wide array of epithelia and this will surely be a goal of investigators over the next decade.

Acknowledgments The authors would like to thank Patrick Thibodeau (University of Pittsburgh) for the modeling of KCa3.1 (Fig. 20.7) and David Hess for the trafficking model (Fig. 20.8). DCD has been supported by National Institutes of Health grants (HL092157), and KLH has been supported by a UORG grant and an OSMS Strategic Research award from the University of Otago.

References

Ackerman SJ, Liu L, Kwatia MA, Savage MP, Leonidas DD, Swaminathan GJ, Acharya KR (2002) Charcot-Leyden crystal protein (galectin-10) is not a dual function galectin with lysophospholipase activity but binds a lysophospholipase inhibitor in a novel structural fashion. J Biol Chem 277:14859–14868

Adelman JP, Maylie J, Sah P (2012) Small-conductance Ca^{2+}-activated K^+ channels: form and function. Annu Rev Physiol 75:245–269

Aggarwal SK, MacKinnon R (1996) Contribution of the S4 segment to gating charge in the Shaker K^+ channel. Neuron 16:1169–1177

Akabas MH, Stauffer DA, Xu M, Karlin A (1992) Acetylcholine receptor channel structure probed in cysteine-substitution mutants. Science 258:307–310

Akabas MH, Kaufmann C, Archdeacon P, Karlin A (1994) Identification of acetylcholine receptor channel-lining residues in the entire M2 segment of the alpha subunit. Neuron 13:919–927

Albaqumi M, Srivastava S, Li Z, Zhdnova O, Wulff H, Itani O, Wallace DP, Skolnik EY (2008) KCa3.1 potassium channels are critical for cAMP-dependent chloride secretion and cyst growth in autosomal-dominant polycystic kidney disease. Kidney Int 74:740–749

Al-Hazza A, Linley JE, Aziz Q, Maclennan KA, Hunter M, Sandle GI (2012) Potential role of reduced basolateral potassium (IKCa3.1) channel expression in the pathogenesis of diarrhoea in ulcerative colitis. J Pathol 226:463–470

Allen D, Fakler B, Maylie J, Adelman JP (2007) Organization and regulation of small conductance Ca^{2+}-activated K^+ channel multiprotein complexes. J Neurosci 27:2369–2376

Almassy J, Won JH, Begenisich TB, Yule DI (2012) Apical Ca^{2+}-activated potassium channels in mouse parotid acinar cells. J Gen Physiol 139:121–133

Amasheh S, Barmeyer C, Koch CS, Tavalali S, Mankertz J, Epple HJ, Gehring MM, Florian P, Kroesen AJ, Zeitz M, Fromm M, Schulzke JD (2004) Cytokine-dependent transcriptional down regulation of epithelial sodium channel in ulcerative colitis. Gastroenterology 126: 1711–1720

Anderson PA, Greenberg RM (2001) Phylogeny of ion channels: clues to structure and function. Comp Biochem Physiol B Biochem Mol Biol 129:17–28

Ashton N, Argent BE, Green R (1991) Characteristics of fluid secretion from isolated rat pancreatic ducts stimulated with secretin and bombesin. J Physiol (Lond) 435:533–546

Babbey CM, Ahktar N, Wang E, Chen CC, Grant BD, Dunn KW (2006) Rab10 regulates membrane transport through early endosomes of polarized Madin-Darby canine kidney cells. Mol Biol Cell 17:3156–3175

Bae H, Lee HJ, Kim K, Kim JH, Kim T, Ko JH, Bang H, Lim I (2014) The stimulating effects of nitric oxide on intermediate conductance Ca^{2+}-activated K^{+} channels in human dermal fibroblasts through PKG pathways but not PKA pathways. Chin J Physiol 57:137–151

Bailey MA, Grabe M, Devor DC (2010) Characterization of the PCMBS-dependent modification of KCa3.1 channel gating. J Gen Physiol 136:367–387

Balut CM, Gao Y, Luke C, Devor DC (2010a) Immunofluorescence-based assay to identify modulators of the number of plasma membrane KCa3.1 channels. Future Med Chem 2: 707–713

Balut CM, Gao Y, Murray SA, Thibodeau PH, Devor DC (2010b) ESCRT-dependent targeting of plasma membrane localized KCa3.1 to the lysosomes. Am J Physiol Cell Physiol 299: C1015–C1027

Balut CM, Loch CM, Devor DC (2011) Role of ubiquitylation and USP8-dependent deubiquitylation in the endocytosis and lysosomal targeting of plasma membrane KCa3.1. FASEB J 25:3938–3948

Barmeyer C, Rahner C, Yang Y, Sigworth FJ, Binder HJ, Rajendran VM (2010) Cloning and identification of tissue-specific expression of KCNN4 splice variants in rat colon. Am J Physiol Cell Physiol 299:C251–C263

Basalingappa KM, Rajendran VM, Wonderlin WF (2011) Characteristics of Kcnn4 channels in the apical membranes of an intestinal cell line. Am J Physiol Gastrointest Liver Physiol 301: G905–G911

Begenisich T, Nakamoto T, Ovitt CE, Nehrke K, Brugnara C, Alper SL, Melvin JE (2004) Physiological roles of the intermediate conductance, Ca^{2+}-activated potassium channel KCNN4. J Biol Chem 46:47681–47687

Bernard K, Bogliolo O, Soriani J, Ehrenfeld J (2003) Modulation of chloride secretion by basolateral SK4-like channels in a human bronchial cell line. J Membr Biol 196:15–31

Bertuccio CA, Lee S-L, Wu G, Butterworth MB, Hamilton KL, Devor DC (2014) Anterograde trafficking of KCa3.1 in polarized epithelia is Rab1- and Rab8-dependent and recycling endosome-independent. PLoS One 9, e92013

Bi K, Tanaka Y, Coudronniere N, Sugie K, Hong S, van Stipdonk MJ, Altman A (2001) Antigen-induced translocation of PKC-theta to membrane rafts is required for T cell activation. Nat Immunol 2:556–563

Bildl W, Strassmaier T, Thurm H, Andersen J, Eble S, Oliver D, Knipper M, Mann M, Schulte U, Adelman JP, Fakler B (2004) Protein kinase CK2 is coassembled with small conductance Ca^{2+}-activated K^{+} channels and regulates channel gating. Neuron 43:847–858

Bolanz KA, Hediger MA, Landowski CP (2008) The role of TRPV6 in breast carcinogenesis. Mol Cancer Ther 7:271–279

Browning JG, Hardcastle J, Hardcastle PT, Sanford PA (1977) The role of acetylcholine in the regulation of ion transport by rat colon mucosa. J Physiol 272:737–754

Burckhardt B-C, Gogelein H (1992) Small and maxi K^+ channels in the basolateral membrane of isolated crypts from rat distal colon: single-channel and slow whole-cell recordings. Pflugers Arch 420:54–60

Cai X, Srivastava S, Sun Y, Li Z, Wu H, Zuvela-Jelaska L, Li J, Salamon RS, Backer JM, Skolnik EY (2011) Tripartite motif containing protein 27 negatively regulates CD4 T cells by ubiquitinating and inhibiting the class II PI3K-C2beta. Proc Natl Acad Sci U S A 108:20072–20077

Cancino J, Torrealba C, Soza A, Yuseff MI, Gravotta D, Henklein P, Rodriguez-Boulan E, Gonzalez A (2007) Antibody to AP1B adaptor blocks biosynthetic and recycling routes of basolateral proteins at recycling endosomes. Mol Biol Cell 18:4872–4884

Catalan MA, Pena-Munzenmayer G, Melvin JE (2014) Ca^{2+}-dependent K^+ channels in exocrine salivary glands. Cell Calcium 55:362–368

Chang D, Dawson DC (1988) Digitonin-permeabilized colonic cell layers: demonstration of calcium-activated basolateral K^+ and Cl^- conductances. J Gen Physiol 92:281–306

Cowley EA, Linsdell P (2002) Characterization of basolateral K^+ channels underlying anion secretion in the human airway cell line Calu-3. J Physiol 538:747–757

Cuthbert AW (2001) Assessment of CFTR chloride channel openers in intact normal and cystic fibrosis murine epithelia. Br J Pharmacol 132:659–668

Cuthbert AW, Hickman ME, Thorn P, MacVinish LJ (1999) Activation of Ca^{2+}- and cAMP-sensitive K^+ channels in murine colonic epithelia by 1-ethyl-2-benzimidazolone. Am J Physiol Cell Physiol 277:C111–C120

DeCoursey TE, Chandy KG, Gupta S, Cahalan MD (1984) Voltage-gated K^+ channels in human T lymphocytes: a role in mitogenesis? Nature 307:465–468

Deutsch C (2002) Potassium channel ontogeny. Annu Rev Physiol 64:19–46

Devor DC, Duffey ME (1992) Carbachol induces K^+, Cl^- and nonselective cation conductances in T84 cells: a perforated patch-clamp study. Am J Physiol Cell Physiol 263:C780–C787

Devor DC, Frizzell RA (1993) Calcium-mediated agonists activate an inwardly rectified K^+ channel in colonic secretory cells. Am J Physiol Cell Physiol 265:C1271–C1280

Devor DC, Frizzell RA (1998) Modulation of K^+ channels by arachidonic acid in T84 cells. I. Inhibition of the Ca^{2+}-dependent K^+ channel. Am J Physiol Cell Physiol 274:C138–C148

Devor DC, Simasko SM, Duffey ME (1990) Carbachol induces oscillations of membrane potassium conductance in a colonic cell line, T84. Am J Physiol Cell Physiol 258:C318–C326

Devor DC, Ahmed Z, Duffey ME (1991) Cholinergic stimulation produces oscillations of cytosolic Ca^{2+} in a secretory epithelial cell line, T84. Am J Physiol Cell Physiol 260:C598–C608

Devor DC, Sekar MC, Frizzell RA, Duffey ME (1993) Taurodeoxycholate activates K^+ and Cl^- conductances via an IP_3-mediated release of Ca^{2+} from intracellular stores. J Clin Invest 92: 2173–2181

Devor DC, Singh AK, Bridges RJ, Frizzell RA (1996a) Modulation of Cl^- secretion by benzimidazolones. II. Coordinate regulation of apical G_{Cl} and basolateral G_K. Am J Physiol Lung Cell Mol Physiol 271:L785–L795

Devor DC, Singh AK, Frizzell RA, Bridges RJ (1996b) Modulation of Cl^- secretion by benzimidazolones. I. Direct activation of a Ca^{2+}-dependent K^+ channel. Am J Physiol Lung Cell Mol Physiol 271:L775–L784

Devor DC, Singh AK, Gerlach AC, Frizzell RA, Bridges RJ (1997) Inhibition of intestinal Cl^- secretion by clotrimazole: direct effect on basolateral membrane K^+ channels. Am J Physiol Cell Physiol 273:C531–C540

Devor DC, Singh AK, Lambert LC, DeLuca A, Frizzell RA, Bridges RJ (1999) Bicarbonate and chloride secretion in Calu-3 human airway epithelial cells. J Gen Physiol 113:743–760

Devor DC, Bridges RJ, Pilewski JM (2000) Pharmacological modulation of ion transport across wild-type and ΔF508 CFTR-expressing human bronchial epithelia. Am J Physiol Cell Physiol 279:C461–C479

Dharmsathaphorn K, Pandol SJ (1986) Mechanism of chloride secretion induced by carbachol in a colonic epithelial cell line. J Clin Invest 77:348–354

Dharmsathaphorn K, McRoberts JA, Mandel KG, Tisdale LD, Masui H (1984) A human colonic tumor cell line that maintains vectorial electrolyte transport. Am J Physiol Gastrointest Liver Physiol 246:G204–G208
Dharmsathaphorn K, Mandel KG, Masui H, McRoberts JA (1985) Vasoactive intestinal polypeptide-induced chloride secretion by a colonic epithelial cell line: direct participation of a basolaterally localized Na^+, K^+, Cl^- cotransport system. J Clin Invest 75:462–471
Dharmsathaphorn K, Cohn J, Beuerlein G (1989) Multiple calcium-mediated effector mechanisms regulate chloride secretory responses in T84 cells. Am J Physiol Cell Physiol 256:C1224–C1230
Dhennin-Duthille I, Gautier M, Faouzi M, Guilbert A, Brevet M, Vaudry D, Ahidouch A, Sevestre H, Ouadid-Ahidouch H (2011) High expression of transient receptor potential channels in human breast cancer epithelial cells and tissues: correlation with pathological parameters. Cell Physiol Biochem 28:813–822
Di L, Srivastava S, Zhdanova O, Sun Y, Li Z, Skolnik EY (2010) Nucleoside diphosphate kinase B knock-out mice have impaired activation of the K^+ channel KCa3.1, resulting in defective T cell activation. J Biol Chem 285:38765–38771
Dong C, Yang L, Zhang X, Gu H, Lam ML, Claycomb WC, Xia H, Wu G (2010) Rab8 interacts with distinct motifs in α_{2B}- and β_2-adrenergic receptors and differentially modulates their transport. J Biol Chem 285:20369–20380
Doyle DA, Cabral JM, Pfuetzner RA, Kuo A, Gulbis JM, Cohen SL, Chait BT, MacKinnon R (1998) The structure of the potassium channel: molecular basis of K^+ conduction and selectivity. Science 280:69–77
Dunn PM (1998) The action of blocking agents applied to the inner face of Ca^{2+}-activated K^+ channels from human erythrocytes. J Membr Biol 165:133–143
Ellgaard L, Molinari M, Helenius A (1999) Setting the standards: quality control in the secretory pathway. Science 286:1882–1888
Falke JJ, Dernburg AF, Sternberg DA, Zalkin N, Milligan DL, Koshland DE Jr (1988) Structure of a bacterial sensory receptor. A site-directed sulfhydryl study. J Biol Chem 263:14850–14858
Fanger CM, Ghanshani S, Logsdon NJ, Rauer H, Kalman K, Zhou J, Beckingham K, Chandy KG, Cahalan MD, Aiyar J (1999) Calmodulin mediates calcium-dependent activation of the intermediate conductance KCa channel, IKCa1. J Biol Chem 274:5746–5754
Flores CA, Melvin JE, Figueroa CD, Sepulveda FV (2007) Abolition of Ca^{2+}-mediated intestinal anion secretion and increased stool dehydration in mice lacking the intermediate conductance Ca^{2+}-dependent K^+ channel KCNN4. J Physiol (London) 583:705–717
Florey HW, Wright RD, Jennings MA (1941) The secretions of the intestine. Physiol Rev 21: 36–69
Fölsch H, Ohno H, Bonifacino JS, Mellman I (1999) A novel clathrin adaptor complex mediates basolateral targeting in polarized epithelial cells. Cell 99:189–198
Foskett JK, Gunter-Smith PJ, Melvin JE, Turner RJ (1989) Physiological localization of an agonist-sensitive pool of Ca^{2+} in parotid acinar cells. Proc Natl Acad Sci U S A 86:161–171
Fuller SD, Bravo R, Simons K (1985) An enzymatic assay reveals that proteins destined for the apical or basolateral domains of an epithelial cell line share the same late Golgi compartments. EMBO J 4:297–307
Furness JB, Robbins HL, Selmer IS, Hunne B, Chen MX, Hicks GA, Moore S, Neylon CB (2003) Expression of intermediate conductance potassium channel immunoreactivity in neurons and epithelial cells of rat gastrointestinal tract. Cell Tissue Res 314:179–189
Futter CE, Gibson A, Allchin EH, Maxwell S, Ruddock LJ, Odorizzi G, Domingo D, Trowbridge IS, Hopkins CR (1998) In polarized MDCK cells basolateral vesicles arise from clathrin-gamma-adaptin-coated domains on endosomal tubules. J Cell Biol 141:611–623

Gao L, Yankaskas JR, Fuller CM, Sorscher EJ, Matalon S, Forman HJ, Venglarik CJ (2001) Chlorzoxazone or 1-EBIO increases Na^+ absorption across cystic fibrosis airway epithelial cells. Am J Physiol Lung Cell Mol Physiol 281:L1123–L1129

Gao Y, Chotoo CK, Balut CM, Sun F, Bailey MA, Devor DC (2008) Role of S3 and S4 transmembrane domain charged amino acids in channel biogenesis and gating of KCa2.3 and KCa3.1. J Biol Chem 283:9049–9059

Gao Y, Balut CM, Bailey MA, Patino-Lopez G, Shaw S, Devor DC (2010) Recycling of the Ca^{2+}-activated K^+ channel, KCa2.3, is dependent upon RME-1, Rab35/EPI64C, and an N-terminal domain. J Biol Chem 285:17938–17953

Garneau L, Klein H, Banderali U, Longpré-Lauzon A, Parent L, Sauvé R (2009) Hydrophobic interactions as key determinants to the KCa3.1 channel closed configuration. An analysis of KCa3.1 mutants constitutively active in zero Ca^{2+}. J Biol Chem 284:389–403

Gerlach AC, Gangopadhyay NN, Devor DC (2000) Kinase-dependent regulation of the intermediate conductance, calcium-dependent potassium channel, hIK1. J Biol Chem 275:585–598

Gerlach AC, Syme CA, Giltinan L, Adelman JP, Devor DC (2001) ATP-dependent activation of the intermediate conductance, Ca^{2+}-activated K^+ channel, hIK1, is conferred by a C-terminal domain. J Biol Chem 276:10963–10970

Ghiani CA, Eisen A, Yuan X, DePinho RA, McBain CJ, Gallo V (1999) Neurotransmitter receptor activation triggers $p27^{Kip1}$ and $p21^{CIP1}$ accumulation and G1 cell cycle arrest in oligodendrocyte progenitors. Development 126:1077–1090

Grantham JJ, Chapman AB, Torres VE (2006a) Volume progression in autosomal dominant polycystic kidney disease: the major factor determining clinical outcomes. Clin J Am Soc Nephrol 1: 148–157

Grantham JJ, Torres VE, Chapman AB, Guay-Woodford LM, Bae KT, King BF Jr, Wetzel LH, Baumgarten DA, Kenney PJ, Harris PC, Klahr S, Bennet WM, Hirscvhman GN, Meyers CM, Zhang X, Zhu F, Miller JP, Investigators CRISP (2006b) Volume progression in polycystic kidney disease. N Engl J Med 354:2122–2130

Gray MA, Pollard CE, Harris A, Coleman L, Greenwell JR, Argent BE (1990) Anion selectivity and block of the small conductance chloride channel on pancreatic duct cells. Am J Physiol Cell Physiol 259:C752–C761

Greger R, Schlatter E (1984a) Mechanism of NaCl secretion in the rectal gland of spiny dogfish (*Squalus acanthias*). I. Experiments in isolated in vitro perfused rectal gland tubules. Pflügers Arch 402:63–75

Greger R, Schlatter E (1984b) Mechanism of NaCl secretion in the rectal gland of spiny dogfish (*Squalus acanthias*). II. Effects of inhibitors. Pflugers Arch 402:364–375

Greger R, Schlatter E, Wang F, Forrest JN Jr (1984) Mechanism of NaCl secretion in rectal gland tubules of spiny dogfish (*Squalus acanthias*). III. Effects of stimulation of secretion by cyclic AMP. Pflugers Arch 402:376–384

Greig ER, Boot-Handford RP, Mani V, Sandle GI (2004) Decreased expression of apical Na+ channels and basolateral Na+, K+-ATPase in ulcerative colitis. J Pathol 204:84–92

Grygorczyk R, Schwarz W (1985) Ca^{2+}-activated K^+ permeability in human erythrocytes: modulation of single-channel events. Eur Biophys J 12:57–65

Gutman GA, Chandy KG, Adelman JP, Aiyar J, Bayliss DA, Clapham DE, Covarriubias M, Desir GV, Furuichi K, Ganetzky B, Garcia ML, Grissmer S, Jan LY, Karschin A, Kim D, Kuperschmidt S, Kurachi Y, Lazdunski M, Lesage F, Lester HA, McKinnon D, Nichols CG, O'Kelly I, Robbins J, Robertson GA, Rudy B, Sanguinetti M, Seino S, Stuehmer W, Tamkun MM, Vandenberg CA, Wei A, Wulff H, Wymore RS, International Union of Pharmacology (2003) International Union of Pharmacology. XLI. Compendium of voltage-gated ion channels: potassium channels. Pharmacol Rev 55:583–586

Hamilton KL, Meads L, Butt AG (1999) 1-EBIO stimulates Cl^- secretion by activating a basolateral K^+ channel in the mouse jejunum. Pflugers Arch 439:158–166
Hamilton KL, Syme CA, Devor DC (2003) Molecular localization of the inhibitory arachidonic acid binding site to the pore of hIK1. J Biol Chem 278:16690–16697
Hardcastle PT, Eggenton J (1973) The effect of acetylcholine on the electrical activity of intestinal epithelial cells. Biochim Biophys Acta 298:95–100
Hawker PC, McKay JS, Turnberg LA (1980) Electrolyte transport across colonic mucosa from patients with inflammatory bowel disease. Gastroenterology 79:508–511
Hayashi M, Wang J, Hede SE, Novak I (2012) An intermediate-conductance Ca^{2+}-activated K^+ channel is important for secretion in pancreatic duct cells. Am J Physiol Cell Physiol 303: C151–C159
Hazelton B, Mitchell B, Tupper J (1979) Calcium, magnesium, and growth control in the WI-38 human fibroblast cell. J Cell Biol 83:487–498
Hede SE, Amstrup J, Klaerke DA, Novak I (2005) $P2Y_2$ and $P2Y_4$ receptors regulate pancreatic Ca^{2+}-activated K^+ channel differently. Pflugers Arch 450:429–436
Hirschberg B, Maylie J, Adelman JP, Marrion NV (1998) Gating of recombinant small-conductance Ca-activated K^+ channels by calcium. J Gen Physiol 111:565–581
Holmgren M, Liu Y, Xu Y, Yellen G (1996) On the use of thiol-modifying agents to determine channel topology. Neuropharmacology 35:797–804
Hoque KM, Woodward OM, van Rossum DB, Zachos NC, Chen L, Leung GP, Guggino WB, Guggino SE, Tse CM (2010) Epac1 mediates protein kinase A-independent mechanism of forskolin-activated intestinal chloride secretion. J Gen Physiol 135:43–58
Howarth M, Takao K, Hayashi Y, Ting AY (2005) Targeting quantum dots to surface proteins in living cells with biotin ligase. Proc Natl Acad Sci U S A 102:7583–7588
Huang C, Shen S, Ma Q, Chen J, Gill A, Pollock CA, Chen XM (2013) Blockade of KCa3.1 ameliorates renal fibrosis through the TGF-β1/Smad pathway in diabetic mice. Diabetes 62: 2923–2934
Huang C, Pollock CA, Chen XM (2014) Role of potassium channel KCa3.1 in diabetic nephropathy. Clin Sci (Lond) 127:423–433
Hubel KA (1976) Intestinal ion transport: effect of norepinephrine, pilocarpine, and atropine. Am J Physiol 231:252–257
Hubel KA (1977) Effects of bethanechol on intestinal ion transport in the rat. Proc Soc Exp Bio Med 154:41–44
Isaacs PET, Corbett CL, Riley AK, Hawker PC, Turnberg LA (1976) *In vitro* behavior of human intestinal mucosa. J Clin Invest 58:535–542
Ishiguro H, Steward MC, Naruse S, Ko SBH, Goto H, Case RM, Kondo T, Yamamoto A (2009) CFTR functions as a bicarbonate channel in pancreatic duct cells. J Gen Physiol 133:315–326
Ishii TM, Silvia C, Hirschberg B, Bond CT, Adelman JP (1997) A human intermediate conductance calcium-activate potassium channel. Proc Natl Acad Sci U S A 94:11651–11656
Jacobowitz D (1965) Histochemical studies of the autonomic innervation of the gut. J Pharmacol Exp Ther 149:358–364
Jager H, Dreker T, Buck A, Giehl K, Gress T, Grissmer S (2004) Blockage of intermediate-conductance Ca^{2+} activated K^+ channels inhibit human pancreatic cancer cell growth in vitro. Mol Pharmacol 65:630–638
Jensen BS, Strobaek D, Christophersen P, Jorgensen TD, Hansen C, Silahtaroglu A, Olesen SP, Ahring PK (1998) Characterization of the cloned human intermediate-conductance Ca^{2+}-activated K^+ channel. Am J Physiol Cell Physiol 275:C848–C856
Jensen BS, Strobaek D, Olesen SP, Christophersen P (2001) The Ca^{2+}-activated K^+ channel of intermediate conductance: a molecular target for novel treatments? Curr Drug Targets 2: 401–422
Joiner WJ, Wang L-Y, Tang MD, Kaczmarek LK (1997) hSK4, a member of a novel subfamily of calcium-activated potassium channels. Proc Natl Acad Sci U S A 94:11013–11018

Joiner WJ, Khanna R, Schlichter LC, Kaczmarek LK (2001) Calmodulin regulates assembly and trafficking of SK4/IK1 Ca^{2+}-activated K^+ channels. J Biol Chem 276:37980–37985

Joiner WJ, Basavappa S, Vidyasagar S, Nehrke K, Krishnan S, Binder HJ, Boulpaep EL, Rajendran VM (2003) Active K^+ secretion through multiple K_{Ca}-type channels and regulation by IK_{Ca} channels in rat proximal colon. Am J Physiol Liver Physiol 285:G185–G196

Jones HM, Hamilton KL, Papworth GD, Syme CA, Watkins SC, Bradbury NA, Devor DC (2004) Role of the NH_2 terminus in the assembly and trafficking of the intermediate conductance Ca^{2+}-activated K^+ channel hIK1. J Biol Chem 279:15531–15540

Jones HM, Hamilton KL, Devor DC (2005) Role of an S4-S5 linker lysine in the trafficking of the Ca^{2+}-activated K^+ channels IK1 and SK3. J Biol Chem 280:37257–37265

Jones HM, Bailey MA, Baty CJ, MacGregor GG, Syme CA, Hamilton KL, Devor DC (2007) An NH2-terminal multi-basic RKR motif is required for the ATP-dependent regulation of hIK1. Channels (Austin) 2:80–91

Jung SR, Kim K, Hille B, Nguyen TD, Koh DS (2006) Pattern of Ca^{2+} increase determines the type of secretory mechanism activated in dog pancreatic duct epithelial cells. J Physiol 576: 163–178

Karlin A, Akabas MH (1998) Substituted-cysteine accessibility method. Methods Enzymol 293: 123–145

Keen JE, Khawaled R, Farrens DL, Neelands T, Rivard A, Bond CT, Janowsky A, Fakler B, Adelman JP, Maylie J (1999) Domains responsible for constitutive and Ca^{2+}-dependent interactions between calmodulin and small conductance Ca^{2+}-activated potassium channels. J Neurosci 19:8830–8838

Khanna R, Chang MC, Joiner WJ, Kaczmarek LK, Schlichter LC (1999) hSK4/hIK1, a calmodulin-binding KCa channel in human T lymphocytes. Roles in proliferation and volume regulation. J Biol Chem 274:14838–14849

Klein H, Garneau L, Banderali U, Simoes M, Parent L, Sauvé R (2007) Structural determinants of the closed KCa3.1 channel pore in relation to channel gating: results from a substituted cysteine accessibility analysis. J Gen Physiol 129:299–315

Klein H, Garneau L, Trinh NT, Prive A, Dionne F, Goupil E, Thuringer D, Parent L, Brochiero E, Sauve R (2009) Inhibition of the KCa3.1 channels by AMP-activated protein kinase in human airway epithelial cells. Am J Physiol Cell Physiol 296:C285–C295

Kobe B, Deisenhofer J (1994) The leucine-rich repeat: a versatile binding motif. Trends Biochem Sci 19:415–421

Kohler R, Hirschberg B, Bond CT, Kinzie JM, Marrion NV, Maylie J, Adelman JP (1996) Small-conductance, calcium-activated potassium channels from mammalian brain. Science 273: 1709–1714

Lallet-Daher H, Roudbaraki M, Bavencoffe A, Mariot P, Gackière F, Bidaux G, Urbain R, Gosset P, Delcourt P, Fleurisse L, Slomianny C, Dewailly E, Mauroy B, Bonnal JL, Skryma R, Prevarskaya N (2009) Intermediate-conductance Ca^{2+}-activated K^+ channels (IKCa1) regulate human prostate cancer cell proliferation through a close control of calcium entry. Oncogene 28:1792–1806

Ledoux J, Bonex AD, Nelson MT (2008) Ca^{2+}-activated K^+ channels in murine endothelial cells: block by intracellular calcium and magnesium. J Gen Physiol 131:125–135

Lee WS, Ngo-Anh TJ, Bruening-Wright A, Maylie J, Adelman JP (2003) Small conductance Ca^{2+}-activated K^+ channels and calmodulin: cell surface expression and gating. J Biol Chem 278:25940–25946

Lee S-Y, Lee A, Chen J, MacKinnon R (2005) Structure of the KvAP voltage-dependent K^+ channel and its dependence on the lipid domain. Proc Natl Acad Sci U S A 102:15441–15446

Leinders T, van Kleef RG, Vijverberg HP (1992) Single Ca^{2+}-activated K^+ channels in human erythrocytes: Ca^{2+} dependence of opening frequency but not of open lifetimes. Biochim Biophys Acta 1112:67–74

Li H, Findlay IA, Sheppard DN (2004) The relationship between cell proliferation, Cl^- secretion, and renal cyst growth: a study using CFTR inhibitors. Kidney Int 66:1926–1938

Li H, Yao J, Tong X, Guo Z, Wu Y, Sun L, Pan N, Wu H, Xu T, Ding J (2007) Interaction sites between the *Slo1* pore and the NH_2 terminus of the β_2 subunit, probed with a three-residue sensor. J Biol Chem 282:17720–17728

Li W, Halling DB, Hall AW, Aldrich RW (2009) EF hands at the N-lobe of calmodulin are required for both SK channel gating and stable SK-calmodulin interaction. J Gen Physiol 134: 281–293

Linsdell P, Tabcharani JA, Hanrahan JW (1997) Multi-ion mechanism for ion permeation and block in the cystic fibrosis transmembrane conductance regulator chloride channel. J Gen Physiol 110:365–377

Logsdon NJ, Kang J, Togo JA, Christian EP, Aiyar J (1997) A novel gene, hKCa4, encodes the calcium-activated potassium channel in human T lymphocytes. J Biol Chem 272:32723–32726

Lohrmann E, Burhoff I, Nitschke RB, Lang HJ, Mania D, Englert HC, Hropot M, Warth R, Rohm W, Bleich M, Greger R (1995) A new class of inhibitors of cAMP-mediated Cl secretion in rabbit colon, acting by the reduction of cAMP-activated K conductance. Pfluegers Arch 429:517–530

Loo DDF, Kaunitz JD (1989) Ca^{2+} and cAMP activate K^+ channels in the basolateral membrane of crypt cells isolated from rabbit distal colon. J Membr Biol 110:19–28

Lu Y, Xiong X, Helm A, Kimani K, Bragin A, Skach WR (1998) Co- and posttranslational translocation mechanisms direct cystic fibrosis transmembrane conductance regulator N terminus transmembrane assembly. J Biol Chem 273:568–576

MacVinish LJ, Keogh J, Cuthbert AW (2001) EBIO, an agent causing maintained epithelial chloride secretion by co-ordinate actions at both apical and basolateral membranes. Pflugers Arch 443: S127–S131

Madara JL, Stafford J, Dharmsathaphorn K, Carlson S (1987) Structural analysis of a human intestinal epithelial cell line. Gastroenterology 92:1133–1145

Mall M, Gonska T, Thomas J, Schreiber R, Seyedwitz HH, Kuehr J, Brandis M, Kunzelmann K (2003) Modulation of Ca^{2+}-activated Cl^- secretion by basolateral K^+ channels in human normal and cystic fibrosis airway epithelia. Pediatr Res 53:608–618

Mandell KG, Dharmsathaphorn K, McRoberts JA (1986) Characterization of a cAMP-activated Cl^- transport pathway in the apical membrane of a human colonic epithelial cell line. J Biol Chem 261:704–712

Mangos S, Lam PY, Zhao A, Liu Y, Mudumana S, Vasilyev A, Liu A, Drummond IA (2010) The ADPKD genes pkd1a/b and pkd2 regulate extracellular matrix formation. Dis Model Mech 3:354–365

Martinez O, Goud B (1998) Rab proteins. Biochim Biophys Acta 1404:101–112

Maruyama Y, Petersen OH (1984) Control of K^+ conductance by cholecystokinin and Ca^{2+} in single pancreatic acinar cells studied by the patch-clamp technique. J Membr Biol 79:293–300

Maruyama Y, Gallacher DV, Petersen OH (1983a) Voltage and Ca^{2+}-activated K^+ channel in basolateral acinar cell membranes of mammalian salivary glands. Nature 302:827–829

Maruyama Y, Petersen OH, Flanagan P, Pearson GT (1983b) Quantification of Ca^{2+}-activated K^+ channels under hormonal control in pig pancreas acinar cells. Nature 305:228–232

McCann JD, Welsh MJ (1990) Basolateral K^+ channels in airway epithelia. II. Role in Cl^- secretion and evidence for two types of K^+ channel. Am J Physiol Lung Cell Mol Physiol 258:L343–L348

McCann JD, Matsuda J, Garcia M, Kaczorowski G, Welsh MJ (1990) Basolateral K^+ channels in airway epithelia I. Regulation by Ca^{2+} and block by charybdotoxin. Am J Physiol Lung Cell Mol Physiol 258:L334–L342

Millership J, Devor DC, Hamilton KL, Balut CM, Bruce JIE, Fearson IM (2011) Calcium activated K^+ channels increase cell proliferation independent of K^+ conductance. Am J Physiol Cell Physiol 300:C792–C802

Morales P, Garneau L, Klein H, Lavoie MF, Parent L, Sauvé R (2013) Contribution of the KCa3.1 channel-calmodulin interactions to the regulation of the KCa3.1 gating process. J Gen Physiol 142:37–60

Morris AP, Gallacher DV, Lee JAC (1986) A large conductance, voltage-and calcium-activated K^+ channel in the basolateral membrane of rat enterocytes. FEBS Lett 206:87–92

Muth TR, Caplan MJ (2003) Transport protein trafficking in polarized cells. Annu Rev Cell Dev Biol 19:333–366

Myers MP, Khanna R, Lee EJ, Papazian DM (2004) Voltage sensor mutations differentially target misfolded K^+ channel subunits to proteasomal and non-proteasomal disposal pathways. FEBS Lett 568:110–116

Nakamoto T, Srivastava A, Romanenko VG, Ovitt CE, Perez-Cornejo P, Arreola J, Begenisich T, Melvin JE (2007) Functional and molecular characterization of the fluid secretion mechanism in human parotid acinar cells. Am J Physiol Regul Integr Comp Physiol 292:R2380–R2390

Nehrke K, Quinn CC, Begenisich T (2003) Molecular identification of Ca^{2+}-activated K^+ channels in parotid acinar cells. Am J Physiol Cell Physiol 284:C535–C546

Nguyen TD, Moody MW (1998) Calcium-activated potassium conductances on cultured non-transformed dog pancreatic duct epithelial cells. Pancreas 17:348–358

Nicolaou SA, Neumeier L, Peng Y, Devor DC, Conforti L (2007) The Ca^{2+}-activated K^+ channel KCa3.1 compartmentalizes in the immunological synapse of human T lymphocytes. Am J Physiol Cell Physiol 292:C1431–C1439

Nilius B, Wohlrab W (1992) Potassium channels and regulation of proliferation of human melanoma cells. J Physiol 445:537–548

O'Hara B, de la Rosa DA, Rajendra VM (2014) Multiple mineralocorticoid response elements localized in different introns regulate intermediate conductance K^+ (KCNN4) channel expression in the rat distal colon. PLoS One 9, e98695

Ohno H, Aguilar RC, Yeh D, Taura D, Saito T (1998) The medium subunits of adaptor complexes recognize distinct but overlapping sets of tyrosine-based sorting signals. J Biol Chem 273: 25915–25921

Ohya S, Niwa S, Yanagi A, Fukuyo Y, Yamamura H, Imaizumi Y (2011) Involvement of dominant-negative spliced variants of the intermediate conductance Ca^{2+}-activated K^+ channel, $K_{Ca}3.1$, in immune function of lymphoid cells. J Biol Chem 286:16940–16952

Orzech E, Cohen S, Weiss A, Aroeti B (2000) Interactions between the exocytic and endocytic pathways in polarized Madin–Darby canine kidney cells. J Biol Chem 275:15207–15219

Ouadid-Ahidouch H, Roudbaraki M, Delcourt P, Ahidouch A, Joury N, Prevarskaya N (2004) Functional and molecular identification of intermediate-conductance Ca^{2+}-activated K^+ channels in breast cancer cells: association with cell cycle progression. Am J Physiol Cell Physiol 287:C125–C134

Papazian DM (1999) Potassium channels: some assembly required. Neuron 23:7–10

Papazian DM, Shao XM, Seoh S-A, Mock AF, Huang Y, Wainstock DH (1995) Electrostatic interactions of S4 voltage sensor in Shaker K^+ channel. Neuron 14:1293–1301

Parihar AS, Coghlan MJ, Gopalakrishnan M, Shieh C-C (2003) Effects of intermediate-conductance Ca^{2+}-activated K^+ channel modulators on human prostate cancer cell proliferation. Eur J Pharmacol 417:157–164

Pedersen KA, Schroder RL, Skaaning-Jensen B, Strobeak D, Olesen SP, Christophersen P (1999) Activation of the human intermediate-conductance Ca^{2+}-activated K^+ channel by 1-ethyl-2-benzimidazolinone is strongly Ca^{2+}-dependent. Biochim Biophys Acta 1420:231–240

Reddy GR, Kotlyarevska K, Ransom RF, Menon RK (2008) The podocyte and diabetes mellitus: is the podocyte the key to the origins of diabetic nephropathy? Curr Opin Nephrol Hypertens 17:32–36

Reuss L, Reinach P, Weinman SA, Grady TP (1983) Intracellular ion activities and Cl^- transport mechanisms in bullfrog corneal epithelium. Am J Physiol Cell Physiol 244:C336–C347
Rimele TJ, O'Dorisio MS, Gaginella TS (1981) Evidence for muscarinic receptors on rat colonic epithelial cells: binding of [^{3}H]quinuclidinyl benzilate. J Pharmacol Exp Ther 218:426–434
Roberts DD, Lewis SD, Ballou DP, Olson ST, Shafer JA (1986) Reactivity of small thiolate anions and cysteine-25 in papain toward methyl methanethiosulfonate. Biochemistry 25:5595–5601
Romanenko V, Nakamoto T, Srivastava A, Melvin JE, Begenisich T (2006) Molecular identification and molecular roles of parotid acinar maxi-K channels. J Physiol 281:27964–27972
Romanenko VG, Nakamoto T, Srivastava A, Begenisich T, Melvin JE (2007) Regulation of membrane potential and fluid secretion by Ca^{2+}-activated K^+ channels in mouse submandibular glands. J Physiol 581:801–817
Romanenko VG, Roser KS, Melvin JE, Begenisich T (2009) The role of cholesterol and the cytoskeleton in the interaction between IK1 and maxi-K channels. Am J Physiol Cell Physiol 296:C878–C888
Roth EK, Hirtz S, Duerr J, Wenning D, Eichler I, Seydwitz HH, Amaral MD, Mall MA (2011) The K^+ channel opener 1-EBIO potentiates residual function of mutant CFTR in rectal biopsies from cystic fibrosis patients. PLoS One 6, e24445
Rufo PA, Jiang L, Moe SJ, Brugnara C, Alper SL, Lencer WI (1996) The antifungal antibiotic, clotrimazole, inhibits Cl^- secretion by polarized monolayers of human colonic epithelial cells. J Clin Invest 98:2066–2075
Rufo PA, Merlin D, Riegler M, Ferguson-Maltzman MH, Dickinson BL, Brugnara C, Alper SL, Lencer WI (1997) The antifungal antibiotic, clotrimazole, inhibits chloride secretion by human intestinal T84 cells via blockade of distinct basolateral K^+ conductances. Demonstration of efficacy in intact rabbit colon and in an in vivo mouse model of cholera. J Clin Invest 100: 3111–3120
Sadlish H, Skach WR (2004) Biogenesis of CFTR and other polytopic membrane proteins: new roles for the ribosome-translocon complex. J Membr Biol 202:115–126
Sandle GI (1998) Salt and water absorption in the human colon: a modern appraisal. Gut 43: 294–299
Sandle GI, Wills NK, Alles W, Binder HJ (1986) Electrophysiology of the human colon: evidence of segmental heterogeneity. Gut 27:999–1005
Sandle GI, Higgs N, Crowe P, Marsh MM, Venkatesan S, Peters TJ (1990) Cellular basis for defective electrolyte transport in inflamed human colon. Gastroenterology 99:97–105
Sato Y, Sakaguchi M, Goshima S, Nakamura T, Uozumi N (2002) Integration of Shaker-type K^+ channel, KAT1, into the endoplasmic reticulum membrane: synergistic insertion of voltage-sensing segments, S3-S4, and independent insertion of pore-forming segments, S5-P-S6. Proc Natl Acad Sci U S A 99:60–65
Sato Y, Sakaguchi M, Goshima S, Nakamura T, Uozumi N (2003) Molecular dissection of the contribution of negatively and positively charged residues in S2, S3, and S4 to the final membrane topology of the voltage sensor in the K^+ channel, KAT1. J Biol Chem 278: 13227–13234
Schreiber R (2005) Ca^{2+} signaling, intracellular pH and cell volume in cell proliferation. J Membr Biol 205:129–137
Schumacher MA, Rivard AF, Bachinger HP, Adelman JP (2001) Structure of the gating domain of a Ca^{2+}-activated K^+ channel complexed with Ca^{2+}/calmodulin. Nature 410:1120–1124
Schumacher MA, Crum M, Miller MC (2004) Crystal structures of apocalmodulin and an apocalmodulin/SK potassium channel gating domain complex. Structure 12:849–860
Schwab A, Wulf A, Schulz C, Kessler W, Nechyporuk-Zloy V, Romer M, Reinhardt J, Weinhold D, Dieterich P, Stock C, Hebert SC (2006) Subcellular distribution of calcium-sensitive potassium channels (IK1) in migrating cells. J Cell Physiol 206:86–94

Schwab A, Nechyporuk-Zloy V, Gassner B, Schulz C, Kessler W, Mally S, Romer M, Stock C (2012) Dynamic redistribution of calcium sensitive potassium channels ($hK_{Ca}3.1$) in migrating cells. J Cell Physiol 227:686–696

Seoh S-A, Sigg D, Papazian DM, Bezanilla F (1996) Voltage-sensing residues in the S2 and S4 segments of the Shaker K^+ channel. Neuron 16:1159–1167

Sepulveda FV, Mason WT (1985) Single channel recordings obtained from basolateral membranes of isolated rabbit enterocytes. FEBS Lett 191:87–91

Sheikh IA, Koley H, Chakrabarti MJ, Hoque KM (2013) The EZpac1 signaling pathway regulates Cl^- secretion via modulation of apical KCNN4c channels in diarrhea. J Biol Chem 288: 20404–20415

Simms LA, Doecke JD, Roberts RL, Fowler EV, Zhao ZZ, McGuckin MA, Huang N, Hayward NK, Webb PM, Whiteman DC, Cavanaugh JA, McCallum R, Florin TH, Barclay ML, Gearry RB, Merriman TR, Montgomery GW, Radford-Smith GL (2010) KCNN4 gene variant is associated with ileal Crohn's Disease in the Australian and New Zealand population. Am J Gastroenterol 105:2209–2217

Simoes M, Garneau L, Klein H, Banderali U, Hobeila F, Roux B, Parent L, Sauvé R (2002) Cysteine mutagenesis and computer modeling of the S6 region of an intermediate conductance IKCa channel. J Gen Physiol 120:99–116

Singh S, Syme CA, Singh AK, Devor DC, Bridges RJ (2001) Benzimidazolone activators of chloride secretion: potential therapeutics for cystic fibrosis and chronic obstructive pulmonary disease. J Pharmacol Exp Ther 296:600–611

Singh SK, O'Hara B, Talukder JR, Rajendran VM (2012) Aldosterone induces active K^+ secretion by enhancing mucosal expression of Kcnn4c and Kcnma1 channels in rat distal colon. Am J Physiol Cell Physiol 302:C1353–C1360

Skalhegg BS, Tasken K, Hansson V, Huitfeldt HS, Jahnsen T, Lea T (1994) Location of cAMP-dependent protein kinase type I with the TCR-CD3 complex. Science 263:84–87

Smith PL, Frizzell RA (1984) Chloride secretion by canine tracheal epithelium IV. Basolateral membrane K permeability parallels secretion rate. J Membr Biol 77:187–199

Smith JJ, Welsh MJ (1992) cAMP stimulates bicarbonate secretion across normal, but not cystic fibrosis airway epithelia. J Clin Invest 89:1148–1153

Srivastava S, Li Z, Lin L, Liu G, Ko K, Coetzee WA, Skolnik EY (2005) The phosphatidylinositol 3-phosphate phosphatase myotubularin- related protein 6 (MTMR6) is a negative regulator of the Ca^{2+}-activated K^+ channel KCa3.1. Mol Cell Biol 25:3630–3638

Srivastava S, Choudhury P, Li Z, Liu G, Nadkarni V, Ko K, Coetzee WA, Skolnik EY (2006) Phosphatidylinositol 3-phosphate indirectly activates KCa3.1 via 14 amino acids in the carboxy terminus of KCa3.1. Mol Biol Cell 17:146–154

Srivastava S, Zhdanova O, Di L, Li Z, Albaqumi M, Wulff H, Skolnik EY (2008) Protein histidine phosphatase 1 negatively regulates CD4 T cells by inhibiting the K^+ channel KCa3.1. Proc Natl Acad Sci U S A 105:14442–14446

Strobl JS, Wonderlin WF, Flynn DC (1995) Mitogenic signal transduction in human breast cancer cells. Gen Pharmacol 26:1643–1649

Sukhareva M, Hackos DH, Swartz KJ (2003) Constitutive activation of the Shaker Kv channel. J Gen Physiol 122:541–556

Suzuki K, Petersen OH (1985) The effects of Na^+ and Cl^- removal and of loop diuretics on acetylcholine-evoked membrane potential changes in mouse lacrimal acinar cells. Q J Exp Physiol 70:437–445

Syme CA, Gerlach AC, Singh AK, Devor DC (2000) Pharmacological activation of cloned intermediate- and small-conductance Ca^{2+}-activated K^+ channels. Am J Physiol Cell Physiol 278:C570–C581

Syme CA, Hamilton KL, Jones HM, Gerlach AC, Giltinan L, Papworth GD, Watkins SC, Bradbury NA, Devor DC (2003) Trafficking of the Ca^{2+}-activated K^+ channel, hIK1, is dependent upon a C-terminal leucine zipper. J Biol Chem 278:8476–8486

Thompson J, Begenisich T (2006) Membrane-delimited inhibition of maxi-K channel activity by the intermediate conductance Ca^{2+}-activated K channel. J Gen Physiol 127:159–169

Thompson J, Begenisich T (2009) Mechanistic details of BK channel inhibition by the intermediate conductance, Ca^{2+}-activated K channel. Channels 3:194–204

Tidball CS (1961) Active chloride transport during intestinal secretion. Am J Physiol 200:309–312

Tiwari-Woodruff SK, Schulteis CT, Mock AF, Papazian DM (1997) Electrostatic interactions between transmembrane segments mediate folding of Shaker K^+ channel subunits. Biophys J 72:1489–1500

Tiwari-Woodruff SK, Lin MA, Schulteis CT, Papazian DM (2000) Voltage-dependent structural interactions in the Shaker K^+ channel. J Gen Physiol 115:123–138

Trautmann A, Marty A (1984) Activation of Ca-dependent K channels by carbamylcholine in rat lacrimal glands. Proc Natl Acad Sci U S A 81:611–615

Tu L, Wang J, Helm A, Skach WR, Deutsch C (2000) Transmembrane biogenesis of Kv1.3. Biochemistry 39:824–836

Vandorpe DH, Shmukler BE, Jiang L, Lim B, Maylie J, Adelman JA, de Franceschi L, Cappellini MD, Brugnara C, Alper SL (1998) cDNA cloning and functional characterization of the mouse Ca^{2+}-gated K^+ channel, mIK1. Role in regulatory volume decrease and erythroid differentiation. J Biol Chem 273:21542–21553

Wallace DP, Grantham JJ, Sullivan LP (1996) Chloride and fluid secretion by cultured human polycystic kidney cells. Kidney Int 50:1327–1336

Wallner M, Meera P, Toro L (1999) Molecular basis of fast inactivation in voltage and Ca^{2+}-activated K^+ channels: a transmembrane β-subunit homolog. Proc Natl Acad Sci U S A 96: 4137–4142

Wang S, Melkoumian Z, Woodfork KA, Cather C, Davidson AG, Wonderlin WF, Strobl SJ (1998) Evidence for an early G1 ionic event necessary for cell cycle progression and survival in the MCF-7 human breast carcinoma cell line. J Cell Physiol 176:456–464

Wang J, Kristian A, Novak I (2013) Purinergic regulation of CFTR and Ca^{2+}-activated Cl^- channels and K^+ channels in human pancreatic duct epithelium. Am J Physiol Cell Physiol 304:C673–C684

Warth R, Hamm K, Bleich M, Kunzelmann K, von Hahn T, Schreiber R, Ullrich E, Mengel M, Trautmann N, Kindle P, Schwab A, Greger R (1999) Molecular and functional characterization of the small Ca^{2+}-regulated K^+ channel (rSK4) of colonic crypts. Pflugers Arch 438:437–444

Wasserman SI, Barrett KI, Huott PA, Beuerlein G, Kagnoff MF, Dharmsathaphorn K (1988) Immune-related intestinal Cl^- secretion. I. Effect of histamine on the T84 cell line. Am J Physiol Cell Physiol 254:C53–C62

Welsh MJ, McCann JD (1985) Intracellular calcium regulates basolateral potassium channels in a chloride-secreting epithelium. Proc Natl Acad Sci U S A 82:8823–8826

Welsh MJ, Smith PL, Frizzell RA (1982) Chloride secretion by canine tracheal epithelium. II. The cellular electrical potential profile. J Membr Biol 70:227–238

Weymer A, Huott P, Liu W, McRoberts JA, Dharmsathaphorn K (1985) Chloride secretory mechanism induced by prostaglandin E_1 in a colonic epithelial cell line. J Clin Invest 76:1828–1836

Wilson SM, Brown SG, McTavish N, McNeill RP, Husband EM, Inglis SK, Olver RE, Clunes MT (2006) Expression of intermediate-conductance, Ca^{2+}-activated K^+ channel (KCNN4) in H441 human distal airway epithelial cells. Am J Physiol Lung Cell Mol Physiol 291:L957–L965

Wonderlin WF, Strobl JS (1996) Potassium channels, proliferation and G1 progression. J Membr Biol 154:91–107

Wonderlin WF, Woodfork KA, Strobl JS (1995) Changes in membrane potential during the progression of MCF-7 human mammary tumor cells through the cell cycle. J Cell Physiol 165:177–185

Wright RD, Jennings MA, Florey HW, Lium R (1940) The influence of nerves and drugs on secretion by the small intestine and an investigation of the enzymes in intestinal juice. Q J Exp Physiol 30:73–120

Wu ZC, Kisslinger SD, Gaginella TS (1982) Functional evidence for the presence of cholinergic nerve endings in the colonic mucosa of the rat. J Pharmacol Exp Ther 221:664–669

Wulf A, Schwab A (2002) Regulation of a calcium-sensitive K^+ channel (cIK1) by protein kinase C. J Membr Biol 187:71–79

Wulff H, Castle NA (2010) Therapeutic potential of KCa3.1 blockers: recent advances and promising trends. Expert Rev Clin Pharmacol 3:385–396

Wulff H, Zhorov BS (2008) K^+ channel modulators for the treatment of neurological disorders and autoimmune diseases. Chem Rev 108:1744–1773

Wulff H, Miller MJ, Hansel W, Grissmer S, Cahalan MD, Chandy KG (2000) Design of a potent and selective inhibitor of the intermediate-conductance Ca^{2+}-activated K^+ channel: a potential immunosuppressant. Proc Natl Acad Sci U S A 97:8151–8156

Wulff H, Gutman GA, Cahalan MD, Chandy KG (2001) Delineation of the clotrimazole/TRAM-34 binding site on the intermediate conductance calcium-activated potassium channel, IKCa1. J Biol Chem 276:32040–32045

Xia XM, Fakler B, Rivard A, Wayman G, Johnson-Pais T, Keen JE, Ishii T, Hirschberg B, Bond CT, Lutsenko S, Maylie J, Adelman JP (1998) Mechanism of calcium gating in small-conductance calcium-activated potassium channels. Nature 395:503–507

Xia XM, Ding JP, Lingle CJ (1999) Molecular basis for the inactivation of Ca^{2+}- and voltage-dependent BK channels in adrenal chromaffin cells and rat insulinoma tumor cells. J Neurosci 19:5255–5264

Ye M, Grantham JJ (1993) The secretion of fluid by renal cysts from patients with autosomal dominant polycystic kidney disease. N Engl J Med 329:310–311

Yi BA, Lin YF, Jan YN, Jan LY (2001) Yeast screen for constitutively active mutant G protein-activated potassium channels. Neuron 29:657–667

Zhang Z, Zhou Y, Ding JP, Xia XM, Lingle CJ (2006) A limited access compartment between the pore domain and cytosolic domain of the BK channel. J Neurosci 26:11833–11843

Zhang X, Wang G, Dupre DJ, Feng Y, Robitaille M, Lazartigues E, Feng YH, Hebert TE, Wu G (2009) Rab1 GTPase and dimerization in the cell surface expression of angiotensin II type 2 receptor. J Pharmacol Exp Ther 330:109–117

Zhou W, Vergara L, Konig R (2004) T cell receptor induced intracellular redistribution of type I protein kinase A. Immunology 113:453–459

Zhu Y, Ye J, Huizinga JD (2007) Clotrimazole-sensitive K^+ currents regulate pacemaker activity in interstitial cells of Cajal. Am J Physiol Gastrointest Liver Physiol 292:G1715–G1725

Zimmerman TW, Binder HJ (1982) Muscarinic receptors on rat isolated colonic epithelial cells: a correlation between inhibition of [^{3}H]quinuclidinyl benzilate binding and alteration in ion transport. Gastroenterology 83:1244–1251

Chapter 21
BK Channels in Epithelia

Donghai Wen, Ryan J. Cornelius, and Steven C. Sansom

Abstract Large, Ca^{2+}-activated K^+ channels (BK) are comprised of the pore-forming α subunit (BK-α) and one of four β subunits (β1–4), which bestows functional diversity. BK channels are localized in the distal nephron of kidneys and a variety of extrarenal, fluid-secreting epithelia. In the kidneys, flow activates BK, which are located in two cell types of the distal nephron. The BK-α/β1 resides in the connecting tubule cells, and the BK-α/β4 is localized in intercalated cells, where they mediate K^+ secretion in mice on a high K^+, alkaline diet.

BK channels mediate K^+ secretion from goblet cells of the colon to generate fluid volume, as part of a neurogenic-stimulated response to infection or bacterial imbalance. In the ductal cells of the exocrine glands and pancreas, BK mediate K^+ secretion in conjunction with HCO_3^- to generate an alkaline volume in order to neutralize the acidic contents of the mouth and stomach. In pulmonary epithelia, BK mediate secretion of K^+, as a counter cation to Cl^-, to drive secretion of fluid volume.

BK also reside in the basolateral membranes of some epithelial cells, where they have a role to drive Cl^- secretion or Na^+ absorption by serving to recycle cellular K^+ that entered via Na^+-K^+-ATPase. In summary, BK-mediated K^+ secretion, which contributes to K^+ homeostasis in response to aldosterone in renal distal tubules and colon, serves to drive high volumes of fluid in response to muscarinic and/or adrenergic stimulation in the colon and other extrarenal epithelia.

Keywords Maxi K^+ • Distal nephron • Colon • Exocrine glands • K^+ secretion

21.1 Introduction

Large, Ca^{2+}- and voltage-activated K^+ channels (BK) are expressed in epithelial cells lining the gastrointestinal track, exocrine glands, and renal distal tubules. The pore-forming channels (BK-α) have a single channel conductance ranging from

D. Wen • R.J. Cornelius • S.C. Sansom (✉)
Department of Cellular and Integrative Physiology, University of Nebraska Medical Center, Omaha, NE 68198-5850, USA
e-mail: ssansom@unmc.edu

K.L. Hamilton, D.C. Devor (eds.), *Ion Channels and Transporters of Epithelia in Health and Disease*, Physiology in Health and Disease,
DOI 10.1007/978-1-4939-3366-2_21

120 to 250 pS and are encoded by a single gene (*KCNMA1*). Two splice variants have been prevalently studied: stress regulated exon (STREX) with 58 amino acid inserts and ZERO, with no amino acid inserts. The BK-α is associated with at least four different ancillary (β) subunits (BK-β1–4). A new gamma family of subunits, the leucine-rich repeat-containing protein (LRRC), has recently been described as a modulator of BK (Yan and Aldrich 2010, 2012). The LRRC26, in particular, yields a very large (−140 mV) shift in activation of BK-α/γ and might explain the increased open probability and function of BK in physiological in vivo conditions (Yan and Aldrich 2010).

The BK-β1 subunit is predominantly expressed with BK-α in smooth muscle cells, and the BK-β4 is predominantly expressed with BK-α in the brain (Brenner et al. 2000). However, any of the four β subunits can be associated with BK-α in the different types of epithelial cells. Although the BK-α is able to conduct K^+ current in the absence of a β subunit in expression systems, it is unclear whether the BK-α functions solo in epithelial cells in vivo. The variable roles of the β subunits include: bestowing increased sensitivity of the BK-α to Ca^{2+} and voltage, promoting the membrane insertion of BK-α, and modifying the sensitivity to toxins (Gan et al. 2008).

BK can reside in either the apical or basolateral membranes of epithelial cells. In the basolateral membranes, BK participate in the governance of the cell membrane potential. In the apical membrane of epithelial cells, BK mediate K^+ secretion. Although quiescent at resting membrane potentials, BK are activated by a combination of increased intracellular [Ca^{2+}] and depolarizing cell membrane potentials. In addition, the ZERO variant is activated by protein kinase A phosphorylation (Tian et al. 2001, 2003). When apical BK are activated, K^+ is driven from the cell to the lumen down its electrochemical gradient. The basolateral Na^+-K^+-ATPase is the primary driving force that increases K^+ above electrochemical equilibrium but also sets up gradients for Na^+, Cl^-, and K^+ to enter the cell via the Na^+-K^+-$2Cl^-$ cotransporter (NKCC) in the basolateral membrane. After elevating above electrochemical equilibrium, Cl^- exerts a depolarizing force across the apical membrane for K^+ to exit the cell.

This chapter will report studies of BK as a K^+ secretory channel in epithelial cells that secrete a K^+-rich fluid. Although other K^+ channels have been implicated, BK mediate K^+ secretion in all K^+-secreting epithelial cells. There are many differences and similarities in the mechanisms of regulating K^+ secretion in the renal and extrarenal epithelia. BK of some of the extrarenal cells respond to sympathetic and parasympathetic nerve stimulation, with increases in cAMP and intracellular Ca^{2+}, respectively. However, such adrenergic and cholinergic stimulation of K^+-secreting renal cells has not been established. On the other hand, much like BK in the distal tubule, BK channels in the colon respond to aldosterone (Sorensen et al. 2008).

21.2 BK-Mediated K^+ Secretion in Renal Epithelia

Microelectrode experiments revealed that the rabbit cortical collecting duct (CCD) possessed a large apical membrane K^+ conductance, which was enhanced in magnitude when the animals were given a synthetic mineralocorticoid (Sansom and O'Neil 1985). In 1984, BK were the first ion-selective channels functionally identified in mammalian renal epithelial cells using the patch clamp technique applied to split-open isolated rabbit CCD (Hunter et al. 1984). Subsequently, transcript for BK was found in the distal nephron segments using RT-PCR (Morita et al. 1997), and BK-α protein was identified in the connecting tubule (CNT) and CCD using immunohistochemistry (IHC) (Holtzclaw et al. 2010a). The discovery of BK in the cortical collecting ducts indicated their role to mediate K^+ secretion.

The function of BK to secrete K^+ became uncertain when the apical BK of rat CCD served to extrude K^+ as a volume regulator during exposure to hypo-osmotic solutions (Stoner and Morley 1995). However, renal BK-mediated K^+ secretion was subsequently demonstrated by raising the K^+ concentration of the solution bathing the isolated *Ambystoma* (amphibian) collecting duct (Stoner and Viggiano 1998). Several years later, the isolated rabbit cortical collecting duct displayed BK-mediated K^+ secretion when perfused at flows of 6 nl/min but not at flows of 0.5 nl/min (Woda et al. 2001). The mechanism for flow-stimulated BK was partly due to a shear stress-induced increase in intracellular [Ca^{2+}] (Woda et al. 2002).

The availability of mice with genetically ablated BK-α (BK-α-KO) and BK-β subunits enabled in vivo studies of BK-mediated K^+ secretion and the roles of the subunits. As the first in vivo study, a mouse with a genetic ablation of the BK-β1 subunit (β1-KO) exhibited a diminished capacity to excrete K^+ when flow was increased by volume expansion (Pluznick et al. 2003b). That the BK-β1 subunit was localized to the apical membrane of the principal cells of the mouse connecting tubule suggested a secretory role for the BK-α/β1 under high flow conditions (Pluznick et al. 2005). BK-α-KO also exhibited defective K^+ secretion when flow was increased by pharmacological blockade of V2 vasopressin receptors (Rieg et al. 2007). BK-α-KO excreted K^+ at a rate similar to WT, but the plasma [aldosterone] was very high, presumably to compensate with a higher expression of Na^+-K^+-ATPase and epithelial Na^+ channels (ENaC). In the absence of BK, the renal outer medullary K^+ channel (ROMK; Kir1.1), also identified in the CNT/CCD (Frindt and Palmer 2004; Mennitt et al. 1997), upregulates and secretes K^+ in high K^+ conditions (Sansom and Welling 2007).

A micropuncture study of ROMK-KO mice also indicated that high distal flow stimulates BK-mediated K^+ secretion (Bailey et al. 2006). A primary role of ROMK is to mediate apical K^+ recycling in the thick ascending limb. Therefore, ROMK-KO display Bartter's syndrome with increased tubular flow due to a dysfunctional concentrating mechanism. ROMK-KO mice exhibited increased iberiotoxin (IBTX)-sensitive K^+ secretion in the late distal tubule of mice. This was additional in vivo evidence that high flow stimulates BK-mediated K^+ secretion. It was notable

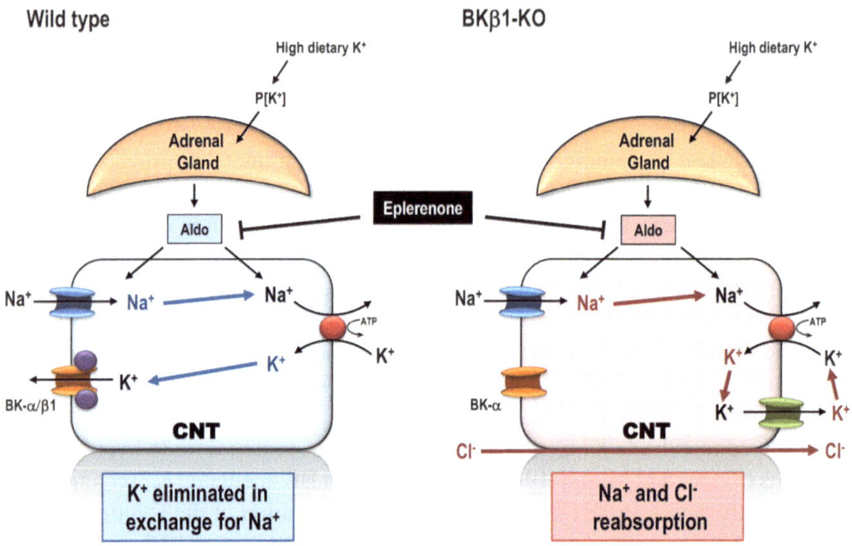

Fig. 21.1 Pathophysiology of hypertension in β1-KO. High K^+ diet induces adrenal secretion of aldosterone, which enhances ENaC-mediated Na^+ reabsorption which drives K^+ secretion through BK-α/β1 in the CNT of wild-type mice. In β1-KO, the Na^+/K^+ exchange is decoupled because of the defective K^+ secretion. Instead, the Na^+ reabsorption is linked with Cl^- absorption through the paracellular pathway. In β1-KO, retention of Na^+ and Cl^- causes hypertension, which can be alleviated by eplerenone, an aldosterone receptor antagonist (Modified from Grimm et al. 2009b)

that the micropuncture study also showed that high K^+ consumption induced BK-mediated K^+ secretion (Bailey et al. 2006).

Subsequent studies revealed that consuming a high K^+ diet resulted in increased mRNA and protein expression of BK-α and its subunits (Holtzclaw et al. 2011; Najjar et al. 2005). β1-KO mice exhibited normal K^+ homeostasis when placed on a regular diet, but when placed on a high K^+ diet, the plasma [K^+] increased, plasma [aldo] increased, and mean arterial blood pressure increased by 20 mmHg (Grimm et al. 2009b). The failure to excrete K^+ resulted in elevated plasma [K^+] and a compensating increase in ENaC-mediated Na^+ reabsorption that was accompanied with Cl^- reabsorption instead of K^+ secretion. Accumulating Na^+ and Cl^- reabsorption led to volume expansion and hypertension (see Fig. 21.1).

BK-β1 was localized in the apical membrane of CNT cells (the equivalent of principal cells of the CCD), and β1-KO mice clearly exhibited a K^+ secretory defect when given a high K^+ diet; however, both patch clamping and IHC (Fig. 21.2) localized the preponderance of BK-α to the intercalated cells (IC) of CNT and CCD (Pacha et al. 1991). Patch clamp experiments also showed a high density of BK in IC of rats (Pacha et al. 1991). IC are enriched with mitochondria and are the acid/base-transporting cells of the distal nephron. The α-IC are the acid-secreting cells and the β-IC are the base (HCO_3^-)-secreting cells. Unlike the principal cells, which have the BK-β1 subunit, the BK-β4 subunit is expressed with the BK-α in IC (Grimm et al. 2007). The localization of BK to IC was interesting because the PC

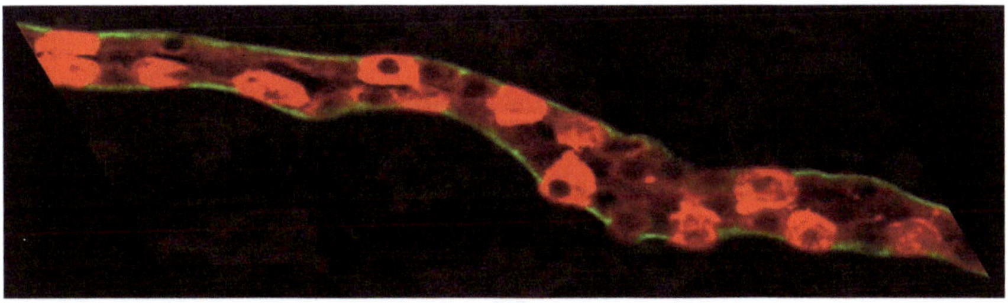

Fig. 21.2 Localization of BK-α in the CCD cells. Immunostaining showed that BK-α (*red*) is predominantly localized in the cytoplasm of intercalated cells, as identified by the absence of basolateral aquaporin 3 (*green*), which identifies PC

were considered the Na^+ and K^+ transporting cells with an abundance of basolateral Na^+-K^+-ATPase that increased with a high K^+ diet. Intercalated cells contain an abundance of H^+-ATPase but nearly undetectable Na^+-K^+-ATPase. The PC and IC are not physically coupled; there are no reports of gap junctions in the CCD. However, the two cell types are linked by the high luminal electronegative field (transepithelial potential; V_{te}), which influences electrical transport of both PC and IC cells.

β4-KO mice exhibited attenuated K^+ secretion when placed on a high K^+ diet, revealing that BK-α/β4 mice also could mediate K^+ secretion under certain conditions. However, the defective K^+ excretion was demonstrated only when the diet was high in K^+ and alkaline (Cornelius et al. 2012). When the high K^+ diet was acidic, both WT and β4-KO mice exhibited impaired K^+ homeostasis, with plasma [K^+] approaching 7 mM. Thus, secreting K^+ via BK-α/β4 mice in the distal nephron is linked to HCO_3^- secretion. In pancreatic duct cells, HCO_3^- and BK-mediated K^+ secretion are also linked (see below) (Venglovecz et al. 2011). HCO_3^- secretion could supply a counter-anion for K^+ secretion. Moreover luminal HCO_3^- is more resistant than Cl^- to back flux (from lumen to plasma) through tight junctions, thereby enhancing the electronegative lumen potential that drives K^+ secretion.

Additional studies revealed the separate roles of a high K^+ diet and alkalinity on BK channel expression in IC cells. The high K^+ load, regardless of accompanying acid or base, increased the expression level of BK-α in IC; however, the BK-β4 subunit was required for the BK-α to traffic to the apical membrane (Fig. 21.3). The BK-β4 expression is increased by alkalinity and prevents the lysosomal breakdown of BK-α, thereby increasing its time in the apical membrane (Wen et al. 2013, 2014b). These results explained the attenuated K^+ excretion of β4-KO mice on the high K^+, alkaline diet.

It was concluded that BK-α/β4 is enhanced in the apical membrane of IC and mediates K^+ secretion when animals consume a high K^+, alkaline diet. However, the driving force is supplied by the aldosterone-regulated Na^+-K^+-ATP, localized in the basolateral membrane of PC. This could hypothetically be explained by a mechanism whereby K^+ is secreted from the IC, with K^+ delivery through a basolateral NKCC1, the secretory isoform of the Na^+-K^+-$2Cl^-$ cotransporter.

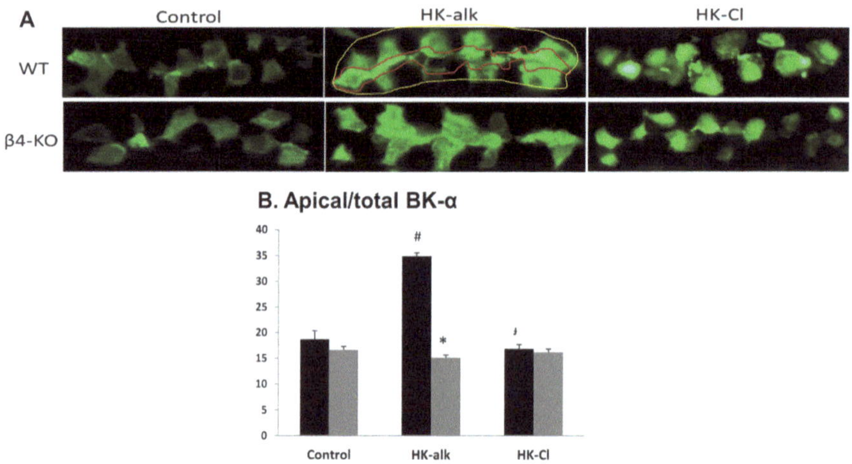

Fig. 21.3 IHC analysis of CCD for BK-α in wild type (WT) and β4KO on control, high K^+ alkaline (HK-alk) and high K^+ chloride (HK-Cl) diets. (**a**) Representative stainings reveal BK-α in the CCD of WT and β4-KO on control, HK-alk and HK-Cl diets. The outlines in *upper middle panel* show how fluorescence intensity was measured. The areas occupied by the *red* and *yellow lines* represent apical and total BK-α staining, respectively. (**b**) Summary bar plots of quantitated fluorescence for proportion of apical/total BK-α expression levels. Values are mean ± SEM. $*P < 0.01$ vs. WT; $^{\#}P < 0.01$ vs. control; $^{J}P < 0.01$ vs. HK-alk; $n \geq 9$ in each group (Adapted from Wen et al. 2013)

The recently described transporter Na^+-dependent HCO_3^-/Cl^- exchanger, localized in IC (Leviel et al. 2010), may be involved as a way to recycle the Na^+ and secrete the bicarbonate (Liu et al. 2011) (Fig. 21.4).

Although BK was clearly increased when mice were fed high K^+, it was not certain that aldosterone was the regulator. The role of aldosterone in BK-mediated K^+ secretion was best demonstrated by mice given a low Na^+, high K^+ diet (LNaHK). In these mice, the aldosterone levels increase to more than 3000 pg/ml, which is tenfold the plasma aldosterone (P[aldo]) of mice on a regular diet. When LNaHK mice were adrenalectomized, the P[K^+] was greater than 9 mM and they survived only 1 day. When LNaHK mice were adrenalectomized and placed on a low dose of aldosterone to achieve P[aldo] of approximately 500 pg/ml, the BK expression was near basal levels and the urinary K^+ excretion and flow rates only increased to 25 % of sham values (Cornelius et al. 2014).

The BK-α/β4 is also found in the mouse medullary collecting duct cells where high distal flow activates BK-α/β4 and reduces cellular K^+ content and volume (Holtzclaw et al. 2010a). The shear stress from high flow also reduces the size of MDCK-C11 cells (Holtzclaw et al. 2010b), which are a subclone that predominantly express an IC phenotype (Gekle et al. 1994). The K^+ exiting the cell may have been a counter cation for ATP release from connexins in the apical membrane (Holtzclaw et al. 2010b). ATP activates BK via purinergic receptors, thereby completing a positive (feedforward) loop (Bugaj et al. 2012). In support of this notion, β4-KO mice, unlike WT mice, do not exhibit cell volume reduction during

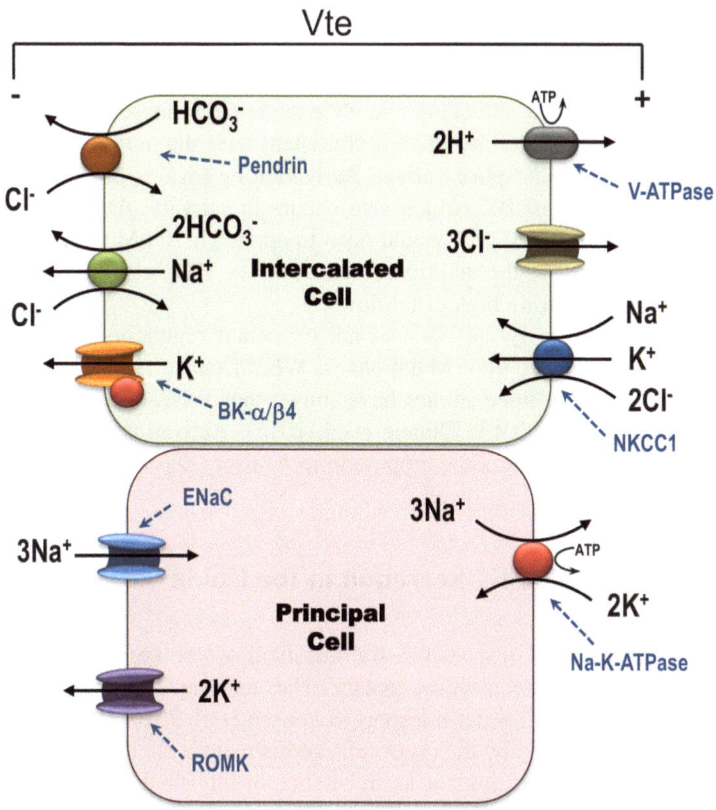

Fig. 21.4 Previously defined transporters of the cortical collecting duct explain how the Na^+/K^+-ATPase of the principal cells (PC) drive BK-α/β4-mediated K^+ secretion in the intercalated cells (IC). HCO_3^- is secreted in exchange for Cl^- via pendrin raising the luminal [HCO_3^-] and forcing Cl^- through the IC instead of the paracellular pathway. The resistance of the tight junction pathway increases, augmenting the electronegative driving force for K^+ secretion. The large plasma to lumen Na gradient and increased intracellular HCO_3^- create a driving force for both K^+ and $NaHCO_3$ secretion from IC. The Na^+ recycles to generate a higher ratio of K^+ secreted per Na^+ reabsorbed (Modified from Wen et al. 2014a)

high urinary flows generated from mice on high K^+ diets (Holtzclaw et al. 2010a). Moreover, when incorporating BK-β4 siRNA, MDCK-C11 cells fail to exhibit a shear stress-induced loss of volume (Holtzclaw et al. 2010b).

The BK-β1 was detected in the apical membrane of the CNT; however, BK-α was only faintly detected in the same cells by IHC, leading to the speculation that BK-α in PC was a different splice variant from that of IC. Unlike MDCK-C11, which contain BK-α/β4, the PC-like MDCK-C7 cells contain BK-β1 but not BK-α (Holtzclaw et al. 2010b). The absence of endogenous BK-α in MDCK-C7 makes it difficult to study its regulation in vitro. The split-open patch technique revealed that PC contained BK, which were quiescent unless the MAPK pathway was inhibited

(Li et al. 2006). Moreover, PGE2 inhibits BK of PC by inhibiting MAPK (Jin et al. 2007).

In isolated perfused rabbit CCDs, BK was constitutively inhibited by protein kinase A (PKA) but activated at high flow via the protein kinase C (PKC) pathway (Liu et al. 2009). Inhibition by PKA is consistent with the notion that the BK of CCDs are of the STREX splice variant. Activation by PKC is consistent with the muscarinic activation of BK, which also occurs in salivary glands (see below). Because the intracellular [Ca^{2+}] would need to approach 10 μM to activate BK at negative cell potentials, the additional activation by PKC can explain how BK opens to secrete K^+ during high distal flows.

The WNK (with no lysine) kinases are important regulators of Na^+ and K^+ balance in the distal nephron. Mutations in WNK4 cause familial hypertensive hyperkalemia. At least three studies have shown that WNK4 inhibits BK (Wang et al. 2013; Yue et al. 2013; Zhuang et al. 2011). Activation of the lysosomal breakdown pathway of BK is the proposed mechanism (Zhuang et al. 2011).

21.3 BK-Mediated K^+ Secretion in the Colon

The colon is primarily responsible for intestinal water absorption, a process achieved by the surface cells, which contain ENaC that mediate the Na^+-absorption that drives transepithelial water transport (Sorensen et al. 2010a). However, water secretion is accomplished by the crypt cells and is mainly elicited as a pathophysiological response to infections or an imbalance of intestinal bacteria. The fluid-secreting intestinal crypt cells are comprised of two cell types: the columnar cells and the goblet cells. The columnar cells secrete water driven by cystic fibrosis transmembrane conductance regulator (CFTR)-mediated Cl^- secretion. The goblet cells secrete a mucous solution driven by BK-mediated K^+ secretion (see Fig. 21.5) (Linley et al. 2014).

Organisms eliminate intestinal infection by a rapid clearing response that is performed by enteric nervous discharge. The cholinergic innervation activates primarily CFTR-mediated Cl^- secretion but also K^+ secretion. These ions create the electrogenic and osmotic forces to drive water secretion (Rechkemmer et al. 1996). Colonic pseudo-obstruction produces a predominant response to a sympathetic β-adrenergic increase in cAMP that initiates K^+ secretion and produces intestinal K^+ concentrations of near 150 mM (Van Dinter et al. 2005). Persistent intestinal diarrhea often results in severe hypokalemia (Field 2003) that can be corrected by intestinal resection (Yusuf et al. 1999).

BK channels are the primary mediators of intestinal K^+ secretion. A model of ulcerative colitis presents with enhanced colonic K^+ secretion (Kanthesh et al. 2013) and BK expression (Kanthesh et al. 2013; Sandle et al. 2007). Specific BK inhibitors, such as IBTX and paxilline, show that BK are the exclusive secretors of K^+ in the distal colon (Matos et al. 2007). Although BK have been reported in the apical membranes of surface cells (Hay-Schmidt et al. 2003), most studies find BK

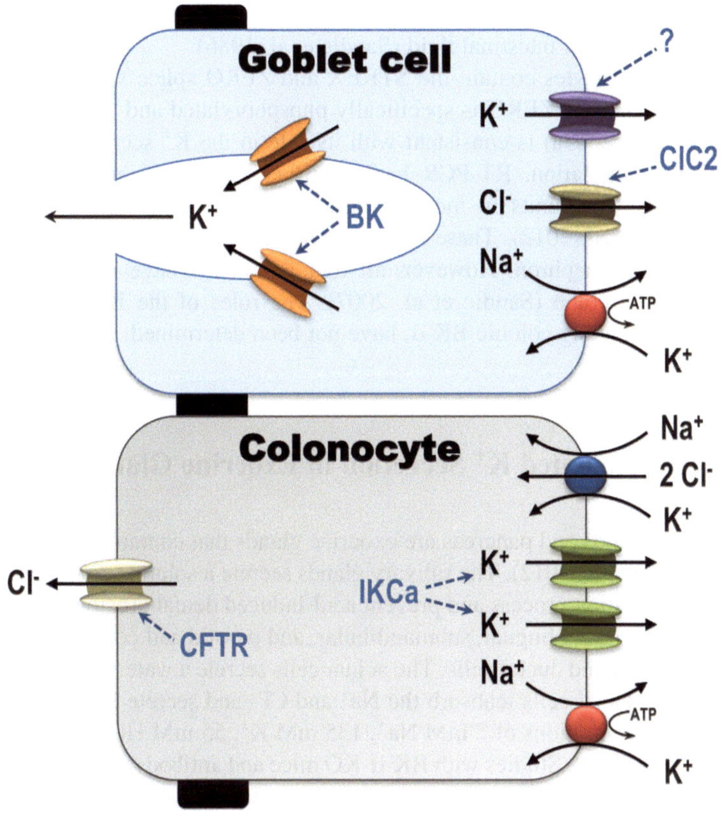

Fig. 21.5 Mechanism of BK-mediated K^+ secretion in the colon. The BK-mediated K^+ secretion in the apical membrane of goblet cells is linked to the CFTR-mediated Cl^- secretion in the apical membrane of colonocytes (Modified from Linley et al. 2014)

only in the crypt cells (Sausbier et al. 2006; Sorensen et al. 2010a). BK activity is enhanced in crypt cells under various conditions such as end-stage renal disease (ESRD) (Mathialahan et al. 2005), secretory diarrhea (Simon et al. 2008), and ulcerative colitis (Sandle et al. 2007). BK-mediated K^+ secretion was demonstrated in vivo (Sausbier et al. 2006; Sorensen et al. 2010a) when BK-α-KO exhibited a substantially reduced K^+ secretory response to α-adrenergic stimulation (Sorensen et al. 2010b).

Under normal physiological conditions, the gastrointestinal tract is responsible for approximately 10 % of K^+ excretion and can have a role to maintain K^+ homeostasis. Intestinal resection does not result in hyperkalemia because the kidneys will elevate their capacity to secrete K^+. However, despite having a minimal role to maintain K^+ homeostasis, the colonic BK are still regulated by aldosterone, which increases BK-α expression (Sorensen et al. 2010a) and BK-mediated K^+ secretion in the distal colon (Sorensen et al. 2008). Moreover, patients with end-stage renal disease (ESRD) exhibit enhanced gastrointestinal K^+

secretion (Sandle et al. 1987), and their plasma K^+ concentration can be reduced by stimulating a K^+-rich intestinal fluid (Sandle et al. 1986).

Colonic enterocytes contain the STREX and ZERO splice variants (Kanthesh et al. 2013). That the ZERO is specifically phosphorylated and activated by PKA (Pluznick et al. 2003a) is consistent with its role in the K^+ secretory response to sympathetic stimulation. RT-PCR has identified the BK-β1 and BK-β4 as the predominating β subunits in mouse (Flores et al. 2007) and guinea pig distal colon (Zhang et al. 2012). These are the same subunits associated with BK-α in the mouse distal nephron. However, all four β subunits have been described in human sigmoid colon (Sandle et al. 2007). The roles of the BK subunits, with respect to modulating colonic BK-α, have not been determined.

21.4 BK-Mediated K^+ Secretion in Exocrine Glands

The salivary glands and pancreas are exocrine glands that contain similar transport processes (Lee et al. 2012). The salivary glands secrete a solution rich in HCO_3^- to initiate the digestion process and prevent acid-induced dental erosion. The salivary glands include the sublingual, submandibular, and parotid and comprise two major cell types: acinar and ductal cells. The acinar cells secrete a watery fluid rich in Na^+ and Cl^-. The ductal cells reabsorb the Na^+ and Cl^- and secrete K^+ and HCO_3^- to yield final concentrations of 2 mM Na^+, 135 mM K^+, 55 mM HCO_3^-, and 78 mM Cl^- (Lee et al. 2012). Studies with BK-α-KO mice and antibody staining reveal that apical BK channels mediate K^+ secretion in striated and excretory ductal cells of submandibular glands (Nakamoto et al. 2008). Both the parasympathetic nerves, via muscarinic M3 receptors, and the sympathetic nerves, via β-adrenergic generation of cAMP, have roles in both the acinar and ductal secretory mechanisms.

The pancreas is an exocrine gland, which also contains acinar and ductal cells. The ductal cells secrete HCO_3^- in order to neutralize the gastric acid in the small intestine. Immunostaining revealed BK-α in the apical membrane of the ductal cells (see Fig. 21.6; Venglovecz et al. 2011). HCO_3^- secretion is attenuated when BK are blocked by apically applied IBTX, indicating a link between HCO_3^- secretion and BK-mediated K^+ secretion (Venglovecz et al. 2011). In ductal cells, the bile acid chenodeoxycholate (CDC) activates BK from the apical side in low concentrations (100 μM) via IP3-mediated release of $[Ca^{2+}]$ from intracellular stores (Venglovecz et al. 2011). It was speculated that CDC-induced BK-mediated fluid secretion in the pancreas was only evoked during a "hypersecretion" condition, activated to pre-emptively combat the consequences of disease resulting from compromised bile secretion. (Interested readers are directed to Chap. 6 for further information about HCO_3^- secretion in epithelia.)

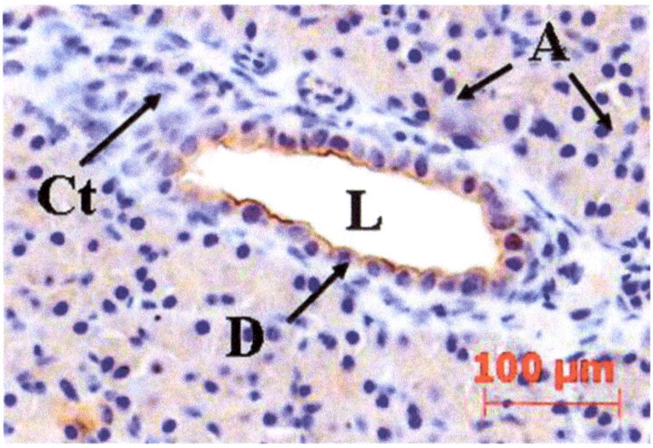

Fig. 21.6 IHC with polyclonal antibody raised against amino acids 1097–1196 of the BK-α. The immunostaining revealed apical expression of BK channels in the intralobular/interlobular pancreatic ducts. Staining is undetectable in basolateral membranes of ductal cells or acinar cells (Adapted from Venglovecz et al. 2011)

21.5 Pulmonary Epithelia

The dynamics of pulmonary mucociliary clearance depends on ionic secretion that drives transepithelial water transport, much like the colon and exocrine glands. Pulmonary BK channels were first identified in cultured pulmonary cells (A549) using PCR (Leroy et al. 2006) and then patch clamp techniques (Ridge et al. 1997). Although water secretion driven by CFTR-mediated Cl secretion has been well studied, BK-mediated K^+ secretion has been only recently shown to promote the pulmonary water for proper mucociliary beating. In normal human bronchial epithelial (NHBE) cells, ATP activates BK-mediated K^+ secretion (Manzanares et al. 2011), with K serving as a counter cation for Cl^- secretion. Inhibition of BK reduces the volume of the liquid surface (Manzanares et al. 2011). The cytokine, IFN shown to decrease pulmonary fluid secretion, increases mRNA of BK-β2 and decreases BK-β4. The recently identified LRR26, which associates with BK-α in NHBE, is dissociated with IFN-γ treatment (Manzanares et al. 2014).

21.6 Role of BK in the Basolateral Membranes of Epithelia

BK is also localized in basolateral membranes of some epithelia, where it influences Na^+ absorption or Cl^- secretion. BK mediate recycling of the K^+ entering the cell via Na^+-K^+-ATPase across the basolateral membrane, thereby creating a negative cellular potential that drives either lumen to cell Na^+ uptake or cell to lumen Cl^- extrusion. In the mouse colon, recycling K^+ via basolateral BK augments Cl^- secretion in the proximal colon and Na^+ absorption in the surface cells of the distal colon (Puntheeranurak et al. 2007). However, in nasal epithelia, outward basolateral BK currents support both Na^+ absorption and Cl^- secretion in the same cell (Clarke et al. 1992). Basolateral BK-mediated K^+ recycling and apical CFTR-mediated Cl^- secretion are initiated by muscarinic ligands, such as acetylcholine and bradykinin (Clarke et al. 1992). Activation of phospholipase C with the generation of IP3 induces release of stores of Ca^{2+}, which activate BK.

The BK-β1 subunit was also described with IHC in the basolateral membrane of connecting tubule cells when animals were subjected to a low Na^+ diet (Grimm et al. 2009a). Compared with wild-type mice, the genetically ablated BK-β1 became extremely dehydrated and Na^+ and volume depleted. These results indicated that the BK-α/β1 trafficked to the basolateral membrane of CNT cells to mediate K^+ recycling, hyperpolarize the basolateral membrane, and support ENaC-mediated Na^+ reabsorption while preventing K^+ secretion. The basolateral localization of BK-β1 with a low Na^+ diet contrasted with the localization of BK-β1 in the apical membrane of principal cells in mice on a high K^+ diet. However, the hormonal response to the two diets is different. Angiotensin II, which is elevated with the low Na diet but not with the high K^+ diet, is a potential regulator of BK-α/β1 trafficking to the basolateral membrane.

At the luminal sides of epithelial cells, depolarizing potentials and flow are cooperative activators of apical BK. However, BK in the basolateral membranes are presented with hyperpolarizing membrane potentials. A study of BK channels in pancreatic acinar cells, in which BK reside in the basolateral membrane, addressed the issue of BK quiescence at resting membrane potentials (Romanenko et al. 2010). When expressed in heterologous expression systems or neuronal cells at a cell $[Ca^{2+}]$ of 500 nM, BK-α requires a membrane potential of more than 75 mV to initiate activation to an open probability >0.1; however, with the same 500 nM intracellular $[Ca^{2+}]$, native BK in parotid cells are active at cell potentials of −25 mV. Therefore, in native epithelial cells, BK may be activated by components not found in neural cells or heterologous expression systems. The association with BK-α of subunits BK-β1 or BK-β4, found in parotid cells (Nehrke et al. 2003), can shift the potential gating in high intracellular Ca^{2+} by 60 mV, a value still less than the 100 mV shift in activation of BK of the parotid BK. It remains to be determined if the newly discovered γ subunit LRRC 26, which confers even greater voltage/Ca^{2+} sensitivity to BK, is associated with BK-α and BK-β1 in the basolateral membranes of epithelial cells.

The details on the role of basolateral BK channels in Na^+ and Cl^- transport, beyond the original observational studies, have not been forthcoming. Basolateral channels are difficult to study by patch clamping applied to a basement membrane barrier. In addition, studies with BK knockout mice have indicated considerable redundancy of basolateral K^+ channels, with other K^+ channels upregulating in the absence of BK to control membrane potential. Indeed, most studies show that the intermediate conductance K^+ channels (IK, KCa2.3), which are activated by Ca^{2+} and not membrane potential, are the predominant channels that mediate recycling of K^+ across the basolateral membrane (Linley et al. 2014). Nevertheless, the evidence for an important role for basolateral BK to support Na^+ reabsorption and Cl^- secretion has been substantial.

21.7 Relevance of Flow Stimulation of BK-Mediated K^+ Secretion

High luminal flows are associated with activation of BK in a variety of epithelia. In the intestines, a high volume of secreted water is driven by K^+. However, as shown for the distal tubule BK, high flow also activates BK by shear stress-induced increases in intracellular [Ca^{2+}] concentration. The relevance of flow-induced K^+ secretion in the kidney relates to the diet of the animal. The amount of the isotonic K^+ concentration that can be achieved by secretion, according to the transtubular K gradient (TTKG) of approximately 25 in mice on a low Na^+, high K^+ diet (Wen et al. 2014a), is limited to approximately 125 mM in the lumen of the cortical collecting ducts. Therefore, more volume is necessary to enhance the K^+ secretory rate. Indeed, when mice consume 12 times the normal diet of K^+, the lumen to plasma [K^+] gradient increases by threefold, and the flow increases by fourfold so that the output matches the input of K^+.

21.8 Summary

A variety of epithelial cells secrete a K^+-rich fluid-utilizing BK channels placed in the apical membrane. BK-mediated K secretion is electrogenic and will drive the transport of ions and fluid. At the same time, secondary active transport of other ions, such as Na reabsorption and bicarbonate secretion, can drive electrogenic BK-mediated K^+ secretion.

In the colon, exocrine glands, and lungs, BK-mediated K^+ secretion drives the secretion of water. Unlike the distal nephron of the kidneys, colonic BK-mediated K^+ secretion is independent of ENaC-mediated Na^+ absorption. Na^+ absorbs via ENaC in surface cells while K^+ is secreted via BK channels in goblet cells. Aldosterone enhances the expressions of colonic ENaC, BK, and Na^+-K^+-ATPase (Zemanova and Pacha 1998), indicating that the colon has a role to maintain K^+ homeostasis. However, in the extrarenal cells, Na^+ absorption reduces volume, whereas K^+ secretion stimulates luminal volume. Colonic BK-mediated K^+ secretion is cholinergically and adrenergically stimulated, separately from ENaC-mediated Na^+ absorption. K^+ secretion generates a high volume in reaction to infection in the colon and associates with HCO_3^- secretion to generate an alkaline environment in the exocrine glands.

Unlike the colon and exocrine glands, the Na^+-reabsorbing and K^+-secreting cells are intermixed in the distal nephron of the kidneys. Potassium is either secreted from the same (principal) cell as ENaC-mediated Na^+ reabsorption or from an adjacent intercalated cell. However, ENaC-mediated Na^+ absorption will create a transtubular electric field that influences BK-mediated K^+ secretion from PC or IC. Treatment of mice on a low Na^+, high K^+ diet with amiloride, which blocks ENaC, decreases BK-mediated K^+ secretion and urinary flow

(Wen et al. 2014a, b). Therefore, for animals on a high K^+ diet, BK-mediated K^+ secretion may generate a high flow in the distal nephron like other K^+-secreting epithelia. High BK-mediated K^+ secretion with minimal Na^+ reabsorption can be accomplished as proposed in Fig. 21.4 whereby the driving force of Na^+-K^+-ATPase of the PC and the H^+-ATPase of the IC are coordinating a high rate of K^+ secretion with Na^+ reabsorption and HCO_3^- secretion. Although plasma to lumen flux of Na^+ has been demonstrated in the isolated CCD, future studies will determine whether the pathways are as hypothesized in Fig. 21.4.

References

Bailey MA, Cantone A, Yan Q et al (2006) Maxi-K channels contribute to urinary potassium excretion in the ROMK-deficient mouse model of Type II Bartter's syndrome and in adaptation to a high-K diet. Kidney Int 70:51–59

Brenner R, Jegla TJ, Wickenden A et al (2000) Cloning and functional characterization of novel large conductance calcium-activated potassium channel beta subunits, hKCNMB3 and hKCNMB4. J Biol Chem 275:6453–6461

Bugaj V, Sansom SC, Wen D et al (2012) Flow-sensitive K^+-coupled ATP secretion modulates activity of the epithelial Na^+ channel in the distal nephron. J Biol Chem 287:38552–38558

Clarke LL, Paradiso AM, Mason SJ et al (1992) Effects of bradykinin on Na^+ and Cl^- transport in human nasal epithelium. Am J Physiol Cell Physiol 262:C644–C655

Cornelius RJ, Wen D, Hatcher LI et al (2012) Bicarbonate promotes BK-alpha/beta4-mediated K excretion in the renal distal nephron. Am J Physiol Renal Physiol 303:F1563–F1571

Cornelius RJ, Wen D, Sansom SC (2014) Aldosterone enhances urine flow in mice on a low Na, high K diet. FASEB J (1 Supplement):892.7 (Abstract)

Field M (2003) Intestinal ion transport and the pathophysiology of diarrhea. J Clin Invest 111:931–943

Flores CA, Melvin JE, Figueroa CD et al (2007) Abolition of Ca^{2+}-mediated intestinal anion secretion and increased stool dehydration in mice lacking the intermediate conductance Ca^{2+}-dependent K^+ channel Kcnn4. J Physiol 583:705–717

Frindt G, Palmer LG (2004) Apical potassium channels in the rat connecting tubule. Am J Physiol Renal Physiol 287:F1030–F1037

Gan G, Yi H, Chen M et al (2008) Structural basis for toxin resistance of beta4-associated calcium-activated potassium (BK) channels. J Biol Chem 283:24177–24184

Gekle M, Wunsch S, Oberleithner H et al (1994) Characterization of two MDCK-cell subtypes as a model system to study principal cell and intercalated cell properties. Pflugers Arch 428:157–162

Grimm PR, Foutz RM, Brenner R et al (2007) Identification and localization of BK-beta subunits in the distal nephron of the mouse kidney. Am J Physiol Renal Physiol 293:F350–F359

Grimm PR, Irsik DL, Liu L et al (2009a) Role of BKbeta1 in Na+ reabsorption by cortical collecting ducts of Na^+-deprived mice. Am J Physiol Renal Physiol 297:F420–F428

Grimm PR, Irsik DL, Settles DC et al (2009b) Hypertension of Kcnmb1−/− is linked to deficient K secretion and aldosteronism. Proc Natl Acad Sci U S A 106:11800–11805

Hay-Schmidt A, Grunnet M, Abrahamse SL et al (2003) Localization of Ca^{2+} activated big-conductance K^+ channels in rabbit distal colon. Pflugers Arch 446:61–68

Holtzclaw JD, Grimm PR et al (2010a) Intercalated cell BK-alpha/beta4 channels modulate sodium and potassium handling during potassium adaptation. J Am Soc Nephrol 21:634–645

Holtzclaw JD, Liu L, Grimm PR et al (2010b) Shear stress-induced volume decrease in C11-MDCK cells by BK-{alpha}/{beta}4. Am J Physiol Renal Physiol 299:F507–F516

Holtzclaw JD, Grimm PR, Sansom SC (2011) Role of BK channels in hypertension and potassium secretion. Curr Opin Nephrol Hypertens 20:512–517

Hunter M, Lopes AG, Boulpaep EL et al (1984) Single channel recordings of calcium-activated potassium channels in the apical membrane of rabbit cortical collecting tubules. Proc Natl Acad Sci U S A 81:4237–4239

Jin Y, Wang Z, Zhang Y et al (2007) PGE2 inhibits apical K channels in the CCD through activation of the MAPK pathway. Am J Physiol Renal Physiol 293:F1299–F1307

Kanthesh BM, Sandle GI, Rajendran VM (2013) Enhanced K^+ secretion in dextran sulfate-induced colitis reflects upregulation of large conductance apical K^+ channels (BK; Kcnma1). Am J Physiol Cell Physiol 305:C972–C980

Lee MG, Ohana E, Park HW et al (2012) Molecular mechanism of pancreatic and salivary gland fluid and HCO_3 secretion. Physiol Rev 92:39–74

Leroy C, Prive A, Bourret JC et al (2006) Regulation of ENaC and CFTR expression with K^+ channel modulators and effect on fluid absorption across alveolar epithelial cells. Am J Physiol Lung Cell Mol Physiol 291:L1207–L1219

Leviel F, Hubner CA, Houillier P et al (2010) The Na^+-dependent chloride-bicarbonate exchanger SLC4A8 mediates an electroneutral Na^+ reabsorption process in the renal cortical collecting ducts of mice. J Clin Invest 120:1627–1635

Li D, Wang Z, Sun P et al (2006) Inhibition of MAPK stimulates the Ca^{2+}-dependent big-conductance K channels in cortical collecting duct. Proc Natl Acad Sci U S A 103:19569–19574

Linley J, Loganathan A, Kopanati S et al (2014) Evidence that two distinct crypt cell types secrete chloride and potassium in human colon. Gut 63(3):472–479

Liu W, Wei Y, Sun P et al (2009) Mechanoregulation of BK channel activity in the mammalian cortical collecting duct: role of protein kinases A and C. Am J Physiol Renal Physiol 297: F904–F915

Liu W, Schreck C, Coleman RA et al (2011) Role of NKCC in BK channel-mediated net K^+ secretion in the CCD. Am J Physiol Renal Physiol 301:F1088–F1097

Manzanares D, Gonzalez C, Ivonnet P et al (2011) Functional apical large conductance, Ca^{2+}-activated, and voltage-dependent K^+ channels are required for maintenance of airway surface liquid volume. J Biol Chem 286:19830–19839

Manzanares D, Srinivasan M, Salathe ST et al (2014) IFN-gamma reduction of large conductance, Ca^{2+} activated, voltage-dependent K^+ (BK) channel activity in airway epithelial cells leads to mucociliary dysfunction. Am J Physiol Lung Cell Mol Physiol 306:L453–L462

Mathialahan T, MacLennan KA, Sandle LN et al (2005) Enhanced large intestinal potassium permeability in end-stage renal disease. J Pathol 206:46–51

Matos JE, Sausbier M, Beranek G et al (2007) Role of cholinergic-activated KCa1.1 (BK), KCa3.1 (SK4) and KV7.1 (KCNQ1) channels in mouse colonic Cl^- secretion. Acta Physiol (Oxf) 189:251–258

Mennitt PA, Wade JB, Ecelbarger CA et al (1997) Localization of ROMK channels in the rat kidney. J Am Soc Nephrol 8:1823–1830

Morita T, Hanaoka K, Morales MM et al (1997) Cloning and characterization of maxi K^+ channel alpha-subunit in rabbit kidney. Am J Physiol Renal Physiol 273:F615–F624

Najjar F, Zhou H, Morimoto T et al (2005) Dietary K^+ regulates apical membrane expression of maxi-K channels in rabbit cortical collecting duct. Am J Physiol Renal Physiol 289:F922–F932

Nakamoto T, Romanenko VG, Takahashi A et al (2008) Apical maxi-K (KCa1.1) channels mediate K^+ secretion by the mouse submandibular exocrine gland. Am J Physiol Cell Physiol 294:C810–C819

Nehrke K, Quinn CC, Begenisich T (2003) Molecular identification of Ca^{2+}-activated K^+ channels in parotid acinar cells. Am J Physiol Cell Physiol 284:C535–C546

Pacha J, Frindt G, Sackin H, Palmer LG (1991) Apical maxi K channels in intercalated cells of CCT. Am J Physiol Renal Physiol 261:F696–F705

Pluznick JL, Kudlacek PE, Padanilam BJ et al (2003a) Identification of BKCa-beta subunit in human glomerular mesangial cells (HMC) in culture. FASEB J 17(5):A1227 (Abstract)
Pluznick JL, Wei P, Carmines PK et al (2003b) Renal fluid and electrolyte handling in BKCa-beta1 −/− mice. Am J Physiol Renal Physiol 284:F1274–F1279
Pluznick JL, Wei P, Grimm PR et al (2005) BK-{beta}1 subunit: immunolocalization in the mammalian connecting tubule and its role in the kaliuretic response to volume expansion. Am J Physiol Renal Physiol 288:F846–F854
Puntheeranurak S, Schreiber R, Spitzner M et al (2007) Control of ion transport in mouse proximal and distal colon by prolactin. Cell Physiol Biochem 19:77–88
Rechkemmer G, Frizzell RA, Halm DR (1996) Active potassium transport across guinea-pig distal colon: action of secretagogues. J Physiol 493(Pt 2):485–502
Ridge FP, Duszyk M, French AS (1997) A large conductance, Ca^{2+}-activated K^+ channel in a human lung epithelial cell line (A549). Biochim Biophys Acta 1327:249–258
Rieg T, Vallon V, Sausbier M et al (2007) The role of the BK channel in potassium homeostasis and flow-induced renal potassium excretion. Kidney Int 72:566–573
Romanenko VG, Thompson J, Begenisich T (2010) Ca^{2+}-activated K channels in parotid acinar cells: the functional basis for the hyperpolarized activation of BK channels. Channels (Austin) 4:278–288
Sandle GI, Gaiger E, Tapster S et al (1986) Enhanced rectal potassium secretion in chronic renal insufficiency: evidence for large intestinal potassium adaptation in man. Clin Sci (Lond) 71:393–401
Sandle GI, Gaiger E, Tapster S et al (1987) Evidence for large intestinal control of potassium homoeostasis in uraemic patients undergoing long-term dialysis. Clin Sci (Lond) 73:247–252
Sandle GI, Perry MD, Mathialahan T et al (2007) Altered cryptal expression of luminal potassium (BK) channels in ulcerative colitis. J Pathol 212:66–73
Sansom SC, O'Neil RG (1985) Mineralocorticoid regulation of apical cell membrane Na^+ and K^+ transport of the cortical collecting duct. Am J Physiol Renal Physiol 248:F858–F868
Sansom SC, Welling PA (2007) Two channels for one job. Kidney Int 72:529–530
Sausbier M, Matos JE, Sausbier U et al (2006) Distal colonic K^+ secretion occurs via BK channels. J Am Soc Nephrol 17:1275–1282
Simon M, Duong JP, Mallet V et al (2008) Over-expression of colonic K^+ channels associated with severe potassium secretory diarrhoea after haemorrhagic shock. Nephrol Dial Transplant 23:3350–3352
Sorensen MV, Matos JE, Sausbier M et al (2008) Aldosterone increases KCa1.1 (BK) channel-mediated colonic K^+ secretion. J Physiol 586:4251–4264
Sorensen MV, Matos JE, Praetorius HA et al (2010a) Colonic potassium handling. Pflugers Arch 459:645–656
Sorensen MV, Sausbier M, Ruth P et al (2010b) Adrenaline-induced colonic K^+ secretion is mediated by KCa1.1 (BK) channels. J Physiol 588:1763–1777
Stoner LC, Morley GE (1995) Effect of basolateral or apical hyposmolarity on apical maxi K channels of everted rat collecting tubule. Am J Physiol Renal Physiol 268:F569–F580
Stoner LC, Viggiano SC (1998) Environmental KCl causes an upregulation of apical membrane maxi K and ENaC channels in everted Ambystoma collecting tubule. J Membr Biol 162:107–116
Tian L, Duncan RR, Hammond MS et al (2001) Alternative splicing switches potassium channel sensitivity to protein phosphorylation. J Biol Chem 276:7717–7720
Tian L, Coghill LS, MacDonald SH et al (2003) Leucine zipper domain targets cAMP-dependent protein kinase to mammalian BK channels. J Biol Chem 278:8669–8677
Van Dinter TGJ, Fuerst FC, Richardson CT et al (2005) Stimulated active potassium secretion in a patient with colonic pseudo-obstruction: a new mechanism of secretory diarrhea. Gastroenterology 129:1268–1273

Venglovecz V, Hegyi P, Rakonczay Z Jr et al (2011) Pathophysiological relevance of apical large-conductance Ca^{2+}-activated potassium channels in pancreatic duct epithelial cells. Gut 60:361–369

Wang Z, Subramanya AR, Satlin LM et al (2013) Regulation of large-conductance Ca^{2+}-activated K^+ channels by WNK4 kinase. Am J Physiol Cell Physiol 305:C846–C853

Wen D, Cornelius RJ, Yuan Y et al (2013) Regulation of BK-alpha expression in the distal nephron by aldosterone and urine pH. Am J Physiol Renal Physiol 305:F463–F476

Wen D, Cornelius RJ, Rivero-Hernandez D et al (2014a) Relation between BK-alpha/beta4-mediated potassium secretion and ENaC-mediated sodium reabsorption. Kidney Int 86 (1):139–145

Wen D, Cornelius RJ, Sansom SC (2014b) Interacting influence of diuretics and diet on BK channel-regulated K homeostasis. Curr Opin Pharmacol 15C:28–32

Woda CB, Bragin A, Kleyman TR et al (2001) Flow-dependent K^+ secretion in the cortical collecting duct is mediated by a maxi-K channel. Am J Physiol Renal Physiol 280:F786–F793

Woda CB, Leite M Jr, Rohatgi R et al (2002) Effects of luminal flow and nucleotides on [Ca(2+)] (i) in rabbit cortical collecting duct. Am J Physiol Renal Physiol 283:F437–F446

Yan J, Aldrich RW (2010) LRRC26 auxiliary protein allows BK channel activation at resting voltage without calcium. Nature 466:513–516

Yan J, Aldrich RW (2012) BK potassium channel modulation by leucine-rich repeat-containing proteins. Proc Natl Acad Sci U S A 109:7917–7922

Yue P, Zhang C, Lin DH et al (2013) WNK4 inhibits Ca^{2+}-activated big-conductance potassium channels (BK) via mitogen-activated protein kinase-dependent pathway. Biochim Biophys Acta 1833:2101–2110

Yusuf TE, Soemijarsih M, Arpaia A et al (1999) Chronic microscopic enterocolitis with severe hypokalemia responding to subtotal colectomy. J Clin Gastroenterol 29:284–288

Zemanova Z, Pacha J (1998) Localization of Na, K-ATPase activity in developing rat distal colon: role of corticosteroids. Mech Ageing Dev 101:129–143

Zhang J, Halm ST, Halm DR (2012) Role of the BK channel (KCa1.1) during activation of electrogenic K^+ secretion in guinea pig distal colon. Am J Physiol Gastrointest Liver Physiol 303:G1322–G1334

Zhuang J, Zhang X, Wang D et al (2011) WNK4 kinase inhibits Maxi K channel activity by a kinase-dependent mechanism. Am J Physiol Renal Physiol 301:F410–F419

Chapter 22
Orai Channels

Trevor J. Shuttleworth

Abstract The highly calcium-selective ion channels formed by the Orai proteins represent the principal route for the agonist-induced entry of extracellular calcium necessary for the generation of the calcium signals involved in the initiation and regulation of a multitude of diverse responses in non-excitable cells. In marked contrast to the voltage-gated calcium channels of excitable cells, the molecular components of these channels (the Orai proteins) and their activation and regulation were only identified less than 10 years ago. Consequently, we are just beginning to understand the details of their unique biophysical properties, modes of activation, and functional roles.

Keywords Calcium release-activated calcium (CRAC) channel • Arachidonate-regulated calcium (ARC) channel • Stromal interacting molecule (STIM) • Calcium signals • Calcium oscillations

22.1 Introduction

The origins of our understanding of agonist-activated calcium entry pathways in non-excitable cells actually began some 20 years before the discovery of the Orai proteins, with the characterization of the so-called capacitative entry of calcium by Putney (Putney 1986). In this, it was proposed that the agonist-induced entry of calcium was specifically initiated as a result of a depletion of endoplasmic reticulum calcium stores. Initially, it was proposed that the entering calcium passed directly into the empty stores situated close to the entry sites in the plasma membrane. However, it was subsequently shown that the calcium actually passed directly into the cytosol (Kwan and Putney 1990; Putney 1990), and the mechanism became known as "store-operated calcium entry." Again, it took a further 3 years before the basic biophysical features of the conductance responsible, named I_{CRAC} (calcium release-activated calcium), were first described (Hoth and Penner 1992,

T.J. Shuttleworth (✉)
Department of Pharmacology and Physiology, University of Rochester Medical Center, 601 Elmwood Avenue, Rochester, NY 14642, USA
e-mail: trevor_shuttleworth@urmc.rochester.edu

K.L. Hamilton, D.C. Devor (eds.), *Ion Channels and Transporters of Epithelia in Health and Disease*, Physiology in Health and Disease,
DOI 10.1007/978-1-4939-3366-2_22

1993; Zweifach and Lewis 1993). Over the subsequent 13 years, considerable information was obtained on the properties and behavior of these channels, but the molecular basis of their composition and activation remained unknown. Only in 2005 was the protein known as stromal-interacting molecule 1 (STIM1, also originally known as GOK) identified as the essential sensor of ER calcium store depletion and regulator of the CRAC channel activity (Roos et al. 2005; Liou et al. 2005; Zhang et al. 2005), followed a year later by the discovery of the novel protein Orai1 (together with its close relatives Orai2 and Orai3) as the actual channel-forming molecule (Feske et al. 2006; Vig et al. 2006b; Zhang et al. 2006; Prakriya et al. 2006; Yeromin et al. 2006).

Consequently, the "Orai channel" field is still very young, and there is much yet to be learnt about this important family of calcium channels that clearly play key roles in a diverse range of critical cellular activities. In particular, it is now clear that the Orai proteins can form additional endogenous channels that are distinct from the store-operated CRAC channels and which act in entirely unique ways to modulate agonist-induced calcium signals in cells. Therefore, this chapter will review the studies describing the properties, structure, and regulation of the well-known store-operated Orai channels (CRAC channels), as well as those of the less extensively studied *store-independent* Orai channels such as the arachidonic acid-regulated (ARC) channels. In addition, the evidence indicating the specific roles of these channels in both the normal physiological functions, as well as those in pathological conditions, of epithelial tissue will be discussed.

22.2 Store-Operated CRAC Channels

22.2.1 Biophysical Properties

The biophysical properties of the CRAC channels have been comprehensively studied over the past 20–25 years and will only be briefly summarized here. More detailed information can be found in numerous excellent reviews (Parekh and Putney 2005; Prakriya 2009; McNally and Prakriya 2012; Derler et al. 2012). As already noted, the initial biophysical characterization of the store-operated CRAC channels occurred several years before their molecular identity was eventually revealed. As such, these early studies established their essential biophysical nature as highly calcium-selective, small-conductance channels whose gating is not membrane voltage dependent. Thus, in whole-cell patch-clamp studies, these channels display marked inward rectification, with inward currents of around 0.6–1.2 pA/pF at −80 mV, depending on cell type, but with no significant outward currents at membrane potentials up to +60 mV. Brief pulses to negative potentials reveal a significant calcium-dependent fast inactivation (Hoth and Penner 1993; Zweifach and Lewis 1995; Fierro and Parekh 1999). Unlike voltage-gated calcium channels, permeability to other divalent cations particularly Ba^{2+} and Sr^{2+} is significantly lower than that for Ca^{2+}. Moreover, the high selectivity for calcium of these channels is further demonstrated by the fact that replacing all extracellular Na^+ and K^+ with $NMDG^+$ results in only a very minor reduction in the inward current, indicating an approximately 1000-fold selectivity for calcium ions over sodium ions. In addition, the currents are effectively blocked by low concentrations of lanthanum (Hoth and Penner 1993), nickel (Zweifach and Lewis 1993), or gadolinium (Broad et al. 1999). Similar to voltage-gated calcium channels, complete removal of external divalent cations results in the appearance of inward monovalent currents that are some eight- to tenfold larger than the Ca^{2+} currents in the same cells which decay over the subsequent 30s or so and display an approximate eightfold selectivity for Na^+ over Cs^+ (Prakriya and Lewis 2002). With regard to their single-channel calcium conductance, it is clear that it is exceptionally small and can only be estimated by noise analysis of the larger monovalent currents measured in the absence of external divalent ions. Such approaches result in an estimated unitary CRAC channel Ca^{2+} conductance of ~20 fS (Zweifach and Lewis 1993; Prakriya and Lewis 2002), a value that is some 500-fold smaller than that for voltage-gated calcium channels.

22.2.2 STIM1 and CRAC Channel Activation

Initially, a variety of diverse "agents" and/or mechanisms were proposed as the essential signal linking depletion of the ER calcium stores and the activation of the CRAC channels in the plasma membrane. These included the involvement of a "second" or "third" messenger (e.g., InsP4), a diffusible calcium influx factor ("CIF"), or via a direct interaction with the InsP3 receptors in the ER. However, it is now clear that STIM1 located in the ER membrane is the key molecule linking store depletion with the activation of the CRAC channels. Thus, the normal store-operated activation of the CRAC channels begins with the loss of bound calcium from an EF-hand domain in the luminal N-terminal portion of STIM1 in the ER membrane as these calcium stores become depleted (Zhang et al. 2005; Liou et al. 2005). This induces the oligomerization of STIM1 molecules (Stathopulos et al. 2006; Liou et al. 2007), followed by their translocation, within the ER membrane domain, to regions that are located close to the plasma membrane (ER-PM "junctions") (Wu et al. 2006) (Fig. 22.1). Here, the STIM1 oligomers interact with, and trap, the Orai1 molecules that form the CRAC channel in the plasma membrane—a process that can be visualized by the formation of discrete puncta representing the co-localization of fluorescently tagged STIM1 and Orai1 molecules at the cell surface (Luik et al. 2006; Xu et al. 2006). The actual induction of CRAC channel activity depends on a region in the extensive cytosolic portion of the STIM1 molecule generally known as the "CRAC activation domain" (CAD, residues 342–448) (Park et al. 2009) that interacts with two distinct regions of the Orai1 subunits—first with a coiled-coil domain in the C-terminal region of Orai1 and second with a sequence in the cytosolic N-terminal portion of Orai1 (McNally et al. 2013).

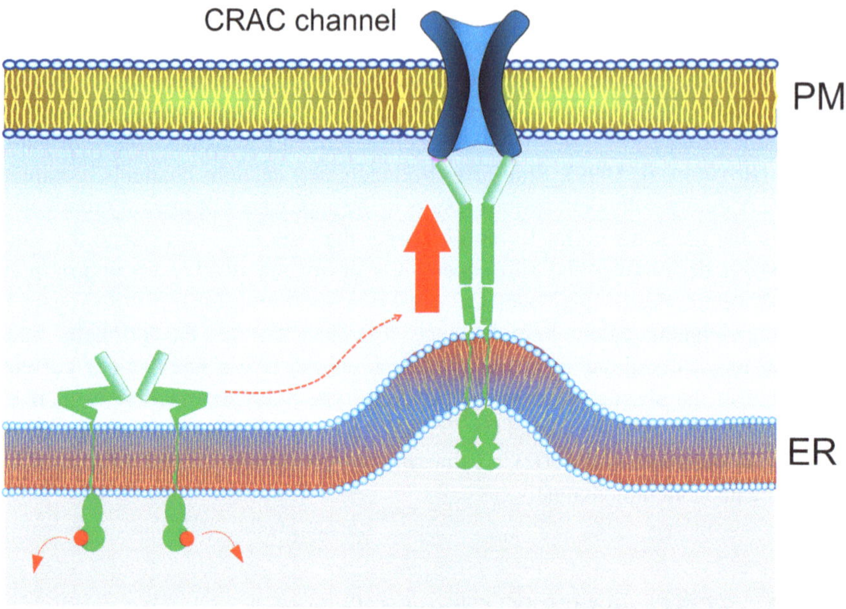

Fig. 22.1 Diagram illustrating the mechanism underlying the activation of the CRAC channels following depletion of endoplasmic reticulum (ER) calcium stores. A fall in calcium ion concentration within the ER store results in the loss of calcium ions (*red*) from the EF-hand domains of STIM1 (*green*) located in the ER membrane. This induces the oligomerization of the STIM1 molecules and transformation into an extended conformation, followed by the translocation of the STIM1 oligomers within the ER membranes to regions that lie close to the plasma membrane (PM). Here, the STIM1 molecules interact with and trap the CRAC channels in the plasma membrane, resulting in their activation

22.2.3 Molecular Basis of CRAC Channel Structure

Prior to the discovery of the Orai proteins in 2006, various molecular entities had been proposed as forming the low-conductance, highly calcium-selective channels responsible for agonist-induced calcium entry in non-excitable cells. Principal among these were members of the TRPC family, an idea largely based on the belief that the sustained portion of the PLC-dependent responses to light in *Drosophila* eyes was induced as a result of calcium store depletion and that these were impaired in animals bearing the *trp* mutations (Hardie and Minke 1992). Subsequent studies however revealed that the activation of the *Drosophila* TRP channels was not dependent on calcium store depletion (Hardie et al. 2001). Despite this, the suggestion that the vertebrate TRPC family of channel proteins represents at least some store-operated calcium channels still persists, although this notion remains controversial (DeHaven et al. 2009). More importantly, it is clear that the biophysical properties of the channels formed from the TRP proteins are entirely distinct from those of the true highly calcium-selective store-operated CRAC channels—a fact that was confirmed by the subsequent discovery of the Orai family of proteins.

As already discussed, the discovery of Orai1 in 2006 definitively revealed that this protein formed the store-operated CRAC channels. The key studies in this long-awaited development involved combinations of a rigorous gene mapping of a family displaying immunodeficiency that was attributed to the absence of functional CRAC channel activity (Feske et al. 2006) and whole-genome screening of *Drosophila* S2 cells (Feske et al. 2006; Vig et al. 2006b; Zhang et al. 2006). Although these studies revealed that the Orai protein family includes two additional members (Orai2 and Orai3), and all three members are known to exist in mammals, it is clear that the endogenous CRAC channel is formed exclusively from Orai1 subunits. Importantly, although studies involving the overexpression of Orai2 and Orai3 have indicated that these can form store-operated conductances, they display biophysical and pharmacological features that are entirely distinct from those seen with endogenous CRAC channels (Vig et al. 2006a; Lis et al. 2007).

Various approaches have been utilized to examine how the Orai1 subunits assemble to form the functional endogenous CRAC channels. For example, we used an approach involving the expression of concatenated assemblies of different numbers of Orai1 subunits, either with or without the co-expression of a dominant-negative Orai1(E106Q) mutant monomer (Mignen et al. 2008b). The anticipation was that, if any of the expressed concatemers required the incorporation of additional subunits to form a functional CRAC channel, the presence of the co-expressed dominant-negative mutant would result in significantly reduced store-operated currents. Thus, although expression of each of the concatemers alone resulted in conductances showing all the features of endogenous CRAC channels, co-expression of the dominant-negative Orai1 monomer significantly reduced each of these currents except those seen with the concatenated Orai1 tetramer (Mignen et al. 2008b). This tetrameric stoichiometry of the functional channel was supported by a variety of additional studies involving single-molecule photobleaching (Ji et al. 2008; Penna et al. 2008), high-resolution electron microscopy (Maruyama et al. 2009), and single-molecule brightness analysis (Madl et al. 2010), although other studies have suggested that the functionally active tetrameric channel may be formed from Orai1 dimers only upon activation (Penna et al. 2008; Demuro et al. 2011). In marked contrast to these findings, a recent report examining the crystal structure of the purified *Drosophila* Orai channel indicated a hexameric stoichiometry (Hou et al. 2012). In this, the functional analysis of this hexameric channel was limited to the demonstration that (a) a constitutively active mutant version incorporated into liposomes allowed the passage of Na^+ ions but not $NMDG^+$ in the absence of external divalent cations, (b) this was absent in the K163W mutant (equivalent to the R91W SCID mutant in mammalian Orai1), and (c) this could be blocked by high concentrations of Gd^{3+} (0.3 mM). However, subsequent examination of a concatenated Orai1 hexameric channel demonstrated that this construct does not result in a store-operated conductance with properties that match endogenous CRAC channels (Thompson and Shuttleworth 2013a). Specifically, such a channel failed to duplicate the high Ca^{2+} selectivity of the CRAC channels,

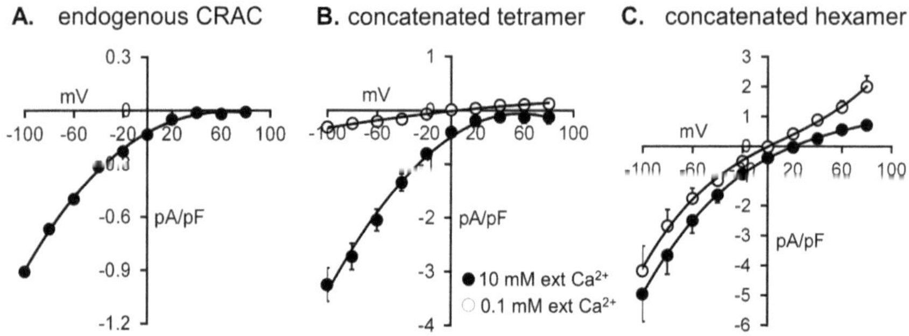

Fig. 22.2 Representative current–voltage relationship of endogenous CRAC channels (**a**) and those of a concatenated Orai1 tetramer (**b**) or a concatenated Orai1 hexamer (**c**). Whole-cell patch-clamp current values were recorded over the voltage range of −100 mV to +80 mV and data obtained with an external medium containing 10 mM calcium (*filled circles*) and, in the case of the concatenated constructs, with an external medium containing 0.1 mM calcium (*open circles*). Note that the concatenated hexamer differs from both the concatenated tetramer and the endogenous CRAC channels by displaying significant outward (Cs^+) currents at voltages above a markedly left-shifted reversal potential (~+20 mV). In addition, reducing external calcium to 0.1 mM essentially eliminates currents through the expressed Orai1 tetramer while having minimal effect on inward currents through the expressed hexamer and actually increasing outward Cs^+ currents (Adapted from Thompson and Shuttleworth 2013a)

displaying marked permeability to Na^+ (and, to a somewhat lesser extent, Cs^+) even in the presence of high external Ca^{2+} (10 mM) (Fig. 22.2). Of course, it should always be remembered that the concatenated construct approach is not without its own problems, perhaps most critically as they relate to the nature of the sequences inserted in the construct to link the subunits. However, even given such potential concerns, it would still seem problematic to explain why, if the CRAC channel truly is a hexamer, the store-operated Ca^{2+} currents seen on expression of a concatenated Orai1 tetramer are completely unaffected by the co-expression of a dominant-negative Orai1(E106Q) construct (Mignen et al. 2008b) and that the expressed Orai1 tetramer duplicates essentially all the key biophysical properties of the endogenous CRAC channel, while a similarly concatenated Orai1 hexamer clearly does not (Thompson and Shuttleworth 2013a).

Despite this obvious controversy, the crystal structure certainly provides useful information regarding the features of the channel pore that are likely important for the selectivity and permeation of Ca^{2+} ions. Most particular of these are the positioning of the critical glutamate residue (E106) at the apical end of the first transmembrane domain, to form the inner part of the selectivity filter, and a somewhat surprising finding of a relatively basic region at the cytosolic end of the pore that may undergo a flexing and/or expansion internally to residue R91 as part of the actual gating process to allow Ca^{2+} ions to pass (Hou et al. 2012).

22.3 Store-Independent ARC Channels

The idea that CRAC channel activity might not be able to explain all modes of agonist-induced cytosolic calcium entry in non-excitable cells really began with a consideration of the underlying mechanism inherent in the store-operated model of calcium entry. In particular, the essential dependence on an InsP3-induced depletion of the ER calcium store to subsequently induce the activation of the calcium entry pathway represented a strict sequence of events that might impose some temporal constraints on the activation of calcium entry. Specifically, in this system, the InsP3-induced release of calcium from the ER calcium stores must precede the onset of calcium entry. Initial studies designed to examine this involved the application of NaF and $AlCl_3$ to induce a gradual generation of the G-protein activating fluoroaluminate ion (AlF_4^-), thereby slowing the rate of InsP3 generation. Curiously, these experiments revealed a clear increased entry of calcium that significantly preceded any detectable release of calcium from intracellular stores (Shuttleworth 1990). Further studies showed that the agonist-induced rate of Ca^{2+} entry was independent of the cyclical emptying and refilling of the agonist-sensitive Ca^{2+} pool during $[Ca^{2+}]_i$ oscillations and that the relevant Ca^{2+} entry pathway remained activated for extended periods following the inhibition of oscillations under conditions in which agonist-sensitive stores could be shown to be full (Shuttleworth and Thompson 1996b). Taken together, these findings were difficult to reconcile with the release of stored Ca^{2+} being the *necessary prerequisite* for the induction of Ca^{2+} entry, as discussed in more detail elsewhere (Shuttleworth 1999; Shuttleworth and Mignen 2003). Of course, one possibility is that the Ca^{2+} entry process is activated by a minor subset of the cellular ER store whose depletion cannot be readily detected, as was indeed suggested at the time (Ribeiro and Putney 1996; Parekh et al. 1997; Huang and Putney 1998). Nevertheless, the above findings certainly raised the possibility of an agonist-activated Ca^{2+} entry pathway distinct from the store-operated, CRAC channel-mediated pathway. The search for the identity of such a store-independent Ca^{2+} entry channel, and its mode of activation, eventually lead to the discovery of the arachidonic acid-regulated “ARC channel” (Shuttleworth 1996; Shuttleworth and Thompson 1998; Mignen and Shuttleworth 2000). Interestingly, like the CRAC channel, the ARC channel subsequently proved to be a new, but distinct, member of the Orai channel family.

22.3.1 *Biophysical Properties*

Biophysically, the basic properties of the store-independent Orai channels known as the ARC channels are very similar to those noted above for the CRAC channels. They show the same markedly inwardly rectifying I/V curve with negligible outward currents at voltages up to +60 mV, and their whole-cell current magnitudes are generally similar to that of the CRAC channels when measured in the same cells. In addition, like the CRAC channels, calcium currents through the ARC channels are profoundly inhibited by low micromolar concentrations (<5 μM) of gadolinium (Mignen et al. 2003).

Despite the close similarity in biophysical properties between the store-operated CRAC channels and the store-independent ARC channels, some differences do exist however. For example, ARC channel currents do not display the clear calcium-dependent fast inactivation seen in CRAC (Mignen and Shuttleworth 2000). Moreover, recordings of the monovalent currents in ARC channels as measured in the absence of external divalent cations indicate that these display a modest permeability to Cs^+ ions that is typically not seen in CRAC channel currents under the same conditions (Mignen et al. 2003). Although the above differences shown between the CRAC channel and ARC channel currents indicate that they may result from the existence of two distinct channel entities, the fact that they are rather subtle could potentially reflect the activity of the same channel under different modes of regulation. Critically, however, it has been shown that provided cytosolic calcium levels are maintained at approximately resting levels (i.e., ~100 nM), the two currents are strictly additive within the same cell, demonstrating that these two biophysically very similar conductances are indeed entirely separate entities—a fact subsequently confirmed by their distinct molecular compositions (see Sect. 22.3.3 below).

22.3.2 STIM1 and ARC Channel Activation

As already noted, the molecular characterization of CRAC channel-mediated calcium entry began with the identification of STIM1 as the essential sensor for the depletion of ER calcium stores. Given this, the clear indication was that this molecule would likely play no part in the activation of the strictly *storebin-dependent* ARC channels. Surprisingly however, it was shown that STIM1 was equally essential for the activation of the ARC channels, although such activation depended not on the pool of STIM1 in the ER, but rather on the minor pool of STIM1 located in the plasma membrane (Mignen et al. 2007)—a finding reconciled by the realization that, as noted above, STIM1 was originally identified as a PM-resident protein (Manji et al. 2000). Nevertheless, this finding raised the question as to how STIM1 in the plasma membrane operated to induce the activation of the ARC channels in the same membrane. Most immediately obvious were the implications of the location in the extracellular environment of the N-terminal portion of STIM1 with its Ca^{2+}-binding EF-hand. Here, given its measured affinity for Ca^{2+} of ~0.2–0.3 mM (Stathopulos et al. 2006; Brandman et al. 2007; Luik et al. 2008), the EF-hand would never normally lose its bound Ca^{2+}, an event that is the essential initiating step in the activation of the CRAC channels on store depletion. This suggested that, in marked contrast to the CRAC channels, the N-terminal portion of STIM1 would be unlikely to have any significant role in the activation of the ARC channels. Consistent with this, expression of a construct comprising just the cytosolic region of STIM1 anchored to the inner face of the plasma membrane via an N-terminally attached sequence based on the SH4 domain of Lck kinase (named the Lck-STIM1-C construct) enabled the exclusive activation of the ARC channels by an appropriate agonist (e.g., carbachol) that would normally be able to activate both ARC and CRAC channels (Thompson and Shuttleworth 2013b). Moreover, again in marked contrast to the situation with CRAC channels where STIM1 in the ER transiently interacts with the channels in the PM, the STIM1 located in the PM is apparently constitutively associated with the ARC channels even under resting conditions (Thompson and Shuttleworth 2013b). Clearly then, although both CRAC and ARC channels share the requirement for STIM1 for their activation, they rely on distinct pools of the total cellular STIM1, and the underlying mechanism of activation is entirely different (Fig. 22.3).

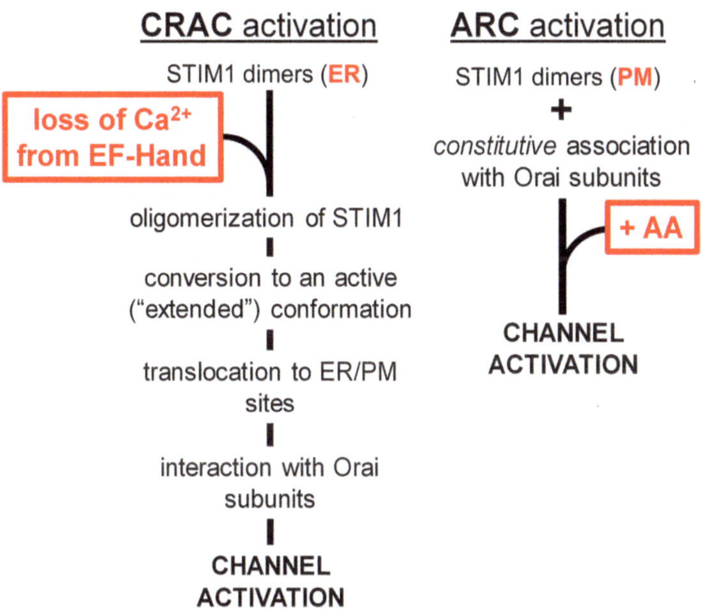

Fig. 22.3 Diagrams comparing the sequence of events involved in the STIM1-mediated activation of CRAC channels versus ARC channels. Endogenous STIM1 located in both the membrane of the endoplasmic reticulum (ER) and in the plasma membrane (PM) is thought to exist as dimers under resting conditions. Activation of the CRAC channels by ER-located STIM1 involves a complex series of discrete events that eventually lead to its interaction with the CRAC channel in the PM. In contrast, STIM1 in the PM is constitutively associated with the ARC channels and only requires the generation/addition of low concentrations of intracellular arachidonic acid to induce channel activation (See Thompson and Shuttleworth 2013b for details)

22.3.3 Molecular Basis of ARC Channel Structure

Given the close biophysical similarity of currents through ARC channels and CRAC channels, together with their common dependence on STIM1 for activation—albeit via an entirely different mechanism—the finding that the ARC channels are also formed from Orai proteins was not surprising. Thus, co-expression of combinations of the three Orai subtypes, followed by examination for the appearance of AA-activated currents with properties consistent with endogenous ARC channels, revealed a requirement for both Orai1 and Orai3, but apparently not Orai2, in forming functional ARC channels (Mignen et al. 2008a). Based on our understanding that CRAC is a tetramer, we used the concatenated approach to show that all possible Orai1/Orai3 combinations in a tetrameric assembly gave predominantly store-operated currents, rather than AA-activated currents. Expanding this same approach to examine Orai1/Orai3 pentamers revealed that expression of a pentamer uniquely comprised of three Orai1 subunits and two Orai3 subunits gave large AA-activated currents that duplicated all the essential biophysical and pharmacological properties of the endogenous ARC channels (Mignen et al. 2009). Finally, subsequent experiments involving the incorporation of individual Orai1/Orai3 "chimeric" subunits into the concatenated pentamer showed that AA-activated ARC channel activity depended specifically on the presence of a minimum of two copies of the N-terminal Orai3 region within the pentameric assembly (Thompson et al. 2010).

22.4 Other Orai Channels

22.4.1 Biophysical Properties

As noted above, several studies have shown that the expression of different Orai proteins in cells can result in the appearance of conductances with a variety of unique features. Principal among these are the conductances seen on the overexpression of Orai3, where the resulting channels display the unique ability to be activated either in a store-dependent manner by STIM1 or in a store-independent manner by the CRAC channel inhibitor 2-aminoethoxydiphenyl borate (2-APB) (Lis et al. 2007). Curiously, these distinct modes of activation result in profound differences in selectivity of the Orai3 channel, with STIM1 activation producing a Ca^{2+}-selective conductance, while activation by 2-APB results in a nonselective conductance (DeHaven et al. 2008; Lis et al. 2007; Schindl et al. 2008; Zhang et al. 2008; Peinelt et al. 2008). Although studies on these "Orai3 channels" have provided important biophysical information on the molecular basis of the gating and selectivity of Orai channels (Yamashita et al. 2011; Amcheslavsky et al. 2013; Yamashita and Prakriya 2014), in the context of the discussion here, it is important to note that the unique features of these "Orai3 channels" have yet to be seen in any *endogenous* conductances. Consequently, their relevance in any physiological context is currently unclear.

More relevant in any physiological context are certain few examples where Orai3 has been shown to be a component of endogenous conductances that appear to display features distinct from the ARC channels described above. Perhaps most notable of these are the store-independent "LRC channels" whose activation depends on the arachidonic acid metabolite leukotriene C_4 (Gonzalez-Cobos et al. 2013; Zhang et al. 2013). The basic biophysical features of these channels are apparently indistinguishable from those of the ARC channels, and they share the inability to be inhibited by the rather promiscuous drug 2-APB (Zhang et al. 2013). However, the reported whole-cell current magnitudes through these LRC channels are extremely small (Zhang et al. 2014), equivalent to only ~2–3 % of the corresponding currents through ARC channels recorded under similar conditions (Mignen et al. 2003).

22.5 Physiological Roles of Orai Channels

Although the original characterization of the store-operated CRAC channels, and the identification of Orai1 as the protein forming their molecular composition, was based almost exclusively on their critical role in various immune responses (e.g., severe combined immunodeficiency/SCID, etc.) (Shaw and Feske 2012), numerous subsequent studies have revealed these channels are also involved in a variety of additional clinical conditions including allergic responses (Di Capite et al. 2011), platelet activation and thrombus formation (Braun et al. 2009), cardiac hypertrophy (Hulot et al. 2011), and smooth muscle proliferation (Trebak 2012). However, it is perhaps worth noting that the fundamental dependence of ER Ca^{2+} store depletion for the activation of the CRAC channels would suggest that an essential role of these channels is likely to be related to the maintenance of the Ca^{2+} content of these stores. Such a role is clearly of paramount importance for cellular viability, as any sustained depletion will result in the disruption of normal protein folding and covalent modification, lipid biosynthesis, etc. in the ER leading to the induction of the "ER stress response" and, ultimately, apoptosis. Obviously, such considerations could certainly explain the widespread expression of these channels in diverse cell types. However, although less comprehensively examined, evidence indicates that the store-independent ARC channels are also present in a variety of different cell types, including various exocrine secretory cells such as salivary and pancreatic acinar cells (Mignen et al. 2005) and avian nasal gland cells (Shuttleworth 1996), as well as the insulin-secreting pancreatic β-cells (Yeung-Yam-Wah et al. 2010), endothelial cells (Fiorio Pla and Munaron 2001), astrocytes (Yang et al. 2005), and in a variety of different cell lines, including HEK cells (Mignen and Shuttleworth 2000), HeLa, COS and RBL cells (Mignen et al. 2003), SH-SY5Y cells (Yoo et al. 2000), and K562 cells (Li et al. 2008).

In any discussion of the physiological roles of the Orai channels, it is important to remember that the activation of the CRAC channels versus the ARC channels depends on two discrete and spatially distinct pools of STIM1. Consequently, this enables these two biophysically similar and molecularly related conductances to operate in entirely separate ways to independently induce distinct physiological responses. However, in general terms, it is clear that the principal role of both the CRAC and ARC channels is in the agonist-induced generation of the cytosolic calcium signals that are essential for the initiation and regulation of a wide range of critical processes, including gene transcription; the activation of certain adenylyl cyclases, kinases, and phosphatases; as well as the regulation of the cell cycle, cell proliferation, and apoptosis. With regard to the actual nature of these agonist-activated Ca^{2+} signals, it is clear that maximal activation of the relevant receptors typically results in a sustained elevation of cytosolic Ca^{2+} concentration. However, because such sustained elevations in cytosolic Ca^{2+} above a certain value can typically induce cell death, such "amplitude-based" signals provide only a limited ability to generate signals able to discriminate between the various appropriate Ca^{2+}-dependent effector systems in cells. With the development of systems enabling examination of responses in single cells, it became clear that physiologically relevant signals generally took the form of a series of Ca^{2+} oscillations (Berridge 1990). This, together with the uncovering of various downstream systems capable of responding to such frequency-based signals, revealed that such responses are able to generate distinct "digital" signals over an almost infinite range (Thomas et al. 1996). As for the specific role of Orai channels in such signals, it was originally thought that the CRAC channels, with their dependence on Ca^{2+} store depletion for activation, were required to replenish Ca^{2+} lost from the cell and recharge the ER stores between each oscillation. However, several lines of evidence were difficult to reconcile with such an effect, and subsequent studies indicated that Ca^{2+} entry also played a key role in actually "driving" the intrinsic oscillatory mechanism itself. For example, oscillatory Ca^{2+} signals typically cease almost immediately on removing extracellular Ca^{2+}, yet intracellular Ca^{2+} stores are not obviously depleted under these conditions (Shuttleworth and Thompson 1996a). Moreover, the oscillations return without any appreciable delay on the restoration of external Ca^{2+} (Martin and Shuttleworth 1994; Morgan and Jacob 1996). Such findings clearly raised questions as to how such activity could be achieved by the store-operated CRAC channels (Shuttleworth 1999), although this has been challenged (Bird and Putney 2005; Bird et al. 2009). More direct evidence for an essential role of the store-independent ARC channels in providing the predominant pathway for agonist-induced Ca^{2+} entry during oscillatory signals was revealed, at least in certain cell types (Shuttleworth 1996; Mignen et al. 2001; Mignen et al. 2005). Nevertheless, the question remained as to how such Ca^{2+} entry could actually act to drive and/or modulate such oscillatory responses. Subsequent studies revealed that one way in which this is achieved involves the specific, and highly local, activation of a member of the PLCδ family (PLCδ3) by the Ca^{2+} entering via ARC channels (Thompson and Shuttleworth 2011). The ARC channel-induced activation of the PLCδ results in the generation of additional InsP3, resulting in

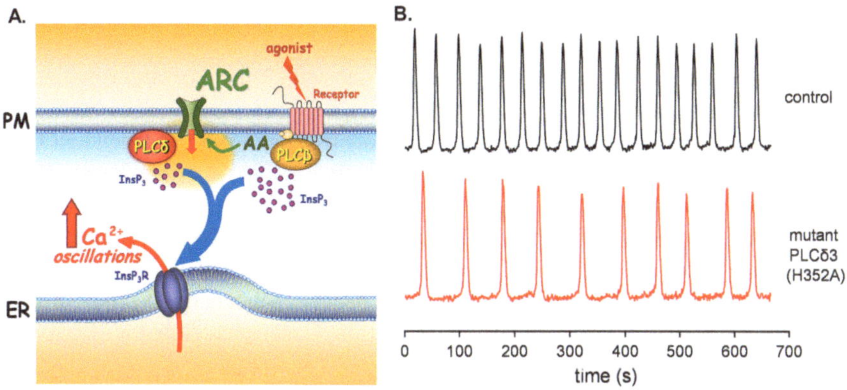

Fig. 22.4 (**a**) Diagram illustrating how the local stimulation of PLCδ by calcium entering exclusively via the ARC channels might act to modulate the frequency of agonist-induced calcium oscillations. The ARC channel-mediated, calcium-dependent activation of PLCδ results in the generation of $InsP_3$ that adds to that produced by the direct agonist-induced activation of PLCβ, thereby effectively raising total $InsP_3$ levels and increasing the oscillatory behavior of the $InsP_3$ receptors on the ER. (**b**) Figure showing the effect of expression of a catalytically dead mutant PLCδ3 (H352A) on agonist-induced ARC channel-mediated calcium oscillations in HEK cells (See Thompson and Shuttleworth 2011 for further details)

an increase in the frequency of the Ca^{2+} oscillations and, presumably, the modulation of the intensity and/or nature of the downstream effects (Fig. 22.4). Importantly, similar activation of Ca^{2+} entry via the CRAC channels failed to duplicate this activity.

Of more direct relevance in the context of this volume with its focus on epithelial tissues, it should be noted that most of the original studies that led to the concept of store-operated Ca^{2+} entry (or, as it was originally described, "capacitative Ca^{2+} entry") were carried out using the epithelial secretory cells of the lachrymal gland (Parod and Putney 1978; Putney 1986; Putney et al. 1998). As additional evidence of the important role of Orai channels in epithelial cell function are the extensive reports of the involvement of these channels in various epithelial cancers—particularly breast, prostate, cervical, and colorectal cancers (Yang et al. 2009; Flourakis et al. 2010; Chen et al. 2011). Finally, there are also examples where an impaired Orai channel function can impact epithelial function in an indirect way. For example, the knockdown of STIM1 (and STIM2) and consequent loss of Orai channel function has been shown to impact salivary gland exocrine function via the induction of the autoimmune condition known as Sjögren's syndrome, resulting in the destruction of the secretory tissue (Cheng et al. 2012).

An important point of note is that the highly polarized nature of epithelial tissues provides the opportunity for entirely discrete signaling process at the opposite ends of the cells, many of which involve Orai channels. An example of this is the generation of distinct apical versus basal Ca^{2+} signals by activation of store-operated Ca^{2+} channels (presumably CRAC channels) by purinergic receptors in human nasal epithelial monolayers (Paradiso et al. 1995). In a similar way, the expression of distinct Ca^{2+} entry channel types in the opposite membrane domains of such polarized tissues can result in distinct responses (Ko et al. 1996), and apical agonist receptors can activate a store-operated Ca^{2+} entry that is exclusively located in the basolateral membrane (Bahra et al. 2004).

As to the distinct respective roles of store-operated and store-independent Ca^{2+} entry, via the CRAC and ARC channels, respectively, this is perhaps most clearly revealed in the acinar cells of the pancreas. Here, the role of cytosolic Ca^{2+} signals in both physiological (Petersen and Tepikin 2008) and pathological (Kruger et al. 2000; Petersen and Sutton 2006; Criddle et al. 2007; Petersen et al. 2009) responses is well established, with both types of response being dependent on the entry of Ca^{2+} (Yule and Gallacher 1988; Raraty et al. 2000). Consistent with the discussion above, the physiologically relevant secretory responses in these cells are driven by oscillatory increases in cytosolic Ca^{2+} signals. However, these signals typically originate in, and are often confined to, the extreme apical region of the cell (Kasai et al. 1993; Thorn et al. 1993). In contrast, the principal site for any agonist-induced Ca^{2+} entry would seem to be in the more basal regions of the cells, where the relevant agonist receptors are located. For a mechanism dependent on store depletion to activate such entry channels would require that activated STIM1 located in the ER membrane of the depleted apical portion of the store could somehow translocate through the ER in regions where such depletion has not occurred while still retaining its ability to interact with and activate the store-dependent CRAC channels located in the plasma membrane of the basal regions of the cell (Fig. 22.5a). Moreover, the entering Ca^{2+} would then have to traverse the cell, without significant buffering, to interact with InsP3 receptors in the apical ER to impact the frequency of Ca^{2+} oscillations in this region. In contrast, if the relevant source of Ca^{2+} entry was via the store-independent ARC channels, their agonist-induced activation could generate additional InsP3 via the PLCδ-mediated mechanism outlined above, which could then readily diffuse to the apical region of the cell without significant degradation to act on the activity of the InsP3 receptors and modulate the resulting oscillatory Ca^{2+} signals responsible for controlling secretion (Fig. 22.5b). Consistent with such a role is the demonstration that, indeed, it is the store-independent ARC channels that provide the predominant source of receptor-induced Ca^{2+} entry at the low agonist concentrations that result in oscillatory Ca^{2+} signals in these cells (Mignen et al. 2005). In contrast to this specific role of the ARC channels in regulating the physiologically relevant oscillatory Ca^{2+} signals associated with secretion, the sustained Ca^{2+} signals associated with activation of the CRAC channels following the depletion of ER Ca^{2+} stores have been shown to result in the premature activation of trypsinogen, and subsequent cellular disruption and necrosis, that is associated with the induction of pancreatitis (Gerasimenko et al. 2013)—an example of where CRAC channel activity is uniquely associated with pathological rather than physiological responses.

An additional example of an important, although perhaps unexpected, role of Orai channels in epithelial function comes from studies investigating the basis for the reported gender differences in the disease cystic fibrosis. Here, females display a markedly poorer general prognosis with a 70–30 % difference in the appearance

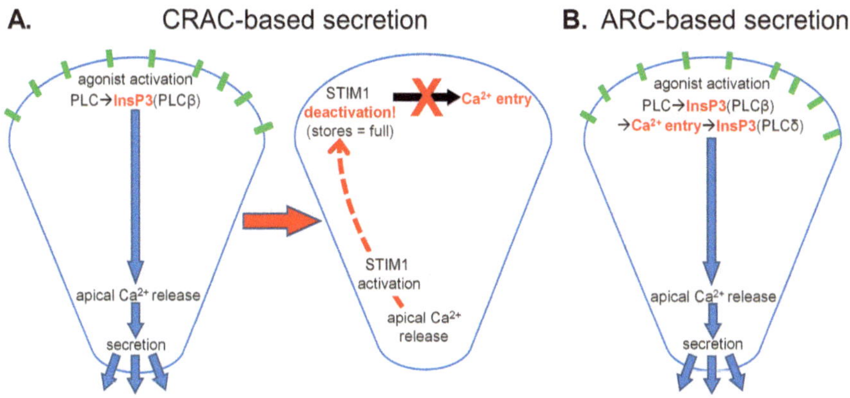

Fig. 22.5 Diagrams comparing the potential regulation of exocrine secretory activity in a pancreatic acinar cell by CRAC channel-mediated and ARC channel-mediated pathways. The CRAC channel pathway (**a**) depends on depletion of ER calcium stores, which typically occurs *only* in the apical part of the cell. STIM1 activated in this region of the ER would then have to translocate to the more basal regions in order to activate CRAC channels—but ER calcium stores would unlikely be depleted in this region, potentially inducing the *deactivation* of the translocated STIM1. In contrast, the *direct* agonist activation of the ARC channels (**b**) would generate $InsP_3$ which would diffuse to the apical ER stores, inducing the release of calcium to drive the secretory process

of early onset COPD (chronic obstructive pulmonary disease), an increased risk of asthma, and a significantly decreased average life-span (Rosenfeld et al. 1997; Carey et al. 2007). Examination of the possible basis for such gender differences in cystic fibrosis revealed that estrogen, acting at the ESR1 receptor, impairs normal STIM1 function in the bronchial epithelial cells as a result of a dephosphorylation of a serine residue (S575) in the cytosolic portion of the molecule (Coakley et al. 2008). This dephosphorylation inhibits the normal Ca^{2+} store depletion-induced oligomerization of STIM1 and its ability to interact with Orai1, leading to a reduced CRAC channel activity. As a result, the normal activation of transepithelial secretion of chloride ions through Ca^{2+}-activated chloride channels is impaired, leading to a reduced fluid secretion by the bronchial epithelium, increasing the susceptibility for lung infections (Coakley et al. 2008).

Finally, it should be noted that attempts to apply pharmacological approaches to delineate the physiological roles of Orai channels have met with only limited success. Thus, essentially all of the agents currently available and most frequently used to inhibit these channels—namely, SKF-96365, 2-aminoethoxydiphenyl borate (2-APB), and pyrazole derivatives (YM-58483, BTP2, Pyr2)—suffer from poor specificity, displaying a variety of effects on other channels particularly members of the various TRP families including TRPC3/5/6/7, TRPV1/2/3, and TRPM4/6/7. More importantly, there is a critical but almost universally overlooked problem in any attempt to develop appropriate drugs that target these channels. Thus, the current emphasis on CRAC channels together with our recent knowledge of the detailed molecular nature of the Orai1 molecules that form these channels and the STIM1 that regulates their activity clearly raises the opportunity to directly target these molecules. Indeed, several major pharmaceutical companies are actively involved in this task. However, such approaches overlook a fundamental and critical problem—namely, that both Orai1 and STIM1 are also essential components of the coexisting, but *functionally entirely distinct*, ARC channels. Consequently, any attempts to target these proteins would not be selective for CRAC channel activity—or, indeed, ARC channel activity—alone. True selectivity for either of these Orai channels will require a far more subtle and incisive approach, most likely based on a detailed understanding of the distinct modes of activation of these coexisting channels.

22.6 Conclusions

It should be clear from the above that the channels encoded by the Orai family of proteins represent a vitally important component of cellular calcium signaling in a diverse range of cell types. While the critical role of the CRAC channels in a variety of immunological and allergic reactions has essentially driven the majority of studies to date, it is becoming increasingly clear that these, as well as other endogenous Orai channels, are widely expressed in a variety of different cell types including those in epithelial tissues. Consequently, although the study of their roles in such tissues is somewhat complicated by their characteristically small whole-cell currents, the demonstrated importance of agonist-induced calcium signals in these cell types indicates that they are likely to be critical. Clearly, there is still much to be learnt about their activities in such tissues under both physiological and pathophysiological conditions.

Acknowledgments The studies reported here from the author's laboratory were supported by NIH grant GM040557 to TJS. The excellent technical assistance of Ms. Jill Thompson is also acknowledged.

References

Amcheslavsky A, Safrina O, Cahalan MD (2013) Orai3 TM3 point mutation G158C alters kinetics of 2-APB-induced gating by disulfide bridge formation with TM2 C101. J Gen Physiol 142:405–412

Bahra P, Mesher J, Li S et al (2004) P2Y2-receptor-mediated activation of a contralateral, lanthanide-sensitive calcium entry pathway in the human airway epithelium. Br J Pharmacol 143:91–98

Berridge MJ (1990) Calcium oscillations. J Biol Chem 265:9583–9586

Bird GS, Hwang SY, Smyth JT et al (2009) STIM1 is a calcium sensor specialized for digital signaling. Curr Biol 19:1724–1729

Bird GS, Putney JW Jr (2005) Capacitative calcium entry supports calcium oscillations in human embryonic kidney cells. J Physiol 562:697–706

Brandman O, Liou J, Park WS, Meyer T (2007) STIM2 is a feedback regulator that stabilizes basal cytosolic and endoplasmic reticulum Ca^{2+} levels. Cell 131:1327–1339

Braun A, Varga-Szabo D, Kleinschnitz C et al (2009) Orai1 (CRACM1) is the platelet SOC channel and essential for pathological thrombus formation. Blood 113:2056–2063

Broad LM, Cannon TR, Taylor CW (1999) A non-capacitative pathway activated by arachidonic acid is the major Ca^{2+} entry mechanism in rat A7r5 smooth muscle cells stimulated with low concentrations of vasopressin. J Physiol 517:121–134

Carey MA, Card JW, Voltz JW et al (2007) It's all about sex: gender, lung development and lung disease. Trends Endocrinol Metab 18:308–313

Chen YF, Chiu WT, Chen YT et al (2011) Calcium store sensor stromal-interaction molecule 1-dependent signaling plays an important role in cervical cancer growth, migration, and angiogenesis. Proc Natl Acad Sci U S A 108:15225–15230

Cheng KT, Alevizos I, Liu X et al (2012) STIM1 and STIM2 protein deficiency in T lymphocytes underlies development of the exocrine gland autoimmune disease, Sjogren's syndrome. Proc Natl Acad Sci U S A 109:14544–14549

Coakley RD, Sun H, Clunes LA et al (2008) 17beta-Estradiol inhibits Ca^{2+}-dependent homeostasis of airway surface liquid volume in human cystic fibrosis airway epithelia. J Clin Invest 118:4025–4035

Criddle DN, McLaughlin E, Murphy JA et al (2007) The pancreas misled: signals to pancreatitis. Pancreatology 7:436–446

DeHaven WI, Jones BF, Petranka JG et al (2009) TRPC channels function independently of STIM1 and Orai1. J Physiol 587:2275–2298

DeHaven WI, Smyth JT, Boyles RR et al (2008) Complex actions of 2-aminoethyldiphenyl borate on store-operated calcium entry. J Biol Chem 283:19265–19273

Demuro A, Penna A, Safrina O et al (2011) Subunit stoichiometry of human Orai1 and Orai3 channels in closed and open states. Proc Natl Acad Sci U S A 108:17832–17837

Derler I, Schindl R, Fritsch R et al (2012) Gating and permeation of Orai channels. Front Biosci 17:1304–1322

Di Capite JL, Bates GJ, Parekh AB (2011) Mast cell CRAC channel as a novel therapeutic target in allergy. Curr Opin Allergy Clin Immunol 11:33–38

Feske S, Gwack Y, Prakriya M et al (2006) A mutation in Orai1 causes immune deficiency by abrogating CRAC channel function. Nature 441:179–185

Fierro L, Parekh AB (1999) Fast calcium-dependent inactivation of calcium release-activated calcium current (CRAC) in RBL-1 cells. J Membr Biol 168:9–17

Fiorio Pla A, Munaron L (2001) Calcium influx, arachidonic acid, and control of endothelial cell proliferation. Cell Calcium 30:235–244

Flourakis M, Lehen'kyi V, Beck B et al (2010) Orai1 contributes to the establishment of an apoptosis-resistant phenotype in prostate cancer cells. Cell Death Dis 1, e75

Gerasimenko JV, Gryshchenko O, Ferdek PE et al (2013) Ca^{2+} release-activated Ca^{2+} channel blockade as a potential tool in antipancreatitis therapy. Proc Natl Acad Sci U S A 110:13186–13191

Gonzalez-Cobos JC, Zhang X, Zhang W et al (2013) Store-independent Orai1/3 channels activated by intracrine leukotriene C4: role in neointimal hyperplasia. Circ Res 112:1013–1025

Hardie RC, Minke B (1992) The trp gene is essential for a light-activated Ca^{2+} channel in Drosophila photoreceptors. Neuron 8:643–651

Hardie RC, Raghu P, Moore S et al (2001) Calcium influx via TRP channels is required to maintain PIP2 levels in Drosophila photoreceptors. Neuron 30:149–159

Hoth M, Penner R (1992) Depletion of intracellular calcium stores activates a calcium current in mast cells. Nature 355:353–356

Hoth M, Penner R (1993) Calcium release-activated calcium current in rat mast cells. J Physiol 465:359–386

Hou X, Pedi L, Diver MM et al (2012) Crystal structure of the calcium release-activated calcium channel Orai. Science 338:1308–1313

Huang Y, Putney JW Jr (1998) Relationship between intracellular calcium store depletion and calcium release-activated calcium current in a mast cell line (RBL-1). J Biol Chem 273:19554–19559
Hulot JS, Fauconnier J, Ramanujam D et al (2011) Critical role for stromal interaction molecule 1 in cardiac hypertrophy. Circulation 124:796–805
Ji W, Xu P, Li Z et al (2008) Functional stoichiometry of the unitary calcium-release-activated calcium channel. Proc Natl Acad Sci U S A 105:13668–13673
Kasai H, Li YX, Miyashita Y (1993) Subcellular distribution of Ca^{2+} release channels underlying Ca^{2+} waves and oscillations in exocrine pancreas. Cell 74:669–677
Ko WH, Chan HC, Wong PY (1996) Anion secretion induced by capacitative Ca^{2+} entry through apical and basolateral membranes of cultured equine sweat gland epithelium. J Physiol 497:19–29
Kruger B, Albrecht E, Lerch MM (2000) The role of intracellular calcium signaling in premature protease activation and the onset of pancreatitis. Am J Pathol 157:43–50
Kwan CY, Putney JW Jr (1990) Uptake and intracellular sequestration of divalent cations in resting and methacholine-stimulated mouse lacrimal acinar cells. Dissociation by Sr^{2+} and Ba^{2+} of agonist-stimulated divalent cation entry from the refilling of the agonist-sensitive intracellular pool. J Biol Chem 265:678–684
Li YS, Wu P, Zhou XY et al (2008) Formyl-peptide receptor like 1: a potent mediator of the Ca^{2+} release-activated Ca^{2+} current ICRAC. Arch Biochem Biophys 478:110–118
Liou J, Fivaz M, Inoue T et al (2007) Live-cell imaging reveals sequential oligomerization and local plasma membrane targeting of stromal interaction molecule 1 after Ca^{2+} store depletion. Proc Natl Acad Sci U S A 104:9301–9306
Liou J, Kim ML, Heo WD et al (2005) STIM is a Ca^{2+} sensor essential for Ca^{2+}-store-depletion-triggered Ca^{2+} influx. Curr Biol 15:1235–1241
Lis A, Peinelt C, Beck A et al (2007) CRACM1, CRACM2, and CRACM3 are store-operated Ca^{2+} channels with distinct functional properties. Curr Biol 17:794–800
Luik RM, Wang B, Prakriya M et al (2008) Oligomerization of STIM1 couples ER calcium depletion to CRAC channel activation. Nature 454:538–542
Luik RM, Wu MM, Buchanan J et al (2006) The elementary unit of store-operated Ca^{2+} entry: local activation of CRAC channels by STIM1 at ER-plasma membrane junctions. J Cell Biol 174:815–825
Madl J, Weghuber J, Fritsch R et al (2010) Resting state Orai1 diffuses as homotetramer in the plasma membrane of live mammalian cells. J Biol Chem 285:41135–41142
Manji SS, Parker NJ, Williams RT et al (2000) STIM1: a novel phosphoprotein located at the cell surface. Biochim Biophys Acta 1481:147–155
Martin SC, Shuttleworth TJ (1994) Ca^{2+} influx drives agonist-activated $[Ca^{2+}]_i$ oscillations in an exocrine cell. FEBS Lett 352:32–36
Maruyama Y, Ogura T, Mio K et al (2009) Tetrameric Orai1 is a teardrop-shaped molecule with a long, tapered cytoplasmic domain. J Biol Chem 284:13676–13685
McNally BA, Prakriya M (2012) Permeation, selectivity and gating in store-operated CRAC channels. J Physiol 590:4179–4191
McNally BA, Somasundaram A, Jairaman A et al (2013) The C- and N-terminal STIM1 binding sites on Orai1 are required for both trapping and gating CRAC channels. J Physiol 591:2833–2850
Mignen O, Shuttleworth TJ (2000) I_{ARC}, a novel arachidonate-regulated, noncapacitative Ca^{2+} entry channel. J Biol Chem 275:9114–9119
Mignen O, Thompson JL, Shuttleworth TJ (2001) Reciprocal regulation of capacitative and arachidonate-regulated noncapacitative Ca^{2+} entry pathways. J Biol Chem 276:35676–35683
Mignen O, Thompson JL, Shuttleworth TJ (2003) Ca^{2+} selectivity and fatty acid specificity of the noncapacitative, arachidonate-regulated Ca^{2+} (ARC) channels. J Biol Chem 278:10174–10181

Mignen O, Thompson JL, Shuttleworth TJ (2007) STIM1 regulates Ca^{2+} entry via arachidonate-regulated Ca^{2+}-selective (ARC) channels without store depletion or translocation to the plasma membrane. J Physiol 579:703–715

Mignen O, Thompson JL, Shuttleworth TJ (2008a) Both Orai1 and Orai3 are essential components of the arachidonate-regulated Ca^{2+}-selective (ARC) channels. J Physiol 586:185–195

Mignen O, Thompson JL, Shuttleworth TJ (2008b) Orai1 subunit stoichiometry of the mammalian CRAC channel pore. J Physiol 586:419–425

Mignen O, Thompson JL, Shuttleworth TJ (2009) The molecular architecture of the arachidonate-regulated Ca^{2+}-selective ARC channel is a pentameric assembly of Orai1 and Orai3 subunits. J Physiol 587:4181–4197

Mignen O, Thompson JL, Yule DI et al (2005) Agonist activation of arachidonate-regulated Ca^{2+}-selective (ARC) channels in murine parotid and pancreatic acinar cells. J Physiol 564:791–801

Morgan AJ, Jacob R (1996) Ca^{2+} influx does more than provide releasable Ca^{2+} to maintain repetitive spiking in human umbilical vein endothelial cells. Biochem J 320:505–517

Paradiso AM, Mason SJ, Lazarowski ER et al (1995) Membrane-restricted regulation of Ca^{2+} release and influx in polarized epithelia. Nature 377:643–646

Parekh AB, Fleig A, Penner R (1997) The store-operated calcium current I_{CRAC}: nonlinear activation by $InsP_3$ and dissociation from calcium release. Cell 89:973–980

Parekh AB, Putney JW Jr (2005) Store-operated calcium channels. Physiol Rev 85:757–810

Park CY, Hoover PJ, Mullins FM et al (2009) STIM1 clusters and activates CRAC channels via direct binding of a cytosolic domain to Orai1. Cell 136:876–890

Parod RJ, Putney JW Jr (1978) The role of calcium in the receptor mediated control of potassium permeability in the rat lacrimal gland. J Physiol 281:371–381

Peinelt C, Lis A, Beck A et al (2008) 2-Aminoethoxydiphenyl borate directly facilitates and indirectly inhibits STIM1-dependent gating of CRAC channels. J Physiol 586:3061–3073

Penna A, Demuro A, Yeromin AV et al (2008) The CRAC channel consists of a tetramer formed by Stim-induced dimerization of Orai dimers. Nature 456:116–120

Petersen OH, Sutton R (2006) Ca^{2+} signalling and pancreatitis: effects of alcohol, bile and coffee. Trends Pharmacol Sci 27:113–120

Petersen OH, Tepikin AV (2008) Polarized calcium signaling in exocrine gland cells. Annu Rev Physiol 70:273–299

Petersen OH, Tepikin AV, Gerasimenko JV et al (2009) Fatty acids, alcohol and fatty acid ethyl esters: toxic Ca^{2+} signal generation and pancreatitis. Cell Calcium 45:634–642

Prakriya M (2009) The molecular physiology of CRAC channels. Immunol Rev 231:88–98

Prakriya M, Feske S, Gwack Y et al (2006) Orai1 is an essential pore subunit of the CRAC channel. Nature 443:230–233

Prakriya M, Lewis RS (2002) Separation and characterization of currents through store-operated CRAC channels and Mg^{2+}-inhibited cation (MIC) channels. J Gen Physiol 119:487–507

Putney JW Jr (1986) A model for receptor-regulated calcium entry. Cell Calcium 7:1–12

Putney JW Jr (1990) Capacitative calcium entry revisited. Cell Calcium 11:611–624

Putney JW Jr, Huang Y, Bird GS (1998) Calcium signalling in lacrimal acinar cells. Adv Exp Med Biol 438:123–128

Raraty M, Ward J, Erdemli G et al (2000) Calcium-dependent enzyme activation and vacuole formation in the apical granular region of pancreatic acinar cells. Proc Natl Acad Sci U S A 97:13126–13131

Ribeiro CM, Putney JW (1996) Differential effects of protein kinase C activation on calcium storage and capacitative calcium entry in NIH 3T3 cells. J Biol Chem 271:21522–21528

Roos J, DiGregorio PJ, Yeromin AV et al (2005) STIM1, an essential and conserved component of store-operated Ca^{2+} channel function. J Cell Biol 169:435–445

Rosenfeld M, Davis R, FitzSimmons S et al (1997) Gender gap in cystic fibrosis mortality. Am J Epidemiol 145:794–803

Schindl R, Bergsmann J, Frischauf I et al (2008) 2-Aminoethoxydiphenyl borate alters selectivity of Orai3 channels by increasing their pore size. J Biol Chem 283:20261–20267

Shaw PJ, Feske S (2012) Physiological and pathophysiological functions of SOCE in the immune system. Front Biosci 4:2253–2268
Shuttleworth TJ (1990) Fluoroaluminate activation of different components of the calcium signal in an exocrine cell. Biochem J 269:417–422
Shuttleworth TJ (1996) Arachidonic acid activates the noncapacitative entry of Ca^{2+} during $[Ca^{2+}]_i$ oscillations. J Biol Chem 271:21720–21725
Shuttleworth TJ (1999) What drives calcium entry during $[Ca^{2+}]_i$ oscillations?–challenging the capacitative model. Cell Calcium 25:237–246
Shuttleworth TJ, Mignen O (2003) Calcium entry and the control of calcium oscillations. Biochem Soc Trans 31:916–919
Shuttleworth TJ, Thompson JL (1996a) Ca^{2+} entry modulates oscillation frequency by triggering Ca^{2+} release. Biochem J 313:815–819
Shuttleworth TJ, Thompson JL (1996b) Evidence for a non-capacitative Ca^{2+} entry during $[Ca^{2+}]$ oscillations. Biochem J 316:819–824
Shuttleworth TJ, Thompson JL (1998) Muscarinic receptor activation of arachidonate-mediated Ca^{2+} entry in HEK293 cells is independent of phospholipase C. J Biol Chem 273:32636–32643
Stathopulos PB, Li GY, Plevin MJ et al (2006) Stored Ca^{2+} depletion-induced oligomerization of stromal interaction molecule 1 (STIM1) via the EF-SAM region: an initiation mechanism for capacitive Ca^{2+} entry. J Biol Chem 281:35855–35862
Thomas AP, Bird GS, Hajnoczky G et al (1996) Spatial and temporal aspects of cellular calcium signaling. FASEB J 10:1505–1517
Thompson JL, Mignen O, Shuttleworth TJ (2010) The N-terminal domain of Orai3 determines selectivity for activation of the store-independent ARC channel by arachidonic acid. Channels 4:398–410
Thompson JL, Shuttleworth TJ (2011) Orai channel-dependent activation of phospholipase C-δ: a novel mechanism for the effects of calcium entry on calcium oscillations. J Physiol 589:5057–5069
Thompson JL, Shuttleworth TJ (2013a) How many Orai's does it take to make a CRAC channel? Sci Rep 3:1961
Thompson JL, Shuttleworth TJ (2013b) Molecular basis of activation of the arachidonate-regulated Ca^{2+} (ARC) channel, a store-independent Orai channel, by plasma membrane STIM1. J Physiol 591:3507–3523
Thorn P, Lawrie AM, Smith PM et al (1993) Local and global cytosolic Ca^{2+} oscillations in exocrine cells evoked by agonists and inositol trisphosphate. Cell 74:661–668
Trebak M (2012) STIM/Orai signalling complexes in vascular smooth muscle. J Physiol 590:4201–4208
Vig M, Beck A, Billingsley JM et al (2006a) CRACM1 multimers form the ion-selective pore of the CRAC channel. Curr Biol 16:2073–2079
Vig M, Peinelt C, Beck A et al (2006b) CRACM1 is a plasma membrane protein essential for store-operated Ca^{2+} entry. Science 312:1220–1223
Wu MM, Buchanan J, Luik RM et al (2006) Ca^{2+} store depletion causes STIM1 to accumulate in ER regions closely associated with the plasma membrane. J Cell Biol 174:803–813
Xu P, Lu J, Li Z et al (2006) Aggregation of STIM1 underneath the plasma membrane induces clustering of Orai1. Biochem Biophys Res Commun 350:969–976
Yamashita M, Prakriya M (2014) Divergence of Ca^{2+} selectivity and equilibrium Ca^{2+} blockade in a Ca^{2+} release-activated Ca^{2+} channel. J Gen Physiol 143:325–343
Yamashita M, Somasundaram A, Prakriya M (2011) Competitive modulation of Ca^{2+} release-activated Ca^{2+} channel gating by STIM1 and 2-aminoethyldiphenyl borate. J Biol Chem 286:9429–9442
Yang KT, Chen WP, Chang WL et al (2005) Arachidonic acid inhibits capacitative Ca^{2+} entry and activates non-capacitative Ca^{2+} entry in cultured astrocytes. Biochem Biophys Res Commun 331:603–613

Yang S, Zhang JJ, Huang XY (2009) Orai1 and STIM1 are critical for breast tumor cell migration and metastasis. Cancer Cell 15:124–134
Yeromin AV, Zhang SL, Jiang W et al (2006) Molecular identification of the CRAC channel by altered ion selectivity in a mutant of Orai. Nature 443:226–229
Yeung-Yam-Wah V, Lee AK, Tse FW et al (2010) Arachidonic acid stimulates extracellular Ca^{2+} entry in rat pancreatic β cells via activation of the noncapacitative arachidonate-regulated Ca^{2+} (ARC) channels. Cell Calcium 47:77–83
Yoo AS, Cheng I, Chung S et al (2000) Presenilin-mediated modulation of capacitative calcium entry. Neuron 27:561–572
Yule DI, Gallacher DV (1988) Oscillations of cytosolic calcium in single pancreatic acinar cells stimulated by acetylcholine. FEBS Lett 239:358–362
Zhang SL, Kozak JA, Jiang W et al (2008) Store-dependent and -independent modes regulating Ca^{2+} release-activated Ca^{2+} channel activity of human Orai1 and Orai3. J Biol Chem 283:17662–17671
Zhang SL, Yeromin AV, Zhang XH et al (2006) Genome-wide RNAi screen of Ca^{2+} influx identifies genes that regulate Ca^{2+} release-activated Ca^{2+} channel activity. Proc Natl Acad Sci U S A 103:9357–9362
Zhang SL, Yu Y, Roos J et al (2005) STIM1 is a Ca^{2+} sensor that activates CRAC channels and migrates from the Ca^{2+} store to the plasma membrane. Nature 437:902–905
Zhang X, Gonzalez-Cobos JC, Schindl R et al (2013) Mechanisms of STIM1 activation of store-independent leukotriene C4-regulated Ca^{2+} channels. Mol Cell Biol 33:3715–3723
Zhang X, Zhang W, Gonzalez-Cobos JC et al (2014) Complex role of STIM1 in the activation of store-independent Orai1/3 channels. J Gen Physiol 143:345–359
Zweifach A, Lewis RS (1993) Mitogen-regulated Ca^{2+} current of T lymphocytes is activated by depletion of intracellular Ca^{2+} stores. Proc Natl Acad Sci U S A 90:6295–6299
Zweifach A, Lewis RS (1995) Rapid inactivation of depletion-activated calcium current (I_{CRAC}) due to local calcium feedback. J Gen Physiol 105:209–226

Chapter 23
ADPKD Channels: The Polycystins

Bonnie L. Blazer-Yost

Abstract The initial members of the polycystin family were first described and characterized as the proteins mutated in a common genetic disease, autosomal dominant polycystic kidney disease (ADPKD). Over the past few decades, these proteins, polycystins 1 and 2 (PC1, PC2), have been found to have a variety of cellular functions in multiple intracellular locations. In addition, other homologous proteins have been identified and linked to the transient receptor potential family of channels. PC1 and PC2 or homologous family members form mechanosensitive Ca^{2+} channels in the primary cilia. Elsewhere in the cell, the PCs may form Ca^{2+} channels that are gated by other stimuli and each of the PCs appears to play a variety of functional roles in multiple cellular pathways. While much is currently known about the polycystins and their roles in ADPKD initiation and progression, the tissue-specific binding partners and the interactions that modulate the normal physiological functions of all the polycystins are still emerging.

Keywords Ciliopathies • Primary cilium • Calcium channels • Transient receptor potential channels

23.1 Polycystic Kidney Disease

Polycystins are the proteins that are mutated in autosomal dominant polycystic kidney disease (ADPKD). ADPKD is the most common renal genetic disease in the human population with an estimated incidence of 1:400–1:1000 (Harris and Torres 2009). ADPKD is not limited to a specific race or gender. There is also an autosomal recessive (ARPDK) form which is relatively rare, affects children in the neonatal period, and is characterized by tubular dilation (Harris and Torres 2014). In contrast, the more common form, ADPKD, is characterized by closed, fluid-filled cysts which form predominately in the kidney and liver bile ducts. Progressive cyst growth and associated fibrosis ultimately affect renal function

---

B.L. Blazer-Yost (✉)
Department of Biology, Indiana University, Purdue University Indianapolis, 723 West Michigan Street, Indianapolis, IN 46202, USA
e-mail: bblazer@iupui.edu

K.L. Hamilton, D.C. Devor (eds.), *Ion Channels and Transporters of Epithelia in Health and Disease*, Physiology in Health and Disease,
DOI 10.1007/978-1-4939-3366-2_23

leading to end-stage renal disease in approximately 50 % of patients (Chapman 2008; Chang and Ong 2008). While patients are born with numerous small cysts, expansion progresses slowly, and renal function is typically not compromised until the fifth decade of life, but then the decline is precipitous often requiring dialysis or transplantation within 5–10 years (Chapman 2008).

In ADPKD, the progression to renal failure in mid-life is often preceded by patient discomfort and pain as the cysts enlarge until the kidneys are ultimately about 10 times normal size. The renal pathology is the primary cause of morbidity and mortality. The most common extrarenal manifestation of PKD is liver cysts that arise from the cholangiocytes lining the hepatic bile duct, causing discomfort and some degree of organ damage. There is also an increased risk of pancreatic disease as well as cardiovascular abnormalities, most commonly intracranial aneurysms and hypertension (Harris and Torres 2014).

Current therapy is limited to renal cyst aspiration to relieve pain and, ultimately, renal transplantation after organ failure. Several pharmaceutical agents are progressing through clinical trials, but thus far none have been approved by the FDA. The disease has been estimated to account for approximately 4 % of dialysis patients and represents a significant cost to the healthcare systems both before and after renal demise (Lentine et al. 2010).

It is widely accepted that secretion of ions and fluid into the lumen by the cyst-lining epithelial cells contributes to cyst expansion. These high-resistance epithelia form a barrier that isolates and maintains the cyst fluid in a compartment separate from the extracellular milieu. Studies performed nearly two decades ago indicate that renal cyst growth in ADPKD is driven by Cl^- secretion (Ye and Grantham 1993; Grantham et al. 1995; Davidow et al. 1996; Hanaoka et al. 1996), and more recent studies have shown a remarkably similar profile in freshly isolated bile duct epithelia (Muchatuta et al. 2009). The cystic fibrosis transmembrane conductance regulator (CFTR) is the major Cl^- channel responsible for secretion into both kidney and biliary cysts (Ye and Grantham 1993; Grantham et al. 1995; Davidow et al. 1996; Hanaoka et al. 1996; Muchatuta et al. 2009). Abnormal growth and proliferation contribute to the overall expansion of the cysts. Molecules that target either the secretory pathway or one of the multiple proliferative pathways are under investigation for possible drug intervention (Watnick and Germino 2010; Pei 2010; Harris and Torres 2014).

The causative genes for ADPKD were identified between 1994 and 1996 and are termed *PKD1* (16p13.3) and *PKD2* (4q22). The proteins encoded by these genes were called polycystin 1 (PC1) or polycystin 2 (PC2) (The European Polycystic Kidney Disease Consortium 1994; The American PKD1 Consortium 1995; The International Polycystic Kidney Disease Consortium 1995; Hughes et al. 1995; Mochizuki et al. 1996). PC1 is involved in approximately 78 % of ADPKD patients, another 13 % involve PC2, and 9 % of cases have undocumented mutations (Rossetti et al. 2007; Audrezet et al. 2012; Harris and Torres 2014).

23.2 Polycystins: Nomenclature

Polycystins 1 and 2 are also termed polycystic kidney disease proteins 1 and 2 (PKD1 and PKD2) and form the prototypic members of two distinct families of proteins. The rest of the members of these two groups have been termed polycystin 1- or 2-like (PKD1L or PKD2L). There are four additional PC1 homologues, PKD1L1, PKD1L2, and PKD1L3 (Yuasa et al. 2002, 2004; Li et al. 2003a) and PKDREJ (receptor for egg jelly) (Hughes et al. 1999). There are two additional members to the PC2 family—PKD2L1 and PKD2L2 (Nomura et al. 1998; Veldhuisen et al. 1999; Guo et al. 2000).

Some of the polycystins, particularly the PC2 family, have been found to be cation channels. Because of the similarity in structure and function, the whole family has been characterized as a branch of the transient receptor potential (TRP) channels. With respect to its channel characteristics, PC2 is also known as TRPP2, the founding member of the TRPP subcategory, while the other two members of this family are known as TRPP3 and TRPP5 (Delmas 2005; Giamarchi et al. 2006; Nilius and Owsianik 2011). The PC1 family of proteins does not show structural similarity to TRP channels. While some investigators have termed PC1 TRPP1, this as well as TRPP nomenclature for the other PC1 family members is not widely used.

23.3 Localization

After the elucidation of full-length *PKD1* cDNA and the production of antibodies to the expressed protein, it was found that PC1 was localized to the epithelial cells of the distal nephron and vascular endothelial cells in both maturing fetal nephrons and normal adult kidney (Retailleau and Duprat 2014; Ibraghimov-Beskrovnaya et al. 1997). The PC1 protein was also detected in ductal epithelia of the liver and pancreas and in smooth muscle cells (Ward et al. 1996; Geng et al. 1996, 1997; Ibraghimov-Beskrovnaya et al. 1997). PC1 is a large transmembrane protein whose subcellular localization has subsequently been demonstrated in the endoplasmic reticulum, plasma membrane, and membranes of the primary cilium, a hairlike structure on the apical membrane (Scheffers et al. 2000; Boletta et al. 2001; Newby et al. 2002).

In the plasma membrane, PC1 plays a role in cell-cell interactions in barrier epithelial cells. In cultured epithelial cells, the PC1 was localized to the lateral cell junctions (Ibraghimov-Beskrovnaya et al. 1997; Bukanov et al. 2002). The subsequent expression of the full-length protein in Madin-Darby canine kidney (MDCK) cell line substantiated the cell-cell junction localization (Boletta et al. 2001), and antibodies raised against the Ig-like domains, an area of PC1 thought to mediate protein-protein interactions, disrupt the cell-cell interactions. These data suggest homophilic interactions via the Ig-like domains (Ibraghimov-Beskrovnaya et al. 2000). In transgenic kidney cell lines generated from the newborn mouse kidney, PC1 was found in the lateral cell borders where it formed a complex with PC2 (Newby et al. 2002).

Within the intracellular junctions, PC1 has been found in desmosomes where it contributes to intercellular adhesion (Scheffers et al. 2000; Streets et al. 2003). In ADPKD cysts, the usual co-localization between desmosomal proteins and PC1 is lost and both proteins appeared in a punctate intracellular localization (Russo et al. 2005). These proteins also localize to adherens junction (Huan and van Adelsberg 1999; Wilson et al. 1999; Geng et al. 2000) where they bind to distinct proteins contributing to the junctional complexes.

PC2 was first found in the premedial Golgi and ER membranes (Cai et al. 1999; Koulen et al. 2002; Li et al. 2005) and this is still considered the predominant localization. Subsequently, several groups have established the presence of PC2 on the cell membrane (Himpens et al. 1993; Gonzalez-Perrett et al. 2001; Luo et al. 2003) as well as in the primary cilium (Yoder et al. 2002; Nauli et al. 2003).

Localization of both PC1 and PC2 in the primary cilia is important for their function in ADPKD (Yoder et al. 2002; Pazour et al. 2002; Nauli et al. 2003). Transport of PC1 into the primary cilia, but not the plasma membrane per se, requires binding to the proteins of the BBSome (Su et al. 2014). The BBSome, a large complex of proteins that appears to gate protein trafficking into the cilia, is named for Bardet-Biedl syndrome, a severe ciliopathy involving these proteins. Specifically, PC1 binds BBS1, BBS4, BBS5, and BBS8 of the components that make up the BBSome. The N-terminal sequence motif $R_6V_7xP_8$ is required for PC2 localization to the cilia (Geng et al. 2006).

The other PC isoforms are more widely distributed (review, Delmas 2005). For example, PKD1L1 is found in the fetal and adult heart and in the Leydig cells of the mouse testis (Yuasa et al. 2002). PKD2L2 and PKDREJ are also found in testis. Interestingly PKD1L1 and PKD2L1 have been identified in primary cilia where they form a Ca^{2+} channel analogous to the channel formed by PC1/PC2 (DeCaen et al. 2013; Delling et al. 2013).

In summary, the polycystins have been localized to the primary cilia, plasma membrane, cytoplasmic vesicles, endoplasmic reticulum, and cell-cell and cell-extracellular matrix contacts. This diversity of cellular and subcellular localization suggests a variety of functional roles that are likely dependent on an equally diverse array of binding partners for each of the polycystins.

23.4 Structure

PC1 is a large glycoprotein which contains 4302 amino acids and has a molecular weight of approximately 600 kDa when glycosylated. The N-terminal extracellular portion contains approximately 75 % of the amino acids including two leucine-rich repeats, a C-type lectin receptor, LDL-A domain, 16 Ig-like (or PKD) domains, a domain homologous to the sea urchin sperm receptor for jelly (REJ), and a G protein-coupled receptor proteolytic site (GPS) (Hughes et al. 1995; Sanford et al. 1997; Ibraghimov-Beskrovnaya et al. 1997; Bycroft et al. 1999). There are 11 transmembrane domains and a short C-terminus cytoplasmic domain which contains a coiled-coil and a heterotrimeric G protein-binding domain that can activate purified G proteins in vitro (Parnell et al. 1998). This protein has multiple binding partners via both the extracellular and intracellular domains (Retailleau and Duprat 2014). One of the most important interactions occurs via the coiled-coil intracellular domains (Qian et al. 1997; Tsiokas et al. 1997).

PC2 is a 968 polypeptide of approximately 110 kDa (Mochizuki et al. 1996; Hanaoka et al. 2000; Newby et al. 2002). The protein has six membrane spanning domains and both the N and C termini are cytosolic (Mochizuki et al. 1996). The C terminus contains a coiled coil, an ER retention signal, and EF-hand motif (Mochizuki et al. 1996; Tsiokas et al. 1999; Anyatonwu and Ehrlich 2004). The N terminus has a ciliary sorting motif.

The additional four members of the PC1 family, PKD1L1–3, and PKDREJ all have a large extracellular N-terminal domain, 11 predicted transmembrane-spanning segments, an REJ domain, and a GPS site (Hughes et al. 1999; Yuasa et al. 2002). Only PKD1L1 shares the intracellular C-terminal coiled-coil domain with PC1 (Delmas 2005).

PKD2L1 and PKD2L2 are structurally similar to PC2 (Nomura et al. 1998; Chen et al. 1999; Veldhuisen et al. 1999). All have the six membrane-spanning domains, a pore domain, and a coiled-coil in the carboxyl-terminus. Only PKD2L1 shares the PC2 EF-hand motif that binds Ca^{2+} (Delmas 2005).

23.5 Binding Partners

Multiple binding partners have been identified for both polycystins. For example, a proteomic analysis identified 79 PC2-associated proteins in arterial myocytes (Sharif-Naeini et al. 2009), and a review by Retailleau and Duprat (2014) details 37 PC1 partners and 32 PC2 partners, most of which are characterized by the PC interacting domains, localization of the interaction, and proposed functional role. Given these daunting numbers, this section will only review the most well characterized of the PC/binding protein interactions.

The exact localization of each of the polycystins appears to be dependent on binding partners. With regard to ADPKD, one of the primary binding partners of PC1 is PC2 (Qian et al. 1997; Tsiokas et al. 1997; Hanaoka et al. 2000). For example, PC1 seems to be necessary for the trafficking of PC2 to the plasma membrane (Hanaoka et al. 2000). Once it is in the plasma membrane, phosphorylation by glycogen synthase kinase 3 (GSK3) is important for maintaining PC2 on the lateral membrane. Inhibitors of GSK3 cause the internalization of membrane-bound but not cilia-localized PC2 (Streets et al. 2006).

A coiled-coil domain in the C-terminus domain of the PC1 protein was shown to specifically interact with an analogous structure in the C terminus of PC2 and naturally occuring mutations disrupt this interaction (Qian et al. 1997). In transgenic and nontransgenic kidney cell lines and in the human adult kidney, PC1 and PC2 were shown to form a stable complex in the lateral membrane. That this interaction takes place via the C-terminal tails is substantiated by the finding that truncated forms of PC1 lacking the C-terminal 205 amino acids were unable to bind PC2 (Newby et al. 2002). PC1 and PC2 also interact in the apical plasma membrane and the ER (Somlo and Ehrlich 2001; Koulen et al. 2002; Luo et al. 2003; Delmas et al. 2004).

Given the similarity of ADPKD in patients with PC1 mutations compared to those with PC2 mutations, the co-localization and binding of the two PCs is likely to be important for disease progression. The PC2/PC1 complex also interacts with fibrocystin, the protein that is mutated in autosomal recessive PKD (Qian et al. 1997; Wu et al. 2006; Kim et al. 2008). All three of these disease-associated proteins have been localized to the primary cilium. PC2 interacts with PDK1L1 in the primary cilium and this interaction is thought to be important in establishing left-right asymmetry during development (Vogel et al. 2010; Field et al. 2011).

PC1 binds four of the seven proteins of the BBSome complex (Su et al. 2014). This protein complex localizes at the basal body and, via G protein interactions, is thought to control protein flux into the cilia (Berbari et al. 2008).

The polycystins can form heterodimers or heteromultimers not only with each other but with other members of the TRP family proteins, particularly those of the TRPC family involved in G protein-coupled receptor activation. PC2 interacts with TRPC1 in transfected cells and in vitro via a C-terminal sequence as well as a a domain located within the transmembrane segments S2 and S5 (Tsiokas et al. 1999). The heteromeric PC2/TRPC1 channels are found in the primary cilia and form unique channels that are distinct from the channels formed by either protein alone (Bai et al. 2008).

Several of the demonstrated binding partners indicate a role for PC1 and PC2 in the functions of maintenance of cellular structure and cell-cell interactions. The large external N-terminal domain of PC1binds to extracellular matrix proteins including laminin, fibronectin, and collagens I, II, and IV (Weston et al. 2001; Malhaus et al. 2002). Binding to extracellular substrates not only serves a structural role but also leads to intracellular changes and recruitment of both signaling intermediates and structural proteins.

In aortic vascular smooth muscle cells, both PC1 and PC2 were found in dense plaques that are involved in the cell-extracellular matrix interactions (Qian et al. 2003b). This finding suggests an analogous function in this cell type, but the consequences of this finding remain to be explored.

In cultured human fetal collecting duct cells, a variety of techniques showed that PC1 associates with focal adhesion proteins, talin, vinculin, p130Cas, focal adhesion kinase pp125^{FAK}, α-actinin, paxillin, and pp60c-src in subconfluent cultures (Wilson et al. 1999; Geng et al. 2000). The binding complex changed to include the cell-cell adherens junction proteins, E-cadherin and β- and γ-catenins when the cells attained confluency. Some components of the multiprotein complex were altered by tyrosine phosphorylation and Ca^{2+} concentration (Geng et al. 2000). In addition to PC1 binding to well-characterized focal adhesion complex proteins, PC2 binds to CD2AP, an adapter protein that functions in the assembly of focal adhesion complexes (Lehtonen et al. 2000). The interaction between PC2 and CD2AP is via the C-terminal portions of both proteins suggesting that the binding takes place in an intracellular compartment.

Adherens junctions connecting epithelial cells at apical junctions are important for the maintenance of cellular polarity, structure, and function. The adherens junction is linked to the cortical actin network, thus connecting it to the cellular interior. PC1 is part of a complex containing the junctional proteins, E-cadherin and α-, β-, and γ-catenins in human fetal kidneys (Huan and van Adelsberg 1999). In addition, PC1 binds annexin A5, a Ca^{2+}- and lipid-binding protein that is important for the initial recruitment of E-cadherin to the membrane (Markoff et al. 2007; Streets et al. 2009).

In desmosomes PC1 is localized with desmoplakins I and II (Scheffers et al. 2000). It has been suggested that certain proteins within the desmosome complex are mislocalized in cells isolated from ADPKD cysts making these structures fragile and more sensitive to sheer stress (Silberberg et al. 2005).

In cultured cells, the cytoplasmic tail of PC1 binds the intermediate filament proteins, vimentin, cytokeratins 8 and 18, and desmin (Xu et al. 2001). PC2 connects to the actin cytoskeleton by binding Hax-1 (Gallagher et al. 2000), tropomyosin-1 (Li et al. 2003b), filamin A (Sharif-Naeini et al. 2009), α-actinin (Li et al. 2005), and mDia1 (Rundle et al. 2004). The junctional complex connection to the actin cytoskeleton and PC proteins has led to the hypothesis that it is important for sensing pressure and flow, cell-cell adhesion, and cell morphogenesis and proliferation.

Using PC1 as bait, a yeast two-hybrid screen identified Pacsin 2 (protein kinase C and casein kinase 2 substrate in neurons), a ubiquitously expressed member of a family of proteins that localize to sites of high actin turnover (Yao et al. 2014). The complex containing these two proteins was found to contain N-Wasp (neural Wiskott-Aldrich syndrome protein) protein. The complex formation was necessary for N-Wasp interaction with Arp3, an actin filament nucleation protein. The authors suggest that complex formation is required for establishing and maintaining the actin cytoskeleton in tubular epithelial cells (Yao et al. 2014).

PC2 also binds to other proteins with functional consequences. Notably it interacts with both the inositol 1,4,5-trisphosphate receptor (IP3R) in the ER and the ryanodine receptor in the sarcoplasmic reticulum of cardiac myocytes, thereby further modulating intracellular Ca^{2+} (Li et al. 2005; Anyatonwu et al. 2007; Santoso et al. 2011; Mekahli et al. 2012). The etiology of ADPKD progression is thought to involve changes in intracelluar Ca^{2+}, and therefore, this aspect of PC function is important for understanding disease progression.

PC2 trafficking between the ER, Golgi, and plasma membrane is dependent on an acidic cluster in the carboxy-terminal domain that binds phosphofurin acidic cluster proteins (PACS-1 and PASC-2) whose binding to PC2 is, in turn, regulated by casein kinase 2 phosphorylation (Kottgen et al. 2005). These proteins are important for retrieval of PC2 to the trans-Golgi and the ER.

A protein allied with maintaining PC2 in the Golgi is PIGEA-14 (polycystin-2 interactor, Golgi, and endoplasmic reticulum-associated protein with a molecular mass of 14 kDa), a small protein containing a coiled-coil domain (Hidaka et al. 2004). This protein was identified as a PC2 binding protein on a yeast two-hybrid screen. On a subsequent screen, the protein was found to bind GM130, a component of the cis-Golgi apparatus. An N-terminal RVxP sequence of the PC2 protein is thought to be responsible for its localization to the cilia (Geng et al. 2006).

Taken together these results suggest that the binding partners for PC1 and PC2 are varied and may influence function, hypothetically in a tissue-specific manner. In this regard, much research remains to elucidate the binding partner roles in PC trafficking, localization, and function.

23.6 Function

The proposed functions for PC1 and PC2 are almost as varied as the identified binding partners. Mechanosensation via the primary cilium is thought to be key in the enhanced cell division and misaligned polarity found in ADPKD. Kinesin-2 a ciliary heterotrimer that is important for ciliary growth and cell cycle regulation interacts with both PC2 and fibrocystin. Mutations in the proteins of this complex all cause cystic disease (Lin et al. 2003; Li et al. 2006; Wu et al. 2006).

PC1/PC2 form a calcium permeable cation channel on the cell membrane and in the primary cilia (Hanaoka et al. 2000; Yoder et al. 2002; Nauli et al. 2003). Based on data showing that the primary cilium functions as an osmotic or flow sensor (Praetorius and Spring 2001), various groups have explored whether the polycystin complex can sense changes in flow and respond with an increase in Ca^{2+} influx. Nauli et al. (2003) were the first to demonstrate that renal cells isolated from PC1 null transgenic mice did not increase Ca^{2+} influx in response to fluid flow and that blocking antibodies raised against PC2 abolished the normal response to flow. They postulated that PC1 is the mechanosensor, while PC2 is the cation channel. Other studies suggested that the increased Ca^{2+} influx tiggers the release of additional Ca^{2+} from IP3-dependent stores, a process that is amplified by nucleotide release (Praetorius and Leipziger 2009).

Using a unique single cell imaging technique of LLC-PK (pig kidney) cells grown on microwires, Jin et al. (2013) demonstrated that fluid-shear stress can initiate Ca^{2+} signaling in the primary cilium via a PC2-dependent mechanism. These authors suggested that the Ca^{2+} signaling can spread to the cytoplasm or can be contained within the cilium depending on the nature of the stimulus. The fluid shear-induced increase in ciliary Ca^{2+} is amplified in the cytoplasm via the ryanodine receptor, whereas chemical stimuli do not result in a change in cytosolic Ca^{2+} (Jin et al. 2013).

The hypothesis that the increased Ca^{2+} influx into the cilium can spread to the cytoplasm has been challenged by the group of Clapham (DeCaen et al. 2013;

Delling et al. 2013). Using an expressed calcium sensor, patch clamp recording, and fluorescently labeled cilia in a cultured human retina pigmented epithelial cell line, these investigators found that a heterodimeric Ca^{2+} channel formed by PKD1L1 and PKD2L1 mediated an increase in ciliary Ca^{2+} (Delling et al. 2013). While the authors were able to confirm transcripts for PC1 and 2 channels, only siRNA for PKD1L1 and PKD2L1 reduced the Ca^{2+} current. The channels were activated by external nucleotides but were not mechanosensitive (DeCaen et al. 2013). In contrast to previous studies, the Ca^{2+} wave within the cilia did not change the cytoplasmic Ca^{2+} levels. In addition, the cilioplasmic Ca^{2+} level was, at rest, maintained at a higher level than the cytoplasmic level, indicating a separation between the two and suggesting that the cilium is a functionally distinct compartment with regard to ionic composition (Delling et al. 2013). Bai et al. (2008) have descibed another channel in the cilia composed of a PC2/TRPC1 multimer with properties that are distinct from the PC1/PC2 channel and have postulated this to be the entity that is responsible for mechanosensitivity.

The polycystins also appear to directly regulate intracellular Ca^{2+} release. When stably expressed in LLC-PK1 kidney cells, PC2 behaved as an intracellular Ca^{2+} release channel augmenting InsP3-mediated Ca^{2+} release (Koulen et al. 2002). The expressed channel was found in the ER membrane and patch clamp analysis showed the protein to be a Ca^{2+}-activated, high-conductance channel that is permeable to divalent cations (Koulen et al. 2002).

In the oocyte expression system, exogenous PC2 and the type I IP_3R were shown to interact and modulate intracellular Ca^{2+} signaling (Li et al. 2005). In more detailed studies, it has been shown that both PC1 and PC2 interact with the IP_3R. Interestingly PC2 stimulates IP_3-dependent increases in intracellular Ca^{2+}, while PC1 inhibits the same process (Li et al. 2005).

The findings in the oocyte expression system were verified by expressing full-length PC1 in MDCK cells. In the MDCK cells, the PC1 overexpression inhibited intracellular Ca^{2+} release in response to ATP by reducing the interaction of PC2 and IP_3R (Santoso et al. 2011). PC1 bound the stromal interaction molecule 1 (STIM1) which helps maintain its position in the ER membrane while inhibiting store-operated Ca^{2+} entry (SOCE) (Woodward et al. 2010). Based on inhibitor studies in the PC1 overexpressing MDCK cells, it has been proposed that the PI3K/Akt pathway is at least partially rsponsible for modulating the competing effects of the PC1/PC2/STIM1/IP_3R complex (Santoso et al. 2011).

A further verification of an ER-localized PC1/PC2 complex was confirmed in conditionally immortalized human proximal tubule epithelial cells where one or both of the PCs were knocked down with miRNA (Mekahli et al. 2012). In experiments that appear to contradict the ying/yang action of PC2/PC1 noted above, these investigators found that both polycystins were necessary for IP_3-induced Ca^{2+} release under fixed Ca^{2+} concentrations. Neither polycystin by itself was able to modulate intracellular Ca^{2+} concentrations (Mekahli et al. 2012).

PC2 has also been shown to interact with the ryanodine receptor of the sarcoplasmic reticulum and inhibit its Ca^{2+} channel activity in cardiac myocytes (Anyatonwu et al. 2007). Interestingly, an increased PC2-mediated Ca^{2+} influx was blocked by inhibitors of the ryanodine receptor suggesting a close interaction with influx and intracellular release. However, the flow-induced Ca^{2+} flux was not blocked by inhibitors of G proteins, phospholipase C, or IP_3R (Nauli et al. 2003).

PC1 and PC2 have been localized at sites of cell-matrix interactions, in cell-cell junctions, and in cell-substrate interactions. The binding partners suggest roles in extracellular matrix interactions, cell adhesion, and actin binding as well as the myriad intracellular pathways that are responsive to changes in these types of interactions. It is intruiging to speculate that the placement of the proteins suggests that they may be involved in force transmission and the sensing of changes in mechanical stress on the cells. How this type of signaling may interact with, or contribute to, the Ca^{2+}-induced concentration changes remains speculative. While many functions of the PCs within the cytoskeleton have been proposed, there is scant experimental evidence to support these hypotheses, particularly in in vivo systems. The functional consequences of changes in these aspects of the PC1/PC2 function and their roles in the development of ADPKD remain the topic of ongoing studies.

PC2 may have other functions that are not as well characterized as those outlined above. In proximal tubule cells, PC2 regulates the activity of piezo-1-dependent stretch-activated channels that are important in the response to luminal pressure (Peyronnet et al. 2013). PC2 has also been shown to be an epidermal growth factor-activated channel on the plasma membrane of LLC-PK1 kidney cells over-expressing PC2 (Ma et al. 2005). PC1 has been implicated in growth and resistance to apoptosis, and furthermore, the protein can induce spontaneous tubulogenesis in MDCK cells (Boletta et al. 2000). PC1 also regulates the cell cycle (Bhunia et al. 2002).

In summary, the PC1/PC2 complex has been postulated to form a Ca^{2+} channel in the apical membrane, a Ca^{2+} release channel in the ER, and a mechanosensitive channel in the primary cilium. In addition, other roles such as cell-cell and cell-matrix interaction proteins and alternative channel regulators have been suggested.

23.7 Mechanisms Underlying Functional Effects

Despite the differences in protein structure and function, mutations in either PC1 or PC2 result in autosomal dominant polycystic kidney disease. On average, the PC1 mutations lead to end-stage renal disease (ESRD) at age 54, while patients with aberrant PC2 experience ESRD at an approximate age of 74 (Hateboer et al. 1999). However, there is substantial patient to patient variability that is masked by reporting simple averages. What is more striking is the similarity of slow disease progression stretching over many decades which suggests analogous or overlapping pathways mediated by these two proteins.

While the exact mechanism is a matter of debate, it is clear that PC1 and PC2 have roles in regulating intracellular homeostatic Ca^{2+} balance and the importance of this function is underscored by the changes that are documented in ADPKD.

Disease-associated decreases in intracellular Ca^{2+} concentrations modulate intracellular signaling pathways including those regulating cAMP. Renal principal cells contain a calcium-/calmodulin-stimulated phosphodiesterase 1 as well as an adenylyl cyclase 6 (AC6) which is inhibited by calcium (Wang et al. 2010; Torres and Harris 2014; Harris and Torres 2014). A chronic decrease in intracellular Ca^{2+} would, therefore, increase the amount of cAMP. Using single and double targeted knockouts of collecting duct PC1 and AC6, Rees et al. (2014) showed that ablation of AC6 mitigated the cytogenesis and renal failure caused by a knockout of PC1.

In animal models of cystic disease, cAMP levels are increased in a wide variety of tissues. The laboratory of Grantham first showed an increase in the kidney (Yamaguchi et al. 1997) and this was later substantiated by other laboratories (Gattone et al. 2003; Smith et al. 2006; Starremans et al. 2008). In ADPKD, cyst formation in the bile duct cholangiocytes is documented in humans and animal models, and increased intracellular cAMP has been found in these cells as well (Masyuk et al. 2007). In addition, increased levels have been found in vascular smooth muscle cells (Kip et al. 2005) and the choroid plexus (Banizs et al. 2007). One interesting caveat of this generalized finding is that the animals used represent a variety of orthologous and non-orthologous cystic models indicating the generalized finding of increased cAMP in cystic disease.

In renal distal tubular cells, the site of most cyst formation in humans, the increased cAMP is generally envisioned as arising from the basolateral plasma membrane AC6 that is complexed to the basolateral vasopressin (V2) receptor and somatostatin receptors via the G proteins G_s and G_i, respectively (Harris and Torres 2014). In addition to the AC6 that is coupled to basolateral hormone receptors, a signaling complex containing a scaffolding protein AKAP150, AC5/6, protein kinase A (PKA), phosphodiesterase 4CC, and PC2 has been identified in cilia, putting these signaling components on the opposite side of the cell from the peptide hormone receptors in polarized epithelia (Choi et al. 2011).

Regardless of the localization of the adenylyl cyclase complex, chronic elevation of cAMP stimulates PKA, a hub protein for downstream signaling leading to both increased secretory channel stimulation and enhanced cellular proliferation (Yamaguchi et al. 2003; Torres and Harris 2014), and has led to the suggestion that strategies targeting cAMP signaling would be quite effective for the treatment of ADPKD. Interestingly, targeting the increased cAMP with Tolvaptan, a V2 receptor antagonist, not only inhibits vasopressin-induced Cl^- secretion but also decreases cell proliferation in human ADPKD cells (Reif et al. 2011). Tolvaptan was also remarkably successful in inhibiting cyst growth in rodent models (Gattone et al. 2003) and has, in recent years, been tested extensively in clinical trials. While phase III clinical trials have shown significant results in retarding cyst growth, the side effects including frequent urination, dehydration, and increased liver enzymes may limit long-term compliance. In phase III tolvaptan trials, these side effects contributed to a 23 % discontinuation rate (Torres et al. 2012).

Other agents in clinical trials target both epithelial cell proliferation and intracellular pathways controlling growth or Cl^- channel activity. These chemotherapeutic and immunosuppressive drugs have multiple side effects and may also not be ideal for long-term therapy (Watnick and Germino 2010; Pei 2010; Walz et al. 2010; Serra et al. 2010; Chang and Ong 2012).

Interestingly cAMP has different effects on normal and cystic cells. Increased cAMP stimulates growth in cells derived from human ADPKD patients but has the opposite effect in normal human kidney cells (Yamaguchi et al. 2000, 2003; Hanaoka and Guggino 2000). The mechanism for this divergent functional effect has been linked to cAMP-dependent B-Raf signaling to MEK, a kinase that stimulates signal-regulated kinase (ERK) in cells obtained from cystic kidneys but has the opposite effect in normal human kidney cells and in the M-1 cell line derived from normal mouse principal cells of the cortical collecting duct (Yamaguchi et al. 2003). The key appears to be a Ca^{2+}-induced increase in Akt kinase activity that suppresses B-Raf signaling in normal cells. Indeed, increases in intracellular Ca^{2+} in ADPKD cells can restore normal anti-mitogenic response to increases in cAMP (Yamaguchi et al. 2006), while Ca^{2+} restriction can switch normal cells to a cystic phenotype in which cAMP activates B-Raf/MEL/ERK and cellular proliferation (Yamaguchi et al. 2004).

There are a number of other pathways that have been shown to incorporate the polycystins and are, therefore, targets for inhibiting ADPKD progression. PC2 can be activated by epidermal growth factor (EGF) receptor in a phospholipase C/PIP2-dependent mechanism (Ma et al. 2005). In the primary cilium, EGF receptor activation causes an activation of PLC and a subsequent decrease in PIP2. The PIP2 lipid is postulated to inhibit PC2 channel activity (Ma et al. 2005).

PC1 regulates the tuberous sclerosis complex 2 (TSC2) tumor suppressor, a part of the mechanistic target of the rapamycin (mTOR) signaling pathway (Kleymenova et al. 2001; Shillingford et al. 2006; Distefano et al. 2009). The C terminus regulates the localization of TSC2 by sequestering it at the membrane, thereby preventing an Akt phosphorylation event and allowing the repression of mTOR (Dere et al. 2010). The finding that the immunosuppressive drug rapamycin that inhibits the mTOR pathway prevents cyst growth in rodent models (Tao et al. 2005) has led to preclinical and clinical studies directed toward the use of this drug as a treatment option for ADPKD (Chang and Ong 2008; Pei 2010; Watnick and Germino 2010; Harris and Torres 2014). Unfortunately to date, these have not been successfully translated to human patients and FDA approval.

The C-terminal tail of PC1 is important for a variety of signaling cascades in which this protein appears to be a player. These include modulation of the Wnt signaling pathway (Kim et al. 1999a, b), the AP-1 transcription factor complex (Le et al. 2004, 2005), Janus kinase-signal transduction transducer and activator of transcription (JAK-STAT) pathway (Bhunia et al. 2002), and STAT6-dependent gene expression (Low et al. 2006). This cleavage of PC1 appears to be required for a number of functions and multiple cleavage sites have been described.

Regulation of the JAK-STAT pathway requires the cleavage of PC1 at a G protein cleavage site (GPS) that is located immediately following the REJ domain (Qian et al. 2002; Wei et al. 2007; Yu et al. 2007). After cleavage, most of the 370 kDa N-terminal fragment (NTF) remains non-covalently tethered to the 150 kDa C-terminal fragment (CTF) (Qian et al. 2002). The majority of the uncleaved PC1 resides in the ER. The cleaved but associated NTF/CTF and the dissociated NTF traffic through the secretory pathway to the plasma membrane (Kurbegovic et al. 2014).

A different cleavage reaction that is initiated in response to mechanical stimuli produces a C-terminal tail (CTT) of approximately 200 amino acids that enters the nucleus (Chauvet et al. 2004; Low et al. 2006). The CTT is released by the action of γ-secretase (Merrick et al. 2012). This cleavage product associates with β-catenin and disrupts canonical Wnt signaling (Lal et al. 2008). The Wnt pathway is important for fate determination, proliferation, and planar cell polarity as well as renal development and morphology (Lancaster and Gleeson 2010), connecting the pathway to the abnormalities inherent in cyst formation.

Nuclear CTT activates the signal transducer and activator of transcription-6 (STAT6) and STAT3 (Low et al. 2006; Talbot et al. 2011) as well as the common transcriptional co-activator p300 (Merrick et al. 2012). Animals with a genetic inactivation of STAT6 or after treatment with a STAT6 inhibitor showed a decrease in cyst growth (Olsan et al. 2011). The nuclear CTT appears to regulate the proliferative aspect of cyst growth.

A third type of PC1 cleavage plays a role in the function of PC1 in the ER membrane. This type of cleavage produces a 100 kDa fragment thought to be derived from cleavage in the third intracellular loop and, therefore, contains the final six transmembrane domains and the C terminus (Li et al. 2009; Woodward et al. 2010). This protein fragment named P100 interacts with stromal interaction molecule 1 (STIM1). STIMs are a family of ER resident proteins that bind Ca^{2+} and act as a sensor of ER stores (Hogan and Rao 2007). Under conditions of Ca^{2+} depletion, STIM1 interacts with and activates a plasma membrane protein, Orai, a subunit of the store-operated Ca^{2+} channels (Prakriya et al. 2006). P100 binding to STIM1 causes an inhibition of store-operated channels (Woodward et al. 2010). This fragment, also called QIF38, associates with the IP3 receptor to inhibit Ca^{2+} release for ER stores after IP3 stimulation (Li et al. 2009). Taken together, the results suggest that PC1 provides a brake for transient Ca^{2+} responses.

The intricacies of the multiple pathways that are modulated in response to the decreases in intracellular Ca^{2+} in ADPKD cells have been reviewed in additional detail (Mekahli et al. 2013). Due to differences in experimental systems, several questions of relevance of all the pathways remain unanswered. How many of these pathways are important in vivo and the tissue localization of most of them remain poorly resolved issues.

23.8 Channel Characteristics

PC2 can form nonselective cation channels in artificial lipid bilayers (Gonzalez-Perrett et al. 2001; Koulen et al. 2002; Zhang et al. 2009). In vivo, PC2 interacts with a variety of binding partners, some of which alter the nature of the channel making it very difficult to provide the exact channel characteristics and adding the theoretical possibility of multiple channel types in the same cell. In addition, channel activators can vary from mechanical or osmotic stress to Ca^{2+} concentrations.

In renal tissue, polycystins in the primary cilia have been shown to be important flow sensors (Nauli et al. 2003; Xu et al. 2007; Kottgen et al. 2008; Jin et al. 2013). In vascular tissue these proteins are important for pressure sensing (Sharif-Naeini et al. 2009; Narayanan et al. 2013), for flow sensing (Nauli et al. 2008), and in stretch-induced injury of blood-brain barrier epithelial cells (Berrout et al. 2012).

Single channel studies have indicated that PC2 is a divalent cation channel (Gonzalez-Perrett et al. 2001). In the ER, PC2 was characterized as a high-conductance, voltage-dependent channel that is activated by increases in cytosolic Ca^{2+} within the physiololgical range and subsequently releases Ca^{2+} from intracellular stores (Koulen et al. 2002).

PC1 is the protoypical binding partner for PC2 and appears to be important for channel activity (Hanaoka et al. 2000; Delmas et al. 2004). However, like many of the TRP channels, PC2 can form homodimers or heterodimers with other TRP channels, specifically TRPC1, TRPC4, and TRPV4 (Tsiokas et al. 1999; Bai et al. 2008; Kottgen et al. 2008).

TRPV4 plays a role in vertebrate mechanosensation and hypo-osmotic stress (Alessandri-Haber et al. 2003). However, $TRPV4^{-/-}$ mice do not become cystic (Liedtke and Friedman 2003). TRPV4/PC2 heteromeric channels are localized to the primary cilia where they form a mechano- and thermosensitive channel in MDCK cells (Kottgen et al. 2008). Inhibitors as well as knockdown experiments were used to show that TRPV4 is an essential component of the ciliary flow sensor. Atomic force microscopy has indicated that both the TRPC1/PC2 and TRPV4/PC2 channels have a 2:2 stoichiometry with an alternating subunit arrangement (Kobori et al. 2009; Stewart et al. 2010).

In PCK rats, defective in fibrocystin and, therefore, an orthologous model of autosomal recessive PKD, pharmacological activation of TRPV4 reduced the hyperproliferative phenotype of the cholangiocytes which line the bile ducts and are prone to cystic development (Gradilone et al. 2010). In isolated renal tubules from the same model, the cystic cells demonstrated impaired TRPV4 activity (Zaika et al. 2013). Fibrocystin is expressed in the primary cilium where it interacts with PC2 (Wang et al. 2007; Kim et al. 2008), presumably forming a complex that includes TRPV4. Patch clamp studies from collecting duct-derived cysts that were compared with normal split-open collecting duct epithelia from the same animals indicated that the cystic cells have impaired TRPV4 responses to an agonist, GSK1016790A, and impaired flow-mediated increases in intracellular Ca^{2+}. Chronic treatment of the slowly progressing PCK cystic rat model with a TRPV4 agonist attenuated cyst growth in vivo. Inexplicably, TRPV4 was localized predominately to the apical membrane in cystic cells compared with a more intracellular localization in normal renal cells from the same animal (Zaika et al. 2013).

TRPC1/PC2 heteromeric channels exhibit single channel conductance, amiloride sensitivity, and ion permeability that is distinct from that shown by either homomeric channel alone (Bai et al. 2008). PC2 and TRPC1 are implicated in stretch-induced injury of the blood-brain barrier (Berrout et al. 2012). PC2 contributes to pressure sensing in vascular smooth mucle cells and also plays a role in myogenic tone in both mesenteric arteries and cerebral arteries (Sharif-Naeini et al. 2009; Narayanan et al. 2013). PC2 alters the mechanosensitivity of a variety of stretch-activated potassium channels as well as nonselective cation channels (Peyronnet et al. 2012; Sharif-Naeini et al. 2010).

Consistent with its homology to PC2, PKD2L1 was shown to be a Ca^{2+}-regulated cation channel (Chen et al. 1999). DeCaen et al. (2013), using patch clamp experiments of primary cilia from cells isolated from transgenic mice with a cilia-specific fluorescent label, described an outwardly rectifying Ca^{2+} channel formed by a PKD1L1/PKD2L1 heteromeric channel. The channels were activated by extracellular uridine and adenosine phosphates and inhibited by Gd^{3+} and ruthenium red. However, the channel lacked the pressure sensitivity of a mechanosensitive channel (DeCaen et al. 2013).

In summary, the characterisitcs of the channels formed by the polycystins vary according to the nature of the often heterologous subunits as well as the location and other binding partners of the channel.

23.9 Gene Dosage and Developmental Switches

PC1 is considered important in development, particularly in tubular morphogenesis (Ibraghimov-Beskrovnaya et al. 1997; Arnould et al. 1998). The effect of PC1 on renal tubular morphogenesis is mediated by GTP-binding proteins and PKCα resulting in the activation of the transcription factor AP-1 that modulates growth responses and apoptosis (Arnould et al. 1998). PC1 has been found to associate with Par3 faciliating the formation of aPar3/aPKC complex that is important for cell polarity and oriented cell migration (Castelli et al. 2013). PC1 may also be required for the maturation and/or stability of angiogenically derived vascular cells (Kim et al. 2000). Using knockin of a hypomorphic disease variant in mice, Hopp and colleagues found that disease development and severity correlated with functional PC1 dosage (Hopp et al. 2012). A targeted deletion of *PKD1* results in death in utero with edema, focal vascular leaks, and, ultimately, massive hemorrhage suggesting that PC1 is necessary for structural integrity of the blood vessels (Kim et al. 2000).

Two groups have used inducible kidney-specific knockout models to show that the timing of the loss of PC1 is critical for disease severity. Loss of PC1 before postnatal day 13 results in rapid onset and severe cystic disease. Loss of PC1 after day 14 results in a more slowly progressing, mild cystic disease (Piontek et al. 2007; Lantinga-van Leeuwen et al. 2007) suggesting that there is a developmental switch in the importance of PC1 expression at the terminal end of renal maturation. The two groups found little difference in cellular proliferation rates in slowly progressing adult-onset cystic disease compared to control animals, suggesting that growth regulation was likely not the key defect in the later stages of cyst growth (Piontek et al. 2007; Lantinga-van Leeuwen et al. 2007).

PC2 localized to the cilia appears to be involved in the process that leads to left-right asymmetry in mice. Specifically the PC is involved in sensing asymmetric Ca^{2+} signaling (McGrath et al. 2003). *PKD2* null mice die in utero with structural defects in cardiac septation as well as cyst formation in maturing nephrons and pancreatic ducts (Wu et al. 2000).

23.10 Polycystins in Cardiovascular Tissues

Because of the association with ADPKD, the role of the polycystins has been most extensively studied in epithelial cells. However the association of cardiovascular abnormalities, including aneurysms in coronary and cerebral blood vessels in ADPKD patients (Schievink et al. 1992; Gabow 1993), and the demonstration of death from cardiovascular defects in both PC1 and PC2 null mice (Kim et al. 2000; Wu et al. 2000) suggest major roles for these proteins in non-epithelial tissues.

Early on PC1 was detected in both endothelial cells (Ibraghimov-Beskrovnaya et al. 1997) and elastic artery vascular smooth muscle cells from normal and PKD patients (Griffin et al. 1997). Subsequently the creation of a mouse expressing a *PKD1* with a lacZ reporter gene showed that PC1 was involved in both vascular and skeletal development and was expressed at high levels in vascular smooth muscle, endothelial cells, cardiac valves, and myocytes (Boulter et al. 2001; Volk et al. 2003).

In endothelial cells, PC1 is localized predominately to the primary cilia where it acts as a putative mechanosensor facilitating the release of the vasodilator nitric oxide (Nauli et al. 2008). Likewise in microvessel endothelial cells in the brain, mechanical injury induces an influx of Ca^{2+} and the synthesis of nitric oxide followed by actin rearrangement. Both TRPC1 and PC2 are crucial for this response which protects the blood-brain barrier (Berrout et al. 2012).

Haploinsufficiency resulting in a decrease in PC2 in vascular smooth muscle cells caused intracranial vascular abnormalities under hypertensive conditions. These findings were coincident with lowered intracellular Ca^{2+} and a decreased total sarcoplasma reticulum Ca^{2+} store due to a decreased activity of the store-activated Ca^{2+} channel (Qian et al. 2003a). Consistent with the regulation of intracellular Ca^{2+} stores, PC2 localized to the SR membrane in vascular smooth muscle cells (Qian et al. 2003b). Decreases in smooth muscle Ca^{2+} have also been shown to increase cellular proliferation and apoptosis (Kip et al. 2005).

In cardiac myocytes, PC2 binds to and inhibits the ryanodine receptor (RyR2) leading to decreased Ca^{2+} release from the SR stores (Anyatonwu et al. 2007). In arterial myocytes the PC1/PC2 ratio has been shown to be important for the cellular response to intraluminal pressure changes. PC2 alone inhibits the activity of stretch-activated channels, while the PC1/PC2 complex reverses this inhibition. Proteomic analysis revealed potential PC1/PC2 complex binding partners, filamin A and actin, as potential intermediates in this effect (Sharif-Naeini et al. 2009).

PC2 has also been found in non-tubular renal tissue. In glomerular mesangial cells, angiotensin II stimulation results in an increased expression of membrane-bound PC2 and facilitates its interaction with TRPC1 and TRPC4 forming channel complexes that mediate the agonist-stimulated Ca^{2+} responses that are important for mesangial cell regulation of glomerular filtration (Du et al. 2008).

23.11 Summary

The first members of the polycystin family of proteins, PC1 and PC2, came to the fore when they were shown to be the proteins that were mutated in ADPKD, one of the most common genetic diseases in the human population. There is little doubt that the PC2 family of TRP channels are bona fide cation channels. The role of PC1 in this regard is less clear—this protein could be considered a channel subunit or accessory protein required for channel targeting or activity. However, the similarity in disease progression and severity when either PC1 or PC2 is aberrant suggests a close relationship between at least some aspects of these proteins. While the exact mechanisms causing the initiation of cyst growth are not firmly established, it is well accepted that cystic cells have lowered intracellular Ca^{2+} and that this contributes in a meaningful way to cyst growth by altering the pathways involved in Cl^- secretion and proliferation. The most successful therapy in clinical trials, tolvaptan, targets the pathway initiated by the lowered intracellular Ca^{2+}.

Like other channel proteins, notably CFTR, the polycystins appear to have multiple functions in a variety of cell types. However, the plethora of reported binding partners and roles for the polycystins does suggest a note of caution. Many of the studies rely on over-, under-, and knockout expression systems. In some of the cultured cell models, the manipulations are performed in cells that do not normally express detectable levels of the introduced proteins. There are many orthologous and non-orthologous rodent models of polycystic kidney disease, but often the non-orthologous models more closely resemble the human disease progression than the orthologous models, and there is no agreement in the field as to which models most accurately reflect the various disease components. The next phase of research regarding the polycystins should, therefore, emphasize the role of physiologically relevant levels of proteins expressed in cells, organs, and organisms where they have been shown to have a functional role.

References

Alessandri-Haber N, Yeh JJ, Boyd AE et al (2003) Hypotonicity induces TRPV4-mediated nociception in rat. Neuron 39:497–511

Anyatonwu GI, Ehrlich BE (2004) Calcium signaling and polycystin 1. Biochem Biophys Res Commun 322:1364–1373
Anyatonwu GI, Estrada M, Tian X et al (2007) Regulation of ryanodine receptor-dependent calcium signaling by polycystin-2. Proc Natl Acad Sci U S A 104:6454–6459
Arnould T, Kim E, Tsiokas L et al (1998) The polycystic kidney disease 1 gene product mediates protein kinase C α-dependent and c-Jun N-terminal kinase-dependent activation of the transcription factor AP-1. J Biol Chem 273:6013–6018
Audrezet MP, Gall JM, Chen JM et al (2012) Autosomal dominant polycystic kidney disease: comprehensive mutation analysis of PKD1 and PKD2 in 700 unrelated patients. Hum Mutat 33:1239–1250
Bai C-X, Giamarchi A, Rodat-Despoix L et al (2008) Formation of a new receptor-operated channel by heteromeric assembly of TRPP2 and TRPC1 subunits. EMBO Rep 9:472–479
Banizs B, Komlosi P, Bevensee MO et al (2007) Altered pH(i) regulation and Na^+/HCO_3^- transporter activity in choroid plexus of cilia defective Tg737(^{orpk}) mutant mouse. Am J Physiol Cell Physiol 292:C1409–C1416
Berbari NF, Lewis JS, Bishop GA et al (2008) Bardet-Biedl syndrome proteins are required for the localization of G-protein coupled receptors to primary cilia. Proc Natl Acad Sci U S A 105: 4242–4246
Berrout J, Jin M, O'Neil RG (2012) Critical role of TPPP2 and TRPC1 channels in stretch-induced injury of blood-brain barrier endothelial cells. Brain Res 1436:1–12
Bhunia AK, Piontek K, Boletta A et al (2002) PKD1 induces $p21^{waf1}$ and regulation of the cell cycle via direct activation of the JAK-STAT signaling pathway in a process requiring PKD2. Cell 109:157–168
Boletta A, Qian F, Onuchic LF et al (2000) Polycystin-1, the gene product of *PKD1*, induces resistance to apoptosis and spontaneous tubulogenesis in MDCK cells. Mol Cell 6:1267–1273
Boletta A, Qian F, Onuchic LF et al (2001) Biochemical characterization of bona fide polycystin-1 in vitro and in vivo. Am J Kidney Dis 38:1421–1429
Boulter C, Mulroy S, Webb S et al (2001) Cardiovascular, skeletal, and renal defects in mice with a targeted disruption of the Pkd1 gene. Proc Natl Acad Sci U S A 98:12174–12179
Bukanov NO, Husson H, Dackowski WR et al (2002) Functional polycystin-1 expression is developmentally regulated during epithelial morphogenesis in vitro: downregulation and loss of membrane localization during cytogenesis. Hum Mol Genet 11:923–936
Bycroft M, Bateman A, Clarke J et al (1999) The structure of a PKD domain from polycystin-1: implications for polycystic kidney disease. EMBO J 18:297–305
Cai Y, Maeda Y, Cedzick A et al (1999) Identification and characterization of polycystin-1, the PKD2 gene product. J Biol Chem 274:28557–28565
Castelli M, Boca M, Chiaravalli M et al (2013) Polycystin-1 binds Par3/aPKC and controls convergent extension during renal tubular morphogenesis. Nat Commun 4:2658. doi:10.1038/ncomms3658
Chang MY, Ong AC (2008) Autosomal dominant polycystic kidney disease: recent advances in pathogenesis and treatment. Nephron Physiol 108:1–7
Chang MY, Ong ACM (2012) Mechanism-based therapeutics for autosomal dominant polycystic kidney disease: recent progress and future prospects. Nephron Clin Pract 120:25–35
Chapman AB (2008) Approaches to testing new treatments in autosomal dominant polycystic kidney disease: insights from the CRISP and HALT-PKD studies. Clin J Soc Nephrol 3: 1197–1204
Chauvet V, Tian X, Husson H et al (2004) Mechanical stimuli induce cleavage and nuclear translocation of the polycystin-1 C terminus. J Clin Invest 114:1433–1443
Chen XZ, Vassilev PM, Basora N et al (1999) Polycystin-L is a calcium-regulated cation channel permeable to calcium ions. Nature 401:383–386
Choi Y-H, Suzuki A, Hajarnis S et al (2011) Polycystin-2 and phosphodiesterase 4C are components of a ciliary A-kinase anchoring protein complex that is disrupted in cystic diseases. Proc Natl Acad Sci U S A 108:10679–10684

Davidow CJ, Maser RL, Rome LA et al (1996) The cystic fibrosis transmembrane conductance regulator mediates transepithelial fluid secretion by human autosomal dominant polycystic kidney disease epithelium in vitro. Kidney Int 50:208–218
DeCaen PG, Delling M, Vien TN et al (2013) Direct recording and molecular identification of the calcium channel of primary cilia. Nature 504:315–318
Delling M, DeCaen PG, Doerner JF et al (2013) Primary cilia are specialized calcium signaling organelles. Nature 504:311–314
Delmas P (2005) Polycystins: polymodal receptor/ion-channel cellular sensors. Pflugers Arch Eur J Physiol 451:264–276
Delmas P, Nauli SM, Li X et al (2004) Gating of the polycystin ion channel signaling complex in neurons and kidney cells. FASEB J 18:740–742
Dere R, Wilson PD, Sanford RN et al (2010) Carboxy terminal tail of polycystin-1 regulates localization of TSC2 to repress mTOR. PLoS One 5, e9239. doi:10.1371/journal.pone.0009239
Distefano G, Boca M, Rowe I et al (2009) Polycystin-1 regulates extracellular signal-regulated kinase-dependent phosphorylation of tuberin to control cell size through nTOR and its downstream effectors S6K and 4EBP1. Mol Cell Biol 29:2359–2371
Du J, Ding M, Sours-Brothers S et al (2008) Mediation of angiotensin II-induced Ca^{2+} signaling by polycystin 2 in glomerular mesangial cells. Am J Physiol Renal Physiol 294:F909–F918
Field S, Riley K-L, Grimes DT et al (2011) Pkd1L1 establishes left-right asymmetry and physically interacts with Pkd2. Development 138:1131–1142
Gabow PA (1993) Autosomal dominant polycystic kidney disease. N Engl J Med 329:332–342
Gallagher AR, Cedzich A, Gretz N et al (2000) The polycystic kidney disease protein PKD2 interacts with Hax-1, a protein associated with the actin cytoskeleton. Proc Natl Acad Sci U S A 97: 4017–4022
Gattone VH II, Wang X, Harris PC et al (2003) Inhibition of renal cystic disease development and progression by a vasopressin V2 receptor antagonist. Nat Med 9:1323–1326
Geng L, Segal Y, Peissel B et al (1996) Identification and localization of polycystin, the PKD1 gene product. J Clin Invest 98(2674–2682):1996
Geng L, Segal Y, Pavlova A et al (1997) Distribution and developmentally regulated expression of murine polycystin. Am J Physiol Renal Physiol 272:F451–F459
Geng L, Burrow CR, Li H-P et al (2000) Modification of the composition of polycystin-1 multiprotein complexes by calcium and tyrosine phosphorylation. Biochim Biophys Acta 1533:21–35
Geng L, Okuhara D, Yu Z et al (2006) Polycystin-2 traffics to cilia independently of polycystin-1 by using an N-terminal RVxP motif. J Cell Sci 119:1383–1395
Giamarchi A, Padilla B, Coste B et al (2006) The versatile nature of the calcium-permeable cation channel TRPP2. EMBO Rep 7:787–793
Gonzalez-Perrett S, Kim K, Ibarra I et al (2001) Polycystin-2, the protein mutated in autosomal dominant polycystic kidney disease (ADPKD), is a Ca^{2+}-permeable nonselective cation channel. Proc Natl Acad Sci U S A 98:1182–1187
Gradilone SA, Masyuk TV, Huang BQ et al (2010) Activation of Trpv4 reduces the hyperproliferative phenotype of cystic cholangiocytes from an animal model of ARPKD. Gastroenterology 139:304–331
Grantham JJ, Ye M, Gattone VH II et al (1995) In vitro fluid secretion by epithelium from polycystic kidneys. J Clin Invest 95:195–202
Griffin MD, Torres VE, Grande JP et al (1997) Vascular expression of polycystin. J Am Soc Nephrol 8:616–626
Guo L, Schreiber TH, Weremowicz S et al (2000) Identification and characterization of a novel polycystin family member, polycystin-L2, in mouse and human sequence expression, alternative splicing and chromosomal localization. Genomics 64:241–251
Hanaoka K, Guggino WB (2000) cAMP regulates cell proliferation and cyst formation in autosomal polycystic kidney disease cells. J Am Soc Nephrol 11:1179–1187

Hanaoka K, Devuyst O, Schwiebert EM et al (1996) A role for CFTR in human autosomal dominant polycystic kidney disease. Am J Physiol Cell Physiol 270:C389–C399
Hanaoka K, Qian F, Boletta A et al (2000) Co-assembly of polycystin-1 and -2 produces unique cation permeable currents. Nature 408:990–994
Harris PC, Torres VE (2009) Polycystic kidney disease. Annu Rev Med 60:321–337
Harris PC, Torres VE (2014) Genetic mechanisms and signaling pathways in autosomal dominant polycystic kidney disease. J Clin Invest 124:2315–2322
Hateboer N, Dijk MAV, Bogdanovk N et al (1999) Comparison of phenotypes of polycystic kidney disease types 1 and 2. Lancet 353:103–107
Hidaka S, Konecke V, Osten L et al (2004) PIGEA-14, a novel coiled coil protein affecting the intracellular distribution of polycystin-1. J Biol Chem 279:35009–35016
Himpens B, DeSmedt H, Casteels R (1993) Intracellular Ca^{2+} signaling induced by vasopressin, ATP, and epidermal growth factor in epithelial LLC-PK1 cells. Am J Physiol Cell Physiol 265: C966–C975
Hogan PG, Rao A (2007) Dissecting ICRAC, a store-operated calcium current. Trends Biochem Sci 32:235–245
Hopp K, Ward CJ, Hommerding CJ et al (2012) Functional polycystin-1 dosage governs autosomal dominant polycystic kidney disease severity. J Clin Invest 122:4257–4273
Huan Y, van Adelsberg J (1999) Polycystin-1, the PKD1 gene product, is in a complex containing E-cadherin and the catenins. J Clin Invest 104:1459–1468
Hughes J, Ward C, Peral B (1995) The polycystic kidney disease 1 (PKD1) gene encodes a novel protein with multiple cell recognition domains. Nat Genet 10:151–160
Hughes J, Ward CJ, Aspinwall R et al (1999) Identification of a human homologue of the sea urchin receptor for egg jelly: a polycystic kidney disease-like protein. Hum Mol Genet 8: 543–549
Ibraghimov-Beskrovnaya O, Dackowski WR, Foggensteiner L et al (1997) Polycystin: in vitro synthesis, in vivo tissue expression, and subcellular localization identifies a large membrane-associated protein. Proc Natl Acad Sci U S A 94:6397–6402
Ibraghimov-Beskrovnaya O, Bukanov NO, Donohue LC (2000) Strong hemophilic interactions of the Ig-like domains of polycystin-1, the protein product of an autosomal dominant polycystic kidney disease gene, PKD1. Hum Mol Genet 9:1641–1649
Jin X, Mohieldin AM, Muntean BS et al (2013) Cilioplasm is a cellular compartment for calcium signaling in response to mechanical and chemical stimuli. Cell Mol Life Sci 71:2165–2178
Kim E, Arnould T, Sellin L et al (1999a) Interaction between RGS7 and polycystin. Proc Natl Acad Sci U S A 96:6371–6376
Kim E, Arnould T, Sellin LK et al (1999b) The polycystic kidney disease I gene product modulates Wnt signaling. J Biol Chem 274:4947–4953
Kim K, Drummond I, Ibraghimov-Beskrovnaya O et al (2000) Polycystin 1 is required for the structural integrity of blood vessels. Proc Natl Acad Sci U S A 97:1731–1736
Kim I, Li C, Liang D et al (2008) Polycystin-2 expression is regulated by a PC2-binding domain in the intracellular portion of fibrocystin. J Biol Chem 283:31559–31566
Kip SN, Hunter LLW, Ren Q et al (2005) $[Ca^{2+}]_i$ reduction increases cellular proliferation and apoptosis in vascular smooth muscle cells: relevance to the ADPKD phenotype. Circ Res 96: 873–880
Kleymenova E, Ibraghimov-Beskrovnaya O, Kugoh H et al (2001) Tuberin-dependent membrane localization of polycystin-1: a functional link between polycystic kidney disease and the TSC2 tumor suppressor gene. Mol Cell 7:823–832
Kobori T, Smith GD, Sanford R et al (2009) The transient receptor potential (TRP) channels TRPP2 and TRPC1 form a heterotetramer with a 2:2 stoichiometry and an alternating subunit arrangement. J Biol Chem 284:33507–35513
Kottgen M, Benzing T, Simmen T et al (2005) Trafficking of TRPP2 by PACS proteins represents a novel mechanism of ion channel regulation. EMBO J 24:705–716

Kottgen M, Buchholz B, Garcia-Gonzalez MA et al (2008) TRPP2 and TRPV4 form a polymodal sensory channel complex. J Cell Biol 182:437–447

Koulen P, Cai Y, Geng L et al (2002) Polycystin-2 is an intracellular calcium release channel. Nat Cell Biol 4:191–197

Kurbegovic A, Kim H, Xu H et al (2014) Novel functional complexity of polycystin-1 by GPS cleavage in vivo: role in polycystic kidney disease. Mol Cell Biol 34:3341–3353

Lal M, Song X, Pluznick JL et al (2008) Polycystin-1 C-terminal tail associates with β-catenin and inhibits canonical Wnt signaling. Hum Mol Genet 17:3105–3117

Lancaster MA, Gleeson JG (2010) Cystic kidney disease: the role of Wnt signaling. Trends Mol Med 6:349–360

Lantinga-van Leeuwen IS, Leonhard WN, van der Wal A et al (2007) Kidney-specific inactivation of the Pkd1 gene induces rapid cyst formation in developing kidneys and a slow onset of disease in adult animals. Hum Mol Genet 16:3188–31964

Le NH, van der Bent P, Huls G et al (2004) Aberrant polycystin-1 expression results in modifications of activator protein-1 activity, whereas Wnt signaling remains unaffected. J Biol Chem 279:27472–27481

Le NH, van der Wal A, van der Bent P et al (2005) Increased activity of activator protein-1 transcription factor components ATF2, cJun, and cFos in human and mouse autosomal dominant polycystic kidney disease. J Am Soc Nephrol 16:2724–2731

Lehtonen S, Ora A, Olkkonen VM et al (2000) In vivo interaction of the adapter protein CD2-associated protein with the type 2 polycystic kidney disease protein, polycystin-2. J. Biol Chem 275:32888–32893

Lentine KL, Xiao H, Machnicki G et al (2010) Renal function and healthcare costs in patients with polycystic kidney disease. Clin J Am Soc Nephrol 5:1471–1479

Li A, Tian X, Sung SW et al (2003a) Identification of two novel polycystic kidney disease-1-like genes in human and mouse genomes. Genomics 81:596–608

Li Q, Dai Y, Guo L et al (2003b) Polycystin-2 associates with tropomyosin-1, an actin microfilament component. J Mol Biol 325:949–962

Li Y, Wright JM, Qian F et al (2005) Polycystin 2 interacts with type I inositol 1,4,5-trisphosphate receptor to modulate intracellular Ca^{2+} signaling. J Biol Chem 280:41298–41306

Li Q, Montalbetti N, Wu Y et al (2006) Polycystin-2 cation channel function is under the control of microtubular structures in primary cilia of renal epithelial cells. J Biol Chem 281:37566–37575

Li Y, Santoso NG, Yu S et al (2009) Polycystin-1 interacts with inositol 1,4,5-trisphosphate receptor to modulate intracellular Ca^{2+} signaling with implications for polycystic kidney disease. J Biol Chem 284:36431–36441

Liedtke W, Friedman JM (2003) Abnormal osmotic regulation in trpv4-/- mice. Proc Natl Acad Sci U S A 100:13698–13703

Lin F, Hiesberger T, Cordes K et al (2003) Kidney-specific inactivation of the KIF3A subunit of kinesin-II inhibits renal cystogenesis and produces polycystic kidney disease. Proc Natl Acad Sci U S A 100:5286–5291

Low SH, Vasanth S, Larson CH et al (2006) Polycystin-1, STAT6 and P100 function in a pathway that transduces ciliar mechanosensation and is activated in polycystic kidney disease. Dev Cell 10:57–68

Luo Y, Vassilev PM, Li X et al (2003) Native polycystin 2 functions as a plasma membrane Ca^{2+}-permeable cation channel in renal epithelia. Mol Cell Biol 23:2600–2607

Ma R, Li W-P, Rundle D et al (2005) PKD2 functions as an epidermal growth factor-activated plasma membrane channel. Mol Cell Biol 25:8285–8298

Malhaus AN, Abuknsha RA, Price RG (2002) Interaction of the leucine-rich repeats of polycystin-1 with extracellular matrix proteins: possible role in cell proliferation. J Am Soc Nephrol 13: 19–26

Markoff A, Bogdanova N, Knope M et al (2007) Annexin A5 interacts with polycystin-1 and interferes with the polycystin-1 stimulated recruitment of E-cadherin into adherens junctions. J Mol Biol 369:954–966

Masyuk TV, Masyuk AI, Torres VE et al (2007) Octreotide inhibits hepatic cystogenesis in a rodent model of polycystic liver disease by reducing cholangiocyte adenosine 3′,5′-cyclic monophosphate. Gastroenterology 132:1104–1116
McGrath J, Somlo S, Makova S et al (2003) Two populations of mode monocilia initiate left-right asymmetry in the mouse. Cell 114:61–73
Mekahli D, Sammels E, Luyten T et al (2012) Polycystin-1 and polycystin-2 are both required to amplify inositol-trisphosphate-induced Ca^{2+} release. Cell Calcium 51:452–458
Mekahli D, Parys JB, Bultynck G et al (2013) Polycystins and cellular Ca^{2+} signaling. Cell Mol Life Sci 70:2697–2712
Merrick D, Chapin H, Baggs JE et al (2012) The γ-secretase cleavage product of polycystin-1 regulates TCF and CHOP-mediated transcriptional activation through a p300-dependent mechanism. Dev Cell 22:197–210
Mochizuki T, Wu G, Hayashi T et al (1996) PKD2, a gene for polycystic kidney disease that encodes an integral membrane protein. Science 272:1339–1342
Muchatuta M, Gattone V, Witzmann F et al (2009) Structural and functional analysis of liver cysts from the BALB/c-cpk mouse model of PKD. Exp Biol Med 234:17–27
Narayanan D, Bulley S, Leo MD et al (2013) Smooth muscle cell transient receptor potential polycystin (TRPP2) channels contribute to the myogenic response in cerebral arteries. J Physiol 591:50131–55046
Nauli SM, Alenghat FJ, Luo Y et al (2003) Polycystins 1 and 2 mediate mechanosensation in the primary cilium of kidney cells. Nat Genet 33:129–137
Nauli SM, Kawanabe Y, Kaminiski JJ et al (2008) Endothelial cilia are fluid shear sensors that regulate calcium signaling and nitric oxide production through polycystin-1. Circulation 117: 1161–1171
Newby LJ, Streets AJ, Zhao Y (2002) Identification, characterization, and localization of a novel kidney polycystin-1-polycystin-2 complex. J Biol Chem 277:20763–20773
Nilius B, Owsianik G (2011) The transient receptor potential family of ion channels. Genome Biol 12:218–229
Nomura H, Turco AE, Pei Y et al (1998) Identification of PKDL, a novel polycystic kidney disease 2-like gene whose murine homologue is deleted in mice with kidney and retinal defects. J Biol Chem 273:25967–25973
Olsan EE, Mukherjee S, Wulkersdorfer B et al (2011) Signal transducer and activator of transcription-6 (STAT6) inhibition suppresses renal cyst growth in polycystic kidney disease. Proc Natl Acad Sci U S A 108:18067–18072
Parnell SC, Magenheimer BS, Maser RL et al (1998) The polycystic kidney disease-1 protein, polycystin-1, binds and activates heterotrimeric G-proteins in vitro. Biochem Biophys Res Commun 251:625–631
Pazour GJ, San Augustin FF, Follit JA et al (2002) Polycystin-2 localizes to kidney cilia and the ciliary level is elevated in orpk mice with polycystic kidney disease. Curr Biol 12:R378–R380
Pei Y (2010) Of mice and men: therapeutic mTOR inhibition in polycystic kidney disease. J Am Soc Nephrol 21:383–394
Peyronnet R, Sharif-Naeini R, Folgering JHA et al (2012) Mechanoprotection by polycystins against apoptosis is mediated through the opening of stretch-activated K_{2P} channels. Cell Rep 1:241–250
Peyronnet R, Martins JR, Duprat F (2013) Piezol-dependent stretch-activated channels are inhibited by polycystin-2 in renal tubular epithelial cells. EMBO Rep 14:1143–1148
Piontek K, Menezes LF, Garcia-Gonzalez MA et al (2007) A critical developmental switch defines the kinetic of kidney cyst formation after loss of Pkd1. Nat Med 13:1490–1495
Praetorius HA, Leipziger J (2009) Released nucleotides amplify the cilium-dependent flow-induced $[Ca^{2+}]i$ response in MDCK cells. Acta Physiol 197:241–251
Praetorius HA, Spring KR (2001) Bending the MDCK cell primary cilium increases intracellular calcium. J Membr Biol 184:71–79

Prakriya M, Feske S, Gwack Y et al (2006) Orai is an essential pore subunit of the CRAC channel. Nature 443:230–233

Qian F, Germino FJ, Cai Y et al (1997) PKD1 interacts with PKD2 through a probable coiled-coil domain. Nat Genet 16:179–183

Qian F, Boletta A, Bhunia AK et al (2002) Cleavage of polycystin-1 requires the receptor for egg jelly domain and is disrupted by human autosomal-dominant polycystic kidney disease 1-associated mutations. Proc Natl Acad Sci U S A 99:16981–16986

Qian Q, Hunter LW, Li M et al (2003a) Pkd2 haploinsufficiency alters intracellular calcium regulation in vascular smooth cells. Hum Mol Genet 12:1875–1880

Qian Q, Li M, Cai Y et al (2003b) Analysis of the polycystins in aortic vascular smooth muscle cells. J Am Soc Nephrol 14:2280–2287

Rees S, Kittikulsuth W, Roos K et al (2014) Adenylyl cyclase 6 deficiency ameliorates polycystic kidney disease. J Am Soc Nephrol 25:232–237

Reif GA, Yamaguchi T, Nivens E et al (2011) Tolvaptan inhibits ERK-dependent cell proliferation, Cl^- secretion, and in vitro cyst growth of human ADPKD cell stimulated by vasopressin. J Physiol Renal Physiol 301:F1005–F1013

Retailleau K, Duprat F (2014) Polycystins and partners: proposed role in mechanosensitivity. J Physiol 592:2453–2471

Rossetti S, Consugar MB, Chapman AB et al (2007) Comprehensive molecular diagnostics in autosomal dominant polycystic kidney disease. J Am Soc Nephrol 18:2143–2160

Rundle DR, Gorbsky G, Tsiokas L (2004) PKD2 interacts and co-localizes with mDia1 to mitotic spindles of dividing cells: role of nDia1 in PKD2 localization to mitotic spindles. J Biol Chem 279:29728–29739

Russo RJ, Husson H, Joly D et al (2005) Impaired formation of desmosomal junctions in ADPKD epithelia. Histochem Cell Biol 124:487–497

Sanford R, Sgotto B, Aparicio S et al (1997) Comparative analysis of the polycystic kidney disease 1 (PKD1) gene reveals an integral membrane glycoprotein with multiple evolutionary conserved domains. Hum Mol Genet 9:1483–1489

Santoso NG, Cabotaru L, Guggino WB (2011) Polycystin-1, 2 and STIM1 interact with IP_3R to modulate ER Ca^{2+} release through the PI3K/Akt pathway. Cell Physiol Biochem 27:715–726

Scheffers MS, van der Bent P, Prins F et al (2000) Polycystin-1, the product of the polycystic kidney disease 1 gene, co-localizes with desmosomes in MDCK cells. Hum Mol Genet 9: 2743–2750

Schievink WI, Torres VE, Piepgras DG et al (1992) Saccular intracranial aneurysms in autosomal dominant polycystic kidney disease. J Am Soc Nephrol 3:88–95

Serra AL, Poster D, Kistler AD et al (2010) Sirolimus and kidney growth in autosomal dominant polycystic kidney disease. N Engl J Med 363:820–829

Sharif-Naeini R, Folgering JH, Bichet D et al (2009) Polycystin-1 and -2 dosage regulates pressure sensing. Cell 139:587–596

Sharif-Naeini R, Folgering JH, Bichet D et al (2010) Sensing pressure in the cardiovascular system: Gq-coupled mechanoreceptors and TRP channels. J Mol Cell Cardiol 48:83–89

Shillingford JM, Murcia NS, Larson CH et al (2006) The mTOR pathway is regulated by polycystin-1, and its inhibition reverses renal cystogenesis in polycystic kidney disease. Proc Natl Acad Sci U S A 103:5466–5471

Silberberg M, Charron AJ, Bacallao R et al (2005) Mispolarization of desmosomal proteins and altered intercellular adhesion in autosomal dominant polycystic kidney disease. Am J Physiol Renal Physiol 288:F1153–F1163

Smith LA, Bukanaov NO, Husson H et al (2006) Development of polycystic kidney disease in juvenile cystic kidney mice: insights into pathogenesis, ciliary abnormalities and common features with human disease. J Am Soc Nephrol 17:2821–2831

Somlo S, Ehrlich B (2001) Human disease: calcium signaling in polycystic kidney disease. Curr Biol 11:R356–R360

Starremans PG, Li X, Finnerty PE (2008) A mouse model for polycystic disease through a somatic in-frame deletion in the 5′ end of Pkd1. Kidney Int 73:1394–1405

Stewart AP, Smith GD, Sandford RN et al (2010) Atomic force microscopy reveals the alternating submit arrangement of the TRPP2-TRPV4 heterotetramer. Biophys J 99:790–797
Streets AJ, Newby LJ, O'Hare MJ et al (2003) Functional analysis of PKD1 transgenic lines reveals a direct role for polycystin-1 in mediating cell-cell adhesion. J Am Soc Nephrol 14: 1804–1815
Streets AJ, Moon DJ, Kane ME et al (2006) Identification of an N-terminal glycogen synthase kinase 3 phosphorylation site which regulates the functional localization of polycystin-2 in vivo and in vitro. Hum Mol Genet 15:1465–1473
Streets AJ, Wagner BE, Harris PC et al (2009) Homophilic and heterophilic polycystin 1 interactions regulate E-cadherin recruitment and junction assembly in MDCK cells. J Cell Sci 122: 1410–1417
Su X, Driscoll K, Yao G et al (2014) Bardet-Biedl syndrome proteins 1 and 3 regulate the ciliary trafficking of polycystic kidney disease 1 protein. Hum Mol Genet 23:5441–5451
Talbot JJ, Shillingford JM, Vasanth S et al (2011) Polycystin-1 regulates STAT activity by a dual mechanism. Proc Natl Acad Sci U S A 108:7985–7990
Tao Y, Kin J, Schrier RW et al (2005) Rapamycin markedly slows disease progression in a rat model of polycystic kidney disease. J Am Soc Nephrol 16:46–51
The American PKD1 consortium (1995) Analysis of the genomic sequence for the autosomal dominant polycystic kidney disease (PKD1) predicts the presence of a leucine rich repeat. Hum Mol Genet 4:575–582
The European Polycystic Kidney Disease Consortium (1994) The polycystic kidney disease gene encodes a 14 kb transcript and lies within a duplicated region on chromosome 16. Cell 77:881–894
The International Polycystic Kidney Disease Consortium (1995) Polycystic kidney disease: the complete structure of the PKD1 gene and its protein. Cell 81:289–298
Torres VE, Harris PC (2014) Strategies targeting cAMP signaling in the treatment of polycystic kidney disease. J Am Soc Nephrol 25:18–32
Torres VE, Chapman AB, Devuyst O et al (2012) Tolvaptan in patients with autosomal dominant polycystic kidney disease. N Engl J Med 367:2407–2418
Tsiokas L, Kim E, Arnould T et al (1997) Homo- and heterodimeric interactions between the gene products of PKD1 and PKD2. Proc Natl Acad Sci U S A 94:6965–6970
Tsiokas L, Arnould T, Zhu C et al (1999) Specific association of the gene product of PKD2 with the TRPC1 channel. Proc Natl Acad Sci U S A 96:3934–3939
Veldhuisen B, Spruit L, Dauwerse HG et al (1999) Genes homologous to the autosomal dominant polycystic kidney disease genes (PKD1 and PKD2). Eur J Hum Genet 7:860–872
Vogel P, Read R, Hansen GM et al (2010) Situs inversus in Dpcd/Poll$^{-/-}$, Nme7$^{-/-}$. And Pkd1l1$^{-/-}$ mice. Vet Pathol 47:120–131
Volk T, Schwoerer AP, Thiessen S et al (2003) A polycystin-2-like large conductance cation channel in rat left ventricular myocytes. Cardiovasc Res 58:76–88
Walz G, Budde K, Mannaa M et al (2010) Everolimus in patients with autosomal dominant polycystic kidney disease. N Engl J Med 363:830–840
Wang S, Zhang J, Nauli SM et al (2007) Fibrocystin/polyductin, found in the same protein complex with polycystin-2, regulates calcium responses in kidney epithelia. Mol Cell Biol 27:3241–3252
Wang X, Ward CJ, Harris PC et al (2010) Cyclic nucleotide signaling in polycystic kidney disease. Kidney Int 77:129–140
Ward C, Turley H, Ong AC et al (1996) Polycystin, the polycystic kidney-disease-1 protein, is expressed by epithelial cells in fetal, adult and polycystic kidney. Proc Natl Acad Sci U S A 93:1524–1528
Watnick T, Germino GG (2010) mTOR inhibitors in polycystic kidney disease. N Engl J Med 363: 879–881
Wei W, Hackmann K, Xu H et al (2007) Characterization of cis-autoproteolysis of polycystin-1, the product of human polycystic kidney disease 1 gene. J Biol Chem 282:21729–21737

Weston BS, Bagneris C, Price RG et al (2001) The polycystin-1 C type lectin domain binds carbohydrate in a calcium-dependent manner, and interacts with extracellular matrix proteins in vitro. Biochim Biophys Acta 1536:161–176

Wilson PD, Geng L, Li X et al (1999) The PKD1 gene product, "polycystin-1" is a tyrosine-phosphorylated protein that colocalizes with alpha2beta1-integrin in focal clusters in adherent renal epithelia. Lab Invest 79:1311–1323

Woodward OM, Li Y, Yu S et al (2010) Identification of a polycystin-1 cleavage product, P100r, that regulates store operated Ca^{2+} entry through interactions with STIM1. PLoS One 5, e12305. doi:10.1371/journal.pone.0012305

Wu G, Markowitz GS, Li L et al (2000) Cardiac defects and renal failure in mice with targeted mutations in Pkd2. Nat Genet 24:75–78

Wu Y, Dai X-Q, Li Q et al (2006) Kinesin-2 mediates physical and functional interactions between polycystin-2 and fibrocystin. Hum Mol Genet 15:3280–3292

Xu GM, Sikaneta T, Sullivan BM (2001) Polycystin-1 interacts with intermediate filaments. J Biol Chem 276:46544–46552

Xu C, Rossetti S, Jiang L et al (2007) Human ADPKD primary cyst epithelial cells with a novel, single codon deletion in the PKD1 gene exhibit defective ciliary polycystin localization and loss of flow-induced Ca^{2+} signaling. Am J Physiol Renal Physiol 292:F930–F945

Yamaguchi T, Nagao S, Kasahara M et al (1997) Renal accumulation and excretion of cyclic adenosine monophosphate in a murine model of slowly progressive polycystic kidney disease. Am J Kidney Dis 30:703–709

Yamaguchi T, Pelling JC, Ramaswamy NT et al (2000) cAMP stimulates the in vitro proliferation of renal cyst epithelial cells by activating the extracellular signal-regulated kinase pathway. Kidney Int 57:1460–1471

Yamaguchi T, Nagao S, Wallace DP (2003) cAMP activates B-raf and ERK in cyst epithelial cells from autosomal dominant polycystic kidneys. Kidney Int 63:1983–1994

Yamaguchi T, Wallace DP, Magenheimer BS et al (2004) Calcium restriction allows cAMP activation of the B-Raf/ERK pathway, switching cells to a cAMP-dependent growth-stimulated phenotype. J Biol Chem 279:40419–40430

Yamaguchi T, Hempson SJ, Reif GA et al (2006) Calcium restores a normal proliferation phenotype in human polycystic kidney disease epithelial cells. J Am Soc Nephrol 17:178–187

Yao G, Su X, Nguyen V et al (2014) Polycystin-1 regulates actin cytoskeleton organization and directional cell migration through a novel PC1-pacsin 2-N-Wasp complex. Hum Mol 23: 2769–2779

Ye M, Grantham JJ (1993) The secretion of fluid by renal cysts from patients with autosomal dominant polycystic kidney disease. N Engl J Med 329:310–313

Yoder BK, Hou X, Guay-Woodford LM (2002) The polycystic kidney disease proteins, polycystin-1, polycystin-2, polaris and cystin, are co-localized in renal cilia. J Am Soc Nephrol 13:2508–2516

Yu S, Hackmann K, Gao J et al (2007) Essential role of cleavage of polycystin-1 at G protein-coupled receptor proteolytic site for kidney tubular structure. Proc Natl Acad Sci U S A 104: 18688–18693

Yuasa T, Venugopal B, Weremowicz S et al (2002) The sequence, expression and chromosomal localization of a novel polycystic kidney disease 1-like gene, PKD1L1, in human. Genomics 79:376–386

Yuasa T, Takakura A, Denker BM et al (2004) Polycystin-1L2 is a novel G-protein-binding protein. Genomics 84:126–138

Zaika O, Mamenko M, Berrout J et al (2013) TRPV4 dysfunction promotes renal cystogenesis in autosomal recessive polycystic kidney disease. J Am Soc Nephrol 24:604–616

Zhang P, Luo Y, Chasan B et al (2009) The multimeric structure of polycystin-2 (TRPP2): structural-functional correlates of homo- and hetero-multimers with TRPC1. Hum Mol Genet 18:1238–1251

Chapter 24
CFTR and Cystic Fibrosis: A Need for Personalized Medicine

Neil A. Bradbury

Abstract Cystic fibrosis (CF), a common lethal genetic disease, is caused by mutations in the cystic fibrosis transmembrane conductance regulator (*cftr*) gene, which codes for an epithelial anion channel. Since the identification of CF as a genetic disease, patients have been treated exclusively with drugs that address the symptoms of the disease (antibiotics for airway disease, pancreatic supplements to replace digestive enzymes, and anti-inflammatories to reduce airway inflammation) rather than treating the basic defect. Nevertheless, this approach has resulted in a marked improvement in the survival of patients with CF over the last few decades, such that the median predicted survival is now around 40 years of age. Since the discovery of the gene encoding CFTR and the identification of mutations underlying the defects in CFTR function, the hunt has been on to discover drugs that would correct the basic defect in the CFTR protein. Although almost 2000 CFTR mutations have been described in patients, they mostly fall into a few broad categories of disruption. Given the disparate ways in which CFTR mutations affect the CFTR protein however, it has become apparent that a single drug regimen will not be effective in treating all patients. Thus, a patient's genotype would have to be taken into account when deciding which drug would be appropriate to treat individual patients with CF. Individual or personalized medicine as a concept has been increasingly highlighted in recent years, yet with CF, personalized medicine is not a mere academic exercise but rather a necessity in order to effectively treat all the mutations that are found in CF patients. This chapter presents the current CFTR mutation classifications and shows how such classification is essential in the establishment of a mutation-specific targeted drug therapy for each individual with CF.

Keywords CFTR • Potentiator • Corrector • HBE • Genotype • Mutation class • High-throughput screening

N.A. Bradbury (✉)
Department of Physiology and Biophysics, Chicago Medical School, Rosalind Franklin University of Medicine and Science, 3333 Green Bay Rd, North Chicago, IL 60064, USA
e-mail: neil.bradbury@rosalindfranklin.edu

K.L. Hamilton, D.C. Devor (eds.), *Ion Channels and Transporters of Epithelia in Health and Disease*, Physiology in Health and Disease,
DOI 10.1007/978-1-4939-3366-2_24

24.1 Introduction

Cystic fibrosis (CF) is a severe autosomal recessive disease affecting some 1:2500–4500 newborns among Caucasian populations (Davis et al. 1996). Although not completely absent from other ethnicities, the penetrance of CF occurs at a much lower rate within these racial groups. The overall pathology of CF is characterized by abnormal epithelial electrolyte transport, which results in the major clinical manifestations of elevated sweat chloride concentrations, abnormal chloride and bicarbonate transport in the airways and pancreas, pancreatic insufficiency (predominantly exocrine pancreas but also endocrine in some patients), recurrent lung infections, chronic bronchiectasis, and ultimately death from respiratory failure (Boat et al. 1989). One of the cardinal symptoms of CF is the presence of high levels of salt in the sweat, a feature which has been associated with infantile mortality for centuries. Medieval folklore avowed "woe to the child, who when kissed on the forehead tastes salty; that child is bewitched and soon must die" (Quinton 1999; Alonzo De Los Ruyzes De Fonteca 1606). CF was first comprehensively described in the 1930s by Dorothy Hansine Anderson (1938), when she noted pancreatic lesions and congested airways in young children who had purportedly died from coeliac disease. Subsequently, working with Dr. Paul di Sant'Agnese, the discovery was made that children whom Andersen had identified as having "cystic fibrosis of the pancreas" displayed markedly elevated sweat electrolyte levels, a finding that was to become a hallmark for the disease and the gold standard for diagnosis. Pioneering work by Dr. Paul Quinton in the late 1970s and early 1980s established that CF patients had elevated sweat electrolytes due to the absence of a chloride reabsorptive pathway in CF sweat ducts (Quinton 1983), although the identity of this pathway would remain unknown for the next few years. During the late 1980s, a major effort into identifying the genetic basis for CF culminated in the discovery of the cystic fibrosis transmembrane conductance regulator (*CFTR*) gene in 1989 (Riordan et al. 1989; Kerem et al. 1989), and the identification of a 3 base-pair deletion that was present in most patients with CF. The international effort marked the first time that a gene had been identified without any knowledge of the protein for which it coded.

Until recently, clinical care for CF patients focused exclusively on the delivery of therapies to relieve the symptoms associated with the disease. Pancreatic enzyme replacement therapy and chest physiotherapy were important early improvements in CF care (Levy et al. 1986), while the use of systemic antibiotics directed against *Pseudomonas aeruginosa* and *Staphylococcus aureus* has further extended patient survival (Ratjen 2001; Davis et al. 1996). Recent developments in the use of inhaled tobramycin, recombinant human DNase, and hypertonic saline have also played their role in improving patient quality of life (Elkins et al. 2006). Although these therapies only treated symptoms, they nonetheless heralded an increase in life expectancy from around 5 years in the 1970s to about 35–42 years of age today (Fig. 24.1). Indeed, a newborn diagnosed with CF today can expect to live well into their fifth or sixth decade of life (Dodge et al. 2007). Although there is continued

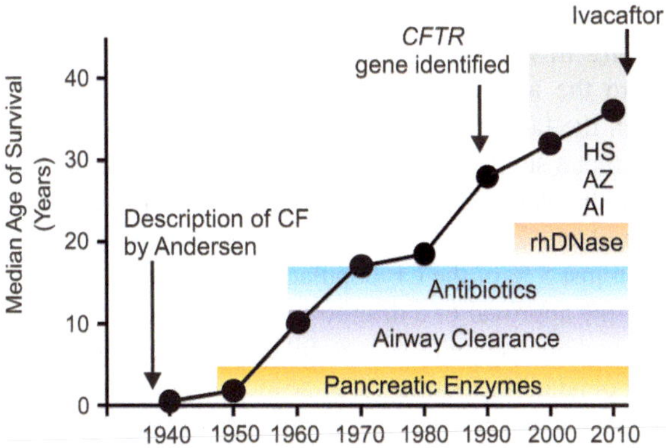

Fig. 24.1 Median survival in cystic fibrosis (CF). CF survival over time, associated with milestones (*arrows*) and CF therapies (*boxes*). Steady increases in median survival have followed the introduction of new therapies and changes in care delivery. *AI* inhaled aztreonam, *AZ* azithromycin, *CFTR* cystic fibrosis transmembrane conductance regulator, *HS* hypertonic saline

development of many exciting symptomatic therapies to treat patients with CF, this chapter will focus on strategies that impact the basic defect in CF, i.e., normalizing the function of mutant CFTR protein.

Strategies to treat the basic defect in CF can be classified into two main categories, targeting the genetic abnormality and pharmacological correction of the protein abnormality. The first is a genetic strategy aimed at inserting a wt CFTR gene in mutant cells. Following the identification of the *CFTR* gene, there was great excitement among CF patients, parents, and caregivers with the hope that a gene therapy “cure” would be on the immediate horizon. Despite the initial enthusiasm for this approach, the ability to deliver efficiently the gene to appropriate target tissues has remained elusive (Armstrong et al. 2014; Griesenbach and Alton 2012). Although the interest in gene therapy has waned in recent years, the new advances in gene-editing technologies, which have the potential to correct directly the aberrant genomic DNA sequences, will likely renew enthusiasm for a genetic therapy. The second strategy to treat patients with CF is to use pharmacological tools to restore the activity of the mutant protein; it is this latter approach that is the focus of this chapter. In 1998, the Cystic Fibrosis Foundation (CFF) initiated a groundbreaking collaboration with pharmaceutical companies to support drug discovery efforts for correcting the basic defect in CF. So far, the CFF has invested more than $300 million to leverage early-phase CF-related drug discovery programs, an approach that culminated in the FDA approval of ivacaftor (Kalydeco, Vertex Pharmaceuticals) as the first treatment for the basic defect in CF.

Since the sequencing of the human genome (an effort led at the National Institutes of Health by Francis Collins, one of the codiscoverers of the CFTR gene), we have now moved into the post-genomic era in biology. This has ushered in the prospect of individualized medicine, that is, drug treatments based on knowledge of an individual patient’s genome. While this is still as yet an unattainable goal for many diseases, individualized therapeutic approaches are a necessity for patients with CF. To date around 1900 mutations have been identified in *CFTR*, resulting in five described classes of cellular mechanisms by which mutations lead to the absence of CFTR protein and/or function in the plasma membrane. Given the fact that CF can be caused by at least five different mechanisms, it is clear that a single pharmacological approach will not be effective in all patients. Indeed, the first FDA-approved drug to treat the basic defect in CF, Kalydeco (McPhail and Clancy 2013), partially corrects only one class of mutation and is effective in just 5 % of the CF population who bear the appropriate mutation (s). Nonetheless, the approval of a drug to treat the basic defect in CF was a major accomplishment and represented a successful and unique collaboration between the Cystic Fibrosis Foundation, industry, and academia. Continuing research in pharmaceutical companies and university research labs is aimed at identifying novel drug candidates to treat all classes of mutations. Identifying a patient's particular mutation class will become increasingly important as class-specific drugs become available to clinicians, bringing CF care into the era of bespoke medicine.

24.2 Biology of CFTR

CF is a monogenic disorder affecting individuals predominantly of Caucasian descent. In Europe, around 1:3500 neonates are diagnosed annually with CF, while slightly less in the USA, with an incidence of around 1:2000–3000 live births (CF Foundation 2011). Since CF is an autosomal recessive disease, a patient must inherit two mutant *CFTR* genes (alleles), one from each parent. The *CFTR* gene is located on the long arm of chromosome 7 at position q31.2 and is comprised of 27 exons, spanning over 190 kb. Following RNA splicing, the resultant *CFTR* mRNA is 6.5 kb, and all exons are required for normal CFTR function. The CFTR protein is composed of 1480 amino acids, which form a membrane-integrated anion channel localized predominantly to the apical membrane of epithelial cells (the sweat duct expresses CFTR in both the apical and basolateral membranes). CFTR is a member of a large adenosine triphosphate (ATP)-binding cassette (ABC) transporter family, which are generally responsible for the transport of small molecules in an ATP-dependent fashion. For example, P-glycoprotein (MDR1) confers resistance to chemotherapeutic drugs by "pumping" the drugs out of cells across the plasma membrane (Choi and Yu 2014). Similarly, the ABC transport protein ALDP (*ABCD1*) is present in peroxisomal membranes and works to transport fatty acids into peroxisomes for degradation. Mutations in ALDP are associated with X-linked adrenoleukodystrophy, a condition brought to the attention of many through the Hollywood movie "Lorenzo's Oil." As with all ABC transporters, CFTR is composed of two membrane spanning domains (MSDs, each composed of six individual transmembrane segments), which for CFTR forms the anion pore, and two nucleotide binding domains (NBDs). Unique to CFTR is the presence of a regulatory or R domain, linking the two halves (Fig. 24.2). It is now known that the NBDs do not act independently, but rather dimerize so that each NBD confers two half ATP binding pockets.

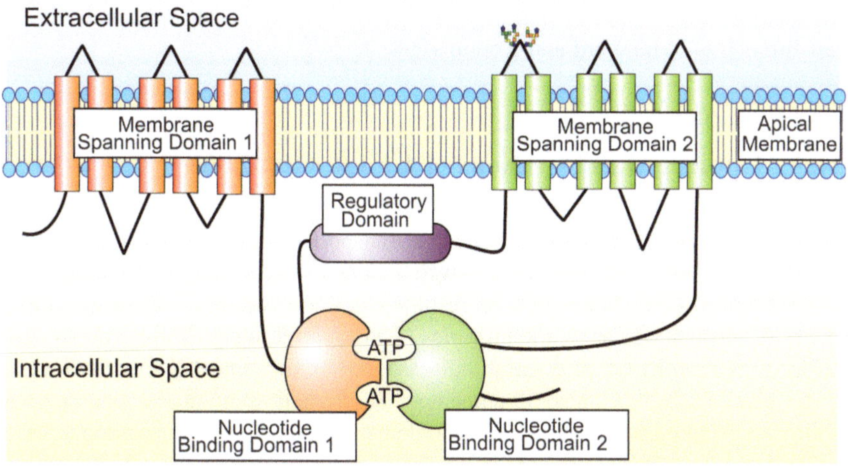

Fig. 24.2 Domain organization of CFTR in the plasma membrane. The cystic fibrosis transmembrane conductance regulator (CFTR) is composed of two repeated motifs, each containing a six member helical membrane spanning domain (MSD) and a nucleotide binding domain (NBD). The regulatory (R) domain links the two halves and serves as a regulatory domain containing multiple phosphorylation consensus sites. Two N-linked carbohydrate moieties are added to the fourth extracellular loop. Activation of CFTR occurs upon phosphorylation of the R domain and binding of ATP to the NBDs

Once the DNA encoding CFTR has been translated into mRNA, the mRNA leaves the nucleus and interacts with ribosomes that initiate the translation of CFTR within the endoplasmic reticulum (ER). During its synthesis, CFTR is co-translationally modified such that a core carbohydrate group is added to the fourth extracellular loop on each of two asparagine residues (designated Band B CFTR). Within the ER, the complex co-translational folding process utilizes multiple accessory proteins, including membrane-associated calnexin, calreticulin, and other chaperone proteins (Harada et al. 2006; Rosser et al. 2008) to cause CFTR to achieve its final 3D structure. During the folding process, appropriate interactions between the NBDs and cytoplasmic loops from the MSDs are required to help stabilize the final 3D conformation and yield an ER-export competent protein. This process is remarkably inefficient, with more than half of all CFTR protein synthesized failing to achieve the final conformation. Misfolded CFTR is targeted to the ER-associated degradation (ERAD) system, primarily involving the ubiquitin proteasome pathway. For those CFTR molecules that fold properly and exit the ER, final processing of CFTR occurs within the Golgi apparatus, with the conversion of the carbohydrate adduct from a mannose-enriched chain to a mature complex oligosaccharide (Band C CFTR). After exiting the Golgi, CFTR is trafficked to the apical plasma membrane. In the plasma membrane, CFTR turns over through clathrin-mediated endocytosis at a rate of 10 % per minute (Bradbury et al. 1994), followed by fairly efficient recycling back to the cell surface (Picciano et al. 2003). Another layer of quality control exists for CFTR at the plasma membrane, where poorly functioning or unfolded CFTR is ubiquitinated and targeted for degradation by lysosomes (Sharma et al. 2004). Overall, CFTR appears to have a half-life of 12–24 h (Ward and Kopito 1994).

24.3 Clinical Manifestations of CFTR Mutations

CFTR is involved in fluid and electrolyte transport across epithelial cells (Welsh and Ramsey 2001). The absence or dysfunction of CFTR results in altered ionic composition of epithelial secretions resulting in the secretion of thick dehydrated mucus or inspissated pancreatic enzymes (Welsh and Ramsey 2001). Shortly after the identification of CF as a unique disease, the association of CF with elevated sweat electrolytes was also documented. In normal individuals, the chloride (followed by sodium) present in the initial sweat produced by the secretory coil is reabsorbed by the duct cells in a CFTR-dependent manner, resulting in the appearance of a hypotonic fluid on the skin surface; evaporation of fluid from the cell surface helping to cool the body. In CF individuals, this reabsorptive process is absent or significantly impaired leading to the appearance of elevated salt levels in the sweat. Note sweat can be produced in the secretory coil in a CFTR-independent manner. The majority of patients diagnosed with CF have sweat chloride levels >60 mmol/L. Although a cutoff of >60 mmol/L is considered diagnostic of CF, many patients have significantly higher sweat chloride levels. A comparison of sweat chloride with predicted residual CFTR function based upon the specific mutation carried by a patient (*see below*) reveals a clear relation between genotype and phenotype (Fig. 24.3). Unaffected individuals are presumed to have 100 % CFTR function and show a mean sweat chloride of ~20 mmol/L. Carriers with one mutant allele show a small increase in sweat chloride (~26 mmol/L) compared to normal subjects. CF patients carrying one severe mutant allele and one mild allele show a sweat chloride of ~80 mmol/L, whereas CF patients with two severe mutant alleles have sweat chlorides >100 mmol/L (Rowe et al. 2007).

Exocrine pancreatic insufficiency is thought to occur in >90 % of CF patients. Defective secretion of pancreatic enzymes into the small intestine results in fat malabsorption, leading to steatorrhea and failure to thrive. Although not initially affected, the endocrine function of the pancreas can also be compromised leading to CF-related diabetes (CFRD), with around 25 % of patients over the age of 25 displaying symptoms of CF-related diabetes (CF Foundation 2011). At present, the mechanism for this endocrine disruption in CF is not fully known. The most significant tissues, in terms of morbidity and mortality in CF, are the lungs and airways. CFTR is expressed in submucosal glands and in the apical surface of ciliated epithelial cells (Engelhardt et al. 1992; Kreda et al. 2005). Normal clearance and maintenance of sterility of the airways occurs through the interaction of cilia with the airway surface liquid (ASL), a pericilliary layer of aqueous fluid and mucus that lines the respiratory tract (Boucher 2004). The correct depth of the ASL is achieved through the net effect of osmotic gradients generated by the efflux of

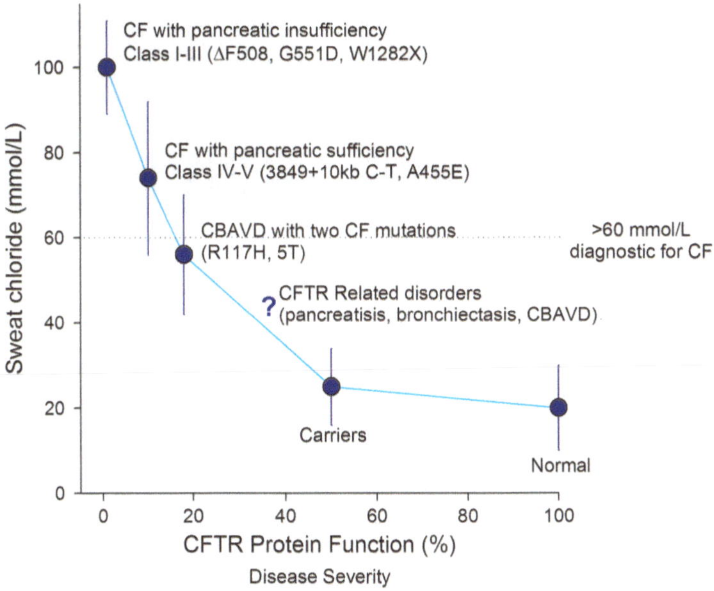

Fig. 24.3 Correlation of CFTR protein/function with sweat chloride levels. Patients with class I–III mutations, have absent or minimal CFTR results in severe CF disease, with pancreatic insufficiency and highly elevated sweat electrolytes. Residual function mutations, class IV–V, are associated with milder disease, generally pancreatic sufficient and sweat electrolytes at (above or below) the diagnostic cutoff of 60 mmol/L. CF-related disorders with some CFTR dysfunction, for example congenital bilateral absence of the vas deferens (CBAVD), present with intermediate or normal sweat electrolyte levels. Heterozygous individuals with one normal and one mutant *CFTR* allele and homozygous normal individuals are asymptomatic and have low sweat electrolyte levels. Adapted from Accurso et al. (2014)

chloride ions through CFTR, coupled to a small influx of sodium ions through the epithelial sodium channel (ENaC). Absence or dysfunction of CFTR in the airways reduces chloride efflux, which, coupled to increased sodium influx, leads to an osmotic imbalance such that the ASL becomes markedly dehydrated (i.e., reduced). This in turn leads to flattening of the cilia against the cell surface preventing their beating, which coupled to increased mucus viscosity impairs the normal clearance of the airways by the mucocilliary escalator (Fig. 24.4). Over time this causes plugging of the airways with thick mucus and bacterial infections leading to inflammation and eventually bronchiectasis (Rowe et al. 2005).

24.4 Classification of CFTR Mutations

Although some 1900 sequence variations have thus far been described in *CFTR* (www.cftr2.org), they nonetheless fall into two broad categories: those affecting protein production and those affecting protein function. Given this division of

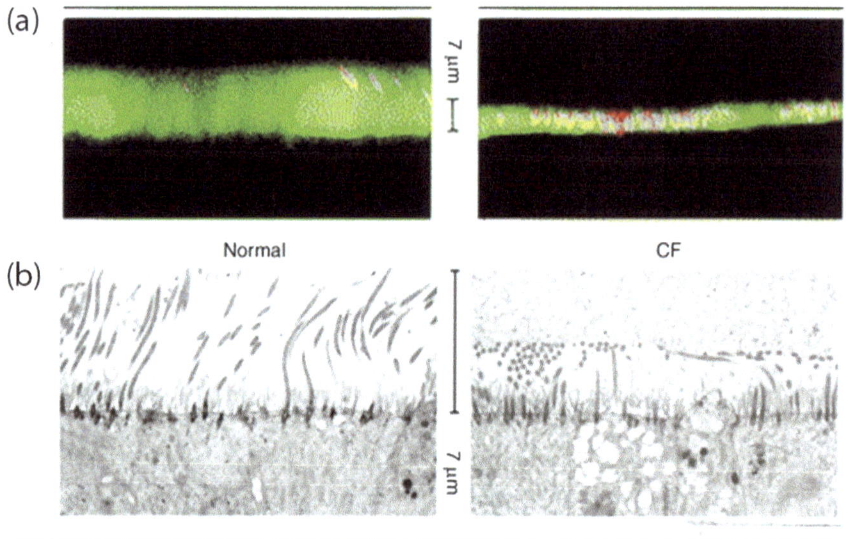

Fig. 24.4 Normal mucus clearance is abolished in the CF airway. (**a**) Confocal images of airway surface liquid (ASL) covering normal and CF airway surfaces. The aqueous component of the ASL is labeled with a green fluorescent dextran and the mucus layer with red fluorescent microspheres. The normal airway cells maintain an appropriate ASL height above the culture, whereas all available ASL has been absorbed from CF cultures, and the microsphere-labeled mucus layer has invaded the periciliary region. (**b**) Perfluorocarbon-osmium fixed cultures. In the normal cultures, a distinct periciliary layer is seen, with cilia fully extended. In CF cultures, the cilia are flattened, and the mucus later has invaded the periciliary space and adhered to the ciliary surfaces. Reproduced with permission from Matsui et al. (1998a)

mutational consequences, it is not surprising that different categories of drugs will likely be required to treat patients with CF, depending upon their unique genetic makeup. Moreover, despite the fact that this broad division in mutational consequences can accommodate most mutations, several mutations (including the common ΔF508 mutation) affect both protein production and function. As a refinement to the classification of CFTR mutations, Welsh and Smith (Welsh and Smith 1993) broke down mutations into five major subclasses (Fig. 24.5). Of the >1900 *CFTR* mutations thus far identified, only ~20 mutations are present at a frequency of >0.1 % (Rogan et al. 2011). Indeed, excluding the common ΔF508 variant (which has an allelic frequency of ~90 %, CF Genetic Analysis Consortium 1990), only four *CFTR* mutations, G551D, W1282X, G542X, and N1301K, have a global prevalence of around 1–3 % each (Rogan et al. 2011; Clancy and Jain 2012).

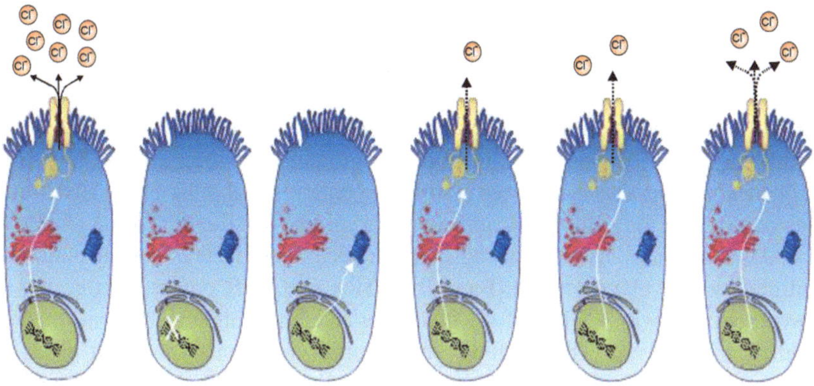

Normal	Class I	Class II	Class III	Class IV	Class IV
Defect	Defective Synthesis	Defective processing or maturation	Defective regulation	Defective conductance	Reduced synthesis and stability
Therapy	Readthrough	Correctors (+ potentiators)	Potentiators	Potentiators	Potentiators Spicing modulators
Mutations (examples)	**G542X** W1282X R553X, E882X 621 + 1G → T	**ΔF508** N1303K ΔI507 R1066C	**G551D** G551S G178R G1244E	R117H R334W R347P R1070W	3272 6A→G A455E D565G 3849+1kb C→T

Fig. 24.5 Classification of *CFTR* mutations according to protein production and function. Each of the five classes of mutations is described. An epithelial cell illustration depicts the localization of mutant CFTR protein relative to normal (wild-type) CFTR

24.4.1 Class I Mutations Prevent the Production of Full-Length CFTR

Class I mutations are typically characterized by a reduction in the quantity of CFTR produced and result in the complete or partial absence of CFTR protein. Such loss of CFTR protein typically arises from frameshift mutations (insertions or deletions), nonsense mutations (premature termination codons, PTC), or mRNA splicing defects (Rowntree and Harris 2003). For example, G542X and W1282X are mutations generating PTC, resulting in premature termination of translation and the production of a truncated protein. Mutations generating PTC can also lead to a marked loss in CFTR mRNA levels though nonsense-mediated decay (NMD) which degrades abnormal PTC containing mRNA species (Maquat 1995). The G542X mutation is the most prevalent mutation of its class worldwide, being expressed on at least one allele of up to 4 % of CF patients. Although the W1282X mutation is less common than G542X worldwide, it can reach a significantly higher frequency in specific populations, due to founder effects. Thus, 48 % of all mutant CFTR alleles in Ashkenazi Jews in Israel carry the W1282X mutation (Shoshani et al. 1992). Similarly, 24 % of CF patients in the French Reunion Island bear the Y122X mutation (Dugueperoux et al. 2003).

24.4.2 Class II Mutations Alter the Intracellular Processing of CFTR

Class II mutations are associated with defective protein processing, arising from misfolding and/or decreased stability of the protein structure. Class II mutations result in a protein that is retained within the endoplasmic reticulum (ER), retro-translocated into the cytoplasm, and subject to proteosomal degradation (Jensen et al. 1995; Ward et al. 1995). During initial translation, CFTR is co-translationally modified by core glycosylation (Band B CFTR). As CFTR progresses through the biosynthetic pathway, CFTR is fully glycosylated in the Golgi to a higher molecular weight Band C form. This shift in molecular weight can be assayed easily by western blot analysis and provides a simple indication of the mutation type and progress of CFTR through the cell. Depending upon the particular class II mutation, either a partial reduction (L206W) or a complete absence (Δ F508) of mature Band C CFTR is seen. Indeed, the most common mutation, occurring on ~70 % of all mutant alleles and on ~90 % of all CF patients, is the ΔF508 mutation. Initially thought to be purely a protein folding problem, subsequent crystallization of wt and ΔF508 NBD1 revealed little difference in overall protein structure (Lewis et al. 2004). The current model argues for an energetic and kinetic instability in ΔF508, causing a destabilized structure and an inability of NBD1 to interact appropriately with the fourth intracellular loop (ICL4) (Protasevich et al. 2010; Cui et al. 2007; He et al. 2010). This combination of defects has clear clinical implications, as correction of both NBD1 kinetics and NBD1/ICL4 surface stability will likely be required to restore ΔF508 processing (Mendoza et al. 2012; Rabeh et al. 2012).

24.4.3 Class III Mutations Alter CFTR Channel Regulation

Class III mutations are usually located within the ATP binding domains of CFTR (NBD1, NBD2) and are referred to as *Gating* mutations. These are missense mutations that have little or no impact on protein processing (mutant proteins being inserted efficiently into the plasma membrane) but rather are resistant to normal activation and gating as a result of protein kinase phosphorylation (or can cause inappropriately rapid dephosphorylation and inactivation). The dominant mutation in this class is G551D, which results in a protein with an open probability (P_o) ~100-fold lower than that of wt CFTR.

24.4.4 Class IV Mutations Alter CFTR Channel Conductance

Class IV mutations generally reside within those membrane spanning domains proposed to be part of the pore-forming region. Class IV missense mutations produce a protein that is exported from the ER to the cell surface, the channel can be activated by kinase phosphorylation, but the channel has a reduced capacity to conduct anions through the pore. Since such mutations reduce the ability of the channel to conduct anions, rather than abolish conductance, and in other aspects the protein is "normal," it can be anticipated that class IV mutations are generally associated with milder forms of CF. In this class, the R117H mutation is the best characterized, yielding a protein that is properly processed and forms a cAMP-/PKA-dependent chloride channel with reduced conductance (Reddy and Quinton 2001; Sheppard et al. 1993). It should be noted, however, that the R117H protein also exhibits a lower open probability compared to wild-type CFTR, also engendering some class III characteristics to the channel (Reddy and Quinton 2001; Sheppard et al. 1993).

24.4.5 Class V Mutations Alter the Amount of Functional CFTR Protein

Class V mutations result in the generation of a fully functional CFTR channel; however the level of protein expression is considerably reduced compared to wild type. Typically these mutations occur in introns, close to splice sites, decreasing the efficiency of intron excision. Such splice mutations can result in complete or partial loss of an exon depending on how inefficient the pre-mRNA splicing becomes. If an exon is inefficiently incorporated into the mRNA, lower mRNA levels will be generated, yielding lower protein levels. The classic example of this kind of mutation is that leading to the skipping of exon 10. The incorporation or exclusion of this exon is correlated with a polymorphism in the poly-T tract upstream of the splice acceptor site. Thus a poly-T tract of 7T or 9T results in up to 90 % splicing efficiency (i.e., normal amounts of CFTR mRNA), whereas a poly-T tract of 5T reduces splicing efficiency to ~10–40 % (i.e., only 10–40 % of wt CFTR mRNA levels are generated) (Chu et al. 1992). Such mutations may not give rise to clinical CF, but may lead to other complications such as male infertility, as seen in CBAVD patients. A sixth class of mutation that decreases the stability of otherwise normal CFTR at the plasma membrane has been proposed (Silvis et al. 2003; Haardt et al. 1999). However, these mutations are now incorporated into class V mutations causing reductions in the amount of functional CFTR.

24.5 Multi-class Mutations

Although many mutations do in fact fall neatly within these five classes, some mutations cause multiple distinct defects. Indeed, this is the case for the common ΔF508 mutation, where not only does the loss of phenylalanine 508 lead to inefficient processing and exit of ΔF508 from the ER, but even for mutant CFTR that does reach the cell surface, ΔF508 displays channel defects characterized by a reduced open probability (P_o) and reduced stability of the protein at the cell surface (Cheng et al. 1990; Dalemans et al. 1991). The R117H mutation, although initially described as a class IV mutation with reduced channel conductance (Reddy and Quinton 2001; Sheppard et al. 1993), also exhibits a reduction in open probability, giving its characteristics associated with class III mutations. As discussed, CFTR conducts both chloride and bicarbonate ions and several mutations differentially affect the ability of CFTR to conduct chloride and bicarbonate. For example, the D648V mutation has little effect on chloride transport but reduces bicarbonate transport by 50–65 % (Choi et al. 2001), which would put the mutation in the class IV conduction category. Interestingly however, the D648V mutation also affects splicing of exon 13 by altering an exonic splicing enhancer (Aznarez et al. 2003), thus reducing mRNA levels which would also put the mutation also in class IV.

Although each affected allele generally harbors one mutation, some CF alleles possess two distinct mutations, referred to as complex alleles. As might be expected, the combined effects of the mutations lead to a more severe phenotype than would be expected from either mutation alone (Clain et al. 2001). For example, the effects of the R117H mutation can be modified by *cis* acting 5T/7T/9T polypyrimidine tract elements in exon 9. The R117H-7T genotype is associated with a milder phenotype even to the extent of a clinical presentation of male infertility, rather than cystic fibrosis. In contrast, the R117H-5T genotype causes elevated sweat electrolytes and clinical CF, which can be severe (Trujilliano et al. 2013).

24.6 Different Problems Need Different Solutions

As can be appreciated, CFTR mutations and their clinical phenotype are very complex. Although all CF patients have mutations in their CFTR gene, some patients have very severe disease, whereas others have a milder disease presentation. Complicating this picture even further is the concept of modifier genes, which are genes that are different from CFTR, yet whose activity can modify the severity of the disease (Cutting 2010; Blackman et al. 2013). From the various mechanisms by which CFTR function can be compromised, it is clear that a single pharmacological approach is unlikely to be viable. A drug which is effective in activating class V mutations will have little efficacy on class I or II mutations. Therefore,

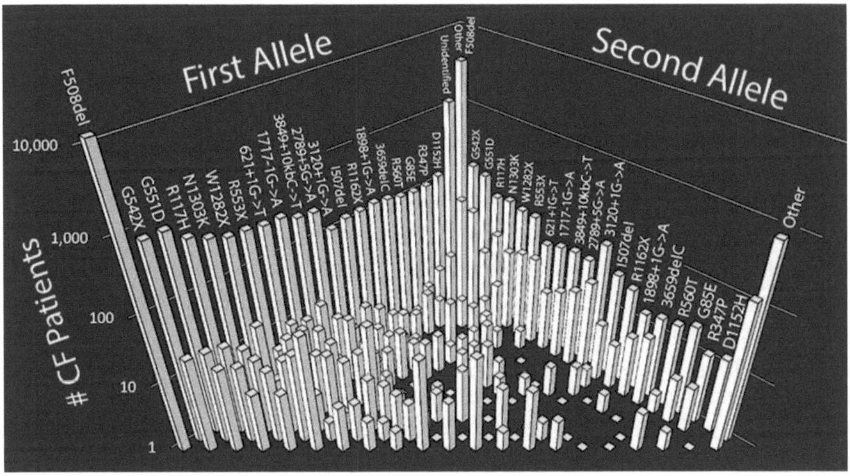

Fig. 24.6 Allelic variation among CF patients. Data is taken from 25,976 CF patients with at least one allele recorded in the US CF Foundation Patient Registry (2012). Note that the Y-axis is on a logarithmic scale

minimally knowing a patient's genotype (i.e., the class of mutation expressed in that patient) will provide information on whether a particular drug treatment will be warranted in that patient. While some patients harbor the same mutation on each allele, this is not the case for all CF patients. Shown in Fig. 24.6 is an analysis of data taken from nearly 26,000 CF patients, where at least one mutant CF allele is known (US Cystic Fibrosis Foundation 2012 Patient Registry). This highlights the complexity of the problem in treating all patients with CF. While ΔF508 clearly predominates, it is also apparent that many patients have different mutations on each allele. Whether each allelic mutation will be amenable to the same pharmacotherapy or whether CFTR expressed from each allele will require a different drug again emphasizes the need to know the CFTR genetic makeup of each patient and tailoring of drug treatments to that individual patient.

24.7 How to Find a CFTR Drug?

The ideal drug to treat patients with CF would be one that normalizes multiple aspects of CFTR dysfunction, including defective protein folding, restricted ER exit, plasma membrane targeting, and surface stability, restores normal channel activation and gating, has no or minimal off-target effects, and can be taken as a single daily oral dose. Although remarkable advances in drug development for CF have been made in the past few years, such an ideal drug has not yet been realized. Nonetheless drugs have been identified that restore some aspects of CFTR dysfunction. The high-resolution 3D structures of wild-type (wt) or mutant CFTR have not yet been determined. However, structures of NBD-1 based upon crystal structure and NMR spectroscopy have been generated (Lewis et al. 2004, 2010), and full-length models of CFTR based on homology modeling (Mendoza and Thomas 2007; Serohijos et al. 2008) have been developed. Such models have been used for structure-based virtual screening of compounds, in an attempt to identify chemicals that would bind to mutant CFTR and restore folding and/or function. Using in silico screening based upon homology models, Kalid and colleagues (Kalid et al. 2010) identified three putative binding sites at the interfaces between the cytoplasmic domains that would be amenable to drug binding. Computer-based analysis, docking a virtual library of 496 candidates, has generated 15 potential compounds, although none has yet reached clinical trials.

High-throughput screening (HTS) for CF drugs initially relied on fluorescence methodologies, including the halide-sensitive yellow fluorescent protein (YFP), developed by Verkman (Rowe and Verkman 2014), and a membrane potential-sensitive dye pair, developed by Molecular Devices Inc for use on an FLIPR III plate reader (Gonzalez et al. 1999) and employed by Vertex Pharmaceuticals. Initial screening protocols relied on evaluating the effects of compounds not on all CFTR mutations, but rather on the most prevalent, i.e., ΔF508 and G551D, CFTR expressed in heterologous systems. Although useful, such cell models do not entirely recapitulate the target tissues in a CF patient, i.e., the airway epithelium. The development and implementation of primary airway epithelial cells as a model for drug screening have made an immense impact on the search for drugs to treat patients with CF. This advance has become particularly important, as several promising compounds which were effective in correcting CFTR in heterologous systems subsequently turned out to be ineffective when tested on airway cells, the major target tissue in CF patients. Airway epithelial cells, derived either from a CF patient at the time of lung transplant or from cadaveric wt tissues, are expanded in tissue culture prior to plating on permeable supports and grown at an air-liquid interface to recapitulate the native airway environment. Such cell monolayers develop an electrically tight epithelial layer containing ciliated epithelial cells. The advantage of this approach is that it facilitates the implementation of highly sensitive electrical measurements of CFTR function. Two dominant culture methods are at present utilized, one developed by the University of North Carolina (Matsui et al. 1998b) and one developed by Vertex Pharmaceuticals (Van Goor et al. 2009). These two different culture methods do give rise to differences in the electrical properties of each monolayer, though which approach (if either) most faithfully recapitulates the native airway environment is not known. Nonetheless, it is clear that such models have proven to be invaluable in CF drug discovery and formed the basis of assays leading to the current FDA-approved drug ivacaftor and drugs currently in clinical trial. Outlined below are the three mechanisms of actions for repairing CFTR protein/function and the main clinical candidates for therapy at this time (Table 24.1, Fig. 24.7).

Table 24.1 CFTR therapies based on patient genotype

Mutation class	Protein defect	Genotype examples	Correction strategy	Drugs in development
I	Protein production	G542X, W1282X	PTC suppressor	Ataluren
II	Protein trafficking	ΔF508	Corrector + potentiator	VX-809 + ivacaftor (approved by FDA 2015, marketed as Orkambi)
III	Channel gating	G551D	Potentiator	Ivacaftor (approved by FDA)
IV	Channel conductance	R117H	Potentiator?	No data
V	Low mRNA levels	3849 + 10 kb C → T	Potentiator?	No data

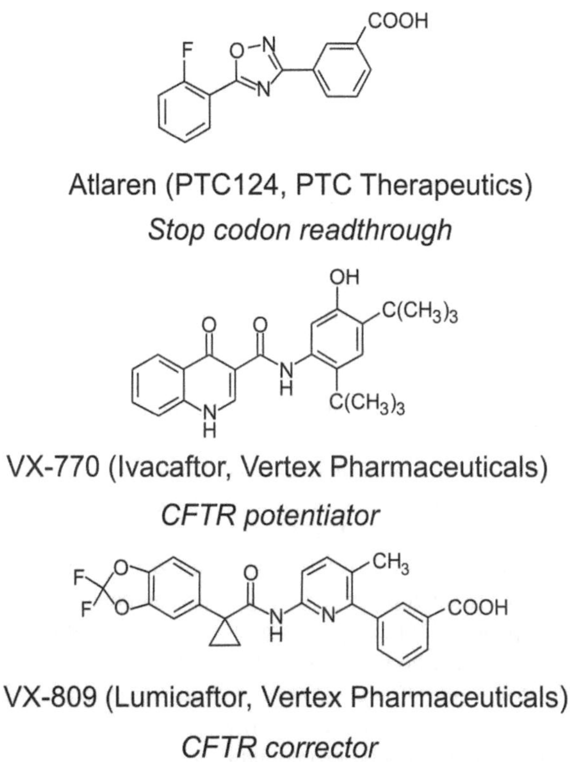

Fig. 24.7 Three CFTR modulators are now in clinical trials or approved by the United States Federal Drug Administration (FDA). (**a**) Ataluren, a PTC read-through compound currently in phase III trials for patients aged 6 years and above who have nonsense mutations. (**b**) VX-770 (Ivacaftor, marketed as Kalydeco) is a CFTR potentiator used to treat patients with at least one copy of the G551D allele. VX-770 is FDA approved for patient use. Evaluation of Kalydeco for other class III mutations is underway. (**c**) A combination of VX-809 (lumacaftor) and XV-770 (ivacaftor) is marketed as Orkambi and is FDA approved to treat patients with the ΔF508 mutation. Whether potentiators will be of use in class IV and V mutations is still uncertain

24.8 Suppressors of Premature Termination: Making More CFTR

For organisms employing the standard genetic code, including humans, translational termination occurs when one of the three STOP codons—UAA, UGA, or UAG—enters the ribosomal A site (Frischmeyer and Dietz 1999). Premature termination codons (PTC) arise when a single base-pair mutation generates an erroneous stop codon within the open reading frame of a gene. This can arise either as a nonsense mutation, frameshift deletions or insertions, or from aberrant mRNA splicing that results in truncated reading frames. Approximately 10 % of patients with CF exhibit a PTC mutation on at least one allele. The antimicrobial activities of aminoglycoside antibiotics arise from their ability to inhibit bacterial protein synthesis at high doses. At low doses, however, aminoglycosides bind inefficiently to eukaryotic ribosomes resulting in a translational read-through by the insertion of an amino acid at the stop codon (Martin et al. 1989; Fan-Minogue and Bedwell 2008). The potential for aminoglycosides in the treatment of PTC mutations (Class I mutations) in CFTR was first recognized in the late 1990s when studies by Bedwell (Bedwell et al. 1997) and Howard (Howard et al. 1996), using transient expression systems, demonstrated that full-length CFTR could be detected following aminoglycoside treatment of cells expressing R1162X, G542X, and R553X CFTR constructs. Moreover, not only was full-length protein restored, but cAMP-activated chloride currents could also be observed. Studies using a *G542X-hCFTR* mouse (human G542X CFTR cDNA on a murine *Cftr*-null background) demonstrated that exposure to the aminoglycoside gentamycin increased the expression of full-length CFTR protein in the gut (the site of CF-related mortality in the mouse model) and was associated with the appearance of a cAMP-activated chloride conductance (Du et al. 2002). Proof-of-principle studies in patients with PTC in CFTR have been able to demonstrate increased CFTR channel activity, as measured by nasal PD (from basal potential difference 45 ± 8 to 34 ± 11 mV, $p = 0.005$) and by immunofluorescence staining of nasal cells using an antibody recognizing the carboxyl terminus of CFTR (Wilschanski et al. 2000, 2003). These studies applied topical (nasal) gentamicin for 2 weeks and resulted in a statistically significant improvement in CFTR function in treated groups compared to control. No effect was observed in patients homozygous for the ΔF508 mutation. In studies with larger patient populations, however, a relative lack of efficacy of gentamicin has been reported (Rowe and Clancy 2009). Although the search for other more effective aminoglycosides is underway (Prayle et al. 2010), a lack of potency, and renal and otic toxicity may be problematic.

High-throughput screening (HTS) of compounds that promote read-through of PTC mutations in a luciferase-based reporter system led to the development of PTC124 (Ataluren, PTC Therapeutics, South Plainfield New Jersey). Ataluren, a 1,2,4-oxadiazole benzoic acid (Fig. 24.7), causes PTC read-through but appears to interact with mammalian ribosomes in a manner distinct from aminoglycosides (Welch et al. 2007); this may explain why ataluren does not possess any antibacterial properties. At present, ataluren represents the only clinical trial using an oral compound to treat PTC mutations in CF patients. Initial preclinical data argued that ataluren was 4- to 15-fold more potent at causing PTC read-through than control compound. Three phase II clinical trials have reported mixed results: two studies reporting improvements in CFTR function (as assessed by nasal PD) and one study reporting no change in function (Sermet-Gaudelus et al. 2010; Kerem et al. 2008; Clancy et al. 2006). All studies however relied on a small patient population, a problem that is being addressed in a larger phase III clinical trial (www.cff.org/research/DrugDevelopmentPipeline/#CFTRMODULATION).

Potential problems and limitations to extending the use of compounds to treat PTC mutations include (1) specificity of read-through agents for mutant vs. authentic stop codons, (2) low levels of mRNA transcripts due to nonsense-mediated decay (NMD), (3) potential ototoxicity of aminoglycosides, and (4) generation of novel mutations through random amino acid insertion. The inappropriate read-through of authentic stop codons is likely more a theoretical than an actual problem. Structural differences between authentic stop codons and PTC mutations argue that read-through is limited to PTC mutations. For example, the 3′ untranslated region (UTR) is a strong stimulus for translation termination at appropriate stop codons and is absent from PTC stop codons (Amrani et al. 2004). mRNA transcripts for PTC mutations tend to be expressed at low levels due to NMD (Maquat 2002); however increased mRNA stabilization has been seen with administration of aminoglycosides in vitro (Bedwell et al. 1997). Ototoxicity has been linked to long-term use of aminoglycosides such as gentamycin, tobramycin, and kanamycin (Al-Malky et al. 2014). No long-term studies using ataluren have yet been performed, so it is not known if similar problems with hearing loss will be associated with chronic ataluren use. Read-through of PTC mutations obligatorily inserts an amino acid in the place of the mutant stop codon. The amino acid inserted appears to be somewhat random [with a preference for glutamine, tyrosine, cysteine, arginine, or tryptophan, at least in yeast (Blanchet et al. 2014)]. Thus there is the generation of a pool of translated full-length protein bearing at least one amino acid substation. These proteins will have either normal or limited function, depending on where in the protein the PTC was present. For example, a tyrosine substation in the amino terminus of CFTR leads to a channel with normal function, but a markedly increased endocytic rate and mild clinical CF (Silvis et al. 2003).

24.9 CFTR Potentiators: Opening a Closed Channel

Potentiator drugs work on class III and class IV mutations that result in channels with defective gating or conductance, respectively (Rogan et al. 2011). Molecules that increase the activity of mutant CFTR have been known since shortly after the identification of the *CFTR* gene, when the phytoestrogens, genistein and isoflavone (Hwang et al. 1997), were identified as compounds increasing the open probability of CFTR. The mechanisms of action of genistein are still unclear, but may include

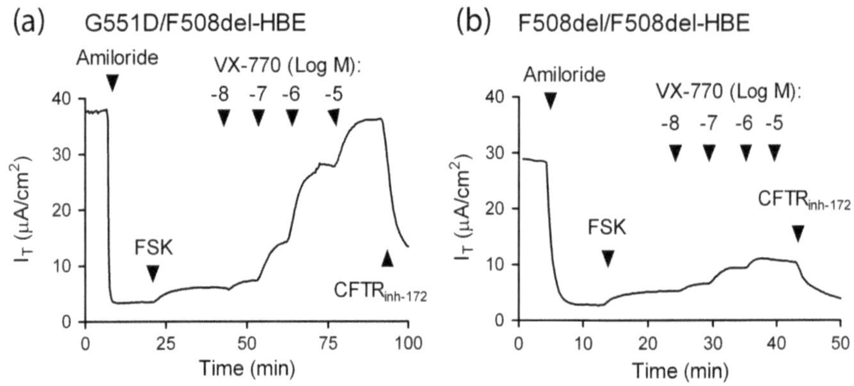

Fig. 24.8 VX-770 potentiation of CFTR-mediated chloride secretion in primary cultures of G551D/DF508 HBE. The Ussing chamber technique was used to record transepithelial currents (I_T) resulting from CFTR-mediated chloride secretion. Currents were measured under a basolateral to apical chloride gradient, amiloride (30 μM) to block ENaC sodium currents, and forskolin (10 μM, FSK) to activate CFTR. The addition of $CFTR_{inh-172}$ (20 μM) to the apical surface was used to confirm that the I_T was CFTR mediated. (**a**) Current trace from G551D/ΔF508 HBE cells showing efficacy of VX-770 on increasing CFTR currents. (**b**) Current trace from ΔF508/ΔF508 HBE showing minimal effect of VX-770 potentiator alone on chloride secretion. Reproduced with permission from Van Goor et al. (2009)

tyrosine kinase inhibition, inhibition of topoisomerase, or direct interaction with CFTR (Weinreich et al. 1997; Illek et al. 1995). Unfortunately, genistein also increases ENaC-mediated sodium absorption (Devor et al. 2000), which, along with the multiple known targets of genistein, make it an unlikely clinical candidate for CF therapy. In an exciting collaboration between the CF Foundation and the pharmaceutical company Vertex Pharmaceuticals (Cambridge, MA), a high-throughput screen (HTS) was initiated utilizing a library of 228,000 compounds, leading to the discovery of VX-770, ivacaftor (trade name Kalydeco) (Fig. 24.7). Initial HTS assays were confirmed using single channel patch-clamp analysis in Fischer rat thyroid (FRT) cells expressing G551D CFTR. 10 μM ivacaftor increased the P_o of G551D CFTR by approximately sixfold ($p < 0.05$), findings subsequently validated in primary human bronchial epithelial cells (HBE) from G551D patients (Fig. 24.8). In addition, ciliary beat frequency in HBE cells increased to near wild-type levels, due to the increase in ASL height accompanying increased CFTR activity. Based on these preclinical data, one phase II (Accurso et al. 2010) and three phase III (Ramsey et al. 2011) clinical trials were initiated. In all studies, ivacaftor led to a marked and persistent improvement in lung function (as measure by forced expiratory volume, FEV_1) and other markers such as a reduction in sweat chloride and a movement of nasal PD out of the CF range. Excitingly in the phase III trial, FEV_1 increased by greater than 10 %, accompanied by a weight increase of >2.5 kg. On January 31, 2012, the US Food and Drug Administration (FDA) approved ivacaftor to treat patients with the G551D mutation (Van Goor et al. 2009).

The molecular mechanism underlying the effectiveness of ivacaftor in increasing channel open probability remains elusive, though initial work by Hwang and colleagues (Jih and Hwang 2013) argues that ivacaftor works by stabilizing a post-hydrolytic open state, leading to an uncoupling of ATP hydrolysis from the gating cycle and causing an increased residence time in the open state. Further work identifying the precise molecular details of how potentiators work will likely lead to the development of more potent potentiators. Although the G551D mutation is the second most prevalent CFTR mutation, it is only present in approximately 4–5 % of all CF patients, and therefore, many patients will not benefit from ivacaftor alone. Initially only approved for the treatment of patients with the G551D mutation, ivacaftor also appears to work on several other gating mutations, including G178R, S549R, G551S, G970R, G1244E, S1251N, S1255P, and G1349D (Yu et al. 2012); clinical trials for patients bearing these mutations are in progress. Interestingly, evidence also suggests that potentiators may also be of benefit in increasing the P_o of class II mutations, including ΔF508 (Van Goor et al. 2009), that have been brought to the cell surface. As a monotherapy in ΔF508 patients, ivacaftor modestly reduced sweat chloride levels relative to baseline and placebo (<3 mM, $p < 0.05$), but did not improve lung function outcomes (Flume et al. 2012). While ivacaftor has provided a proof of concept that small molecules can interact with CFTR in a therapeutically beneficial manner, it is nonetheless only effective in a small number of patients. Indeed, since potentiators only work on CFTR protein that is present at the cell surface, it is clear that potentiators alone will be of no benefit to patients with either class I or class II mutations.

24.10 CFTR Correctors: Getting CFTR to the Cell Surface

As discussed above, the ΔF508 mutation disrupts the appropriate folding of nascent CFTR causing its premature degradation in the ER (Rowe et al. 2005), resulting in little or no CFTR reaching the plasma membrane. CFTR "correctors" work by increasing the stability of ΔF508 in the ER, facilitating its ER exit and trafficking to the cell surface. Correction of ΔF508 trafficking was first observed in vitro by culturing cells at 27 °C (Denning et al. 1992), providing proof of principle that ΔF508 could be "corrected." Subsequently, chemical chaperones such as glycerol were found to cause small increases in surface ΔF508 CFTR (Sato et al. 1996); however an open parallel group study in CF subjects using topical glycerol failed to show any effect on nasal PD (Chadwick et al. 1998). Butyric acid and its analogues were also shown to increase CFTR mRNA in heterologous ΔF508-expressing cells and to increase the expression of ΔF508 at the cell surface. Interest in the potential use of the butyric acid analogue sodium 4-phenylbutyrate (4-PBA), which also increases ΔF508 CFTR expression, arose because the compound was already approved for the clinical treatment of patients with urea cycle disorders (Iannitti and Palmieri 2011). In primary human nasal epithelial cells expressing ΔF508/

W1282X CFTR, forskolin-induced CFTR currents were increased in the presence of 4-PBA (Rubenstein et al. 1997). In early pilot studies of 4-PBA in CF patients, a small change in nasal PD measurement, consistent with an increase in CFTR function, was observed, although no changes in sweat electrolytes were noted (Rubenstein and Zeitlin 1998). Although phase I/II trials demonstrated a modest correction in ΔF508 function, later studies revealed that the high doses of 4-PBA required to see these effects produced intolerable adverse effects (Rubenstein and Zeitlin 1998; Rubenstein et al. 1997).

Curcumin, derived from the spice turmeric, is a SERCA pump inhibitor that reduces ER calcium concentrations. Treatment of cells expressing ΔF508 CFTR with curcumin showed a modest increase in ΔF508 expression, and ΔF508 mice exposed to curcumin displayed a modest correction in nasal PD (Egan et al. 2004). However, subsequent evaluation of curcumin using human airway epithelial cells failed to replicate these findings (Grubb et al. 2006; Loo et al. 2004). Moreover, it appears that the doses of curcumin required to achieve ΔF508 correction in early cell models are not physiologically achievable in human subjects (Song et al. 2004).

The early promise and subsequent disappointment in compounds, such as 4-PBA and curcumin, to display any efficacy in patients with CF heightened the need for identifying more potent correctors with clinically acceptable pharmacokinetic profiles. Utilizing HTS approaches, KM11096, corr-2b, corr-4a, VRT-325, VX-809, and VX-661 were identified as potential clinical candidates (Dormer et al. 2005; Poschet et al. 2007; Loo et al. 2005; Van Goor et al. 2011) (http://investors.vrtx.com/releasedetail.cfm?releaseid=563453).

In FRT cells expressing ΔF508, VX-809 incubation (48 h) (Fig. 24.8) improved ΔF508 trafficking by approximately sevenfold ($EC_{50} = 81$ nM) and increased CFTR channel activity by approximately fivefold. With a slightly lower EC_{50} in primary human airway epithelial cells cultured from patients homozygous for ΔF508-CFTR ($EC_{50} = 350$ nM), VX-809 nonetheless increased CFTR maturation by eightfold and increased CFTR-mediated chloride transport (measured by short circuit current, *Isc*) by fourfold. Upon VX-809 treatment of ΔF508 HBE cells, the CFTR-dependent chloride currents were determined to be ~14 % of that observed in non-CF HBE cells (Van Goor et al. 2011). In clinical trials, VX-809, lumacaftor (Vertex Pharmaceuticals, Cambridge, MA) has been assessed by multicenter, randomized, double-blind, placebo controlled trials in adults homozygous for the ΔF508 mutation. The safety profile of lumacaftor was similar to that of placebo and showed that lumacaftor was well tolerated (Clancy et al. 2012). Moreover, lumacaftor treatment resulted in a modest, but significant, dose-dependent reduction in sweat chloride levels (8 mM drop). However, no significant improvements were observed in lung function (FEV_1) or in CFTR-dependent nasal PD. A second CFTR corrector, VX-661, entered clinical trials for CF patients homozygous and heterozygous for the ΔF508 mutation (http://clinicaltrials.gov/ct2/show/NCT01531673). In a dose-escalation monotherapy, VX-661 caused a decrease in sweat chloride and a relative change in FEV_1, consistent with an improvement in lung function (~9 %) (Donaldson et al. 2013). Importantly, FEV_1 returned to pretreatment levels following cessation of VX-661 treatment. Mechanistic studies suggest that correctors interact directly with ΔF508-CFTR and/or CFTR-associated proteins during biogenesis to facilitate ER exit of ΔF508 (Ren et al. 2013; Grove et al. 2009). Studies by Cyr and colleagues have argued that VX-809 suppresses folding defects in ΔF508 CFTR by binding to membrane spanning domain 1 (MSD1) to enhance and stabilize the interaction between the first nucleotide binding domain (NBD1) and the MSD1 and MSD2 domains (Ren et al. 2013).

24.11 If One Is Good, Two Must Be Better: Combination of Treatments for CF

Since correctors allow ΔF508 to reach the cell surface and potentiators increase CFTR activity, it seems reasonable that a combination therapy of correctors and potentiators would amplify the effects of each modulator. A recent phase II clinical trial has investigated the effects of combining lumacaftor with ivacaftor in the treatment of patients with the ΔF508 mutation (Boyle et al. 2012). A significant drop in sweat electrolytes (9.1 mM, $p < 0.001$) was observed following twice daily ivacaftor (250 mg) with once daily lumacaftor (150 mg). For patients receiving only 150 mg ivacaftor, no drop in sweat electrolytes was observed. Curiously, an improvement of FEV_1 of 3.5 % was observed in patients taking 150 mg ivacaftor but not in those receiving 250 mg ivacaftor. What the source of this apparent contradiction would be is not entirely clear. Recent in vitro studies have, however, raised concern about the combined use of ivacaftor with lumacaftor (Cholon et al. 2014; Veit et al. 2014), which argued that chronic dosage of potentiators, such as ivacaftor, could inhibit the activity of correctors such as lumacaftor, a finding that may not be apparent in short-term clinical trials. Thus, the reduced correction efficiency of ΔF508 CFTR, as well as other processing mutants, in the presence of potentiators may abrogate any benefit from the use of correctors. Whether this finding is true for all potentiator/corrector pairs is not known; however studies from other groups argue that it may be a lumacaftor/ivacaftor phenomenon and that other potentiator/corrector pairs may not be susceptible to competing actions. Moreover, it is likely that patients would receive a single medication containing both correctors and potentiators, rather than single doses of corrector with multiple potentiator dosing.

An alternative strategy to combining a corrector drug with a potentiator drug is to combine both mechanisms of action into a single molecule. These compounds are called dual-acting CFTR modulators. Compared to multicomponent therapies, dual-acting modulators have several advantages. Pharmacokinetic/pharmacodynamic studies for single drugs are much simpler to conduct and more straightforward in terms of clinical trials (Morphy and Rankovic 2005). Joining a corrector compound (derived from CFTR corrector Corr-4a) with a potentiator moiety (derived from the phenylglycine potentiator PG01) using a hydrolysable ethylene glycol spacer, Mills and colleagues generated a hybrid dual-acting CFTR modulator (Mills et al. 2010). At least in FRT cells expressing ΔF508 CFTR, such an approach appears to have validity. However, because the two components are linked by a single cleavable linker, the compound does in fact break down into its constituent components and is therefore not a true dual-acting modulator. Screening over 100,000 compounds for a true dual-acting molecule, Verkman and colleagues identified six commercially available compounds that exhibited dual correcting and potentiating properties (Phuan et al. 2011). Although these compounds, especially CoPo-22, have an EC_{50} in the low micromolar range, their efficacy is still below that of other single-acting compounds. Whether a truly efficacious dual-acting compound will be identified remains to be determined.

24.12 How Much Is Enough?

A critical unanswered question in CF therapy is "how much correction do you need for long-term benefit?" On the surface, this seems a simple question, yet is surprisingly difficult to answer. The assumption is made that obligate heterozygotes (i.e., parents of children with CF) have half the amount of CFTR relative to noncarriers, yet this has not been extensively investigated. The other assumption made is that heterozygotes (even with half the amount of CFTR as non-CF carriers) are symptomless. This for the most part appears to be true, although recent studies suggest there may be differences (Aeffner et al. 2013). Certainly heterozygotes and homozygous wt individuals appear to have similar CFTR function, as monitored by sweat electrolytes and nasal PD. Conversely patients with two severe mutations, e. g., homozygous for ΔF508, have little or no detectable CFTR function or protein and have sweat electrolytes ~100 mM (Fig. 24.1). Increasing CFTR protein/function, associated with milder mutations, is seen in pancreatic sufficient CF patients, with near-normal protein/function possibly in patients with idiopathic pancreatitis (Wilschanski et al. 2006). If sweat electrolyte levels are used as a biomarker, a value of ~70 mM would place a patient in a pancreatic sufficient-like state [known to be associated with a decade or so of increased life expectancy (Davis et al. 2004)] and ~60 mM essentially normal. However, sweat electrolytes, not normally a pathologic concern, appear to show little correlation with important clinical manifestations such as lung function (as measured by FEV_1). Even for those patients with residual CFTR function, lung disease including bronchiectasis with sinopulmonary infections and endocrine problems associated with CF-related diabetes can have devastating effects. Thus it is clear that the present CFTR modulators will still necessitate treatment for residual disease.

24.13 Future Perspectives

Until recently, a patient's CFTR genotype was more a research issue than a clinical issue, though the separation into severe and mild mutations would certainly impact prognosis. However, a patient's genotype had little impact upon treatment regimens. With the marketing of ivacaftor, that all changed. Now, it is clear that a physician must have a detailed knowledge of a patient's CFTR genotype in order to determine the most appropriate treatment; admittedly at present this comes down to whether a patient's mutations will benefit from potentiators such as ivacaftor or not. As correctors and other potentiators arrive in the clinician's armamentarium, the need to know a patient's genotype will only increase. In a short period, we have therefore moved from a roughly one-size-fits-all approach to treating CF patients to a bespoke treatment based on a patients DNA. The era of personalized medicine for CF has arrived.

Acknowledgments The author wishes to thank Dr Ashvani Singh (Principle Scientist, Abbvie) for helpful discussions and manuscript review. Work in the author's lab is funded by NIH and the Cystic Fibrosis Foundation

References

Accurso FJ, Rowe SM, Clancy JP, Boyle MP, Dunitz JM, Durie PR, Sagel SD, Hornick DB, Konstan MW, Donaldson SH, Moss RB, Pilewski JM, Rubenstein RC, Uluer AZ, Aitken ML, Freedman SD, Rose LM, Mayer-Hamblett N, Dong Q, Zha J, Stone AJ, Olson ER, Ordonez CL, Campbell PW, Ashlock MA, Ramsey BW (2010) Effect of VX-770 in persons with cystic fibrosis and the G551D-CFTR mutation. N Engl J Med 363:1991–2003

Accurso FJ, Van Goor F, Zha J, Stone AJ, Dong Q, Ordonez CL, Rowe SM, Clancy JP, Konstan MW, Hoch HE, Heltshe SL, Ramsey BW, Campbell PW, Ashlock MA (2014) Sweat chloride as a biomarker of CFTR activity: proof of concept and ivacaftor clinical trial data. J Cyst Fibros 13:139–147

Aeffner F, Abdulrahman B, Hickman-Davis JM, Janssen PM, Amer A, Bedwell DM, Sorscher EJ, Davis IC (2013) Heterozygosity for the F508del mutation in the cystic fibrosis transmembrane conductance regulator anion channel attenuates influenza severity. J Infect Dis 208:780–789

Al-Malky G, Suri R, Sirimanna T, Dawson SJ (2014) Normal hearing in a child with the m.1555A>G mutation despite repeated exposure to aminoglycosides. Has the penetrance of this pharmacogenetic interaction been overestimated? Int J Pediatr Otorhinolaryngol 78:969–973

Alonso y de los Ruyzes de Fontecha J (1606) Diez Previlegios para Mgeres Preñadas. Alcalá de Henares, Henares, Spain, p 212

Amrani N, Ganesan R, Kervestin S, Mangus DA, Ghosh S, Jacobson A (2004) A faux 3′-UTR promotes aberrant termination and triggers nonsense-mediated mRNA decay. Nature 432:112–118

Anderson D (1938) Cystic fibrosis of the pancreas: clinical and pathological study. Am J Dis Child 56:344

Armstrong DK, Cunningham S, Davies JC, Alton EW (2014) Gene therapy in cystic fibrosis. BMJ 99:465–468

Aznarez I, Chan EM, Zielenski J, Blencowe BJ, Tsui LC (2003) Characterization of disease-associated mutations affecting an exonic splicing enhancer and two cryptic splice sites in exon 13 of the cystic fibrosis transmembrane conductance regulator gene. Hum Mol Genet 12:2031–2040

Bedwell DM, Kaenjak A, Benos DJ, Bebok Z, Bubien JK, Hong J, Tousson A, Clancy JP, Sorscher EJ (1997) Suppression of a CFTR premature stop mutation in a bronchial epithelial cell line. Nat Med 3:1280–1284

Blackman SM, Commander CW, Watson C, Arcara KM, Strug LJ, Stonebraker JR, Wright FA, Rommens JM, Sun L, Pace RG, Norris SA, Durie PR, Drumm ML, Knowles MR, Cutting GR (2013) Genetic modifiers of cystic fibrosis-related diabetes. Diabetes 62:3627–3635

Blanchet S, Cornu D, Argentini M, Namy O (2014) New insights into the incorporation of natural suppressor tRNAs at stop codons in Saccharomyces cerevisiae. Nucleic Acids Res 42:10061–10072

Boat TF, Welsh MJ, Beadudet AL, Scriver CR (1989) Cystic fibrosis. In: Scriver CR, Beadudet AL, Valle D, Sly WS (eds) The metabolic basis of inherited disease, 8th edn. McGraw-Hill, New York, NY

Boucher RC (2004) New concepts of the pathogenesis of cystic fibrosis lung disease. Eur Respir J 23:146–158

Boyle MP, Bell S, Konstan M, McColley SA, Kang L, Patel N (2012) The investigational CFTR corrector, VX-809 (lumacaftor) co-administered with the oral potentiator ivacaftor improved CFTR and lung function in F508DEL homozygous patients: Phase II study results. Pediatr Pulmonol 47(Suppl 35):315

Bradbury NA, Cohn JA, Venglarik CJ, Bridges RJ (1994) Biochemical and biophysical identification of cystic fibrosis transmembrane conductance regulator chloride channels as components of endocytic clathrin-coated vesicles. J Biol Chem 269:8296–8302

CF Foundation (2011) Patient Registry. Annual data report. CF Foundation, Bethesda, MD

CF Genetic Analysis Consortium (1990) Worldwide survey of the DF508 mutation – report from the cystic fibrosis genetic analysis consortium. Am J Hum Genet 47:354–359

Chadwick S, Browning JE, Stern M (1998) Nasal application of glycerol in DF508 cystic fibrosis patients. Thorax 53:A60

Cheng SH, Gregory RJ, Marshall J, Paul S, Souza DW, White GA, O'riordan CR, Smith AE (1990) Defective intracellular transport and processing of CFTR is the molecular basis of most cystic fibrosis. Cell 63:827–834

Choi YH, Yu AM (2014) ABC transporters in multidrug resistance and pharmacokinetics, and strategies for drug development. Curr Pharm Des 20:793–807

Choi JY, Muallem D, Kiselyov K, Lee MG, Thomas PJ, Muallem S (2001) Aberrant CFTR-dependent HCO_3^- transport in mutations associated with cystic fibrosis. Nature 410:94–97

Cholon DM, Quinney NL, Fulcher ML, Esther CR Jr, Das J, Dokholyan NV, Randell SH, Boucher RC, Gentzsch M (2014) Potentiator ivacaftor abrogates pharmacological correction of DeltaF508 CFTR in cystic fibrosis. Sci Transl Med 6:246ra96

Chu CS, Trapnell BC, Curristin SM, Cutting G, Crystal RG (1992) Extensive posttranscriptional deletion of the coding sequences for part of nucleotide-binding fold 1 in respiratory epithelial mRNA transcripts of the cystic fibrosis transmembrane conductance regulator gene is not associated with the clinical manifestations of cystic fibrosis. J Clin Invest 90:785–790

Clain J, Fritsch J, Lehmann-Che J, Bali M, Arous M, Goossens M, Edelman A, Fanen P (2001) Two mild cystic fibrosis-associated mutations result in severe cystic fibrosis when combined in cis and reveal a residue important for cystic fibrosis transmembrane conductance regulator processing and function. J Biol Chem 276:9045–9049

Clancy JP, Jain M (2012) Personalized medicine in cystic fibrosis: dawning of a new era. Am J Respir Crit Care Med 186:593–597

Clancy JP, Konstan MW, Rowe SM, Accurso FJ, Zeitlin PL, Hirawat S (2006) A phase II study of PTC124 in CF patients harboring premature stop mutations. North American Cystic Fibrosis Conference, A269

Clancy JP, Rowe SM, Accurso FJ, Aitken ML, Amin RS, Ashlock MA, Ballmann M, Boyle MP, Bronsveld I, Campbell PW, de Boeck K, Donaldson SH, Dorkin HL, Dunitz JM, Durie PR, Jain M, Leonard A, Mccoy KS, Moss RB, Pilewski JM, Rosenbluth DB, Rubenstein RC, Schechter MS, Botfield M, Ordonez CL, Spencer-Green GT, Vernillet L, Wisseh S, Yen K, Konstan MW (2012) Results of a phase IIa study of VX-809, an investigational CFTR corrector compound, in subjects with cystic fibrosis homozygous for the F508del-CFTR mutation. Thorax 67:12–18

Cui L, Aleksandrov L, Chang XB, Hou YX, He L, Hegedus T, Gentzsch M, Aleksandrov A, Balch WE, Riordan JR (2007) Domain interdependence in the biosynthetic assembly of CFTR. J Mol Biol 365:981–994

Cutting GR (2010) Modifier genes in Mendelian disorders: the example of cystic fibrosis. Ann N Y Acad Sci 1214:57–69

Dalemans W, Barbry P, Champigny G, Jallat S, Dott K, Dreyer D, Crystal RG, Pavirani A, Lecocq JP, Lazdunski M (1991) Altered chloride ion channel kinetics associated with the delta F508 cystic fibrosis mutation. Nature 354:526–528

Davis PB, Drumm ML, Konstan MW (1996) Cystic fibrosis. Am J Respir Crit Care Med 1545:1229–1256

Davis PB, Schluchter MD, Konstan MW (2004) Relation of sweat chloride concentration to severity of lung disease in cystic fibrosis. Pediatr Pulmonol 38:204–209

Denning GM, Anderson MP, Amara JF, Marshall J, Smith AE, Welsh MJ (1992) Processing of mutant cystic fibrosis transmembrane conductance regulator is temperature-sensitive. Nature 358:761–764

Devor DC, Bridges RJ, Pilewski J (2000) Pharmacological modulation of ion transport across wild-type and DeltaF508 CFTR-expressing human bronchial epithelia. Am J Physiol 279: C461–C479

Dodge JA, Lewis PA, Stanton M, Wilsher J (2007) Cystic fibrosis mortality and survival in the UK: 1947–2003. Eur Respir J 29:522–526

Donaldson SH, Pilewski J, Griese M (2013) VX-661, an investigational CFTR corrector, in combination with ivacaftor, a CFTR potentiator, in patients with CF homozygous for the F508del-CFTR mutation, interim analysis. European Cystic Fibrosis Conference, Lisbon, Portugal

Dormer RL, Harris CM, Clark Z, Pereira MM, Doull IJ, Norez C, Becq F, Mcpherson MA (2005) Sildenafil (Viagra) corrects DeltaF508-CFTR location in nasal epithelial cells from patients with cystic fibrosis. Thorax 60:55–59

Du M, Jones JR, Lanier J, Keeling KM, Lindsey JR, Tousson A, Bebok Z, Whitsett JA, Dey CR, Colledge WH, Evans MJ, Sorscher EJ, Bedwell DM (2002) Aminoglycoside suppression of a premature stop mutation in a Cftr-/- mouse carrying a human CFTR-G542X transgene. J Mol Med 80:595–604

Dugueperoux I, Bellis G, Lesure JF, Renouil M, Flodrops H, de Braekeleer M (2003) Cystic fibrosis at the Reunion Island (France): spectrum of mutations and genotype-phenotype for the Y112X mutation. J Cyst Fibros 3:185–188

Egan ME, Pearson M, Weiner SA, Rajendran V, Rubin D, Glockner-Pagel J, Canny S, Du K, Lukacs GL, Caplan MJ (2004) Curcumin, a major constituent of turmeric, corrects cystic fibrosis defects. Science 304:600–602

Elkins MR, Robinson M, Rose BR, Harbour C, Moriarty CP, Marks GB, Belousova EG, Xuan W, Bye PT (2006) A controlled trial of long-term inhaled hypertonic saline in patients with cystic fibrosis. N Engl J Med 354:229–240

Engelhardt JF, Yankaskas JR, Ernst SA, Yang Y, Marino CR, Boucher RC, Cohn JA, Wilson JM (1992) Submucosal glands are the predominant site of CFTR expression in the human bronchus. Nat Genet 2:240–248

Fan-Minogue H, Bedwell DM (2008) Eukaryotic ribosomal RNA determinants of aminoglycoside resistance and their role in translational fidelity. RNA 14:148–157

Flume PA, Liou TG, Borowitz DS, Li H, Yen K, Ordonez CL, Geller DE (2012) Ivacaftor in subjects with cystic fibrosis who are homozygous for the F508del-CFTR mutation. Chest 142:718–724
Frischmeyer PA, Dietz HC (1999) Nonsense-mediated mRNA decay in health and disease. Hum Mol Genet 8:1893–1900
Gonzalez JE, Oades K, Leychkis Y, Harootunian A, Negulescu PA (1999) Cell-based assays and instrumentation for screening ion-channel targets. Drug Discov Today 4:431–439
Griesenbach U, Alton EW (2012) Progress in gene and cell therapy for cystic fibrosis lung disease. Curr Pharm Des 18:642–662
Grove DE, Rosser MF, Ren HY, Naren AP, Cyr DM (2009) Mechanisms for rescue of correctable folding defects in CFTRDelta F508. Mol Biol Cell 20:4059–4069
Grubb BR, Gabriel SE, Mengos A, Gentzsch M, Randell SH, Van Heeckeren AM, Knowles MR, Drumm ML, Riordan JR, Boucher RC (2006) SERCA pump inhibitors do not correct biosynthetic arrest of deltaF508 CFTR in cystic fibrosis. Am J Respir Cell Mol Biol 34:355–363
Haardt M, Benharouga M, Lechardeur D, Kartner N, Lukacs GL (1999) C-terminal truncations destabilize the cystic fibrosis transmembrane conductance regulator without impairing its biogenesis. A novel class of mutation. J Biol Chem 274:21873–21877
Harada K, Okiyoneda T, Hashimoto Y, Ueno K, Nakamura K, Yamahira K, Suhgahara T, Shuto T, Wada I, Suico MA, Kaoi H (2006) Calreticulin negatively regulates the cell surface expression of cystic fibrosis transmembrane conductance regulator. J Biol Chem 281:12841–12848
He L, Aleksandrov LA, Cui L, Jensen TJ, Nesbitt KL, Riordan JR (2010) Restoration of domain folding and interdomain assembly by second-site supressors of the delta F508 mutation in CFTR. EMBO J 24:3103–3112
Howard M, Frizzell RA, Bedwell DM (1996) Aminoglycoside antibiotics restore CFTR function by overcoming premature stop mutations. Nat Med 2:467–469
Hwang TC, Wang F, Yang IC, Reenstra WW (1997) Genistein potentiates wild-type and delta F508-CFTR channel activity. Am J Physiol Cell Physiol 273:C988–C998
Iannitti T, Palmieri B (2011) Clinical and experimental applications of sodium phenylbutyrate. Drugs R D 11:227–249
Illek B, Fischer H, Santos GF, Widdicombe JH, Machen TE, Reenstra WW (1995) cAMP-independent activation of CFTR Cl channels by the tyrosine kinase inhibitor genistein. Am J Physiol Cell Physiol 268:C886–C893
Jensen TJ, Loo MA, Pind S, Williams DB, Goldberg AL, Riordan JR (1995) Multiple proteolytic systems, including the proteosome, contribute to CFTR processing. Cell 83:129–135
Jih KY, Hwang TC (2013) Vx-770 potentiates CFTR function by promoting decoupling between the gating cycle and ATP hydrolysis cycle. Proc Natl Acad Sci U S A 110:4404–4409
Kalid O, Mense M, Fischman S, Shitrit A, Bihler H, Ben-Zeev E, Schutz N, Pedemonte N, Thomas PJ, Bridges RJ, Wetmore DR, Marantz Y, Senderowitz H (2010) Small molecule correctors of F508del-CFTR discovered by structure-based virtual screening. J Comput Aided Mol Des 24:971–991
Kerem B, Rommens JM, Buchanan JA, Markiewicz D, Cox TK, Chakravarti A, Buchwald M, Tsui LC (1989) Identification of the cystic fibrosis gene: genetic analysis. Science 245:1073–1080
Kerem E, Hirawat S, Armoni S, Yaakov Y, Shoseyov D, Cohen M, Nissim-Rafinia M, Blau H, Rivlin J, Aviram M, Elfring GL, Northcutt VJ, Miller LL, Kerem B, Wilschanski M (2008) Effectiveness of PTC124 treatment of cystic fibrosis caused by nonsense mutations: a prospective phase II trial. Lancet 372:719–727
Kreda SM, Mall M, Mengos A, Rochelle L, Yankaskas JR, Riordan JR, Boucher RC (2005) Characterization of wild-type and deltaF508 cystic fibrosis transmembrane conductance regulator in human respiratory epithelia. Mol Biol Cell 16:2154–2167
Levy L, Durie P, Pencharz P, Corey M (1986) Prognostic factors associated with patient survival during nutritional rehabilitation in malnourished children and adolescents with cystic fibrosis. J Pediatr Gastroenterol 5:97–102

Lewis HA, Buchanan SG, Burley SK, Conners K, Dickey M, Dorwart M, Fowler R, Gao X, Guggino WB, Hendrickson WA, Hunt JF, Kearins MC, Lorimer D, Maloney PC, Post KW, Rajashankar KR, Rutter ME, Sauder JM, Shriver S, Thibodeau PH, Thomas P, Zhang M, Zhao X, Emtage S (2004) Structure of nucleotide-binding domain 1 of the cystic fibrosis transmembrane conductance regulator. EMBO J 23:282–293

Lewis DF, Ito Y, Lake BG (2010) Quantitative structure-activity relationships (QSARs) for inhibitors and substrates of CYP2B enzymes: importance of compound lipophilicity in explanation of potency differences. J Enzyme Inhib Med Chem 25:679–684

Loo TW, Bartlett MC, Clarke DM (2004) Thapsigargin or curcumin does not promote maturation of processing mutants of the ABC transporters, CFTR, and P-glycoprotein. Biochem Biophys Res Commun 325:580–585

Loo TW, Bartlett MC, Clarke DM (2005) Rescue of DeltaF508 and other misprocessed CFTR mutants by a novel quinazoline compound. Mol Pharm 2:407–413

Maquat LE (1995) When cells stop making sense: effects of nonsense codons on RNA metabolism in vertebrate cells. RNA 1:453–465

Maquat LE (2002) Nonsense-mediated mRNA decay. Curr Biol 12:R196–R197

Martin R, Mogg AE, Heywood LA, Nitschke L, Burke JF (1989) Aminoglycoside suppression at UAG, UAA and UGA codons in Escherichia coli and human tissue culture cells. Mol Gen Genet 217:411–418

Matsui H, Grubb BR, Tarran R, Randell SH, Gatzy JT, Davis CW, Boucher RC (1998a) Evidence for periciliary liquid layer depletion, not abnormal ion composition, in the pathogenesis of cystic fibrosis airways disease. Cell 95:1005–1015

Matsui H, Randell SH, Peretti SW, Davis CW, Boucher RC (1998b) Coordinated clearance of periciliary liquid and mucus from airway surfaces. J Clin Invest 102:1125–1131

Mcphail GL, Clancy JP (2013) Ivacaftor: the first therapy acting on the primary cause of cystic fibrosis. Drugs Today (Barc) 49:253–260

Mendoza JL, Thomas PJ (2007) Building an understanding of cystic fibrosis on the foundation of ABC transporter structures. J Bioenerg Biomembr 39:499–505

Mendoza JL, Schmidt A, Li Q, Nuvaga E, Barrett T, Bridges RJ, Feranchak AP, Brautigam CA, Thomas PJ (2012) Requirements for efficient correction of DF508 CFTR revealed by analyses of evolved sequences. Cell 148:164–174

Mills AD, Yoo C, Butler JD, Yang B, Verkman AS, Kurth MJ (2010) Design and synthesis of a hybrid potentiator-corrector agonist of the cystic fibrosis mutant protein DeltaF508-CFTR. Bioorg Med Chem Lett 20:87–91

Morphy R, Rankovic Z (2005) Designed multiple ligands. An emerging drug discovery paradigm. J Med Chem 48:6523–6543

Phuan PW, Yang B, Knapp JM, Wood AB, Lukacs GL, Kurth MJ, Verkman AS (2011) Cyanoquinolines with independent corrector and potentiator activities restore DeltaPhe508-cystic fibrosis transmembrane conductance regulator chloride channel function in cystic fibrosis. Mol Pharm 80:683–693

Picciano JA, Ameen N, Grant BD, Bradbury NA (2003) Rme-1 regulates the recycling of the cystic fibrosis transmembrane conductance regulator. Am J Physiol Cell Physiol 285:C1009–C1018

Poschet JF, Timmins GS, Taylor-Cousar JL, Ornatowski W, Fazio J, Perkett E, Wilson KR, Yu HD, de Jonge HR, Deretic V (2007) Pharmacological modulation of cGMP levels by phosphodiesterase 5 inhibitors as a therapeutic strategy for treatment of respiratory pathology in cystic fibrosis. Am J Physiol Lung Cell Mol Physiol 293:L712–L719

Prayle A, Watson A, Fortnum H, Smyth A (2010) Side effects of aminoglycosides on the kidney, ear and balance in cystic fibrosis. Thorax 65:654–658

Protasevich I, Yang Z, Wang C, Atwell S, Zhao X, Emtage S, Wetmore DR, Hunt JF, Brouillette CG (2010) Thermal unfolding studies show the disease causing F508del mutation in CFTR thermodynamically destabilizes nucleotide-binding domain 1. Prot Sci 19:1917–1931

Quinton PM (1983) Chloride impermeability in cystic fibrosis. Nature 301:421–422

Quinton PM (1999) Physiological basis of cystic fibrosis: a historical perspective. Physiol Rev 79: S3–S22
Rabeh WM, Bossard F, Xu H, Okiyoneda T, Bagdany M, Mulvihill CM, Du K, di Bernardo S, Liu Y, Konermann L, Roldan A, Lukacs GL (2012) Correction of both NBD1 energetics and domain interface is required to restore ΔF508 CFTR folding and function. Cell 148:150–163
Ramsey BW, Davies J, Mcelvaney NG, Tullis E, Bell SC, Drevinek P, Griese M, Mckone EF, Wainwright CE, Konstan MW, Moss R, Ratjen F, Sermet-Gaudelus I, Rowe SM, Dong Q, Rodriguez S, Yen K, Ordonez C, Elborn JS (2011) A CFTR potentiator in patients with cystic fibrosis and the G551D mutation. N Engl J Med 365:1663–1672
Ratjen F (2001) Changes in strategies for optimal antibacterial therapy in cystic fibrosis. Int J Antimicrob Agents 17:93–96
Reddy MM, Quinton PM (2001) Selective activation of cystic fibrosis transmembrane conductance regulator Cl^- and HCO_3^- conductances. JOP 2:212–218
Ren HY, Grove DE, de la Rosa O, Houck SA, Sopha P, Van Goor F, Hoffman BJ, Cyr DM (2013) VX-809 corrects folding defects in cystic fibrosis transmembrane conductance regulator protein through action on membrane-spanning domain 1. Mol Biol Cell 24:3016–3024
Riordan JR, Rommens JM, Kerem B, Alon N, Rozmahel R, Grzelczak Z, Zielenski J, Lok S, Plavsic N, Chou JL et al (1989) Identification of the cystic fibrosis gene: cloning and characterization of complementary DNA. Science 245:1066–1073
Rogan MP, Stoltz DA, Hornick DB (2011) Cystic fibrosis transmembrane conductance regulator intracellular processing, trafficking, and opportunities for mutation-specific treatment. Chest 139:1480–1490
Rosser MF, Grove DE, Chen L, Cyr DM (2008) Assembly and misassembly of cystic fibrosis transmembrane conductance regulator: folding defects caused by deletion of F508 occur before and after the calnexin-dependent association of membrane spanning domain (MSD) 1 and MSD2. Mol Biol Cell 19:4570–4579
Rowe SM, Clancy JP (2009) Pharmaceuticals targeting nonsense mutations in genetic diseases: progress in development. BioDrugs 23:165–174
Rowe SM, Verkman AS (2014) Cystic fibrosis transmembrane regulator correctors and potentiators. In: Riordan JR, Boucher RC, Quinton PM (eds) Cold Spring Harbor perspectives in medicine. Cold Spring Harbor Laboratory Press, Cold Spring Harbor, NY
Rowe SM, Miller S, Sorscher E (2005) Cystic fibrosis. N Engl J Med 352:1992–2001
Rowe SM, Accurso FJ, Clancy JP (2007) Detection of cystic fibrosis transmembrane conductance regulator activity in early phase clinical trials. Proc Am Thorac Soc 4:387–398
Rowntree RK, Harris A (2003) The phenotypic consequences of CFTR mutations. Am Hum Genet 67:471–485
Rubenstein RC, Zeitlin PL (1998) A pilot clinical trial of oral sodium 4-phenylbutyrate (Buphenyl) in deltaF508-homozygous cystic fibrosis patients: partial restoration of nasal epithelial CFTR function. Am J Respir Crit Care Med 157:484–490
Rubenstein RC, Egan ME, Zeitlin PL (1997) In vitro pharmacologic restoration of CFTR-mediated chloride transport with sodium 4-phenylbutyrate in cystic fibrosis epithelial cells containing delta F508-CFTR. J Clin Invest 100:2457–2465
Sato S, Ward CL, Krouse ME, Wine JJ, Kopito RR (1996) Glycerol reverses the misfolding phenotype of the most common cystic fibrosis mutation. J Biol Chem 271:635–638
Sermet-Gaudelus I, Boeck KD, Casimir GJ, Vermeulen F, Leal T, Mogenet A, Roussel D, Fritsch J, Hanssens L, Hirawat S, Miller NL, Constantine S, Reha A, Ajayi T, Elfring GL, Miller LL (2010) Ataluren (PTC124) induces cystic fibrosis transmembrane conductance regulator protein expression and activity in children with nonsense mutation cystic fibrosis. Am J Respir Crit Care Med 182:1262–1272
Serohijos AW, Hegedus T, Aleksandrov AA, He L, Cui L, Dokholyan NV, Riordan JR (2008) Phenylalanine-508 mediates a cytoplasmic-membrane domain contact in the CFTR 3D structure crucial to assembly and channel function. Proc Natl Acad Sci U S A 105:3256–3261

Sharma M, Pampinella F, Nemes C, Benharouga M, So J, Du K, Bache KG, Papsin B, Zerangue N, Stenmark H, Lukacs GL (2004) Misfolding diverts CFTR from recycling to degradation: quality control at early endosomes. J Cell Biol 164:923–933

Sheppard DN, Rich DP, Ostedgaard LS, Gregory RJ, Smith AE, Welsh MJ (1993) Mutations in CFTR associated with mild-disease form Cl^- channels with altered pore properties. Nature 362:160–164

Shoshani T, Augarten A, Gazit E, Bashan N, Yahav Y, Rivlin Y, Tal A, Seret H, Yaar L, Kerem E, Bat-Sheva K (1992) Association of a nonsense mutation (W1282X), the most common mutation in the Ashkenazi Jewish cystic fibrosis patients in Israel, with presentation of severe disease. Am J Hum Genet 50:222–228

Silvis MR, Picciano JA, Bertrand C, Weixel K, Bridges RJ, Bradbury NA (2003) A mutation in the cystic fibrosis transmembrane conductance regulator generates a novel internalization sequence and enhances endocytic rates. J Biol Chem 278:11554–11560

Song Y, Sonawane ND, Salinas D, Qian L, Pedemonte N, Galietta LJ, Verkman AS (2004) Evidence against the rescue of defective DeltaF508-CFTR cellular processing by curcumin in cell culture and mouse models. J Biol Chem 279:40629–40633

Trujillano D, Ramos MD, González J, Tornador C, Sotillo F, Escaramis G, Ossowski S, Armengol L, Casals T, Estivill X (2013) Next generation diagnostics of cystic fibrosis and CFTR-related disorders by targeted multiplex high-coverage resequencing of CFTR. J Med Genet 50:455–462

Van Goor F, Hadida S, Grootenhuis PD, Burton B, Cao D, Neuberger T, Turnbull A, Singh A, Joubran J, Hazlewood A, Zhou J, Mccartney J, Arumugam V, Decker C, Yang J, Young C, Olson ER, Wine JJ, Frizzell RA, Ashlock M, Negulescu P (2009) Rescue of CF airway epithelial cell function in vitro by a CFTR potentiator, VX-770. Proc Natl Acad Sci U S A 106:18825–18830

Van Goor F, Hadida S, Grootenhuis PD, Burton B, Stack JH, Straley KS, Decker CJ, Miller M, Mccartney J, Olson ER, Wine JJ, Frizzell RA, Ashlock M, Negulescu PA (2011) Correction of the F508del-CFTR protein processing defect in vitro by the investigational drug VX-809. Proc Natl Acad Sci U S A 108:18843–18848

Veit G, Avramescu RG, Perdomo D, Phuan PW, Bagdany M, Apaja PM, Borot F, Szollosi D, Wu YS, Finkbeiner WE, Hegedus T, Verkman AS, Lukacs GL (2014) Some gating potentiators, including VX-770, diminish DeltaF508-CFTR functional expression. Sci Transl Med 6:246ra97

Ward CL, Kopito RR (1994) Intracellular turnover of cystic fibrosis transmembrane conductance regulator. Inefficient processing and rapid degradation of wild-type and mutant proteins. J Biol Chem 269:25710–25718

Ward CL, Omura S, Kopito RR (1995) Degradation of CFTR by the ubiquitin-proteasome pathway. Cell 83:121–127

Weinreich F, Wood PG, Riordan JR, Nagel G (1997) Direct action of genistein on CFTR. Pflugers Arch 434:484–491

Welch EM, Barton ER, Zhuo J, Tomizawa Y, Friesen WJ, Trifillis P, Paushkin S, Patel M, Trotta CR, Hwang S, Wilde RG, Karp G, Takasugi J, Chen G, Jones S, Ren H, Moon YC, Corson D, Turpoff AA, Campbell JA, Conn MM, Khan A, Almstead NG, Hedrick J, Mollin A, Risher N, Weetall M, Yeh S, Branstrom AA, Colacino JM, Babiak J, Ju WD, Hirawat S, Northcutt VJ, Miller LL, Spatrick P, He F, Kawana M, Feng H, Jacobson A, Peltz SW, Sweeney HL (2007) PTC124 targets genetic disorders caused by nonsense mutations. Nature 447:87–91

Welsh M, Ramsey B (2001) Cystic fibrosis. In: Scriver C, Beaudet A, Valle D (eds) The metabolic and molecular basis of inherited disease. McGraw-Hill, New York, NY

Welsh MJ, Smith AE (1993) Molecular mechanisms of CFTR chloride channel dysfunction in cystic fibrosis. Cell 73:1251–1254

Wilschanski M, Famini C, Blau H, Rivlin J, Augarten A, Avital A, Kerem B, Kerem E (2000) A pilot study of the effect of gentamicin on nasal potential difference measurements in cystic fibrosis patients carrying stop mutations. Am J Respir Crit Care Med 161:860–865

Wilschanski M, Yahav Y, Yaacov Y, Blau H, Bentur L, Rivlin J, Aviram M, Bdolah-Abram T, Bebok Z, Shushi L, Kerem B, Kerem E (2003) Gentamicin-induced correction of CFTR function in patients with cystic fibrosis and CFTR stop mutations. N Engl J Med 349:1433–1441

Wilschanski M, Dupuis A, Ellis L, Jarvi K, Zielenski J, Tullis E, Martin S, Corey M, Tsui LC, Durie P (2006) Mutations in the cystic fibrosis transmembrane regulator gene and in vivo transepithelial potentials. Am J Respir Crit Care Med 174:787–794

Yu H, Burton B, Huang CJ, Worley J, Cao D, Johnson JP Jr, Urrutia A, Joubran J, Seepersaud S, Sussky K, Hoffman BJ, Van Goor F (2012) Ivacaftor potentiation of multiple CFTR channels with gating mutations. J Cyst Fibros 11:237–245

Chapter 25
Renal Aquaporins in Health and Disease

Marleen L.A. Kortenoeven, Emma T.B. Olesen, and Robert A. Fenton

Abstract Aquaporins (AQPs) are a large family of membrane proteins that act as semipermeable channels. The majority of AQPs are permeable to water, but a subset of the family can also transport glycerol, urea, and other small solutes. Currently, 13 AQP homologues have been identified in mammals, termed AQP0–12. These aquaporins are highly abundant in epithelial cells and non-epithelial cells in various tissues including the kidney, brain, liver, lungs, and salivary glands. In this chapter we focus on AQPs expressed in kidney epithelial cells. We summarize the current knowledge with respect to their localization and function within the kidney tubule and their critical role in mammalian water homeostasis. We describe a number of water balance disorders resulting from altered AQP function and provide an overview of some of the treatment strategies for these disorders.

Keywords Urinary-concentrating mechanism • Nephrogenic diabetes insipidus • NDI • Hyponatremia • Vasopressin • Hypernatremia • Polyuria • Knockout mouse

25.1 Introduction

Aquaporins (AQPs) are a large family of small membrane proteins (approximately 30 kDa) that act as semipermeable channels. To date, 13 aquaporins have been identified in mammals, named AQP0–12, which are expressed in various tissues such as the kidney, brain, liver, lungs, and salivary glands (Ishibashi et al. 2009).

M.L.A. Kortenoeven • E.T.B. Olesen
Department of Biomedicine, Aarhus University, Aarhus, Denmark

Center for Interactions of Proteins in Epithelial Transport (InterPrET), Aarhus University, Aarhus, Denmark

R.A. Fenton (✉)
Center for Interactions of Proteins in Epithelial Transport (InterPrET), Aarhus University, Aarhus, Denmark

Department of Biomedicine, Aarhus University, Wilhelm Meyers Alle 3, BLDG 1234, Aarhus 8000, Denmark
e-mail: rofe@biomed.au.dk

K.L. Hamilton, D.C. Devor (eds.), *Ion Channels and Transporters of Epithelia in Health and Disease*, Physiology in Health and Disease,
DOI 10.1007/978-1-4939-3366-2_25

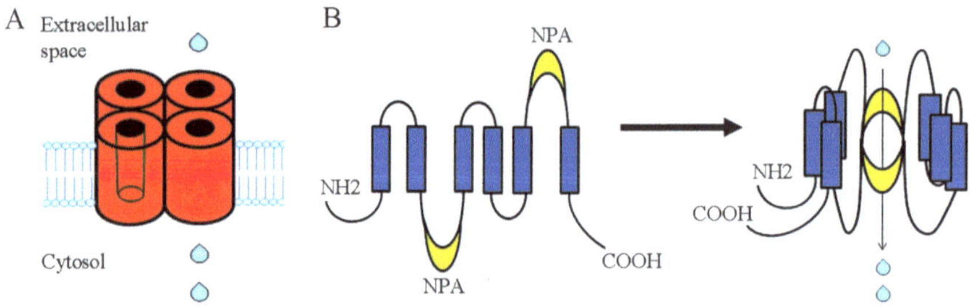

Fig. 25.1 Schematic representation of the structural organization of aquaporins. (**a**) Four AQP monomers assemble to form tetramers, which are the functional units in the membrane. Each AQP monomer has a central water-transporting pore. (**b**) A single monomer of an AQP channel consists of six membrane-spanning alpha-helices and has short intracellular amino and carboxyl termini. Almost all AQPs have two highly conserved Asn-Pro-Ala (NPA) motifs in each monomer. The NPA sequences have been proposed to interact in the membrane to form the pathway for translocation of water across the plasma membrane. Modified from Moeller HB, Fenton RA (2012) Cell biology of vasopressin-regulated aquaporin-2 trafficking. Pflugers Arch 464:133–144

This chapter specifically focuses on renal AQPs. Aquaporins have also been identified and cloned from amphibians, plants, yeast, bacteria, and various lower organisms, where they likely fulfill a variety of functions. A single monomer of an AQP channel consists of six membrane-spanning alpha-helices that have a central water-transporting pore (see Fig. 25.1) (Murata et al. 2000; Sui et al. 2001). AQPs have short intracellular amino and carboxyl termini, which contain major regulatory domains for channel function. Almost all AQPs have two highly conserved Asn-Pro-Ala (NPA) motifs in each monomer. Structural analysis of AQP1 has shown that these motifs reside on opposite sides of the AQP1 monomer and are important for water selective pore formation (Murata et al. 2000). The two most recently identified mammalian AQPs, AQP11 and AQP12, have a conserved C-terminal NPA motif, but the N-terminal NPA motif is deviated to an Asn-Pro-Cys (NPC) motif for AQP11 and to an Asn-Pro-Thr (NPT) motif for AQP12 (Ishibashi et al. 2014). The function of AQP12 is still unknown, but AQP11 has been shown to have water channel activity, despite the incompletely conserved NPA, although this is relatively low (Yakata et al. 2007, 2011). Four AQP monomers assemble to form tetramers, which are the functional units in the membrane. Precisely where these tetramers form is unknown, and most likely formation occurs differently for each AQP family member. For one of the mammalian AQPs, AQP4, these tetramers further assemble into supramolecular square arrays (Yang et al. 1996), which may further influence the activity and molecular regulation of AQP4 (Fenton et al. 2010; Silberstein et al. 2004).

Water, if given enough time, can cross the majority of lipid bilayers from biological membranes by simple diffusion. However, this process is relatively slow and energetically unfavorable. The water permeability of the lipid membrane is greatly enhanced by insertion of AQPs, which provide a semi-selective pore through which molecules can rapidly translocate. Although termed AQPs based on their initial characterization, besides water, a subset of the mammalian AQPs, the aquaglyceroporins AQP3, AQP7, AQP9, and AQP10, can transport glycerol and urea [(Ishibashi et al. 1994, 1997a, 2002; Ko et al. 1999), reviewed in Rojek et al. (2008)], while both AQP7 and AQP9 are permeable for arsenite (Liu et al. 2002). In addition, AQP9 has been shown to be permeable to a wide range of non-charged solutes, like mannitol, sorbitol, purines, and pyrimidines (Tsukaguchi et al. 1998). AQP6 may function primarily as an anion transporter (Ikeda et al. 2002; Yasui et al. 1999a). AQPs have also been proposed to transport other small molecules and gases, including carbon dioxide, ammonia, nitric oxide, and hydrogen peroxide (Miller et al. 2010; Musa-Aziz et al. 2009; Wang and Tajkhorshid 2010). Besides functioning as channels for water and small solutes, AQPs may also play other roles in the cell, including a structural role in, e.g., cell-to-cell adhesion. For example, AQP0, a lens-specific AQP molecule, forms single-layered as well as double-layered crystals, each layer in an opposing membrane, suggesting that AQP0 has an adhesive role in the ocular lens fiber cells (Fotiadis et al. 2000). In addition, AQP2 interacts with integrins to promote renal epithelial cell migration, contributing to the structural and functional integrity of the mammalian kidney (Chen et al. 2012).

25.2 Renal Aquaporins

Nine of the known AQPs are expressed in the kidney, with various cellular and subcellular distributions (Fig. 25.2). AQP1 is constitutively localized in the plasma membranes of proximal tubule cells, descending thin limbs of Henle cells, and endothelial cells from the descending vasa recta (Nielsen et al. 1993b, 1995b; Zhai et al. 2007). AQP2, AQP3, and AQP4 are expressed in cells from the connecting tubule, collecting duct principal cells, and inner medulla collecting duct cells. Whereas AQP2 is predominantly expressed in the apical plasma membrane and membranes of intracellular vesicles (it traffics between the two compartments; see later), AQP3 and AQP4 are constitutively localized to the basolateral plasma membrane (Ecelbarger et al. 1995; Fushimi et al. 1993; Nielsen et al. 1993a; Terris et al. 1995). AQP5 is expressed in the apical membrane of intercalated cells in the connecting tubule and the cortical part of the collecting duct (Procino et al. 2011b), cells which are predominantly involved in acid-base balance rather than water transport. AQP7 is localized in the apical plasma membrane of cells from the S3 segment of proximal tubules (Ishibashi et al. 2000; Nejsum et al. 2000). In contrast to these AQPs, AQP6, AQP8, and AQP11 are not detected in the plasma membrane but are localized to intracellular membranes only. AQP6 is localized in membrane-bound vesicles within intercalated cells of the connecting tubule and collecting duct (Ohshiro et al. 2001; Yasui et al. 1999b). AQP8 and AQP11 are both localized intracellularly in proximal tubule cells, while AQP8 has also been detected in the collecting ducts (Elkjaer et al. 2001; Morishita et al. 2005).

In addition to removal of various waste substances from the body, maintaining acid-base balance and tight control of plasma electrolyte concentrations, the kidney plays a critical role in regulation of body water content. Body water balance is

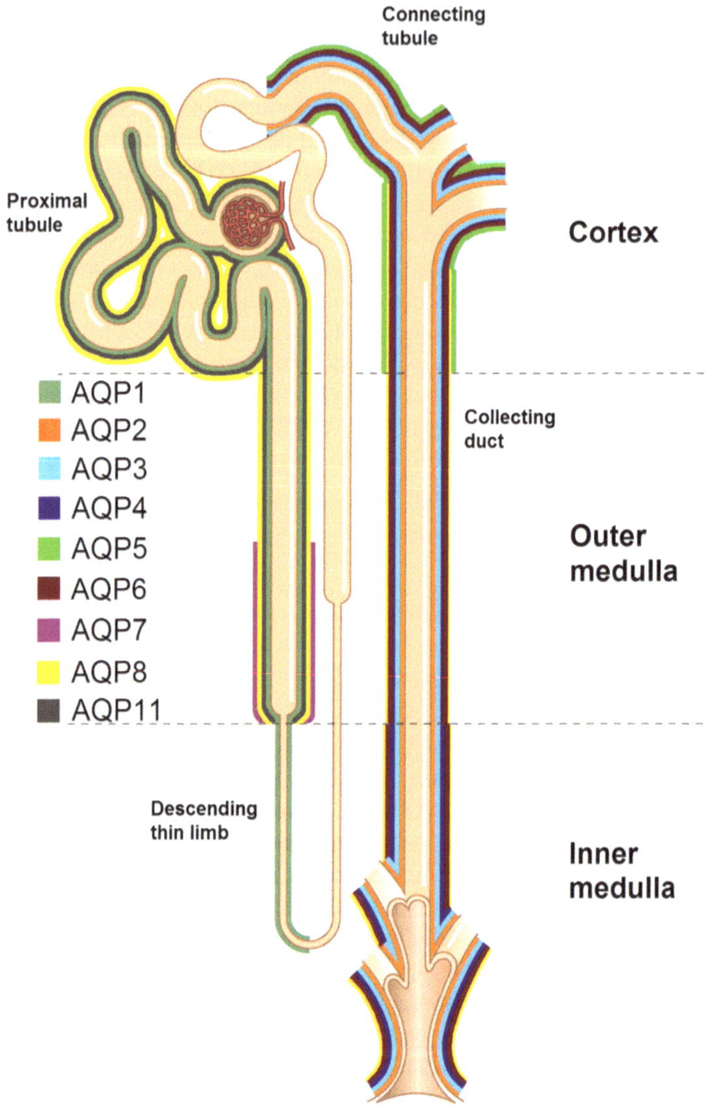

Fig. 25.2 Expression of aquaporins along the nephron. Blood is filtered at the glomerulus, and the filtrate is modified as it travels through the nephron to make the final urine. Most of the glomerular filtrate is reabsorbed through AQP1 in the proximal tubule and descending thin limbs of Henle, although AQP7 is also expressed in the S3 segment of the proximal tubule. AQP1 is also expressed in the descending vasa recta, facilitating the removal of water. In the connecting tubule and collecting duct, AQP2 is mainly expressed at the apical membrane and intracellular vesicles of principal cells, while AQP3 and AQP4 are present at the basolateral membrane, representing exit pathways for water reabsorbed via AQP2. AQP5 is expressed in the apical membrane of intercalated cells in the connecting tubule and the cortical collecting duct. In contrast to these AQPs, AQP6, AQP8, and AQP11 are localized in intracellular membranes only. AQP6 is localized to intercalated cells of the collecting duct and connecting tubule, AQP8 is expressed in proximal tubules and weakly in collecting ducts, while AQP11 is localized to proximal tubules. Modified

tightly controlled by regulating both water intake and urinary water excretion. In the kidney, 180 L of plasma is filtered by the human glomeruli each day, of which less than 1 % is finally excreted in the urine (Stanton and Koeppen 1998). Approximately 67 % of the filtered water is reabsorped in the proximal tubule and 15 % in the descending thin limb of Henle, which are both constitutive processes. Depending on the body's needs, the remaining fluid can be reabsorbed in the connecting tubule and collecting duct, defining the final urine concentration. The processes of water reabsorption are tightly regulated by a variety of hormones and systemic and local factors. A major regulator of body water balance is the peptide hormone arginine vasopressin (AVP), the effects of which allow the body to adapt to periods of water load or water restriction (Stanton and Koeppen 1998). Of the nine AQPs expressed in the kidney, only AQP1–4 and possibly AQP7 have a role in the process of urine concentration and maintenance of body water balance. In the following section we provide an overview of each of the renal AQPs, focusing on their role in renal epithelial cell and kidney function. Readers are directed toward an extensive catalogue of reviews that provide more details of individual AQP function and regulation (Badaut et al. 2014; Delporte 2014; Ishibashi et al. 2009, 2014; Rojek et al. 2008; Schey et al. 2014; Verkman 2008).

25.2.1 Aquaporin 1

AQP1 is the first molecularly identified aquaporin and was originally cloned from erythrocytes. AQP1 is expressed in the apical and basolateral membrane of epithelial cells in the proximal tubule (Nielsen et al. 1993b), where the majority of fluid filtered by the glomerulus is reabsorbed by an active near-isosmolar transport mechanism. AQP1 is also expressed in the descending thin limb of long-looped nephrons, but not short-looped nephrons, and the descending vasa recta (Nielsen et al. 1995b; Zhai et al. 2007). Recently, it was shown that AQP1 is also expressed in the thick ascending limb of the loop of Henle (TAL) of mice and rats, where it is localized both intracellularly and on the basolateral plasma membrane (Cabral and Herrera 2012). Although AQP1 expression is constitutively high, it can be modulated by hypertonicity and angiotensin II (Bouley et al. 2009). The antidiuretic hormone AVP does not regulate AQP1 expression in the proximal tubule, descending thin limb or vasa recta; however, the expression in the TAL was shown to be increased in rats by chronic infusion of the synthetic vasopressin analogue dDAVP (1-desamino-8-D-arginine vasopressin, desmopressin), which could be either direct or secondary to an increased interstitial osmolality (Cabral and Herrera 2012; Terris et al. 1996).

Fig. 25.2 (continued) from Kortenoeven ML, Fenton RA (2014) Renal aquaporins and water balance disorders. Biochim Biophys Acta 1840:1533–1549

AQP1 knockout mice have a reduced urine osmolality and increased urine volume compared with wild-type mice, and the osmolality is not increased in response to injection of dDAVP (Ma et al. 1998; Schnermann et al. 1998). In addition, AQP1 knockout mice also do not increase urine osmolality in response to water deprivation and become severely dehydrated and lethargic, altogether showing that AQP1 plays an essential role in maintaining body water balance. AQP1 deletion produces a 78 % decrease in osmotic water permeability across the proximal tubule epithelium, indicating that the major pathway for osmotically driven water transport in the proximal tubule is transcellular and mediated by AQP1 water channels (Schnermann et al. 1998). In addition, the rate of proximal fluid reabsorption in AQP1 knockout mice is reduced by approximately 50 % relative to control mice. Osmotic water permeabilities of microperfused thin descending limb of Henle and outer medullary descending vasa recta are reduced in AQP1 knockout mice as well (Chou et al. 1999; Pallone et al. 2000). However, micropuncture flow measurements of distal nephron segments showed that the distal fluid delivery is normal in AQP1 knockout mice, because of a decrease in the single-nephron glomerular filtration rate, probably caused by activation of the tubuloglomerular feedback mechanism (Schnermann et al. 1998). The increased urinary output, despite normal distal delivery, suggests that the urine-concentrating defect seen in AQP1 knockout mice results primarily from reduced fluid absorption in the collecting duct. The decreased water permeability in the thin limbs and vasa recta is predicted to reduce countercurrent multiplication and counter current exchange, respectively, which will prevent the formation of a hypertonic medullary interstitium. This conclusion is supported by the absence of an increase in urine osmolality in AQP1 knockout mice after dDAVP stimulation of collecting duct water permeability (Ma et al. 1998). Consistent with the phenotype of AQP1 knockout mice, humans with a loss-of-function mutation in AQP1 are unable to maximally concentrate their urine when challenged by water deprivation (King et al. 2001).

AQP1 in the TAL is unlikely involved in the process of urine concentration, as the TAL is impermeable to water, indicating an absence of water channels at the apical membrane. However, TAL cells rapidly shrink and/or swell in response to acute changes in basolateral osmolality. Compared with wild-type mice, the rate of water flux across the basolateral side of TALs was reduced by around 50 % in tubules isolated from AQP1 knockout mice (Cabral and Herrera 2012). AQP1 could play an important role in the regulation of TAL cell volume during changes in interstitial osmolality, such as during a high-salt diet or water deprivation.

25.2.2 Aquaporin 2

AQP2 is abundantly expressed in the plasma membrane and intracellular membrane vesicles of the principal cells along the whole connecting tubule and collecting duct (Fushimi et al. 1993; Nielsen et al. 1993a). Although AQP2 is mainly associated with the apical plasma membrane, AQP2 is also found in the basolateral plasma membrane, especially in the connecting tubule and inner medulla collecting duct (Christensen et al. 2003). AQP2 localization and abundance are regulated by several hormones and signaling molecules, including AVP. During hypernatremia, hypovolemia, or hypotension, AVP is released from the posterior pituitary gland into the bloodstream (Bankir 2001; Baylis 1989; Voisin and Bourque 2002). AVP regulates the body's retention of water, increasing both the osmotic driving force for water reabsorption and the transcellular route for water transport, causing the kidneys to concentrate the urine. The transcellular route of regulation mainly occurs via modulating plasma membrane expression of AQP2. AVP interacts with the vasopressin type-2 receptor (V2R), present in the basolateral membrane of renal connecting tubule and collecting duct principal cells (Fenton et al. 2007; Mutig et al. 2007). This initiates a signal transduction cascade, involving Gs protein-mediated activation of adenylate cyclase, a rise in intracellular cAMP, activation of protein kinases, and subsequent phosphorylation of AQP2 water channels. This results in the redistribution of AQP2 from intracellular storage vesicles to the apical membrane (Katsura et al. 1997; Kuwahara et al. 1995; Nielsen et al. 1995a), greatly increasing the rate-limiting water permeability of the apical plasma membrane (see Fig. 25.3) (Flamion and Spring 1990; Strange and Spring 1987). Water will now enter the principal cells through AQP2 driven by the transcellular osmotic gradient of sodium and (in the medulla) urea and will leave the cells through AQP3 and AQP4 in the basolateral membrane (Ecelbarger et al. 1995; Terris et al. 1995), leading to a concentration of the urine. Once the water balance is restored, reduced AVP levels will result in AQP2 internalization, which returns the apical membrane water permeability to basal levels. AQP2 is retrieved to intracellular vesicles, resulting in either lysosomal degradation or intracellular storage and relocation to the plasma membrane upon restimulation with AVP (Kamsteeg et al. 2006; Katsura et al. 1996).

A number of posttranslational modifications of AQP2 are required for the trafficking of AQP2 to/from the apical plasma membrane [for extensive review, see Moeller et al. (2011)]. AQP2 can be phosphorylated at five residues, Thr244, Ser256, Ser261, Ser264, and Ser269 (Thr269 in human AQP2; see Fig. 25.4) (Hoffert et al. 2006, 2012). The role of Thr244 phosphorylation is unknown. Phosphorylation of Ser256, Ser264, and Ser269 are increased in abundance upon dDAVP stimulation, while Ser261 phosphorylation is decreased (Fenton et al. 2008; Hoffert et al. 2007, 2008). Phosphorylated Ser261-AQP2 is mainly found in intracellular vesicles and seems to be absent from the plasma membrane (Hoffert et al. 2007). Ser261 phosphorylation is thought to stabilize AQP2 ubiquitination (Tamma et al. 2011). Phosphorylated Ser256-AQP2 and Ser264-AQP2 reside in intracellular vesicles and the plasma membrane, and upon stimulation with dDAVP, there is a change in distribution (or increase in abundance) to the plasma membrane (Christensen et al. 2000; Fenton et al. 2008). The function of Ser264-AQP2 phosphorylation is currently unknown. Ser256 is phosphorylated by protein kinase A (PKA) upon AVP stimulation and phosphorylation at this site is essential for translocation to the plasma membrane (Fushimi et al. 1997; Katsura et al. 1997;

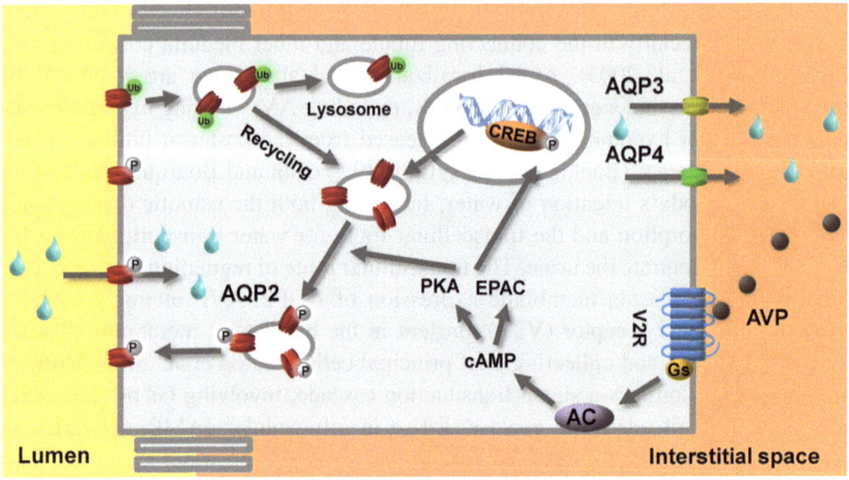

Fig. 25.3 Regulation of AQP2-mediated water reabsorption. AVP binds to the vasopressin type-2 receptor (V2R) at the basolateral membrane of collecting duct and collecting duct principal cells. This induces a signaling cascade, involving Gs protein-mediated activation of adenylate cyclase (AC), a rise in intracellular cAMP, activation of protein kinase A (PKA), and possibly the exchange factor directly activated by cAMP (EPAC) and subsequent phosphorylation of AQP2. This results in the redistribution of AQP2 from intracellular vesicles to the apical membrane. Driven by the transcellular osmotic gradient, water will then enter principal cells through AQP2 and leave the cell via AQP3 and AQP4, which are expressed in the basolateral membrane, resulting in concentrated urine. In the long term, vasopressin also increases AQP2 expression via phosphorylation of the cAMP responsive element-binding protein (CREB), which stimulates transcription from the AQP2 promoter. Once the water balance is restored, AVP levels drop and AQP2 is internalized via ubiquitination. Internalized AQP2 can either be targeted to recycling pathways or to degradation via lysosomes. Modified from Kortenoeven ML, Fenton RA (2014) Renal aquaporins and water balance disorders. Biochim Biophys Acta 1840:1533–1549

Van Balkom et al. 2002). A study using oocytes as a model system for exogenous AQP2 expression showed that Ser256 phosphorylation of at least three of four monomers of an AQP2 tetramer is needed to redistribute AQP2 tetramers from storage vesicles to the apical plasma membrane (Kamsteeg et al. 2000). Phosphorylated Ser269-AQP2 is solely found in the apical plasma membrane (Moeller et al. 2009). Ser269 phosphorylation is thought to be involved in apical plasma membrane retention of AQP2 (Hoffert et al. 2008) and interaction of AQP2 with proteins participating in the endocytic machinery (Moeller et al. 2010). In addition to phosphorylation, AQP2 is also modified by ubiquitination at Lys270 (Kamsteeg et al. 2006). Ubiquitination of AQP2 is believed to mediate the endocytosis of AQP2 from the plasma membrane upon AVP removal. Recently it was shown that, although AQP2 Ser269 phosphorylation and Lys270 ubiquitination can occur in parallel, the effects of phosphorylation can override the effect of ubiquitination on AQP2 endocytosis, possibly by decreasing the interaction of AQP2 with proteins of the endocytic machinery (Moeller et al. 2014).

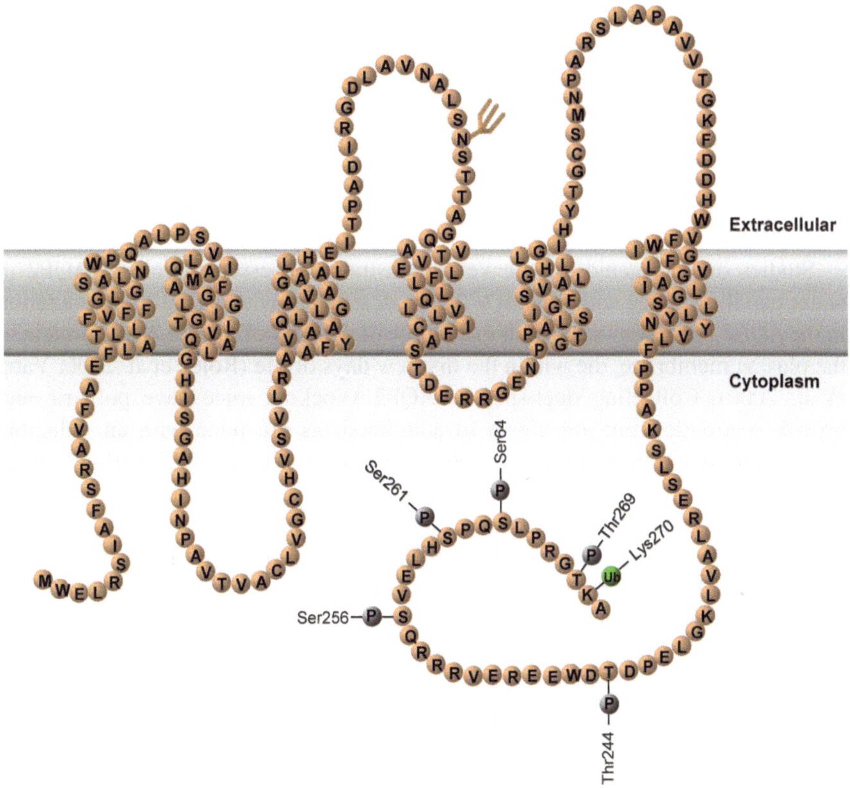

Fig. 25.4 Schematic representation of the human AQP2 protein. AQP2 is a 271-amino acid protein that is predicted to have six transmembrane domains with an intracellular N- and C-terminus. The known phosphorylation (P) and ubiquitination (Ub) sites are indicated. Modified from Kortenoeven ML, Fenton RA (2014) Renal aquaporins and water balance disorders. Biochim Biophys Acta 1840:1533–1549

In addition to playing an acute role in AQP2 translocation, the AVP-mediated increase in cAMP also increases AQP2 abundance via PKA-mediated phosphorylation of the cAMP responsive element-binding protein (CREB), which stimulates transcription from the AQP2 promoter (Hozawa et al. 1996; Matsumura et al. 1997; Yasui et al. 1997). The PKA-CREB pathway is involved in the initial increase in AQP2 abundance after AVP stimulation, but not in the long-term effect of AVP (Kortenoeven et al. 2012b). Instead, activation of the exchange protein directly activated by cAMP (EPAC) levels may be involved in the long-term regulation of AQP2 (Kortenoeven et al. 2012b). EPAC might also play a role in the regulation of AQP2 localization, as perfusion of mouse inner medullary collecting ducts with an EPAC activator resulted in intracellular Ca^{2+} mobilization and translocation of AQP2 to the apical membrane (Yip 2006).

Several other factors besides AVP stimulate AQP2 expression or translocation, for example, extracellular tonicity or insulin or long-term aldosterone stimulation

(Bustamante et al. 2005; Hasler et al. 2003, 2005), while other factors decrease AVP-induced AQP2 abundance or trafficking, like extracellular purines, prostaglandin E2, and dopamine [Boone et al. 2011; Nejsum et al. 2005; Wildman et al. 2009; Zelenina et al. 2000; for review, see Boone and Deen (2008)]. Although prostaglandin E2 antagonizes AVP-induced water permeability, prostaglandin E2 or selective prostaglandin E2 receptor agonists increase the translocation and abundance of AQP2 in the absence of high levels of AVP (Li et al. 2009; Olesen et al. 2011).

Studies on transgenic mice have demonstrated an essential role of AQP2 in water handling. Mice with total *AQP2* gene deletion or with a Thr126Met mutation in the *AQP2* gene, shown to result in a failure of delivery of mature AQP2 protein to the plasma membrane, die within the first few days of life (Rojek et al. 2006; Yang et al. 2000). Collecting duct-specific AQP2 knockout mice have polyuria and growth retardation but are viable to adulthood, as are mice with an inducible *AQP2* gene deletion resulting in >95 % reduction of renal AQP2 (Rojek et al. 2006; Yang et al. 2006a). Both mouse models have severe polyuria, accompanied by a very low urine osmolality. In these mice, the urine osmolality is not increased after water deprivation. Mice with a specific deletion of AQP2 in the connecting tubules have defective renal water handling under basal conditions, with a 1.5-fold higher urine volume and reduced urine osmolality (Kortenoeven et al. 2013a), indicating that AQP2 in the connecting tubules also plays an important role in regulating body water balance. However, in contrast to the collecting duct-specific knockout, these mice are able to concentrate their urine under conditions of high circulating AVP, suggesting that the connecting tubule plays a minor, yet still significant contribution to water conservation, and the collecting duct has the major role in AVP-mediated antidiuresis. Alternatively, increased water reabsorption in the collecting duct can compensate for dysfunction of the connecting tubule. Although a loss-of-function mutation in the *AQP2* gene is not lethal in humans, it does lead to a severe urinary-concentrating defect known as nephrogenic diabetes insipidus (NDI), confirming the importance of AQP2 in renal water handling (Robben et al. 2006a).

25.2.3 Aquaporin 3

AQP3 is constitutively expressed in the basolateral membrane of the connecting tubule and collecting duct principal cells (Coleman et al. 2000; Ecelbarger et al. 1995; Ishibashi et al. 1997b) where it functions as an exit pathway for water reabsorbed via AQP2. There is no evidence for short-term regulation of AQP3 by AVP-induced trafficking. However, water deprivation or AVP infusion, resulting in long-term increased AVP levels, increases AQP3 protein and mRNA levels both in the cortex and medulla (Ecelbarger et al. 1995; Ishibashi et al. 1997b; Murillo-Carretero et al. 1999; Terris et al. 1996).

AQP3 knockout mice have an increased urine volume and a lower urine osmolality. In addition, osmotic water permeability of the cortical collecting duct basolateral membrane is more than threefold reduced by AQP3 deletion, showing the importance of AQP3 for the urine-concentrating mechanism (Ma et al. 2000). After water deprivation or dDAVP injection, AQP3 knockout mice are however able to concentrate their urine partially to around 30 % of that in wild-type mice, which might be due to the remaining presence of basolateral AQP4 or AQP2 in the principal cells. In agreement with this, AQP3/AQP4 double-knockout mice have a greater impairment of urinary-concentrating ability than AQP3 single-knockout mice (Ma et al. 2000).

Besides water, AQP3 can transport both urea and glycerol. The glycerol-transporting function of AQP3 has been shown to play an important role in skin hydration (Rojek et al. 2008), and AQP3 knockout mice develop hypotriglyceridemia, which may be related to the glycerol-transporting function of AQP3 in the kidney (Ma et al. 2000). The physiological role of renal AQP3-mediated glycerol or urea transport has however not been delineated in detail.

25.2.4 Aquaporin 4

AQP4 is localized in the basolateral membrane of connecting tubule and collecting duct principal cells, with highest levels in the inner medullary collecting duct, where it represents an additional pathway for water exit from the principal cells into the interstitium (Coleman et al. 2000; Mobasheri et al. 2007; Terris et al. 1995). AQP4 has been shown to be expressed in the basolateral plasma membrane of proximal tubule S3 segments of mice as well, which may be species specific, as no AQP4 expression was detected in the S3 segment of humans or rats (Van Hoek et al. 2000). Six alternatively spliced isoforms of AQP4 have been identified in rats. Three of these isoforms have been demonstrated to transport water: the short M23 consisting of 301 amino acids, M1 consisting of 323 amino acids, and the long Mz consisting of 364 amino acids (Moe et al. 2008). The M23 isoform gives rise to formation of supramolecular square arrays (Furman et al. 2003). Long-term increased AVP levels do not increase AQP4 protein abundance in the renal inner medulla, where AQP4 is constitutively highly expressed (Terris et al. 1995, 1996). However, AQP4 mRNA was shown to be increased in the cortex, where unstimulated AQP4 expression is low, and this increase could partly be prevented by treatment with a V2R antagonist, suggesting that AVP regulates AQP4 expression (Murillo-Carretero et al. 1999). This was further confirmed by a study in AVP-deficient Brattleboro rats, showing that chronic dDAVP administration increases AQP4 protein abundance in the connecting tubule, cortical collecting duct, and outer part of the inner medullary collecting duct while decreasing AQP4 abundance in the middle and inner segments of the inner medullary collecting duct (Poulsen et al. 2013). In addition, long-term AVP infusion of Brattleboro rats increases the short AQP4 splice form M23 and decreases the long isoform M1 (Van Hoek et al. 2009).

Transepithelial osmotic water permeability in microperfused, vasopressin-stimulated, inner medullary collecting ducts of AQP4 knockout mice is fourfold lower than in wild-type mice, indicating that AQP4 is the dominant basolateral membrane water channel in inner medullary collecting duct (Chou et al. 1998). However, under basal conditions, AQP4 knockout mice do not have a difference in urine osmolality compared with wild-type mice (Ma et al. 1997). After water deprivation, urine osmolality is significantly reduced in AQP4 knockout mice compared to controls, and this is not further increased by injection of dDAVP, showing that AQP4 knockout mice have a mild urinary-concentrating defect. The large decrease in inner medullary water permeability and the absence of a more profound defect in urinary-concentrating ability in mice lacking AQP4 suggest that the majority of collecting duct water reabsorption occurs in segments proximal of the inner medullary collecting duct. This is in agreement with micropuncture studies showing that under antidiuretic conditions, the amount of water reabsorbed in the connecting tubule and initial collecting duct is much greater than that absorbed in the medullary collecting duct (Lassiter et al. 1961).

25.2.5 Aquaporin 5

Recently, Procino et al. showed that AQP5 is expressed in rat, mice, and human kidneys (Procino et al. 2011b), which contrasts with previous studies that have failed to localize AQP5 in kidney. AQP5 was detected in the apical membrane of type-B intercalated cells in the connecting tubule and the cortical part of the collecting duct, where it colocalized with the sodium-independent chloride/iodide transporter pendrin. AQP5 was absent from intercalated cells in the medullary part of the collecting duct (Procino et al. 2011b). The absence of a detectable AQP in the basolateral membranes of type-B intercalated cells suggests that AQP5 does not mediate net transepithelial water reabsorption in this cell type. Instead, AQP5 has been speculated to be acting as an osmosensor in these cells, regulating the cell volume when exposed to the luminal hypotonicity produced by solute reabsorption (in the absence of water reabsorption) from the thick ascending limb and distal convoluted tubule. Chronic potassium depletion in mice, a treatment that elicits metabolic alkalosis, resulted in a strong reduction in both pendrin and AQP5 expression in type-B intercalated cells. Simultaneously, both proteins shifted distribution from the apical cell membrane to an intracellular compartment (Procino et al. 2013), suggesting that pendrin and AQP5 might be subject to a common regulatory pathway. It has been proposed that AQP5-dependent water flux could be coupled to pendrin-mediated Cl^- influx. Supporting the localization and role of AQP5 in type-B intercalated cells, functional AQP5 plasma membrane expression can be induced by extracellular GTP in M1-CCD cells, a cell line that shows an intermediate phenotype between principal and intercalated cells (Mancinelli et al. 2014).

Additional evidence for a role of AQP5 in kidney has been proposed in studies of a renal collecting duct/connecting tubule-specific knockout model of Dot1l, a histone H3 K79 methyltransferase (Wu et al. 2013). In wild-type control mice, AQP5 is barely detectable. However, in the Dot1a knockout mice AQP5 was robustly expressed in renal intercalated cells and principal cells. In the principal cells from knockout mice, AQP5 colocalized with AQP2. AQP5 can be co-immunoprecipitated with AQP2 in AQP2-expressing cell lines, and AQP5 appears to impair the cell surface expression of AQP2. Despite this recent evidence for AQP5 expression in renal epithelial cells, the precise subcellular localization and function of AQP5 in the kidney need further investigation.

25.2.6 Aquaporin 6

AQP6 is an unusual AQP, in that it possesses anion permeability that is activated by low pH or $HgCl_2$. When expressed in *Xenopus laevis* oocytes, AQP6 exhibits low basal water permeability; however, when exposed to Hg^{2+} or low pH, the water permeability of AQP6 oocytes rises (Yasui et al. 1999a). In addition, AQP6 is permeated by anions in response to acidic pH or Hg^{2+} activation, with selectivity for nitrate (Ikeda et al. 2002; Yasui et al. 1999a). Each subunit of the AQP6 tetramer has an individual ion-conducting pore (Ikeda et al. 2002). AQP6 is expressed exclusively in the kidney (Ma et al. 1996; Ohshiro et al. 2001) where, in contrast to most other AQPs, it is localized only in intracellular membranes. Localization of AQP6 in these intracellular membranes is dependent on the N-terminus of AQP6 (Beitz et al. 2006). AQP6 has a relatively broad distribution in the kidney. It is detected in type-A intercalated cells of the renal connecting tubule and cortical and medullary collecting duct, where it colocalizes with the V-type H-ATPase in intracellular vesicles (Ohshiro et al. 2001; Yasui et al. 1999a). In addition, AQP6 has been detected in membrane vesicles within podocyte cell bodies and foot processes and within the subapical compartments of cells from segments 2 and 3 of the proximal tubule, although this localization is debated (Ohshiro et al. 2001; Yasui et al. 1999b). AQP6 expression is increased in a rat model of chronic alkalosis, but the cellular distribution of AQP6 is not altered (Promeneur et al. 2000). In addition, water loading or lithium-induced nephrogenic diabetes insipidus increases AQP6 expression, while water dehydration decreases the expression of AQP6 (Ohshiro et al. 2001; Promeneur et al. 2000). Although the physiological role of AQP6 is still undefined, its location and anion permeability suggest a role in the acidification of intracellular vesicles. As nitrate can inhibit the V-type H-ATPase, AQP6 may be involved in the regulation of H-ATPase activity in acid-secreting type-A intercalated cells (Ikeda et al. 2002).

25.2.7 Aquaporin 7

AQP7, a member of the aquaglyceroporins that can transport both water and glycerol, is expressed in the apical brush border of cells from the S3 segment of the proximal tubule (Ishibashi et al. 2000; Nejsum et al. 2000). Although AQP7 is a functional water channel, as emphasized by the reduced osmotic water permeabilities in apical membrane vesicles isolated from the proximal tubule of AQP7 knockout mice (Sohara et al. 2005), AQP7 knockout mice do not have impaired renal water handling or urinary-concentrating ability. This might be because in the S3 segment of the proximal tubule, AQP1 is abundantly expressed, making AQP7 water transport redundant. This hypothesis was confirmed in AQP1–7 double-knockout mice, which had a significantly greater urine volume accompanied by a proportional decrease in urine osmolality compared with AQP1 knockout mice, suggesting that the amount of water reabsorbed through AQP7 in the proximal straight tubules is physiologically substantial. In contrast to the absence of a defect in water handling, AQP7 knockout mice show significantly increased renal glycerol excretion (Sohara et al. 2005); thus the main physiological role of AQP7 in the kidney might therefore be related to glycerol, rather than water, reabsorption.

25.2.8 Aquaporin 8

AQP8 is permeable to water, but it can also facilitate ammonia transport (Holm et al. 2005; Liu et al. 2006). Ammonia is produced predominantly from the metabolism of glutamine in the mitochondria of the proximal tubule and is secreted into the urine as a result of specialized transport processes in several nephron segments (Knepper et al. 1989). The regulation of renal ammonium excretion plays a central role in the maintenance of systemic acid-base balance (Knepper et al. 1989). Immunohistochemistry has demonstrated strong AQP8 expression along the length of the proximal tubule (Elkjaer et al. 2001), with weaker expression in the collecting ducts. In proximal tubule cells, labeling was confined to the cytoplasm, with no labeling of the apical brush border and an absence of, or very weak, labeling of basolateral plasma membrane domains. Similarly, in collecting duct cells, the weak labeling of AQP8 was associated with intracellular structures. Intracellularly, AQP8 has been demonstrated to be located in the inner mitochondrial membrane in the rat kidney, which is similar to its localization in rat hepatocytes (Lee et al. 2005). Despite its high water conductance, AQP8 does not contribute greatly to the facilitated transport of water across the mitochondrial membrane (Calamita et al. 2006; Yang et al. 2006c). Instead, AQP8 efficiently facilitates the transport of ammonia through mitochondrial membranes (Soria et al. 2010). Comparative studies demonstrated that the AQP8-mediated movement of ammonia across mitochondrial membranes was markedly higher than that of water (Soria et al. 2010), suggesting that ammonia diffusional transport is a major function for mitochondrial AQP8. In addition, AQP8 is expressed in the mitochondria of a human renal proximal tubule cell line, and knockdown of its expression induced a decrease in the rate of ammonia released in the culture medium, either from cells grown at neutral or acidic pH. AQP8 expression was also upregulated in this cell line in response to metabolic acidosis, together with a greater ammonia excretion rate (Molinas et al. 2012). In vivo, ammonia-loaded rats, which are a model of metabolic acidosis, have increased AQP8 expression in the kidney cortex (Molinas et al. 2012). Taken together, these data suggest that AQP8 could play an important role in the adaptive response of proximal tubule cells to acidosis, possibly by facilitating mitochondrial ammonia transport. However, a major physiological role of AQP8 in renal ammonia handling has been questioned, since AQP8 knockout mice show defects in renal ammonia excretion only in response to acidosis by chronic ammonium loading (Yang et al. 2006b). Since ammonia transport is critical to maintain the systemic acid-base balance, it cannot be ruled out that other compensatory ammonia transport pathways may exist and compensate in the AQP8 knockout mice.

An additional role of AQP8 in renal water handling has been dismissed based on studies in knockout mice. Urine osmolalities were not different in wild-type vs. AQP8 knockout mice at baseline or after water deprivation (Yang et al. 2005). Furthermore, because a subtle effect of AQP8 might be detected after deletion of the major proximal nephron water channel AQP1, urine osmolality was also measured in AQP1 knockout and AQP8/AQP1 double-knockout mice. However, urine osmolality was also not significantly different in AQP1 knockout mice vs. mice lacking AQP8 and AQP1 together (Yang et al. 2005), providing evidence against a significant role of AQP8 in the urinary-concentrating ability in mice.

25.2.9 Aquaporin 11

AQP11, in contrast to most other AQPs that have two highly conserved short "NPA" (asparagine-proline-alanine) sequences that form loops directed into the membrane constituting a pore, has unique "NPC" motifs distinct from those of other AQPs (Morishita et al. 2005). Experiments using reconstituted AQP11 in proteoliposomes or Sf9 cell membrane vesicles revealed that AQP11, despite the incompletely conserved NPA motif and low homology with other AQP family proteins, has water channel activity, but this is relatively low (Yakata et al. 2007, 2011). AQP11 is also permeable to glycerol (Madeira et al. 2014). Changing the "NPC" motif of AQP11 to "NPA" reduced the water permeability and oligomerization of AQP11 (Ikeda et al. 2011). In mice kidneys, AQP11 is expressed selectively in cells along the proximal tubule. AQP11 is localized intracellularly, confined to the region around the nucleus, and is absent at the apical brush border membrane (Inoue et al. 2014; Morishita et al. 2005). Thus far, AQP11 antibodies that are suitable for subcellular localization studies of AQP11, e.g., immunogold electron microscopy have not been developed, hampering studies to localize AQP11 to a specific organelle. However, in CHO-K1 cells and in vivo in mice expressing GFP or HA-tagged AQP11, AQP11 localizes to the endoplasmic reticulum (ER) (Ikeda et al. 2011; Inoue et al. 2014; Morishita et al. 2005).

The majority of AQP11 knockout mice die before weaning due to advanced renal failure from polycystic kidneys. The cyst formation in these mice is preceded by epithelial cell swelling with intracellular vacuolization of the proximal tubule (Morishita et al. 2005). The cysts in the AQP11 knockout mice originate mainly from the proximal tubules (Inoue et al. 2014; Morishita et al. 2005). Although these cysts are absent in the kidneys of newborn mice, they are apparent after 3 weeks when functional renal failure is evident. Remarkably, the cyst epithelia contain vacuoles in the cytoplasm, which are already present at birth. These vacuoles are limited to the proximal tubule, where AQP11 is expressed, suggesting that the dysfunction of AQP11 is the direct cause of the vacuole formation. As vacuolization precedes the cyst formation, the proximal tubule cysts may be caused by the vacuolization. The vacuoles strongly stain for an ER marker and ribosomes are attached to some of the vacuoles, suggesting that they originate from the ER (Morishita et al. 2005). A similar phenotype to the total AQP11 knockout mice is observed in another mouse model with a spontaneous amino acid substitution (Cys227Ser) in the predicted E-loop region of AQP11. These mice also present with proximal tubule injury, with formation of giant vacuoles in the renal cortex and die from renal failure at an early age (Tchekneva et al. 2008). Examining the total AQP11 knockout mouse model, it is unclear if the kidney failure is caused by functional insufficiency of the proximal tubule in the absence of AQP11 or by abnormal development of the nephron and structural malformation of the kidney in the absence of AQP11. Recently, an inducible AQP11 knockout mouse model has been developed, where *AQP11* gene disruption can be temporally controlled by injection of tamoxifen (Nielsen et al. 2014). Microscopic analyses of the proximal tubules after induction of *AQP11* gene deletion in adult mice (after the completion of kidney development) revealed no cellular damage of proximal tubule cells or cyst formation for up to 4 months after disruption of the *AQP11* gene. This suggests that continuous proximal tubule injury during the neonatal period (as in total AQP11 knockout mice) results in abnormal cell development and cyst formation. However, the exact function of AQP11 in proximal tubule cells remains unknown. As proximal tubule cells exhibit ER vacuolization in AQP11 knockout mouse kidneys, and AQP11 knockout mice show increased ER stress response and oxidative stress, AQP11 may play an important role in the homeostasis of the ER (Atochina-Vasserman et al. 2013; Okada et al. 2008).

The kidneys from AQP11 knockout mice exhibit increased protein expression levels of polycystin-1 (PC-1) and decreased protein expression levels of polycystin-2 (PC-2). Despite the increased PC-1 levels, PC-1 is abnormally N-glycosylated and less is localized in the plasma membrane (Inoue et al. 2014). These differences in PC-1 and PC-2 could represent a key mechanism of cyst formation in the AQP11 knockout mice. PC-1 and PC-2 are usually located in the plasma membrane and primary cilia of renal epithelia, where they form part of a complex that acts as a flow sensor on the cilium. Mutations in the genes coding for PC-1 and PC-2 are the cause of autosomal-dominant polycystic kidney disease (ADPKD), a disease characterized by cyst formation in the kidney that in many cases ultimately results in end-stage renal disease. Autosomal recessive polycystic kidney disease (ARPKD) can be caused by mutations in the gene coding for fibrocystin, which is in the same complex as PC-2 within the primary cilia. The precise function of fibrocystin is at present unknown, but it may mediate its affects through PC-2 (Harris and Torres 2009).

Although in PKD, lesions are localized to collecting ducts, murine models of ARPKD demonstrate an early phase of proximal tubular cystic involvement that disappears shortly after birth, but is followed by a phase of collecting duct cyst formation (Nakanishi et al. 2000). Similarly, in mice homozygous for Pkd1 mutations (the gene coding for PC-1), renal cyst formation begins in the proximal tubules, followed by cyst formation in the collecting duct (Lu et al. 1997). In humans, ARPKD has a transient phase of proximal tubular cyst formation during fetal development, with a gradual shift in the site of cystic nephron involvement, from proximal tubules to collecting ducts during the fetal period (Nakanishi et al. 2000). This suggests that AQP11 knockout mice and polycystic kidney disease share a common pathogenic mechanism for cyst formation.

25.3 Renal Aquaporins in Disease

25.3.1 Water Balance Disorders Associated with Hyponatremia and Increased Aquaporin Levels

Disturbances of water balance are frequently occurring and are characterized by hyponatremia, hypernatremia, or polyuria. Several of these pathologies result from too high or too low levels of renal AQPs, mainly AQP2. Hyponatremia with high AQP2 levels can be caused by either osmoregulation disorders or, most frequently, by diseases with low effective circulating blood volume. In these latter conditions, osmoregulation is overruled by the body's regulation of volume, leading to the development of hyponatremia as a consequence of the body's goal to maintain sufficient circulating volume.

25.3.1.1 Congestive Heart Failure

Congestive heart failure (CHF) patients have high blood levels of AVP that contribute to hyponatremia and an increased extracellular volume (Schrier 2006). AQP2 excretion is increased in the urine of CHF patients (Pedersen et al. 2003). AQP2 has been shown to be excreted in the urine in exosomes via an apical pathway, rather than whole cell shedding (Wen et al. 1999). Exosomes are formed from multivesicular bodies (MVBs) within the cell, which, under certain conditions, dock and fuse with the plasma membrane and empty their contents (exosomes) into the urinary space. The urinary excretion of AQP2 in exosomes directly correlates with the amount of renal AQP2 at the apical membrane (Wen et al. 1999) and thus can be used as a marker for the expression of AQP2 in the kidney. Rat models of CHF show increased renal AQP2 trafficking and total expression (Lin et al. 2011; Nielsen et al. 1997; Xu et al. 1997), whereas AQP1 and AQP3 levels are not increased, indicating that the effect on AQP2 in CHF is selective (Nielsen et al. 1997). The increased AQP2 levels are most likely the result of increased circulating levels of AVP. However, studies have shown that rats with CHF have increased protein levels of AQP2 even during conditions with unchanged circulating levels of AVP (Brond et al. 2013). It was shown that in these rats, the V2 receptor obtained an increased ability to recycle to the plasma membrane and consequently collecting ducts showed an increased accumulation of cAMP in response to AVP. This indicates increased AVP sensitivity in the collecting ducts, which might contribute to the increased AQP2 levels and hyponatremia seen in CHF (Brond et al. 2013). Administration of the V2R antagonist Mozavaptan to CHF rats inhibited the AQP2 upregulation and trafficking and led to an increase in urine output and a rise in plasma osmolality, suggesting a major role for AQP2 in CHF-induced water retention (Xu et al. 1997). In agreement with these animal data, V2R antagonists have been shown to increase urine volume and correct hyponatremia in CHF patients (Aronson et al. 2011; Schrier et al. 2006).

25.3.1.2 Hepatic Cirrhosis

Hepatic cirrhosis is a chronic disease associated with water retention, hyponatremia, and increased levels of AVP (Schrier 2006). The development of hyponatremia in cirrhosis patients is thought to be caused by systemic vasodilation, leading to increased secretion of AVP and consequently water retention and hyponatremia. The role of AQP2 in hepatic cirrhosis is not clear as, unlike for CHF, the changes in AQP2 levels vary considerably between different studies. In some studies of rat models of hepatic cirrhosis induced by carbon tetrachloride, AQP2 protein and mRNA levels were increased (Asahina et al. 1995; Fujita et al. 1995). The mRNA expression of renal AQP2 was furthermore found to correlate with the volume of ascites, suggesting that AQP2 plays a crucial role in abnormal water retention followed by the development of ascites in hepatic cirrhosis (Asahina et al. 1995). In another study, an increase in plasma membrane expression of AQP2 was observed, although total AQP2 abundance was not changed, while total expression of AQP1 and AQP3 were increased as well (Fernandez-Llama et al. 2000). In contrast, AQP2, AQP3, and AQP4 levels were decreased, and AQP1 levels unchanged, in a rat model of liver cirrhosis induced by chronic common bile duct ligation, suggesting that an increase in renal aquaporins is not essential for the development of excessive water retention with hyponatremia (Fernandez-Llama et al. 1999).

Human studies have demonstrated an increased excretion of AQP2 in the urine of cirrhosis patients compared with control subjects (Chung et al. 2010; Ivarsen et al. 2003). AQP2 levels correlated with the clinical severity of the cirrhosis and were highest in patients with ascites, suggesting a critical role for AQP2 in cirrhosis-induced water retention. However, other studies have demonstrated no increase, or even a decrease, in urinary AQP2 in cirrhosis patients (Esteva-Font et al. 2006; Pedersen et al. 2003). An explanation for the differences between these studies is at present lacking.

25.3.1.3 Syndrome of Inappropriate Secretion of Antidiuretic Hormone

The syndrome of inappropriate secretion of antidiuretic hormone (SIADH) results from abnormally increased AVP levels, leading to excessive renal water reabsorption and compromised osmoregulation, which ultimately can result in life-threatening hyponatremia (Palmer 2003). The most common causes of SIADH are neoplasia (notably small-cell lung cancer), neurological diseases, lung diseases, and a wide variety of drugs, particularly psychoactive drugs and chemotherapy (Baylis 2003). AQP2 levels are markedly increased in SIADH rats, but can be reduced using V2R antagonists, a process that correlates closely with a marked diuresis and a normalization of serum sodium levels (Fujita et al. 1995). In agreement with these animal data, V2R antagonists have been shown to induce diuresis and correct hyponatremia in SIADH patients (Schrier et al. 2006; Soupart et al. 2006; Wong et al. 2003). SIADH patients are normovolemic or slightly hypervolemic and often show excessive release of AVP that is not related to plasma osmolality, e.g., ectopic AVP production by neoplasms (Hoorn et al. 2014). Other patients continue to regulate water excretion normally, but do so around a lower plasma osmolality set point compared to normal. The underlying mechanism of this is unclear (Baylis 2003).

25.3.1.4 Nephrogenic Syndrome of Inappropriate Antidiuresis

Another cause of normovolemic hyponatremia is the nephrogenic syndrome of inappropriate antidiuresis (NSIAD), a rare disorder caused by a mutation in the *AVPR2* gene, encoding the V2R. Although the clinical presentations of NSIAD resemble those of SIADH, i.e., hyponatremia and lack of urinary dilution, in NSAID patients the plasma AVP levels are undetectable or very low (Levtchenko and Monnens 2010). The *AVPR2* mutation usually results in a change in amino acid 137 from an arginine into a cysteine or a leucine (Arg137Cys/Leu) (Feldman et al. 2005), although a Phe229Val mutation was observed in an infant with NSIAD (Carpentier et al. 2012). The mutation of the *AVPR2* results in constitutive activation of the V2R, resulting in increased basal cAMP production compared to the wild-type V2R (Carpentier et al. 2012; Feldman et al. 2005; Tenenbaum et al.

2009). This constitutive activation of the V2R will result in enhanced AQP2 abundance and membrane targeting, which explains the inappropriate antidiuresis. In contrast, mutation of amino acid 137 into a histidine (Arg137His) results in increased V2R endocytosis which causes impaired agonist binding leading to reduced AQP2 abundance and trafficking, ultimately causing the opposite disease phenotype and a loss of the kidney's ability to concentrate urine (Barak et al. 2001). As the *AVPR2* is located on the X-chromosome, NSIAD is an X-linked genetic disease affecting predominantly males. Although female carriers of the mutation are clinically asymptomatic, a proportion demonstrated an impaired ability to dilute urine during a water-load test (Decaux et al. 2007; Ranchin et al. 2010).

25.3.1.5 Treatment of Pathologies Resulting from Increased Aquaporin Levels

Euvolemic or hypervolemic chronic hyponatremia, caused either by CHF, liver cirrhosis, SIADH, or NSIAD, is usually treated via simple fluid restriction (Elhassan and Schrier 2011). However, not all patients comply with this regime and are subsequently supplemented with oral intake of urea (to induce an osmotic diuresis) or with loop diuretics, e.g., furosemide. In addition to this, newer more effective treatments are becoming more common, including the use of either demeclocycline or V2R antagonists.

Demeclocycline

Demeclocycline is a bacteriostatic antibiotic of the tetracycline group, use of which causes water diuresis and subsequently NDI (Forrest et al. 1974; Singer and Rotenberg 1973). Demeclocycline is one of the options currently used to treat sustained hyponatremia in patients with SIADH (Sherlock and Thompson 2010). In SIADH patients, demeclocycline restores sodium plasma concentration to normal levels, permitting unrestricted water intake in these patients (Forrest et al. 1978). Demeclocycline selectively inhibits water reabsorption in the distal part of the nephron (Wilson et al. 1973) and AVP-induced osmotic water flow in the toad urinary bladder (a model system of mammalian collecting duct) (Feldman and Singer 1974; Hirji and Mucklow 1991; Singer and Rotenberg 1973). In a mouse collecting duct cell line (mpkCCD), demeclocycline decreases dDAVP-induced cAMP generation and adenylate cyclase 3 and 5/6 abundances (Kortenoeven et al. 2013b), leading to reduced *AQP2* gene transcription and subsequently reduced AQP2 protein abundance. In a rat model of SIADH, demeclocycline increases urine volume, decreases urine osmolality, and corrects the hyponatremia, a result of decreased AQP2 and adenylate cyclase 5/6 abundances in the renal inner medulla. Thus, demeclocycline attenuates hyponatremia by attenuating the AVP-induced signaling cascade and AQP2 abundance in the inner medullary collecting duct (Kortenoeven et al. 2013b).

V2R Antagonists

Vasopressin-receptor antagonists, or vaptans, have become promising drugs for the treatment of AQP2-mediated hyponatremia. OPC-31260 (mozavaptan) increased the urinary flow rate, lowered the urinary osmolality, and attenuated the hyponatremia in rat models of SIADH, CHF, and liver cirrhosis (Fujisawa et al. 1993; Tsuboi et al. 1994; Xu et al. 1997). OPC-31260 treatment also results in water diuresis and lessens the hyponatremia in SIADH patients (Saito et al. 1997a). Four other vaptans, the V2R antagonists tolvaptan, lixivaptan, and satavaptan and the V2R/V1a antagonist conivaptan increase diuresis and correct hyponatremia in patients with euvolemic or hypervolemic hyponatremia due to CHF, liver cirrhosis, and SIADH (Aronson et al. 2011; Ghali et al. 2006; Gines et al. 2008; Schrier et al. 2006; Soupart et al. 2006; Wong et al. 2003).

25.3.2 Water Balance Disorders Associated with Polyuria and Decreased Aquaporin Levels

Polyuria and polydipsia are symptoms of a range of diseases and disturbances to homeostasis, which can originate at extrarenal, intrarenal, or postrenal sites. Given the tight hormonal regulation of AQP2, it is not surprising that polyuria is often accompanied by altered levels of plasma AVP, which are either pathologically decreased, as seen in central and gestational diabetes insipidus, or compensatory increased in polyuria from other causes, such as diabetes mellitus.

25.3.2.1 Central DI

Diabetes insipidus (DI) is characterized by impaired renal water reabsorption, leading to polyuria and hypotonic urine. DI patients can have hypernatremia, but most often have a normal plasma osmolality, provided the thirst mechanism is normal and there is adequate access to fluid. Central DI, also known as neurohypophysial DI, is caused by impaired AVP release from the posterior pituitary gland. Under normal circumstances, AVP is produced by the magnocellular neurons of the supraoptic nucleus of the hypothalamus, which project axons into the posterior pituitary gland, from which they release AVP into the bloodstream upon activation (Voisin and Bourque 2002). Central DI can be congenital but is usually acquired as a result of local inflammatory or autoimmune disease, vascular disease, Langerhans cell histiocytosis, sarcoidosis, intracranial tumors, or trauma resulting from surgery or an accident (Di et al. 2012). It is treated by administration of the synthetic AVP analogue dDAVP, which drastically decreases urine output in central DI patients. The Brattleboro rat is an animal model of central diabetes insipidus, which has undetectable levels of plasma AVP due to a random genetic mutation (Valtin et al.

1962) and displays decreased expression of AQP2, which can be reversed by chronic AVP infusion (Digiovanni et al. 1994). In agreement with this, central DI patients excrete decreased amounts of AQP2 in the urine compared with healthy controls and AQP2 excretion remains unchanged after hypertonic saline infusion, but increases after exogenous AVP administration (Saito et al. 1997b, 1999). AVP regulates a variety of proteins involved in urine concentration, such as AQP3 and AQP4 in connecting tubules and collecting ducts, urea transporters in inner medullary collecting ducts and the loop of Henle, and the Na–K–2Cl cotransporter 2 (NKCC2) in the thick ascending limb, which may be hypothesized to also be affected in central DI patients.

The rarest form of central DI is familial neurohypophysial DI (FNDI), which is characterized by severe polyuria and polydipsia, and in almost all cases is caused by an inherited mutation in the *AVP* gene. The *AVP* gene encodes a 164 kDa long precursor peptide, consisting of a signal peptide, AVP, neurophysin II and copeptin. This precursor is enzymatically cleaved to form active AVP (Land et al. 1982). At least 70 different mutations resulting in a defective precursor peptide and a deficiency of active AVP have been identified in FNDI. The majority of these mutations show an autosomal dominant pattern of inheritance [http://www.hgmd.org (Arima and Oiso 2010)] and most of the mutations are located in either the signal peptide or the neurophysin II moiety. The carriers display a normal phenotype at birth, but typically between 1 and 6 years of age, symptoms of compulsive drinking and polyuria appear, even though they carry one normal allele. FNDI is hypothesized to result from accumulation of mutant AVP precursor in the ER, resulting in an aberrant endoplasmic morphology and possible cell dysfunction and death (Arima and Oiso 2010; Ito and Jameson 1997; Russell et al. 2003). Mutant precursors form heterodimers with the wild-type AVP precursor, thereby impairing intracellular trafficking of both forms and reducing the bioavailability of active AVP by means of a dominant negative effect (Arima and Oiso 2010; Ito et al. 1999).

Patients with an autosomal recessive instead of dominant pattern of inheritance have a missense mutation in the region encoding the AVP domain resulting in reduced biological activity of the mutant AVP peptide (Abu et al. 2010; Willcutts et al. 1999). In contrast to other forms of central DI, circulating AVP levels in these patients are high. In one family with autosomal-dominant FNDI, no mutations in the coding region, introns, or the promoter of the *AVP* gene were found (Ye et al. 2005). It was found that the gene(s) responsible for FNDI in this family was/were located on chromosome 20, implying mutations in other genes than the *AVP* gene can cause FNDI as well.

25.3.2.2 Gestational DI

A transient form of DI can occur during pregnancy. Gestational DI is due to an abnormal increase in an AVP-degrading enzyme produced by the placenta termed vasopressinase, leading to AVP deficiency (Ananthakrishnan 2009). This form of DI may be associated with preeclampsia and HELLP syndrome (consisting of hemolysis, elevated liver enzymes, and low platelets), possibly because the decreased hepatic function in these conditions leads to a decrease in vasopressinase degradation (Hadi et al. 1985; Katz and Bowes 1987; Nasrat et al. 1997). dDAVP is resistant to degradation by vasopressinase and is an effective treatment for gestational DI (Ananthakrishnan 2009).

25.3.2.3 Nephrogenic DI

Nephrogenic DI (NDI) is a collective term for congenital or acquired polyuric conditions where the kidney is unable to normalize urine concentration in response to AVP stimulation.

Congenital NDI

Congenital NDI can be divided into X-linked, autosomal recessive, and autosomal dominant NDI. The urine-concentrating defect is present from birth, and symptoms usually arise during the first weeks of life. X-linked congenital NDI is the cause of disease in more than 90 % of all congenital NDI patients and occurs mainly in males; however, due to skewed X-chromosome inactivation, some heterozygous females have variable degrees of clinical NDI symptoms (Arthus et al. 2000; Satoh et al. 2008). It is caused by loss-of-function mutations in the *AVPR2* gene, encoding the V2R (Lolait et al. 1992), and over 220 such gene mutations have been described to cause NDI (http://www.hgmd.org). Mutations in the *AVPR2* gene interfere with receptor signaling, thus making the principal cells of the collecting duct insensitive to AVP, resulting in a severe urinary concentration defect. The most common mechanism underlying this insensitivity is misfolding of the V2R protein caused by a missense mutation and consequently retention in the ER. Other mechanisms include mRNA instability, splicing errors, frameshifts, or nonsense mutations resulting in truncation of the receptor, diminished binding of the Gs protein, reduced affinity for AVP, and misrouting of the V2R to different organelles in the cell (Robben et al. 2006a).

Approximately 10 % of patients diagnosed with NDI have mutations in the *AQP2* gene, which is an autosomal recessive disorder in more than 90 % of cases. To date, 52 AQP2 mutations have been described, of which 44 are involved in recessive NDI (http://www.hgmd.org). The majority of these comprise missense mutations. Typically, mutations are found in between the first and last transmembrane domains of AQP2, which form the channels' water pore. Nearly all mutations cause misfolding of the protein, which is consequently trapped in the ER and rapidly degraded by the proteasome (Robben et al. 2006a). AQP2 could not be detected in the urine of recessive NDI patients, presumably due to extensive degradation of the protein (Deen et al. 1996). It was shown that several AQP2 mutants are able to function as water channels when overexpressed in oocytes, where a fraction of the AQP2 mutants escape the cell's quality control system and reach the plasma membrane (Canfield et al. 1997; Marr et al. 2001, 2002; Mulders et al. 1997). Thus, in these cases the phenotype is caused exclusively by ER retention of the protein. The healthy parents of patients with recessive NDI express both mutant and wild-type AQP2 proteins. However, since these AQP2 mutants are not able to form heterotetramers with wild-type AQP2, this leaves only the formation of the functional wild-type AQP2 homotetramers, which likely explains the autosomal recessive pattern of inheritance (Kamsteeg et al. 1999).

Currently, eight different AQP2 mutations have been described to cause autosomal dominant NDI, which is the rarest form of the disease. The identified mutations comprise missense mutations and small nucleotide deletion or insertions. All mutations are found in the carboxyl-terminal tail of AQP2, which is essential for correct intracellular routing of the protein, and the mutations cause missorting of the protein to other cellular organelles than the apical plasma membrane, e.g., late endosomes/lysosomes and the basolateral plasma membrane (Kamsteeg et al. 1999; Robben et al. 2006a). Dominant AQP2 mutants are able to form heterotetramers with wild-type AQP2 causing misrouting of both forms of the protein and consequently severely compromised collecting duct water reabsorption. Six of the mutations introduce a missorting signal whereas the remaining two, Arg254Leu and Arg254Gln, interfere with Ser256 phosphorylation of AQP2 and prevent the translocation of AQP2 to the plasma membrane, leading to impaired water reabsorption (De Mattia et al. 2005; Savelkoul et al. 2009). Heterozygous dominant NDI patients display generally milder symptoms than recessive NDI patients, suggesting that some wild-type AQP2 homotetramers are formed that are able to reach the apical plasma membrane (Robben et al. 2006a).

Although mutations in the carboxyl tail of AQP2 usually lead to dominant NDI, one mutation, Pro262Leu, causes recessive NDI. Like dominant NDI mutants, Pro262Leu-AQP2 is missorted to intracellular vesicles and is able to form heterotetramers with wild-type AQP2 (De Mattia et al. 2004). However, when expressed together with wild-type AQP2, wild-type/Pro262Leu-AQP2 heterotetramers are targeted to the apical membrane of MDCK cells. This indicates that, in contrast to AQP2 mutants in dominant NDI, the apical sorting signal of wild-type AQP2 overrules the sorting signal in Pro262Leu-AQP2.

Few subjects have been identified with loss-of-function mutations in AQPs other than AQP2. Humans lacking functional AQP1 do not have polyuria and no obvious clinical phenotype under normal conditions. However, when challenged by water deprivation, they have an impaired ability to maximally concentrate their urine (King et al. 2001). Under nonstressed conditions, AQP3 null individuals do not suffer from any clear clinical abnormality (Roudier et al. 2002). Mutations in AQP5 in humans have been shown to cause autosomal dominant diffuse nonepidermolytic palmoplantar keratoderma (Blaydon et al. 2013). Three unrelated children with a homozygous Gly264Val mutation in the *AQP7* gene, which previously has been shown in oocytes to result in a lack of both water and glycerol permeability, were shown to have psychomotor retardation of unknown etiology, a mild platelet secretion defect, and pronounced renal glycerol loss (Goubau et al. 2013; Kondo et al. 2002). However, the renal phenotype of individuals with defective AQP3,

AQP5, or AQP7 has not been studied in detail. Human gene mutations abolishing the function of other renal AQPs have not been reported.

Acquired NDI

Electrolyte Disorders

Acquired forms of NDI are much more common than the rare hereditary disorders. Hypokalemia and hypercalcemia are common electrolyte disorders, which cause NDI by mechanisms that are not completely understood (Gill and Bartter 1961; Rubini 1961). In rat models of hypokalemia, the condition is associated with decreased AQP2 mRNA and protein expression (Jeon et al. 2007; Marples et al. 1996) alongside decreased expression of renal urea and sodium transporters in the distal nephron. This may contribute to the urinary-concentrating defect by reducing interstitial tonicity (Elkjaer et al. 2002; Jeon et al. 2007; Jung et al. 2003).

Hypercalcemia decreases AQP2 protein abundance in rats, but does not affect AQP2 mRNA levels; indicating that the effect is independent of AQP2 transcription but may result from increased AQP2 degradation or decreased AQP2 translation. Hypercalcemia also decreases AQP2 apical membrane accumulation (Earm et al. 1998). Moreover, hypercalcemic rats have decreased AQP1 and AQP3 levels, while AQP4 levels remain unchanged (Wang et al. 2002b). In addition, hypercalcemia has been shown to decrease the expression of renal sodium transporters such as NKCC2, decrease NaCl reabsorption in the thick ascending limb, and reduce medullary tonicity, likely contributing to the decreased urine-concentrating ability (Levi et al. 1983; Peterson 1990; Wang et al. 2002a, 2004).

In humans, it was shown that in children with enuresis, hypercalciuria is associated with a decrease in nocturnal AQP2 levels in the urine and that decreasing urine calcium excretion by means of a low calcium diet is associated with an increase in urine AQP2 and decreased nocturnal urine volume in these children (Valenti et al. 2000, 2002). In another study a higher baseline urine AQP2 excretion was found in hypercalciuric subjects. In contrast to normocalciurics (Procino et al. 2012), dDAVP administration did not increase AQP2 excretion in these subjects, suggesting a disturbed AQP2 regulation. In addition, prolonged bed rest has been shown to induce bone demineralization and transient increase in urinary calcium followed by a transient decrease in AQP2 excretion, while the return of calciuria to baseline is followed by a recovery of AQP2 excretion, indicating that urinary calcium modulates AQP2 expression (Tamma et al. 2014).

The effects of hypercalcemia on AQP2 have been attributed to activation of the apical calcium-sensing receptor (CaSR), which might be a defense mechanism to reduce the risk of calcium renal stone formation in states of high calcium in the tubular lumen. CaSR is a G protein-coupled receptor which is activated by interaction with calcium at the extracellular domain (Alfadda et al. 2014). It is thought to couple to Gq and activates intracellular signaling pathways including phospholipase C and mitogen-activated protein kinases (Tfelt-Hansen et al. 2003). Sands et al. showed that the CaSR is expressed at the apical border of cells in the terminal inner medullary collecting duct and that purified subapical endosomes contain both CaSR and AQP2 (Sands et al. 1998). Activation of the CaSR by high extracellular calcium antagonizes forskolin-induced AQP2 trafficking to the apical plasma membrane and decreases AVP-induced AQP2 expression in cultured cells (Bustamante et al. 2008; Procino et al. 2004, 2012), suggesting that hypercalcemia-induced activation of the CaSR in vivo may have a similar effect resulting in subsequent urine concentration defects. Alternatively, high calcium may activate the protease calpain, which is expressed in the inner medullary collecting ducts, leading to degradation of AQP2 (Puliyanda et al. 2003).

Obstruction of the Urinary Tract

A relatively common condition associated with long-term impairment of urinary-concentrating ability is obstruction of the urinary tract, a serious clinical condition and a common cause of end-stage renal disease. In adults, obstruction is mostly caused by stones, enlargement of the prostate, or urinary tract neoplasms, whereas in children, it is usually due to congenital abnormalities (Iyasere et al. 2012). During the recovery phase after release of the urinary tract obstruction, the response of the collecting duct to AVP and water reabsorption in the collecting duct is decreased, resulting in polyuria (Klahr et al. 1988). Experimental bilateral or unilateral obstruction of the ureters is associated with markedly reduced expression of AQP1, AQP2, AQP3, and AQP4, as well as of renal sodium and urea transporters (Frokiaer et al. 1996; Li et al. 2001, 2003a–c, 2004). Both AQP2 protein and mRNA levels are decreased (Frokiaer et al. 1997; Stodkilde et al. 2011). After obstruction, AQP2 redistributes to early endosomes and lysosomes, suggesting degradation of AQP2 (Stodkilde et al. 2011). Following release of the obstruction, there is a marked polyuria during which period AQP1, AQP2, and AQP3 levels remain decreased, providing an explanation at the molecular level for postobstructive polyuria (Li et al. 2001).

Bilateral obstruction in rats is associated with increased renal cyclooxygenase-2 (COX-2) expression and increased prostaglandin E_2 (PGE_2) excretion in the urine (Cheng et al. 2004; Norregaard et al. 2005). Several prostaglandins are increased in the inner medulla, and to a lesser extent in the cortex, of rats with ureteral obstruction. In these conditions, COX-2 is the predominant isoform involved in prostaglandin synthesis, with minor, but significant, contributions from COX-1 (Norregaard et al. 2010). Ureteral obstruction did not significantly decrease AQP2 and AQP3 protein levels in the cortex and outer medulla of COX-2 knockout mice, in contrast to wild-type mice, although a decrease in inner medullary AQP2 expression remained (Nilsson et al. 2012), suggesting that COX-2-derived prostaglandins may contribute in part to downregulation of AQPs in the collecting duct in ureteral obstruction. Treatment of rats with COX-2 inhibitors prevents the increase in PGE_2, whereas the downregulation of AQP2 and the development of polyuria are alleviated in some studies (Cheng et al. 2004; Norregaard et al. 2005), but have no effect or are even augmented in others (Jensen et al. 2010). Thus, the effect of COX-2 inhibition in the model is not completely understood [for discussion see Olesen and Fenton (2013)].

Lithium-Induced NDI

The most common cause of NDI is lithium therapy. Lithium is a frequently prescribed drug used by 1 in 1000 of the western population (Manji et al. 2001; Timmer and Sands 1999) as a mood-stabilizing drug in such conditions as bipolar disorder, schizoaffective disorder, and recurrent depression. Unfortunately, more than 50 % of patients undergoing lithium treatment develop impaired urinary-concentrating ability and in 20 % of patients this leads to the development of clinical NDI (Boton et al. 1987). Plasma lithium levels must be kept within a narrow therapeutic range and lithium-NDI patients are at risk for dehydration-induced lithium toxicity. Moreover, prolonged lithium treatment may result in development of renal microcysts and end-stage renal disease (Farres et al. 2003; Timmer and Sands 1999). However, since lithium is still regarded as the most powerful mood-stabilizing drug available, substitution of lithium therapy with one of the newer drugs is not always an option.

In the short term, lithium-NDI coincides with AQP2 and AQP3 downregulation and natriuresis in rats, without gross changes in renal morphology (Laursen et al. 2004; Marples et al. 1995; Mu et al. 1999). The lithium-induced natriuresis is suggested to be due to the reduced expression of salt-transporting proteins, including the NaCl cotransporter (NCC), the β- and γ-subunits of ENaC, and the Na-K-ATPase (Kwon et al. 2000; Laursen et al. 2004; Nielsen et al. 2003). A study using connecting tubule-specific AQP2 knockout mice suggested that lithium does not severely affect AQP2 in the connecting tubule, in contrast to the collecting duct (Kortenoeven et al. 2013a). Recently, it was shown in rats that already 4 hours after lithium exposure, urine output was increased and phosphorylation of both mitogen-activated protein kinases (MAPK) ERK1/2 and p38, as well as of Ser261-AQP2, was increased. Pretreatment with MAPK inhibitors reversed the increased Ser261-AQP2 phosphorylation, suggesting that ERK1/2 and p38 are early targets of lithium and may play a role in the onset of lithium-induced polyuria (Trepiccione et al. 2014).

Chronic lithium treatment also leads to a severe decrease in the fraction of principal cells and an increase in the fraction of intercalated cells, which are involved in acid-base balance regulation (Christensen et al. 2004, 2006), suggesting that part of the lithium-induced NDI might be explained by epithelial remodeling. Recently, it was shown that lithium treatment initiates proliferation of principal cells but that a significant percentage of these cells are arrested in the late G2 phase of the cell cycle, which might explain the reduced principal/intercalated cell ratio (de Groot et al. 2014). In agreement with the studies in rats, a study in humans showed that 4-week therapy with lithium carbonate leads to a reduced urinary AQP2 excretion (Walker et al. 2005).

Lithium exerts its effects by entering principal cells through the epithelial sodium channel ENaC, which has a higher permeability for lithium than for sodium (Kellenberger et al. 1999). In the mouse collecting duct cell lines $mCCD_{c11}$ and mpkCCD, co-incubation of lithium with the ENaC blockers amiloride or benzamil prevented the lithium-induced AQP2 downregulation and decreased the intracellular lithium concentration (Kortenoeven et al. 2009), showing that ENaC is the major cellular entry pathway for lithium. These cell studies were confirmed by experiments in rats, where simultaneous treatment with amiloride and lithium decreased urine output and increased urine osmolality relative to lithium alone (Kortenoeven et al. 2009). Dual treatment also attenuated AQP2 downregulation and completely prevented the reduction of principal/intercalated cell ratios. Christensen et al. determined that collecting duct-specific αENaC knockout mice do not demonstrate polyuria or a reduction in urine osmolality when treated with lithium, and AQP2 protein levels are only decreased in the inner medulla of these mice, confirming that ENaC is the entry pathway for lithium (Christensen et al. 2011). In agreement with these animal studies, it has been shown in lithium-NDI patients that blocking ENaC with amiloride significantly reduces urine volume and increases urine osmolality (Batlle et al. 1985; Bedford et al. 2008; Kosten and Forrest 1986).

After entry into the principal cell, lithium does not affect the pathway activated by AVP, as indicated by data in mpkCCD mouse collecting duct cells, showing that lithium does not affect AVP-induced cAMP generation, PKA-dependent AQP2 phosphorylation, or the phosphorylation of the AQP2 transcription factor CREB (Li et al. 2006). Furthermore, in AVP-deficient Brattleboro rats with clamped blood dDAVP levels, dDAVP-induced cAMP generation is not affected by lithium treatment (Li et al. 2006).

Instead, lithium leads to inactivation of intracellular glycogen synthase kinase (Gsk) 3β, which is temporally related to increased COX-2 expression in the kidney and in cultured principal and interstitial cells (Kortenoeven et al. 2009, 2012a; Rao et al. 2004, 2005). Consistent with the role of COX-2 in prostaglandin production, lithium increased PGE_2 excretion in the urine of rats and mice (Kotnik et al. 2005; Rao et al. 2005). As PGE_2 reduces AVP-stimulated water reabsorption in perfused collecting ducts (Hebert et al. 1990; Nadler et al. 1992), this suggests a possible role for PGE_2 in lithium-induced NDI development. This theory is supported by studies showing that blocking prostaglandin production by indomethacin reduces the urine output of lithium-treated rats (Kim et al. 2008) as well as of patients with lithium-induced NDI (Allen et al. 1989; Weinstock and Moses 1990). In mpkCCD cells, prostaglandins do not affect *AQP2* gene transcription, but increase lysosomal degradation of AQP2 (Kortenoeven et al. 2012a), suggesting that prostaglandins might have a role in lithium-induced NDI by decreasing AQP2 protein stability.

In mpkCCD cells, lithium decreases AQP2 protein abundance as well as AQP2 transcription independently of endogenous prostaglandin production. The pathway leading to a decreased AQP2 transcription is not known, but might involve one of the pathways driven by Gsk3β, as not only lithium but also other Gsk3 inhibitors lead to a downregulation of AQP2 expression in mpkCCD cells (Kortenoeven et al. 2009), and, in agreement with a Gsk3β regulation of AQP2, collecting duct-specific Gsk3β knockout mice are polyuric and have reduced AQP2 mRNA and protein levels after water deprivation compared with wild-type mice (Rao et al. 2010).

25.3.2.4 Acute Tubulointerstitial Nephritis

Tubular dysfunction resulting in reduced water and salt reabsorption occurs as a result of inflammatory processes in the kidney tubules and/or interstitium as a complication of a variety of pathological conditions including rabdomyolysis, septicemia, leptospirosis, and renal vascular occlusion. This causes acute tubulointerstitial nephritis (ATIN) leading to acute renal failure of intrarenal origin. Diuresis is frequently maintained despite reduction in GFR, resulting in non-oliguric renal failure (Raghavan and Eknoyan 2014). In several of these conditions, kidney aquaporin expression has been studied in patients as well as animal models.

Leptospirosis-induced ATIN is typically non-oliguric and accompanied by a urinary-concentrating defect (Ooi et al. 1972; Seguro and Andrade 2013). Downregulation of AQP1 in proximal tubules coupled with upregulation of AQP2 has been observed in autopsies after lethal outcome of the disease (Araujo et al. 2010), but animal models have suggested that AQP2 is in fact downregulated in the acute phase. In one such study, AVP resistance in the inner medullary collecting duct was observed (Magaldi et al. 1992) and AQP2 downregulation occurs in response to injections of glycolipoprotein, which is thought to be the virulence factor of leptospirosis (Cesar et al. 2012).

Acute renal failure in septicemia can be pre- or intrarenal, and a urinary-concentrating defect is frequently observed in patients despite elevated plasma AVP levels (Bagshaw et al. 2006; Jochberger et al. 2009). Lipopolysaccharide (LPS) is a known virulence factor, which induces a urinary-concentrating defect alongside acute downregulation of AQP2 in experimental animal models (Grinevich et al. 2004; Olesen et al. 2009), and its effects have been ascribed in part to a direct molecular link between the pro-inflammatory mediator NFκB and the AQP2 promoter (Hasler et al. 2008).

Loss of renal concentrating ability in patients suffering from acute renal ischemia is well known, and downregulation of renal AQP1, AQP2, and AQP3 has been observed in animal models of ischemia-reperfusion (Kwon et al. 1999). Downregulation of renal aquaporins was prevented by the anti-inflammatory agent αMSH, which strongly indicates that the concentrating defect is induced by an inflammatory process.

Taken together, these studies suggest that inflammatory mediators cause acute, vasopressin-independent (dys)regulation of renal aquaporins and that this process plays a role in the pathophysiology of ATIN and explains maintained diuresis in non-oliguric acute renal failure.

25.3.2.5 Diabetes Mellitus

A major cause of polyuria is diabetes mellitus (DM). Especially type-1 DM may present as frequent urination, and feared complications of the disease are ketoacidosis and the hyperglycemic hyperosmolar state (Maletkovic and Drexler

2013). As the name suggests, the latter is a state of excessive water loss and increased plasma osmolality. The regulation of AQPs in patients during this medical emergency has not been investigated, but animal models have revealed a dramatic upregulation of AQP2, AQP2 phosphorylated at Ser256 and AQP3 (Nejsum et al. 2001). This occurs in the Brattleboro rat as well and is thus independent from increases in plasma AVP levels (Kim et al. 2004a), but is blunted in adrenelectomized rats (Klein et al. 2006), indicating a role for the renin-angiotensin-aldosterone system, although candesartan, an angiotensin-II type-1 receptor antagonist, failed to prevent upregulation of renal AQPs (Blount et al. 2008).

Thus, polyuria in DM is caused exclusively by osmotic diuresis due to hyperglycemia and ketonuria and not by downregulation of renal AQPs. On the contrary, these display a compensatory upregulation by an unknown mechanism, which is assumed to alleviate the acute hyperosmolar state which may develop in DM patients (Satake et al. 2010).

25.3.2.6 Treatment of Polyuria

Administration of dDAVP can successfully treat central DI (Robinson 1976). It is also an effective treatment for gestational DI, since dDAVP is resistant to degradation by vasopressinase (Ananthakrishnan 2009). dDAVP can be administered either as an intranasal spray, tablet, or sublingual melting formulation. For the majority of NDI patients, treatment with dDAVP is however not effective. The main strategy for treating NDI is a sufficient supply of fluid to replace the urinary water losses, usually combined with a low-solute diet, which probably, by causing a slight hypovolemia and activation of the RAAS system, increases the proximal reabsorption of water and thereby decreases the urine volume (Beaser 1947; Blalock et al. 1977). Additionally, thiazide diuretics, alone or in combination with amiloride or indomethacin, can efficiently decrease the urine volume in NDI patients, although usually not to normal levels and not in all patients.

Diuretics

Currently, thiazide diuretics are used to reduce polyuria in NDI patients. Several studies have shown that in patients with congenital NDI, as well as lithium-induced NDI, thiazide diuretics effectively decrease urine output and increase urine osmolality (Alon and Chan 1985; Crawford et al. 1960; Forrest et al. 1974; Jakobsson and Berg 1994). The precise mechanism by which thiazide diuretics cause their paradoxical antidiuretic effect is largely unknown. The thiazide-sensitive NaCl cotransporter, NCC, is expressed in the renal distal convoluted tubule, a segment which is water impermeable; therefore, the water-preserving effect of thiazides is unlikely related to a direct effect on this part of the nephron. However, the thiazide-mediated inhibition of NCC will cause an increased renal sodium excretion, which will induce a decrease in extracellular volume, leading to a decreased glomerular filtration rate (GFR), increased activation of the renin-angiotensin-aldosterone system, and a compensatory increase in proximal sodium and water reabsorption. Consequently, less water is delivered to the collecting duct, and less urine is produced. Recently it has been shown that thiazide blocks not only NCC but also the Na^{+}-dependent Cl^{-}/HCO_3^{-} exchanger NDCBE in the intercalated cells of the collecting duct, which might further contribute to this mechanism (Leviel et al. 2010), and in addition, hydrochlorothiazide inhibits carbonic anhydrases (Matsumoto et al. 1989). Treatment of NCC knockout mice with lithium and hydrochlorothiazide revealed a significantly reduced urine volume and increased cortical AQP2 abundance compared with NCC knockout mice treated with lithium alone (Sinke et al. 2014), indicating that part of the antidiuretic effect of thiazide in lithium-induced NDI is NCC independent, which may involve a reduction in GFR due to thiazide-mediated proximal tubular carbonic anhydrase inhibition or inhibition of NDCBE in the collecting duct (Sinke et al. 2014). However, it has also been shown that thiazide directly enhances water permeability of isolated collecting ducts (Cesar and Magaldi 1999) and that thiazide treatment increases the expression of AQP2 in rats with lithium-induced NDI as well as in the mpkCCD mouse collecting duct cell line (Kim et al. 2004b; Sinke et al. 2014), suggesting a direct effect of thiazide on the collecting duct.

Combining treatment of hydrochlorothiazide with either a prostaglandin synthesis inhibitor, e.g., indomethacin, or with the ENaC blocker amiloride is more effective in reducing urine volume than thiazide alone (Alon et al. 1985; Alon and Chan 1985; Jakobsson and Berg 1994; Kirchlechner et al. 1999; Konoshita et al. 2004; Monnens et al. 1984; Rascher et al. 1987). Besides this, amiloride also prevents hypokalemia and metabolic alkalosis, common side effects of thiazide therapy (Alon et al. 1985; Konoshita et al. 2004).

Future Therapies

In recent years many promising therapies for NDI have been identified, e.g., treatment with cell-permeable non-peptide V2R antagonists. Most V2R mutations associated with X-linked NDI result in ER retention of the receptor due to misfolding, but the retained receptor may still be functional. It was shown that a group of compounds known as "chemical chaperones," including glycerol and DMSO, could rescue the plasma membrane expression these mutant V2R, probably by stabilization of the receptors conformation (Robben et al. 2006b). Non-peptide V2R antagonists have been tested for their efficacy as chaperones as well, to increase specificity of the treatment and to reduce toxicity. Once the antagonist-bound V2R reaches the plasma membrane, the antagonist needs to be replaced by AVP or dDAVP to allow activation of the receptor (Fig. 25.5). Several cell-permeable V2R antagonists have indeed been shown to stabilize ER-retained V2R mutants in cultured cells, inducing the receptor to escape from the ER and reach the plasma membrane (Morello et al. 2000; Robben et al. 2007). Subsequent

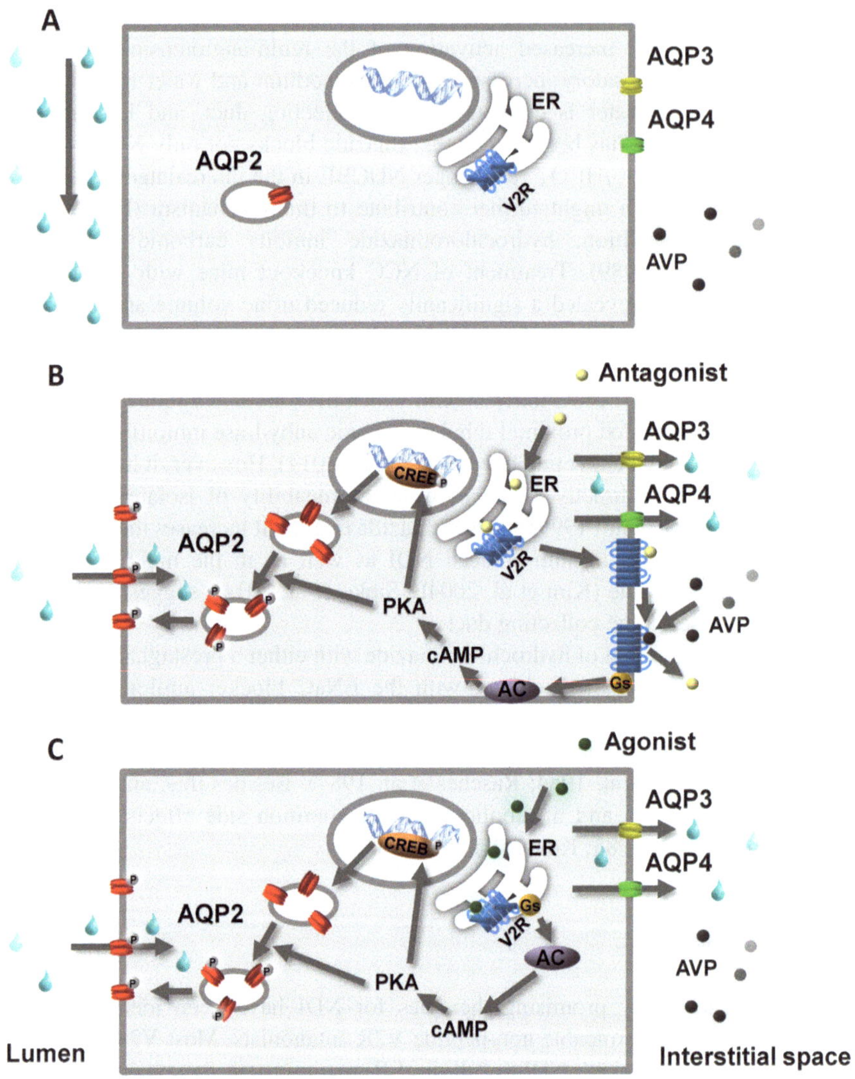

Fig. 25.5 Possible treatment strategy for X-linked NDI. (**a**) X-linked NDI is caused by mutations in the gene encoding the V2R. Most mutations are retained in the endoplasmic reticulum (ER) caused by misfolding of the protein. (**b**) Rescue of plasma membrane expression of the V2R by V2R antagonists. Antagonists bind to misfolded V2R mutants in the ER. This aids the correct folding of the V2R and allows the V2R to escape the ER and reach the plasma membrane. In the plasma membrane, the antagonist is displaced by AVP to allow activation of the receptor, leading to increased AQP2 at the plasma membrane. (**c**) Activation of misfolded V2R mutants by cell-permeable agonists. The agonists reach the misfolded V2R in the ER. This allows signaling to occur from ER-retained receptors, leading to increased AQP2 trafficking. Other cell-permeable agonists also aid proper folding of the V2R, leading to increased expression of V2R at the plasma membrane. Modified from Kortenoeven ML, Fenton RA (2014) Renal aquaporins and water balance disorders. Biochim Biophys Acta 1840:1533–1549

AVP administration leads to the generation of cAMP in these cells, showing a functional rescue of the receptors (Robben et al. 2007). Cell membrane expression of V2R mutants can also be rescued in cultured cells using the V1a receptor antagonist SR49059 (relcovaptan, with moderate affinity for V2R). Furthermore, SR49059 significantly reduces the urine output in X-linked NDI patients, proving in principle that non-peptide antagonists can rescue V2R mutant activity in vivo (Bernier et al. 2006). A limitation of non-peptide antagonists is that their efficiency may depend on the type and location of the specific V2R mutation, meaning that different mutations may require different antagonists to achieve V2R rescue and some mutants may be insensitive to all antagonists (Los et al. 2010).

Another possible treatment for X-linked NDI is the use of cell-permeable non-peptide V2R agonists (Fig. 25.5). The non-peptide V2R agonists OPC51803, VA999088, and VA999089 have been shown to induce a cAMP response in polarized renal cells, eventually leading to an increase in AQP2 trafficking to the apical plasma membrane in six out of seven tested dDAVP-insensitive V2R mutants (Robben et al. 2009). Interestingly, the non-peptide agonists did not affect ER localization of the mutant receptors and failed to induce receptor maturation, indicating that the observed cAMP signal was derived from V2R still retained in the ER. Recently, three structurally related non-peptide agonists MCF14, MCF18, and MCF57 were demonstrated to increase cAMP concentrations of intracellularly retained V2R mutants as well. However, this coincided with an increased plasma membrane expression and increased maturation of the V2R mutants (Jean-Alphonse et al. 2009), indicating a different mechanism of rescue.

In other forms of NDI than X-linked NDI or X-linked NDI where V2R mutants are not functional or insensitive to rescue, alternative treatments might be applied which bypass the V2R signaling cascade and increase trafficking or abundance of AQP2 via activation of another pathway. One option is to stimulate PGE_2 receptors. PGE_2 reduces AVP-induced water permeability; however, PGE_2 increases collecting duct water permeability in the absence of high levels of AVP, probably via activation of the PGE_2 receptors EP2 and EP4 (Hebert et al. 1990; Nadler et al. 1992; Sakairi et al. 1995). As mentioned above, inhibiting general prostaglandin synthesis by indomethacin has been shown to reduce urine volume in NDI patients. However, specific EP2 or EP4 agonists might be more effective, since no activation is expected of prostaglandin receptors involved in decreasing collecting duct water permeability. The EP2 agonist butaprost and the EP4 agonist CAY10580 were shown to increase AQP2 apical membrane accumulation in MDCK cells and butaprost increased AQP2 membrane accumulation in slices of kidney tissue (Olesen et al. 2011). The EP4 agonist ONO-AE-329 has been shown to decrease urine output, increase urine osmolality, and increase AQP2 abundance in a mouse model for X-linked NDI, while the EP2 agonist butaprost has been shown to reduce urine flow rate and increase urine osmolality in a rat model of NDI, suggesting that specific EP2 or EP4 agonists could be promising treatment strategies for NDI (Li et al. 2009; Olesen et al. 2011).

Another possible treatment option for NDI is the use of statins or HMG-CoA reductase inhibitors. Statins are a class of drugs that are used in the treatment of hypercholesterolemia, because they competitively inhibit the rate-limiting enzyme HMG-CoA in the biosynthesis of cholesterol. Fluvastatin, lovastatin, and simvastatin have been shown to increase the apical plasma membrane expression of AQP2 in cultured cells by inhibiting AQP2 endocytosis (Li et al. 2011; Procino et al. 2010, 2011a), which has been suggested to be due to a decrease in active Rho involved in regulation of the cytoskeleton and AQP2 translocation (Li et al. 2011). In vivo studies have shown that simvastatin, fluvastatin, and atorvastatin can decrease urine volume, increase urine osmolality, and increase AQP2 expression (Danilovic et al. 2012; Li et al. 2011; Procino et al. 2011a). Future studies have to determine if statins can indeed be used as a therapy for NDI.

An alternative method to bypass the V2R-pathway is increasing intracellular cGMP concentrations. An increase in cGMP, mediated by atrial natriuretic peptide or nitric oxide, has been shown to increase AQP2 abundance at the apical plasma membrane (Boone et al. 2010; Bouley et al. 2000; Wang et al. 2006). In both transfected cells and in AVP-deficient Brattleboro rats, preventing cGMP degradation by the PDE5 inhibitor sildenafil results in an increased plasma membrane expression of AQP2 (Bouley et al. 2005). In addition, sildenafil has been shown to reduce urine volume and increase total membrane AQP2 levels in rats with lithium-induced NDI (Sanches et al. 2012), indicating that sildenafil could be a potential treatment strategy for NDI.

Currently there are no effective therapies to treat or prevent sepsis-induced acute renal failure, because the pathogenesis of acute renal failure in sepsis remains incompletely understood. In a sepsis mouse model, inhibiting NFκB by pyrrolidine dithiocarbamate or small interfering RNA attenuated the sepsis-induced downregulation of AQP2 and V2R and ameliorated the hypotension and acute renal failure, suggesting that inhibition of NFκB could be a possible treatment strategy (Hocherl et al. 2010).

25.4 Conclusion

Nine of the twelve known AQPs are expressed in the kidney. Of these aquaporins, only AQP1–4 and possibly AQP7 have a role in the process of urine concentration, while others are involved in diverse functions such as acidification of intracellular vesicles and ammonia transport across mitochondrial membranes. Regulation of renal aquaporins is critical to osmoregulation and the maintenance of body water homeostasis. Considering the important role of aquaporins in water transport, it is not surprising that a number of aquaporins, especially AQP2, are involved in diseases associated with disturbed water homeostasis. In the last decade, our understanding of the molecular mechanisms of these disorders has increased enormously, which has opened up several possible treatment strategies. However, the exact molecular mechanisms which regulate the different aquaporins in different conditions are still largely unknown. Future basic and translational studies are needed to fully understand and treat various water balance disorders.

Acknowledgments Research in the authors' laboratory is supported by the Danish Medical Research Council, the Lundbeck Foundation, the Novo Nordisk Foundation, and the Carlsberg Foundation.

References

Abu LA, Levy-Khademi F, Abdulhadi-Atwan M, Bosin E, Korner M, White PC, Zangen DH (2010) Autosomal recessive familial neurohypophyseal diabetes insipidus: onset in early infancy. Eur J Endocrinol 162:221–226

Alfadda TI, Saleh AM, Houillier P, Geibel JP (2014) Calcium-sensing receptor 20 years later. Am J Physiol Cell Physiol 307:C221–C231

Allen HM, Jackson RL, Winchester MD, Deck LV, Allon M (1989) Indomethacin in the treatment of lithium-induced nephrogenic diabetes insipidus. Arch Intern Med 149:1123–1126

Alon U, Chan JC (1985) Hydrochlorothiazide-amiloride in the treatment of congenital nephrogenic diabetes insipidus. Am J Nephrol 5:9–13

Alon U, Wellons MD, Chan JC (1985) Hydrochlorothiazide, amiloride and tolmetin in the treatment of diabetes insipidus of Brattleboro rats. Med Biol 63:117–122

Ananthakrishnan S (2009) Diabetes insipidus in pregnancy: etiology, evaluation, and management. Endocr Pract 15:377–382

Araujo ER, Seguro AC, Spichler A, Magaldi AJ, Volpini RA, De BT (2010) Acute kidney injury in human leptospirosis: an immunohistochemical study with pathophysiological correlation. Virchows Arch 456:367–375

Arima H, Oiso Y (2010) Mechanisms underlying progressive polyuria in familial neurohypophysial diabetes insipidus. J Neuroendocrinol 22(7):754–757

Aronson D, Verbalis JG, Mueller M, Krum H (2011) Short- and long-term treatment of dilutional hyponatraemia with satavaptan, a selective arginine vasopressin V2-receptor antagonist: the DILIPO study. Eur J Heart Fail 13:327–336

Arthus MF, Lonergan M, Crumley MJ, Naumova AK, Morin D, De Marco LA, Kaplan BS, Robertson GL, Sasaki S, Morgan K, Bichet DG, Fujiwara TM (2000) Report of 33 novel AVPR2 mutations and analysis of 117 families with X-linked nephrogenic diabetes insipidus. J Am Soc Nephrol 11:1044–1054

Asahina Y, Izumi N, Enomoto N, Sasaki S, Fushimi K, Marumo F, Sato C (1995) Increased gene expression of water channel in cirrhotic rat kidneys. Hepatology 21:169–173

Atochina-Vasserman EN, Biktasova A, Abramova E, Cheng DS, Polosukhin VV, Tanjore H, Takahashi S, Sonoda H, Foye L, Venkov C, Ryzhov SV, Novitskiy S, Shlonimskaya N, Ikeda M, Blackwell TS, Lawson WE, Gow AJ, Harris RC, Dikov MM, Tchekneva EE (2013) Aquaporin 11 insufficiency modulates kidney susceptibility to oxidative stress. Am J Physiol Renal Physiol 304:F1295–F1307

Badaut J, Fukuda AM, Jullienne A, Petry KG (2014) Aquaporin and brain diseases. Biochim Biophys Acta 1840:1554–1565

Bagshaw SM, Langenberg C, Bellomo R (2006) Urinary biochemistry and microscopy in septic acute renal failure: a systematic review. Am J Kidney Dis 48:695–705

Bankir L (2001) Antidiuretic action of vasopressin: quantitative aspects and interaction between V1a and V2 receptor-mediated effects. Cardiovasc Res 51:372–390

Barak LS, Oakley RH, Laporte SA, Caron MG (2001) Constitutive arrestin-mediated desensitization of a human vasopressin receptor mutant associated with nephrogenic diabetes insipidus. Proc Natl Acad Sci U S A 98:93–98

Batlle DC, von Riotte AB, Gaviria M, Grupp M (1985) Amelioration of polyuria by amiloride in patients receiving long-term lithium therapy. N Engl J Med 312:408–414

Baylis PH (1989) Regulation of vasopressin secretion. Baillieres Clin Endocrinol Metab 3:313–330

Baylis PH (2003) The syndrome of inappropriate antidiuretic hormone secretion. Int J Biochem Cell Biol 35:1495–1499

Beaser SB (1947) Renal excretory function and diet in diabetes insipidus. Am J Med Sci 213:441–449

Bedford JJ, Weggery S, Ellis G, McDonald FJ, Joyce PR, Leader JP, Walker RJ (2008) Lithium-induced nephrogenic diabetes insipidus: renal effects of amiloride. Clin J Am Soc Nephrol 3:1324–1331

Beitz E, Liu K, Ikeda M, Guggino WB, Agre P, Yasui M (2006) Determinants of AQP6 trafficking to intracellular sites versus the plasma membrane in transfected mammalian cells. Biol Cell 98:101–109

Bernier V, Morello JP, Zarruk A, Debrand N, Salahpour A, Lonergan M, Arthus MF, Laperriere A, Brouard R, Bouvier M, Bichet DG (2006) Pharmacologic chaperones as a potential treatment for x-linked nephrogenic diabetes insipidus. J Am Soc Nephrol 17:232–243

Blalock T, Gerron G, Quiter E, Rudman D (1977) Role of diet in the management of vasopressin-responsive and -resistant diabetes insipidus. Am J Clin Nutr 30:1070–1076

Blaydon DC, Lind LK, Plagnol V, Linton KJ, Smith FJ, Wilson NJ, McLean WH, Munro CS, South AP, Leigh IM, O'Toole EA, Lundstrom A, Kelsell DP (2013) Mutations in AQP5, encoding a water-channel protein, cause autosomal-dominant diffuse nonepidermolytic palmoplantar keratoderma. Am J Hum Genet 93:330–335

Blount MA, Sands JM, Kent KJ, Smith TD, Price SR, Klein JD (2008) Candesartan augments compensatory changes in medullary transport proteins in the diabetic rat kidney. Am J Physiol Renal Physiol 294:F1448–F1452

Boone M, Deen PM (2008) Physiology and pathophysiology of the vasopressin-regulated renal water reabsorption. Pflugers Arch 456:1005–1024

Boone M, Kortenoeven M, Robben JH, Deen PM (2010) Effect of the cGMP pathway on AQP2 expression and translocation: potential implications for nephrogenic diabetes insipidus. Nephrol Dial Transplant 25:48–54

Boone M, Kortenoeven ML, Robben JH, Tamma G, Deen PM (2011) Counteracting vasopressin-mediated water reabsorption by ATP, dopamine, and phorbol esters: mechanisms of action. Am J Physiol Renal Physiol 300:F761–F771

Boton R, Gaviria M, Batlle DC (1987) Prevalence, pathogenesis, and treatment of renal dysfunction associated with chronic lithium therapy. Am J Kidney Dis 10:329–345

Bouley R, Breton S, Sun T, McLaughlin M, Nsumu NN, Lin HY, Ausiello DA, Brown D (2000) Nitric oxide and atrial natriuretic factor stimulate cGMP-dependent membrane insertion of aquaporin 2 in renal epithelial cells. J Clin Invest 106:1115–1126

Bouley R, Pastor-Soler N, Cohen O, McLaughlin M, Breton S, Brown D (2005) Stimulation of AQP2 membrane insertion in renal epithelial cells in vitro and in vivo by the cGMP phosphodiesterase inhibitor sildenafil citrate (Viagra). Am J Physiol Renal Physiol 288:F1103–F1112

Bouley R, Palomino Z, Tang SS, Nunes P, Kobori H, Lu HA, Shum WW, Sabolic I, Brown D, Ingelfinger JR, Jung FF (2009) Angiotensin II and hypertonicity modulate proximal tubular aquaporin 1 expression. Am J Physiol Renal Physiol 297:F1575–F1586

Brond L, Mullertz KM, Torp M, Nielsen J, Graebe M, Hadrup N, Nielsen S, Christensen S, Jonassen TE (2013) Congestive heart failure in rats is associated with increased collecting duct vasopressin sensitivity and vasopressin type 2 receptor reexternalization. Am J Physiol Renal Physiol 305:F1547–F1554

Bustamante M, Hasler U, Kotova O, Chibalin AV, Mordasini D, Rousselot M, Vandewalle A, Martin PY, Feraille E (2005) Insulin potentiates AVP-induced AQP2 expression in cultured renal collecting duct principal cells. Am J Physiol Renal Physiol 288:F334–F344

Bustamante M, Hasler U, Leroy V, de Seigneux S, Dimitrov M, Mordasini D, Rousselot M, Martin PY, Feraille E (2008) Calcium-sensing receptor attenuates AVP-induced aquaporin-2 expression via a calmodulin-dependent mechanism. J Am Soc Nephrol 19:109–116

Cabral PD, Herrera M (2012) Membrane-associated aquaporin-1 facilitates osmotically driven water flux across the basolateral membrane of the thick ascending limb. Am J Physiol Renal Physiol 303:F621–F629

Calamita G, Gena P, Meleleo D, Ferri D, Svelto M (2006) Water permeability of rat liver mitochondria: a biophysical study. Biochim Biophys Acta 1758:1018–1024

Canfield MC, Tamarappoo BK, Moses AM, Verkman AS, Holtzman EJ (1997) Identification and characterization of aquaporin-2 water channel mutations causing nephrogenic diabetes insipidus with partial vasopressin response. Hum Mol Genet 6:1865–1871

Carpentier E, Greenbaum LA, Rochdi D, Abrol R, Goddard WA III, Bichet DG, Bouvier M (2012) Identification and characterization of an activating F229V substitution in the V2 vasopressin receptor in an infant with NSIAD. J Am Soc Nephrol 23:1635–1640

Cesar KR, Magaldi AJ (1999) Thiazide induces water absorption in the inner medullary collecting duct of normal and Brattleboro rats. Am J Physiol 277:F756–F760

Cesar KR, Romero EC, de Braganca AC, Blanco RM, Abreu PA, Magaldi AJ (2012) Renal involvement in leptospirosis: the effect of glycolipoprotein on renal water absorption. PLoS One 7, e37625

Chen Y, Rice W, Gu Z, Li J, Huang J, Brenner MB, Van HA, Xiong J, Gundersen GG, Norman JC, Hsu VW, Fenton RA, Brown D, Lu HA (2012) Aquaporin 2 promotes cell migration and epithelial morphogenesis. J Am Soc Nephrol 23:1506–1517

Cheng X, Zhang H, Lee HL, Park JM (2004) Cyclooxygenase-2 inhibitor preserves medullary aquaporin-2 expression and prevents polyuria after ureteral obstruction. J Urol 172:2387–2390

Chou CL, Ma TH, Yang BX, Knepper MA, Verkman S (1998) Fourfold reduction of water permeability in inner medullary collecting duct of aquaporin-4 knockout mice. Am J Physiol 43:C549–C554

Chou CL, Knepper MA, Hoek AN, Brown D, Yang B, Ma T, Verkman AS (1999) Reduced water permeability and altered ultrastructure in thin descending limb of Henle in aquaporin-1 null mice. J Clin Invest 103:491–496

Christensen BM, Zelenina M, Aperia A, Nielsen S (2000) Localization and regulation of PKA-phosphorylated AQP2 in response to V(2)-receptor agonist/antagonist treatment. Am J Physiol Renal Physiol 278:F29–F42

Christensen BM, Wang W, Frokiaer J, Nielsen S (2003) Axial heterogeneity in basolateral AQP2 localization in rat kidney: effect of vasopressin. Am J Physiol Renal Physiol 284:F701–F717

Christensen BM, Marples D, Kim YH, Wang W, Frokiaer J, Nielsen S (2004) Changes in cellular composition of kidney collecting duct cells in rats with lithium-induced NDI. Am J Physiol Cell Physiol 286:C952–C964

Christensen BM, Kim YH, Kwon TH, Nielsen S (2006) Lithium treatment induces a marked proliferation of primarily principal cells in rat kidney inner medullary collecting duct. Am J Physiol Renal Physiol 291:F39–F48

Christensen BM, Zuber AM, Loffing J, Stehle JC, Deen PM, Rossier BC, Hummler E (2011) alphaENaC-mediated lithium absorption promotes nephrogenic diabetes insipidus. J Am Soc Nephrol 22:253–261

Chung SH, Jun DW, Kim KT, Chae JD, Park EK, Son BK, Kim SH, Jo YJ, Park YS (2010) Aquaporin-2 urinary excretion in cirrhosis: relationship to vasopressin and nitric oxide. Dig Dis Sci 55:1135–1141

Coleman RA, Wu DC, Liu J, Wade JB (2000) Expression of aquaporins in the renal connecting tubule. Am J Physiol Renal Physiol 279:F874–F883

Crawford JD, Kennedy GC, HILL LE (1960) Clinical results of treatment of diabetes insipidus with drugs of the chlorothiazide series. N Engl J Med 262:737–743

Danilovic A, Lopes RI, Sanches TR, Shimizu MH, Oshiro FM, Andrade L, Denes FT, Seguro AC (2012) Atorvastatin prevents the downregulation of aquaporin-2 receptor after bilateral ureteral obstruction and protects renal function in a rat model. Urology 80:485.e15–485.e20

De Groot T, Alsady M, Jaklofsky M, Otte-Holler I, Baumgarten R, Giles RH, Deen PM (2014) Lithium causes G2 arrest of renal principal cells. J Am Soc Nephrol 25:501–510

De Mattia F, Savelkoul PJ, Bichet DG, Kamsteeg EJ, Konings IB, Marr N, Arthus MF, Lonergan M, van Os CH, van der Sluijs P, Robertson G, Deen PM (2004) A novel mechanism in recessive nephrogenic diabetes insipidus: wild-type aquaporin-2 rescues the apical membrane expression of intracellularly retained AQP2-P262L. Hum Mol Genet 13:3045–3056

De Mattia F, Savelkoul PJ, Kamsteeg EJ, Konings IB, van der Sluijs P, Mallmann R, Oksche A, Deen PM (2005) Lack of arginine vasopressin-induced phosphorylation of aquaporin-2 mutant AQP2-R254L explains dominant nephrogenic diabetes insipidus. J Am Soc Nephrol 16:2872–2880

Decaux G, Vandergheynst F, Bouko Y, Parma J, Vassart G, Vilain C (2007) Nephrogenic syndrome of inappropriate antidiuresis in adults: high phenotypic variability in men and women from a large pedigree. J Am Soc Nephrol 18:606–612

Deen PMT, van Aubel RA, van Lieburg AF, van Os CH (1996) Urinary content of aquaporin 1 and 2 in nephrogenic diabetes insipidus. J Am Soc Nephrol 7:836–841

Delporte C (2014) Aquaporins in salivary glands and pancreas. Biochim Biophys Acta 1840:1524–1532

Di Iorgi N, Napoli F, Allegri AE, Olivieri I, Bertelli E, Gallizia A, Rossi A, Maghnie M (2012) Diabetes insipidus–diagnosis and management. Horm Res Paediatr 77:69–84

Digiovanni SR, Nielsen S, Christensen EI, Knepper MA (1994) Regulation of collecting duct water channel expression by vasopressin in Brattleboro rat. Proc Natl Acad Sci U S A 91:8984–8988

Earm JH, Christensen BM, Frokiaer J, Marples D, Han JS, Knepper MA, Nielsen S (1998) Decreased aquaporin-2 expression and apical plasma membrane delivery in kidney collecting ducts of polyuric hypercalcemic rats. J Am Soc Nephrol 9:2181–2193

Ecelbarger CA, Terris J, Frindt G, Echevarria M, Marples D, Nielsen S, Knepper MA (1995) Aquaporin-3 water channel localization and regulation in rat kidney. Am J Physiol 38:F663–F672

Elhassan EA, Schrier RW (2011) Hyponatremia: diagnosis, complications, and management including V2 receptor antagonists. Curr Opin Nephrol Hypertens 20:161–168

Elkjaer ML, Nejsum LN, Gresz V, Kwon TH, Jensen UB, Frokiaer J, Nielsen S (2001) Immunolocalization of aquaporin-8 in rat kidney, gastrointestinal tract, testis, and airways. Am J Physiol Renal Physiol 281:F1047–F1057

Elkjaer ML, Kwon TH, Wang W, Nielsen J, Knepper MA, Frokiaer J, Nielsen S (2002) Altered expression of renal NHE3, TSC, BSC-1, and ENaC subunits in potassium-depleted rats. Am J Physiol Renal Physiol 283:F1376–F1388

Esteva-Font C, Baccaro ME, Fernandez-Llama P, Sans L, Guevara M, Ars E, Jimenez W, Arroyo V, Ballarin JA, Gines P (2006) Aquaporin-1 and aquaporin-2 urinary excretion in cirrhosis: relationship with ascites and hepatorenal syndrome. Hepatology 44:1555–1563

Farres MT, Ronco P, Saadoun D, Remy P, Vincent F, Khalil A, Le Blanche AF (2003) Chronic lithium nephropathy: MR imaging for diagnosis. Radiology 229:570–574

Feldman HA, Singer I (1974) Comparative effects of tetracyclines on water flow across toad urinary bladders. J Pharmacol Exp Ther 190:358–364

Feldman BJ, Rosenthal SM, Vargas GA, Fenwick RG, Huang EA, Matsuda-Abedini M, Lustig RH, Mathias RS, Portale AA, Miller WL, Gitelman SE (2005) Nephrogenic syndrome of inappropriate antidiuresis. N Engl J Med 352:1884–1890

Fenton RA, Brond L, Nielsen S, Praetorius J (2007) Cellular and subcellular distribution of the type-2 vasopressin receptor in the kidney. Am J Physiol Renal Physiol 293:F748–F760

Fenton RA, Moeller HB, Hoffert JD, Yu MJ, Nielsen S, Knepper MA (2008) Acute regulation of aquaporin-2 phosphorylation at Ser-264 by vasopressin. Proc Natl Acad Sci U S A 105:3134–3139

Fenton RA, Moeller HB, Zelenina M, Snaebjornsson MT, Holen T, MacAulay N (2010) Differential water permeability and regulation of three aquaporin 4 isoforms. Cell Mol Life Sci 67:829–840

Fernandez-Llama P, Turner R, Dibona G, Knepper MA (1999) Renal expression of aquaporins in liver cirrhosis induced by chronic common bile duct ligation in rats. J Am Soc Nephrol 10:1950–1957

Fernandez-Llama P, Jimenez W, Bosch-Marce M, Arroyo V, Nielsen S, Knepper MA (2000) Dysregulation of renal aquaporins and Na-Cl cotransporter in CCl4-induced cirrhosis. Kidney Int 58:216–228

Flamion B, Spring KR (1990) Water permeability of apical and basolateral cell membranes of rat inner medullary collecting duct. Am J Physiol 259:F986–F999

Forrest JN Jr, Cohen AD, Torretti J, Himmelhoch JM, Epstein FH (1974) On the mechanism of lithium-induced diabetes insipidus in man and the rat. J Clin Invest 53:1115–1123

Forrest JN Jr, Cox M, Hong C, Morrison G, Bia M, Singer I (1978) Superiority of demeclocycline over lithium in the treatment of chronic syndrome of inappropriate secretion of antidiuretic hormone. N Engl J Med 298:173–177

Fotiadis D, Hasler L, Muller DJ, Stahlberg H, Kistler J, Engel A (2000) Surface tongue-and-groove contours on lens MIP facilitate cell-to-cell adherence. J Mol Biol 300:779–789

Frokiaer J, Marples D, Knepper MA, Nielsen S (1996) Bilateral ureteral obstruction downregulates expression of vasopressin-sensitive AQP-2 water channel in rat kidney. Am J Physiol 39:F657–F668

Frokiaer J, Christensen BM, Marples D, Djurhuus JC, Jensen UB, Knepper MA, Nielsen S (1997) Downregulation of aquaporin-2 parallels changes in renal water excretion in unilateral ureteral obstruction. Am J Physiol 42:F213–F223

Fujisawa G, Ishikawa S, Tsuboi Y, Okada K, Saito T (1993) Therapeutic efficacy of non-peptide ADH antagonist OPC-31260 in SIADH rats. Kidney Int 44:19–23

Fujita N, Ishikawa S, Sasaki S, Fujisawa G, Fushimi K, Marumo F, Saito T (1995) Role of water channel AQP-CD in water retention in SIADH and cirrhotic rats. Am J Physiol 38:F926–F931

Furman CS, Gorelick-Feldman DA, Davidson KG, Yasumura T, Neely JD, Agre P, Rash JE (2003) Aquaporin-4 square array assembly: opposing actions of M1 and M23 isoforms. Proc Natl Acad Sci U S A 100:13609–13614

Fushimi K, Uchida S, Hara Y, Hirata Y, Marumo F, Sasaki S (1993) Cloning and expression of apical membrane water channel of rat kidney collecting tubule. Nature 361:549–552

Fushimi K, Sasaki S, Marumo F (1997) Phosphorylation of serine 256 is required for cAMP-dependent regulatory exocytosis of the aquaporin-2 water channel. J Biol Chem 272:14800–14804

Ghali JK, Koren MJ, Taylor JR, Brooks-Asplund E, Fan K, Long WA, Smith N (2006) Efficacy and safety of oral conivaptan: a V1A/V2 vasopressin receptor antagonist, assessed in a randomized, placebo-controlled trial in patients with euvolemic or hypervolemic hyponatremia. J Clin Endocrinol Metab 91:2145–2152

Gill JR Jr, Bartter FC (1961) On the impairment of renal concentrating ability in prolonged hypercalcemia and hypercalciuria in man. J Clin Invest 40:716–722

Gines P, Wong F, Watson H, Milutinovic S, del Arbol LR, Olteanu D (2008) Effects of satavaptan, a selective vasopressin V(2) receptor antagonist, on ascites and serum sodium in cirrhosis with hyponatremia: a randomized trial. Hepatology 48:204–213

Goubau C, Jaeken J, Levtchenko EN, Thys C, Di MM, Martens GA, Gerlo E, De VR, Buyse GM, Goemans N, Van GC, Freson K (2013) Homozygosity for aquaporin 7 G264V in three unrelated children with hyperglyceroluria and a mild platelet secretion defect. Genet Med 15:55–63

Grinevich V, Knepper MA, Verbalis J, Reyes I, Aguilera G (2004) Acute endotoxemia in rats induces down-regulation of V2 vasopressin receptors and aquaporin-2 content in the kidney medulla. Kidney Int 65:54–62

Hadi HA, Mashini IS, Devoe LD (1985) Diabetes insipidus during pregnancy complicated by preeclampsia. A case report. J Reprod Med 30:206–208

Harris PC, Torres VE (2009) Polycystic kidney disease. Annu Rev Med 60:321–337

Hasler U, Mordasini D, Bianchi M, Vandewalle A, Feraille E, Martin PY (2003) Dual influence of aldosterone on AQP2 expression in cultured renal collecting duct principal cells. J Biol Chem 278:21639–21648
Hasler U, Vinciguerra M, Vandewalle A, Martin PY, Feraille E (2005) Dual effects of hypertonicity on aquaporin-2 expression in cultured renal collecting duct principal cells. J Am Soc Nephrol 16:1571–1582
Hasler U, Leroy V, Jeon US, Bouley R, Dimitrov M, Kim JA, Brown D, Kwon HM, Martin PY, Feraille E (2008) NF-kappaB modulates aquaporin-2 transcription in renal collecting duct principal cells. J Biol Chem 283:28095–28105
Hebert RL, Jacobson HR, Breyer MD (1990) PGE2 inhibits AVP-induced water flow in cortical collecting ducts by protein kinase C activation. Am J Physiol 259:F318–F325
Hirji MR, Mucklow JC (1991) Transepithelial water movement in response to carbamazepine, chlorpropamide and demeclocycline in toad urinary bladder. Br J Pharmacol 104:550–553
Hocherl K, Schmidt C, Kurt B, Bucher M (2010) Inhibition of NF-kappaB ameliorates sepsis-induced downregulation of aquaporin-2/V2 receptor expression and acute renal failure in vivo. Am J Physiol Renal Physiol 298:F196–F204
Hoffert JD, Pisitkun T, Wang G, Shen RF, Knepper MA (2006) Quantitative phosphoproteomics of vasopressin-sensitive renal cells: regulation of aquaporin-2 phosphorylation at two sites. Proc Natl Acad Sci U S A 103:7159–7164
Hoffert JD, Nielsen J, Yu MJ, Pisitkun T, Schleicher SM, Nielsen S, Knepper MA (2007) Dynamics of aquaporin-2 serine-261 phosphorylation in response to short-term vasopressin treatment in collecting duct. Am J Physiol Renal Physiol 292:F691–F700
Hoffert JD, Fenton RA, Moeller HB, Simons B, Tchapyjnikov D, McDill BW, Yu MJ, Pisitkun T, Chen F, Knepper MA (2008) Vasopressin-stimulated increase in phosphorylation at Ser269 potentiates plasma membrane retention of aquaporin-2. J Biol Chem 283:24617–24627
Hoffert JD, Pisitkun T, Saeed F, Song JH, Chou CL, Knepper MA (2012) Dynamics of the G protein-coupled vasopressin V2 receptor signaling network revealed by quantitative phosphoproteomics. Mol Cell Proteomics 11:M111
Holm LM, Jahn TP, Moller AL, Schjoerring JK, Ferri D, Klaerke DA, Zeuthen T (2005) NH3 and NH_4^+ permeability in aquaporin-expressing Xenopus oocytes. Pflugers Arch 450:415–428
Hoorn EJ, Monserez DA, Fenton RA, Overdevest I, Apperloo AJ, Zietse R, Hardillo JA (2014) Olfactory neuroblastoma with hyponatremia. J Clin Oncol. doi:10.1200/JCO.2013.49.1464
Hozawa S, Holtzman EJ, Ausiello DA (1996) cAMP motifs regulating transcription in the aquaporin 2 gene. Am J Physiol 39:C1695–C1702
Ikeda M, Beitz E, Kozono D, Guggino WB, Agre P, Yasui M (2002) Characterization of aquaporin-6 as a nitrate channel in mammalian cells: requirement of pore-lining residue threonine-63. J Biol Chem 277:39873–39879
Ikeda M, Andoo A, Shimono M, Takamatsu N, Taki A, Muta K, Matsushita W, Uechi T, Matsuzaki T, Kenmochi N, Takata K, Sasaki S, Ito K, Ishibashi K (2011) The NPC motif of aquaporin-11, unlike the NPA motif of known aquaporins, is essential for full expression of molecular function. J Biol Chem 286:3342–3350
Inoue Y, Sohara E, Kobayashi K, Chiga M, Rai T, Ishibashi K, Horie S, Su X, Zhou J, Sasaki S, Uchida S (2014) Aberrant glycosylation and localization of polycystin-1 cause polycystic kidney in an AQP11 knockout model. J Am Soc Nephrol. doi:10.1681/ASN.2013060614
Ishibashi K, Sasaki S, Fushimi K, Uchida S, Kuwahara M, Saito H, Furukawa T, Nakajima K, Yamaguchi Y, Gojobori T (1994) Molecular cloning and expression of a member of the aquaporin family with permeability to glycerol and urea in addition to water expressed at the basolateral membrane of kidney collecting duct cells. Proc Natl Acad Sci U S A 91:6269–6273
Ishibashi K, Kuwahara M, Gu Y, Kageyama Y, Tohsaka A, Suzuki F, Marumo F, Sasaki S (1997a) Cloning and functional expression of a new water channel abundantly expressed in the testis permeable to water, glycerol, and urea. J Biol Chem 272:20782–20786

Ishibashi K, Sasaki S, Fushimi K, Yamamoto T, Kuwahara M, Marumo F (1997b) Immunolocalization and effect of dehydration on AQP3, a basolateral water channel of kidney collecting ducts. Am J Physiol 41:F235–F241
Ishibashi K, Imai M, Sasaki S (2000) Cellular localization of aquaporin 7 in the rat kidney. Exp Nephrol 8:252–257
Ishibashi K, Morinaga T, Kuwahara M, Sasaki S, Imai M (2002) Cloning and identification of a new member of water channel (AQP10) as an aquaglyceroporin. Biochim Biophys Acta 1576:335–340
Ishibashi K, Hara S, Kondo S (2009) Aquaporin water channels in mammals. Clin Exp Nephrol 13:107–117
Ishibashi K, Tanaka Y, Morishita Y (2014) The role of mammalian superaquaporins inside the cell. Biochim Biophys Acta 1840:1507–1512
Ito M, Jameson JL (1997) Molecular basis of autosomal dominant neurohypophyseal diabetes insipidus – cellular toxicity caused by the accumulation of mutant vasopressin precursors within the endoplasmic reticulum. J Clin Invest 99:1897–1905
Ito M, Yu RN, Jameson JL (1999) Mutant vasopressin precursors that cause autosomal dominant neurohypophyseal diabetes insipidus retain dimerization and impair the secretion of wild-type proteins. J Biol Chem 274:9029–9037
Ivarsen P, Frokiaer J, Aagaard NK, Hansen EF, Bendtsen F, Nielsen S, Vilstrup H (2003) Increased urinary excretion of aquaporin 2 in patients with liver cirrhosis. Gut 52:1194–1199
Iyasere O, Xu G, Harris K (2012) Urinary tract obstruction. Br J Hosp Med (Lond) 73:696–700
Jakobsson B, Berg U (1994) Effect of hydrochlorothiazide and indomethacin treatment on renal function in nephrogenic diabetes insipidus. Acta Paediatr 83:522–525
Jean-Alphonse F, Perkovska S, Frantz MC, Durroux T, Mejean C, Morin D, Loison S, Bonnet D, Hibert M, Mouillac B, Mendre C (2009) Biased agonist pharmacochaperones of the AVP V2 receptor may treat congenital nephrogenic diabetes insipidus. J Am Soc Nephrol 20:2190–2203
Jensen AM, Bae EH, Norregaard R, Wang G, Nielsen S, Schweer H, Kim SW, Frokiaer J (2010) Cyclooxygenase 2 inhibition exacerbates AQP2 and pAQP2 downregulation independently of V2 receptor abundance in the postobstructed kidney. Am J Physiol Renal Physiol 298:F941–F950
Jeon US, Han KH, Park SH, Lee SD, Sheen MR, Jung JY, Kim WY, Sands JM, Kim J, Kwon HM (2007) Downregulation of renal TonEBP in hypokalemic rats. Am J Physiol Renal Physiol 293: F408–F415
Jochberger S, Dorler J, Luckner G, Mayr VD, Wenzel V, Ulmer H, Morgenthaler NG, Hasibeder WR, Dunser MW (2009) The vasopressin and copeptin response to infection, severe sepsis, and septic shock. Crit Care Med 37:476–482
Jung JY, Madsen KM, Han KH, Yang CW, Knepper MA, Sands JM, Kim J (2003) Expression of urea transporters in potassium-depleted mouse kidney. Am J Physiol Renal Physiol 285: F1210–F1224
Kamsteeg EJ, Wormhoudt TA, Rijss JPL, van Os CH, Deen PMT (1999) An impaired routing of wild-type aquaporin-2 after tetramerization with an aquaporin-2 mutant explains dominant nephrogenic diabetes insipidus. EMBO J 18:2394–2400
Kamsteeg EJ, Heijnen I, van Os CH, Deen PMT (2000) The subcellular localization of an aquaporin-2 tetramer depends on the stoichiometry of phosphorylated and nonphosphorylated monomers. J Cell Biol 151:919–930
Kamsteeg EJ, Hendriks G, Boone M, Konings IB, Oorschot V, van der Sluijs P, Klumperman J, Deen PM (2006) Short-chain ubiquitination mediates the regulated endocytosis of the aquaporin-2 water channel. Proc Natl Acad Sci U S A 103:18344–18349
Katsura T, Ausiello DA, Brown D (1996) Direct demonstration of aquaporin-2 water channel recycling in stably transfected LLC-PK1 epithelial cells. Am J Physiol 39:F548–F553

Katsura T, Gustafson CE, Ausiello DA, Brown D (1997) Protein kinase A phosphorylation is involved in regulated exocytosis of aquaporin-2 in transfected LLC-PK1 cells. Am J Physiol 41:F816–F822
Katz VL, Bowes WA Jr (1987) Transient diabetes insipidus and preeclampsia. South Med J 80:524–525
Kellenberger S, Gautschi I, Schild L (1999) A single point mutation in the pore region of the epithelial Na+ channel changes ion selectivity by modifying molecular sieving. Proc Natl Acad Sci U S A 96:4170–4175
Kim D, Sands JM, Klein JD (2004a) Role of vasopressin in diabetes mellitus-induced changes in medullary transport proteins involved in urine concentration in Brattleboro rats. Am J Physiol Renal Physiol 286:F760–F766
Kim GH, Lee JW, Oh YK, Chang HR, Joo KW, Na KY, Earm JH, Knepper MA, Han JS (2004b) Antidiuretic effect of hydrochlorothiazide in lithium-induced nephrogenic diabetes insipidus is associated with upregulation of aquaporin-2, Na-Cl co-transporter, and epithelial sodium channel. J Am Soc Nephrol 15:2836–2843
Kim GH, Choi NW, Jung JY, Song JH, Lee CH, Kang CM, Knepper MA (2008) Treating lithium-induced nephrogenic diabetes insipidus with a COX-2 inhibitor improves polyuria via upregulation of AQP2 and NKCC2. Am J Physiol Renal Physiol 294:F702–F709
King LS, Choi M, Fernandez PC, Cartron JP, Agre P (2001) Defective urinary-concentrating ability due to a complete deficiency of aquaporin-1. N Engl J Med 345:175–179
Kirchlechner V, Koller DY, Seidl R, Waldhauser F (1999) Treatment of nephrogenic diabetes insipidus with hydrochlorothiazide and amiloride. Arch Dis Child 80:548–552
Klahr S, Harris K, Purkerson ML (1988) Effects of obstruction on renal functions. Pediatr Nephrol 2:34–42
Klein JD, Kozlowski S, Antoun TA, Sands JM (2006) Adrenalectomy blocks the compensatory increases in UT-A1 and AQP2 in diabetic rat kidney. J Membr Biol 212:139–144
Knepper MA, Packer R, Good DW (1989) Ammonium transport in the kidney. Physiol Rev 69:179–249
Ko SB, Uchida S, Naruse S, Kuwahara M, Ishibashi K, Marumo F, Hayakawa T, Sasaki S (1999) Cloning and functional expression of rAOP9L a new member of aquaporin family from rat liver. Biochem Mol Biol Int 47:309–318
Kondo H, Shimomura I, Kishida K, Kuriyama H, Makino Y, Nishizawa H, Matsuda M, Maeda N, Nagaretani H, Kihara S, Kurachi Y, Nakamura T, Funahashi T, Matsuzawa Y (2002) Human aquaporin adipose (AQPap) gene. Genomic structure, promoter analysis and functional mutation. Eur J Biochem 269:1814–1826
Konoshita T, Kuroda M, Kawane T, Koni I, Miyamori I, Tofuku Y, Mabuchi H, Takeda R (2004) Treatment of congenital nephrogenic diabetes insipidus with hydrochlorothiazide and amiloride in an adult patient. Horm Res 61:63–67
Kortenoeven ML, Li Y, Shaw S, Gaeggeler HP, Rossier BC, Wetzels JF, Deen PM (2009) Amiloride blocks lithium entry through the sodium channel thereby attenuating the resultant nephrogenic diabetes insipidus. Kidney Int 76:44–53
Kortenoeven ML, Schweer H, Cox R, Wetzels JF, Deen PM (2012a) Lithium reduces aquaporin-2 transcription independent of prostaglandins. Am J Physiol Cell Physiol 302:C131–C140
Kortenoeven ML, Trimpert C, van den Brand M, Li Y, Wetzels JF, Deen PM (2012b) In mpkCCD cells, long-term regulation of aquaporin-2 by vasopressin occurs independent of protein kinase A and CREB but may involve Epac. Am J Physiol Renal Physiol 302:F1395–F1401
Kortenoeven ML, Pedersen NB, Miller RL, Rojek A, Fenton RA (2013a) Genetic ablation of aquaporin-2 in the mouse connecting tubules results in defective renal water handling. J Physiol 591:2205–2219
Kortenoeven ML, Sinke AP, Hadrup N, Trimpert C, Wetzels JF, Fenton RA, Deen PM (2013b) Demeclocycline attenuates hyponatremia by reducing aquaporin-2 expression in the renal inner medulla. Am J Physiol Renal Physiol 305:F1705–F1718

Kosten TR, Forrest JN (1986) Treatment of severe lithium-induced polyuria with amiloride. Am J Psychiatry 143:1563–1568

Kotnik P, Nielsen J, Kwon TH, Krzisnik C, Frokiaer J, Nielsen S (2005) Altered expression of COX-1, COX-2, and mPGES in rats with nephrogenic and central diabetes insipidus. Am J Physiol Renal Physiol 288:F1053–F1068

Kuwahara M, Fushimi K, Terada Y, Bai L, Marumo F, Sasaki S (1995) cAMP-dependent phosphorylation stimulates water permeability of aquaporin-collecting duct water channel protein expressed in Xenopus oocytes. J Biol Chem 270:10384–10387

Kwon TH, Frøkiaer J, Fernández-Llama P, Knepper MA, Nielsen S (1999) Reduced abundance of aquaporins in rats with bilateral ischemia-induced acute renal failure: prevention by alpha-MSH. Am J Physiol 277:F413–F427

Kwon TH, Laursen UH, Marples D, Maunsbach AB, Knepper MA, Frokiaer J, Nielsen S (2000) Altered expression of renal AQPs and Na(+) transporters in rats with lithium-induced NDI. Am J Physiol Renal Physiol 279:F552–F564

Land H, Schutz G, Schmale H, Richter D (1982) Nucleotide sequence of cloned cDNA encoding bovine arginine vasopressin-neurophysin II precursor. Nature 295:299–303

Lassiter WE, Gottschalk CW, Mylle M (1961) Micropuncture study of net transtubular movement of water and urea in nondiuretic mammalian kidney. Am J Physiol 200:1139–1147

Laursen UH, Pihakaski-Maunsbach K, Kwon TH, Ostergaard JE, Nielsen S, Maunsbach AB (2004) Changes of rat kidney AQP2 and Na, K-ATPase mRNA expression in lithium-induced nephrogenic diabetes insipidus. Nephron Exp Nephrol 97:e1–e16

Lee WK, Bork U, Gholamrezaei F, Thevenod F (2005) Cd(2+)-induced cytochrome c release in apoptotic proximal tubule cells: role of mitochondrial permeability transition pore and Ca^{2+} uniporter. Am J Physiol Renal Physiol 288:F27–F39

Levi M, Peterson L, Berl T (1983) Mechanism of concentrating defect in hypercalcemia. Role of polydipsia and prostaglandins. Kidney Int 23:489–497

Leviel F, Hübner CA, Houillier P, Morla L, El Moghrabi S, Brideau G, Hassan H, Parker MD, Kurth I, Kougioumtzes A, Sinning A, Pech V, Riemondy KA, Miller RL, Hummler E, Shull GE, Aronson PS, Doucet A, Wall SM, Chambrey R, Eladari D (2010) The Na+-dependent chloride-bicarbonate exchanger SLC4A8 mediates an electroneutral Na+ reabsorption process in the renal cortical collecting ducts of mice. J Clin Invest 120:1627–1635

Levtchenko EN, Monnens LA (2010) Nephrogenic syndrome of inappropriate antidiuresis. Nephrol Dial Transplant 25:2839–2843

Li C, Wang W, Kwon TH, Isikay L, Wen JG, Marples D, Djurhuus JC, Stockwell A, Knepper MA, Nielsen S, Frokiaer J (2001) Downregulation of AQP1, -2, and -3 after ureteral obstruction is associated with a long-term urine-concentrating defect. Am J Physiol Renal Physiol 281:F163–F171

Li C, Wang W, Knepper MA, Nielsen S, Frokiar J (2003a) Downregulation of renal aquaporins in response to unilateral ureteral obstruction. Am J Physiol Renal Physiol 284:F1066–F1079

Li C, Wang W, Kwon TH, Knepper MA, Nielsen S, Frokiaer J (2003b) Altered expression of major renal Na transporters in rats with bilateral ureteral obstruction and release of obstruction. Am J Physiol Renal Physiol 285:F889–F901

Li C, Wang W, Kwon TH, Knepper MA, Nielsen S, Frokiaer J (2003c) Altered expression of major renal Na transporters in rats with unilateral ureteral obstruction. Am J Physiol Renal Physiol 284:F155–F166

Li C, Klein JD, Wang W, Knepper MA, Nielsen S, Sands JM, Frokiaer J (2004) Altered expression of urea transporters in response to ureteral obstruction. Am J Physiol Renal Physiol 286: F1154–F1162

Li Y, Shaw S, Kamsteeg EJ, Vandewalle A, Deen PM (2006) Development of lithium-induced nephrogenic diabetes insipidus is dissociated from adenylyl cyclase activity. J Am Soc Nephrol 17:1063–1072

Li JH, Chou CL, Li B, Gavrilova O, Eisner C, Schnermann J, Anderson SA, Deng CX, Knepper MA, Wess J (2009) A selective EP4 PGE2 receptor agonist alleviates disease in a new mouse model of X-linked nephrogenic diabetes insipidus. J Clin Invest 119:3115–3126

Li W, Zhang Y, Bouley R, Chen Y, Matsuzaki T, Nunes P, Hasler U, Brown D, Lu HA (2011) Simvastatin enhances aquaporin-2 surface expression and urinary concentration in vasopressin-deficient Brattleboro rats through modulation of Rho GTPase. Am J Physiol Renal Physiol 301:F309–F318

Lin QQ, Lin R, Ji QL, Zhang JY, Wang WR, Yang LN, Zhang KF (2011) Effect of exercise training on renal function and renal aquaporin-2 expression in rats with chronic heart failure. Clin Exp Pharmacol Physiol 38(3):179–185

Liu Z, Shen J, Carbrey JM, Mukhopadhyay R, Agre P, Rosen BP (2002) Arsenite transport by mammalian aquaglyceroporins AQP7 and AQP9. Proc Natl Acad Sci U S A 99:6053–6058

Liu K, Nagase H, Huang CG, Calamita G, Agre P (2006) Purification and functional characterization of aquaporin-8. Biol Cell 98:153–161

Lolait SJ, O'Carroll AM, McBride OW, Konig M, Morel A, Brownstein MJ (1992) Cloning and characterization of a vasopressin V2 receptor and possible link to nephrogenic diabetes insipidus. Nature 357:336–339

Los EL, Deen PM, Robben JH (2010) Potential of nonpeptide (ant)agonists to rescue vasopressin V2 receptor mutants for the treatment of X-linked nephrogenic diabetes insipidus. J Neuroendocrinol 22:393–399

Lu W, Peissel B, Babakhanlou H, Pavlova A, Geng L, Fan X, Larson C, Brent G, Zhou J (1997) Perinatal lethality with kidney and pancreas defects in mice with a targetted Pkd1 mutation. Nat Genet 17:179–181

Ma TH, Yang BX, Kuo WL, Verkman AS (1996) cDNA cloning and gene structure of a novel water channel expressed exclusively in human kidney: Evidence for a gene cluster of aquaporins at chromosome locus 12q13. Genomics 35:543–550

Ma TH, Yang BX, Gillespie A, Carlson EJ, Epstein CJ, Verkman AS (1997) Generation and phenotype of a transgenic knockout mouse lacking the mercurial-insensitive water channel aquaporin-4. J Clin Invest 100:957–962

Ma TH, Yang BX, Gillespie A, Carlson EJ, Epstein CJ, Verkman AS (1998) Severely impaired urinary concentrating ability in transgenic mice lacking aquaporin-1 water channels. J Biol Chem 273:4296–4299

Ma T, Song Y, Yang B, Gillespie A, Carlson EJ, Epstein CJ, Verkman AS (2000) Nephrogenic diabetes insipidus in mice lacking aquaporin-3 water channels. Proc Natl Acad Sci U S A 97:4386–4391

Madeira A, Fernandez-Veledo S, Camps M, Zorzano A, Moura TF, Ceperuelo-Mallafre V, Vendrell J, Soveral G (2014) Human aquaporin-11 is a water and glycerol channel and localizes in the vicinity of lipid droplets in human adipocytes. Obesity. doi:10.1002/oby.20792

Magaldi AJ, Yasuda PN, Kudo LH, Seguro AC, Rocha AS (1992) Renal involvement in leptospirosis: a pathophysiologic study. Nephron 62:332–339

Maletkovic J, Drexler A (2013) Diabetic ketoacidosis and hyperglycemic hyperosmolar state. Endocrinol Metab Clin North Am 42:677–695

Mancinelli R, La Rovere RM, Fulle S, Miscia S, Marchisio M, Pierdomenico L, Lanuti P, Procino G, Barbieri C, Svelto M, Fano-Illic G, Pietrangelo T (2014) Extracellular GTP is a potent water-transport regulator via aquaporin 5 plasma-membrane insertion in M1-CCD epithelial cortical collecting duct cells. Cell Physiol Biochem 33:731–746

Manji HK, Moore GJ, Chen G (2001) Bipolar disorder: leads from the molecular and cellular mechanisms of action of mood stabilizers. Br J Psychiatry Suppl 41:s107–s119

Marples D, Christensen S, Christensen EI, Ottosen PD, Nielsen S (1995) Lithium-induced downregulation of aquaporin-2 water channel expression in rat kidney medulla. J Clin Invest 95:1838–1845

Marples D, Frokiaer J, Dorup J, Knepper MA, Nielsen S (1996) Hypokalemia-induced downregulation of aquaporin-2 water channel expression in rat kidney medulla and cortex. J Clin Invest 97:1960–1968

Marr N, Kamsteeg EJ, Van Raak M, van Os CH, Deen PMT (2001) Functionality of aquaporin-2 missense mutants in recessive nephrogenic diabetes insipidus. Pflugers Arch 442:73–77

Marr N, Bichet DG, Hoefs S, Savelkoul PJ, Konings IB, De Mattia F, Graat MP, Arthus MF, Lonergan M, Fujiwara TM, Knoers NVAM, Landau D, Balfe WJ, Oksche A, Rosenthal W,

Muller D, van Os CH, Deen PMT (2002) Cell-biologic and functional analyses of five new aquaporin-2 missense mutations that cause recessive nephrogenic diabetes insipidus. J Am Soc Nephrol 13:2267–2277
Matsumoto K, Miyazaki H, Fujii T, Hashimoto M (1989) Binding of sulfonamides to erythrocytes and their components. Chem Pharm Bull (Tokyo) 37:1913–1915
Matsumura Y, Uchida S, Rai T, Sasaki S, Marumo F (1997) Transcriptional regulation of aquaporin-2 water channel gene by cAMP. J Am Soc Nephrol 8:861–867
Miller EW, Dickinson BC, Chang CJ (2010) Aquaporin-3 mediates hydrogen peroxide uptake to regulate downstream intracellular signaling. Proc Natl Acad Sci U S A 107:15681–15686
Mobasheri A, Marples D, Young IS, Floyd RV, Moskaluk CA, Frigeri A (2007) Distribution of the AQP4 water channel in normal human tissues: protein and tissue microarrays reveal expression in several new anatomical locations, including the prostate gland and seminal vesicles. Channels (Austin) 1:29–38
Moe SE, Sorbo JG, Sogaard R, Zeuthen T, Petter OO, Holen T (2008) New isoforms of rat Aquaporin-4. Genomics 91:367–377
Moeller HB, Knepper MA, Fenton RA (2009) Serine 269 phosphorylated aquaporin-2 is targeted to the apical membrane of collecting duct principal cells. Kidney Int 75:295–303
Moeller HB, Praetorius J, Rutzler MR, Fenton RA (2010) Phosphorylation of aquaporin-2 regulates its endocytosis and protein-protein interactions. Proc Natl Acad Sci U S A 107:424–429
Moeller HB, Olesen ET, Fenton RA (2011) Regulation of the water channel aquaporin-2 by posttranslational modification. Am J Physiol Renal Physiol 300:F1062–F1073
Moeller HB, Aroankins TS, Slengerik-Hansen J, Pisitkun T, Fenton RA (2014) Phosphorylation and ubiquitylation are opposing processes that regulate endocytosis of the water channel aquaporin-2. J Cell Sci 127:3174–3183
Molinas SM, Trumper L, Marinelli RA (2012) Mitochondrial aquaporin-8 in renal proximal tubule cells: evidence for a role in the response to metabolic acidosis. Am J Physiol Renal Physiol 303:F458–F466
Monnens LAH, Jonkman A, Thomas C (1984) Response to indomethacin and hydrochlorothiazide in nephrogenic diabetes insipidus. Clin Sci 66:709–715
Morello JP, Salahpour A, Laperriere A, Bernier V, Arthus MF, Lonergan M, Petaja-Repo U, Angers S, Morin D, Bichet DG, Bouvier M (2000) Pharmacological chaperones rescue cell-surface expression and function of misfolded V2 vasopressin receptor mutants. J Clin Invest 105:887–895
Morishita Y, Matsuzaki T, Hara-chikuma M, Andoo A, Shimono M, Matsuki A, Kobayashi K, Ikeda M, Yamamoto T, Verkman A, Kusano E, Ookawara S, Takata K, Sasaki S, Ishibashi K (2005) Disruption of aquaporin-11 produces polycystic kidneys following vacuolization of the proximal tubule. Mol Cell Biol 25:7770–7779
Mu J, Johansson M, Hansson GC, Lundgren O (1999) Lithium evokes a more pronounced natriuresis when administered orally than when given intravenously to salt-depleted rats. Pflugers Arch 438:159–164
Mulders SM, Knoers NVAM, van Lieburg AF, Monnens LAH, Leumann E, Wuhl E, Schober E, Rijss JPL, van Os CH, Deen PMT (1997) New mutations in the AQP2 gene in nephrogenic diabetes insipidus resulting in functional but misrouted water channels. J Am Soc Nephrol 8:242–248
Murata K, Mitsuoka K, Hirai T, Walz T, Agre P, Heymann JB, Engel A, Fujiyoshi Y (2000) Structural determinants of water permeation through aquaporin-1. Nature 407:599–605
Murillo-Carretero MI, Ilundain AA, Echevarria M (1999) Regulation of aquaporin mRNA expression in rat kidney by water intake. J Am Soc Nephrol 10:696–703
Musa-Aziz R, Chen LM, Pelletier MF, Boron WF (2009) Relative CO2/NH3 selectivities of AQP1, AQP4, AQP5, AmtB, and RhAG. Proc Natl Acad Sci U S A 106:5406–5411
Mutig K, Paliege A, Kahl T, Jons T, Muller-Esterl W, Bachmann S (2007) Vasopressin V2 receptor expression along rat, mouse, and human renal epithelia with focus on TAL. Am J Physiol Renal Physiol 293:F1166–F1177

Nadler SP, Zimpelmann JA, Hebert RL (1992) PGE2 inhibits water permeability at a post-cAMP site in rat terminal inner medullary collecting duct. Am J Physiol 262:F229–F235
Nakanishi K, Sweeney WE Jr, Zerres K, Guay-Woodford LM, Avner ED (2000) Proximal tubular cysts in fetal human autosomal recessive polycystic kidney disease. J Am Soc Nephrol 11:760–763
Nasrat H, Wali FN, Warda A (1997) Diabetes insipidus, a rare complication of HELLP syndrome. Report of local experience and review of the literature. J Obstet Gynaecol 17:64–65
Nejsum LN, Elkjaer M, Hager H, Frokiaer J, Kwon TH, Nielsen S (2000) Localization of aquaporin-7 in rat and mouse kidney using RT-PCR, immunoblotting, and immunocytochemistry. Biochem Biophys Res Commun 277:164–170
Nejsum LN, Kwon TH, Marples D, Flyvbjerg A, Knepper MA, Frokiaer J, Nielsen S (2001) Compensatory increase in AQP2, p-AQP2, and AQP3 expression in rats with diabetes mellitus. Am J Physiol Renal Physiol 280:F715–F726
Nejsum LN, Zelenina M, Aperia A, Frokiaer J, Nielsen S (2005) Bidirectional regulation of AQP2 trafficking and recycling: involvement of AQP2-S256 phosphorylation. Am J Physiol Renal Physiol 288:F930–F938
Nielsen S, Digiovanni SR, Christensen EI, Knepper MA, Harris HW (1993a) Cellular and subcellular immunolocalization of vasopressin-regulated water channel in rat kidney. Proc Natl Acad Sci U S A 90:11663–11667
Nielsen S, Smith BL, Christensen EI, Knepper MA, Agre P (1993b) CHIP28 water channels are localized in constitutively water-permeable segments of the nephron. J Cell Biol 120:371–383
Nielsen S, Chou CL, Marples D, Christensen EI, Kishore BK, Knepper MA (1995a) Vasopressin increases water permeability of kidney collecting duct by inducing translocation of aquaporin-CD water channels to plasma membrane. Proc Natl Acad Sci U S A 92:1013–1017
Nielsen S, Pallone T, Smith BL, Christensen EI, Agre P, Maunsbach AB (1995b) Aquaporin-1 water channels in short and long loop descending thin limbs and in descending vasa recta in rat kidney. Am J Physiol 268:F1023–F1037
Nielsen S, Terris J, Andersen D, Ecelbarger C, Frokiaer J, Jonassen T, Marples D, Knepper MA, Petersen JS (1997) Congestive heart failure in rats is associated with increased expression and targeting of aquaporin-2 water channel in collecting duct. Proc Natl Acad Sci U S A 94:5450–5455
Nielsen J, Kwon TH, Praetorius J, Kim YH, Frokiaer J, Knepper MA, Nielsen S (2003) Segment-specific ENaC downregulation in kidney of rats with lithium-induced NDI. Am J Physiol Renal Physiol 285:F1198–F1209
Nielsen S, Rojek A, Andreasen A (2014) Aquaporin 11 deletion results in induceable proximal tubule injury in response to metabolic challenge (LB837). FASEB J 28(1):Suppl LB837
Nilsson L, Madsen K, Topcu SO, Jensen BL, Frokiaer J, Norregaard R (2012) Disruption of cyclooxygenase-2 prevents downregulation of cortical AQP2 and AQP3 in response to bilateral ureteral obstruction in the mouse. Am J Physiol Renal Physiol 302:F1430–F1439
Norregaard R, Jensen BL, Li C, Wang W, Knepper MA, Nielsen S, Frokiaer J (2005) COX-2 inhibition prevents downregulation of key renal water and sodium transport proteins in response to bilateral ureteral obstruction. Am J Physiol Renal Physiol 289:F322–F333
Norregaard R, Jensen BL, Topcu SO, Wang G, Schweer H, Nielsen S, Frokiaer J (2010) Urinary tract obstruction induces transient accumulation of COX-2-derived prostanoids in kidney tissue. Am J Physiol Regul Integr Comp Physiol 298:R1017–R1025
Ohshiro K, Yaoita E, Yoshida Y, Fujinaka H, Matsuki A, Kamiie J, Kovalenko P, Yamamoto T (2001) Expression and immunolocalization of AQP6 in intercalated cells of the rat kidney collecting duct. Arch Histol Cytol 64:329–338
Okada S, Misaka T, Tanaka Y, Matsumoto I, Ishibashi K, Sasaki S, Abe K (2008) Aquaporin-11 knockout mice and polycystic kidney disease animals share a common mechanism of cyst formation. FASEB J 22:3672–3684
Olesen ET, Fenton RA (2013) Is there a role for PGE2 in urinary concentration? J Am Soc Nephrol 24:169–178

Olesen ET, de Seigneux S, Wang G, Lutken SC, Frokiaer J, Kwon TH, Nielsen S (2009) Rapid and segmental specific dysregulation of AQP2, S256-pAQP2 and renal sodium transporters in rats with LPS-induced endotoxaemia. Nephrol Dial Transplant 24:2338–2349

Olesen ET, Rutzler MR, Moeller HB, Praetorius HA, Fenton RA (2011) Vasopressin-independent targeting of aquaporin-2 by selective E-prostanoid receptor agonists alleviates nephrogenic diabetes insipidus. Proc Natl Acad Sci U S A 108:12949–12954

Ooi BS, Chen BT, Tan KK, Khoo OT (1972) Human renal leptospirosis. Am J Trop Med Hyg 21:336–341

Pallone TL, Edwards A, Ma T, Silldorff EP, Verkman AS (2000) Requirement of aquaporin-1 for NaCl-driven water transport across descending vasa recta. J Clin Invest 105:215–222

Palmer BF (2003) Hyponatremia in patients with central nervous system disease: SIADH versus CSW. Trends Endocrinol Metab 14:182–187

Pedersen RS, Bentzen H, Bech JN, Nyvad O, Pedersen EB (2003) Urinary aquaporin-2 in healthy humans and patients with liver cirrhosis and chronic heart failure during baseline conditions and after acute water load. Kidney Int 63:1417–1425

Peterson LN (1990) Vitamin D-induced chronic hypercalcemia inhibits thick ascending limb NaCl reabsorption in vivo. Am J Physiol 259:F122–F129

Poulsen SB, Kim YH, Frokiaer J, Nielsen S, Christensen BM (2013) Long-term vasopressin-V2-receptor stimulation induces regulation of aquaporin 4 protein in renal inner medulla and cortex of Brattleboro rats. Nephrol Dial Transplant 28:2058–2065

Procino G, Carmosino M, Tamma G, Gouraud S, Laera A, Riccardi D, Svelto M, Valenti G (2004) Extracellular calcium antagonizes forskolin-induced aquaporin 2 trafficking in collecting duct cells. Kidney Int 66:2245–2255

Procino G, Barbieri C, Carmosino M, Rizzo F, Valenti G, Svelto M (2010) Lovastatin-induced cholesterol depletion affects both apical sorting and endocytosis of aquaporin-2 in renal cells. Am J Physiol Renal Physiol 298:F266–F278

Procino G, Barbieri C, Carmosino M, Tamma G, Milano S, De Benedictis L, Mola MG, Lazo-Fernandez Y, Valenti G, Svelto M (2011a) Fluvastatin modulates renal water reabsorption in vivo through increased AQP2 availability at the apical plasma membrane of collecting duct cells. Pflugers Arch 462:753–766

Procino G, Mastrofrancesco L, Sallustio F, Costantino V, Barbieri C, Pisani F, Schena FP, Svelto M, Valenti G (2011b) AQP5 is expressed in type-B intercalated cells in the collecting duct system of the rat, mouse and human kidney. Cell Physiol Biochem 28:683–692

Procino G, Mastrofrancesco L, Tamma G, Lasorsa DR, Ranieri M, Stringini G, Emma F, Svelto M, Valenti G (2012) Calcium-sensing receptor and aquaporin 2 interplay in hypercalciuria-associated renal concentrating defect in humans. An in vivo and in vitro study. PLoS One 7, e33145

Procino G, Milano S, Tamma G, Dossena S, Barbieri C, Nicoletti MC, Ranieri M, Di MA, Nofziger C, Svelto M, Paulmichl M, Valenti G (2013) Co-regulated pendrin and aquaporin 5 expression and trafficking in Type-B intercalated cells under potassium depletion. Cell Physiol Biochem 32:184–199

Promeneur D, Kwon TH, Yasui M, Kim GH, Frokiaer J, Knepper MA, Agre P, Nielsen S (2000) Regulation of AQP6 mRNA and protein expression in rats in response to altered acid-base or water balance. Am J Physiol Renal Physiol 279:F1014–F1026

Puliyanda DP, Ward DT, Baum MA, Hammond TG, Harris HW Jr (2003) Calpain-mediated AQP2 proteolysis in inner medullary collecting duct. Biochem Biophys Res Commun 303:52–58

Raghavan R, Eknoyan G (2014) Acute interstitial nephritis – a reappraisal and update. Clin Nephrol 82:149–162

Ranchin B, Boury-Jamot M, Blanchard G, Dubourg L, Hadj-Aissa A, Morin D, Durroux T, Cochat P, Bricca G, Verbavatz JM, Geelen G (2010) Familial nephrogenic syndrome of inappropriate antidiuresis: dissociation between aquaporin-2 and vasopressin excretion. J Clin Endocrinol Metab 95:E37–E43

Rao R, Hao CM, Breyer MD (2004) Hypertonic stress activates glycogen synthase kinase 3beta-mediated apoptosis of renal medullary interstitial cells, suppressing an NFkappaB-driven cyclooxygenase-2-dependent survival pathway. J Biol Chem 279:3949–3955

Rao R, Zhang MZ, Zhao M, Cai H, Harris RC, Breyer MD, Hao CM (2005) Lithium treatment inhibits renal GSK-3 activity and promotes cyclooxygenase 2-dependent polyuria. Am J Physiol Renal Physiol 288:F642–F649

Rao R, Patel S, Hao C, Woodgett J, Harris R (2010) GSK3beta mediates renal response to vasopressin by modulating adenylate cyclase activity. J Am Soc Nephrol 21:428–437

Rascher W, Rosendahl W, Henrichs IA, Maier R, Seyberth HW (1987) Congenital nephrogenic diabetes insipidus-vasopressin and prostaglandins in response to treatment with hydrochlorothiazide and indomethacin. Pediatr Nephrol 1:485–490

Robben JH, Knoers NV, Deen PM (2006a) Cell biological aspects of the vasopressin type-2 receptor and aquaporin 2 water channel in nephrogenic diabetes insipidus. Am J Physiol Renal Physiol 291:F257–F270

Robben JH, Sze M, Knoers NV, Deen PM (2006b) Rescue of vasopressin v2 receptor mutants by chemical chaperones: specificity and mechanism. Mol Biol Cell 17:379–386

Robben JH, Sze M, Knoers NV, Deen PM (2007) Functional rescue of vasopressin V2 receptor mutants in MDCK cells by pharmacochaperones: relevance to therapy of nephrogenic diabetes insipidus. Am J Physiol Renal Physiol 292:F253–F260

Robben JH, Kortenoeven ML, Sze M, Yae C, Milligan G, Oorschot VM, Klumperman J, Knoers NV, Deen PM (2009) Intracellular activation of vasopressin V2 receptor mutants in nephrogenic diabetes insipidus by nonpeptide agonists. Proc Natl Acad Sci U S A 106:12195–12200

Robinson AG (1976) DDAVP in the treatment of central diabetes insipidus. N Engl J Med 294:507–511

Rojek A, Fuchtbauer EM, Kwon TH, Frokiaer J, Nielsen S (2006) Severe urinary concentrating defect in renal collecting duct-selective AQP2 conditional-knockout mice. Proc Natl Acad Sci U S A 103:6037–6042

Rojek A, Praetorius J, Frokiaer J, Nielsen S, Fenton RA (2008) A current view of the mammalian aquaglyceroporins. Annu Rev Physiol 70:301–327

Roudier N, Ripoche P, Gane P, Le Pennec PY, Daniels G, Cartron JP, Bailly P (2002) AQP3 deficiency in humans and the molecular basis of a novel blood group system, GIL. J Biol Chem 277:45854–45859

Rubini ME (1961) Water excrtion in potassium-deficient man. J Clin Invest 40:2215–2224

Russell TA, Ito M, Ito M, Yu RN, Martinson FA, Weiss J, Jameson JL (2003) A murine model of autosomal dominant neurohypophyseal diabetes insipidus reveals progressive loss of vasopressin-producing neurons. J Clin Invest 112:1697–1706

Saito T, Ishikawa S, Abe K, Kamoi K, Yamada K, Shimizu K, Saruta T, Yoshida S (1997a) Acute aquaresis by the nonpeptide arginine vasopressin (AVP) antagonist OPC-31260 improves hyponatremia in patients with syndrome of inappropriate secretion of antidiuretic hormone (SIADH). J Clin Endocrinol Metab 82:1054–1057

Saito T, Ishikawa SE, Sasaki S, Nakamura T, Rokkaku K, Kawakami A, Honda K, Marumo F (1997b) Urinary excretion of aquaporin-2 in the diagnosis of central diabetes insipidus. J Clin Endocrinol Metab 82:1823–1827

Saito T, Ishikawa S, Ito T, Oda H, Ando F, Higashiyama M, Nagasaka S, Hieda M, Saito T (1999) Urinary excretion of aquaporin-2 water channel differentiates psychogenic polydipsia from central diabetes insipidus. J Clin Endocrinol Metab 84:2235–2237

Sakairi Y, Jacobson HR, Noland TD, Breyer MD (1995) Luminal prostaglandin E receptors regulate salt and water transport in rabbit cortical collecting duct. Am J Physiol 269:F257–F265

Sanches TR, Volpini RA, Massola Shimizu MH, Braganca AC, Oshiro-Monreal F, Seguro AC, Andrade L (2012) Sildenafil reduces polyuria in rats with lithium-induced NDI. Am J Physiol Renal Physiol 302:F216–F225

Sands JM, Flores FX, Kato A, Baum MA, Brown EM, Ward DT, Hebert SC, Harris HW (1998) Vasopressin-elicited water and urea permeabilities are altered in IMCD in hypercalcemic rats. Am J Physiol 274:F978–F985

Satake M, Ikarashi N, Ichikawa Y, Maniwa A, Toda T, Ito K, Ochiai W, Sugiyama K (2010) The role of renal aquaporin 2 in the alleviation of dehydration associated with diabetic polyuria in KKAy mice. Life Sci 87:475–480

Satoh M, Ogikubo S, Yoshizawa-Ogasawara A (2008) Correlation between clinical phenotypes and X-inactivation patterns in six female carriers with heterozygote vasopressin type 2 receptor gene mutations. Endocr J 55:277–284

Savelkoul PJ, De Mattia F, Li Y, Kamsteeg EJ, Konings IB, van der Sluijs P, Deen PM (2009) p. R254Q mutation in the aquaporin-2 water channel causing dominant nephrogenic diabetes insipidus is due to a lack of arginine vasopressin-induced phosphorylation. Hum Mutat 30: E891–E903

Schey KL, Wang Z, Wenke L, Qi Y (2014) Aquaporins in the eye: expression, function, and roles in ocular disease. Biochim Biophys Acta 1840:1513–1523

Schnermann J, Chou CL, Ma T, Traynor T, Knepper MA, Verkman AS (1998) Defective proximal tubular fluid reabsorption in transgenic aquaporin-1 null mice. Proc Natl Acad Sci U S A 95:9660–9664

Schrier RW (2006) Water and sodium retention in edematous disorders: role of vasopressin and aldosterone. Am J Med 119:S47–S53

Schrier RW, Gross P, Gheorghiade M, Berl T, Verbalis JG, Czerwiec FS, Orlandi C (2006) Tolvaptan, a selective oral vasopressin v2-receptor antagonist, for hyponatremia. N Engl J Med 355:2099–2112

Seguro AC, Andrade L (2013) Pathophysiology of leptospirosis. Shock 39(Suppl 1):17–23

Sherlock M, Thompson CJ (2010) The syndrome of inappropriate antidiuretic hormone: current and future management options. Eur J Endocrinol 162(Suppl 1):S13–S18

Silberstein C, Bouley R, Huang Y, Fang P, Pastor-Soler N, Brown D, Van Hoek AN (2004) Membrane organization and function of M1 and M23 isoforms of aquaporin-4 in epithelial cells. Am J Physiol Renal Physiol 287:F501–F511

Singer I, Rotenberg D (1973) Demeclocycline-induced nephrogenic diabetes insipidus. In-vivo and in-vitro studies. Ann Intern Med 79:679–683

Sinke AP, Kortenoeven ML, de Groot T, Baumgarten R, Devuyst O, Wetzels JF, Loffing J, Deen PM (2014) Hydrochlorothiazide attenuates lithium-induced nephrogenic diabetes insipidus independently of the sodium-chloride cotransporter. Am J Physiol Renal Physiol 306:F525–F533

Sohara E, Rai T, Miyazaki J, Verkman AS, Sasaki S, Uchida S (2005) Defective water and glycerol transport in the proximal tubules of AQP7 knockout mice. Am J Physiol Renal Physiol 289:F1195–F1200

Soria LR, Fanelli E, Altamura N, Svelto M, Marinelli RA, Calamita G (2010) Aquaporin-8-facilitated mitochondrial ammonia transport. Biochem Biophys Res Commun 393:217–221

Soupart A, Gross P, Legros JJ, Alfoldi S, Annane D, Heshmati HM, Decaux G (2006) Successful long-term treatment of hyponatremia in syndrome of inappropriate antidiuretic hormone secretion with satavaptan (SR121463B), an orally active nonpeptide vasopressin V2-receptor antagonist. Clin J Am Soc Nephrol 1:1154–1160

Stanton BA, Koeppen BM (1998) Solute and water transport along the nephron: tubular function. In: Berne RM, Levy MN (eds) Physiology, 4th edn. Mosby, St. Louis, MO, pp 699–714

Stodkilde L, Norregaard R, Fenton RA, Wang G, Knepper MA, Frokiaer J (2011) Bilateral ureteral obstruction induces early downregulation and redistribution of AQP2 and phosphorylated AQP2. Am J Physiol Renal Physiol 301:F226–F235

Strange K, Spring KR (1987) Cell membrane water permeability of rabbit cortical collecting duct. J Membr Biol 96:27–43

Sui H, Han BG, Lee JK, Walian P, Jap BK (2001) Structural basis of water-specific transport through the AQP1 water channel. Nature 414:872–878

Tamma G, Robben JH, Trimpert C, Boone M, Deen PM (2011) Regulation of AQP2 localization by S256 and S261 phosphorylation and ubiquitination. Am J Physiol Cell Physiol 300:C636–C646

Tamma G, Di MA, Ranieri M, Svelto M, Pisot R, Bilancio G, Cavallo P, De Santo NG, Cirillo M, Valenti G (2014) A decrease in aquaporin 2 excretion is associated with bed rest induced high calciuria. J Transl Med 12:133

Tchekneva EE, Khuchua Z, Davis LS, Kadkina V, Dunn SR, Bachman S, Ishibashi K, Rinchik EM, Harris RC, Dikov MM, Breyer MD (2008) Single amino acid substitution in aquaporin 11 causes renal failure. J Am Soc Nephrol 19:1955–1964

Tenenbaum J, Ayoub MA, Perkovska S, Adra-Delenne AL, Mendre C, Ranchin B, Bricca G, Geelen G, Mouillac B, Durroux T, Morin D (2009) The constitutively active V2 receptor mutants conferring NSIAD are weakly sensitive to agonist and antagonist regulation. PLoS One 4, e8383

Terris J, Ecelbarger CA, Marples D, Knepper MA, Nielsen S (1995) Distribution of aquaporin-4 water channel expression within rat kidney. Am J Physiol 38:F775–F785

Terris J, Ecelbarger CA, Nielsen S, Knepper MA (1996) Long-term regulation of four renal aquaporins in rats. Am J Physiol 40:F414–F422

Tfelt-Hansen J, Schwarz P, Brown EM, Chattopadhyay N (2003) The calcium-sensing receptor in human disease. Front Biosci 8:s377–s390

Timmer RT, Sands JM (1999) Lithium intoxication. J Am Soc Nephrol 10:666–674

Trepiccione F, Pisitkun T, Hoffert JD, Poulsen SB, Capasso G, Nielsen S, Knepper MA, Fenton RA, Christensen BM (2014) Early targets of lithium in rat kidney inner medullary collecting duct include p38 and ERK1/2. Kidney Int. doi:10.1038/ki.2014.107

Tsuboi Y, Ishikawa S, Fujisawa G, Okada K, Saito T (1994) Therapeutic efficacy of the non-peptide AVP antagonist OPC-31260 in cirrhotic rats. Kidney Int 46:237–244

Tsukaguchi H, Shayakul C, Berger UV, Mackenzie B, Devidas S, Guggino WB, Van Hoek AN, Hediger MA (1998) Molecular characterization of a broad selectivity neutral solute channel. J Biol Chem 273:24737–24743

Valenti G, Laera A, Pace G, Aceto G, Lospalluti ML, Penza R, Selvaggi FP, Chiozza ML, Svelto M (2000) Urinary aquaporin 2 and calciuria correlate with the severity of enuresis in children. J Am Soc Nephrol 11:1873–1881

Valenti G, Laera A, Gouraud S, Pace G, Aceto G, Penza R, Selvaggi FP, Svelto M (2002) Low-calcium diet in hypercalciuric enuretic children restores AQP2 excretion and improves clinical symptoms. Am J Physiol Renal Physiol 283:F895–F903

Valtin H, Schroeder HA, Benirschke K, Sokol HW (1962) Familial hypothalamic diabetes insipidus in rats. Nature 196:1109–1110

Van Balkom BWM, Savelkoul PJ, Markovich D, Hofman E, Nielsen S, van der Sluijs P, Deen PMT (2002) The role of putative phosphorylation sites in the targeting and shuttling of the aquaporin-2 water channel. J Biol Chem 277:41473–41479

Van Hoek AN, Ma T, Yang B, Verkman AS, Brown D (2000) Aquaporin-4 is expressed in basolateral membranes of proximal tubule S3 segments in mouse kidney. Am J Physiol Renal Physiol 278:F310–F316

Van Hoek AN, Bouley R, Lu Y, Silberstein C, Brown D, Wax MB, Patil RV (2009) Vasopressin-induced differential stimulation of AQP4 splice variants regulates the in-membrane assembly of orthogonal arrays. Am J Physiol Renal Physiol 296:F1396–F1404

Verkman AS (2008) Dissecting the roles of aquaporins in renal pathophysiology using transgenic mice. Semin Nephrol 28:217–226

Voisin DL, Bourque CW (2002) Integration of sodium and osmosensory signals in vasopressin neurons. Trends Neurosci 25:199–205

Walker RJ, Weggery S, Bedford JJ, McDonald FJ, Ellis G, Leader JP (2005) Lithium-induced reduction in urinary concentrating ability and urinary aquaporin 2 (AQP2) excretion in healthy volunteers. Kidney Int 67:291–294

Wang Y, Tajkhorshid E (2010) Nitric oxide conduction by the brain aquaporin AQP4. Proteins 78:661–670
Wang W, Kwon TH, Li C, Frokiaer J, Knepper MA, Nielsen S (2002a) Reduced expression of Na-K-2Cl cotransporter in medullary TAL in vitamin D-induced hypercalcemia in rats. Am J Physiol Renal Physiol 282:F34–F44
Wang W, Li C, Kwon TH, Knepper MA, Frokiaer J, Nielsen S (2002b) AQP3, p-AQP2, and AQP2 expression is reduced in polyuric rats with hypercalcemia: prevention by cAMP-PDE inhibitors. Am J Physiol Renal Physiol 283:F1313–F1325
Wang W, Li C, Kwon TH, Miller RT, Knepper MA, Frokiaer J, Nielsen S (2004) Reduced expression of renal Na^+ transporters in rats with PTH-induced hypercalcemia. Am J Physiol Renal Physiol 286:F534–F545
Wang W, Li C, Nejsum LN, Li H, Kim SW, Kwon TH, Jonassen TE, Knepper MA, Thomsen K, Frokiaer J, Nielsen S (2006) Biphasic effects of ANP infusion in conscious, euvolumic rats: roles of AQP2 and ENaC trafficking. Am J Physiol Renal Physiol 290:F530–F541
Weinstock RS, Moses AM (1990) Desmopressin and indomethacin therapy for nephrogenic diabetes insipidus in patients receiving lithium carbonate. South Med J 83:1475–1477
Wen H, Froklaer J, Kwon TH, Nielsen S (1999) Urinary excretion of aquaporin-2 in rat is mediated by a vasopressin-dependent apical pathway. J Am Soc Nephrol 10:1416–1429
Wildman SS, Boone M, Peppiatt-Wildman CM, Contreras-Sanz A, King BF, Shirley DG, Deen PM, Unwin RJ (2009) Nucleotides downregulate aquaporin 2 via activation of apical P2 receptors. J Am Soc Nephrol 20:1480–1490
Willcutts MD, Felner E, White PC (1999) Autosomal recessive familial neurohypophyseal diabetes insipidus with continued secretion of mutant weakly active vasopressin. Hum Mol Genet 8:1303–1307
Wilson DM, Perry HO, Sams WM Jr, Dousa TP (1973) Selective inhibition of human distal tubular function by demeclocycline. Curr Ther Res Clin Exp 15:737–740
Wong F, Blei AT, Blendis LM, Thuluvath PJ (2003) A vasopressin receptor antagonist (VPA-985) improves serum sodium concentration in patients with hyponatremia: a multicenter, randomized, placebo-controlled trial. Hepatology 37:182–191
Wu H, Chen L, Zhang X, Zhou Q, Li JM, Berger S, Borok Z, Zhou B, Xiao Z, Yin H, Liu M, Wang Y, Jin J, Blackburn MR, Xia Y, Zhang W (2013) Aqp5 is a new transcriptional target of Dot1a and a regulator of Aqp2. PLoS One 8, e53342
Xu DL, Martin PY, Ohara M, Stjohn J, Pattison T, Meng XZ, Morris K, Kim JK, Schrier RW (1997) Upregulation of aquaporin-2 water channel expression in chronic heart failure rat. J Clin Invest 99:1500–1505
Yakata K, Hiroaki Y, Ishibashi K, Sohara E, Sasaki S, Mitsuoka K, Fujiyoshi Y (2007) Aquaporin-11 containing a divergent NPA motif has normal water channel activity. Biochim Biophys Acta 1768:688–693
Yakata K, Tani K, Fujiyoshi Y (2011) Water permeability and characterization of aquaporin-11. J Struct Biol 174:315–320
Yang BX, Brown D, Verkman AS (1996) The mercurial insensitive water channel (AQP-4) forms orthogonal arrays in stably transfected Chinese hamster ovary cells. J Biol Chem 271:4577–4580
Yang B, Gillespie A, Carlson EJ, Epstein CJ, Verkman AS (2000) Neonatal mortality in an aquaporin-2 knock-in mouse model of recessive nephrogenic diabetes insipidus. J Biol Chem 276:2775–2779
Yang B, Song Y, Zhao D, Verkman AS (2005) Phenotype analysis of aquaporin-8 null mice. Am J Physiol Cell Physiol 288:C1161–C1170
Yang B, Zhao D, Qian L, Verkman AS (2006a) Mouse model of inducible nephrogenic diabetes insipidus produced by floxed aquaporin-2 gene deletion. Am J Physiol Renal Physiol 291: F465–F472

Yang B, Zhao D, Solenov E, Verkman AS (2006b) Evidence from knockout mice against physiologically significant aquaporin 8-facilitated ammonia transport. Am J Physiol Cell Physiol 291:C417–C423
Yang B, Zhao D, Verkman AS (2006c) Evidence against functionally significant aquaporin expression in mitochondria. J Biol Chem 281:16202–16206
Yasui M, Zelenin SM, Celsi G, Aperia A (1997) Adenylate cyclase-coupled vasopressin receptor activates AQP2 promoter via a dual effect on CRE and AP1 elements. Am J Physiol 41:F443–F450
Yasui M, Hazama A, Kwon TH, Nielsen S, Guggino WB, Agre P (1999a) Rapid gating and anion permeability of an intracellular aquaporin. Nature 402:184–187
Yasui M, Kwon TH, Knepper MA, Nielsen S, Agre P (1999b) Aquaporin-6: an intracellular vesicle water channel protein in renal epithelia. Proc Natl Acad Sci U S A 96:5808–5813
Ye L, Li X, Chen Y, Sun H, Wang W, Su T, Jiang L, Cui B, Ning G (2005) Autosomal dominant neurohypophyseal diabetes insipidus with linkage to chromosome 20p13 but without mutations in the AVP-NPII gene. J Clin Endocrinol Metab 90:4388–4393
Yip KP (2006) Epac-mediated Ca^{2+} mobilization and exocytosis in inner medullary collecting duct. Am J Physiol Renal Physiol 291:F882–F890
Zelenina M, Christensen BM, Palmer J, Nairn AC, Nielsen S, Aperia A (2000) Prostaglandin E(2) interaction with AVP: effects on AQP2 phosphorylation and distribution. Am J Physiol Renal Physiol 278:F388–F394
Zhai XY, Fenton RA, Andreasen A, Thomsen JS, Christensen EI (2007) Aquaporin-1 is not expressed in descending thin limbs of short-loop nephrons. J Am Soc Nephrol 18:2937–2944

Part III
Pharmacology of Potassium Channels

Chapter 26
Recent Developments in the Pharmacology of Epithelial Ca^{2+}-Activated K^+ Channels

Antonio Nardi, Søren-Peter Olesen, and Palle Christophersen

Abstract Calcium (Ca^{2+})-activated potassium (K^+) channels (K_{Ca}) in epithelia serve important functions in fluid and salt secretion and may be attractive targets for drug development for epithelial disorders, such as cystic fibrosis, diarrhoea, COPD, polycystic kidney disease, and glaucoma. Two very different types of K_{Ca} channels are generally found in epithelia: the big conductance, Ca^{2+}-activated K^+ channel (BK, $K_{Ca}1.1$), and the intermediate conductance, Ca^{2+}-activated K^+ channel (IK, $K_{Ca}3.1$). These channels are differentially expressed in various cells and tissues and serve different physiological and potentially also pathophysiological functions in epithelia.

The present chapter aims at giving a brief review of the physiology and function of BK and IK channels, a description of the classical pharmacology of these channels, and an in-depth overview of the status of drug discovery and development programmes in the pharmaceutical industry as well as academia, based on both literature and patent references, and it ends with a discussion of recent advances in the understanding of the molecular pharmacology of these channels. The review is comprehensive and the focus is throughout ion channel pharmacology with exemplary cases from the epithelial as well as non-epithelial fields.

Keywords MaxiK • $K_{Ca}1.1$ • IK channel • $K_{Ca}3.1$ • Large conductance Ca^{2+}-activated K^+ channel • $K_{Ca}1.1$ • Intermediate conductance Ca^{2+}-activated K+ channel • $K_{Ca}3.1$ • Iberiotoxin

A. Nardi (✉)
Grünenthal GmbH, Zieglerstraße 6, 52078 Aachen, Germany
e-mail: antonio.nardi@grunenthal.com

S.-P. Olesen
Panum Institute, University of Copenhagen, Blegdamsvej 3, 2200 Copenhagen N, Denmark

P. Christophersen
Saniona A/S, Baltorpvej 154, 2750 Ballerup, Denmark

K.L. Hamilton, D.C. Devor (eds.), *Ion Channels and Transporters of Epithelia in Health and Disease*, Physiology in Health and Disease,
DOI 10.1007/978-1-4939-3366-2_26

Abbreviation List

BK channel	Big conductance Ca^{2+}-activated K^+ channel, $K_{Ca}1.1$
Cam	Calmodulin
ChTx	Charybdotoxin
COPD	Chronic obstructive pulmonary disorder
CFTR	Cystic fibrosis transmembrane conductance regulator
CYP	Cytochrome P450
FLIPR	Fluorescent imaging plate reader
IbTx	Iberiotoxin
IK channel	Intermediate conductance Ca^{2+}-activated K^+ channel $K_{Ca}3.1$
IV relation	Current through the ion channel depicted as function of the membrane potential
K_{Ca}	Ca^{2+}-activated K^+ channel including all SK IK, and BK channels (IUPHAR nomenclature)
k_{off}	Kinetic off-rate of ligand (inhibitor or activator) binding to the channel (s^{-1})
k_{on}	Kinetic on-rate of ligand binding to the channel ($M^{-1} \times s^{-1}$)
K_v	Voltage-gated K channel (IUPHAR nomenclature)
KCNN	Potassium intermediate/small conductance calcium-activated channel subfamily N (gene name)
KCNMA	Potassium large conductance calcium-activated channel subfamily M, alpha member 1 (gene name)
KCNMB	Potassium large conductance calcium-activated channel subfamily M, beta member 1 (gene name)
P2Y	Purinergic 2Y receptors
pS	Conductance though a single ion channel measured by the unit 10^{-12} Siemens
RKC domain	Regulator of K^+ conductance domain
SK channel	Small conductance Ca^{2+}-activated K^+ channel ($K_{Ca}2.1$, $K_{Ca}2.2$, $K_{Ca}2.3$)
TEA	Tetraethylammonium
TM	Transmembrane domain
TMEM16A	Transmembrane protein member 16A, Ca^{2+}-activated Cl^- channel
V_m	Membrane potential (mV)

26.1 Introduction

26.1.1 K_{Ca} Channel Expression and Function in Epithelia

The presence of Ca^{2+}-activated K^+ channels has been demonstrated in most epithelia, either by physiological or expression studies. SK ($K_{Ca}2.1$), IK ($K_{Ca}3.1$), and

BK ($K_{Ca}1.1$) channels have been reported, but the most comprehensive expression across different types of epithelia is of the IK and BK channels with physiological roles varying from one epithelium to another. IK channels are broadly expressed in both secretory epithelia like the colonic crypts and exocrine glands as well as in nonsecretory epithelia like the skin and the bladder epithelium (Thompson-Vest et al. 2006). In secretory epithelia, IK channels are often expressed on the basolateral membrane, where their activation hyperpolarises the cell and thereby—in cooperation with other K^+ channels—helps maintain a large driving force for Cl^- secretion across the luminal membrane via chloride channels, such as the cAMP-activated CFTR and the Ca^{2+}-activated TMEM16A (Schroeder et al. 2008; Caputo et al. 2008). Pharmacological activation of IK channels by compounds such as 1-EBIO (for full chemical names, please see the figure legends associated with a detailed description of each pharmacological agent below) and chlorzoxazone was initially suggested as a way to increase Cl^- and water secretion across epithelia (Devor et al. 1996) and even as a possible therapeutic principle for restoration of epithelial transport in patients suffering from cystic fibrosis (Mall et al. 2003; Roth et al. 2011). Rationalised by a similar line of arguments, IK channel inhibition was shown by use of IK channel knockout mice and pharmacological blockers to be a realistic opportunity for treatment of diarrhoea (Rufo et al. 1997). Physiologically, activation of epithelial IK channels is coupled to various secretagogues, such as ATP acting via P2Y receptors (Dutta et al. 2009; Hayashi et al. 2011; Bardou et al. 2009) or acetylcholine acting via muscarinic receptors (Matos et al. 2007). In non-secretory epithelia, the role of IK channels is less clear but they may participate in epithelial cell volume regulation and thereby protect, for example, bladder and kidney epithelial cells against osmotic stress, or they may be involved in the controlled proliferation, repair, and differentiation of stratified epithelia (Denda et al. 2007; Manaves et al. 2004; Koegel and Alzheimer 2001; Koegel et al. 2003).

BK channels are also found in secretory epithelia and are typically expressed luminally where they contribute to K^+ secretion in parallel with Cl^-. In the airways, epithelial BK channels can be activated by secreted ATP via P2Y receptors on the luminal membrane, which probably helps preserve a well-hydrated surface liquid and a low viscosity, which is important for normal cilia beat rate and function (Manzanares et al. 2011). BK channels also play an important role for global K^+ homeostasis via K^+ excretion into the intestinal lumen of the distal colon. That is, upon feeding a K^+-rich diet to mice, a BK splice variant sensitive to cAMP regulation was specifically up-regulated by aldosterone and gave rise to an increased adrenaline-dependent K^+ secretory response (Sorensen et al. 2008, 2010). BK channels are also functionally expressed in the ciliary nonpigmented epithelium of the eye, where they may be important for the regulation of aqueous humour secretion (Ryan et al. 1998; Cullinane et al. 2002).

26.1.2 Basic Properties of BK and IK Channels

BK channels are named according to their large (big) unitary conductance of approximately 250 pS as measured electrophysiologically in symmetric K^+ solutions using patch-clamp or lipid bilayer techniques. Despite their large single channel conductance, BK channels are highly K^+ selective, with the reversal potential following closely the Nernst potential for potassium ion when Na^+ and K^+ concentration gradients are varied. In contrast to many other potassium channels, the single channel IV relation in symmetric K^+ is linear over a broad voltage span ($\pm$100 mV) and only exhibits weak outward rectification ("constant field-like rectification") with varied Na^+/K^+ ratios in the extracellular phase. BK channel gating is basically controlled by two factors: the intracellular Ca^{2+} concentration and the membrane voltage (V_m). At "resting" levels of Ca^{2+} and V_m, the BK channels are effectively closed, whereas the combination of increased intracellular Ca^{2+} and depolarised membrane potential activates the channel. A systematic analysis reveals that the open-state probability dependency of V_m follows classic Boltzmann equation curves as most K_v channels, which are shifted gradually in the hyperpolarising direction by increasing intracellular Ca^{2+} concentrations (Barrett et al. 1982; Latorre et al. 1982).

BK channels are encoded by the *KCNMA1* gene, and the human protein is 1182 amino acids long (Dworetzky et al. 1994). This is assembled as a homotetramer in the membrane and constitutes the functional BK channel with the central K^+ selective pore created at the interface of the subunits. In analogy to other K^+ channels, the reentrant P-loop between TM5 and TM6 contains the GYG sequence that forms the main selectivity filter of the pore. The extremely long C-termini are clearly exposed to the intracellular side of the membrane and contain a pair of homology RCK domains regulating the K^+ conductance. These latter domains create a "gating ring" structure (Wu et al. 2010) upon which Ca^{2+} binding occurs directly at Ca^{2+} binding "bowls" (Schreiber and Salkoff 1997), i.e., molecular regions that are rich in negatively charged amino acids. Selective binding of Ca^{2+} shortens the C-terminal-RKC1 linker which seems to be central for opening of the channel gate (Javaherian et al. 2011), which itself is positioned in the pore region (Chen et al. 2014). In contrast to the C-terminus, it was debated whether the N-terminus is extracellular or intracellular. It is now clear that the BK channel is unique among potassium channels in being a 7 TM channel with an extra transmembrane helix called TM0 preceding the equivalent TM1–4 in K_v channels. In analogy with K_v channels, the voltage-sensing part of the BK channel is constituted by TM0–TM4, with the highly charged TM4 likely playing a central role in voltage sensing, but with additionally important characteristics imposed by the participation of TM0 (Pantazis et al. 2010a, b; Niu et al. 2013).

Understanding the nature of the cooperation of the voltage- and Ca^{2+}-sensing parts of the protein represents a major challenge for the molecular biophysics of BK channels and is also a major challenge for rationalising the pharmacology of the gating process. The TM0 of the BK channel alpha subunit also plays a unique role in the co-assembly and interactions with the beta subunits (beta1–4; encoded by *KCNMB*1–4) (Wallner et al. 1996), which can influence both the gating and pharmacology of the channel (see later). Adding to BK heterogeneity is the fact that a number of splice variants exist as well (McCobb et al. 1995).

IK channels are so named because their single channel conductance is intermediate (approximately 30 pS at 0 mV and symmetric K^+ concentrations) between the high conductance of BK channels and the low conductance of SK channels. As for the BK channel, the IK channel pore is highly selective for K^+ over Na^+, but in contrast to BK channels the single channel IV curve reveals a prominent inward rectification in symmetric K^+. Also, in contrast to BK channels, the gating of IK channels is not influenced by V_m; rather it is solely dependent on the intracellular Ca^{2+} concentration: At "resting" Ca^{2+} levels the channel is closed, but it is sensitive to much smaller increases (nM) in intracellular Ca^{2+} concentration than the BK channel (μM, at normal ranges of membrane potentials) and may even be more sensitive to low Ca^{2+} fluctuations than SK channels (Ishii et al. 1997).

The IK channel belongs to the same gene family as the SK channels (the *KCNN* family, encoded by *KCNN4*) with which it shares many functional and physiological traits. The IK channel alpha subunit (427 aa) (Ishii et al. 1997) is a 6 TM protein only remotely related to the BK channel except for sharing the overall K^+ channel topology, at least around the pore region. The IK alpha subunit does not have any binding site for Ca^{2+}, but relies on constitutively bound calmodulin (Cam) at the C-terminal, which acts as the Ca^{2+} sensor. Upon binding of Ca^{2+} to calmodulin, a conformational change is transferred from the C-terminal to the TM5-P-TM6 regions, where the physical gate is positioned in IK and SK channels (Garneau et al. 2014; Bruening-Wright et al. 2007). The understanding of this transformation is pivotal to the understanding of gating pharmacology (selectivity, site, and mode of actions) of both IK and SK channels (see later section).

26.1.3 Basic Pharmacology of BK and IK Channels

The pivotal description of BK channels in 1981–1982 by the groups of Karl Magleby and Ramon Latorre initiated decades of extensive characterisations at the single channel level, including detailed descriptions of inhibition by inorganic ions, like Cs^+ and Ba^{2+} as well as small organic ions like TEA. The main conclusions from this purely biophysical (nonstructural) approach to BK channel pharmacology were that the positively charged inhibitors mimicked the K-ion and were able to pass some way into the permeation pore, where they were immobilised at a selectivity filter and thereby prevented the normal flow of K-ions through the channel. In other words, these inhibitors are all blockers of the open channel. Whereas easily comprehended for the monovalent alkali metal ions, Ba^{2+} might intuitively have been thought of as a competitive Ca^{2+} analogue preventing channel opening at the Ca^{2+} binding site. This was, however, not the case (Miller 1987; Neyton and Miller 1988). Being "substrate analogues" for the K-ion, it is not very surprising that the blocking mechanism of these ions gave important insights to specific binding sites positioned deep in the pore in the electrical field generated by the membrane potential. Remarkably good quantitative descriptions of different blocking phenomena, such as voltage-dependent block (increased binding affinity with increased potential), "fast" and "slow" blockers (*apparent* reduction in current vs. *apparent* change of gating; high vs. low blocker k_{off}), as well as sidedness of action (different blocker affinities from inside vs. outside of channel), were achieved (Kehl 1996). In addition to the blocker ions being biophysical "probes", the studies clearly pointed towards the existence of unique pharmacological sites in the open BK channel pore region, since both TEA added from the outside and Ba^{2+} added from the inside exhibited much more potent block of BK channels than of most other K^+ channels including other Ca^{2+}-activated K^+ channels (Brown et al. 1988; Lang and Ritchie 1990; Rae et al. 1990).

The first described high-affinity peptide blocker of the BK channel was charybdotoxin, a constituent from the venom of the scorpion *Leiurus quinquestriatus hebraeus* (Miller et al. 1985). Charybdotoxin (ChTx) was shown to bind to the outer pore mouth, thus causing blocked periods lasting several seconds in the single channel record. Closer analysis revealed a simple bimolecular reaction mechanism with k_{on} proportional to blocker concentration and a concentration-independent k_{off} (Anderson et al. 1988). Despite its high affinity, charybdotoxin proved not to be a specific blocker of BK channels, since several K_v channels and also the IK channel were blocked with very high affinity. Chemically modified charybdotoxin was nonetheless very important in the early days for defining BK channel pharmacology, since a binding assay—developed by scientists at Merck—based on the ^{125}I-labelled peptide picked up BK modulators with radically different chemical structures and modulatory properties, including certain secondary metabolites of "rye grass staggers", fungi which lie behind outbreaks of frequent tremors and convulsions to livestock cattle in New Zeeland and Australia. These tremorgenic indole alkaloids, which are steroid-like structures, interacted in complex ways with ^{125}I-ChTx binding to smooth muscle membranes and became the first potent class of non-peptide BK channel inhibitors identified (Knaus et al. 1994). Interestingly, the functional modes of actions of these inhibitors were more complex than the simple blocking mechanism shown by the peptide inhibitors, since they clearly influenced gating parameters (Strobaek et al. 1996; Sanchez and McManus 1996) and are thus better categorised as negative gating modulators than pore blockers. Another class of molecules also displacing ^{125}I-ChTx was the dehydrosoyasaponin-1 (DHS-1), a secondary metabolite of the medical herb *Desmodium adscendens*. DHS-1 was, in contrast to the indole alkaloids, identified in functional assays as a very potent activator of BK channels (McManus et al. 1993) acting exclusively from the inside of the membrane and only in the presence of the beta1 subunit. Shortly thereafter, the first synthetic molecule (NS1619; see later) with BK-activating properties was described (Olesen et al. 1994). In contrast to DHS-1, NS1619 easily crosses biological membranes, works independently of whether it is added from the outside or inside of the membrane, and does not require the beta subunit. Despite its much lower potency,

NS1619 became the most widely used tool compound for investigation of positive modulation of BK channels for many years and also inspired several lead optimisation programmes in the pharmaceutical industry. These original positive modulators did not make it into the clinical development phases and have also largely been replaced nowadays by more adequate research compounds as will be discussed in more details in the section to follow.

Whereas the BK channel was only recognised after the development of single channel recording techniques, the description of Ca^{2+}-activated K^+ fluxes and hyperpolarisations mediated by the erythrocyte IK channel (Hoffman et al. 2003; Lew and Ferreira 1976; Heinz and Passow 1980; Vestergaard-Bogind et al. 1985) was studied for decades before its characterisation at the single channel level. Therefore it is not particularly surprising that various low-affinity compounds were identified more or less by serendipity during the years. One important example is cetiedil, a drug with vasodilative properties, which was shown to counteract sickling of erythrocytes from sickle cell patients and also to inhibit the Ca^{2+}-activated K^+ fluxes (Berkowitz and Orringer 1981). Despite the fact that cetiedil entered clinical trials for sickle cell anaemia that were later discontinued, it was the identification of two other compound classes as IK channel inhibitors that became the guides for further optimisation towards clinical useful IK channel blockers: the triarylmethane (TRAM) class of antimycotics (Alvarez et al. 1992) and the dihydropyridine class of antihypertensives (Ellory et al. 1994) (see later). Potent class representatives are clotrimazole and nitrendipine. Despite its limited usefulness in vivo, clotrimazole was shown to reduce Cl^- secretion from colon epithelia cells by inhibiting the basolateral K^+ conductance and to inhibit cholera toxin-induced diarrhoea in mice (Rufo et al. 1997).

In contrast to the inhibitor field, the multiple erythrocyte K^+ transport studies did not reveal clear leads for development of activators of IK channels, with propranolol as a possible (and still unexplored) exception (Schwarz et al. 1989). The pivotal observation in that area rather came from the study of basic epithelial transport. Daniel Devor investigated the action of NS1619-like molecules (NS1619 per se does *not* work; Balut et al. 2012), i.e., benzimidazolones on Cl^- secretion across colon and tracheal cell layers, and found compelling evidence for a directly activating effect of the compound 1-EBIO on the basolateral Ca^{2+}-activated K^+ conductance and thus on overall transepithelial Cl^- secretion (Devor et al. 1996). The biophysical properties of the epithelial 1-EBIO-sensitive channel were indistinguishable from the cloned IK channel (Ishii et al. 1997) which was soon after also shown to be sensitive to 1-EBIO (Jensen et al. 1998). Interestingly, several clinically used drugs with an overall structure like 1-EBIO additionally activated the IK channel and stimulated epithelial Cl^- secretion (Singh et al. 2000; Syme et al. 2000). These compounds also activate SK channels, albeit with approximately 10-fold lower potency.

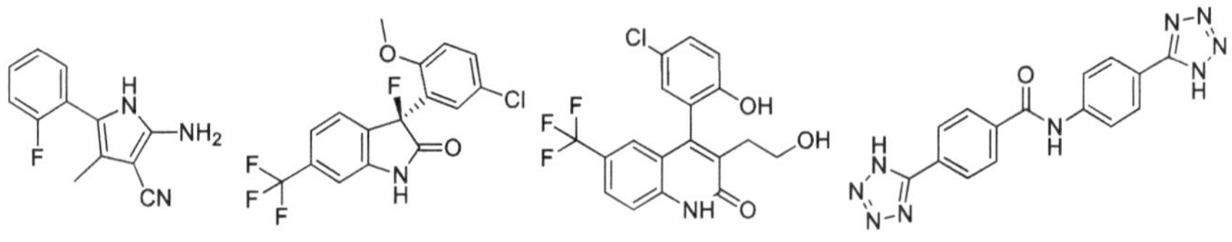

Fig. 26.1 Chemical structures of NS8, BMS-204352, BMS-223131—BK channel openers progressed to clinical development and discontinued. Andolast is a topical and supposedly a non-selective BK channel opener, currently in phase III for respiratory diseases. **NS-8**: 2-amino-5-(2-fluorophenyl)-4-methyl-1H-pyrrole-3-carbonitrile. **BMS-204352**: (S)-3-(5-chloro-2-methoxyphenyl)-3-fluoro-6-(trifluoromethyl)indolin-2-one. **BMS-223131**: 4-(5-chloro-2-hydroxyphenyl)-3-(2-hydroxyethyl)-6-(trifluoromethyl)quinolin-2(1H)-one. **Andolast**: N-(4-(1H-tetrazol-5-yl)phenyl)-4-(1H-tetrazol-5-yl)benzamide

26.2 BK Channel Modulator Chemistry

26.2.1 BK Channel Activators

Consistent with the wealth of publications highlighted in the previous sections suggesting that the pharmacological modulation of BK channels is a clear opportunity to develop new therapeutics, several research efforts have been directed towards the identification of BK channel activators and inhibitors (Nardi and Olesen 2008). As a result, a variety of pharmacological agents have been identified, although very little success has been made in bringing these forward to clinical settings, neither in epithelia-related disorders nor in other diseases. With regard to BK activators, the attrition rate has been exceptionally high and it culminated in only a few development compounds, i.e., andolast, NS8, BMS-204352, BMS-223131 (Fig. 26.1), and TA1702 (chemical structure undisclosed) (Nardi and Olesen 2008). Out of these, only one compound, namely, andolast, is still currently under active development (phase III).

NS8 (Fig. 26.1) (Parihar et al. 2003; La Fuente et al. 2014) was a pyrrole development compound designed by Nippon Shinyaku. Notwithstanding its low potency at the channel and functional effects in the micromolar range, the in vivo pharmacological profile in preclinical settings has prompted the European licensee Apogepha Arzneimittel to develop it first up to phase II for the potential treatment of overactive bladder and to later discontinue it due to a lack of efficacy at the expected therapeutic dosage in a proof-of-concept clinical trial (Akada et al. 2002; Nakamura et al. 1996, 1999; Malysz et al. 2004; Tanaka et al. 2003). Further structural optimisation of NS8 either via expansion of the 5-membered ring into a 6-membered ring (Harada et al. 2003) or via its fusion with other rings (Turner et al. 2003, 2004) did afford more potent derivative compounds but they did not emerge up to the clinical stage.

BMS-204352 (Fig. 26.1) (MaxiPost™) is a relatively non-selective BK channel activator whose efficacy as a neuroprotective agent in in vivo pharmacological

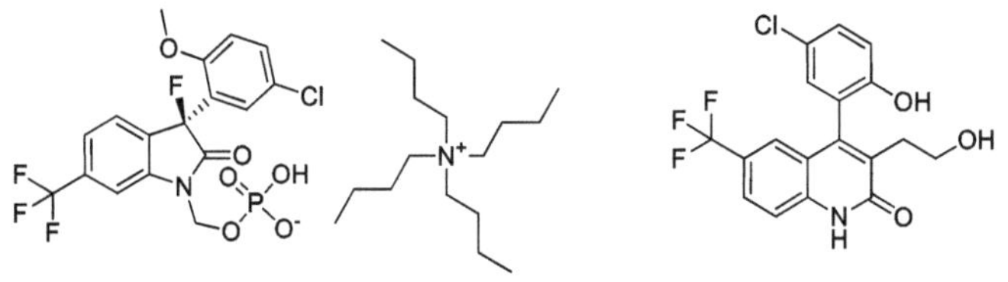

Fig. 26.2 BMS-554216 and BMS-223131, chemical entities structurally related to BMS-204352 (Fig. 26.1). **BMS-554216**: tetrabutylammonium (S)-(3-(5-chloro-2-methoxyphenyl)-3-fluoro-2-oxo-6-(trifluoromethyl)indolin-1-yl)methyl hydrogenphosphate. **BMS-223131**: 4-(5-chloro-2-hydroxyphenyl)-3-(2-hydroxyethyl)-6-(trifluoromethyl)quinolin-2(1H)-one

models of stroke was demonstrated in a series of studies (Hewawasam et al. 1997; Gribkoff et al. 2001; Mackay 2001; Krishna et al. 2002a–c; Jensen 2002; Zhang et al. 2005). Congruous with this experimental evidence and the rationale that a BK channel activator might prevent the ischemic cell depolarisation and excitotoxicity, Bristol-Myers Squibb progressed the compound in phase III for the treatment of acute ischemic stroke. The drug failed to show superiority over placebo, a result that was disappointingly also confirmed in a magnetic resonance imaging sub-study comparing the effect of 1 mg/kg of MaxiPost with placebo in more than 100 patients (Warach et al. 2002). The poor aqueous solubility of BMS-204352 has considerably limited the administrable dose by intravenous infusion, and this has been claimed as one of the potential explanations for lack of efficacy. Prodrugs of BMS-204352 bearing diverse solubilising moieties on the amidic nitrogen (Gillman et al. 2003) (Gillman and Bocchino 2004; Starrett et al. 2005), including BMS-554216 (Fig. 26.2), were subsequently explored. The latter compound has been reported as being more water soluble (11 mg/ml), able to induce higher brain concentration in preclinical species, and to be efficacious in the same species in reducing oedema after traumatic brain injury (Starrett 2004). This compound may have entered the clinical phase at some point, yet its further development has not been reported recently, and it is presumed to be halted. Additional efforts around the same chemical space of BMS-204352 capitalised more structurally related BK openers that were evaluated preclinically and clinically by the same company as new therapeutics for urinary incontinence, erectile dysfunction and bowel disorders. Among several others, BMS-223131 was synthesised and profiled (Hewawasam et al. 2003, 2004, 2005; Boy et al. 2004, 2005a, b; Boy 2005), and notwithstanding the modest ability to increase the BK current at 20 μM by about fourfold, it showed iberiotoxin-sensitive efficacy in an in vivo rat model of erectile function (Hewawasam et al. 2004) and reduced stress-induced colonic motility and visceral nociception (Sivarao et al. 2005). The same compound was later acquired in an in-licensing deal by Xention (XEN-D0401) and, albeit initially reported in the company clinical pipeline for the treatment of overactive bladder, no further developments have been announced. One development issue that was clearly identified and reported in the public domain was the inhibitory action on the cytochrome P450 (CYP) 2C9 isoform (IC_{50} of 1.7 μM), so that a medicinal chemistry effort aimed at reducing the CYP inhibitory properties whilst preserving the BK channel activity has been attempted (Vrudhula et al. 2005). Additional complications to this series of molecules may also be their lack of selectivity between different K-channels, as demonstrated for BMS-204352 with respect to K_v7 channels (Schroder et al. 2001) and later confirmed in a number of papers.

TA1702 (chemical structure not disclosed) is a BK activator that GlaxoSmithKline (GSK) acquired from Tanabe Seiyaku and was in clinical development (phase I, 2005) for the treatment of overactive bladder (Nardi and Olesen 2008). A couple of years later, with no public disclosures on the scientific grounds for this decision, the compound was again withdrawn both from the product development pipeline of Mitsubishi Tanabe Pharma Corporation and GSK (Nardi and Olesen 2008).

Andolast® (or CR2039) (Fig. 26.1), as dry powder inhaled formulation, is currently developed (phase III) by Rottapharm, a multinational pharmaceutical group based in Italy, for the potential treatment of bronchial asthma, allergic rhinitis, and chronic obstructive pulmonary disease (COPD) (Nardi and Olesen 2008). This small and simple molecule is claimed in the company website to act as "a calcium (Ca^{2+})-activated potassium (K^+) channel" opener and it is believed that the BK-opening activity is an important mechanism of action for its efficacy in preclinical models (Rovati et al. 2012) in defiance of diverse pharmacological properties and action at different cellular levels (Makovec et al. 1990, 1992; Malerba et al. 2009).

Contradictory to the poor clinical success of selective BK channel openers, a wealth of very diverse late clinical or marketed compounds have been reported to possess BK channel activation as an add-on or ancillary mechanism, including puerarin or kakonein (phase II, treatment of alcohol abuse; Sun et al. 2007), cilostazol (launched, intermittent claudication, Pletal®; Wu et al. 2004), Zonisamide (launched, Excegran®, epilepsy; Huang et al. 2007), riluzole (launched, amyotrophic lateral sclerosis, Rilutek®; Wang et al. 2008), unoprostone (launched, glaucoma, Rescula®; Cuppoletti et al. 2012), acetazolamide (launched, epilepsy, etc., Diamox®; Tricarico et al. 2004), a number of cox-2 inhibitors, such as rofecoxib (launched, arthritis, etc., Celebrex®; Aihara et al. 2004), and vinpocetine (launched, cognitive disorders, etc., Cavinton®; Wu et al. 2001). Moreover, activation of BK channels can be produced by a multiplicity of endogenous molecules, including polyunsaturated arachidonic acid, metabolites of lipoxygenase (Duerson et al. 1996), and cytochrome P450 epoxygenase (Morin et al. 2007; Hercule et al. 2007; Lu et al. 2001; Benoit et al. 2001; Fukao et al. 2001), which have also been employed as starting points for further chemical derivatisation in the synthesis of new epoxyeicosatrienoic acid derivatives (Yang et al. 2007; Falck et al. 2003, 2011; Lauterbach et al. 2002; Schunck et al. 2010; Lucas et al. 2010; Hercule et al. 2007) (Fig. 26.3). The finding that 17β-estradiol (Fig. 26.3), an endogenous sex hormone, is a direct and β-subunit-dependent activator of BK channels (De Wet et al. 2006; Valverde et al. 1999) prompted additional studies leading to the discovery of the BK opening properties of a number of estrogenic compounds without steroid structure, including tamoxifen (Coiret et al. 2007; Dick

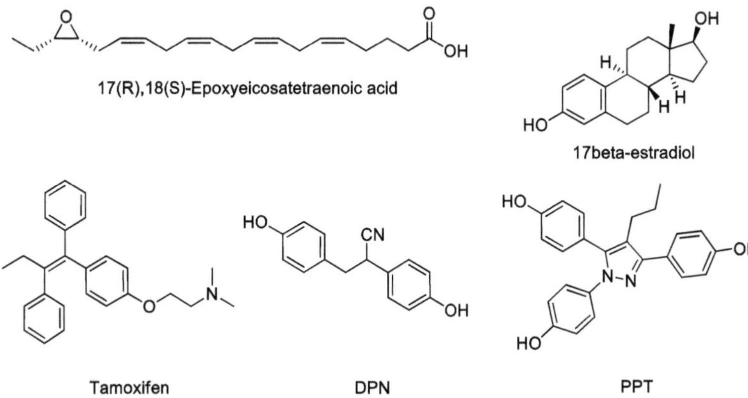

Fig. 26.3 Endogenous BK-channel modulators and structural analogues: 17(R),18(S)-epoxyeicosatetraenoic acid, 17β-estradiol, tamoxifen, DPN, PPT. **17(R),18(S)-Epoxyeicosatetraenoic acid**: (5Z,8Z,11Z,14Z)-16-((2R,3S)-3-ethyloxiran-2-yl)hexadeca-5,8,11,14-tetraenoic acid. **17beta-estradiol**: (8R,9S,13S,14S,17S)-13-methyl-7,8,9,11,12,13,14,15,16,17-decahydro-6H-cyclopenta[a]phenanthrene-3,17-diol. **Tamoxifen**: (Z)-2-(4-(1,2-diphenylbut-1-en-1-yl) phenoxy)-N,N-dimethylethanamine. **DPN**: 2,3-bis(4-hydroxyphenyl)propanenitrile. **PPT**: 4,4′,4-(4-propyl-1H-pyrazole-1,3,5-triyl)triphenol

et al. 2001, 2002; Dick and Sanders 2001) and close analogues (Sha et al. 2005), DPN and PPT (Fig. 26.3; Wang et al. 2007) as well as steroid structures with or without hormonal activity, such as lithocholate (Bukiya et al. 2007) and synthetic analogues (Bukiya et al. 2013), testosterone, and corticosterone (King et al. 2006; Unemoto et al. 2007). Nitric oxide (NO) is one more example of an endogenous mediator exerting BK activating properties directly and indirectly (Carrier et al. 1997; Gragasin et al. 2004; Archer et al. 1994; Haburcak et al. 1997; Bolotina et al. 1994; Lee et al. 2006) and it theoretically opens the opportunity to endow any drug with a donor of nitric oxide to induce "physiological" BK channel activation (Vaali et al. 1998).

Natural products have also been demonstrated to be a rich source of BK channel openers, which has been the objective of comprehensive reviews (Nardi et al. 2003; Nardi and Olesen 2008). Polyphenolic compounds, such as puerarin (Sun et al. 2007), mallotoxin (Zakharov et al. 2005), trans-resveratrol (Wu 2003; Sheu et al. 2008), and several other flavonoids or terpene compounds, such as pimaric acid (PiMA; Imaizumi and Ohwada 2002; Imaizumi et al. 2002) and dehydroabietic acid (Ohwada et al. 2003), are clearly privileged scaffolds whose promiscuous pharmacological profile has led to several synthetic efforts to improve upon both potency and selectivity (Ohwada et al. 2003, 2007; Li et al. 2003c; Rittenhouse et al. 1997; Tashima et al. 2006; Cui et al. 2008) and also the well-profiled Cym04 (Fig. 26.4; Cui et al. 2010; Gessner et al. 2012).

Whereas pharmacological selectivity and potency have certainly been the limit for preclinical studies of endogenous substances, natural compounds, and the original prototype phenolic BK activators (NS004, NS1619, NS1608; Nardi and Olesen 2008), a more potent and more selective pharmacological tool NS11021

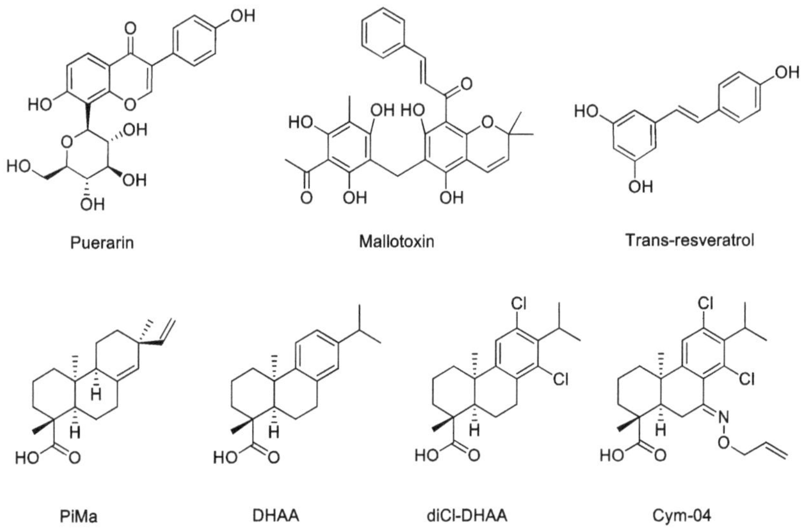

Fig. 26.4 Natural BK channel activators puerarin, mallotoxin, trans-resveratrol, PiMa, DHAA and the synthetic derivatives diCl-DHAA and Cym-04. **Puerarin**: 7-hydroxy-3-(4-hydroxyphenyl)-8-((2S,3R,4R,5S,6R)-3,4,5-trihydroxy-6-(hydroxymethyl)tetrahydro-2H-pyran-2-yl)-4H-chromen-4-one. **Mallotoxin**: (E)-1-(6-(3-acetyl-2,4,6-trihydroxy-5-methylbenzyl)-5,7-dihydroxy-2,2-dimethyl-2H-chromen-8-yl)-3-phenylprop-2-en-1-one. **Trans-resveratrol:** (E)-5-(4-hydroxystyryl)benzene-1,3-diol. **PiMa**: (1R,4aS,4bS,7S,10aR)-1,4a,7-trimethyl-7-vinyl-1,2,3,4,4a,4b,5,6,7,9,10,10a-dodecahydrophenanthrene-1-carboxylic acid. **DHAA**: (1R,4aR,10aR)-7-isopropyl-1,4a-dimethyl-1,2,3,4,4a,9,10,10a-octahydrophenanthrene-1-carboxylic acid. **diCl-DHAA**: (1R,4aR,10aR)-6,8-dichloro-7-isopropyl-1,4a-dimethyl-1,2,3,4,4a,9,10,10a-octahydrophenanthrene-1-carboxylic acid. **Cym-04**: (1R,4aR,10aR,E)-9-((allyloxy)imino)-6,8-dichloro-7-isopropyl-1,4a-dimethyl-1,2,3,4,4a,9,10,10a-octahydrophenanthrene-1-carboxylic acid

(Bentzen et al. 2007), equipped also with its corresponding negative control NS13558 (Bentzen et al. 2010), has recently been disclosed and is available commercially by several vendors (Tocris, Sigma; Fig. 26.5). NS11021, originally discovered at NeuroSearch, currently appears to be the pharmacological BK channel activating tool of choice (EC_{50}: 400 nM) given the widespread use among different research groups (Wulf-Johansson et al. 2010; Wojtovich et al. 2013; Venglovecz et al. 2011; Liu et al. 2014; Layne et al. 2010; Kun et al. 2009; Kiraly et al. 2013; Borchert et al. 2013; Bentzen et al. 2009; Bednarczyk et al. 2013; Aon et al. 2010). A common feature for these latter molecules, as well as for most of the known BK activators, is the presence of an acidic functionality, especially carboxylic acids. This plethora includes, for instance, CTBIC (Fig. 26.5; Gormemis et al. 2005), a compound able to reduce bladder voiding, relax bladder smooth muscle, and inhibit intestinal motility (Dela Pena et al. 2009a, b) and whose carboxylic acid group has been suggested to be critical for its binding mode (Lee et al. 2012).

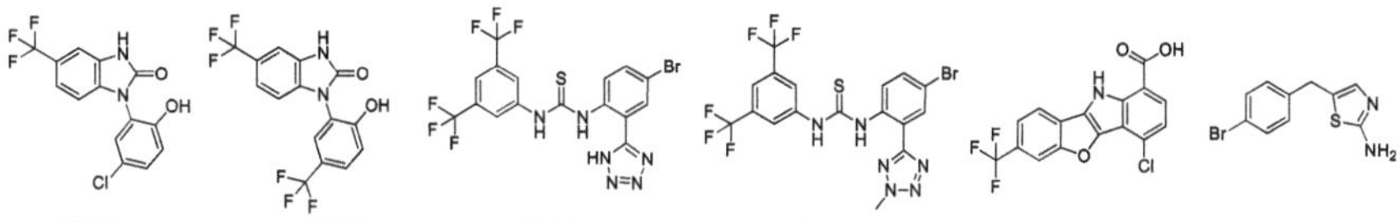

Fig. 26.5 BK activator tool compounds: the former and outclassed NS004 and NS1619, the current NS11021 and the newly identified NS19504. **NS004**: 1-(5-chloro-2-hydroxyphenyl)-5-(trifluoromethyl)-1H-benzo[d]imidazol-2(3H)-one. **NS1619**: 1-(2-hydroxy-5-(trifluoromethyl)phenyl)-5-(trifluoromethyl)-1H-benzo[d]imidazol-2(3H)-one. **NS11021**: 1-(3,5-bis(trifluoromethyl)phenyl)-3-(4-bromo-2-(1H-tetrazol-5-yl)phenyl)thiourea. **NS13558**: 1-(3,5-bis(trifluoromethyl)phenyl)-3-(4-bromo-2-(2-methyl-2H-tetrazol-5-yl)phenyl)thiourea. **CTBIC**: 4-chloro-7-(trifluoromethyl)-10H-benzofuro[3,2-b]indole-1-carboxylic acid. **NS19504**: 5-(4-bromobenzyl)thiazol-2-amine

Carboxylic acids are however regarded with caution in current drug development as they may act as acylators either via glucuronidation or acyl-CoA formation, which may constitute bioactivation pathways that lead to toxicity (Sallustio et al. 2000; Li et al. 2003a, b). Tetrazole moieties, present, for instance, in NS11021, are common bioisostere for carboxylate in view of their planar structure and ability to ionise at physiological pH (Pinter et al. 2011; Biot et al. 2004; Song et al. 2013). Whereas tetrazoles may offer several advantages over the carboxylic acids, such as the ability to circumvent some of the classical phase II biotransformation of carboxylic acids or the higher lipophilicity leading to a higher permeability, they may nonetheless suffer from similar undesirable properties, such as high protein binding and limited extravascular tissue distribution (Maillard et al. 2000; Treiber et al. 2003; Davi et al. 2000; Laeis et al. 2001). With this in mind, NeuroSearch recently launched a high-throughput screening campaign employing a new FLIPR-based Tl^+-flux assay as a primary screen and an automated QPatch assay as a secondary screen in order to identify and profile small-molecule BK activators without any acidic functionalities present in the molecule (Nausch et al. 2014). These efforts materialised in a low-molecular-weight and structurally different commercial molecule, namely, NS19504 (Fig. 26.5), which activated the BK channel with an EC_{50} value of 11.0 ± 1.4 μM in the FLIPR assay, produced activation from a concentration of 0.3 μM, and left-shifted the voltage activation curve by 60 mV at 10 μM, in manual and automated (QPatch) patch-clamp assays. Furthermore, NS19504 (1 μM) showed an iberiotoxin-sensitive reduction of spontaneous phasic contractions of guinea pig urinary bladder strips, providing further support for the role of BK channels in urinary bladder function. This completely distinct chemotype may now offer a new pharmacological tool with very diverse chemical physical properties and a new starting point for a lead optimisation campaign.

Finally, in the last few years, and since the most comprehensive review on BK channel modulators (Nardi and Olesen 2008), relatively few small molecule openers have been identified. Among these are a set of novel anthraquinone analogues (named the GoSlo-SR family; Roy et al. 2012), structurally similar to compounds previously reported as purinoceptor antagonists (Weyler et al. 2008; Baqi et al. 2009). This set of molecules was tested using the inside-out patch-clamp technique in rabbit bladder smooth muscle cells whilst holding patches at −60 mV. The best compound emerging ($EC_{50} = 2.30$ μM) was 5r or GoSlo-SR-5-44, able to shift the voltage-activation curve by more than −100 mV in a concentration-dependent manner and in the presence of a low concentration of calcium ions (100 nM). GoSlo-SR-5-69 is from the same chemical class of GoSlo-SR-5-44, both appearing in the same patent applications (Hollywood et al. 2012a, b), but fully disclosed only later (Roy et al. 2014) as an optimised analogue with an EC_{50} of 251 nM. Other more recent compounds still characterised by a carboxylic acid moiety were divulged by Amgen and Merck. The Merck tetrahydroquinoline 36 was able to produce an approximately sevenfold increase of the maximum current observed compared to control at 3 μM compound concentration, whereas the Amgen tetrahydroquinoline Z displayed an EC_{50} of 109 nM. These compounds were able to block spontaneous neuronal firing in an electrophysiological model of migraine (Gore et al. 2010) and to induce smooth muscle relaxation (Ponte et al. 2012). Finally, more carboxylic acid derivatives were identified by Japanese groups from the University of Tokyo (no biological data; Ohwada et al. 2009) and from Mitsubishi Tanabe Pharma (compound 1 and compound 2, Fig. 26.6). The BK activator ($EC_{50} = 0.37$ μM) described by this latter group did not display the cyclooxygenase-inhibitory activity ($IC_{50} > 30$ μM) common for this chemical

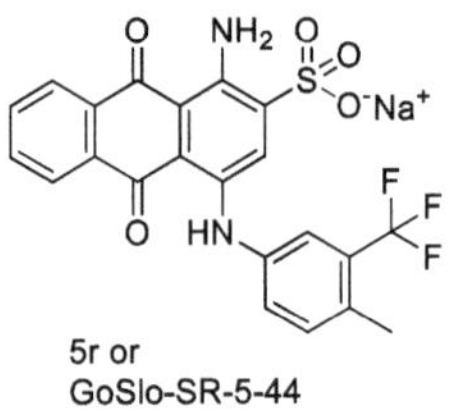

5r or
GoSlo-SR-5-44

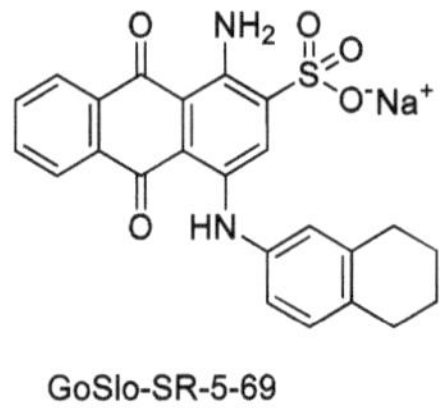

GoSlo-SR-5-69

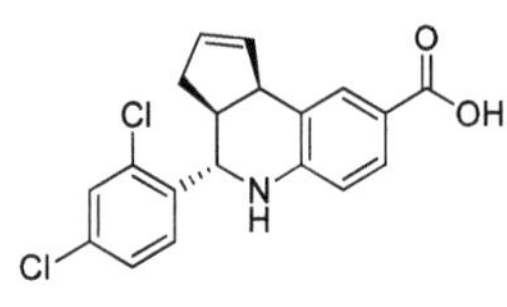

Compound 36

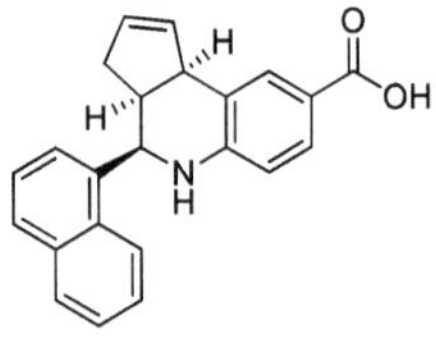

Z

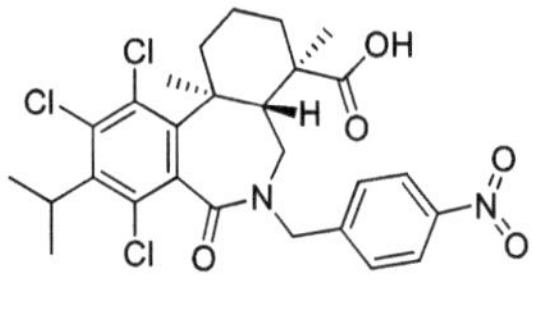

1

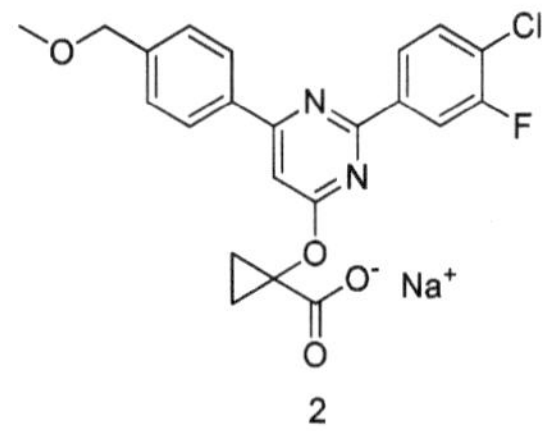

2

Fig. 26.6 Recently disclosed BK activators. **5r or GoSlo-SR-5-44**: sodium 1-amino-4-((4-methyl-3-(trifluoromethyl)phenyl)amino)-9,10-dioxo-9,10-dihydroanthracene-2-sulfonate. **GoSlo-SR-5-69**: sodium 1-amino-9,10-dioxo-4-((5,6,7,8-tetrahydronaphthalen-2-yl)amino)-9,10-dihydroanthracene-2-sulfonate. **Compound 36**: (3aS,4S,9bR)-4-(2,4-dichlorophenyl)-3a,4,5,9b-tetrahydro-3H-cyclopenta[c]quinoline-8-carboxylic acid. **Z**: (3aS,4R,9bR)-4-(naphthalen-1-yl)-3a,4,5,9b-tetrahydro-3H-cyclopenta[c]quinoline-8-carboxylic acid. **1**: (4R,4aR,11bS)-8,10,11-trichloro-9-isopropyl-4,11b-dimethyl-6-(4-nitrobenzyl)-7-oxo-2,3,4,4a,5,6,7,11b-octahydro-1H-dibenzo[c,e] azepine-4-carboxylic acid. **2**: sodium 1-((2-(4-chloro-3-fluorophenyl)-6-(4-(methoxymethyl) phenyl)pyrimidin-4-yl)oxy)cyclopropanecarboxylate

template and, administered at 10 mg/kg i.v. in rats, it significantly (>50 %) reduced the frequency of rhythmic bladder contractions induced by iberiotoxin (Tsuzuki and Hirai 2009).

26.2.2 BK Channel Inhibitors

BK channel inhibition has been shown to be a secondary effect of a number of marketed drugs in the same manner as is the case for BK channel activators. The most noteworthy of these inhibitors are probably verapamil (Harper et al. 2001), ketamine (Hayashi et al. 2011; Klockgether-Radke et al. 2005; Denson et al. 1994; Denson and Eaton 1994), simvastatin (Seto et al. 2007), clotrimazole (Wu et al. 1999), and ketoconazole (Wu et al. 1999). Ketoconazole may be particularly interesting, since this molecule (in contrast to clotrimazole) does not potently inhibit IK channels (Jensen et al. 1999). Furthermore, synthetic analogues and partial structures of ketoconazole have interestingly afforded compounds with either activating or inhibiting BK channel activity (Power et al. 2006), confirming the idea that known drugs may serve well as useful template structures.

The most valuable and used inhibiting tool compounds for in vitro and ex vivo pharmacological studies are however BK inhibitors from natural sources. Of these, the most frequently used is certainly iberiotoxin (Kunze et al. 1994; Kozlowski et al. 1996; Haghdoost-Yazdi et al. 2008), a 37-amino acid peptide isolated from the venom of the scorpion *Buthus tamulus* (Galvez et al. 1990), with affinity (K_i) at the channel down to the single-digit nanomolar range. Blockade of BK channels by iberiotoxin is promoted by a very selective binding to the external mouth of the channel that physically blocks the conduction pathway (Candia et al. 1992). A remarkable selectivity feature of iberiotoxin is that it is inactive on BK channels associated with the neuronal beta4 subunit (Wang et al. 2006). Several other peptidic scorpion toxins have been reported to similarly block the BK channel, although with different degrees of selectivity and mechanisms of action, including charybdotoxin (Tang et al. 2010), slotoxin (Garcia-Valdes et al. 2001), BmP09 (Yao et al. 2005), and martentoxin (Shi et al. 2008; Ji et al. 2003; Wang et al. 2013; Tao et al. 2012).

Indole-diterpenes of natural origin are among the most selective and potent non-peptidic BK channel inhibitors. This group includes paxilline (IC_{50}: 1.9 nM) (Sanchez and McManus 1996; Strobaek et al. 1996), verruculogen, paspalicine, paspalitrem C, and penitrem A (DeFarias et al. 1996; Gribkoff et al. 1996; Knaus et al. 1994). Structurally similar tremorgenic indole alkaloids have been claimed to be useful as a topical treatment of glaucoma and intraocular hypertension via blockage of potassium channel conductance in the ciliary epithelium (Adorante et al. 1996). Their ability to reduce intraocular pressure in proof of concept studies in rabbit comparably to the β-blocker timolol catalysed the interest for scientists at Merck to establish a fluorescence resonance energy transfer (FRET)/voltage ion probe reader (VIPR) assay to identify small-molecule inhibitors of BK channels

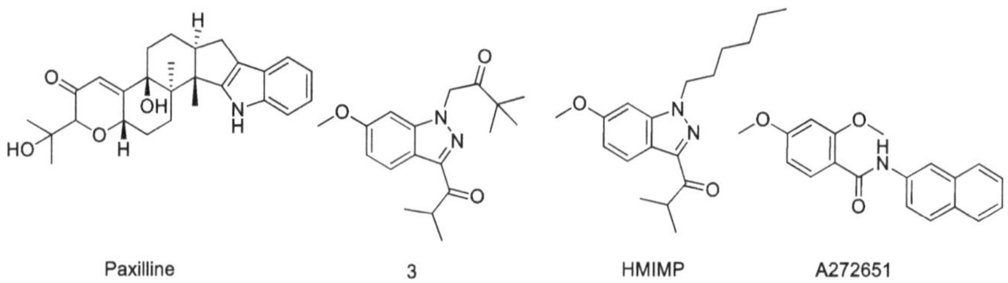

Fig. 26.7 Naturally occurring (paxilline) and synthetic small-molecule (compound 3, HMIMP, A272651) BK-channel inhibitors. **Paxilline**: (4bS,6aS,12bS,12cR,14aS)-4b-hydroxy-2-(2-hydroxy propan-2-yl)-12b,12c-dimethyl-5,6,6a,7,12,12b,12c,13,14,14a-decahydro-2H-chromeno[5′,6′:6,7] indeno[1,2-b]indol-3(4bH)-one. **3**: 1-(3-isobutyryl-6-methoxy-1H-indazol-1-yl)-3,3-dimethylbutan-2-one. **HMIMP**: 1-(1-hexyl-6-methoxy-1H-indazol-3-yl)-2-methylpropan-1-one. **A272651**: 2,4-dimethoxy-N-(naphthalen-2-yl)benzamide

with potency in the low, double-digit nanomolar range. This assay has been presumably used to identify and/or profile a large array of compounds and similar diterpene frameworks lacking tremorgenic liability whilst retaining high affinity for BK channels. Such compounds have been subsequently patented for the same therapeutic benefit in an impressive number of patents (Goetz et al. 2003; Goetz and Kaczorowski 2004; Garcia et al. 2001; Doherty et al. 2004a, b, 2007, 2008a, b; Gao and Shen 2006; Doherty and Shen 2006; Chen et al. 2004a, 2005a, b; Boyd et al. 2005). One particular compound from these studies (compound 3, Fig. 26.7) has later been the specific object of optimisation for ophthalmic formulation (Miller and Yu 2008), leading to the supposition that this compound might have been the frontrunner emerging from this drug discovery programme. It is not currently known whether this compound was further advanced to clinical settings prior to the supposed discontinuation of the entire research project.

Furthermore, other synthetic small molecules have been identified. The small-molecule A272651 (Fig. 26.7; Shieh et al. 2007), disclosed by Abbott, was to our knowledge the first small-molecule BK channel inhibitor to be identified and fully described in a scientific article. However, with an IC_{50} of about 5 μM, a potency >1000-fold lower than that of iberiotoxin and paxilline, its value as a pharmacological tool compound appears questionable, a conclusion that is reinforced by the absence of any additional studies or publications with this very compound in PubMed/Medline. Novel indole, indazole, imidazole, benzofuran, benzothiophene, benzimidazole, oxoquinoline, 1,2-dihydroquinolin-2-one, 1,2-dihydroquinoxalin-2-one, and 1,2-dihydronaphthyridin-2-one derivatives as BK inhibitors had been earlier reported in the patent literature by Merck, but poorly described from a pharmacological standpoint (Doherty et al. 2007; Chen et al. 2003, 2004b, 2005a, b; Doherty and Shen 2005, 2006; Gao and Shen 2006). One such small molecule, HMIMP (Fig. 26.7), has been interestingly profiled in a study carried out later by GlaxoSmithKline and identified as a highly potent ($IC_{50} = 2$ nM) and beta-subunit independent BK channel blocker (Zeng et al. 2008). Whilst the potency is comparable to iberiotoxin, the mechanism of action is surely very different, with the small molecule acting allosterically and able to block the BK channel from either side of the membrane (Zeng et al. 2008).

26.3 IK Channel Modulator Chemistry

26.3.1 IK Channel Activators

Whereas small-molecule BK channel inhibitors and, as described in the next section, IK channel inhibitors usually display a high degree of selectivity towards the other subtypes of calcium-activated potassium channels, IK small-molecule openers usually possess only a low degree of selectivity and they are typically embedded with SK opening properties. One such molecule is the benzimidazolone 1-EBIO (Devor et al. 1996; Singh et al. 2000), with an EC_{50} of 30 μM (Wulff et al. 2007) at IK and an EC_{50} of 589 μM on SK2 in a whole-cell patch-clamp assay (Hougaard et al. 2009; Fig. 26.8). This compound was originally suggested to have potential therapeutic benefit for disease characterised by mucus-congested airways such as cystic fibrosis and chronic obstructive pulmonary disease (Singh et al. 2001) as well as for disease characterised by altered neuronal excitability (Garduno et al. 2005; Anderson et al. 2006; Kobayashi et al. 2008). Improvement of potency and selectivity of 1-EBIO has been the object of a number of medicinal chemistry campaigns. DC-EBIO (Singh et al. 2000; Fig. 26.8) is one analogue identified in a structure-activity relationship study, namely, the 5,6-dichloro-1-ethyl analogue, displaying an improved (EC_{50}: 750 nM) potency profile over other differently decorated benzimidazolones. Furthermore, installation of an oxime moiety in the corresponding indolin-2-one nucleus and retention of a dichloro pattern (6,7)

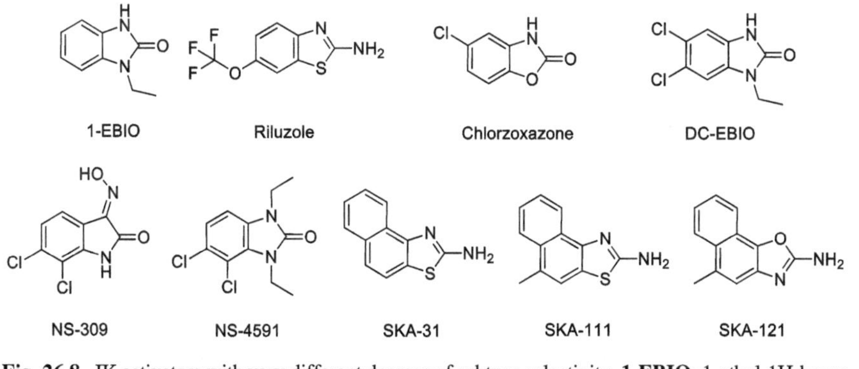

Fig. 26.8 IK activators with very different degrees of subtype selectivity. **1-EBIO**: 1-ethyl-1H-benzo[d]imidazol-2(3H)-one. **Riluzole**: 6-(trifluoromethoxy)benzo[d]thiazol-2-amine. **Chlorzoxazone**: 5-chlorobenzo[d]oxazol-2(3H)-one. **DC-EBIO**: 5,6-dichloro-1-ethyl-1H-benzo[d]imidazol-2(3H)-one. **NS-309**: (E)-6,7-dichloro-3-(hydroxyimino)indolin-2-one. **NS-4591**: 4,5-dichloro-1,3-diethyl-1H-benzo[d]imidazol-2(3H)-one. **SKA-31**: naphtho[1,2-d]thiazol-2-amine. **SKA-111**: 5-methylnaphtho[1,2-d]thiazol-2-amine. **SKA-121**: 5-methylnaphtho[2,1-d]oxazol-2-amine

afforded NS309 (Fig. 26.8; Strobaek et al. 2004), which demonstrated to be some 30-fold more potent than DC-EBIO in a back-to-back comparison. NS309, although featuring poor metabolic stability in in vivo settings, due to its high potency and excellent penetration into various tissues, has served as a valuable tool compound for demonstrating participation of IK/SK channels in many in vitro and ex vivo experimental setups. Recently, it has been used to advocate the involvement of IK/SK channels in the regulation of detrusor smooth muscle contractions (Parajuli et al. 2013), for demonstration of the pivotal role of IK/SK channels in epithelial and endothelial-dependent smooth muscle relaxation (Kroigaard et al. 2012; Brondum et al. 2010) and for establishing SK channel activation for stabilisation of cerebellar Purkinje cell activity a neuroprotective and anti-ataxic principle in the cerebellum (Kasumu et al. 2012). Based on these observations, a new quest for novel positive modulators of SK/IK channels with improved bioavailability led to the identification of NS4591 (Fig. 26.8). In rats, the bioavailability of NS4591 after oral dosing was estimated to be 97 % (Hougaard et al. 2009). In whole-cell patch-clamp experiments, NS4591 doubled IK-mediated currents at a concentration of 45 nM, whereas 530 nM was required for doubling of SK3-mediated currents under the same experimental conditions. Furthermore, in ex vivo settings, NS4591 inhibited the firing of action potentials in primary bladder afferent DRG neurons and also reduced carbachol-induced contractions of bladder rings. Consistent with this in vitro and ex vivo profile, NS4591 alleviated in vivo bladder overactivity induced by capsaicin in awake rats as well as by acetic acid in anesthetised cats, further supporting the hypothesis that SK/IK channels fill an important role in bladder function and suggest that positive modulators of SK/IK channels may be of benefit for the therapeutic treatment of overactive bladder.

Analogue small molecules, such as methylxanthines (theophylline, IBMX, and caffeine; Schroder et al. 2000), the marketed glutamate-release inhibitor riluzole (Grunnet et al. 2001; Cao et al. 2001), and the marketed chlorzoxazone (Syme et al. 2000; Fig. 26.8), characterised by relatively simple bicyclic chemical frameworks, have also exhibited similar and poor level of potency and specificity for the IK channel opening. The diversity of therapeutic indications addressed by riluzole underlies the wide range of mechanisms of action embedded in this molecule: First launched in 1996, it is approved as a treatment for amyotrophic lateral sclerosis and it is currently in clinical development for additional indications as diverse as bipolar disorders (phase II), multiple sclerosis (phase II), malignant melanoma (phase I), ataxia (phase II), etc.

Improvement of potency and selectivity of riluzole towards IK channels has been attempted and achieved in the form of SKA-31 (Fig. 26.8; Sankaranarayanan et al. 2009). After screening 55 riluzole analogues, scientists at the University of California identified this latter molecule as a more potent (10-fold, EC_{50}: 250 nM) and more selective IK channel opener, including 10-fold selectivity over SK and other ion channels. This in vitro profile, combined with an acceptable pharmacokinetic profile, has made it possible to use it as a tool compound in several studies and to probe the therapeutic potential of IK channel opening, including the potential preservation of coronary flow in diabetic myocardium (Mishra et al. 2014).

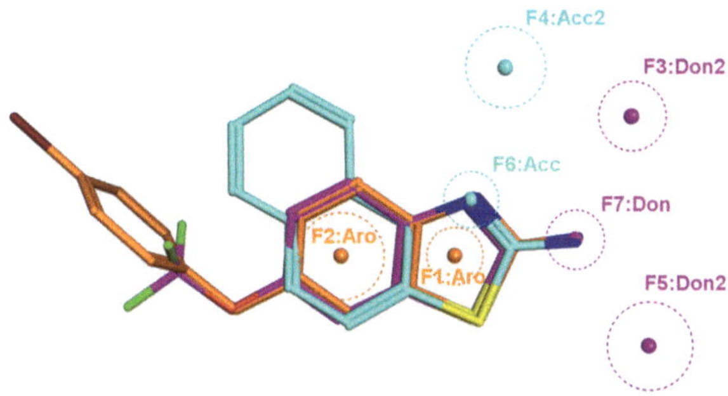

Fig. 26.9 Standard modeling techniques have been applied to generate, overlay, and visualise the pharmacophore features of the compounds NS19504 (*orange*), riluzole (*magenta*), and SKA-31 (*cyan*) as implemented in MOE2013.08 (Molecular Operating Environment 2014, CCG, Montreal, Canada). Figure generated by Dr. Achim Kless (Grünenthal). **NS19504**: 5-(4-bromobenzyl) thiazol-2-amine. **Riluzole**: 6-(trifluoromethoxy)benzo[d]thiazol-2-amine. **SKA-31**: naphtho [1,2-d]thiazol-2-amine

Nonetheless, the need for a yet more potent and selective IK (over SK) remains in order to investigate the potential of the IK channel as target for novel endothelial antihypertensives and to further dissect the functional effects of IK channel activation from those of SK channel activation in the above as well as in other studies (Radtke et al. 2013). Analogues of SKA-31, namely, SKA-111 and SKA-121 (Fig. 26.8), have been very recently identified as better in vitro/ex vivo tool IK activators (EC_{50}: 100 nM for both; Coleman et al. 2014). Surprisingly, SKA-111 showed a 120-fold selectivity for IK over SK, despite the difference of only a methyl substituent when compared to SKA-31. SKA-121 carries instead an additional structural difference of the core, with the thiazole being replaced by an oxazole ring. In common for all (SKA-31, SKA-111, SKA-121), the retention of a primary amino group at the terminal and right-hand site is an evidently important pharmacophoric feature. Interesting, such a primary amino functionality is present in both IK activators (riluzole, SKA-31, SKA-111, SKA-121) and BK activators (NS19504), suggesting the intriguing possibility of this group serving as an anchor point across different calcium-activated potassium channels and the possibility of fine-tuning the specificity via molecular decoration and the molecular diversification of the left-hand site of the molecules. This is clearly represented in Fig. 26.9, where NS19504 (orange), riluzole (magenta), and SKA-31 (cyan) have been overlaid and the potential common pharmacophoric features (aromatics, accepton/ hydrogen bond donors) have been highlighted.

26.3.2 IK Channel Inhibitors

A number of peptide toxins, encompassing maurotoxin, charybdotoxin, and OSK1, isolated from venoms of animals such as scorpions have demonstrated nanomolar potency as IK inhibitors (Wei et al. 2005; Mouhat et al. 2005; Jaravine et al. 1997). These toxins, however, not additionally featured with the specificity that is ideally required for chemical tools, have allowed small molecules to take centre stage in most of IK pharmacological studies. TRAM-34 (Wulff et al. 2001; Fig. 26.10), designed at the University of California Irvine (IC_{50}: 9 nM, patch-clamp assay; Strobaek et al. 2013; Wulff et al. 2000), and the significantly less potent antifungal imidazole, clotrimazole (IC_{50}: 3 μM, in perforated-patch whole-cell recording experiments; Wu et al. 1999; Brugnara et al. 1996; McNaughton-Smith et al. 2008) originally developed at Bayer, are the prototype small-molecule agents that currently serve as gold reference standards for the pharmacological blockade of IK channels. TRAM-34 and clotrimazole share a number of common or independent shortcomings, including important effects on a number of human and rat CYP isoforms (Agarwal et al. 2013), suggesting caution in their use in vivo. On the opposite end of the spectrum of small-molecule IK inhibitor agents is ICA-17043 (Senicapoc, Fig. 26.10; IC_{50}: 11 nM, rubidium efflux assay; Stocker et al. 2003; Ataga and Stocker 2009; Ataga et al. 2006, 2008, 2011), a drug candidate that has been advanced to clinical phase III as an orally available drug for the chronic treatment of sickle cell disease (Brugnara and de Franceschi 2006). This Icagen compound, in which the heteroaromatic rings of clotrimazole and TRAM-34 have been replaced by a carboxamide ($CONH_2$) function, still belongs to the TRAM class of compounds, thus emphasising the limited chemical diversity of available IK inhibitors. The triarylmethane moiety was postulated by Wulff et al. (2001) to serve as a propeller to fit "snugly into the internal cavity just below the selectivity filter with each of the phenyl rings locked in place by interactions with Thr250 and Val275 from three of the four IK subunits", supporting and reinforcing the idea that structural diversification taking place in other molecular regions would probably deliver more easily pharmacologically active agents. In fact, a patent application from NeuroSearch illustrates a large series of di- and triarylmethane derivatives (an exemplary compound 4 is reported in Fig. 26.11; Demnitz et al. 2003), with the most preferred compounds inhibiting IK channels in electrophysiology experiments in the 'low micromolar and nanomolar range". In one more follow-up patent application, NeuroSearch again presented a structurally related diarylmethane

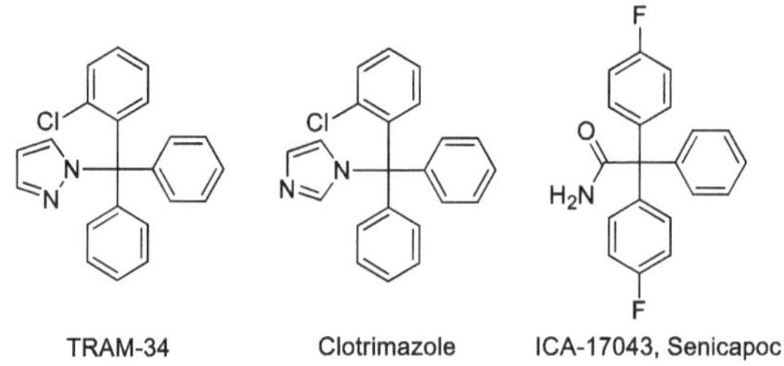

Fig. 26.10 Traditional IK channel inhibitors. **TRAM-34**: 1-((2-chlorophenyl)diphenylmethyl)-1H-pyrazole. **Clotrimazole**: 1-((2-chlorophenyl)diphenylmethyl)-1H-imidazole. **ICA-17043, Senicapoc**: 2,2-bis(4-fluorophenyl)-2-phenylacetamide

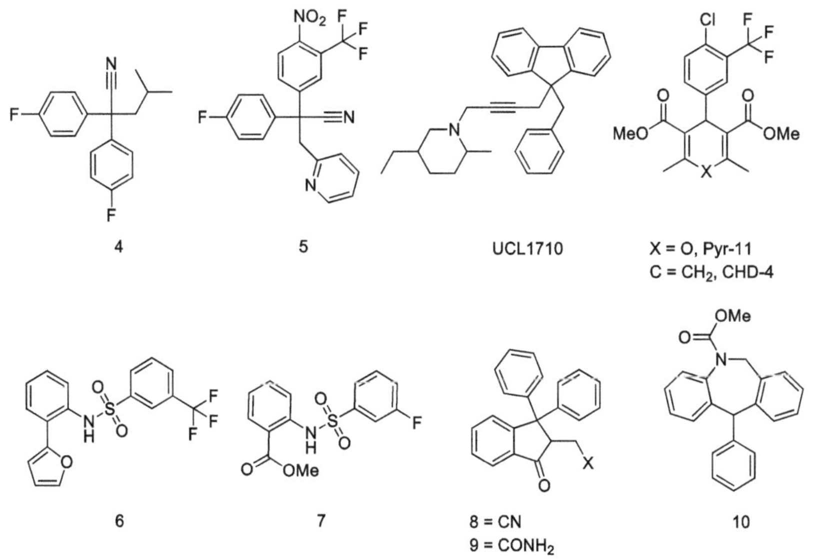

Fig. 26.11 IK channel inhibitors characterised by the traditional TRAM motif (*top*) as well as structurally diversified molecular frameworks. **4**: 2,2-bis(4-fluorophenyl)-4-methylpentanenitrile. **5**: 2-(4-fluorophenyl)-2-(4-nitro-3-(trifluoromethyl)phenyl)-3-(pyridin-2-yl)propanenitrile. **UCL1710**: 1-(4-(9-benzyl-9H-fluoren-9-yl)but-2-yn-1-yl)-5-ethyl-2-methylpiperidine. **Pyr-11**: dimethyl 4-(4-chloro-3-(trifluoromethyl)phenyl)-2,6-dimethyl-4H-pyran-3,5-dicarboxylate. **CHD-4**: dimethyl 4′-chloro-3,5-dimethyl-3′-(trifluoromethyl)-1,4-dihydro-[1,1′-biphenyl]-2,6-dicarboxylate. **6**: N-(2-(furan-2-yl)phenyl)-3-(trifluoromethyl)benzenesulfonamide. **7**: methyl 2-(3-fluorophenylsulfonamido)-benzoate. **8**: 2-(3-oxo-1,1-diphenyl-2,3-dihydro-1H-inden-2-yl)acetonitrile. **9**: 2-(3-oxo-1,1-diphenyl-2,3-dihydro-1H-inden-2-yl)acetamide. **10**: methyl 11-phenyl-6,11-dihydro-5H-dibenzo[b,e]azepine-5-carboxylate

derivative (in Fig. 26.11 the exemplary compound 5), with a K_i value for block of hIK channels of 65 nM (Demnitz et al. 2005). These and other analogues have been the subject of other recent excellent reviews, and the reader is referred to these reviews for further detailed background (Wulff and Castle 2010; Wulff et al. 2007).

The primary endpoint of the clinical study for senicapoc, named ASSERT (A Stratified Sickle Events Randomized Trial), was vaso-occlusive crisis rate. In a press release dated 3 April 2007, Icagen announced termination of the trial because of the low probability of achieving a reduction in crisis rate, the primary endpoint (Nardi et al. 2008). However, Senicapoc reemerged later with a growing body of data linking IK channels to immune cell activation and cellular proliferation as well as efficacy data from a study that demonstrated the ability of this compound to reverse antigen-induced changes in late-phase airway resistance and airway hyper-reactivity in a sheep model of asthma (Van Der Velden et al. 2013). This triggered the same company to initiate a phase IIa clinical trial in exercise-induced asthma, which failed to meet the new primary endpoints (decrease in FEV1, etc.), but succeeded to spark a high interest for IK channels as drug target for allergic and inflammatory diseases. This interest has recently resulted in reinvigorated efforts towards the identification of structurally different small-molecule ligands. Molecular diversification of these pharmacological tools would allow the scientific community to appraise whether the chemical space represented by these triarylmethane compounds is the reason for failure in delivering successful drugs as well as to potentially elude a number typical pharmaceutical and drugability issues for this class of compounds, e.g., limited freedom to operate still available, low plasma exposure upon oral dosing, lack of brain penetration, etc.

Medicinal chemists had actually witnessed in the past some efforts in this direction. From experimental work based on the pioneer IK channel inhibitor cetiedil (Berkowitz and Orringer 1981), University College London identified compounds such as UCL1710 (Fig. 26.11; Roxburgh et al. 2001; Narenjkar et al. 2004), with inhibitory potency of 300 nM. A close inspection of the chemical structures described so far and depicted in Figs. 26.10 and 26.11 does, however, highlight a common motif preserved, i.e., two or three, frequently electron-deficient, aromatic rings compactly arranged around a "tight core" of one or two carbon atoms, with a fourth substituent being slightly more "versatile", as in UCL1710.

Furthermore, the observation that the L-type Ca^{2+} channel blocker nifedipine inhibited the IK channel in the low micromolar range (Kaji 1990) also inspired efforts culminating in the 4H-pyran (Pyr-11; Urbahns et al. 2003) and the 1,4-cyclohexadiene (CHD-4; Nardi et al. 2008; Fig. 26.11), able to block IK with IC_{50} values of 8 and 1.5 nM, respectively. Additionally, Icagen has disclosed a series of novel diarylsulfonamide IK inhibitors for use in glaucoma, sickle cell disease, abnormal cell proliferation, and inflammation (Atkinson et al. 2004). A common feature of these compounds is the left-hand aniline-derived aryl group usually substituted by one or more groups, of which one is invariably either a methyl ester or a presumably isosteric five-membered heterocycle ortho to the sulfonyl nitrogen. On the right-hand sulfonyl aryl ring, one or more electronegative halogen or trifluoromethyl substituents imply the necessity for electron deficiency. Although no discrete potencies are quoted in the patent, compounds in Fig. 26.11 are two examples (compounds 6 and 7) of more potent derivatives as determined in $^{86}Rb^{+}$-efflux assays from red blood cells ($IC_{50} < 500$ nM). Finally, also the Brugnara group demonstrated the switching from the "classical" triarylmethane chemotype to more restricted diphenyl indanones (Bellott et al. 1999b; Alper et al. 1999b); Compounds 8 and 9, Fig. 26.11) or 11-phenyl-dibenzazepines (Alper et al. 1999a; Bellott et al. 1999a; exemplary compound 10, Fig. 26.11) that yield new classes of IK blockers, which can be considered conformationally restricted triarylmethanes.

More recently, NeuroSearch (nowadays Saniona A/S) has significantly addressed the need of identification of new chemical scaffolds by describing a 384-well high-throughput Tl^{+} influx assay that was capable of screening rapidly more than 200,000 compounds (Jorgensen et al. 2013) and to single out several new chemical series of inhibitors as well as activators of IK channels, including the inhibitor NS6180 (Fig. 26.12), making its first appearance in the public domain in February 2012, at the 56th Annual Biophysical Society Meeting in San Diego and

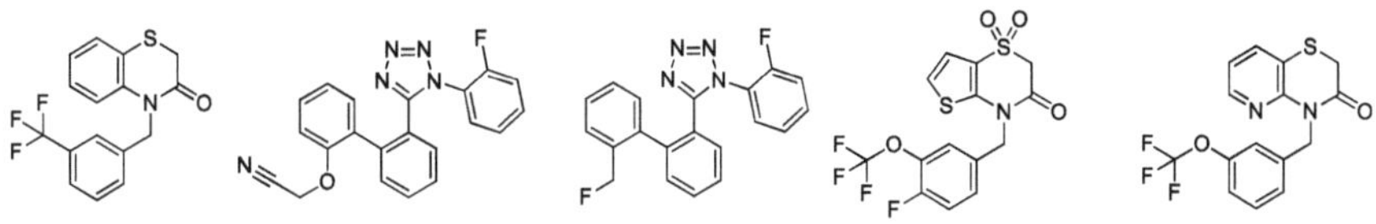

NS-6180 11 12 13 14

Fig. 26.12 Innovative molecular frameworks, such as the thiazinone compounds identified at NeuroSearch (Saniona) and Boehringer Ingelheim Pharma as IK channel inhibitors (*top*). Below are chemical structures of tetrazole IK inhibitors, built on a novel molecular backbone with pharmacophoric features resembling the traditional TRAM motif. **NS-6180**: 4-(3-(trifluoromethyl)benzyl)-2H-benzo[b][1,4]thiazin-3(4H)-one. **11**: 2-((2′-(1-(2-fluorophenyl)-1H-tetrazol-5-yl)-[1,1′-biphenyl]-2-yl)oxy)acetonitril. **12**: 5-(2′-(fluoromethyl)-[1,1′-biphenyl]-2-yl)-1-(2-fluorophenyl)-1H-tetrazole. **13**: 4-(4-fluoro-3-(trifluoromethoxy)benzyl)-2H-thieno [3,2-b][1,4]thiazin-3(4H)-one 1,1-dioxide. **14**: 4-(3-(trifluoromethoxy)benzyl)-2H-pyrido[3,2-b] [1,4]thiazin-3(4H)-one

being fully disclosed in a publication shortly thereafter (Strobaek et al. 2013). This compound showed anti-inflammatory actions in a model of inflammatory bowel disease, possibly by a local action in the gut. NS6180 (Strobaek et al. 2013) is a commercial benzothiazinone inhibitor that is equipotent to TRAM-34 (IC_{50}: 10–20 nM) and that has no structural resemblance to TRAM analogues or previous IK inhibitors. Irrespective of the chemical diversity, the presence in the binding site of the same amino acid residues highlighted above (Thr250 and Val275) is also highly critical for the pharmacological effect of NS6180, suggesting nonetheless a similar binding mode in the inner vestibule of the channel. One more chemical series by Saniona is represented by the tetrazole derivatives (Demnitz and Jorgensen 2014) exemplified by compounds 11 and 12 in Fig. 26.12 with IC_{50} of 16 nM and 17 nM, respectively. From these, it may be concluded that the tetrazole ring may serve a double role, both as bioisosteric replacement of the carboxamide moiety of TRAM-34 and as a platform to correctly position the three classic aromatic rings in the binding space.

Contextually with its commercial origin, either the benzothiazinone compound NS6180 or its molecular framework currently is not yet included in any published patent application. Similar fused thiazin-3-ones have recently been claimed by Boehringer Ingelheim Pharma GmbH & Co (Priority date Jun 21, 2012) as IK inhibitors, also identified in a FLIPR assay (Burke et al. 2013). Two of the exemplified compounds (compounds 13, 14; Fig. 26.12) had in such an assay an IC_{50} value of 39 nM and 52 nM, respectively, and they have been reported to be potentially useful for several inflammatory and autoimmune diseases, including psoriasis, asthma, inflammatory bowel disease, etc. It remains to be seen whether these minor structural changes are able to exhibit advantages over NS6180, such as the extremely low bioavailability upon oral dosing in preclinical settings described for this latter compound (Strobaek et al. 2013).

Finally, even more recently, a medium-throughput assay has been developed using the QPatch-16 system for secondary profiling of inhibitors, which appears very helpful for confirmation of hits from high-throughput screening campaigns and appreciation of their real potency without the classical right shift for lipophilic compounds (Jenkins et al. 2013).

26.4 Interaction Sites for BK/IK Channel Modulators

BK channel activators interact with several different sites on BK channels as seen from their various dependencies on beta-subunit co-expression and splice variants (Saito et al. 1997). A recent analysis revealed that Cym04 (Fig. 26.4) and NS1619 (Fig. 26.5), which both act on the BK alpha subunit per se, are inactive on a splice variant comprising the linker between the S6 and the RKC1 domain in the C-terminal (Gessner et al. 2012). Since mutational analysis revealed that Cym04 specifically lost its activating effect on BK channels upon deletion of two amino acids in the distal linker (Phe336 and Ser335 or Ser337), whereas amino acid substitutions were ineffective, it was suggested that the effect of Cym04 was to shorten the S6-RKC1 linker, thereby facilitating force transduction from the C-terminal "gating ring" gate, much in line with the suggested mechanism for gating by Ca^{2+}. However, the actual binding site for Cym04 was not finally revealed by these studies. It is noteworthy that mallotoxin (Fig. 26.4), another BK channel activator, acted independently of the S6-RKC linker in the same study, alluding to the existence of other sites for positive modulators. One example of this is the benzofuroindole CTBIC (Fig. 26.5) site positioned at the extracellular side of the channel between two adjacent alpha subunits (Lee et al. 2012). It is very intriguing that pharmacological gating modulation can occur at such remotely placed sites and may well reflect that the physiological gating process involves many parts of the channel protein (Chen et al. 2014). In contrast to the BK channel activators, no information is yet available for the site of actions of the inhibitors of the tremorgenic indole alkaloid class as well as for most of the additional activators described in the literature.

The binding site for the triarylmethane class (clotrimazole, a.o.) of IK channel inhibitors has been delineated by a combined chimeric and site-directed mutagenesis approach, taking advantage of the more than 1000-fold affinity difference between the very closely related IK and SK channels (Wulff et al. 2001). Replacement of just two amino acids, Thr(250) and Val(275), in the IK channel with the corresponding residues in SK3 completely abolished activity. Conversely, these two amino acids also yielded full TRAM sensitivity when incorporated in SK channels. Thr(250) and Val(275) are positioned in extracellular end of S6 and in the pore helix, respectively, just below the K^+ selectivity filter and are therefore only accessible from the inside of the channel as shown by membrane impermeable TRAMs. Benzothiazinones, a novel structural class of high potency IK inhibitors exemplified by the compound NS6180, are also obligatory dependent on these two amino acids (Strobaek et al. 2013), whereas it is noteworthy that nifedipine and nitrendipine act independently of this site, which probably also counts for the derived cyclohexadienes and phenylpyrans (Urbahns et al. 2003). A resent abstract pinpoints the dihydropyridine interaction site on IK channels to the fenestration between S5 and S6 (Nguyen et al. 2015). Interestingly, the "TRAM site" on the IK channel was later shown to correspond exactly to the negative gating modulator site on SK channels defined by the 5-aminobenzimidazole compounds, such as NS8593 (Jenkins et al. 2011), raising the question whether TRAMs should also be considered gating modulators rather than pore blockers. Functional evidence for this notion is still lacking.

Use of 1-EBIO in a mutational approach founded the notion that the activator interaction site was in the C-terminal of IK and SK channels (Pedarzani et al. 2001). This has recently been confirmed by the successful co-crystallisation of several activators with the SK C-terminal and Cam, defining an activator site comprised both by SK channel and Cam residues in a narrow pocket between the two proteins (Zhang et al. 2012, 2013). Importantly, point mutations guided by the crystal revealed that SK/IK channel selectivity could be rationalised for both SK and IK selective compounds (CyPPA and NS309), which proves that the identified CamBD/Cam site is an activator site on the IK channel as well. In analogy with the SK channels, which have a common interaction site for positive and negative modulators in the transmembrane regions (TM5; Hougaard et al. 2009, 2012), other positive modulator sites probably also exist on IK channels, although the hitherto identified SK channel modulators working at this site only exhibit low affinity for the IK channel.

26.5 Conclusions and Perspectives

The preceding paragraphs have outlined the established classical pharmacology of IK and BK channels and described the various attempts done in both academia and the pharmaceutical industry to expand the chemical scope within this field, both with the aim of target validation and drug development. Target validating type of experiments where tool compounds have been used in combination with knockout and transgenic animal studies points towards a yet unexploited potential for use of both activators and inhibitors of KCa channels for therapeutic uses in serious epithelial diseases, such as IK channel inhibitors for treating cAMP-dependent secretion and cyst growth in autosomal-dominant polycystic kidney disease (Albaqumi et al. 2008), BK channel inhibitors for glaucoma, and IK activators for cystic fibrosis. In spite of this, it is noteworthy that a relatively low number of realised industrial drug discovery/development programmes appear to have been specifically targeted against correcting pathological conditions via interaction with the epithelial KCa channels. An outstanding exception is probably the BK channel inhibitor programme from Merck against glaucoma. On the other hand, several of these industrial programmes have been directed towards conditions such as asthma, chronic obstructive pulmonary disease, or inflammatory bowel disease, where the active participation of the epithelium in the complex disease process cannot be excluded, although the primary therapeutic actions may well be via KCa channels expressed in other cells, for example, direct smooth muscle relaxation in overactive bladder (i.e., BK channel activators) or direct inhibition of immune cells mediating the inflammatory responses in asthma and IBD (i.e., IK channel blockers). Thus, for a number of both novel and classic BK and IK channel modulators, we are still awaiting profiling of the direct effects on the epithelial channels as well as the corresponding change in epithelial physiology in the relevant disease models.

As the review describes, the pharmaceutical efforts within the K_{Ca} channel field have resulted in a wealth of structures with direct, or in some cases indirect, actions on the channels. Understanding the nature of how this wealth of molecular frameworks ultimately interacts with the channel proteins or with their associated subunits in order to produce unique modulatory effects is only beginning to be uncovered. Similarly, knowledge of the long-term effects on K_{Ca} channel expression and trafficking is only sporadically known, although epithelial pharmacology has delivered a clear example of downregulation of IK channels in response to activators (Manaves et al. 2004; Koegel et al. 2003). The little body of knowledge of the structural determinants underlying ligand-receptor interaction within the K_{Ca} channels as opposed to other families of ion channel targets or potassium channel types has surely hampered the original drug design of truly innovative K_{Ca} modulators by means of structural biology, virtual screening, and/or fragment-based approaches. This outstanding evidence, together with lack, until recently, of established high-throughput screening assays formats and the omnipresence of large chemical libraries often optimised for classical targets such as G-protein-coupled receptor or kinases, may have represented a key deterrent for several pharmas and big pharmas to enter the field and, by consequence, may have limited the full exploitation of the field itself. This becomes, for instance, very evident when considering the overall limited chemical diversity that has emerged so far, especially in the field of BK and IK channel activators, where several drug discovery programmes appear too often guided by a patent busting approach and with the resulting molecules displaying very similar structural features and similar drugability hurdles. As more information on structural biology is emerging (Yuan et al. 2010), high-throughput screening assay formats for both acute and lasting modulatory effects are being published (Jorgensen et al. 2013; Balut et al. 2010), and molecular pharmacology approaches based on mutations are made available (Strobaek et al. 2013; Nausch et al. 2014), we would expect the scenario to significantly change in the short-term future. It is with this in mind that we look forward to a larger exploration of the chemical space of IK and BK channel modulators and the disclosure of more potent, specific, and drugable pharmacological agents to allow new exciting insight into the therapeutic potential of epithelial BK and IK channels in the patient population.

References

Adorante JS, Woldemussie E, Ruiz G (1996) Method for reducing intraocular pressure in the mammalian eye by administration of potassium channel blockers. WO 1996033719

Agarwal JJ, Zhu Y, Zhang QY, Mongin AA, Hough LB (2013) TRAM-34, a putatively selective blocker of intermediate-conductance, calcium-activated potassium channels, inhibits cytochrome P450 activity. PLoS One 8(5), e63028. doi:10.1371/journal.pone.0063028

Aihara H, Hosaka T, Kashiwagi T, Kohnomi S, Kono RTCC (2004) Calcium-activated K channel activator. EP 1400243

Akada K, Izumi S, Yasuura H (2002) Sustained release oral preparations. WO 2001070221

Albaqumi M, Srivastava S, Li Z, Zhdnova O, Wulff H, Itani O, Wallace DP, Skolnik EY (2008) $K_{Ca}3.1$ potassium channels are critical for cAMP-dependent chloride secretion and cyst growth in autosomal-dominant polycystic kidney disease. Kidney Int 74(6):740–749. doi:10.1038/ki.2008.246

Alper S, Bellot EM, Brugnara C, Clifford JJ, Froimowitz M, Gao YD, Haidar RM, Halperin J, Kelleher EW, Kher FM (1999a) Use of substituted 11-phenyl-dibenzazepine compounds for the treatment or prevention of sickle cell disease, inflammatory diseases characterized by abnormal cell proliferation, diarrhea and scour. WO 1999026628

Alper S, Bellot EM, Brugnara C, Clifford JJ, Froimowitz M, Gao YD, Haidar RM, Halperin J, Kelleher EW, Kher FM (1999b) Use of substituted diphenyl indanone, indane and indole compounds for the treatment or prevention of sickle cell disease, inflammatory diseases characterized by abnormal cell proliferation, diarrhea and scours. WO1999026624

Alvarez J, Montero M, Garcia-Sancho J (1992) High affinity inhibition of Ca^{2+}-dependent K^+ channels by cytochrome P-450 inhibitors. J Biol Chem 267(17):11789–11793

Anderson CS, MacKinnon R, Smith C, Miller C (1988) Charybdotoxin block of single Ca^{2+}-activated K^+ channels. Effects of channel gating, voltage, and ionic strength. J Gen Physiol 91 (3):317–333

Anderson NJ, Slough S, Watson WP (2006) In vivo characterisation of the small-conductance K_{Ca} (SK) channel activator 1-ethyl-2-benzimidazolinone (1-EBIO) as a potential anticonvulsant. Eur J Pharmacol 546(1–3):48–53. doi:10.1016/j.ejphar.2006.07.007

Aon MA, Cortassa S, Wei AC, Grunnet M, O'Rourke B (2010) Energetic performance is improved by specific activation of K^+ fluxes through K_{Ca} channels in heart mitochondria. Biochim Biophys Acta 1797(1):71–80. doi:10.1016/j.bbabio.2009.08.002

Archer SL, Huang JM, Hampl V, Nelson DP, Shultz PJ, Weir EK (1994) Nitric oxide and cGMP cause vasorelaxation by activation of a charybdotoxin-sensitive K channel by cGMP-dependent protein kinase. Proc Natl Acad Sci U S A 91(16):7583–7587

Ataga KI, Stocker J (2009) Senicapoc (ICA-17043): a potential therapy for the prevention and treatment of hemolysis-associated complications in sickle cell anemia. Expert Opin Investig Drugs 18(2):231–239. doi:10.1517/13543780802708011

Ataga KI, Orringer EP, Styles L, Vichinsky EP, Swerdlow P, Davis GA, Desimone PA, Stocker JW (2006) Dose-escalation study of ICA-17043 in patients with sickle cell disease. Pharmacotherapy 26(11):1557–1564. doi:10.1592/phco.26.11.1557

Ataga KI, Smith WR, De Castro LM, Swerdlow P, Saunthararajah Y, Castro O, Vichinsky E, Kutlar A, Orringer EP, Rigdon GC, Stocker JW, Investigators ICA (2008) Efficacy and safety of the Gardos channel blocker, senicapoc (ICA-17043), in patients with sickle cell anemia. Blood 111(8):3991–3997. doi:10.1182/blood-2007-08-110098

Ataga KI, Reid M, Ballas SK, Yasin Z, Bigelow C, James LS, Smith WR, Galacteros F, Kutlar A, Hull JH, Stocker JW, Investigators ICAS (2011) Improvements in haemolysis and indicators of erythrocyte survival do not correlate with acute vaso-occlusive crises in patients with sickle cell disease: a phase III randomized, placebo-controlled, double-blind study of the Gardos channel blocker senicapoc (ICA-17043). Br J Haematol 153(1):92–104. doi:10.1111/j.1365-2141.2010.08520.x

Atkinson RN, Mcnaughton-Smith GA, Reed AD (2004) Sulfonamides as potassium channel blockers. WO 2004016221

Balut CM, Gao Y, Luke C, Devor DC (2010) Immunofluorescence-based assay to identify modulators of the number of plasma membrane $K_{Ca}3.1$ channels. Future Med Chem 2 (5):707–713. doi:10.4155/fmc.10.182

Balut CM, Hamilton KL, Devor DC (2012) Trafficking of intermediate ($K_{Ca}3.1$) and small ($K_{Ca}2.x$) conductance, Ca^{2+}-activated K^+ channels: a novel target for medicinal chemistry efforts? ChemMedChem 7(10):1741–1755. doi:10.1002/cmdc.201200226

Baqi Y, Weyler S, Iqbal J, Zimmermann H, Muller CE (2009) Structure-activity relationships of anthraquinone derivatives derived from bromaminic acid as inhibitors of ectonucleoside triphosphate diphosphohydrolases (E-NTPDases). Purinergic Signal 5(1):91–106. doi:10.1007/s11302-008-9103-5

Bardou O, Trinh NT, Brochiero E (2009) Molecular diversity and function of K^+ channels in airway and alveolar epithelial cells. Am J Physiol Lung Cell Mol Physiol 296(2):L145–L155. doi:10.1152/ajplung.90525.2008

Barrett JN, Magleby KL, Pallotta BS (1982) Properties of single calcium-activated potassium channels in cultured rat muscle. J Physiol 331:211–230

Bednarczyk P, Koziel A, Jarmuszkiewicz W, Szewczyk A (2013) Large-conductance Ca^{2+}-activated potassium channel in mitochondria of endothelial EA.hy926 cells. Am J Physiol Heart Circ Physiol 304(11):H1415–H1427. doi:10.1152/ajpheart.00976.2012

Bellott EM, Brugnara C, Clifford JJ, Fluckiger R, Gao YD, Haidar RM, Halperin J, Kelleher EW, Lombardy RJ, Moussa AM (1999a) Substituted 11-phenyl-dibenzazepine compounds useful for the treatment or prevention of diseases characterized by abnormal cell proliferation. WO 1999026929

Bellott EM, Brugnara C, Clifford JJ, Fluckiger R, Gao YD, Haidar RM, Halperin J, Kelleher EW, Lombardy RJ, Moussa AM (1999b) Substituted diphenyl indanone, indane and indole compounds and analogues thereof useful for the treatment or prevention of diseases characterized by abnormal cell proliferation. WO 1999026611

Benoit C, Renaudon B, Salvail D, Rousseau E (2001) EETs relax airway smooth muscle via an EpDHF effect: BK(Ca) channel activation and hyperpolarization. Am J Physiol Lung Cell Mol Physiol 280(5):L965–L973

Bentzen BH, Nardi A, Calloe K, Madsen LS, Olesen SP, Grunnet M (2007) The small molecule NS11021 is a potent and specific activator of Ca^{2+}-activated big-conductance K^+ channels. Mol Pharmacol 72(4):1033–1044. doi:10.1124/mol.107.038331

Bentzen BH, Osadchii O, Jespersen T, Hansen RS, Olesen SP, Grunnet M (2009) Activation of big conductance Ca^{2+}-activated K^+ channels (BK) protects the heart against ischemia-reperfusion injury. Pflugers Archiv Eur J Physiol 457(5):979–988. doi:10.1007/s00424-008-0583-5

Bentzen BH, Andersen RW, Olesen SP, Grunnet M, Nardi A (2010) Synthesis and characterisation of NS13558: a new important tool for addressing $K_{Ca}1.1$ channel function ex vivo. Naunyn Schmiedebergs Arch Pharmacol 381(3):271–283. doi:10.1007/s00210-009-0456-2

Berkowitz LR, Orringer EP (1981) Effect of cetiedil, an in vitro antisickling agent, on erythrocyte membrane cation permeability. J Clin Invest 68(5):1215–1220

Biot C, Bauer H, Schirmer RH, Davioud-Charvet E (2004) 5-Substituted tetrazoles as bioisosteres of carboxylic acids. Bioisosterism and mechanistic studies on glutathione reductase inhibitors as antimalarials. J Med Chem 47(24):5972–5983. doi:10.1021/jm0497545

Bolotina VM, Najibi S, Palacino JJ, Pagano PJ, Cohen RA (1994) Nitric oxide directly activates calcium-dependent potassium channels in vascular smooth muscle. Nature 368 (6474):850–853. doi:10.1038/368850a0

Borchert GH, Hlavackova M, Kolar F (2013) Pharmacological activation of mitochondrial BK (Ca) channels protects isolated cardiomyocytes against simulated reperfusion-induced injury. Exp Biol Med 238(2):233–241. doi:10.1177/1535370212474596

Boy K (2005) Such as N-[4-[[4-(5-chloro-2-hydroxyphenyl)-1,2-dihydro-2-oxo-6-(trifluoromethyl)-3-quinolinyl]thio]phenyl]-4-morpholineacetamide; potassium channel modulators. US 20050124611

Boy KM, Guernon JM, Sit SY, Xie K, Hewawasam P, Boissard CG, Dworetzky SI, Natale J, Gribkoff VK, Lodge N, Starrett JE Jr (2004) 3-Thio-quinolinone maxi-K openers for the treatment of erectile dysfunction. Bioorg Med Chem Lett 14(20):5089–5093. doi:10.1016/j.bmcl.2004.07.080

Boy K, Hewawasam P, Sit SY (2005a) 3-Thia-4-arylquinolin-2-one derivatives. US 20050080105
Boy K, Hewawasam P, Sit SY, Xie K (2005b) Administering a therapeutically effective amount of 3-thia-4-(2-oxy/hydroxyphenyl)quinolin-2-one derivative to a patient to treat erectile dysfunction or irritable bowel syndrome. US 20050176763
Boyd EA, Fisher MH, Garcia ML, Kaczorowski GJ, Meinke PT, Parsons WH, Price S, Stibbard J (2005) Ophthalmic compositions for treating ocular hypertension. WO 2004087051
Brondum E, Kold-Petersen H, Simonsen U, Aalkjaer C (2010) NS309 restores EDHF-type relaxation in mesenteric small arteries from type 2 diabetic ZDF rats. Br J Pharmacol 159 (1):154–165. doi:10.1111/j.1476-5381.2009.00525.x
Brown PD, Loo DD, Wright EM (1988) Ca^{2+}-activated K^+ channels in the apical membrane of Necturus choroid plexus. J Membr Biol 105(3):207–219
Bruening-Wright A, Lee WS, Adelman JP, Maylie J (2007) Evidence for a deep pore activation gate in small conductance Ca^{2+}-activated K^+ channels. J Gen Physiol 130(6):601–610. doi:10.1085/jgp.200709828
Brugnara C, de Franceschi L (2006) Clinical trials of new therapeutic pharmacology for sickle cell disease. Sante 16(4):263–268
Brugnara C, Gee B, Armsby CC, Kurth S, Sakamoto M, Rifai N, Alper SL, Platt OS (1996) Therapy with oral clotrimazole induces inhibition of the Gardos channel and reduction of erythrocyte dehydration in patients with sickle cell disease. J Clin Invest 97(5):1227–1234. doi:10.1172/JCI118537
Bukiya AN, Liu J, Toro L, Dopico AM (2007) Beta1 (KCNMB1) subunits mediate lithocholate activation of large-conductance Ca^{2+}-activated K^+ channels and dilation in small, resistance-size arteries. Mol Pharmacol 72(2):359–369. doi:10.1124/mol.107.034330
Bukiya AN, McMillan JE, Fedinec AL, Patil SA, Miller DD, Leffler CW, Parrill AL, Dopico AM (2013) Cerebrovascular dilation via selective targeting of the cholane steroid-recognition site in the BK channel beta1-subunit by a novel nonsteroidal agent. Mol Pharmacol 83 (5):1030–1044. doi:10.1124/mol.112.083519
Burke MJ, Mckibben B, Tschantz MA (2013) Fused thiazin-3-ones as $K_{Ca}3.1$ inhibitors. WO 2013191984
Candia S, Garcia ML, Latorre R (1992) Mode of action of iberiotoxin, a potent blocker of the large conductance Ca^{2+}-activated K^+ channel. Biophys J 63(2):583–590. doi:10.1016/S0006-3495(92)81630-2
Cao Y, Dreixler JC, Roizen JD, Roberts MT, Houamed KM (2001) Modulation of recombinant small-conductance Ca^{2+}-activated K^+ channels by the muscle relaxant chlorzoxazone and structurally related compounds. J Pharmacol Exp Ther 296(3):683–689
Caputo A, Caci E, Ferrera L, Pedemonte N, Barsanti C, Sondo E, Pfeffer U, Ravazzolo R, Zegarra-Moran O, Galietta LJ (2008) TMEM16A, a membrane protein associated with calcium-dependent chloride channel activity. Science 322(5901):590–594. doi:10.1126/science.1163518
Carrier GO, Fuchs LC, Winecoff AP, Giulumian AD, White RE (1997) Nitrovasodilators relax mesenteric microvessels by cGMP-induced stimulation of Ca-activated K channels. Am J Physiol 273(1 Pt 2):H76–H84
Chen MH, Doherty JB, Liu L, Meinke PT, Natarajan R, Parsons WH, Shen DM, Shu M, Stelmach JE, Wisnoski D (2003) Ophthalmic compositions for treating ocular hypertension. WO 2003105847
Chen MH, Doherty JB, Liu L, Natarajan SR, Shen DM, Tynebor RM (2004a) Ophthalmic compositions for treating ocular hypertension. WO 2004043354
Chen MH, Doherty JB, Liu L, Natarajan SR, Tynebor RM (2004b) Ophthalmic compositions for treating ocular hypertension. WO 2004043933
Chen MH, Doherty JB, Liu L, Natarajan S, Tynebor RM (2005a) Ophthalmic compositions for treating ocular hypertension. WO 2005026128
Chen MH, Doherty JB, Liu L, Natarajan SR, Shen DM, Shu M (2005b) Ophthalmic compositions for treating ocular hypertension. WO 2005002520

Chen X, Yan J, Aldrich RW (2014) BK channel opening involves side-chain reorientation of multiple deep-pore residues. Proc Natl Acad Sci U S A 111(1):E79–E88. doi:10.1073/pnas.1321697111

Coiret G, Borowiec AS, Mariot P, Ouadid-Ahidouch H, Matifat F (2007) The antiestrogen tamoxifen activates BK channels and stimulates proliferation of MCF-7 breast cancer cells. Mol Pharmacol 71(3):843–851. doi:10.1124/mol.106.028290

Coleman N, Brown BM, Olivan-Viguera A, Singh V, Olmstead MM, Valero MS, Kohler R, Wulff H (2014) New Positive K_{Ca} Channel Gating Modulators with Selectivity for $K_{Ca}3.1$. Mol Pharmacol. doi:10.1124/mol.114.093286

Cui YM, Yasutomi E, Otani Y, Yoshinaga T, Ido K, Sawada K, Ohwada T (2008) Design, synthesis and characterization of podocarpate derivatives as openers of BK channels. Bioorg Med Chem Lett 18(19):5197–5200. doi:10.1016/j.bmcl.2008.08.081

Cui YM, Yasutomi E, Otani Y, Ido K, Yoshinaga T, Sawada K, Ohwada T (2010) Design, synthesis, and characterization of BK channel openers based on oximation of abietane diterpene derivatives. Bioorg Med Chem 18(24):8642–8659. doi:10.1016/j.bmc.2010.09.072

Cullinane AB, Leung PS, Ortego J, Coca-Prados M, Harvey BJ (2002) Renin-angiotensin system expression and secretory function in cultured human ciliary body non-pigmented epithelium. Br J Ophthalmol 86(6):676–683

Cuppoletti J, Malinowska DH, Tewari KP, Chakrabarti J, Ueno R (2012) Unoprostone isopropyl and metabolite M1 activate BK channels and prevent ET-1-induced $[Ca^{2+}]i$ increases in human trabecular meshwork and smooth muscle. Invest Ophthalmol Vis Sci 53(9):5178–5189. doi:10.1167/iovs.11-9046

Davi H, Tronquet C, Miscoria G, Perrier L, DuPont P, Caix J, Simiand J, Berger Y (2000) Disposition of irbesartan, an angiotensin II AT1-receptor antagonist, in mice, rats, rabbits, and macaques. Drug Metab Dispos 28(1):79–88

De Wet H, Allen M, Holmes C, Stobbart M, Lippiat JD, Callaghan R (2006) Modulation of the BK channel by estrogens: examination at single channel level. Mol Membr Biol 23(5):420–429. doi:10.1080/09687860600802803

DeFarias FP, Carvalho MF, Lee SH, Kaczorowski GJ, Suarez-Kurtz G (1996) Effects of the K^+ channel blockers paspalitrem-C and paxilline on mammalian smooth muscle. Eur J Pharmacol 314(1–2):123–128

Dela Pena IC, Yoon SY, Kim SM, Lee GS, Park CS, Kim YC, Cheong JH (2009a) Inhibition of intestinal motility by the putative BK(Ca) channel opener LDD175. Arch Pharm Res 32(3):413–420. doi:10.1007/s12272-009-1315-x

Dela Pena IC, Yoon SY, Kim SM, Lee GS, Ryu JH, Park CS, Kim YC, Cheong JH (2009b) Bladder-relaxant properties of the novel benzofuroindole analogue LDD175. Pharmacology 83(6):367–378. doi:10.1159/000218739

Demnitz J, Jorgensen S (2014) Novel tetrazole derivatives and their use as potassium channel modulators. WO 2014001363

Demnitz J, Gouliaev AH, Sloek FA, Teuber L (2003) Potassium channel modulators. WO 2003059873

Demnitz J, Madsen LS, Olesen SP, Stroebaek D (2005) Diarylmethyl derivatives as potassium channel modulators. WO 2005003094

Denda M, Tsutsumi M, Inoue K, Crumrine D, Feingold KR, Elias PM (2007) Potassium channel openers accelerate epidermal barrier recovery. Br J Dermatol 157(5):888–893. doi:10.1111/j.1365-2133.2007.08198.x

Denson DD, Eaton DC (1994) Ketamine inhibition of large conductance Ca^{2+}-activated K^+ channels is modulated by intracellular Ca^{2+}. Am J Physiol 267(5 Pt 1):C1452–C1458

Denson DD, Duchatelle P, Eaton DC (1994) The effect of racemic ketamine on the large conductance Ca^{2+}-activated potassium (BK) channels in GH3 cells. Brain Res 638(1–2):61–68

Devor DC, Singh AK, Frizzell RA, Bridges RJ (1996) Modulation of Cl- secretion by benzimidazolones. I. Direct activation of a Ca^{2+}-dependent K^+ channel. Am J Physiol 271 (5 Pt 1):L775–L784

Dick GM, Sanders KM (2001) (Xeno)estrogen sensitivity of smooth muscle BK channels conferred by the regulatory beta1 subunit: a study of beta1 knockout mice. J Biol Chem 276 (48):44835–44840. doi:10.1074/jbc.M106851200

Dick GM, Rossow CF, Smirnov S, Horowitz B, Sanders KM (2001) Tamoxifen activates smooth muscle BK channels through the regulatory beta 1 subunit. J Biol Chem 276(37):34594–34599. doi:10.1074/jbc.M104689200

Dick GM, Hunter AC, Sanders KM (2002) Ethylbromide tamoxifen, a membrane-impermeant antiestrogen, activates smooth muscle calcium-activated large-conductance potassium channels from the extracellular side. Mol Pharmacol 61(5):1105–1113

Doherty JB, Shen DM (2005) Ophthalmic compositions for treating ocular hypertension. WO 2005020917

Doherty JB, Shen DM (2006) Ophthalmic compositions for treating ocular hypertension. WO 2006020003

Doherty J, Chen MH, Liu L, Natarajan S, Tynebor R (2004a) Ophthalmic compositions for treating ocular hypertension. US 20040097575

Doherty JB, Chen MH, Liu L, Natarajan SR, Tynebor RM (2004b) Ophthalmic compositions for treating ocular hypertension. WO 2004043932

Doherty JB, Dong-Ming S, Min S, Fengqi Z (2007) Ophthalmic compositions for treating ocular hypertension. WO 2007108968

Doherty JB, Dong-Ming S, Min S (2008a) Ophthalmic compositions for treating ocular hypertension. WO 2008030390

Doherty JB, Natarajan SR, Shen DM, Zhang F (2008b) Ophthalmic compositions for treating ocular hypertension. WO 2007146136

Duerson K, White RE, Jiang F, Schonbrunn A, Armstrong DL (1996) Somatostatin stimulates BKCa channels in rat pituitary tumor cells through lipoxygenase metabolites of arachidonic acid. Neuropharmacology 35(7):949–961

Dutta AK, Khimji AK, Sathe M, Kresge C, Parameswara V, Esser V, Rockey DC, Feranchak AP (2009) Identification and functional characterization of the intermediate-conductance Ca^{2+}-activated K^+ channel (IK-1) in biliary epithelium. Am J Physiol Gastrointest Liver Physiol 297 (5):G1009–G1018

Dworetzky SI, Trojnacki JT, Gribkoff VK (1994) Cloning and expression of a human large-conductance calcium-activated potassium channel. Brain Res Mol Brain Res 27(1):189–193

Ellory JC, Culliford SJ, Smith PA, Wolowyk MW, Knaus EE (1994) Specific inhibition of Ca-activated K channels in red cells by selected dihydropyridine derivatives. Br J Pharmacol 111(3):903–905

Falck JR, Reddy LM, Reddy YK, Bondlela M, Krishna UM, Ji Y, Sun J, Liao JK (2003) 11,12-Epoxyeicosatrienoic acid (11,12-EET): structural determinants for inhibition of TNF-alpha-induced VCAM-1 expression. Bioorg Med Chem Lett 13(22):4011–4014

Falck JR, Wallukat G, Puli N, Goli M, Arnold C, Konkel A, Rothe M, Fischer R, Muller DN, Schunck WH (2011) 17(R),18(S)-epoxyeicosatetraenoic acid, a potent eicosapentaenoic acid (EPA) derived regulator of cardiomyocyte contraction: structure-activity relationships and stable analogues. J Med Chem 54(12):4109–4118. doi:10.1021/jm200132q

Fukao M, Mason HS, Kenyon JL, Horowitz B, Keef KD (2001) Regulation of BK(Ca) channels expressed in human embryonic kidney 293 cells by epoxyeicosatrienoic acid. Mol Pharmacol 59(1):16–23

Galvez A, Gimenez-Gallego G, Reuben JP, Roy-Contancin L, Feigenbaum P, Kaczorowski GJ, Garcia ML (1990) Purification and characterization of a unique, potent, peptidyl probe for the high conductance calcium-activated potassium channel from venom of the scorpion Buthus tamulus. J Biol Chem 265(19):11083–11090

Gao YD, Shen DM (2006) Ophthalmic compositions for treating ocular hypertension. WO 2006044232

Garcia M, Kaczorowski G, McManus O (2001) Method for treating ocular hypertension. US 20010047025

Garcia-Valdes J, Zamudio FZ, Toro L, Possani LD (2001) Slotoxin, alphaKTx1.11, a new scorpion peptide blocker of MaxiK channels that differentiates between alpha and alpha+beta (beta1 or beta4) complexes. FEBS Lett 505(3):369–373

Garduno J, Galvan E, Fernandez de Sevilla D, Buno W (2005) 1-Ethyl-2-benzimidazolinone (EBIO) suppresses epileptiform activity in in vitro hippocampus. Neuropharmacology 49 (3):376–388. doi:10.1016/j.neuropharm.2005.03.021

Garneau L, Klein H, Lavoie MF, Brochiero E, Parent L, Sauve R (2014) Aromatic-aromatic interactions between residues in KCa3.1 pore helix and S5 transmembrane segment control the channel gating process. J Gen Physiol 143(2):289–307. doi:10.1085/jgp.201311097

Gessner G, Cui YM, Otani Y, Ohwada T, Soom M, Hoshi T, Heinemann SH (2012) Molecular mechanism of pharmacological activation of BK channels. Proc Natl Acad Sci U S A 109 (9):3552–3557. doi:10.1073/pnas.1114321109

Gillman K, Bocchino D (2004) Useful in therapy of disorders which are responsive to the opening of potassium channels such as ischemia, stroke, convulsions, asthma, epilepsy, irritable bowel syndrome, migraine, traumatic brain injury, spinal cord injury, sexual dysfunction. US 20040152646

Gillman KW, Hewawasam P, Schmitz WD, Lopez OD, Starrett JE, Provencal DP (2003) Phosphate prodrugs of fluorooxindoles. WO 2003080047

Goetz MA, Kaczorowski GJ (2004) Compositions and methods for treating glaucoma and ocular hypertension. WO 2003077845

Goetz MA, Kaczorowski GJ, Monaghan RL, Strohl WR, Tkacz JS (2003) Novel maxi-k channel blockers, methods of use and process for making the same. WO 2003105868

Gore VK, Ma VV, Yin R, Ligutti J, Immke D, Doherty EM, Norman MH (2010) Structure-activity relationship (SAR) investigations of tetrahydroquinolines as BKCa agonists. Bioorg Med Chem Lett 20(12):3573–3578. doi:10.1016/j.bmcl.2010.04.125

Gormemis AE, Ha TS, Im I, Jung KY, Lee JY, Park CS, Kim YC (2005) Benzofuroindole analogues as potent BK(Ca) channel openers. ChemBioChem 6(10):1745–1748. doi:10.1002/cbic.200400448

Gragasin FS, Michelakis ED, Hogan A, Moudgil R, Hashimoto K, Wu X, Bonnet S, Haromy A, Archer SL (2004) The neurovascular mechanism of clitoral erection: nitric oxide and cGMP-stimulated activation of BKCa channels. FASEB J 18(12):1382–1391. doi:10.1096/fj.04-1978com

Gribkoff VK, Lum-Ragan JT, Boissard CG, Post-Munson DJ, Meanwell NA, Starrett JE Jr, Kozlowski ES, Romine JL, Trojnacki JT, McKay MC, Zhong J, Dworetzky SI (1996) Effects of channel modulators on cloned large-conductance calcium-activated potassium channels. Mol Pharmacol 50(1):206–217

Gribkoff VK, Starrett JE Jr, Dworetzky SI, Hewawasam P, Boissard CG, Cook DA, Frantz SW, Heman K, Hibbard JR, Huston K, Johnson G, Krishnan BS, Kinney GG, Lombardo LA, Meanwell NA, Molinoff PB, Myers RA, Moon SL, Ortiz A, Pajor L, Pieschl RL, Post-Munson DJ, Signor LJ, Srinivas N, Taber MT, Thalody G, Trojnacki JT, Wiener H, Yeleswaram K, Yeola SW (2001) Targeting acute ischemic stroke with a calcium-sensitive opener of maxi-K potassium channels. Nat Med 7(4):471–477. doi:10.1038/86546

Grunnet M, Jespersen T, Angelo K, Frokjaer-Jensen C, Klaerke DA, Olesen SP, Jensen BS (2001) Pharmacological modulation of SK3 channels. Neuropharmacology 40(7):879–887

Haburcak M, Wei L, Viana F, Prenen J, Droogmans G, Nilius B (1997) Calcium-activated potassium channels in cultured human endothelial cells are not directly modulated by nitric oxide. Cell Calcium 21(4):291–300

Haghdoost-Yazdi H, Janahmadi M, Behzadi G (2008) Iberiotoxin-sensitive large conductance Ca^{2+} -dependent K^+ (BK) channels regulate the spike configuration in the burst firing of cerebellar Purkinje neurons. Brain Res 1212:1–8. doi:10.1016/j.brainres.2008.03.030

Harada H, Watanuki S, Takuwa T, Kawaguchi K, Okazaki T, Hirano Y, Saitoh C (2003) Medicine comprising dicyanopyridine derivative. US 20030232860

Harper AA, Catacuzzeno L, Trequattrini C, Petris A, Franciolini F (2001) Verapamil block of large-conductance Ca-activated K channels in rat aortic myocytes. J Membr Biol 179 (2):103–111

Hayashi Y, Kawaji K, Sun L, Zhang X, Koyano K, Yokoyama T, Kohsaka S, Inoue K, Nakanishi H (2011) Microglial Ca^{2+}-activated K^+ channels are possible molecular targets for the analgesic effects of S-ketamine on neuropathic pain. J Neurosci 31(48):17370–17382. doi:10.1523/JNEUROSCI.4152-11.2011

Heinz A, Passow H (1980) Role of external potassium in the calcium-induced potassium efflux from human red blood cell ghosts. J Membr Biol 57(2):119–131

Hercule HC, Salanova B, Essin K, Honeck H, Falck JR, Sausbier M, Ruth P, Schunck WH, Luft FC, Gollasch M (2007) The vasodilator 17,18-epoxyeicosatetraenoic acid targets the pore-forming BK alpha channel subunit in rodents. Exp Physiol 92(6):1067–1076. doi:10.1113/expphysiol.2007.038166

Hewawasam P, Meanwell NA, Gribkoff VK (1997) Useful in reducing neuronal damage during ischemic stroke. US 5602169

Hewawasam P, Fan W, Ding M, Flint K, Cook D, Goggins GD, Myers RA, Gribkoff VK, Boissard CG, Dworetzky SI, Starrett JE Jr, Lodge NJ (2003) 4-Aryl-3-(hydroxyalkyl)quinolin-2-ones: novel maxi-K channel opening relaxants of corporal smooth muscle targeted for erectile dysfunction. J Med Chem 46(14):2819–2822. doi:10.1021/jm030005h

Hewawasam P, Fan W, Cook DA, Newberry KS, Boissard CG, Gribkoff VK, Starrett J, Lodge NJ (2004) 4-Aryl-3-(mercapto)quinolin-2-ones: novel maxi-K channel opening relaxants of corporal smooth muscle. Bioorg Med Chem Lett 14(17):4479–4482. doi:10.1016/j.bmcl.2004.06.051

Hewawasam P, Sit SY, Starrett J (2005) Potassium channel modulators. US 20050080082

Hoffman JF, Joiner W, Nehrke K, Potapova O, Foye K, Wickrema A (2003) The hSK4 (KCNN4) isoform is the Ca^{2+}-activated K^+ channel (Gardos channel) in human red blood cells. Proc Natl Acad Sci U S A 100(12):7366–7371. doi:10.1073/pnas.1232342100

Hollywood MA, Mchale N, Roy S, Sergeant GP, Thornbury K (2012a) Anthraquinone compounds and their uses. WO 2012035122

Hollywood MA, Thornbury K, Sergeant GP, Mchale N, Roy S (2012b) Anthraquinone compounds and their uses. EP 2439200

Hougaard C, Fraser MO, Chien C, Bookout A, Katofiasc M, Jensen BS, Rode F, Bitsch-Norhave J, Teuber L, Thor KB, Strobaek D, Burgard EC, Ronn LC (2009) A positive modulator of $K_{Ca}2$ and $K_{Ca}3$ channels, 4,5-dichloro-1,3-diethyl-1,3-dihydro-benzoimidazol-2-one (NS4591), inhibits bladder afferent firing in vitro and bladder overactivity in vivo. J Pharmacol Exp Ther 328(1):28–39. doi:10.1124/jpet.108.143123

Hougaard C, Hammami S, Eriksen BL, Sorensen US, Jensen ML, Strobaek D, Christophersen P (2012) Evidence for a common pharmacological interaction site on $K_{Ca}2$ channels providing both selective activation and selective inhibition of the human $K_{Ca}2.1$ subtype. Mol Pharmacol 81(2):210–219. doi:10.1124/mol.111.074252

Huang CW, Huang CC, Wu SN (2007) Activation by zonisamide, a newer antiepileptic drug, of large-conductance calcium-activated potassium channel in differentiated hippocampal neuron-derived H19-7 cells. J Pharmacol Exp Ther 321(1):98–106. doi:10.1124/jpet.106.116954

Imaizumi Y, Ohwada T (2002) Potassium channel opener. WO 2002087559

Imaizumi Y, Sakamoto K, Yamada A, Hotta A, Ohya S, Muraki K, Uchiyama M, Ohwada T (2002) Molecular basis of pimarane compounds as novel activators of large-conductance Ca^{2+}-activated K^+ channel alpha-subunit. Mol Pharmacol 62(4):836–846

Ishii TM, Silvia C, Hirschberg B, Bond CT, Adelman JP, Maylie J (1997) A human intermediate conductance calcium-activated potassium channel. Proc Natl Acad Sci U S A 94 (21):11651–11656

Jaravine VA, Nolde DE, Reibarkh MJ, Korolkova YV, Kozlov SA, Pluzhnikov KA, Grishin EV, Arseniev AS (1997) Three-dimensional structure of toxin OSK1 from Orthochirus scrobiculosus scorpion venom. Biochemistry 36(6):1223–1232. doi:10.1021/bi9614390

Javaherian AD, Yusifov T, Pantazis A, Franklin S, Gandhi CS, Olcese R (2011) Metal-driven operation of the human large-conductance voltage- and Ca^{2+}-dependent potassium channel (BK) gating ring apparatus. J Biol Chem 286(23):20701–20709. doi:10.1074/jbc.M111.235234

Jenkins DP, Strobaek D, Hougaard C, Jensen ML, Hummel R, Sorensen US, Christophersen P, Wulff H (2011) Negative gating modulation by (R)-N-(benzimidazol-2-yl)-1,2,3,4-tetrahydro-1-naphthylamine (NS8593) depends on residues in the inner pore vestibule: pharmacological evidence of deep-pore gating of K(Ca)2 channels. Mol Pharmacol 79(6):899–909. doi:10.1124/mol.110.069807

Jenkins DP, Yu W, Brown BM, Lojkner LD, Wulff H (2013) Development of a QPatch automated electrophysiology assay for identifying $K_{Ca}3.1$ inhibitors and activators. Assay Drug Dev Technol 11(9–10):551–560. doi:10.1089/adt.2013.543

Jensen BS (2002) BMS-204352: a potassium channel opener developed for the treatment of stroke. CNS Drug Rev 8(4):353–360

Jensen BS, Strobaek D, Christophersen P, Jorgensen TD, Hansen C, Silahtaroglu A, Olesen SP, Ahring PK (1998) Characterization of the cloned human intermediate-conductance Ca^{2+}-activated K^+ channel. Am J Physiol 275(3 Pt 1):C848–C856

Jensen BS, Odum N, Jorgensen NK, Christophersen P, Olesen SP (1999) Inhibition of T cell proliferation by selective block of Ca^{2+}-activated K^+ channels. Proc Natl Acad Sci U S A 96 (19):10917–10921

Ji YH, Wang WX, Ye JG, He LL, Li YJ, Yan YP, Zhou Z (2003) Martentoxin, a novel K^+-channel-blocking peptide: purification, cDNA and genomic cloning, and electrophysiological and pharmacological characterization. J Neurochem 84(2):325–335

Jorgensen S, Dyhring T, Brown DT, Strobaek D, Christophersen P, Demnitz J (2013) A high-throughput screening campaign for detection of Ca^{2+}-activated K^+ channel activators and inhibitors using a fluorometric imaging plate reader-based Tl^+-influx assay. Assay Drug Dev Technol 11(3):163–172. doi:10.1089/adt.2012.479

Kaji DM (1990) Nifedipine inhibits calcium-activated K transport in human erythrocytes. Am J Physiol 259(2 Pt 1):C332–C339

Kasumu AW, Hougaard C, Rode F, Jacobsen TA, Sabatier JM, Eriksen BL, Strobaek D, Liang X, Egorova P, Vorontsova D, Christophersen P, Ronn LC, Bezprozvanny I (2012) Selective positive modulator of calcium-activated potassium channels exerts beneficial effects in a mouse model of spinocerebellar ataxia type 2. Chem Biol 19(10):1340–1353. doi:10.1016/j.chembiol.2012.07.013

Kehl SJ (1996) Block of BK (maxi K) channels of rat pituitary melanotrophs by Na^+ and other alkali metal ions. Pflugers Archiv 432(4):623–629

King JT, Lovell PV, Rishniw M, Kotlikoff MI, Zeeman ML, McCobb DP (2006) Beta2 and beta4 subunits of BK channels confer differential sensitivity to acute modulation by steroid hormones. J Neurophysiol 95(5):2878–2888. doi:10.1152/jn.01352.2005

Kiraly I, Pataricza J, Bajory Z, Simonsen U, Varro A, Papp JG, Pajor L, Kun A (2013) Involvement of large-conductance Ca^{2+} -activated K^+ channels in both nitric oxide and endothelium-derived hyperpolarization-type relaxation in human penile small arteries. Basic Clin Pharmacol Toxicol 113(1):19–24. doi:10.1111/bcpt.12059

Klockgether-Radke AP, Huneck S, Meyberg S, Neumann P, Hellige G (2005) Ketamine enantiomers differentially relax isolated coronary artery rings. Eur J Anaesthesiol 22 (3):215–221

Knaus HG, McManus OB, Lee SH, Schmalhofer WA, Garcia-Calvo M, Helms LM, Sanchez M, Giangiacomo K, Reuben JP, Smith AB III et al (1994) Tremorgenic indole alkaloids potently inhibit smooth muscle high-conductance calcium-activated potassium channels. Biochemistry 33(19):5819–5828

Kobayashi K, Nishizawa Y, Sawada K, Ogura H, Miyabe M (2008) K^+-channel openers suppress epileptiform activities induced by 4-aminopyridine in cultured rat hippocampal neurons. J Pharmacol Sci 108(4):517–528

Koegel H, Alzheimer C (2001) Expression and biological significance of Ca^{2+}-activated ion channels in human keratinocytes. FASEB J 15(1):145–154. doi:10.1096/fj.00-0055com

Koegel H, Kaesler S, Burgstahler R, Werner S, Alzheimer C (2003) Unexpected down-regulation of the hIK1 Ca^{2+}-activated K^+ channel by its opener 1-ethyl-2-benzimidazolinone in HaCaT keratinocytes. Inverse effects on cell growth and proliferation. J Biol Chem 278(5):3323–3330. doi:10.1074/jbc.M208914200

Kozlowski ES, Johnson G, Dischino DD, Dworetzky SI, Boissard CG, Gribkoff VK (1996) Synthesis and biological evaluation of an iodinated iberiotoxin analogue, [mono-iodo-Tyr5, Phe36]-iberiotoxin. Int J Pept Protein Res 48(2):194–199

Krishna R, Palme H, Zeng J, Srinivas N (2002a) Effect of dose and input rate on the brain penetration of BMS-204352 following intravenous administration to rats. Biopharm Drug Dispos 23(6):227–231. doi:10.1002/bdd.317

Krishna R, Shah VR, Mantha S, Vachharajani NN, Srinivas N (2002b) Pharmacokinetics and dose proportionality of BMS-204352 after intravenous administration to dogs. Biopharm Drug Dispos 23(2):83–86

Krishna R, Shah VR, Srinivas N (2002c) Pharmacokinetics and dose proportionality of BMS-204352 after intraarterial administration to rats. Biopharm Drug Dispos 23 (6):233–237. doi:10.1002/bdd.318

Kroigaard C, Dalsgaard T, Nielsen G, Laursen BE, Pilegaard H, Kohler R, Simonsen U (2012) Activation of endothelial and epithelial $K_{Ca}2.3$ calcium-activated potassium channels by NS309 relaxes human small pulmonary arteries and bronchioles. Br J Pharmacol 167 (1):37–47. doi:10.1111/j.1476-5381.2012.01986.x

Kun A, Matchkov VV, Stankevicius E, Nardi A, Hughes AD, Kirkeby HJ, Demnitz J, Simonsen U (2009) NS11021, a novel opener of large-conductance Ca^{2+}-activated K^+ channels, enhances erectile responses in rats. Br J Pharmacol 158(6):1465–1476. doi:10.1111/j.1476-5381.2009.00404.x

Kunze WA, Bornstein JC, Furness JB, Hendriks R, Stephenson DS (1994) Charybdotoxin and iberiotoxin but not apamin abolish the slow after-hyperpolarization in myenteric plexus neurons. Pflugers Archiv 428(3–4):300–306

La Fuente JM, Fernandez A, Cuevas P, Gonzalez-Corrochano R, Chen M, Angulo J (2014) Stimulation of large-conductance calcium-activated potassium channels inhibits neurogenic contraction of human bladder from patients with urinary symptoms and reverses acetic acid-induced bladder hyperactivity in rats. Eur J Pharmacol. doi:10.1016/j.ejphar.2014.03.060

Laeis P, Puchler K, Kirch W (2001) The pharmacokinetic and metabolic profile of olmesartan medoxomil limits the risk of clinically relevant drug interaction. J Hypertens Suppl 19(1):S21–S32

Lang DG, Ritchie AK (1990) Tetraethylammonium blockade of apamin-sensitive and insensitive Ca^{2+}-activated K^+ channels in a pituitary cell line. J Physiol 425:117–132

Latorre R, Vergara C, Hidalgo C (1982) Reconstitution in planar lipid bilayers of a Ca^{2+}-dependent K^+ channel from transverse tubule membranes isolated from rabbit skeletal muscle. Proc Natl Acad Sci U S A 79(3):805–809

Lauterbach B, Barbosa-Sicard E, Wang MH, Honeck H, Kargel E, Theuer J, Schwartzman ML, Haller H, Luft FC, Gollasch M, Schunck WH (2002) Cytochrome P450-dependent eicosapentaenoic acid metabolites are novel BK channel activators. Hypertension 39(2 Pt 2):609–613

Layne JJ, Nausch B, Olesen SP, Nelson MT (2010) BK channel activation by NS11021 decreases excitability and contractility of urinary bladder smooth muscle. Am J Physiol Regul Integr Comp Physiol 298(2):R378–R384. doi:10.1152/ajpregu.00458.2009

Lee JE, Kwak J, Suh CK, Shin JH (2006) Dual effects of nitric oxide on the large conductance calcium-activated potassium channels of rat brain. J Biochem Mol Biol 39(1):91–96

Lee BC, Lim HH, Kim S, Youn HS, Lee Y, Kim YC, Eom SH, Lee KW, Park CS (2012) Localization of a site of action for benzofuroindole-induced potentiation of BKCa channels. Mol Pharmacol 82(2):143–155. doi:10.1124/mol.112.078097

Lew VL, Ferreira HG (1976) Variable Ca sensitivity of a K-selective channel in intact red-cell membranes. Nature 263(5575):336–338

Li C, Grillo MP, Benet LZ (2003a) In vitro studies on the chemical reactivity of 2,4-dichlorophenoxyacetyl-S-acyl-CoA thioester. Toxicol Appl Pharmacol 187(2):101–109

Li C, Olurinde MO, Hodges LM, Grillo MP, Benet LZ (2003b) Covalent binding of 2-phenylpropionyl-S-acyl-CoA thioester to tissue proteins in vitro. Drug Metab Dispos 31 (6):727–730

Li Y, Johnson G, Romine JL, Meanwell NA, Martin SW, Dworetzky SI, Boissard CG, Gribkoff VK, Starrett JE Jr (2003c) Novel openers of Ca^{2+}-dependent large-conductance potassium channels: symmetrical pharmacophore and electrophysiological evaluation of bisphenols. Bioorg Med Chem Lett 13(8):1437–1439

Liu H, Xin T, He W, Li F, Su ZQ (2014) Myelinated Ah-type trigeminal ganglion neurons in female rats: neuroexcitability, chemosensitivity to histamine, and potential clinical impact. Neurosci Lett 567:74–79. doi:10.1016/j.neulet.2014.03.039

Lu T, Katakam PV, VanRollins M, Weintraub NL, Spector AA, Lee HC (2001) Dihydroxyeicosatrienoic acids are potent activators of Ca^{2+}-activated K^{+} channels in isolated rat coronary arterial myocytes. J Physiol 534(Pt 3):651–667

Lucas D, Goulitquer S, Marienhagen J, Fer M, Dreano Y, Schwaneberg U, Amet Y, Corcos L (2010) Stereoselective epoxidation of the last double bond of polyunsaturated fatty acids by human cytochromes P450. J Lipid Res 51(5):1125–1133. doi:10.1194/jlr.M003061

Mackay KB (2001) BMS-204352 (Bristol Myers Squibb). Curr Opin Investig Drugs 2(6):820–823

Maillard MP, Rossat J, Brunner HR, Burnier M (2000) Tasosartan, enoltasosartan, and angiotensin II receptor blockade: the confounding role of protein binding. J Pharmacol Exp Ther 295 (2):649–654

Makovec F, Peris W, Rovati AL (1990) Derivatives of n-phenylbenzamide with anti-ulcer and anti-allergy activity and a method for their preparation. WO 1990009989

Makovec F, Peris W, Revel L, Giovanetti R, Redaelli D, Rovati LC (1992) Antiallergic and cytoprotective activity of new N-phenylbenzamido acid derivatives. J Med Chem 35 (20):3633–3640

Malerba M, Mennuni L, Piepoli T, Caselli G, Makovec F, Rovati LC, D' Amato M, Ferrari F (2009) Andolast acts at different cellular levels to inhibit immunoglobulin E synthesis. Int J Immunopathol Pharmacol 22(1):85–94

Mall M, Gonska T, Thomas J, Schreiber R, Seydewitz HH, Kuehr J, Brandis M, Kunzelmann K (2003) Modulation of Ca^{2+}-activated Cl^{-} secretion by basolateral K^{+} channels in human normal and cystic fibrosis airway epithelia. Pediatr Res 53(4):608–618. doi:10.1203/01.PDR.0000057204.51420.DC

Malysz J, Buckner SA, Daza AV, Milicic I, Perez-Medrano A, Gopalakrishnan M (2004) Functional characterization of large conductance calcium-activated K^{+} channel openers in bladder and vascular smooth muscle. Naunyn Schmiedebergs Arch Pharmacol 369(5):481–489. doi:10.1007/s00210-004-0920-y

Manaves V, Qin W, Bauer AL, Rossie S, Kobayashi M, Rane SG (2004) Calcium and Vitamin D increase mRNA levels for the growth control hIK1 channel in human epidermal keratinocytes but functional channels are not observed. BMC Dermatol 4:7. doi:10.1186/1471-5945-4-7

Manzanares D, Gonzalez C, Ivonnet P, Chen RS, Valencia-Gattas M, Conner GE, Larsson HP, Salathe M (2011) Functional apical large conductance, Ca^{2+}-activated, and voltage-dependent

K^+ channels are required for maintenance of airway surface liquid volume. J Biol Chem 286 (22):19830–19839. doi:10.1074/jbc.M110.185074
Matos JE, Sausbier M, Beranek G, Sausbier U, Ruth P, Leipziger J (2007) Role of cholinergic-activated $K_{Ca}1.1$ (BK), $K_{Ca}3.1$ (SK4) and $K_V7.1$ (KCNQ1) channels in mouse colonic Cl^- secretion. Acta Physiol 189(3):251–258. doi:10.1111/j.1748-1716.2006.01646.x
McCobb DP, Fowler NL, Featherstone T, Lingle CJ, Saito M, Krause JE, Salkoff L (1995) A human calcium-activated potassium channel gene expressed in vascular smooth muscle. Am J Physiol 269(3 Pt 2):H767–H777
McManus OB, Harris GH, Giangiacomo KM, Feigenbaum P, Reuben JP, Addy ME, Burka JF, Kaczorowski GJ, Garcia ML (1993) An activator of calcium-dependent potassium channels isolated from a medicinal herb. Biochemistry 32(24):6128–6133
McNaughton-Smith GA, Burns JF, Stocker JW, Rigdon GC, Creech C, Arrington S, Shelton T, de Franceschi L (2008) Novel inhibitors of the Gardos channel for the treatment of sickle cell disease. J Med Chem 51(4):976–982. doi:10.1021/jm070663s
Miller C (1987) Trapping single ions inside single ion channels. Biophys J 52(1):123–126. doi:10.1016/S0006-3495(87)83196-X
Miller O, Yu R (2008) Topical ophthalmic formulations. WO 2008027341
Miller C, Moczydlowski E, Latorre R, Phillips M (1985) Charybdotoxin, a protein inhibitor of single Ca^{2+}-activated K^+ channels from mammalian skeletal muscle. Nature 313 (6000):316–318
Mishra RC, Wulff H, Cole WC, Braun AP (2014) A pharmacologic activator of endothelial K_{Ca} channels enhances coronary flow in the hearts of type 2 diabetic rats. J Mol Cell Cardiol 72:364–373. doi:10.1016/j.yjmcc.2014.04.013
Morin C, Sirois M, Echave V, Gomes MM, Rousseau E (2007) Functional effects of 20-HETE on human bronchi: hyperpolarization and relaxation due to BKCa channel activation. Am J Physiol Lung Cell Mol Physiol 293(4):L1037–L1044. doi:10.1152/ajplung.00145.2007
Mouhat S, Visan V, Ananthakrishnan S, Wulff H, Andreotti N, Grissmer S, Darbon H, De Waard M, Sabatier JM (2005) K^+ channel types targeted by synthetic OSK1, a toxin from Orthochirus scrobiculosus scorpion venom. Biochem J 385(Pt 1):95–104. doi:10.1042/BJ20041379
Nakamura A, Tanaka M, Tsuda M (1996) Pyrrole derivatives and medicinal composition. WO 1996040634
Nakamura A, Tanaka M, Tsuda M (1999) Potassium channel activators. WO 1999036068
Nardi A, Olesen SP (2008) BK channel modulators: a comprehensive overview. Curr Med Chem 15(11):1126–1146
Nardi A, Calderone V, Chericoni S, Morelli I (2003) Natural modulators of large-conductance calcium-activated potassium channels. Planta Med 69(10):885–892. doi:10.1055/s-2003-45095
Nardi A, Demnitz D, Garcia ML, Polosa R (2008) Potassium channels as drug targets for therapeutic intervention in respiratory diseases. Expert Opin Ther Pat 18(12):1361–1384
Narenjkar J, el Assem SK, Ganellin CR (2004) Inhibition of the antigen-induced activation of RBL-2H3 cells by cetiedil and some of its analogues. Eur J Pharmacol 483(2–3):107–116
Nausch B, Rode F, Jorgensen S, Nardi A, Korsgaard MP, Hougaard C, Brown WD, Bonev AD, Dyhring T, Strobaek D, Olesen SP, Christophersen P, Grunnet M, Nelson MT, Ronn LC (2014) NS19504: a novel BK channel activator with relaxing effect on bladder smooth muscle spontaneous phasic contractions. J Pharmacol Exp Therap. doi:10.1124/jpet.113.212662
Neyton J, Miller C (1988) Discrete Ba^{2+} block as a probe of ion occupancy and pore structure in the high-conductance Ca^{2+}-activated K^+ channel. J Gen Physiol 92(5):569–586
Nguyen HM, Wulff H, Varov-Varovoy V (2015) Validation of $K_{Ca}3.1$ channel nifedipine interaction site predicted by Rosetta modeling method. Biophysical J 108(2):582a–583a
Niu X, Liu G, Wu RS, Chudasama N, Zakharov SI, Karlin A, Marx SO (2013) Orientations and proximities of the extracellular ends of transmembrane helices S0 and S4 in open and closed BK potassium channels. PLoS One 8(3), e58335. doi:10.1371/journal.pone.0058335

Ohwada T, Nonomura T, Maki K, Sakamoto K, Ohya S, Muraki K, Imaizumi Y (2003) Dehydroabietic acid derivatives as a novel scaffold for large-conductance calcium-activated K^+ channel openers. Bioorg Med Chem Lett 13(22):3971–3974

Ohwada T, Sakamaki Y, Kabasawa Y, Yasutomi E (2009) Potassium channel opener. WO 2009110468

Olesen SP, Munch E, Moldt P, Drejer J (1994) Selective activation of Ca^{2+}-dependent K^+ channels by novel benzimidazolone. Eur J Pharmacol 251(1):53–59

Owada T, Tajima T, Chokai Y, Akaha S (2007) Potassium channel opening agent. JP 2007186480

Pantazis A, Gudzenko V, Savalli N, Sigg D, Olcese R (2010a) Operation of the voltage sensor of a human voltage- and Ca^{2+}-activated K^+ channel. Proc Natl Acad Sci U S A 107(9):4459–4464. doi:10.1073/pnas.0911959107

Pantazis A, Kohanteb AP, Olcese R (2010b) Relative motion of transmembrane segments S0 and S4 during voltage sensor activation in the human BK(Ca) channel. J Gen Physiol 136 (6):645–657. doi:10.1085/jgp.201010503

Parajuli SP, Hristov KL, Soder RP, Kellett WF, Petkov GV (2013) NS309 decreases rat detrusor smooth muscle membrane potential and phasic contractions by activating SK3 channels. Br J Pharmacol 168(7):1611–1625. doi:10.1111/bph.12049

Parihar AS, Groebe DR, Scott VE, Feng J, Zhang XF, Warrior U, Gopalakrishnan M, Shieh CC (2003) Functional analysis of large conductance Ca^{2+}-activated K^+ channels: ion flux studies by atomic absorption spectrometry. Assay Drug Dev Technol 1(5):647–654. doi:10.1089/154065803770381002

Pedarzani P, Mosbacher J, Rivard A, Cingolani LA, Oliver D, Stocker M, Adelman JP, Fakler B (2001) Control of electrical activity in central neurons by modulating the gating of small conductance Ca^{2+}-activated K^+ channels. J Biol Chem 276(13):9762–9769. doi:10.1074/jbc.M010001200

Pinter T, Jana S, Courtemanche RJ, Hof F (2011) Recognition properties of carboxylic acid bioisosteres: anion binding by tetrazoles, aryl sulfonamides, and acyl sulfonamides on a calix[4]arene scaffold. J Org Chem 76(10):3733–3741. doi:10.1021/jo200031u

Ponte CG, McManus OB, Schmalhofer WA, Shen DM, Dai G, Stevenson A, Sur S, Shah T, Kiss L, Shu M, Doherty JB, Nargund R, Kaczorowski GJ, Suarez-Kurtz G, Garcia ML (2012) Selective, direct activation of high-conductance, calcium-activated potassium channels causes smooth muscle relaxation. Mol Pharmacol 81(4):567–577. doi:10.1124/mol.111.075853

Power EC, Ganellin CR, Benton DC (2006) Partial structures of ketoconazole as modulators of the large conductance calcium-activated potassium channel (BK(Ca)). Bioorg Med Chem Lett 16 (4):887–890. doi:10.1016/j.bmcl.2005.11.001

Radtke J, Schmidt K, Wulff H, Kohler R, de Wit C (2013) Activation of $K_{Ca}3.1$ by SKA-31 induces arteriolar dilatation and lowers blood pressure in normo- and hypertensive connexin40-deficient mice. Br J Pharmacol 170(2):293–303. doi:10.1111/bph.12267

Rae JL, Dewey J, Rae JS, Cooper K (1990) A maxi calcium-activated potassium channel from chick lens epithelium. Curr Eye Res 9(9):847–861

Rittenhouse AR, Parker C, Brugnara C, Morgan KG, Alper SL (1997) Inhibition of maxi-K currents in ferret portal vein smooth muscle cells by the antifungal clotrimazole. Am J Physiol 273(1 Pt 1):C45–C56

Roth EK, Hirtz S, Duerr J, Wenning D, Eichler I, Seydewitz HH, Amaral MD, Mall MA (2011) The K^+ channel opener 1-EBIO potentiates residual function of mutant CFTR in rectal biopsies from cystic fibrosis patients. PLoS One 6(8), e24445. doi:10.1371/journal.pone.0024445

Rovati LC, Giordani A, Caselli G (2012) N-phenylbenzamide derivatives as drugs for the treatment of copd. CA 2514373

Roxburgh CJ, Ganellin CR, Athmani S, Bisi A, Quaglia W, Benton DC, Shiner MA, Malik-Hall M, Haylett DG, Jenkinson DH (2001) Synthesis and structure-activity relationships of cetiedil analogues as blockers of the Ca^{2+}-activated K^+ permeability of erythrocytes. J Med Chem 44(20):3244–3253

Roy S, Morayo Akande A, Large RJ, Webb TI, Camarasu C, Sergeant GP, McHale NG, Thornbury KD, Hollywood MA (2012) Structure-activity relationships of a novel group of large-conductance Ca^{2+}-activated K^+ (BK) channel modulators: the GoSlo-SR family. ChemMedChem 7(10):1763–1769. doi:10.1002/cmdc.201200321

Roy S, Large RJ, Akande AM, Kshatri A, Webb TI, Domene C, Sergeant GP, McHale NG, Thornbury KD, Hollywood MA (2014) Development of GoSlo-SR-5-69, a potent activator of large conductance Ca^{2+}-activated K^+ (BK) channels. Eur J Med Chem 75:426–437. doi:10.1016/j.ejmech.2014.01.035

Rufo PA, Merlin D, Riegler M, Ferguson-Maltzman MH, Dickinson BL, Brugnara C, Alper SL, Lencer WI (1997) The antifungal antibiotic, clotrimazole, inhibits chloride secretion by human intestinal T84 cells via blockade of distinct basolateral K^+ conductances. Demonstration of efficacy in intact rabbit colon and in an in vivo mouse model of cholera. J Clin Invest 100 (12):3111–3120. doi:10.1172/JCI119866

Ryan JS, Tao QP, Kelly ME (1998) Adrenergic regulation of calcium-activated potassium current in cultured rabbit pigmented ciliary epithelial cells. J Physiol 511(Pt 1):145–157

Saito M, Nelson C, Salkoff L, Lingle CJ (1997) A cysteine-rich domain defined by a novel exon in a slo variant in rat adrenal chromaffin cells and PC12 cells. J Biol Chem 272(18):11710–11717

Sallustio BC, Sabordo L, Evans AM, Nation RL (2000) Hepatic disposition of electrophilic acyl glucuronide conjugates. Curr Drug Metab 1(2):163–180

Sanchez M, McManus OB (1996) Paxilline inhibition of the alpha-subunit of the high-conductance calcium-activated potassium channel. Neuropharmacology 35(7):963–968

Sankaranarayanan A, Raman G, Busch C, Schultz T, Zimin PI, Hoyer J, Kohler R, Wulff H (2009) Naphtho[1,2-d]thiazol-2-ylamine (SKA-31), a new activator of $K_{Ca}2$ and $K_{Ca}3.1$ potassium channels, potentiates the endothelium-derived hyperpolarizing factor response and lowers blood pressure. Mol Pharmacol 75(2):281–295. doi:10.1124/mol.108.051425

Schreiber M, Salkoff L (1997) A novel calcium-sensing domain in the BK channel. Biophys J 73 (3):1355–1363. doi:10.1016/S0006-3495(97)78168-2

Schroder RL, Jensen BS, Strobaek D, Olesen SP, Christophersen P (2000) Activation of the human, intermediate-conductance, Ca^{2+}-activated K^+ channel by methylxanthines. Pflugers Archiv 440(6):809–818

Schroder RL, Jespersen T, Christophersen P, Strobaek D, Jensen BS, Olesen SP (2001) KCNQ4 channel activation by BMS-204352 and retigabine. Neuropharmacology 40(7):888–898

Schroeder BC, Cheng T, Jan YN, Jan LY (2008) Expression cloning of TMEM16A as a calcium-activated chloride channel subunit. Cell 134(6):1019–1029. doi:10.1016/j.cell.2008.09.003

Schunck WH, Wallukat G, Fischer R, Schmidt C, Müller DN, Puli N, Falck JR (2010) Novel eicosanoid derivatives. WO 2010081683

Schwarz W, Keim H, Fehlau R, Fuhrmann GF (1989) Modulation of the Ca^{2+}- or Pb^{2+}-activated K^+-selective channels in human red cells. I. Effects of propranolol. Biochim Biophys Acta 978 (1):32–36

Seto SW, Au AL, Lam TY, Chim SS, Lee SM, Wan S, Tjiu DC, Shigemura N, Yim AP, Chan SW, Tsui SK, Leung GP, Kwan YW (2007) Modulation by simvastatin of iberiotoxin-sensitive, Ca^{2+}-activated K^+ channels of porcine coronary artery smooth muscle cells. Br J Pharmacol 151(7):987–997. doi:10.1038/sj.bjp.0707327

Sha Y, Tashima T, Mochizuki Y, Toriumi Y, Adachi-Akahane S, Nonomura T, Cheng M, Ohwada T (2005) Compounds structurally related to tamoxifen as openers of large-conductance calcium-activated K^+ channel. Chem Pharm Bull 53(10):1372–1373

Sheu SJ, Bee YS, Chen CH (2008) Resveratrol and large-conductance calcium-activated potassium channels in the protection of human retinal pigment epithelial cells. J Ocul Pharmacol Ther 24(6):551–555. doi:10.1089/jop.2008.0013

Shi J, He HQ, Zhao R, Duan YH, Chen J, Chen Y, Yang J, Zhang JW, Shu XQ, Zheng P, Ji YH (2008) Inhibition of martentoxin on neuronal BK channel subtype (alpha+beta4): implications for a novel interaction model. Biophys J 94(9):3706–3713. doi:10.1529/biophysj.107.122150

Shieh CC, Turner SC, Zhang XF, Milicic I, Parihar A, Jinkerson T, Wilkins J, Buckner SA, Gopalakrishnan M (2007) A-272651, a nonpeptidic blocker of large-conductance Ca^{2+}-activated K^+ channels, modulates bladder smooth muscle contractility and neuronal action potentials. Br J Pharmacol 151(6):798–806. doi:10.1038/sj.bjp.0707278

Singh AK, Devor DC, Gerlach AC, Gondor M, Pilewski JM, Bridges RJ (2000) Stimulation of Cl^- secretion by chlorzoxazone. J Pharmacol Exp Ther 292(2):778–787

Singh S, Syme CA, Singh AK, Devor DC, Bridges RJ (2001) Benzimidazolone activators of chloride secretion: potential therapeutics for cystic fibrosis and chronic obstructive pulmonary disease. J Pharmacol Exp Ther 296(2):600–611

Sivarao DV, Newberry K, Langdon S, Lee AV, Hewawasam P, Plym MJ, Signor L, Myers R, Lodge NJ (2005) Effect of 4-(5-chloro-2-hydroxyphenyl)-3-(2-hydroxyethyl)-6-(trifluoromethyl)-quinolin-2(1H)-one (BMS-223131), a novel opener of large conductance Ca^{2+}-activated K^+ (maxi-K) channels on normal and stress-aggravated colonic motility and visceral nociception. J Pharmacol Exp Ther 313(2):840–847. doi:10.1124/jpet.104.079285

Song WH, Liu MM, Zhong DW, Zhu YL, Bosscher M, Zhou L, Ye DY, Yuan ZH (2013) Tetrazole and triazole as bioisosteres of carboxylic acid: discovery of diketo tetrazoles and diketo triazoles as anti-HCV agents. Bioorg Med Chem Lett 23(16):4528–4531. doi:10.1016/j.bmcl.2013.06.045

Sorensen MV, Matos JE, Sausbier M, Sausbier U, Ruth P, Praetorius HA, Leipziger J (2008) Aldosterone increases $K_{Ca}1.1$ (BK) channel-mediated colonic K+ secretion. J Physiol 586 (Pt 17):4251–4264. doi:10.1113/jphysiol.2008.156968

Sorensen MV, Sausbier M, Ruth P, Seidler U, Riederer B, Praetorius HA, Leipziger J (2010) Adrenaline-induced colonic K^+ secretion is mediated by $K_{Ca}1.1$ (BK) channels. J Physiol 588 (Pt 10):1763–1777. doi:10.1113/jphysiol.2009.181933

Starrett JEJG, Gillmsan KW, Bocchino DM et al (2004) Maxi-K openers as neuroprotective agents. 227th ACS Nattional Meet, Anaheim, March 28–April 1

Starrett J, Lopez O, Hewawasam P, Ding M (2005) N-substituted prodrugs of fluorooxindoles. US 20050203089

Stocker JW, De Franceschi L, McNaughton-Smith GA, Corrocher R, Beuzard Y, Brugnara C (2003) ICA-17043, a novel Gardos channel blocker, prevents sickled red blood cell dehydration in vitro and in vivo in SAD mice. Blood 101(6):2412–2418. doi:10.1182/blood-2002-05-1433

Strobaek D, Christophersen P, Holm NR, Moldt P, Ahring PK, Johansen TE, Olesen SP (1996) Modulation of the Ca^{2+}-dependent K^+ channel, hslo, by the substituted diphenylurea NS 1608, paxilline and internal Ca^{2+}. Neuropharmacology 35(7):903–914

Strobaek D, Teuber L, Jorgensen TD, Ahring PK, Kjaer K, Hansen RS, Olesen SP, Christophersen P, Skaaning-Jensen B (2004) Activation of human IK and SK Ca^{2+}-activated K^+ channels by NS309 (6,7-dichloro-1H-indole-2,3-dione 3-oxime). Biochim Biophys Acta 1665(1–2):1–5. doi:10.1016/j.bbamem.2004.07.006

Strobaek D, Brown DT, Jenkins DP, Chen YJ, Coleman N, Ando Y, Chiu P, Jorgensen S, Demnitz J, Wulff H, Christophersen P (2013) NS6180, a new $K_{Ca}3.1$ channel inhibitor prevents T-cell activation and inflammation in a rat model of inflammatory bowel disease. Br J Pharmacol 168(2):432–444. doi:10.1111/j.1476-5381.2012.02143.x

Sun XH, Ding JP, Li H, Pan N, Gan L, Yang XL, Xu HB (2007) Activation of large-conductance calcium-activated potassium channels by puerarin: the underlying mechanism of puerarin-mediated vasodilation. J Pharmacol Exp Ther 323(1):391–397. doi:10.1124/jpet.107.125567

Syme CA, Gerlach AC, Singh AK, Devor DC (2000) Pharmacological activation of cloned intermediate- and small-conductance Ca^{2+}-activated K^+ channels. Am J Physiol Cell Physiol 278(3):C570–C581

Tanaka M, Sasaki Y, Kimura Y, Fukui T, Hamada K, Ukai Y (2003) A novel pyrrole derivative, NS-8, suppresses the rat micturition reflex by inhibiting afferent pelvic nerve activity. BJU Int 92(9):1031–1036

Tang QY, Zhang Z, Xia XM, Lingle CJ (2010) Block of mouse Slo1 and Slo3 K^{+} channels by CTX, IbTX, TEA, 4-AP and quinidine. Channels 4(1):22–41

Tao J, Shi J, Liu ZR, Ji YH (2012) Martentoxin: a unique ligand of BK channels. Sheng li xue bao [Acta physiologica Sinica] 64(4):355–364

Tashima T, Toriumi Y, Mochizuki Y, Nonomura T, Nagaoka S, Furukawa K, Tsuru H, Adachi-Akahane S, Ohwada T (2006) Design, synthesis, and BK channel-opening activity of hexahydrodibenzazepinone derivatives. Bioorg Med Chem 14(23):8014–8031. doi:10.1016/j.bmc.2006.07.042

Thompson-Vest N, Shimizu Y, Hunne B, Furness JB (2006) The distribution of intermediate-conductance, calcium-activated, potassium (IK) channels in epithelial cells. J Anat 208 (2):219–229. doi:10.1111/j.1469-7580.2006.00515.x

Treiber A, Van Giersbergen PL, Dingemanse J (2003) In vivo and in vitro disposition profile of tezosentan, an intravenous dual endothelin receptor antagonist, in humans. Xenobiotica 33 (4):399–414. doi:10.1080/0049825021000061624

Tricarico D, Barbieri M, Mele A, Carbonara G, Camerino DC (2004) Carbonic anhydrase inhibitors are specific openers of skeletal muscle BK channel of K^{+}-deficient rats. FASEB J 18(6):760–761. doi:10.1096/fj.03-0722fje

Tsuzuki Y, Hirai M (2009) Dérivés de pyrimidine, pyridine et triazine en tant qu'ouvreurs de canaux maxi-k. WO 2009125870

Turner SC, Carroll WA, White TK, Gopalakrishnan M, Coghlan MJ, Shieh CC, Zhang XF, Parihar AS, Buckner SA, Milicic I, Sullivan JP (2003) The discovery of a new class of large-conductance Ca^{2+}-activated K^{+} channel opener targeted for overactive bladder: synthesis and structure-activity relationships of 2-amino-4-azaindoles. Bioorg Med Chem Lett 13 (12):2003–2007

Turner S, White T, Carroll W (2004) Pyrrolopyridine potassium channel openers. US 20040192697

Unemoto T, Matsushita M, Tamura K, Tanaka Y, Koike K, Kogo H (2007) Role of BK channels in testosterone-induced relaxation of the aorta in spontaneously hypertensive rats. Biol Pharm Bull 30(8):1477–1480

Urbahns K, Horvath E, Stasch JP, Mauler F (2003) 4-Phenyl-4H-pyrans as IK(Ca) channel blockers. Bioorg Med Chem Lett 13(16):2637–2639

Vaali K, Li L, Paakkari I, Vapaatalo H (1998) Relaxing effects of NO donors on guinea pig trachea in vitro are mediated by calcium-sensitive potassium channels. J Pharmacol Exp Ther 286 (1):110–114

Valverde MA, Rojas P, Amigo J, Cosmelli D, Orio P, Bahamonde MI, Mann GE, Vergara C, Latorre R (1999) Acute activation of Maxi-K channels (hSlo) by estradiol binding to the beta subunit. Science 285(5435):1929–1931

Van Der Velden J, Sum G, Barker D, Koumoundouros E, Barcham G, Wulff H, Castle N, Bradding P, Snibson K (2013) $K_{Ca}3.1$ channel-blockade attenuates airway pathophysiology in a sheep model of chronic asthma. PLoS One 8(6), e66886. doi:10.1371/journal.pone.0066886

Venglovecz V, Hegyi P, Rakonczay Z Jr, Tiszlavicz L, Nardi A, Grunnet M, Gray MA (2011) Pathophysiological relevance of apical large-conductance Ca^{2+}-activated potassium channels in pancreatic duct epithelial cells. Gut 60(3):361–369. doi:10.1136/gut.2010.214213

Vestergaard-Bogind B, Stampe P, Christophersen P (1985) Single-file diffusion through the Ca^{2+}-activated K^{+} channel of human red cells. J Membr Biol 88(1):67–75

Vrudhula VM, Dasgupta B, Boissard CG, Gribkoff VK, Santone KS, Dalterio RA, Lodge NJ, Starrett JE Jr (2005) Analogs of a potent maxi-K potassium channel opener with an improved inhibitory profile toward cytochrome P450 isozymes. Bioorg Med Chem Lett 15 (19):4286–4290. doi:10.1016/j.bmcl.2005.06.056

Wallner M, Meera P, Toro L (1996) Determinant for beta-subunit regulation in high-conductance voltage-activated and Ca^{2+}-sensitive K^{+} channels: an additional transmembrane region at the N terminus. Proc Natl Acad Sci U S A 93(25):14922–14927

Wang B, Rothberg BS, Brenner R (2006) Mechanism of beta4 subunit modulation of BK channels. J Gen Physiol 127(4):449–465. doi:10.1085/jgp.200509436
Wang YJ, Lin MW, Wu SN, Sung RJ (2007) The activation by estrogen receptor agonists of the BK(Ca)-channel in human cardiac fibroblasts. Biochem Pharmacol 73(9):1347–1357. doi:10.1016/j.bcp.2006.12.029
Wang YJ, Lin MW, Lin AA, Wu SN (2008) Riluzole-induced block of voltage-gated Na^+ current and activation of BKCa channels in cultured differentiated human skeletal muscle cells. Life Sci 82(1–2):11–20. doi:10.1016/j.lfs.2007.10.015
Wang J, Qian W, Zhu Q, Chen J, Huan F, Gao R, Xiao H (2013) Martentoxin, a large-conductance Ca^{2+}-activated K^+ channel inhibitor, attenuated TNF-alpha-induced nitric oxide release by human umbilical vein endothelial cells. J Biomed Res 27(5):386–393. doi:10.7555/JBR.27.20120080
Warach S, Hacke W, Hsu C, Luby M, Sullivan M, Noonan T, Lin CY, Fernandes L, Brunell R, Bozik M (2002) Effect of MaxiPost on ischemic lesions in patients with acute stroke: the POST-010 MRI substudy. Stroke 33(1):383.
Wei AD, Gutman GA, Aldrich R, Chandy KG, Grissmer S, Wulff H (2005) International Union of Pharmacology. LII. Nomenclature and molecular relationships of calcium-activated potassium channels. Pharmacol Rev 57(4):463–472. doi:10.1124/pr.57.4.9
Weyler S, Baqi Y, Hillmann P, Kaulich M, Hunder AM, Muller IA, Muller CE (2008) Combinatorial synthesis of anilinoanthraquinone derivatives and evaluation as non-nucleotide-derived P2Y2 receptor antagonists. Bioorg Med Chem Lett 18(1):223–227. doi:10.1016/j.bmcl.2007.10.082
Wojtovich AP, Nadtochiy SM, Urciuoli WR, Smith CO, Grunnet M, Nehrke K, Brookes PS (2013) A non-cardiomyocyte autonomous mechanism of cardioprotection involving the SLO1 BK channel. Peer J 1, e48. doi:10.7717/peerj.48
Wu SN (2003) Large-conductance Ca^{2+}-activated K^+ channels: physiological role and pharmacology. Curr Med Chem 10(8):649–661
Wu SN, Li HF, Jan CR, Shen AY (1999) Inhibition of Ca^{2+}-activated K^+ current by clotrimazole in rat anterior pituitary GH3 cells. Neuropharmacology 38(7):979–989
Wu SN, Li HF, Chiang HT (2001) Vinpocetine-induced stimulation of calcium-activated potassium currents in rat pituitary GH3 cells. Biochem Pharmacol 61(7):877–892
Wu SN, Liu SI, Huang MH (2004) Cilostazol, an inhibitor of type 3 phosphodiesterase, stimulates large-conductance, calcium-activated potassium channels in pituitary GH3 cells and pheochromocytoma PC12 cells. Endocrinology 145(3):1175–1184. doi:10.1210/en.2003-1430
Wu Y, Yang Y, Ye S, Jiang Y (2010) Structure of the gating ring from the human large-conductance Ca^{2+}-gated K^+ channel. Nature 466(7304):393–397. doi:10.1038/nature09252
Wulff H, Castle NA (2010) Therapeutic potential of KCa3.1 blockers: recent advances and promising trends. Expert Rev Clin Pharmacol 3(3):385–396. doi:10.1586/ecp.10.11
Wulff H, Miller MJ, Hansel W, Grissmer S, Cahalan MD, Chandy KG (2000) Design of a potent and selective inhibitor of the intermediate-conductance Ca^{2+}-activated K^+ channel, IKCa1: a potential immunosuppressant. Proc Natl Acad Sci U S A 97(14):8151–8156
Wulff H, Gutman GA, Cahalan MD, Chandy KG (2001) Delineation of the clotrimazole/TRAM-34 binding site on the intermediate conductance calcium-activated potassium channel, IKCa1. J Biol Chem 276(34):32040–32045. doi:10.1074/jbc.M105231200
Wulff H, Kolski-Andreaco A, Sankaranarayanan A, Sabatier JM, Shakkottai V (2007) Modulators of small- and intermediate-conductance calcium-activated potassium channels and their therapeutic indications. Curr Med Chem 14(13):1437–1457
Wulf-Johansson H, Amrutkar DV, Hay-Schmidt A, Poulsen AN, Klaerke DA, Olesen J, Jansen-Olesen I (2010) Localization of large conductance calcium-activated potassium channels and their effect on calcitonin gene-related peptide release in the rat trigemino-neuronal pathway. Neuroscience 167(4):1091–1102. doi:10.1016/j.neuroscience.2010.02.063

Yang W, Holmes BB, Gopal VR, Kishore RV, Sangras B, Yi XY, Falck JR, Campbell WB (2007) Characterization of 14,15-epoxyeicosatrienoyl-sulfonamides as 14,15-epoxyeicosatrienoic acid agonists: use for studies of metabolism and ligand binding. J Pharmacol Exp Ther 321 (3):1023–1031. doi:10.1124/jpet.107.119651

Yao J, Chen X, Li H, Zhou Y, Yao L, Wu G, Chen X, Zhang N, Zhou Z, Xu T, Wu H, Ding J (2005) BmP09, a "long chain" scorpion peptide blocker of BK channels. J Biol Chem 280 (15):14819–14828. doi:10.1074/jbc.M412735200

Yuan P, Leonetti MD, Pico AR, Hsiung Y, MacKinnon R (2010) Structure of the human BK channel Ca^{2+}-activation apparatus at 3.0 A resolution. Science 329(5988):182–186. doi:10.1126/science.1190414

Zakharov SI, Morrow JP, Liu G, Yang L, Marx SO (2005) Activation of the BK (SLO1) potassium channel by mallotoxin. J Biol Chem 280(35):30882–30887. doi:10.1074/jbc.M505302200

Zeng H, Gordon E, Lin Z, Lozinskaya IM, Willette RN, Xu X (2008) 1-[1-Hexyl-6-(methyloxy)-1H-indazol-3-yl]-2-methyl-1-propanone, a potent and highly selective small molecule blocker of the large-conductance voltage-gated and calcium-dependent K^+ channel. J Pharmacol Exp Ther 327(1):168–177. doi:10.1124/jpet.108.139733

Zhang D, Krishna R, Wang L, Zeng J, Mitroka J, Dai R, Narasimhan N, Reeves RA, Srinivas NR, Klunk LJ (2005) Metabolism, pharmacokinetics, and protein covalent binding of radiolabeled MaxiPost (BMS-204352) in humans. Drug Metab Dispos 33(1):83–93. doi:10.1124/dmd.104.001412

Zhang M, Pascal JM, Schumann M, Armen RS, Zhang JF (2012) Identification of the functional binding pocket for compounds targeting small-conductance Ca^{2+}-activated potassium channels. Nat Commun 3:1021. doi:10.1038/ncomms2017

Zhang M, Pascal JM, Zhang JF (2013) Unstructured to structured transition of an intrinsically disordered protein peptide in coupling Ca^{2+}-sensing and SK channel activation. Proc Natl Acad Sci U S A 110(12):4828–4833. doi:10.1073/pnas.1220253110

Chapter 27
Small-Molecule Pharmacology of Epithelial Inward Rectifier Potassium Channels

Sujay V. Kharade and Jerod S. Denton

Abstract Inward rectifier potassium (Kir) channels play fundamental roles in epithelial fluid and electrolyte transport in several organ systems and appear, in some cases, to represent novel drug targets for human diseases. The discovery that loss-of-function mutations in genes encoding Kir1.1 and Kir4.1 give rise to renal tubule phenotypes that resemble those in patients taking diuretics has motivated drug discovery efforts aimed at developing novel classes of diuretics targeting these channels. In many cases, however, even a rudimentary understanding of the functions of Kir channels in epithelia is lacking due, in part, to the lack of precise pharmacological tools for specifically modulating the activity of Kir channels in complex biological cells and tissues. This barrier is gradually disappearing with the development of Kir channel-specific modulators that can be used alone or in combination to explore the physiology of Kir channels. The goals of this chapter are threefold: (1) to provide an overview of the expression patterns and known functions of Kir channels in epithelial cells, (2) provide a summary of the small-molecule pharmacology of Kir channels expressed in epithelial cells to help guide the design of pharmacological experiments for dissecting their functions, and (3) discuss in detail the ongoing efforts to develop small-molecule antagonists of Kir1.1 (ROMK) for use as diuretics to serve as a model for Kir channel-directed drug discovery.

S.V. Kharade
Department of Anesthesiology, Vanderbilt University Medical Center, 1161 21st Avenue South, Nashville, TN 37232, USA
e-mail: sujay.v.kharade@vanderbilt.edu

J.S. Denton (✉)
Department of Anesthesiology, Vanderbilt University Medical Center, 1161 21st Avenue South, Nashville, TN 37232, USA

Department of Pharmacology, Vanderbilt University Medical Center, 1161 21st Avenue South, Nashville, TN 37232, USA

Institute of Chemical Biology, Vanderbilt University Medical Center, 1161 21st Avenue South, Nashville, TN 37232, USA

Institute for Global Health, Vanderbilt University Medical Center, 1161 21st Avenue South, Nashville, TN 37232, USA
e-mail: jerod.s.denton@vanderbilt.edu

K.L. Hamilton, D.C. Devor (eds.), *Ion Channels and Transporters of Epithelia in Health and Disease*, Physiology in Health and Disease,
DOI 10.1007/978-1-4939-3366-2_27

Keywords ROMK • Kir4.1 • Kir7.1 • Drug discovery • Thallium flux • Medicinal chemistry

27.1 Overview of Inward Rectifier Potassium Channel Structure and Function

The Kir channel family is comprised of 16 genes (denoted *KCNJx*) in humans that encode seven distinct subfamilies of proteinaceous subunits (denoted Kir1–7), many of which are expressed in epithelial cells of various organs (Table 27.1). The simplest functional unit of Kir channels is a tetramer, which is formed by the assembly of four membrane-spanning subunits around an aqueous pore (Fig. 27.1). In most cases, a clear understanding of the functions, regulatory pathways, and therapeutic potential of Kir channels is unknown due to the lack of selective pharmacological tools that can discriminate between individual family members. This challenge is exacerbated by the fact that some Kir channels are heteromeric assemblies of nonidentical subunits (e.g., Ki4.1/Kir5.1—see below), which increases the functional, regulatory, and pharmacological diversity of the channel family. Thus, an important, albeit challenging, goal in the field that will help elucidate the integrative physiology and druggability of Kir channels will be development of sharp pharmacological tools that can discriminate between the myriad Kir channel subtypes in epithelia and other tissues.

Under almost all physiological conditions, K^+ ions follow an outwardly directed electrochemical gradient created by the Na^+-K^+-ATPase. As will be discussed below, the movement of K^+ ions through inward rectifier K^+ channels helps energize substrate-coupled or electrophysiologically coupled transport processes in epithelia that contribute to the tissue, organ, and systemic fluid and electrolyte balance (Fig. 27.2). The term "inward rectification" refers to the electrophysiological behavior of Kir channels under voltage-clamp conditions. By measuring current at voltages more positive and more negative than the Nernst electrochemical potential for K^+ (E_K), one observes larger *inward* current amplitude (i.e., at voltages more negative than E_K) than *outward* current amplitude (i.e., at voltages more positive than E_K). This is in contrast to the behavior of voltage-gated K^+ (K_v) channels, which typically carry only outward current following channel opening at depolarized potentials. The mechanism underlying inward rectification in Kir channels has been studied extensively (Ficker et al. 1994; Lu and MacKinnon 1994; Nichols et al. 1994; Wible et al. 1994; Fakler et al. 1995; Lopatin et al. 1995; Nishida and MacKinnon 2002; Pegan et al. 2005; Kurata et al. 2007; Hibino et al. 2010) and is fundamentally different from that giving rise to outward rectification in K_v channels (Hille 1992). In Kir channels, at voltages more positive than E_K, intracellular polyvalent cations such as Mg^{2+} and polyamines (e.g., spermine) enter the pore in a steeply voltage-dependent manner and physically occlude the ion-conduction pathway. Negatively charged residues that mediate these electrostatic interactions with blocking cations have been identified through

Table 27.1 Kir channel distribution in mammalian epithelial cells

Kir channel	Organ (cell type)	Reference
Kir1.1 (ROMK)	Kidney (TAL, DCT, CCD)	Boim et al. (1995), Hebert et al. (2005)
Kir2.1	Mammary gland (mammary secretory)	Kamikawa and Ishikawa (2014)
Kir2.3	Kidney (CCD)	Welling (1997), Millar et al. (2006)
Kir4.1	Kidney (DCT, CCD)	Lourdel et al. (2002), Lachheb et al. (2008), Zaika et al. (2013), Scholl et al. (2009), Paulais et al. (2011)
	Cochlea (stria vascularis)	Rozengurt et al. (2003), Hibino et al. (1997)
	Stomach (parietal cells)	Fujita et al. (2002)
Kir5.1	Kidney (DCT)	Lachheb et al. (2008), Lourdel et al. (2002), Paulais et al. (2011), Zaika et al. (2013)
Kir6.1/ SUR2B	Lung (alveolar type II)	Leroy et al. (2004), Monaghan et al. (1997), Sakuma et al. (1998)
Kir7.1	Brain (choroid plexus)	Doring et al. (1998), Krapivinsky et al. (1998)
	Retina (RPE)	Nakamura et al. (1999), Yang et al. (2008), Yasuda et al. (2003), Shimura et al. (2001)
	Kidney (PT, TAL, CD, DCT)	Derst et al. (2001), Nakamura et al. (1999), Ookata et al. (2000)
	Intestine	Partiseti et al. (1998), Nakamura et al. (1999)
	Stomach	Nakamura et al. (1999)
	Thyroid (follicular cells)	Nakamura et al. (1999)

PT proximal tubule, *TAL* thick ascending limb, *DCT* distal convoluted tubule, *CCD* cortical collecting duct

mutagenesis and crystallography experiments. The identity and location of these so-called rectification residues will be revisited later in the chapter, since some of these appear to contribute to selective small-molecule binding sites in Kir channels.

27.2 The Renal Outer Medullary Potassium Channel (ROMK, Kir1.1, *KCNJ1*)

27.2.1 Overview of ROMK Expression and Functions in Renal Tubule Epithelia

Kir1.1, which is commonly referred to as the renal outer medullary K^+ (ROMK) channel, is the founding member of the Kir channel family. The complementary DNA (cDNA) of the gene encoding ROMK (i.e., *KCNJ1*) was cloned by two independent groups more than 20 years ago. Dr. Steven Hebert's group cloned splice variant 1 (i.e., ROMK1 or Kir1.1a) of the channel from the inner stripe of the

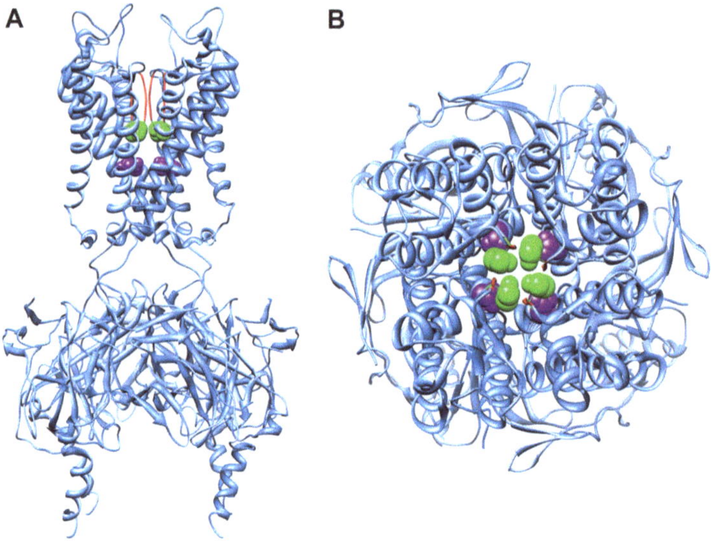

Fig. 27.1 Kir channel structure and small-molecule binding sites. (**a**) Side and (**b**) top-down view of a ribbon diagram of chicken Kir2.2 crystal structure (Tao et al. 2009). The selectivity filter has been colored *red*. Amino acids corresponding to those implicated in small-molecule binding in ROMK (N171, *purple*) and Kir4.1 (E158, *purple*; T128, *green*) are shown as space-filling models for clarity. The image was generated from protein data bank file PDB 3JYC using Chimera software (Pettersen et al. 2004). See text for details

outer medulla of rat kidneys and showed that it forms a barium (Ba^{2+})-inhibitable K^+ channel when heterologously expressed in *Xenopus* oocytes (Ho et al. 1993). Dr. Larry Palmer's group subsequently reported cloning of the ROMK2 splice variant, which lacks the first 19 N-terminal amino acids found in ROMK1 (Zhou et al. 1994). A ROMK3 splice variant was later reported by Hebert and colleagues and contains a unique stretch of 26 N-terminal amino acids (Boim et al. 1995). All three splice variants express robust, weakly rectifying K^+ currents that are functionally indistinguishable. We will therefore not differentiate between the splice variants and simply refer to them collectively as ROMK for the remainder of the chapter.

ROMK is abundantly expressed in the thick ascending limb (TAL) of Henle's loop and collecting duct (CD) of the nephron, but is also expressed at lower levels in the distal convoluted tubule (DCT) and connecting tubule (CNT) (Boim et al. 1995). In the TAL, ROMK forms an ~30 picoSiemens (pS) channel in the luminal membrane that underlies to the so-called secretory K^+ (SK) channel, first identified with patch-clamp electrophysiology (Wang et al. 1990). The activity of ROMK is essential for NaCl reabsorption in the TAL, as well as distal nephron water reabsorption required to generate concentrated urine (Fig. 27.2). The first step in NaCl reabsorption in the TAL is catalyzed by the apical membrane Na^+-K^+-$2Cl^-$ cotransporter, NKCC2, which is the molecular target for the loop diuretics furosemide and bumetanide. Because the concentration of K^+ in the luminal fluid is lower

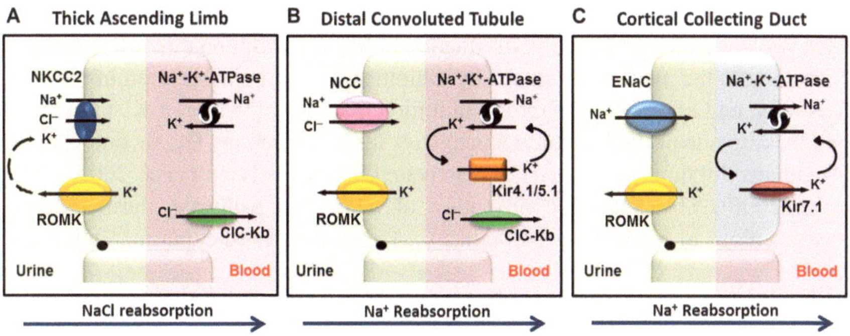

Fig. 27.2 Role of renal Kir channels in renal solute transport. Schematic of different nephron segments where Kir channels are thought to play an important role in transepithelial Na^+ reabsorption and K^+ secretion. (**a**) Thick ascending limb: Loop diuretic target Na^+-K^+-$2Cl^-$ cotransporter (NKCC2) is the major pathway for the reabsorption of NaCl across the apical membrane. However, since luminal K^+ concentration is relatively low compared to Na^+ and Cl^- concentrations, K^+ must be recycled in the opposite direction through the renal outer medullary potassium (ROMK) channel to maintain the robust activity of NKCC2 and efficient NaCl reabsorption. On the basolateral membrane, Na^+ is taken back in the blood by Na^+-K^+-ATPase pump, while Cl^- makes its way back through ClC-Kb. (**b**) Distal convoluted tubule: The thiazide-sensitive Na^+-Cl^- cotransporter (NCC) mediates NaCl entry across the apical membrane, whereas K^+ secretion is mediated by ROMK. On the basolateral membrane K^+ recycling via Kir4.1/5.1 maintains optimal activity of the Na^+-K^+-ATPase since the extracellular K^+ concentration is low and helps Na^+ reabsorption. Kir7.1 (not shown) is thought to be also involved in the basolateral K^+ recycling. ClC-Kb mediates Cl^- exit across the basolateral membrane. (**c**) Cortical collecting duct: The amiloride-sensitive epithelial sodium channel (ENaC) is responsible for apical Na^+ entry; however, to negate the resulting electropositivity and to maintain its activity, ROMK mediates the K^+ secretion. On the basolateral side the Na^+ is taken back by Na^+-K^+-ATPase, and Kir7.1 is thought to be responsible for the necessary K^+ recycling in this part of the nephron

than that of Na^+ and Cl^-, K^+ must be continually recycled across the luminal membrane by ROMK to maintain the activity of NKCC2 (Fig. 27.2). Inhibition of ROMK with the nonspecific blocker Ba^{2+} reduces NaCl reabsorption in the TAL (Greger and Schlatter 1981), essentially by starving NKCC2 of substrate K^+ ions needed for its activity. This observation provided some of the first evidence that pharmacological inhibition of ROMK in the TAL may represent a target for a novel class of loop diuretic by indirectly inhibiting NKCC2.

In the aldosterone-sensitive distal nephron, ROMK comprises the major constitutive secretory pathway for K^+ (Giebisch 1995; Wang et al. 1997; Welling and Ho 2009). K^+ secretion in principal cells of the CD occurs by a two-step process (Fig. 27.2). First, K^+ is transported from blood across the basolateral membrane by the Na^+-K^+-ATPase. K^+ ions subsequently move down their electrochemical gradient and cross the apical membrane via ROMK channels. K^+ secretion through ROMK makes the apical membrane potential more negative, thereby increasing the driving force for Na^+ reabsorption by the epithelial Na^+ channel (ENaC). Consistent with this model, inhibition of ROMK with Ba^{2+} depolarizes the apical membrane potential to limit the ENaC-mediated Na^+ reabsorption (Koeppen et al. 1983). This observation suggests that inhibition of ROMK activity in the distal nephron may have two consequences: Firstly, a reduction in Na^+ reabsorption through ENaC may contribute to the natriuresis caused by inhibition of NaCl reabsorption in the TAL. Secondly, and importantly, ROMK inhibition may limit the loss of K^+ in the urine by blocking one of the two known pathways for K^+ secretion. The second pathway is mediated by flow-stimulated Ca^{2+}-activated K^+ channels (Liu et al. 2007; Bailey et al. 2006). The relative contribution of these two pathways during chronic pharmacological ROMK inhibition has yet to be determined experimentally.

27.2.2 Genetic Validation of ROMK as a Diuretic Target

The first genetic validation of ROMK as a diuretic target came from the discovery by Dr. Richard Lifton and colleagues of autosomal recessive loss-of-function (LOF) mutations in *KCNJ1* in patients with type II Bartter syndrome (Simon et al. 1996). The clinical manifestations of type II Bartter syndrome include symptoms that would be expected from loss of ROMK activity in the nephron, namely, hypernatremia, hypokalemic alkalosis, polyuria, and low blood pressure (Bartter et al. 1962). In 2008, Lifton and colleagues found that heterozygous carriers of *KCNJ1* mutations in the Framingham Heart Cohort have lower blood pressure, are protected from development of hypertension with age, and, importantly, exhibit no overt clinical symptoms of type II Bartter syndrome (Ji et al. 2008). Taken together, these data suggest that partial ROMK inhibition may lead to favorable clinical outcomes, but that complete inhibition may lead to pathological salt and water wasting. However, given that a majority of *KCNJ1*$^{-/-}$ mice exhibit hydronephrosis and renal insufficiency (Lu et al. 2002), it is conceivable that a genetic loss of ROMK function may result in abnormal kidney development and function that would not occur in genetically normal individuals taking a medication that inhibits ROMK activity. With the recent development of drug-like inhibitors of ROMK (see below), the answer to this question may soon be revealed.

27.2.3 ROMK Drug Discovery

A discussion of the small-molecule pharmacology for ROMK begins with the efforts of Clark and colleagues, who first demonstrated that administration to rats of glyburide, a sulfonylurea K_{ATP} channel inhibitor, induced a dose-dependent stimulation of renal Na^+ excretion without affecting K^+ excretion (Clark et al. 1993). Wang and colleagues later showed that this was associated with a reduction in Na^+ and K^+ transport in the TALH and CD via inhibition of ROMK (Wang et al. 1995). However, whether the high doses of glyburide used in these experiments altered renal tubule transport through inhibition of other transport mechanisms is unclear. Assessing whether specific inhibition of ROMK function will elicit natriuresis and diuresis without inducing kaliuresis would require development of more potent and selective small-molecule inhibitors.

In 1996, our group at Vanderbilt began developing a high-throughput screening (HTS) platform to enable the discovery of novel inhibitors of ROMK. Because single-cell, voltage-clamp electrophysiology is too slow and laborious to enable robust drug discovery, we set out to develop an assay that can be performed in 384-well plates to dramatically improve throughput. The assay we chose takes advantage of the permeability of the ROMK channel pore to the K^+ congener thallium (Tl^+) and utilizes an intracellular dye whose fluorescence emission increases dose-dependently with Tl^+ concentration. There are Tl^+-sensitive dyes that are commercially available, including BTC, FluoZin-2, FluxOR (all from Life Technologies), and Thallos (TEFLabs). We have developed robust assays using FluoZin-2 (Lewis et al. 2009; Raphemot et al. 2011; Raphemot et al. 2013) and Thallos (Raphemot, Swale, and Denton, unpublished data), but have recently begun using Thallos for all of our Kir channel assays because of its lower cost. The Tl^+ flux assay is performed using T-REx-HEK-293 (Life Technologies) cells that express ROMK from a tetracycline-inducible promoter. The use of an inducible system serves at least two important purposes. First, inducible ROMK expression limits cell line degeneration and loss of assay performance that can be caused by the constitutive overexpression of some Kir channels. Secondly, an inducible expression provides a convenient way to test compounds for activity toward endogenous Tl^+ flux pathways, such as the Na^+-K^+-ATPase and voltage-gated K^+ channels expressed in HEK-293 cells. The assay is technically straightforward and performed essentially as follows. After loading cells with the acetoxymethyl ester form of Thallos, the cells are preincubated with small molecules (one compound per well) for 20 min, and the plate is moved to the stage of a kinetic imaging plate reader capable of measuring fluorescence from each well of the plate simultaneously (e.g., Fig. 27.3a). The assay is initiated by adding a Tl^+ stimulus buffer to every well of the plate, while monitoring fluorescence emission. As shown in Fig. 27.3b, extracellular Tl^+ addition stimulates robust Tl^+ flux in cells induced overnight with tetracycline, but not in uninduced cells not expressing ROMK or in induced cells that have been preincubated with the peptide ROMK inhibitor tertiapin-Q (TPNQ, Fig. 27.3b). In a HTS campaign, ROMK modulators are identified by monitoring the slope of fluorescence change over time: inhibitors decrease the slope by inhibiting ROMK-dependent Tl^+ flux (e.g., Fig. 27.3a, well N17), whereas activators increase Tl^+ flux. A step-by-step description of how to develop and validate a Kir channel Tl^+ flux assay for HTS has been described elsewhere (Raphemot et al. 2013).

Using the Tl^+ flux assay, we performed a screen of approximately 225,000 compounds from the NIH Molecular Libraries Small Molecule Repository (MLSMR) for pharmacological modulators of ROMK (Lewis et al. 2009). The results of the screen have been deposited in a searchable format in the National Center for Biotechnology Information PubChem Bioassay database (https://pubchem.ncbi.nlm.nih.gov). Compounds were screened at a nominal concentration of 10 μM and selected as "hits" if they were associated with a change in a well's

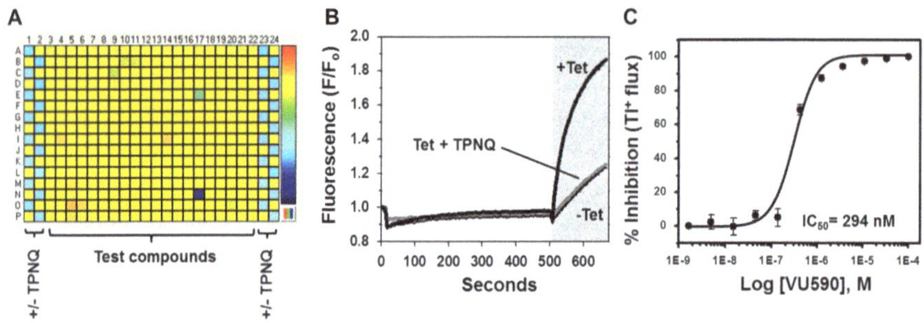

Fig. 27.3 Tl^+ flux assay of ROMK function for HTS and inhibition by VU590. (**a**) Representative fluorescence "heat map" recorded from a 384-well plate containing 20,000 T-REx-HEK293-ROMK cells/well induced overnight with tetracycline. Fluorescence intensity has been pseudocolored using the color scale shown to the right of the plate, with cooler colors (*blues*) indicating lower fluorescence and hotter colors (*yellow* and *red*) indicating high fluorescence and enhanced Tl^+ flux. Every other well of columns 1–2 and 23–24 has been treated with the ROMK inhibitor TPNQ to provide a measure of unblocked and blocked ROMK-mediated Tl^+ flux for quality control purposes. Each well in columns 3–22 has been treated with 10 μM test compound (one per well). Note that column N17 received an inhibitory small molecule, leading to lower rate of Tl^+ flux and fluorescence. This compound would be identified as a "hit" and picked for subsequent analysis. (**b**) Timecourse of Tl^+-induced (*stippled box*) changes in fluorescence in T-REx-HEK293-ROMK cells that were cultured overnight without tetracycline (−Tet) or with tetracycline (+Tet), or with tetracycline and pretreated with TPNQ (Tet + TPNQ). Cells that were left uninduced or induced and treated with TPNQ exhibit low background Tl^+ flux, whereas those that were induced to express ROMK and left unblocked exhibit robust Tl^+ flux. (**c**) Dose-dependent inhibition of ROMK-mediated Tl^+ flux by the small-molecule inhibitor VU590. (**b**) and (**c**) are reproduced with permission from Lewis et al. (2009)

fluorescence that differed by greater than 3 standard deviations from the mean fluorescence of control wells. The hits were picked for confirmation testing in tetracycline-induced and uninduced cells, and those exhibiting activity in uninduced cells were excluded from further analysis because they were likely to also affect endogenous Tl^+ flux pathways. ROMK-dependent modulators were reordered as dry powders, dissolved freshly in DMSO, and used for testing in concentration-response experiments against ROMK and other Kir channels.

The first publically disclosed small-molecule inhibitor of ROMK was 7,13-*bis* [(4-nitrophenyl)methyl]-1,4,10-trioxa-7,13-diazacyclopentadecane, which we termed VU590 (Lewis et al. 2009), and consists of two nitrophenyl groups flanking a nitrogen- and oxygen-rich crown ether moiety. It was almost rejected as a "false-positive" because crown ethers are well known to chelate alkali metals such as K^+, raising the possibility that the activity observed in the Tl^+ flux assay was due to chelation of Tl^+ and not frank inhibition of ROMK. However, patch-clamp experiments showed that VU590 is indeed a bona fide inhibitor of ROMK that dose-dependently inhibits the channel with a 50 % inhibition concentration (IC_{50}) of approximately 300 nM (Fig. 27.3c; Lewis et al. 2009). Moreover, VU590 was moderately selective. At a concentration of 10 μM, VU590 had no effect on the closely related Kir channels Kir2.1 or Kir4.1, providing at least a 30-fold selectivity window for ROMK over these channels. Additional experiments revealed that VU590 was not completely selective for ROMK, as it inhibited Kir7.1 with an IC_{50} of ~8 μM. This presented a problem for the development of a small-molecule probe that could be used to investigate the effects of pharmacological inhibition of ROMK on renal tubule electrolyte transport, because both ROMK and Kir7.1 are expressed in K^+-secreting principal cells of the collecting duct (see below for discussion of Kir7.1). Thus, any effects of VU590 on renal tubule function could be attributed to inhibition of ROMK, Kir7.1, or both Kir channels.

The pharmacological activity of a small molecule can be refined by making changes to its chemical structure and hence ability to engage its primary target potently and specifically. This is often done by synthesizing a library of structurally distinct analogs and testing them for activity against the primary target (e.g., ROMK), as well as off-targets that are to be avoided (e.g., Kir7.1). Establishing so-called structure-activity relationships (SAR) can reveal which fragments of a compound form are important for ligand-receptor interactions and help guide subsequent steps in an iterative cycle of lead optimization. In the case of VU590, however, we found that virtually any change to its chemical structure led to a dramatic loss of potency toward ROMK. For example, migrating both nitro groups from the 4-position to the 3- or 2-position led to a complete loss of ROMK activity. Replacing the nitro groups with 4-chlorine, 4-thiomethyl, or 4-methyl afforded only weak ROMK inhibitors with IC_{50} values ranging between 10 and 30 μM. Despite our ability to improve the potency of VU590, this limited round of chemistry revealed that the presence and position of the nitro groups are essential for the potency of VU590 toward ROMK.

Having developed an understanding of the structure-activity relationships (SAR) between VU590 and ROMK, we turned our attention to another bis-nitro compound identified in the primary screen: 1,3-bis(5-nitro-1*H*-benzo[*d*]imidazol-2-yl)propane , (BNBI, Fig. 27.4). , ,Although BNBI is relatively weak inhibitor of ROMK (IC_{50} ~ 8 μM), it exhibits no activity toward Kir7.1 at concentrations up to 100 μM. In an effort to improve the potency of BNBI while maintaining selectivity over Kir7.1, we focused on the linker region, which differs from that of VU590 by the notable absence of an ether oxygen (Table 27.2). Remarkably, the addition of a single ether oxygen improved the potency of BNBI by ~30-fold (IC_{50} = 240 nM). This BNBI analog was termed VU591.

The selectivity of VU591 was characterized in a series of patch-clamp experiments on HEK-293 cells transfected with plasmids encoding mostly human inward rectifier K^+ channels, including Kir2.1, Kir2.3, Kir3.1/3.2, Kir4.1, Kir6.2/SUR1, and Kir7.1, as well as the voltage-gated K^+ channels Kv1.3 and Kv11.1 and the voltage- and calcium-activated, large-conductance "BK" K^+ channel Slol/β1. Kir2.3, Kir4.1, Kv1.3, and Slo1/β1 were important off-targets to consider because these channels contribute to K^+ transport in the nephron. The selectivity of VU591 was also characterized in a series of radioligand-displacement assays against 68 neuronal and cardiac ion channels, transporters, and receptors. With the exception of Kir6.2/SUR1, hERG, dopamine $D_{4.2}$ receptor, $GABA_A$ receptor, dopamine

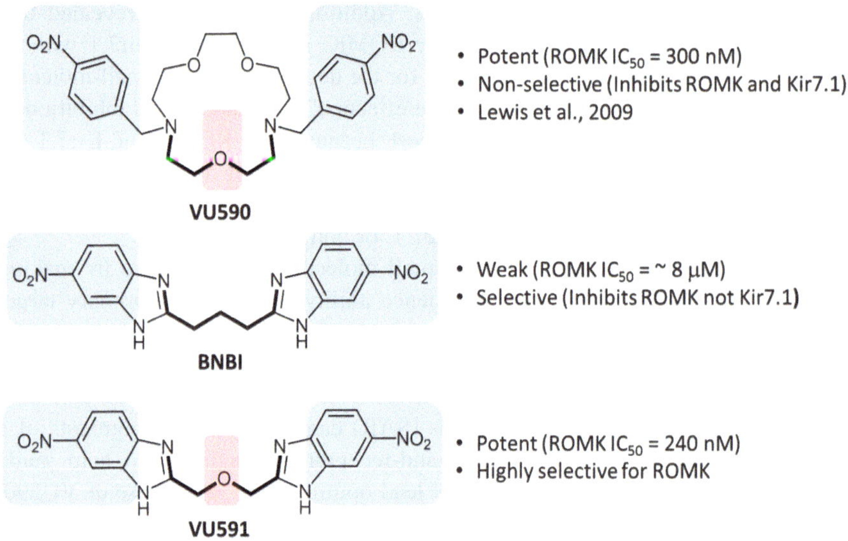

Fig. 27.4 Chemical structures and pharmacological properties of VU590, BNBI, and VU591. Nitro groups and ether oxygen, which are critical for ROMK inhibitory activity, are highlighted with *gray* and *red boxes*, respectively. See text for details

transporter, and norepinephrine transporter, all of which were only weakly sensitive to VU591, the ancillary pharmacology of VU591 was clean (Bhave et al. 2011).

A number of studies have suggested that ROMK exists as part of a multi-protein complex in the renal tubule epithelial cells (Yoo et al. 2004; Ruknudin et al. 1998; Tanemoto et al. 2000; McNicholas et al. 1996), raising the possibility that protein-protein interactions could obscure or alter the binding site for inhibitory small molecules such as VU591. Because all of the discovery and characterization work on VU591 had been performed on heterologously expressed ROMK channels in HEK293 fibroblasts, we tested if VU591 inhibits native ROMK channels expressed in the collecting duct. Single rat collecting duct tubules were isolated and perfused with a buffer containing DMSO (solvent control) or 10 μM VU591 for 10 min, after which the concentration of Na^+, K^+, and Cl^- in the perfusate was analyzed so that the rate of ion flux could be calculated. Over a 10-min collection period, VU591 inhibited net K^+ secretion by approximately 50 %, which is consistent with inhibition of ROMK function in this nephron segment (Bhave et al., 2011).

The metabolic stability of VU591 was assessed in vitro by measuring its rate of degradation in the presence of human or rat liver microsomes. These studies revealed that VU591 was more stable than the clinically used drugs propanolol and nifedipine, suggesting it may escape first-pass metabolism in the liver after administration in animal models. VU591 exhibits high serum protein binding (98 % and 99 % in rat and human, respectively), which could limit its ability to engage and inhibit ROMK in vivo (Bhave et al. 2011). Studies are underway to assess the effect of VU591 on renal function in vivo.

Table 27.2 Inhibitors of epithelial Kir channels

Compound	Structure	Kir target	IC_{50}/EC_{50} (max)	Selectivity	References
3	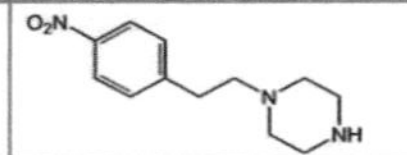	Kir1.1	NA	NA	Tang et al. (2012)
5	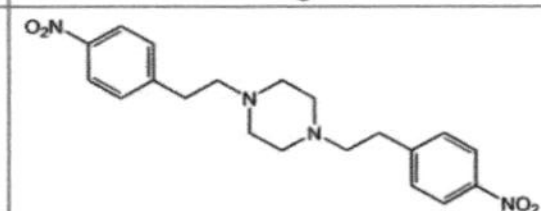	Kir1.1	52 nM	≫Kir2.1, 2.3, Kir7.1 hERG blocker	Tang et al. (2012)
13	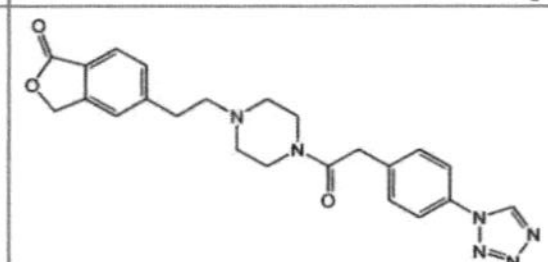	Kir1.1	60 nM	≫hERG	Tang et al. (2012)
18	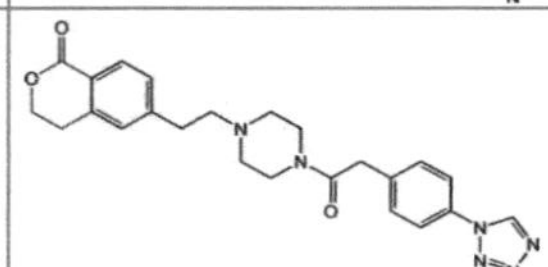	Kir1.1	60 nM	≫hERG	Tang et al. (2012)
30	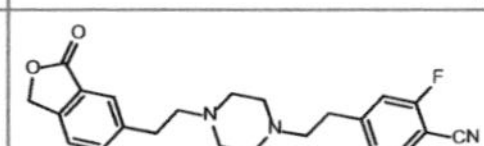	Kir1.1	30–49[a] nM	>Kir2.1, 2.3 ≫Kir4.1, Kir7.1 hERG	Tang et al. (2012)
VU590	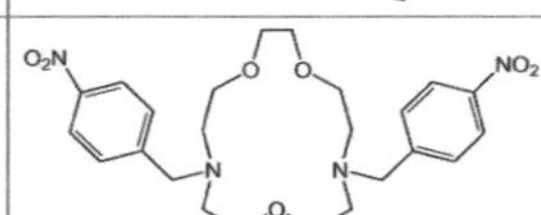	Kir1.1	0.29 μM[b] (100 %)	>Kir7.1 (IC_{50} ~ 8 μM)	Lewis et al. (2009)
VU591	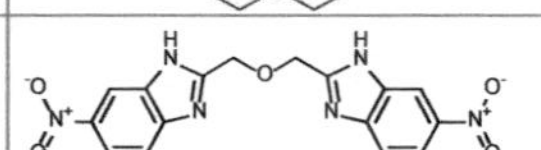	Kir1.1	0.30 μM[b] (100 %)	>Kir2.1, Kir2.3, Kir4.1	Bhave et al. (2011)

(continued)

Table 27.2 (continued)

Compound	Structure	Kir target	IC_{50}/EC_{50} (max)	Selectivity	References
Pregnenolone sulfate	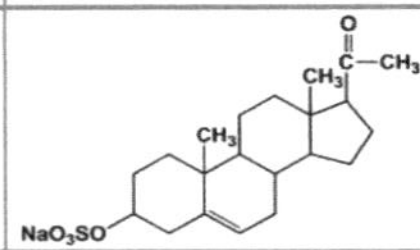	Kir2.3	~15 μM (60 %)	>Kir2.1/2.3, Kir2.2/2.3	Kobayashi et al. (2009)
VU717	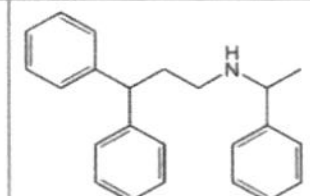	Kir4.1	~6.0 μM	NA	Raphemot et al. (2013)
Nortriptyline	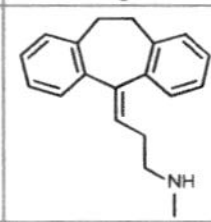	Kir4.1	38 μM	>Kir1.1, 2.1	Su et al. (2007)
Fluoxetine	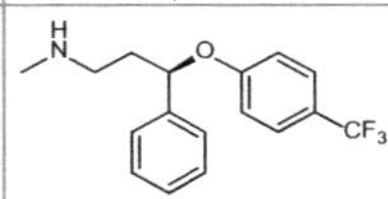	Kir3	7–13[c] μM (58–74 %)	>Kir1.1, Kir2.1 Known SSRI	Kobayashi et al. (2004), Ohno et al. (2007)
		Kir4.1	15.2 μM (100 %)		
		Kir4.1	16–38 μM; voltage dependent (100 %)		
VU573	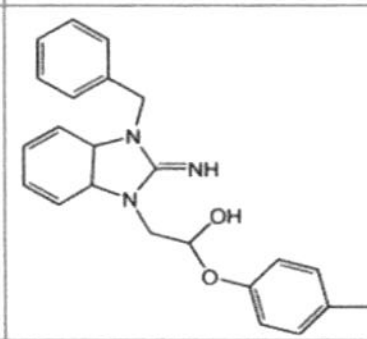	Kir2.3	0.8 μM (70 %)	>Kir1.1, Kir2.1	Raphemot et al. (2011)
		Kir3.x	1.3 μM (80 %)		
		Kir7.1	0.9 μM (70 %)		

[a]IC_{50}/EC_{50} values accompanied by percent maximal inhibition/activation observed in parentheses
[b]Results obtained from Tl^+ flux experiments
[c]Results from experiments on Kir channels heterologously expressed in *Xenopus* oocytes, which have been reported to yield significantly higher IC_{50} values when compared to mammalian cells (Matsuda et al. 2006)

Merck Research Laboratories screened approximately 1.5 million compounds from their internal small-molecule library for inhibitors of ROMK. The group employed a fluorescence resonance energy transfer (FRET) assay that reports changes in membrane potential using the dye pair DiSBAC2(3) and CC2-DMPE, which are commercially available from Life Technologies. They selected one compound, termed compound **3**, for development due to its moderate activity toward ROMK ($IC_{50} \sim 5$ μM) and lack of activity toward Kir2.1 and Kir2.3 at doses up to 100 μM in radiolabeled rubidium uptake or Tl^+ flux assays. Interestingly, purification of **3** by HPLC led to a complete loss of ROMK inhibitory activity. A detailed analysis of the sample by mass spectrometry revealed a minor sample impurity with a molecular weight that was larger than expected for the screening hit, raising the possibility that the active compound was actually a symmetrical, bis-nitro derivative of **3**. The suspected compound was synthesized and shown to inhibit ROMK with an $IC_{50} = 52$ nM and exhibits greater than 1900-fold selectivity over Kir2.1 and Kir2.3. A liability of the new compound, termed **5**, was that it inhibits the cardiac K^+ channel Kv11.1, more commonly known as hERG (for human EAG-related gene) with single-nanomolar affinity ($IC_{50} = 5$ nM). Because inhibition of hERG can delay repolarization of ventricular action potentials leading to cardiac arrhythmias and sudden death, Merck set out to develop an analog of **5** with reduced potency toward hERG.

Similar to what was observed for VU590 (Lewis et al. 2009), the initial lead optimization campaign on **5** revealed that the nitro groups extending from the 4′ position of the phenyl ring are critically important for potency toward ROMK. Removing one or both nitro groups or replacing them with chloride, methoxy, trifluoromethyl, or methylsulfone led to a loss of inhibitory activity. Interestingly, shortening the distance between the nitro groups by one carbon resulted in a complete loss of activity, indicating that a precise geometrical configuration between the nitro groups and binding site in ROMK is required for high-affinity block.

After extensive medicinal chemistry efforts, the Merck team was ultimately successful in replacing the nitro groups and developing a compound, termed **30** (Table 27.1), that preferentially inhibits ROMK ($IC_{50} = 55$ nM) over hERG ($IC_{50} = 170$ nM) and the inward rectifiers Kir2.1, Kir2.3, Kir4.1, and Kir7.1 ($IC_{50} > 100$ μM). Compound **30** also exhibits pharmacokinetic properties that should make it useful for in vivo testing, although no such studies have been published to date. In a subsequent paper (Tang et al. 2012), the Merck group reported a series of **5** derivatives (termed **13** and **18**, Table 27.2) exhibiting even better selectivity for ROMK ($IC_{50} \sim 60$ nM) over hERG ($IC_{50} = 5$–20 μM). Importantly, oral administration of either compound to volume-loaded Sprague–Dawley rats led to a significant natriuresis and diuresis that matched or exceeded that observed with the control diuretic hydrochlorothiazide. This was the first demonstration that acute ROMK inhibition with a pharmacological antagonist could induce urinary salt and water excretion.

A fundamental question, however, is whether or not ROMK inhibition will lead to an increased K^+ excretion during diuresis. Indeed, as noted earlier, a genetic loss of ROMK in Bartter syndrome is associated with hypokalemia due to enhanced urinary K^+ excretion, a phenotype that is also observed in ROMK knockout mice. The residual pathway for K^+ secretion in the absence of ROMK appears to be mediated by calcium-activated BK channels, which are stimulated by increased delivery of fluid from the TAL to the distal nephron. To begin to address this issue, the Merck group treated normotensive rats and dogs with compound **13** and measured urinary volume and electrolyte concentrations over a 4-h period (Garcia et al. 2013). Once again, the effects of compound **13** on renal function were comparable to those observed with hydrochlorothiazide. An important distinction, however, is that, unlike hydrochlorothiazide, **13** did not stimulate K^+ excretion over the study period. Taken together, these data strongly support the idea that ROMK represents a molecular target for a novel class of diuretic. Looking ahead, it will be important to examine the effects of chronic ROMK inhibition on fluid and electrolyte (particularly K^+) balance and interactions with commonly used diuretics. These studies will reveal which nephron segments are targeted in vivo by ROMK inhibitors, how potential remodeling of renal tubule ion channels and transporters influences the long-term efficacy of ROMK inhibitors, and help inform the best combination of diuretics to prescribe clinically to meet target blood pressure in hypertensive patients.

27.2.4 ROMK Inhibitor Mechanisms of Action

Unlike many drug targets, Kir channels are relatively small (~350 amino acids) and architecturally simple proteins that lack modulatable gating mechanisms found in many voltage-gated K^+ channels that are common targets of inhibitory toxins and small molecules. From a screen of 225,000 compounds at Vanderbilt, VU590 was the only sub-micromolar inhibitor identified. Merck reported that only a "few" tractable hits were identified in their 1.5 million compound screen. Taken together, these low hit rates support the idea that Kir channels are difficult targets for developing potent small-molecule inhibitors. Furthermore, Kir channels share 40–60 % amino acid identity with each other, raising the question of whether a selective small-molecule inhibitor of ROMK could be developed. Our development of VU591 and Merck's development of their class of inhibitor show that it is indeed possible to develop sharp pharmacological tools for Kir channels, but it requires screening a large number of compounds and luck.

It is interesting to note that the lead compounds identified at Vanderbilt (i.e., VU590) and Merck (compound **5**) are both symmetrical and contain nitrophenyl groups flanking a nitrogen-containing linker (Table 27.2). The similarity between the compounds begs the question of whether they interact with the same binding sites in ROMK.

The first clues to the location of a potential binding site of VU590 arose from a series of patch-clamp experiments in which we observed that the onset of block was associated with increased current rectification of ROMK (Fig. 27.5a). As noted

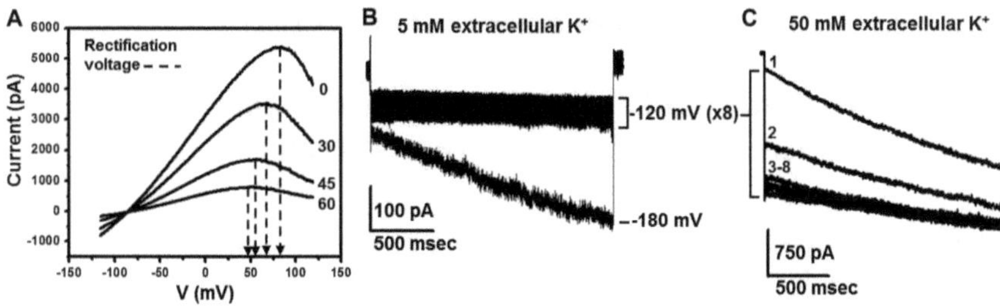

Fig. 27.5 Knockoff of VU590 from the ROMK channel pore. (**a**) Representative ROMK current traces evoked from voltage ramps before and during the onset of block by 10 μM VU590. The time (in sec) after VU590 application is shown next to each current trace. The *arrows* are drawn at the point of current inflection to indicate the rectification voltage. (**b**) ROMK current traces recorded from a cell incubated with 5 mM K^+ and 10 μM VU590 after achieving maximal block. Eight repetitive pulses to −120 mV had virtually no effect on current amplitude. However, a step to −180 mV led to time-dependent unblock of the channel. (**c**) Elevation of extracellular K^+ to 50 mM progressively displaces VU590 during repetitive steps (labeled 1–8) to −120 mV, indicating that the VU590 binding site is likely to be in the ion-conduction pathway. Reproduced with permission from Lewis et al. (2009)

earlier, rectification is caused by block of outward current by intracellular Mg^{2+} and polyamines. The increased rectification in the presence of VU590 suggested that VU590 was being swept into the pore at positive voltages and inducing rectification. As a more direct test of pore block, we determined if inhibition of ROMK exhibits voltage dependence. Under a physiological K^+ gradient, where E_K is approximately −80 mV, ROMK was completely inhibited by VU590 at every voltage except −180 mV, where a small, time-dependent inward current was observed (Fig. 27.5b). A much larger current could be evoked at −120 mV after increasing the extracellular K^+ concentration and hence electrochemical driving force for K^+ entry (Fig. 27.5c). These observations are consistent with a classical "knockoff" effect, whereby VU590 occupying a binding site in the ion-conduction pathway can be displaced by inwardly directed K^+ ions. Similar results were observed for VU591, suggesting both Vanderbilt inhibitors are pore blockers (Bhave et al. 2011; Lewis et al. 2009).

Compound **13** showed no evidence of voltage-dependent block, but the Merck team nevertheless set out to test if its binding site is located within the pore. Mutation of only one residue, asparagine 171 (N171), which is located in the membrane-spanning pore immediately below the selectivity filter (Fig. 27.1), weakened the inhibition of ROMK by **13** (Tang et al. 2012). It will be of interest to determine if the bis-nitro parent compound, **5**, interacts with the same site as **13**, VU590, and VU591. Furthermore, given the structural similarities between **5**, VU590, and VU591, it will be important to determine if N171 and other amino acids are important for block of ROMK by the Vanderbilt inhibitors. We are currently employing molecular modeling, in silico docking, and site-directed mutagenesis to comprehensively assess potential binding for VU590 and VU591 in ROMK.

27.3 Kir4.1 (*KCNJ10*) and Kir4.1/5.1 (*KCNJ16*)

27.3.1 Are Basolateral Kir4.1-Containing Channels Basolateral Membrane Diuretic Targets?

Loop (i.e., furosemide, bumetanide) and thiazide (i.e., hydrochlorothiazide) diuretics work by inhibiting their respective molecular targets in the luminal membrane of the TAL and DCT, respectively. In order to reach their target sites, the diuretics must first be secreted into the tubule fluid by proximal tubule cells following their transepithelial transport via organic anion transporters and multidrug resistance proteins. Competitive interactions between loop and thiazide diuretics with certain antibiotics, nonsteroidal anti-inflammatory drugs, antiviral drugs, as well as organic acids in the setting of renal failure can limit the secretion of these diuretics into the tubule fluid and hence their efficacy and predictability at lowering blood volume and pressure. Diuretics acting on basolateral targets that can be reached directly from the blood would presumably avoid this limitation and provide clinicians with more options for managing blood pressure in the setting of other comorbidities. A growing body of genetic and physiological data suggests that Kir4.1 homomeric and/or Kir4.1/5.1 heteromeric channels may represent such a target.

Soon after the cDNA encoding Kir4.1 was cloned (Takumi et al. 1995; Bond et al. 1994), Ito and colleagues (Ito et al. 1996) demonstrated that the Kir4.1 protein was immunolocalized to the basolateral membrane of what was probably the DCT, connecting tubule, and principal cells of the CD. This was the first time that any molecularly defined K^+ channel was shown to exhibit polarized expression in the basolateral membrane of renal tubule epithelial cells. Tucker and colleagues (Tucker et al. 2000) subsequently showed that Kir5.1, which is electrically silent when expressed alone, is also localized to the distal nephron. This raised the intriguing notion, which was subsequently borne out by a series of elegant physiological and genetic studies, that Kir4.1/Kir5.1 heteromeric channels play critical roles in the regulation of electrolyte transport in the nephron. These studies also provided genetic validation for Kir4.1/5.1 channels as putative diuretic targets.

27.3.2 Genetic Validation of Kir4.1 as a Diuretic Target

The first piece of genetic evidence that Kir4.1 is essential for urine concentration by the kidney came from the discovery by two independent groups that inactivating mutations in the Kir4.1-encoding gene, *KCNJ10*, give rise to severe salt and water wasting in patients with EAST or SeSAME syndrome (Bockenhauer et al. 2009; Scholl et al. 2009). The disease also presents with severe neurological symptoms, including seizures, ataxia, and intellectual disability, which reflects the loss of homomeric Kir4.1 expression in astrocytes and glia of the central nervous system. The renal phenotype associated with EAST/SeSAME syndrome is thought to reflect loss of Kir4.1 activity primarily in the DCT; however, recent evidence suggests that the loss of channel activity in the CCD may also contribute to the pathophysiology of the disease. In the DCT, NaCl is reabsorbed from the tubule fluid across the luminal plasma membrane via the electroneutral NaCl cotransporter, NCC. The steep Na^+ concentration gradient across the luminal membrane is maintained by the activity of the basolateral Na^+-K^+-ATPase, which pumps three Na^+ ions out in exchange for two K^+ ions in, during each catalytic cycle. Chloride ions entering the cell across the luminal membrane move down their electrochemical gradient and exit the basolateral membrane via ClC-Kb (Fig. 27.2). Homomeric Kir4.1 and/or heteromeric Kir4.1/5.1 channels facilitate net NaCl reabsorption in the DCT in two important ways. First, they recycle K^+ ions across the basolateral membrane to help sustain robust activity of the Na^+-K^+-ATPase. Second, they hyperpolarize the basolateral membrane potential and thus increase the electrochemical driving force for Cl^- exit. Thus, a loss of Kir4.1 expression in SeSAME/EAST syndrome would reduce NaCl reabsorption in the DCT, leading to a loss of Na^+ and water in the urine. In principle, pharmacological inhibition of Kir4.1-containing channels in the DCT should recapitulate the renal consequences of thiazide diuretics, which directly inhibit NCC.

A potential caveat of extrapolating from genetic studies to justify a molecular target for drug discovery is that the phenotypes may be associated with mechanisms that are triggered to compensate for the loss of the gene product during development. It is therefore beneficial to have independent verification using, for example, acute manipulation of target function in genetically wild-type animals, to further validate the target for drug discovery. Recently, Zaika et al. (2013) showed that homomeric Kir4.1 and heteromeric Kir4.1/5.1 channels are functionally coupled to a D2 dopaminergic signaling pathway in principal cells of the CCD. Cell-attached patch-clamp recordings from the basolateral membrane revealed that acute dopamine application led to a rapid inhibition of Kir4.1 homomers and Kir4.1/5.1 heteromers and depolarization of the basolateral membrane. This in turn would reduce the electrochemical driving force for Na^+ reabsorption through ENaC in the luminal membrane and potentially explain the natriuretic actions of dopamine in the CCD. Taken together, these data support the notion that Kir4.1-containing channels in the basolateral membrane of the DCT and CCD represent unexploited drug targets for a novel class of diuretic.

27.3.3 Kir4.1 Pharmacology

There are currently no nanomolar-affinity Kir4.1 or Kir4.1/5.1 small-molecule inhibitors available for exploring the therapeutic potential of these channels for the treatment of hypertension. However, homomeric Kir4.1 is blocked by the selective serotonin reuptake inhibitor (SSRI) fluoxetine ($IC_{50}=15$ μM; Ohno et al. 2007) and tricyclic antidepressants (TCA) nortriptyline ($IC_{50}=28$ μM; (Su et al. 2007) (Table 27.2). These compounds appear to be somewhat selective for Kir4.1 within the Kir channel family since they exhibit only weak inhibitory activity toward Kir1.1 and Kir2.1 (Ohno et al. 2007; Su et al. 2007). Similar to what was observed for VU590- and VU591-dependent block of ROMK (Bhave et al. 2011; Lewis et al. 2009), fluoxetine and nortriptyline exhibit voltage-dependent knockoff from the Kir4.1 channel pore. Mutation of T128 and E158 weakens block of Kir4.1 by fluoxetine and nortriptyline, suggesting these residues participate in the pore blocking mechanism (Furutani et al. 2009). Fluoxetine sensitivity can be conferred to ROMK by mutating N171, which is equivalent to E158 in Kir4.1, to an E or a D (N171D/E), providing further support for E158 playing a role in the fluoxetine mechanism of action. Computational modeling simulations suggest that both drugs can interact with E158 and T128 on diagonally apposed subunits of the tetramer, raising the possibility that these drugs may be specific for homomeric Kir4.1 vs. heteromeric Kir4.1/5.1 channels since Kir5.1 has an asparagine residue (N161) at the equivalent position of Kir4.1–E158.

In an effort to develop more potent and selective inhibitors of Kir4.1, we recently screened a library of 3655 compounds using a Tl^+ flux assay similar to the one developed for ROMK (discussed above). Sixteen confirmed Kir4.1 inhibitors were identified in the screen, the most potent of which was 3,3-diphenyl-N-(1-phenylethyl)propan-1-amine, which we termed VU717 (Raphemot et al. 2013). An automated patch-clamp electrophysiology assay was developed around the IonFlux HT benchtop platform (Molecular Devices), which enables 32 patch-clamp recordings to be made simultaneously, to improve throughput and facilitate compound characterization. In both Tl^+ flux and electrophysiology assays, VU717 inhibited Kir4.1 with an IC_{50} of ~6 μM, which represents a modest improvement in potency over fluoxetine and nortriptyline. The continued development of improved inhibitors targeting Kir4.1 and Kir4.1/5.1 channels will provide critically needed tools for exploring the therapeutic potential of these channels in hypertension and other disorders.

27.4 Kir7.1 (*KCNJ13*)

27.4.1 Overview of Kir7.1 Expression and Function

The gene encoding Kir7.1 (i.e., *KCNJ13*) was originally cloned from human fetal brain by the laboratory of Dr. David Clapham in 1998 (Krapivinsky et al. 1998) and subsequently identified in epithelial cells of the eye, intestine, stomach, thyroid, and kidney (Table 27.1), where it is thought to play important roles in regulation of epithelial salt and water homeostasis (Kusaka et al. 2001; Nakamura et al. 1999).

The most comprehensive understanding of Kir7.1 physiology has come from studies of retinal pigment epithelial (RPE) cells of the eye, which are polarized epithelial cells that carry out several important supportive functions required for vision (Strauss 2005). Kir7.1 activity is critical for electrolyte and water transport across the RPE. Water produced by metabolically active neurons and photoreceptors must be continuously transported from the subretinal space, across RPE cells, and into the blood. Water transport is driven primarily by a transepithelial Cl^- gradient created by apical membrane Na^+-K^+-$2Cl^-$ cotransporter, which is energized by an inwardly directed Na^+ gradient established by the Na^+-K^+-ATPase. Unlike most polarized epithelial cells, the Na^+-K^+-ATPase of RPE cells is localized to the apical membrane and helps maintain robust transport activity of the Na^+-K^+-$2Cl^-$ cotransporter by pumping Na^+ ions into the subretinal space. Kir7.1 is also expressed in the apical membrane and recycles substrate K^+ across the apical membrane to help sustain the activity of the Na^+-K^+-$2Cl^-$ cotransporter. Consistent with playing an important role in RPE transport functions, a LOF mutation (R162W) in *KCNJ13* has been associated with the rare disorder snowflake vitreoretinal degeneration (Hejtmancik et al. 2008; Pattnaik et al. 2013).

In the kidney, Kir7.1 exhibits an interesting expression pattern and regulated expression, suggesting the channel could play an important role in renal function. Kir7.1 is expressed on the basolateral membrane of multiple nephron segments, including DCT, CCD, and to a lesser extent, TAL (Ookata et al. 2000). The functions of Kir7.1 in the nephron are purely speculative at this point, but may be analogous to and perhaps overlapping with that of Kir4.1/5.1. Thus, in the DCT, Kir7.1 may help energize the Na^+-K^+-ATPase and in turn maintain a favorable driving force for Na^+ reabsorption by the NaCl cotransporter. In the CCD, Kir7.1 may contribute to a favorable electrochemical gradient for K^+ excretion and Na^+ reabsorption (Fig. 27.2). Indeed the observation that renal expression of Kir7.1 is increased with dietary K^+ is consistent with a role of this channel in regulating K^+ excretion (Ookata et al. 2000). Due to its expression in the DCT and CCD, it is intriguing to speculate that pharmacological inhibitors of the channel may recapitulate some of clinical symptoms of thiazide and K^+-sparing diuretics by inhibiting Kir7.1 in these segments, respectively. However, progress toward developing the pharmacology of Kir7.1 thus far has been limited.

27.4.2 Kir7.1 Pharmacology

At present, the small-molecule pharmacology of Kir7.1 is rudimentary and limited to two inhibitors developed by our group (Table 27.2). As noted earlier, the ROMK antagonist VU590 exhibits moderately potent activity toward Kir7.1 (IC_{50} ~8 μM; Lewis et al. 2009). VU590 was recently used to help uncover a role of Kir7.1 in regulation of uterine muscle tone during pregnancy and implicates Kir7.1 as a novel target for dystocia and postpartum hemorrhage (McCloskey et al. 2014). The other inhibitor, termed VU573, was discovered as a weak inhibitor of ROMK in the

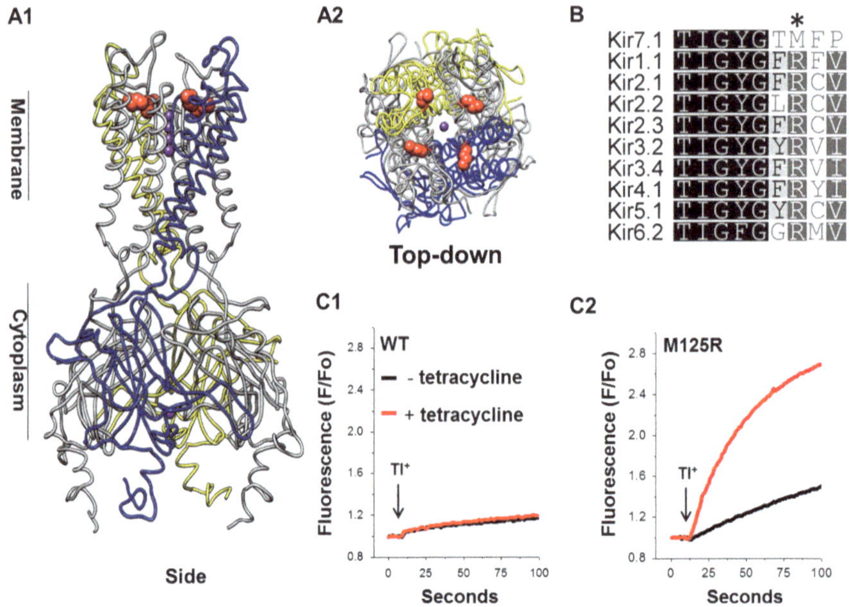

Fig. 27.6 Development of Kir7.1 (M125R) Tl^+ flux assay. (**a1**) Side and (**a2**) top-down views of a Kir7.1 model with M125 highlighted in *red*. (**b**) Alignment showing M125 is unique to Kir7.1. (**c1** and **c2**) M125R but not wild-type (WT) Kir7.1 mediates robust Tl^+ flux in tetracycline-induced cells. See text for details

225,000 compound HTS and shown in secondary assays to be a preferential inhibitor of Kir2.3 (IC_{50} ~ 1 μM), Kir3.1/3.2 (IC_{50} ~ 1.3 μM), and Kir7.1 (IC_{50} ~ 1.7 μM; Raphemot et al. 2011). A lead optimization campaign was initiated to improve the potency and selectivity of VU573 for any of the three channels. A library of 26 VU573 analogs was synthesized and tested in the hope of identifying compounds with improved pharmacology. To improve the throughput of compound testing, Tl^+ flux assays were developed for Kir2.3 and Kir3.1/3.2 using standard methods. However, despite several attempts using several different polyclonal and monoclonal cell lines and experimental conditions, we were unable to convincingly demonstrate tetracycline-inducible (i.e., Kir7.1 channel dependent) Tl^+ flux. Unlike other members of the Kir channel family, Kir7.1 contains a methionine (M) residue at position 125 (M125) near the extracellular pore mouth (Fig. 27.6a). This amino acid confers unique functional properties to the channel, including reduced sensitivity to extracellular K^+ and Ba^{2+} and lower unitary conductance (Doring et al. 1998; Krapivinsky et al. 1998). We considered the possibility that mutation of M125 to an arginine (M125R), which is found at the equivalent position in every other family member (Fig. 27.6b), would increase Tl^+ flux through the channel. Indeed, as shown in Fig. 27.6c, Kir7.1 (M125R) exhibited robust tetracycline-inducible Tl^+ flux that could be readily measured in 384-well format (Raphemot et al. 2011). The tetracycline-inducible assay (Fig. 27.7b) reports dose-dependent

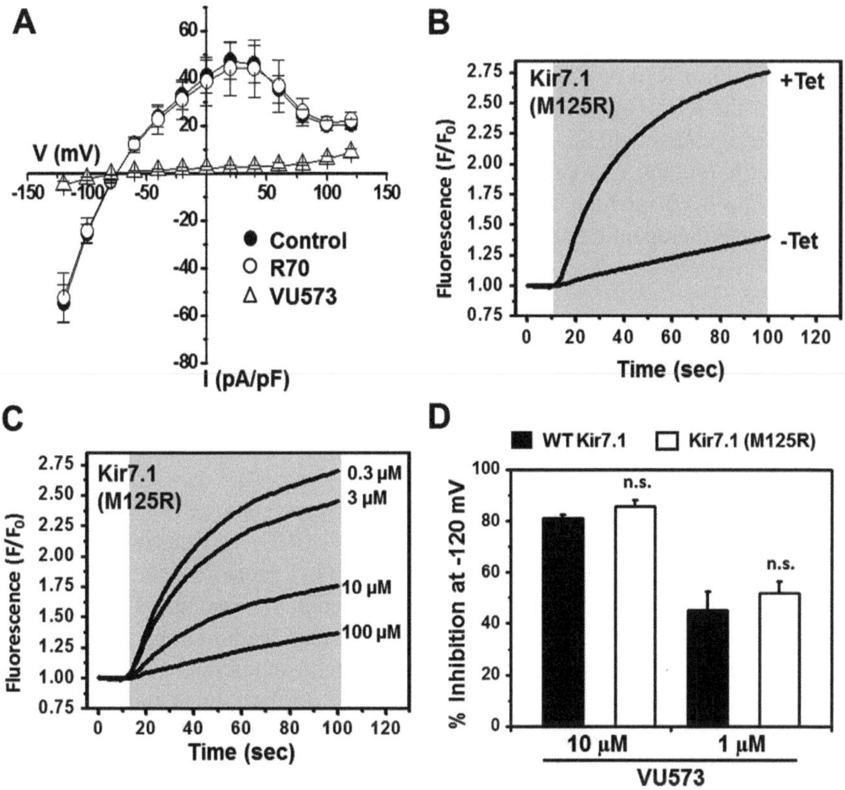

Fig. 27.7 Development of the VU573 series of Kir channel antagonists. (**a**) Kir7.1 current traces recorded in the absence or presence of 10 μM R70 or VU573 suggest that R70 is an inactive analog. (**b**) Tl^+ flux-dependent FluoZin-2 fluorescence recorded from polyclonal T-REx-HEK-293-Kir7.1 (M125R) cells cultured overnight in the absence (−Tet) or presence (+Tet) of tetracycline. The fluorescence emission was recorded before and after the addition of extracellular Tl^+ (*shaded box*). (**c**) Pretreatment of induced cells expressing Kir7.1 with VU573 shows dose-dependent reduction in the FluoZin-2 fluorescence. (**d**) VU573 causes equal pharmacological inhibition of the Kir7.1 wild type and the M125R mutant. Modified from Raphemot et al. (2011) with permission

inhibition of the channel (Fig. 27.7c), enabling rapid assessment of compound potency to support lead optimization. Furthermore, the M125R mutation has no effect on the pharmacological sensitivity of Kir7.1 compared to the wild-type channel (Fig. 27.7d).

Using these assays, we found that most of the VU573 analogs retained potency toward Kir2.3, Kir3.1/3.2, and Kir7.1. No improvements in potency were observed. However, and importantly, we identified two analogs, termed R70 (Fig. 27.7a) and R5C, that were completely inactive toward Kir7.1 (Raphemot et al. 2011). Thus, in principle, it may be possible to use a combination of the Kir channel inhibitors described in this chapter to tease out the functional expression of Kir7.1 in epithelial cells of the CCD, for example, where Kir1.1, Kir2.3, Kir4.1, and Kir7.1 are expressed. For example, in the presence of the Kir1.1 inhibitor VU591 and Kir2.3 inhibitor R70 or R5C, a residual current that is blocked by VU573 should represent the Kir7.1-dependent component of the whole-cell inward rectifier current. This of course will have to be verified experimentally. We are currently using the M125R mutant as a surrogate for the wild-type channel in HTS assays for finding improved modulators of Kir7.1 that can ultimately be used to probe the function and therapeutic potential of Kir7.1 in vivo.

27.5 Kir2.3 (*KCNJ4*)

27.5.1 A Basolateral Channel with No Known Functions

Dr. Paul Welling cloned and heterologously expressed a cDNA encoding the inward rectifier Kir2.3 (IRK3) from the mouse M1 CCD cell line and showed that it forms a K^+ current with biophysical and regulatory properties similar to that of the native small-conductance K^+ channel expressed in the basolateral membrane of the CCD (Welling 1997). When expressed in polarized Madin-Darby canine kidney epithelial cells, Kir2.3 is directed to the basolateral membrane via a C-terminal PDZ-interacting domain (Le Maout et al. 2001) further suggesting that Kir2.3 contributes to the basolateral K^+ conductance in the CCD. Millar et al. (2006) used Ba^{2+} as a Kir channel probe to establish a pharmacological "fingerprint" of the macroscopic K^+ currents expressed in principal cells of the mouse CCD. The authors concluded that the voltage-dependent and kinetic properties of Ba^{2+} block of the current are consistent with the corresponding properties of Kir2.3 (Millar et al. 2006). However, there are at least four distinct Kir channels (Kir1.1, Kir2.3, Kir4.1, and Kir7.1) expressed in principal cells of the CCD (Table 27.1). Studies aimed at determining if Kir2.3 is indeed functionally expressed in the CCD and exploring its physiological functions in the nephron will benefit from the judicious use of sharper pharmacological tools capable of discriminating Kir2.3 from other Kir channels.

27.5.2 Kir2.3 Pharmacology

As is the case with most members of the Kir channel family, the small-molecule pharmacology of Kir2.3 is still rudimentary. However, one could envision experiments in which VU591 and the VU573 series described earlier in the chapter could be used to identify Kir2.3-dependent currents in patch-clamp experiments, as follows. As noted above, VU573 inhibits the known CCD Kir channels with the rank-order potency: Kir2.3 = Kir7.1 ($IC_{50} \sim 1$ μM) > ROMK ($IC_{50} \sim 15$ μM) ≫

Kir4.1 ($IC_{50} > 30$ μM). In the presence of the ROMK inhibitor VU591, which should preserve currents carried by Kir2.3, Kir4.1, and Kir7.1, application of VU573 will preferentially inhibit Kir2.3 and Kir7.1. The relative contributions of Kir2.3 (and Kir7.1) to the VU573-inhibitable component could be separated using R70 or R5C (Table 27.2), the VU573 analog that inhibits Kir2.3, but has no apparent activity toward Kir7.1 (discussed above; Raphemot et al. 2011). In principle, a response to R5C in the presence of VU591 should provide pharmacological evidence of functional Kir2.3 channels in the CCD.

Another tool compound that could be used to identify functional Kir2.3 channels in the distal nephron is pregnenolone sulfate (PREGS; Table 27.2), a neurosteroid that also acutely potentiates currents through Kir2.3 at micromolar concentrations (Kobayashi et al. 2009). Although PREGS has no apparent effect on ROMK, its activity toward Kir4.1, Kir4.1/5.1, and Kir7.1 has not yet been reported and should be evaluated before utilizing in the nephron to study Kir2.3.

27.6 Conclusions and Future Perspectives

Since ROMK was cloned more than 20 years ago, the rate of discoveries implicating Kir channels in fundamentally important physiological processes in epithelial and other tissues has far outpaced the development of sharp pharmacological tools for exploring the biology and potential druggability of inward rectifiers. With the recent development of small-molecule modulators of Kir channels expressed in various epithelial tissues, the pharmacological toolkit for dissecting the relative contributions of Kir channels to epithelial cell physiology is expanding. However, while these success stories demonstrate proof-of-concept that potent and highly selective small-molecule modulators can be developed using conventional drug discovery approaches, there is still a lot of work to be done. Notably, for studies of epithelial Kir channel physiology, the development of safe, efficacious, nanomolar-affinity inhibitors of homomeric Kir4.1, heteromeric Kir4.1/5.1, and Kir7.1 will provide critically needed tools for examining the therapeutic potential of these channels for the treatment of hypertension and edema. Furthermore, and importantly, the development of potent and selective inhibitors of ROMK has created unprecedented opportunities for finally assessing whether ROMK represents a viable therapeutic target for disorders of excessive fluid balance.

References

Bailey MA, Cantone A, Yan Q, MacGregor GG, Leng Q, Amorim JB, Wang T, Hebert SC, Giebisch G, Malnic G (2006) Maxi-K channels contribute to urinary potassium excretion in the ROMK-deficient mouse model of Type II Bartter's syndrome and in adaptation to a high-K diet. Kidney Int 70(1):51–59

Bartter FC, Pronove P, Gill JR Jr, Maccardle RC (1962) Hyperplasia of the juxtaglomerular complex with hyperaldosteronism and hypokalemic alkalosis. A new syndrome. Am J Med 33:811–828

Bhave G, Chauder BA, Liu W, Dawson ES, Kadakia R, Nguyen TT, Lewis LM, Meiler J, Weaver CD, Satlin LM, Lindsley CW, Denton JS (2011) Development of a selective small-molecule inhibitor of Kir1.1, the renal outer medullary potassium channel. Mol Pharmacol 79(1):42–50

Bockenhauer D, Feather S et al (2009) Epilepsy, ataxia, sensorineural deafness, tubulopathy, and KCNJ10 mutations. N Engl J Med 360(19):1960–1970

Boim MA, Ho K, Shuck ME, Bienkowski MJ, Block JH, Slightom JL, Yang Y, Brenner BM, Hebert SC (1995) ROMK inwardly rectifying ATP-sensitive K^+ channel. II. Cloning and distribution of alternative forms. Am J Physiol 268(6 Pt 2):F1132–F1140

Bond CT, Pessia M, Xia XM, Lagrutta A, Kavanaugh MP, Adelman JP (1994) Cloning and expression of a family of inward rectifier potassium channels. Receptors Channels 2(3): 183–191

Clark MA, Humphrey SJ, Smith MP, Ludens JH (1993) Unique natriuretic properties of the ATP-sensitive K^+-channel blocker glyburide in conscious rats. J Pharmacol Exp Ther 265(2):933–937

Derst C, Hirsch JR, Preisig-Muller R, Wischmeyer E, Karschin A, Doring F, Thomzig A, Veh RW, Schlatter E, Kummer W, Daut J (2001) Cellular localization of the potassium channel Kir7.1 in guinea pig and human kidney. Kidney Int 59(6):2197–2205

Doring F, Derst C, Wischmeyer E, Karschin C, Schneggenburger R, Daut J, Karschin A (1998) The epithelial inward rectifier channel Kir7.1 displays unusual K^+ permeation properties. J Neurosci 18(21):8625–8636

Fakler B, Brandle U, Glowatzki E, Weidemann S, Zenner HP, Ruppersberg JP (1995) Strong voltage-dependent inward rectification of inward rectifier K^+ channels is caused by intracellular spermine. Cell 80(1):149–154

Ficker E, Taglialatela M, Wible BA, Henley CM, Brown AM (1994) Spermine and spermidine as gating molecules for inward rectifier K^+ channels. Science 266(5187):1068–1072

Garcia ML, Priest BT et al (2013) Pharmacologic inhibition of the renal outer medullary potassium channel causes diuresis and natriuresis in the absence of kaliuresis. J Pharmacol Exp Ther 348(1):153–164

Fujita A, Horio Y, Higashi K, Mouri T, Hata F, Takeguchi N, Kurachi Y (2002) Specific localization of an inwardly rectifying K^+ channel, Kir4.1, at the apical membrane of rat gastric parietal cells; its possible involvement in K^+ recycling for the H^+-K^+-pump. J Physiol 540(Pt 1):85–92

Furutani K, Ohno Y, Inanobe A, Hibino H, Kurachi Y (2009) Mutational and in silico analyses for antidepressant block of astroglial inward-rectifier Kir4.1 channel. Mol Pharmacol 75(6): 1287–1295

Giebisch G (1995) Renal potassium channels: an overview. Kidney Int 48:1004–1009

Greger R, Schlatter E (1981) Presence of luminal K^+, a prerequisite for active NaCl transport in the cortical thick ascending limb of Henle's loop of rabbit kidney. Pflugers Arch 392(1):92–94

Hebert SC, Desir G, Giebisch G, Wang W (2005) Molecular diversity and regulation of renal potassium channels. Physiol Rev 85(1):319–371

Hejtmancik JF, Jiao X, Li A, Sergeev YV, Ding X, Sharma AK, Chan CC, Medina I, Edwards AO (2008) Mutations in KCNJ13 cause autosomal-dominant snowflake vitreoretinal degeneration. Am J Hum Genet 82(1):174–180

Hibino H, Horio Y, Inanobe A, Doi K, Ito M, Yamada M, Gotow T, Uchiyama Y, Kawamura M, Kubo T, Kurachi Y (1997) An ATP-dependent inwardly rectifying potassium channel, K_{AB}-2 (Kir4. 1), in cochlear stria vascularis of inner ear: its specific subcellular localization and correlation with the formation of endocochlear potential. J Neurosci 17(12):4711–4721

Hibino H, Inanobe A, Furutani K, Murakami S, Findlay I, Kurachi Y (2010) Inwardly rectifying potassium channels: their structure, function, and physiological roles. Physiol Rev 90(1): 291–366

Hille B (1992) Ionic channels of excitable membranes. Sunderland Associates, Sunderland, MA

Ho K, Nichols CG, Lederer WJ, Lytton J, Vassilev PM, Kanazirska MV, Hebert SC (1993) Cloning and expression of an inwardly rectifying ATP-regulated potassium channel. Nature 362(6415):31–38

Ito M, Inanobe A, Horio Y, Hibino H, Isomoto S, Ito H, Mori K, Tonosaki A, Tomoike H, Kurachi Y (1996) Immunolocalization of an inwardly rectifying K^+ channel, K_{AB}-2 (Kir4.1), in the basolateral membrane of renal distal tubular epithelia. FEBS Lett 388(1):11–15

Ji W, Foo JN, O'Roak BJ, Zhao H, Larson MG, Simon DB, Newton-Cheh C, State MW, Levy D, Lifton RP (2008) Rare independent mutations in renal salt handling genes contribute to blood pressure variation. Nat Genet 40(5):592–599

Kamikawa A, Ishikawa T (2014) Functional expression of a Kir2.1-like inwardly rectifying potassium channel in mouse mammary secretory cells. Am J Physiol Cell Physiol 306(3): C230–C240

Kobayashi T, Washiyama K, Ikeda K (2004) Modulators of G protein-activated inwardly rectifying K^+ channels: potentially therapeutic agents for addictive drug users. Ann N Y Acad Sci 1025:590–594

Kobayashi T, Washiyama K, Ikeda K (2009) Pregnenolone sulfate potentiates the inwardly rectifying K channel Kir2.3. PLoS One 4(7):e6311

Koeppen BM, Biagi BA, Giebisch GH (1983) Intracellular microelectrode characterization of the rabbit cortical collecting duct. Am J Physiol 244(1):F35–F47

Krapivinsky G, Medina I, Eng L, Krapivinsky L, Yang Y, Clapham DE (1998) A novel inward rectifier K^+ channel with unique pore properties. Neuron 20(5):995–1005

Kurata HT, Cheng WW, Arrabit C, Slesinger PA, Nichols CG (2007) The role of the cytoplasmic pore in inward rectification of Kir2.1 channels. J Gen Physiol 130(2):145–155

Kusaka S, Inanobe A, Fujita A, Makino Y, Tanemoto M, Matsushita K, Tano Y, Kurachi Y (2001) Functional Kir7.1 channels localized at the root of apical processes in rat retinal pigment epithelium. J Physiol 531(Pt 1):27–36

Lachheb S, Cluzeaud F, Bens M, Genete M, Hibino H, Lourdel S, Kurachi Y, Vandewalle A, Teulon J, Paulais M (2008) Kir4.1/Kir5.1 channel forms the major K^+ channel in the basolateral membrane of mouse renal collecting duct principal cells. Am J Physiol Renal Physiol 294(6):F1398–F1407

Le Maout S, Welling PA, Brejon M, Olsen O, Merot J (2001) Basolateral membrane expression of a K^+ channel, Kir 2.3, is directed by a cytoplasmic COOH-terminal domain. Proc Natl Acad Sci U S A 98(18):10475–10480

Leroy C, Dagenais A, Berthiaume Y, Brochiero E (2004) Molecular identity and function in transepithelial transport of K(ATP) channels in alveolar epithelial cells. Am J Physiol Lung Cell Mol Physiol 286(5):L1027–L1037

Lewis LM, Bhave G, Chauder BA, Banerjee S, Lornsen KA, Redha R, Fallen K, Lindsley CW, Weaver CD, Denton JS (2009) High-throughput screening reveals a small-molecule inhibitor of the renal outer medullary potassium channel and Kir7.1. Mol Pharmacol 76(5):1094–1103

Liu W, Morimoto T, Woda C, Kleyman TR, Satlin LM (2007) Ca^{2+} dependence of flow-stimulated K secretion in the mammalian cortical collecting duct. Am J Physiol Renal Physiol 293(1): F227–F235

Lopatin AN, Makhina EN, Nichols CG (1995) The mechanism of inward rectification of potassium channels: "Long-pore plugging" by cytoplasmic polyamines. J Gen Physiol 106(5): 923–955

Lourdel S, Paulais M, Cluzeaud F, Bens M, Tanemoto M, Kurachi Y, Vandewalle A, Teulon J (2002) An inward rectifier K^+ channel at the basolateral membrane of the mouse distal convoluted tubule: similarities with Kir4-Kir5.1 heteromeric channels. J Physiol 538(Pt 2): 391–404

Lu Z, MacKinnon R (1994) Electrostatic tuning of Mg^{2+} affinity in an inward-rectifier K^+ channel. Nature 371(6494):243–246

Lu M, Wang T, Yan Q, Yang X, Dong K, Knepper MA, Wang W, Giebisch G, Shull GE, Hebert SC (2002) Absence of small conductance K^+ channel (SK) activity in apical membranes of thick ascending limb and cortical collecting duct in ROMK (Bartter's) knockout mice. J Biol Chem 277(40):37881–37887

Matsuda T, Ito M, Ishimaru S, Tsuruoka N, Saito T, Iida-Tanaka N, Hashimoto N, Yamashita T, Tsuruzoe N, Tanaka H, Shigenobu K (2006) Blockade by NIP-142, an antiarrhythmic agent, of carbachol-induced atrial action potential shortening and GIRK1/4 channel. J Pharmacol Sci 101(4):303–310

McCloskey C, Rada C, Bailey E, McCavera S, Van den Berg H, Atia J, Rand DA, Shmygol A, Chan Y-W, Quenby S, Brosens JJ, Vatish M, Zhang J, Denton JS, Taggart MJ, Kettleborough C, Tickle D, Jerman J, Wright P, Dale T, Kanumilli S, Trezise DJ, Thornton S, Brown P, Catalano R, Lin N, England SK, Blanks AM (2014) The inwardly rectifying K^+ channel Kir7.1 controls uterine excitability throughout pregnancy. EMBO Mol Med 6(9):1161–1174

McNicholas CM, Guggino WB, Schwiebert EM, Hebert SC, Giebisch G, Egan ME (1996) Sensitivity of a renal K^+ channel (ROMK2) to the inhibitory sulfonylurea compound glibenclamide is enhanced by coexpression with the ATP-binding cassette transporter cystic fibrosis transmembrane regulator. Proc Natl Acad Sci U S A 93(15):8083–8088

Millar ID, Taylor HC, Cooper GJ, Kibble JD, Robson L (2006) A Kir2.3-like K^+ conductance in mouse cortical collecting duct principal cells. J Membr Biol 211(3):173–184

Monaghan AS, Baines DL, Kemp PJ, Olver RE (1997) Inwardly rectifying K^+ currents of alveolar type II cells isolated from fetal guinea-pig lung: regulation by G protein- and Mg^{2+}-dependent pathways. Pflugers Arch 433(3):294–303

Nakamura N, Suzuki Y, Sakuta H, Ookata K, Kawahara K, Hirose S (1999) Inwardly rectifying K+ channel Kir7.1 is highly expressed in thyroid follicular cells, intestinal epithelial cells and choroid plexus epithelial cells: implication for a functional coupling with Na+, K+-ATPase. Biochem J 342(Pt 2):329–336

Nichols CG, Ho K, Hebert S (1994) Mg^{2+}-dependent inward rectification of ROMK1 potassium channels expressed in Xenopus oocytes. J Physiol 476(3):399–409

Nishida M, MacKinnon R (2002) Structural basis of inward rectification: cytoplasmic pore of the G protein-gated inward rectifier GIRK1 at 1.8 A resolution. Cell 111(7):957–965

Ohno Y, Hibino H, Lossin C, Inanobe A, Kurachi Y (2007) Inhibition of astroglial Kir4.1 channels by selective serotonin reuptake inhibitors. Brain Res 1178:44–51

Ookata K, Tojo A, Suzuki Y, Nakamura N, Kimura K, Wilcox CS, Hirose S (2000) Localization of inward rectifier potassium channel Kir7.1 in the basolateral membrane of distal nephron and collecting duct. J Am Soc Nephrol 11(11):1987–1994

Partiseti M, Collura V, Agnel M, Culouscou JM, Graham D (1998) Cloning and characterization of a novel human inwardly rectifying potassium channel predominantly expressed in small intestine. FEBS Lett 434(1–2):171–176

Pattnaik BR, Tokarz S, Asuma MP, Schroeder T, Sharma A, Mitchell JC, Edwards AO, Pillers DA (2013) Snowflake vitreoretinal degeneration (SVD) mutation R162W provides new insights into Kir7.1 ion channel structure and function. PLoS One 8(8):e71744

Paulais M, Bloch-Faure M, Picard N, Jacques T, Ramakrishnan SK, Keck M, Sohet F, Eladari D, Houillier P, Lourdel S, Teulon J, Tucker SJ (2011) Renal phenotype in mice lacking the Kir5.1 (Kcnj16) K^+ channel subunit contrasts with that observed in SeSAME/EAST syndrome. Proc Natl Acad Sci U S A 108(25):10361–10366

Pegan S, Arrabit C, Zhou W, Kwiatkowski W, Collins A, Slesinger PA, Choe S (2005) Cytoplasmic domain structures of Kir2.1 and Kir3.1 show sites for modulating gating and rectification. Nat Neurosci 8(3):279–287

Pettersen EF, Goddard TD, Huang CC, Couch GS, Greenblatt DM, Meng EC, Ferrin TE (2004) UCSF Chimera-a visualization system for exploratory research and analysis. J Comput Chem 25(13):1605–1612

Raphemot R, Lonergan DF, Nguyen TT, Utley T, Lewis LM, Kadakia R, Weaver CD, Gogliotti R, Hopkins C, Lindsley CW, Denton JS (2011) Discovery, characterization, and structure-activity relationships of an inhibitor of inward rectifier potassium (Kir) channels with preference for Kir2.3, Kir3.x, and Kir7.1. Front Pharmacol 2:75

Raphemot R, Kadakia RJ, Olsen ML, Banerjee S, Days E, Smith SS, Weaver CD, Denton JS (2013) Development and validation of fluorescence-based and automated patch clamp-based functional assays for the inward rectifier potassium channel kir4.1. Assay Drug Dev Technol 11(9–10):532–543

Rozengurt N, Lopez I, Chiu CS, Kofuji P, Lester HA, Neusch C (2003) Time course of inner ear degeneration and deafness in mice lacking the Kir4.1 potassium channel subunit. Hear Res 177(1–2):71–80

Ruknudin A, Schulze DH, Sullivan SK, Lederer WJ, Welling PA (1998) Novel subunit composition of a renal epithelial KATP channel. J Biol Chem 273(23):14165–14171

Sakuma T, Takahashi K, Ohya N, Nakada T, Matthay MA (1998) Effects of ATP-sensitive potassium channel opener on potassium transport and alveolar fluid clearance in the resected human lung. Pharmacol Toxicol 83(1):16–22

Scholl UI, Choi M, Liu T, Ramaekers VT, Hausler MG, Grimmer J, Tobe SW, Farhi A, Nelson-Williams C, Lifton RP (2009) Seizures, sensorineural deafness, ataxia, mental retardation, and electrolyte imbalance (SeSAME syndrome) caused by mutations in KCNJ10. Proc Natl Acad Sci U S A 106(14):5842–5847

Shimura M, Yuan Y, Chang JT, Zhang S, Campochiaro PA, Zack DJ, Hughes BA (2001) Expression and permeation properties of the K^+ channel Kir7.1 in the retinal pigment epithelium. J Physiol 531(Pt 2):329–346

Simon DB, Karet FE, Rodriguez-Soriano J, Hamdan JH, DiPietro A, Trachtman H, Sanjad SA, Lifton RP (1996) Genetic heterogeneity of Bartter's syndrome revealed by mutations in the K^+ channel, ROMK. Nat Genet 14(2):152–156

Strauss O (2005) The retinal pigment epithelium in visual function. Physiol Rev 85(3):845–881

Su S, Ohno Y, Lossin C, Hibino H, Inanobe A, Kurachi Y (2007) Inhibition of astroglial inwardly rectifying Kir4.1 channels by a tricyclic antidepressant, nortriptyline. J Pharmacol Exp Ther 320(2):573–580

Takumi T, Ishii T, Horio Y, Morishige K, Takahashi N, Yamada M, Yamashita T, Kiyama H, Sohmiya K, Nakanishi S et al (1995) A novel ATP-dependent inward rectifier potassium channel expressed predominantly in glial cells. J Biol Chem 270(27):16339–16346

Tanemoto M, Vanoye CG, Dong K, Welch R, Abe T, Hebert SC, Xu JZ (2000) Rat homolog of sulfonylurea receptor 2B determines glibenclamide sensitivity of ROMK2 in Xenopus laevis oocyte. Am J Physiol Renal Physiol 278(4):F659–F666

Tang H, Walsh SP, Yan Y, de Jesus RK, Shahripour A, Teumelsan N, Zhu Y, Ha S, Owens KA, Thomas-Fowlkes BS, Felix JP, Liu J, Kohler M, Priest BT, Bailey T, Brochu R, Alonso-Galicia M, Kaczorowski GJ, Roy S, Yang L, Mills SG, Garcia ML, Pasternak A (2012) Discovery of selective small molecule ROMK inhibitors as potential new mechanism diuretics. ACS Med Chem Lett 3(5):367–372

Tao X, Avalos JL, Chen J, MacKinnon R (2009) Crystal structure of the eukaryotic strong inward-rectifier K^+ channel Kir2.2 at 3.1 A resolution. Science 326(5960):1668–1674

Tucker SJ, Imbrici P et al (2000) pH dependence of the inwardly rectifying potassium channel, Kir5.1, and localization in renal tubular epithelia. J Biol Chem 275(22):16404–16407

Wang WH, White S, Geibel J, Giebisch G (1990) A potassium channel in the apical membrane of rabbit thick ascending limb of Henle's loop. Am J Physiol 258(2 Pt 2):F244–F253

Wang T, Wang WH, Klein-Robbenhaar G, Giebisch G (1995) Effects of glyburide on renal tubule transport and potassium-channel activity. Ren Physiol Biochem 18(4):169–182

Wang W, Hebert SC, Giebisch G (1997) Renal K^+ channels: structure and function. Annu Rev Physiol 59:413–436

Welling PA (1997) Primary structure and functional expression of a cortical collecting duct Kir channel. Am J Physiol 273(5 Pt 2):F825–F836

Welling PA, Ho K (2009) A comprehensive guide to the ROMK potassium channel: form and function in health and disease. Am J Physiol Renal Physiol 297(4):F849–F863

Wible BA, Taglialatela M, Ficker E, Brown AM (1994) Gating of inwardly rectifying K^+ channels localized to a single negatively charged residue. Nature 371(6494):246–249

Yang D, Zhang X, Hughes BA (2008) Expression of inwardly rectifying potassium channel subunits in native human retinal pigment epithelium. Exp Eye Res 87(3):176–183

Yasuda K, Shimura M, Nakazawa T, Sato H, Tomita H, Sugano E, Tamai M (2003) Expression and functional properties of unique inward rectifier K^+ channel Kir7.1 in the porcine iris and retinal pigment epithelium. Curr Eye Res 27(5):279–287

Yoo D, Flagg TP, Olsen O, Raghuram V, Foskett JK, Welling PA (2004) Assembly and trafficking of a multiprotein ROMK (Kir 1.1) channel complex by PDZ interactions. J Biol Chem 279(8):6863–6873

Zaika O, Mamenko M, Palygin O, Boukelmoune N, Staruschenko A, Pochynyuk O (2013) Direct inhibition of basolateral Kir4.1/5.1 and Kir4.1 channels in the cortical collecting duct by dopamine. Am J Physiol Renal Physiol 305(9):F1277–F1287

Zhou H, Tate SS, Palmer LG (1994) Primary structure and functional properties of an epithelial K channel. Am J Physiol 266(3 Pt 1):C809–C824

Part IV
Diseases in Epithelia

Chapter 28
Diseases of the Pancreas Involving Channels/Transporters

David C. Whitcomb

Abstract The pancreas is a digestive gland that is functionally divided into the exocrine pancreas producing digestive enzymes and the endocrine pancreas producing insulin and other metabolic hormones. Diseases of the exocrine pancreas are primarily inflammatory, such as acute pancreatitis and chronic pancreatitis. These diseases are symptom-defined clinical syndromes reflecting the final pathologic pathway of multiple complex gene × gene and gene × environment interactions. Diseases of the pancreas originating with channel–transporter dysfunction are dominated by dysfunction in the cystic fibrosis transmembrane conductance regulator (CFTR). Risk for chronic pancreatitis is also linked to claudin-2, a tight junction protein that forms a cation–water channel between cells. The dysfunction, however, is unrelated to channel function. New genetic discoveries of a unique class of CFTR variants that specifically alter CFTR bicarbonate permeability and conductance are discussed, along with their implications for human disease.

Keywords Cystic fibrosis transmembrane conductance regulator, CFTR • Claudin-2, CLDN2 • Cationic trypsinogen, PRSS1 • Pancreatitis • Pancreatic duct • Bicarbonate • Cystic fibrosis • CFTR-related disorders

28.1 Introduction

The pancreas is a digestive gland located in the retroperitoneum of the abdomen. The pancreas has two functional components: the endocrine pancreas secreting hormones into the bloodstream and the exocrine pancreas secreting juices into the proximal intestine just downstream of the stomach. The endocrine pancreas consists

D.C. Whitcomb (✉)
Department of Medicine, Cell Biology and Molecular Physiology, University of Pittsburgh and UPMC, Pittsburgh, PA 15213, USA

Department of Cell Biology and Molecular Physiology, University of Pittsburgh and UPMC, Pittsburgh, PA 15213, USA

Department of Human Genetics, University of Pittsburgh and UPMC, Pittsburgh, PA 15213, USA
e-mail: whitcomb@pitt.edu

K.L. Hamilton, D.C. Devor (eds.), *Ion Channels and Transporters of Epithelia in Health and Disease*, Physiology in Health and Disease,
DOI 10.1007/978-1-4939-3366-2_28

of clusters of islet cells that form about a million islets of Langerhans, suspended in a sea of exocrine pancreas cells. Four major and one minor type of islet cells each produce different hormones including alpha cells (glucagon), beta cells (insulin), delta cells (somatostatin), and PP/gamma cells (pancreatic polypeptide), as well as scarce epsilon cells (ghrelin). These cells release their respective hormones into the bloodstream in a highly regulated way to regulate intermediate metabolism. The exocrine pancreas consists of the acinar cells, which make pancreatic digestive pro-enzymes (zymogens), and duct cells that produce bicarbonate-rich fluid to alkalinize and flush the zymogens into the duodenum through the pancreatic duct system. The exocrine pancreas responds to neurohormonal stimuli through multilevel, integrated, and well-regulated control systems (Noble and Liddle 2005; Deng and Whitcomb 1998; Mussa and Verberne 2013).

Different types of medical and surgical specialists manage human diseases of the endocrine and exocrine pancreas, even though there is significant interaction between the endocrine and exocrine pancreas. This chapter will focus on the exocrine pancreas and diseases associated with the acinar and duct cell pathobiology linked to ion channels.

28.2 Human Pancreatic Diseases

The majority of the pancreatic mass is composed of acinar cells (~80 %) with only a small percent of duct cells (~5 %). Compared to other organs, the exocrine pancreas is relatively simple. The acinar cells are polarized epithelial cells that serve one primary function—to manufacture precursors to pancreatic digestive enzymes (zymogens), package them into vesicles, and secrete them into the pancreatic duct system on demand by exocytosis. The duct cells form an arborizing network of ducts linking the acinar cells to the intestine and secrete fluid as a result of ion transport through specific channels and transporters. The microanatomy of the exocrine pancreas is the acinus, which is a cluster of acinar cells joined at the center and connected by tight junction proteins that allow the apical surface to form a vestibule where secreted zymogens accumulate (Fig. 28.1). The vestibule is connected to the duct system forming a cul-de-sac. Within the pocket are duct-like cells called centroacinar cells. These cells connect with duct cells to form the pancreatic ductal system.

The primary diseases of the exocrine pancreas from a medical perspective are acquired inflammatory disorders (pancreatitis) and neoplastic disorders (e.g., pancreatic ductal adenocarcinoma). The inflammatory diseases are broadly divided into acute pancreatitis and chronic pancreatitis, which are complex syndromes defined by clinical criteria. Acute pancreatitis is defined as sudden inflammation of the pancreas with two of three criteria: (1) typical pain, (2) elevated concentrations of pancreatic digestive enzymes such as amylase or lipase in the serum (greater than three times normal values), and/or (3) evidence of pancreatic inflammation on abdominal imaging. Chronic pancreatitis is a complex syndrome of pancreatic

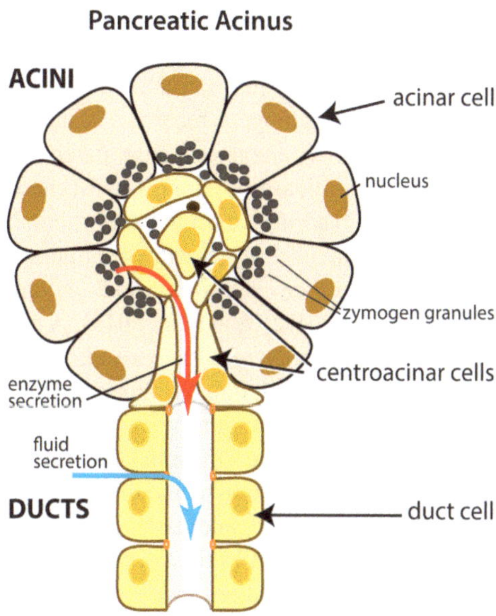

Fig. 28.1 Microanatomy of the exocrine pancreas. The functional unit of the exocrine pancreas is the pancreatic acinus, composed of acinar cells and duct cells organized to form a cul-de-sac linked to an upstream duct. The acinar cells are polarized epithelial cells with the basolateral membrane away from the duct and the apical membrane facing the lumen of the cul-de-sac. The acinar cells are protein factories, rich in endoplasmic reticulum and mitochondria to produce digestive enzyme zymogens. The zymogens are concentrated in zymogen granules at the apical membrane (*black circles*). With stimulation of the acinar cells on the basolateral side, there is calcium-mediated excitation–secretion coupling with zymogen exocytosis into the lumen (*red arrow*). The duct cells begin as centroacinar cells and then form a duct that links with ducts from other acini. Duct cells transport ions (chloride and/or bicarbonate) from the basolateral membrane to the apical membrane where high ionic concentrations are generated. Water is drawn into the lumen by osmotic forces (*blue arrow*) resulting in hydrostatic pressure which forces the pancreatic juice (ionic solution and proteins) through the ducts and into the duodenum

injury and inflammation resulting in recurrent and persistent pathologic features which include pathologic fibrosis, atrophy, calcifications, pain, diabetes mellitus (from islet destruction), and increased risk of pancreatic cancer (Sarles et al. 1989; Sarner and Cotton 1984; Witt et al. 2007). Thus, these pancreatic diseases have been defined by symptoms rather than pathogenic mechanisms. Indeed, as late as 1995, chronic pancreatitis was considered an "enigmatic process of uncertain pathogenesis, unpredictable clinical course, and unclear treatment" (Steer et al. 1995). However, studies over the subsequent 20 years demonstrate that pancreatitis is a complex set of gene–gene and gene–environment interactions that predispose the pancreas to injury, inflammation, and its complications, including cancer (Chen and Ferec 2012; Whitcomb 2013).

Recent studies demonstrate that inflammatory disorders are primarily linked to risk factors that cause unregulated trypsin activity within the pancreas (Chen and Ferec 2012; Whitcomb 2013). Trypsin is the master catalysts of zymogen activation. Activation of inactive trypsinogen to active trypsin results in an activation cascade with activation of additional trypsinogen as well as other zymogens to become active digestive enzymes. When activation occurs inside the pancreas rather than being delayed until the zymogens reach the lumen of the small intestine, the prematurely activated digestive enzymes cause parenchymal digestion with pancreatic injury signaling [e.g., release of damage-associated molecular patterns, DAMPs (Hoque et al. 2012)], and stimulate acute inflammation. Other mechanisms of immune system activation include acinar stress signaling from the unfolded protein response and likely other stressors, and autoimmune pancreatitis (Rosendahl et al. 2008; Beer et al. 2012; Witt et al. 2013). A number of rare genetic and developmental disorders of the pancreas have also been identified.

Unraveling the complex genetics of human exocrine pancreatic diseases has progressed rapidly because the singular purpose of the exocrine pancreas is to manufacture, store, and secrete digestive enzymes. The mechanisms of zymogen synthesis, packaging, and secretion are known, and the mechanism of duct cell fluid secretion is also known. This allows for candidate gene approaches to understand human genetics, in contrast to "unbiased" statistical approaches, such as genome-wide association studies, to identify disease-associated or clearly pathogenic genetic variants. To date, about 20 genetic variants have been linked to pancreatitis, but only two are an ion channel or transporter—the cystic fibrosis transmembrane conductance transporter gene (*CFTR*) and claudin-2 gene (*CLDN2*).

The common pathway linking genetic defects to human pancreatic inflammatory diseases is in genes linked to regulation of and protection from trypsin within the pancreas. The strongest pathogenic genetic variant causes autosomal hereditary pancreatitis. The common genetic defect was *gain-of-function* mutations in the cationic trypsinogen gene (*PRSS1*) (Whitcomb et al. 1996; Gorry et al. 1997). Premature activation of mutant trypsinogen to trypsin results in a persistently active protease that cannot be degraded by the calcium-regulated trypsin auto-degradation process (Fig. 28.2).

The molecular self-destruction protective mechanism regulating the trypsin molecule has two components: hydrolysis of the autolysis site at R122 by a second trypsin molecule and hydrolysis of the trypsin molecule at Leu81–Glu82 peptide bond by another trypsin-activated enzyme, serine protease, chymotrypsin C (CTRC) (Szabo and Sahin-Toth 2012). Loss-of-function mutations in *CTRC* also increase the risk of chronic pancreatitis (Beer et al. 2012; Rosendahl et al. 2008; LaRusch et al. 2015), although the risk is much lower that with *PRSS1* mutations and is usually pathogenic only in the context of additional risk factors. The trypsin autodigestion protective mechanism only works in solutions with low levels of calcium, and therefore exists within the acinar cell under normal conditions, but not in high calcium concentrations as exist in the pancreatic duct due to the action of calcium pumps transferring calcium from the acinar cell to the duct lumen. Fortunately, there are several other protective mechanisms if the trypsin autodigestion protective mechanism fails (Whitcomb 2013).

The pancreatic acinar cells respond to acute inflammation with upregulation of a trypsin inhibitor, serine protease inhibitor Kazal type 1 gene (*SPINK1*), also known as pancreatic secretory trypsin inhibitor (PSTI) (Pfutzer et al. 2000; Witt et al. 2000). SPINK1 is a specific inhibitor of the two major isoforms of trypsin,

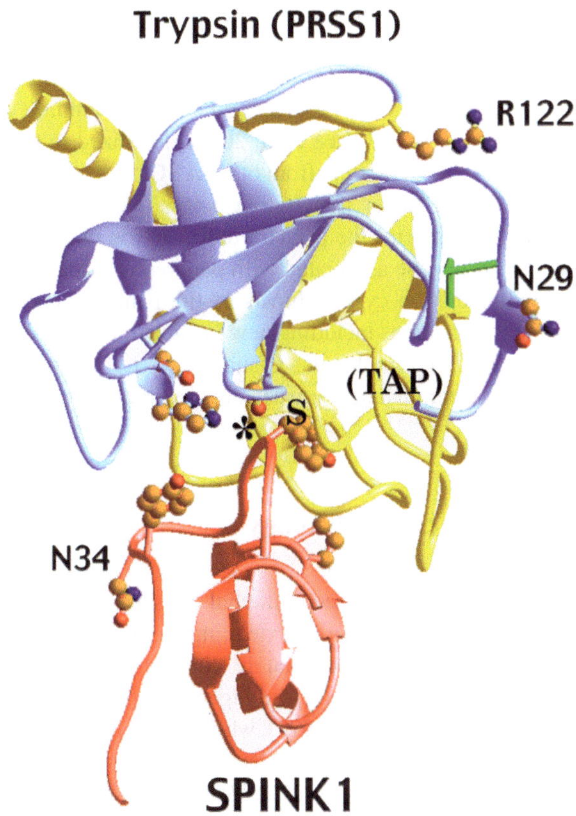

Fig. 28.2 Trypsin–SPINK1 complex. Ribbon diagram of trypsin (*yellow* and *blue*) and SPINK1 (*red*) showing the interaction of the two molecules. Trypsin is painted *yellow* and *blue* to illustrate the two globular domains attached by the connecting chain (*top*). Arginine 122 (numbered R117 using a protein numbering system) is the target of auto-degradation by another trypsin molecule when the side chain flexes away from the molecule. Mutations R122H and N29I are associated with altered activity and risk of pancreatitis. N34 is the site of a SPINK1 (pancreatic secretory trypsin inhibitor) variant (N34S) that is linked to risk of pancreatitis, although the functional variant is likely in a regulatory element since the amino acid substitution does not affect protein function. TAP indicates the location of the trypsinogen activation peptide, an eight-amino-acid sequence that is cleaved by trypsin or enterokinase to change inactive trypsinogen to active trypsin. *Asterisk* indicates active catalytic site of trypsin. S, indicates the specificity pocket of the serene protease that selects the amino acid side chain for cleavage. Note that this molecule is a derivative of trypsin and SPINK1 to mimic chymotrypsin binding to stabilize the protein complex for crystallography (Image courtesy of William Furry PhD)

PRSS1 (cationic trypsin) and PRSS2 (anionic trypsin). Pathogenic variants in *SPINK1* are loss-of-function or loss of expression types. *SPINK1* mutations are associated with recurrent acute pancreatitis and/or chronic pancreatitis linked to multiple other etiologies (Aoun et al. 2008). SPINK1 is synthesized in the acinar cell with PRSS1 and PRSS2 so that the utility of SPINK1 as a trypsin inhibitor is seen both in the acinar cell and in the duct system.

The duct cells are important for the production of a bicarbonate-rich juice that flushes the zymogens out of the pancreas and into the duodenum where zymogens are normally activated and digest nutrients as they leave the stomach. CFTR is critical to normal duct function. CFTR is a regulated ATP-binding cassette (ABC) transporter-type protein [ABCC7 (Lee et al. 2012)] localized to the apical plasma membrane of epithelial cells. It differs from other ABC transporters in that it acts as a regulated anion channel rather than a transporter (Bear et al. 1992). Loss of function of the CFTR represents the primary defect in ion channel/transport function that has been shown to be associated with pancreatic disease in humans.

This chapter will focus on cystic fibrosis of the pancreas and cystic fibrosis-related disorders of the pancreas including new insights into the role of bicarbonate secretion through the CFTR molecule. Claudin-2, a tight junction protein that functions as a cation–water channel between cells, will also be discussed.

28.3 Pancreatic Duct Cell Biology

The pancreatic duct cells are polarized epithelial cells that form an arborizing tree of cells with slightly different functions extending from every acinus to the intestine. The intercalated ducts are immediately connected to the acinar cells and centroacinar cells, and these cells have the highest level of CFTR expression by immunohistochemistry (Marino et al. 1991), while CFTR is expressed at high levels in the small, intercalated ducts and at lower levels in the interlobular ducts by in situ hybridization to CFTR mRNA (Strong et al. 1994).

The duct cells are linked by tight junctions that do not allow fluid or molecules to pass between the cells when the pancreas is at rest (Gunzel and Yu 2013). We believe, but it has not been proven, that there is an exchange of claudin-3 and/or claudin-4 for claudin-2 in the small ducts with stimulation so that sodium and water can pass between the cells and join bicarbonate to mediate sodium bicarbonate secretion (Gunzel and Yu 2013; Amasheh et al. 2002; Rahner et al. 2001; LaRusch and Whitcomb 2012). As the intercalated ducts from adjacent acini coalesce, they form interlobular ducts, then interlobar ducts, and finally the main pancreatic duct. Progressively larger ducts display changes in anatomic features as well as epithelial cell function.

28.4 Ion Transport Across the Pancreatic Ductal Epithelial Cell

Ion transport across lipid membranes is facilitated by ion channels and exchangers and is dependent on the electrochemical gradient across the membrane. CFTR, as an anion-specific channel, specifically facilitates conductance of chloride (Cl^-), bicarbonate (HCO_3^-), and other anions (Bear et al. 1992). The direction of ion

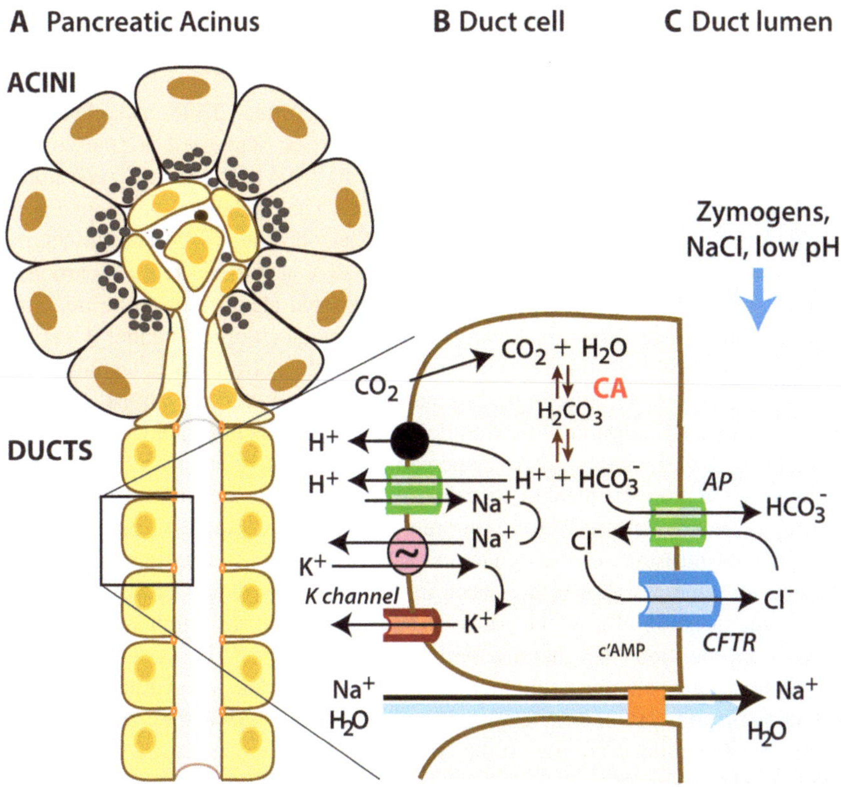

Fig. 28.3 The Case–Argent pancreatic duct model. One of the first, and most influential, hypothetical models of pancreatic duct physiology after the discovery of CFTR and before the sodium bicarbonate cotransporter (NBC). CFTR was believed to be a selective chloride (Cl^-) channel (no significant HCO_3^- conductance), so bicarbonate (HCO_3^-) secretion was linked to a chloride–bicarbonate antiporter (AP). The role of CFTR was believed to actively conduct Cl^- out of the duct cell keeping intracellular chloride levels low enough to all Cl^-/HCO_3^- exchanger to occur—a hypothesis that influenced subsequent models. The HCO_3^- was believed to be generated within the duct cell through the effects of carbonic anhydrase (CA)

movement across the membrane is determined by the concentration of the ion on each side of the membrane and the electrical potential. From a physiologic perspective, ions can only be secreted or absorbed across epithelial cell barrier if there are channels or transporters capable of moving a specific ion across both the apical (luminal) and basolateral (plasma) membranes. Interestingly, the human body uses CFTR to *secrete* Cl^- in the lung, *secrete* HCO_3^- into the pancreatic duct, and *absorb* Cl^- in the sweat gland. The differences are all related to cell organization, membrane potential, and electrochemical gradients.

Potential mechanisms used by the pancreas to secrete bicarbonate have been tested using multiple experimental, theoretical, and mathematical models. The first important conceptual model was developed by Case and Argent [see review (Case and Argent 1993)]. The model (Fig. 28.3) depicted entry of CO_2 into the duct cell from the blood by diffusion, transformation of $CO_2 + H_2O$ into H^+ and HCO_3^- within the duct cell (facilitated by carbonic anhydrase; CA), and then secretion of HCO_3^- into the duct lumen by an antiporter in exchange for Cl^-. The intracellular Cl^- levels were maintained at a low level by opening of the CFTR channel to allow Cl^- to exit by diffusion. The electrical properties suggested a transcellular $V_t = 5$ mV (duct negative), a basolateral potential of $V_{bl} = -35$ mV, and apical potential of $V_a = -35$ mV (intracellular negative). The model was developed as an adaptation of models in the urinary bladder (Stetson et al. 1985) and guinea pig gallbladder (Stewart et al. 1989), with the rationale that if there were a bicarbonate channel, the relative high permeability of the known channels to chloride, compared to bicarbonate, would make any anion channel a chloride channel (Case and Argent 1993). However, upon experimental and mathematical model testing, it was not capable of generating high luminal bicarbonate concentrations (Steward et al. 2005) (see discussion of pancreatic duct cell models in Chap. 6).

CFTR is clearly important to pancreatic duct bicarbonate secretion, but the channel is far more permeable to chloride than bicarbonate, indicating that the process of bicarbonate secretion is more complex than chloride secretion. Direct experimental measures of the dynamics of bicarbonate secretion in situ are challenging, so alternate approaches, such as mathematical modeling, were used to answer the question of how the pancreas secretes a juice with high sodium bicarbonate concentrations that increase with stimulation.

The first major attempt at mathematical simulation of a pancreatic duct cell at rest and with stimulation was developed by Sohma et al. (1996). This mathematical model, built in collaboration with the Argent group, tested the Case–Argent hypothetical model. Variables included intracellular $[Na^+]$, $[K^+]$, $[Cl^-]$, and $[HCO_3^-]$, membrane potential, and cell volume, plus extracellular fluids with or without a junctional pathway separating plasma from lumen. An equation for single diffusive pathway was included for Cl^- on the apical membrane (e.g., to include CFTR-related conductance), but it did not include conductance of HCO_3^- through CFTR. Dynamic simulation of fluid secretion indicated that it was not possible to generate high HCO_3^- concentrations in the duct lumen with the proposed conceptual model. Furthermore, HCO_3^- flux *decreased* as luminal HCO_3^- concentrations increased, pH was a problem, and the question of how opening of the CFTR channel resulted in brisk secretion of a sodium bicarbonate fluid, as seen in vivo, was not adequately addressed.

The discovery of the pancreatic sodium bicarbonate cotransporter (NBC_p) and the possibility of a HCO_3^- conductance pathway on the apical membrane resulted in another set of mathematical modeling experiment, published by Sohma et al. (2000). This elaborate model required a series of duct cells with different properties extending from the acinus toward the duct. Secretion began with Cl^- conductance through CFTR, and then the pancreatic juice was modified with Cl^-/HCO_3^- exchange, until a high HCO_3^- concentration was reached. However, simulation of fluid secretion with this model did not *generate* high luminal HCO_3^- concentrations and sustained, high volume fluid secretion. Furthermore, the HCO_3^- would likely fall, rather than rise with CFTR channel opening and fluid volume increasing during stimulated secretion (Steward et al. 2005) (see Chap. 6).

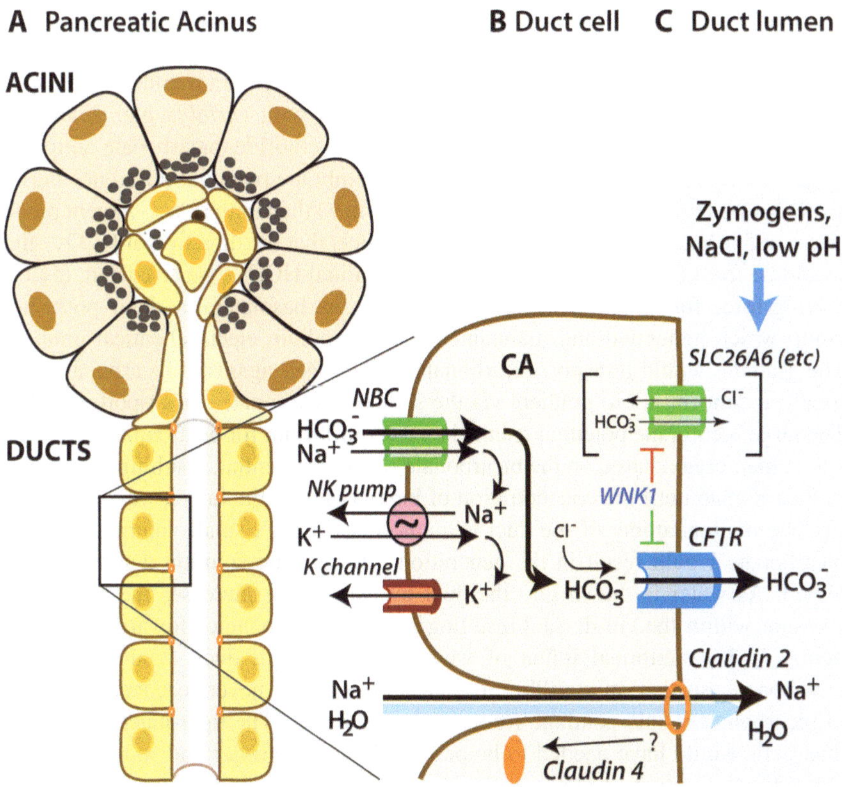

Fig. 28.4 Minimal model for a bicarbonate-secreting duct system. (**a**) Pancreatic acinus including the acinar cells and duct cells as a functional unit. (**b**) Proximal duct cell. (**c**) Luminal fluid beginning with acinar cell secretion of zymogens (pre-activated pancreatic digestive enzymes), a NaCl-rich fluid. Bicarbonate secretion is driven by a basolateral sodium–potassium pump generating a sodium gradient sufficient to bring bicarbonate into the cell via a sodium bicarbonate cotransporter (NBD). Bicarbonate exits the duct cell down its electrochemical gradient through CFTR. Bicarbonate flux though CFTR is much greater than chloride when (1) the basolateral membrane is impermeable to chloride so that equilibrates across the apical membrane, and (2) WNK1 acts on CFTR to increase bicarbonate permeability. The model predicts that sodium and water follow bicarbonate through a paracellular pathway facilitated by the insertion of claudin-2 and withdrawal of claudin-4 from the tight junction. With active bicarbonate secretion the chloride–bicarbonate antiporters such as SLC26A6 are inhibited, but it becomes important when CFTR closes and the cell returns to equilibrium. Carbonic anhydrase (CA) is important in protecting the duct cell from wide pH changes (based on Whitcomb and Ermentrout 2004)

An alternative model was proposed that was conceptually different on several levels (Whitcomb 1999). The model was tested with the Whitcomb–Ermentrout equation-based mathematical modeling and secretion simulations (Whitcomb and Ermentrout 2004) and published in 2004. The Whitcomb–Ermentrout model addressed multiple conceptual differences from the Case–Argent models that were believed to be important in pancreatic physiology (Fig. 28.4): (1) The microanatomy of the pancreas included the duct with a small volume so that ion secretion could change the ion concentration in the duct. (2) The basolateral membrane needed to be impermeable to chloride to prevent transepithelial transport of chloride. (3) The CFTR channel was modeled with *variable permeability of bicarbonate* compared to chloride, initially 1:5. A chloride–bicarbonate antiporter (e.g., SLC26A6) was also included. (4) The membrane potential difference during secretion was predicted to be higher (e.g., −60 mV) than in the Case–Argent model (e.g., −35 mV), since the Nernst equation predicts that the direction of HCO_3^- flux would be from the lumen *into* the cell when luminal HCO_3^- was elevated. (5) The driving force for secretion was predicted to be the basolateral sodium–potassium pump which generated and maintained a large sodium electrochemical gradient. The gradient would transport bicarbonate into the cell against a bicarbonate electrical (± concentration) gradient via the sodium bicarbonate cotransporter (NBC). The large membrane potential results in a disequilibrium for bicarbonate across the apical membrane, based on the bicarbonate reversal potential, resulting in an efflux of bicarbonate until the concentration of bicarbonate in the lumen was >150 mM. (6) The microanatomy of the duct is a cul-de-sac with a small volume. Efflux of bicarbonate would result in the generation a hypertonic solution within the duct lumen. As water is drawn into the duct by osmosis, it increases the hydrostatic pressure within the cul-de-sac, resulting in brisk flow of juice toward the duodenum. (7) The continual influx of water into the small ducts would dilute the bicarbonate, so that an equilibrium concentration was never reached, and efflux of bicarbonate would continue from the duct cells. (8) The tight junction between duct cells would have needed to be permeable to both water and sodium, but not chloride, to facilitate sodium entry into the duct lumen following an electrical gradient generated by bicarbonate secretion. Water would follow the sodium and bicarbonate through the paracellular pathway and/or via a transcellular pathway linked to aquaporins. (9) The roll of CFTR could be central to pancreatic bicarbonate secretion if it were permeable to both chloride and bicarbonate. Its central question is *not the relative permeability* of CFTR to chloride or bicarbonate but on the reversal potential (linked to the concentration of the anions on either side of the apical membrane) compared to the membrane potential during secretion, resulting in *anion flux*. If there were no pathway for chloride to cross the basolateral membrane, then with the opening of CFTR, the chloride concentrations would equilibrate across the luminal membrane based on membrane potential. In contrast, bicarbonate is continually pulled across the basolateral membrane into the duct cell via NBC, so that there is continued disequilibrium across the apical membrane and outward anion flux. Thus, the CFTR would conduct bicarbonate rather than chloride since bicarbonate, but not chloride, has an outward driving force.

To test this conceptual model, pancreatic duct secretion was simulated using the Whitcomb–Ermentrout model (Whitcomb and Ermentrout 2004), applying the parameters of Sohma (Sohma et al. 2000) with the exception of *CFTR* which was made permeable to bicarbonate in variable ratios to chloride. This simple model generated high concentrations of luminal bicarbonate (>140 mM) with high fluid output that was initiated by opening the CFTR channel and ended with closing the

CFTR channel. The model also demonstrated that CFTR acted as a Cl^- and/or HCO_3^- channel depending on the conductance pathways on the basolateral membrane.

The Whitcomb–Ermentrout model also suggested that when conditions were optimized for maximal luminal bicarbonate secretion, several predictions could be made: (1) The basolateral membrane of duct cells from species that secreted high concentrations of bicarbonate in pancreatic fluid would be nearly impermeable to chloride. (2) The apical chloride–bicarbonate antiporter was not necessary for bicarbonate secretion. In fact, the apical chloride–bicarbonate antiporter actually ran *backwards* (with low flux) from predicted direction of anion flux required by the Case–Argent model (bicarbonate entering the cell in exchange for chloride). The model suggested that excluding the antiporter was only important after the CFTR channel was closed, since it played a role in the equilibration of chloride within the cell to resting levels (Fig. 28.4, SLC26A6 in parentheses). (3) Bicarbonate secretion was dependent on conductance through CFTR. When bicarbonate conductance, but not chloride conductance, was minimized, bicarbonate conductance and pancreatic fluid secretion rates dropped proportionally and independent of the antiporters. Thus, the model predicted that mutations that alter CFTR bicarbonate secretion, but not chloride secretion, would increase the risk of pancreatitis, since duct cell function and zymogen flushing are critical for protection of the pancreas. (4) There needed to be a paracellular pathway between the duct cells that was restricted to cations (e.g., sodium) but impermeable to chloride and bicarbonate to cause directional flow of cations and reduce the transcellular electrical potential (Fig. 28.4). (5) The model predicted that chloride would be washed out of the pancreas, and chloride concentrations within the duct and the lumen would drop to zero unless there were a small chloride leak or chloride entering the duct from upstream acinar cells.

The other major prediction of the model is that if bicarbonate was secreted through CFTR, then the pancreatic model of bicarbonate secretion could be modified to focus on the proximal duct, rather than a series of duct cells that first secrete chloride proximally and then exchange chloride for bicarbonate distally (Sohma et al. 2000; Lee et al. 2012). The minimal model of Whitcomb and Ermentrout required testing, but over a decade would elapse before the critical information was available. The prediction that *CFTR* mutations selectively affecting bicarbonate would predispose to pancreatitis, and the unique features of claudin-2, found to be present in the small pancreatic ducts (Rahner et al. 2001; Lee et al. 2011; LaRusch and Whitcomb 2012), is discussed below.

28.5 Duct Function and Pancreatitis

Pancreatic duct physiology is largely regulated dependent on CFTR function. The details of the intracellular mechanism regulating CFTR opening are considered elsewhere (Chap. 6). However, there appear to be several mechanisms that regulate CFTR activity that are important in protecting the pancreas from pancreatitis.

There are multiple extracellular hormones, neurotransmitters, and receptors that are linked to duct function and specifically CFTR state. The first is secretin, the first hormone discovered in 1902 and the major challenge to the Pavlov perspective that the visceral organs were solely regulated by neural connections. Secretin is released from "S" cells in the duodenum mucosa into the bloodstream following luminal acidification. The pancreatic duct cells express secretin receptors, as demonstrated by receptor autoradiography of rat ducts (Ulrich et al. 1998) and later human tissues (Korner et al. 2005). The possibility that secretin also acts on afferent vagal fibers or interneurons has not been excluded. Secretin receptors are seven-transmembrane G-protein-coupled receptor family that increases intracellular cyclic AMP, as does the neuropeptide vasoactive intestinal peptide (VIP). Secretin stimulation of the duct results in CFTR-mediated fluid secretion.

Collection of human duct cells from fresh pancreas without disease remains a challenging obstacle to mechanistic research. Some insight can be gained by evaluation of well-differentiated duct cell lines that are capable of forming polarized monolayers with tight junctions or electrophysiology studies. Wang and Novak (2013) recently publish a series of studies using human-derived Capan-1 cells. They confirmed transepithelial anion transport in human pancreatic duct epithelium with Ussing chambers in response to secretin and VIP via cAMP but also documented responses to acetylcholine (muscarinic receptors) and adenosine, adenosine 5′-triphosphate (ATP), uridine 5′-triphosphate (UTP), and other pharmacological agents via purinergic P2 receptors and calcium signaling pathways (Wang and Novak 2013). These experiments demonstrate multiple pathways converging on the duct cell to stimulate secretion.

From the perspective of integrated physiology and systems biology, stimulation of the duct cells appears to occur under two conditions, meal stimulation and danger/inflammation. Multiple mechanisms are associated with the danger/inflammation cell activation. Extracellular ATP and UTP activate transepithelial anion secretion when applied to either the apical (luminal) or basolateral membrane (Wang and Novak 2013). ATP and UTP are normally intracellular molecules. These purines may be released by acinar cells during secretion, thus providing paracrine activation of local duct cell secretion (Novak 2008; Haanes et al. 2014) by binding to luminal receptors. Basolateral receptors may serve a different function. High concentrations are generated within the regional interstitial spaces with cell injury membrane disruption. Thus, ATP and UTP, by acting on the basolateral membrane of epithelial cells, on inflammatory cells, on neurons, or on other cells responding to injury, represent DAMP molecules (Gombault et al. 2012).

The protease-activated receptors PAR2 (coagulation factor II (thrombin) receptor-like 1; F2RL1) and PAR4 (coagulation factor II (thrombin) receptor-like 3; F2RL3) are expressed in the pancreas and may be important for regulating duct secretion, inflammation, and pancreatic pain (Ceppa et al. 2011). PAR2 can be either protective or pathogenic in pancreatitis, depending on the model (Laukkarinen et al. 2008). PAR2 is located on the apical membrane of the guinea pig and human duct cells, is sensitive to trypsin activation, and appeared to inhibit CFTR-mediated pancreatic secretion (Pallagi et al. 2011). The role of PARs in pancreatitis is likely important, but more studies are required to determine their role or effects under different circumstances.

Calcium-sensing receptor (CaSR) is a member of the G-protein-coupled receptor (GPCR) superfamily (Brown et al. 1993). CaSR plays an important role in the body's overall calcium homeostasis and is expressed by cells of the parathyroid gland and renal tubules involved in the calcium metabolism. CaSR was identified in both human pancreatic acinar and ductal cells, as well as in various non-exocrine cells (Bruce et al. 1999), although its functional significance in the pancreas has not yet been well studied. On the other hand, mutations in *CASR* that are classified as loss-of-function variants are associated with chronic pancreatitis in patients with *SPINK1* mutations (Felderbauer et al. 2003; Pidasheva et al. 2004; Muddana et al. 2008). This is a hypothesis-generating observation that CaSRs protect the pancreas from acute pancreatitis with high intraductal calcium concentrates favoring trypsin activation and protection from inhibition. *CASR* variants alone are not sufficient to cause chronic pancreatitis, but when combined with another major risk factor, such as a pathogenic *SPINK1* variant, the combined risk is sufficient to cause recurrent acute pancreatitis and eventually chronic pancreatitis.

28.6 Genetics of Human *CFTR* Variants

The biology of the CFTR channel is reviewed in Chap. 16. In humans, the primary organ affected by severe recessive *CFTR* mutations is the human pancreas. Indeed, the term "cystic fibrosis of the pancreas" refers to the chronic inflammation fibrosis and cyst that occur in infants who died of maldigestion from pancreatic exocrine failure and failure-to-thrive prior to pancreatic enzyme replacement therapy. The fact that *CFTR* mutations have such devastating effects on the pancreas, and that no other disease of ion transporters affects the pancreas to cause a Mendelian genetic syndrome, demonstrates the central role of CFTR in pancreatic physiology.

Over 2000 mutations and/or sequence variants in *CFTR* have been identified in human subjects worldwide. The genetic variants are not equivalent in functional effect. The various consequences of pathogenic sequence variants on CFTR function have been divided into several classification systems. One of the clinical uses of classifications is to predict the consequence of the CFTR variant with complete loss of function resulting in a "severe" phenotype, reduced but residual function resulting in a "mild-variable" phenotype, and CFTR variants with no functional consequences being classified as "benign" (Zielenski 2000; Rowntree and Harris 2003). As an inherited disease, cystic fibrosis is an autosomal recessive genetic disorder and the severity or phenotype of the disease correlates with the amount of dysfunction in the *milder* mutation. In general loss of a single *CFTR* allele does not result in human disease—up to 3 % of the population may carry a single copy of the common *CFTR* F508del variant and be asymptomatic. However, as will be discussed below, even a 50 % reduction in CFTR expression (e.g., one allele with a severe pathogenic *CFTR* variant and the other allele normal) makes the pancreas more susceptible to dysfunction when stressed. *CFTR* mutations classified as mild-variable variants typically retain at least 10 % of normal function. In most cases, a 75–80 % reduction in overall *CFTR* function (e.g., one allele with a severe mutations and the other with a mild-variable mutation) makes organs susceptible to injury from even moderate environmental or genetic stressors. With the retention of some function, the phenotype is milder than with no CFTR function, and some organs may function nearly normally while others become diseased. These milder clinical syndromes are classified as either atypical cystic fibrosis or as a CFTR-associated disorder (Kerem 2006; Bombieri et al. 2011). Patients without full-blown cystic fibrosis may have recurrent acute pancreatitis or chronic pancreatitis as part of the CFTR-related syndrome (Bombieri et al. 2011; Cohn et al. 1998; Sharer et al. 1998).

28.6.1 CFTR Bicarbonate-Defective Variance

The Whitcomb–Ermentrout model predicted that *CFTR* variants that specifically affected bicarbonate secretion would increase the risk of pancreatitis and disease in other organs using CFTR for bicarbonate conductance but not the risk of disease in organs that used CFTR for chloride secretion, such as the sweat gland and lungs. This is a challenging prediction in humans, since *CFTR* functional testing is primarily done using sweat chloride testing, and the physician specialists caring for patients with cystic fibrosis are pulmonary physicians.

To test this hypothesis we evaluated the CFTR gene in 984 patients from the North American Pancreatitis Study 2 [NAPS2 (Whitcomb et al. 2008)] in which patients underwent extensive phenotyping during ascertainment by experts from over 20 major pancreas centers (LaRusch et al. 2014). To improve the pretest probability that any *CFTR* variant was truly associated with pancreatic disease, we conducted a literature review to identify all *CFTR* variants that have been reported at least twice in previous chronic pancreatitis case-control genetic studies, plus common severe, cystic fibrosis-causing *CFTR* variants ($CFTR^{CF}$). Using a panel of 81 *CFTR* variants, we identified 43 variants in pancreatitis patients, including nine seen multiple times that were previously classified as benign for classic cystic fibrosis (*CFTR* R74Q, R75Q, R117H, R170H, L967S, L997F, D1152H, S1235R, and D1270N) (LaRusch et al. 2014).

The nine candidate bicarbonate-defective *CFTR* variants ($CFTR^{BD}$) were individually cloned into normal CFTR vectors and expressed in HEK-293T cells for electrophysiological analysis. The goal was to test the hypothesis that the $CFTR^{BD}$ variants were specifically associated impairment of the known WNK1-SPAK (With-No-Lysine 1; Ste20-related proline/alanine-rich) pathway-stimulated increase in CFTR bicarbonate permeability (Park et al. 2010; LaRusch et al. 2014). WNK1 is member of the "with-no-K" (Lys) kinases that serves as an intracellular sensor of osmolality, chloride concentration, and other factors and respond by activating additional kinases linked to a variety of ion channels and exchangers, including CFTR (Anselmo et al. 2006; Richardson and Alessi 2008; Park et al. 2010). Wild-type (WT) CFTR responds to WNK1-SPAK activation with an increase in CFTR HCO_3^- permeability and conductance to reach that of chloride (Park et al. 2010). In contrast, the candidate $CFTR^{BD}$ variants, which had normal chloride permeability and conductance, generally failed to respond to WNK1-SPAK activation with increased bicarbonate permeability or conductance (Fig. 28.5). Treatment with the CFTR inhibitor $CFTR_{inh}$-172 (20 μM) inhibited >90 % of the HCO_3^- currents indicating that CFTR mediates most of the HCO_3^- currents rather than affecting other pathways.

The nine $CFTR^{BD}$ variants all result in amino acid substitutions, and the specific amino acids, whose code they alter, are scattered across the gene. Three-dimensional models revealed that the location of the variant amino acids was clustered in four locations: L997F and D1152H projected into the lumen of the anion channel causing a reduction in functional diameter to block bicarbonate conductance; R74, R75, R170, L967, and R1162L [a candidate that did not fully meet the predefined criteria (LaRusch et al. 2014)] are at the hinge region that modulates the collective movements of the nucleotide-binding domains (NBDs) with respect to membrane-spanning domains (MSDs); D1270 and S1235 were on the apex surface of NBD2; and R117H projects into the extracellular space (Fig. 28.6a, b).

Molecular dynamics (MD) simulations, based on homology-modeled structures of ABC transporters, were used to examine the effect of $CFTR^{BD}$ variants on the structure and dynamics of the channel. The channel diameters of the WT and mutants L997F and D1152H (Fig. 28.6c, d) demonstrate that the channel diameter is observed to narrow down from an average value of 10.3–7.5 Å (standard deviation, $\sigma = 0.5$ Å) at the pore region, near the L997F amino acid substitution (Fig. 28.6e), and from an average of 9.9 Å to 4.3 Å ($\sigma = 1.1$ Å) for the $CFTR^{BD}$ mutant D1152H (Fig. 28.6f). On the other hand, L967, L997, D1152, and R1162 appear to serve as anchors for dynamic movements between NBD1 and NBD2, which affects channel gating (LaRusch et al. 2014).

CFTR is utilized in multiple tissues for anion transport, including several organs in which CFTR is linked to bicarbonate secretion. To determine if $CFTR^{BD}$ variants are associated with disease in other organs, the prevalence of sinusitis and male infertility was compared between pancreatitis cases with and without $CFTR^{BD}$ variants (LaRusch et al. 2014). Risk of sinusitis was significantly increased among carriers of $CFTR^{BD}$ ($p = 0.001$; OR 2.60; CI 1.43–4.60), cystic fibrosis-associated variants ($CFTR^{CF}$) ($p = 0.01$; OR 2.47; CI 1.18–4.91), or either $CFTR^{BD}$ or $CFTR^{CF}$ variant allele ($p = 0.0001$; OR 2.55; CI 1.55–4.15) (LaRusch et al. 2014). Likewise, risk of male infertility was in patients with pancreatitis similar to controls, but the significantly increased risk was associated with either $CFTR^{BD}$ or $CFTR^{CF}$ alleles ($p = 0.023$; OR 10.7; CI 1.03–536) or as a recessive genotype ($p = 1.2 \times 10^{-7}$; OR 303; CI 23–15,783) (LaRusch et al. 2014). Together, these data provide compelling evidence of disease in humans associated with defects in CFTR bicarbonate conductance.

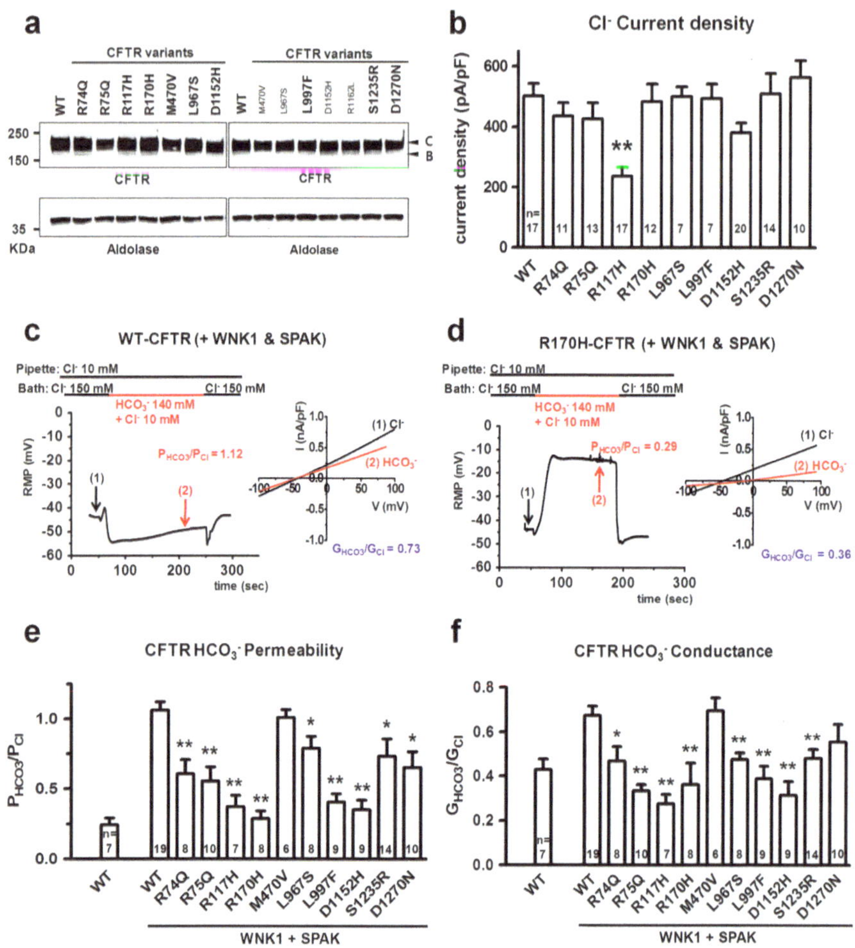

Fig. 28.5 Functional characteristics of the nine $CFTR^{BD}$ variants (LaRusch et al. 2014). (**a**) Wild-type (WT) and variant CFTR proteins were expressed in HEK 293T cells and immunoblotted with anti-CFTR and anti-aldolase antibodies. Replicate lanes are in small font. Band B, expected size of immature ER core-glycosylated CFTR; band C, mature complex-glycosylated CFTR. (**b**) Whole-cell Cl_2 currents were measured in WT and variant CFTR-expressing HEK 293T cells, as described in Methods. (**c**) Whole-cell currents of WT-CFTR were measured in HEK 293T cells co-expressed with WNK1 and SPAK using patch pipette contained a low concentration of Cl_2 (10 mM). A representative trace of reversal potential measurement is shown in the *left panel*. The permeability ratio P_{HCO3}/P_{Cl} was calculated according to the Goldman–Hodgkin–Katz equation. I–V relationships at the indicated points are presented in the accompanying graph. The conductance ratio G_{HCO3}/G_{Cl} was calculated by measuring each outward current (i.e., slope between E_{rev} and $E_{rev}+25$ mV). *RMP* resting membrane potential. (**d**) Whole-cell currents of R170H-CFTR were measured in HEK 293T cells using the same protocol shown in (**c**) and (**e**). A summary of the P_{HCO3}/P_{Cl} values obtained from WT-CFTR in the standard state (*left*) compared to WT-CFTR and the nine CFTRBD variants with WNK1 + SPAK activation (*right*, *underlined*). (**f**) A summary of the G_{HCO3}/G_{Cl} values in the standard state (*left*) with WNK1 + SPAK activation (*right*). Values throughout are means 6 SEM. *p, 0.05, **p, 0.01: difference from WT in cells co-expressed with WNK1 and SPAK

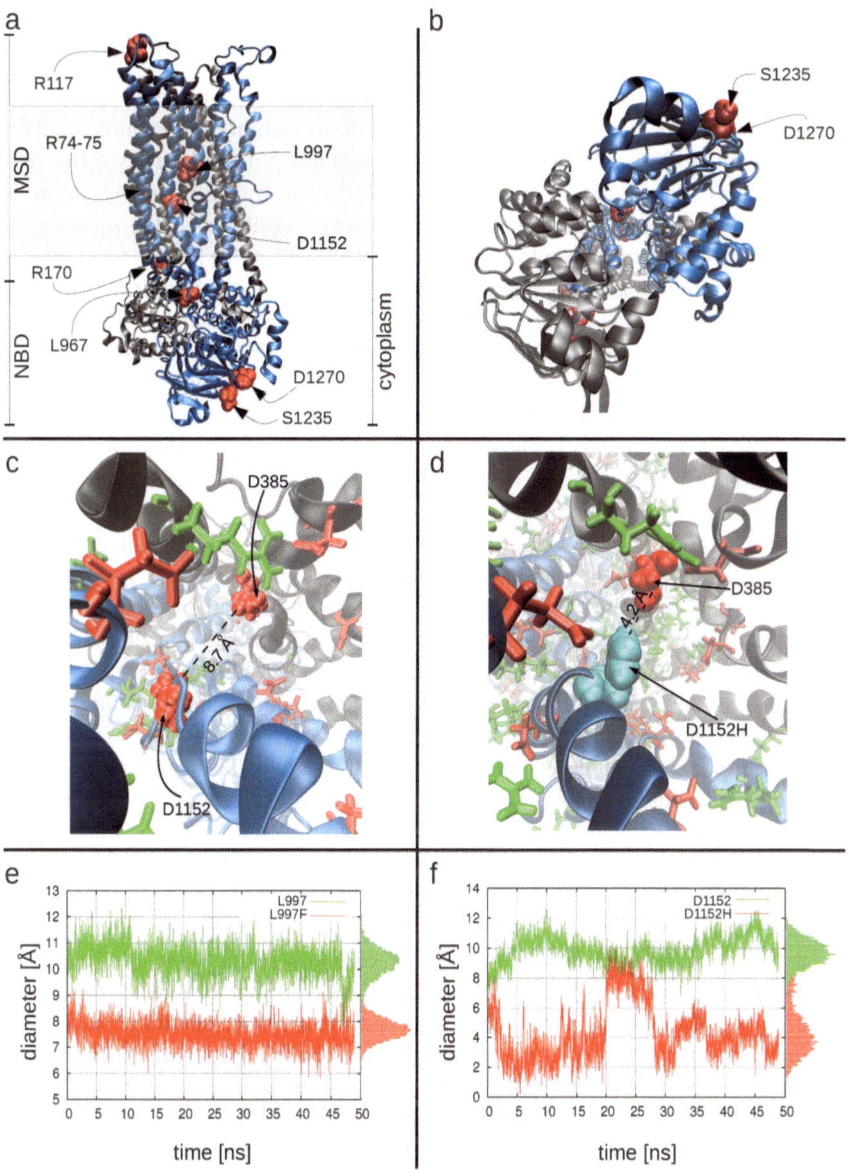

Fig. 28.6 Molecular modeling and simulations of WT-CFTR and variants (LaRusch et al. 2014). (**a**) and (**b**) display the side and bottom views, respectively, of the WT-CFTR dimer, where the two nucleotide-binding domains and the two membrane-spanning domains are labeled as NBD and MSD. The *shaded region* indicates the location of the lipid bilayer. Color key: *black*, subunit 1 of CFTR, with residues 1–859; *blue*, subunit 2, residues 860–1480; *red* CFTR variants studied. (**c**) shows the charge distribution around D1152H: This negatively charged residue (*left*; shown in *red* space-filling representation) is surrounded by several positively charged residues (*green*), especially on its side of the cavity, creating an attractive force that keeps the residue from extending into the cavity. Also shown are other negatively charged residues (*red stick* or space-filling representation), including D385, diametrically opposite to D1152. (**d**) shows the corresponding scene for the variant residue, D1152H (*cyan*), which can move toward the center of the cavity, thus

28.7 Claudin-2 in the Pancreas

The claudins are family of at least 27 different tight junction proteins in mammals with a molecular range of 20–27 kDa and four membrane-spanning domains (Angelow et al. 2008; Krause et al. 2008; Gunzel and Yu 2013). They differ from other tight junction proteins such as the occludins and tricellulin as they form the "gasket" that seals the paracellular space between epithelial cells, separate the apical from basolateral membranes, mediate cell–cell adhesion, and polymerize to form tight junction fibrils (Angelow et al. 2008). The most extensive work on claudins has been done in the kidney where various claudins are expressed in the different tubules and ducts. Although claudins have not been as extensively studied in the pancreas, several claudins, including claudin-1, claudin-2, claudin-3, and claudin-4, are expressed in mouse and human pancreas (Lee et al. 2011).

Genetic variants in the claudin-2 gene (*CLDN2*) locus confer a very high risk for pancreatitis, especially in alcoholics (Whitcomb et al. 2012). The finding was especially interesting since *CLDN2* is located on the X chromosome, and since men have one X chromosome, the risk is conferred in a dominant inheritance pattern, whereas women, who have two X chromosomes, have risk in a recessive pattern (Whitcomb et al. 2012). The risk variant was not in the coding region and appeared to increase inflammation response through aberrant localization on acinar cells. The functional consequences appear to reflect abnormal expression of claudin-2 on the acinar cell (or partially dedifferentiated acinar cell) after injury. The effect could be gain of function and appears recessive in women as one of the two X chromosome alleles is silenced, but this has not been demonstrated yet. Although the pathobiology is not known, these findings brought attention to the potential role of the claudins in normal pancreatic biology.

Claudin-2 is a highly regulated tight junction protein normally expressed between cells of the pancreatic ducts, kidney proximal tubules, gall bladder hepatobiliary system, and intestinal epithelium (Lee et al. 2011; Aung et al. 2006). Claudin-2 differs from other claudins in the pancreas because it forms a low-resistance cation-selective ion and water channel (Van Itallie et al. 2008; Amasheh et al. 2002). In the pancreatic duct claudin-2 may allow sodium and water to freely enter the duct lumen where it mixes with the bicarbonate that is actively secreted through the duct cells via a CFTR-dependent pathway to form pancreatic juice (Fig. 28.4). The expression of claudin-2 in human pancreas has recently been confirmed in pancreatic samples by quantitative PCR of cDNA,

---

Fig. 28.6 (continued) leading to a constriction in the channel diameter. Channel diameter at the location of variant residues: (**e**) shows the diameter of the channel at the location of L997, as a function of time, both for the WT (L997, *green curve*) and the variant (F997, *red curve*), based on closest interatomic distance between L997/F997 and D385. On (**f**), the same information for the WT and variant D1152H is shown. In both plots, the pore diameter in the WT is larger than that stabilized in the mutants. The histograms of channel sizes are shown along the right ordinate

by Western blot, by immunofluorescence, and by immunohistochemistry (Whitcomb et al. 2012).

Claudin-2 and other tight junction proteins are continually cycled by insertion into the plasma membrane and removal by endocytosis (Dukes et al. 2012). The *CLDN2* promoter contains an NFκB-binding site (Sakaguchi et al. 2002), and expression is upregulated by tumor necrosis factor alpha (TNFα) (Mankertz et al. 2009), interleukin-6 (IL-6) (Suzuki et al. 2011), and the Wnt/LEF-1/beta-catenin pathway (Mankertz et al. 2004) and downregulated by epidermal growth factor (Ikari et al. 2011) by endocytosis linked to the endosomal sorting complex required for transport (ESCRT) mechanism (Dukes et al. 2012). Expression of claudin-4, which is also expressed in the pancreas but differs from claudin-2 by forming high resistance barrier to ions and water (Van Itallie et al. 2001; Angelow et al. 2008), appears to be regulated in a reciprocate manner. In the gallbladder, for example, the immunohistochemical staining of claudin-2 was weak in normal tissue but became intense with inflammation, while claudin-4 staining shifted from strong to faint (Laurila et al. 2007). Madin-Darby canine kidney (MDCKII) cells respond to EGF with ERK1/2-mediated downregulation of claudin-2 and upregulation of claudin-1, claudin-3, and claudin-4 (Singh and Harris 2004), possibly through STAT3 and Src (Garcia-Hernandez et al. 2014). Both claudin-2 and claudin-4 are phosphorylated by WNK4 (Yamauchi et al. 2004), and possibly WNK1 (Ohta et al. 2006), although the importance of this observation has not been applied to pancreatic duct cells. The dynamic changes may also be important during stimulated pancreatic secretion, which is not usually seen in human tissues since samples coming from humans are typically linked to surgical operations after the patient has been fasting for some time.

28.8 Summary and Conclusions

Diseases of the exocrine pancreas related to ion transporters are currently limited to cystic fibrosis and recurrent acute and chronic pancreatitis. While there may be other diseases, the clinical severity has not been sufficient to be fully recognized or appreciated. Clinical recognition of pathology is generally secondary to the increased susceptibility to intrapancreatic trypsin activation, which causes significant inflammatory responses that damage the pancreas. Ion transport dysfunction that reduces ductal fluid secretion and flushing of trypsin out of the duct results in pancreatic injury and disease. CFTR is used by the pancreas for bicarbonate secretion. A new finding of *CFTR* variants that specifically affect regulated bicarbonate conductance in response to WNK1 activation provides new insights into ion channel biology and models of pancreatic duct physiology and has implications for multiple related organ systems.

References

Amasheh S, Meiri N, Gitter AH, Schoneberg T, Mankertz J, Schulzke JD, Fromm M (2002) Claudin-2 expression induces cation-selective channels in tight junctions of epithelial cells. J Cell Sci 115:4969–4976

Angelow S, Ahlstrom R, Yu AS (2008) Biology of claudins. Am J Physiol Renal Physiol 295: F867–F876. doi:10.1152/ajprenal.90264.2008

Anselmo AN, Earnest S, Chen W, Juang YC, Kim SC, Zhao Y, Cobb MH (2006) WNK1 and OSR1 regulate the Na^+, K^+, $2Cl^-$ cotransporter in HeLa cells. Proc Natl Acad Sci U S A 103:10883–10888. doi:10.1073/pnas.0604607103

Aoun E, Chang CC, Greer JB, Papachristou GI, Barmada MM, Whitcomb DC (2008) Pathways to injury in chronic pancreatitis: decoding the role of the high-risk SPINK1 N34S haplotype using meta-analysis. PLoS One 3, e2003

Aung PP, Mitani Y, Sanada Y, Nakayama H, Matsusaki K, Yasui W (2006) Differential expression of claudin-2 in normal human tissues and gastrointestinal carcinomas. Virchows Arch 448:428–434. doi:10.1007/s00428-005-0120-2

Bear CE, Li CH, Kartner N, Bridges RJ, Jensen TJ, Ramjeesingh M, Riordan JR (1992) Purification and functional reconstitution of the cystic fibrosis transmembrane conductance regulator (CFTR). Cell 68:809–818

Beer S, Zhou J, Szabo A, Keiles S, Chandak GR, Witt H, Sahin-Toth M (2012) Comprehensive functional analysis of chymotrypsin C (CTRC) variants reveals distinct loss-of-function mechanisms associated with pancreatitis risk. Gut 62:1616–1624. doi:10.1136/gutjnl-2012-303090

Bombieri C, Claustres M, De Boeck K, Derichs N, Dodge J, Girodon E, Sermet I, Schwarz M, Tzetis M, Wilschanski M, Bareil C, Bilton D, Castellani C, Cuppens H, Cutting GR, Drevinek P, Farrell P, Elborn JS, Jarvi K, Kerem B, Kerem E, Knowles M, Macek M Jr, Munck A, Radojkovic D, Seia M, Sheppard DN, Southern KW, Stuhrmann M, Tullis E, Zielenski J, Pignatti PF, Ferec C (2011) Recommendations for the classification of diseases as CFTR-related disorders. J Cyst Fibros 10(Suppl 2):S86–S102. doi:10.1016/S1569-1993(11)60014-3

Brown EM, Gamba G, Riccardi D, Lombardi M, Butters R, Kifor O, Sun A, Hediger MA, Lytton J, Hebert SC (1993) Cloning and characterization of an extracellular Ca^{2+}-sensing receptor from bovine parathyroid. Nature 366:575–580

Bruce JI, Yang X, Ferguson CJ, Elliott AC, Steward MC, Case RM, Riccardi D (1999) Molecular and functional identification of a Ca^{2+} (polyvalent cation)-sensing receptor in rat pancreas. J Biol Chem 274:20561–20568

Case RM, Argent BE (1993) Pancreatic duct cell secretion. Control and mechanism of transport. In: Go VLW et al (eds) The pancreas: biology, pathobiology and disease, 2nd edn. Raven Press, New York, NY, pp 301–350

Ceppa EP, Lyo V, Grady EF, Knecht W, Grahn S, Peterson A, Bunnett NW, Kirkwood KS, Cattaruzza F (2011) Serine proteases mediate inflammatory pain in acute pancreatitis. Am J Physiol Gastrointest Liver Physiol 300:G1033–G1042. doi:10.1152/ajpgi.00305.2010

Chen JM, Ferec C (2012) Genetics and pathogenesis of chronic pancreatitis: the 2012 update. Clin Res Hepatol Gastroenterol 36:334–340. doi:10.1016/j.clinre.2012.05.003

Cohn JA, Friedman KJ, Noone PG, Knowles MR, Silverman LM, Jowell PS (1998) Relation between mutations of the cystic fibrosis gene and idiopathic pancreatitis. N Engl J Med 339:653–658

Deng X, Whitcomb DC (1998) Neurohormonal control of the exocrine pancreas. Curr Opin Gastroenterol 14:362–368

Dukes JD, Whitley P, Chalmers AD (2012) The PIKfyve inhibitor YM201636 blocks the continuous recycling of the tight junction proteins Claudin-1 and Claudin-2 in MDCK cells. PLoS One 7, e28659. doi:10.1371/journal.pone.0028659

Felderbauer P, Hoffmann P, Einwachter H, Bulut K, Ansorge N, Schmitz F, Schmidt WE (2003) A novel mutation of the calcium sensing receptor gene is associated with chronic pancreatitis in a family with heterozygous SPINK1 mutations. BMC Gastroenterol 3:34

Garcia-Hernandez V, Flores-Maldonado C, Rincon-Heredia R, Verdejo-Torres O, Bonilla-Delgado J, Meneses-Morales I, Gariglio P, Contreras RG (2014) EGF regulates claudin-2 and -4 expression through STAT3 and Src in MDCK cells. J Cell Physiol 230:105–115. doi:10.1002/jcp.24687

Gombault A, Baron L, Couillin I (2012) ATP release and purinergic signaling in NLRP3 inflammasome activation. Front Immunol 3:414. doi:10.3389/fimmu.2012.00414

Gorry MC, Gabbaizedeh D, Furey W, Gates LK Jr, Preston RA, Aston CE, Zhang Y, Ulrich C, Ehrlich GD, Whitcomb DC (1997) Mutations in the cationic trypsinogen gene are associated with recurrent acute and chronic pancreatitis. Gastroenterology 113:1063–1068

Gunzel D, Yu AS (2013) Claudins and the modulation of tight junction permeability. Physiol Rev 93:525–569. doi:10.1152/physrev.00019.2012

Haanes KA, Kowal JM, Arpino G, Lange SC, Moriyama Y, Pedersen PA, Novak I (2014) Role of vesicular nucleotide transporter VNUT (SLC17A9) in release of ATP from AR42J cells and mouse pancreatic acinar cells. Purinergic Signal 10:431–440. doi:10.1007/s11302-014-9406-7

Hoque R, Malik AF, Gorelick F, Mehal WZ (2012) Sterile inflammatory response in acute pancreatitis. Pancreas 41:353–357. doi:10.1097/MPA.0b013e3182321500

Ikari A, Takiguchi A, Atomi K, Sugatani J (2011) Epidermal growth factor increases clathrin-dependent endocytosis and degradation of claudin-2 protein in MDCK II cells. J Cell Physiol 226:2448–2456. doi:10.1002/jcp.22590

Kerem E (2006) Atypical CF and CF related diseases. Paediatr Respir Rev 7(Suppl 1):S144–S146

Korner M, Hayes GM, Rehmann R, Zimmermann A, Friess H, Miller LJ, Reubi JC (2005) Secretin receptors in normal and diseased human pancreas: marked reduction of receptor binding in ductal neoplasia. Am J Pathol 167:959–968. doi:10.1016/S0002-9440(10)61186-8

Krause G, Winkler L, Mueller SL, Haseloff RF, Piontek J, Blasig IE (2008) Structure and function of claudins. Biochim Biophys Acta 1778:631–645. doi:10.1016/j.bbamem.2007.10.018

LaRusch J, Whitcomb DC (2012) Genetics of pancreatitis with a focus on the pancreatic ducts. Minerva Gastroenterol Dietol 58:299–308

LaRusch J, Jung J, General IJ, Lewis MD, Park HW, Brand RE, Gelrud A, Anderson MA, Banks PA, Conwell D, Lawrence C, Romagnuolo J, Baillie J, Alkaade S, Cote GA, Gardner TB, Amann ST, Slivka A, Sandhu B, Aloe A, Kienholz ML, Yadav D, Barmada MM, Bahar I, Lee MG, Whitcomb DC (2014) Mechanisms of CFTR functional variants that impair regulated bicarbonate permeation and increase risk for pancreatitis but not for cystic fibrosis. PLoS Genet 10, e1004376

LaRusch J, Lozano-Leon A, Stello K, Moore A, Muddana V, O'Connell M, Diergaarde B, Yadav D, Whitcomb D (2015) The common chymotrypsinogen C (CTRC) variant G60G (C.180T) increases risk of chronic pancreatitis but not recurrent acute pancreatitis in a North American population. Clin Transl Gastroenterol 6:e68

Laukkarinen JM, Weiss ER, van Acker GJ, Steer ML, Perides G (2008) Protease-activated receptor-2 exerts contrasting model-specific effects on acute experimental pancreatitis. J Biol Chem 283:20703–20712. doi:10.1074/jbc.M801779200

Laurila JJ, Karttunen T, Koivukangas V, Laurila PA, Syrjala H, Saarnio J, Soini Y, Ala-Kokko TI (2007) Tight junction proteins in gallbladder epithelium: different expression in acute acalculous and calculous cholecystitis. J Histochem Cytochem 55:567–573. doi:10.1369/jhc.6A7155.2007

Lee JH, Kim KS, Kim TJ, Hong SP, Song SY, Chung JB, Park SW (2011) Immunohistochemical analysis of claudin expression in pancreatic cystic tumors. Oncol Rep 25:971–978. doi:10.3892/or.2011.1132

Lee MG, Ohana E, Park HW, Yang D, Muallem S (2012) Molecular mechanism of pancreatic and salivary gland fluid and HCO_3 secretion. Physiol Rev 92:39–74. doi:10.1152/physrev.00011.2011

Mankertz J, Hillenbrand B, Tavalali S, Huber O, Fromm M, Schulzke JD (2004) Functional crosstalk between Wnt signaling and Cdx-related transcriptional activation in the regulation of the claudin-2 promoter activity. Biochem Biophys Res Commun 314:1001–1007
Mankertz J, Amasheh M, Krug SM, Fromm A, Amasheh S, Hillenbrand B, Tavalali S, Fromm M, Schulzke JD (2009) TNFalpha up-regulates claudin-2 expression in epithelial HT-29/B6 cells via phosphatidylinositol-3-kinase signaling. Cell Tissue Res 336:67–77. doi:10.1007/s00441-009-0751-8
Marino CR, Matovcik LM, Gorelick FS, Cohn JA (1991) Localization of the cystic fibrosis transmembrane conductance regulator in pancreas. J Clin Invest 88:712–716
Muddana V, Lamb J, Greer JB, Elinoff B, Hawes RH, Cotton PB, Anderson MA, Brand RE, Slivka A, Whitcomb DC (2008) Association between calcium sensing receptor gene polymorphisms and chronic pancreatitis in a US population: role of serine protease inhibitor Kazal 1type and alcohol. World J Gastroenterol 14:4486–4491
Mussa BM, Verberne AJ (2013) The dorsal motor nucleus of the vagus and regulation of pancreatic secretory function. Exp Physiol 98:25–37. doi:10.1113/expphysiol.2012.066472
Noble MD, Liddle RA (2005) Neurohormonal control of exocrine pancreatic secretion. Curr Opin Gastroenterol 21:531–537
Novak I (2008) Purinergic receptors in the endocrine and exocrine pancreas. Purinergic Signal 4:237–253. doi:10.1007/s11302-007-9087-6
Ohta A, Yang SS, Rai T, Chiga M, Sasaki S, Uchida S (2006) Overexpression of human WNK1 increases paracellular chloride permeability and phosphorylation of claudin-4 in MDCKII cells. Biochem Biophys Res Commun 349:804–808. doi:10.1016/j.bbrc.2006.08.101
Pallagi P, Venglovecz V, Rakonczay Z Jr, Borka K, Korompay A, Ozsvari B, Judak L, Sahin-Toth M, Geisz A, Schnur A, Maleth J, Takacs T, Gray MA, Argent BE, Mayerle J, Lerch MM, Wittmann T, Hegyi P (2011) Trypsin reduces pancreatic ductal bicarbonate secretion by inhibiting CFTR Cl− channels and luminal anion exchangers. Gastroenterology 141:2228–2239.e2226. doi:10.1053/j.gastro.2011.08.039
Park HW, Nam JH, Kim JY, Namkung W, Yoon JS, Lee JS, Kim KS, Venglovecz V, Gray MA, Kim KH, Lee MG (2010) Dynamic regulation of CFTR bicarbonate permeability by $[Cl^-]_i$ and its role in pancreatic bicarbonate secretion. Gastroenterology 139:620–631. doi:10.1053/j.gastro.2010.04.004, [pii] S0016-5085(10)00550-0
Pfutzer RH, Barmada MM, Brunskill AP, Finch R, Hart PS, Neoptolemos J, Furey WF, Whitcomb DC (2000) SPINK1/PSTI polymorphisms act as disease modifiers in familial and idiopathic chronic pancreatitis. Gastroenterology 119:615–623
Pidasheva S, D'Souza-Li L, Canaff L, Cole DE, Hendy GN (2004) CASRdb: calcium-sensing receptor locus-specific database for mutations causing familial (benign) hypocalciuric hypercalcemia, neonatal severe hyperparathyroidism, and autosomal dominant hypocalcemia. Hum Mutat 24:107–111
Rahner C, Mitic LL, Anderson JM (2001) Heterogeneity in expression and subcellular localization of claudins 2, 3, 4, and 5 in the rat liver, pancreas, and gut. Gastroenterology 120:411–422
Richardson C, Alessi DR (2008) The regulation of salt transport and blood pressure by the WNK-SPAK/OSR1 signalling pathway. J Cell Sci 121:3293–3304. doi:10.1242/jcs.029223, [pii] 121/20/3293
Rosendahl J, Witt H, Szmola R, Bhatia E, Ozsvari B, Landt O, Schulz HU, Gress TM, Pfutzer R, Lohr M, Kovacs P, Bluher M, Stumvoll M, Choudhuri G, Hegyi P, te Morsche RH, Drenth JP, Truninger K, Macek M Jr, Puhl G, Witt U, Schmidt H, Buning C, Ockenga J, Kage A, Groneberg DA, Nickel R, Berg T, Wiedenmann B, Bodeker H, Keim V, Mossner J, Teich N, Sahin-Toth M (2008) Chymotrypsin C (CTRC) variants that diminish activity or secretion are associated with chronic pancreatitis. Nat Genet 40:78–82
Rowntree RK, Harris A (2003) The phenotypic consequences of CFTR mutations. Ann Hum Genet 67:471–485
Sakaguchi T, Gu X, Golden HM, Suh E, Rhoads DB, Reinecker HC (2002) Cloning of the human claudin-2 5′-flanking region revealed a TATA-less promoter with conserved binding sites in

mouse and human for caudal-related homeodomain proteins and hepatocyte nuclear factor-1alpha. J Biol Chem 277:21361–21370. doi:10.1074/jbc.M110261200
Sarles H, Adler G, Dani R, Frey C, Gullo L, Harada H, Martin E, Norohna M, Scuro LA (1989) The pancreatitis classification of Marseilles, Rome 1988. Scand J Gastroenterol 24:641–642
Sarner M, Cotton PB (1984) Definitions of acute and chronic pancreatitis. Clin Gastroenterol 13:865–870
Sharer N, Schwarz M, Malone G, Howarth A, Painter J, Super M, Braganza J (1998) Mutations of the cystic fibrosis gene in patients with chronic pancreatitis. N Engl J Med 339:645–652
Singh AB, Harris RC (2004) Epidermal growth factor receptor activation differentially regulates claudin expression and enhances transepithelial resistance in Madin-Darby canine kidney cells. J Biol Chem 279:3543–3552. doi:10.1074/jbc.M308682200
Sohma Y, Gray MA, Imai Y, Argent BE (1996) A mathematical model of the pancreatic ductal epithelium. J Membr Biol 154:53–67
Sohma Y, Gray MA, Imai Y, Argent BE (2000) HCO_3 transport in a mathematical model of the pancreatic ductal epithelium. J Membr Biol 176:77–100
Steer ML, Waxman I, Freedman S (1995) Chronic pancreatitis. N Engl J Med 332:1482–1490
Stetson DL, Beauwens R, Palmisano J, Mitchell PP, Steinmetz PR (1985) A double-membrane model for urinary bicarbonate secretion. Am J Physiol Renal Physiol 249:F546–F552
Steward MC, Ishiguro H, Case RM (2005) Mechanisms of bicarbonate secretion in the pancreatic duct. Annu Rev Physiol 67:377–409
Stewart CP, Winterhager JM, Heintze K, Petersen KU (1989) Electrogenic bicarbonate secretion by guinea pig gallbladder epithelium: apical membrane exit. Am J Physiol Cell Physiol 256: C736–C749
Strong TV, Boehm K, Collins FS (1994) Localization of cystic fibrosis transmembrane conductance regulator mRNA in the human gastrointestinal tract by in situ hybridization. J Clin Invest 93:347–354. doi:10.1172/JCI116966
Suzuki T, Yoshinaga N, Tanabe S (2011) Interleukin-6 (IL-6) regulates claudin-2 expression and tight junction permeability in intestinal epithelium. J Biol Chem 286:31263–31271. doi:10.1074/jbc.M111.238147
Szabo A, Sahin-Toth M (2012) Determinants of chymotrypsin C cleavage specificity in the calcium-binding loop of human cationic trypsinogen. FEBS J 279:4283–4292. doi:10.1111/febs.12018
Ulrich CD II, Wood P, Hadac EM, Kopras E, Whitcomb DC, Miller LJ (1998) Cellular distribution of secretin receptor expression in rat pancreas. Am J Physiol Gastrointestinal Liver Physiol 275:G1437–G1444
Van Itallie C, Rahner C, Anderson JM (2001) Regulated expression of claudin-4 decreases paracellular conductance through a selective decrease in sodium permeability. J Clin Invest 107(10):1319–1327. doi:10.1172/JCI12464
Van Itallie CM, Holmes J, Bridges A, Gookin JL, Coccaro MR, Proctor W, Colegio OR, Anderson JM (2008) The density of small tight junction pores varies among cell types and is increased by expression of claudin-2. J Cell Sci 121:298–305. doi:10.1242/jcs.021485
Wang J, Novak I (2013) Ion transport in human pancreatic duct epithelium, Capan-1 cells, is regulated by secretin, VIP, acetylcholine, and purinergic receptors. Pancreas 42:452–460. doi:10.1097/MPA.0b013e318264c302
Whitcomb DC (1999) Pancreatic bicarbonate secretion: the role of CFTR and the sodium-bicarbonate cotransporter. Gastroenterology 117:275–277
Whitcomb DC (2013) Genetic risk factors for pancreatic disorders. Gastroenterology 144:1292–1302. doi:10.1053/j.gastro.2013.01.069
Whitcomb DC, Ermentrout GB (2004) A mathematical model of the pancreatic duct cell generating high bicarbonate concentrations in pancreatic juice. Pancreas 29:E30–E40
Whitcomb DC, Gorry MC, Preston RA, Furey W, Sossenheimer MJ, Ulrich CD, Martin SP, Gates LK Jr, Amann ST, Toskes PP, Liddle R, McGrath K, Uomo G, Post JC, Ehrlich GD (1996)

Hereditary pancreatitis is caused by a mutation in the cationic trypsinogen gene. Nat Genet 14:141–145

Whitcomb DC, Yadav D, Adam S, Hawes RH, Brand RE, Anderson MA, Money ME, Banks PA, Bishop MD, Baillie J, Sherman S, DiSario J, Burton FR, Gardner TB, Amann ST, Gelrud A, Lo SK, DeMeo MT, Steinberg WM, Kochman ML, Etemad B, Forsmark CE, Elinoff B, Greer JB, O'Connell M, Lamb J, Barmada MM, North American Pancreatic Study G (2008) Multicenter approach to recurrent acute and chronic pancreatitis in the United States: the North American Pancreatitis Study 2 (NAPS2). Pancreatology 8:520–531. doi:10.1159/000152001

Whitcomb DC, Larusch J, Krasinskas AM, Klei L, Smith JP, Brand RE, Neoptolemos JP, Lerch MM, Tector M, Sandhu BS, Guda NM, Orlichenko L, Albert MS, Albin RL, Apostolova LG, Arnold SE, Baldwin CT, Barber R, Barnes LL, Beach TG, Beecham GW, Beekly D, Bennett DA, Bigio EH, Bird TD, Blacker D, Boxer A, Burke JR, Buxbaum JD, Cairns NJ, Cantwell LB, Cao C, Carney RM, Carroll SL, Chui HC, Clark DG, Cribbs DH, Crocco EA, Cruchaga C, Decarli C, Demirci FY, Dick M, Dickson DW, Duara R, Ertekin-Taner N, Faber KM, Fallon KB, Farlow MR, Ferris S, Foroud TM, Frosch MP, Galasko DR, Ganguli M, Gearing M, Geschwind DH, Ghetti B, Gilbert JR, Gilman S, Glass JD, Goate AM, Graff-Radford NR, Green RC, Growdon JH, Hakonarson H, Hamilton-Nelson KL, Hamilton RL, Harrell LE, Head E, Honig LS, Hulette CM, Hyman BT, Jicha GA, Jin LW, Jun G, Kamboh MI, Karydas A, Kaye JA, Kim R, Koo EH, Kowall NW, Kramer JH, Kramer P, Kukull WA, Laferla FM, Lah JJ, Leverenz JB, Levey AI, Li G, Lin CF, Lieberman AP, Lopez OL, Lunetta KL, Lyketsos CG, Mack WJ, Marson DC, Martin ER, Martiniuk F, Mash DC, Masliah E, McKee AC, Mesulam M, Miller BL, Miller CA, Miller JW, Montine TJ, Morris JC, Murrell JR, Naj AC, Olichney JM, Parisi JE, Peskind E, Petersen RC, Pierce A, Poon WW, Potter H, Quinn JF, Raj A, Raskind M, Reiman EM, Reisberg B, Reitz C, Ringman JM, Roberson ED, Rosen HJ, Rosenberg RN, Sano M, Saykin AJ, Schneider JA, Schneider LS, Seeley WW, Smith AG, Sonnen JA, Spina S, Stern RA, Tanzi RE, Trojanowski JQ, Troncoso JC, Tsuang DW, Valladares O, Van Deerlin VM, Van Eldik LJ, Vardarajan BN, Vinters HV, Vonsattel JP, Wang LS, Weintraub S, Welsh-Bohmer KA, Williamson J, Woltjer RL, Wright CB, Younkin SG, Yu CE, Yu L, Alkaade S, Amann ST, Anderson MA, Baillie J, Banks PA, Conwell D, Cote GA, Cotton PB, Disario J, Farrer LA, Forsmark CE, Johnstone M, Gardner TB, Gelrud A, Greenhalf W, Haines JL, Hartman DJ, Hawes RA, Lawrence C, Lewis M, Mayerle J, Mayeux R, Melhem NM, Money ME, Muniraj T, Papachristou GI, Pericak-Vance MA, Romagnuolo J, Schellenberg GD, Sherman S, Simon P, Singh VP, Slivka A, Stolz D, Sutton R, Weiss FU, Wilcox CM, Zarnescu NO, Wisniewski SR, O'Connell MR, Kienholz ML, Roeder K, Barmada MM, Yadav D, Devlin B, Whitcomb DC, Larusch J, Krasinskas AM, Klei L, Smith JP, Brand RE, Neoptolemos JP, Lerch MM, Tector M, Sandhu BS, Guda NM, Orlichenko L, Albert MS, Albin RL, Apostolova LG, Arnold SE, Baldwin CT, Barber R, Barnes LL, Beach TG, Beecham GW, Beekly D, Bennett DA, Bigio EH, Bird TD, Blacker D, Boxer A, Burke JR, Buxbaum JD, Cairns NJ, Cantwell LB, Cao C, Carney RM, Carroll SL, Chui HC, Clark DG, Cribbs DH, Crocco EA, Cruchaga C, Decarli C, Demirci FY, Dick M, Dickson DW, Duara R, Ertekin-Taner N, Faber KM, Fallon KB, Farlow MR, Ferris S, Foroud TM, Frosch MP, Galasko DR, Ganguli M, Gearing M, Geschwind DH, Ghetti B, Gilbert JR, Gilman S, Glass JD, Goate AM, Graff-Radford NR, Green RC, Growdon JH, Hakonarson H, Hamilton-Nelson KL, Hamilton RL, Harrell LE, Head E, Honig LS, Hulette CM, Hyman BT, Jicha GA, Jin LW, Jun G, Kamboh MI, Karydas A, Kaye JA, Kim R, Koo EH, Kowall NW, Kramer JH, Kramer P, Kukull WA, Laferla FM, Lah JJ, Leverenz JB, Levey AI, Li G, Lin CF, Lieberman AP, Lopez OL, Lunetta KL, Lyketsos CG, Mack WJ, Marson DC, Martin ER, Martiniuk F, Mash DC, Masliah E, McKee AC, Mesulam M, Miller BL, Miller CA, Miller JW, Montine TJ, Morris JC, Murrell JR, Naj AC, Olichney JM, Parisi JE, Peskind E, Petersen RC, Pierce A, Poon WW, Potter H, Quinn JF, Raj A, Raskind M, Reiman EM, Reisberg B, Reitz C, Ringman JM, Roberson ED, Rosen HJ, Rosenberg RN, Sano M, Saykin AJ, Schneider JA, Schneider LS, Seeley WW, Smith AG, Sonnen JA, Spina S, Stern RA, Tanzi RE, Trojanowski JQ, Troncoso JC, Tsuang DW, Valladares O, Van Deerlin VM, Van Eldik LJ, Vardarajan BN,

Vinters HV, Vonsattel JP, Wang LS, Weintraub S, Welsh-Bohmer KA, Williamson J, Woltjer RL, Wright CB, Younkin SG, Yu CE, Yu L, Alkaade S, Amann ST, Anderson MA, Baillie J, Banks PA, Conwell D, Cote GA, Cotton PB, Disario J, Farrer LA, Forsmark CE, Johnstone M, Gardner TB, Gelrud A, Greenhalf W, Haines JL, Hartman DJ, Hawes RA, Lawrence C, Lewis M, Mayerle J, Mayeux R, Melhem NM, Money ME, Muniraj T, Papachristou GI, Pericak-Vance MA, Romagnuolo J, Schellenberg GD, Sherman S, Simon P, Singh VP, Slivka A, Stolz D, Sutton R, Weiss FU, Wilcox CM, Zarnescu NO, Wisniewski SR, O'Connell MR, Kienholz ML, Roeder K, Barmada MM, Yadav D, Devlin B (2012) Common genetic variants in the CLDN2 and PRSS1-PRSS2 loci alter risk for alcohol-related and sporadic pancreatitis. Nat Genet 44:1349–1354. doi:10.1038/ng.2466

Witt H, Luck W, Hennies HC, Classen M, Kage A, Lass U, Landt O, Becker M (2000) Mutations in the gene encoding the serine protease inhibitor, Kazal type 1 are associated with chronic pancreatitis. Nat Genet 25:213–216

Witt H, Apte MV, Keim V, Wilson JS (2007) Chronic pancreatitis: challenges and advances in pathogenesis, genetics, diagnosis, and therapy. Gastroenterology 132:1557–1573

Witt H, Beer S, Rosendahl J, Chen JM, Chandak GR, Masamune A, Bence M, Szmola R, Oracz G, Macek M Jr, Bhatia E, Steigenberger S, Lasher D, Buhler F, Delaporte C, Tebbing J, Ludwig M, Pilsak C, Saum K, Bugert P, Masson E, Paliwal S, Bhaskar S, Sobczynska-Tomaszewska A, Bak D, Balascak I, Choudhuri G, Nageshwar Reddy D, Rao GV, Thomas V, Kume K, Nakano E, Kakuta Y, Shimosegawa T, Durko L, Szabo A, Schnur A, Hegyi P, Rakonczay Z Jr, Pfutzer R, Schneider A, Groneberg DA, Braun M, Schmidt H, Witt U, Friess H, Algul H, Landt O, Schuelke M, Kruger R, Wiedenmann B, Schmidt F, Zimmer KP, Kovacs P, Stumvoll M, Bluher M, Muller T, Janecke A, Teich N, Grutzmann R, Schulz HU, Mossner J, Keim V, Lohr M, Ferec C, Sahin-Toth M (2013) Variants in CPA1 are strongly associated with early onset chronic pancreatitis. Nat Genet 45:1216–1220. doi:10.1038/ng.2730

Yamauchi K, Rai T, Kobayashi K, Sohara E, Suzuki T, Itoh T, Suda S, Hayama A, Sasaki S, Uchida S (2004) Disease-causing mutant WNK4 increases paracellular chloride permeability and phosphorylates claudins. Proc Natl Acad Sci U S A 101:4690–4694. doi:10.1073/pnas.0306924101

Zielenski J (2000) Genotype and phenotype in cystic fibrosis. Respiration 67:117–133

Chapter 29
Secretory Diarrhea

Nadia Ameen, Sascha Kopic, Md. Kaimul Ahsan, and Dmitri V. Kravtsov

Abstract Secretory diarrhea (SD) remains a major cause of morbidity and mortality globally particularly among infants and children. Enterotoxin-mediated diarrhea due to cholera, *Escherichia coli* (*E. coli*), and rotavirus remain major causes of SD in developing countries, while genetic diseases that produce defects in intestinal ion transporters and motor proteins are increasingly implicated in diarrheal diseases in developed countries. In recent years, significant advances have been made in understanding various factors that contribute to the underlying pathogenesis of SD. These advances are the result of research progress in the fields of genetics, cell biology, and physiology and animal models that have culminated in the development of new therapeutic agents for treating SD. The major advances in pathophysiology and treatment of enterotoxin and genetic causes of SD in childhood will be reviewed.

Keywords Secretory diarrhea • Microvillus inclusion disease • Cystic fibrosis transmembrane conductance regulator • Congenital chloride diarrhea • Downregulated in adenoma • Congenital sodium diarrhea • CFTR high expresser cell

29.1 Introduction

The molecular physiology of ion transporters is discussed in detail in other chapters. This chapter will focus on advances in the underlying pathophysiology of enterotoxin-mediated and childhood secretory diarrhea. The roles of specific intestinal ion transporters relevant to SD pathogenesis will be reviewed, and efforts to

N. Ameen (✉) • Md.K. Ahsan • D.V. Kravtsov
Department of Pediatrics, Yale University School of Medicine, 333 Cedar St, LCI-409C, New Haven, CT 06520, USA
e-mail: nadia.ameen@yale.edu; mdkaimul.ahsan@yale.edu; Dmitri.Kravtsov@yale.edu

S. Kopic
Department of Surgery and Cellular and Molecular Physiology, Yale University School of Medicine, New Haven, CT, USA
e-mail: Sascha.Kopic@yale.edu

K.L. Hamilton, D.C. Devor (eds.), *Ion Channels and Transporters of Epithelia in Health and Disease*, Physiology in Health and Disease,
DOI 10.1007/978-1-4939-3366-2_29

develop new therapies based on research advances in intestinal transporter physiology will be discussed.

The proximal small intestine is a major site for fluid absorption and secretion and the target site for enterotoxin-mediated secretory diarrhea (Cohen et al. 1989; Field and Semrad 1993). Under healthy conditions, more than 90 % of the 7 L of fluid that is secreted into the intestine each day is absorbed. Fluid secretion is driven largely by active secretion of chloride into the lumen and is necessary to maintain fluidity of intestinal contents during various stages of digestion, allowing for diffusion of enzymes and nutrients to preserve digestive functions. However in disease states, dysregulation of intestinal transport mechanisms alters the balance between absorptive and secretory processes such that secretion predominates and results in diarrhea (Barrett 2000).

In the small intestine, mature columnar enterocytes line villi or fernlike structures that project into the lumen and increase the absorptive surface area. Immature enterocytes line the crypts that are positioned between villi and extend in the opposite direction toward the submucosa. Enterocytes possess polarized distribution of plasma membrane that is divided into two defined regions, the apical brush border membrane (BBM) and basolateral membrane (BLM). This division is functionally significant because segregation of specific membrane transport proteins to either the apical or basolateral membrane facilitates transepithelial transport processes (absorption and secretion) of electrolytes. Villus enterocytes are tall columnar epithelial cells endowed with a well-developed brush border and express major absorptive transport proteins such as Na^+/H^+ exchanger (NHE3) and sodium glucose transporter 1 (SGLT1) on the apical BBM. Crypt cells are shorter and more immature compared to villus cells, possess a less developed brush border, and express abundant apical secretory (e.g., cystic fibrosis transmembrane conductance regulator, CFTR) and basolateral transporters (sodium potassium chloride co-transporter, NKCC1) to regulate anion secretion (Binder 2012; Binder and Reuben 2012) (Fig. 29.2).

In the healthy intestine, absorption and secretion occur simultaneously, but net absorption of electrolytes dominates. The geographic separation of villi and crypt supported separation of absorptive and secretory functions that were previously ascribed to villus and crypt, respectively. Recent data, however, suggest that both processes occur simultaneously indicating significant functional plasticity along enterocytes of the crypt-villus axis (Jakab et al. 2011). Over the last decade, new intestinal ion transporters have been identified, and significant advances made in understanding their modes of regulation both in cultured cells and the native epithelium. This new knowledge has had significant impact on new understanding of the normal ion transport processes, the resulting pathophysiological changes that lead to diarrhea, and design of targeted therapies (Binder 2012).

Secretory diarrhea (SD) results when bacterial or viral enterotoxins signal enterocytes to increase intracellular second messengers (cAMP, cGMP, calcium). This process leads to downstream activation of membrane transporters that result in increased anion secretion (chloride, Cl^-, and bicarbonate, HCO_3^-) that occurs with or without inhibition of sodium (Na^+) absorption. The increase in solute transport across the gut epithelium is associated with water transport into the lumen (Barrett and Keely 2000; Barrett 2000).

29.2 Clinical Relevance and Burden of Disease

SD is characterized by a large volume of watery stools that is present despite food withdrawal and most commonly follows infections by a host of bacterial, viral, and parasitic pathogens. In developing countries infectious pathogens are the most common cause of SD, and this remains a major cause of mortality in children less than 5 years of age (Kopic and Geibel 2010). The first estimates of the global burden of diarrhea in children became available in the 1980s. Those data indicated that diarrheal illnesses accounted for about 4.6 million deaths in children each year (Kopic and Geibel 2010). While the introduction of oral rehydration solution (ORS) by the WHO has significantly reduced the number of deaths attributable to diarrheal disease in the developing world, diarrhea and its sequelae (malnutrition, stunting, cognitive delay) remain a major cause of morbidity and mortality in those less than 5 years of age. In the developed world, death due to SD is rare and associated with noninfectious and genetic diseases of childhood such as microvillus inclusion disease (MVID) and congenital chloride diarrhea (CCD) (Sherman et al. 2004).

29.3 Infectious SD

Vibrio cholerae (*V. cholerae*) and *Escherichia coli* (*E. coli*) are the two most common bacterial pathogens responsible for SD. Of all enteric pathogens responsible for SD in humans, *V. cholerae* induces the most severe SD affecting humans and results in a rapidly fatal condition (Kopic and Geibel 2010). Significant advances have been made in understanding the mechanisms by which intestinal ion transporters contribute to the underlying pathogenesis of *V. cholerae* and *E. coli*-elicited SD. Translation of the research advances is currently being tapped for design of new therapies, some of which are in clinical trials for treatment of SD (Donowitz et al. 2012). Rotavirus is another common cause of SD and an important contributor to mortality in children. There has been recent advances in understanding rotavirus pathogenesis; however less is understood of the mechanisms underlying rotavirus-induced diarrhea compared to bacterial secretory enterotoxins (Lorrot and Vasseur 2007).

29.4 Channels and Transporters Relevant to SD

29.4.1 Cystic Fibrosis Transmembrane Conductance Regulator

CFTR chloride channels represent the primary apical exit pathway for chloride secretion from the enterocyte BBM. Detailed discussion of CFTR is provided in

Chap. 17. CFTR is the gene product responsible for the genetic disease cystic fibrosis (CF). It is also central to the underlying pathogenesis of *V. cholerae*- and *E. coli*-induced diarrhea. In CF, mutations in *CFTR* lead to trafficking defects that result in reduced or absent functional CFTR on the enterocyte BBM. The net result is failure to secrete chloride and bicarbonate, constipation, and intestinal obstruction. In SD, intracellular second messenger (cyclic nucleotides) signaling leads to increased CFTR abundance and anion secretion on the BBM (Golin-Bisello et al. 2005). Long before the CFTR gene was discovered, CF-affected populations are noted to be protected against cholera epidemics and SD, but the underlying explanation for these observations came only after the CFTR gene was cloned and functional characterization conducted in vivo (Riordan et al. 1989; Field and Semrad 1993). Indeed, studies in genetically engineered mice confirmed that levels of expression of functional CFTR on the enterocyte brush border correlated with fluid secretory responses and diarrhea following exposure to cholera toxin (CTX). $CFTR^{-/-}$ mice fail to secrete fluid similar to CF-affected humans, while heterozygote mice expressing 50 % of wild-type (WT) CFTR secreted 50 % of the normal chloride and fluid following CTX exposure. These studies confirm the selective advantage of the CF heterozygote to cholera and SD (Gabriel et al. 1994).

29.4.1.1 CFTR Localization in the Intestine: Relevance to SD

Chloride is taken up from the bloodstream across the basolateral membrane via NKCC1 that is driven secondarily by low intracellular sodium concentrations established by the active sodium/potassium ATPase pump (Na^+/K^+-ATPase) on the basolateral membrane. This arrangement allows chloride to accumulate in the cell above its electrochemical equilibrium. When apical chloride channels are opened, chloride flows out of the cell down this electrochemical gradient resulting in chloride secretion (Kopic and Geibel 2010). Early electrophysiological studies indicated that CFTR was confined to the apical membrane, consistent with a model of channel opening as the only mechanism regulating fluid secretion (Field and Semrad 1993). This model presumed that CFTR was confined to crypt epithelial cells, consistent with the paradigm of functional segregation of secretion and absorptive functions along the crypt-villus axis. However, the generation of superior specific anti-CFTR antibodies enabled detailed visualization of the subcellular distribution and mapping of CFTR distribution and its expression gradient along the proximal-distal and crypt-villus axis of the intestine (Jakab et al. 2011; Ameen et al. 1995). CFTR is present in the apical domain of epithelial cells of the stomach (Weis et al. 2013), Brunner glands of the proximal duodenum (Jakab et al. 2011), and crypt and villus enterocytes (Ameen et al. 2000), and very high levels are present in a subpopulation of crypt and villus enterocytes, called CFTR high expresser (CHE) in the duodenum and jejunum (Ameen et al. 1995; Jakab et al. 2013). CFTR is absent from goblet cells. Highest levels of CFTR are observed in the proximal small intestine and lowest levels in the colon (Jakab et al. 2011).

Surprisingly, localization studies revealed that while CFTR was indeed present on the apical membranes of enterocytes as predicted, it was not confined to the apical membrane. CFTR is present in subapical vesicles and on the apical membrane in all CFTR-expressing epithelial cells in the intestine (Ameen et al. 2000). Ultrastructural localization studies in the native intestine revealed that surprisingly about one third of the total CFTR in the enterocyte was localized to the apical membrane, while the majority of channels reside on the membranes of endosomal vesicles within the subapical cytoplasm of both crypt and villus enterocytes (Ameen et al. 2000). This subcellular distribution in the native intestine failed to reconcile with the existing paradigm of a singular regulation of CFTR by cyclic nucleotide-dependent phosphorylation on the apical membrane as suggested by electrophysiological studies. Indeed, further studies in the native intestine confirmed that this subcellular distribution in native enterocytes was critical for acute regulation of CFTR and fluid secretory response to enterotoxins under conditions of SD (Jakab et al. 2012; Golin-Bisello et al. 2005). CFTR distribution in the intestine resembled that of other transporters such as the insulin-stimulated glucose transporter 4 (Glut4). Localization and ultrastructural studies demonstrated that in adipose tissues endogenously expressing this transporter, Glut4 acutely and tightly controls glucose transport by shuttling the transporter from subapical vesicles into and out of the plasma membrane, thereby regulating the number and function of transporters on the plasma membrane (Rowland et al. 2011). Since then, immunolocalization approaches have been employed and confirmed that similar mechanisms of regulation are at work for other ion transporters in vivo (Jakab et al. 2012). Enterotoxin-dependent regulation of CFTR-mediated fluid secretion by traffic is discussed in the sections dedicated to *V. cholera* and *E. coli* pathogenesis.

29.4.1.2 CFTR and Bicarbonate

Clinicians have long noted defects in intestinal alkalization in the human CF intestine (Gelfond et al. 2013; Pratha et al. 2000). In CF, increased acidity due to absent CFTR on the enterocyte BBM contributes to bile acid-induced diarrhea, malabsorption, maldigestion, bacterial overgrowth, poor motility, and accumulation of thick inspissated mucus (Baxter et al. 1988). In addition to chloride, CTX and *E. coli* heat-stable enterotoxin (STa) elicit significant bicarbonate secretion and contribute to metabolic acidosis in affected individuals (De Jonge 1975b). Until recently, chloride secretion was central to CFTR function on the apical membrane of epithelia. But increasingly, the importance of CFTR to bicarbonate secretion is recognized, and more attention is given to this aspect of its function following the identification of chloride/bicarbonate exchangers on the brush border membrane in the intestine. In the intestine, CFTR-mediated bicarbonate secretion is functionally linked to members of the SLC26 family of apical chloride/bicarbonate exchangers that are predominantly found on the villus epithelium (SLC26A3, DRA, downregulated in adenoma; SLC26A6, PAT1, putative anion transporter 1; SLC4A2, anion exchanger AE2) (Singh et al. 2010). Apical bicarbonate secretion is dependent on basolateral HCO_3^- entry transporters such as NBC, sodium bicarbonate co-transporter. Mutations in the apical SLC26A3 (DRA) are directly linked to an important cause of childhood diarrhea referred to as congenital chloride diarrhea (CCD) that is discussed later in this section (Hoglund et al. 1996b; Makela et al. 2002). Immunolocalization studies were critical in mapping specific cellular localization sites of bicarbonate/chloride entry and exit pathways in the native intestine. These studies contributed to unraveling an important role for villus enterocytes of the proximal small intestine in CFTR-mediated bicarbonate secretion and provide a plausible explanation for the observed increase in duodenal luminal acidity in the CF human intestine (Jakab et al. 2012).

29.4.2 Sodium/Hydrogen Exchanger

The fundamentals of sodium absorption in epithelia are discussed in Chap. 2. In the intestine, the NHE3 plays a critical role in electroneutral sodium chloride absorption and contributes to the pathogenesis of SD (Singh et al. 2014; Binder 2012). NHE3 is present on the apical domain of villus enterocytes of the small intestine and surface cells of the colon (Jakab et al. 2011). NHE3 function on the apical membrane is linked to CFTR and Cl^-/HCO_3^- exchanger (Singh et al. 2010). Like CFTR, NHE3 is distributed on the apical membrane and in subapical vesicles of enterocytes and regulated by traffic into and out of the plasma membrane (Jakab et al. 2012). Under conditions of enterotoxin-induced SD, cyclic nucleotides (cAMP, cGMP) and second messengers maximize fluid secretion by simultaneously regulating traffic of NHE3 and CFTR in opposing directions on the enterocyte brush border (Fig. 29.2). Cyclic nucleotide signaling leads to rapid traffic of endosomal-associated CFTR into the brush border membrane to increase CFTR abundance and function, while simultaneously inhibiting sodium absorption by blocking NHE3 internalization, leading to increased NHE3 abundance on the BBM (Jakab et al. 2012).

29.5 CFTR High Expresser Cells

In 1995, our group identified a small (2 %) subpopulation of enterocytes expressing very high levels of CFTR in the human and rat intestine (Ameen et al. 1995). These cells, called CFTR high expresser (CHE) cells, are present in the crypt and villus of the duodenum and jejunum, but are more abundant in villus epithelium. CHE cells face the lumen between the long fernlike villi where they are positioned to respond to luminal milieu (Jakab et al. 2013). Immunoelectron microscopic studies reveal that villus CHE cells possess a prominent cytoskeleton, but express lower levels of the BBM myosin Myosin Ia (Myo1a), and possess slightly shorter microvilli, endocytic and exocytic trafficking machinery, and a large pool of subapical endosomes. In distinct contrast to villus enterocytes of the small intestine, CHE cells possess unique anti-absorptive features, in that they lack expression for absorptive apical proteins including NHE3 and sucrose isomaltase (Jakab et al. 2013; Ameen et al. 1995). CHE also lack apical chloride/bicarbonate exchangers that regulate bicarbonate secretion in conjunction with CFTR/NHE3. Interestingly, CHE cells possess higher levels of the basolateral NKCC1. Moreover, cAMP agonists including vasoactive intestinal peptide (VIP) and CTX trigger robust exocytosis of CFTR from the subapical endosomal pool into the BBM of CHE cells that can be reversed upon washout (Ameen et al. 1999), supporting a pro-secretory behavior in CHE cells.

In the absence of direct evidence to demonstrate chloride secretion originating from CHE cells, morphological characterization supports a role for robust CFTR-mediated fluid secretion from villus CHE cells following cyclic nucleotide stimulation, suggesting that these cells contribute to enterotoxin-mediated diarrhea.

29.6 Motors/Binding Partners in Brush Border Membrane CFTR Traffic: Relevance to SD

CFTR has a long half-life (>72 h) and efficiently maintains its surface abundance in the intestine by apical endocytosis, recycling, and return to the plasma membrane through the exocytic pathway (Silvis et al. 2009). This constitutive recycling has been observed in a number of CFTR-expressing epithelia including the intestine and requires no second messenger stimulation (Ameen et al. 2007). In the intestine, CFTR also undergoes robust regulated trafficking in enterocytes following second messenger signaling (cAMP, cGMP) (Golin-Bisello et al. 2005; Ameen et al. 1999). This is a unique feature of its behavior that is prominent in the proximal small intestine and appears to be critical for acute responses to bacterial enterotoxins in eliciting the massive increase in fluid secretion that results in SD. Enterocytes possess a prominent cytoskeleton support network rich in actin and microtubule motors that facilitates movement or traffic of CFTR into and out of the BBM. The role of molecular motors in CFTR traffic is reviewed more comprehensively in (Kravtsov and Ameen 2013). Both cAMP (CTX, heat-labile *E. coli* toxin) and cGMP (heat-stable *E. coli* toxin) mediated enterotoxins that utilize motors to redistribute CFTR into and out of the BBM in the native intestine. While much is left to understand about the roles of motors in CFTR traffic, the molecular motors do not appear to distinguish between second messengers, as available data indicate that cAMP and cGMP agonists employ similar motors to move CFTR. Cell-type specificity is a recognized characteristic of CFTR traffic that is linked to its function in native epithelial cells and dictated largely by specific adaptor signaling (Silvis et al. 2009; Guggino and Stanton 2006). In addition to cell-specific adaptors, CFTR traffic into and out of the BBM in the intestine is powered differentially by specific actin motors in crypt vs. villus enterocytes (Kravtsov and Ameen 2013). The role of actin motors and adaptors in SD (both cAMP and cGMP) pathogenesis will be discussed briefly.

The actin-binding motor, Myosin VI (Myo6), is present in many epithelial cells including crypt and villus enterocytes (Ameen and Apodaca 2007). Myo6 is a unique minus end-directed motor that regulates clathrin-mediated endocytosis of a specific class of receptors (low-density lipoprotein receptor, LDLR) from the plasma membrane (Buss et al. 2004; Roberts et al. 2004; Hasson 2003). Studies in transgenic mice lacking Myo6 provided important insights into the physiologic role that Myo6 plays in apical endocytosis of CFTR from the BBM and its contribution to STa-elicited diarrhea pathogenesis (Ameen and Apodaca 2007). Receptors for STa binding (guanylyl cyclase, GCC) are enriched on the apical membranes of villus enterocytes in the jejunum but also present in the crypt. STa-elicited fluid secretion is exclusively mediated by CFTR (Vaandrager et al. 1997; De Jonge 1975a). Absence of Myo6 in mice leads to defective apical endocytosis and accumulation of CFTR on the surface of crypt and villus enterocytes in the jejunum and increased STa-mediated fluid secretion, indicating the importance of this minus end-directed actin motor in regulating the enterocyte response to enterotoxin-elicited diarrhea (Ameen and Apodaca 2007) (Fig. 29.1; Myo6, *middle panel*).

LDLR receptor family members utilize Myo6 and the endocytic adaptor, Dab2, to facilitate clathrin-mediated endocytosis (Roberts et al. 2004; Buss et al. 2001; Morris et al. 2002). However, in contrast to airway epithelial cells, in the intestine, Dab2 does not play a direct role in CFTR endocytosis (Cihil et al. 2012; Collaco et al. 2010; Madden and Swiatecka-Urban 2012). Instead, the specific binding partner, AP2-alpha, directly binds to CFTR, while Dab2 is an indirect binding partner in directing Myo6-dependent endocytosis from the enterocyte BBM (Collaco et al. 2010).

A key step in enterotoxin-mediated increase in CFTR abundance and fluid secretion on the BBM is cAMP- or cGMP-regulated exocytosis of endosomal-associated CFTR into the BBM of crypt and villus enterocytes. CFTR exocytosis requires molecular motors. The specific exocytic motors and binding proteins responsible for CFTR exocytosis in the crypt have not been identified to date, but are likely to be a combination of microtubule- and actin-based motors. CFTR exocytosis into the villus enterocyte BBM of the proximal small intestine was recently shown to be dependent on the exocytic actin-binding motor Myosin Ia (Myo1a), a BBM myosin family member (Kravtsov et al. 2012). Myo1a is expressed on the apical domain of mature villus enterocytes of the small intestine, but is not in the crypt as it is critical for BBM development. Transgenic mice lacking Myo1a fail to traffic CFTR from subapical endosomes into the BBM under steady-state conditions but also following cAMP agonists. These findings are intriguing and may be crucial for the pathogenesis of CTX-mediated bicarbonate secretion in light of the new studies indicating the importance of CFTR in villus bicarbonate secretion in the proximal small intestine (Jakab et al. 2011; Kravtsov et al. 2012) (Fig. 29.1; Myo1a, *right panel*).

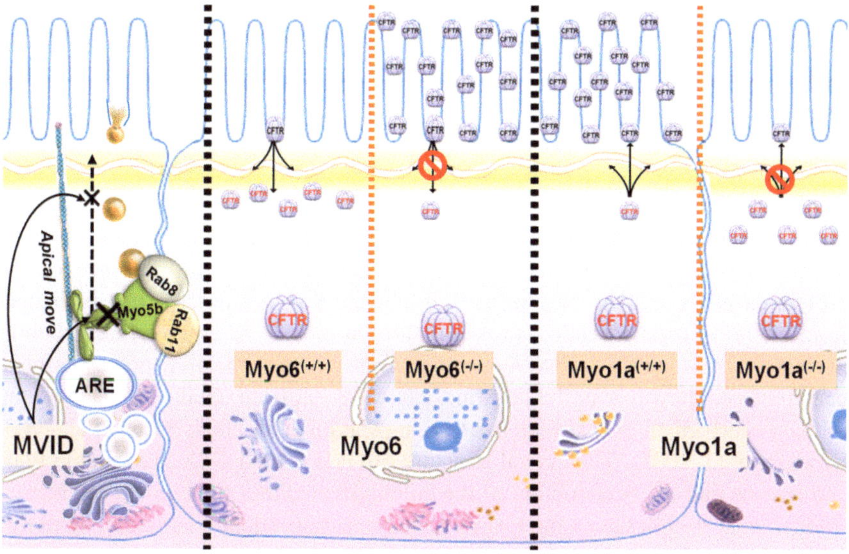

Fig. 29.1 Schematic diagram of myosin motor transport system-mediated recycling. Myo5b moves exocytic vesicles along actin filaments in the intestinal epithelial cells (*left panel*). Myo5b is mutated in MVID, leading to defective ARE-mediated recycling and exocytosis of apical proteins. Myo6 facilitates apical endocytosis of CFTR in the intestine. Deficiency of Myo6 leads to apical accumulation of CFTR (*middle panel*). Myo1a is important for the exocytosis and tethering of BB membrane-inserted CFTR in the villus enterocyte. Deficiency of Myo1a results in the accumulation of CFTR in the subapical compartment (*right panel*)

These observations suggest that targeting CFTR trafficking (either endocytosis or exocytosis) mechanisms into the enterocyte brush border membrane may be key to reducing fluid secretion in secretory diarrhea.

29.7 The Pathophysiology of Cholera Toxin-Mediated Diarrhea

Cholera still remains a major healthcare concern. According to a WHO report, over 190,000 cases of cholera were reported in 2008. As a result of the feco-oral transmission route, cholera outbreaks are typically associated with poor sanitary conditions occurring in the wake of natural or humanitarian disasters. The disease is characterized by its fulminant disease progression with water losses of up to 1 L/h resulting in severe and potentially fatal dehydration within hours (Kopic and Geibel 2010).

Cholera is caused by an enterotoxin secreted by the bacterium *V. cholera*. CTX belongs to the family of AB_5 toxins. Of note, the heat-labile enterotoxin LT-I secreted by ETEC is a member of the same toxin family (see below). AB_5 toxins are large multimeric proteins consisting of an enzymatically active alpha-subunit and five identical beta-subunits which are assembled in a ringlike fashion. The alpha-subunit is further divided into an A1 and an A2 domain, the latter of which is responsible for linking the entire alpha-subunit to the pentameric beta-subunit ring (Gill et al. 1981; Sixma et al. 1993).

Following secretion the toxin attaches to GM1-type membrane gangliosides. Gangliosides are glycosphingolipids in the plasma membranes of cells. GM1 is particularly enriched in so-called caveolae of the BBM (Parton 1994). Caveolae are distinct plasma membrane invaginations characterized by the presence of the protein caveolin and can be internalized into the cell in a process that is distinct from clathrin-mediated endocytosis (Singh et al. 2003). Alternative, i.e., clathrin dependent and clathrin and caveolin independent, internalization processes have been reported; however in the intestine, caveolae-mediated uptake seems to be the primary mechanism of toxin internalization (Massol et al. 2004; Singh et al. 2003; Kirkham et al. 2005). The crystal structure of GM1-bound CTX is available to us and reveals that the beta-subunit of the holotoxin mediates the binding interaction (Merritt et al. 1994). Given the toxin's pentameric beta-subunit structure, up to five GM1 gangliosides can be bound by the holotoxin.

Once internalized the toxin undergoes complex intracellular trafficking before exerting its hypersecretory activity. The holotoxin is first transported via endosomes to the trans-Golgi network (Pelkmans et al. 2004; Richards et al. 2002). From the trans-Golgi network, the holotoxin enters the ER, where it is further processed (Majoul et al. 1998; Richards et al. 2002). Here the protein is unfolded upon which the alpha1 subunit dissociates from the toxin. This process is catalyzed by the enzyme protein disulfide isomerase (PDI) (Tsai et al. 2001). Alternatively, it has been suggested that PDI is only responsible for the dissociation process of A1 and that the unfolding takes place spontaneously without enzymatic catalysis (Taylor et al. 2011). The alpha1 subunit then exits the ER (endoplasmic reticulum), most likely utilizing the ER-associated degradation system which physiologically translocates misfolded proteins from the ER into the cytosol for subsequent degradation (Hazes and Read 1997; Bernardi et al. 2008; Massey et al. 2009; Pande et al. 2007; Taylor et al. 2010). Misfolded proteins bound for degradation are usually ubiquitinated; however the alpha1 subunit escapes this fate given its lack of lysine residues, which are necessary for attachment of the ubiquitin tag (Hazes and Read 1997; Rodighiero et al. 2002). Once released into the cytosol, the alpha1 subunit exerts its toxic activity by transferring an ADP-ribose moiety from NAD+ to the alpha-subunit of a G-protein (Moss and Richardson 1978; Lee et al. 1991). The activated G-protein then stimulates adenylate cyclase, resulting in increased cAMP levels, PKA activation, and subsequent CFTR opening.

CTX-induced increase in cAMP levels drives robust traffic of endosomal-associated CFTR into the apical membrane to increase its abundance by at least twofold that is accompanied by a fourfold increase in fluid secretion in the native intestine. PKA phosphorylation is necessary for CFTR traffic from subapical vesicles into the enterocyte BBM, as well as channel opening on the plasma membrane (Golin-Bisello et al. 2005) (Fig. 29.2). However the contribution of

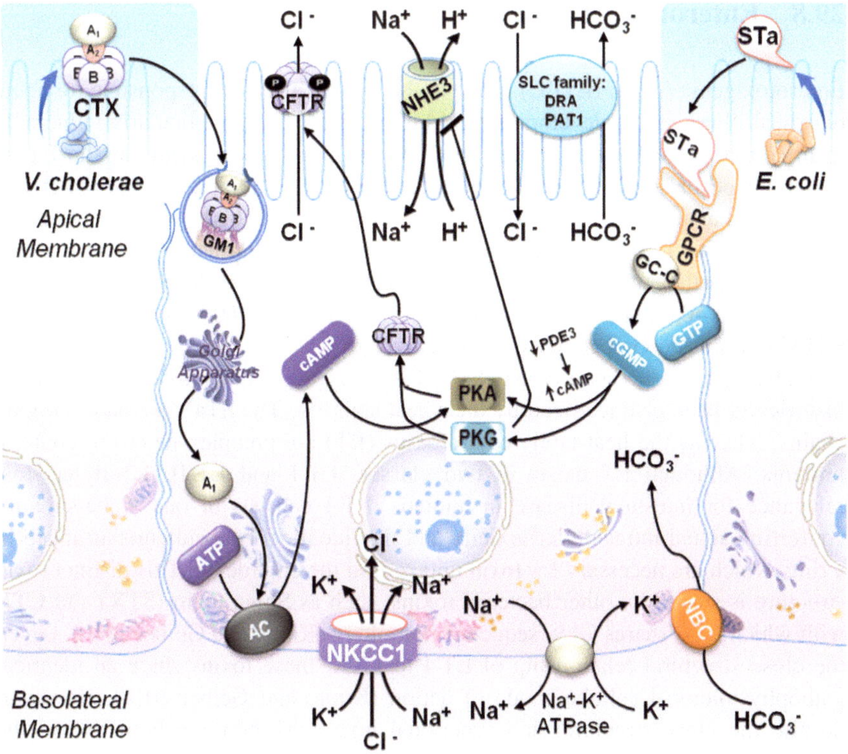

Fig. 29.2 Schematic illustration of ion transporters and cellular pathways relevant to secretory diarrhea pathogenesis. On the *left*, CTX binds to the surface ganglioside GM1 f.b. internalization, Golgi processing, and release of alpha-subunit of the toxin. The released alpha-subunit stimulates adenylate cyclase (AC), leading to increased intracellular levels of cAMP. On the *right*, heat-stable enterotoxin (STa) binds to G-protein coupling receptor (GPCR) and activates guanylyl cyclase C (GC-C), leading to increased intracellular levels of cGMP. Increased intracellular cAMP/cGMP increases intestinal fluid secretion by regulating apical trafficking of CFTR and/or NHE3 through activation of PKG or PKA signaling pathway, resulting in secretory diarrhea

endosomal traffic vs. phosphorylation of resident BBM CFTR channels to chloride secretion remains to be defined. The preferential abundance of CFTR in endosomes within enterocytes of the small intestine coupled with robust PKA-dependent traffic into the enterocyte BBM suggests that regulated traffic is critical to cholera-induced fluid secretion and secretory diarrhea pathogenesis.

29.8 Enterotoxigenic *E. coli* in SD

Enterotoxigenic *E. coli* (ETEC) is one of the major pathogens responsible for acute diarrheal illness in the developing world. For infants being particularly vulnerable to infection, it has been estimated that 280 million cases of symptomatic ETEC infections occur annually in children under the age of 5 (Kopic and Geibel 2010). Infection is associated with a significant mortality of 370,000 deaths per year, the majority of which are children (Kopic and Geibel 2010). Furthermore, ETEC is one of the main pathogens responsible for the development of traveler's diarrhea. Out of an estimated 40,000 daily cases of traveler's diarrhea, 50–60 % are attributable to ETEC (Kopic and Geibel 2010).

ETEC can produce two molecularly distinct types of enterotoxins, which have historically been characterized by their heat stability. The STa's are short peptide chains, whereas the heat-labile enterotoxins (LT) are complex hetero-oligomeric proteins. Although LT exists in two classes, LT-I and LT-II, LT-II has low relevance for intestinal disease in humans. LT-I consists of one alpha-subunit, conferring actual intracellular toxicity and five identical beta-subunits arranged in a ring, which are necessary for toxin uptake into the enterocyte. This distinct toxin structure is shared by other bacterial toxins, such as Shiga toxin (STX) and CTX with which LT-I shares 78 % sequence homology (Kopic and Geibel 2010). Given the close structural relationship of LT-I to CTX, these toxins share an identical pathophysiological cellular mode of action (Kopic and Geibel 2010). However, despite the close structural and functional homology between both toxins, the disease progression of CTX is in general more severe, which has been attributed to a more stable linkage between the alpha- and beta-subunits of CTX, thereby increasing the toxin effectiveness (Rodighiero et al. 1999). The pathophysiological mode of action of LT-I is further described in the section on CTX (Sect. 29.5).

The heat-stable enterotoxin of ETEC is produced in two variations: STa and STb. Both toxins can be experimentally discerned by their methanol solubility. STb is mostly associated with diarrheal disease in piglets and is not relevant for human pathology (Alderete and Robertson 1978; Takeda et al. 1979; Burgess et al. 1978). STa consists of a short 18 or 19 amino acid (AA) chain and can be further classified into STIa and STIb depending on the respective AA sequence (Staples et al. 1980; Aimoto et al. 1982; So and McCarthy 1980; Moseley et al. 1983). The STa peptide contains three characteristic disulfide bonds, which confer the characteristic heat stability to the toxin. The six involved cysteine residues are part of a 13AA sequence called the toxic domain, which is essential for eliciting a hypersecretory response in the enterocyte (Gariepy and Schoolnik 1986; Takao et al. 1985b; Gariepy et al. 1986; Aimoto et al. 1983). Analogous to LT-I and CTX, similar toxin architectures can be found in the heat-stable toxins from *V. cholerae* O1/non-O1 and *Yersinia enterocolitica* (Takao et al. 1985a, b; Yoshino et al. 1993).

The differences between LT-I and STa are not restricted to structural aspects but also predictably extend to divergent pathophysiological modes of action. Rather than relying on internalization and intracellular processing, STa exerts its toxic effects by binding to a receptor localized on the BBM of the enterocyte. First indications for the involvement of receptor binding emerged in the 1980s when radioisotope labeling of STa revealed characteristic binding kinetics including competitive inhibition (Giannella et al. 1983). A few years earlier, it had also been observed that STa exposure is associated with an increase in intracellular cGMP levels, which had been attributed to increased guanylate cyclase (GC) activity (Guerrant et al. 1980; Field et al. 1978). Today, we know that that the long elusive STa receptor is in fact a membrane-bound GC, which had initially been characterized as receptors of atrial natriuretic peptide (isoforms A and B) and were only later identified (isoform C) as STa receptors (Chinkers et al. 1989; Lowe et al. 1989; Schulz et al. 1990). The central role of GC-C in ETEC pathophysiology is effectively underlined by GC-C-deficient animals, which are immune to the hypersecretory effects of STa (Mann et al. 1997; Schulz et al. 1997). Interestingly, receptor levels are reported to decrease with age, which may explain higher disease burden in children (Cohen et al. 1988). Attempts at characterizing the binding site of STa to GC-C have been made utilizing receptor mutagenesis and photoaffinity labeling (Wada et al. 1996; Hasegawa et al. 1999). In lack of a precise crystal structure of GC-C, homology modeling with GC-A has also been conducted (Hasegawa et al. 2005; Lauber et al. 2009).

GC-C also has physiological ligands. Uroguanylin and guanylin are produced by local intestinal cells and are endogenous activators of GC-C, thereby serving as physiological secretagogues (Currie et al. 1992; de Sauvage et al. 1992; Nakazato 2001; Lorenz et al. 2003; Joo et al. 1998). STa shares significant structural homology to both peptides, including the characteristic disulfide bonds (Skelton et al. 1994).

GC-C possesses an uneven distribution pattern along the intestine. Its density was shown to decrease from the proximal to distal small intestine (Krause et al. 1994). Expression also seems to be heterogeneous along the vertical crypt-villus axis; however only inconclusive evidence about expression characteristics is available (Cohen et al. 1992; Krause et al. 1994).

STa binding to GC-C results in increased catalytic activity of the intracellular cyclase domain of the receptor, thereby elevating intracellular cGMP levels. cGMP in turn activates two intracellular signaling pathways: cGMP-dependent protein kinase isoform II (cGKII aka PKG) pathway and PKA pathway (Vaandrager et al. 2000). Preferential activation of these pathways varies along the intestine (the PKA pathway seems to be more specific to the colon); however a functional interplay between both pathways is most likely (Vaandrager et al. 2000). cGKII activation causes increased chloride secretion through CFTR and concomitant inhibition of sodium absorption through NHE, thereby increasing luminal osmolarity and water loss (Chao et al. 1994; Field et al. 1978; Pfeifer et al. 1996) (Fig. 29.2). In line with this hypothesis, cGKII-deficient animals demonstrate an impaired secretory response to STa exposure (Vaandrager et al. 2000; Pfeifer et al. 1996). The PKA-mediated pathway is less well characterized. It has been suggested that activation of cGMP leads to increased phosphodiesterase activity, thereby leading to an accumulation of cAMP and PKA activation (Vaandrager et al. 2000).

29.9 Treatment of SD

The role of ORS and current treatment approaches to SD can be found in a number of excellent reviews (Farthing 1994, 2001). ORS has remained the primary therapeutic modality for fluid replacement and rehydration for SD since its development in the 1970s. ORS formulation takes advantage of glucose-dependent sodium absorption and the stoichiometry of SGLT1 transport from the BBM of enterocytes. While ORS does not directly address the underlying pathophysiology of SD and significantly reduce stool volume and frequency, its implementation has saved millions of lives particularly in the developing world. A very recent study examined the reasons for continued fluid secretion and diarrhea with ORS. Interestingly, Yin et al. found that glucose increases intracellular calcium and stimulates active chloride secretion in the small intestine (Yin et al. 2014). These new studies have implications for current ORS therapy and suggest that modification to the current formulation may be possible to help reduce diarrhea and fluid secretion.

Being the common molecular endpoint of various SD entities, CFTR has been identified as a potential pharmacological target. To date several specific, nonabsorbable small-molecule CFTR inhibitors have been discovered with the aim of abrogating the hypersecretion of chloride occurring during SD (Ma et al. 2002; Muanprasat et al. 2004; Sonawane et al. 2005). Despite their excellent experimental efficacy of reducing chloride and water flux through CFTR, clinical applications have not been successful in the light of toxicity issues due to impairment of mitochondrial function (Kelly et al. 2010). Crofelemer is a new antisecretory agent that has had more success in progress toward clinical use (Crutchley et al. 2010). This compound is a purified proanthocyanidin oligomer extracted from the latex of a South American plant that has been used by locals for many years to treat diarrhea. Crofelemer has been shown to inhibit both CFTR and calcium-activated chloride channels (CaCCs) at the luminal membranes in enterocytes (Tradtrantip et al. 2010). This compound is currently in clinical trials for treatment of infectious secretory diarrhea including cholera and diarrhea-predominant irritable bowel syndrome and was recently approved for HIV diarrhea (Tradtrantip et al. 2010; Cottreau et al. 2012; Crutchley et al. 2010; Yeo et al. 2013).

29.10 CFTR/NHE3 in Secretory Diarrhea

CTX-triggered elevations of cAMP levels further inhibit Na^+ uptake on the villus enterocyte brush border membranes by NHE3, thereby increasing the osmotic driving force for fluid secretion (Donowitz and Welsh 1986). Net fluid secretion results from the simultaneous coordinated increase in CFTR activation and inhibition of NHE3-mediated Na^+ absorption through rapid internalization of NHE3 from the BBM that is facilitated by signaling through the cytoskeleton, Na^+/H^+ exchange regulatory cofactor, and adaptors (Singh et al. 2014). CFTR and NHE3 are expressed in the villus epithelium of the proximal small intestine although NHE3 is more abundant than CFTR with the exception of the CHE subpopulation (Jakab et al. 2011).

While the evidence is undisputable that CFTR and NHE3 are major mediators of CTX-induced diarrhea, the role/contribution of specific cell types, crypt and villus enterocytes, in CTX-induced secretion is not clear. CTX elicits secretion of mucus, chloride, and bicarbonate in the intestine. Mucus is secreted by goblet cells, but it is unclear whether CFTR is present in this cell type. How CTX activates chloride/bicarbonate secretion, whether specific cell types are associated with chloride versus bicarbonate secretion, and the details of how these are functionally linked to mucus secretion and NHE3 function are emerging but not fully understood. The crypt is a major site of CFTR expression and presumed target for CTX and chloride secretion. However, early studies in native epithelium revealed that CTX exerts its action first by elevating cAMP levels in villus enterocytes and it elicits fluid secretion from both villus and crypt cells (De Jonge 1975b). The observation that villus enterocytes of the proximal small intestine express CFTR and chloride/bicarbonate exchangers on the apical membrane and contribute significantly to cAMP-activated bicarbonate secretion supports a model whereby CTX elicits chloride secretion predominantly from crypt cells, while villus cells secrete bicarbonate and inhibit sodium absorption via NHE3 (Jakab et al. 2012). Moreover, this model is consistent with recent data indicating that CFTR-mediated bicarbonate secretion from enterocytes is critical for mucus release from neighboring goblet cells (Yang et al. 2013). This scenario contrasts with conditions where CFTR is absent from the apical membranes of enterocytes, as in CF. Absence of CFTR in the enterocyte brush border is associated with thick acidic adherent mucus, lack of fluid secretion, and intestinal obstruction (Eggermont 1996). Thus a picture is unfolding in support of cell-specific functions in CTX-induced diarrhea, but the details will require further studies aimed at mapping cellular sites along the crypt-villus epithelium using intact epithelium. In addition to its direct effects on enterocytes, *V. cholera* stimulates fluid secretion indirectly through its effects on neural pathways by stimulation of enterochromaffin cells to release serotonin, which in turn activates release of VIP from local enteric neurons (Camilleri et al. 2012; Navaneethan and Giannella 2008). Through its release of cAMP, VIP in turn activates signaling and insertion of CFTR associated with endosomes into the enterocyte brush border to increase anion secretion (Ameen et al. 1999).

29.11 Rotavirus

Rotavirus is the most common cause of gastroenteritis and diarrhea in young children. In healthy children, the disease is self-limiting; however, rotavirus is associated with significant morbidity. The number of reported rotavirus-associated deaths in young children was 453,000 in 2008 (Bohles et al. 2014; Srivastava et al. 1999). Morbidity and mortality is increased under conditions of poor sanitation, absence of clean water supplies, overcrowding, and malnutrition. Rotavirus is a non-enveloped virus that is stable against environmental influences. Rotavirus targets enterocytes of the upper two thirds of the villus epithelium in the small intestine, where it produces an enterotoxin called NSP4 (Lorrot and Vasseur 2007; Lorrot et al. 2006). Rotavirus-induced diarrhea was previously thought to be primarily anti-absorptive, as the virus is known to induce functional defects in villus BBM disaccharidase activities, glucose and sodium transport that contributes to massive water loss into the lumen without histologic damage. Moderate net chloride secretion coincident with diarrhea has been reported, but the increase in luminal chloride following rotavirus is much lower compared to that in pure SDs elicited by bacterial enterotoxins of *V. cholerae* or *E. coli*. Other studies reported both chloride influx and efflux across the villus BBM following rotavirus infection (Lorrot et al. 2006).

The precise source of chloride secretion in rotavirus diarrhea remains unclear. Rotavirus is able to elicit diarrhea in mice lacking CFTR, suggesting that alternate chloride transport pathways are implicated. CaCCs and a chloride channel ClC-2 are suggested as alternate chloride efflux pathways to CFTR in the intestine. With regard to rotavirus pathogenesis, existing data indicate that the mechanisms are quite different compared to bacterial enterotoxins that cause a pure secretory diarrhea. Recently, studies have focused on CaCCs. The rotavirus nonstructural protein NSP4 and the synthetic peptide $NSP4_{114\text{-}135}$ were shown to elicit diarrhea in young rodents (Ousingsawat et al. 2011). Infected intestinal cells release NSP4 into the lumen where it acts as a viral toxin and induces calcium-dependent chloride secretion and inhibition of SGLT1. New data using the synthetic peptide $NSP4_{114\text{-}135}$ suggest that rotavirus may elicit diarrhea by activating the calcium-activated chloride channel TMEM16A, while inhibiting the epithelial Na^+ channel ENaC and the SGLT1 (Ousingsawat et al. 2011). However, the role of CaCCs in the native intestine remains controversial and their contribution to secretory diarrhea pathogenesis is currently unclear. More detailed discussions of CaCCs can be found in excellent reviews (Barrett and Keely 2000; Pedemonte and Galietta 2014) and in Chap. 18 of this book.

29.11.1 *Treatment for Rotavirus*

The development of effective rotavirus vaccines has already markedly reduced diarrhea, morbidity, and mortality from rotavirus gastroenteritis in affected children worldwide (Gray 2011). In addition to vaccines, a number of groups are investigating the role of CaCCs in rotavirus-induced diarrhea with a view to therapeutic intervention.

29.12 Congenital Diarrhea

This section reviews the secretory diarrheal diseases that are not induced by any infectious organisms, but rather have the following features in common: they all manifest early in life, often immediately after birth, and they are a result of inborn genetic defects and often are notoriously difficult to treat. Such entities are collectively called congenital diarrheas.

29.12.1 *Congenital Chloride Diarrhea*

Gamble et al. and Darrow (Gamble et al. 1945; Darrow 1945) first described infants with diarrhea from birth, metabolic alkalosis, stool chloride levels in excess of stool sodium values, and failure to thrive. They termed the disorder "congenital alkalosis with chloride diarrhea." It was later renamed CCD (McReynolds et al. 1974), an autosomal recessive form of severe chronic diarrhea characterized by secretion of large amounts of watery stool containing high levels of chloride, leading to dehydration, hypokalemia, and metabolic alkalosis. The electrolyte disorder resembles the renal disorder, Bartter syndrome, except that CCD is not associated with calcium abnormalities. Hyponatremia, hypochloremia, and metabolic alkalosis are the notable biochemical abnormalities in affected children and infants (Choi et al. 2009). Voluminous watery stools are present from birth, and polyhydramnios is a typical feature, presumably due to intrauterine diarrhea (Holmberg et al. 1975). The disease is most common in Finnish (incidence 1:20,000), Polish (incidence 1:200,000), and Arabian populations (incidence up to 1:5000) (Kere et al. 1999). Defective Cl^-/HCO_3^- exchange was proposed as the pathophysiologic mechanism as early as 1970s (Holmberg et al. 1975; Holmberg 1978), but it was not until 1990 that a gene defect was implicated in CCD. Linkage disequilibrium analysis in Finnish families (Kere et al. 1993) positioned the candidate gene on chromosome 7 close to but distinct from the CFTR gene, and more detailed genetic and physical mapping studies suggested a candidate gene known as SLC26A3 (Hoglund et al. 1995, 1996a, b). Each of the three high-incidence populations has its own founder mutation: in Finland, the V317del mutation is present in all but one

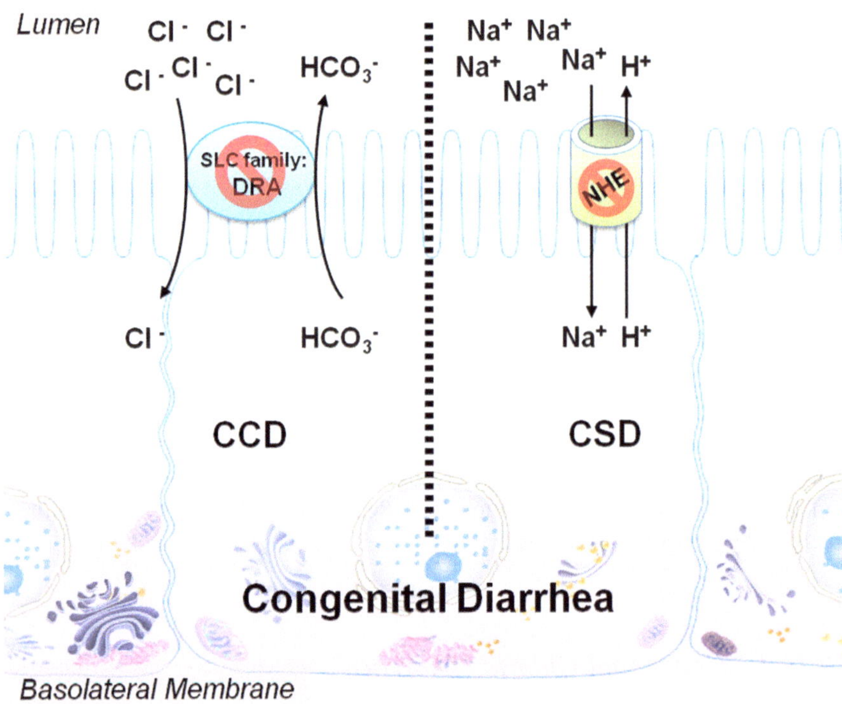

Fig. 29.3 Schematic diagram of pathogenesis of congenital chloride (CCD) and sodium diarrhea (CSD). DRA (SLC26A3) is a member of the SLC26 sulfate permease/anion transporter family that is primarily expressed on the apical brush border membrane of the intestinal (ileum, colon) epithelium and regulates Na^+-independent Cl^-/HCO_3^- (or OH^-) exchange. Defect or dysfunction of DRA leads to accumulation of Cl^- in the lumen resulting in large volume of watery Cl^- rich diarrhea (*CCD*). Na^+/H^+ exchangers (NHEs) regulate the Na^+/H^+ exchange across the plasma membrane. Mutations in NHEs are thought to result in malabsorption of Na^+ manifesting as Na^+-rich watery diarrhea (*CSD*)

chromosome studied to date (98 %); in Saudi Arabia and Kuwait, the G187X mutation is present in 17 of 18 chromosomes (94 %); and in Poland the most common mutation I675–676 accounts for 16 of 34 disease chromosomes (47 %) (Kere et al. 1999). DRA (SLC26A3) encodes a Na^+-independent Cl^-/HCO_3^- (or OH^-) exchanger. It is a member of the SLC26 sulfate permease/anion transporter family, and it is expressed mainly in the apical brush border of the intestinal epithelium. The basic defect in CCD is loss of SLC26A3-mediated transport in the surface epithelium of ileum and colon. This results in defective intestinal absorption of Cl^- and secretion of HCO_3^-. Secondarily the coupled Na^+/H^+ transport through the Na^+/H^+ exchangers (NHE2 or NHE3) is defective, leading to loss of both sodium and chloride and watery Cl^--rich diarrhea (Fig. 29.3; CCD, *left panel*). The only extraintestinal tissues that express SLC26A3 are eccrine sweat glands and seminal vesicles (Makela et al. 2002).

Interestingly, DRA is the exit pathway for SO_4^{2-} (Whittamore et al. 2013; Tyagi et al. 2001) as well as HCO_3^-; DRA was shown to be upregulated in the colon of mice lacking NHE3 exchanger (Melvin et al. 1999). Taken together with the reports that CFTR markedly activates Cl^- and OH^-/HCO_3^- by DRA, SLC26A6, and pendrin (Ko et al. 2002), it may be safe to postulate that the intestinal machinery responsible for Cl^- and HCO_3^- transport is regulated as a functional unit.

29.12.1.1 Treatment for CCD

Treatment of CCD consists of replacement with a high chloride intake to prevent volume depletion. Determining the optimal replacement dose is challenging because inadequate salt substitution paradoxically decreases diarrhea volume, and excessive salt administration also increases diarrhea by osmotic mechanisms (Wedenoja et al. 2010). In patients with CCD, oral intake of chloride, sodium, and potassium must exceed fecal output (i.e., there must be a positive gastrointestinal balance) so that obligatory losses in sweat can be replaced. A positive balance can best be ensured by a high intake of chloride, even though it exacerbates diarrhea.

Aichbichler and colleagues successfully used a proton-pump inhibitor, omeprazole, for treatment of CCD in a 34-year-old male. Omeprazole reduced the number of stools from 6–12 per day down to 2–4 per day and prevented episodes of fecal incontinence. However, it did not reduce the need for careful monitoring of dietary intake, serum electrolyte concentrations, and urinary chloride excretion (Aichbichler et al. 1997).

Canani and co-workers examined butyrate as a possible treatment for CCD in 2004, concluding that butyrate could be effective in treating CCD (Canani et al. 2004). It is easily administered useful in preventing severe dehydration episodes and may be a promising therapeutic approach for a long-term treatment in this rare and severe condition (Canani et al. 2004). Wedenoja et al. reported on the controversies associated with butyrate treatment (Wedenoja et al. 2008). Recently, Canani and colleagues revisited their study of butyrate efficiency, demonstrating its dependence on genotype in seven CCD children; the effect of butyrate was related in part to a different modulation of the expression of the two main apical membrane Cl^- exchangers of epithelial cells, both members of the SLC26 anion family (Canani et al. 2013). To improve the early diagnosis and prevent unnecessary surgical interventions for conditions like CCD that mimic meconium ileus (Krzemien et al. 2013), genetic and ultrasonographic methods are useful for prenatal diagnosis.

29.12.2 Congenital Sodium Diarrhea

Congenital sodium diarrhea (CSD) was initially described in patients with a diarrhea closely resembling CCD, but with stool electrolyte and other studies excluding this possibility, and rather suggesting a defect in sodium/proton exchange (Booth et al. 1985; Holmberg and Perheentupa 1985). Clinical features reported included a net Na^+ secretory state (stool Na^+ up to 145 mEq/L) by jejunal perfusion studies, defective Na^+/H^+ exchange, and high bicarbonate content (Fig. 29.3; CSD, *right panel*). Polyhydramnios was reported as a complication of pregnancy, watery diarrhea began immediately after birth, and serum studies showed metabolic acidosis and hyponatremia (Booth et al. 1985).

The specific member of the NHE family that is directly responsible for the disease remains to be identified. Muller et al. conducted analysis of clinical and laboratory data from five affected patients and concluded that CSD is an autosomal recessive disorder, but is not related to mutations in the NHE1, NHE2, NHE3, and NHE5 genes encoding for currently known sodium/proton exchangers (Muller et al. 2000). Recently, NHE8, another member of the NHE family, was cloned and its expression was found to be higher in the neonatal intestine compared to adult, suggesting that it may be implicated in CSD. However, genetic studies of five patients failed to identify mutations in NHE8 (Baum et al. 2011).

Several variants of the disease have been reported (Keller et al. 1990; Fell et al. 1992), including the variant associated with morphological malformations, such as anal atresia and double kidney (Bird et al. 2007; Heinz-Erian et al. 2009). The latter group applied a genome-wide SNP scan and identified a homozygous c.593-1G→A splicing mutation in SPINT2, encoding a Kunitz-type serine-protease inhibitor, in one extended kindred with syndromic CSD. The same mutation and four distinct, homozygous, or compound heterozygous mutations (p.Y163C, c.1A→T, c.337+2T→C, c.553+2T→A) were identified in all syndromic patients. No SPINT2 mutations were found in classic CSD patients, leading the authors to delineate syndromic CSD as a distinct disease entity caused by SPINT2 loss-of-function mutations (Heinz-Erian et al. 2009).

Treatment of CSD remains supportive in the absence of more detailed understanding of the mechanisms leading to disease.

29.12.3 Microvillus Inclusion Disease

MVID is a life-threatening rare autosomal recessive disorder of childhood that presents with severe intractable diarrhea. Without total parenteral nutritional (TPN) support or intestinal transplantation, death from diarrhea and dehydration is inevitable. Fecal sodium and chloride concentration approximate those of serum, and the volume of stool loss often exceeds that seen in cholera. In contrast to CCD, oligohydramnios or prenatal diarrhea is not a feature of MVID, and diarrhea is noted most often after feeding in the first hours and days of life (Cutz et al. 1989; Sherman et al. 2004). MVID has been identified in infants from multiple ethnic backgrounds, is associated with consanguinity, and clusters in certain populations including infants of Navajo Indian descent (Pohl et al. 1999). Two variants of MVID are recognized: an early and late onset. Late-onset MVID can present within the first month of life and is usually less severe than the early variant (Croft et al. 2000).

In 1978, Davidson and co-workers first recognized MVID in a report of five infants with persistent severe diarrhea from birth and marked abnormalities of absorption with failure to thrive. The diarrhea eventually led to death in four out of five children due to failure to control and correct fluid and nutritional abnormalities. Ion flux studies demonstrated a net secretory state that accounts for the voluminous diarrhea. Marker perfusion studies of the proximal jejunum also revealed abnormal glucose absorption and a blunted response of Na^+ absorption. Net secretion of Na^+ and H_2O was also reported in MVID patients (Davidson et al. 1978).

Initial unifying morphological findings in the small intestine of patients suffering from MVID included villus atrophy, crypt hypoplasia without an increase in mitoses or inflammatory cell infiltrate in the lamina propria and absence of a brush border in villus enterocytes, increase in lysosome-like inclusions, and autophagocytosis (Davidson et al. 1978). Follow-up reports confirmed the ultrastructural presence of microvilli decorating intracytoplasmic inclusions (microvillus inclusions, MIs) in the enterocytes, leading to the designation MVID and the recognition of this feature as a pathognomonic hallmark of the disease.

In the following decade, several other cases were identified (Phillips et al. 1985; Cutz et al. 1989) and characterized. Cutz's group concluded that MVID might very well represent the most common cause of severe refractory diarrhea in the neonatal period. First notions were made of the possible pathogenesis of the disease, implicating defective intracellular transport leading to aberrant assembly or differentiation of the components of the enterocyte brush border. Rectal biopsy was proposed as a dependable and relatively easy method for early diagnosis (Cutz et al. 1989).

The clear definition of the clinical presentation and diagnostic features facilitated further analysis of the pathogenesis of MVID. Rhoads and co-workers conducted a study on the transport abnormalities of the MVID small intestine (Rhoads et al. 1991). They analyzed proximal and distal tract outputs via a proximal jejunostomy placement in one patient showing that less fluid was lost from mouth to jejunum (60 mL/kg than from jejunum to anus, 100 mL/kg/day). They also performed Na^+,Cl^- and conductance analysis in the piece of jejunum excised during jejunostomy and reported that compared with the intestine of normal infants, this infant's epithelium showed transmural conductance and unidirectional ion fluxes that were only 30 % of normal. With respect to both Na^+ and Cl^-, the excised jejunum was in a net secretory state.

Phillips and Schmitz surveyed 23 cases of MVID and described the pattern of the diarrhea, defining congenital and late-onset cases, and the volume and ion content of stools in affected patients (Phillips and Schmitz 1992). Late-onset cases appeared to have a better prognosis, which was still very grim overall. The authors also reported that the presence of inclusions is characteristic of the mature enterocytes, while in the less differentiated cells in the crypts, a counterpart of the inclusions would be the abundant "secretory granules." Electron micrographs presented in their paper demonstrated that there are at least three different and distinctive appearances of the proximal small intestinal mucosa in MVID and that not all enterocytes completely lack microvilli, but enterocytes possessing mature brush borders were not abundant. These authors documented MIs attached to the apical plasma membrane and to the apical cell-cell contact areas (Phillips and Schmitz 1992).

The presence of intracellular vacuolar apical compartment with microvilli was noted previously in the cultured MDCK cell line following the disruption of cell-cell contacts (Vega-Salas et al. 1988). Chronic treatment of cultured cells with microtubular transport inhibitors such as vincristine, vinblastine, and cytochalasin resulted in the formation of basolateral inclusions laden with microvilli and the collapse of apical polarity (Carruthers et al. 1986; Ellinger and Pavelka 1986; Achler et al. 1989). By the mid-1990s MVID was considered among the disorders resulting from alterations in epithelial polarity (Fish and Molitoris 1994). During this period, intestinal transplantation emerged as a recognized treatment for MVID (Oliva et al. 1994; Bunn et al. 2000; Randak et al. 1998).

Attempts at more detailed characterization of the molecular defect leading to diarrhea in MVID began to appear in the late 1990s to early 2000s. Michail et al. (1998) observed reduction in the expression levels of NHE2, NHE3, and SGLT1, but not NHE1 in the intestinal samples from MVID patients. They concluded that the patients with MVID have defects in apical but not basolateral membrane transport systems and proposed that these defects are related to the pathogenesis of the disease (Michail et al. 1998). In 2000, Ameen and Salas executed a large-scale comparative analysis of protein distribution in the MVID intestine, comparing apical (CFTR, actin, NHE3, sucrase-isomaltase, alkaline phosphatase, and cGKII) and basolateral (Na^+/K^+-ATPase) protein localization patterns. They demonstrated that all apical proteins examined were mis-localized in MVID enterocytes and redistributed from the BB area into the cytosol and MIs. The effect was more pronounced in mature villus enterocytes. Distribution of the basolateral marker Na^+/K^+-ATPase was not affected (Ameen and Salas 2000), pointing to an apical-specific trafficking defect; however, the latter discovery could be case specific, as subsequent studies identified redistribution of basolateral transferrin receptor as well as Na^+/K^+-ATPase in MVID tissues (Dong et al. 2012; Thoeni et al. 2014b).

Increased paracellular macromolecular transport and reduced glucose uptake in duodenal biopsies of patients with MVID were demonstrated by Bijlsma et al. (2000). Reinshagen and colleagues used immunogold electron microscopy to demonstrate that labeled ovalbumin appears in the inclusions after apical treatment, concluding that MIs are the result of autophagocytosis of the apical membrane in MVID (Reinshagen et al. 2002). Studies of small GTP-binding protein

Rab8 in the knockout mouse model by Sato et al. initially seemed very promising for deciphering the MVID pathogenesis, because the absence of Rab8 leads to the formation of BB-decorated inclusions of KO mice resembling those of MIs in the human MVID intestine (Sato et al. 2007). However, Muller and co-workers identified the gene of Myosin Vb (Myo5b), a 214 kDa molecular motor, as the target gene responsible for MVID (Muller et al. 2008). This was consistent with previous observations of a 200 kDa band disappearance from intestinal immunoblots by Carruthers and co-workers (Carruthers et al. 1985). The group also demonstrated that the absence of Myo5b affects cell polarity (Muller et al. 2008). In 2010, Ruemmelle et al. confirmed that Myo5b was implicated in MVID pathogenesis and identified 15 novel Myo5b mutations in MVID patients. They also provided the first in vitro cellular model of MVID based on siRNA silencing of Myo5b expression (Ruemmele et al. 2010).

Current research focuses on the characterization of apical transport pathways connecting apical recycling endosomes (ARE) to the apical plasma membrane. MVID is assigned to the category of diseases associated with defects in "enterocyte differentiation and polarization" (Canani and Terrin 2011). Myo5b binding partners including Rab8 and Rab11 are uniformly accepted as small GTP-ases that are involved in the pathogenesis of MVID (Dhekne et al. 2014; Talmon et al. 2012; Dong et al. 2012) (Fig. 29.1; MVID, *left panel*). Recently, aberrations in the phosphorylation of ezrin and defects in apical PDK1 signaling were identified giving an insight into the mechanism of apical BB abnormalities and defects in apical sodium transport in the human MVID intestine and an intestinal Caco2 cell model (Kravtsov et al. 2014; Dhekne et al. 2014). In 2014, Wiegerinck et al. provided an insight into the early- and late-onset variants of MVID by demonstrating that the latter is the result of mutations in the Syntaxin 3 gene (Wiegerinck et al. 2014). The involvement of a different gene may explain the difference in the patient presentation in the variant form of MVID.

Despite the significant progress in understanding the genetic and molecular mechanisms involved in the pathogenesis of MVID over the past 5 years, several important questions remain unanswered. The underlying reasons for the severity of the secretory diarrhea remain baffling. In addition, although Myo5b is expressed in many epithelia outside the intestinal tract (Rodriguez and Cheney 2002), the reason for specificity of clinical disease manifestation in the small intestine is not understood. In fact, there is a single report of two cases of extraintestinal pathology in MVID describing the development of renal Fanconi syndrome in patients on parenteral nutrition. The authors were unable to establish a correlation of Myo5b mutations with aberrant apical plasma membrane morphology or altered ARE organization in renal tubular epithelial cells (Golachowska et al. 2012).

29.12.3.1 Treatment MVID

In 1994, the first successful intestinal transplantation was performed for MVID (Oliva et al. 1994). Since then, there have been no advances in therapeutic modalities largely because of lack of understanding of the underlying pathogenesis of the secretory diarrhea. Thus TPN and intestinal transplantation are the only currently available treatment options. While supportive, these are not ideal therapeutic modalities. Long-term use of TPN is associated with many complications including growth retardation, septicemia, cirrhosis, and cholelithiasis, all of which contribute to impaired survival in MVID patients. Intestinal transplantation is expensive, available only in specialized tertiary case centers, and associated with high mortality.

29.12.4 Tufting Enteropathy

Congenital tufting enteropathy (CTE) or intestinal epithelial dysplasia is a rare autosomal recessive disease affecting multiple epithelial tissues. CTE presents with intractable diarrhea of infancy characterized by villous atrophy, absence of inflammation, and focal epithelial tufts involving the small intestine and colon (Goulet et al. 1995). CTE presents early in life with chronic watery diarrhea and failure to thrive, and most affected individuals require parenteral nutrition for normal growth and development (Sivagnanam et al. 2008).

The disease was first described by Reifen et al. (1994) in a group of three children, each of whom presented early in life with protracted watery diarrhea with volumes in excess of 1500 mL/day. The diarrhea was refractory to a variety of diets, necessitating the use of TPN. Characteristic histopathologic findings included partial villous atrophy accompanied by crypt hyperplasia and an increased number of mitotic figures in the crypts. The most striking findings in the small intestine of the three children were observed in the surface epithelium, which showed focal epithelial "tufts." Tufts were composed of closely packed enterocytes with apical rounding of the plasma membrane, resulting in a teardrop configuration to the cells. Approximately 80–90 % of epithelial surface contained tufts, versus <10 % in the normal jejunum. Colonic biopsies showed no histopathologic abnormalities: specifically, no tufting lesions were observed (Reifen et al. 1994).

Although CTE was initially considered a defect of a2b1 integrin (Goulet et al. 1995; Patey et al. 1997), recent genotyping studies by Sivagnanam et al., supported by direct sequencing of genes on chromosome 2p21, discovered a homozygous mutation in epithelial cell adhesion molecule (EpCAM) of affected patients (Sivagnanam et al. 2008). Immunohistochemistry and immunoblotting of patient intestinal tissue revealed decreased expression of EpCAM. The authors concluded that EpCAM mutations are responsible for CTE and presented further evidence for EpCAM as the gene for congenital tufting enteropathy by identifying a novel homozygous C>T substitution at c.412 of exon 3 of EpCAM resulting in amino acid change from arginine (CGA) to a stop codon (UGA) (Sivagnanam et al. 2010b).

As more cases of CTE were identified, a variant form of the disease associated with various malformations (most consistently keratitis and choanal atresia) (Abely et al. 1998; El-Matary et al. 2007; Bird et al. 2007; Al-Mayouf et al. 2009) was described and found to be caused by mutations of SPINT2, a gene associated with a variant form of the CSD (Sivagnanam et al. 2010a). In the review of 57 cases of CTE, an EpCAM mutation was involved in 41 patients (73 %) who mainly displayed isolated digestive symptoms. Twelve other patients (21 %) carried mutations in SPINT2 and were phenotypically characterized by morphological malformations. Finally, four patients (7 %) with isolated digestive symptoms had no detectable EpCAM or SPINT2 mutation (Salomon et al. 2014). The precise mechanism of diarrhea following the loss of EpCAM is still unknown. Dysfunctional epithelial barrier is one of the possible explanations (Thoeni et al. 2014a), and EpCAM can possibly interact with the tight junction protein Claudin7 in the human intestine (Mueller et al. 2014). Treatment of CTE thus relies on TPN and bowel transplantation as major modes of life support.

29.13 Summary

SD remains a major global health problem that contributes to significant morbidity and mortality in children. Until recently, no new therapies were developed for treating SD, as pharmaceutical industry resources have been focused on developing therapies for diseases affecting the developed world. Fortunately, over the last decade, there have been significant advances in elucidating the underlying pathophysiology of infectious and genetic causes of secretory diarrhea. Genetic advances contributed to identification of several new intestinal ion transporters and molecules such as molecular motors relevant for infectious and inherited SD pathogenesis. Cell biological approaches combined with conventional electrophysiologic ion transport methods have elucidated novel cell-specific mechanisms (endocytosis, exocytosis, adaptors, molecular motors) that regulate intestinal ion transporters and contribute to SD. These advances have led to the development of new therapies such as crofelemer that is currently approved and in use for HIV-related diarrhea. The identification of CFTR as the major apical exit pathway for chloride in the intestine led to the development of inhibitors that are the pipeline for cholera and *E. coli*-based SD. The rapid pace of recent research advances in the pathogenesis of congenital diarrheas including CCD, CSD, and MVID will undoubtedly lead to the development of new and specific therapies in the coming years. This will prove a relief for the many affected families of inherited childhood diarrheas since intestinal transplant remains the only hope of survival, but is far from ideal.

Acknowledgments The authors wish to acknowledge Anne Collaco for assistance and technical support and the NIH-R01 DK 077065 to NA.

References

Abely M, Hankard GF, Hugot JP et al (1998) Intractable infant diarrhea with epithelial dysplasia associated with polymalformation. J Pediatr Gastroenterol Nutr 27(3):348–352

Achler C, Filmer D, Merte C et al (1989) Role of microtubules in polarized delivery of apical membrane proteins to the brush border of the intestinal epithelium. J Cell Biol 109(1):179–189

Aichbichler BW, Zerr CH, Santa Ana CA et al (1997) Proton-pump inhibition of gastric chloride secretion in congenital chloridorrhea. N Engl J Med 336(2):106–109

Aimoto S, Takao T, Shimonishi Y et al (1982) Amino-acid sequence of a heat-stable enterotoxin produced by human enterotoxigenic Escherichia coli. Eur J Biochem 129(2):257–263

Aimoto S, Watanabe H, Ikemura H et al (1983) Chemical synthesis of a highly potent and heat-stable analog of an enterotoxin produced by a human strain of enterotoxigenic Escherichia coli. Biochem Biophys Res Commun 112(1):320–326

Alderete JF, Robertson DC (1978) Purification and chemical characterization of the heat-stable enterotoxin produced by porcine strains of enterotoxigenic Escherichia coli. Infect Immun 19(3):1021–1030

Al-Mayouf SM, Alswaied N, Alkuraya FS et al (2009) Tufting enteropathy and chronic arthritis: a newly recognized association with a novel EpCAM gene mutation. J Pediatr Gastroenterol Nutr 49(5):642–644

Ameen N, Apodaca G (2007) Defective CFTR apical endocytosis and enterocyte brush border in myosin VI-deficient mice. Traffic 8(8):998–1006

Ameen NA, Salas PJ (2000) Microvillus inclusion disease: a genetic defect affecting apical membrane protein traffic in intestinal epithelium. Traffic 1(1):76–83

Ameen NA, Ardito T, Kashgarian M et al (1995) A unique subset of rat and human intestinal villus cells express the cystic fibrosis transmembrane conductance regulator. Gastroenterology 108(4):1016–1023

Ameen NA, Martensson B, Bourguinon L et al (1999) CFTR channel insertion to the apical surface in rat duodenal villus epithelial cells is upregulated by VIP in vivo. J Cell Sci 112(Pt 6): 887–894

Ameen NA, van Donselaar E, Posthuma G et al (2000) Subcellular distribution of CFTR in rat intestine supports a physiologic role for CFTR regulation by vesicle traffic. Histochem Cell Biol 114(3):219–228

Ameen N, Silvis M, Bradbury NA (2007) Endocytic trafficking of CFTR in health and disease. J Cyst Fibros 6(1):1–14

Barrett KE (2000) New insights into the pathogenesis of intestinal dysfunction: secretory diarrhea and cystic fibrosis. World J Gastroenterol 6(4):470–474

Barrett KE, Keely SJ (2000) Chloride secretion by the intestinal epithelium: molecular basis and regulatory aspects. Annu Rev Physiol 62:535–572

Baum M, Martin MG, Booth IW et al (2011) Nucleotide sequence of the Na^+/H^+ exchanger-8 in patients with congenital sodium diarrhea. J Pediatr Gastroenterol Nutr 53(5):474–477

Baxter PS, Dickson JA, Variend S et al (1988) Intestinal disease in cystic fibrosis. Arch Dis Child 63(12):1496–1497

Bernardi KM, Forster ML, Lencer WI et al (2008) Derlin-1 facilitates the retro-translocation of cholera toxin. Mol Biol Cell 19(3):877–884

Bijlsma PB, Kiliaan AJ, Scholten G et al (2000) Increased paracellular macromolecular transport and subnormal glucose uptake in duodenal biopsies of patients with microvillus inclusion disease. Comparisons to other chronic diarrhea patients and to nondiarrhea patients. Ann NY Acad Sci 915:267–269

Binder HJ (2012) Intestinal fluid and electrolyte movement. In: Boron WF, Boulpaep EL (eds) Medical physiology a cellular and molecular approach, updated 2nd edn, vol 1. Saunders Elsevier, Philadelphia, PA, p 933

Binder HJ, Reuben A (2012) Nutrient digestion and absorption. In: Boron WF, Boulpaep EL (eds) Medical physiology a cellular and molecular approach, updated 2nd edn, vol 1. Saunders Elsevier, Philadelphia, PA, p 949

Bird LM, Sivagnanam M, Taylor S et al (2007) A new syndrome of tufting enteropathy and choanal atresia, with ophthalmologic, hematologic and hair abnormalities. Clin Dysmorphol 16(4):211–221

Bohles N, Busch K, Hensel M (2014) Vaccines against human diarrheal pathogens: current status and perspectives. Hum Vaccin Immunother 10(6)

Booth IW, Stange G, Murer H et al (1985) Defective jejunal brush-border Na^+/H^+ exchange: a cause of congenital secretory diarrhoea. Lancet 1(8437):1066–1069

Bunn SK, Beath SV, McKeirnan PJ et al (2000) Treatment of microvillus inclusion disease by intestinal transplantation. J Pediatr Gastroenterol Nutr 31(2):176–180

Burgess MN, Bywater RJ, Cowley CM et al (1978) Biological evaluation of a methanol-soluble, heat-stable Escherichia coli enterotoxin in infant mice, pigs, rabbits, and calves. Infect Immun 21(2):526–531

Buss F, Arden SD, Lindsay M et al (2001) Myosin VI isoform localized to clathrin-coated vesicles with a role in clathrin-mediated endocytosis. EMBO J 20(14):3676–3684

Buss F, Spudich G, Kendrick-Jones J (2004) Myosin VI: cellular functions and motor properties. Annu Rev Cell Dev Biol 20:649–676

Camilleri M, Nullens S, Nelsen T (2012) Enteroendocrine and neuronal mechanisms in pathophysiology of acute infectious diarrhea. Dig Dis Sci 57(1):19–27

Canani RB, Terrin G (2011) Recent progress in congenital diarrheal disorders. Curr Gastroenterol Rep 13(3):257–264

Canani RB, Terrin G, Cirillo P et al (2004) Butyrate as an effective treatment of congenital chloride diarrhea. Gastroenterology 127(2):630–634

Canani RB, Terrin G, Elce A et al (2013) Genotype-dependency of butyrate efficacy in children with congenital chloride diarrhea. Orphanet J Rare Dis 8:194

Carruthers L, Phillips AD, Dourmashkin R et al (1985) Biochemical abnormality in brush border membrane protein of a patient with congenital microvillus atrophy. J Pediatr Gastroenterol Nutr 4(6):902–907

Carruthers L, Dourmashkin R, Phillips A (1986) Disorders of the cytoskeleton of the enterocyte. Clin Gastroenterol 15(1):105–120

Chao AC, de Sauvage FJ, Dong YJ et al (1994) Activation of intestinal CFTR Cl^- channel by heat-stable enterotoxin and guanylin via cAMP-dependent protein kinase. EMBO J 13(5): 1065–1072

Chinkers M, Garbers DL, Chang MS et al (1989) A membrane form of guanylate cyclase is an atrial natriuretic peptide receptor. Nature 338(6210):78–83

Choi M, Scholl UI, Ji W et al (2009) Genetic diagnosis by whole exome capture and massively parallel DNA sequencing. Proc Natl Acad Sci U S A 106(45):19096–19101

Cihil KM, Ellinger P, Fellows A et al (2012) Disabled-2 protein facilitates assembly polypeptide-2-independent recruitment of cystic fibrosis transmembrane conductance regulator to endocytic vesicles in polarized human airway epithelial cells. J Biol Chem 287(18): 15087–15099

Cohen MB, Guarino A, Shukla R et al (1988) Age-related differences in receptors for Escherichia coli heat-stable enterotoxin in the small and large intestine of children. Gastroenterology 94(2): 367–373

Cohen MB, Thompson MR, Giannella RA (1989) Differences in jejunal and ileal response to E. coli enterotoxin: possible mechanisms. Am J Physiol Gastrointest Liver Physiol 257(1): G118–G123

Cohen MB, Mann EA, Lau C et al (1992) A gradient in expression of the Escherichia coli heat-stable enterotoxin receptor exists along the villus-to-crypt axis of rat small intestine. Biochem Biophys Res Commun 186(1):483–490

Collaco A, Jakab R, Hegan P et al (2010) Alpha-AP-2 directs myosin VI-dependent endocytosis of cystic fibrosis transmembrane conductance regulator chloride channels in the intestine. J Biol Chem 285(22):17177–17187
Cottreau J, Tucker A, Crutchley R et al (2012) Crofelemer for the treatment of secretory diarrhea. Expert Rev Gastroenterol Hepatol 6(1):17–23
Croft NM, Howatson AG, Ling SC et al (2000) Microvillous inclusion disease: an evolving condition. J Pediatr Gastroenterol Nutr 31(2):185–189
Crutchley RD, Miller J, Garey KW (2010) Crofelemer, a novel agent for treatment of secretory diarrhea. Ann Pharmacother 44(5):878–884
Currie MG, Fok KF, Kato J et al (1992) Guanylin: an endogenous activator of intestinal guanylate cyclase. Proc Natl Acad Sci U S A 89(3):947–951
Cutz E, Rhoads JM, Drumm B et al (1989) Microvillus inclusion disease: an inherited defect of brush-border assembly and differentiation. N Engl J Med 320(10):646–651
Darrow DC (1945) Congenital alkalosis with diarrhea. J Pediatr 26:519–532
Davidson GP, Cutz E, Hamilton JR et al (1978) Familial enteropathy: a syndrome of protracted diarrhea from birth, failure to thrive, and hypoplastic villus atrophy. Gastroenterology 75(5): 783–790
De Jonge HR (1975a) The localization of guanylate cyclase in rat small intestinal epithelium. FEBS Lett 53(2):237–242
De Jonge HR (1975b) The response of small intestinal villous and crypt epithelium to choleratoxin in rat and guinea pig. Evidence against a specific role of the crypt cells in choleragen-induced secretion. Biochem Biophys Acta 381(1):128–143
de Sauvage FJ, Keshav S, Kuang WJ et al (1992) Precursor structure, expression, and tissue distribution of human guanylin. Proc Natl Acad Sci U S A 89(19):9089–9093
Dhekne HS, Hsiao NH, Roelofs P et al (2014) Myosin Vb and Rab11a regulate phosphorylation of ezrin in enterocytes. J Cell Sci 127(Pt 5):1007–1017
Dong W, Chen X, Chen P et al (2012) Inactivation of MYO5B promotes invasion and motility in gastric cancer cells. Dig Dis Sci 57(5):1247–1252
Donowitz M, Welsh MJ (1986) Ca^{2+} and cyclic AMP in regulation of intestinal Na, K and Cl transport. Annu Rev Physiol 48:135–150
Donowitz M, Alpers DH, Binder HJ et al (2012) Translational approaches for pharmacotherapy development for acute diarrhea. Gastroenterology 142(3):e1–e9
Eggermont E (1996) Gastrointestinal manifestations in cystic fibrosis. Eur J Gastroenterol Hepatol 8(8):731–738
Ellinger A, Pavelka M (1986) Colchicine-induced tubular, vesicular and cisternal organelle aggregates in absorptive cells of the small intestine of the rat. II. Endocytosis studies. Biol Cell 58(1):31–41
El-Matary W, Dalzell AM, Kokai G et al (2007) Tufting enteropathy and skeletal dysplasia: is there a link? Eur J Pediatr 166(3):265–268
Farthing MJ (1994) Oral rehydration therapy. Pharmacol Ther 64(3):477–492
Farthing MJ (2001) Treatment of gastrointestinal viruses. Novartis Found Symp 238:289–300
Fell JM, Miller MP, Finkel Y et al (1992) Congenital sodium diarrhea with a partial defect in jejunal brush border membrane sodium transport, normal rectal transport, and resolving diarrhea. J Pediatr Gastroenterol Nutr 15(2):112–116
Field M, Semrad CE (1993) Toxigenic diarrheas, congenital diarrheas, and cystic fibrosis: disorders of intestinal ion transport. Annu Rev Physiol 55:631–655
Field M, Graf LH Jr, Laird WJ et al (1978) Heat-stable enterotoxin of Escherichia coli: in vitro effects on guanylate cyclase activity, cyclic GMP concentration, and ion transport in small intestine. Proc Natl Acad Sci U S A 75(6):2800–2804
Fish EM, Molitoris BA (1994) Alterations in epithelial polarity and the pathogenesis of disease states. N Engl J Med 330(22):1580–1588
Gabriel SE, Brigman KN, Koller BH et al (1994) Cystic fibrosis heterozygote resistance to cholera toxin in the cystic fibrosis mouse model. Science 266(5182):107–109

Gamble JL, Fahey KR, Appleton J et al (1945) Congenital alkalosis with diarrhea. J Pediatr 26(6): 509–518
Gariepy J, Schoolnik GK (1986) Design of a photoreactive analogue of the Escherichia coli heat-stable enterotoxin STIb: use in identifying its receptor on rat brush border membranes. Proc Natl Acad Sci U S A 83(2):483–487
Gariepy J, Lane A, Frayman F et al (1986) Structure of the toxic domain of the Escherichia coli heat-stable enterotoxin ST I. Biochemistry 25(24):7854–7866
Gelfond D, Ma C, Semler J et al (2013) Intestinal pH and gastrointestinal transit profiles in cystic fibrosis patients measured by wireless motility capsule. Dig Dis Sci 58(8):2275–2281
Giannella RA, Luttrell M, Thompson M (1983) Binding of Escherichia coli heat-stable enterotoxin to receptors on rat intestinal cells. Am J Physiol Gastrointest Liver Physiol 8(4):G492–G498
Gill DM, Clements JD, Robertson DC et al (1981) Subunit number and arrangement in Escherichia coli heat-labile enterotoxin. Infect Immun 33(3):677–682
Golachowska MR, van Dael CM, Keuning H et al (2012) MYO5B mutations in patients with microvillus inclusion disease presenting with transient renal Fanconi syndrome. J Pediatr Gastroenterol Nutr 54(4):491–498
Golin-Bisello F, Bradbury N, Ameen N (2005) STa and cGMP stimulate CFTR translocation to the surface of villus enterocytes in rat jejunum and is regulated by protein kinase G. Am J Physiol Cell Physiol 289(3):C708–C716
Goulet O, Kedinger M, Brousse N et al (1995) Intractable diarrhea of infancy with epithelial and basement membrane abnormalities. J Pediatr 127(2):212–219
Gray J (2011) Rotavirus vaccines: safety, efficacy and public health impact. J Intern Med 270(3): 206–214
Guerrant RL, Hughes JM, Chang B et al (1980) Activation of intestinal guanylate cyclase by heat-stable enterotoxin of Escherichia coli: Studies of tissue specificity, potential receptors, and intermediates. J Infect Dis 142(2):220–228
Guggino WB, Stanton BA (2006) New insights into cystic fibrosis: molecular switches that regulate CFTR. Nat Rev Mol Cell Biol 7(6):426–436
Hasegawa M, Hidaka Y, Matsumoto Y et al (1999) Determination of the binding site on the extracellular domain of guanylyl cyclase C to heat-stable enterotoxin. J Biol Chem 274(44): 31713–31718
Hasegawa M, Matsumoto-Ishikawa Y, Hijikata A et al (2005) Disulfide linkages and a three-dimensional structure model of the extracellular ligand-binding domain of guanylyl cyclase C. Protein J 24(5):315–325
Hasson T (2003) Myosin VI: two distinct roles in endocytosis. J Cell Sci 116(Pt 17):3453–3461
Hazes B, Read RJ (1997) Accumulating evidence suggests that several AB-toxins subvert the endoplasmic reticulum-associated protein degradation pathway to enter target cells. Biochemistry 36(37):11051–11054
Heinz-Erian P, Muller T, Krabichler B et al (2009) Mutations in SPINT2 cause a syndromic form of congenital sodium diarrhea. Am J Hum Genet 84(2):188–196
Hoglund P, Sistonen P, Norio R et al (1995) Fine mapping of the congenital chloride diarrhea gene by linkage disequilibrium. Am J Hum Genet 57(1):95–102
Hoglund P, Haila S, Scherer SW et al (1996a) Positional candidate genes for congenital chloride diarrhea suggested by high-resolution physical mapping in chromosome region 7q31. Genome Res 6(3):202–210
Hoglund P, Haila S, Socha J et al (1996b) Mutations of the Down-regulated in adenoma (DRA) gene cause congenital chloride diarrhoea. Nat Genet 14(3):316–319
Holmberg C (1978) Electrolyte economy and its hormonal regulation in congenital chloride diarrhea. Pediatr Res 12(2):82–86
Holmberg C, Perheentupa J (1985) Congenital Na^+ diarrhea: a new type of secretory diarrhea. J Pediatr 106(1):56–61
Holmberg C, Perheentupa J, Launiala K (1975) Colonic electrolyte transport in health and in congenital chloride diarrhea. J Clin Invest 56(2):302–310

Jakab RL, Collaco AM, Ameen NA (2011) Physiological relevance of cell-specific distribution patterns of CFTR, NKCC1, NBCe1, and NHE3 along the crypt-villus axis in the intestine. Am J Physiol Gastrointest Liver Physiol 300(1):G82–G98

Jakab RL, Collaco AM, Ameen NA (2012) Cell-specific effects of luminal acid, bicarbonate, cAMP, and carbachol on transporter trafficking in the intestine. Am J Physiol Gastrointest Liver Physiol 303(8):G937–G950

Jakab RL, Collaco AM, Ameen NA (2013) Characterization of CFTR High Expresser cells in the intestine. Am J Physiol Gastrointest Liver Physiol 305(6):G453–G465

Joo NS, London RM, Kim HD et al (1998) Regulation of intestinal Cl^- and HCO_3^- secretion by uroguanylin. Am J Physiol Gastrointest Liver Physiol 274(4 Pt 1):G633–G644

Keller KM, Wirth S, Baumann W et al (1990) Defective jejunal brush border membrane sodium/proton exchange in association with lethal familial protracted diarrhoea. Gut 31(10): 1156–1158

Kelly M, Trudel S, Brouillard F et al (2010) Cystic fibrosis transmembrane regulator inhibitors CFTR(inh)-172 and GlyH-101 target mitochondrial functions, independently of chloride channel inhibition. J Pharmacol Exp Ther 333(1):60–69

Kere J, Sistonen P, Holmberg C et al (1993) The gene for congenital chloride diarrhea maps close to but is distinct from the gene for cystic fibrosis transmembrane conductance regulator. Proc Natl Acad Sci U S A 90(22):10686–10689

Kere J, Lohi H, Hoglund P (1999) Genetic disorders of membrane transport. III. Congenital chloride diarrhea. Am J Physiol Gastrointest Liver Physiol 276(1):G7–G13

Kirkham M, Fujita A, Chadda R et al (2005) Ultrastructural identification of uncoated caveolin-independent early endocytic vehicles. J Cell Biol 168(3):465–476

Ko SB, Shcheynikov N, Choi JY et al (2002) A molecular mechanism for aberrant $CFTR^-$dependent HCO_3^- transport in cystic fibrosis. EMBO J 21(21):5662–5672

Kopic S, Geibel JP (2010) Toxin mediated diarrhea in the 21 century: the pathophysiology of intestinal ion transport in the course of ETEC. V. Cholerae and rotavirus infection. Toxins 2(8):2132–2157

Krause WJ, Cullingford GL, Freeman RH et al (1994) Distribution of heat-stable enterotoxin/guanylin receptors in the intestinal tract of man and other mammals. J Anat 184(Pt 2):407–417

Kravtsov DV, Ameen NA (2013) Molecular motors and apical CFTR traffic in epithelia. Int J Mol Sci 14(5):9628–9642

Kravtsov DV, Caputo C, Collaco A et al (2012) Myosin Ia is required for CFTR brush border membrane trafficking and ion transport in the mouse small intestine. Traffic 13(8):1072–1082

Kravtsov D, Mashukova A, Forteza R et al (2014) Myosin 5b loss of function leads to defects in polarized signalling: implication for microvillus inclusion disease pathogenesis and treatment. Am J Physiol Gastrointest Liver Physiol 307(10):G992–G1001

Krzemien G, Szmigielska A, Jankowska K et al (2013) Congenital chloride diarrhea mimicking meconium ileus in newborn. Med Wieku Rozwoj 17(4):320–323

Lauber T, Tidten N, Matecko I et al (2009) Design and characterization of a soluble fragment of the extracellular ligand-binding domain of the peptide hormone receptor guanylyl cyclase-C. Protein Eng Des Sel 22(1):1–7

Lee CM, Chang PP, Tsai SC et al (1991) Activation of Escherichia coli heat-labile enterotoxins by native and recombinant adenosine diphosphate-ribosylation factors, 20-kD guanine nucleotide-binding proteins. J Clin Invest 87(5):1780–1786

Lorenz JN, Nieman M, Sabo J et al (2003) Uroguanylin knockout mice have increased blood pressure and impaired natriuretic response to enteral NaCl load. J Clin Invest 112(8): 1244–1254

Lorrot M, Vasseur M (2007) How do the rotavirus NSP4 and bacterial enterotoxins lead differently to diarrhea? Virol J 4:31

Lorrot M, Benhamadouche-Casari H, Vasseur M (2006) Mechanisms of net chloride secretion during rotavirus diarrhea in young rabbits: do intestinal villi secrete chloride? Cell Physiol Biochem 18(1–3):103–112

Lowe DG, Chang MS, Hellmiss R et al (1989) Human atrial natriuretic peptide receptor defines a new paradigm for second messenger signal transduction. EMBO J 8(5):1377–1384

Ma T, Thiagarajah JR, Yang H et al (2002) Thiazolidinone CFTR inhibitor identified by high-throughput screening blocks cholera toxin-induced intestinal fluid secretion. J Clin Invest 110(11):1651–1658

Madden DR, Swiatecka-Urban A (2012) Tissue-specific control of CFTR endocytosis by Dab2: cargo recruitment as a therapeutic target. Commun Integr Biol 5(5):473–476

Majoul I, Sohn K, Wieland FT et al (1998) KDEL receptor (Erd2p)-mediated retrograde transport of the cholera toxin A subunit from the Golgi involves COPI, p23, and the COOH terminus of Erd2p. J Cell Biol 143(3):601–612

Makela S, Kere J, Holmberg C et al (2002) SLC26A3 mutations in congenital chloride diarrhea. Hum Mutat 20(6):425–438

Mann EA, Jump ML, Wu J et al (1997) Mice lacking the guanylyl cyclase C receptor are resistant to STa-induced intestinal secretion. Biochem Biophys Res Commun 239(2):463–466

Massey S, Banerjee T, Pande AH et al (2009) Stabilization of the tertiary structure of the cholera toxin A1 subunit inhibits toxin dislocation and cellular intoxication. J Mol Biol 393(5): 1083–1096

Massol RH, Larsen JE, Fujinaga Y et al (2004) Cholera toxin toxicity does not require functional Arf6- and dynamin-dependent endocytic pathways. Mol Biol Cell 15(8):3631–3641

McReynolds EW, Roy S III, Etteldorf JN (1974) Congenital chloride diarrhea. Am J Dis Child 127(4):566–570

Melvin JE, Park K, Richardson L et al (1999) Mouse down-regulated in adenoma (DRA) is an intestinal Cl^-/HCO_3^- exchanger and is up-regulated in colon of mice lacking the NHE3 Na^+/H^+ exchanger. J Biol Chem 274(32):22855–22861

Merritt EA, Sarfaty S, Van den Akker F et al (1994) Crystal structure of cholera toxin B-pentamer bound to receptor G(M1) pentasaccharide. Protein Sci 3(2):166–175

Michail S, Collins JF, Xu H et al (1998) Abnormal expression of brush-border membrane transporters in the duodenal mucosa of two patients with microvillus inclusion disease. J Pediatr Gastroenterol Nutr 27(5):536–542

Morris SM, Arden SD, Roberts RC et al (2002) Myosin VI binds to and localises with Dab2, potentially linking receptor-mediated endocytosis and the actin cytoskeleton. Traffic 3(5): 331–341

Moseley SL, Hardy JW, Hug MI et al (1983) Isolation and nucleotide sequence determination of a gene encoding a heat-stable enterotoxin of Escherichia coli. Infect Immun 39(3):1167–1174

Moss J, Richardson SH (1978) Activation of adenylate cyclase by heat-labile Escherichia coli enterotoxin. Evidence for ADP-ribosyltransferase activity similar to that of choleragen. J Clin Invest 62(2):281–285

Muanprasat C, Sonawane ND, Salinas D et al (2004) Discovery of glycine hydrazide pore-occluding CFTR inhibitors: mechanism, structure-activity analysis, and in vivo efficacy. J Gen Physiol 124(2):125–137

Mueller JL, McGeough MD, Pena CA et al (2014) Functional consequences of EpCam mutation in mice and men. Am J Physiol Gastrointest Liver Physiol 306(4):G278–G288

Muller T, Wijmenga C, Phillips AD et al (2000) Congenital sodium diarrhea is an autosomal recessive disorder of sodium/proton exchange but unrelated to known candidate genes. Gastroenterology 119(6):1506–1513

Muller T, Hess MW, Schiefermeier N et al (2008) MYO5B mutations cause microvillus inclusion disease and disrupt epithelial cell polarity. Nat Genet 40(10):1163–1165

Nakazato M (2001) Guanylin family: new intestinal peptides regulating electrolyte and water homeostasis. J Gastroenterol 36(4):219–225

Navaneethan U, Giannella RA (2008) Mechanisms of infectious diarrhea. Nat Clin Pract Gastroenterol Hepatol 5(11):637–647

Oliva MM, Perman JA, Saavedra JM et al (1994) Successful intestinal transplantation for microvillus inclusion disease. Gastroenterology 106(3):771–774

Ousingsawat J, Mirza M, Tian Y et al (2011) Rotavirus toxin NSP4 induces diarrhea by activation of TMEM16A and inhibition of Na^+ absorption. Pflugers Arch 461(5):579–589
Pande AH, Scaglione P, Taylor M et al (2007) Conformational instability of the cholera toxin A1 polypeptide. J Mol Biol 374(4):1114–1128
Parton RG (1994) Ultrastructural localization of gangliosides; GM1 is concentrated in caveolae. J Histochem Cytochem 42(2):155–166
Patey N, Scoazec JY, Cuenod-Jabri B et al (1997) Distribution of cell adhesion molecules in infants with intestinal epithelial dysplasia (tufting enteropathy). Gastroenterology 113(3): 833–843
Pedemonte N, Galietta LJ (2014) Structure and function of TMEM16 proteins (anoctamins). Physiol Rev 94(2):419–459
Pelkmans L, Burli T, Zerial M et al (2004) Caveolin-stabilized membrane domains as multifunctional transport and sorting devices in endocytic membrane traffic. Cell 118(6):767–780
Pfeifer A, Aszodi A, Seidler U et al (1996) Intestinal secretory defects and dwarfism in mice lacking cGMP-dependent protein kinase II. Science 274(5295):2082–2086
Phillips AD, Schmitz J (1992) Familial microvillous atrophy: a clinicopathological survey of 23 cases. J Pediatr Gastroenterol Nutr 14(4):380–396
Phillips AD, Jenkins P, Raafat F et al (1985) Congenital microvillous atrophy: specific diagnostic features. Arch Dis Child 60(2):135–140
Pohl JF, Shub MD, Trevelline EE et al (1999) A cluster of microvillous inclusion disease in the Navajo population. J Pediatr 134(1):103–106
Pratha VS, Hogan DL, Martensson BA et al (2000) Identification of transport abnormalities in duodenal mucosa and duodenal enterocytes from patients with cystic fibrosis. Gastroenterology 118(6):1051–1060
Randak C, Langnas AN, Kaufman SS et al (1998) Pretransplant management and small bowel-liver transplantation in an infant with microvillus inclusion disease. J Pediatr Gastroenterol Nutr 27(3):333–337
Reifen RM, Cutz E, Griffiths AM et al (1994) Tufting enteropathy: a newly recognized clinicopathological entity associated with refractory diarrhea in infants. J Pediatr Gastroenterol Nutr 18(3):379–385
Reinshagen K, Naim HY, Zimmer KP (2002) Autophagocytosis of the apical membrane in microvillus inclusion disease. Gut 51(4):514–521
Rhoads JM, Vogler RC, Lacey SR et al (1991) Microvillus inclusion disease. In vitro jejunal electrolyte transport. Gastroenterology 100(3):811–817
Richards AA, Stang E, Pepperkok R et al (2002) Inhibitors of COP-mediated transport and cholera toxin action inhibit simian virus 40 infection. Mol Biol Cell 13(5):1750–1764
Riordan JR, Rommens JM, Kerem B et al (1989) Identification of the cystic fibrosis gene: cloning and characterization of complementary DNA. Science 245(4922):1066–1073
Roberts R, Lister I, Schmitz S et al (2004) Myosin VI: cellular functions and motor properties. Philos Trans R Soc Lond B Biol Sci 359(1452):1931–1944
Rodighiero C, Aman AT, Kenny MJ et al (1999) Structural basis for the differential toxicity of cholera toxin and Escherichia coli heat-labile enterotoxin: construction of hybrid toxins identifies the A2-domain as the determinant of differential toxicity. J Biol Chem 274(7): 3962–3969
Rodighiero C, Tsai B, Rapoport TA et al (2002) Role of ubiquitination in retro-translocation of cholera toxin and escape of cytosolic degradation. EMBO Rep 3(12):1222–1227
Rodriguez OC, Cheney RE (2002) Human myosin-Vc is a novel class V myosin expressed in epithelial cells. J Cell Sci 115(Pt 5):991–1004
Rowland AF, Fazakerley DJ, James DE (2011) Mapping insulin/GLUT4 circuitry. Traffic 12(6): 672–681

Ruemmele FM, Muller T, Schiefermeier N et al (2010) Loss-of-function of MYO5B is the main cause of microvillus inclusion disease: 15 novel mutations and a CaCo-2 RNAi cell model. Hum Mutat 31(5):544–551

Salomon J, Goulet O, Canioni D et al (2014) Genetic characterization of congenital tufting enteropathy: epcam associated phenotype and involvement of SPINT2 in the syndromic form. Hum Genet 133(3):299–310

Sato T, Mushiake S, Kato Y et al (2007) The Rab8 GTPase regulates apical protein localization in intestinal cells. Nature 448(7151):366–369

Schulz S, Green CK, Yuen PS et al (1990) Guanylyl cyclase is a heat-stable enterotoxin receptor. Cell 63(5):941–948

Schulz S, Lopez MJ, Kuhn M et al (1997) Disruption of the guanylyl cyclase-C gene leads to a paradoxical phenotype of viable but heat-stable enterotoxin-resistant mice. J Clin Invest 100(6):1590–1595

Sherman PM, Mitchell DJ, Cutz E (2004) Neonatal enteropathies: defining the causes of protracted diarrhea of infancy. J Pediatr Gastroenterol Nutr 38(1):16–26

Silvis MR, Bertrand CA, Ameen N et al (2009) Rab11b regulates the apical recycling of the cystic fibrosis transmembrane conductance regulator in polarized intestinal epithelial cells. Mol Biol Cell 20(8):2337–2350

Singh RD, Puri V, Valiyaveettil JT et al (2003) Selective caveolin-1-dependent endocytosis of glycosphingolipids. Mol Biol Cell 14(8):3254–3265

Singh AK, Riederer B, Chen M et al (2010) The switch of intestinal Slc26 exchangers from anion absorptive to HCOFormula secretory mode is dependent on CFTR anion channel function. Am J Physiol Cell Physiol 298(5):C1057–C1065

Singh V, Yang J, Chen TE et al (2014) Translating molecular physiology of intestinal transport into pharmacologic treatment of diarrhea: stimulation of Na^+ absorption. Clin Gastroenterol Hepatol 12(1):27–31

Sivagnanam M, Mueller JL, Lee H et al (2008) Identification of EpCAM as the gene for congenital tufting enteropathy. Gastroenterology 135(2):429–437

Sivagnanam M, Janecke AR, Muller T et al (2010a) Case of syndromic tufting enteropathy harbors SPINT2 mutation seen in congenital sodium diarrhea. Clin Dysmorphol 19(1):48

Sivagnanam M, Schaible T, Szigeti R et al (2010b) Further evidence for EpCAM as the gene for congenital tufting enteropathy. Am J Med Genet A 152A(1):222–224

Sixma TK, Kalk KH, Van Zanten BAM et al (1993) Refined structure of Escherichia coli heat-labile enterotoxin, a close relative of cholera toxin. J Mol Biol 230(3):890–918

Skelton NJ, Garcia KC, Goeddel DV et al (1994) Determination of the solution structure of the peptide hormone guanylin: observation of a novel form of topological stereoisomerism. Biochemistry 33(46):13581–13592

So M, McCarthy BJ (1980) Nucleotide sequence of the bacterial transposon Tn1681 encoding a heat-stable (ST) toxin and its identification in enterotoxigenic Escherichia coli strains. Proc Natl Acad Sci U S A 77(7):4011–4015

Sonawane ND, Muanprasat C, Nagatani R Jr et al (2005) In vivo pharmacology and antidiarrheal efficacy of a thiazolidinone CFTR inhibitor in rodents. J Pharm Sci 94(1):134–143

Srivastava R, Zinman HM, McKenzie MD et al (1999) Rotavirus vaccine. Pediatrics 104(6): 1419–1420

Staples SJ, Asher SE, Giannella RA (1980) Purification and characterization of heat-stable enterotoxin produced by a strain of E. coli pathogenic for man. J Biol Chem 255(10): 4716–4721

Takao T, Shimonishi Y, Kobayashi M et al (1985a) Amino acid sequence of heat-stable enterotoxin produced by Vibrio cholerae non-01. FEBS Lett 193(2):250–254

Takao T, Tominaga N, Yoshimura S et al (1985b) Isolation, primary structure and synthesis of heat-stable enterotoxin produced by Yersinia enterocolitica. Eur J Biochem 152(1):199–206

Takeda Y, Takeda T, Yano T et al (1979) Purification and partial characterization of heat-stable enterotoxin of enterotoxigenic Escherichia coli. Infect Immun 25(3):978–985
Talmon G, Holzapfel M, DiMaio DJ et al (2012) Rab11 is a useful tool for the diagnosis of microvillous inclusion disease. Int J Surg Pathol 20(3):252–256
Taylor M, Navarro-Garcia F, Huerta J et al (2010) Hsp90 is required for transfer of the cholera toxin A1 subunit from the endoplasmic reticulum to the cytosol. J Biol Chem 285(41):31261–31267
Taylor M, Banerjee T, Ray S et al (2011) Protein-disulfide isomerase displaces the cholera toxin A1 subunit from the holotoxin without unfolding the A1 subunit. J Biol Chem 286(25): 22090–22100
Thoeni C, Amir A, Guo C et al (2014a) A novel nonsense mutation in the EpCAM gene in a patient with congenital tufting enteropathy. J Pediatr Gastroenterol Nutr 58(1):18–21
Thoeni CE, Vogel GF, Tancevski I et al (2014b) Microvillus inclusion disease: loss of Myosin vb disrupts intracellular traffic and cell polarity. Traffic 15(1):22–42
Tradtrantip L, Namkung W, Verkman AS (2010) Crofelemer, an antisecretory antidiarrheal proanthocyanidin oligomer extracted from Croton lechleri, targets two distinct intestinal chloride channels. Mol Pharmacol 77(1):69–78
Tsai B, Rodighiero C, Lencer WI et al (2001) Protein disulfide isomerase acts as a redox-dependent chaperone to unfold cholera toxin. Cell 104(6):937–948
Tyagi S, Kavilaveettil RJ, Alrefai WA et al (2001) Evidence for the existence of a distinct SO_4^{--}-OH^- exchange mechanism in the human proximal colonic apical membrane vesicles and its possible role in chloride transport. Exp Biol Med 226(10):912–918
Vaandrager AB, Bot AG, De Jonge HR (1997) Guanosine 3′,5′-cyclic monophosphate-dependent protein kinase II mediates heat-stable enterotoxin-provoked chloride secretion in rat intestine. Gastroenterology 112(2):437–443
Vaandrager AB, Bot AG, Ruth P et al (2000) Differential role of cyclic GMP-dependent protein kinase II in ion transport in murine small intestine and colon. Gastroenterology 118(1): 108–114
Vega-Salas DE, Salas PJ, Rodriguez-Boulan E (1988) Exocytosis of vacuolar apical compartment (VAC): a cell-cell contact controlled mechanism for the establishment of the apical plasma membrane domain in epithelial cells. J Cell Biol 107(5):1717–1728
Wada A, Hirayama T, Kitaura H et al (1996) Identification of ligand recognition sites in heat-stable enterotoxin receptor, membrane-associated guanylyl cyclase C by site-directed mutational analysis. Infect Immun 64(12):5144–5150
Wedenoja S, Holmberg C, Hoglund P (2008) Oral butyrate in treatment of congenital chloride diarrhea. Am J Gastroenterol 103(1):252–254
Wedenoja S, Hoglund P, Holmberg C (2010) Review article: the clinical management of congenital chloride diarrhoea. Aliment Pharmacol Ther 31(4):477–485
Weis VG, Sousa JF, LaFleur BJ et al (2013) Heterogeneity in mouse spasmolytic polypeptide-expressing metaplasia lineages identifies markers of metaplastic progression. Gut 62(9): 1270–1279
Whittamore JM, Freel RW, Hatch M (2013) Sulfate secretion and chloride absorption are mediated by the anion exchanger DRA (Slc26a3) in the mouse cecum. Am J Physiol Gastrointest Liver Physiol 305(2):G172–G184
Wiegerinck CL, Janecke AR, Schneeberger K et al (2014) Loss of syntaxin 3 causes variant microvillus inclusion disease. Gastroenterology 147(1):65–68, e10
Yang N, Garcia MA, Quinton PM (2013) Normal mucus formation requires cAMP-dependent $HCO3^-$ secretion and Ca^{2+}-mediated mucin exocytosis. J Physiol 591:4581–4593
Yeo QM, Crutchley R, Cottreau J et al (2013) Crofelemer, a novel antisecretory agent approved for the treatment of HIV-associated diarrhea. Drugs Today 49(4):239–252
Yin L, Vijaygopal P, MacGregor GG et al (2014) Glucose stimulates calcium-activated chloride secretion in small intestinal cells. Am J Physiol Cell Physiol 306(7):C687–C696
Yoshino K, Miyachi M, Takao T et al (1993) Purification and sequence determination of heat-stable enterotoxin elaborated by a cholera toxin-producing strain of Vibrio cholerae O1. FEBS Lett 326(1–3):83–86

Index

K.L. Hamilton, D.C. Devor (eds.), *Ion Channels and Transporters of Epithelia in Health and Disease*, Physiology in Health and Disease,
DOI 10.1007/978-1-4939-3366-2

D

F

G

H

MIX
Papier aus verantwortungsvollen Quellen
Paper from responsible sources
FSC® C105338
FSC
www.fsc.org

If you have any concerns about our products,
you can contact us on
ProductSafety@springernature.com

In case Publisher is established outside the EU,
the EU authorized representative is:
Springer Nature Customer Service Center GmbH
Europaplatz 3, 69115 Heidelberg, Germany

Printed by Libri Plureos GmbH
in Hamburg, Germany